Márton Veress • Szabolcs Leél-Őssy
Editors

Cave and Karst Systems of Hungary

Springer

Editors
Márton Veress
SEK, Department of Geography
Eötvös Loránd University
Szombathely, Hungary

Szabolcs Leél-Őssy
Department of Physical and Applied Geology
Eötvös Loránd University
Budapest, Hungary

ISSN 2364-4591 ISSN 2364-4605 (electronic)
Cave and Karst Systems of the World
ISBN 978-3-030-92959-6 ISBN 978-3-030-92960-2 (eBook)
https://doi.org/10.1007/978-3-030-92960-2

This Springer imprint is published by the registered company Springer Nature Switzerland AG
The registered company address is: Gewerbestrasse 11, 6330 Cham, Switzerland

In memory of Jakucs László

Cave and Karst Systems of the World

Series Editor

James W. LaMoreaux, P.E.LaMoreaux and Associates, Tuscaloosa, AL, USA

Márton Veress • Szabolcs Leél-Őssy
Editors

Cave and Karst Systems of Hungary

Springer

Editors
Márton Veress
SEK, Department of Geography
Eötvös Loránd University
Szombathely, Hungary

Szabolcs Leél-Őssy
Department of Physical and Applied Geology
Eötvös Loránd University
Budapest, Hungary

ISSN 2364-4591 ISSN 2364-4605 (electronic)
Cave and Karst Systems of the World
ISBN 978-3-030-92959-6 ISBN 978-3-030-92960-2 (eBook)
https://doi.org/10.1007/978-3-030-92960-2

This Springer imprint is published by the registered company Springer Nature Switzerland AG
The registered company address is: Gewerbestrasse 11, 6330 Cham, Switzerland

In memory of Jakucs László

Preface

This book describes Hungarian karst areas in a versatile and complex way from an analytical point of view. It may not be unfounded to say that it is the most detailed and most comprehensive work that has ever been done also including the latest research results.

Hungarian karst areas do not constitute uniform karst regions such as the karst areas of the Alps and China. Disregarding some exceptions, karst features are not large, some features are absent or their occurrence is rare (collapse dolines, karren features of bare surfaces). Despite this, karst features and karst processes are diverse. However, Hungarian karsts are rich in caves (more than 4000 caves are known), hypogene caves being rich in hydrothermal crystals are significant. The surfaces of these karsts may be good examples of temperate karsts which developed on the peneplain parts of former tropical karsts simultaneously with powerful fluvial surface formation, being affected by hydrothermal impacts in some places.

Forty-five authors participated in the preparation of this book who are experts of various disciplines, many of them are internationally acknowledged researchers. They work at different places such as at universities, in museums, research institutes, and national parks. The age of the authors is also various, there are retired researches, but Ph.D. students may also occur among them.

This book has 21 chapters. Two chapters deal with the geological–hydrological background, one chapter is about paleokarst, three is about the surface of karsts, one chapter focuses on the general description of Hungarian karsts and one concentrates on the history of karst and cave research. Seven chapters deal with caves, three chapters are about the biology and ecology of karsts, and two chapters deal with the concretions of the karst. We are thankful to the authors of the chapters and those who had technical contribution to this book.

Here, we would like to draw the reader's attention to the fact that karst areas have dual or different names in the various chapters. Names of some karst areas used in international publications on geology can also be found in some chapters. Thus, "Hills" occurs at Szalonna Karst (or at Esztramos) as well as at the Buda Mountains, Gerecse Mountains, and Pilis Mountains, while "Range" is used at the Transdanubian Mountains. However, these are mountains and plateaus according to both their geological (structure, built-up) and geographical (morphology, altitude) characteristics and to their name as well, though it is without doubt that their elevation may be partly below 500 m and parts of hill type also occur in their areas.

This book was written with the moral support and the contribution of several members of the *Hungarian Speleological Society*.

Szombathely, Hungary
Budapest, Hungary

Márton Veress
Szabolcs Leél-Őssy

Contents

Contributors

Gáspár Albert Department of Cartography and Geoinformatics, Eötvös Loránd University, Budapest, Hungary

Gergely Balázs Behavioural Ecology Group, Department of Systematic Zoology and Ecology, Eötvös Loránd University, Budapest, Hungary

Ilona Bárány Kevei Department of Climatology and Landscape Ecology, University of Szeged, Szeged, Hungary

Károly Barta Department of Geoinformatics, Physical and Environmental Geography, University of Szeged, Szeged, Hungary

Zoltán Bátori Department of Ecology, University of Szeged, Szeged, Hungary

Norbert Bauer Department of Botany, Hungarian Natural History Museum, Budapest, Hungary

Ágnes Berentés Institute for Geological and Geochemical Research, Research Centre for Astronomy and Earth Sciences, Eötvös Loránd Research Network, Budapest, Hungary

Barbara Bódai Department of Physical Geography, Eötvös Loránd University, Budapest, Hungary

Tamás Budai Division of Geology, Supervisory Authority of Regulatory Affairs, Budapest, Hungary;
Department of Geology and Meteorology, University of Pécs, Pécs, Hungary

Katalin Csondor Department of Geology, ELTE Eötvös Loránd University, Budapest, Hungary

Gábor Csorba Hungarian Natural History Museum, Budapest, Hungary

György Czuppon Institute for Geological and Geochemical Research, Research Centre for Astronomy and Earth Sciences, Eötvös Loránd Research Network, Budapest, Hungary

Attila Demény Institute for Geological and Geochemical Research, Research Centre for Astronomy and Earth Sciences, Eötvös Loránd Research Network, Budapest, Hungary

Péter Dobosy Institute of Aquatic Ecology, Centre for Ecological Research, Eötvös Loránd Research Network, Budapest, Hungary

Csaba Egri Department of National Parks and Landscape Protection, Deputy State Secretariat for Nature Conservation, Ministry of Agriculture, Budapest, Hungary

László Erdős Department of Ecology, University of Debrecen, Debrecen, Hungary

Anita Erőss Department of Geology, ELTE Eötvös Loránd University, Budapest, Hungary

Péter Gruber Aggtelek National Park Directorate, Jósvafõ, Hungary

Attila Hevesi Institute of Geography and Geoinformatics, University of Miskolc, Miskolc-Egyetemváros, Hungary

Veronika Iván Department of Geosciences, University of Padova, Padova, Italy

Sándor Kele Institute for Geological and Geochemical Research, Research Centre for Astronomy and Earth Sciences, Budapest, Hungary

Zsolt Kercsmár Division of Geology, Supervisory Authority of Regulatory Affairs, Budapest, Hungary

Márton Kiss Centre for Ecological Research, Institute of Ecology and Botany, Vácrátót, Hungary

Gyula Konrád University of Pécs, Pécs, Hungary

László Kordos Savaria University Centre, ELTE SEK Eötvös Lóránd University, Szombathely, Hungary

Richard Kovács Duna-Ipoly National Park Directorate, Budapest, Hungary

György Less Faculty of Earth Science & Engineering, Institute of Mineralogy and Geology, University of Miskolc, Miskolc-Egyetemváros, Hungary

Szabolcs Leél-Őssy Department of Physical and Applied Geology, Eötvös Loránd University, Budapest, Hungary;
Department of Geology, Eötvös Loránd University, Budapest, Hungary

László Lénárt Department of Hydrogeology and Engineering Geology, University of Miskolc, Miskolc, Hungary

László Mari Department of Physical Geography, Eötvös Loránd University, Budapest, Hungary

Andrea Mindszenty Department of Geology, Eötvös Loránd University, Budapest, Hungary

Judit Mádl-Szőnyi Department of Geology, ELTE Eötvös Loránd University, Budapest, Hungary

János Móga Department of Physical Geography, Eötvös Loránd University, Budapest, Hungary

Mihály Óvári Nuclear Security Department, Centre of Energy Research, Eötvös Loránd Research Network, Budapest, Hungary

Péter Paulovics Juhász Gyula Faculty of Education, University of Szeged, Szeged, Hungary

Krisztina Sebe Department of Geology and Meteorology, University of Pécs, Pécs, Hungary

Ildikó Selmeczi Division of Geology, Supervisory Authority of Regulatory Affairs, Budapest, Hungary

József Stieber Stieber Environmental Ltd, Budapest, Hungary

Csaba Szinetár Department of Biology, Savaria University Centre, ELTE SEK, Eötvös Loránd University, Szombathely, Hungary

Gábor Szunyogh University of Óbuda, Budapest, Hungary

Kinga Székely Institute of Speleology, Budapest, Hungary

Tamás Telbisz Department of Physical Geography, Eötvös Loránd University, Budapest, Hungary

Ádám Tóth Department of Geology, ELTE Eötvös Loránd University, Budapest, Hungary

Márton Veress Department of Geography, Savaria University Centre, ELTE SEK Eötvös Lóránd University, Szombathely, Hungary

Márton Veress Department of Geography, Eötvös Lóránd University, Szombathely, Hungary

Szilárd Vetési-Foith University of Pécs, Pécs, Hungary

Magdolna Virág Department of Geology, Eötvös Loránd University, Budapest, Hungary

András Vojtkó Department of Botany, Eszterházy Károly University of Applied Sciences, Eger, Hungary

A General Description of Karsts in Hungary

Márton Veress

Abstract

The general characteristics of the karsts in Hungary are surveyed. The main properties of karst areas are described: their position, elevation, expansion, division, geomorphology and karst types.

Keywords

Carpathian Basin • Karst area of Hungary • Karst type

1 Introduction

Hungary (its area is 93011 km^2) is situated in Central Europe, in the Carpathian Basin, between the latitudes of 45° 48′ N and 48° 35′ N, in the temperate zone, in the catchment area of River Danube. Its crust developed from microplates with a different origin (Géczy 1973; Haas et al. 1999). Among them, the most important is the Alpaca Macrostructural Unit and the Tisza Macrostructural Unit (Fülöp 1989; Nagy 1971; Fodor and Csontos 1998; Haas 2015). The Alpaca Macrostructural Unit includes the Transdanubian Mountains and a part of the North Hungarian Mountains, in the south it originates from the environment of the Southern Alp, and from the southern margin of the Tethys sea arm. The Tisza Macrostructural Unit which originates from the northern margin of the Tethys sea arm involves the Great Hungarian Plain and a part of Transdanubia southwards from Lake Balaton (Laubscher 1971; Nagy 1971; Haas et al. 1999). Thus, the collision of the Tethys sea arm took place by the inversion of the positions of the two microplates. As a result of this, some karst areas being close to each other have a mainly Haas.

M. Veress (✉)
Department of Geography, Savaria University Centre, ELTE SEK Eötvös Lóránd University, Károlyi Gáspár tér 4, Szombathely, 9700, Hungary
e-mail: veress.marton@sek.elte.hu

During the collision of the oceanic arms of the Carpathians and the Dinarides, frictional heat was generated to the effect of the subducting oceanic crust which resulted in the development of mantle plume (Stegena et al. 1975). The flow of the mantle plume has thinned out the lower part of the crust since the Middle Miocene, which resulted in the isostatic subsidence of the crust, the formation of tensile structures, high heat flux and basalt volcanism (until the end of Pliocene) (Stegena et al. 1975; Haas 2015). The Carpathian Basin is a sialic basin according to plate tectonics (Stegena et al. 1975), within which the area of Hungary is separated into larger and smaller basins of Miocene age or younger than this (Kőrössy 1963, 1970). Between the basins, there are low-elevation mountains with a small expansion, which bear the karst areas of Hungary.

2 A General Description of the Karstification of the Karsts of Hungary

Karst occurs in the following regions (mountains) in Hungary: Aggtelek-Rudabánya Mountains, Bükk Region, Nézsa-Csővár Hills (North Hungarian Mountains), Bakony Region, Vértes-Velence Mountains, the Gerecse Region (Gerecse Hills), Buda Mountains (Buda Hills), Pilis (Hills) Mountains (Transdanubian Mountains) and the Mecsek Region (Western Mecsek), and southwards from it the Villány Mountains and the Beremend block.

The proportion of karst areas in the country's total area is not significant. Karst covers 1.45% of the area of the country, and their total expansion is 1350 km^2 (Jakucs 1977). Hungary's karsts are of small expansion, and the expansion of homogenous karst areas is between some 10 and 115 km^2 (Bükk Plateau). According to expansion, Leél-Őssy (1959) distinguished karst plateaus, karstic block plateaus and karst blocks. The surface elevation of karst areas is below 900 m, and their elevation as compared to the local base level of erosion is some 100 m at most. The small

M. Veress and Sz. Leél-Őssy (eds.), *Cave and Karst Systems of Hungary*,
Cave and Karst Systems of the World, https://doi.org/10.1007/978-3-030-92960-2_1

proportion of the karst in the total area of the country can be ascribed to the characteristics of lithogenesis and to the significant expansion of basin areas. However on its low karsts with small expansion, the karst features are diverse, but there are significant differences between various karst areas regarding the size and density of the features. According to Hevesi (1991a), the present karstification in the karst areas of Hungary can be traced back to the Upper Sarmatian.

Non-karstic rocks often occur in Hungarian karst areas and in their environs. According to Leél-Őssy (1959), there are karst areas which are independent karst micro regions (as can be seen below, there is also a microregion group among them; for example the Buda Hills), the area of which is homogenous karst. In addition, there are karst areas within other areas built up of non-karstic rocks (for example the Tapolca Basin).

The karst areas are significantly separated and dissected. This did not favour the development of expanded, homogenous karst systems. The nature of separation is manifested in the fact that the karst areas are surrounded and separated by basins and basin structures. The karst areas are dissected by stream valleys, the wedged-in non-karstic rocks and the blocky structure (Láng 1952). This latter character is also present even in karst areas with nappe structure. The blocky structure causes a dissected character because the blocks with a lower elevation being covered by non-karstic rocks favour the separation and limited local development of karstic systems within a karst area.

Solitary horsts with the small area may rise over their non-karstic environment. There is no significant surface karstification on them, at most subsoil karren occur. The Villány Mountains are a good example of this, where a rounded head of bed remnants (bed head karren) was formed below the soil by dissolution along the bedding plane.

Leél-Őssy (1959) classified Hungarian karst areas according to their altitude. He distinguished more elevated medium mountain karst (with an elevation of 600–950 m). Good examples are the Bükk Plateau and some areas of the Transdanubian Mountains (for example some parts of the Bakony Mountains), lower medium mountain karst (with an elevation of 300–600 m) and karst of basin floor (for example the karst in the environs of Tapolca in the Tapolca Basin).

Although there are karst areas with a fold and nappe structure (Fodor and Csontos 1998) (Bükk Mountains, Aggtelek Karst), they are basically of blocky structure. This is particularly the characteristic of the Transdanubian Mountains, but the Aggtelek Karst also developed into such a karst.

Karst areas are surrounded by the so-called deep karsts that are thickly covered and buried by impermeable beds (Alföldi and Lorberer 1976). There is water transmission (Alföldi and Lorberer 1976; Erőss 2010) and thus, dissolved material transmission between karst areas and the deep karsts of their environment. Partially water flow takes place from the karst into the sediments of the basins (Sárváry 1971; Mádlné Szőnyi and Tóth 2015), but karst areas receive water from their environment too. The heating up water of the basin sediments or surface streams, e.g., the water of the River Danube, is added to the karstwater of the Transdanubian Mountains (Papp 1942; Bodor et al. 2014).

The karstwater or waters of other origin are heated up because of low reciprocal gradient (its value is 20–22 m/°C), which can be explained by the upper mantle of high elevation and flowing upwards from a great depth it is mixed with the karstwater of karst areas. Thus, the hot water effect is significant in several karst areas (e.g. Buda Hills) in karstification.

In some karst areas of Hungary, volcanic rocks are also present mainly in the form of Pliocene basalt volcanism triggered by mantle plume (Lóczy 1913). Therefore some karstic terrains (e.g. Kab Mountain) were covered by basalt to various degrees.

Karstification took place several times on the karsts of Hungary during the history of geology. Its features can be mainly studied in the area of the Transdanubian Mountains and also in the karsts of the North Hungarian Mountains and on the karst of the Mecsek (see chapter "Paleokarst in Hungary"). According to Hevesi (1991a), its tropical karst landforms were destroyed due to the lack of adequate (consolidated rocks) caprock. The terrain of karst areas is mostly exhumed Late Cretaceous tropical karst peneplain or younger abrasion surface and pediment. Features that developed during older karstification may also affect recent karstification. Thus, above covered karstic mounds, where the superficial deposit is thin (Veress 2009, 2016) or in karstic depressions (Veress and Zentai 2009; Veress 2016). In the former case, karstification is favoured by the thin superficial deposit, while in the latter, the water accumulates in the depressions. Non-karstic rocks (volcanic rocks, gravel, loess etc.) survived in a significant distribution in the recent past, but partly today too because of the low surface elevation of its karst areas. However, the presence of non-karstic rocks was also contributed by the young evolution of some of them (e.g. loess, gravel). Therefore, karst areas are fluviokarsts or karsts that developed from fluviokarsts. However, among the karsts of Transdanubian type and of Aggtelek type, only those of Aggtelek type (see below) can be regarded as typical fluviokarst. The karst areas of Transdanubian type are specific fluviokarsts since there are no ponors with blind valleys (except Kab Mountain) in their areas. Among karst types, soil-covered karsts are of the greatest expansion, but the distribution of concealed karsts is also significant (Bakony Mountains and Mecsek Mountains, Veress 2010). Buried karst also occurs here (some blocks of the Bakony Mountains and the southern part of Aggtelek karst). Bare karst is absent except in some larger and smaller patches (walls of gorges, doline slopes). In case of adequate slope conditions, mixed autogenic-allogenic karsts (Aggtelek Karst, Bükk Mountains, Kab Mountain) also developed at non-karstic rock intercalations and caprocks.

The karstic rocks of karst areas are often vertically separated from each other by non-karstic intercalations too (e.g. the Bakony Mountains). This favours the development of local flow systems and local cavity formation.

The climate of karst areas slightly differs because of the short distances between them, but it is different to a larger and smaller extent from the climate of their environment as a result of their larger elevation. According to the Trewartha system, the climate of Hungary belongs to the continental climate subtype (Mezősi 2011, 2016). According to the Köppen climate classification, the northern part of the Transdanubia (Transdanubian Mountains) belongs to the uniformly warm-moderate climate with long and cool summers (fb), The North Hungarian Mountains and thus their karsts have a uniformly wet-boreal climate with long summer periods (Dfb), and the southern part of Transdanubia with the Mecsek Mountains has a uniformly wet, warm-temperate climate with long and hot summers (Cfa) (Bartoly and Pongrácz 2013). The annual average temperature of the karst areas is 8–11 °C, but in the Bükk Mountains it is below 8 °C (Mezősi 2011, 2016). However, during the summer months the temperature may also be below 0 °C in the dolines of the Bükk plateau at night (Kevei Bárány 2018). The amount of precipitation is 550–800 mm in the karst areas, but this value may be above 800 mm in the northern part of the Bakony Mountains and in the middle part of the Bükk Mountains (Mezősi 2011, 2016). The vegetation in karst areas are oak, hornbeam-oak and beech (Zólyomi 1989); their soil is brown forest soil and its black variety, the rendzina soil (Bulla 1964).

Karst areas are cultural landscapes where human activity has been present for a long time. For example, in the Bakony Region this is proved by graves and fortresses from the Bronz Age (Éri et al. 1969, Dax et al. 1972). However, the original forests of some karst areas were preserved for a long time. Thus, during the Hungarian conquest, there was a continuous forest cover in the Hungarian mountains (Bartha and Orbán 1995). The first significant deforestation took place during the times of Ottoman conquests because of castle buildings (Medzihradszky 1996). In the Bakony Region, the disintegration and shrinking of untouched forests can be connected to the glasshouses. The operation of glasshouses began in about 1715 in the mountains and lasted until 1859 with various sites (Éri 1966). The denudation of the superficial deposit may have become more intensive as a result of deforestation and later agriculture (from the beginning and more intensively the middle of the twentieth century). This may have resulted in the increased infilling and transformation of certain depressions, for example in Bakony Karst (Veress 1987).

A karstification of larger and smaller degree occurs in all macro regions of the country (the Great Hungarian Plain, the Little Hungarian Plain, the Transdanubian Hills, Alpokalja, North Hungarian Mountains, Transdanubian Mountains Fig. 1), except the two plain areas. According to Jakucs (1977), the karst areas are the following (Fig. 2):

- The karst landscapes of the Transdanubian Mountains are the Keszthely Mountains, the Balaton Uplands, the Tapolca Basin with its Sarmatian limestone terrain, the Northern Bakony Mountains and the Southern Bakony Mountains, the Vértes Mountains, the Gerecse Region (with the Zsámbék Basin), the Pilis Mountains (with the Pomáz calcareous tufa plateau) and the Buda Hills (with the Tétény Plateau and the Pest Plateau with its Lajta Limestone patches).
- The karstic inselbergs of Transdanubia such as the Lajta limestone karst of the shore of Lake Fertő (Alpokalja), the karst of the Mecsek Mountains, the karst of the Villány Mountains and the karst block of Beremend (the latter are the inselbergs of the Transdanubian Hills). Among them, in the area of the shore of Lake Fertő, the depressions are scars of quarrying. Only some cavities have depressions, which developed by collapse; thus, these areas are not dealt with here. The lack of karst features is also characteristic of the Villány Mountains as well as Szokolya Basin and Törökmező; thus, they are not described in other chapters of the book either.
- The karsts of the North-Hungarian Mountain Region, such as the South-Börzsöny Lajta limestone karst (Szokolya Basin, Törökmező), the limestone blocks of Cserhát Region (Naszály, Romhány Mountain, Csővár Mountain), the karsts of the Bükk Mountains with the Uppony block, the Aggtelek Karst Region with the karsts of the Rudabánya block, the Szalonna block and the Szendrő block.

The Kőszeg Mountains (Alpokalja) can also be mentioned where there are larger and smaller solution features in the greenschist with high calcareous content and in calcareous phyllite on the cliff walls and larger and smaller cavities also occur (Veress et al. 2015).

Among the listed areas, the karstification of the Aggtelek Karst (Aggtelek-Rudabánya Mountains), the Bükk Mountains (Bükk Region), Northern Bakony and Southern Bakony (the Bakony Region), and the Mecsek Mountains (Western Mecsek) is more significant as compared to the other areas. In brackets the presently used names of regions

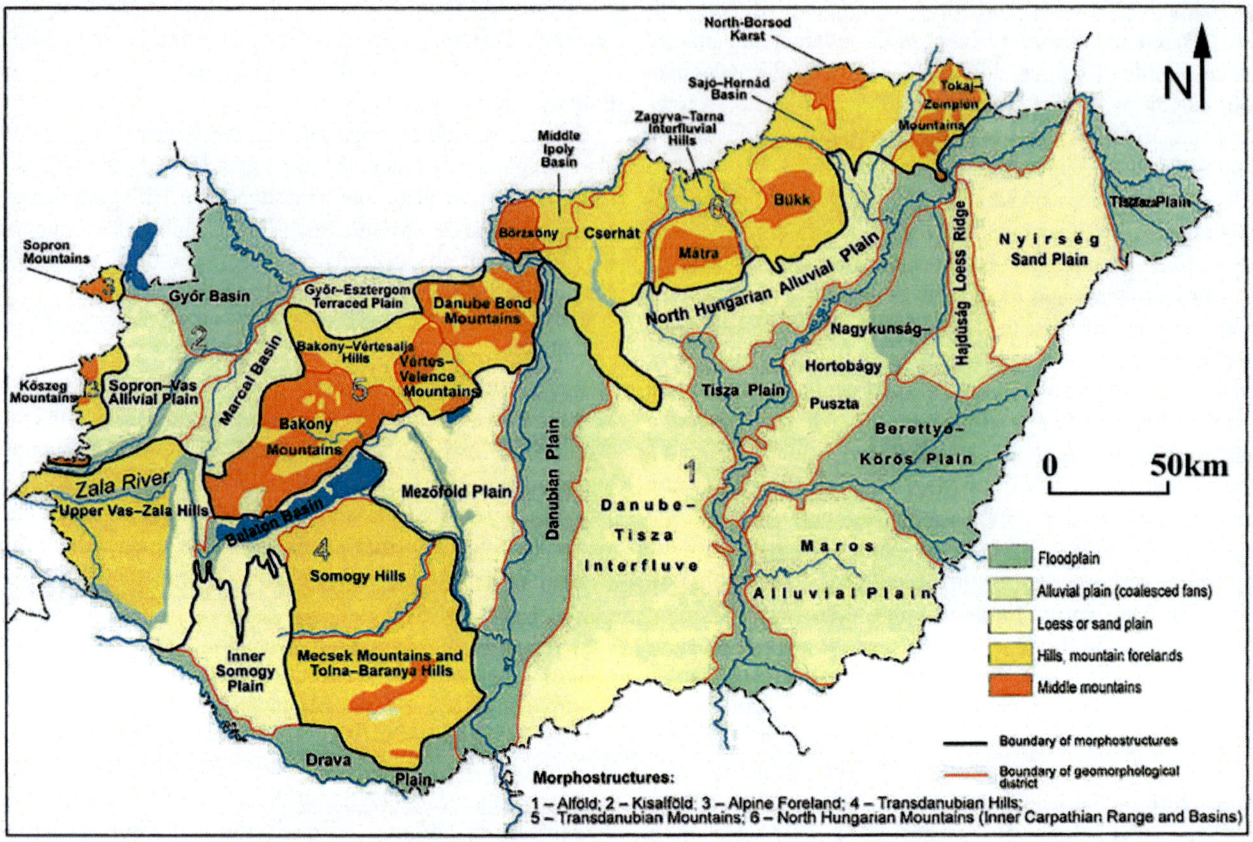

Fig. 1 Major geomorphological districts of Hungary (Lóczy 2015)

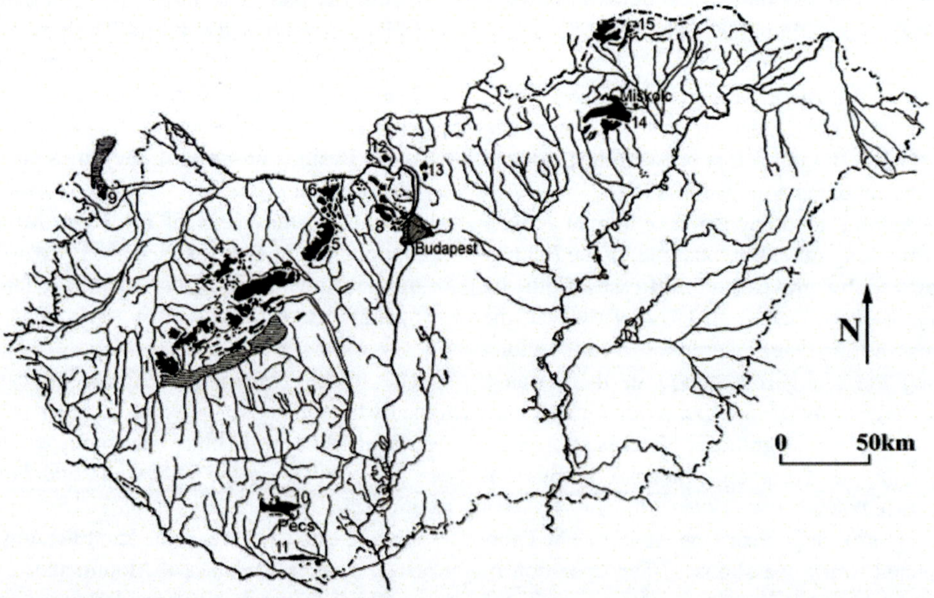

Fig. 2 Karst areas of Hungary (Jakucs 1977). Legend: 1. Keszthely Mountains, 2. Balaton Uplands with Tapolca Basin, 3. Southern Bakony, 4. Northern Bakony, 5. Vértes Mountains, 6. Gerecse Region (with the Zsámbék Basin), 7. Pilis Mountains (with the Pomáz calcareous tufa plateau), 8. Buda Mountains (with the Tétény Plateau), 9. karst near Lake Fertő, 10. Mecsek Mountains, 11. Villány Mountains and the Beremend karst block, 12. South-Börzsöny Lajta Limestone karst, 13. limestone blocks of the Cserhát Region, 14. Bükk Mountains (with the Uppony block), 15. Aggtelek Karst Region (with the limestone areas of Rudabánya block, Szalonna block and the Szendrő block)

and a more exact region classification can be found. If we estimate the proportion of surface karstification and subsurface karstification as compared to each other among the karst areas, we can see that this proportion is similar to the case of the Aggtelek-Rudabánya Mountains, the Bükk Region and the Bakony Region. In the Mecsek Mountains, surface karst is dominant, while in the Buda Hills and on the karst block of Beremend, subsurface karst is significant. The degree of karstification of the above-mentioned areas as compared to each other can also be estimated. Thus, based on the estimated value of calcareous tufa formed at karst springs, the most karstified mountains are the Bükk Region, which are followed by the Aggtelek-Rudabánya Mountains (though, the larger amount of calcareous tufa of the former can be attributed to the larger mass of the mountains), the Western Mecsek and the Bakony Region. The Gerecse Region and the Buda Mountains also have a significant amount of calcareous tufa. However, their calcareous tufa is older and are not connected or only partially connected to the present karstification of the mountains (Schweitzer 2002; Leél-Őssy 2012).

Leél-Őssy (1959, 1960) and, then elaborating his theory, Jakucs () put Hungarian karsts into two categories from an evolutionary point of view taking their karstification characteristics (karst features) into consideration: the Transdanubian type and the Aggtelek type. The former includes the karst areas of the Transdanubian Mountains, while the karsts of the North Hungarian Mountains (Aggtelek Karst and Bükk Mountains) belong to the latter. Although the karst of the Mecsek Mountains is situated in Transdanubia, it was regarded as a transitional and classified as a member of the Aggtelek type, while although the limestone blocks of the Cserhát Region are located in the North Hungarian Mountains they were put into the Transdanubian type. He drew attention to the fact that although there is a short distance between the karsts of the two types, they are significantly different regarding their karstification characteristics. According to him, there is a lack of karren, dolines (solution dolines), ponors, erosion caves in the area of the Transdanubian type, but the crumbling (physical weathering) of dolomite triggered by the effect of hydrothermal water is characteristic. In addition, caves of hydrothermal origin or those being exposed to hydrothermal effect are widespread. On the contrary, on the karsts of Aggtelek type, there are karren, well-developed (solution) dolines, doline systems, uvalas, ponors with blind valleys and erosion caves. (We also mention that spring caves are also characteristic in addition to erosion caves on the latter, which cannot be found on karsts of the Transdanubian type or their occurrence is only subordinate.)

Taking into consideration the karst built up of non-karstic and karstic rocks and also non-karstic rock covering the karst, Hevesi (1986a) classified karst and developed a nomenclature for the different types. Elaborating Jakucs's interpretation on the landforms of the karsts of the Aggtelek and the Transdanubian character (Hevesi 1991a), and then adopting his classification he classified the karst types occurring on the karst areas of Aggtelek and Transdanubia (Hevesi 1991b).

If we consider the characteristics of Aggtelek karst and Transdanubian karst, we can state that mixed autogenic-allogenic karsts are widespread on the Aggtelek karsts, while concealed karsts with subsidence dolines are specific of the Transdanubian karsts, which are mainly of suffosion origin. On Transdanubian karsts there are not only concealed karsts, but also soil-covered karst and mixed autogenic-allogenic karst. Mixed autogenic-allogenic karsts can only be found in the Bakony Mountains (Kab Mountain) and (according to Hevesi 1991b) in the Pilis Mountains. In addition, buried karst occurs on Kab Mountain (Veress and Unger 2015), on Fekete Mountain (Southern Bakony), and in the Gerecse and the Pilis Mountains in patches. Thus, on karsts of the Transdanubian type, in the above-mentioned areas, ponors with blind valleys and caprock dolines also occur (the latter for example on basalt, sandstone and marl).

Concealed karst also occurs on Aggtelek karst subordinately in larger karst depressions for example in dolines filled with superficial deposit or in ponors (Veress and Zentai 2009; Veress 2016). Although soil-covered karst, as it has already been mentioned, also occurs on the karsts of the Transdanubian type, there is an almost complete lack of solution dolines. The karst of the Mecsek Mountains is a transition between the two types because concealed karst with landforms specific of concealed karst developed on the terrain dissected by karst landforms which are characteristic of the karst of Aggtelek type. Considering the karsts of the two types, further distinctions can be described. Thus, they are not only different regarding the quality (type) of the features, but also the quantitative characteristics of the forms. On the karsts of the Aggtelek type, the density of depressions is probably larger by order(s) of magnitude than on the karsts of the Transdanubian type. The size of the features is also larger in case of the karsts of Aggtelek type. While disregarding DSDs (see chapter "The Surface Morphology of Karsts in Hungary"), the diameter of the features is some metres or maybe some 10 m on the karsts of the Transdanubian type, depressions with a diameter exceeding 100 m often occur on karsts of the Aggtelek type. If we compare the complexity of landforms, we can state that the depressions of the karsts of Aggtelek type are more complex (karst features are built up of several partial features and larger features contain smaller forms). The size of caves and thus the quantity of dissolved material is greater on the karsts of Aggtelek type. Here, disregarding the hydrothermal caves of the Buda Hills and the cave system of Tapolca, there are several caves the size of which exceeds the size of the caves of the karsts of Transdanubian type. However, present cave explorations indicate that on karsts of the Transdanubian type, the length, and mainly the depth, of caves that developed in the vadose zone does not lag behind the karsts of the Aggtelek type. However, undoubtedly, influent caves of erosional origin and through caves are specific of karsts of the Aggtelek type. However, the number and density of caves are not smaller on all karsts of Transdanubian type. Thus, for example in the Bakony Region, more than 700 caves occur though with those of non-karstic origin together (Eszterhás 2013). This can be attributed to the fact that larger and smaller phreatic cavity systems were dissected into several short cave-remains by downcutting streams (Veress 2000a, b, 2016).

However, there is also a difference in the extent to which the karst areas are dissected. While the karsts of Aggtelek type are less dissected and thus, they constitute plateaus of great expansion, the karsts of Transdanubian type are mountain tops of small expansion (Leél-Őssy 1959; Hevesi 1991a). Therefore, on the latter, despite the former existing cover, no stream and valley network could develop on the blocks with small expansion either (Hevesi 1991a).

According to Jakucs (1977), the reason for the differences regarding the landforms is that the karsts of Aggtelek type have been continually karstified since the end of the Cretaceous period, while karstification was interrupted on the karsts of Transdanubian type as a result of their burial and this process could only continue after they became exposed. However, the karsts of Aggtelek type became also covered several times at least partially (Láng 1955; Vass et al. 1994; Hevesi 1978, 1991a; Sásdi 1990; Császár 1997; Gyuricza and Sásdi 2009) thus, and their continuous karstification at least in their whole area has not been able to take place since the end of the Cretaceous. However, it is indisputable that the covering of the karsts of Transdanubian type (terrains built up of Mesozoic rocks) was of larger extent and duration and also the fact that their terrains built up of Mesozoic rocks became covered with consolidated rocks (limestone, sandstone) to a greater degree than the karsts of the Aggtelek type. It is also unquestionable that the denudation of their superficial deposits is of a smaller degree even today than that of the karsts of Aggtelek type.

However, the covering of various degrees of the karsts of the two types is hardly sufficient to interpret morphological differences. On the karsts of the Transdanubian type, there is a lack of drawdown dolines disregarding two exceptions. These two areas are the Tapolca karst and the Devecser karst (Móga et al. 2011; Veress and Vetési-Foith 2019a, b). But just these areas are built up of young limestones (Tapolca karst Sarmatian limestone, Devecser karst Lajta limestone). Thus, solution dolines can not only develop on older areas built up of limestones, that is on old surfaces which have been karstifying for a long time, but also on young surfaces too. However, morphological differences cannot be explained by the young uplift mentioned by Láng (1958) because it would follow that there is no solution dolines on terrain sections of low elevation. However, as it has already been mentioned, solution dolines occur on such surface (the karst around Tapolca in the Bakony Mountains) where the altitude of the surface is low (120–180 m).

The lack of solution dolines was also contributed by the fact that the inclination of the surface is large on the soil-covered karsts of the more elevated blocks with an incomplete superficial deposit of the Transdanubian Mountains. It was established that 85% of the dolines develop on terrains with a dip smaller than 12° in the Miroč Mountains in Serbia (Telbisz et al. 2007). However, on lower surfaces with a small inclination where there is an opportunity for doline development, the permeable superficial deposit (loess) survived. Although there are shafts below this permeable superficial deposit (Veress 2000a, 2018) they occur one by one probably due to the effect of the cover (Veress 2000a). Since the loess is washed in into the karst, the epikarst is probably poorly developed and it is different from the epikarst on typical soil-covered karsts. This does not favour the development of solution dolines. This fact is supported by the experience that on the gypsum karsts of Italy drawdown (solution) dolines are not characteristic (Waele et al. 2017) which is explained by the lack or the limited development of the epikarst (Waele et al. 2017; Ferrarese et al. 2002; Sauro 2003).

The covering of various types on the karsts of Transdanubia and Aggtelek may give an explanation for the differences in older karst landforms (paleokarst). Since consolidated caprocks are more widespread in the Transdanubian Mountains, the features of older karstification(s) could have been preserved. Some of the paleokarstic features (Cretaceous-Eocene) on karsts of Aggtelek type may have been completely or partially destroyed (Hevesi 1991a), probably during the denudation of loose superficial deposit (which is thus more easily destroyable). However, the depressions at (and around) which no denudation took place, on the contrary they become filled in or even covered, could have survived such as the Dász doline in Aggtelek karst or the depressions bearing the ponors of Baradla cave (Veress 2017).

The various distribution of concealed karst and its features on the karsts of the two types can be attributed to the presence of loose superficial deposits (mainly loess) of different degrees which can be ascribed to the different elevation of the karst surfaces of low inclination of the karsts of Aggtelek and Transdanubian type according to us. (Unconsolidated, cohesive superficial deposit being characteristic of concealed karst only occurs subordinately.) The development of loess has an elevation limit which is 400 m (Mezősi 2011), but loess can even be found over 400 m on the Tés

Plateau (Bakony Region). However, while for example the most karstic terrain can be found above 450 m (Kis Plateau) in the Bükk Mountains, and the Nagy-Plateau is located above 650 mm, the majority of the blocks of the Transdanubian Mountains with a small inclination are situated below 500 m (see chapter "The Surface Morphology of Karsts in Hungary"). Thus, a significant proportion of the blocks of the Transdanubian Mountains is a surface of a low inclination which belongs to the zone of loess development. Loess could have survived on these especially if the block is of large expansion and because streams are absent and the mounds of the bedrock hinder denudation, ensuring the development of concealed karst and its features. The survival of loess (and in general of the superficial deposit) was also contributed by the fact that the uplift of the Bakony Mountains, which is the most peculiar concealed karst area of the Transdanubian Mountains, is young (Láng 1958); therefore there has not been enough time for the denudation of the superficial deposit. Here, we mention that since concealed karst features mainly developed on loess, loess formation affects not only their distribution but their age too. For example, the loess is of Würmian age in the Bakony Region (Pécsi 1982; Sás 1987). Older features could only be formed during the reworking of the older superficial deposit (gravel cover) or above bedrock passages that had developed during an earlier karstification process. On karsts of Aggtelek type, concealed karst is more subordinate probably because on more elevated terrains of the Bükk Mountains there are only patches of slope loess (Hevesi 1978). However, loess (Tóth 1986) and its mixed variety is mentioned in karstic depressions (Veress and Zentai 2009). On Aggtelek karst it is subordinate and red earth is more widespread (Zámbó 1970) which is less permeable and this does not favour concealed karstification. The close relation between the survival of loess and the development of surface karst features is proved by the features of Mecsek karst that developed on abrasional terraces (Lovász 1981) and which belongs to Aggtelek karsts (Hevesi 1991b). The low elevation levels of this karst (250–380 m and 450–470 m) favoured loess development, while the hardly dissected nature of the surface favoured the survival of loess. Regarding their areas, the dolines occur on surfaces with an inclination of 2°–7° in the greatest proportion (Lippmann et al. 2008). The features of concealed karst, the subsidence dolines occur on terrains dissected by older solution dolines (Szabó 1968) and covered with loess in significant distribution and in the greatest density as regards Hungary.

The Aggtelek and Transdanubian karsts are not individual karst types since considering all their characteristics other non-Hungarian karst areas cannot be classified as these karst types. However, taking some of their characteristics into consideration, these, especially the karst areas of the Transdanubian type, can also be put into several karst types (karst of mountains of medium height, soil-covered karst, concealed karst, autogenic karst etc.) They cannot be ranked among evolutionary types as only some of their characteristics can be explained by different evolutionary events. Thus, it is better to use the 'of…character' expression used by Hevesi (1991a, b) instead of the term 'type'. However, creating these two larger units, Hungarian karst areas can be compared and distinguished according to their features. In our view, the differences of Transdanubian and Aggtelek karsts cannot be explained by only one cause. The presence or absence of some characteristics (for example the presence or lack of mixed autogenic-allogenic karst and the features related to that) can be traced back to evolutionary characteristics. Others, like the features of concealed karst or soil-covered karst deduced from the characteristics of the epikarst. The reason for the latter distinction is that the constituting blocks (plateaus) are covered to various extents, which can be explained by their different elevation, expansion, and by the inclination of the surface. These differences can be explained by tectonic reasons. The differences regarding hydrothermal nature can be interpreted by the difference in the distance between the karst and the magma below the surface. There may be several reasons for the lack of erosional caves on Transdanubian karsts (thus, the small area of blocks and the small non-karstic catchment area, the young erosional denudation of short duration, the nature of the superficial deposit, the erosional denudation of karstic cavities in the gorges etc.).

3 The Parts, Morphology and Karst Types of Hungarian Karst Areas

At the division of karst areas, the works of 'the Inventory of microregions in Hungary' (Marosi and Somogyi 1990) and 'The Physical Geography of Hungary' (Bulla 1964; Mezősi 2011, 2016) and some articles (Ádám 1983, 1984) were taken into consideration. During the categorization of the landscapes macro regions, meso regions, micro region groups and micro regions are distinguished. The description of the distribution of karst areas follows this categorization.

Hungarian karst areas occur in three macro regions: North Hungarian Mountains, Transdanubian Mountains and Transdanubian Hills.

3.1 The North Hungarian Mountains and Their Karsts

The North Hungarian Mountains are a macro region situated in the north-eastern part of Hungary between the River Ipoly and the River Bodrog, north from the Great Hungarian Plain, but on their southwestern part they reach beyond the River

Danube (Visegrád Mountains). The majority of their area is constituted by mountains which belong to the inner volcanic zone of the Carpathian Mountains (Visegrád Mountains, Börzsöny, Cserhát Region, Mátra Region, Tokaj-Zemplén Mountains). In their eastern part, two karst areas are wedged in between the volcanic mountains, the Bükk Region and the Aggtelek-Rudabánya Mountains (Fig. 3). In the west, some smaller karstic blocks occur in the Cserhát Region.

The Aggtelek-Rudabánya Mountains (Aggtelek-Rudabánya Hills)

They are a meso region surrounded by the valleys of Bódva, Sajó and Csermoslya, and they are separated into two micro region groups: the Aggtelek karst and the Rudabánya-Szalonna Mountains. Westwards from the River Bodva the Aggtelek Karst and eastwards from it the Rudabánya Mountains and the Szalonna Karst (Estramos or Szalonna Mountains) are situated. The latter is called Szalonna Plateau or Estramos Hills. However, according to the latest region classifications, the Szalonna Karst does not belong to the Aggtelek-Rudabánya Mountains. (This meso region, the parts of which called Szilice Plateau and Kecső Plateau reach into the area of Slovakia, is also called Gömör Torna Karst and Northern Borsod Karst.) The Aggtelek Karst micro region group is separated into two micro regions: the Aggtelek Mountains and the Alsó-hegy Plateau. The Aggtelek Mountains is surrounded by the valleys of Sajó, Bódva and Ménes, the area of which is 135 km^2 and its elevation is 350–500 m, but some mounds may be more elevated than 500 m (Nagyoldal 604 m).

The Aggtelek Mountains are separated into plateaus which are the following (Fig. 4): in the north, the Haragistya Plateau, the Jósvafő Plateau, the Nagyoldal Plateau and the Szinpetri Plateau, southwards from these the Aggtelek Plateau and the Galyaság. The latter is separated into four plateaus or parts by Móga (2002a). These are the following from W to E: the western part of Galyaság, the plateau around Mount Pitics, the Teresztenye Plateau and the eastern part of Galyaság. The plateaus are Late Cretaceous tropical karst peneplains (Mezősi 1984), while according to other opinion (Zámbó 1998) pediment parts elevated to various elevations. The area of Alsó-hegy Plateau is 50 km^2, its elevation is 400–500 m, its western part is the Derenk Plateau and its eastern part is the Szilasi Plateau, but the Dusa Plateau (Páska-Bükk) also belongs to this area (Móga 2002b).

In the area of Aggtelek Karst, the plateaus are separated from each other by valleys having permanent streams. Mass movements occur frequently in the valleys. On the plateaus there are epigenetic, erosion valleys and in some places grabens with creeks and gullies occur in the thick cover. Its plateaus are separated into two levels: a more elevated level with mounds and a lower, less dissected level. According to its development, its karst is renewed allogenic covered karst (Veress 2016). This type develops from cryptokarst in a way that the karst becomes separated from its non-karstic sediment sources or from its non-karstic environment in an

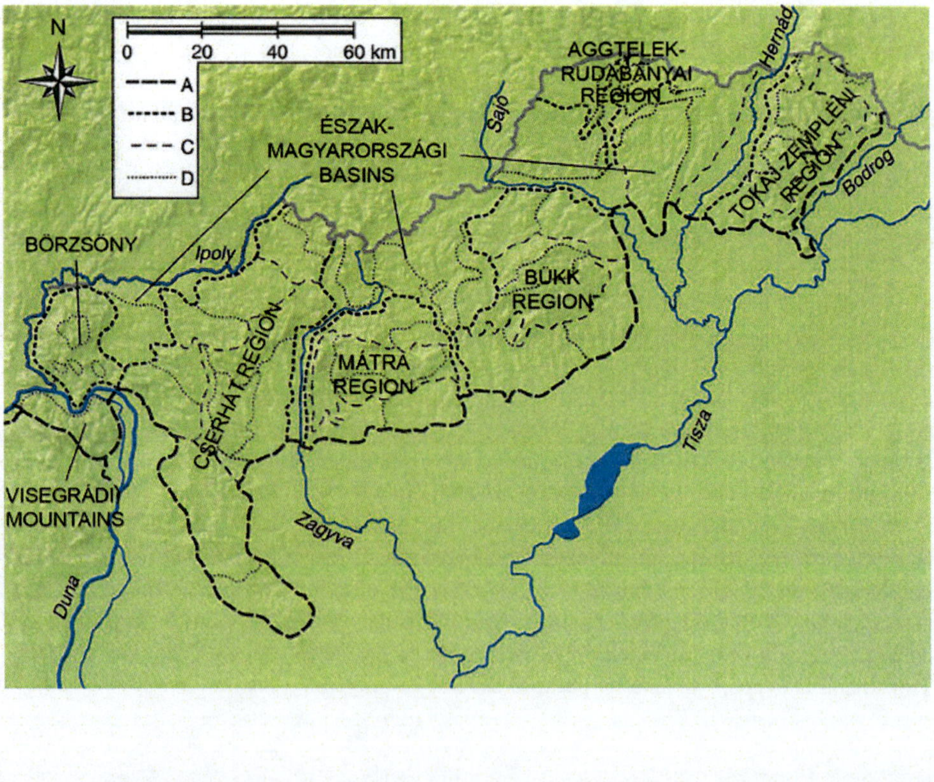

Fig. 3 Landscape geographical division of the North Hungarian Mountains (Mezősi 2011). Legend: A. border of macro region, B. border of meso region, C. border of micro region group, D. border of micro region

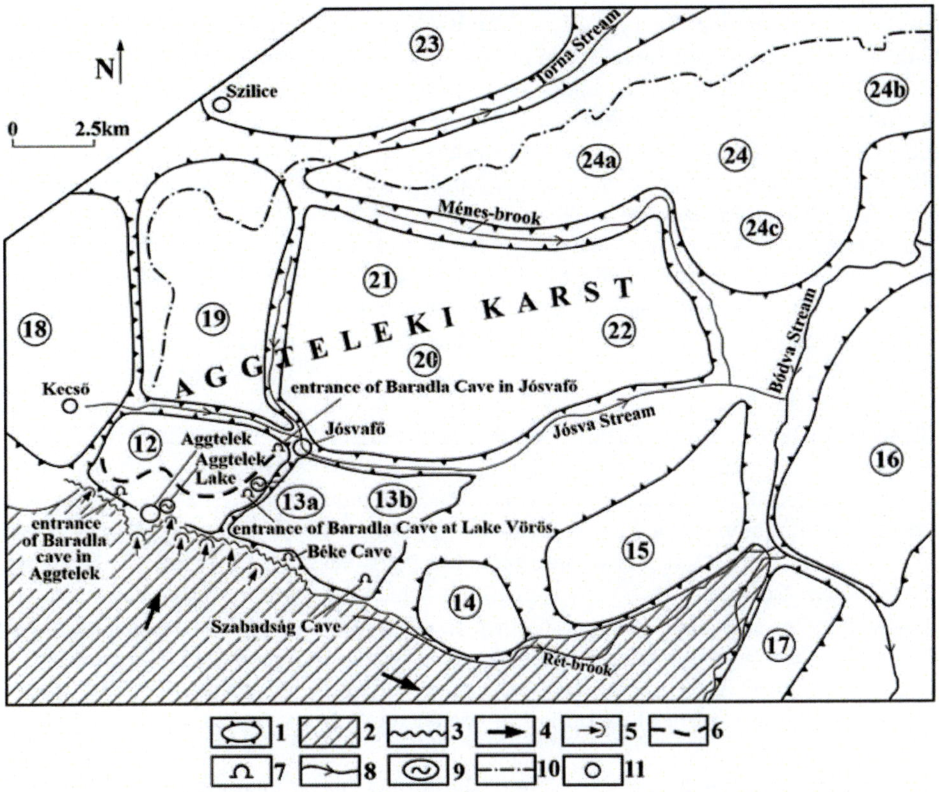

Fig. 4 Aggtelek-Rudabánya Mountains (Veress and Unger 2015). Legend: 1. karst plateau, 2. buried karst, 3. rock boundary, 4. dip of buried karst, 5. ponor, 6. Baradla Cave, 7. cave entrance, 8. stream, 9. doline lake, 10. national border, 11. settlement, 12. Aggtelek Plateau, 13a. Western Galyaság, 13b. Mount Pitics, 14. Teresztenye Plateau, 15. Eastern Galyaság, 16. Szalonna Plateau (Estramos), 17. Rudabánya Mountains, 18. Kecső Plateau, 19. Haragistya, 20. Jósvafő Plateau, 21. Nagyoldal, 22. Szinpetri Plateau, 23. Szilice Plateau, 24. Alsó-hegy Plateau (24a. Derenk Plateau, 24b. Szilas Plateau, 24c. Dusa Plateau)

erosional or tectonical way. Subsequently, the karstification is renewed and modified during the partial denudation of the superficial deposit. The Aggtelek Karst became separated from its northern, non-karstic environment by the development of the Torna valley and has been separated into three zones (Veress 2014). The northern zone is separated into two subzones: an autogenic karst, which the Alsó-hegy Plateau belongs to. Here, soil-covered karst is dominant, but concealed karst and mixed autogenic-allogenic karst also occur locally. The southern part of this zone can be found southwards between the Ménes valley and Jósva-valley (the area of Haragistya plateau and Szinpetri plateau), on its autogenic karst plateaus there is soil-covered karst, but concealed karst is also more expanded. The middle zone reaches from Jósva valley to the valley of the Rét Stream. Its western part was renewed and was transformed into a mixed autogenic-allogenic karst. On its autogenic karst plateaus (Aggtelek Plateau, Mount Pitics, Teresztenye Plateau, the eastern part of Galyaság) soil-covered karst and concealed karst occur. Southwards from the Rét Stream buried karst can be found. Karst features being characteristic of this karst type are widespread on the above mentioned soil-covered karsts. The area of Rudabánya Mountain (micro region) is 20 km^2, and its elevation is 158–376 m. It is buried karst and concealed karst dissected by gullies. The Szalonna Mountain (micro region) is an exhumed horst, its area is 50 km^2 and its elevation is 160–520 m. Its karst is concealed karst (Hevesi 1991b, soil-covered karst), but its karst landforms are not significant. Its subsurface karst shows warm water effects.

The Bükk Region

The Bükk Region (meso region) is surrounded by the River Tarna in the west, by the River Sajó in the north and east and by the Great Hungarian Plain in the south (Borsodi-Mezőség). Its area is 1795 km^2. The elevation of the surface is the largest here among Hungarian karsts. Eleven mounds are situated above an elevation of 950 m, but none of them reaches 1000 m.

The area of the mountains was denuded into a tropical peneplain (Pinczés 1968), during the Lower Eocene (Hevesi 1978), from the Upper Cretaceous-Middle Cretaceous to the Eocene (Pinczés 1980), and between the Upper Cretaceous and the Miocene (Tóth and Fejes 1984). Taking into consideration the elevation distribution of spring caves, Tóth and Fejes (1984) described two denudation levels on the Bükk Plateau, the level consuming the tropical peneplain and a level where the tropical peneplain develops (an older and a younger one of Pliocene age). The Bükk Region is separated into three micro region groups: the Központi Bükk (Central Bükk), the Bükkalja and the Bükklába (Fig. 5). The micro regions of the Central Bükk are the Bükk Plateau, the

Northern Bükk and the Southern Bükk. The Bükk Plateau with an expansion of more than 20 km in NE-SW direction and with an area of 115 km^2 is separated into two areas by the valley of Garadna Stream downcutting into non-karstic rock. One area is the Small Plateau with an elevation of 450–500 m and situated north from the Garadna Stream and the other is the Great Plateau with an elevation of 650–850 m and situated in the south. The separation of the surface into two levels is particularly characteristic of the Great Plateau: the upper, more elevated which is not uniform, but is separated into mounds and the lower which is slightly dissected by epigenetic valleys.

The Northern Bükk (its area is 100 km^2) is a pediment dissected by gullies and interfluves bordering the Small Plateau. Here, a smaller karstic area (Kemesnye Mountain) can be mentioned. The Southern Bükk is a micro region situated south from the Great Plateau, its area is 335 km^2 and its elevation is 210–790 m. It is separated into two parts: the Southwestern Bükk and the Southeastern Bükk, which are separated by the Hór Valley from each other. In the area of the Southeastern Bükk the karst is less widespread, while it is more widespread in the area of the Southeastern Bükk (in the latter area the superficial expansion of karstic rocks is larger).

Bükkalja is separated into the Tárkány Basin, and the Eger Bükkalja and Miskolc Bükkalja micro regions. The Tárkány Basin (its area is 30 km^2) is a pediment dissected by erosion-derasion valleys and derasion valleys. No karstification occurs in its area (buried karst). The Eger Bükkalja (its area is 480 km^2) is also a pediment transformed in an erosion way (hilly region) without karstification. The Miskolc Bükkalja (its area is 240 km^2) is a pediment dissected by valleys and built up of non-karstic rocks.

The northern part of the mountains is the Bükklába which is separated into the Tardona Hills (its area is 260 km^2) and the Uppony Mountains (its area is 240 km^2). The Tardona Hills are a micro region which is a pediment dissected in an erosional way. No karstification occurs in its area. The Uppony Mountains is a micro region dissected by erosion valleys and karstification is present in its area.

The uplift of the Bükk Region took place by upwarping (Moldvay 1969). Its central part uplifted mostly and the earliest. Therefore, it is upwarping karst according to development (Veress 2000b) where karstification gradually decreases towards the margins and probably it is rejuvenated too. Its karst is most sharply developed on the Bükk Plateau (both surface and subsurface karst) where karst features are present at both levels, but they are particularly prevailing at the lower. Its Small Plateau is allogenic karst, autogenic karst, mixed autogenic-allogenic karst, with soil-covered karst on its autogenic karst sections (Hevesi 1991b). The Great Plateau is autogenic karst, mixed autogenic-allogenic karst, with soil-covered karst on its autogenic karst sections (Hevesi 1991b) and in patches with concealed karst (Veress

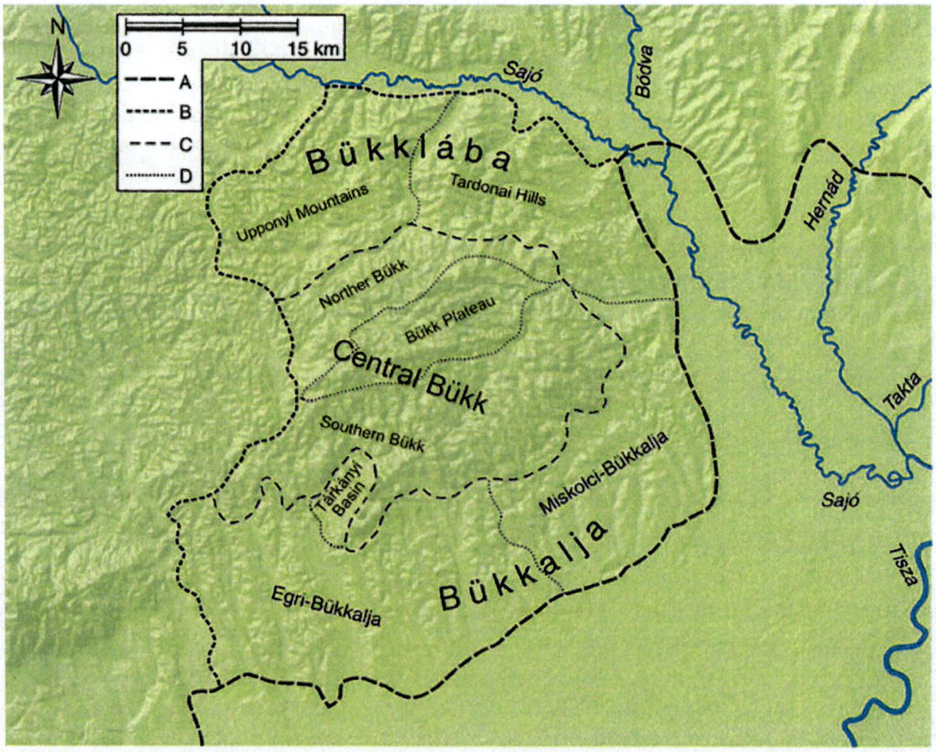

Fig. 5 Geographical regions of the Bükk Region (Mezősi 2011). Legend: A. border of macro region, B. border of meso region, C. border of micro region group, D. border of micro region

and Zentai 2009). In the area of the Southern Bükk, the limestone is disrupted by non-karstic intercalations. The non-karstic rocks prevail especially in the area of the Southwestern Bükk (intercalating and as a superficial deposit too). Consequently, the karstification of the Southeastern Bükk is more expanded. The Southeastern Bükk is autogenic karst, mixed autogenic-allogenic karst and buried karst to a small extent (Hevesi 1986b, 1991b), with soil-covered karst on its autogenic karst. The Southwestern Bükk is autogenic karst, mixed autogenic-allogenic karst and partially buried karst, with soil-covered karst on its autogenic karst sections (Hevesi 1986c, 1991b).

The Nézsa-Csővár Hills

They are a micro region belonging to the western part of the Cserhát Region (Western Cserhát), with an elevation of 155–442 m, and an area of 120 km^2, 15% of which is a block with horsts of the mountain range of medium height character. Its blocks are the Romhány Mountain, the Naszály Mountain and the Csővár Mountain. The area of the mountains is predominantly thus covered (crypto) by karst. Its surface karst features are not widespread, and its caves are more considerable. Its caves are affected by hydrothermal water (Ozoray and Láng 1957). Although these blocks do not belong to the Transdanubian Mountains geographically (the River Danube separates them from it), they do belong to it from a tectonic and historical geological point of view.

3.2 The Transdanubian Mountains and Their Karsts

The Transdanubian Mountains are a macro region, which is a block mountains with NE-SW direction, its expansion is 160 km in this direction and its area is 6810 km^2. The mountains are surrounded by the River Danube, the Great Hungarian Plain, the Little Hungarian Plain and the Transdanubian Hills. The mountains are separated into meso regions by faults with NW–SE direction. These are the Bakony Region, the Vértes-Velence Mountains and the Dunazug Mountains (Fig. 6). Their mountains are built up of blocks with different elevation and development which were formed by the dissection of the Late Cretaceous peneplain (Pécsi 1980, 1983).

The Bakony Region

The Bakony Region is the southwestern part of the Transdanubian Mountains. It is bordered by the Little Hungarian Plain in NW, by the Transdanubian Hills in W and SW, by

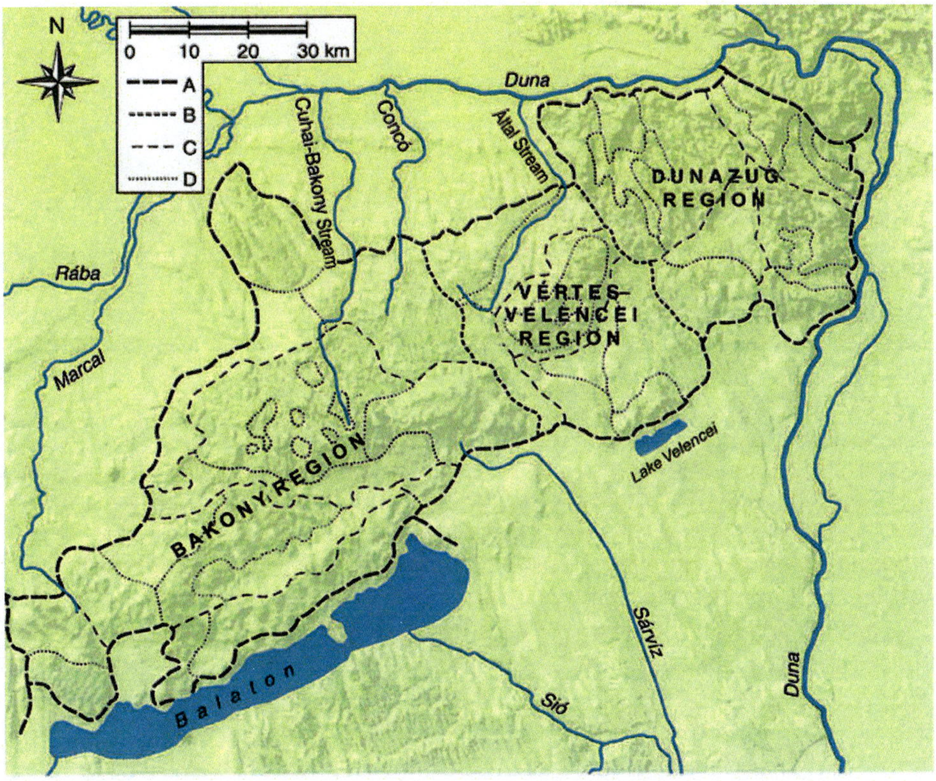

Fig. 6 Landscape geographical division of the Transdanubian Mountains (Mezősi 2011). Legend: A. border of macro region, B. border of meso region, C. border of micro region group, D. border of micro region

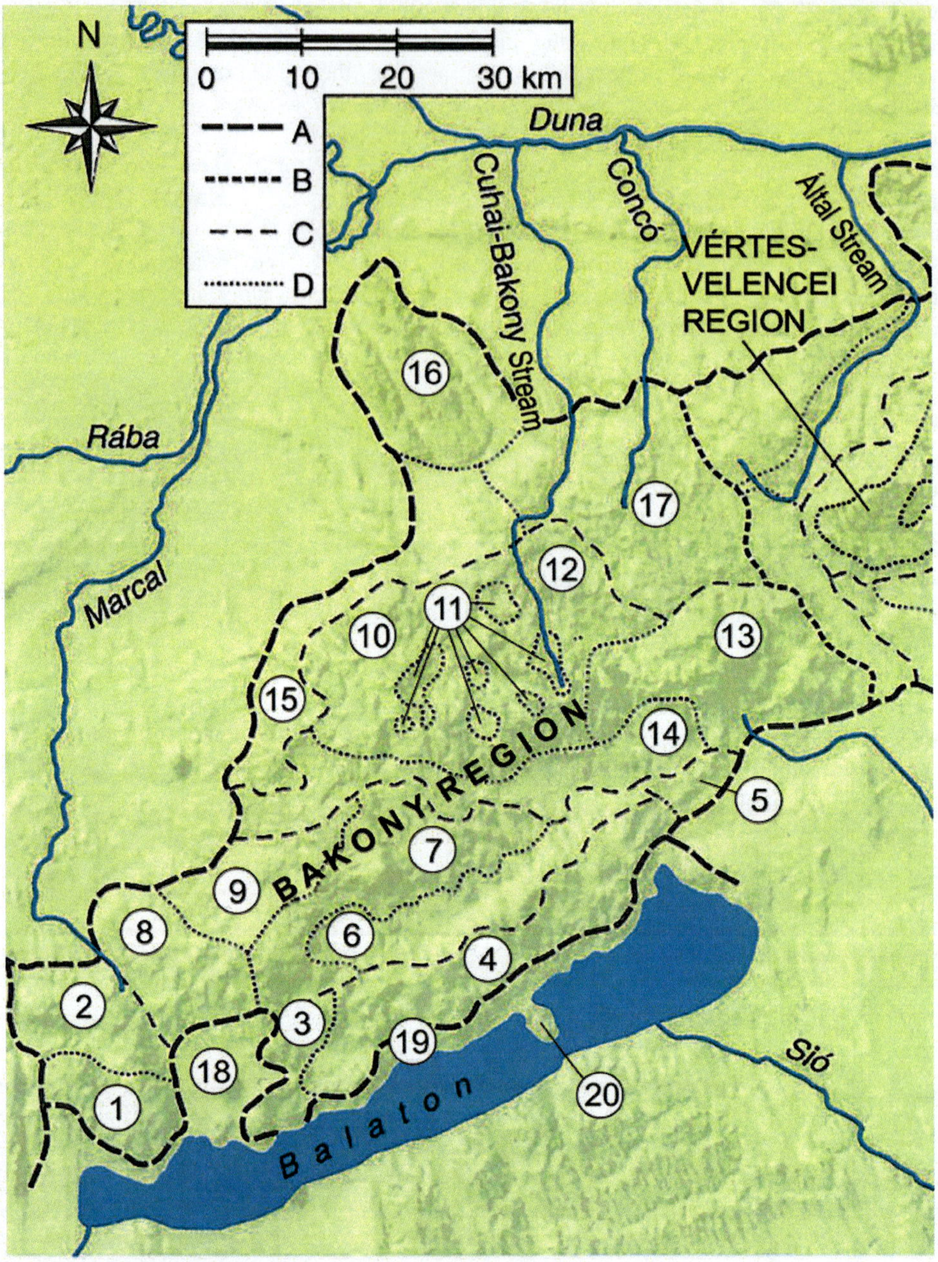

Fig. 7 Division of the Bakony Region (Mezősi 2011). Legend: A. border of macro region, B. border of meso region, C. border of micro region group, D. border of micro region, 1. Keszthely Plateau, 2. Tátika Group, 3. Badacsony-Gulács Group, 4. Balaton Uplands, 5. Vilonya Mountains, 6. Veszprém-Nagyvázsony Basin, 7. Mount Kab-Mount Agár Group, 8. Sümeg-Tapolca Ridge, 9. Devecser-Bakonyalja, 10. Öreg Bakony (Old Bakony), 11. Small basins of Bakony, 12. Sűrű Mountains Group, 13. Tés Plateau, 14. Veszprém-Devecser Graben, 15. Pápai-Bakonyalja, 16. Pannonhalma Hills, 17. Súri-Bakonyalja, 18. Tapolca Basin, 19. Balaton Riviera, 20. Tihany Peninsula

Lake Balaton, the micro regions of the Balaton Basin (Balaton-Riviera, the Tapolca Basin), by the Great Hungarian Plain (Mezőföld) in SE and by the Vértes Mountains (and the Mór Graben with a graben structure) in NE. The elevation of the Bakony Region is 150–700 m and its area is 4300 km^2. It is constituted by the Bakony Mountains (2200 km^2) and the surrounding regions with a lower elevation. It is separated into five micro region groups: Keszthely Mountains, Balaton Uplands, Northern Bakony, Southern Bakony and Bakonyalja (Fig. 7). The Northern Bakony and the Southern Bakony are separated from each other by the Séd Stream and the Veszprém-Devecser graben. The Northern Bakony is separated into micro regions called Old Bakony and Eastern Bakony by the Cuha Stream with NS direction.

The geomorphology of the mountains is constituted by horst surfaces with various elevation which are Late Cretaceous tropical peneplains (Pécsi 1980), by fault scarps, while at the margin of the mountains, in the basins there are abrasion platforms of Miocene age and pediments of Middle-Oligocene, Middle-Miocene, Pliocene and Pleistocene age (Pécsi 1963, 1980, 1983, 1991). In addition, mountains with basalt cover and buttes (Mezősi 2011, 2016), erosion valleys being inherited to a various degree also occur, on the loess cover, creeks and gullies (Veress 2000a), and at the margin of the mountains, alluvial cones (Juhász 1983) occur.

According to its evolution, the karst of the mountains is karst with horsts (Veress 2016) since they are built of blocks with different evolution and elevation (see chapter "An Interpretation of Karstification and Karst Processes by Hungarian Researchers"), but the Kab Mountain is mantled allogenic-covered karst (where mixed allogenic-autogenic karst, concealed karst and buried karst occur). In case of this karst type, the consolidated impermeable rock covers the karst in large expansion (this cover is basalt on the Kab Mountain). On karst of horst type, karst type may be different on various blocks and thus, karstification may be of different character and extent on the various blocks (Veress 2016; Veress and Vetési-Foith 2019a). The subsurface karst of its southern part was also affected by hot water. The karstification of the mountains is in an initial phase.

The Keszthely Mountains is separated into the Keszthely Plateau and the Tátika group micro regions. (The western margin of the Keszthely Mountains is the Keszthely Riviera which belongs to the Basin of Lake Balaton.) The Keszthely Mountains (their area is 150 km^2, and their elevation is 350–450 m) are constituted by a horst group, at their margin there are fault scarps and its surface is dissected by epigenetic valleys. Their subsurface karstification is more significant than its surface karstification. They are characterized by soil-covered karst (without features that are specific of this karst type) and in patches by covered (concealed) karst. In the area of the Tátika group (its area is 200 km^2), there are volcanic features and no karstification takes place here.

The parts (micro regions) of Balaton Uplands are the Badacsony-Gulács group, the Balaton Uplands and their small basins and the Vilonya Mountains. In the area of the Badacsony-Gulács group (its area is 75 km^2), volcanic features are characteristic, where no karstification takes place. In the area of the Balaton Uplands and in their small basins (with an area of 320 km^2, and an altitude of 200–400 m) horsts, volcanic features, intermountain basins, erosion and derasion valleys and boulder fields alternate. Their karstic terrains are soil-covered karsts (without features that are specific of the karst type), without a significant feature development. The karren landforms are more specific, developed on the sandstone of boulder fields (Kál Basin). In the area of the Vilonya Mountains (with an area of 55 km^2) volcanic features and horsts of low elevation, mainly horst with dolomite, occur. The karstification of dolomite surfaces (soil-covered karst) is not significant. The Tapolca Basin with buttes and the Balaton Riviera, a part of which is the Tihany Peninsula, belong to the Balaton Basin (the Balaton Basin belongs to the Transdanubian Hills). The Tapolca karst is situated at the margin of the Tapolca Basin, where both surface (soil-covered karst and concealed karst) and subsurface karst are significant.

The parts (micro regions) of the Southern Bakony are the Veszprém-Nagyvázsony Basin (with an area of 250 km^2) the Kab Mountain-Agártető group (with an area of 340 km^2), the Sümeg-Tapolca ridge (its area is 180 km^2) and the Devecseri-Bakonyalja (its area is 180 km^2). In the area of the Kab Mountain-Agártető group, there are horsts, erosion valleys, volcanic cones and pediments. Surface karstification is significant on the Kab Mountain and in its environs. As already mentioned, its karst types are allogenic karst, mixed allogenic-autogenic karst (at the margin of the basalt cover and inside it at limestone outcrops), cryptokarst, concealed karst and buried karst. On the Sümeg-Tapolcai Ridge, mainly low, exhumed horsts are characteristic, the surface of which is exhumed, former tropical karstic peneplain (the environs of Nyirád).

The Devecseri-Bakonyalja is a pediment which is dissected by different, but generally low horsts, erosion valleys and alluvial cones; its karst is soil-covered karst (with features characteristic of this type) and some concealed karst patches.

The micro regions of the Northern Bakony are the Old Bakony, the small basins of Bakony, the Eastern Bakony and the Veszprém-Devecser graben. The horsts of Old Bakony (its area is 700 km^2) are of various elevation, but those with an elevation above 500–600 m are also characteristic such as the Kőris Mountain, the Som Mountain, the Hajag etc. In its area there are erosion valleys (the valley sections between the basins are gorges or of gorge nature), and loess and other loose sediments creeks also can be found, while at the margin of the blocks, glacis, pediment and abrasion platforms occur. It is one of the characteristics of concealed karst micro regions of the mountains.

In the area of the small basins of the Bakony Mountains (Bakonybél basin, Porva Basin,, Zirc Basin, Hárskút Basin, Lókút Basin, Pénzesgyőr Basin, Csehbánya Basin and the Dudar basin belonging to the Eastern Bakony) an erosion valley network is predominant. The area of the basins is buried karst, but in the basins with a higher elevation (Hárskút Basin, Lókút Basin, Porva Basin) both concealed and soil-covered karst (without features specific of soil-covered karst) are present.

More significant horst groups of the Eastern Bakony (with an area of 370 km^2) are the Sűrű mountain group and the Tés Plateau. In the area of the former there are erosion valleys with intermittent or permanent streams (which have sections with gorges on blocks of higher elevation). Its karst is soil-covered karst (without features specific of the type) and some concealed karst patches. Abrasion platforms, pediments and glacis are also characteristic at the plateau margins of the landforms of the Eastern Bakony. The area of the Tés Plateau is only dissected by some epigenetic dry valleys. Surface (but also subsurface) karstification is the most significant here in the mountains. The most widespread concealed karst of the Eastern Bakony (but of the whole Bakony Region too) can be found here.

In the area of the Veszprém-Devecser graben (its area is 300 km^2, its elevation is 100–400 m) remnants of tropical karst peneplains occur at the surface (landforms with cones) and buried. Its geomorphology is determined by abrasion platforms, pediments, erosion and derasion valleys. On its non-buried parts, soil-covered karst is characteristic.

The micro regions of Bakonyalja are the Pápa-Bakonyalja, the Pannonhalma Hills and the Súr-Bakonyalja. In the Pápa-Bakonyalja alluvial cones and pediments occur. The area of this micro region is one of the most important resurgence sites of the karstwater of the mountains (Tapolcafő). The areas of the Pannonhalma Hills and the Súr-Bakonyalja are dissected by erosion and derasion valleys. No karstification occurs in the area of these micro regions.

The Vértes-Velence Mountains

They are separated into Vértesalja Hills, the Vértes Mountains, the Velence Mountains and its environs micro region group. Micro regions of the Vértes Mountains are the Vértes Plateau, the margin area of Vértes and the Gánt Basin. The Vértes Plateau can be regarded as a karst area. The Vértes is separated from the Bakony Region by the Mór Graben from Gerecse by Vál valley and Tatabánya Basin, while in the northwest it is surrounded by the Little Hungarian Plain and in the southeast by the Zámoly Basin.

Because of its young elevation, the margins of Vértes Plateau are bordered by fault scarps in each side. Its area is 130 km^2 and its average elevation is 400 m; only some blocks have an altitude of 500 m (e.g. Körtvélyes 480 m). Although its exhumed horsts are of various elevation, it is plateau-like; only some basins deepen into its area, for example Gánt Basin or Kőbányapuszta Basin (Bulla 1964). Its valleys are epigenetic valleys that developed along structural lines. Pediments commonly occur at the margin of Vértes.

Soil-covered karst is widespread on Vértes Plateau (without features that are specific of the karst type) and some smaller concealed karst patches occur. On dolomite there are also bare karstic patches (Csillag and Fodor 2008). Neither its surface karstification nor its subsurface karstification is significant.

The Dunazug Mountains

The Dunazug Mountains are separated into the Gerecse Region, the Bicske-Zsámbék Basin, the Buda Hills and the Pilis Mountains. These are micro region groups.

The Gerecse Region is separated from Vértes by the Tatabánya Basin. In the north it is bordered by the River Danube, while in the east it is separated from Pilis by the Dorog Graben. Its area is 650 km^2. The elevation of its surface is 300–600 m, but a block with an altitude of over 600 m also occurs (Gerecse 634 m). The Gerecse Region is also separated into blocks and block groups of various elevations which developed by the dividing up of a Late Cretaceous tropical karst peneplain. Its parts (micro regions) are Western-Gerecse, Central-Gerecse, Eastern-Gerecse and the small basins of Gerecse (Fig. 8).

The surface morphology of Western-Gerecse is constituted by fault scarps, terraces, erosion valleys, erosion-derasion valleys, derasion valleys, abrasion platforms, pediments and features of mass movement. The

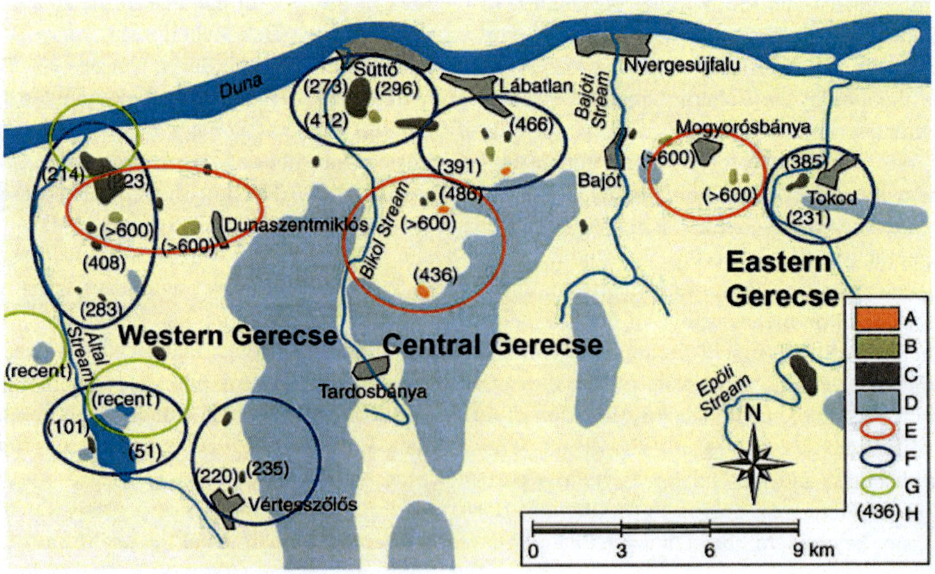

Fig. 8 Parts and calcareous tufa of the Gerecse Region (Mezősi 2011). Legend: A: Upper Pannonian springs and calcareous tufa, B: Upper Pliocene springs and calcareous tufa, C: Quaternary springs and calcareous tufa, D: Surficial distribution of Mesozoic Calcareous rocks, E: zones of spring outlets older than 600 thousand years, F: Middle and Upper Pleistocene zones of spring outlets, Holocene zones of spring outlets, H: elevation of spring

landforms of Central-Gerecse are fault scarps, erosion valleys (with sections of gorges), abrasion platforms, pediments and features of mass movement. Eastern-Gerecse is dissected by rows of horsts and basins with grabens. The slopes of horsts are surrounded by fault scarps, erosion and erosion valleys with terraces often occur. The small basins of Gerecse are dissected by erosion valleys; there is no karstification in their area (buried karst). The northern margin of Gerecse is lined by Danube terraces protected by calcareous sinter (Schweitzer 2002).

Its karst is soil-covered karst (without features specific of this karst type), concealed karst, and in the area of the basins, buried karst. Its karstification, mainly its surface karstification, is not significant, but regarding its degree, it comes after the Bakony Region in the Transdanubian Mountains. Its surface karst (concealed karst) mostly occurs in Central-Gerecse and Eastern-Gerecse. The karstification of the surface is in an initial phase, and its larger and smaller patches are far from each other. Its subsurface karst was affected by warm water.

The Buda Mountain-Group is situated among the Danube Valley, Mezőföld (Great Hungarian Plain) and the Pilisvörösvár Graben. Their area is 500 km^2 and elevation is 250–500 m. They are separated into blocks, block groups and basins. Their parts (micro regions) are the Buda Mountains, the Tétény Plateau, the Budaőrs Basin and the Budakeszi Basin. The parts of the Buda Mountains are the group of Hármashatár Mountain (between the Hűvös Valley and the Pilisvörösvár Graben), the Szabadság Mountain (between the Hűvös Valley and the Budaőrs Basin), the Nagy-Szénás group (between the Nagykovács Basin and the Pilisvörösvár Graben), the Nagy-Kopasz (between the Nagykovács Basin and the Budakeszi Basin).

The area of the Budai Mountain-Group consists of horst assemblages of various elevation, which got into different altitude following the dissection of the Late Cretaceous tropical karstic peneplain. The landforms of its surface are diverse: fault scarps and erosion valleys occur in the mountains and area of blocks (basins) with a lower elevation, and there are valleys and pediments with different width and glacis. Regarding their karst, caves with a hydrothermal effect (Hármashatár Mountain) and freshwater limestones covering river terraces are predominant (Schweitzer 2002).

The Pilis is situated between the Pillisvörösvár Graben and the Visegrád Mountains (the latter belongs to the North Hungarian Mountains because it is built up of volcanic rock). Its micro regions are the Pilis Mountain-Group (130 km^2) and the Pilis Basins (120 km^2). The elevation of the surface of the Pilis Mountain-Group is 450–500 m, but some blocks with an elevation of over 700 m also occur. It is separated into horst groups by its erosion valleys such as the Kevély Mountain Group, the Hosszú Mountain, the Fekete-Kétágú Mountain Group and the Pilis Block. The roof levels of these mountains are also the elevated exhumed surfaces of the Late Cretaceous tropical karstic peneplain. Dry valleys commonly occur on its dolomite surfaces and there are alluvial cones at the feet of fault scarps. Its subsurface karst is more significant, and the cave development was influenced by hydrothermal effect. Its surface karst is soil-covered karst (without features specific of the karst type) and with some concealed karst patches and cryptokarst patches.

3.3 The Mecsek Region (Western Mecsek)

The Mecsek Region is a micro region group, which is part of the meso region of the Mecsek-Tolna-Baranya Hills. (The micro region groups of the meso region are the Mecsek Region, the Tolna Hills, the Baranya Hills and Zselic.) The meso region belongs to the Transdanubian Hills macro region. The Mecsek Region is separated into the Mecsek Mountains and the Baranya-Hegyhát micro regions. Karst can be found in the Mecsek Mountains. The Mecsek Mountains are enclosed by the Great Hungarian Plain (Sárköz), the South-Baranya Hills, the Zselic and the Völgység. The Villány Mountains (micro region) is a block of small expansion being separated from the Mecsek and is a part of the Baranya Hills.

The area of the Mecsek Mountains is 350 km^2; it is a system of peneplains with SSW–NNE direction and their elevation is 300–600 m. The Western Mecsek is a series of horsts with peneplains, while the Eastern Mecsek is built up of rows of horsts radiating from the Dobogó-Zengő group. The features of the mountains are the pediment half-planes, alluvial cones and erosion valleys. Their karst developed in the northern part of the Western Mecsek area with an expansion of 30 km^2. Western Mecsek is an anticline with a WE direction. This area is a pediment of Carpathian age. It is separated into two parts: a more elevated part (450–470 m) which is an abrasion platform of Middle Miocene age, and a less elevated part (250–380 m). This latter is also an abrasion platform of Pannonian age and dissected tectonically (Lovász 1971; Hevesi 2001). Its karst is separated into three parts (Fig. 9): the Abaliget karst west from the Szuadó valley, the Orfű karst in the east and the Misina Tubes karst area eastwards from the latter (Lovász 1977).

In the south, its karst is mixed autogenic-allogenic karst (Barta and Tarnai 1997), but in its northern part the autogenic part (Veress 2016) was transformed into concealed karst (Hevesi 2001). The area of the Misina Tubes is a karst, the more elevated part of which was autogenic karst according to Czigány and Lovász's data (2006), while its lower part was probably covered which turned into allogenic karst, but eventually losing its impermeable cover it became concealed karst. Therefore, these karst areas (as already mentioned) are regarded to be transitional between the karsts of

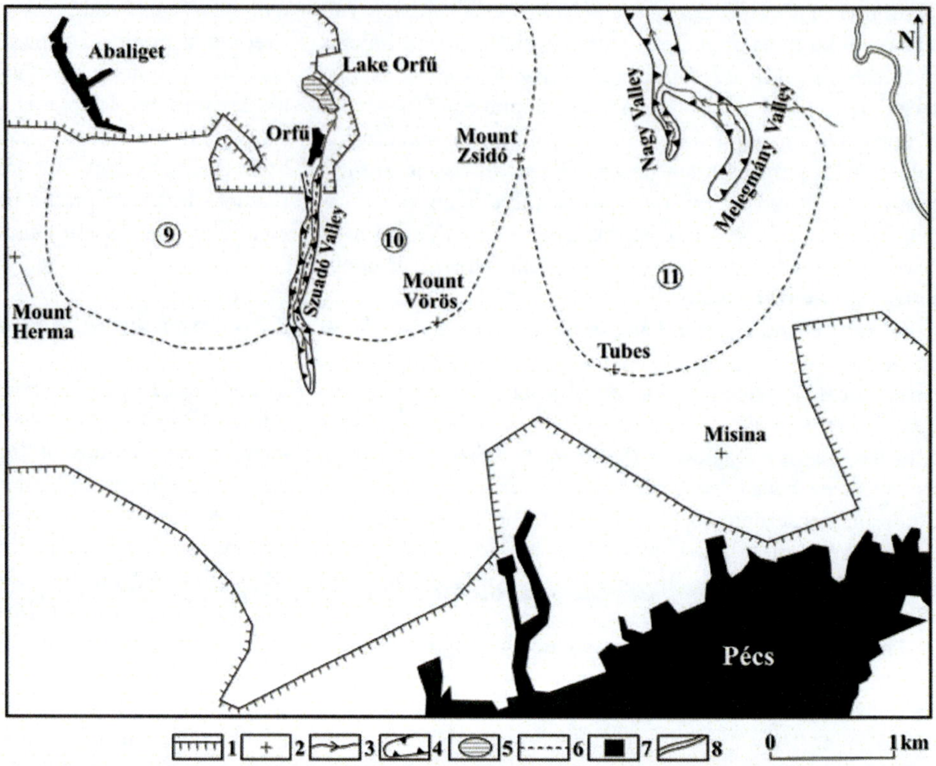

Fig. 9 Parts of Western Mecsek karst (Mezősi 2011). Legend 1. margin of mountains, 2. mountain, 3. stream, 4. valley, 5. lake, 6. border of karstic part area, 7. settlement, 8. road, 9. Abaliget Karst, 10. Orfű Karst, 11. Misina-Tubes Karst

Aggtelek and Transdanubian character. According to its evolution, it is semi-allogenic-covered karst. In such karst, the valleys of the non-karstic terrain surrounding the karst are the epigenetic valleys going through the karst, the streams of which transported or transport the superficial deposit of the karst or the rocks of the bordering non-karstic terrain (Veress 2016).

The Villány Mountains is (with an area of 50 km^2) surrounded by a pediment half-plane, above which there is an abrasion platform and above this are block sections with an elevation of 350–400 m. In a southern direction from the mountains, there are covered blocks of lower elevation, among which the most important is the Beremend block because of its subsurface karst. On the latter, a cave development of hydrothermal origin also played a role. The karst of the Villány Mountains is not significant. Their karst type is soil-covered karst (without features specific of this karst type), while bare karst occurs in some sites on its side slopes as a result of the denudation of the soil.

4 Conclusions

The main characteristics of Hungarian karsts were described. These are the following: their geological structure may be different (as a result of their various former development environment), they have a small area, they are dissected, they have a reversible hydrological relation with their environment, there is a strong hydrothermal effect in their area and paleokarsts are widespread. On its karsts, the fluvial surface formation was determined not only in the past but it is also important in the present. According to their karst landforms, Hungarian karsts can be put into two groups, the karst of Aggtelek group and the karst of Transdanubian group, but as a result of the presence of non-karstic rocks, several karst types are present in various karst areas. The karst types are of small expansion, partly because several types occur in the same karst area and partly because they are surrounded by non-karstic areas. Soil-covered karst, mixed autogenic-allogenic karst and concealed karst are the most widespread. According to their constituting rocks, they are mainly limestone karsts, but dolomite karsts are also widespread particularly in the Transdanubian Mountains. The climate, vegetation and soil of karst areas are not significantly different.

In Hungary, karst occurs in three macro regions: areas of the North Hungarian Mountains, Aggtelek-Rudabánya Mountains and Bükk Region (and on some smaller blocks). They are also found in the areas of the Transdanubian Hills, Mecsek Mountains and Villány Mountains (with the Beremend block); Transdanubian Mountains, Bakony Region, Vértes-Velence Mountains (Vértes Plateau), Dunazug Mountains (Gerecse, Buda Mountains and Pilis). The parts, expansion, elevation, geomorphology and karst types of these karst areas were also studied.

References

Alföldi L, Lorberer Á (1976) A karsztos kőzetek háromdimenziós áramlásának vizsgálata kútadatok alapján (A study of the three-dimensional flow of karst rocks based on well data). Hidrológiai Közlöny 10:433–443 (in Hungarian)

Ádám L (1983) A Dunántúli-középhegység alakrajzi jellemzése (A morphographic description of the Transdanubian Mountains). Földr Ért XXXII(3–4):413–420 (in Hungarian)

Ádám L (1984) Az Észak-magyarországi-hegyvidék alakrajzi jellemzése (A morphographic description of the North Hungarian Mountains). Földr Ért XXXIII(4):321–345 (in Hungarian)

Barta K, Tarnai T (1997) Karsztkutatás az orfűi Vízfő-forrás vízgyűjtő területén (Karst research on the catchment of the Vízfő spring, Orfű). Karszt és Barlang I–II:12–19 (in Hungarian)

Bartha D, Orbán S (1995) Magyar erdők (Hungarian forests). In: Járainé Komlódi M (ed) Pannon enciklopédia. Magyarország növényvilága. Dunakanyar Kiadó, Budapest, pp 222–230 (in Hungarian)

Bartholy J, Pongrácz R (2013) A légkör földrajza (Geography of the atmosphere). In: Szabó J (ed) Általános természeti földrajz I. ELTE, Eötvös Kiadó, Budapest, pp 47–114 (in Hungarian)

Bodor P, Erőss Á, Mádlné Szőnyi J, Czuppon Gy (2014) A Duna és a felszíni viszonyok kapcsolata a Rózsa-dombi megcsapolás területén (Interaction between the groundwater and the Danube at the Rózsadomb discharge area). Karsztfejlődés XIX:63–75 (in Hungarian)

Bulla B (1964) Magyarország természeti földrajza (Physical Geography of Hungary). Tankönyvkiadó, Budapest, 424 pp (in Hungarian)

Császár G (szerk) (1997) Lithostratigraphical units of Hungary. MÁFI Kiadvány, Budapest 114 pp

Csillag G, Fodor L (2008) Geomorfológia (Geomorphology). In: Budai T, Fodor L (eds) A Vértes-hegység földtana. Magyarázó a Vértes-hegység földtani térképéhez (1:50000) (Explanatory book to the geological map of the Vértes Hills (1:50000). Magyar Állami Földtani Intézet, Budapest, pp 135–144 (in Hungarian)

Czigány Sz, Lovász Gy (2006) A mecseki karszt térképezésének újabb eredményei (New results of the mapping on the karstic area of Mecsek Mountains). Közlemények a Pécsi Tudományegyetem Földrajzi Intézetének Természetföldrajzi Tanszékéről 26:3–14 (in Hungarian)

Dax M, Éri I, Mithay S, Palágyi Sz, Torma I (1972) Veszprém megye régészeti topográfiája, a Pápai és Zirci járás (Archaeological topography of Veszprém County, the Pápa and Zirc districts), 4. Akadémia Kiadó, Budapest, 330 pp (in Hungarian)

Éri I (1966) Adatok a bakonyi üveghuták történetéhez (Data on the history of the glasshouses of the Bakony Mountains). A Veszprém Megyei Múzeumok Közleményei 5:143–180 (in Hungarian)

Eszterhás I (2013) Bazaltbarlangok a Bakonyban (Basalt Caves in the Bakony Mountains, Hungary). Karsztfejlődés XVIII:183–204 (in Hungarian)

Erőss A (2010) Characterization of fluids and evaluation of their effects on karst development at the Rózsadomb and Gellért Hill, Buda Thermal Karst Hungary. Doktori Disszertáció, ELTE, 171 pp

Éri I, Kelemen M, Németh P, Torma I (1969) Veszprém megye régészeti topográfiája, a Veszprémi járás 2 (Archaeological topography of Veszprém County, the Veszprém district 2). Akadémia Kiadó, Budapest, 340 pp (in Hungarian)

Ferrarese F, Macaluso T, Madonia G, Palmeri A, Sauro U (2002) Solution and recrystallization processes and associated landforms in gypsum outcrops of Sicily. Geomorphology 49(1–2):25–43. https://doi.org/10.1016/S0169-555X(02)00159-9

Fodor L, Csontos L (1998) Magyarországi szerkezetföldtani kutatások és ezek legfontosabb eredményei (Structural geological research in Hungary and their most significant results). Földtani Közlöny 128 (1):123–143 (in Hungarian)

Fülöp J (1989) Bevezetés Magyarország geológiájába (Introduction to Hungary's geology). Akadémia Kiadó, Budapest, 246 pp (in Hungarian)

Géczy B (1973) Lemeztektonika és paleogeográfia a kelet-mediterrán mezozóos térségben (Plate tectonics and palaeography in the Eastern Mediterranean Mesozoic area)—MTA X. Osztály Közleményei 5(3–4):219–225 (in Hungarian)

Gyuricza Gy, Sásdi L (2009) A Baradla-barlangrendszer kialakulásának kérdései a tágabb környezet földtani fejlődésének tükrében (Theory of the evolution of the Baradla Cave as mirrored in the geological evolution of its neighbourhood). Földtani Közlöny 139(1):83–92 (in Hungarian)

Haas J (2015) Geological and tectonic background. In: Lóczy D (ed) Landscapes and landforms of Hungary. Springer, Cham, pp 7–27. https://doi.org/10.1007/978-3-319-08997-3

Haas J, Hámor G, Korpás E (1999) Geological setting and tectonic evolution of Hungary. Geol Hung 24:179–196

Hevesi A (1978) A Bükk szerkezet- és felszínfejlődésének vázlata (A sketch of the structure and landscape evolution of the Bükk Mountains). Földr Ért XXVII(2):169–203 (in Hungarian)

Hevesi A (1986a) Hideg vizek létrehozta karsztok osztályozása (The classification of karsts created by cold waters). Földr Ért XXXV(3–4):231–254 (in Hungarian)

Hevesi A (1986b) A Déli-Bükk karsztja I. rész: Délkeleti-Bükk (Karst of the South Bükk Part I: South-Eastern Bükk). Karszt és Barlang I:3–14 (in Hungarian)

Hevesi A (1986c) A Déli-Bükk karsztja II. rész: Délnyugati-Bükk (Karst of the South Bükk Part II: South-Western Bükk). Karszt és Barlang II:87–94 (in Hungarian)

Hevesi A (1991a) Magyarország karsztvidékeinek kialakulása és formakincse, I. rész. (Development and landforms of karst regions in Hungary I). Földrajzi Közlemények 1–2:25–35 (in Hungarian)

Hevesi A (1991b) Magyarország karsztvidékeinek kialakulása és formakincse, II. rész. (Development and landforms of karst regions in Hungary II). Földrajzi Közlemények CXV(3–4):99–120 (in Hungarian)

Hevesi A (2001) A Nyugat-Mecsek felszíni karsztosodásának kérdései (Questions of the surface karstification of the Western Mecsek). Karsztfejlődés VI:103–111 (in Hungarian)

Jakucs L (1977) A magyarországi karsztok fejlődéstörténeti típusai (Evolutionary types of Hungarian karsts). Karszt és Barlang I–II:1–16 (in Hungarian)

Juhász Á (1983) Az Északi-Bakony előtere és a Pannonhalmi-dombság domborzata (Geomorphology of the northern foreland of the Bakony Mountains and the Pannonhalma Hills). Földr Ért XXXII (3–4):421–432 (in Hungarian)

Kevei Bárány I (2018) Karsztdolinákról másképpen (Geoökológiai értelmezés) (From karst dolines otherwise (Geoecological interpretation). Karsztfejlődés XXIII:163–185. https://doi.org/10.17701/18.163-185 (in Hungarian)

Kőrössy L (1963) Magyarország medenceterületeinek összehasonlító földtani szerkezete (Comparison between the geological structure of the basin regions of Hungary). Földtani Közlöny 89(2):115–124 (in Hungarian)

Kőrössy L (1970) Entwicklungsgeschichte der neogenen Becken in Ungarn. Acta Geol Sci Hung 14:421–429

Laubscher HP (1971) Das Alpen-Dinariden-Problem und die Palinspastik der südlichen Tethys. Geol Rundsch 60(3):813–833

Láng S (1952) Geomorfológiai-karsztmorfológiai kérdések (Geomorphological-karst morphological questions). Földr Ért I (1):120–126 (in Hungarian)

Láng S (1955) Geomorfológiai tanulmányok az Aggteleki karszton (Geomorphological studies on Aggtelek karst). Földr Ért 4(1):1–16 (in Hungarian)

Láng S (1958) A Bakony geomorfológiai képe (Geomorphological picture of the Bakony Mountains). Földrajzi Közlemények 6:325–343 (in Hungarian)

Leél-Őssy S (1959) Magyarország karsztvidékei (Karst regions of Hungary). Karszt és Barlangkutatás, pp 79–88 (in Hungarian)

Leél-Őssy S (1960) Magyarország karsztvidékei (Karst regions of Hungary). Földr Ért 9(1–4):490–495 (in Hungarian)

Leél-Őssy S (2012) A Budai Vár-hegy geológiai viszonyai (Geological conditions of the Buda Castle Hill). Karsztfejlődés XVII: 5–23 (in Hungarian)

Lippmann L, Kiss K, Móga J (2008) Az Abaligeti-Orfűi karszt karsztos felszínformáinak vizsgálata térinformatikai módszerekkel (GIS investigation of the karst landforms in the Abaliget-Orfű Karst). Karsztfejlődés XIII:151–166 (in Hungarian)

Lóczy L (1913) A Balaton környékének geológiai képződményei és ezeknek vidékek szerinti telepedése (The geological features of the surroundings of Lake Balaton and their regional distribution). A Balaton tudományos tanulmányozásának eredményei, I. K. Budapest, 617 pp (in Hungarian)

Lóczy D (2015) Geomorphological Regions. In: Lóczy (ed) Landscapes and landforms of Hungary. Springer, Cham, Heidelberg, New York, pp 39–43

Lovász Gy (1971) Adatok az Abaligeti-karszt geomorfológiai és hidrológiai jellemzéséhez (Data for the geomorphological and hydrological description of the Abaligetien karst). Földr Ért XX (3):283–295 (in Hungarian)

Lovász Gy (1977) Geomorfológiai körzetek (Geomorphological districts). In: Lovász Gy (ed) Baranya megye természetföldrajza. Baranya megyei Levéltár, Pécs, pp 62–68 (in Hungarian)

Lovász Gy (1981) A Baranyai-dombság, a Mecsek és a Villányi-hegység (The Baranya Hills, the Mecsek and the Villány Mountains). In: Pécsi M (ed) A Dunántúli-dombság (Dél-Dunántúl). Akadémiai Kiadó, Budapest, pp 124–136 (in Hungarian)

Marosi S, Somogyi S (1990) Magyarország kistájainak katasztere (Inventory of microregions in Hungary). MTA Földrajztudományi Kutató Intézet, Budapest, 1023 pp (in Hungarian)

Mádlné Szőnyi J, Tóth Á (2015) Basin-scale conceptual groundwater flow model for an unconfined and confined thick carbonate region. Hydrogeol J 23(7):1359–1380. https://doi.org/10.1007/s1040-015-1274-x

Medzihradszky Zs (1996) A magyarországi erdők rövid története (A short story of Hungarian forests). Földrajzi Közlemények CXX(2–3):181–186 (in Hungarian)

Mezősi G (1984) A Sajó-Bódva köz felszínfejlődése (The geomorphic evolution of the interfluve of the Sajo and Bodva rivers). Földr Ért XXIII(3):181–206 (in Hungarian)

Mezősi G (2011) Magyarország természetföldrajza (The physical geography of Hungary). Akadémia Kiadó, Budapest, 393 pp (in Hungarian)

Mezősi G (2016) The physical geography of Hungary. Springer, Cham, 334 pp. https://doi.org/10.1007/978-3-319-45133-1

Móga J (2002a) Felszínalaktani vizsgálatok a Galyaság területén (Geomorphological investigations in the area of Galyaság). Karsztfejlődés VII:173–186 (in Hungarian)

Móga J (2002b) A tornai Alsó-hegy (The Alsó Hill of Torna). Karszt és Barlang 1998–1999 évf. (I–II):95–104 (in Hungarian)

Móga J, Kiss K, Szabó M, Kériné Borsodi A, Kéri A, Mari L, Knáb M, Iván V (2011) Természeti és antropogén hatások vizsgálata a Tapolcai-karszt epikarszt rendszereiben (Investigations of natural and anthropogenic effects in the epikarst systems of Tapolca Karst). Karsztfejlődés XVI:183–201 (in Hungarian)

Moldvay L (1969) A neotektonikus felszínalakulás jelenségei a magyarországi középhegységekben (Phenomena of neotectonic surface development in Hungarian Mountains). Magyar Állami Földtani Intézet Évi Jelentése 1969-ről, Budapest, pp 587–637 (in Hungarian)

Nagy E (1971) A lábai fázis jelentősége a Dunántúli szerkezetfejlődés szempontjából (Significance of the Lábán phase in view of the structural development of Transdanubia). A Magyar Állami Földtani Intézet Évi Jelentése 1969 évről, Budapest pp 583–586 (in Hungarian)

Ozoray Gy, Láng G (1957) Karsztjelenségek a csővári rögcsoportban (Karst phenomena in the block groups of Csővár). Karszt- és Barlangkutatási Tájékoztató 36–37 (in Hungarian)

Papp F (1942) Dunántúl karsztvizei és a feltárás lehetősége Budapesten (The karstwaters of Transdanubia and the opportunity of exploration in Budapest). Hidrológiai Közlöny 21(7–12):146–196 (in Hungarian)

Pécsi M (1963) Hegylábi (pediment) felszínek a magyarországi középhegységekben (Pediment surfaces in Hungarian central mountains). Földrajzi Közlemények IX(3):195–212 (in Hungarian)

Pécsi M (1980) A Pannóniái-medence morfogenetikája (The Morphogenesis of the Pannonian basin). Földr Ért 29:105–127 (in Hungarian)

Pécsi M (1982) Quaternary studies in Hungary – Elmélet – Módszer – Gyakorlat. Akadémia Kiadó, Budapest, 313 pp

Pécsi M (1983) A Dunántúli-középhegység sasbérc felszínei és a denudációs kronológia (Horsts in the Transdanubian Mountains and denudation chronology). Földrajzi Értesítő XXXII(3–4):504–505 (in Hungarian)

Pécsi M (1991) Geomorfológia és domborzatminősítés (Geomorphology and relief classification). Földrajz Tudományi Kutató Intézet, Budapest, 296 pp (in Hungarian)

Pinczés Z (1968) Herausbildung der Tertiären Oberflächen des Bükk-Gebirges. Acta Geographica Debrecina:189–200

Pinczés Z (1980) Néhány gondolat a Bükk felszínfejlődésével kapcsolatban (Some thoughts on the landscape evolution of the Bükk Mountains). Földr Ért XXIX(1):102–103 (in Hungarian)

Sauro U (2003) Dolines and sinkholes: aspects of evolution and problems of classification. Acta Carsologica 32(2):41–52

Sás L (1987) Földtani alapok (Basics of geology). In: PÉCSI M (szerk.) A Dunántúli-középhegység A. Akadémia kiadó, Budapest, pp 40–131 (in Hungarian)

Sásdi L (1990) Az Aggtelek-Rudabányai-hegység karsztjának földtani fejlődéstörténete (Geological evolution of the karst of the Aggtelek-Rudabánya Mountains). Karszt és Barlang I:3–8 (in Hungarian)

Sárváry I (1971) A természeti tényezőktől függő karsztvízszint süllyedés a Dunántúli középhegységben (Subsidence of karstwater table depending on natural factors in the Transdanubian Mountains). Hidrológiai Közlöny 10:429–484 (in Hungarian)

Schweitzer F (2002) Édesvízi mészkő előfordulások (Freshwater limestone occurrences). Karszt és Barlang 1998–1999 évf. (I–II) (in Hungarian)

Stegena L, Géczy B, Horvath F (1975) Late Cenozoic evolution of the Pannonian Basin. Tectonophysics 26:71–90

Szabó PZ (1968) A magyarországi karsztosodás fejlődéstörténeti vázlata (An evolutionary outline of Hungarian karstification). Dunántúli Tudományos Gyűjtemény, pp 65–102 (in Hungarian)

Telbisz T, Mari L, Kohán B, Jelena C (2007) A szerbiai Miroč-hegység töbreinek térinformatikai GPS-es terepi vizsgálata (A spatial informatics GPS study of the dolines of the Serbian Miroč Mountains). Karsztfejlődés XII:71–90 (in Hungarian)

Tóth G (1986) Adatok a Magas-Bükk nem karsztosodó üledéktakaróinak ismeretéhez (Data on the non-karstic sediment layers of the High-Bükk). Nehézipari Műszaki Egyetem Közleményei I Sorozat, Bányászat 33(1–4):197–202 (in Hungarian)
Tóth G, Fejes P (1984) Idősebb pliocén lepusztulási szint kimutatása a Bükkben (Report on an older Pliocene denudation level in the Bükk mountains). Karszt és Barlang II:65–72 (in Hungarian)
Vass D, Elečko J, Horská A, Petrik F, Barkač Z, Mello J, Vozárová A, Gy R, Dubéci B (1994) Základné črty geológie turnianskej depresie. Geol Páce Správy 99:7–22
Veress M (1987) Karsztos mélyedések működése bakonyi fedett karsztokon (Activity of karst depressions on the covered karst in the Bakony Mountains). Földr Ért XXXVI(1–2):91–114 (in Hungarian)
Veress M (2000a) Covered karst evolution Northern Bakony Mountains, W-Hungary. A Bakony Természettud. Kut. Eredményei 23, Bakonyi Természettudományi Múzeum, Zirc, 167 pp.
Veress M (2000b) Középhegységi karsztok néhány típusa (Some types of middle mountain karsts). Földrajzi Közlemények CXXIV(1–4):1–28 (in Hungarian)
Veress M (2009) Investigation of covered karst form development using geophysical measurements. Zeits F Geomorph 53(4):469–486
Veress M (2010) A magyarországi eltemetett és rejtett karsztos térszínek felszínfejlődése (Development of cryptokarstic and latent karstic surface in Hungary). Földrajzi Közlemények 134(4):373–391 (in Hungarian)
Veress M (2014) Az Aggteleki-karszt karsztos zónái. Kutatások az Aggteleki Nemzeti Park területén II (Researches in Aggtelek National Park and Biosphere Reserve II). Aggteleki Nemzeti Park Igazgatóság, Jósvafő, pp 27–43 (in Hungarian)
Veress M (2016) Covered karst. Springer. https://doi.org/10.1007/978-94-017-7518-2536p
Veress M (2017) A depresszió exhumálódás és a barlangképződés kapcsolata az Aggteleki-karszton (The relationship between depression exhumation and cave development on the Aggtelek Karst). Földrajzi Közlemények 141(2):97–111 (in Hungarian)
Veress M (2018) A Bakonyvidék Genetikai Aknatípusai (The genetic shaft types of the Bakony Region). Karsztfejlődés XXIII:55–84. https://doi.org/10.17701/18.55-84
Veress M, Zentai Z (2009) Karsztjelenségek minősítése a Bükk-hegység néhány mintaterületén a mészkőfekü morfológiájának és a fedőüledékek szerkezetének értékelésével (Classification of Karst phenomena on a few sample areas of the Bükk Mountains by using morphology of the limestone floor and the structure of the sedimentary rock). Karszt és Barlang I–II:37–54 (in Hungarian)
Veress M, Unger Z (2015) Baradla-Domica: large cave system on the Hungarian-Slavak Border. In: Lóczy D (ed) Landscapes and landforms of Hungary. Springer, Dordrecht, Cham, pp. 167–178. https://doi.org/10.1007/978-3-319-08997-3
Veress M, Zentai Z, Unger Z (2015) Hat rock: solution features in metamorphic mountains. In: Lóczy D (ed) Landscapes and landforms of Hungary. Springer, Cham, pp 47–54. https://doi.org/10.1007/978-3-319-08997-3
Veress M, Vetési-Foith Sz (2019a) The Distribution of Surface Karst Features in the Bakony Region (Transdanubian Mountains Hungary). J Geol Res 1(1):21–25
Veress M, Vetési-Foith Sz (2019b) A Bakonyvidék felszíni karsztja (The surface of the Bakony Regions in Hungary). Földrajzi Közlemények 143(2):167–186 (in Hungarian)
Waele JD, Piccini L, Columbu A, Madonia G, Vattano M, Calligaris Ch, D'Angeli I, Parise M, Chiesi M, Sivelli M, Vigna B, Zini L, Chiarini V, Sauro F, Drysdale R, Forti P (2017) Evaporite karst in Italy: a review. Int J Speleol 46(2):137–168
Zámbó L (1970) A vörösagyagok és a felszíni karsztosodás kapcsolata az Aggteleki-karszt délnyugati részén (The relationship between red clays and surface karstification at the southwestern part of Aggtelek karst). Földrajzi Közlemények 94(18):281–293 (in Hungarian)
Zámbó L (1998) Az Aggteleki karszt felszínalaktani jellemzése (Surface morphological characterisation of Aggtelek karst). Földr Ért XLVII(3):359–378 (in Hungarian)
Zólyomi B (1989) Természetes növénytakaró (Natural vegetation). In: Pécsi M (ed) Magyar Nemzeti Atlasz. Cartographia, Budapest, 89 pp (in Hungarian)

History of Karst and Cave Research in Hungary

Kinga Székely, Attila Hevesi, Szabolcs Leél-Őssy, and Tamás Telbisz

Abstract

In Hungary, geographical publications already mentioned special karst objects more than half a thousand years ago. About 250 years ago, some authors explicitly dealt with limestone mountains including their spectacular large cave entrances and their famous gorges. The caves of Hungary are widely known in the world based on their natural, cultural and historical values as well as their role in scientific research, despite the fact that there are no Hungarian caves among the longest or deepest caves in the world. In many cases, the hypogenic caves in Hungary served as study areas even for foreign researchers. The "Bükk Culture" is an internationally acknowledged denomination for a prehistoric culture named after the artefacts found in the caves of the Bükk Mountains. Likewise, the Middle Pliocene "Estramontinum' vertebrate biostratigraphic stage refers to the exploration site at Esztramos Hill. The work of Hungarian cave researchers contributed to solving the problems of several special karstic phenomena (e.g. the formation of helictites, the effect of tidal phenomena on the discharge of karst springs). Furthermore, the contribution of Hungarian cave researchers was also important in establishing new scientific fields (e.g. biospeleology) or raising them to an international level (e.g. speleotherapy). Our oldest-known, best-researched cave system is Baradla, which is one of the longest dripstone caves in the temperate zone with an active stream, and one of the earliest tourist caves in the world. In the nineteenth century, it was recorded as the second-longest cave in the world after Mammoth Cave (USA). The international reputation of our researchers and the universal value of our caves are proved by the fact that Hungary hosted the 10th International Speleological World Congress in 1989, and the caves of Aggtelek Karst (together with Slovak Karst) are the only objects listed in the Natural Category of World Heritage sites in Hungary.

Keywords

Surface karst phenomena and landforms • Archaeological findings • Cave exploration • Cave mapping • Cave protection

1 Introduction

The scientific study of karsts in Hungary started only from the beginning of the nineteenth century, or rather from the second half of this century. The research issues included surface karst morphology, the scientific study of caves as well as cave explorations. Within the second, the archaeological excavations in caves had an outstanding role. An overview of this research history is given in this chapter, with a special focus on the history of the Hungarian Karst and Cave Research Society and its legal predecessor.

In the second half of the nineteenth century, karst research in Hungary took place within the framework of the Austro-Hungarian Monarchy due to the lack of independence of the country. For this reason, German (Austrian) and Hungarian karst research had a strong impact on each other. It is noteworthy that after Hungary lost two-thirds of its territory after the First World War as a result of the Treaty of

K. Székely (✉)
Institute of Speleology, Budapest, Hungary

A. Hevesi
Institute of Geography and Geoinformatics, University of Miskolc, Miskolc-Egyetemváros, 3515, Hungary
e-mail: ecoheves@uni-miskolc.hu

S. Leél-Őssy
Department of Physical and Applied Geology, Eötvös Loránd University, Pázmány Péter sétány 1/C, 1117 Budapest, Hungary
e-mail: leel-ossy.szabolcs@ttk.elte.hu

T. Telbisz
Department of Physical Geography, Eötvös Loránd University, Pázmány Péter sétány 1/C, 1117 Budapest, Hungary
e-mail: tamas.telbisz@ttk.elte.hu

M. Veress and Sz. Leél-Őssy (eds.), *Cave and Karst Systems of Hungary*, Cave and Karst Systems of the World, https://doi.org/10.1007/978-3-030-92960-2_2

Trianon (1920), a number of karst areas became part of the neighbouring countries. Thus, it is understandable that Hungarian researchers working before Trianon played an important role in the examination and investigation of these karst lands, and even after the detachment of these karst terrains, many of them remained at the forefront of the interest of certain Hungarian karst researchers.

Caves, as special gifts of nature, have always attracted people, but their perceptions have changed with age. The more than a thousand-year-old history of Hungarian cave research can be divided into a few, well-distinguished, large periods. In the first stage, which lasted for about 500 years from the time of the Hungarian conquest of the Carpathian Basin (895 AD), we cannot speak about cave description or cave research, but the places referring to caves (Odvaskő, Kőlyuk, Likaskő and Kőpest) were well-known landmarks, thus they were used in donation letters and charters. The earliest of such royal charter, a letter of donation issued by the first Hungarian king, St. Stephen (ca. 975–1038) to the Benedictine abbey in Bakonybél, dates from 1037. The "Odvaskő" (Oduaskw) mentioned in this charter, which is a rocky hill still known by this name, is the first written cave-related geographical name in Hungary. In a charter from the Anjou period, issued in 1355, the name "Likaskw" probably refers to Kis-Kevély Hill (in Buda Mountains), which hides the Kis-kevély Cave that was inhabited by prehistoric people. In 1535, Oláh Miklós (1493–1568) archbishop of Esztergom published his work describing Hungary. This publication mentions the large karst springs of Buda and the rock cavities under the buildings of the Castle Hill in Buda (Dénes 2015).

2 Research of Surface Karst Features in the Carpathians and Carpathian Basin

Naturally, at first, the large or strange cavities and rock gates found in the limestone mountains caught the attention of naturalists, and of course the high, steep-walled, cavity dotted rock gorges, such as the Torda Gorge (Cheile Turzii) in Transylvania. Simultaneously with the documents about the above-mentioned features, medieval records also report the "stone-forming" springs from certain locations (i.e. they recognized the formation of freshwater limestone). One has to go back to the end of the fifteenth century to find the origin of these records. The Italian-born Petrus Ransanus, who lived in the court of King Matthias (1458–1490) and Queen Beatrix between 1488 and 1492, already mentioned the ice cave in the southern neighbourhood of Szepesváralja (Spišské Podhradie, Slovakia) in his work on Hungary (1489). This is not a real underground cave, but a pseudo-cave created by the tilting of the split freshwater limestone blocks of Pleistocene and Holocene origin as a result of slow gravitational slope movements (Schönviszky 1968; Hevesi 1999). The book by György Wernher (Georg Wernher—1487–1556) on the "wonderful waters" of Hungary (1549, 1551) describes the freshwater limestones of Ruschbach, Szepesvár (Spišský hrad), Szentjános (Liptovský Ján) and Szentandrás (Liptovský Ondrej; all in Slovakia today). Wernher also mentions the Drevnyik (Drevenik) Mountain Ice Cave, which was later described by Speer (1683) and Szentiványi (1689, 1709), too.

In the work of Bertalanffy (1757), the author of the first completely Hungarian-language geography book, we can read about surface karst phenomena around Lake Czirkniczi (Cerknica, Slovenia), without, of course, the technical term "polje".

In his book presenting the mountains of Hungary (1714), István Csiba mentions in connection with the Torda Gorge that large underground cavities can rupture as a result of earthquakes. The first truly scientific explanation for the formation of Torda Gorge and similar gorges can be found in the work of János Kászonújfalvi Szabó (1803; cited by Tulogdy 1943 and Hevesi 1976). He pointed to the varied erosional resistance of the surface-building rocks as the most important reason behind the formation of large limestone gorges. This view was repeated by Téglási Ercsei (1837), who added that "limestone mountains are rich in cavities and spacious caves, and consequently they are so weak in certain parts that they cannot withstand forces beyond their strength and therefore collapse" (as cited by Tulogdy 1943).

We can read about the subsurface and surface formations occurring in the limestone mountains already in the works of Apáczai Csere (1653), Mitterpacher (1789),Varga (1809) and Katona (1824). Nevertheless, karst landforms smaller than limestone gorges, such as dolines, even if they are remarkable, appear on Hungarian maps for the first time only during the First Military Survey (1785–1788), namely on the Aggtelek-Jósvafő section, and on the manuscript sketches of the Bükk Nagy-Fennsík and Cserépváralja (Hevesi 2002a, b, c, 2007). On the other hand, the word "töbör", the Hungarian term for "doline" already appears in the form of "teber" in a 1599 field boundary description of Jólész (Jovice), a village in Gömör-Torna-Karszt (Dénes 2002).

The first map published in print, which can be called "karst morphological map", was drawn by Keresztély Raisz in 1802. With our current knowledge, the linearly arranged valley dolines, the dolines near the mountain summits, as well as the funnel-like sinkholes and doline lakes are all well recognizable (Fig. 1). He was also the first to scientifically describe the relationship between water drained at the bottom of sinkholes (the so-called "ravaszlyuk" in Hungarian), flowing in the cave streams and leaving the cave across the springs. In his work published in German in 1807, the words "töbör" (i.e. doline) and "ravaszlyuk" (i.e. sink point) appear in Hungarian.

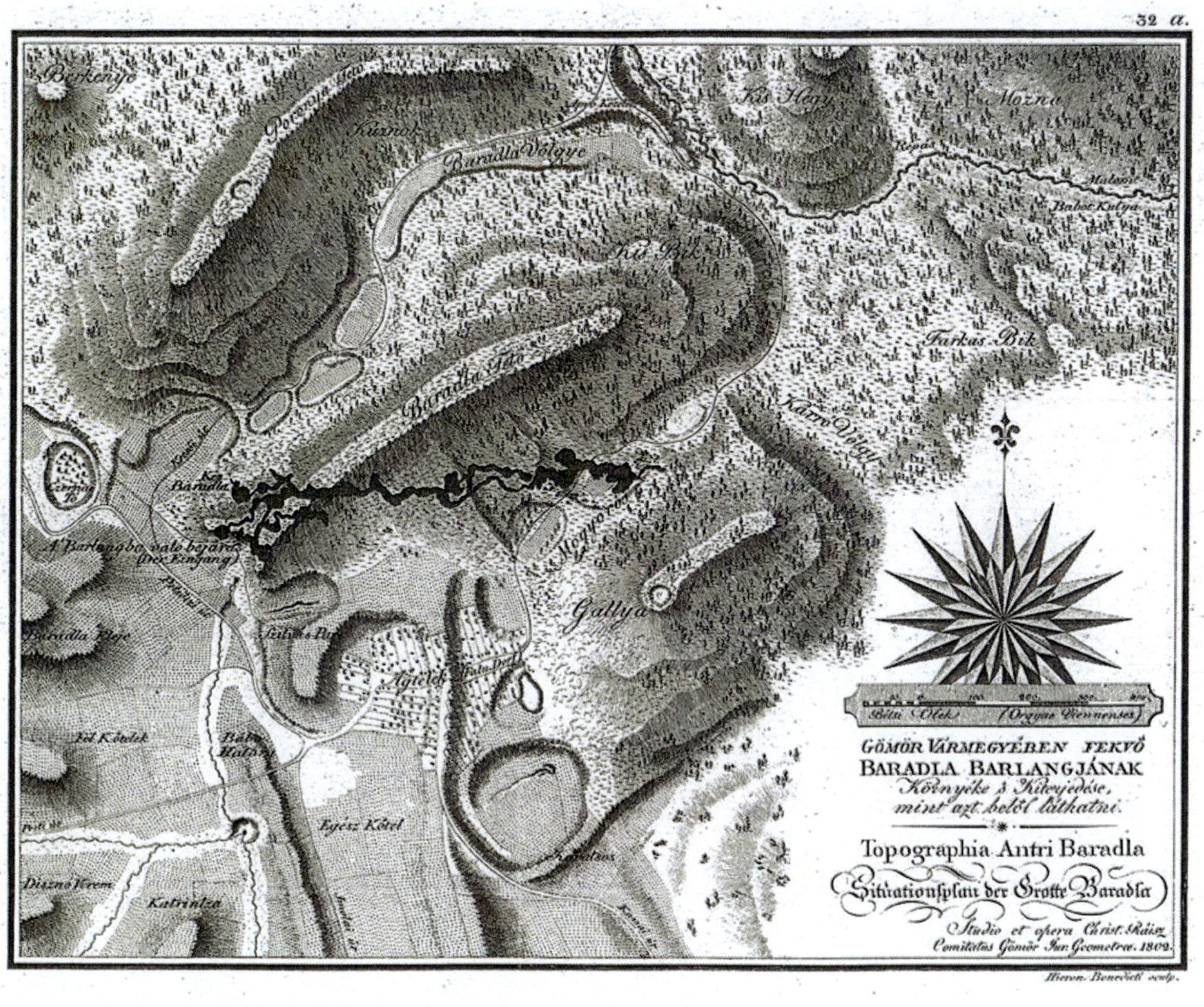

Fig. 1 The map of Baradla Cave by József Sartory (1794)

Following the karst geomorphological map of Raisz (1802), Imre Vass, who drew an even more precise and detailed map about the area, described Aggtelek Cave and issued the first Hungarian karst geomorphological study about Aggtelek Karst (Vass 1831a, b). He considered it natural that both dolines and caves were created by waters infiltrating through "secret pathways". Not only did he indicate the different types of dolines on a map, but he also described their partly different origin.

In his most important book (1863–1865), which is a milestone in the history of Hungarian geography, János Hunfalvy provided a comprehensive presentation on the karst phenomena of almost all limestone mountains in the then Hungary and Croatia (Hunfalvy 1863). The doline-dotted terrains and the travertine accumulations, where they are really characteristic or special, are presented in more detail (e. g. Bihar Mts, Torna Mts; Bükk Mts; Northern Carpathians in general). With regard to the Classical Karst, he also drew the attention to the fact that its "Unforested, bare nature… is not its original property" (Vol. II, p. 562).

Following János Hunfalvy, Géza Czirbusz gave a general description of karst phenomena in the Aranyos (Arieş) Valley in Transylvania (1890), and also dealt with the origin of limestone gorges in the framework "gap valleys" (1892).

Cholnoky (1926, 1930) summarized the experiences in karst geomorphology accumulated by the first half of the twentieth century, supplemented by his own and his student Gábor Strömpl's studies. Both of his works, more precisely the third volume of the latter work, contain figures showing karstification and polje types, the vast majority of which can still be used with confidence, even at the beginning of the twenty-first century. He classified poljes according to their origin on the basis of thorough geological studies and using the research results of Cvijić (1925) (Rybár and Veress

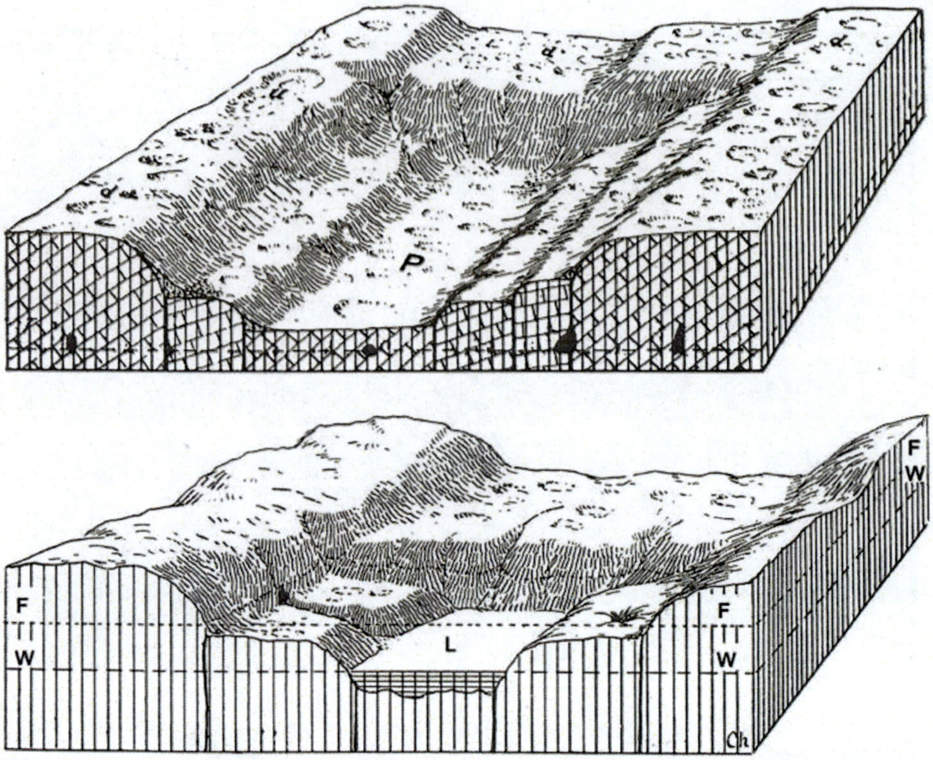

Fig. 2 **a** Block diagram of a dry, tectonic polje (P: polje, Cholnoky 1926). **b** Block diagram of a permanently water-covered polje (L: lake, F: flood, W: small water, Cholnoky 1926)

2020). Unfortunately, he also widely used the word "doline", which had become an international term by that time, even though it is misleading in some way. His genetical polje classification (Figs. 2 and 3) is still used in the international literature, and he was one of the first researchers to recognize the genetical connections between the cave and river terraces (1932). He proved the cave rupture origins of several spectacular limestone gorges with a number of unequivocal examples that are still valid today. He used examples not only from the territory of historical Hungary and the Dinaric Mountains, but also from Algeria, for example, Rummel Gorge. In this summary work, he only partially touches on the morphology of karren. Nonetheless, in his work we can read for the first time in the Hungarian karst literature that the CO_2 and H_2CO_3 content of the infiltrating rainwater increases under the influence of vegetation and soil.

Among Cholnoky's direct and "indirect" students are the already mentioned Strömpl (1912, 1913, 1914, 1923, 1933, 1935, 1944, 1945), Jaskó (1933), Kerekes (1936, 1938), Láng (1937, 1943, 1949, 1954, 1955) and Leél-Őssy (1954, 1960), who dealt mainly with karst geomorphological studies, the formation and morphology of gorges, dolines and poljes.

By the middle of the twentieth century, the general interpretation of karstification had fundamentally changed. In addition to CO_2 from the air, the CO_2 enrichment of water seeping through vegetation and soil was thoroughly studied according to climate zones (Lehmann 1948, 1954, 1955, 1956, 1959, 1960, 1962; Bögli 1954, 1956, 1957, 1963). All these new results lead to the necessary reworking of previous theories about karst morphogenesis.

Regarding the formation of surface karst forms, Jakucs (1950, 1951, 1953, 1956, 1968, 1978) and Balázs (1959, 1960, 1961, 1962, 1963a, 1964, 1965), who visited humid tropical karsts in many places of the world, should be highlighted among other Hungarian karst scientists. The "Morphogenetics of karst regions: variants of karst evolution" book of Jakucs (1971) was published in both English (Jakucs 1977) and Russian language (besides Hungarian), and is still one of the summary guides to karst research in Hungary and abroad.

The modern studies on the conditions of sinkhole formation are connected to the name of Kevei Bárány (1980, 1991, 1998), Kevei Bárány, Jakucs (1984)—a student and later a colleague of László Jakucs—and László Zámbó. The former mainly studied the microclimate, soil and vegetation conditions of the sinkholes, whereas the latter studied the seasonal, and even the daily changes in the amount of CO_2, H_2CO_3 and humic acid seeping into the dolines (1970, 1986, Zámbó and Telbisz 2000, Zámbó and Tompa 2001, Zámbó and Ford 2003). Móga (1999, 2011) and Szeremley (1972) correctly explained the connections between the formation of dolines, stream sinks and cave passages.

Theories about doline formation were further enriched with new data by Hevesi (1986) and Telbisz (2001a, 2001b).

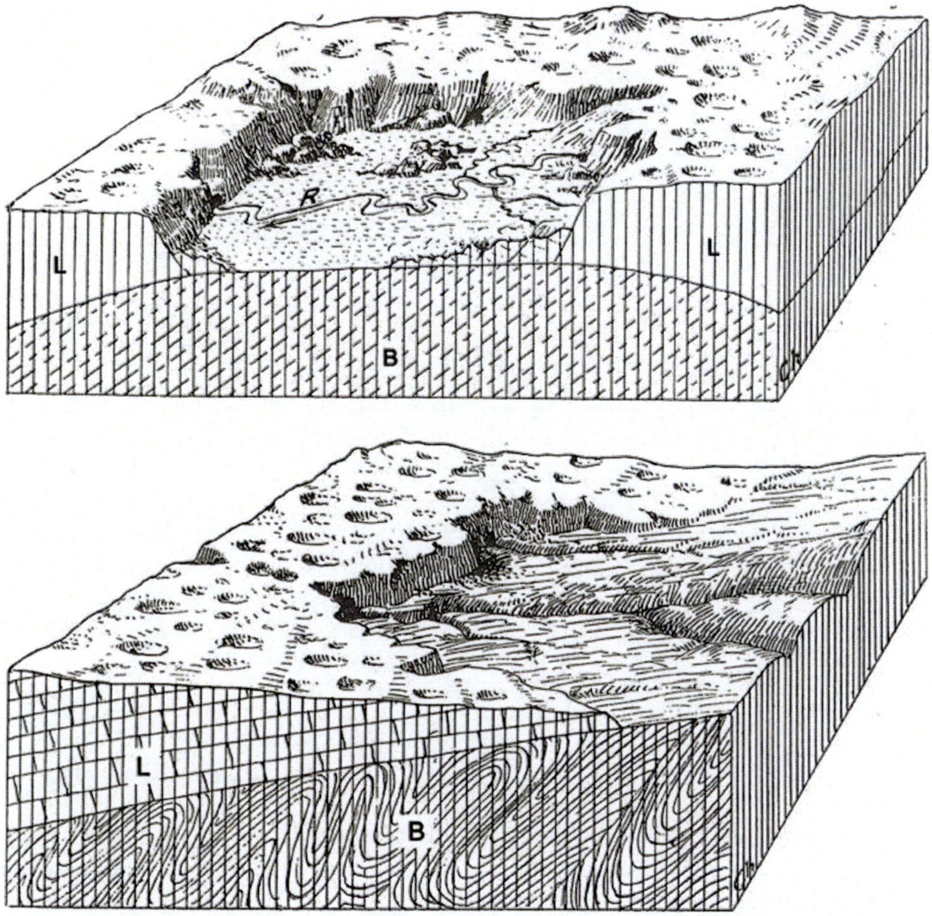

Fig. 3 **a** Schematic block diagram of a denudation polje (L: limestone, B: bedrock, R: river, Cholnoky 1926). **b** Schematic block diagram of a border polje (L: limestone, B: non-karstic rock, Cholnoky 1926)

Taking into account their different formation conditions, Veress (1999, 2000) classified these "diagnostic" karst landforms in more detail. Attila Hevesi introduced the concept of "blind valley stream captured by karst margin sinkholes and flowing in hidden passages" (2002). The characterization of karst margin and karst inland sinks was previously studied in detail by Leél-Őssy (1959).

The student of László Jakucs, Márton Veress, started the modern, internationally significant study of the formation of the most superficial smaller scale karst forms, the karren (runnels or "devil's ploughs"). In this respect, he examined almost all the mountains built up of (at least partly) karstic rocks in Hungary; he also carried out expeditions to the humid tropical zone karsts and even to the high mountains. In particular, he successfully studied the development of karren and depressions by comparing field and laboratory experiments. For years, he repeatedly returned to the on-site examination of karren in Austria, the Totes Gebirge and the Dachstein Mountains (Veress 1995, 1998, 2004a, b, 2008, 2010, 2011, 2013, 2014, 2015, 2016, 2017) and summarized the results of studies on karstification processes almost to the present (Veress 2004a, b).

The freshwater limestone formations of the Carpathians and the Carpathian Basin were previously studied by Scheuer and Schweitzer (1973a, 1973b, 1981, 1983). In the 2000s, Sándor Kele continued to research Hungarian travertine deposits using modern methods (Kele 2009; Kele et al. 2009, 2011). As Jenő Cholnoky examined the correlations between the formation of cave and valley terraces, the above authors and Márton Pécsi correlated the river-terrace levels and the elevation levels of travertine deposits (1984). Their research also yielded archaeological results of international significance (e.g. in the case of Vértesszőlős).

Following the work of Boros (1956, 1968), the role of algae and mosses in travertine formation was the most thoroughly studied and measured in Hungary by Hevesi (1970, 1971, 1972). The process of travertine cave formation by the creation of a limestone tuff ceiling was described by Jakucs (1971) and supported by several examples from the Bükk Mountains by Hevesi (1972, 2002a, b, c).

From the second half of the twentieth century to the beginning of the twenty-first century, there was a significant increase in Hungarian karst geomorphological mapping. The maps of the most characteristic karst areas of the Bükk and

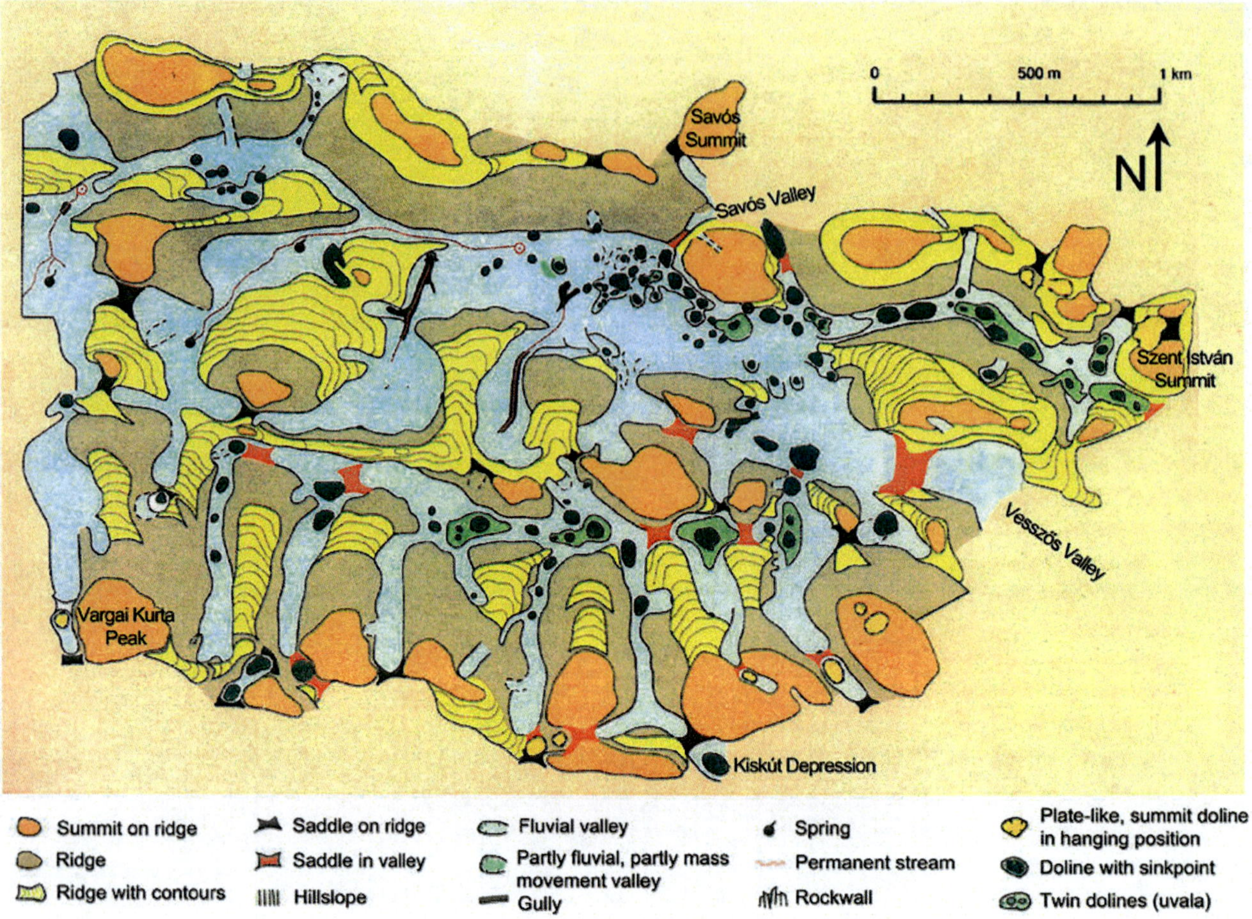

Fig. 4 Geomorphological map of the Létrási karst depression in the Bükk Mountains (Hevesi 1997)

the Mecsek Mountains were drawn by Hevesi (1986, 2002; Fig. 4). Móga (1998) made similarly comprehensive maps of the Gömör-Torna Karst (Fig. 5).

From the middle of the twentieth century, several researchers undertook the general characterization of the most important Hungarian karst terrains. The karst morphology of Aggtelek and its surroundings was presented by Jaskó (1935),Jakucs (1960, 1964), Sásdi (1998), Zámbó (1998), Bakony Mountains by Veress (1999), Vértes Mountains by Sásdi (2016), Bükk Mountains by Strömpl (1914), Tóth (1983), Hevesi (1986a, b, c, 1990, 2002a, b, c), and Sásdi (1997), Mecsek Mountains by Szabó (1953, 1956a, b),Bartha and Tarnai (1997), and Hevesi (2001). Summary reviews of all Hungarian karst terrains were prepared by Jakucs (1957, 1960, 1961), Leél-Őssy (1960), Szabó (1968), Hevesi (1991, 1997) and Veress (2011). The effect of the impervious rock cover on karstification and the consequences of its denudation were analyzed by Dénes (1974),Hevesi (1986, 2002c) and Veress (2016).

3 Cave Research and Explorations

3.1 Early Period

If only the territory of present-day Hungary is taken into consideration, whose borders were established by the Treaty of Trianon in 1920, and other areas belonging to Hungary before the First World War are ignored, then the first book mentioning a concrete cave was published by György Wernher (Georgius Wernherus 1497–1567) in Basel in 1549. This publication with the title "*De admirandis Hungariae aquis hypomonemation*" provides some information about Baradla Cave, though the location of the cave is incorrect. This location error was repeated in later editions in Vienna, Basel, Köln and Strassbourg during the 1600s. The correction was made only by Matthias Bel (1684–1749), who published the first detailed geographical description of the then Hungary under the title of "*Notitia Hungariae*

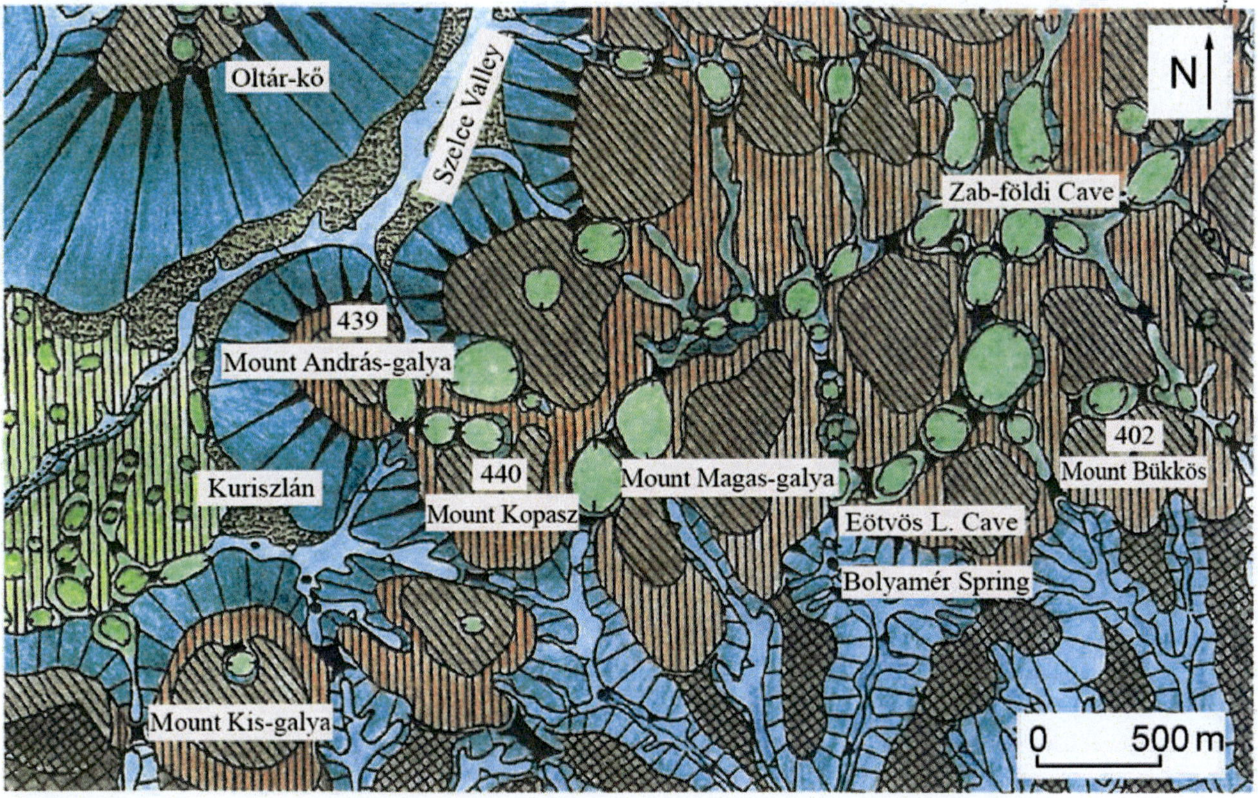

Fig. 5 Doline series in hanging valleys at the left side of Jósva Stream, NE from Jósvafő village (Móga 1998)

novae historico geographica" in 1742. The knowledge-seeking, naturalist approach in caving, the creation of cave maps and the thorough description of cave phenomena started in the 1700s, and Matthias Bel was one of the pioneers who intended to extensively present the caves of Hungary. At his request, György Buchholtz (1688–1737), a naturalist of his age, visited, described and mapped several caves, but all of them are beyond the present country borders of Hungary. Although Buchholtz visited Baradla Cave in 1723, it was not documented, and when Bel finished his manuscript about Gömör County, he noted that Baradla Cave has not yet been scientifically explored (Bél 1992).

3.2 Nineteenth Century

In the territory of present-day Hungary, the first caving activity with purely scientific purposes took place in 1794, when János Farkas and József Sartory visited Baradla Cave in order to scientifically describe its features. Farkas published the first detailed Hungarian language description of the cave, whereas Sartory created maps about the then known sections (Fig. 6). Although both original works have been lost since then, a copy of their map was discovered in 1962 (Bendefy 1962). Cave historians consider it as the first cave map created by an engineer, but in fact it is rather a draft based on some direction and distance measurements.

From this time on, it became more common to describe visits to caves, to report geological, mineralogical, climatological and biological observations, and to survey passages. In the first period, the target was mostly Baradla Cave, and later Abaliget Cave, and the description and popularization of this knowledge were done not only by Hungarians but by foreign naturalists as well. As a result of German-language publications, these two caves became part of the international cave literature at that time. Domonkos Teleki (1773–1798), the son of Sámuel Teleki the chancellor of Transylvania, visited Baradla Cave in 1793 during his travel in Hungary (Fig. 7). His travel diary was published in Hungarian (1796) and in German (1805), too. In the same year, Robert Townson (1762–1827), an English naturalist, also visited the cave. His notes about his travel were published in London (1797), Paris (1799, 1803), Leipzig (1800) and Haga (1800, 1801). He was the first to carry out temperature measurements in Baradla Cave. Furthermore, he also reported the existence of shaft caves on Alsó-hegy plateau in Aggtelek Karst (Hadobás 1992).

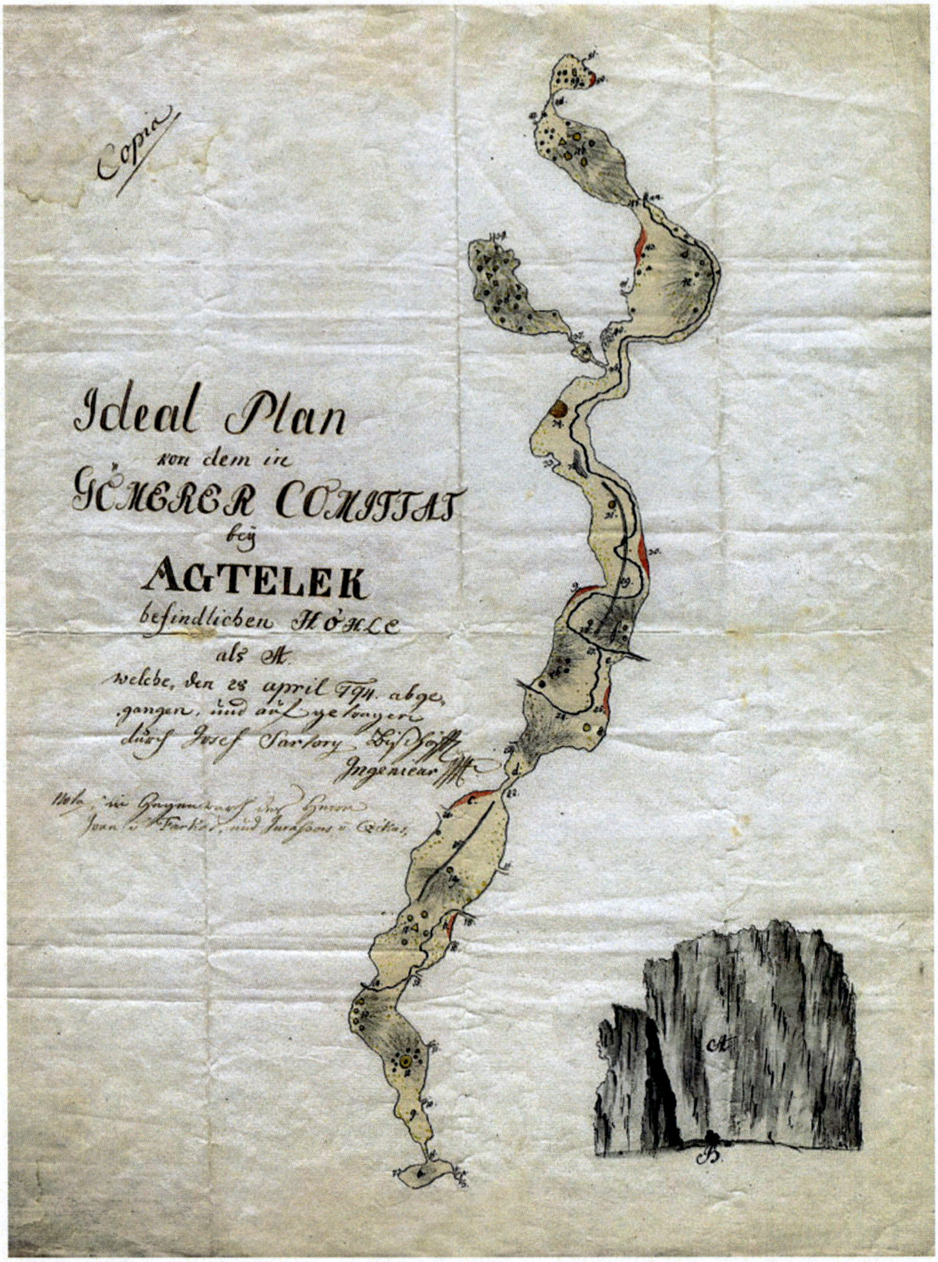

Fig. 6 The copy of the Baradla Cave map of József Sartory József (1794)

Stanisław Staszic (1755–1826), the father of Polish geology visited Baradla Cave in 1799. He presented several details of the cave in his book published in Warsaw. As an appendix to his book, he also presented the first known geological map of Hungary (1815), in which he marked the location of Baradla Cave. The famous Russian poet, Fyodor Nikolaevich Glinka (1786–1880) returned from the Napoleonic war as a military officer, and on his way, he visited Baradla Cave in 1805. In his book published in Moscow (1815) he provided an illustrative description about the cave and he also appended the Russian language version of Keresztély Raisz's cave map. Francois-Surpice Beudant (1787–1850), a French geologist visited several caves during his nine-month-long staying in Hungary. He presented Baradla Cave in his book published in Paris (1822) citing Raisz's work, but his personal visit to the cave has not been approved. John Paget (1808–1892) an English physician and George Edwards Hering (1805–1879) a landscape painter visited Baradla in 1835 during their one-and-a-half-year-long travel in Hungary. Paget reported his experiences in a book, which had several editions (London 1839, 1849, 1850, 1855, Leipzig 1842, 1845),

Fig. 7 The entrance of Baradla Cave in the book of Domonkos Teleki (1796)

while Hering published an album of engravings, in which one image is about Baradla (Fig. 8), and it is the first printed illustration about the cave (1838).

The first scientifically important step was made by Keresztély Raisz (1766–1849), who was a land surveyor in Gömör County. He was the first to create a plan view and longitudinal map of Baradla Cave with scale, labels, orientation, and symbols related to slope conditions and cave deposits. He was also the first in combining the surface profile and the underground passages in the same map (Fig. 1). These maps and drawings were published several times after the first edition (Görög 1802), for instance in the Hungarian, Latin, German and French language works of Bertuch (1813). Raisz also created a German language description of the cave connecting the underground features to surface karst landforms. He established the nomenclature of the cave, fixed the folk names and his own denominations were used for different features and parts of the cave (1807). He also explored new parts in the cave by going beyond the previously known endpoint of Baradla. He also mentioned the lower cave of Baradla, though he had some erroneous ideas about it. Furthermore, he tried to explain the origin of the large number of human bones found in the cave. Nonetheless, he insisted on his views about the origin of Baradla by fire disregarding the opinion of many of his contemporaries, who had more correct ideas about the origin of caves. Raisz mentioned the black, "lava-looking" gravels found in the passages as proves of the cave's origin by fire. Because of these misinterpretations, later ages honoured only his map, but his cave description and explorations were underestimated for long.

The successor of Raisz, Imre Vass (1795–1863) also a land surveyor, systematically visited all parts of the cave for several years, and in 1825 he successfully crossed the

Fig. 8 The Danse Room of Baradla Cave in the hand-coloured lithography of George Hering

Vaskapu (Iron Gate), which was the then known endpoint of the cave and explored new parts of the main branch in a length of 4.5 km. He thoroughly mapped the newly discovered parts and created plan view and longitudinal section drawings (Fig. 6 in Chapter 10). Based on previous geological studies, he described the geological and hydrological settings of the area, the history of the cave exploration, and the formations found in the cave. He presented his ideas about the origin of the cave and the dripstones. He observed that the cave temperature did not follow outside air temperature changes. Based on his measurements at several points of the cave, he stated that there is a constant 10 °C in the cave. He was also interested in the caves of other countries and compared them with Baradla. At that time, Baradla was recorded as the second-longest cave on Earth. The maps of Imre Vass (1829) were created by engineering accuracy and artistic design. Furthermore, the scientifically based description of the cave was published in Hungarian and German languages (1831c). Therefore, Baradla had a leading role in the history of cave research, as there had not been similar publications and surveys about Mammoth Cave (which was and still is the longest known cave on Earth) to that time. The original manuscript of the map was probably lost during the Second World War in 1944, but the original description is still preserved; thus it is hypothesized that it is the oldest preserved cave monograph manuscript in the world (Székely 2012).

In the first half of the nineteenth century, the second most significant location was Abaliget Cave from the viewpoint of map survey and scientific documentation. The cave is situated near the eponymous village in the Mecsek mountains. In 1786, the miller József Mattenheim already penetrated the cave behind the spring, which operated the watermill, but we do not have any detailed data about his exploratory activity. It was only in 1819, that Vince Kölesi (1780–1847) a nearby farmer, after long preparation also entered the cave through the spring risking his own life and reached the main branch. He not only discovered but also surveyed the cave and described its formations, the cave deposits and even the cave life-forms. His work was published already in the coming year (1820), but his cave map remained unpublished. However, it has been preserved as a manuscript and a low-quality copy of the map (Fig. 9) was published by Adolf Schmidl in 1863.

In Hungary, the first geographer who scientifically dealt with the origin, morphology and deposits of caves was Adolf Schmidl (1802–1863), who is also known as the father of scientific cave research in Hungary. He was a professor of geography at the Budapest Joseph Polytechnicum from 1857 until his death. His main study areas included the karst terrains of Austria (1858) and Slovenia (1854), as well as the Bihor Mountains (1863), which are today in Romania, but belonged to the area of Hungary at that time. Moreover, he studied the Ice Cave of Silica (today in Slovakia), Baradla Cave (1856), Abaliget Cave (1863) and thermal springs of Buda. His research carried out in Baradla was also presented in the Academy of Vienna. In the printed versions of his presentations (1856, 1857) he summarized the knowledge of his predecessors and presented the results of his own observations and measurements comparing them with data and experiences from Slovenian and Austrian caves. He was the person who pronounced that Baradla is the second-longest cave on Earth after Mammoth Cave (USA) and he also outlined that the beauty and richness of the dripstone formations in Baradla are only slightly less than those of Postojna Cave. During the presentation of the surface terrains, he mentioned dolines, stream sinks, lakes and watercourses. He explained the formation of the entrance and passages of Baradla, described the movement of water and air in the cave, the characteristics of the enclosing rocks, and compared the extent of the cave with the dimensions of other European caves. Analysing the climatic conditions, he also reported the data of his predecessors as well as his own measurements. He described the colour, shape and frequency of speleothems. He presented the flora of the cave and the latest results of fauna research. With the breakthrough of a flowstone, he found the bones of a cave bear, but he could not examine in detail these finds. Based on Vass's work and his own observations in Baradla Cave, he presented eight side branches of the cave. He also explored a new part when he penetrated about 400 m deeper in the Retek-ág (Radish branch) than his predecessors (Hadobás 1988).

From the middle of the nineteenth century, the more specialized scientific examination of the caves also began. Among the scientific works, biological research stands out, which is mainly related to the name of the zoologist János

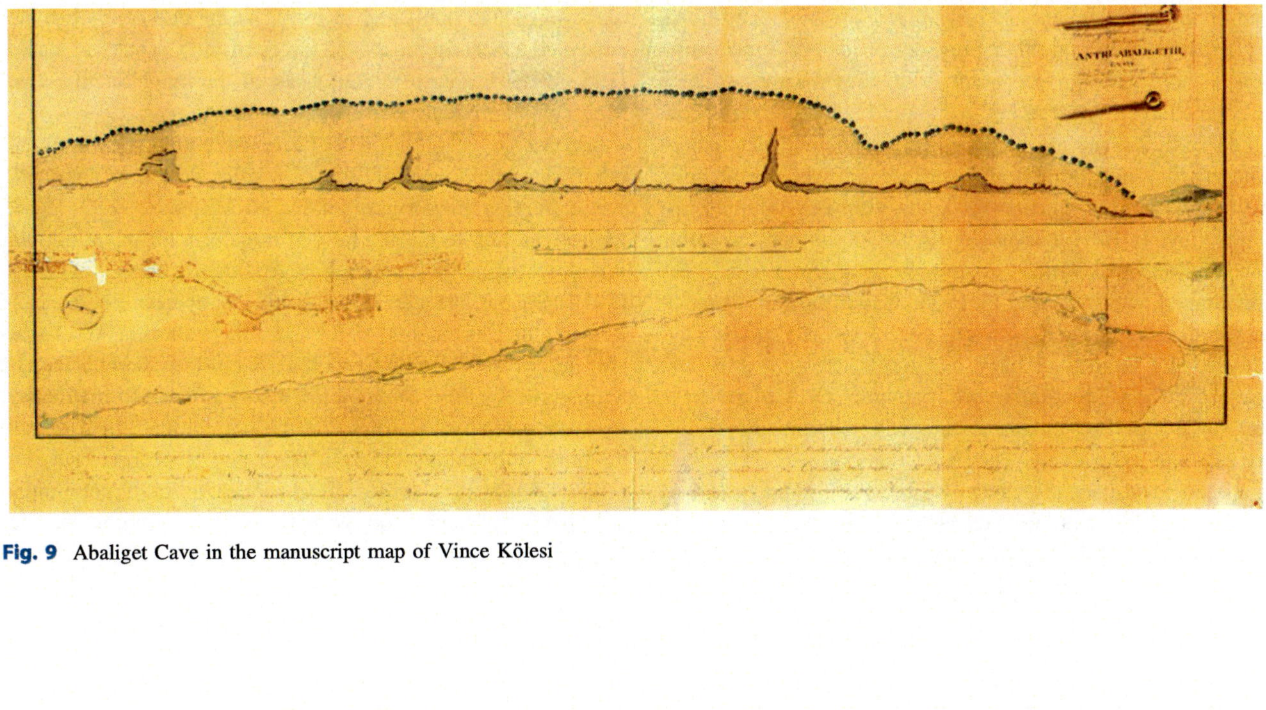

Fig. 9 Abaliget Cave in the manuscript map of Vince Kölesi

Frivaldszky (1822–1895), a member of the Hungarian Academy of Sciences. He was the first to process the fauna of Hungarian caves based on his own collection (1865). János Petényi Salamon (1799–1855), the founder of Hungarian ornithology and palaeontology, made a palaeontological description of the cavities of the Beremend limestone quarry (Kubinyi 1864). Baron Jenő Nyáry (1836–1914), an archaeologist and royal chamberlain, was of the opinion that significant traces of the prehistory of mankind were preserved in caves (in addition to river valleys), thus he made excavations in Baradla three times in 1876–1877. His book (1881) on the excavation, which contains the exact drawings of 335 finds, is the first significant Hungarian cave excavation documentation. Its value is shown by the fact that Lajos Kossuth (former governor of Hungary who lived in exile after the War of Independence in 1848–49), who was familiar with natural sciences as well, presented and analysed the contents of Nyáry's book in a 40-page study(1883). In 1869, József Szabó (1822–1894), the founder of the Hungarian geological school, organized an "expedition" on behalf of the Geological Society to explore and scientifically study the Csörgő-lyuk Cave (Fig. 10) in Mátra Mountains (Szabó 1890). He was the first to visit the longest cave of non-karst origin in Hungary, and in addition to its description, he also provided information on its formation, the enclosing rocks and the method of the survey (Síkhegyi and Székely 2014).

3.3 Twentieth Century

Although significant results were achieved already in the nineteenth century, the most important cave explorations and studies occurred in the twentieth century, thanks to deliberate, regular and organized work. At the turn of the nineteenth to the twentieth century, there was no institutional cave research, and no official register of underground values was established. For the number of known caves—which may not have even reached a hundred—we can only conclude from the reports published in dailies and magazines. The gradual development in cave research was significantly influenced by two events. First, the discovery of the Pál-völgy Cave in 1904 marked the start of exploratory research; second, the excavation of the Szeleta Cave (in Bükk mountains), which began in 1906, was a milestone in scientific research. Thanks to the successes, the caves have aroused the interest of more and more scientists and cave tourists. It has become more common to explore new, unknown cave systems and to study the already known caves more extensively. People of this era also recognized the need to protect the caves, while at the same time they worked a lot to popularize caves and make them accessible for a wider public. In the history of cave research, the political and economic events of the twentieth century caused large ups and downs. However, really significant changes took place only after the Second World War, when the reviving cave research, both the scientific and exploratory activities, began to move in new directions.

3.4 Scientific Research

When Ottokár Kadić (1876–1957), the father of Hungarian cave research, began excavating the Szeleta Cave (Bükk Mountains) in November 1906, at the suggestion of naturalist Otto Herman (1835–1914), he did not think that his activity would determine the direction of scientific cave research for about four decades. During this period, the acquisition of new knowledge about caves was mainly related to palaeontological and archaeological excavations. In 1907, Kadić found traces of the prehistoric man in the Szeleta Cave (1915). During the excavation of Jenő Hillebrand (1884–1950) in 1909, Neolithic children's bones were found in the Balla Cave, also in Bükk (1911). Although the authenticity of the finds was initially questioned by many, eventually the Archaeological Congress held in Tübingen in 1911 confirmed the authenticity of the Palaeolithic finds in the Szeleta Cave. Several prominent scientists were involved in the works, such as the anatomist Mihály Lenhossék (1863–1937), palaeontologist and archaeologist Tivadar Kormos (1881–1946), palaeontologist and ornithologist Kálmán Lambrecht (1889–1936) and Mária Mottl (1906–1980) geologist and palaeontologist. Thanks to regular state support, excavations took place not only in the caves of Bükk (Fig. 11) but also in the well-known, large cave entrances of Pilis,Gerecse, Bakony and Aggtelek Karst. In connection with the excavations, Kadić visited, surveyed and described all the caves in the area. His reports on his work were also published in print. At the end of his life, he compiled a description of the caves of the Carpathian Basin (1952). These data provided most of the input for the cave registry setup decades after his death.

Of course, in addition to the excavation type research, other long-standing professional results were also obtained. For example, the geographer Rezső Milleker (1887–1945) figured out before his age how to transform a camera into an instrument suitable for the photogrammetric survey of caves (1914). Rezső Maucha (1884–1962), a hydrobiologist, carried out measurements of the Baradla waters in 1930 that still serves as a reference (1931). Geologist Ferenc Pávai-Vajna (1886–1964) founded the theory of cave formation by thermal water by his article on the corrosional effect of hot vapours and gases (1931). Hydrologist Hubert Kessler (1907–1994) set up a theoretical model for the development of Alsó-hegy shaft (pothole)caves (1932). Geologists and geographers such as Jenő Cholnoky (1870–1950), József

Fig. 10 Profile of Csörgőlyuk Cave at Ágasvár (Mátra Mountains) in the drawing of Lajos Tanhoffer (1870)

Kerekes (1911–1957) and Gábor Strömpl (1885–1945) made karst morphology including speleology widely accepted as a science. Biologists, mainly Elemér Bokor (1887–1928), Antal Gebhardt (1887–1972) and Endre Dudich (1895–1971) significantly contributed to the spread of the science of biospeleology in Europe (Dudich 1932) (Fig. 12).

After the Second World War, the scientific life slowly rearranged, and for several decades hydrological studies (discharge measurements, correlation analysis etc.) had a leading role in karst research, which greatly contributed to the exploration of active stream cave systems. In the 1960s, the study of the climatic components, the healing effect and the indicator role of the caves came to the fore. Since the turn of the millennium, mathematical modelling of cave development has become a significant new approach.

In the 1950s, a view was strongly prevalent among certain geologists in Hungary, according to which the caves are uninteresting topics for science, so there was a significant lag in the geological-morphological study of caves compared to other countries. The change was brought about by cave researchers entering higher education, thanks to whom "karst and cave research schools" were formed: thanks to László Jakucs (1926–2001) and Ilona Kevei Bárány in Szeged, Sándor Leél-Őssy (1924–1998), and later his student, László Zámbó (1936–2017) and his son, Szabolcs Leél-Őssy (1960–) in Budapest, to Márton Veress in Szombathely, and to Attila Hevesi and László Lénárt in Miskolc. As a result, in recent decades, many university students have chosen a topic related to karsts and caves, and their results have drawn attention to the wide range of research opportunities in connection with caves. Due to the development of cave mapping methods and in some cases the personal craftsmanship of certain speleologists, cave maps made in the last 30 years, which are the basis of further research, provide information not only on the size and direction of passages, but also faithfully reflect the

Fig. 11 Ottokár Kadić and Mária Mottl in front of Suba-lyuk Cave, with the people working at the excavation (photo by L. Hajdú 1932)

Fig. 12 Endre Dudich during biological collection work in Baradla Cave

nature and morphological characteristics of passages, the cave deposits, and help to more deeply understand cave development.

In addition to articles published in journals, the shear amount of research results is also reflected in the collection of reports of more than a thousand volumes, which summarize the annual activities of cave researchers and speleoclubs. Since 1975, the most accurate reports are regularly awarded by the common tender of the Hungarian Karst and Cave Research Society and the Nature Conservation Office. The tender named after Jenő Cholnoky was created to stimulate the quality documentation of cave research and surveys.

3.5 Exploratory Research

By the turn of the twentieth century, mountain tourism was already a popular pastime. More and more people set out to explore the unknown caves, whose entrances were observed during the hiking tours. Moreover, vertical caves with deep shafts could be visited, thanks to the improving gears used for mountaineering.

The first significant success in the exploratory research of this period was the discovery of the Pálvölgyi Cave in 1904. The first cave expedition with rope ladders and manila ropes started in 1911 (Fig. 13) to descend into the vertical caves of Alsó-hegy plateau (Aggtelek Karst). However, the deepest caves were not explored until 1927. By the 1930s, explorations were taking place in almost all karst areas, and many caves with natural entrances or artificially opened by human activity were under exploration. Significant previously unknown sections were explored, and entirely new parts or systems were discovered by removing sediment fillings from the passages. The most significant results of this period were: penetrating the then endpoint of the Baradla and discovering a new 500 m passage in 1922, which provided an opportunity to create an artificial exit from the cave at Jósvafő village. In 1932, by penetrating the siphon of the Styx stream, the connection of Baradla and Domica caves was proved (Fig. 14). In the same year, the length of Baradla Cave was increased by about 1,000 m by exploring further sections of the Retek-ág (Radish branch). The caves discovered in the capital (1930: Szemlő-hegy Cave, 1932: Ferenc-hegy Cave) provided new evidences for the theory of cave formation by thermal waters. Meanwhile, in order to explore the large cave systems of the Bükk Mountains, the researchers used all their technical knowledge to mechanize the removal of sediments filling the caves.

The first significant result of the post-Second World War cave explorations was the discovery of the Sátorkőpuszta Cave (1946), a thermal water origin cave, which is richly decorated with gypsum formations. In 1948, the cave researchers of BETE club explored the "Centenáris" section of the Mátyás-hegy Cave, increasing the length of the cave to 1540 m. In the 1950s, researchers set out to explore cave systems that were previously unknown. At that time, the discoverers had some professionally grounded ideas about the probable location of the unknown caves based on water tracing experiments and other hydrological or geological arguments. Explorers systematically tried to penetrate these unknown caves from sinkholes or from sources and achieved remarkable successes. First the large active stream caves of the Aggtelek Karst (Béke Cave 1952, Vass Imre Cave, Szabadság Cave 1954, Kossuth Cave 1956, Meteor Cave 1961), then the large stream sinkhole caves of Bükk (Bolhási-víznyelő Cave, Jávorkúti-víznyelő Cave 1953, Szepesi Cave, Létrási-vizes Cave 1963, Istvánlápai Cave 1964, Szamentu Cave 1967) were explored.

From the seventies onwards, cave explorations received a new impetus again (Vecsembükk Pothole Cave in 1971, Hajnóczy Cave in 1972, Borókás-teber Stream Sink Caves #2 and #4 in 1975, Pál-völgyi Cave in 1980, Baradla Rövid-alsóbarlang in 1982, Danca Cave 1983, József-hegy Cave in 1984, Beremend Crystal Cave in 1984,

Fig. 13 The first explorers of the shaft caves of Alsó-hegy plateau (1911, photo by Imre Gábor Bekey)

Fig. 14 Hubert Kessler when the Baradla-Domica connecting passage was discovered

Nagyharsány Crystal Cave in 1994). Furthermore, as a result of the increasing explorations in Bakony, the previous opinion that these mountains are unfavourable for the formation of a large cave system due to the geological settings was overturned (Alba Regia Cave 1975).

By the 1990s, in addition to exploratory research based on the extraction of cave filling sediments, new methods had become more common. Explorers tried to increase the size of caves by climbing up in high chimneys (e.g. in the Baradla Cave), or by looking for connection passages between neighbouring caves in order to make up interconnected systems. The actually longest cave system in Hungary (Pál-völgy Cave System; 32,000 m) was created by connecting nine previously independent caves. The third longest cave system (Ariadne; 19,300 m) consists of six caves, while the fourth longest cave system (Tapolca; 14,930 m) consists of three individual caves.

In 1910, when the first cave register was made, 230 caves were recorded. In 1977, there were 1231 caves in the database. Today, the number of registered caves exceeds 4100. In 1977, there were 21 caves, whose length exceeded 1 km. In 2002, this number was 29, and today we know 39 caves longer than 1 km. In 1977, the depth of 24 caves exceeded 100 m, while in 2002 this number was 30, and today 38. As a result of recent exploration, the current deepest cave in Hungary is the Bányász Cave (Bükk) with a depth of 300 m.

3.6 Protection of Caves

The idea of cave protection appeared as early as the nineteenth century. Naturally, according to the age, the protection in this period covered only certain elements, such as dripstones and bats, the management of caves and the prevention of the extraction of cave sediments (especially guano). The first draft law on this topic was prepared by the Hungarian Cave Research Society in 1929 under the direction of Ottokár Kadić. The proposal of five chapters and 20 paragraphs has never become final. The legal regulation was first implemented in 1935, within the framework of the Forest Act, in the chapter containing the tasks of nature conservation. This regulation made it possible to protect caves and their surface areas. It prescribed that exploratory research and activities involving the change of status need permissions. It required the compulsory reporting of cave discoveries. It specified the rules of the management and utilization of caves, the possibility of revoking the utilization rights in case of privately owned caves but did not settle the issues related to the ownership of caves, and lacked to express the need to establish a professional institution related to caves. Under the law, the first declaration of a cave placed under protection took place in 1940, and by 1958 a total of 34 caves had been declared protected.

Hungary's first independent nature protection legislation (1961) placed all natural caves longer than 2 m under *ex lege* protection. In 1982, the possibility of putting caves under high protection was introduced. Exclusive state ownership was declared by Act No. LIII on Nature Conservation in 1996. According to this, a cave cannot be sold or privatized. The state is represented by the territorially competent national park directorate. This law, which is still current today, deals with the protection of caves in a separate chapter. The previous regulations have been left in force, but as a step forward, this law explicitly states that the protection of caves extends to its entrance, the whole system of passages, the host rock, the formations, the speleothems, the cave fill in any physical condition, the natural flora and fauna and the artificially constructed passages serving as entrance or communicating between different parts of the cave. It prescribes the designation of surface areas for the protection of caves, the registration of caves and their protection zone in the Land Registry, and the establishment of a National Cave Register.

In 1975, the Nature Conservation Office established the Institute of Speleology to perform state tasks related to caves. Since then, this institution has been reorganized several times and has lost its independence, but within the framework of the Deputy State Secretariat for Nature Protection it continues to prepare legal regulations, coordinate and control the cave-related tasks of national parks, and manage the cave register (Egri et al. 2015).

3.7 Cave Research Organizations

The first scientific organization of cave research, the Cave Research Committee, was established in 1910 within the framework of the Hungarian Geological Society, and in 1913 it developed into a department. Its primary task was to increase knowledge and to register caves. The Hungarian Cave Research Society, as an independent organization, was founded in 1926 by professionals, cave explorers and the interested public. The Society continued the activities of its predecessor organization, the registration of caves and the publication of the journal Barlangkutatás (Cave Research), which has been published since 1912, and also launched a new popular journal entitled Barlangvilág (Cave World, Fig. 15). The Society has played a leading role in all areas of cave research: in cave explorations, in the collection of biological data, in the morphological description of karst areas and in the constructions related to show caves. With the direction of the Society the caves under the Buda Castle were explored and transformed into a show cave to welcome visitors. The Society was also the tour operator, thus the incomes contributed to the budget of the Society. Lecture meetings were held regularly, where members reported on their achievements. Ottokár Kadić with his popularizing articles and lectures not only brought the world of caves closer to the public but also put some higher political leaders at the service of cave research with his convincing power. The operation of the Society ceased during the bombing of Budapest in 1944 when the results of about four decades of work were destroyed in minutes when the office was hit by a bomb (Székely 2010a, b).

The Hungarian Karst and Cave Research Society was reorganized in 1958 and has been operating continuously since then. The Society coordinates the activities of individuals and groups engaged in the scientific and exploratory research of caves. In addition to regular lecture meetings, it organizes the annual National Cave Day as well as a professional meeting for cave researchers. Furthermore, the Society has organized several international conferences, mainly on cave genetics, cave morphology, cave hydrology, cave climate, speleotherapy, lamp flora and cave research history. In addition to several occasional and conference volumes, the Society has published a journal entitled Karszt és Barlang (Karst and Cave) since 1961. Besides, it maintains a cave professional library, organizes study trips to foreign caves and runs courses for the training of cave tour guides and research leaders according to legal regulations (Hazslinszky and Fleck 2010).

4 Conclusions

In Hungary, the very first papers on karst morphological objects (such as rock gorges, rock arches, ice-filled pseudo-caves, springs depositing freshwater limestone) were published at the end of the Middle Ages. From the eighteenth century, a textbook on geography was preserved, which deals specifically with surface karst phenomena. Since the nineteenth century, the main aim of scientists was to find explanations for the origin of different karst landforms. At first, caves were in the focus, but since the middle of the twentieth century, studies on surface karst phenomena have also gained momentum.

Of our caves, the oldest preserved records dating from the sixteenth to seventeenth centuries refer to Baradla cave. The first cave maps were also made about this cave in the late eighteenth century and early nineteenth century. In this century, cave exploration and research were already scientifically based activities, which resulted in studies on the morphology and speleogenesis of caves.

However, really extensive cave explorations and scientific research (including archaeological excavations) took place already in the twentieth century in Hungary. The vast majority of our known caves were discovered during this century, and modern, often instrumental scientific research was only started at that time. Organizations dealing with research and cave protection as well as karst and cave journals were also founded in this century.

BARLANGVILÁG

NÉPSZERŰ BARLANGTANI FOLYÓIRAT

SZERKESZTI:

KADIĆ OTTOKÁR

IX. KÖTET 1–2 FÜZET

MEGJELENIK NÉGYSZER ÉVENKÉNT

ELŐFIZETÉSI ÁR EGY ÉVRE 3 P—EGY FÜZET ÁRA 1 P

KIADJA:

A MAGYAR BARLANGKUTATÓ TÁRSULAT

BUDAPEST, 1939. MÁJUS.

Fig. 15 Cover page of Barlangvilág (Cave World) magazine, which was published between 1926 and 1944

References

Apáczai Csere J (1653) Magyar encyclopedia (Hungarian Encyclopedia), Utrecht (in Hungarian)

Balázs D (1959) A trópusi karszt terminológiai problémái (Terminology problems of tropical karsts). Karszt- és Barlangkutatási Tájékoztató 1959 november (in Hungarian)

Balázs D (1960) A Délkínai-karsztvidék vízrajza (Hydrography of the Southern Chinese Karst). Hidrológiai Közlemények 1960(6):485–492 (in Hungarian)

Balázs D (1961) A Délkínai-karsztvidék természeti földrajza (Physical Geography of the Southern Chinese Karst). Földrajzi Közlemények 9(4):327–346 (in Hungarian)

Balázs D (1962) 1962: A karsztok elterjedésének azonális és zonális feltételei (Azonal and zonal conditions of karst geographic distributions). Karszt és Barlang 1962(I):19–24 (in Hungarian)

Balázs D (1963) Karsztgenetikai problémák (Karst genetical problems). Földrajzi Értesítő 1963(4):487–494 (in Hungarian)

Balázs D (1964) A vegetáció és a karsztkorrózió kapcsolata (The relation of vegetation and karst corrosion). Karszt és Barlang 1964 (I):13–16

Balázs D (1965) A karsztkorrózió általános kémiai vonatkozásai (The general chemical aspects of karst corrosion). Karszt és Barlang. 1965(II):51–60 (in Hungarian)

Barta K, Tarnai T (1997) Karsztkutatás az Orfűi Vízfő-forrás vízgyűjtő területén (Karst research in the area of Vízfő Spring, Orfű). Karszt és Barlang 1997(I-II):12–19 (in Hungarian)

Bél M (1742) Notitia Hungariae novae historico geograpfica I-IV Wien pp 17

Bél M (1992) Gömör vármegye leírása (1749-ből) (Description of Gömör County from 1749). Gömör Néprajza 35:13–14 (in Hungarian)

Bendefy L (1962) Sartory József bányamérnök 1794. évbeli térképe az aggteleki Baradla-barlangról (The map of József Sartory cave engineer about Baradla Cave at Aggtelek created in 1794). Karszt és Barlang 1962(I):5–9 (in Hungarian)

Bertalanffy P (1757) Világnak két rendbéli Rövid Ismerete (Short introduction of the world from two aspects). Nagyszombat. (in Hungarian)

Bertuch J. F (1813) Bilderbuch zum Nutzen und Vergnügen der Jugend Novus orbis pictus. Wien

Beudant F. S (1822) Voyage mineralogique et geologique en Hongrie. Páris, II:90

Boros Á (1956) Magyarország mohaföldrajza (Moss geography of Hungary). Doctoral dissertation, Manuscript ELTE Budapest (in Hungarian)

Boros Á (1968) Bryogeographie und Byoflora Ungarns. Budapest

Bögli A (1954) Das Verhalten von Karbonaten in der Natur. Die Höhle 1954(3–4):36–54

Bögli A (1956) Der Chemismus der Lösungprozesse und der Einfluss der Geteinsbeschaffenheit auf die Entwicklung des Karstes. Report of the Commission on Karst Phenomena IGU, New York

Bögli A (1957) Die Phasen der Kalklösung. Verh. Schweitz. Naturforsch. Ges., Neuenburg

Bögli A (1963) Adatok a karsztbarlangok keletkezéséhez (Some data to the formation of limestone caves). Karszt és Barlang 1963(I):83–86 (in Hungarian)

Czirbusz G (1890) Karsztjelenségek az Aranyos középső szakaszában (Karst phenomena in the middle valley of Aranyos river). Földrajzi Közlemények XVIII:429–446 (in Hungarian)

Czirbusz G (1892) Az áttört völgyek. Erdély I (Water gaps. Transylvania I). (in Hungarian)

Csiba I (1714) Dissertatio Historico-Physica de Montibus Hungariae. Nagyszombat (Tyrnaviae)

Cholnoky J (1926) A földfelszín formáinak ismerete (Presentation of surface landforms). Királyi Magyar Egyetemi Nyomda, Budapest (in Hungarian)

Cholnoky J (1930) A napsugár diadala (The Triumph of Sunshine). Singer és Wolfner Irodalmi Intézet, Budapest (in Hungarian)

Cholnoky J (1932) Dolgozatok. A barlangok és a folyóvölgyek összefüggése (Studies. The relationship of caves and river valleys). Barlangvilág II(1–2):1–10 (in Hungarian)

Dénes Gy (1974) A fokozatosan lepusztuló vízzáró takaró szerepe az exhumálódó karszt morfológia fejlődésében (The role of gradually eroding impervious rock cover in the morphological evolution of exhuming karsts). Karszt és Barlang 1971(I):5–8 (in Hungarian)

Dénes Gy (2002) Magyar karsztföldrajzi szókincsünk középkori elemei (Middle Age Elements of the Hungarian Karst Geographic Vocabulary). Karsztfejlődés VII:243–251 (in Hungarian)

Dudich E (1932) Biologie der Aggteleker Tropfsteinhöhle "Baradla" in Ungarn. Wien

Egri Cs, Székely K, Takácsné BK (2015) A Barlangok védelméről (About cave protection). Túravezetői ismeretek 218–234 (in Hungarian)

Frivaldszky J (1865) Adatok a magyarhoni barlangok faunájához (Some data to the fauna of Hungarian caves). Mathematikai És Természettudományi Közlemények 3:17–53 (in Hungarian)

Glinka F Ny (1815) Piszma ruszkago oficera o Polsi, Avsztijszkih vladnijah, Prusszii i Francii o prodrobnüm opiszaniem pohoda Rosszijam protivu Francuzov, v 1805 i 1806. Moszkva, I:152–165

Görög D (1802–1811) Magyar átlás az az Magyar, Horvát és Tót országoknak vármegyéji (Counties of Hungary, Croatia and Tót countries). Vienne (in Hungarian)

Hadobás S (1988) Schmidl Adolf (1802–1863). Karszt és Barlang 1988 (I):37–42 (in Hungarian)

Hadobás S (1992) Passages concerning caves from Robert Townson Hungarian travelbook. Karszt és Barlang Special Issue, pp 33–36

Hazslinszky T, Fleck N (2010) A II. világháború után újjáalakult társulat. 100 éves a szervezett magyar barlangkutatás (The Society Recreated After WWII. 100 Years of Organised Hungarian Cave Research). Budapest Hungarian Speleological Society, pp 43–71 (in Hungarian)

Hering G (1838) Sketches on the Danube, in Hungary and Transylvania. London

Hevesi A (1970) Az algák és mohák szerepe a bükki forrásmészkő képződésében (The role of algae and mosses in the formation of travertines in the Bükk Mountains). Botanikai Közlemények 57 (3):233–244 (in Hungarian)

Hevesi A (1971) The role the algae and mosses in the formation of the Bükk spring-water limestone. Abstracta Botanica operum ex Instituto Taxonomiae-Oikologiae Plantarum Univ. Sci De L Eötvös I:22–30

Hevesi A (1972) Forrásmészkő-képződés a Bükkben (Travertine formation in the Bükk Mountains). Földrajzi Értesítő 21(2–3):187–206 (in Hungarian)

Hevesi A (1976) Kászonújfalvi Szabó János/1767–1858/ ályája és földrajzi munkássága (Carrier and Geographic Results of Kászonújfalvi Szabó János/1767–1858/). Földrajzi Értesítő XXV(2–4):417–430 (in Hungarian)

Hevesi A (1986a) A Déli-Bükk karsztja I.-II. (The karst of the Southern Bükk Mountains I-II.) Karszt és Barlang 1986a(I):3–14; and 1986a (2):87–94 (in Hungarian)

Hevesi A (1986b) Hidegvizek létrehozta karsztok osztályozása (Classification of karst terrains formed by cold waters). Földrajzi Értesítő 35(3–4):231–254 (in Hungarian)

Hevesi A (1986c) A Bükk hegység felszínfejlődése és karsztja (Landform evolution and karst of Bükk Mountains). MTA Földrajztudományi Kutatóintézet Manuscript PhD thesis (in Hungarian)
Hevesi A (1990) A Bükk felszínfejlődése különös tekintettel a karsztosodásra (Landform evolution of Bükk Mountains with special emphasis on karstification). Magyar Földrajzi Társaság, Budapest (in Hungarian)
Hevesi A (1991) Magyarország karsztvidékeinek kialakulása és formakincse I-II (Formation and geomorphology of Hungarian Karst Terrains I-II). Földrajzi Közlemények 39(1–2):25–35 and 39 (3–4):99–120 (in Hungarian)
Hevesi A (1997) A Bükk. Kertek 2000, Budapest, pp 331–345 (in Hungarian)
Hevesi A (1999) A szepesváraljai Drevnyik-hegy édesvízi-mészköve (The freshwater limestone of Drevenik Hill at Spišské Podhradie). Karsztfejlődés III:71–77 (in Hungarian)
Hevesi A (2001) A Nyugati-Mecsek felszíni karsztosodásának kérdései (Some questions related to the surface karstification of the Western Mecsek Mountains). Karsztfejlődés VI:103–111
Hevesi A (2002a) About the development of limestone Gorge. Geographica Croatica 35:57–66 (in Hungarian)
Hevesi A (2002b) Felszínalaktani jellemzés, karsztformakincs (Geomorphological characterization, karst landforms). In: A Bükki Nemzeti Park (Bükk National Park, ed.: Baráz Csaba), Bükki Nemzeti Park Igazgatóság, Mezőgazda Eger, pp 109–148 (in Hungarian)
Hevesi A (2002c) A magyarországi karsztok fejlődéstörténet és felszínalaktan szerinti csoportosítása; az „Aggteleki jellegű" karsztok felszíni formakincse (The classification of Hungarian karst terrains by landform evolution and morphology; the surface karst morphology of „Aggtelek type" karsts). Akadémiai doktori értekezés, Manuscript Miskolc (in Hungarian)
Hevesi A (2009) Töbör szavunkról, mint szakszóról (Our word „töbör" as a scientific term). Karszt és Barlang 2007(I-II):55–65 (in Hungarian)
Hillebrand J (1911) Az ősember csontjai a Bükkhegység Balla barlangjában (Bones of prehistoric man in Balla Cave, Bükk Mountains). Földtani Közlöny 41(92–95):209–210 (in Hungarian)
Hunfalvy J (1863) A Magyar Birodalom természeti viszonyainak leírása I-II (The natural characteristics of the Hungarian Realm I-II) Pest, pp 539–689 (in Hungarian)
Jakucs L (1951) Vízföldtani megfigyelések a Gömöri-karszton (Hydrogeological observations in the Gömör Karst). Földtani Közlöny 71:464–467 (in Hungarian)
Jakucs L (1953) Adatok néhány Bükk-hegységi karsztforrás ismeretéhez (Some data to the knowledge about certain karst springs in Bükk Mountains). Földtani Intézet Évi Jelentése 1950:49–58 (in Hungarian)
Jakucs L (1957) Aggtelek és vidéke (Aggtelek and its surroundings). Úti kalauz. Budapest (in Hungarian)
Jakucs L (1960) Általános karsztgenetikai, morfológiai és hidrográfiai problémák vizsgálata az Aggteleki-karszton (Investigation of some general karst genetical, morphological and hydrographic problems in Aggtelek Karst). PhD thesis Budapest (in Hungarian)
Jakucs L (1964) Geomorfológiai problémák az Észak-borsodi karsztvidéken (Geomorphological problems in the Észak-Borsod karst terrains). Borsodi Földrajzi Évkönyv, Miskolc V, pp 12–23 (in Hungarian)
Jakucs L (1968) Szempontok a karsztos tájak denudációs folyamatainak és morfogenetikájának értelmezéséhez (Some viewpoints to the interpretation of denudation processes and morphogenetics of karst terrains). Földrajzi Értesítő XVI I(1):17–46 (in Hungarian)
Jakucs L (1971) A karsztok morfogenetikája (The morphogenetics of karst regions). Akadémiai Kiadó Budapest (in Hungarian)
Jakucs L (1977) Morphogenetics of karst regions. Variants of karst evolution. Hilger, Bristol
Jakucs L (1978) A magyarországi karsztok fejlődéstörténeti típusai (Landform evolution types of Hungarian karst terrains). Karszt és Barlang 1977 (I–II):1–16 (in Hungarian)
Jaskó S (1933) Morfológiai megfigyelések és problémák a Gömör--Tornai karsztvidék délkeleti részében (Morphological observations and problems in the southeastern part of Gömör-Torna Karst. Földrajzi Közlemények 9–10:245–258 (in Hungarian)
Jaskó S (1935) A Jósva patak felső völgyének geológiai leírása (Geological description of the upper Jósva Valley). Földtani Közlöny 55:291–300 (in Hungarian)
Kadić O (1915) A Szeleta-barlang kutatásának eredményei (Results of Szeleta Cave research). Magyar Királyi Földtani Intézet Évkönyve 23(4):147–278 (in Hungarian)
Kadić O (1952) A Kárpát medence barlangjai. Magyarország, Csehszlovákia és Románia területén előforduló barlangok ismertetése (Caves of the Carpathian Basin. The presentation of caves in Hungary, Romania and Czechoslovakia). Magyar Állami Földtani Intézet 846 (in Hungarian)
Kászonújfalvi Szabó János (1803) A' Sz.irás, és a' természet' szava a' Földnek, és az emberi nemnek némely főbb változásairól (The words of the Bible and Nature about the major changes of Earth and mankind). Kolozsvár (in Hungarian)
Katona M (1824) Közönséges természeti földleírás (Common physical geographic description). Pest 1824 (in Hungarian)
Kele S (2009) Édesvízi mészkövek vizsgálata a Kárpát-medencéből: paleoklimatológiai és szedimentológiai elemzések (Study of freshwater limestones from the Carpathian Basin: paleoclimatological and sedimentological implications) PhD Theses, Eötvös Loránd University & Hungarian Academy of Sciences, Budapest (in Hungarian)
Kele S, Gy S, Demény A, Shen C-C, Chiang H-W (2009) U-series dating and isotope geochemical study of the Gellért Hill (Budapest) travertine. Central Eur Geol 52(3–4):199–224
Kele S, Scheuer Gy, Demény A, Shen C-C, Chiang H-W (2011) A Rózsadomb (Budapest) édesvízi mészköveinek U/Th sorozatos kormeghatározása és stabilizotóp-geokémiai vizsgálata (Uranium-series dating and geochemical study of the travertines located on the Rózsadomb, Budapest) Földtani Közlöny (Bull. Hung. Geol. Soc.) 141(3):293–312 (in Hungarian)
Kerekes J (1936) A Tárkányi öböl morfológiája (Morphology of Tárkány-inlet). Földrajzi Közlemények LXIV 6–7:80–97 (in Hungarian)
Kerekes J (1938) Az Eger környéki barlangvidék kialakulása (Formation of caves and landscape around Eger). Barlangkutatás 8(1–2):25–30 (in Hungarian)
Kessler H (1932) A zsombolyok keletkezéséről (About the formation of shaft caves). Barlangvilág 2(3–4):20–22 (in Hungarian)
Kevei Bárány I (1980) Some data about the physical and chemical properties of soil of karst dolines. Acta Geographica Szegedensis 20:34–49
Kevei Bárány I (1991) Distinction of doline types based on ecological and morphometric parameters. Conference ont he karst in Hungary, pp 47–54
Kevei Bárány I (1998) Geoecological system of karsts. Acta Carsologica 27(1):13–24
Kevei Bárány I, Jakucs L (1984) Szempontok a karsztok felszínformáinak rendszerezéséhez, különös tekintettel a dolinák típusaira (Some viewpoints to the systematization of karst landforms with

special emphasis on doline types). Földrajzi Értesítő XXXII I (3):259–265 (in Hungarian)
Kossuth L (1883) Tanulmányok báró Nyáry Jenő az Aggteleki barlang mint őskori temető czimű munkája felett (Studies on the work of Baron Jenő Nyáry, the Aggtelek Cave as a prehistoric cemetery). Franklin T. Könyvnyomda (in Hungarian)
Kölesi V (1820) Az újonnan felfedezett Abaligethi Barlangnak leírása (Description of the recently discovered Abaliget Cave). Tudományos Gyűjtemény 4(10):81–96 (in Hungarian)
Kubinyi F (1864) Petényi Salamon János hátrahagyott munkái (Posthumus works of Petényi Salamon János). Pest (in Hungarian)
Láng S (1937) Felvidéki karsztok (Karsts of Northern Hungary). Földrajzi Közlemények, LXV:145–150 (in Hungarian)
Láng S (1943) Karszthidrológiai megfigyelések a Gömör-Tornai--karsztban (Karst hydrological observations in Gömör-Torna Karst). Hidrológiai Közlöny XXII I(1–6):38–58 (in Hungarian)
Láng S (1949) Geomorfológiai és hidrológiai tanulmányok Gömörben (Geomorphological and hydrological studies in Gömör). Hidrológiai Közlöny, XXIX(1–6):38–58 (in Hungarian)
Láng S (1954) Hidrológiai és morfológiai tanulmányok a Bükkben (Hydrological and morphological studies in Bükk). Hidrológiai Közlöny. XXIV(1–2):70–81 (in Hungarian)
Láng S (1955) Geomorfológiai tanulmányok az aggteleki karsztvidéken (Geomorphological studies in Aggtelek Karst). Földrajzi Értesítő IV (I):1–20 (in Hungarian)
Leél-Őssy S (1954) A Magas Bükk geomorfológiája (Geomorphology of the High Bükk). Földrajzi Értesítő III(3):323–356 (in Hungarian)
Leél-Őssy S (1959) A Bükk víznyelőinek és víznyelőbarlangjainak tanulmányozása (Study of sinkholes and sinkhole caves of Bükk Mountains). Földrajzi Értesítő 8(2):179–190 (in Hungarian)
Leél-Őssy S (1960) Magyarország karsztvidékei (Karst terrains of Hungary). Karszt- És Barlangkutatás I:79–89 (in Hungarian)
Lehmann H (1948) Der tropische Kegelkarst auf den Grossen Antillen. Die Erde 2. Berlin
Lehmann H (1954) Das Karst phänomen in den verschiedenen Klimazonen. Erdkunde, Bd VIII. Bonn
Lehmann H (1956) Der Einfluss des Klimas auf die morphologische Entwicklung des Karstes. Rep. of the Comp on Karst Phenomena. Intern. Geographical, New York
Lehmann H (1959) Studien über Poljen in den Venezianischen Voralpen und im Hochappenin. Erdkunde, 13. Bonn
Lehmann H (1960) La termonologie classique du karst sous l'aspect critique de la morphologie climatique moderne. Revue de Géogr. de Lyon, Vol. XXXV. N°1
Lehmann H (1962) Karstmorfologie (Westermanns Lexikon der Geographie), Braunschweig
Maucha R (1931) Az Aggteleki cseppkőbarlang vizeinek chemiai vizsgálata (Chemical analysis of waters from Aggtelek Dripstone Cave). Hidrológiai Közlöny 10:201–207 (in Hungarian)
Milleker R (1914) Barlangok fotografikus felmérése (Photographic Survey of Caves). Barlangkutatás 2(1):34, 49 (in Hungarian)
Mitterpacher L (1789) Physikalishe Erdbeschreibung, Wien
Móga J (1998) Felszínalaktani megfigyelések a Gömör-Tornai-karszton (Geomorphological observations in Gömör-Torna Karst). Egyetemi doktori értekezés ELTE TTK Manuscript (in Hungarian)
Móga J (1999) Reconstuction of the development history of karstic water networks on the southern part of the Gömör-Torna karst ont he basis of ruined caves and landforms. Acta Carsologica 159–174
Móga J (2011) A Tési-fennsík víznyelőinek felszínalaktani vizsgálata (Investigation of the sinkholes of Tés Plateau). A Miskolci Egyetem Közleményei A Sorozat Bányászat 82:207–214 (in Hungarian)
Nyáry J (1881) Az Aggteleki barlang, mint őskori temető (Aggtelek Cave as a Preshitoric Cemetery). Monumenta Hungariae archaeologica nevi praehistorici (in Hungarian)
Paget J (1839) Hungary and Transylvania with Remarks on their condition, social, political, and economical. London I-II
Pávai-Vajna F (1931) A forró oldatok és gőzök-gázok szerepe a barlangképződésnél (The role of hot solutions, steam and gases in the formation of caves). Hidrológiai Közlöny 10:115–122 (in Hungarian)
Pécsi M, Scheuer Gy, Schweitzer F (1984) Plio-Pleistocene tectonic movements and traverline horizons in Hungarian Mountains. Studia Geomorphologica. Carp.-Balc. 17:19–27
Ransanus P (1489) Epithome rerum Hungarorum (1558), Ford. Kulcsár P 1977, Budapest
Raisz K (1802) Gömör vármegyében fekvő Baradla Barlangjának Környéke s Kiterjedése (Surroundings and extenstion of Baradla Cave in Gömör County)... Topographia Antri Baradla...
Raisz K (1807) Topographische Beschreibung der im Gömörer Komitat bei dem Dorfe Aktelek befidlicher Höhle Baradla. Wien
Sásdi L (1997) Újabb adatok a Bükk-hegység karsztjának fejlődéstörténetéhez (New data to the evolution of Bükk Karst). Karszt és Barlang 1997 (I-II):20–25 (in Hungarian)
Sásdi L (1998) Az Aggtelek – Rudabányai-hegyvidék karsztjának fejlődéstörténete (Evolution of the karst of Aggtelek-Rudabánya Mountains). In: Az Aggteleki Nemzeti Park (ed: Baross G.) Mezőgazda Budapest, pp 154–159 (in Hungarian)
Sásdi L (2016) A Vértes karsztjának fejlődéstörténete (Evolution of Vértes karst). Karszt És Barlang 2012–2014:47–55 (in Hungarian)
Scheuer Gy, Schweitzer F (1973a) A magyarországi travertino összletek képződésének fázisai a negyedkorban (Quaternary phases of travertine formation in Hungary). Földrajzi Közlemények 21:133–144 (in Hungarian)
Scheuer Gy, Schweitzer F (1973b) 1981: A hazai édesvízi mészkő összletek származása és összehasonlító vizsgálata (Origin and comparative study of Hungarian freshwater limestone formations). Földtani Közlöny 111(1):67–97 (in Hungarian)
Schmidl A (1856) Die Baradla-Höhle bei Aggtelek und die Lednica-Höhle bei Szilitze im Gömörer Komitate Ungarn. Sitzungber. mathem.-naturw. Classe Akad Wiss Wien 22:579–621
Schmidl A (1857) Die Baradla Höhle bei Aggtelek und die Lednica Eishöhle bei Szilitze im Gömörer Com. Ungarns, Wien, p 45
Schmidl A (1863) Die Abaligeter Höhle. – Akad. Wissensch. mathem.-naturw. 101:1–5
Schönviszky L (1968) A Kárpát-medence legrégebben ismert jégbarlangja (The Longest Known Ice Cave in the Carpathian Basin). Karszt és Barlang, 1968 (I-II):11–16 (in Hungarian)
Síkhegyi F, Székely K (2014) Expedíció a Mátrában – Szabó József 1869 (Expedition in Mátra – Szabó József 1869). – Hungeo 2014 Magyar Földtudományi szakemberek XII. találkozója, pp 210–213 (in Hungarian)
Speer D (1683) Ungarischer oder Dacianischer Simplicissimus
Staszic S (1815) O ziemiorodztwe Karpatow i innych gor i rownin Polski. Warszawa, pp 213–215
Strömpl G (1912) A Vargyas szurdoka (The Vargyas Gorge). Földrajzi Közlemények 40:223–226 (in Hungarian)
Strömpl G (1913) Barlangok alakulása és pusztulása (Formation and distruction of caves). Uránia 14:292–303 (in Hungarian)
Strömpl G (1914) A borsodi Bükk karsztja (Karst of Borsod Bükk). Földrajzi Közlemények 42(2):79–98 (in Hungarian)
Strömpl G (1923) A Gömör-tornai karszt hidrológiája (Hydrology of Gömör-Torna Karst). Hidrológiai Közlöny 3(1):20–33 (in Hungarian)

Strömpl G (1933) Cholnoky Jenő karsztkutatásai (Karst research of Jenő Cholnoky). Földrajzi Közlemények 63(9–10):391–395 (in Hungarian)

Strömpl G (1944) A Szilicei fennsík (Szilice Plateau). Turista Értesítő 18(8):74–75 (in Hungarian)

Strömpl G (1945) A pelsőci Nagyhegy (Nagyhegy at Pelsőc). A Földgömb 15:110–115 (in Hungarian)

Szabó J (1890) Az ágasvári trachyt-barlang a Mátrában (Ágasvár Trachyt Cave in Mátra). Turisták Lapja 2(6):217–224 (in Hungarian)

Szabó Pál Z (1953) A Mecsek karsztvízrendszere (Karst Hydrology of Mecsek Mountains). Hidrológiai Közlöny 1953(7–8):241–251 (in Hungarian)

Szabó Pál Z (1956a) A karszt, mint klimatikus morfológiai probléma (Karst as a Climatic Geomorphological Problem). Dunántúli Tudományos Gyűjtemény. Pécs (in Hungarian)

Szabó Pál Z (1956b) A magyarországi karsztformák klímatörténeti vonatkozásai (Climate historical relations of Hungarian karst forms). Földrajzi Közlemények (in Hungarian)

Szabó Pál Z (1968) A magyarországi karsztosodás fejlődéstörténeti vázlata (Draft of the evolution of Hungarian karstification). Dunántúli Tudományos Gyűjtemény, Pécs (in Hungarian)

Székely K (2010a) A barlangkutatás első tudományos szervezetei. 100 éves a szervezett magyar barlangkutatás (The first scientific organisations of cave research. 100 years of organised cave research in Hungary). Budapest Hungarian Speleological Society, pp 11–25 (in Hungarian)

Székely K (2010b) Az első önálló barlangkutató szervezet. 100 éves a szervezett magyar barlangkutatás (The first independent cave research society. 100 years of organised cave research in Hungary). Budapest Hungarian Speleological Society, pp 26–42 (in Hungarian)

Székely K (2012) Vass Imre ismeretlen kézirata a Baradla 1825. évi feltárásáról (The unknown manuscript of Imre Vass about the 1825 exploration of Baradla Cave). Gömörország 13(1):17–22 (in Hungarian)

Szentiványi M (1689) Curiosiora et selectiora scientiarum miscellanea in tres partes divisa. I- III., Nagyszombat (Tyrnaviae)

Szeremley Sz (1972) A bükki karsztosodásról (About the karstification of Bükk). Manuscript, Miskolc (in Hungarian)

Téglási-Ercsei J (1837) Utazások nemes Torda vármegye alsó kerületeinek nevezetesebb járásaiban (Travels in the municipalities of Torda County). Nemzeti Társalkodó (in Hungarian)

Telbisz T (2001a) Számítógépes töbör-morfometria az Aggteleki-karszt példáján (Computer-aided doline morphometry by the example of Aggtelek Karst). Földrajz az egész világ, Geográfus Doktoranduszok V. Országos Konferenciája Miskolc, pp 46–52 (in Hungarian)

Telbisz T (2001b) Új megközelítések a töbör-morfológiában az Aggteleki-karszt példáján (New perspectives in doline morphology by the example of Aggtelek Karst) Földrajzi Közlemények 49(1–2):95–108 (in Hungarian)

Teleki D (1796) Egynehány hazai utazások leírása Tót- és Horvátországoknak rövid esmértetésével együtt (Description of some home travels with short presentation of Tót and Croat countries). Béts pp 36–39 (in Hungarian)

Teleki D (1805) Reisen durch Ungern und einige angranzende Lander. Pesth, pp 61–63

Tóth G (1983) A Bükk felszínfejlődése és mai formakincse (Landform evolution and present morphology of Bükk). In: Bükki Nemzeti Park (ed: Sándor A.), Mezőgazda Budapest, 62–106 (in Hungarian)

Townson R (1797) Travels in Hungary, with a short account of Vienna in the year 1793. London

Tulogdy J (1943) A Tordai hasadék keletkezése (Formation of Torda Gorge). Földrajzi Közlemények LXX I(3):193–215 (in Hungarian)

Varga M (1809) A tsillagos égnek és a föld golyóbissának az ő tüneményeivel eggyütt való természeti előadása s megesmértetése (Presentation of the Starry Sky and the Earth Globe with its Phenomena). Nagyvárad (in Hungarian)

Vass I (1831a) Az Aggteleki barlang leírása, fekte területével, talprajzolatával és hosszába való átvágásával, két táblában (Description of Aggtelek Cave with area, planform view and longitudinal profile). Landerer Nyomda Pest (in Hungarian)

Vass I (1831b) Neue Beschreibung der Agteleker Höhle. Landerer, Pescht

Veress M (1995) Karros folyamatok és formák rendszerezése Totes Gebirge-i példák alapján (System of karren forms and processes using examples from Totes Gebirge). Karsztfejlődés I:7–30 (in Hungarian)

Veress M (1998) Adatok karrványúk meanderfejlődéséhez (Some data to meanderkarr development). Karsztfejlődés II:75–90 (in Hungarian)

Veress M (1999) Az Északi-Bakony fedett karsztja (The covered karst of Northern Bakony). Zirc (in Hungarian)

Veress M (2000) Középhegységi karsztok néhány típusa (Some types of middle mountain karsts). Földrajzi Közlemények 48(1–4):1–28 (in Hungarian)

Veress M (2004a) A karszt. Szombathely, Berzsenyi Dániel Főiskola (in Hungarian)

Veress M (2004b) A karrformák I. (Karren forms I) Karszt és Barlang 2004b(1–2):3–12 (in Hungarian)

Veress M (2008) The origin of the Bemaraha tsingy (Madagascar). Int J Speleol 37:131–142

Veress M (2010) Karst enviroments: Karren formation in high mountains. Hannover, Springer Netherlands

Veress M (2011) Recens karsztosodás Magyarországon (Recent karstification in Hungary). A Miskolci Egyetem Közleménye, A sorozat, Bányászat 82:271–279 (in Hungarian)

Veress M (2013) Flow dynamics and shape of rinnenkarren systems. Geomorphology 198:115–127

Veress M (2014) The development of the pinnacales (Lena pillars) along Middle Lena (Sakha Republic, Siberia, Russia). Proc Geol Assoc 125:452–462

Veress M (2015) The effect of slope angle ont he development of type a and type b channels of rinnenkarren with field and laboratory measurements. Geomorphology 228:60–70

Veress M (2016) Covered Karsts. Heidelberg, New York, Springer, Berlin

Veress M (2017) Solution Doline development on Glaciokarst in Alpine and Dinaric area. Earth Sci Rev 173:31–48

Vass I (1831c) Az Aggteleki barlang leírása, fekte területével, talprajzolatával és hosszába való átvágásával (Description of Aggtelek Cave with area, planform view and longitudinal profile). Pest (in Hungarian)

Wernher G (1549) De admirandis Hungariae aqvis hypomnemation. Basel

Zámbó L (1970) A vörösagyagok és a felszíni karsztosodás kapcsolata az Aggteleki-karszt délnyugati részén (Relationship of red clays and surface karstification in the southwestern part of Aggtelek Karst). Földrajzi Közlemények 18(4):281–293 (in Hungarian)

Zámbó L (1986) A talaj-hatás jelentősége a karszt morfológiai fejlődésében (The significance of soil effect in karst morphological development). PhD thesis Manuscript Hungarian Academy of Sciences Budapest (in Hungarian)

Zámbó L (1998) Felszínalaktan, Talajtakaró (Geomorphology, Soli Cover). Az Aggteleki Nemzeti Park (szerk.: Baross G.), Mezőgazda Kiadó, Budapest 70–96 and 97–117 (in Hungarian)

Zámbó L, Telbisz T (2000) A karsztkorróziós talajhatás érvényesülése a karrfejlődésben (The soil effect on karst corrosion in the evolution of karren). Karsztfejlődés 5:103–114 (in Hungarian)

Zámbó L, Tompa K (2001) A karsztkorrózió feltételezett huminsavas oldásfolyamatának vizsgálata (Examination of the hypothesized effect of humic acids in the process of karst corrosion). Karsztfejlődés 6:61–70 (in Hungarian)

Zámbó L, Ford D (2003) Corrosional factors of the epicarst. Soil effect on karst processes, pp 7–18

An Interpretation of Karstification and Karst Processes by Hungarian Researchers

Márton Veress

Abstract

This chapter describes theories by which karst phenomena and processes of Hungarian karst areas were interpreted by Hungarian researchers. The theories are overviewed by various topics which are the following: karst types, CO_2 measurements and karstic denudation, karst hydrology, denudation and geomorphic evolution of karst surfaces, processes and phenomena related to epigenetic valley development, surface karst features, and caves (cold-water caves and warm-water caves are discussed separately).

Keywords

Karst area • Karst phenomenon • Interpretation theory • Hungarian researcher

1 Introduction

This chapter describes theories explaining various phenomena and processes of karstification which were created by Hungarian researchers during studying Hungarian karst areas.

This description of theories does not include international antecedents since they were mostly not dealt with by the cited authors either. These references are absent not only because they were less required in previous works, but also because the latest scientific results were hardly accessible or they were not available at all in countries belonging to the Soviet sphere of influence. In addition, in some cases authors explaining the characteristics of a Hungarian karst may have supposed that the theory had probably no antecedents or actually it may not have had any. An example for a theory without antecedents may be the classification of the karst as allogenic and autogenic where the author intended to explain the characteristics of Aggtelek Karst (Jakucs 1956a) and extended this classification over other karsts only in a later work (Jakucs 1968).

2 Karst Types

Venkovits (1959) classified karsts taking into consideration the expansion of the karst, the position of the karstwater and the coveredness of the karst probably based on international resources. The classification, which is generally valid, distinguishes for example block karst according to expansion (its expansion is greater than 1 km^2, its thickness is larger than 40 m), bed karst (interrupted by non-karstifying rocks), karstschuppen (its expansion is smaller than 1 km^2, its thickness is greater than 40 m), or according to coveredness, open karst (karst passages open to the surface), closed karst (it has impermeable cover) and outcropping karst (it outcrops from covered karst).

Leél-Őssy (1959, 1960), and later Jakucs (1977b) put Hungarian karst areas into two types, into the Aggtelek type and Transdanubian type. The former author's classification was based on structure, while the latter's on morphology. Jakucs (1977b) explained the differences by various evolution (see chapter "A General Description of Karsts in Hungary"). Hevesi (1991a, b) used the "of character" instead of "type". According to the above authors, the difference between the two types is primarily manifested in the fact that on Transdanubian karsts the intensity of karstification is lower and dolines and doline systems are absent.

Taking into consideration the expansion of Hungarian karst areas, Leél-Őssy (1959, 1960) distinguished karst plateau (its expansion is at least 50–100 km^2), karst block plateau (with an expansion smaller than 50–100 km^2) and karst block (tectonic mountain), while according to their elevation, he differentiated mountain karst and basin floor karst.

M. Veress (✉)
Department of Geography, Savaria University Centre, ELTE SEK Eötvös Lóránd University, Károlyi Gáspár tér 4, Szombathely, 9700, Hungary
e-mail: veress.marton@sek.elte.hu

M. Veress and Sz. Leél-Őssy (eds.), *Cave and Karst Systems of Hungary*,
Cave and Karst Systems of the World, https://doi.org/10.1007/978-3-030-92960-2_3

Considering the characteristics of Hungarian karst areas (Aggtelek Mountains, or Aggtelek Plateau) Jakucs (1956a, 1968, 1977a) introduced the notions of autogenic karst and allogenic karst. (The waters of autogenic karst are fed by meteoric water, while the waters of allogenic karst are partly fed by the surface waters of the non-karstic terrain). He described the occurrence of the two karst types in case of various geological conditions and defined their characteristics. He carried out a detailed analysis on the development characteristics of autogenic karst at stable and at changing base level of erosion. He thinks that in the former case, a karst peneplain develops as a result of denudation, while in the case, the solution landforms are renewed and cave levels are formed. At the allogenic type, he distinguished the karst completely covered by impermeable beds (today this is called allogenic karst) and the karst partially covered by non-karstic rock (today this is called mixed allogenic-autogenic karst). At the former, he analysed valley inheritance and the subsequent surface and subsurface feature development. At the latter, he studied the development and characteristics of the cave of purely erosion origin (see below).

Hevesi (1986) improved Jakucs's classification, but he did not connect his own classification to Hungarian karst areas. Thus, at autogenic karst, he distinguished the originally bare variety and the variety that lost its cover. At allogenic karst, he also differentiated non-buried variety and the variety that lost its former cover. His third karst type is the covered karst where the cover can be permeable (concealed karst) and impermeable (cryptokarst). His fourth karst type is the partly covered karst, a part of which was always uncovered or has partly lost its cover recently. He also described the surface and subsurface features of the types and varieties.

Veress (2016) classified covered karsts considering the conditions of coveredness. According to their development, he classified Hungarian karsts as semi-allogenic covered karst, renewed allogenic covered karst, mantled allogenic covered karst and horst covered karst (Fig. 1 in chapter "The Surface Morphology of Karsts in Hungary"). The renewed covered karst is separated from its sediment sources (among Hungarian karsts, the Aggtelek-Rudabánya Mountains belongs to here). On semi-allogenic covered karst, the epigenetic valleys of the karst transported the non-karstic cover out of the karst. He regarded the Mecsek Karst as such karst. On mantled karst, the karst is covered by consolidated impermeable rock (Kab Mountain).

Veress (2016), using Pécsi's (1980, 1991) block classification ranked the Bakony Region as horst covered karst. He states that even adjacent blocks may have various karstification since the adjacent blocks may belong to various block types and thus, even to different karst types. Blocks belonging to various types may be covered to a different degree since blocks of various types have different elevations. The elevation determines whether superficial deposit is present on any block and if it is present, what type and what thickness it has. The superficial deposit is primarily loess. It was either not developed on more elevated blocks (see chapter "A General Description of Karsts in Hungary"), or it could only survive on blocks with lower surface and larger expansion. The block classification is valid for the whole Transdanubian Mountains, thus it can also be applied to their karstification. On the karsts of the Transdanubian Mountains being separated to blocks (horsts), the existence, degree and maturity of karstification, the karst type and the frequency of the features primarily depend on the evolutionary type of the bearing block (Veress 2016).

Therefore, the karst types of the mountain parts of the Transdanubian Mountains can be given with the consideration of each block type. Thus, in the area of the block of cryptopeneplain type there is buried karst, while in the area of the horst elevated to summit position there is concealed karst and soil-covered karst. On blocks of threshold type and on horsts in summit position there may be soil-covered karst, but at places where the patches of permeable superficial deposit survived concealed karst may also occur. Concealed karst patches may also occur on those parts of cryptopeneplains where the impermeable cover was denuded, but the loess cover was not.

3 CO_2 Measurements and Karstic Denudation

Also Hungarian karst researchers realised early the significance of CO_2 of biogenic origin in the process of karstification. Several researchers used measurements when they investigated climatic and biological factors which affect CO_2 production and the daily, or seasonal changes of its quantity.

Balázs (1969) built an experiment device to investigate the effect of CO_2 with biogenic origin. He placed limestone debris and various soils originating from Hungarian karst areas into this. As a result of the experiments he established that the dissolved material content of the water seeping through the limestone debris depends on air temperature, on soil quality and whether there is vegetation on it. It also turned out that subsoil saturation already happens in a depth of some decimetres thus, no dissolution takes place below such depth. The quantity of the dissolved material increased fast during seepage then 1–2 days later this rate decreased and became stabile at a lower level.

Jakucs (1977a) elaborated two methods in order to measure the quantity of CO_2 in karst soils in the field. With the application of these methods, he could determine the CO_2 content of the micro areas of the soil to a 20 cm-depth. Among Hungarian karst areas, he took measurements on

Table 1 Corrosional denudation at various soil (cover) thicknesses on Aggtelek Karst (Zámbó 1998)

Thickness of soil cover (m)	Total corrosional denudation (mm/ka)	Minimal corrosional denudation (mm/ka)
0.2–0.5	7	4
0.5–1.5	9	6.5
1.5–4.5	7.4	4.4
4.5–5.5	6.0	3.5
5.5–7.5	13.5	10.0

Table 2 Denudation rates of sites of various morphologies on Aggtelek Karst (Zámbó 1998)

Morphological environment	Denudation (mm/ka)
Plateaus	7
Steep slopes	7
Slopes of medium inclination	6.5
Slopes of small inclination	6
Marginal part of doline floor	10
Doline floor below a cover with good permeability	13–16
Doline floor covered by impermeable beds	1–2

Aggtelek Karst, in the Bükk Mountains, in the Pilis Mountains, in the Gerecse Mountains, in the Bakony Mountains and in the Mecsek Mountains and determined the CO_2 quantity of soils at several hundred places.

Zámbó (1986) carried out his investigations on Aggtelek Karst for several years. He made measuring shafts at 24 sites on doline slopes and doline floors in the cover. He deepened shafts being shallower than 1 m in the soil, while shafts deeper than 1 m (there were also shafts with a depth of 8 m) in the doline sediments. In the shafts he measured soil (superficial deposit) temperature and humidity, the CO_2 content of pore air, the aggressivity and the dissolved material content of the infiltrating water. With the analysis of his measurements, he outlined the spatial structure of the material circulation of the dolines and the seasonal changes of material circulation. Bárány and Mezősi (1978) investigated the biomass of doline soils in the Bükk Mountains. (Aerobic microbes obtain energy during fermentation under semi-anaerobic circumstances, by which they produce CO_2.) There were differences between the microbe quantities of the slopes of the studied dolines with various exposure. It was established that the aerobic microbe number value is the largest (3×10^6/g–5×10^6/g) in the domain with a humidity of 14–25% and with a temperature of 22.2–24.6 °C.

There have been some experiments on the denudation rate of Hungarian karsts. Zámbó (1986, 1998) measured the CO_2 content of the soils of Aggtelek Karst and calculated the total limestone dissolution (denudation originating from the dissolved material content of the limestone debris of the bedrock and the cover), and the minimal denudation (the denudation of the bedrock only), to various soil thicknesses (Table 1) and to various morphological environments (Table 2). He explained the differences by the various thickness and quality of the cover. Zámbó and Telbisz (2000) also made measurements on dissolution in two grikes on Aggtelek Karst by removing soil from one of the grikes. Thus, they could calculate the quantity of the dissolved material in subsoil features and in features without soil. In case of covered karren its value was 1.3 mm/ka, while at uncovered karren it was 1.0 mm/ka.

Izápy and Maucha (2000) calculated average specific ion-transport values based on the ion content of the springs and karstwater producing wells of three areas (Veszprém, Miskolc, Jósvafő). Based on these, they determined the surface denudation values of Hungarian karsts which were 0.0261 (Veszprém), 0.0216 (Miskolc), and 0.0233 (Jósvafő) mm/year. They did not mention the reason for the differences. The data refer to the fact that in contrast with previous authors' opinion, the karstification of the Transdanubian Mountains is not of lower intensity than that of the Bükk Region or the Aggtelek-Rudabánya Mountains.

4 Karst Hydrology

In the area of the Transdanubian Mountains in Hungary, first Szádeczky-Kardoss (1941), and then Papp (1942) made karstwater level maps with the help of elevation data of karst springs, karstwater marshes and some mine gallery water levels. This represented the karstwater level (main karstwater level) of the whole mountains. Since 1950 karstwater level observation wells have been made, the number of which reached 200 during some decades. Based on the data of the wells, the researchers of VITUKI made karstwater level maps for several decades every year. (Maucha 1989). This was started because karstwater inrushes threatened bauxite and coal mining. Karstwater was pumped out in the environs of the mines in order to sink the water level below the mine levels. (Karstwater level sinking was also contributed by drinking water consumption.) The pumped out water quantity reached a value of 460 m^3/min. So-called depressions developed at the karstwater level of the mountains, the degree of water level decrease exceeded 100 m in some places (Csepregi 2007). Mountain marginal springs, where they were reached by water level decrease, ran dry, in addition, the water flew from the sediments of the surrounding basins towards the mountains.

Jaskó (1959) used the expression "karstwater storey" for the locally developing karstwater separating from the main karstwater in the Transdanubian Mountains. He states that karstwater storeys develop at sites where there are impermeable intercalations in the younger rocks over Triassic rocks (or below them in the older ones).

Alföldi and Lorberer (1976) made a three-dimensional flow system by the analysis of the well data of the Transdanubian Mountains. According to this, infiltrating waters flow towards the buried deep karst and then getting warmer there, they move upwards towards mountain marginal springs where the warm and cold karstwaters are mixed (Fig. 1).

Main areas of tapping and water supplement areas were detected in the area of the Buda thermal karst (Mádl-Szőnyi and Tóth 2015), and supporting previous research results, the presence of the water with basin origin in the karstwater (Erőss et al. 2011; Mádl-Szőnyi and Tóth 2015).

Two flow levels were separated above each other in the area of the Bakony Region (Tóth 2018): a flow level between the surface and an altitude of 250 m from the Balaton Uplands towards the Northern Bakony, and the water movement below the altitude of 250 m from the Northern Bakony towards the Balaton Uplands. It was established that the flow of the karstwater from the Bakony Mountains takes place southwards below Lake Balaton and it reaches the Transdanubian Hills (Tóth and Mádl-Szőnyi 2019). A similar flow, taking place towards the surrounding basin (Dráva Basin) was detected from the Mecsek Mountains and the Villány Mountains (Erőss et al. 2019).

Springs were studied in order to establish the vulnerability of the karst (Aggtelek Karst) with the Weights of Evidence (WofE) method (Iván and Mádl-Szőnyi 2017; Iván et al. 2020).

Böcker (1972) revealed karstwater fluctuation in the karstwater observation wells that is triggered by tide effect. Maucha (1977, 1989, 1997) described the effect of the tide to the karstwater level on Aggtelek karst first in an indirect way by the analysis of the discharge fluctuations of springs. He states that the karst is separated into blocks with an expansion of about 50 × 50 m which are separated by fissures being partly or completely filled with water from each other. At low tide, the blocks come closer to each other, while at high tide they move away from each other. He also described the tide effect directly with measurement in Vass Imre

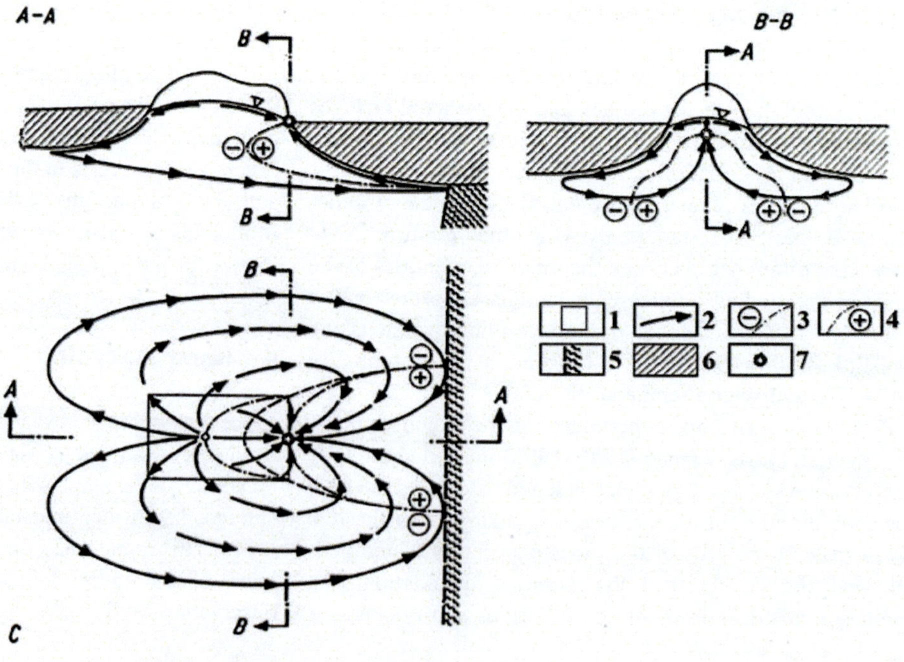

Fig. 1 Three-dimensional flow model of karst hydrothermal waters (Alföldi and Lorberer 1976). 1. karstic rock on the surface, 2. karstwater flow, 3. area of descending water motion (endothermic area), 4. area of ascending water motion (exothermic area), 5. subsurface boundary of karstified carbonate substratum, 6. impermeable and insulating basin sediment, 7. thermal karst spring, A-A′ lateral view, B-B′ front view, C-C′ plan view

Cave (he measured a 0.5–5.0 μm change in passage width) and analysing the discharge of springs, he distinguished width change triggered by tide effect from shifts caused by the infiltrating meteoric water.

With the analysis of the spring discharges of Aggtelek Karst, Maucha (1989) established a hierarchical character of the karstwater flow system (passage system). The karst is separated into blocks in a way that the springs are connected by larger passages, these are connected by smaller passages and the latter are linked by passages with even smaller cross-section. When karstwater empties from the passages, first those with a large cross-section become empty and then the tributaries. At a given time, the passage which is emptying blocks the emptying of all the other passages with smaller cross-section.

During studying the springs of Aggtelek Jakucs (1956b) distinguished two types of karst springs. Type A spring is the outlet of autogenic karst thus, it is in connection with the karstwater of the karst. The discharge of such a karst spring only increases to a small degree and following infiltration in a delayed way, discharge increase is long-lasting, the dissolved material content decreases at most to a small extent during discharge growth. The type B karst spring is the outlet of the mixed allogenic-autogenic karst and taps caves. In rainy season, the discharge growth appears fast, it is of great degree because of the flood developing in the cave and lasts for a short time. The discharge growth takes place simultaneously with the temporary decrease of the dissolved material content since the waters flowing inwards from the non-karstic terrain do not contain dissolved material or they only contain little.

5 The Denudation and Landscape Evolution of Karst Surfaces

Based on his observations on Bükk Plateau, Leél-Őssy (1954) outlined the possibility of the coalescence of karst depressions. The coalescing dolines develop into uvalas and then newer dolines are formed on the floors of uvalas which may also coalesce into uvalas. He mentions such a process from the areas of Zsidó-rét and Nagy-mező which he calls poljes. With the above-mentioned things, the author outlined the karstic denudation of the plateau in several levels.

Jakucs (1977a) sketched the theory of the peneplanation of a karst area. He states that if the karstwater level does not change permanently (because the bearing area does not rise, or the base level of erosion does not subside) the lower limit of the saturation level (this is the zone of doline formation) is shifted lower and lower since dissolution makes the saturation zone thinner. A karstic peneplain develops when the surface of the karst reaches the base level of erosion (Fig. 2).

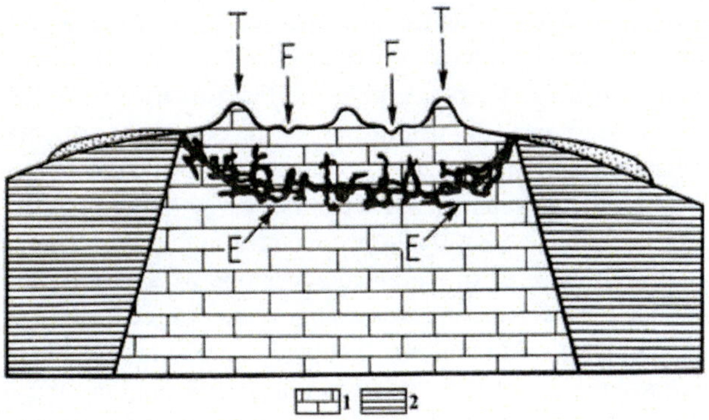

Fig. 2 Peneplanated autogenic karst (Jakucs 1977a). F surface stream, T limestone butte, E infilled phreatic passage, 1. limestone, 2. non-karstic rock

Tóth and Fejes (1984), then Tóth (2001) described the denudation levels of Bükk Plateau by the elevation data of inactive spring cave levels, which represent a base level of erosion belonging to a given denudation level, and by the elevation data of fossil ponors. Particularly, the cave level belonging to the denudation level of 850–870 m (older Pliocene surface) can be revealed as 30 inactive caves are situated at elevations of 800–875 m. (out of 80 spring caves situated at an elevation of 950–500 m 60 caves are above 600 m in the Bükk Region.)

Researches took notice of the rubble beds of dolomite terrains in the Buda Hills early. Leél-Őssy (1952) and Jakucs (1994) explained the physical weathering of the dolomite to warm water effect as follows. Aragonite is deposited from the warm water rising along fractures into the rock texture which is transformed into calcite by temperature decrease, and it causes volume increase and thus, mechanical weathering. However, according to Jakucs (1994), physical weathering also occurs when free sulfuric acid reacts with the calcareous material of the dolomite since in this case it turns into gypsum. The transformation is also accompanied by volume increase and thus, by mechanical weathering.

However, the dolomite is transformed into rubble in cold-water environment too. Jakucs (1977a, b) states that the $CaCO_3$ sticking the dolomite crystals together is dissolved, by which the rock is weathering and rubble develops.

In Hungarian karst areas, but also on other karsts, beds with limestone debris of various thickness develops on limestone. Veress and Péntek (1996) explained this by the partial dissolution of the rock. Solution weathering takes place to a depth at which the infiltrating water becomes saturated. The degree of the thinning of the debris zone because of the dissolution of debris parts coincides with the degree of the downward shift of the saturation level thus, the debris zone becomes thicker from downwards (the debris zone will be in dynamic balance). The degree of the subsidence of the saturation level corresponds to the degree of the lowering of the karst surface.

Investigating the block fields of the Kál Basin, Veress (2003) interpreted the denudation of the sandstone surfaces containing amorphous silica. In case of enough water quantity, the amorphous silica can be solved (large boulder) if the water has a long-duration contact with the rock. This is possible in case of fine-grained sandstone because of slow drainage. (However, the deepening of the karren feature is retarded by the fact that the dissolved material is not able to be transported and thus, it is precipitated.) Therefore, those kamenitzas can primarily deepen which have a decantation runnel, or there is a fracture in the bearing rock since the dissolved material can be transported away in both cases.

6 Processes and Phenomena Related to Epigenetic Valley Development

Jakucs (1956a, 1968, 1977a) described valley inheritance from the non-karstic cover onto the karst with the use of observations of Hungarian karst areas. He (1956a) recognised karstic feature development during inheritance on the Aggtelek Plateau of Aggtelek Karst. He (1968, 1977a) states that infiltration increases on the valley floor part that inherited to the limestone. If the gap system is not mature enough, valley deepening can take place to the karstwater level, by which gorges and gorge-like valleys are formed. If passage development is more significant because of infiltration, the phenomenon of subsurface capture occurs a ponor develops at the rock boundary of the valley floor. Valley development continues below the surface by cave development. The stream entering the phreatic passage system fills the passages, the ceiling of the partially infilled passages is dissolved, and then the passages continue their development in an erosional way (see below).

According to Jakucs (1977a), if no beheading happens in the lack of rock boundary and if the valley floor does not reach the karstwater level, solution dolines develop on the valley floor that deepened into the limestone and thus, row dolines are formed, while, as it has already been mentioned, a ponor develops at the rock boundary of the valley floor by subsurface capture. If newer and newer beheading sites develop towards the valley head, a ponor row is formed. The already developed cave has newer and newer branches. The author emphasises that since the parts outside the valleys are covered during inheritance, the row dolines are always older and larger than solitary dolines that are situated outside the valley. A reason for this is that the development of the latter started later because of the later denudation of the cover than the development of those situated on the valley floor.

Hevesi (1978, 1980) improved the ideas on valley inheritance and the accompanying feature development by his investigations on the karst of the Bükk Mountains. Hevesi (1980) take into consideration the position (elevation) of the karstwater level during inheritance. If the karstwater level is not deeper than 1 m as compared to the valley floor during inheritance, no subsurface capture takes place, the valley deepening continues in the limestone too. The valley may have gorges because the hardness and resistance of the rock favours the survival of steep slopes (Hevesi 2019). If the valley deepening lags behind the subsidence of the karstwater level during the rise of the area, subsurface capture (ponor development) does take place. If the karstwater level is deeper than 2 m as compared to the valley channel, subsurface capture happens early, so early that the stream will have no valley on the karstic rock. (If beheading occurs later, the valley is also inherited onto the limestone.) As a result of the retreat of subsurface capture, newer ponors develop, and the former are transformed into dolines. Through caves may be formed. He thinks that if these (Hevesi 2019) become filled up enough, they develop upwards by dissolution. A gorge develops by the collapse of the cave whose ceiling is thinned out. He states that a large-sized gorge can also develop in a way that several cave levels collapse into each other (Fig. 43 in chapter "The Surface Morphology of Karsts in Hungary"). Younger cave levels can continually develop following the collapse of older caves.

According to Hevesi (1978, 1980), ponors that developed by beheading become partially filled up and they continue their development as dolines, constituting doline rows. On the karst of the Bükk Region, the exposure of the mounds happened earlier than the exposure of the lower valleys with doline rows. Therefore, he thinks that the dolines of the mounds are older and larger than the dolines of the row dolines of the valley floors. At the same time, row doline are suitable for detecting the site of former streams. He thinks that the karsts of Aggtelek character are essentially characterised by valleys with doline rows.

Inheritance shows differences from the above mentioned in the Transdanubian Mountains, particularly in the Bakony Region. According to Veress's (2000a) investigations in the Bakony Region, the valley development is epigenetic-regression here, but in some parts of epigenetic valleys, it is epigenetic-antecedent. The development of epigenetic-antecedent valley parts takes place at sites where a covered block becomes elevated as compared to its environment. In the area of the bearing block, an antecedent valley section (gorge) develops after inheritance if a stream arrives there. A condition of this is present if the block is surrounded by a terrain covered with non-karstic rock (a basin morphologically). It occurs that there is no impermeable intercalation in the carbonate rocks of the rising block particularly if it is exclusively built up of Triassic rocks. The developing gorge reaches the cavities of the karstwater (main karstwater) that developed in the whole area of the mountains. It destroys them completely or if it partly destroys them, it opens them up. It is more common that there are impermeable or partially impermeable beds in the young limestones (Cretaceous, Eocene, maybe Jurassic) on the rising block. Gorge development takes place in this case as well, but the water seeping on the gorge floor creates a local water flow above the impermeable beds. A karstwater storey develops above the main karstwater where local phreatic cavity formation takes place. On the one hand, the phreatic passages promote increased drainage, on the other hand, they also contribute to the increase of the intensity of gorge development because the cavities do not hinder erosion. The remnants of the only partially destroyed cavities survive in the gorge sides, but they promote the denudation of gorge walls by the subsequent denudation of their ceilings. In the gorges of the mountains, where there are impermeable beds below the floor, this process takes place nowadays as well (Veress and Vetési-Foith 2019).

Impermeable intercalations may also occur in the area of epigenetic-regression valley development. The remnants of the phreatic cavities of the karstwater storeys may be exposed during valley development in this case too.

Veress (2000b) states that inheritance does not go with ponor development together in the Bakony Region. The reason for this is at antecedent valley sections that drainage is diffuse and erosion is intensive. As a result of diffuse drainage, no mature passage develop and because of intensive erosion (a relatively large quantity of water arrives at gorge sections, the gradient of gorge floors is large) the passages of the floors are destroyed. However, at epigenetic-regression valleys, the lack of ponor development can be explained by too small discharge.

7 Surface Karst Features

Based on his measurements mainly on Bükk Plateau Jakucs (1977a, b) found that only the surface of the rock is dissolved in dolines (Fig. 3). Taking into consideration the CO_2 content of soils, Jakucs (1977a) states that doline formation takes place to biogenic effect, during which soil development and doline development are processes that strengthen each other (Jakucs 1980). He thinks that doline development is a self-generating process: the soil of the surrounding terrain is washed in into the solution depression that developed on the karst accidentally. Dissolution (the intensity of biogenic CO_2 production will be greater) and thus, deepening also increases below the thickening soil. As a result of deepening, more and more soil is accumulated in the doline and thus, dissolution will be even more intensive.

Jakucs (1977a) recognised the asymmetrical morphology of dolines on Bükk Plateau which is caused by the various inclination of the doline slopes with different exposure. By the exposure of doline sediments he established that the gradient of slopes with various exposure is independent of the inclination of the beds, the dissolution of the bedrock does not modify the inclination of the beds (Fig. 3). According to his and Kevei Bárány's (1986) investigations, microclimatic conditions (temperature, humidity) are different on the slopes of dolines of various exposure. This affects the heat balance and water balance of soils which results in the fact that on the various slopes of the same doline, macroflora and microflora and thus CO_2 production will be different. Because of this, the dissolution intensity and thus, the gradient of the doline slopes with various exposure will be different.

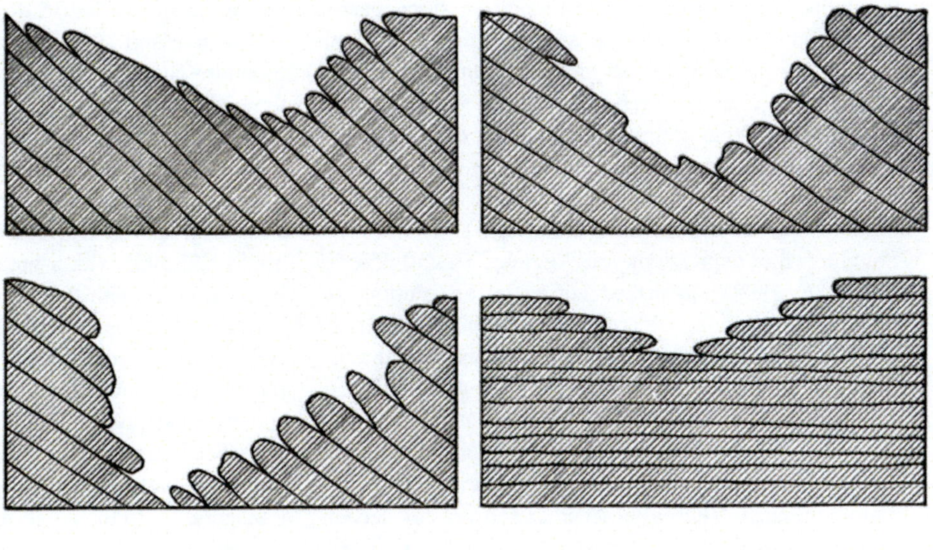

Fig. 3 The unchanged position of the karstic rock beds in exposed dolines prove surface dissolution (Jakucs 1977b)

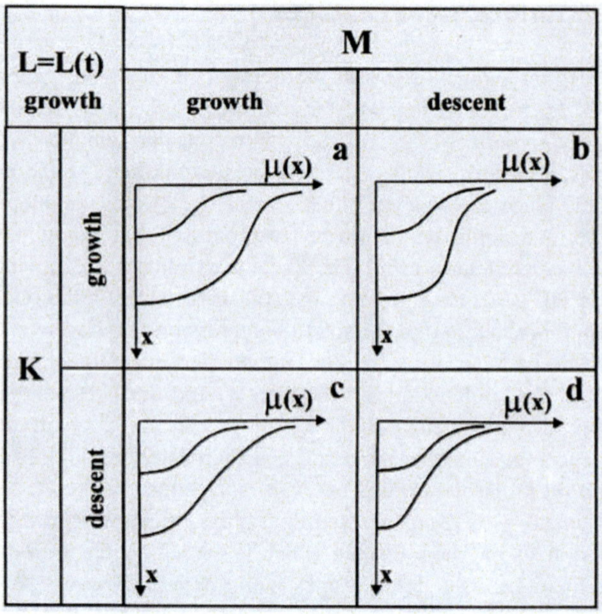

Fig. 4 Solution doline types according to their cross-sections (Péntek et al. 2007). μ(x) function describing the side slope of the doline (meridian function), L > 0 parameter characterising doline depth M < 0 parameter characterising doline radius, K ≥ 1 parameter depending on the value of inflexion point on the meridian function (it expresses the value of concave and convex slope sections as compared to each other)

A function was fitted to the relief draught doline maps (M1:500) made by the use of the data of detailed mapping on Aggtelek Karst, on Bükk Plateau and in the Mecsek Mountains. According to Péntek et al. (2007), generalising the obtained functions, and considering the lengths of the convex and concave parts of their slopes as compared to each other, the dolines are of four kinds (Fig. 4). Doline widening at rim and base (the concave slope part is dominant), doline widening at base (the concave slope part is dominant, but the slope become steep during its development and the floor becomes flat), doline widening at rim (the convex slope becomes longer as compared to the concave slope, while the slope becomes gentle), and doline deepening, but not widening (the length of the convex slope part increases as compared to the length of the concave part, while the slope becomes steeper and steeper).

Studying the depressions of Aggtelek Karst, Zámbó (1970, 1986, 1998) established horizontal dissolution. He states that the infiltrating meteoric waters seep laterally in the fill of the depressions, which is primarily caused by the fact that the material of the fill becomes swollen to the effect of absorption of water. The laterally seeping water reaches the doline wall and exerts its dissolution effect and then moving downwards does not have a solution capacity because it becomes saturated. The dolines widen and do not deepen. Horizontal dissolution also increases the chance of the development of larger karst depressions (uvalas) because the dividing walls between the dolines are destroyed (Fig. 4 in chapter "The Surface Morphology of Karsts in Hungary").

During his investigations in the Bakony Mountains (VES measurements), Veress (2000b, 2008, 2009) studied the cover (thickness, quality and structure) below subsidence dolines. The superficial deposit is primarily loess. He established that the dolines can be syngenetic (the feature that developed in the cover is of the same age as the passage of the bedrock) and postgenetic (the depression is younger than the passage of the bedrock). Out of 43 dolines that he studied by VES measurements in the Bakony Region, 30 dolines were situated above the mound of the bedrock, 11 dolines were above the depression of the bedrock and 2 dolines were above the flat bedrock. Regarding syngenetic subsidence dolines, the cover thickness at their margin is smaller than 3.5 m in case of 9 dolines, it is 3.51–6.0 m at 11 dolines, and it is larger than 6.01 m in case of 4 dolines. (In case of postgenetic subsidence dolines, the cover thickness is more diverse: in case of 2 dolines, it is smaller than 3.5 m, in case of 6 dolines it is 3.51–6.0 m, and in case of 11 dolines, it is larger than 6.01 m.) However, in general in case of a smaller thickness of superficial deposit, there is a greater frequency for the development of dolines (Veress 2000b, 2016). He supposed that postgenetic dolines also develop at a greater thickness of superficial deposit because in case of superficial deposit with a larger thickness material can be transported into the karst through the already developed shafts too. However, there is a smaller chance of the development of syngenetic dolines below thicker superficial deposit since shaft development is less likely.

The majority of dolines develop above the mound of the bedrock because as a matter of course the superficial deposit is thinner at these sites. Surface denudation can also contribute to the thinning out of the superficial deposit. The reason for this is that 15 dolines out of 43 dolines are on the valley floor and 10 dolines are on terrains between exhumed mounds. The superficial deposit is thin below the dolines on terrains between exhumed mounds too. Here, at syngenetic dolines the average thickness of the superficial deposit is 3.85 m. According to data of scientific literature, subsidence dolines develop on thinner and thinner superficial deposit with a greater and greater frequency (Yuan 1987; Chen 1988) and thus, with a greater and greater chance, which is in accordance with the above-mentioned data.

Based on his investigations and observations in the Bakony Mountains, Veress (2005, 2006, 2008, 2009, 2016) recognised a newer covered karst feature, the depression of superficial deposit (DSD). He states that the depression of superficial deposit which bears subsidence dolines (maybe ponors) develops in a way that the superficial deposit is transported into the karst through dolines by pluvial erosion and/or channel erosion. A closed depression (DSD) is formed on the superficial deposit during the local denudation of the superficial deposit. He distinguished real depression of superficial deposit (there is a depression on the bedrock too) and pseudo depression of superficial deposit (there is no depression on the bedrock).

8 Caves

8.1 Cold-Water Caves

During his investigations in Aggtelek, Jakucs (1956a, 1977a) analysed the relation between the geomorphic evolution of the karst and erosional cave development. He distinguished two varieties of allogenic erosional cave development, which are the following.

- The karst is covered, the passage system is underdeveloped in the inheriting valley, and there is no ponor development. Valley deepening can reach the karstwater table.
- The karst is partly covered, ponor development takes place on the floor of the inheriting valley since seepage and subsequent passage development can already take place on karstic terrains partly exposed by valleys. The water flowing into the karst is transmitted into the phreatic passages. It connects them and fills their floors and their upper part becomes flattened due to the effect of rising water table (Fig. 5). The later the ponor development takes place (thus, subsurface capture), the larger the surface valley, the larger height and the wider trapezoid cross-section its subsurface continuation, the cave has (Fig. 6). The trapezoid cross-section (flat roof) develops because since the feeding surface streams increase their catchment areas by retreat, by the time of beheading, the underground creeks will have a discharge large enough to create corridors with wide (flat) ceiling even at the beginning of cave development. The inlet and outlet sections of these caves may have tributary branches (vertical delta formation) as a result of the lower position of ponor sites and the subsidence of the base level of erosion (Fig. 7). During the retreat of the beheading site, a series of passages can also develop which lead onto the cave (Fig. 8). The subsidence of the base level of erosion of great degree does not result in the delta formation of the cave, but it causes the development of several levels (Fig. 9).
- The uncovered part of the karst already rises above its environment at the beginning of its development therefore no valley inheritance takes place on the karst. There are not any epigenetic valleys and row dolines on the surface. The developing cave has a triangular cross-section which can be traced back to the fact that at the beginning of cave development the stream of the ponor is short and its catchment area is small. However, during the deepening of the corridor, the discharge of the underground creek will be larger and larger as a result of the growth of its surface catchment area and thus, the corridor will also be wider and wider towards its present floor.

During his investigations in the Béke Cave (Aggtelek Karst), Jakucs (1956a) discovered a relation between the width of the erosional cave corridor and the size of its belonging catchment area. The value of the quotient of corridor width (in mm) and the expansion of its belonging non-karstic catchment area (in m^2) is almost similar to the values of other caves of Aggtelek Karst at which the bearing rock and the cover are almost identical. He thinks that the reason for this is that more water can flow into the cave from

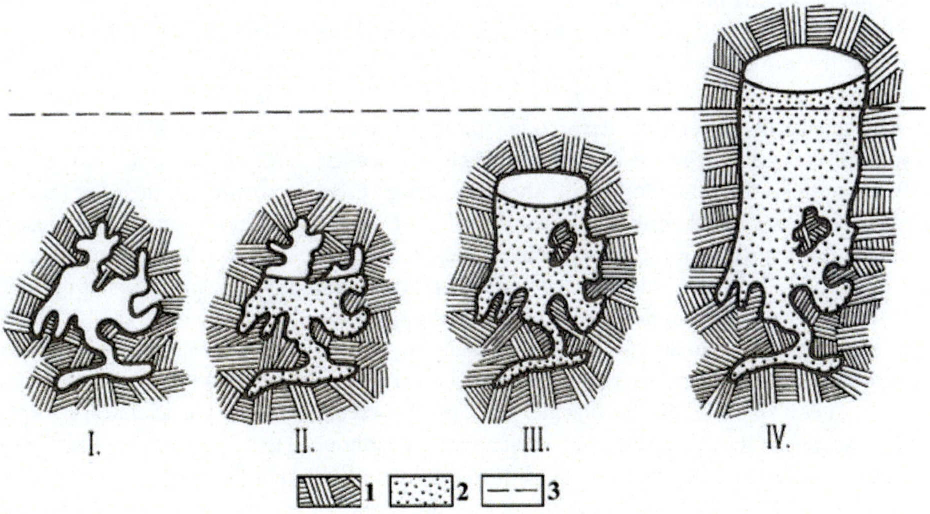

Fig. 5 Upward developing, allogenic erosion cave (Jakucs 1977a). 1. cave wall, 2. filling sediment, 3. level of spring I. phreatic cavity, II. filling effect of the creek entering through a ponor, III.–IV. erosional cave rise

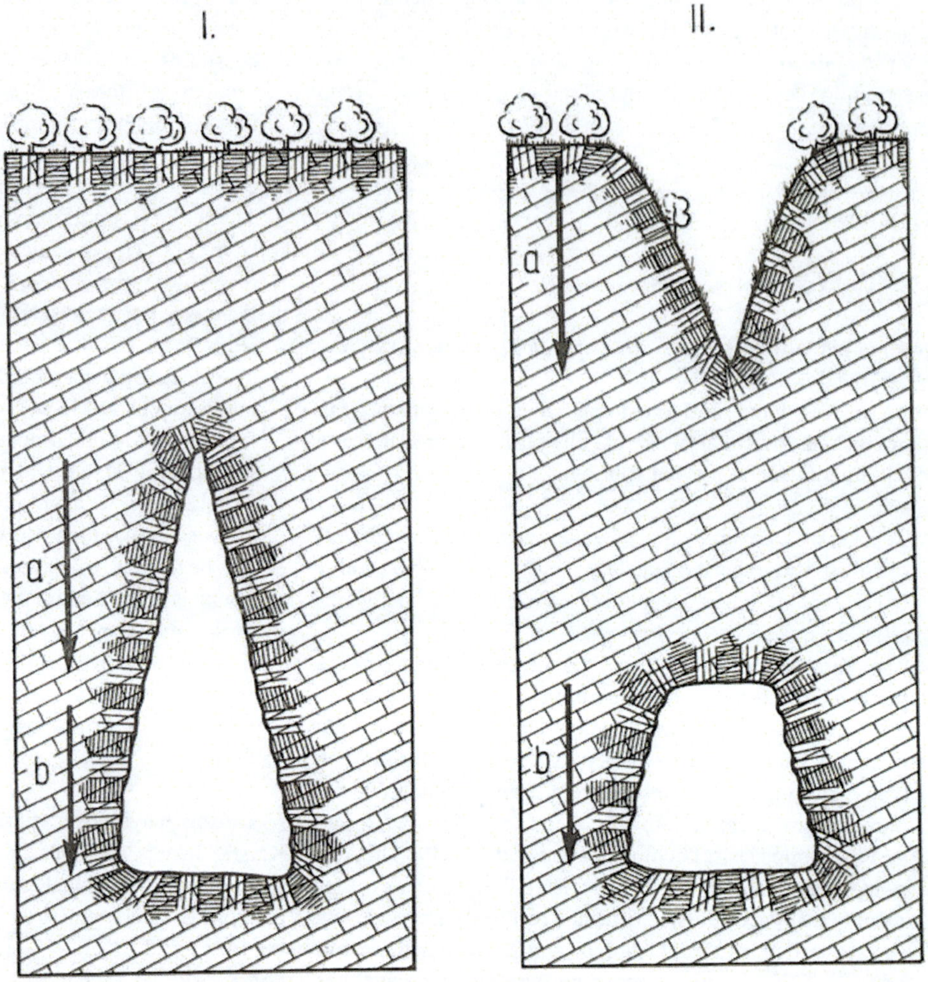

Fig. 6 Erosion cave cross-sections (Jakucs 1977a). I. erosional cave development was not preceded by valley development on the karst, II. cave development was preceded by valley development on the karst, cave part marked with "a" in Fig. 1 is the cave part that coincides with valley development

the larger catchment area and thus, the cave will be wider. This is the average cave width belonging to its catchment area with an expansion of 1 m^2, which he called erosional index number. He thinks that therefore, in the knowledge of the index number and the size of the catchment area, the width of the still unknown erosion caves of the karst can be predicted.

The creeks of the erosion caves often meander, which results in the meandering of the cave passage too. Balázs (1971) studied the meandering of the Szabadság Cave (Aggtelek Karst). He described the features of cave meandering (meandering corridor with asymmetric cross-section, the notch of the concave side, the bench of the convex side and the dead-passage that developed by the budding of the meander passage). He thinks that the meandering of the passage is caused by lateral erosion and corrosion triggered by the lateral shifting of the channel line of the creek. Budding causes the "straightening" (river section without meanders) of the underground creek, but newer meanders (meandering) develop during the deepening of the creek.

Meandering occurs at different elevations as compared to the floor level in Szabadság Cave. He calculated the degree of meandering (sinuosity) of meanders at various elevations by calculating the quotient of the length measured along the channel line of the creek and of the length measured along the centre line of the meander belt width (the stripe between the polygon lines being tangential to the bends). This value is 1.75 at 150 cm, 2.38 at 70 cm and 1.45 at 0 cm. At the 70 cm level, the increase of meandering resulted in the growth and then the budding of the meanders and thus, the development of the presently almost straight cave section and the decrease of sinuosity (Fig. 10).

Epigene caves of the vadose zone, the shafts are common on Hungarian karsts. Their development and characteristics have been studied by several researchers. According to Dénes's (1971) investigations and observations on Aggtelek Karst, the shafts are former inflow caves. They lose their active character when they get far from the margin of non-karstic impermeable cover because of its denudation. As a result of the shifting of the non-karstic cover margin,

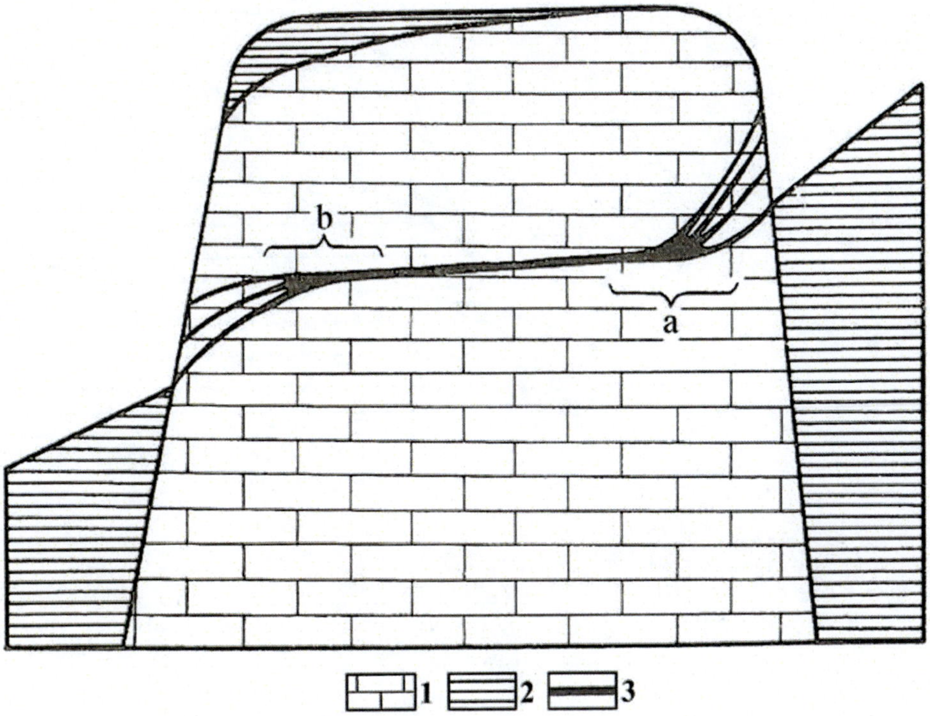

Fig. 7 Delta formation of through cave (Jakucs 1977a). 1. karstic rock, 2. non-karstic rock, 3. cave, a. delta formation at the inlet section, b. delta formation at the outlet section

Fig. 8 Retreat of the beheading site during the shifting of the rock boundary (Jakucs 1977a). 1. karstic rock, 2. non-karstic rock, 3. cave

newer ponors develop which also lose their active character and are transformed into shafts. This author thinks that losing their sediments, the filled shafts become partially active too.

Sárváry (1970) studied the shafts of the plateau of Alsó Hill (Aggtelek-Rudabánya Mountains). He states that the majority of the shafts of the plateau are in dolines, they are widening downwards, and they are mostly built up of shaft parts situated step-like below each other. He thinks that doline development and shaft development are processes stimulating each other. If a fracture taps the still unsaturated waters below the doline floor, the fracture widens and becomes deeper. The so-developing shaft initiative abstracts the waters of the doline floor which have dissolution capacity, the deepening of the floor slows down, and the deepest point of the doline is moved to another place. A newer shaft initiative develops at the developing deepest site. At this time, the deepening of the doline floor slows down here, and the original place will be the deepest point as a result of faster deepening here (thus, the younger shaft part will be of the highest position). The waters of the younger shaft part flow towards the older shaft, the two shaft parts constitute a shaft system by coalescing through waters seeping along bedding planes or fractures.

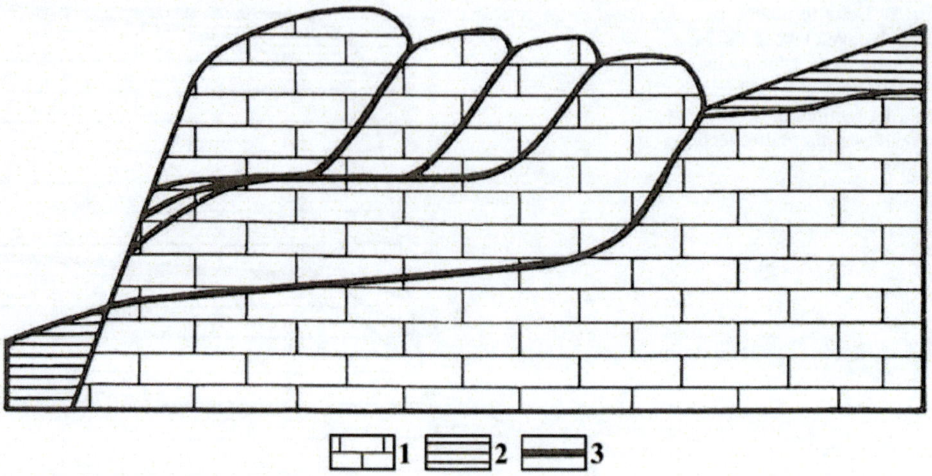

Fig. 9 Development of multilevel cave (Jakucs 1977a). 1. karstic rock, 2. non-karstic rock, 3. cave

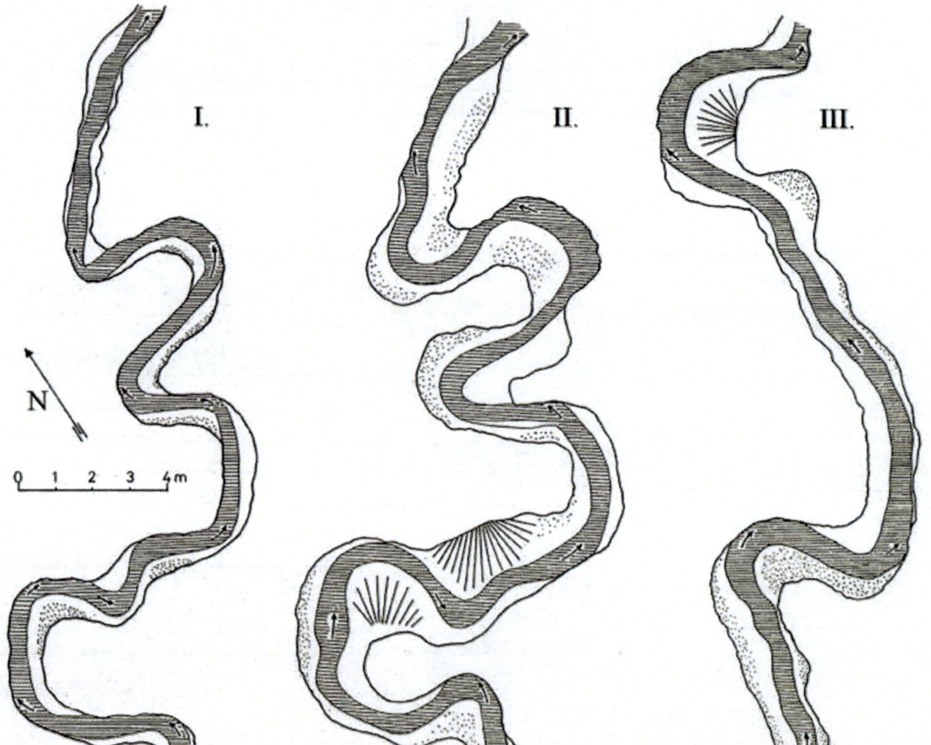

Fig. 10 The meanders of the Szabadság Cave (Aggtelek Karst) at different heights (Balázs 1971). 1. cave meander (supposed at higher levels), 2. bench of concave side, 3. dip direction of floor, 4. stream load, I. cave wall at a height of 150 cm, II. cave wall at a height of 70 cm, III. cave wall at a height of 0 cm

Veress (2018, 2019) investigated the shafts of the Bakony Region. He states that the shafts of the mountains occur below subsidence dolines, ponors and former ponors (today subsidence dolines). According to shaft structure, he distinguished, simple shafts, complex shafts, shaft systems, bifurcating shafts, storeyed shafts and shafts with phreatic passage. In addition to crust structure, the complexity of the shafts is also contributed by tributary shafts that develop as a result of paragenetic effect.

The development of subsidence dolines and the development of their shafts are processes strengthening each other. The presence of superficial deposit is a precondition for shaft formation and development. In the Bakony Region, there is a zone of collapse material below the subsidence dolines in the bedrock (Veress 2000a). The debris zone develops during the collapse of a blind shaft (secondary shaft). The blind shaft develops in the following way: the superficial deposit is washed in and creates a veneer on the wall of the upper part of primary shafts (which develop from grikes). Here, dissolution stops, but below this part it does not stop and thus, the dividing walls between the grikes coalesce by dissolution and a blind shaft develops. Small-sized suffosion dolines already develop during the material transport into the primary shafts. If the collapse of the ceiling of the blind shaft is also inherited onto the cover, a dropout doline develops (particularly, if the cover is cohesive), if it is not inherited, the already existing suffosion doline grows further. It may also occur that a paragenetic tributary shaft develops branching out from the main shaft (Veress 2016). A subsidence doline may also develop above the paragenetic blind shaft either by the collapse of the ceiling of the tributary shaft or above the tributary shaft by suffosion if there are primary shafts there. In this case, the blind tributary shaft is connected to the shafts situated below the subsidence dolines by collapse. In the former case, collapse plays a role in doline development, while in the latter, suffosion and collapse play a role (Fig. 11, Veress 2016).

Veress et al. (1992) described a possible way of phreatic cavity development by using the example of Ördög-lik of the Bakony Mountains. Some parts of the cave developed by the coalescence of spherical cavities. If we consider the development of the spherical cavity with the largest radius as unit duration, its optional radius gives its optional age considered as a unit. Supposing that all the other spherical cavities of the cave grew at a similar speed, their size can also be given to an optional, relative age. In case of an optional, relative age, two adjacent solution pockets connected to each other if the circles symbolising them on the cave map reached each other. The connection of more and more solution pockets into each other resulted in the development of a longer and longer cave part. The coalescences of solution pockets described on the cave map show the already existing cave parts at an optional relative age. Describing the coalesced spherical cavities on the cave map, the history of cave dissolution can be given to various relative ages.

Complex maps 1:100 (describing morphology, hydrology and cave sediments) were made of the large caves of Aggtelek Karst (Baradla Cave, Béke Cave) by traditional and laser scanner surveys lasting for several years (Szunyogh 2002, 2017, 2019; Szunyogh and Kisbán 2004).

8.2 Warm-Water Caves

Summarising his observations of Sátorkő-puszta Cave (Pilis Mountains) and of the Buda Hills Jakucs (1948) states that the spatial pattern of warm-water caves is characterised by the following:

- the cave is vertically dissected and vertically developed,
- downward narrowing and branching, upward widening passages which terminate at the same level at the top,
- they are dissected by large-sized spherical cavities at the top.

Jakucs (1948) states that this cave structure develops at sites where warm water is injected into the cold karstwater from below since the cold water cools down the warm water and its solution capacity increases.

According to Müller (1971), the large size of the caves of the Buda Hills cannot be explained by the CO_2 content of the infiltrating water or by mixing corrosion. Large size requires more CO_2 of metamorphic origin. CO_2 is released during the metamorphosis of the rocks of the sinking Triassic floor of the Great Hungarian Plain.

Kovács and Müller (1980) believe that the karstification of warm water origin of the Buda Hills can be related to two phases. The first phase can be described by the closed cell model because the karst surface was covered by impermeable beds at this time. This is of Miocene age and water motion was triggered by the heat of the subsurface melt. (The water of the cell and the CO_2 of the water originated from here.) The cavities developed along fractures and grikes to hydrothermal effect (Fig. 12).

The second phase is of Late Pliocene-recent age can be described by the open cell model. At this model, the majority of the karst lost its cover, waters infiltrating into the karst reached the floor of the basin that subsided into a great depth. These waters becoming warmer and taking up metamorphic CO_2 flowed upwards and at karst springs they mixed with cold karstwaters. Cavity formation took place to the effects of warm-water dissolution and mixing corrosion. Horizontal, vertically developed cavities were formed (Fig. 13).

Müller (1974) investigated the large-sized solution pockets (spherical cavities) of the caves of the Buda Hills. He explained their evolution by condensed water dissolution. An air space develops above the warm karstwater of the developing cave if the karstwater level sinks. One part of the water evaporates, the vapour from the humid air precipitates on the cold rock wall. The dissolution capacity of the karstwater is ensured by the CO_2 being released from the karstwater and entering the condensed water.

Szunyogh (1982) investigated the condensed water model theoretically. He established how the size of the developing feature depends on the warm air and he gave the growth rate belonging to a given development factor. However, in his later work (Szunyogh 1987) he questioned the merits of the

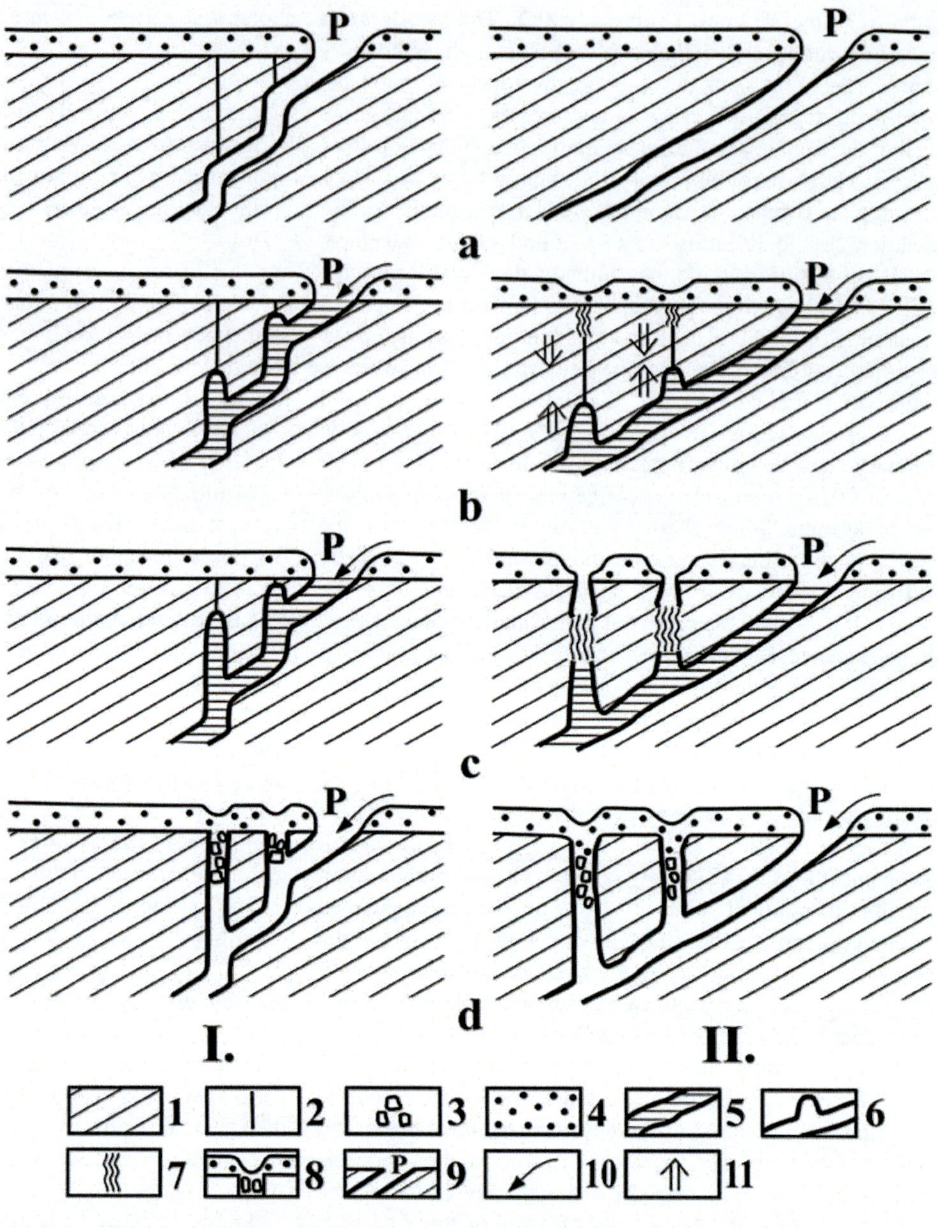

Fig. 11 Development of tributary dolines by the development of the paragenetic blind shaft from the main shaft (I), and the further development of the subsidence doline by the coalescence of the shaft below the doline and the paragenetic blind shaft from the main shaft (II) (Veress 2016, modified). 1. bedding plane, 2. fracture, 3. debris, 4. cover, 5. shaft (pit) temporarily filled with water, 6. paragenetic blind shaft, 7. primary shaft, 8. subsidence doline, 9. doline functioning as ponor, 10. water supply, 11. shaft growth, a. complex shaft of the main doline develops (I), simple shaft of the main doline develops (II), b–c: at complex shaft, upward developing paragenetic blind shafts develop which are transformed into tributary shafts (I), upward developing paragenetic blind shafts (tributary shaft) also develop at simple shaft, in addition shaft close to the surface is formed and doline develops by suffosion (II), d. at sites where blind shafts reach the surface, tributary doline develops by collapse (I), shafts close to the surface and blind shafts of the simple shaft are connected, the tributary doline continues its development by collapse (II)

model since the growth rate of the spherical cavity is too slow. Thus, he thinks that 100,000 years is necessary for the development of a solution pocket with a volume of some m^3.

9 Conclusions

Hungarian karst researchers developed theories in almost all areas of karst phenomena and processes. Particularly, theories on karstwater and karst morphology are significant and long-lasting. This can be partly traced back to the fact that the above-mentioned topics have a large practical significance (karstwater), or because karst areas have characteristics that required new approaches. For example, a significant number of Hungarian karst areas were covered karsts in the geological past or even today. The coveredness made the widespread analysis of the effect of valley inheritance on karstification necessary.

Results related to the latter are globally valid in addition to Hungarian karsts. However, the generalisations of CO_2 investigations, the theories on the development of karst surfaces, on erosional cave development and on the development of subsidence dolines can be regarded as globally valid theories as well. The international spread of some of these theories was probably hindered by the fact that several of them were only published in Hungarian journals.

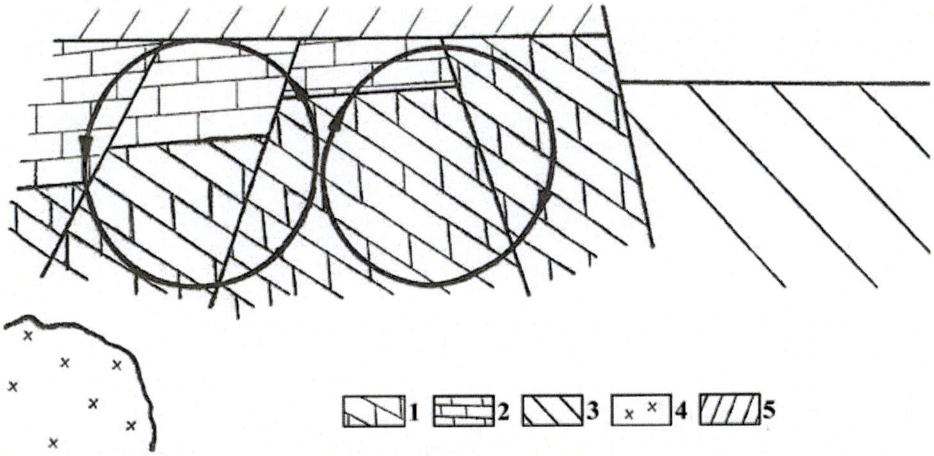

Fig. 12 Closed cell model (Kovács and Müller 1980). 1. Triassic dolomite, 2. Eocene limestone, 3. Oligocene sedimentary rock, 4. andesite magma, 5. former cover

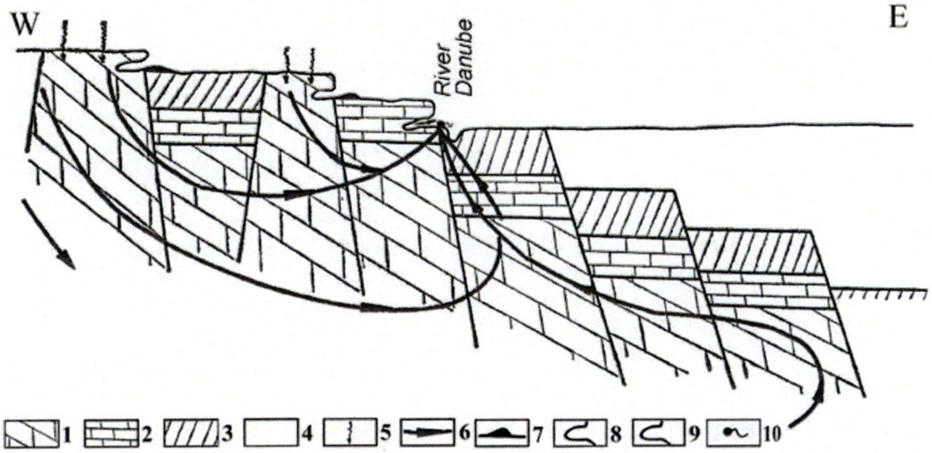

Fig. 13 Open cell model (Kovács and Müller 1980). 1. Triassic dolomite, 2. Eocene limestone, 3. Oligocene clay, 4. Neogene sediment, 5. infiltration, 6. karstwater flow, 7. freshwater limestone, 8. fossil spring cave, 9. active spring cave, 10. thermal spring (with a temperature of 120–60 °C)

References

Alföldi L, Lorberer Á (1976) A karsztos hévizek háromdimenziós áramlásának vizsgálata kút adatok alapján (A study of the three-dimensional flow of karst rocks based on well data). Hidrológiai Közlöny 56(10):433–443 (in Hungarian)

Balázs D (1969) Kísérletek a talaj alatti karsztos korrózióról (Experiments on subsoil karst corrosion). Karszt és Barlang II:57–60 (in Hungarian)

Balázs D (1971) A barlangi meander képződése (The development of cave meander). Karszt és Barlang II:75–80 (in Hungarian)

Bárány I, Mezősi G (1978) Adatok a karsztos dolinák talajökológiai viszonyaihoz (Data on the soil-ecological relations of karst dolines). Földr Ért XXVII(1):65–73 (in Hungarian)

Böcker T (1972) A karsztvizek mozgásviszonyai természetes körülmények között (Karstwater movements under natural conditions). In: Szádeczky-Kardoss E (ed) II. Anyag- és energiaáramlási ankét. Akadémia Kiadó, Budapest, pp 107–121 (in Hungarian)

Chen J (1988) Karst collapses in cities and mining areas, China. Environ Geol Water Sci 12:29–35

Csepregi A (2007) A karsztvíz kiemelés hatása a Dunántúli-Középhegység vízháztartására (The effect of karstwater extraction on the water balance of the Transdanubian Mountains). In: Alföldi L, Kapolyi L (eds) Bányászati karsztvízszint süllyesztés a Dunántúli-Középhegységben. MTA Földrajztudományi Kutatóintézet, Budapest, pp 77–112 (in Hungarian)

Dénes Gy (1971) A fokozatosan lepusztuló vízzáró takaró szerepe az exhumálódó karszt morfológiai fejlődésében (The role of the gradually denuding impermeable cover in the morphological development of exhuming karst). Karszt és Barlang I:5–8 (in Hungarian)

Erőss A, Poros ZS, Mádl-Szőnyi J, Mindszenty A, Molnár F, Ronchi P, Csoma AÉ (2011) Role of karstic and basinal fluids in porosity evolution in the Buda Hills, Hungary. In: AAPG International conference and exhibition 2011: following DaVinci's footsteps to future energy resources: innovations from outcrops to assets AAPG, Paper 1071554

Erőss A, Csondor K, Csobaji L, Zentai-Czauner B, Győri O (2019) Characterization of the regional groundwater flow system in South Transdanubia (Hungary). Geophysical Research Abstracts, EGU2019-14194 21:1–1

Hevesi A (1978) A Bükk szerkezet- és felszínfejlődésének vázlata (A sketch of the structure and landscape evolution of the Bükk Mountains). Földr Ért XXVII(2):169–203 (in Hungarian)

Hevesi A (1980) Adatok a Bükk hegység negyedidőszaki ősföldrajzi képéhez (Data to the quaternary paleogeographical features of the Bükk Mountains). Földtani Közlöny 110(3–4):540–550 (in Hungarian)

Hevesi A (1986) Hideg vizek létrehozta karsztok osztályozása (The classification of karsts created by cold waters). Földr Ért XXXV(3–4):231–254 (in Hungarian)

Hevesi A (1991a) Magyarország karsztvidékeinek kialakulása és formakincse, I. rész (Development and landforms of karst regions in Hungary I). Földrajzi Közlemények XV(1–2):25–35 (in Hungarian)

Hevesi A (1991b) Magyarország karsztvidékeinek kialakulása és formakincse, II. rész (Development and landforms of karst regions in Hungary II). Földrajzi Közlemények CXV(3–4):99–120 (in Hungarian)

Hevesi A (2019) Valamit a mészkőbarlangok és a mészkőszurdokok származási kapcsolatáról (About the relation of the origin of limestone caves and limestone gorges). Karszt és Barlang, Különszám 23–30 (in Hungarian)

Iván V, Mádl-Szőnyi J (2017) State of the art of karst vulnerability assessment: overview, evaluation and outlook. Environ Earth Sci 76:112

Iván V, Stevenazzi S, Pollicino LC, Masetti M, Mádl-Szőnyi J (2020) An enhanced approach to the spatial and statistical analysis of factors influencing spring distribution on a transboundary karst aquifer. Water 12:2133

Izápy G, Maucha L (2000) A magyarországi karsztos denudáció sebességének becslése (Estimation of the denudation rate of Hungarian karst). Karsztfejlődés V:7–20 (in Hungarian)

Jakucs L (1948) A hévforrásos barlangkeletkezés (Cave development of thermal spring origin). Hidrológiai Közlöny XXVIII(1–4):53–58 (in Hungarian)

Jakucs L (1956a) Adatok az Aggteleki-hegység és barlangjainak morfogenetikájához (Some data to the morphogenetics of the mountains and caves of Aggtelek). Földrajzi Közlemények IV:25–38 (in Hungarian)

Jakucs L (1956b) A barlangi árvizekről (On cave floods). Földrajzi Közlemények IV:281–400 (in Hungarian)

Jakucs L (1968) Szempontok a karsztos tájak denudációs folyamatainak és morfológiájának értékeléséhez (Viewpoints to the evaluation of the denudation processes and morphogenesis of karst landscapes). Földr Ért 17:17–46 (in Hungarian)

Jakucs L (1977a) Morphogenetics of Karst Regions. Adam Hilgar, Bristol, 284 pp

Jakucs L (1977b) A magyarországi karsztok fejlődéstörténeti típusai (Evolutionary types of Hungarian karsts). Karszt és Barlang I–II:1–16 (in Hungarian)

Jakucs L (1980) A karszt biológiai produktum (Karst is a biological product). Földrajzi Közlemények XXVIII(4):331–344 (in Hungarian)

Jakucs L (1994) A Budai-hegység hidrotermális karsztja (The hydrothermal karst of the Buda Hills). Földr Ért XLIII(3–4):235–245 (in Hungarian)

Jaskó S (1959) A földtani felépítés és a karsztvíz elterjedésének kapcsolata a Dunántúli Középhegységben (The relation between geological structure and the distribution of karstwater in the Transdanubian Mountains). Hidrológiai Közlöny 39(4):289–297 (in Hungarian)

Kevei Bárány I (1986) Újabb adatok a karsztdolinák képződéséhez (Newer data on the development of karst dolines). A Nehézipari Műszaki Egyetem Közleményei I. sorozat, Bányászat 33(1–4):149–155 (in Hungarian)

Kovács J, Müller P (1980) A Budai-hegyek hévizes tevékenységének kialakulása és nyomai (Origin and traces of hydrothermal activities in the Buda Range). Karszt és Barlang II:93–98

Leél-Őssy S (1952) Karrosodás és karros formák. Hidrológiai Közlöny 32(7–8):298–303 (in Hungarian)

Leél-Őssy S (1954) A Magas Bükk geomorfológiája (The geomorphology of High Bükk). Földr Ért III(2):323–336 (in Hungarian)

Leél-Őssy S (1959) Magyarország karsztvidékei (Karst regions of Hungary). Karszt és Barlangkutatás I:79–88 (in Hungarian)

Leél-Őssy S (1960) Magyarország karsztvidékei (Karst regions of Hungary). Földr Ért 9(1–4):490–495 (in Hungarian)

Maucha L (1977) Study of tidal movements of karstwaters and karstic rocks. Ann Geophys 33(1–2):151–156

Maucha L (1989) A karsztvizek jelentősége és kutatása hazánkban (The significance and research of karstwaters in Hungary). Karszt és Barlang I–II:67–76 (in Hungarian)

Maucha L (1997) A karsztos árapály-jelenség működési mechanizmusa (The mechanism of tidal phenomena in karstified rocks). Karszt és Barlang I–II:40–44 (in Hungarian)

Mádl-Szőnyi J, Tóth Á (2015) Basin-scale conceptual groundwater flow model for an unconfined and confined thick carbonate region. Hydrogeol J 23:1359–1380

Müller P (1971) A metamorf eredetű széndioxid karsztkorróziós hatása (The karst corrosion effect of metamorphic carbon-dioxide). Karszt és Barlang II:53–56 (in Hungarian)

Müller P (1974) A melegvizes barlangok és gömbfülkék keletkezéséről (On the development of warm-water caves and spherical cavities). Karszt és Barlang I:7–11 (in Hungarian)

Papp F (1942) Dunántúl karsztvizei és a feltárás lehetősége Budapesten (The karstwaters of Transdanubia and the opportunity of exploration in Budapest). Hidrológiai Közlöny 21(7–12):146–196 (in Hungarian)

Pécsi M (1980) A Pannóniái-medence morfogenetikája (The Morphogenesis of the Pannonian basin). Földrajzi Értesítő 29:105–127 (in Hungarian)

Pécsi M (1991) Geomorfológia és domborzatminősítés (Geomorphology and relief classification). Földrajz Tudományi Kutató Intézet, Budapest, 296 pp (in Hungarian)

Péntek K, Veress M, Lóczy D (2007) A morphometric classification of solution dolines. Zeits Für Geomorph 51(1):19–30

Sárváry I (1970) A zsombolygenetika kérdéseiről (On the issues of shaft genetics). Karszt és Barlang I:5–14 (in Hungarian)

Szádeczky-Kardoss E (1941) A Dunántúli középhegység karsztvízének néhány problémájáról (On some problems of the karstwater of the Transdanubian Mountains). Hidrológiai Közlöny 21(7–12):67–92 (in Hungarian)

Szunyogh G (1982) A hévizes eredetű gömbfülkék kioldódásának elméleti vizsgálata (A theoretical study of the dissolution of spherical cavities with hydrothermal origin). Karszt és Barlang II:83–88 (in Hungarian)

Szunyogh G (1987) A hévizes eredetű gömbfülkék víztükör alatti kioldódásának elméleti vizsgálata (A theoretical study of the dissolution of spherical cavities with hydrothermal origin below the surface of the water). Karszt és Barlang I–II:29–31 (in Hungarian)

Szunyogh G (2002) A Baradla barlang részletes felmérésének tapasztalata (Experience on the detailed research of Baradla Cave). Karszt és Barlang 1998–1999 (I–II):109–117 (in Hungarian)

Szunyogh G (2017) Gömbfülkék, mennyezeti üstök és vakkürtők a Baradla barlangban (Spherical cavities, ceiling pockets and half tubes). Karsztfejlődés XXII. https://doi.org/10.17701/17.163-177:163-177 (in Hungarian)

Szunyogh G (2019) A barlangtérképezés, mint a dokumentáció alapja (Cave mapping as the basis for documentation). Karszt és Barlang különszám 41–50 (in Hungarian)

Szunyogh G, Kisbán J (2004) Béke barlang. A Komlós patak felszín alatti útja (Béke Cave. The subsurface way of Komlós Stream). Aggteleki Nemzeti Park Igazgatóság, Jósvafő, 140 pp (in Hungarian)

Tóth G (2001) A karsztos résrendszer fejlődésének tényezői a Bükkben (Factors of the development of the karstic joint system in the Bükk Mountains). Karsztfejlődés VI:113–125 (in Hungarian)

Tóth Á (2018) A Balaton-felvidék felszínalatti vizeinek hidraulikai kapcsolata a Bakonnyal és a Balatonnal (Hydraulic connection of

the groundwaters of the Balaton Highland, Bakony Mts. and Lake Balaton). PhD dolgozat, Budapest, p 202
Tóth G, Fejes P (1984) Idősebb pliocén lepusztulási szint kimutatása a Bükkben (Report on an older Pliocene denudation level in the Bükk Mountains). Karszt és Barlang II:65–72 (in Hungarian)
Tóth Á, Mádl-Szőnyi J (2019) Where does the drinking water come from?—Interrelationship between surface water and groundwater in a carbonate area, Hungary. Geophysical Research Abstracts, EGU2019-8902, 21:1–1
Venkovits I (1959) Karsztnevezéktani vita (A debate on karst nomenclature). Karszt és Barlangkutatás I:67–77 (in Hungarian)
Veress M (2000a) Covered karst evolution Northern Bakony Mountains, W-Hungary. A Bakony Természettud. Kut. Eredményei 23, Bakonyi Természettudományi Múzeum, Zirc, 167 pp
Veress M (2000b) Középhegységi karsztok néhány típusa (Some types of middle mountain karsts). Földrajzi Közlemények CXXIV(1–4):1–28 (in Hungarian)
Veress M (2003) Adalékok a homokkő anyagú kőtengerek (Káli-medence) pszeudokarrjainak morfogenetikájához (Data on the morphogenesis of the pseudokarren of block fields with sandstone material (Kál Basin). Folia Musei Historica – Naturalis Bakonyiensis, 2001–2003 Zirc: 7–46 (in Hungarian)
Veress M (2005) Adalékok a Tábla-völgyi-dűlő (Tési-fennsík) fedett karsztosodáshoz (Data on covered karstification of the doline in Tábla Valley, Tés Plateau). Karsztfejlődés X:267–291 (in Hungarian)
Veress M (2006) Adatok a Tési-fennsík két térszínrészletének fedett karsztosodásához (Survey/data on covered karstification of two surface parts of the Tés Plateau). Karsztfejlődés XI:171–184 (in Hungarian)
Veress M (2008) A mészkőfekü morfológiájának hatása a fedett karsztosodásra az Északi-Bakonyban (The effect of the morphology of the limestone bedrock on covered karstification in the Northern Bakony Mountains). Karszt és Barlang 2004–2005. évf.:33–54 (in Hungarian)
Veress M (2009) Investigation of covered karst form development using geophysical measurements. Zeits F Geomorph 53(4):469–486
Veress M (2016) Covered karst. Springer, Berlin, Heidelberg, New York, 536 pp. https://doi.org/10.1007/978-94-017-7518-2
Veress M (2018) A Bakonyvidék genetikai aknatípusai (The genetic shaft types of the Bakony Region). Karsztfejlődés XXIII:55–84 (in Hungarian)
Veress M (2019) Shaft lengths and shaft development types in the vadose zone of the Bakony Region (Transdanubian Mountains, Hungary). J Soil Water Sci 3(1):54–74
Veress M, Péntek K (1996) Theoretical model of surface karstic processes. Z Geomorphol 40(4):461–476
Veress M, Vetési-Foith Sz (2019) The cave opening types of the Bakony region (Transdanubian Mountains Hungary). Int J Curr Res 11605:3546–3562. https://doi.org/10.24941/ijer.34579.05.2019
Veress M, Péntek K, Horváth ET (1992) Evolution of corrosion caverns: Ördöglik Cave, Bakony, Hungary. Cave Sci 19(2):41–50
Yuan D (1987) Environmental and engineering problems of karst geology in China. In: Beck BF, Wilson WL (eds) Karst hydrogeology: engineering and environmental applications. Rotterdam, Balkema, pp 1–11
Zámbó L (1970) A vörösagyagok és a felszíni karsztosodás kapcsolata az Aggteleki-karszt délnyugati részén (The relation of red clays and karstification in the south-western part of Aggtelek Mountains). Földrajzi Közlemények 94(4):281–293 (in Hungarian)
Zámbó L (1986) Karsztvörösagyagok CO_2 termelés és a karsztkorrózió összefüggése (The connection between the CO_2 production of karst red clays and karst corrosion). A Nehézipari Műszaki Egyetem Közleményei I. sorozat Bányászat 33(1–4):125–138 (in Hungarian)
Zámbó L (1998) Az Aggteleki-karszt felszínalaktani jellemzése (Surface morphological characterisation of Aggtelek karst). Földrajzi Értesítő XLVII(3):359–378 (in Hungarian)
Zámbó L, Telbisz T (2000) A karsztkorróziós talajhatás érvényesülése a karsztfejlődésben (The effectiveness of karst corrosion soil impact in karst development). Karsztfejlődés V:103–114 (in Hungarian)

Geology of the Karst Terrains in Hungary

Zsolt Kercsmár, Ildikó Selmeczi, Tamás Budai, György Less, and Gyula Konrád

Abstract

In this chapter we describe the geology of the structural units of Hungary in the Alp-Carpathian orogene system and briefly present the most important stages of their evolution (Sect. 1. Budai T. and Kercsmár Zs.). Geological features of the mountain ranges are summarized according to their tectonic and lithostratigraphic characteristics. The relationship between the stratigraphic units is shown in pre-Cenozoic and Cenozoic lithostratigraphic charts. In the geological characterization of mountains, we deal with rocks suitable for karstification in more detail. In our study we first provide a review on the geology of the Transdanubian Range located NW of the Mid-Hungarian Shear Zone and belonging to the Alcapa Mega-unit (Sect. 2. Kercsmár Zs., Budai T. and Selmeczi I.). We then give an overview of the geological build-up of the Aggtelek–Rudabánya Hills (located next to the Bükk due to nappe tectonics), the Bükk and the Szendrő and Uppony Mountains (located in the shear zone of the Mid-Hungarian Mega-unit) (Sects. 3, 4, 5. Less Gy). The later is followed by the geology of the Mecsek and the Villány Mountains in the southern tectonic unit, i.e. the Tisza Mega-unit (Sect. 6. Konrád Gy).

Keywords

Hungary • Alcapa mega-unit • Tisza mega-unit • Mid-Hungarian mega-unit • Tectonics and geology • Lithostratigraphy • Karst terrains

1 Introduction

Hungary lies in the middle of the Carpathian Basin surrounded by the Alps, the Carpathians and the Dinarides. These mountain ranges belong to the Alpine orogenic system (Fig. 4.1) that was formed by the collision of the European and the African Plate.

The pre-Cenozoic basement of the basin is composed of two different lithospheric mega-units with different geological features (Fig. 4.2). The Alcapa Mega-unit has a highly complex geological build-up and originated from the African Plate, while the Tisza Mega-unit has a European Plate origin. The two tectonic mega-units are attached to each other along the Mid-Hungarian Mega-unit (Mid-Hungarian Shear Zone), which shows close Dinaridic affinity.

The pre-Cenozoic basement of the basin exposed on the surface only in the area of the mountain ranges (Fig. 4.2), while its overwhelming part is covered by several-kilometre-thick sedimentary successions and volcanics formed over the last 19 million years.

The basement of the Alcapa Mega-unit in NW Transdanubia is built up by the tectonic units of the Austroalpine Nappe System. Jurassic metamorphic rocks of the west-Alpine Penninic Unit are exposed in a window below the Austroalpine Nappes in the Kőszeg Mountains while Paleozoic gneiss and schists of the Lower Austroalpine Nappes build up the Sopron Mountains. The Transdanubian Range Unit forms the uppermost nappe of the Austroalpine

Z. Kercsmár (✉) · I. Selmeczi · T. Budai
Division of Geology, Supervisory Authority of Regulatory Affairs, Stefánia u. 14. 1143, Budapest, Hungary
e-mail: zsolt.kercsmar@sztfh.hu

I. Selmeczi
e-mail: ildiko.selmeczi@sztfh.hu

G. Less
Faculty of Earth Science & Engineering, Institute of Mineralogy and Geology, University of Miskolc, Miskolc-Egyetemváros, Miskolc-Egyetemváros, 3515, Hungary
e-mail: foldlgy@uni-miskolc.hu

G. Konrád
University of Pécs, Pécs, Hungary
e-mail: konradgyula@t-email.hu

T. Budai
Department of Geology and Meteorology, University of Pécs, Ifjúság útja 6, 7624 Pécs, Hungary

M. Veress and Sz. Leél-Őssy (eds.), *Cave and Karst Systems of Hungary*,
Cave and Karst Systems of the World, https://doi.org/10.1007/978-3-030-92960-2_4

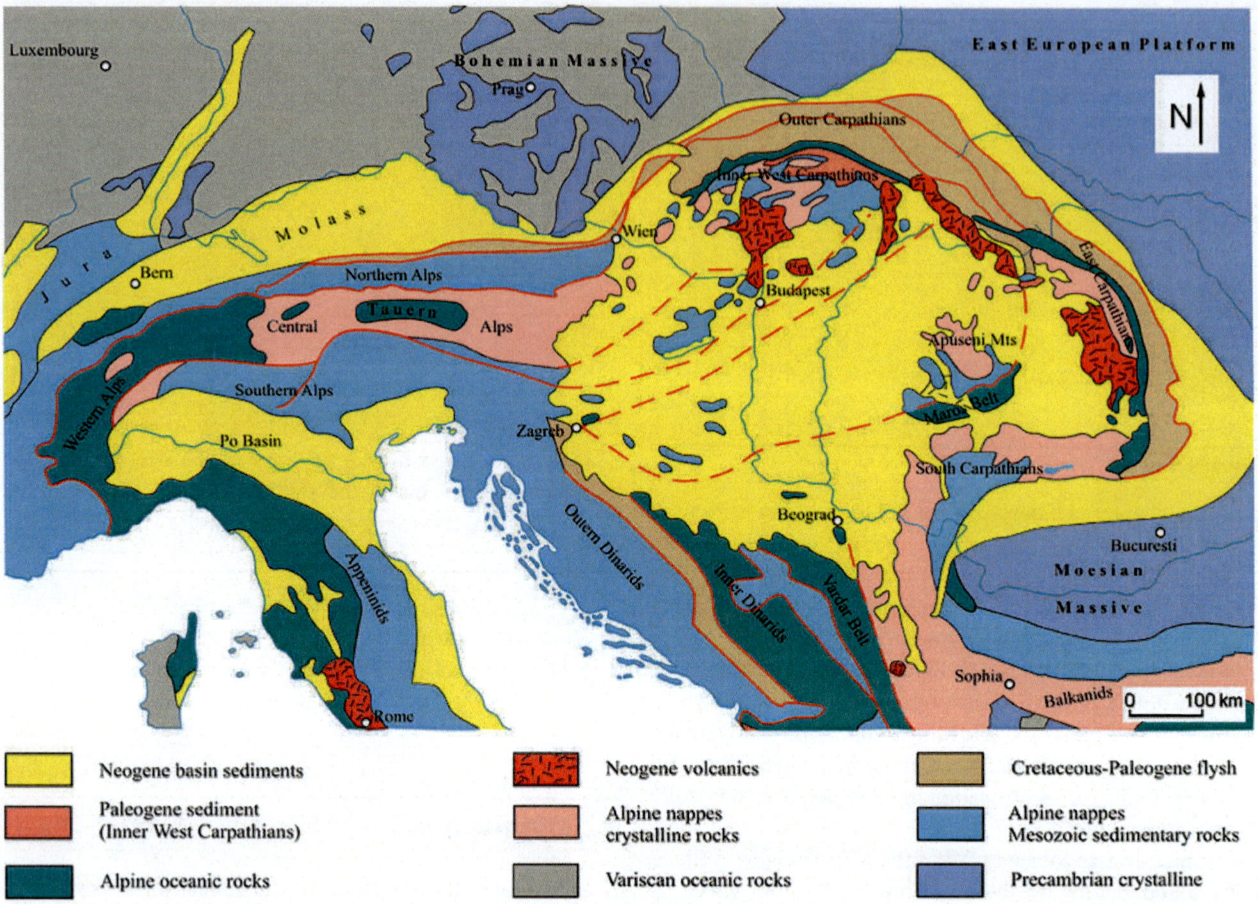

Fig. 4.1 Tectonic position of the Carpathian Basin in the Alpine orogenic system (after Haas et al. 2012 modified)

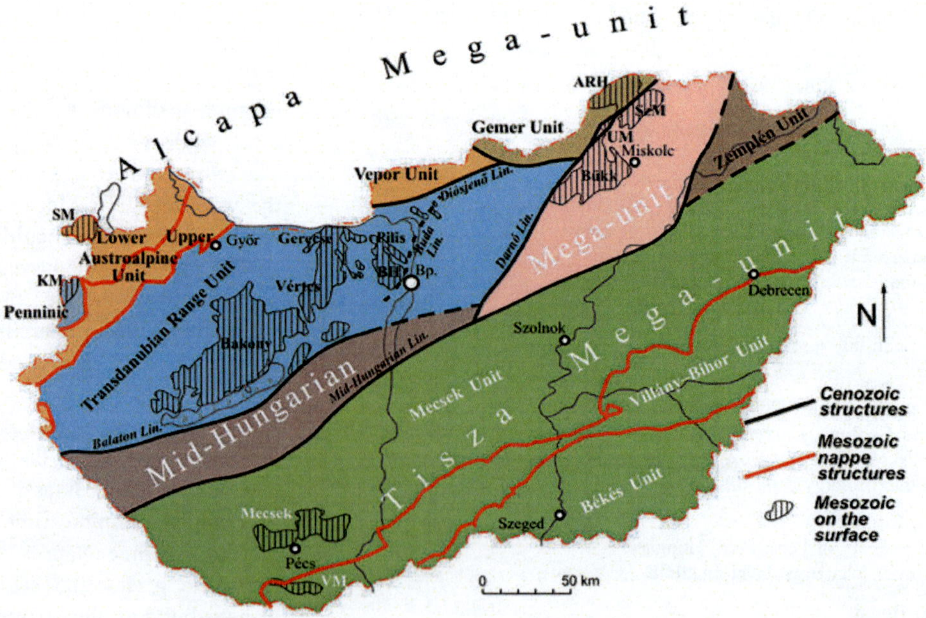

Fig. 4.2 Tectonic structure of the pre-Cenozoic basement of Hungary (after Haas et al. 2010, modified). Dashed area shows the surface extension of the basement. ARM: Aggtelek-Rudabánya Mountains; BH: Buda Hills, KM: Kőszeg Mountains; SM: Sopron Mountains; SzM: Szendrő Mountains; UM: Uppony Mountains; VM: Villány Mountains

Nappe System, which consists of slightly metamorphosed Paleozoic rocks and non-metamorphosed Permian and Mesozoic sedimentary successions. The nappe system of the Aggtelek-Rudabánya Hills of the Gemer Unit is built up mainly of slightly metamorphosed Triassic–Jurassic basin successions and non-metamorphosed Triassic platform carbonates.

The Mid-Hungarian Mega-unit is bordered by the Balaton and the Darnó Line on the NW and by the Mid-Hungarian Lineament on the SE. Its pre-Cenozoic basement crops out to the surface exclusively in the area of the Szendrő–Uppony Mountains and the Bükk Mountains. The slightly metamorphosed succession is built up of Paleozoic siliciclastics and carbonates, whereas most of the Bükk is made up of Permian–Triassic carbonates and Jurassic igneous and siliciclastic rocks.

The Mesozoic nappe system of the basement of the Tisza Mega-unit is exposed in the Mecsek Mountains and the Villány Hills. The pre-Alpine basement of the mega-unit is composed predominantly of polymetamorphic rocks and granites. The several-thousand-meter-thick Carboniferous–Permian siliciclastics are overlain by Triassic to Lower Cretaceous sedimentary sequences with volcanics in the Lower Cretaceous.

The compressional structure of the mega-units started to develop in the Early Cretaceous due to the convergence of the African Plate towards the European Plate. This was followed by uplift and erosion, and new basins were formed in the Late Cretaceous. During the Paleogene the Alcapa Mega-unit moved towards the ENE along the Mid-Hungarian Shear Zone. This was accompanied by andesite volcanism along the Periadriatic Lineament, which bordered the Hungarian Paleogene Basin in the area of the Transdanubian Range and the Bükk Units. At the beginning of the Miocene the main part of the mega-units was a continental terrain, marine sedimentation was limited to the northern sub-basins of the Paleogene Basin. In the middle Miocene the Alcapa and Tisza Mega-units moved into their current position and due to the continuous subduction of the European Plate, an intensive island arc-type andesite volcanism took place. Sea-level rise resulted in carbonate sedimentation in shallow-marine shelf environments, while fine-grained siliciclastic sediments deposited in the basins.

Evolution of the Carpathian Basin during the Late Miocene was determined by thermal subsidence. The central basin of the former Paratethys had closed completely, forming a lake with no runoff. This was Lake Pannon, which gradually filled up by the end of the Miocene due to continuous sediment tansport of rivers flowing down from the Carpathians and from the Alps into the lake basin. During the Pliocene and the Quaternary fluvial sediments deposited in the subsiding areas of the dry-land, while in uplifting mountainous regions denudation went on. The Pliocene Epoch was the main period of mafic basaltic volcanism in the area of the Bakony–Balaton Highland, and it continued even in the Quaternary in the Karancs-Medves area.

2 Transdanubian Range (Transdanubian Mountains)

The Transdanubian Range Unit forms the uppermost nappe of the Austroalpine Nappe System which was not affected by metamorphism during the Alpine cycle (Horváth 1993). Its north-western boundary is the so-called Rába Line which can be interpreted as a Cretaceous thrust that reactivated as a normal fault during the opening of the Pannonian Basin in the Miocene (Haas and Budai eds. 2014). Its prolongation along the Danube is the Diósjenő Line that separates the Transdanubian Range Unit from the crystalline rocks of the Vepor Unit. The south-eastern boundary of the unit is the right-lateral strike-slip zone of the Balaton lineament that belongs to the south-Alpine Periadriatic lineament.

2.1 Tectonics

The pre-Cenozoic structure of the Transdanubian Range Unit is determined by compressional tectonic elements of SW–NE strike (Fig. 4.3). The syncline structure of the unit was formed during the eo-Alpine compressional phases in the middle Cretaceous (Haas et al. 2012). In the axis zone, Jurassic and Lower Cretaceous formations are to be found, while the formations progressively get older towards the flanks, ranging in age from the Late Triassic to the Early Paleozoic. Both sides of the syncline are accompanied by antiforms, in the axes of which Early Paleozoic formations are found. Such an anticline stretches from the Zala Basin through the basement of the Balaton up to the southern foreland of the Buda Hills, as well as in the basement of the Danube Basin, in the north-western foreland of the Bakony Mountains. On the limbs of the syncline overturned and overthrust folds have been formed, among which the most important one is the Litér thrust that stretches over a length of 100 km through the Balaton Highland and through the E Bakony. In the northern part of the unit the strike of the syncline turns gradually to NW–SE direction; in this area the main compressional structure is the Vértessomló–Nagykovácsi line of W–E direction.

2.2 Geological Build-Up

The Early Paleozoic succession of the Transdanubian Range was formed during the Variscan tectonic cycle. Fine-grained siliciclastics siliciclastites (Lovas Slate) were deposited in an open-marine basin from the Ordovician up to the Devonian (Budai et al. 1999). This was occasionally interrupted by basic (Litér Metabasalt) and acidic volcanic activity (Alsóörs Metarhyolite). A shallow-marine carbonate platform was

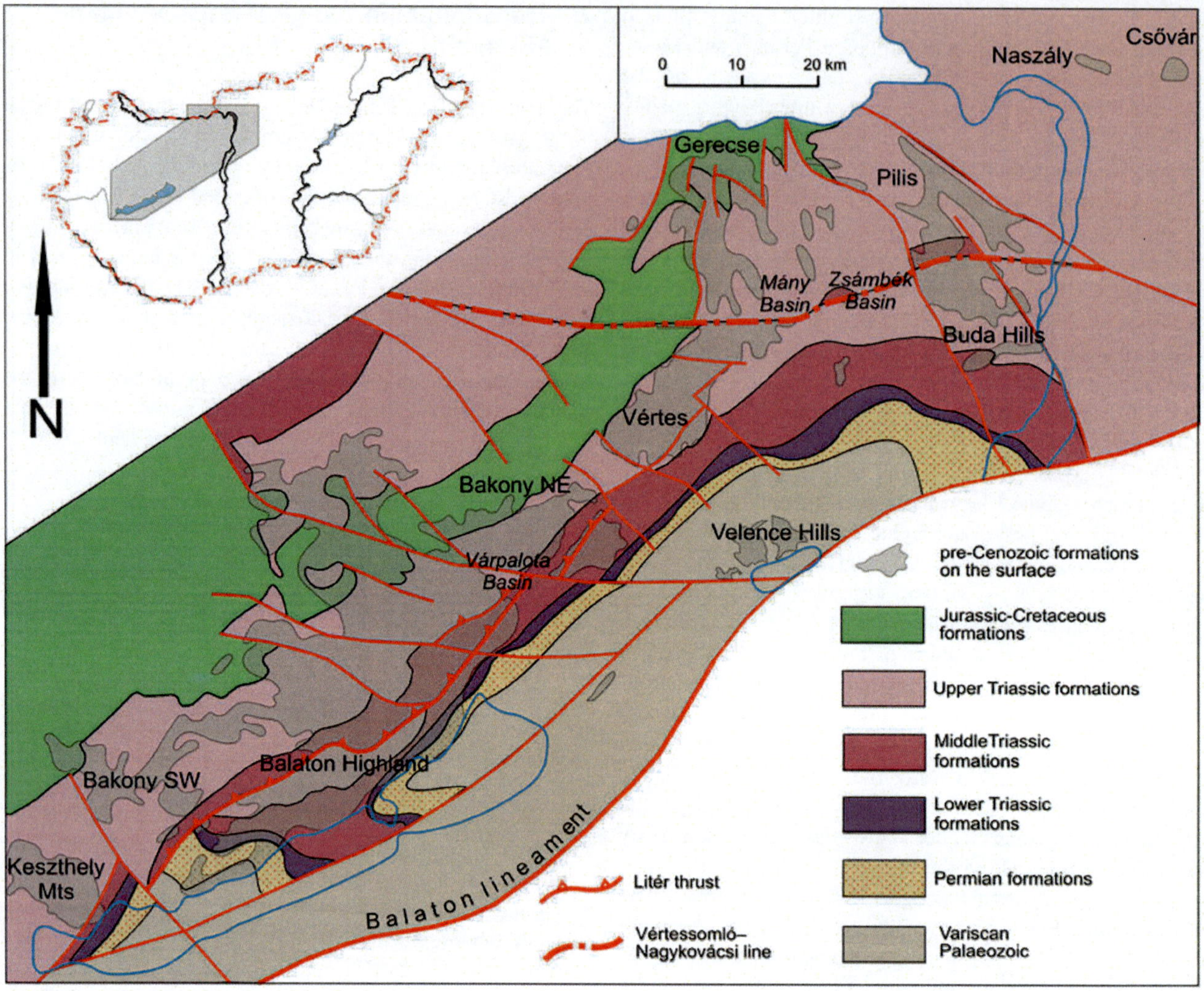

Fig. 4.3 Simplified pre-Cenozoic map of the Transdanubian Range (after Haas et al. 2010, modified)

formed during the Devonian (Polgárdi Limestone). Early Paleozoic formations underwent very low-grade metamorphism during the Variscan orogeny in the Early Carboniferous and their eroded material was deposited during the Late Carboniferous molasse sedimentation (Füle Conglomerate).

The Alpine tectonic cycle began in the Early Permian with the continental rifting of the Neotethys (Fig. 4.4). Post-collision granitoid intrusions are traced along the Balaton lineament (Velence Granite). During the Middle–Late Permian terrestrial sedimentation was going on (Majoros 1983), the dissected denudation surface of the Variscan basement was covered by a cyclic fluvial succession of conglomerate–sandstone–siltstone (Balatonfelvidék Sandstone). The north-eastern part of the Transdanubian Range Unit was flooded by sea during the Late Permian, as a result of which peritidal (Tabajd Evaporite) and shallow lagoonal environments evolved (Dinnyés Dolomite).

Due to the eustatic sea level rise at the beginning of the Triassic, an extensive shallow-marine ramp came into being where siliciclastic and carbonate sedimentation took place (Haas and Budai 1995, 1999). In the north-eastern part of the Transdanubian Range the Lower Triassic limestone and calcareous marl succession (Alcsútdoboz Limestone) overlies the Upper Permian shallow-marine dolomite. In the Balaton Highland the fluvial Upper Permian sandstone is overlain by shallow-marine marl, calcareous marl and limestone (Arács Marl); and in the south-western part of the Balaton Highland by coastal sandy dolomite and sandstone (Köveskál Dolomite). These are covered by shallow-marine red siltstone (Zánka Sandstone) and lagoonal dolomite (Hidegkút Dolomite), whereas the uppermost part of the

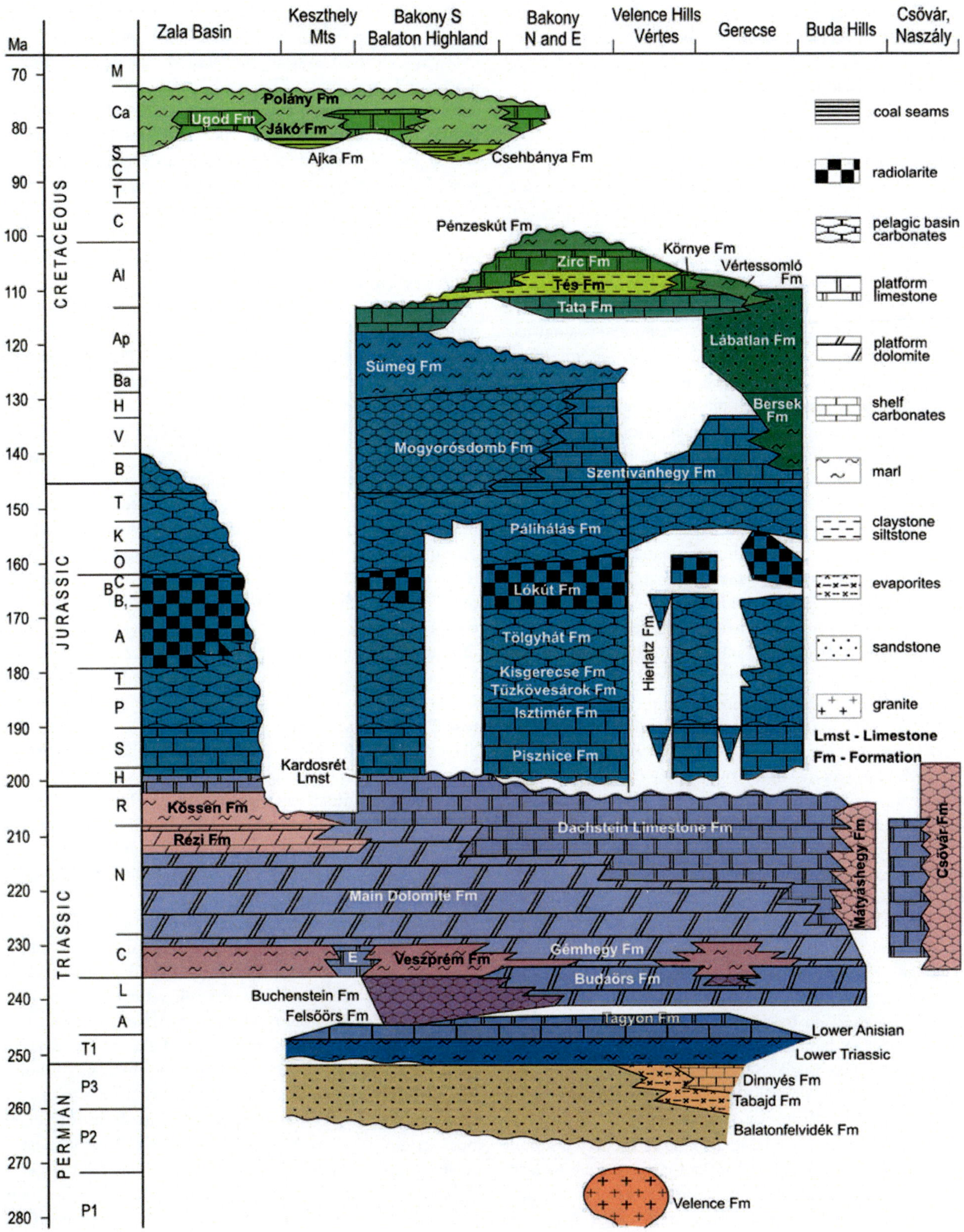

Fig. 4.4 Stratigraphic column of the pre-Cenozoic Alpine formations of the Transdanubian Range Unit (after Haas and Budai eds. 2014)

Lower Triassic sequence is built up of marl and siltstone of basin facies, with intercalating limestone and sandstone (Csopak Fm). The thickness of the whole Lower Triassic succession is about 400 m (Budai et al. 1999).

In the Early Anisian the siliciclastic influx ceased on the shallow-marine shelf and carbonate deposition became prevailing. At the beginning, cellular dolomites were deposited under arid climate (Aszófő Dolomite) then laminitic and bedded, bioturbated limestone (Iszkahegy Limestone) and dolomite (Megyehegy Dolomite) were formed in a subtidal restricted lagoon. The thickness of the whole Lower Anisian shallow-marine carbonate succession may reach 600 m (Budai et al. 1999).

At the beginning of the Middle Anisian extensional tectonics disintegrated the carbonate ramp, as a consequence of the rifting of the Neotethys (Budai and Vörös 1992, 2006). Isolated platforms were formed on the uplifted blocks where cyclic peritidal-lagoonal carbonates deposited (Tagyon Fm). Due to a sea-level drop at the beginning of the Late Anisian these platforms became subaerially exposed but considerable karstification has not happened on them. Coevally, above the down-faulted blocks halfgraben-type basins were opened, characterized by hemipelagic limestone and marl succession (Felsőörs Fm). Deepening of the basins continued up to the end of the Middle Triassic, Ladinian basin successions are dominated by cherty limestones and radiolarites (Buchenstein Fm) with volcanic tuff intercalations (Fig. 4.5). Coevally, huge carbonate platforms evolved in the uplifted north-eastern areas of the Transdanubian Range (Budaörs Dolomite) which prograded against the shallowing basins during the earliest Carnian (Budai et al. 1999). The thickness of the Middle Triassic basin succession can reach 300 m, whereas the maximum thickness of the Ladinian platform carbonate succession is about 1000 m.

Climate changed to more humid at the beginning of the Carnian. This caused increasing siliciclastic influx into the intraplatform basins where clay marl, marl and silty marl (Veszprém Marl) deposited in a thickness up to 800 m (Budai et al. 1999). Coevally, large isolated platforms developed in the south-western (Keszthely Mts) and north-eastern part of the Transdanubian Range (E Bakony, Vértes), characterized by cyclic succession of peritidal and subtidal facies (Ederics and Gémhegy Fm, Fig. 4.6).

The basins were filled up by the Late Carnian and a huge platform started to grow on the levelled topography of the carbonate shelf. Until the Middle Norian a 1–1.5-km-thick cyclic succession of peritidal and shallow lagoonal dolomite was formed (Main Dolomite) on arid climate. Due to more humid climatic conditions the dolomitisation of the sediments of the platform gradually ceased and a maximum 800-m-thick limestone succession (Dachstein Limestone, Fig. 4.7) was formed till the end of the Rhaetian (Haas 2004). Due to extensional tectonic activity intraplatform basins opened in the mid-Carnian characterized by pelagic cherty limestones and dolomites in the Buda Hills (Mátyáshegy Fm) and in the area of the Csővár blocks (Csővár Limestone). The mid-Norian restricted basins formed in the south-western part of the Transdanubian Range Unit (Keszthely Mts and Zala Basin) were represented by high organic content (Rezi Dolomite, Kössen Marl). A similar intraplatform basin is also known in the north-eastern part of the unit (Haas 2002).

In the north-eastern part of the Transdanubian Range Unit the carbonate platform was flooded at the Triassic/Jurassic boundary. However, in some parts of the Bakony Mountains the development of the platform proceeded even in the earliest Jurassic (Kardosrét Limestone) and it was flooded only in the Sinemurian. The huge carbonate platform, formed during the Late Triassic, gradually disintegrated at the beginning of the Jurassic period (Galácz 1988; Vörös and Galácz 1998) due to the extensional tectonics of the rifting Penninic Ocean. On the emerged areas submarine highs were formed characterized by periodic sedimentation. Their tectonically controlled steep slopes were populated by brachiopods and crinoids (Hierlatz Limestone). Coevally, in the surrounding deep basins slow and continuous sedimentation proceeded. The Lower Jurassic succession, lying on the Upper Triassic-Lower Jurassic platform carbonates, comprises thick-bedded (Pisznice Limestone) and cherty limestones (Isztimér Limestone). The upper part of the Lower Jurassic, the Middle Jurassic and the lower part of the Upper Jurassic are made up of clayey, nodular, "ammonitico rosso" type limestones (Tűzkövesárok Limestone, Kisgerecse Marl, Tölgyhát Limestone, Pálihálás Limestone, Fig. 4.8). The pelagic basin facies of the Callovian-Oxfordian is represented by lamellar-siliceous marl and cherty limestone (Lókút Radiolarite). The total thickness of the Jurassic succession exceeds 300 m in the basinal areas of the Bakony Mountains and reaches 20–50 m in the area of the ridges.

Marine sedimentation was continuous during the latest Jurassic and the Early Cretaceous. Thin-bedded, cherty limestone was formed in the deep "maiolica" basin of the south-western part of the Transdanubian Range (Mogyorósdomb Limestone) with a maximum thickness of 150 m. Coevally, the northern part of the unit was dominated by submarine highs where fossiliferous crinoidal limestones were formed (Szentivánhegy Limestone). The upper part of the Lower Cretaceous is made up of pelagic marl in the South Bakony (Sümeg Marl), the thickness of which reaches 250–300 m. It is capped by Aptian-Lower Albian crinoidal limestone (Tata Fm) which overlies older Cretaceous and Jurassic formations with erosional disconformity in the central and north-eastern part of the Transdanubian Range Unit. At the beginning of the Early Albian, Alpine compression led to the development of the syncline

Fig. 4.5 Nodular siliceous Ladinian limestone with volcanic tuff intercalations (Buchenstein Fm) in a quarry near Kádárta (South Bakony Mts) (*Photo* T. Budai)

structure of the Transdanubian Range Unit. As a result of tectonic uplift and subaerial exposure, the overwhelming part of the unit underwent intense denudation (Mindszenty et al. 2001). Triassic and Lower Jurassic carbonates were karstified and in the karstic depressions bauxite accumulated (Alsópere Bauxite). Further on fluvial, lacustrine, swamp and marine clay, clay marl and marl deposited on the erosional surface (Tés Clay Marl) with a maximum thickness of 200 m (Császár 1986). Rudists formed patch-reefs coevally (Környe Limestone) and later (Zirc Limestone, Fig. 4.9) in the shallow-marine lagoon (Császár 2002) and were drowned during the late Albian transgression (Pénzeskút Marl, Vértessomló Siltstone).

Sedimentation pattern of the Gerecse area was significantly different from that of the south-western and central part of the Transdanubian Range during the Early Cretaceous (Budai et al. 2018). Flysch-type, deep basinal siliciclastics were deposited in a thickness of about 600 m (Bersek Marl, Lábatlan Sandstone).

Tectonic movements in the Turonian–Coniacian caused uplifting and subaerial exposure of the Transdanubian Range. Strong karstification took place on the uplifted highs of the dissected surface (Mindszenty et al. 2001) where bauxite was deposited under subtropical climate (Halimba Bauxite). In the area of the basins fluvial (Csehbánya Fm), lacustrine and swamp sediments were deposited (Ajka Coal) during the Santonian. Due to the following sea-level rise carbonate platforms were formed on the highs (Ugod Limestone), while in the basins pelagic marls deposited (Jákó Marl and Polány Marl) with a thickness of 800 m during the Campanian and Maastrichtian (Haas 1999).

After the deposition of the Upper Cretaceous pelagic marls, the Transdanubian terrains became tectonically uplifted as a result of flexural basin evolution in the NW–SE compressional tectonic regime (Tari et al. 1993; Kercsmár 1996, 2003; Kázmér et al. 2003; Kercsmár and Fodor 2005). Karstification and bauxite deposition (Gánt Bauxite) began under tropical climate, and lasted for 10 million years on the

Fig. 4.6 Platform slope succession of mid-Carnian carbonate platform (Gémhegy Dolomite) in the Séd Valley, Veszprém (South Bakony Mts) (*Photo* T. Budai)

uplifted tectonic forebulge. Bauxite occurs in piedmont sediments and karstic dolines (Fig. 4.10). The latter were formed due to the weathering of the Mesozoic carbonates and older bauxites became redeposited into them (Mindszenty et al. 2001).The maximum thickness of the formation is 60 m.

In its role as the starting member of the Middle Eocene transgression cycle (Fig. 4.11), a succession—made up of different sediments—was deposited upon the Paleo–Mesozoic basement or the Gánt Bauxite Fm. It comprises terrestrial piedmont, fluvial, lacustrine, marshy–boggy and paralic sediments, deposited in brackish water environment. On the basis of fossils of the intercalating beds and those of the overlying sediments, the age of the succession is Middle and Late Eocene (Dorog Fm and Kosd Fm). The shallow-marine and deep neritic formations of the Middle Eocene transgression cycle show a continuous transition from the overlying sediments of the coal-bearing formations. In the 10 to 20 (max. 50)-m-thick, shallow-marine marl and calcareous marl, siltstone and sandstone beds there are fossil gastropods, bivalves and corals indicating eutrophic conditions. Frequently, the fossils have been accumulated in a lumachelle-like manner (Csernye Fm). Due to tectonic deepening of the basin the formation gradually passes upward (Kercsmár et al. 2008; Kercsmár 2018) into a homogenous, deep neritic, well-bedded, grey clay marl and marl succession (Csolnok Clay Marl Fm, "Operculina marl"). The foraminifers (Operculina, small nummulites, Assilina, Orthophragminas) indicate oligo– and mesotrophic conditions. The average thickness of the formation in the area of the Transdanubian Range varies between 20 and 120 m. Locally, the clay marl succession with sand-intercalations is overlain by coarse-grained sandstone, which contains exclusively *Nummulites perforatus* (Kecskemétiné-Körmendy 1966; Kercsmár 2005a, b, 2010, 2018). The clayey marl, sandy marl, sand and sandstone succession (Tokod Sandstone Fm) is dominated by sand, subordinately clay, siltstone and clay marl beds with submarine erosional

Fig. 4.7 Peritidal and shallow subtidal Lofer cyclic lagoonal succession of the Norian platform (Dachstein Limestone) in the quarry of Kecske-kő (Gerecse Mts) (*Photo* J. Haas)

horizons. Open-marine silty marl was deposited in the deeper basin environments (Padrag Marl Fm); it was the heteropic facies of the Csolnok Clay Marl Fm. The lower part of the formation consists of planktonic foraminifer-bearing (*Globigerina* sp.) basinal marl ("globigerina marl"), accompanied by thin andesite tuff intercalations according to the intense volcanic activity in the region in the Bartonian and Lutetian (Nadap Andesite, Szentmihály Fm). The upper part of the formation comprises coarse-sandy marl and calcareous, clayey sandstone beds of submarine canyons and fans with large amounts of volcanic grains, accumulated by gravity movements (Báldiné Beke and Báldi 1990; Sztanó and Fodor 1997). The Padrag Marl gradually develops from the glauconite-rich Csolnok Clay Marl, which was deposited in an upper slope environment on the shelf margin. The total thickness of the formation ranges between 100 and 300 m; its age is Bartonian–early Priabonian.

The Eocene carbonate sequences were deposited on shallow-water carbonate ramps; toward the open sea intercalations of basin marls, whereas towards the hinterland intercalations of lagoonal deposits (marly limestones and marl) occur in them. The oldest member of the sequences is the littoral to neritic, and lagoonal early Lutetian–Bartonian Szőc Limestone Fm with a rich fauna of large foraminiferans. The carbonate succession formed in shallow-marine environment (ramp and slope facies) is dominantly made up of yellowish grey limestone, clayey limestone, sandy limestone and calcareous marl (Kercsmár 2015a, b, 2018, Fig. 4.12). It contains fossils indicating normal salinity: primarily large foraminiferans, locally red algae in rock-forming quantity, as well as molluscs, echinoids, decapods, corals and vertebrate remains. The formation stretches over the area of the Transdanubian Range (and the North Hungarian Range), in a form of a carbonate

Fig. 4.8 Lower Jurassic thick-bedded limestone (Pisznice Fm) overlain by Toarcian to Middle Jurassic "ammonitico rosso" marl and limestone (Kisgerecse and Tölgyhát Fm) in the quarry of Pisznice (Gerecse Mts) (*Photo* T. Budai)

succession that becomes younger from SW to NE. In the area of the Transdanubian Range the formation ("Main Nummulitic Limestone") extends up to the blind fold tectonic structure of NE—SW direction (Fodor et al. 1992; Magyari 1996) in the Buda Mts. ("Buda Line", Fig. 4.11). The biostratigraphic division of the limestone beds is confirmed by the large number of Nummulites species (Kopek et al. 1965; Kecskeméti 1980, 1993) and the zone indicator forms of the Orthophragminas (Less 1987).

The youngest member of the Eocene carbonate sequence is the Priabonian Szépvölgy Limestone Fm (Fig. 4.13), which upwards shows a continous transition from the Szőc Limestone in the Velence Hills and in the NW part of the Buda Hills. It contains thin andesite tuff intercalations (Gyalog et al. 2004) due to the volcanism in the Velence Hills (Nadap Andesite). Along the "Buda Line" it overlies the Mesozoic carbonates or the clastic succession of the Kosd Fm. The Priabonian Szépvölgy Limestone Fm is a neritic biogenic limestone with rich flora and fauna, e.g. with corallinacean algae, *Nummulites fabianii*, *Discocyclina papyracaea* (Kázmér 1982; Kecskeméti 1989). Due to rapid subsidence, this shallow-water formation is directly overlain by the pelagic, bathyal Buda Marl Fm (Fig. 4.13). South-eastward, over a short distance, the formation passes into the deep-marine basinal Buda Marl, which, on the margins, overlies it with short temporal transition. As a result of the structural development related to the Eocene basin evolution, the carbonates—formed on the margin of the ramp or accumulated on the upper part of the sedimentary slope—may have undergone plastic deformations and with gravity movements of high density, or in the forms of olistoliths the sediments may have been redeposited into heteropic Priabonian basinal facies (Buda Marl Fm, Fodor et al. 1992, 1994; Magyari 1994a, b, 1996, Kercsmár and Fodor 2005, Fig. 4.14). The Middle Eocene Szőc Limestone is 20–200 m thick, while the thickness of the Upper Eocene Szépvölgy Limestone varies between 30 and 40 m. Allodapic limestone beds and pebbly mudstone intercalations in

Fig.4.9 Karstified surface of the thick-bedded Zirc Limestone (Albian) on Eperkés Hill near Olaszfalu (*Photo* I. Selmeczi)

the Buda Marl Fm suggest an intense subsidence and steep submarine slopes. The light grey, greenish grey (in weathered parts: light brown, yellowish brown), hard, highly silty marl, marly siltstone succession is accompanied by bentonitic and tuffite intercalations of some centimetres or, rarely, a half metre thickness. It contains rich bryozoan assemblages in its lower part (the so called "Bryozoan Marl"), diverse planktonic assemblages in the upper part (P17–18 planktonic foraminiferal Zones and NP 20–21 nannoplankton Zones, Nagymarosy and Báldi-Beke 1988) and thin-shelled *Propeamussium* forms. It is a pelagic siliciclastic succession deposited in water of up to 200 m depth. It has a rapid transition from the heteropic coastal Szépvölgy Limestone, represented by a max. 20-m-thick bioclastic, calcarenitic limestone and calcareous marl succession, containing many bryozoan remains. The age is indicated also by the *Nummulites fabianii* redeposited from the shallow shelf. According to the *Globigerina tapuriensis* planktonic foraminifer, its formation in the north-eastern region extends to the Lower Oligocene. The Late Eocene deposits pass into the Early Oligocene without any break, unlike the developments north-west of the Buda Line (Báldi and Nagymarosy 1976; Fodor et al. 1992), where an early Oligocene hiatus occurs uniformly. The Eocene/Oligocene boundary is characterised here by deepening, in contrast to the development NW of the Buda Line where uplift and erosion took place. W of the Buda Line, as a result of the Oligocene uplift and the following denudation ("infra–Oligocene denudation", Telegdy Roth 1927), the bauxite was redeposited and trapped in the depressions of the karstified basement (Óbarok Bauxite, Fig. 4.15). Several differences exist between the Paleogene sequences NW and SE of the Buda Line (Fig. 4.11). The border between the two depositional types and facies areas is located within a few hundred-metre-wide, very narrow zone, without any sign of transition. This phenomenon led Báldi and Nagymarosy (1976) to the conclusion that the Buda Line may have been a major dislocation zone in post-Paleogene times.

In the area of the Buda Hills, close to the Eocene/Oligocene boundary, the formation gradually passes upward into the Tard Clay, which develops continuously

Fig. 4.10 Eocene karst bauxite (Gánt Bauxite Fm) overlain by Lutetian sediments in the Vértes Hills, Gánt (*Photo* Zs. Kercsmár)

from the Buda Marl, simultaneously with a decreasing carbonate content (Fig. 4.11). The dark grey, laminitic and organic-rich Tard Clay Fm was deposited in an euxinic, subtidal to shallow-bathyal environment, in 30–300 m-deep water E and SE of Buda Line. The light grey, bioturbated Kiscell Clay consists of fine-grained, non-laminitic clayey siltstone. It contains 15–20% carbonate and it shows a continuous transition from the Tard Clay. The formation extends beyond the Buda Line in the west, where it was deposited on the near-shore Hárshegy Sandstone. In the basin it overlies the Tard Clay, while westwards the Hárshegy Sandstone is the heteropic facies of the Kiscell Clay. The characteristically siliciclastic Hárshegy Sandstone Fm dominantly comprises fine-grained and (locally) coarse-grained sandstone, pebbly sandstone, conglomerate, (locally) kaolinic sand and fireclay, with fine sandy, clayey siltstone intercalations deposited in normal-salinity littoral–shallow sublittoral environments (Kercsmár 2015c, Fig. 4.16). The Kiscell Clay is unconformably overlain by the Törökbálint Fm (marine–brackish water and lacustrine clay, claymarl, coal, siltstone, sandstone, conglomerate) which westward interfingers with the heteropic Csatka Fm (terrestrial–fluvial–alluvial clay, sand–sandstone, gravel–conglomerate).

During the Late Oligocene (Chattian and Aquitanian) the main depocentre of the Hungarian Paleogene Basin was shifted to the area between the Buda and Darnó Lines (Fig. 4.11), where the deltaic and shallow-marine Törökbálint Fm interfingers with grey, greenish grey, fine sandy, micaceous clayey siltstone and siltstone of basin facies which consist fine sandstone intercalations (Szécsény Schlier Fm, Selmeczi and Kercsmár 2015, see in Sect. 3).

At the beginning of the Miocene, as a result of uplifting in connection with the subduction of the Magura Ocean, the dominant part of the Transdanubian Range was a dry-land. The Eggenburgian (late Aquitanian–early Burdigalian)

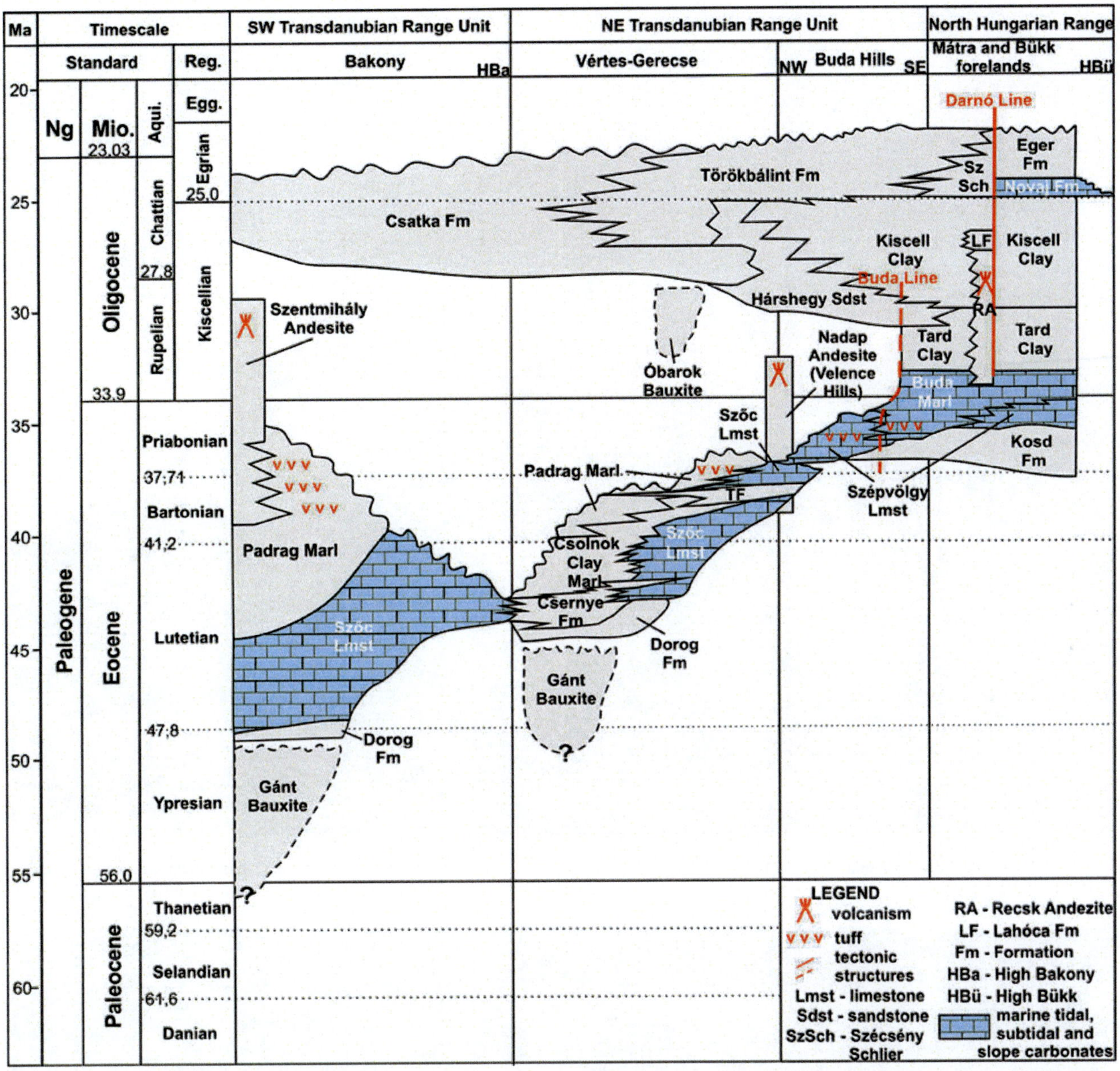

Fig. 4.11 Lithostratigraphy of Paleogene formations of the Transdanubian and North Hungarian Range Units, compiled using data from Sztanó et al. (1998), Haas et al. (1998), Báldi and Sztanó (2000a, b), Kercsmár (2015a, b, 2018, 2005b), Kercsmár et al. (2008, 2019), Less et al. (2008, 2015a, b), Less and Frijia (2016) Less et al. (2016, 2019), Parente and Less (2019), Kercsmár (2010), Kercsmár et al. (2019), unpublished lithostratigraphic charts, Hungarian Stratigraphic Committee

marine sedimentation—upon the remnants of the Hungarian Paleogene Basin—was restricted to the north-easternmost part of the Transdanubian Range, and shallow-marine, sandy–pebbly beds (Budafok Sand) were deposited. During the Early Miocene the area was dominated by terrigenous sedimentation resulting in the deposition of continental-fluvial clastics transected by wells (Somlóvásárhely Fm), and locally reaching a thickness of 400 m. Tuff intercalations observed in some borehole successions may correspond with the "lower rhyolite tuff" (Gyulakeszi Fm, Kókay 1966; Selmeczi 2014). Miocene lithostratigraphic units of the Transdanubian Range are shown by Fig. 4.17.

Karpatian (late Burdigalian) transgression affected the Várpalota Basin (E Bakony) where sediments of the off-shore Garáb Schlier and the heteropic, shoreface Fót Fm (comprising Bryozoa–Balanus bearing calcarenite and pebbly calcareous sandstone) occur (Selmeczi 2015a). A part of this was formerly considered as Ottnangian (middle

Fig. 4.12 Late Lutetian—Bartonian shallow-water carbonate sequence (Szőc Limestone) in the S Gerecse, Tatabánya (*Photo* Zs. Kercsmár)

Fig. 4.13 Priabonian shallow-water carbonate sequence (Szépvölgy Limestone) overlain by deeper-water calcareous, bryozoan marl (at top, right Buda Marl) in the NW Buda Hills, Mátyás Hill quarry (*Photo* Z. Lantos)

Fig. 4.14 Allodapic limestone layers in the Priabonian bathyal marl (Buda Marl), Pusztaszer Street, Buda Hills (*Photo* T. Budai)

Burdigalian, Kókay 1973a, b). Karpatian shoreface sediments are known in the south-eastern foreland of the Vértes (Kókay 1988) and in the Visegrád Mts. (Csillag and Müller 2005). Connections towards the open sea in the Transdanubian Range cannot be traced at the moment (Nagymarosy and Hámor 2012). Subsequently the regression of the Karpatian sea—corresponding with the sea level fall at the Burdigalian-Langhian boundary (Piller et al. 2007)—denudation took place in the area.

As a result of a global sea-level rise and transgression from the Mediterranean (Haq et al. 1988), marine sedimentation started in the early Badenian. Marine formations were deposited in sub-basins located along the rim of the Bakony or in inlets penetrating into the mountains. The sedimentary basins were formed by extensional processes (Balla and Dudko 1989; Kókay 1996). The subsidence of sub-basins along the western margin started in the early period of the Badenian (early Langhian). Transgression from the SW resulted in the formation of siliciclastics and carbonates in the Bakony Mountains. Depending on the hinterland and the coastal morphology, a great variety of sedimentary environments existed close to each other.

Along the shores (with minimum or moderated terrigenous influx) the highly variable lithofacies of the "lower Leithakalk" (Lajta Limestone Fm Öskü and Pécsszabolcs Mbs) were formed (Fig. 4.18). These carbonates of marginal reef facies represent the older, early Badenian Leithakalk in Hungary. The white, greyish-white, frequently loose, porous or hard, compact limestones are rich in fossils, especially in coralline algae, molluscs, corals, echinoderms and bryozoans indicating a normal salinity sedimentary environment in the upper neritic zone. Microfossils (foraminifers) are also frequent. The formation crops out to the surface on the margins of the Bakony, and a patch reef is known also in the Visegrád Mts. It is widespread in borehole successions, too. Over large areas it overlies the Mesozoic basement. The thickness of the lower Badenian limestone is some tens of metres with a maximum of about 100 m (Nyirád, S Bakony). Towards

Fig. 4.15 Redeposited bauxite of Oligocene age (Óbarok Bauxite Fm) in the depression of the Mesozoic karstified basement, Óbarok, S Gerecse (*Photo* Z. Lantos)

the basin centre it laterally interfingers with pelitic sediments of offshore facies representing NN5 nannoplankton Zone (Bence et al. 1999; Selmeczi et al. 2004; Selmeczi 2015b; Kókay 2013).

In case of fluvial influx from the Bakony hinterland, shoreface conglomerates, pebbly sandstones, sandstones and sand were deposited (Pusztamiske Fm). In the western areas, towards the basin centre there is a gradual transition into silty-pelitic sediments (Baden Fm Tekeres Schlier Mb). They locally reach a maximum thickness of 390 m. In an inlet in the W Bakony which was in connection with the sea towards the W in the early Badenian, there is a 160-m-thick pelitic marine succession with a particular faunal assemblage (Herend Fm) above the lower Badenian coal-bearing beds. Volcaniclastic intercalations ("middle rhyolite tuff") are widespread in the Badenian successions.

The oldest Badenian marine successions correspond to NN5 nannoplankton Zone (Selmeczi et al. 2004) which represents late Langhian-early Serravallian age (Hohenegger et al. 2014; Kováč et al. 2018). There are data only from two areas (SW Bakony, Nagygörbő: Báldiné 1983, Jámbor and Korpás 1974; NE Bakony, Várpalota: Kókay 1967, 2013) suggesting earlier Badenian sedimentation.

In the south-eastern foreland of the Vértes, as well as in the Mány and Zsámbék Basins located E of the Gerecse, terrigenous sedimentation took place until the late Badenian (Perbál Fm). The max. 140–150-m-thick succession is made up of variegated clay, clay marl, sand and pebbly sand with tuffaceous and bentonitic clay intercalations corresponding to the "middle rhyolite tuff" (Selmeczi 2008a, 2018; Kercsmár et al. 2020).

The Balaton Highland, the south-eastern foreland of the Vértes up to the Buda Hills, including the Mány and Zsámbék Basins, were invaded by the sea from the S in the late Badenian. The beds of the off-shore clay marl, marl (Baden Fm Szilágy Mb, "Turritella-Corbula Clay Marl") belonging to NN6 nannoplankton Zone (Nagymarosy and Hámor 2012), interfingers with shallow-marine carbonates towards the margins.

The upper Badenian Leithakalk ("upper Leithakalk", Lajta Fm Rákos Mb) is made up of limestone, pebbly limestone, calcareous marl and calcareous sandstone. It occurs on the surface at Budapest and in the southern foreland of the Buda Hills and was transected by wells in the Mány Basin and Zsámbék Basin, in the south-eastern foreland of the Vértes (Alcsútdoboz, Gyúró) and in the southern

Fig. 4.16 Entrance of a tectonic cave in the Oligocene near-shore, littoral–shallow sublittoral sandstone (Hárshegy Sandstone Fm) above the tectonic and karstic cave ("Papp Ferenc cave") in the Mesozoic carbonates in Pilis Mts. (*Photo* Zs. Kercsmár)

part of Balaton Highland where it can be found under the Sarmatian (Selmeczi 2008b). In the western foreland of the Transdanubian Range interbeddings of the limestone occurs in the Szilágy Clay Marl. In some areas the upper Leithakalk discordantly overlaps the basement. Its average thickness is 10–20 m. The formation is rich in coralline algae, molluscs (e.g. Pectinids), echinoderms, Ditrupa worm tubes, as well as foraminifers (e.g. Borelis). In contrast to the early Badenian "lower Leithakalk", corals are almost absent in the late Badenian limestones due to climatic deterioration. The presence of some taxa in the latest Badenian mollusc assemblages indicate seaway connections towards the Eastern Paratethys (Kókay 1985).

Coal formation (Hidas Fm) took place in brackish-water or freshwater swamps in different periods of the Badenian. In the Keszthely Hills (SW Transdanubian Range) and its surroundings kaolin (Cserszegtomaj Kaolin Fm) is found predominantly in karstic dolines located along tectonic lines and at their intersections. Karstic dolines—locally exceeding a depth of 100 m—developed in the Triassic Main Dolomite. The most favourable climatic conditions for kaolin formation existed during the late Karpatian and Badenian (late Burdigalian–Langhian, Bence et al. 1999). "Miocene bauxites" (Vöröstó Fm) were formed as a result of re-sedimentation of older, uncovered bauxites in the Bakony and the Balaton Highland. The red clay with bauxite pebbles and carbonate clasts, bauxitic clay, clayey bauxite deposits usually overlie the karstified terrain of the Mesozoic basement. The max. thickness of the formation exceeds 70 m and is covered by Miocene rocks (Tóth and Varga 2014; Selmeczi 2015b). In the Badenian andesitic volcanic activity took place in the north-easternmost part of the Transdanubian Range producing the mountain ranges at the Danube bend (Selmeczi 2015b).

At the onset of the Sarmatian (late Serravallian)—probably by the uplift of the Dinarids—isolation of the Central Paratethys from the world ocean system began. The highly diverse Badenian biota (e.g. corals, echinoids etc.) vanished;

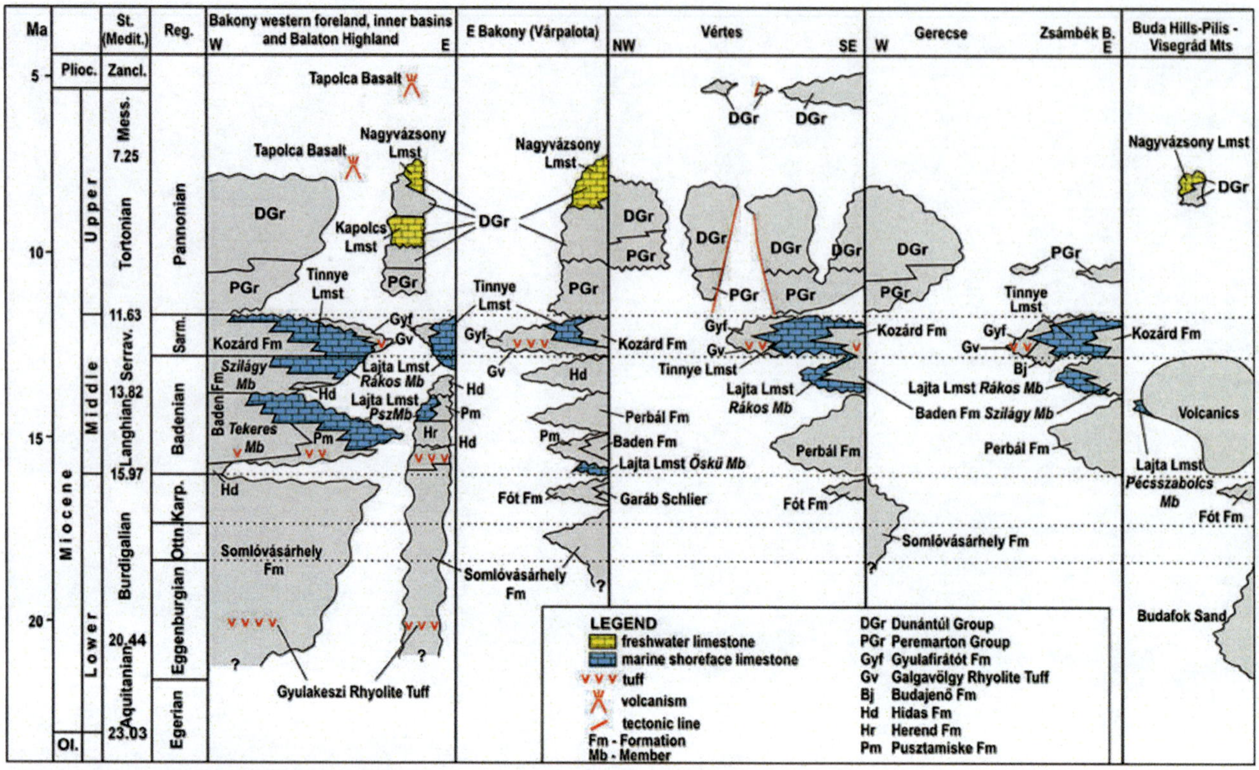

Fig. 4.17 Miocene lithostratigraphic units of the Transdanubian Range

Fig. 4.18 The "lower Leithakalk" (Lajta Limestone Fm Pécsszabolcs Mb) exposed in the railroad-cut between Devecser and Kolontár, at the Sacred Spring (*Photo* I. Selmeczi)

Sarmatian marine successions are characterised by an endemic fauna (Piller et al. 2007; Nagymarosy and Hámor 2012).

In shoreface environments biogenic calcarenite, coarse, porous limestone, oolithic-bioclastic limestone, pebbly calcareous sandstone and locally pea gravel belonging to the Tinnye Limestone Fm were deposited. The limestone is rich in fossil molluscs (e.g. Cardiid bivalves, Ervilia, Sarmatimactra, Venerupis, Cerithium, Calliostoma genera) and in microfossils, especially in foraminifers (in several beds Miliolids are extremely frequent). The limestone has a large areal extent in the Transdanubian Range. The largest, uninterrupted Sarmatian limestone area of Hungary with significant karst formation is located in the Tapolca Basin (Balaton Highland, Fig. 4.19). On the surface the formation occurs in the W Bakony and the Balaton Highland, in the south-eastern foreland of the Vértes, in the Mány and Zsámbék Basins and in the south-western foreland of the Buda Hills, where it has been quarried since the Roman times. Its thickness is of some tens of metres, in the Tapolca Basin it may reach 80–100 m (Bence et al. 1999; Tóth et al. 2010; Selmeczi 2015b; Kercsmár et al. 2020). Towards the basin there is a transition from the Tinnye Limestone into the offshore clay marl, marl and siltstone (Kozárd Fm). Sarmatian formations overlap the Badenian ones in large areas. Sarmatian marine sedimentation took place in brackish-water environments; however, locally in lagoons (e.g. Mány–Zsámbék area and Zánka in the Balaton Highland) hypersaline conditions existed (Jámbor 1976; Görög 1992; Cornée et al. 2009; Selmeczi et al. 2017). Drillings in the Zsámbék Basin penetrated a succession comprising evaporites (Budajenő Fm, Jámbor 1976; Ravasz 1978).

The clay, bentonitic clay, sandstone, gravel, conglomerate beds of the Sarmatian continental succession (Gyulafirátót Fm) are derived from the denudation of uplifted areas in the Transdanubian Range. Sediments can be traced from the W Bakony to the south-eastern foreland of the Vértes, exceeding a thickness of 100 m (Selmeczi 2008b).

Fig.4.19 The the most well-known part of the Tapolca Cave System, i.e. the Lake Cave of Tapolca, formed in the middle Miocene Sarmatian limestone (Tinnye Fm) by the corrosion effect of warm and cold waters (*Photo* S. Horváth)

Rhyolite and dacite tuff and bentonitic tuffite intercalations corresponding to the "upper rhyolite tuff" (Galgavölgy Fm) with a maximum thickness of some metres are characteristic in the lower third of the Sarmatian succession.

In the Late Miocene (Pannonian) seaways were closed and the isolated basin of the Lake Pannon was formed in the area of the relic Central Paratethys sea. It reached its largest areal extent 10–9 million years ago. Siliciclastic-dominated, brackish to freshwater sediments were deposited. The lake became filled up during the Pliocene (Magyar 2010). In the Transdanubian Range the main bulk of the Pannonian (Tortonian-Messinian) succession is represented by siliciclastic formations of the Peremarton and Dunántúl Groups; however, locally, freshwater limestones were also formed (Kapolcs and Nagyvázsony Fm). The Nagyvázsony Limestone precipitated from karst water (Csillag and Sztanó 2015). Basaltic volcanism (Tapolca Basalt Fm) took place in the Balaton Highland and in the S Bakony. Volcanic activity started 8 million years ago in Tihany (Csillag 2015).

During the Pliocene and the Pleistocene this area intensively uplifted and eroded. Quaternary was characterized by terrestrial sedimentation. Thick loess and alluvial successions were added to the sequences in this period. Pleistocene loess is widespread in the forelands of the Transdanubian Range. Locally it may reach a thickness of some tens of metres. Deflation played an important role in relief development during the Quaternary. Loose sediments were affected by wind erosion. In some places drift sand accumulated. The development of terraces along the Danube (which started in the late Pliocene) continued. Travertine formation occurred in connection with spring activity. Incision of the present valley network has started already in the late Pleistocene resulting in the accumulation of thick fluvial successions. In areas built up of carbonate rocks and exposed to meteoric water karstification took place.

3 Bükk Mountains

This review is a brief summary of the geological mapping of the Bükk Mts. resulted in the geological map (Less et al. 2002) and its explanatory booklet (Pelikán et al. 2005) of the territory, both published by the Geological Institute of Hungary. Beside it, the works of several generations (Balogh 1964; Balla 1983; Báldi 1986; Csontos 1999, 2000; Velledits 2000, 2004; Filipovic et al. 2003; Haas et al. 2013; Kövér et al. 2018) have also been used.

The Bükk Mts. lying in NE Hungary is surrounded from SW, SE and NE by young basins below which Bükkian rocks and their structures can be followed. From NW, however, it is bordered by the Darnó zone of sinistral strike-slip character, active till the middle of the Miocene. Paleo-Mesozoic rocks of the Bükk Mts. are of Dinaric affinity (Filipovic et al. 2003), whereas the areas that are located to the NW of the Darnó zone show south-Alpine and Carpathian relations. According to the character of Tertiary rocks and their structures, the Bükk occupied its recent position only in the early Miocene.

3.1 Tectonics and Metamorphism

Paleo-Mesozoic rocks of the Bükk Mts. can be classified into two series (Fig. 4.20). Middle Carboniferous to Jurassic rocks of the Bükk Para-autochton were deposited on continental crust, whereas Triassic-Jurassic rocks of the overthrust Mónosbél-Szarvaskő nappe system indicate deposition either in accretionary prism conditions (Mónosbél nappe) or on oceanic crust in the case of the Szarvaskő nappe (Balla 1983; Csontos 1999, 2000). The age of the earliest overthrust is most probably Late Jurassic. The above two series were buried, metamorphosed and folded most probably in the Cretaceous; this is likely because of the overthrust of higher—maybe Aggtelek-Bódva-type—nappes (see at the Chapter 4.4) which became eroded later almost entirely and recently are represented only in the form of occasional olistoliths. All the known part of the Bükkian Paleo-Mesozoic may have been buried during the whole Cretaceous, Paleocene and most part of the Eocene that caused very low-grade to low-grade metamorphism (1.5–3 but locally up to 5 kbar and 200–350 °C according to Árkai 1983) and also its ductile deformations in at least two phases (one of N-S and the other of E-W compression in recent co-ordinates). These latter finished about 80 million years ago (Dunkl et al. 1994; Árkai et al. 1995) and resulted in the very intensive post-primary nappe folding (North Bükk and Bükkfennsík anticlines, Vesszős syncline, etc.) of southern and south-eastern vergence according to the recent co-ordinates.

3.2 Geological Build-Up

The stratigraphic sequence of the Bükk Para-autochton (Fig. 4.21) starts with a thick (>1000 m), Bashkirian to Lower Moscovian (middle Carboniferous) siliciclastic sequence (Szilvásvárad Fm), interpreted as deep-marine, flysch-like sediments, which are very similar to the youngest Paleozoic rocks of the Szendrő Mountains (Szendrő Phyllite). It can be reasonably supposed that older Paleozoic rocks of the Bükk Para-autochton have also Szendrő-Uppony affinity. It is followed by the Upper Moscovian to Gzhelian (upper Carboniferous) shallow-water carbonate-clastic sequence (Mályinka Fm) in a thickness of 400–500 m. It is very rich in fossils (fusulinids, calcareous algae, corals, brachiopods, etc.). These two formations

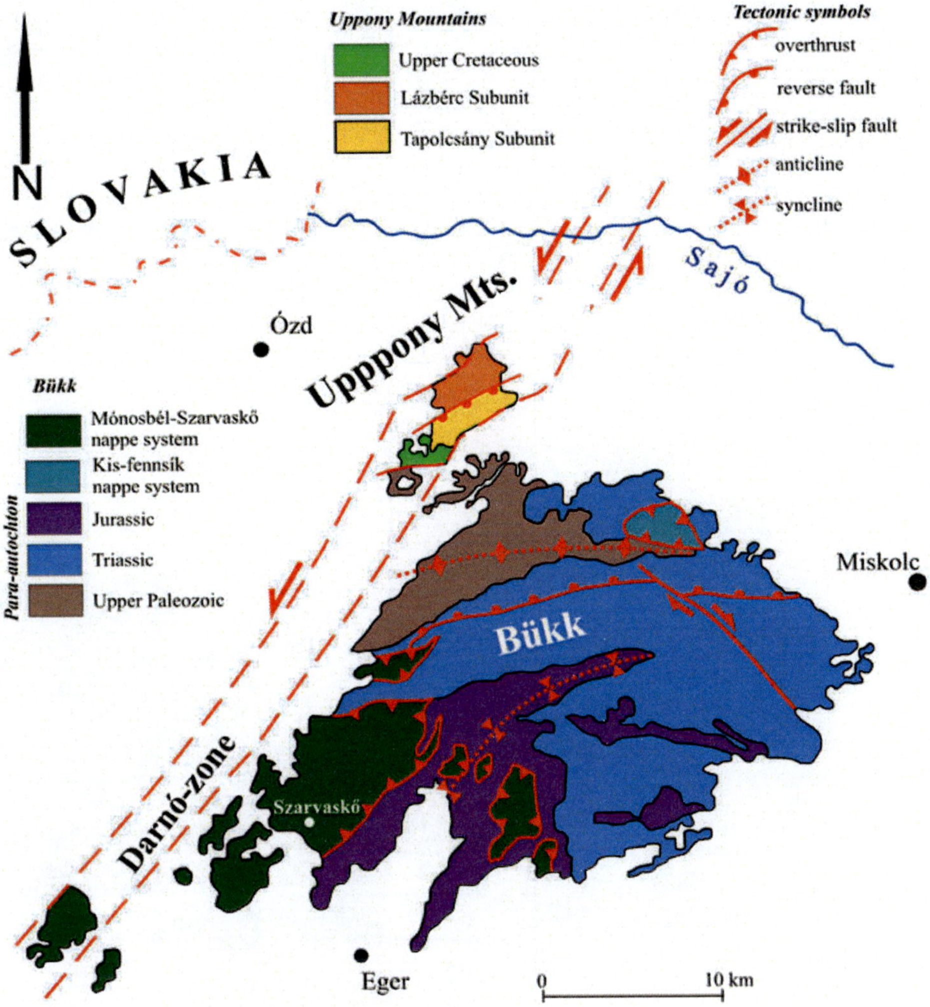

Fig. 4.20 Simplified geological map of the Bükk Mts. and Uppony Mts. (after Haas and Budai eds. 2014, modified)

correspond to the Hochwipfel Flysch and Auernig Group of the Carnic Alps (Ebner et al. 1998). They terminate the Variscan (Hercynian) sedimentary-tectonic cycle. The area was uplifted at the end of the Carboniferous and became dry-land.

After this regression the Alpine cycle starts with continental (variegated sandstone) then hypersaline lagoonal beds (evaporites, dolomites and chloritic clays) of the Middle Permian Szentlélek Fm (corresponding to the south-Alpine Val Gardena/Gröden Fm), about 250 m in thickness. Beside the hiatus, locally, a very small angular unconformity can also be observed on its base. The Paleozoic is terminated by the 250–300-m-thick Upper Permian black algal (*Mizzia* sp. and others) shallow-marine limestones (Nagyvisnyó Limestone corresponding to "Bellerophon Kalk" of the Southern Alps).

One of the peculiarities of the Bükk Mts. is the continuous transition between Permian and Triassic in shallow marine carbonate conditions (Fig. 4.22). The lowermost Triassic is represented by 150–200-m-thick, light-coloured, pinkish-brownish, oolithic carbonates (Gerennavár Limestone), in which karstic phenomena can be found in some places. The rest of the Lower Triassic is made up of the alternation of shallow-marine siliciclastic and carbonate rocks (Ablakoskővölgy Fm, about 300 m in thickness). After the termination of the siliciclastic input the Anisian starts with the approx. 400-m-thick, dark grey Hámor Dolomite representing partly restricted inner platform. Local paleosoils (Sebesvíz Conglomerate) in its top mark a short-term emersion (Velledits 2004) at about the Anisian-Ladinian boundary. They are followed by andesitic and dacitic lavas and tuffs of the max. 350-m-thick Szentistvánhegy Metaandesite deposited both in marine and dry-land conditions. During the Ladinian and Late Triassic sedimentation took place both in carbonate platforms (massive, light grey Fehérkő, Bükkfennsík, Berva and Kisfennsík Limestones, up to 1000 m in thickness, all with common karstic phenomena, Fig. 4.23) and intraplatfom

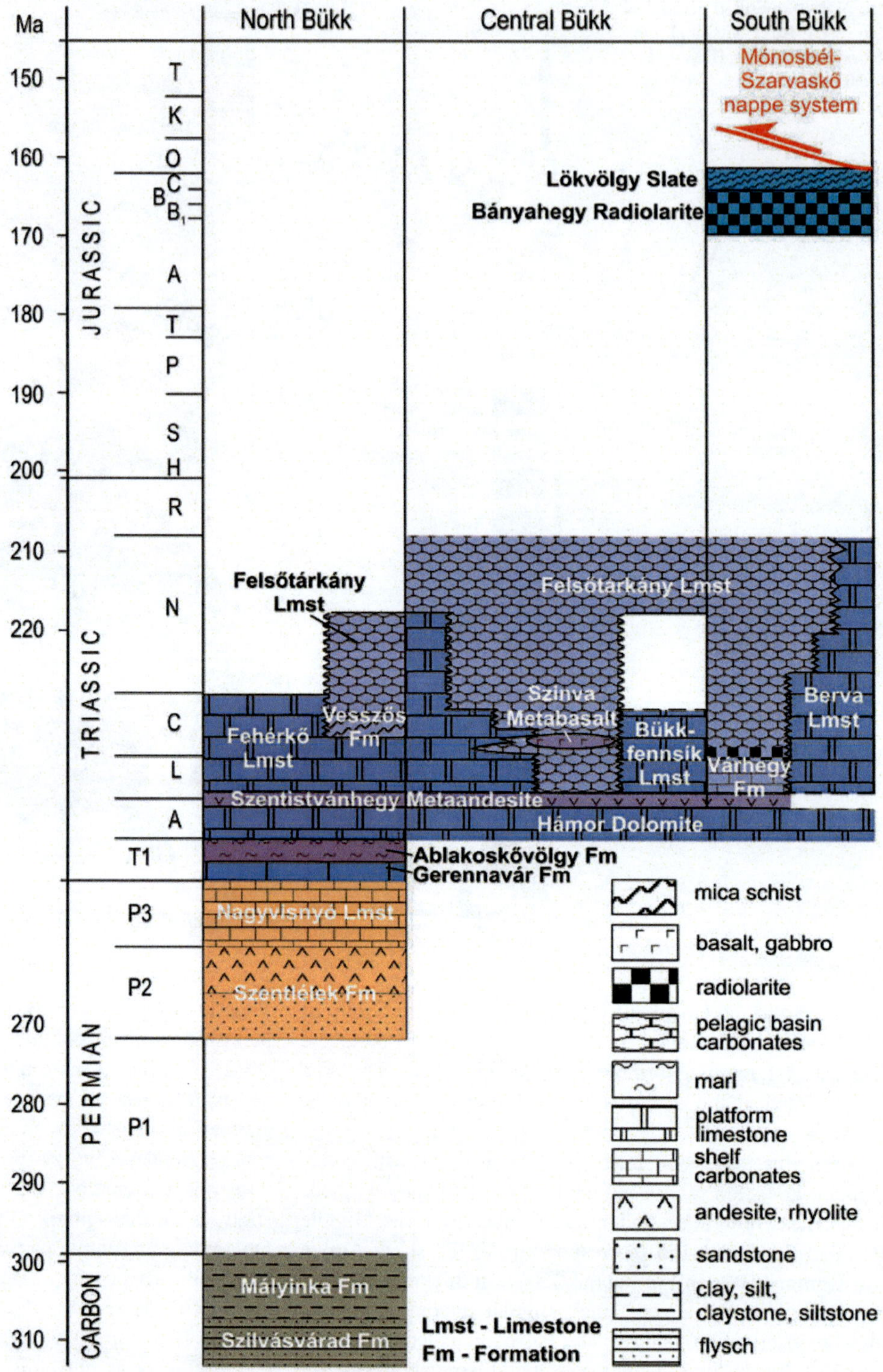

Fig. 4.21 Stratigraphic column of Paleozoic–Mesozoic formations of the Bükk Para-autochton (after Haas and Budai eds. 2014, slightly modified and completed)

basins. The latter are indicated by the 300–500-m-thick grey, cherty Felsőtárkány Limestone and by the radiolarite-bearing siliciclastites of the Várhegy Fm, 20–40 m in thickness (Velledits 2000). Upwards the basins occupied increasingly larger space simultaneously with the platforms having got drowned. The 50–100-m-thick, pinkish and drab, sometimes crinoidal Répáshuta Limestone marks their transition. The carbonate sedimentation was interrupted in the Middle Carnian by a siliciclastic event in the N Bükk (the 150–200-m-thick Vesszős Shale) and quasi-simultaneously by

Fig. 4.22 Marine transition of Permian (P_3 —Nagyvisnyó Limestone) and Triassic (T_1 —Gerennavár Limestone) with "boundary clay" (BCl) between them, N slope of Mt. Bálvány (*Photo* Cs. Péró)

basaltic volcanism in the South and Central Bükk (Szinva and Létrás Metabasalt), 10–50 m in thickness.

All carbonate platforms were drowned at the beginning of the Jurassic, deep-marine conditions developed in the entire area. The carbonate production was, however, so low and potential source areas were so far that practically no sedimentation took place almost during 25 million years. Therefore, very little is known about the Liassic and early Doggerian sedimentation in the Bükk; maybe some red crinoidal limestones represent this time-span. The first, biostratigraphically proven Jurassic sediment is the thin (<30 m) red and green, Bajocian-Bathonian Bányahegy Radiolarite representing under-CCD conditions for this period. The radiolarite locally contains huge olistoliths of the Bükkfennsík and Répáshuta Limestones (Fig. 4.24). Following this, the initial closure of the basin is marked by thick (300–500 m) distal flyschoid carbonate-free dark shales with some turbiditic sandstones (Lökvölgy Slate).

According to Balla (1983) and Csontos (1999, 2000), the para-autochtonous sequence of the Bükk terminates with this shale and all the Mesozoic rocks (in geometrically higher position) belong already to the Mónosbél-Szarvaskő nappe system. However, since biostratigraphic data are very erratic, some authors (Pelikán et al. 2005; Haas et al. 2013) consider the Mónosbél Formation Group is part of the Bükk Para-autochton. Based on their mapping experiences there is a quasi-continuous transition between the Mónosbél Formation Group and the underlying Lökvölgy Slate.

The Mónosbél nappe is built up of the thick (>1000 m) Middle Jurassic Mónosbél Formation Group represented first

Fig. 4.23 Fehérkő Limestone at Miskolc, Felső-Hámor (*Photo* Gy. Less)

by black shales with calcareous intercalations and higher up with olistoliths (Csontos 2000). As a whole, these rocks can be considered as an olistostrome deposited in accretionary prism conditions. The origin of olistoliths is of multiple sources (Haas et al. 2013), among them resedimented oolithic limestones could be originated from the Outer Dinarides, the only carbonate platform of that time. A peculiar single olistolith of about only 10 m in diameter—containing Carnian basalts and Hallstatt Limestone (Fig. 4.25)—may have come from the prolongation of the Bódva depositional area (recently present in the Rudabánya Hills). It might indicate the beginning of the overthrust of the Szarvaskő nappe onto both the Paleo-Mesozoic sequence of the Bükk Para-autochton and the Mónosbél nappe in the late Jurassic.

The Szarvaskő nappe representing the upper part of the Mónosbél-Szarvaskő nappe system is composed predominantly of Middle Jurassic siliciclasts (Vaskapu Fm and others exceeding 300 m in thickness) and of thick (>500 m), slightly atypical Middle-Upper Jurassic ophiolites (Tardos Gabbro and Szarvaskő Basalt, Fig. 4.26) penetrating them with thermal contact. The overthrust of the Mónosbél-Szarvaskő nappe system onto the Bükk para-autochton might have happened still before the end of the Jurassic. In this case these nappes can be considered as huge "megaolistoliths" in the Jurassic sequence of the Bükk Para-autochton.

Cretaceous to Middle Eocene deposits are unknown from the Bükk. The Paleo-Mesozoic sequence of the Bükk Mts. with its structures became exhumed at least until the Late Eocene. Gravitational intra-Bükkian nappes of the Kis-fennsík nappe system might have formed during the process of exhumation. This nappe system is built up of three partial nappes in superposition (Fig. 4.27), i.e. the Harica Valley, Magos-kő and Örvény-kő nappes (their description—though using partly different conception and names—can be found in Forián Szabó and Csontos 2001, and in Velledits et al. 2003).

During the late Eocene a new transgression reached the area, which already became part of the Hungarian Paleogene

Fig. 4.24 Huge olistoliths of uppermost Triassic Bükkfennsík Limestone (above) and Répáshuta Limestone (below) in the Middle Jurassic Bányahegy Radiolarite, Répáshuta, Bánya-hegy (*Photo* Cs. Péró)

Fig. 4.25 A complex olistolith consisting of Upper Triassic Hallstatt Limestone, radiolarite and basalt in the Mónosbél Formation Group (Jurassic), Csipkés-tető, Kavicsos-kilátó, N of Felsőtárkány (*Photo* Cs. Péró)

Basin. This sequence, lying with angular unconformity on the Paleo-Mesozoic sequences, is built up of non-metamorphosed Buda-type Paleogene rocks (Fig. 4.11). After a short period in dry-land conditions, represented by red clays deposited in local depressions (Kosd Fm, up to 100 m in thickness), the first marine transgression reached the southern and north-eastern peripheral parts of the Bükk still in the late Eocene. The 50–100-m-thick Priabonian shallow marine nummulitic (locally karstified) limestone (Szépvölgy Limestone) and Eocene–Oligocene marl (Buda Marl) is overlain by typical, predominantly siliciclastic Oligocene sediments (Tard Clay and Kiscell Clay) of the newly formed, coolish Central Paratethys. They are followed by the calcareous Novaj Fm and finally the regressive Eger Fm, altogether 250–750 m in thickness. However, the central part of the Bükk remained probably uncovered during the whole Paleogene because abrasional lower Egerian (=upper Chattian) sediments of the approx. 25-m-thick Csókás Member of the Novaj Fm—lying directly on the Mesozoic rocks—have been found in the N Bükk (Less 1991; Báldi et al. 1999). This contradicts the data of Dunkl et al. (1994) assuming a 600–1200-m-thick Paleogene cover over the whole Bükk.

The Bükk once again emerged from the very end of the Oligocene and remained dry-land during the Early Miocene. One part of the mountains was intensively karstified whereas other parts served as source areas for the neighbouring basins N of the Bükk. This is also the period when the Bükk moved to its current position. An important volcanic activity of three phases started in the Ottnangian (=Late Burdigalian) with the deposition of the lower, 150–400-m-thick Gyulakeszi Rhyolite Tuff. It was followed by the „middle tuff", i.e. the Tar Dacite Tuff at the Karpatian/Badenian (=Burdigalian/Langhian) boundary (reaching a thickness of 30–50 m), and terminated with the upper, ca. 100-m-thick Galgavölgy Rhyolite Tuff of Sarmatian (=Upper Serravallian) age. These three horizons of tuffs are well exposed in the S Bükk and have been deposited mostly in dry-land conditions. The centre of the volcanic activity was located S of the Bükk. Based on paleomagnetic and stress-field data derived from the above three different tuff horizons Márton and Fodor (1995) demonstrated the history of rotation during the Miocene. The lower tuff shows still a 70–80° counter-clockwise rotation while it is only 30–40° in the middle horizon and finally no rotation could be detected from the upper horizon. Meanwhile some small remnants of Middle Miocene sands in the Central

Fig. 4.26 Pillow lava of the Szarvaskő Basalt, N of Szarvaskő (*Photo* Cs. Péró)

Bükk and also shark teeth from this time-span found in one cave of the karstic plateau of the High Bükk indicate that the marine transgression reached and covered the whole mountain in this time. This is in accord with the fission track data of Dunkl et al. (1994), too. They indicate that the Miocene burial reached or exceeded 1 km. The final, very rapid emersion started 2–3 million years ago and caused the erosion of most of the young sediments.

Fig. 4.27 Contact of two partial nappes of the Kis-fennsík nappe system, Miskolc, Három-kút cave (*Photo* Cs. Péró)

4 Aggtelek–Rudabánya Hills (Aggtelek–Rudabánya Mountains)

Despite of their small dimension, the Aggtelek–Rudabánya Hills, the southernmost element of the Inner Western Carpathians, are geologically among the most complex regions in Hungary. Therefore, there are contradictory views in several aspects that are impossible to discuss here because of the limited extent available. This review is based on the latest geological mapping of the Aggtelek-Rudabánya Hills resulted in the geological map (Less et al. 1988) and its explanatory booklet (Szentpétery et al. 2006). The latter includes results published in several papers (Árkai and Kovács 1986; Árkai et al. 2003; Grill 1988; Hips 1996; Kovács 1986; Kovács et al. 1989; Less 2000; Márton et al. 1989; Réti 1985; Szentpétery 1997). New results including alternative structural model of the area are presented by Kövér et al. (2009a, b, 2018), Velledits et al. (2011) and Fodor (in Haas and Budai eds. 2014).

4.1 Tectonics

The region can be subdivided into two parts, to the Aggtelek Karst and to the Rudabánya Hills (Fig. 4.28). The Aggtelek Karst as part of the Gömör–Torna Karst is clearly the continuation of the Slovak Karst, whereas the Rudabánya Hills are incorporated into the Darnó Zone. They arrived in their recent place from SW at about the Oligocene/Miocene boundary along a sinistral strike-slip (Szentpétery 1997; Less 2000). This means that until the Late Oligocene the Rudabánya Hills were located some tenth of km-s to the S of

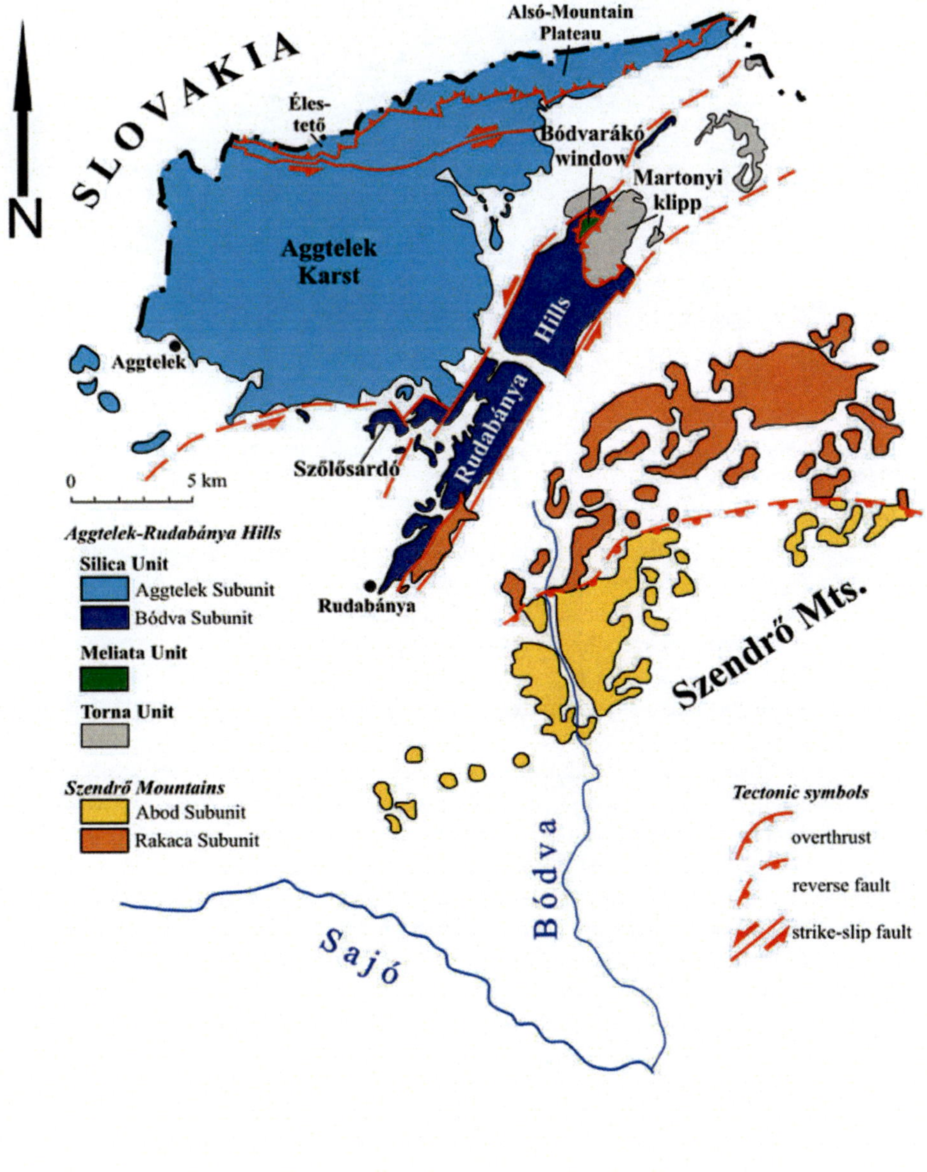

Fig. 4.28 Structural map of the Aggtelek–Rudabánya and Szendrő Mountains (after Haas and Budai eds.. 2014 modified, based data by Kovács et al. 1993; Less et al. 2004)

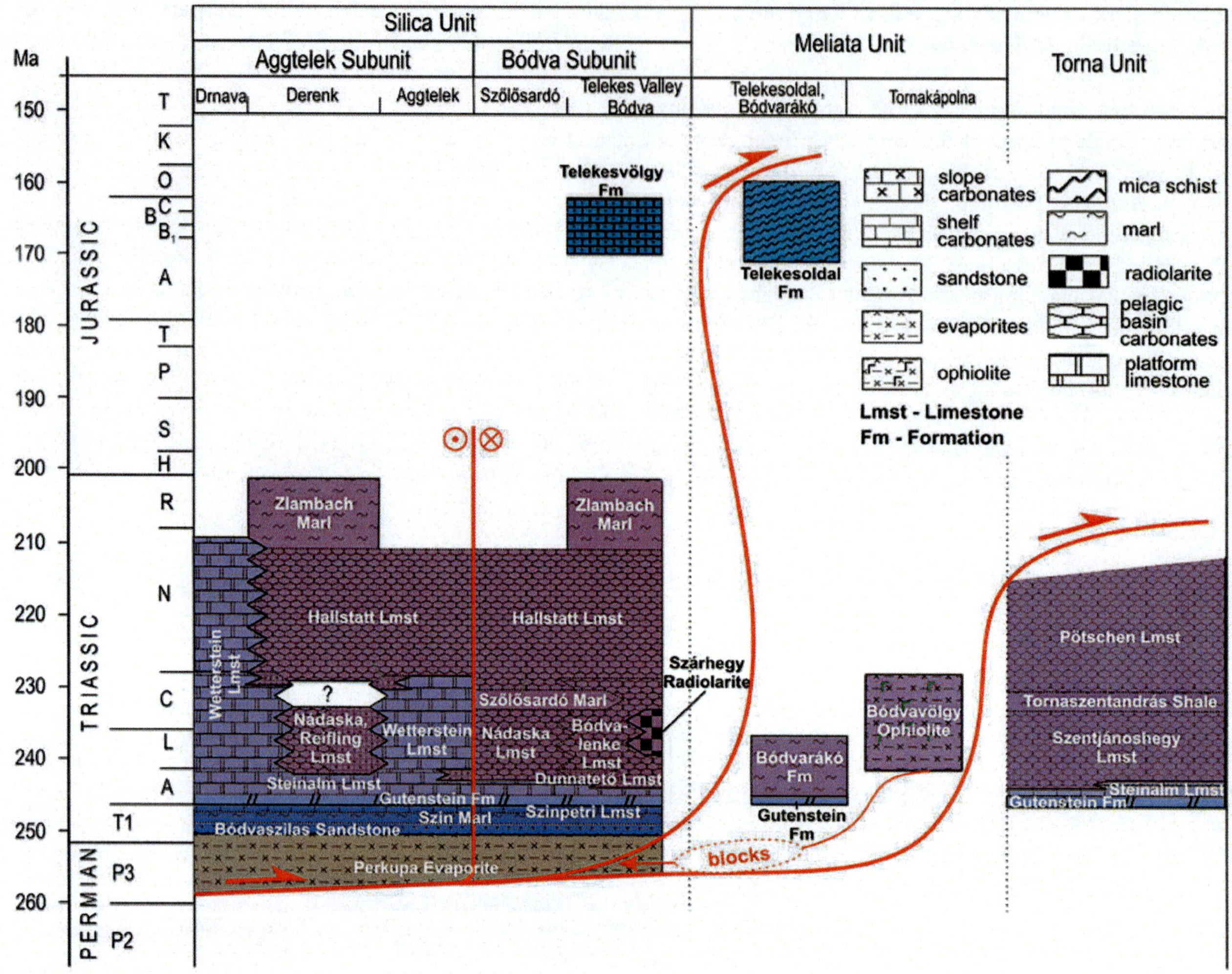

Fig. 4.29 Lithostratigraphic column of Paleozoic–Mesozoic formations of the Aggtelek-Rudabánya Hills (after Haas and Budai eds. 2014, modified and completed)

the Aggtelek Karst. To the SE of the Rudabánya Hills already the Uppony and Szendrő Mountains can be found; their Paleozoic rocks show already a south-Alpine and Dinaric affinity, together with the Paleo-Mesozoic of the Bükk Mts. Thus, the SE margin of the Rudabánya Mts. is once again a sinistral strike-slip delimiting the entire Inner West Carpathians towards SE.

In re-establishing the pre-Miocene structures, the obtained structure is still very complicated. It consists of (in order of superposition from upwards) the neo-allochtonous klippes of Alsó-hegy (Dolný Vrch), Éles-tető (Ostrý Vrch) and Derenk, covering the folded and imbricated structures (of southern vergence in Hungary) of the territory that are superpositioning the primary nappe structure.

This reconstructed primary nappe structure is composed (in order of superposition from upwards) of the Silica, Meliata and Torna (Turňa) Units which are characterized by three different groups of rocks (Fig. 4.29); their metamorphic degrees are also significantly different (Árkai and Kovács 1986; Árkai et al. 2003).

4.2 Geological Build-Up

Non-metamorphosed Uppermost Permian to Jurassic rocks, deposited on continental crust, belong to the Silica series group. They form the Silica nappe system (originally defined by Kozur and Mock 1973), the upper unit of the primary nappe structure of the Aggtelek-Rudabánya Mts. This nappe system is mostly detached from its Paleozoic basement along the plastic Upper Permian/Lowermost Triassic, 250–300-m-thick Perkupa Evaporites (Fig. 4.30) deposited in hypersaline lagoons. The Paleozoic basement of the Silica series group could be in the southern part of the Gemer-Spiš Metalliferous Mts.

Fig. 4.30 Diapyric folds in the Perkupa Evaporite, Alsótelekes (*Photo* Cs. Péró)

The sequences of the Silica series group are partly different in the Aggtelek Karst and Rudabánya Hills. In the first it is called Aggtelek series (with the Aggtelek and Derenk development area) whereas in the Rudabánya Hills its name is Bódva series with the Bódva and Szőlősardó development area. They both start with the Perkupa Evaporite followed by Lower Triassic marine deposits (Hips 1996). This sequence (strongly resembling to the Haselgebirge and Werfen Beds in the N Calcareous Alps) starts with the shallow-marine, siliciclastic Bódvaszilas Sandstone with a thickness of 200–300 m, covered by shallow-marine, first terrigenous but upwards increasingly more calcareous deposits, by the 300–500 m-thick Szin Marl and by the ca. 200-m-thick Szinpetri Limestone. The lower part of the Middle Triassic is built up of shallow-marine carbonates, which were formed first in disoxyc conditions (the mostly dolomitic Gutenstein Fm, 200–300 m in thickness), and later in well-oxygenated platform carbonate conditions (the 250–400-m-thick, intensively karstified Steinalm Limestone).

After the middle Anisian the evolution of the Aggtelek and Bódva series was different. Taking into account that the Rudabánya Hills together with the Bódva series on its top must be pulled back far to the S before the Miocene, the facial distribution of the Upper Anisian-Middle Carnian within the Silica series group suggests a general southward deepening in recent co-ordinates. Therefore, they indicate the position of the Silica depositional area on the northern margin of the Neotethys Ocean (see below).

After or without an intraplatform basinal event (Reifling and/or Schreyeralm Limestones, both 10–50 m in thickness, Velledits et al. 2011) the carbonate platform survived in most part of the Aggtelek series up to the late Carnian. This is the more than 1000-m-thick Wetterstein Limestone in the Aggtelek development area (Fig. 4.31), representing the main body for karstification. However, some intraplatform basins could also exist in the late Ladinian to Carnian interval within the Wetterstein platform (Nádaska Limestone in the Derenk development area, 50–80 m in thickness).

Meanwhile, in the Bódva series no platform carbonates can be found from the middle Anisian onwards. The Szőlősardó development area is characterized by the slope deposits of the Nádaska Limestone (50–120 m in thickness) and by the relatively thick (up to 100 m), terrigenous Szőlősardó Marl marking the middle Carnian Raibl event. The upper Anisian to Carnian succession of the Bódva development area is characterized first by the 10–50-m-thick transitional Dunnatető Limestone then by the deep pelagic Bódvalenke Limestone (10–60 m in thickness, Fig. 4.32) interfingering with the deep marine 25–50-m-thick Szárhegy Radiolarite.

Fig. 4.31 The Wetterstein Limestone at the entrance of the Baradla Cave, Aggtelek (*Photo* Gy. Less)

Fig. 4.32 The type-locality of the Bódvalenke Limestone with red cherts, Bódvalenke (*Photo* Cs. Péró)

After the very diverse Upper Anisian to Middle Carnian, the Upper Carnian and Norian of the Aggtelek and Bódva series became similar: this interval is represented in both series by the same pelagic, reddish Hallstatt and/or grey Pötschen Limestones (both 50–150 m in thickness) covered by the 25–30-m-thick Upper Norian to Rhaetian Zlambach Marl.

The Jurassic in Hungary is known only in the Rudabánya Hills. It has two developments (Grill 1988), both about 500 m in thickness. The Telekesoldal Complex is built up of monotonous black shales then by rhyolitic olistostrome, whereas the Telekesvölgy Complex consists of a lower, variegated marly part and an upper part, which is composed of crinoidal marls and manganese shales. According to Less (2000) the first complex is lying directly upon the Triassic of the Bódva development area, whereas the Triassic basement of the second one is rather uncertain. Meanwhile Kövér et al. (2009b) ranked the Telekesoldal Complex to the Meliata series group and argued for the Bódva type Triassic basement of the Telekesvölgy Complex.

Very weakly (in average) metamorphosed (250–300 °C and 2–3kbar) Triassic and Jurassic rocks—deposited on the oceanic or thinned continental crust of the Neotethys Ocean —are grouped into the Meliata series group. The Tornakápolna series is built up of the Middle Triassic Bódva Valley Ophiolites consisting of serpentinites, gabbros, basalts and deep-marine radiolarites in various thicknesses (up to 250 m, Réti 1985). Jurassic deposits (represented by olistostromes) can also be found, mostly in Slovakia (Mock et al. 1998). Another sequence, the Bódvarákó series was classified into the Meliata series group as well by Less (2000), (according to Kövér et al. 2009a it belongs to the Torna series, see below). This series crops out to the surface in the core of an antiform in the northern part of the Rudabánya Hills and located clearly below the Bódva series of the Silica nappe system. Based on its limited exposures and on borehole data it is built up of the lower Anisian Gutenstein Fm directly (lacking the Steinalm Limestone) covered by the 40–50 m thick middle Anisian to Ladinian deep-marine black, cherty limestone of the Bódvarákó Fm, which is capped by a mysterious olistostrome (with no dating yet) similar to that of the Telekesoldal Complex and from the Jurassic of the Meliata series.

Originally the Meliata series group was located in the Meliata nappe, interpreted as the middle unit in the nappe pile of the Aggtelek-Rudabánya Hills. This original position can only be observed in some small relicts (e.g. the Bódvarákó series), since recently the major part of the series is dismembered and incorporated into the evaporitic basement of the Silica nappe system (Fig. 4.33) due to its later overthusting.

Very weakly and weakly metamorphosed (300–500 °C and 3–5kbar) and intensively deformed Triassic rocks, deposited on continental crust, are grouped into the Torna (Turňa) series forming the lowest known tectonic unit in the primary nappe pile. Primarily it can be found always under the Silica series group in the core of huge antiforms; tectonically dissected blocks in the basal evaporitic layer of the Silica nappe system are not yet known. The Torna series can be secondarily overthrust onto less metamorphosed units, too, as in the case of the Martonyi klipp (Less 2000; Fodor and Koroknai 2000).

The series itself contains only Triassic rocks, its Jurassic part is eroded. The Lower Triassic is unknown from Hungary, these beds in Slovakia (Mello et al. 1997) can be correlated with those of the Silica series group. The Middle–Upper Triassic is most similar to the Szőlősardó development area. In Hungary it starts with the lower Anisian Gutenstein Fm, followed by the max. 400-m-thick middle Anisian, intensively karstified (especially in the Esztramos Hill, near Bódvarákó), shallow-water Steinalm Limestone, and then by the upper Anisian to lower Carnian pelagic Szentjánoshegy Limestone, 25–150 m in thickness. The 30–100-m-thick middle Carnian Tornaszentandrás Shale marks the Raibl event, whereas the Upper Carnian to Norian pelagic Pötschen Limestone (Fig. 4.34), about 100 m in thickness, represents the uppermost member of the Torna series.

The structural evolution of the Aggtelek-Rudabánya Hills is described in detail by Less (2000). It started in the Middle Anisian with rifting and then opening of the Neotethys ocean between the Silica and Torna depositional areas. The oceanic crust subducted northward starting in the Middle Jurassic and simultaneously obducted southward onto the Tornaic continental crust and formed the Meliata nappe. After the collision (at about the Jurassic–Cretaceous boundary) the Silica nappe system was formed in the Early Cretaceous by gravitational gliding to the S, having detached from its Paleozoic basement along thick plastic Upper Permian evaporites. This resulted in the overthrusting of the Silica nappe onto the Torna–Meliata nappe pile and in the reworking of the ophiolites of the Meliata series into the evaporitic basement of the Silica nappe system.

In a later phase (Cretaceous and Paleocene-Eocene rocks are completely missing in the area) folding and imbrication of southern vergence were terminated in forming secondary klippes approximately at the middle of the Cretaceous. The compression was of N–S direction in recent co-ordinates; however, an about 90° counter-clockwise rotation, detected by Márton et al. (1989), must be taken into account in this respect.

The last main phase in the Oligo-Miocene was induced by the Bükk Mts. and Szendrő Mountains having approached far from SW. As a result of this, the Rudabánya Hills moved in three main segments (documented also by Upper Oligocene to Lower Miocene rocks in different facies—

Fig. 4.33 Serpentinite blocks of the Meliata series (dark) tectonically incorporated into the Perkupa Evaporite of the base of the Silica nappe (light), Perkupa (abandoned gypsum mine) (*Photo* G. Nagy)

Szentpétery 1997) from the southern vicinity of the Aggtelek Karst to their eastern neighbourhood along sinistral strike-slips of the Darnó zone. Overthrusting of new secondary klippes (e.g. the Martonyi klippe) and movements along complementary strike-slips of E–W direction (e.g. along the Bódvaszilas–Ménes Valley zone) are also associated with this phase. After the consolidation of the area in the Middle Miocene it became once again a dry-land. In the Late Miocene the Pannonian Lake flooded morphological depressions generated by erosion and brittle faults as it is recorded by the 50–300-m-thick Edelény Variegated Clay containing basal lignite measures. The latest uplift of the area started at the end of the Pliocene, and took place in the foreground of the general emergence of the Western Carpathians. The masses of pebble material arriving from the Carpathians (preserved in the max. 80 m thick Borsod Pebble) played the dominant role in developing the huge cave systems in the previously karstified large limestone bodies.

5 Szendrő–Uppony Mountains

From a geographical point of view, the Szendrő and Uppony Mts. are part of the Bükk and Cserehát; however, based on their structural position and geological structure, they are discussed separately from them.

The Uppony Mountains (Fig. 4.20) are located entirely within the Darnó Fault Zone. They are bordered by the Uppony Fault on the N and NW, separating it from the Triassic series of the Rudabánya Mts. (Fig. 4.35), and by the Nekézseny Fault on the S and SE, separating it from the Bükk Mts.

The Szendrő Mountains (Fig. 4.28)—located NE of the Uppony Mountains, with an areal extent of more than 150 km^2—are bordered by the Darnó Fault Zone on the NW, by the Hernád Line on the E (with the Zemplén Unit on the opposite side) and most probably by the eastern continuation of the Nekézseny Fault to the S against the Bükk Mts. Its northward extension is unknown due to the lack of borehole data.

Fig. 4.34 The intensively folded, cherty Pötschen Limestone from the Torna series, Tornaszentandrás (*Photo* Cs. Péró)

As in the case of the Bükk Mts., both the Uppony and Szendrő Mountains have a strong south-Alpine affinity (Ebner et al. 1998); however, compared to the Bükk Mts., they are built up mostly of older Paleozoic rocks. Theoretically, the sequence of the Bükk Para-autochton may be the stratigraphic cover of the Szendrő–Uppony Mountains. The oldest non-metamorphosed overlying deposits of the metamorphosed Variscan series are made up of Upper Cretaceous conglomerates in the Uppony Mountains and Neogene sediments in the Szendrő Mountains.

Fig. 4.35 Paleozoic rocks of the Uppony Mts. (on the right) bordered by the Middle Triassic Gutenstein Dolomite in the prolongation of the Rudabánya Mts. (on the left) along the sinistral strike-slip of the Darnó zone (at the pylon), S of Uppony (*Photo* Cs. Péró)

5.1 Tectonics and Metamorphism

Two subunits are distinguished in both areas (Fig. 4.36). In the Uppony Mountains these are the Tapolcsány Subunit in the SE overthrust onto the the Lázbérc Subunit in the NW. In the case of the Szendrő Mountains the Abod Subunit in the S is overthrust onto the Rakaca Subunit in the N. Folded and imbricated structures of N-NW vergence (opposite to the general southern trend in both the Aggtelek-Rudabánya Mts. to the N and the Bükk Mts. to the S) are well developed in both areas. Nappe structures, however, have not been detected, yet.

The peculiarity of the Szendrő–Uppony Mountains is the lack of Variscan but the presence of Alpine metamorphism (around 108–118 Ma according to Árkai et al. 1995); its degree, however, is different in the two areas. Low-grade metamorphism (3 kbar and 400–450 °C) is detected in the Szendrő Mountains, whereas it is the boundary between very low and low grade (2.5 kbar and 350 °C) in the case of the Uppony Mountains (Árkai 1983).

5.2 Geological Build-Up

Due to the metamorphism fossil data are rather erratic and uneven from the Szendrő–Uppony Mountains. Pelagic carbonate rocks and olistostromes with pelagic carbonate olistoliths are much better dated than platform carbonate, siliciclastic and siliceous rocks. The age of the latter is based on their stratigraphic position and on south-Alpine analogues. Direct datings are mostly due to the extensive conodont determinations by S. Kovács (mostly in Fülöp 1994). Also, he was the one who elaborated the stratigraphy of Paleozoic rocks (Kovács 1992; Ebner et al. 1998), which is briefly summarized below with some simplifications (Fig. 4.36, see also for the location of particular formations in different subunits).

The oldest rocks are Upper Ordovician to Lower Devonian black, sometimes siliceous shales (Tapolcsány Fm and Irota Fm), 300–500 m in thickness, containing coarse sandstone interbeds. The overlying 60–80-m-thick Strázsahegy Fm is a Lower and/or Middle Devonian olistostrome

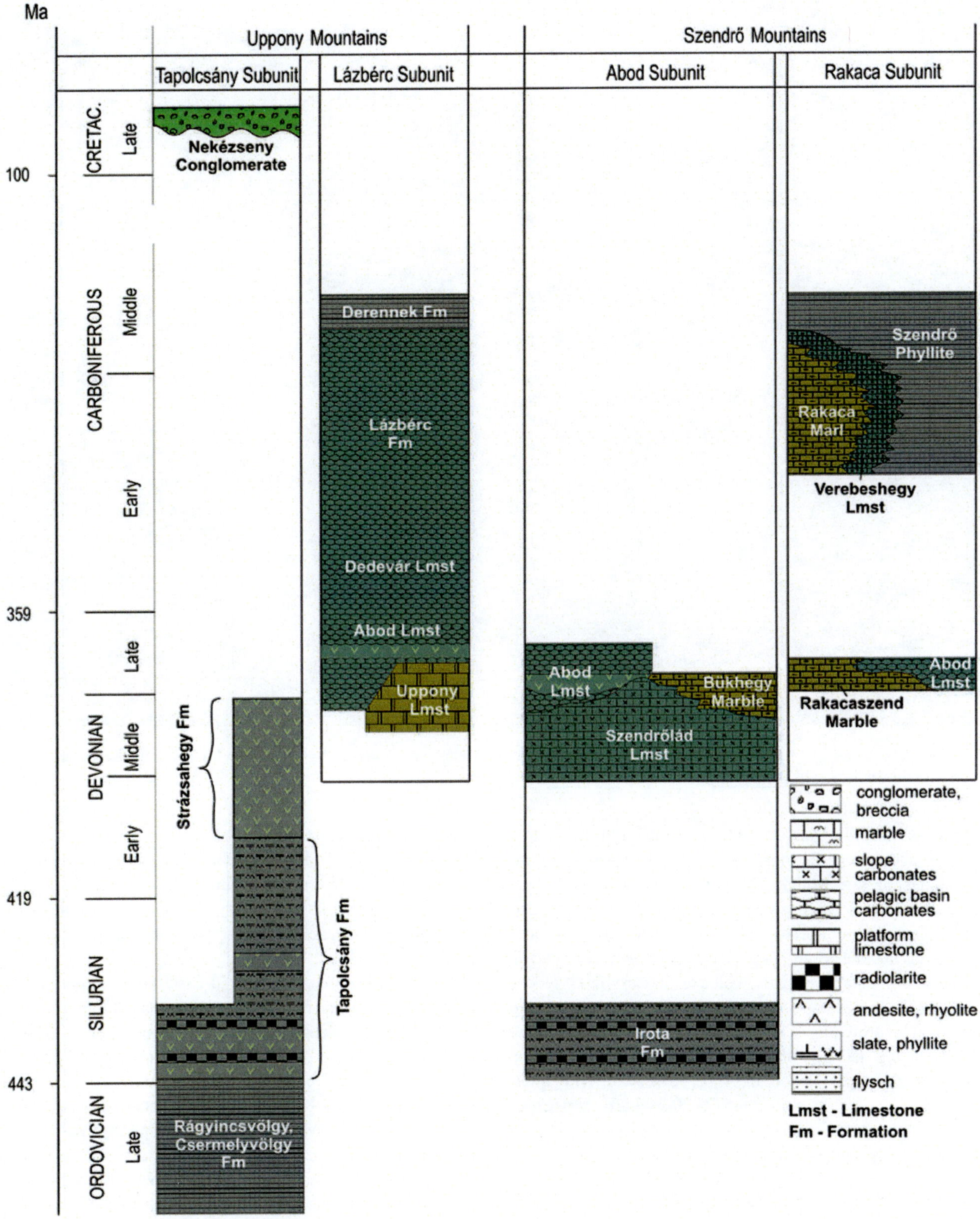

Fig. 4.36 Lithostratigraphic column of Paleozoic–Mesozoic formations of the Uppony and Szendrő Mountains (after Haas and Budai eds. 2014, modified and completed)

Fig. 4.37 The karstified Middle Devonian Uppony Limestone (in the background) overlain by the Upper Devonian tuffitic pelagic Abod Limestone (in the foreground), Uppony gorge, S of Uppony (*Photo* Gy. Less)

with basaltic tuffite matrix containing olistoliths of Upper Silurian and Lower Devonian pelagic limestones. Platform carbonates of late Middle Devonian (Uppony Limestone, Rakacaszend and Bükhegy Marbles) and questionably also of late Early Carboniferous age (Rakaca Marble), each about 200 m in thickness, are intensively karstified (Fig. 4.37). Middle Devonian mixed siliciclastic-carbonate rocks (Szendrőlád Limestone), about 500 m in thickness, are also present. The drowning of carbonate platforms are marked by the overlying pelagic limestones. These are the approx. 200 m thick Upper Devonian Abod Limestone with intercalations of basaltic tuffs, the 1–20-m-thick lowermost Carboniferous Dedevár Limestone with cherts, and the late Early to early Middle Carboniferous Verebeshegy Limestone (Fig. 4.38), 3–30 m in thickness. The increase of terrigenous influx is indicated by lower to middle Carboniferous mixed siliciclastic-carbonate rocks in basinal facies (the 200–300 m thick Lázbérc Fm), by middle Carboniferous calcareous sandstone (Derennek Fm, 20–30 m in thickness) and flyschoid phyllite with olistoliths (the 500–600-m-thick Szendrő Phyllite). This youngest Paleozoic rock of the Szendrő–Uppony Mountains resembles very much the Szilvásvárad Fm of the same age, the oldest Paleozoic rock in the Bükk Mts.

Upper Carboniferous to mid-Cretaceous deposits are unknown; they were most probably eroded during the Late Cretaceous and Paleogene. The non-metamorphosed Upper Cretaceous molasse-like Nekézseny Conglomerate (Fig. 4.39)—formed by relatively deep-marine debris flow (Brezsnyánszky and Haas 1984)—covers older rocks and their structures with angular unconformity in the Uppony Mts. These deposits are also deformed (locally observed in vertical position) and occasionally contain limestone blocks with Hippurites.

Paleogene deposits are unknown in any of the two areas. Both of them were source areas in the early Miocene towards the SE part of the Rudabánya Mts.

Fig. 4.38 Middle Devonian Rakacaszend Limestone of platform facies (lighter, on the left) overlain by the pelagic Lower Carboniferous Verebeshegy Limestone (darker, on the right), Rakacaszend (Szendrő Mountains) (*Photo* Cs. Péró)

6 Mecsek and Villány Mountains

The Pre-Cenozoic sequences of the Mecsek and Villány Mts. are parts of the Tisza Mega-unit. The area of this Mega-unit extends beyond the border of the country, encompassing the Apuseni Mountains and a part of the Slavonian Mountains, as well (Haas and Péró 2004). Its pre-Cenozoic basement consists of three large units: the Mecsek, the Villány–Bihor and the Békés–Codru Nappe Systems (Fig. 4.2). The unit's north-western boundary is the Mid-Hungarian Line. On the South the Tisza Mega-unit borders the Sava-unit, Vardar-zone and Transylvanides (Haas and Budai eds. 2014). The rocks of the basement crop out to the surface only in the area of the Mecsek Mountains in the Mecsek Unit and in the area of the Villány Mts. in the Villány–Bihor Unit.

6.1 Tectonics

The area became attached to the European plate margin by the early Carboniferous Variscan orogeny that formed the crystalline basement (Fig. 4.40). A significant tectonic element of the basement is the Mecsekalja Dislocation Zone (Szederkényi 1977, Fig. 4.41). The rocks of the crystalline basement are unconformably overlain by Upper Carboniferous terrigenous sandstones, and, in the Mecsek, by Permian continental deposits—due to the downlifting motions. After the relatively calm sedimentary period during the Late Permian—Early Triassic, the first signs of tectonic activity, namely listric normal faulting—related to the rifting of the Vardar ocean branch—appeared in the Middle Triassic. From that time onwards, differentiation of the morphology increased, resulting in a different evolution of the Mecsek

Fig. 4.39 Upper Cretaceous Nekézseny Conglomerate. Nekézseny, railway stop (*Photo* Cs. Péró)

and the Villány Mts: sedimentation in the Mecsek was almost continuous until the Early Cretaceous, whereas the Villány area was characterized by long gaps. The area was detached from the European margin by the Magura Ocean, which started rifting in the Middle Jurassic. The Early Cretaceous is characterized by widespread mafic volcanics related to continental rifting (Harangi et al. 1996). Albian clastic sediments in the northern foreland basin of the Villány Mts. already refer to compression-induced denudation. By the Turonian, the microcontinent comprising the Mecsek area collided with the West Carpathian—Austroalpine one (Csontos and Vörös 2004). This tectonic phase caused folding and uplifting, and nappe formation is supposed, as well. The Cretaceous nappes of the Tisza Mega-unit have a NNW vergency, similarly to most of the compression structures within the structural units. The East Mecsek is basically of a syncline structure, whereas the West Mecsek has an eastward dipping perianticlinal one (Fig. 4.41). The imbricate structure of the Villány Mts. is also a result of the Cretaceous orogeny (Figs. 4.41 and 4.42).

Subsequently the Paleogene erosional period the thickest clastic Miocene sequences accumulated in the narrow basins which encircled the mountains and had been formed by strike-slip fault movements (Benkovics 1997) along the renewed Paleozoic Mecsekalja Dislocation Zone (Fig. 4.41). The neotectonic period of the area is characterized by compression/transpression. During this the Mecsek and Villány Mts. were uplifted. The Mecsek Mts. have been pressed up as a large-scale positive flower structure in a broad shear zone.

6.2 Geological Build-Up

The oldest formations of the Pre-Cenozoic basement of the Tisza Mega-unit are assigned to many (pre-Variscan and Variscan) crystalline basement complexes (Ófalu Group, Mórágy, Baksa and Babócsa Complex, Szederkényi 1996). The polimetamorphic basement is covered by the Carboniferous-Triassic continental–shallow-marine sequence. The denudation period following the Variscan orogeny is represented by a 3–4-km-thick Carboniferous (Téseny Fm) and Permian terrestrial molasse which was deposited either on the Carboniferous granite or on the metamorphic Variscan basement. The mostly red fluvial succession of clastic sediments is made up of fining-upward cycles in the Lower Permian (Korpád Sandstone). In the Middle Permian (Cserdi Fm) it

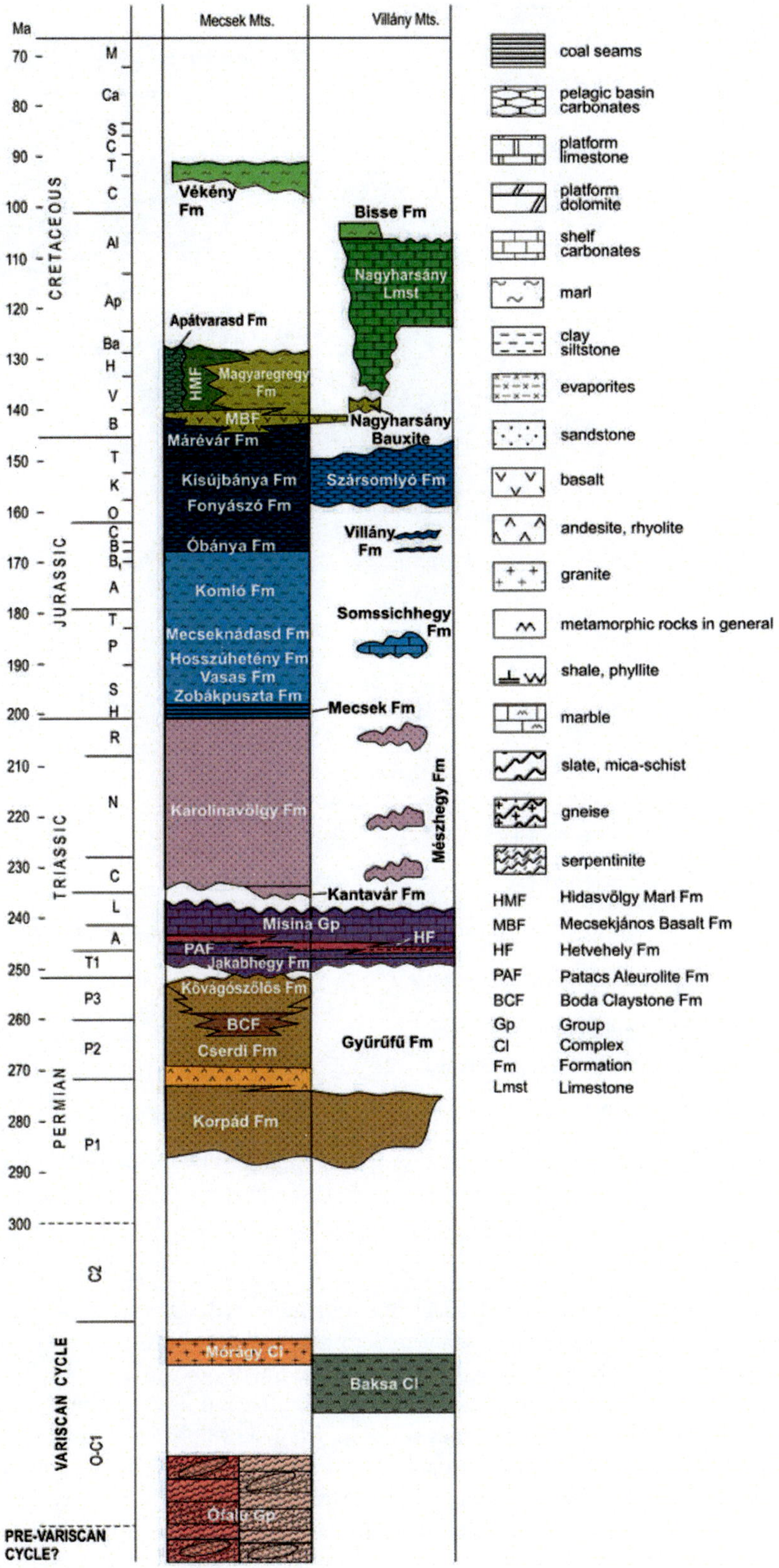

Fig. 4.40 Stratigraphic column of Paleozoic–Mesozoic formations of the Mecsek and Villány Mts. (after Haas and Budai eds. 2014, slightly modified and completed)

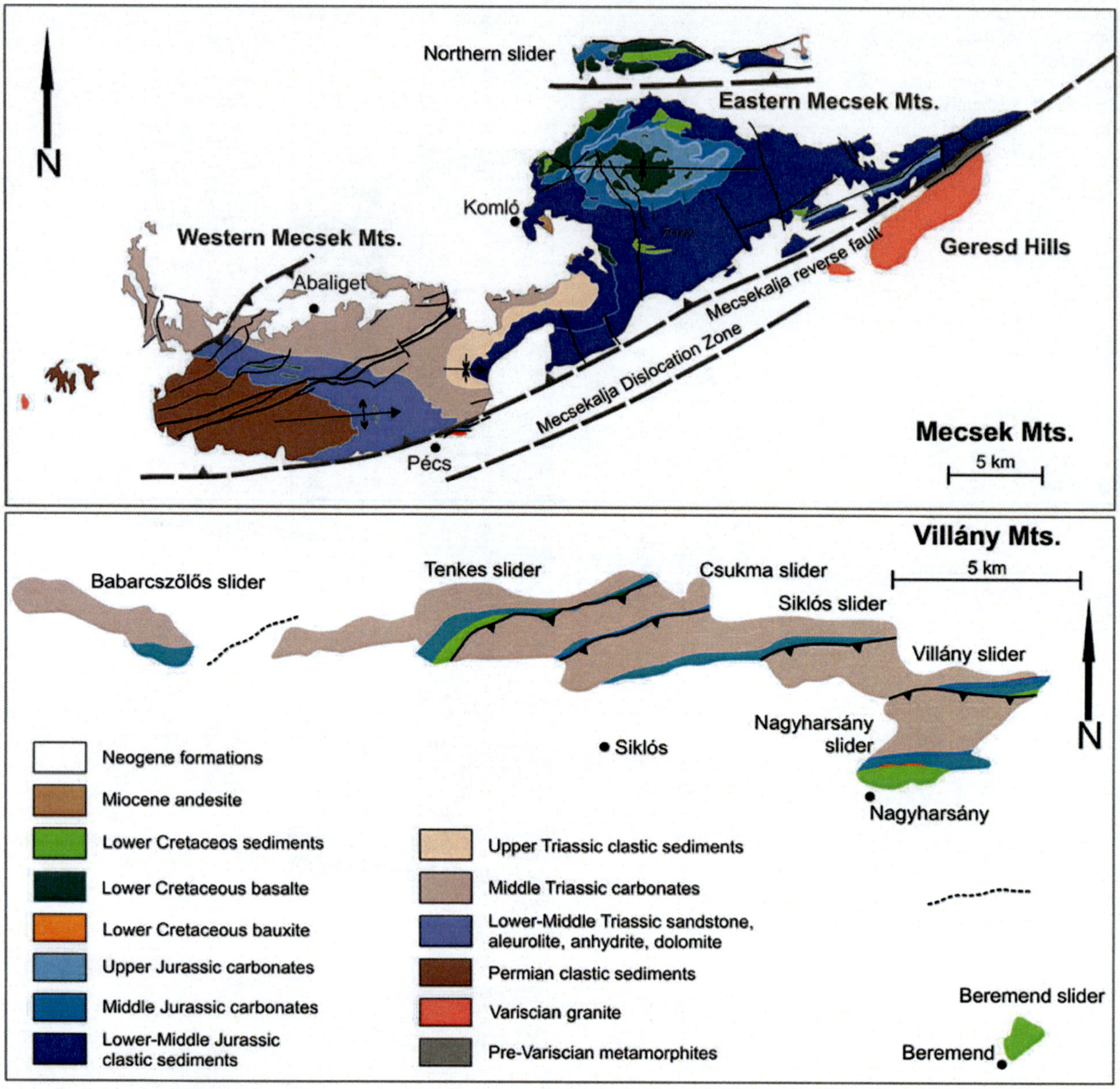

Fig. 4.41 Simplified geological map of the Mecsek and Villány Mts (after Konrád et al. 2010). Legend in the figure below

interfingers with fine-grained lacustrine formations (Boda Claystone). The cyclic fluvial sedimentation continued even in the beginning of the Early Triassic (Kővágószőlős Sandstone). The Boda Claystone is unknown in the Villány Mts. and its surrounding area. The clastic series is known from the whole area of the Tisza Mega-unit in variable thicknesses.

The early continental rifting of the Alpine cycle was accompanied by extensive acidic volcanism. The Gyűrűfű Rhyolite—deposited in the Lower Permian terrestrial molasse succession (Korpád Sandstone)—is made up predominantly of greyish lilac extrusive and sub-volcanic lava rocks, frequently with intercalating tuff, agglomerate and ignimbrite (Szemerédi et al. 2017).

The Triassic of the Tisza Mega-unit shows similarities to the tripartite Germanic epicontinental sequences, in spite of the paleontological and sedimentological evidence (Géczy 1973; Wéber 1990; Török 1993). The coarse conglomerate (Fig. 4.43) formed at the beginning of the Early Triassic indicates the uplift of the denuding background and the start of a new sedimentary cycle. The red, fluvial succession of

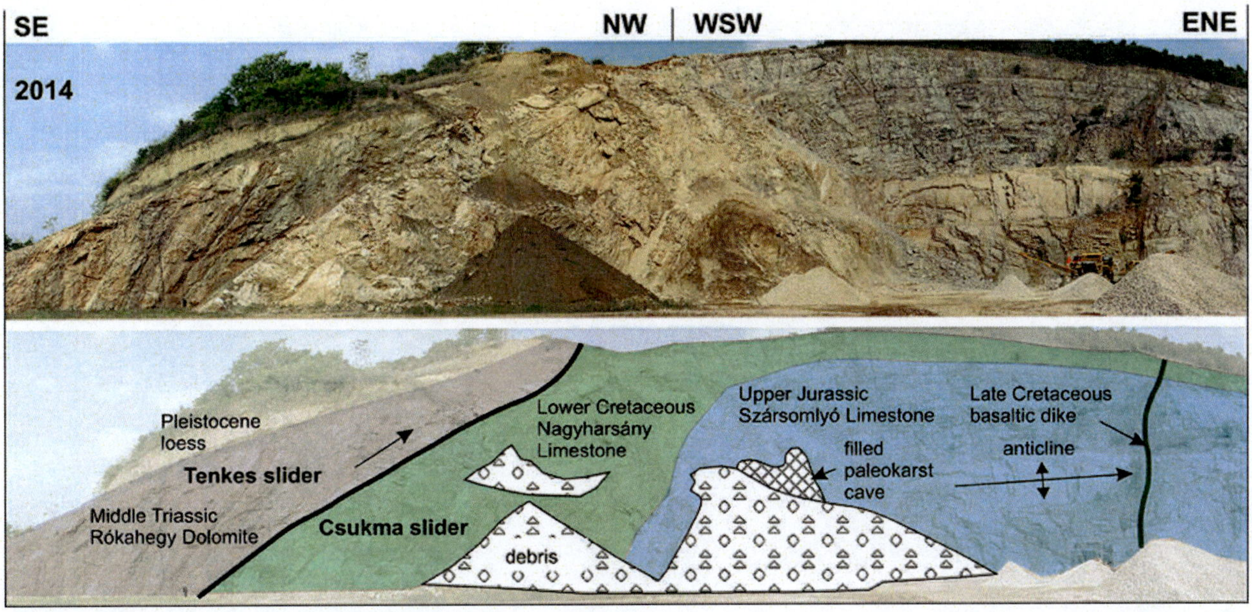

Fig. 4.42 Tilted and folded Jurassic-Cretaceous sequence and upthrust Triassic beds in the Szabolcs Valley quarry, Villány Mts. (*Photo* Gy. Konrád)

Fig. 4.43 Outcrops of the basal conglomerate of the Lower Triassic Jakabhegy Sandstone in the foreground and on side of Jakab Hill (*Photo* Gy. Konrád)

the Jakabhegy Sandstone has a fining-upward character. Over a large area the formation is directly deposited on the metamorphic Variscan basement (Barabás and Barabás--Stuhl 2005).

The continuous Permian–Early Triassic terrigenous clastic sedimentation was followed by the deposition of marine clastics in the late Early Triassic. As a result of the Anisian transgression a shallow marine ramp began to evolve; it was

Fig. 4.44 Lapis and Zuhánya Limestone at Bükkösd in the Bükkösdkő quarry (*Photo* Gy. Konrád)

characterized by the deposition of fine-grained terrigenous clastic sediments (Patacs Siltstone). The formation indicates a continuous transgression and marks the beginning of the transition from a purely siliciclastic to a carbonate ramp. The increase of carbonate content (mainly dolomite), magnesite nodules in the dolomites and enterolithic anhydrites (Hetvehely Fm) in the upper part show a sabkha environment and refer to the climate getting drier.

After the sabkha event a storm-dominated, homoclinal carbonate ramp developed during the Middle Anisian and in the Ladinian (Török 1993). The thickness of the carbonate succession (Misina Group) exceeds 600 m. The succession is represented by dark grey limestones and dolomitized carbonate rocks. The sea level rise resulted in a Wellenkalk-type carbonate sequence (Viganvár and Lapis Limestones). Between the two formations a regression event resulted in the formation of a stromatolitic and pseudostromatolitic dolomite (Rókahegy Dolomite). The varied lithofacies of the Lapis Limestone show primarily inner and mid-ramp deposition mostly above the storm wave base. Proximal areas are characterized by bioclastic tempestite lenses of usually uniform composition, where crinoid ossicles and gastropod and bivalve shells rarely mix. Gutter casts, hummocky cross lamination and sporadic ripple marks also refer to storm-generated flow. More distal tempestites may appear as turbidite floatstones with bivalve shells or as micrograded beds with a scant amount of bioclasts. The thin, wavy limestone layers, separated by clayey films, are distal mid-ramp deposits of quiet (stormless) periods. Beds with preserved ichnofossils show a relatively well-oxygenated seafloor but not very diverse benthic fauna, while the typical nodular limestones (authigenic breccias?) were formed by the rip-up and redeposition of bioturbated beds. At several spots slumps caused syndepositional deformation. The upper part of the formation (Tubes Limestone Mb, Fig. 4.44) is thick-bedded; its crinoidal and ooidal banks and hummocky cross-stratification refer to inner ramp sediments redeposited from sublittoral sand shoals and show a relative sea level fall (Haas et al. 2002).

In the Mecsek and Villány mountains the deepest facies is represented by brachiopod beds (Zuhánya Limestone, Fig. 4.44). Open-marine conditions are indicated by the presence of ammonites and conodonts (Kovács and Rálisch-Felgenhauer 2005). The sedimentological features (slumps, intraclasts) indicates deposition on the outer ramp (Török 1993).

In the upper Anisian significant spatial differences occur in the grade of dolomitization and facies development. In the West Mecsek and the Villány Mountains the carbonates are extensively dolomitized (Csukma Dolomite). These beds were formed in the supratidal to peritidal zone of the inner ramp. In the Villány Mountains, it was overlain by dolomitic marls (Templomhegy Dolomite) and subsequently by Carpathian Keuper-type sandstones (Mészhegy Sandstone). In the central part of the Mecsek, a littoral ooidal, cross-bedded limestone (Kozár Limestone) was deposited. The formation indicates the shallowing of the carbonate ramp. Its sediments were transported by submarine currents from littoral or sublittoral sand shoals to the mid-ramp (Török 1998).

During the Ladinian the Mecsek area was uplifted and an erosional unconformity was formed, only small patches of terrigenous siderite and clay accumulated with a thickness of a few metres. The next sedimentary cycle started with an oncoidal—trigonodus-bearing limestone and continued with lacustrine calcareous marls (Kantavár Calcareous Marl, Fig. 4.45). The monospecific ostracod fauna and the characeae of the lower beds indicate a freshwater lacustrine environment evolved from a completely restricted lagoon (Monostori 1996). The amount of terrigenous clastics gradually increased resulting in the transition into a fluvial, lacustrine, shallow-marine sandstone sequence (Karolinavölgy Sandstone). A coal-bearing paralic sandstone (Mecsek Coal) is known from the Upper Triassic; its deposition continued in the Early Jurassic, as well.

The palaeogeographic picture of the Tisza Mega-unit was diverse in the Late Triassic, and from this time onwards, until the Middle Cretaceous the Villány–Bihor Zone appeared as an uplifted ridge between the rapidly subsiding Mecsek trough on the north and the Codru basins in the south (Fig. 4.40).

In the Mecsek Mts. the thickness of the overlying Upper Triassic (Karolinavölgy Sandstone) and the Lower Jurassic fluvial, as well as deltaic sandstone–siltstone-claystone successions (Mecsek Coal) increases from NE to SW, from 400 to 1800 m. In the area of the East Mecsek the Lower Jurassic succession of pelagic basin facies is made up of sandstone, siltstone and marl (Vasas Marl and Hosszúhetény Marl, Mecseknádasd Sandstone, Óbánya Siltstone, Komló Calcareous Marl). Its thickness increases south-westward where it may attain 2000 m.

In the East Mecsek the pelagic Middle Jurassic is represented by thin-bedded limestone (Óbánya Limestone), the Upper Jurassic by siliceous limestone and radiolarite (Fonyászó Limestone) as well as by the thin-bedded Kisújbánya Limestone and the Upper Jurassic—Lower Cretaceous Márévár Limestone.

The Lower Cretaceous volcanics are represented by alkali basalt, trachybasalt, tephrite and phonolite rocks

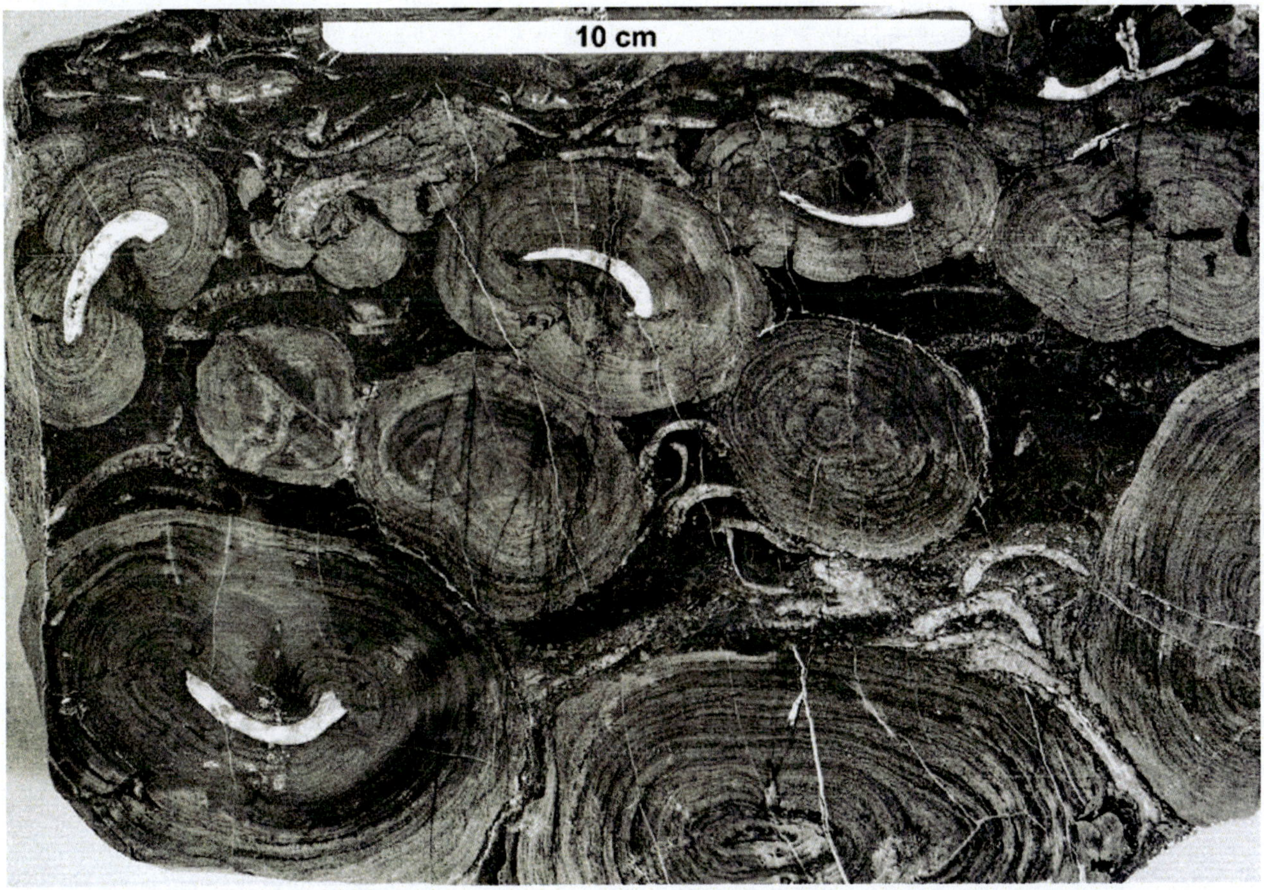

Fig. 4.45 Polished slab of stromatolite oncoids from the basal bed of Kantavár Calcareous Marl. Trigonodus shells are covered by stromatolitic laminae (*Photo* Gy. Konrád)

(Mecsekjános Basalt). Their appearance (pillow lava, hyaloclastite) and characteristic alteration patterns indicate submarine effusion. Early Cretaceous sedimentary rocks were formed in the basin among the volcanic build-ups (Hidasivölgy Marl), on the slopes of volcanoes (Magyaregregy Conglomerate), whereas south of the rift valley shallow-bathyal limestone (Apátvarasd Limestone) accumulated (Császár and Turnšek 1996). Red bathyal basinal marl assigned to the Turonian is known from the northern slivers (Fig. 4.41) of the Mecsek Mountains (Vékény Marl).

In the Villány Mountains the Jurassic sedimentation starts with a transgressive series of Pliensbachian age. The basal conglomerate beds rest unconformably on the eroded top of the Upper Triassic Mészhegy Fm; upwards, with decreasing amount of terrigenous clastics, pure limestone develops; an erosional top surface terminates the sequence. The Pliensbachian Somssichhegy Fm is also restricted to local tectonic depressions. The next, short sedimentary episode took place in the Late Bathonian, when a thin bed of sandy, marine limestone accumulated. It has been preserved also in local lenses. The first widespread Jurassic formation is the famous Callovian stromatolithic, ammonitic bank. This deeper marine Villány Fm is uniformly thin (less than half a metre) and bounded by unconformities from below and above.

From the Oxfordian, the previously episodic and condensed sedimentation switched to continuous accumulation of thick, pure limestones (Szársomlyó Limestone, 300 m, Fig. 4.46) in the Villány Mts. The basal, micritic "pelagic oolite" changes to increasingly shallow-marine limestone up to the Tithonian. The subaerial character of the next regional unconformity in the earliest Cretaceous, is marked by karst bauxites (Harsányhegy Bauxite Fm, Fig. 4.46), overlain by Valanginian–Albian Urgon-type limestones formed on a carbonate platform. The thickness of the Nagyharsány Limestone Fm (Fig. 4.46) may exceed 400 m. The lower part of the formation is characterized by black pebbles and lofer-cycles, the upper part is a thick-bedded grey limestone with rudists.

Younger Cretaceous (Albian to Cenomanian) formations (Bisse Marl) occur only in erosional fragments or in boreholes. The appearance of coarse siliciclastic turbidite formations in the Villány–Bihor Unit suggests significant changes in the Alpine evolution of the Tisza Mega-unit at the end of the Albian. The turbidite formations were probably formed in the foreland basin along the nappe fronts. Turonian pelagic basins evolved in relation with the pushing forward of the nappe fronts.

In the Late Cretaceous and Paleogene a denudation period existed; only small patches of intramontane molasse deposits (Szentlőrinc Fm) are known from the Mecsekalja Dislocation Zone. In the Mecsek Mts. the folded Permian–Mesozoic sequence is overlain by Lower Miocene sediments with a remarkable unconformity. The sedimentation began only in the later Miocene in the surroundings of the Villány Mts.

The Miocene sedimentation started with terrigenous siliciclastic rocks in the Late Eggenburgian (Szászvár Fm), when a mountain range existed south of the current Mecsek Mts (Józsa et al. 2009). During the Badenian the area became a marine sedimentary basin again as part of the Paratethys (Fig. 4.47), so the terrigenous clastic succession and the overlying lacustrine and lagoonal deposits (Kiskunhalas Fm Pécsvárad Mb and Komló Mb) were covered by shallow-marine carbonates (Lajta Fm Pécsszabolcs Mb and Rákos Mb) and siliciclastics (Budafa Fm and Baden Fm Tekeres Mb and Szilágy Mb). During the Sarmatian period the Central Paratethys became isolated from the open sea. Due to uplifting during this phase, sedimentation ceased in the Mecsek Mountains. The Kozárd Clay Marl and the Tinnye Limestone were deposited in the foreland (Fig. 4.48). Following this the freshwater Lake Pannon evolved (Peremarton and Dunántúl Groups, Fig. 4.47). Water coverage was different in the region, the narrow area of the Mecsek and the Villány Mts. emerged above the lake as islands. After the complete upfilling of Lake Pannon the Pliocene is characterized by pedimentation and the formation of red clays began (Tengelic Fm). During the Pleistocene a thick loess succession was added to the sequence.

7 Conclusion

The thickest and the most widespread formations of Hungary that are typically karstic were formed along the broad shallow marine carbonate shelf of the Neotethys Ocean during the Mesozoic. These rocks build up the main part of the Alcapa Mega-Unit (Transdanubian Range Unit and Aggtelek-Rudabánya Unit) and the northern part of the Mid-Hungarian Mega-Unit (Bükk and Szendrő Mountains). The presence of karstic forms in the Silurian–Devonian and Carboniferous carbonate formations in the Paleozoic succession of the Uppony–Szendrő Mountains is not typical. The most intensely karstified formations of the Transdanubian Range, the Bükk Mts and the Aggtelek-Rudabánya Hills are the km thick Triassic limestone series. However, the Cretaceous shallow marine limestones of the Transdanubian Range also show strong karsification. Among the Paleogene-Neogene series, there are only a few Middle-Upper Eocene and Middle Miocene carbonate rock units suitable for karstification, mainly in the area of the Transdanubian Range.

The Tisza Mega-unit, belonging to the European plate margin, forms the basement of the Pannonian Basin south of the Mid-Hungarian Mega-Unit. In the territory of Hungary the pre-Neogene basement of this mega-unit crops out only in two relatively small, isolated areas in South Transdanubia,

Fig. 4.46 Upper Jurassic Szársomlyó Limestone and Lower Cretaceous Nagyharsány Limestone at Nagyharsány in the DDCM quarry. Between the formations the Lower Cretaceous karstic Nagyharsány Bauxite Fm can be seen (*Photo* Gy. Konrád)

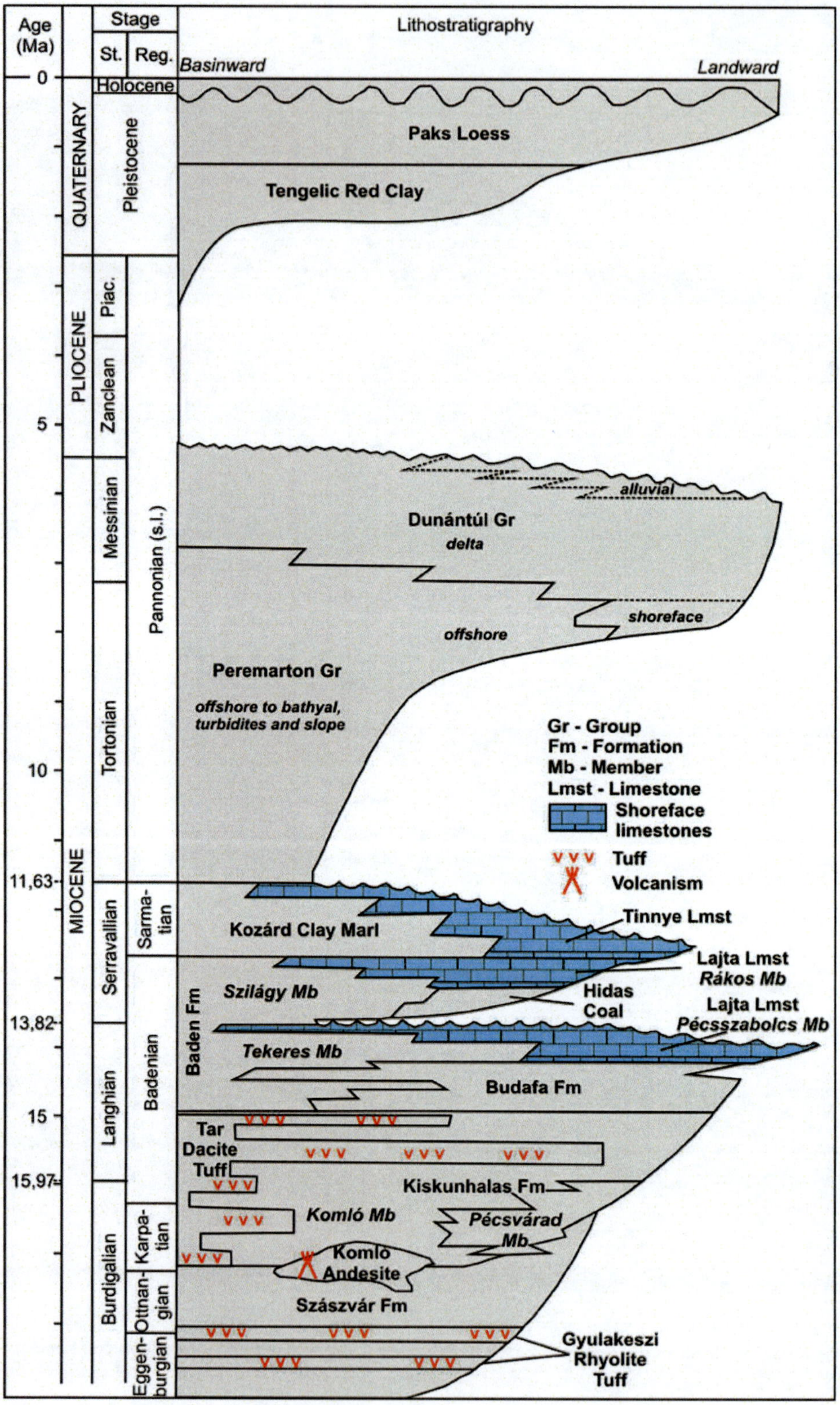

Fig. 4.47 Lithostratigraphy of Cenozoic formations of the Mecsek region, based on Sebe et al. (2019), Sztanó (2015) and Bartha et al. eds (2015).

Fig. 4.48 Outcrop of the Miocene Kozár Limestone at Pécs, Havi Hill. View from the north with the Villány Mts. in the background (*Photo* Gy. Konrád)

i.e. in the Mecsek Mountains and in the Villány Mountains. Considering the pre-Cenozoic successions, which are continuous in the Mecsek and rather discontinuous in the Villány Mountains, the Triassic to Lower Createaceous carbonates are the ones suitable for karstification. Paleogene series are missing in this area. Among the Neogene layers carbonate formations occur only in the Middle Miocene Badenian and Sarmatian successions; however, their karstification is not typical.

References

Árkai P (1983) Very low- and low-grade Alpine metamorphism of the Paleozoic and Mesozoic formations of the Bükkium, NE Hungary. Acta Geol Hung 26(1–2):83–101

Árkai P, Kovács S (1986) Diagenesis and regional metamorphism of the Aggtelek-Rudabánya Mts. Acta Geol Hung 29:349–373

Árkai P, Balogh K, Dunkl I (1995) Timing of low-temperature metamorphism and cooling of the Paleozoic and Mesozoic formations of the Bükkium, innermost Western Carpathians, Hungary. Geol Rundschau 84:334–344

Árkai P, Faryad SW, Vidal O, Balogh K (2003) Very low-grade metamorphism of sedimentary rocks of the Meliata unit, Western Carpathians, Slovakia: implications of phyllosilicate characteristics. Int J Earth Sci 92:68–85

Báldi T, Sztanó O (2000a) Gravitációs tömegmozgásos fáciesek és a vízmélység változásainak jelei a Bükk tengeri oligocén rétegeiben (Gravity mass movements and palaeobathymetric changes in the marine Oligocène deposits of the Bükk Mts.). Földtani Közlöny 130 (3):451–496 (in Hungarian with English abstract)

Báldi T, Sztanó O (2000b) Gravitációs tömegmozgások a Darnó zóna tengeri oligo-miocén üledékeiben: a Dubicsány-31 fúrás értékelése (Gravity mass movements in the Oligocene-Miocene deposits of the Darnó zone: a study based on Dubicsány-31 well. Földtani Közlöny 130(4):673–694 (in Hungarian with English abstract)

Báldi T, Less Gy, Mandic O (1999) Some new aspects of the lower boundary of the stage Egerian (Oligocene, chronostratigraphic scale of the Paratethys area). Abh Geol B-A 56(2):653–668

Báldi T, Nagymarosy A (1976) A hárshegyi homokkő kovásodása és annak hidrotermális eredete. (The silification of Hárshegy sandstone and its hydrotermal origin) Földtani Közlöny 106:257–275 (in Hungarian)

Báldi T (1983) Jelentés a Nagygörbő Ng–1. sz. fúrás miocén szakaszának nannoplankton vizsgálatáról (Report on the nannoplankton studies of the Miocene section of Nagygörbő Ng–1 well). Unpublished report, Archive of the Mining and Geological Survey of Hungary, Budapest (in Hungarian)
Báldi T (1986) Mid-Tertiary stratigraphy and paleogeographic evolution of Hungary. Akad Kiadó, Budapest
Báldiné Beke M, Báldi T (1990) A bakonyi eocén medence süllyedéstörténete (Basin evolution of the Bakony Eocene Basin). Általános Földtani Szemle 25:83–118 (in Hungarian with English abstract)
Balla Z (1983) A szarvaskői szinform rétegsora és tektonikája (Stratigraphy and tectonics of the Szarvaskő synform). Ann Rep Eötvös L Geophys Inst Hung 1982:42–65 (in Hungarian with English abstract)
Balla Z, Dudko A (1989) Large-scale tertiary strike-slip displacements recorded in the structure of the transdanubian range. Geophys Trans 35(1–2):3–63
Balogh K (1964) Die geologische Bildungen des Bükk Gebirges. Ann Hung Geol Inst 48(2):245–719
Barabás A, Barabás-Stuhl Á (2005) Geology of the lower triassic jakabhegy sandstone formation, Hungary, SE Transdanubia. Acta Geol Hung 48:1–47
Bartha I, Kriván Á, Magyar I, Sebe K eds (2015) Neogene of the Paratethyan Region. 6th Workshop on the Neogene of Central and South-Eastern Europe. An RCMNS Interim Colloquium, 31 May–3 June, Orfű, Hungary. Programme, Abstracts, Field Trip Guidebook. – Hungarian Geological Society Budapest:125
Bence G, Budai T, Csillag G, Selmeczi I (1999) Prepannonian miocene. In: Budai T, Csillag G (eds) Geology of the Balaton Highland. Explanation to the geological map of the Balaton Highland, 1:50 000. Geological Institute of Hungary, Occasional Papers 197, pp. 202–207
Benkovics L (1997) Étude structurale et géodynamique des monts Buda, Mecsek et Villány (Hongrie). Ph.D. Thesis, Université des Sciences et Technologies de Lille
Brezsnyánszky K, Haas J (1984) A szenon nekézsenyi konglomerátum formáció sztratotípus szelvényének szedimentológiai és tektonikai vizsgálata (The Nekézseny Conglomerate Formation of Senonian age: a sedimentological and tectonic study of the stratotype section). Földtani Közlöny 114(1):81–100 (in Hungarian with English abstract)
Budai T, Császár G, Csillag G, Dudko A, Koloszár L, Majoros Gy (1999) Geology of the Balaton Highland. Explanation to the Geological Map of the Balaton Highland, 1:50 000. Occasional papers of the geological institute of Hungary 197, Budapest
Budai T, Fodor L, Sztanó O, Kercsmár Zs, Császár G, Csillag G, Gál N, Kele S, Kiszely M, Selmeczi I, Babinszki E, Thamóné Bozsó E, Lantos Z (2018) Geology of the Gerecse mountains. Explanatory book to the geological map of the Gerecse mountains (1:50 000). Mining and Geological Survey of Hungary, Budapest
Budai T, Vörös A (1992) Middle Triassic history of the Balaton Highland: extensional tectonics and basin evolution. Acta Geol Hung 35(3):237–250
Budai T, Vörös A (2006) Middle triassic platform and basin evolution of the Southern Bakony Mountains (Transdanubian Range, Hungary). Rivista Italiana Paleontologica Stratigrafia 112(3):359–371
Cornée J-J, Moisette P, Saintmartin J-P, Dulai A, Tóth E, Görög Á, Kázmér M, Müller P (2009) Marine carbonate systems in the Sarmatian of the Central Paratethys: the Zsámbék Basin of Hungary. Sedimentology 56:728–1750
Császár G (1986) Dunántúli-középhegységi középső-kréta képződmények rétegtana és kapcsolata a bauxitképződéssel (Middle Cretaceous formations of the Transdanubian Central Range: stratigraphy and connection with bauxite genesis). Geologica Hungarica Series Geologica 23:1–295 (in Hungarian with English abstract)
Császár G (2002) Urgon formations in Hungary. Geologica Hungarica Series Geologica 25:1–209
Császár G, Turnšek D (1996) Vestiges of atoll-like formations in the lower cretaceous of the mecsek mountains, Hungary. Cretaceous Res 17(4):381–514
Csillag G, Sztanó O (2015) Upper miocene. In: Kercsmár Zs (ed) Surface geology of Hungary. Explanatory notes to the Geological map of Hungary (1: 500 000). Geological and Geophysical Institute of Hungary, Budapest, pp 45–50
Csillag G, Müller P (2005) Vulkanizmus utáni felszínalakulás Leányfalu környékén (Visegrádi-hegység) (Geomorphic evolution of the Leányfalu area, after the Miocene volcanism, Visegrád Mts., North Hungary). Annual Report of the Geological Institute of Hungary of 2004, pp 261–272 (in Hungarian with English abstract)
Csillag G (2015) Miocene–Pleistocene. In: Kercsmár ZS (ed) Surface geology of Hungary. Explanatory notes to the Geological map of Hungary (1: 500 000). Geological and Geophysical Institute of Hungary, Budapest, p 52
Csontos L, Vörös A (2004) Mesozoic plate tectonic reconstruction of the Carpathian region. Palaeogeogr Palaeoclimatol Palaeoecol 210:1–56
Csontos L (1999) A Bükk hegység szerkezetének főbb vonásai (Structural outline of the Bükk Mts., N Hungary). Földt Közl 129 (4):611–651 (in Hungarian with English abstract)
Csontos L (2000) A Bükk hegység mezozoos rétegtani újraértékelése (Stratigraphic reevaluation of the Bükk Mts., N Hungary). Földt Közl 130(1):95–131 (in Hungarian with English abstract)
Dunkl I, Árkai P, Balogh Kad, Csontos L, Nagy G (1994) A hőtörténet modellezése fission track adatok felhasználásával (Thermal modelling based on apatite fission track dating: the uplift history of the Bükk Mts., Inner Western Carpathians, Hungary). Földt Közl (Budapest) 124(1):1–24 (in Hungarian with English abstract)
Ebner F, Kovács S, Shönlaub P (1998) Stratigraphic and facial correlation of the Szendrő-Uppony Paleozoic (NE Hungary) with the Carnic Alps—South Karawanken and Graz Paleozoic (Southern Alps and Central Eastern Alps): some paleogeographic implications. Acta Geol Hung 41(4):355–388
Filipovic I, Jovanovic D, Sudar M, Pelikán P, Kovács S, Less Gy, Hips K (2003) Comparison of the Variscan-Early Alpine evolution of the Jadar block (NW Serbia) and Bükkium (NE Hungary) terranes; some paleogeographic implications. Slov Geol Mag 9 (1):23–40
Fodor L, Kázmér M, Magyari Á, Fogarasi A (1992) Gravity-flow dominated sedimentation on the Buda paleoslope (Hungary). Record of Late Eocene continental escape of the Bakony Unit. Geol Rundsch 82:695–716
Fodor L, Koroknai B (2000) Ductile deformation and revised lithostratigraphy of the Martonyi subunit (Torna Unit, Rudabánya Mts.), northeastern Hungary. Geol Carp 51(6):355–369
Fodor L, Magyari Á, Fogarasi A, Palotás K (1994) Tercier szerkezetfejlődés és késő paleogén üledékképződés a Budai-hegységben. A Budai-vonal új értelmezése (Tertiary tectonics and Late Palaeogene sedimentation in the Buda Hills, Hungary. A new interpretation of the Buda Line). Földtani Közlöny 124 (2):129–305 (in Hungarian with English abstract)
Forián Szabó M, Csontos L (2001) Structural features of the Kis-Fennsík area (Bükk Mts., NE Hungary). Geol Carp 53 (4):223–234
Fülöp J (1994) Magyarország geológiája. Paleozoikum II (The geology of Hungary. Paleozoic II). Akad Kiadó, Budapest (in Hungarian)
Galácz A (1988) Tectonically controlled sedimentation in the Jurassic of the Bakony Mountains (Transdanubian Central Range, Hungary). Acta Geol Hung 31:313–328

Géczy B (1973) Plate tectonics and paleogeography in the East-Mediterranean Mesozoic. Acta Geol Hung 17:421–428

Görög Á (1992) Sarmatian foraminifera fauna of the Zsámbék Basin, Hungary. Annales Universitatis Scientiarum Budapestinensis De Rolando Eotvos Nominatae, Sectio Geologica 29:31–153

Grill J (1988) A Rudabányai-hegység jura formációi. (Jurassic formations of the Rudabánya Hills). Magy Áll Földt Int Évi Jel 1986:69–103 (in Hungarian with English abstract)

Gyalog L, Horváth I, Daridáné Tichy M, Dudko A, Ódor L, Budai T, Császár G, Cserny T, Csillag G, Kaiser M, Kókay J, Lelkesné Felvári Gy, Less Gy, Ó Kovács L, Selmeczi I, Domos V (eds) (2004) Geology of the Velence Hills and the Balatonfő, Explanatory book of the geological maps of Velence (1:25 000) and Balatonfő–Velence Hills (1:100 000). Regional map series of Hungary, Geologica Institute of Hungary, Budapest

Haas J (1999) Genesis of Late Cretaceous toe-of-slope breccias in the Bakony Mts., Hungary. Sedimentary Geol 128:51–66

Haas J (2002) Origin and evolution of Late Triassic backplatform and intraplatform basins in the Transdanubian Range, Hungary. Geologica Carpathica 53(3):159–178

Haas J (2004) Characteristics of peritidal facies and evidences for subaerial exposures in Dachstein-type cyclic platform carbonates in the Transdanubian Range, Hungary. Facies 5:263–286

Haas J, Budai T (1995) Upper Permian-Triassic facies zones in the transdanubian range. Rivista Italiana Paleontologia Stratigrafia 101 (3):249–266

Haas J, Budai T (1999) Triassic sequence stratigraphy of the Transdanubian Range, Hungary. Geologica Carpathica 50(6):459–475

Haas J, Péró Cs (2004) Mesozoic evolution of the Tisza mega-unit. Int J Earth Sci 93:297–313

Haas J, Juhász Gy, Sztanó O (1998) Merre tart a szedimentológia? Az üledékképződési folyamatok dinamikája (Quo Vadis sedimentology? Dinamics of sedimentary processes). Földtani Közlöny 128 (1):71–91 (in Hungarian with English abstract)

Haas J, Budai T, Hips K, Konrád Gy, Török Á (2002) Magyarországi triász fáciesterületek szekvencia-rétegtani elemzése (Sequence stratigraphy of Triassic faciès areas). Földtani Közlöny 132(1):17–43 (in Hungarian with English abstract)

Haas J, Budai T, Csontos L, Fodor L, Konrád Gy (2010) Pre-Cenozoic geological map of Hungary, 1:500 000. Geological Institute of Hungary, Budapest

Haas J, Pelikán P, Görög Á, Józsa S, Ozsvárt P (2013) Stratigraphy, facies, and geodynamic setting of Jurassic formations in the Bükk Mountains, North Hungary; its relation with the other areas of the Neotethyan realm. Geol Mag 150:18–49

Haas J, Hámor G, Jámbor Á, Kovács S, Nagymarosy A, Szederkényi T (ed) (2012) Geology of Hungary. Regional Geology Reviews, Springer

Haas J, Budai T (eds), Csontos L, Fodor L, Konrád GY, Koroknai B (2014) Geology of the pre-Cenozoic basement of Hungary. Explanatory notes of the "Pre-Cenozoic geological map of Hungary" (1: 500 000). Geological Institute of Hungary, Budapest

Haq BU, Hardenbol J, Vail PR (1988) Mesozoic and Cenozoic chronostratigraphy and cycles of sea level changes. In: Wilgus CK (ed) Sea-level changes—an integrated approach. Society of Economic Paleontologists and Mineralogists, Special Publications 42:71–108

Harangi Sz, Szabó Cs, Józsa S, Szoldán Zs, Árva-Sós E, Balla M, Kubovics I (1996) Mesozoic igneous suites in Hungary: Implications for genesis and tectonic setting in the northwestern part of Tethys. International Geology Review 38(4):336–360

Hips K (1996) Stratigraphic and facies evaluation of the lower triassic formations in the Aggtelek-Rudabánya Mountains, NE Hungary. Acta Geol Hung 39:369–411

Hohenegger J, Coric S, Wagreich M (2014) Timing of the middle miocene badenian stage of the central paratethys. Geol Carpath 65 (1):55–66

Horváth F (1993) Towards a mechanical model for the formation of the Pannonian basin. Tectonophysics 226(1–4):333–357

Jámbor Á (1976) Üledékes kéntelep a Zsámbéki-medence szarmata sorozatában (Sedimentary sulfur deposit in the Sarmatian sequence of the Zsámbék Basin, Transdanubia). Annual Report of the Geological Institute of Hungary of 1974, pp 301–306 (in Hungarian with English abstract)

Jámbor Á, Korpás L (1974) A nagygörbői szerkezetkutató fúrás (Strukturbohrung von Nagygörbő). Annual Report of the Geological Institute of Hungary of 1972, pp 161–166 (in Hungarian with German abstract)

Józsa S, Szakmány Gy, Máthé Z, Barabás A (2009) A Mecsek és környéke miocén konglomerátum összletek felszíni elterjedése és a kavicsanyag összetétele (Occurence of Miocene conglomerates in the Mecsek Mts. and surroundings and composition of theirs gravel material). GeoLitera, Szeged, pp 195–217 (in Hungarian with English abstract)

Kázmér M, Dunkl I, Frisch W, Kuhlemann J, Ozsvárt P (2003) The Paleogene forearc basin of the Eastern Alps and Western Carpathians: subduction erosion and basin evolution. J Geol Soc Lond 160:413–428

Kázmér M (1982) A budai felsőeocén mészkő mikrofácies vizsgálata (Study of the Upper Eocene limestone microfacies in the Buda Hills). Ph.D. Thesis, ELTE Földtani Tanszék, Budapest (in Hungarian)

Kecskeméti T (1989) Bathymetric significance of recent larger foraminifera an example of ap-plication to the Eocene of Hungary. Fragmenta Mineralogica Et Paleontologica 14:73–82

Kecskeméti T (1980) A Bakony-hegységi Nummulites-fauna paleobiogegoráfiai áttekintése (A paleobiographical overview of the Nummulite-fauna of the Bakony Mountains). Földtani Közlöny 110:432–449 (in Hungarian)

Kecskeméti T (1993) Paleokommunitás vizsgálatok a Bakony eocén nagyforaminiferáin (Study of the Eocene large foraminifera paleocommunity in the Bakony Mountains). Őslénytani Viták 39 (in Hungarian with English abstract)

Kecskemétiné-Körmendy A (1966) A Dorog környéki eocén biofácies-vizsgálata (Study of the Eocene biofacies near Dorog). A Magyar Állami Földtani Intézet Évi Jelentése az 1964-es évről:329–337 (in Hungarian with English abstract)

Kercsmár Zs, Fodor L (2005) Syn-sedimentary deformations in the Eocene Tatabanya Basin, Central Hungary. Geolines 19:60–61

Kercsmár Zs, Pálfalvi S, Less Gy, Kordos L (2008) Eocene. In: Budai T, Fodor L (eds) Geology of the Vértes Hills: explanatory book to the geological map of the Vértes Hills, 1:50000. Geological Institute of Hungary, Budapest, pp 58–79

Kercsmár Zs, Budai T, Csillag G, Selmeczi I, Sztanó O (ed) (2015) Surface geology of Hungary. Explanatory notes for the Geological map of Hungary (1:500 000). Geological and Geophysical Institute of Hungary, Budapest. ISBN 978-963-671-302-7

Kercsmár Zs, Ozsvárt P, Less Gy (2019) Trópusi paradicsom Magyarországon (Tropic paradise in Hungary). In: Dulai A, Erdei B, Gasparik M, Matúš H, Kecskeméti T, Kercsmár Zs, Kocsis L, Kövecsi Sz, Less Gy, Ozsvárt P, Szabó M, Kamil Z (eds) (2019) Eocén élővilág a Kárpát-medencében, Üvegház – 22 millió éven át. Magyar Természettudományi Múzeum, Budapest, pp 34–65 (in Hungarian). ISBN 978–615–80345–2–42019

Kercsmár Zs, Budai T, Szurominé Korecz A, Selmeczi I, Musitz B, Lantos Z (2020) A zsámbéki Strázsa-hegy és környékének kainozoos képződményei (Cenozoic formations of the Strázsa Hill at Zsámbék and its surroundings). Bullet Hungarian Geol Soc 150 (1):129–150 (in Hungarian with English abstract)

Kercsmár Zs (1996) Syntectonic sedimentation in the marine Eocene of Tatabánya, Hungary. In: Kurzfassungen der Vorträge und Poster. Sediment '96 11. Sedimentologentreffen 9–15. Mai 1996, Institut für Geologie-Institut für Paläontologie Universitate, Wien, p 74

Kercsmár Zs (2003) Late Lutetian synsediment tectonic activity on the NE part of the Trans-danubian Range (Tatabánya Basin, Vértes Mts., Hungary). In: 22nd IAS Meeting of Sedimentology, abstract book, Opatija, p 94

Kercsmár Zs (2005a) Középső-eocén karbonátos üledékképződési környezetek és egymásra épülésük a Tatabányai-medence DK-i peremén. (Evolution of Middle Eocene carbonate sedimentation on the southeastern part of the Tatabánya Basin) A Magyar Állami Földtani Intézet Évi Jelentése 2004, pp 215–235 (in Hungarian with English abstract)

Kercsmár Zs (2005b) A Tatabányai Eocén Medence földtani felépítésének és fejlődéstörténetének újabb kutatási eredményei, üledékföldtani és tektono-szedimentológiai vizsgálatok alapján (Geology and basin evolution of the Eocene Tatabánya Basin based new research of sedimentology and tectono-sedimentology). Ph.D. thesis, Eötvös Loránd Tudományegyetem Természettudományi Kar, Őslénytani Tanszék, Budapest (in Hungary with English summary)

Kercsmár Zs (2010) Az É-i Gerecse eocén rétegsora a legújabb kutatások tükrében (New results of the research of Eocene sequences in the N Gerecse). In: EMT 12. Bányászati-Kohászati és Földtani Konferencia, Nagyenyed 2010 április 8–11. ISSN 1842–9440, pp 148–153 (in Hungarian)

Kercsmár Zs (2015a) Eocene. In: Kercsmár Zs (ed) Surface geology of Hungary. Explanatory notes for the Geological map of Hungary (1:500 000). Geological and Geophysical Institute of Hungary, Budapest, pp 28–30. ISBN 978-963-671-302-7

Kercsmár Zs (2015b) Eocene-oligocene. In: Kercsmár Zs (ed) Surface geology of Hungary. Explanatory notes for the Geological map of Hungary (1:500 000). Geological and Geophysical Institute of Hungary, Budapest, pp 30–31. ISBN 978–963–671–302–7

Kercsmár Zs (2015c) Oligocene. In: Kercsmár Zs (ed) Surface geology of Hungary. Explanatory notes for the Geological map of Hungary (1:500 000). Geological and Geophysical Institute of Hungary, Budapest, pp 31–33. ISBN 978–963–671–302–7

Kercsmár Zs (2018) Eocene. In: Budai T, Fodor L eds Geology of the Gerecse Mountains. Explanatory book to the geological map of the Gerecse Mountains, 1:50 000 Magyar Bányászati és Földtani Szolgálat kiadványa, Budapest, pp 57–106

Kókay J (1966) A Herend-Márkói barnakőszénterület földtani és őslénytani vizsgálata (Geologische und Palaeontologische Untersuchung des Braunkohlengebietes von Herend-Márkó, Bakony Gebirge, Ungarn). Geologica Hungarica Series Palaeontologica, Fasc 36:1–149 (in Hungarian with shortened German text)

Kókay J (1985) Central and eastern paratethyan interrelations in the light of late Badenian salinity conditions. Geologica Hungarica Series Palaeontologica, Fasc 48:7–95

Kókay J (1988) Tengeri kifejlődésű kárpáti rétegek előfordulása Alcsútdobozon (Marine Karpatian beds at Alcsútdoboz, Hungary). Annual Report of the Geological Institute of Hungary of 1986, pp 263–279 (in Hungarian with English abstract)

Kókay J (2013) Study of the Middle Miocene (Badenian and Sarmatian) formations in the Várpalota Neogene basin. Bullet Hungarian Geol Soc 143(2):145–156

Kókay J (1967) Stratigraphie des Oberhelvets ("Karpatien") von Várpalota (Ungarn). Palaeontographia Italica 63 (n. ser. 33):75–111

Kókay J (1973a) Die Entwicklung der Bántapusztaer Schichtengruppe, M2b–c(d). In: Seneš J (ed) Chronostratigraphie und Neostratotypen. Miozän der zentralen Paratethys Bd. III. M2 Ottnangien. Verlag der Slowakischen Akademie der Wissenschaften, Bratislava, pp 85–88

Kókay J (1973b) Faziostratotypen der Bántapusztaer Schichtengruppe. In: Seneš J (ed) Chronostratigraphie und Neostratotypen. Miozän der zentralen Paratethys, Bd. III. M2 Ottnangien. Verlag der Slowakischen Akademie der Wissenschaften, Bratislava, pp 227–243

Kókay J (1996) A várpalotai neogén medence tektonikai összefoglalója (Tectonic description of the Neogene Várpalota Basin, Bakony Mts, W Hungary). Bullet Hungarian Geol Soc 126(4):417–445 (in Hungarian with English abstract)

Konrád Gy, Sebe K, Halász A, Halmai Á (2010) A Délkelet-Dunántúl földtani fejlődéstörténete – recens analógiák (The major phases in the geological evolution of South-East Transdanubia and their recent analogues). Földrajzi Közlemények 134(3):251–265 (in Hungarian with English abstract)

Kopek G, Kecskeméti T, Dudich E (1965) A Dunántúli-középhegység eocénjének rétegtani kérdései (Questions of the Eocene stratigraphy of Transdanubian Range). A Magyar Állami Földtani Intézet Évi Jelentése 1964:249–263 (in Hungarian with English abstract)

Kováč M, Halásová E, Hudácková N, Holcová K, Hyžný M, Jamrich M, Ruman A (2018) Towards better correlation of the Central Paratethys regional time scale with the standard geological time scale of the Miocene Epoch. Geologica Carpathica 69(3):283–300

Kovács S, Less Gy, Piros O, Réti Zs, Róth L (1989) Triassic formations of the Aggtelek–Rudabánya Mountains (NE Hungary). Acta Geol Hung 32(1–2):31–63

Kovács S, Less Gy, Piros O, Réti Zs, Róth L (1993) Aggtelek–Rudabányai egység (Aggtelek–Rudabánya Unit). In: Haas J (ed) Magyarország litosztratigráfiai alapegységei. Triász. Magy Áll Földt Int, Magy Olaj- és Gáz Rt, Budapest, pp 155–221 (in Hungarian)

Kovács S, Rálisch-Felgenhauer E (2005) Middle Anisian (Pelsonian) platform conodonts from the Triassic of the Mecsek Mts (South Hungary)—their taxonomy and stratigraphic significance. Acta Geol Hung 48(1):69–105

Kovács S (1986) Conodonta-biosztratigráfiai és mikrofácies vizsgálatok a Rudabányai-hegység ÉK-i részén (Conodonta-biostratigraphy and microfacies study of the NE Rudabánya Hills). Magy Áll Földt Int Évi Jel 1984:193–244 (in Hungarian with English abstract)

Kovács S (1989) Geology of North Hungary: Palaeozoic and Mesozoic terranes. In: Kecskeméti T (ed) 21st Eur Micropal Coll Guidebook. Hung Geol Soc, Budapest, pp 15–36

Kovács S (1992) Stratigraphy of the Szendrő–Uppony Paleozoic (Northeastern Hungary). In: Vozár J (ed) Special volume to the problems of the Paleozoic geodynamic domains. Western Carpathians, Eastern Alps, Dinarides (IGCP Project 276). Geol Inst D Štúra Bratislava, pp 93–108

Kövér Sz, Fodor L, Judik K, Németh T, Balogh K, Kovács S (2009a) Deformation history and nappe stacking in Rudabánya Hills (Inner Western Carpathians) unravelled by structural geological, metamorphic petrological and geochronological studies of Jurassic sediments. Geodin Acta 22(1–3):3–29

Kövér SZ, Haas J, Ozsvárt P, Görög Á, Götz AE, Józsa S (2009b) Lithofacies and age data of Jurassic foreslope and basin sediments of Rudabánya Hills (NE Hungary) and their tectonic interpretation. Geol Carp 60(5):351–379

Kövér SZ, Fodor L, Kovács Z, Klötzli U, Haas J, Zajzon N, Szabó CS (2018) Late Triassic acidic volcanic clasts in different Neotethyan sedimentary mélanges: paleogeographic and geodynamic implications. Int J Earth Sci 107(8):2975–2998

Kozur H, Mock R (1973) Zum Alter und zur tektonischen Stellung der Meliata-Serie des Slowakischen Karstes. Geol Zborn Geol Carp 24:225–264

Less Gy (1987) Paleontology and stratigraphy of the European Orthophragminae. Geologica Hungarica Series Palaeontologica 50:1–373

Less Gy (1991) Upper Oligocene larger Foraminifers of the Bükk Mountains (NE Hungary). Ann Rep Geol Inst Hung 1989:411–465

Less Gy (2000) Polyphase evolution of the structure of the Aggtelek-Rudabánya Mountains (NE Hungary), the southernmost element of the Inner Western Carpathians—a review. Slov Geol Mag 6(2–3):260–268

Less Gy, Grill J, Szentpétery I, Róth L, Gyuricza Gy (1988) Az Aggtelek–Rudabányai-hegység fedetlen földtani térképe (Geological map of the Aggtelek–Rudabánya Mts.) 1:25 000. Magy Áll Földt Int, Budapest

Less Gy, Gulácsi Z, Kovács S, Pelikán P, Pentelényi L, Rezessy A, Sásdi L (2002) Geological map of the Bükk Mountains (1:50 000). Geol Inst Hung, Budapest

Less Gy, Mello J, Elečko M, Kovács S, Pelikán P, Pentelényi L, Peregi Zs, Pristaš J, Radócz Gy, Szentpétery I, Vass D, Vozár J, Vozárová A (eds) (2004) Geological map of the Gemer–Bükk area 1:100 000. Geol Inst Hung, Budapest

Less Gy, Báldi-Beke M, Pálfalvi S, Földessy J, Kertész B (2008) New data on the age of the Recsk volcanics and of the adjacent sedimentary rocks. Publications of the University of Miskolc, Series A, Mining 73:57–84

Less Gy, Frijia G, Filipescu S, Holcová K, Mandic O, Sztanó O (2015a) New Sr-isotope stratigraphy (SIS) age-data from the Central Paratethys. In: Gülli E, Piller WE (eds) Abstracts 2nd international congress on stratigraphy strati 2015a 19–23 July 2015, Graz, Austria. ISSN 1608-8166, Berichte des Institutes für Erdwissenschaften Karl-Franzens-Universität Graz 21, p 223

Less Gy, Parente M, Frijia G, Cahuzac B (2015b) New Sr-isotope stratigraphy (SIS) age-data from some European Oligo-Miocene larger benthic foraminiferal sites. In: Gülli E, Piller WE (eds) Abstracts 2nd International Congress on Stratigraphy STRATI 2015b, 19–23 July 2015, Graz, Austria. ISSN 1608-8166, Berichte des Institutes für Erdwissenschaften Karl-Franzens-Universität Graz 21, p 224

Less Gy, Frijia G (2016) Új koradatok a Központi-Paratethysből Sr-izotópvizsgálatok alapján (New age data to the Central-Paratethys based Sr-isotope study). In: A Magyarhoni Földtani Társulat Földtudományi Vándorgyűlése és Kiállítása, Sárospatak, 2016. augusztus 24–27. Magyarhoni Földtani Társulat, Budapest, ISBN: 978–963–8221–62–9, p 57 (in Hungarian)

Less Gy, Sztanó O, Kercsmár Zs (2019) Egyszerűsített paleogén litosztratigráfia a legújabb rétegtani eredmények és értelmezések figyelembevételével (Simplified Paleogene lithostratigraphy based on new geological data). In: Bosnakoff M, Főzy I (eds) 22. Magyar Őslénytani Vándorgyűlés, 2019. május 30–június 1. Döbrönte. Program, Előadáskivonatok, Kirándulásvezető. ISBN 978–963–8221–75–9, Magyarhoni Földtani Társulat, Budapest, pp 23–24 (in Hungarian)

Magyar I (2010) A Pannon-medence ősföldrajza és környezeti viszonyai a késő-miocénben (Paleogeographical and paleoenvironmental conditions of the Pannonian Basin in the Late Miocene). GeoLitera, Szeged (in Hungarian)

Magyari Á (1994a) Késő-eocén hidraulikus breccsásodási jelenségek a Budai-hegység déli részén (Late Eocene hydraulic rebrecciation in the Southern Buda Mountains, Hungary). Földtani Közlöny 124 (1):89–107 (in Hungarian with English abstract)

Magyari Á (1994b) Késő-eocén transzpresszió a Budaörsi-hegyekben (Late Eocene transpression in the Budaörs Hills). In: Fodor L, Magyari Á, Fogarasi A, Palotás K Tercier szerkezetfejlődés és késő paleogén üledékképződés a Budai-hegységben. A Budai-vonal új értelmezése. Földtani Közlöny 124(2):155–173 (in Hungarian with English abstract)

Magyari Á (1996) Eocén szinszediment tektonikai jelenségek és üledékképződésre gyakorolt hatásai a Budai-hegységben (Eocene synsedimentary occurences and these effect of sedimentation in the Buda Mts.). PhD thesis, Eötvös Lóránd Tudomány Egyetem, Természettudományi Kar, Általános és Történeti Földtani Tanszék, Budapest (in Hungarian with English summary)

Majoros Gy (1983) Lithostratigraphy of the permian formations of the transdanubian central mountains. Acta Geol Hung 26(1–2):7–20

Márton E, Fodor L (1995) Combination of paleomagnetic and stress data—a case study from North Hungary. Tectonophysics 242:99–114

Márton E, Márton P, Less Gy (1989) Paleomagnetic evidence of tectonic rotations in the southern margin of the Inner West Carpathians. Phys Earth Planet Int 52:256–266

Mello J, Elečko M, Pristaš J, Reichwalder P, Snopko L, Vass D, Vozárová A, Gaál L, Hanzel V, Hók J, Kováč P, Slavkay M, Steiner A (ed) (1997) Vysvetlivky ku geologickej mape Slovenského Krasu 1:50 000. Vyd. D. Štúra, Bratislava, p 255 (in Slovakian)

Mindszenty A, Csoma A, Török Á, Hips K (2001) Flexura jellegű előtéri deformációhoz köthető karsztbauxitszintek a Dunántúli-középhegységben (Rudistid limestones, bauxites, paleokarst and geodynamics. The case of the Cretaceous of the Transdanubian Central Range). Földtani Közlöny 131(1–2):107–152 (in Hungarian with English abstract)

Mock R, Sýkora M, Aubrecht R, Ožvoldová L, Kronome B, Reichwalder P, Jablonský R (1998) Petrology and stratigraphy of the Meliaticum near the Meliata and Jaklovce Villages, Slovakia. Slovak Geol Mag 4(4):223–260

Monostori M (1996) Ostracods and charophytes from the Triassic Kantavár Formation, Mecsek Mts., Hungary. Acta Geol Hung 39 (3):311–317

Nagymarosy A, Báldi-Beke M (1988) The position of the Paleogene formations of Hungary in the standard nannoplankton zonation. Annales Universitatis Scientiarium Budapestinensis de Rolando Eötvös, Nominatae, Sectio Geologica 28:3–25

Nagymarosy A, Hámor G (2012) Genesis and evolution of the pannonian basin. In: Haas J (ed) Geology of Hungary. Springer, Berlin Heidelberg, pp 149–200

Pelikán P, Budai T, Less Gy, Kovács S, Pentelényi L, Sásdi L (eds) (2005) Geology of the Bükk Mountains. Geol Inst Hung, Budapest

Piller WE, Harzhauser M, Mandic O (2007) Miocene central paratethys stratigraphy. Stratigraphy 4(2–3):151–168

Ravasz Cs (1978) A budajenői kéntartalmú miocén evaporitok ásványkőzettani vizsgálata. (Mineralogical and petrographic study of Miocene sulphur-bearing evaporites at Budajenő, NE Transdanubia). Annual Report of the Geological Institute of Hungary of 1976, pp 177–188 (in Hungarian with English abstract)

Réti Zs (1985) Triassic ophiolite fragments in an evaporitic melange, Northern Hungary. Ofioliti 10:411–422

Sebe K, Selmeczi I, Szuromi-Korecz A, Hably L, Kovács Á, Benkó Zs (2019) Miocene syn-rift lacustrine sediments in the Mecsek Mts. (SW Hungary). - Swiss Journal of Geosciences 112:83–100

Selmeczi I (2014) A Somlóvásárhelyi Formáció (The Somlóvásárhely Formation). Annual Report of the Geological and Geophysical Institute of Hungary 2012–2013:159–166 (in Hungarian with English abstract)

Selmeczi I, Kercsmár Zs (2015) Oligocene-Miocene. In: Kercsmár Zs (ed) Surface geology of Hungary. Explanatory notes for the Geological map of Hungary (1:500 000). Geological and Geophysical Institute of Hungary, Budapest, pp 33–34. ISBN 978–963–671–302–7

Selmeczi I, Bohn-Havas M, Szegő É (2004) Prepannonian Miocene sequences of the SW edge of the Transdanubian Central Range.

Litho- and biostratigraphy. Acta Palaeontologica Romaniae 4:463–466

Selmeczi I, Szurominé Korecz A, Palotás K, Szabadosné Sallay E (2017) Tengerpart a magyar tenger partján (szarmata lagúna Zánkán) (Seashore on the shore of the "Hungarian Sea (Balaton)" (Sarmatian lagoon in Zánka). In: Attila V, Mariann B (eds) 20. Magyar Őslénytani Vándorgyűlés. május 25–27, Tata–Tardos, Program, előadáskivonatok, kirándulásvezető, pp 39–40 (in Hungarian). ISBN 978–963–8221–64–3

Selmeczi I (2008a) Miocene: lower-middle miocene. In: Budai T, Fodor L (eds) Geology of the Vértes Hills: explanatory book to the geological map of the Vértes Hills, 1:50000, Geological Institute of Hungary, Budapest, p 88, 261

Selmeczi I (2008b) Miocene: middle miocene. In: Budai T, Fodor L (eds) Geology of the vértes hills: explanatory book to the geological map of the Vértes Hills, 1:50000, Geological Institute of Hungary, Budapest, pp 88–93, 261–264

Selmeczi I (2015a) Lower miocene. In: Kercsmár ZS (ed) Surface geology of Hungary. Explanatory notes to the Geological map of Hungary (1: 500 000). Geological and Geophysical Institute of Hungary, pp 34–37

Selmeczi I (2015b) Middle miocene. In: Kercsmár ZS (ed) Surface geology of Hungary. Explanatory notes to the Geological map of Hungary (1: 500 000). Geological and Geophysical Institute of Hungary, pp 38–44

Selmeczi I (2018) Miocene: lower-middle miocene. In: Budai T (ed) Geology of the Gerecse Mountains: explanatory book to the geological map of the Gerecse Mountains, 1:50000, Magyar Bányászati és Földtani Hivatal, Budapest, pp 115–116, 330–331

Szederkényi T (1977) Geological evolution of South Transdanubia (Hungary) in Paleozoic time. Acta Miner Petr, Szeged 23:3–14

Szederkényi T (1996) Metamorphic formation and their correlation in the Hungarian part of Tisia Megaunit (Tisia Composite Terrane). Acta Mineralogica Petrographica, Szeged 37:143–160

Szemerédi M, Varga A, Lukács R, Pál-Molnár E (2017) A Gyűrűfűi Riolit Formáció kőzettani vizsgálatának eredményei a Villányi-hegység északi előterében (Petrography of the Gyűrűfű Rhyolite Formation, northern foreland of the Villány Mts, Hungary). Földtani Közlöny 147(4):357–382 (in Hungarian with English abstract)

Szentpétery I, Less Gy, Kovács S, Grill J, Róth L, Gyuricza Gy, Sásdi L, Piros O, Réti Zs, Elsholtz L, Árkai P, Nagy E, Borka Zs, Harnos J, Zelenka T (eds) (2006) Az Aggtelek–Rudabányai-hegység földtana (Geology of the Aggtelek–Rudabánya Mts.). Magyarázó az Aggtelek–Rudabányai-hegység 1988-ban megjelent 1:25 000 méretarányú fedetlen földtani térképéhez (Explanatory book of the geological map (1:25 000) of Aggtelek–Rudabánya Mts.). Magy Áll Földt Int, Budapest (in Hungarian)

Szentpétery I (1997) Sinistral lateral displacement in the Aggtelek–Rudabánya Mts. (North Hungary) based on the facies distribution of oligocene and lower miocene formations. Acta Geol Hung 40 (3):265–272

Sztanó O, Magyari Á, Nagymarosy A (1998) Az Esztergomi-medence oligocén képződményeinek integrált sztratigráfiai vizsgálata: II. Oligocén szekvenciák és értelmezésük (High-resolution stratigraphy in the Esztergom Basin, northeastern Transdanubia, Hungary: II. Oligocène sequences and their interpretation). Földtani Közlöny 128(2–3):455–486 (in Hungarian with English abstract)

Sztanó O, Fodor L (1997) Lejtőüledékek a paleogén medence peremén: a felső-eocén Piszkei Márga (Nyergesújfalu, Sánc-hegy) ülepedési és szerkezeti viszonyai (Bathyal slope deposits in the Paleogene Basin: A case study of the Upper Eocene Piszke Marl (Nyergesújfalu, Sánc Hill, Hungary). Földtani Közlöny 127(3):267–290 (in Hungarian with English abstract)

Sztanó O (2015) Rivers, deltas, turbidite systems: an overview of sedimentation in Lake Pannon. In: Bartha I, Kriván Á, Magyar I, Sebe K (eds) Neogene of the Paratethyan Region. 6th Workshop on the Neogene of Central and South-Eastern Europe. An RCMNS Interim Colloquium, 31 May–3 June, Orfű, Hungary. Programme, Abstracts, Field Trip Guidebook. – Hungarian Geological Society Budapest:89-90

Tari G, Báldi T, Báldi-Beke M (1993) Paleogene retroarc flexural basin beneath the Neogene Pannonian Basin: a geodynamic model. Tectonophysics 226:433–455

Telegdy Roth K (1927) Spuren einer Infraoligozänen Denudation am nordwestichen Rande des Transdanubischen Mittelgebirge. Földtani Közlöny 57:117-128

Török Á (1993) Storm influenced sedimentation in the Hungarian Muschelkalk. In: Hagdorn A, Seilacher A (eds) Muschelkalk. Schöntaler Symp. 1991. Sonderb. Ges. Nat. Württemberg 2, pp 133–142

Török Á (1998) Controls on development of Mid-Triassic ramps: examples from southern Hungary. In: Wright VP, Burchette TP (eds) Carbonate Ramps. Geol. Scol, London, Spec. Publ. 149, pp 339–367

Tóth E, Görög Á, Lécuyer C, Moissette P, Balter V, Monostori M (2010) Palaeoenvironmental reconstruction of the Sarmatian (Middle Miocene) Central Paratethys based on palaeontological and geochemical analyses of foraminifera, ostracods, gastropods and rodents. Geol Mag 147(2):299–314

Tóth K, Varga G (2014) A diszeli bauxit. (The bauxite at Diszel). Bullet Hungarian Geol Soc 144(4):315–328 (in Hungarian with English abstract)

Velledits F (2000) A Berva-völgytől a Hór-völgyig terjedő terület fejlődéstörténete a középső–felső-triászban (Evolution of the area from the Berva Valley to the Hór Valley in the Middle-Upper Triassic). Földt Közl 130(1):47–93 (in Hungarian with English abstract)

Velledits F (2004) Anisian terrestrial sediments in the Bükk Mts. (NE Hungary) and their role in the Triassic rifting of the Vardar-Meliata branch of the Neo-Tethys ocean. Riv It Paleont Strat 110:659–679

Velledits F, Forián-Szabó M, Bérczi-Makk A, Piros O, Józsa S (2003) Stratigraphy and origin of the Kisfennsík nappe (Bükk Mts.; NE Hungary)—is the Silica Unit really present in the Bükk Mts.? Geol Carp 54(3):189–198

Velledits F, Péró Cs, Blau J, Senowbari-Daryan B, Kovács S, Piros O, Pocsai T, Szügyi-Simon H, Dumitrica P, Pálfy J (2011) The oldest Triassic platform margin reef from the Alpine-Carpathian region (Aggtelek, NE Hungary): platform evolution, reefal biota and biostratigraphic framework. Riv Ital Paleont Strat 117(2):221–268

Vörös A, Galácz A (1998) Jurassic palaeogeography of the transdanubian central range (Hungary). Riv Ital Paleontol Stratigr 104(1):69–83

Wéber B (1990) Ladin és felsőtriász rétegek a Ny-Mecsek északi előterében. (Ladinian and Upper Triassic strata in the northern foreland of theWestern Mecsek Mountains). Földtani Közlöny 120:153–180 (in Hungarian with English abstract)

Paleokarst in Hungary

Andrea Mindszenty and Krisztina Sebe

Abstract

Carbonate formations of Mesozoic-Tertiary to Quaternary age of Hungary abound in both early and late paleokarst features often superimposed on each other. *Early (diagenetic) paleokarst* is—as a rule—associated with shallow-water (either marine or freshwater) limestones and/or dolomites. Examples are known from the Transdanubian Range (Transdanubian Mountains), the Northern Highlands (Bükk, Aggtelek, Uppony) and also from South-Hungary (Villány, Mecsek). More important are, however, those *late ("Alpine-type") paleokarst* features which occur at major tectonically controlled subaerial unconformities, and were resulted by the events of the Eoalpine (Cretaceous to Paleogene) and Neoalpine (Miocene to Recent) structural evolution of the Carpatho-Pannonian Region. To separate the effects of the numerous superimposed early and late karst events is often difficult and in addition to thoughtful field-observations requires also sophisticated laboratory methods and the appropriate understanding of the geological evolution of the area concerned. Selected examples of some early (diagenetic) and late "Alpine"-type paleokarsts are presented and the relevant references discussed.

Keywords

Paleokarst • Diagenetic • Unconformity-related • Bauxite-filled • History of research • Selected examples • Transdanubian Range • Mecsek • Villány

A. Mindszenty (✉)
Department of Geology, Eötvös Loránd University, Pázmány P. stny. 1/C, Budapest, 1117, Hungary

K. Sebe
Department of Geology and Meteorology, University of Pécs, Ifjúság str. 6, Pécs, 7624, Hungary
e-mail: sebe@gamma.ttk.pte.hu

1 Introduction

In this chapter, following the definition of James and Choquette (1988), the term *karst* will include all "*macroscopic, microscopic, surface- and subterranean features produced during the chemical dissolution and associated modification of a carbonate sequence on interaction with meteoric (or mixed) waters. Included in this definition are also all subsurface precipitates, and mechanical infills*". Evaporite karst and dissolution-related modification of any other rock types will not be the subject of this discussion. Dissolution processes associated with thermal waters of deep-seated origin (=hypogenic karst) will not be treated here in detail, either (they are the subjects of chapters "Hydrogeology of the Karst Regions in Hungary and The Caves of the Northeastern Part of the Transdanubian Mountains in this volume).

Paleokarst will be defined as *ancient karst* (be it relict or buried/filled) having been produced under conditions different from those prevailing recently (Wright 1982).

In addition to the geomorphological characteristics of classical (Cvijić-type) karst (and what remains of it on burial and fossilization), also *micromorphological features* and *geochemical changes* resulted by intimate interaction of carbonate rocks with meteoric water will be emphasized. In full accordance with Esteban and Klappa (1983), Esteban and Wilson (1993) and many other carbonate sedimentologists, karst is going to be considered here also as "*a diagenetic facies*" of shallow-water carbonates.

2 Paleokarst as a Product of Diagenesis

In shallow-water carbonate environments, the most relevant effects of subaerial exposure—be it *transient, ephemeral* or *long-lasting*—are (a) karstification (i.e. dissolution/cementation of/by $CaCO_3$) and/or (b) soil formation. Products of *transient exposure* are those formed on recurrent, areally more-or-less extensive emersions of the tidal realm. They are

M. Veress and Sz. Leél-Őssy (eds.), *Cave and Karst Systems of Hungary*,
Cave and Karst Systems of the World, https://doi.org/10.1007/978-3-030-92960-2_5

textures and sedimentary structures, well known from textbooks on carbonate sedimentology. Their coalescence in the vertical profile may end up in *ephemeral exposure* typical of shallowing upward eustatic cyclothems. This happens when repeated transient events (such as karstic dissolution and the accumulation of incipient paleosols) take place as a result of high frequency sea-level falls of tens of thousand-years duration. Such coalescence may give rise to textures very similar to those brought about by *long-lasting*, eventually *tectonically controlled* exposure of hundreds of thousands or even million years of duration. This is why sometimes it is extremely difficult to separate paleokarst features formed on "early" karstification from those resulted by "late" processes.

2.1 "Diagenetic" Paleokarst Features Formed as Results of Ephemeral Exposure

Characteristic morphological and micromorphological features formed on *ephemeral exposure*, i.e. *early diagenetic karstification* (according to Esteban and Klappa 1983; Wright 1982; Tucker and Wright 1992; Read and Horbury 1993; Carannante et al. 1987, 1994a,b; Mindszenty and Deák 1999; Mindszenty 2019 and many others) are as follows:

- partly vadose partly phreatic-lens-related, early pervasive microkarst (affecting not only intertidal/supratidal but *also subtidal* lithologies);
- early microcavities/pores, of the micrometer to millimeter size (often of biomoldic origin);
- deposition of internal sediment (vadose silt, infiltrated clay) and vadose to phreatic meteoric cements in the cavities;
- calcrete and/or normally thin, moderately or weakly developed, argillaceous soils blanketing the exposure surface and/or infiltrated into the above-described micropores;
- late underground (phreatic-lens-related or epi-phreatic) karstification + complex karst-fills recording often a long and complex paleohydrological story (including the documents of sea-level rise following the subaerial episode).

Thorough interacton with meteoric and/or mixed waters may leave behind a *geochemical signal* preserved and ready for identification sometimes even after a long and complex burial history (Allen and Matthews 1982; Joachimski 1994; Csoma et al. 2004 and many others). Microsampling and stable isotope analysis of cavitiy-filling $CaCO_3$-cements or the bulk of pervasively cemented portions of carbonate rocks below unconformities or disconformity surfaces are as a rule enriched in ^{12}C. Geochemical studies may help to decipher the details of the paleokarst environment (position of the freshwater lens, boundaries of the vadose-, the phreatic-, and the mixing zone, etc.) even in the case of short-lived exposure. For details, the reader is referred to Allen and Matthews (1982) as cited above and their Fig. 11. Cathodoluminescence of the cavity-filling or pore-filling cement phases may help to understand paleohydrological changes having taken place before, during and after the karst episode (see Csoma 2003).

2.2 "Diagenetic" Paleokarst Features Formed on Long-Lasting Exposure

Surfaces of *long-lasting subaerial exposure* are theoretically underlain by "late" subterranean karst features of the *vadose*, the *epi-phreatic* and the *phreatic* zone. Whether they are all preserved in the stratigraphic record depends on their burial diagenetic and postburial ("telogenetic") history. In addition to burial diagenetic modifications (or complete destruction), they may become overprinted by several subsequent episodes of karstification (see later).

As to these "late" karst features, as a rule, they postdate (sometimes partially erase) earlier generations of well-developed vadose-to-phreatic dissolution-cavities filled by internal sediments, vadose silts and calcite cements. In some cases, they are crosscut by an even later generation of fracture-controlled cavities. The latter are usually filled by complex assemblages of collapse breccia, soil fragments, internal sediments, vadose silts and a late generation of calcite cements, the chemical composition (reflected partly by their cathodoluminescence) and crystal morphology of which suggest deposition from fluctuating oxidizing/reducing meteoric/mixed/marine porewaters.

3 Paleokarst Features as the Fossilized Remnants of Ancient "Exo"-(Surface) Karst Morphology

Surface karst features (sensu Cvijić 1918) may be preserved in the stratigraphic record when (a) filled by soils, soil-derived sediments or mineral precipitates formed and/or accumulated during or immediately after the subaerial period responsible for the karstification or when (b) "sudden" burial takes place at a time when subaerial karst denudation/planation (Grund 1914) has not yet reached its end and thus a recognizable surface relief may be reflected by the basal lithofacies (or the diachronism) of the immediate cover (e.g. Komlóssy 1970; Purdy 1974; Tóth 1980; Rákosi and Tóth 1980; Knauer and Gellai 1978). Attention! Interstratal dissolution and/or cementation may be mistaken for fossilized remnants of ancient exo-karst. Thoughtful considerations based on independent evidence regarding the geological history of the wider surroundings are therefore inevitable for the appropriate distinction.

4 Paleokarst in Hungary—The Geological Framework

The territory of Hungary is 92 000 km^2. It occupies the greater part of the Pannonian Basin. It is a Neogene extensional structure, the basement of which is built up mainly by Paleozoic to Mesozoic and Early Tertiary formations. These are similar to formations known partly from the Eastern Alps, Southern Alps and Dinarids, partly from stable Europe. Within the Pannonian Basin proper, they are covered by thousands of meters thick Neogene sediments. Mesozoic-Tertiary carbonates crop out from underneath the mostly siliciclastic sedimentary cover only in the Transdanubian Range; in the Bükk Mountains, in the Aggtelek Karst and in the Uppony/Rudabánya Hills in the North; and in the Mecsek Mountains; and in the Villány Hills in the South (Fig. 1). The total area of exposed carbonate rocks, potential subjects of karstification, is only cca 1,5% of the territory of Hungary. In addition, underneath about 10–25% of the present land surface, there are carbonate rocks also in the basement, known from boreholes only (Bárdossy and Kordos 1989).

All the above-mentioned carbonates (now either exposed or deeply buried) have been subject to subaerial exposure *cum* karstification several times during the long and complex history of the structural geological evolution of the Pannonian Basin and its basement (see eg. Schmid et al. 2008; Balázs et al. 2016). Shallow marine carbonates, as a rule, show several early diagenetic features of dissolution and precipitation which could be qualified as karstification on *ephemeral* exposure mentioned above. Subsequently, these features became overprinted by burial diagenesis. Then later on, as a result of tectonically controlled uplift many of them were exposed again as part of major regional subaerial unconformities. This way, early (ephemeral) karst features became either destroyed or at least partially or even substantially modified by karstification related to the subsequent *long-lasting* exposure (see above). Often, such—already polygenetic—karst assemblages, partially or completely filled by soil-derived sediments, became covered and buried again and again. To recognize and properly separate the traces of all the individual karst events of such polygenetic assemblages is a difficult, sometimes even an impossible task, as already mentioned above.

Instead of giving a complete inventory of all paleokarst events and their products recognized in Hungary, rather some selected examples will be presented. Detailed regional reviews are available from outstanding scholars like Bárdossy and Kordos in the comprehensive volume on Paleokarst by Bosák (1989) and like L. Korpás who published his monograph of *Hungarian Paleokarst* in 1998 (see later). These reviews include also the grouping of paleokarst phases according to their stratigraphic position and according to their geographic location. Therefore, this time karst features as related to the most important regional unconformities will be discussed and the lateral association of vadose and phreatic karst facies will be illustrated.

Examples are taken from the Transdanubian Range, the largest continuous karst + paleokarst area of Hungary, and from the previously understudied Mecsek and Villány Hills (SW part of the country) from where some interesting new results became available recently.

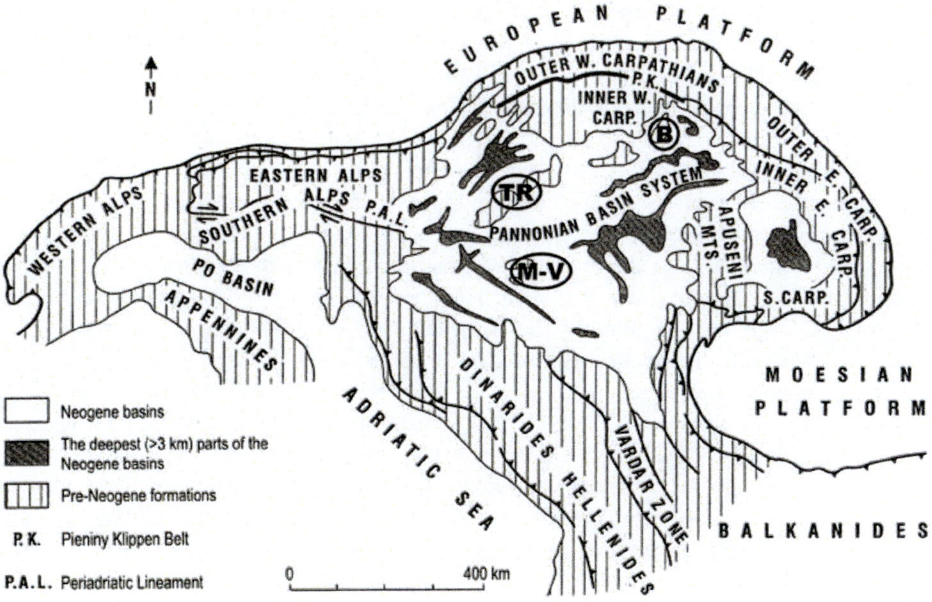

Fig. 1 Outcrops of Mesozoic-Tertiary basement rocks of the Pannonian Basin (base map by Royden & Horváth 1988, modified). **TR:** Transdanubian Range (Pelso Unit); **M-V:** Mecsek-Villány (Tisza Unit, South Hungary); **B:** Bükk (Mid-Hungarian Zone)

5 History of Paleokarst Research in Hungary

In Hungary, the study of buried (i.e. "paleo") karst surfaces began with the discovery of bauxites and the related ancient karst in the early twenties of the last century (e.g. Telegdi Roth 1923, 1927; Vadász 1946; de Weisse 1948 and others). Though the causal relationship between bauxites (important mineral resources) and karstification processes was soon recognized, this has not been a sufficiently important incentive for paleokarst research per se. Buried (or fossil) karst for a long time was considered simply as the equivalent to bauxite-filled old "exo"-karst surface formed under humid tropical conditions (as shown by the overlying bauxite-blanket—the target of exploitation). Fossil, non-bauxitic karst features have not received any particular attention.

Isolated attempts to unfold the sequence of events related to subaerial exposure of carbonate rocks by studying their buried remnants even if they were of non-bauxitic nature began only in the mid-fifties in Hungary. P. Kriván, in a manuscript report on exploration for fire-clay deposits, described buried deep-sinkhole morphologies from the Keszthely Hills (Cserszegtomaj) of the Transdanubian Range (Kriván 1956). In the same year, P. Z. Szabó's paper dedicated to climatic aspects of karst morphology and its *evolution in time* was the first sign of the interest of Hungarian geographers in ancient karst (Szabó 1956). His observations on and interpretation of the bauxite-filled paleokarst of Hungary were correct and provided a sound basis for Hungarian bauxite-geologists to elaborate a terminology of the bauxite deposits properly fit to the geomorphology of the host carbonates and used in the practice of prospecting for karst bauxites even today (doline-filling, sinkhole-filling, canyon-filling and "stratiform" polje-filling deposits). In 1958, Kriván, analyzing the unconformable contact between Triassic and Eocene formations in the Buda Hills, drew the attention to the modifying effects of burial on previously formed karst features (Kriván 1959). E. Végh-Neubrandt (1960) in her monograph on the Triassic of the Gerecse Hills gave a detailed description of textural (also dissolutional) features observed in Late Triassic limestones and described also the repeated alternation of the various lithological types and the intercalated green-clay horizons. Though she speculatively put forward even the idea of astronomically controlled sea-level changes (of Milanković-frequencies) being responsible for the cyclicity observed in the shallow-marine carbonates, 4 years before the advent of Fischer's cyclothemes (Fischer 1964) she was not yet able to realize that what she described was the result of a series of repeated subaerial exposure surfaces (disconformities) with "early" *microkarst features* and paleosols.

The term *paleokarst* in the modern sense of the word turned up in Hungary, and research on paleokarst gained ground more than 20 years later in the mid-eighties. It was a time when in the frames of a research program on Mesozoic-Tertiary Geodynamics of the Periadriatic Region a close collaboration between geologists and geophysicists of Southern Italy and Hungary was started (D'Argenio and Horváth 1987). Inspired by the expertise of the Italian colleagues and by the spectacular paleokarst features of the southern Apennines, description of Hungarian paleokarst phenomena began (Kraus 1988; Bárdossy and Kordos 1989; Müller 1989; Mindszenty 1991, 1992; Török and Szabó-Balog 1990; Korpás and Juhász 1990; Nádor 1991, 1992; Nádor and Sásdi 1991; Korpás et al. 1991; Nádor et al. 1993; Korpás and Juhász 1993; Mindszenty et al. 1994; Juhász et al. 1995; Csoma 1998). The boom was driven also by the famous Paleokarst volume of James and Choquette published by Springer in 1988. It has raised considerable interest towards early diagenetic dissolution phenomena among carbonate sedimentologists and at the same time offered the right approach to researchers interested in ancient karst phenomena. The other volume, the effect of which was likewise beneficial to the development of Hungarian paleokarst research, was P. Bosák's regional review-collection on Paleokarst (Bosák 1989). In it there were included already the first systematic reviews on the paleokarst of Hungary (Bárdossy and Kordos 1989—already mentioned above, and Müller 1989).

In 1998, Korpás, in the introduction of his monograph on paleokarst studies in Hungary, gave an overview on paleokarst research worldwide (Korpás 1998). Recalling the results and statements of Allen and Matthews (1982), Wright (1982), Esteban and Klappa (1983), James and Choquette (1988), Rasmussen and Neumann (1988), Vera et al. (1988) and several others also came to the conclusion that karst features preserved in the stratigraphic record (i.e. paleokarst) are almost always the results of some polygenetic history. Some of the elements of this history belong to the early stages of carbonate diagenesis, while others are late, often tectonically controlled ("telogenetic") changes and their appropriate separation is often difficult if not impossible. He suggested that the superposition of the various karst facies and karst horizons were *"essentially controlled by base-level variations"*. In reviewing "Palaeokarst systems in Hungary", he attempted the integration of the diagenetic and the tectonic approach, however, with only partial success. Focusing on non-bauxitic paleokarst features, he correctly demonstrated that their formation was closely related to phases of carbonate-platform evolution (controlled by climate and tectonics). At the same time, he did not give the appropriate emphasis to those tectonically controlled major, regional unconformities which resulted in long-lasting subaerial exposure at times of humid tropical climate, even though they are outstanding records of paleokarstification in Hungary. His review was organized according to the—often only surmised—stratigraphic position and the geographic location of what he called the "Paleokarst Systems".

A new "wave" of paleokarst studies emerged in the 2000s when Poros et al. (2012), Poros (2011) then Győri et al. (2011, 2014) and Győri (2014), trying to understand the diagenetic history of Mesozoic-Tertiary carbonate sequences of the Buda Hills, came to the conclusion that to unfold the story of the complex set of superimposed geochemical and mineralogical events they observed would require first the understanding of the paleohydrological story behind. Thanks to the parallel research interest of hydrogeologists and speleologists working on the Pleistocene to recent Thermal Karst of the Buda Hills (Erőss et al. 2008, 2011; Pásztor 2016; Virág 2018; Virág et al 2019and others) and to the efforts of geochemists and sedimentologists studying travertines precipitated from the predecessors of thermal karst springs marking the sites of regional *discharge of deep groundwaters* (Kele 2009; Török 2018; Török et al. 2019), the combination of the approach of the different subdisciplines facilitated a better understanding of the whole karst-travertine continuum (see Chaps. 6, 12 and 18 in this volume).

6 Paleokarst in the Transdanubian Range, Hungary—Selected Examples (Fig. 2)

6.1 Examples of Bahamian-Type Karstification Related to Ephemeral Exposure and the Related Early Diagenetic Modification of Peritidal Carbonate Sequences

6.1.1 Cycle-Bound Ephemeral Karstification in the Triassic of the Gerecse Hills

The term *"Bahamian* (or Caribbean) *-type karst*" was introduced informally by Esteban in 1988[1] and adapted also by Bosák 2008. It refers to meso- and microkarst features associated with the "A"-members of Lofer-cyclothems and brought about by high-frequency (IVth order) eustatic falls of sea level in shallow marine carbonate depositional environments (like those known from the Pleistocene/Holocene of the Bahamas).

Such meso- and microkarst features are abundant in many of the cyclically organized Late Triassic carbonate sequences of the Transdanubian Range. They were studied in detail by Haas (2004 by Balog et al. 1997) in the Norian-Rhaetian of the *Bakony Mountains*, by Mindszenty and Deák (1999) and Mindszenty (2019) in the Rhaetian of the *Gerecse Hills* and by Korpás and Juhász (1990) and Juhász et al. (1995) in the *Buda Hills* and in the vicinity of Vác-Naszály. Balog et al. (1997) were focusing on the relationship between the observable early diagenetic changes (dissolution, partial dolomitization, etc.) and the cyclicity of changes of the eustatic sea level. Korpás and Juhász (1990) tried to reconstruct a whole polygenetic Paleokarst System while Mindszenty and Deák (1999) and Mindszenty (2019) having been interested in paleosols occurring at the exposure-surfaces tried to correlate the degree of development of the paleosoils and concomittant karstification with the duration of the exposure. Also—in a somewhat simplified way—they attempted the correlation of the argillaceous versus calcareous nature of paleosols with eventual temporal changes of the climate during deposition of the studied sequence. According to their observations, *argillaceous paleosoils* were accompanied by various *meso- and microkarst features* (like decimeter-scale uneven, sometimes serrate corrosion surfaces, cm-size dissolution-breccia-clasts mixed with the clay (Fig. 3) and—in the substrate—meso- and micro-scale often biomoldic cavities filled sometimes with stalactite-cements, more often with laminated internal sediments and/or with late radiaxial calcites). Infiltration of the soil—material into the substrate, deep below the disconformity surfaces, was also frequently observed. Their interpretation based on the overall similarity with features described and interpreted by Read andHorbury (1993) (from the Middle Ordovician of Virginia) and by D'Argenio et al. (1993, 1997) (from the Cretaceous of the Southern Apennines) was that the observed microkarst probably represented *humid climatic periods*. They also emphasized that the evidence for the humidity of the climate was not the mineralogy of the (apparently airborne and pedogenically modifed) clay phase of the soil but rather the *karstic dissolution* observed along and below the disconformity surface. According to their descriptions, also the *calcareous paleosols(=calcretes)* are accompanied by dissolution phenomena in the substrate, however, these are mostly restricted to meso-scale cracks and fissures somewhat enlarged by dissolution and — instead of clay — filled rather by small, broken-up clasts of the host rock and of the calcareous crust covering the exposure surface. The often laminated, iron-stained (yellowish) and at least partially dolomitized crusts were qualified as *calcretes* and interpreted as formed under slightly

[1] At a workshop of IGCP-287 (Tethyan Bauxites) when discussing the differences between (i) karstification associated with disconformities and affecting yet unlithified carbonate sediments and (ii) those occurring at major regional unconformities obviously affecting already lithified carbonate rocks, M. Esteban used the terms Caribbean (or Bahamian) and Alpine to distinguish the two. The terms have been used informally in paleokarst discussions ever since then.

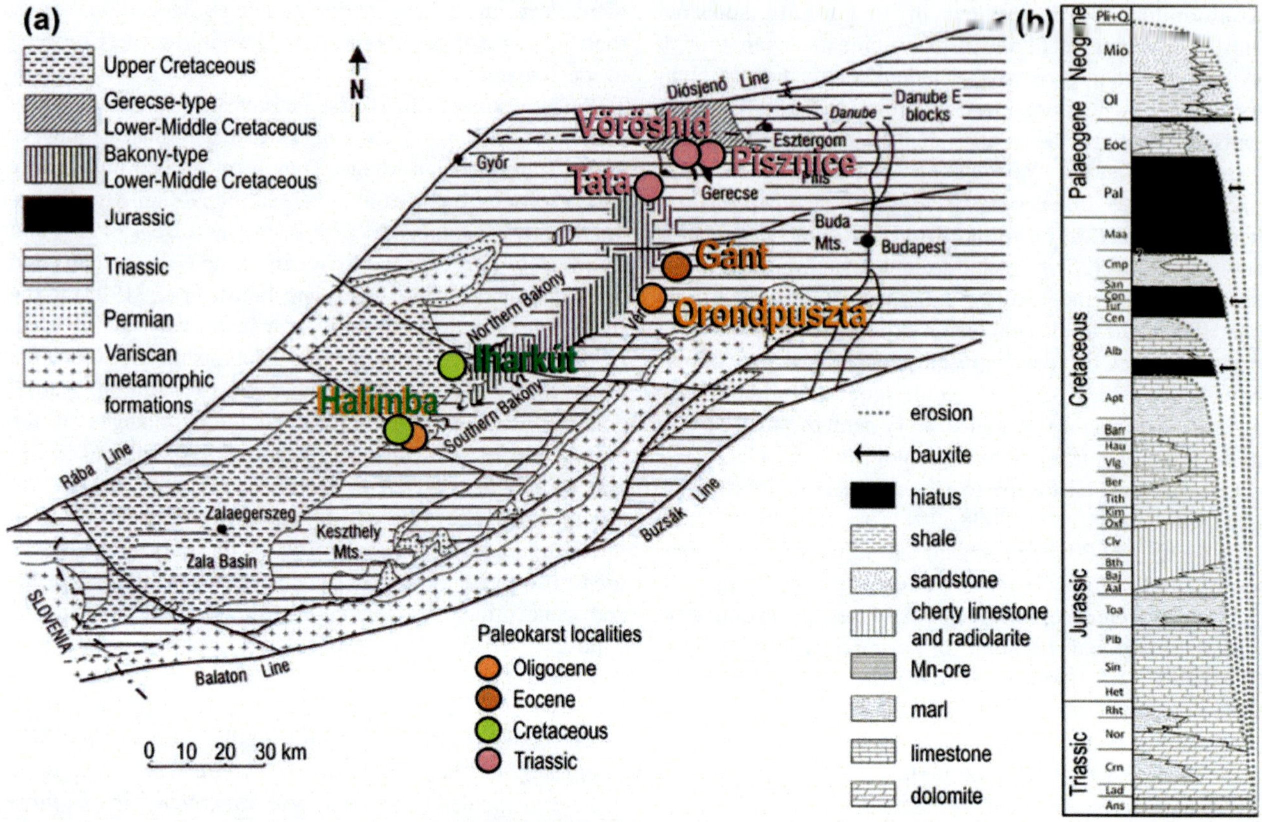

Fig. 2 Pre-Pleistocene paleokarst occurrences of the Transdanubian Range, mentioned in the text (a) and their stratigraphic position (b) (base-map by Haas 2001, stratigraphic column by Mindszenty et al. 2000)1: Halimba (Cretaceous and Tertiary bauxitic paleokarst); 2: Iharkút (Cretaceous, bauxite-filled high-level paleokarst); 3: Orondpuszta (Oligocene dolina-type karst, Vértes Hills); 4: Gánt (Eocene, low-level, polje-type paleokarst); 4: Tata (Late Triassic cycle-bound paleokarst); 5: Gerecse Hills, Vöröshíd and Pisznice (Late Triassic cycle-bound paleokarst)

more evaporative conditions than their argillaceous counterparts. To explain the observed only *moderate to weak karstification* associated with the calcrete-capped exposure surfaces, they suggested that it was the result of only a slight surplus of evaporation above precipitation. As to the duration of the exposure, adapting (and modifying) the terminology developed by Wright and Marriott (1996) for alluvial paleosols they recognized *simple, composite* and *cumulate* paleosols occurring at the cycle-bounding disconformities, each accompanied also by *microkarst features* affecting the underlying subtidal lithologies, as discussed above. They suggested that the duration of exposure in case of simple paleosols would be shorter than in the case of composite ones. The time represented by cumulate (aggrading) paleosols is definitely longer than that represented by either the single or by the composite ones, however, cumulates being the results of a series of episodes of repeated flooding and exposure in a close-to-sea-level (hydromorphic) position, the associated eventual dissolution features cannot be interpreted in terms of conventional karstification (Mindszenty 2019).

6.1.2 Enigmatic Dissolution Phenomena in the Late Triassic of Tata

An ambiguous disconformity surface between Late Triassic and Early Jurassic limestones of Tata demonstrates very well how difficult it may be to recognize eventual imprints of a *karst episode of moderate duration* when subsequent burial has almost completely erased all possibly diagnostic micromorphological and geochemical evidences of it.

In the Tata Geological Garden (=a series of abandoned quarries converted into a geological park), the contact between the cyclically organized Late Triassic (Rhaetian) shallow marine limestone (Dachstein Formation) and the overlying Liassic hemipelagic Ammonitico Rosso can be studied in a superb outcrop. According to Fülöp (1975), Galácz et al. (1985), Haas (1995, 2004), Haas and Hámor (2001) and others, the latest Rhaetian and Early Hettangian are missing here and the contact between the two formations is a bioerosional surface with half-cut and bored Megalodontids. Below this bioeroded (apparently submarine) surface, the last four parasequences of the Dachsten

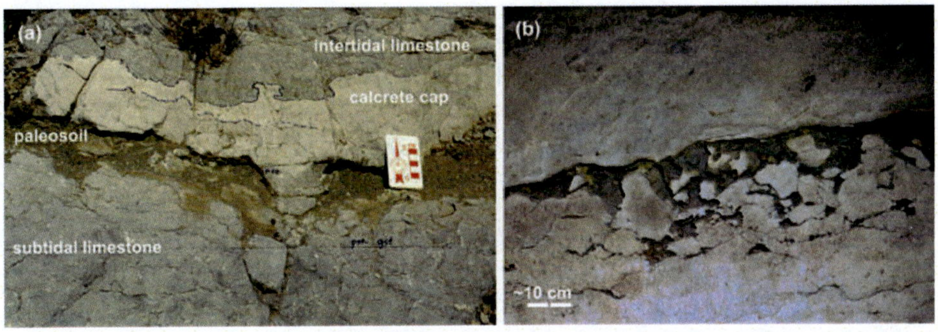

Fig. 3 *"Bahamian-type"* paleokarst. Close-up of the crenulated, mezo-scale karst relief, developed as a result of *ephemeral subaerial exposure* related to the Lofer-cyclicity of Late Triassic platform carbonates in the Gerecse Hills exposed by an abandoned stone-quarry, Vöröshíd (**a**); and by the Pisznice Cave, Nagy-Pisznice (**b**). Note the greenish-grayish, clayey, hydromorphic paleosol blanketing, the karst surface and infiltrating into the underlying, karstified limestone

limestone are full of spectacular dissolutional phenomena affecting the subtidal members of the Lofer cyclothemes *below* the disconformity surfaces (= the "A"-members). Most dissolution phenomena are large biomolds after Megalodontids (Fig. 4), but often also *meter-scale, elongate, bed-parallel cavities*, filled by calcite cements and finely laminated pink calcisiltite do occur (Lantos 2004; Győri 2014). The first cements encrusting the cavity-walls are isopachous radiaxial calcites, and their geochemistry (stable C and O isotopes 1.5 to 3.4 and -3.4 to -0.2, respectively) clearly suggests precipitation from saline (marine) solutions. However, the observed anomalously intense dissolution features and the close relationship between the cavities and the disconformities have called for some episodes of subaerial *exposure cum karstification* longer than those usually related to the "A"-members of the Lofer-cycles. All the same, microsampling and stable isotope analysis of the cavity-filling cements gave *no meteoric signal* and the immediately overlying hemipelagic Ammonitico Rosso, instead of documenting subaerial exposure pointed rather to a swift subsidence having resulted in a drowning unconformity (Type-3, sensu Schlager 1981, 1999). What is more: in the mechanical infill of the dissolution-cavities, no sign of soil-derived material could be identified. The red calcisiltite seems to be identical to the micritic matrix of the overlying pelagic sediments having been infiltrated into the dissolutional cavities through Neptunian dykes crosscutting the Triassic carbonate platform. Therefore, for a long time, the origin of the dissolution-cavities of the Late Triassic of Tata remained the subject of debates.

It wasGyőri (2014) who based on meticulous petrographic and geochemical invetigations, and a thoughtful discussion of all the theoretically possible explanations, concluded that before final drowning, there could have been, indeed, a *longer-than-usual episode of subaerial exposure*. Smaller cavities formed on dissolution related to earlier transient, cycle-bound exposure of the platform might have

Fig. 4 Enigmatic dissolution phenomena in the Late Triassic of Tata: (karstic) dissolution-enlarged cavities of biomoldic origin, some of them having a thin lining of calcite-cement and all filled by red-colored pelagic mud infiltrated from the overlying Early Jurassic Ammonitico Rosso

been enlarged during this later, longer lasting subaerial phase and produced this extraordinary set of *paleokarstic phenomena* preserved in the Late Triassic/Early Jurassic sequence of Tata. As to the timing of the *paleokarstic episode*, the anomalously intense dissolution must have taken place some time during or after the Rhaetian but definitely before the Sinemurian (the small early cavities "preparing the ground for the subsequent intense dissolution" were the products of early diagenesis of the *Triassic* Dachstein Limestone, while the Neptunian dykes filling up the enlarged cavities with the red cacisiltite die out in the *Sinemurian* part of the cover. According to Győri (2014), the reason for the lack of any substantial meteoric geochemical signal in association with the Tata Late Triassic/EarlyJurassic paleokarst may be that after the not very long subaerial period the platform became swiftly and finally drowned, the probably very restricted freshwater lens (if any) formed below the reliefless platform-top faded soon and the still open cavities became immediately filled by seawater which then effectively erased any eventual meteoric signal from the host of the cavities. Radiaxial calcites lining the cavity-walls and characterized by clearly marine stable isotope-geochemistry are already the witnesses of the flooding of the previously formed (and at the time of the flooding still empty) paleokarst.

6.2 Examples of "Alpine"-Type Paleokarst Related to Long-lasting Subaerial Exposure Along Tectonicaly Controlled, Major Regional Unconformities—Karst Bauxite Occurrences and Their Enclosing Karstic Carbonates

The Transdanubian Range is well known for its Cretaceous-Early Tertiary karst bauxites (Fig. 2). They occur at tectonically controlled major regional unconformities of Albian, Santonian, Eocene and Oligocene ages (e.g. Bárdossy 1982; Dudich and Komlóssy 1969; Mindszenty et al. 2000, and others). The apparent stratigraphic gap (vacuity + hiatus sensu Sloss 1963) between their Late Triassic bedrock and the usually Cretaceous to Tertiary cover varies between 35 and 190 Myr. The actual hiatus (=period of non-deposition, erosion and karstification) estimated on the basis of the known minimum gap (at the Csabpuszta locality "only" cca 10–20 Myrs) suggests that the reason for the rather large apparent gap is that much of the missing part of the Triassic-Jurassic Cretaceous and Tertiary sequences fell victim to erosion and karstification during those 10–20 Myrs of subaerial exposure. This substantial erosion was obviously the result of tectonically controlled uplift which brought about the exposure of large former shallow-marine depositional environments. Since most of the exposed area was built up by carbonate sediments (limestones, dolomites and subordinately also calcareous marls) and climatic conditions were favorable for karstification right throughout the Late Cretaceous to Early Paleogene subaerial episodes, a variegated, mostly highly dissected karst relief was formed essentially all over the area now called the Transdanubian Range and beyond. The *style*, the *degree* and the chance of *preservation* of karstification depended on (a) the rate, the style and the amount of uplift of each inidividual sector, (b) the lithology of the uplifted terrain, (c) the resulting/inherited geomorphological position, (d) the time spent on karstification (i.e. the duration of exposure) and (e) the depositional environment of the covering strata.

(a) **Rate, style and amount of uplift/subsidence as the controls of the formation and preservation of the karst landforms**

Slow and continuous uplift results in slow and continuous *downward progress of karstification* in the *vadose zone* until after it reaches the karst water-table (the local base level of erosion). From then on, *lateral corrosion* within the *epi-phreatic and phreatic zone* follows. That means that as long as we find shafts/sinkholes/ponors preserved in the stratigraphic record, extending downwards (vertically or subvertically) from the associated paleosurface, we have good reason to think that these paleokarst features were formed on karstification in the *vadose zone*. On the contrary, bauxite-filled shallow-karst (shallow dolines, poljes or uvalas) would suggest that karstification took place in "close-to-karst water-table" position (*semi-phreatic zone*) mainly by lateral corrosion. The paleo-vertical extent of the preserved karst (=depth of ponors/dolines, amplitude of irregularities of the bottom of bauxite-filled poljes/uvalas, etc.) may facilitate the estimation of the *minimum elevation* of the paleosurface above the paleo-water-table (i.e. above the paleo base level of erosion) at the time of the fossilization of the karst. The study of paleokarstic landforms thus may help to reconstruct the evolution of paleorelief at times of subaerial exposure of carbonate terrains. Inasmuch as relief is the joint result of tectonics + erosion + sedimentation, such reconstructions may give us an insight also into the tectonic evolution of the studied area. Based on this principle, the Eoalpine flexural deformation and the related partial uplift of the Transdanubian Range (Tari 1994; Tari and Linzer 2018) could be convincingly correlated with the established phases of subaerial exposure, karstification and the formation of the well-known Cretaceous/Tertiary bauxite deposits of the area (for details, see Mindszenty et al. 2000).

(b) **Lithological controls of paleokarst morphology**

Among karst morphologists, there is a general consent about the response to karstification of limestones and dolomites being different, and this is reinforced also by hydrogeologists. Indeed, experience shows that in limestone terrains long-lasting karstification results in well developed underground channel systems permitting optimum drainage via open passages (e.g. Palmer 2007 and many others). Dolomites, on the other hand, being harder and more brittle than limestones and therefore usally full of cracks and fissures, do not develop large open passages but provide a likewise excellent drainage through their "diffuse porosity". However

—at least in the humid tropics—this has nothing to do with the *adjoining surface morphology*. The famous tower karst areas (fenglins) of South China, North Vietnam or Indonesia having developed in dolomitic limestones (Jennings 1971; Sweeting 1973; Jakucs 1977; Palmer 2007; and personal observations in N.Vietnam) suggest that in the tropics, instead of pure geochemistry it is rather the physical properties of carbonates which are decisive from the point of view of the intensity of karstification (tower-karst may very well form also in tectonically affected *dolomites* the well-known rigidity of which results in well-developed fissure porosity "preparing" the ground for karstification). This applies also for many of the karst-filling bauxite deposits of the Transdanubian Range. For example, at Iharkút in the North Bakony, sinkholes developed at the crossing of major fault-zones and karstic canyons formed on the coalescence of sinkholes sitting on major fault-lines are the recipients of thick (sometimes > 90 m) industrial-grade bauxite deposits, and their bedrock is almost exclusively the Late Triassic *dolomite* (Szantner et al. 1986; Bárdossy and Mindszenty 2013). Deep sinkholes, likewise formed in dolomite, were reported also from the southern extremes of the Transdanubian Range (from Cserszegtomaj) where the age of karstification was estimated as Miocene (~18 Ma ago) based on U/Pb geochronology-assisted micromineralogical study of the karst-filling material (Kelemen et al. 2020).

(c) **Geomorphological position**

As was already mentioned above, the thickness of any karst bauxite deposit—i.e. the depth of the bauxite-filled karst depressions—is a reliable indicator of the minimum elevation of the karst terrain above the paleokarst water-table at the time of the accumulation of the bauxite (see also in "a"). For example at the Cretaceous Iharkút occurrence (North Bakony), the >90 m deep, bauxite-filled sinkholes show that karstification and accumulation of the karst-fill must have taken place in the *vadose zone,* i.e. obviously above the paleokarst water-table. Since similar to the karst morphology, also the lithofacies particularly the geochemical characteristics (Eh!) of the karst-filling bauxite depends essentially on the topographic position of the karst environment; the lithofacies of bauxites is an additional reliable relief-indicator.

Based on the observed close correlation between the *lithology/geochemistry* of bauxites and the *geometry of the enclosing karst forms,* D'Argenio and Mindszenty (1995) elaborated the lithofacies-concept of karst bauxites. Adapting the terminology of Grubb (1973) and Combes (1969, 1984), they distinguished *high-level* (vadose) and *low-level* (phreatic to semi-phreatic) bauxites which by their morphology *and* lithofacies are indicative of karstification having taken place predominantly high-above or close-to-the base level of erosion at the time of bauxite-formation (Fig. 5).

In the Transdanubian Range, there are some textbook examples of both *high-level* and *low-level* paleokarst terrains each filled with commercial-grade bauxite deposits. *Iharkút* with its Cretaceous deep-sinkhole-filling, fully vadose bauxites is one of the *typical high-karsts*, while the stratiform part of the *Halimba* deposit or the semi-phreatic *Gánt* bauxite, the latter with its large-scale crossbedding and alternating layers of muddy and conglomeratic material, fills a characteristic *low-level karst* (amplitude of the irregularities of the "karren-feld"-like surface of the bedrock in these latter cases is only ~10 to 12 m) (Figs. 6 and 7).

Halimba, with its deep-sinkhole-filling mid-Cretaceous to Santonian bauxites overlain by a large layer-like, crossbedded, conglomeratic Santonian deposit documents an intra-Cretaceous, clearly *tectonically controlled morphological evolution* from high-karst (sinkhole-fillings!) to *low-karst* (stratiform deposit) finally covered by a

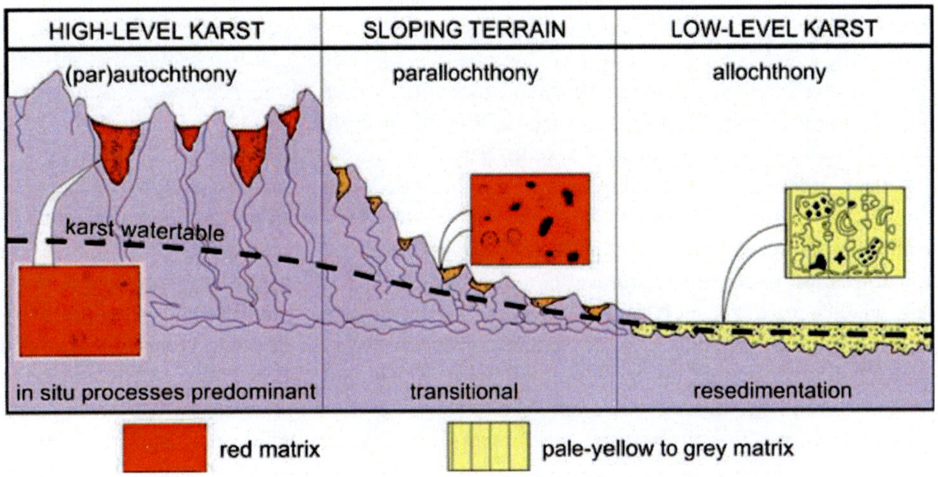

Fig. 5 Cartoon showing the relationship between surface-karst morphology and the lithology of the karst-filling material: (*high-level karst*:—deep-sinkhole morphology—vadose lithofacies of the karst-fill; *low-level karst*: shallow kast-morphology, semi-phreatic lithofacies of the karst-fill, deposit-scale crossbedding)

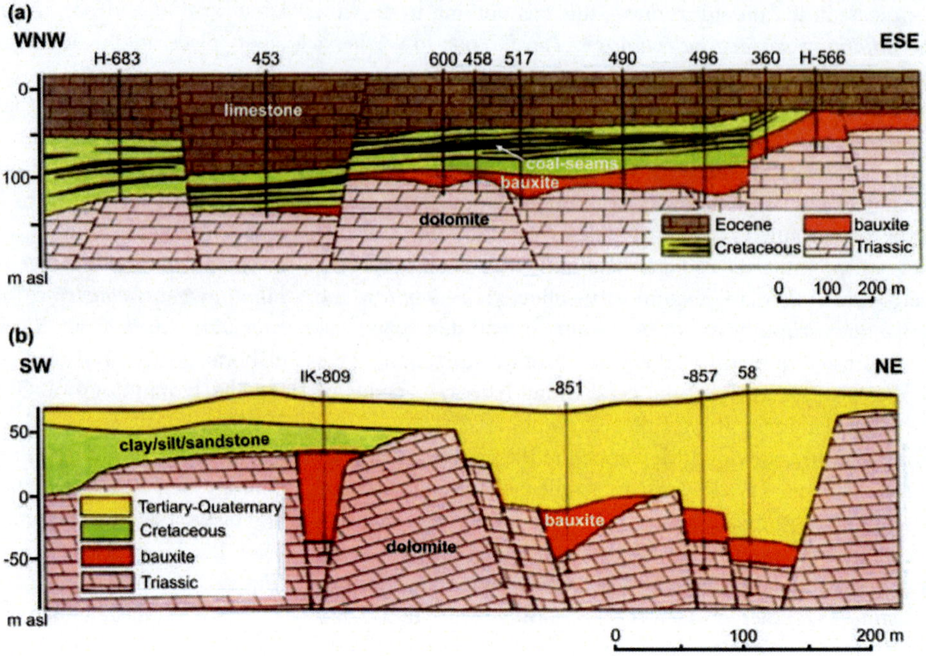

Fig. 6 Typical, *"Alpine-type"*, low-level and high-level karst-relief filled by bauxite, "Stratiform" (polje-filling) deposit of Halimba/South Bakony (**a**); and deep-sinkhole-filling deposit of Iharkút/North Bakony (**b**). Both are of Late Cretaceous age

transgressional sequence, beginning with alluvial siliciclastics, then coal-seams followed by marine marls (Fig. 8).

(d) **The time spent on karstification (the duration of exposure)**

To establish the duration of exposure at any unconformity of disconformity is a difficult, often an impossible task. Degree of karstification, i.e. whether the preserved morphology is juvenile, mature or senile (sensu Cvijić 1918, 1925) may be used as a guideline but considering the often complex history of paleokarst terrains, the resulting estimate is almost always dubious (see also Bosák 2008). There is no better "rule of thumb" than *the longer the duration of karstification, the more complex the resulting karst morphology may be* (see, for example, Ford and Williams 2007; Bosák 2002; Waltham and Fookes 2003; Waltham 2008).

Bauxite-filled karst is exceptional, because the bauxitic nature of the karst-fill facilitates the establishment of at least the minimum duration of the exposure. We know from chronosequences of soil scientists (e.g. Birkeland 1984; Retallack 1990) that the development of ferrallitic pedofeatures (characteristic of bauxites) requires at least several hundreds of thousands or even millions of years spent on soil formation under humid tropical conditions. That means that whence a karst form is filled by bauxite, we have good reason to think that intense karstification must have very probably lasted at least for some million years—fairly sufficient to result in "mature" ore even "senile" and complex karst. Of course, this time-frame, mentioned above, is yet only a "guess", however, nowadays with the advent of FT-dating and of U–Pb geochronology of single (detrital or volcanogenic) zircon crystals (present in the micromineralogical-fraction of many karst bauxites), new tools have been introduced into the age-dating of bauxites (and the enclosing paleokarst). It became possible to establish not only the numerical age of the paleokarst-filling bauxite but also the time-span during which the karst depression was "open" to receive the weathered (pre-bauxitic) material and—with it—any other wind-blown or even volcanogenic mineral grains (Dunkl 1992; Dunkl and Székely 2002; Dunkl et al. 2009; Kelemen et al. 2017, 2020). This way the "guess" mentioned above may (and will) be substantially improved in the future (at least as far as bauxite-filled karst is concerned).

As to the duration of exposure of the bauxite-filled paleokarst surfaces of the Transdanubian Range, it is generally accepted that all four known bauxitic unconformities (the Albian, the Santonian, the Eocene and the Oligocene) must have lasted for several tens of millions of years. Because of the substantial erosion likely preceding the formation of the preserved karst morphology, the apparent stratigraphic gap in which the bauxites occur, is—as a rule—rather large. The deeply karstified bedrock is almost everywhere of Triassic age, only the ages of the cover sequences are different. The only exception is the stratigraphic position of the Csabpuszta Eocene bauxite underlain by the karstified Late Cretaceous Ugod Limestone Formation and covered by an early Mid Eocene transgressional sequence. This is the minimum apparent gap for a bauxite deposit in the

Fig. 7 Examples of bauxite-filled paleokarst morphology in the Transdanubian Range. **a** Opencast mining exposed the once bauxite-filled, "Alpine-type", *high-level* Cretaceous paleokarst terrain developed on the surface of Late Triassic dolomite at Iharkút (Németbánya-XI deposit), North Bakony, TR (max. thickness of the bauxite was ~80 m); **b** Large-scale crossbedding of conglomeratic and pelitomorphous (mudstone-like) layers of allochthonous bauxite (Gánt Bauxite F.), filling the *low-level* Eocene karst, developed on Late Triassic dolomites of the Gánt-Bagoly-hegy area, Vértes Hills (for scale, see the footprints on the scree); **c** Karst pinnacles (formed of Cretaceous limestone) cropping out from below the Eocene bauxite of Csabpuszta, South Bakony (size of hammer: ~30 cm); **d** Karrenfeld formed on the surface of Late Triassic dolomite blanketed by Oligocene bauxite (Óbarok Bauxite F.) at Óbarok, Gerecse Hills (amplitude of the karren: ~1.5–2 m); **e** Oligocene bauxite fills the dolinas developed in Late Triassic dolomite at Orondpuszta, Vértes Hills

Transdanubian Range and even this one is of at least cca 36 million years duration. To refine our concept regarding the time-span necessary for the formation (and partial destruction) of the karst morphology preserved underneath the bauxite deposits of the Transdanubian Range would require additional sophisticated analytical work like the U–Pb geochronology mentioned above.

(e) Depositional environment of the cover

When the deposition of the immediate cover takes place under high-energy conditions, this may be of deteriorating influence for the survival of the underlying paleokarst. Such sediments, like coarse-grained clastics deposited in alluvial channels or breccia-deposits of abrasion shorelines, etc. are

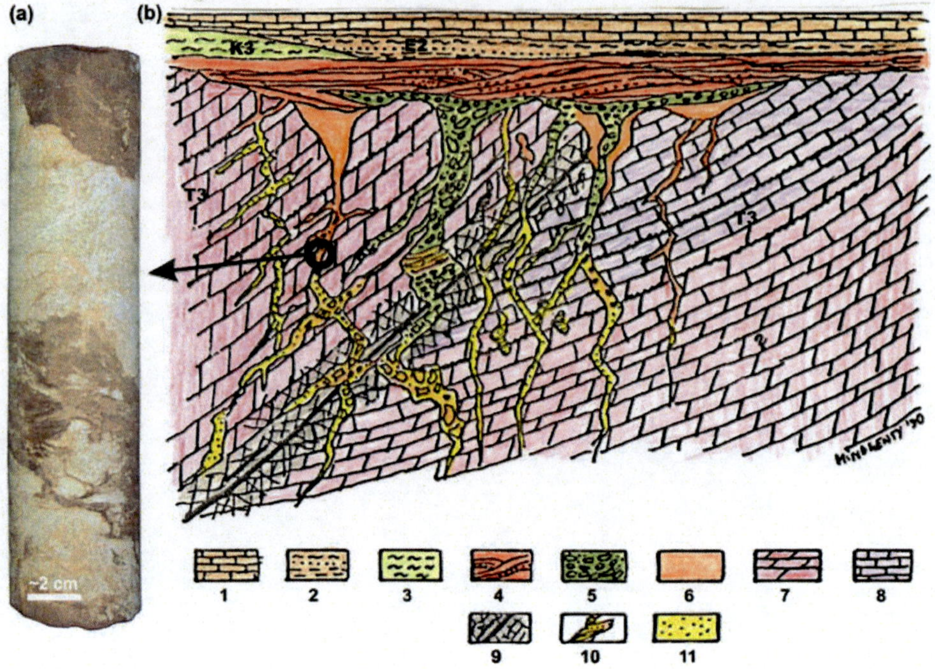

Fig. 8 Superposition of early high-level and later low-level karst fill documenting tectonically controlled uplift, followed by likewise tectonically controlled subsidence of the paleokarst of Halimba, not to scale (Mindszenty et al. 2000). **a** Diamond-core taken from the Triassic dolomite. Note the bauxite-filled paleokarstic cavities! **b** Conceptual profile across the Halimba bauxitic paleokarst. Key: 1: shallow marine limestone (Eocene, Szőc Limestone F.); 2: Lignitiferous clay/silt (Eocene, Darvastó F.); 3: Siltstone/sand/marl (Cretaceous, Csehbánya F.); 4: Allocthonous to parallochthonous, crossbedded bauxite with intercalation of dolomite-pebbles (Late Cretaceous, Halimba Bauxite F.); 5: Dolomite-pebble conglomerate (Late Cretaceous); 6: In situ developed, sinkhole-filling, oolithic bauxite (Late Cretaceous, Nagytárkány Bauxite F.); 7: Dolomite (Late Triassic, Hauptdolomit F.); 8: Limestone (Dachstein F., Late Triassic); 9: Fault-zone; 10: paleokarstic cavities, filled by bauxite/dolomite-silt/calcite cement; 11: Dolomite-silt, from karst-related powderization of dolomite

the indications of intense mechanical- or sometimes bio-erosion (e.g. Rasmussen and Neumann 1988) possibly destroying most of the antecedent karst. On the contrary, fine-grained cover sediments (like overbank sediments of alluvial floodplains or organic-rich muddy layers of freshwater or brackish origin) would facilitate the survival of not only the underlying karst but also the terrestrial material accumulated in it. There is, however, a rather conspicuous effect which such cover sequences may exert on the underlying karst-filling sediments. It is of geochemical/mineralogical nature. Transgression over a previously exposed terrain invariably involves the change of hydrology (from vadose to phreatic). Later on also, the interaction with organic-rich solutions affects both the pH and the Eh of porewaters (e.g. omnipresent iron becomes reduced and mobilized resulting in colorful mottles in the uppermost layers of the doline-filling sediment) (Fig. 9).

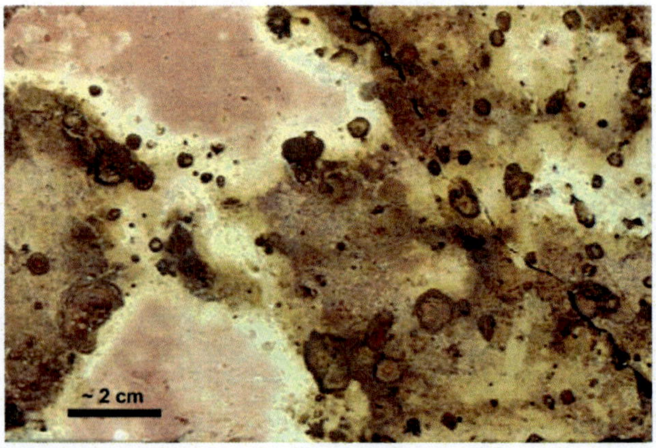

Fig. 9 Redox-mottling in the upper part of the bauxite resulted by the hydrological change associated with the deposition of the coverbeds (hand-specimen, Gánt, Vértes Hills)

A rather particular case is when flooding of the former exo-karst begins from below, via the well-developed subterranean channel system of the karst. This happens when elevated irregular cliffs of the well-developed surface karst hinder the progress of overland transgression, however, the freshwater lens, pushed upwards by the rising sea level, sooner or later intersects the surface of the doline-filling sediment and this way, those dolines will be first soaked then filled by freshwater and—as sea-level rise continues—by saline waters. The result will be a rather characteristic facies-succession overlying the doline-fill. The areal extent of this facies-succession would be limited to the area of or would only slightly overlap the dolines flooded from below. Carannante et al. (1994a, b) and Durn et al. (2003) referring to Bahamian analogues described byRasmussen and Neumann (1988) and Whitaker and Smart (1990) introduced the term "blue-hole" sequence for this facies succession (Fig. 10). For a long time, it has been considered by bauxite-geologists as an indicator-facies heralding the existence of hidden bauxite-filled dolines during the exploration of karst bauxites (Komlóssy 1970, Knauer and Gellai 1978; Rákosi and Tóth 1980). Also, the areal distribution of the basal facies of transgression sequences following a subaerial exposure phase faithfully records all the details of the subaerial relief this way facilitating a paleokarst morphological mapping.

When low-energy transgression (whether from below, or from above) reaches a "clean", i.e. unfilled or only partially filled karstic channel-system, with time, calcite cement and sometimes also mechanical sediments (either as sparse dissolution residue or as infiltrated marine sediments) fill up the open spaces. Whether the cements precipitated from meteoric or marine waters can be established by means of micropetrography, isotope-geochemistry or by fluid inclusion studies (cf. with Csoma et al. 2004). Such analyses were performed on some of the Mesozoic/Tertiary paleokarst areas of the Transdanubian Range, as well, and it became clear that the geochemistry of the studied calcite parageneses filling karstic cavities in Mesozoic limestones reflected faithfully, indeed, the sequence of events from vadose to phreatic environments and from meteoric towards brackish chemistry of the fluids involved (Győri 2014; Győri et al. 2014).

7 Paleokarst in South Hungary (Mecsek Mountains and Villány Hills)

From the Mecsek Mountains, only sporadic occurrences of paleokarst phenomena have been reported so far. To a large part, they are described in manuscripts and in unpublished reports. These paleokarsts all occur in the western part of the mountains, in Middle Triassic limestones. From the few published sites, some were studied from the aspect of karstification (Szabó 1961; Rónaki 1969), the others from other viewpoints, e.g. the Late Pleistocene faunal remains (Kretzoi 1942) or archeological finds were examined (Vértes 1952). In the past few years, research began again, focusing on the relationship of karstification to the uplift history of the area.

The Villány Hills are famous for their numerous relict and/or exhumed paleokarst cavities with abundant Plio-Pleistocene vertebrate fossils (for a summary, see Pazonyi 2009). They have been studied from the early twentieth century (Kormos 1917), mostly from a paleontological viewpoint, and provide type localities for several biozones of the Pliocene and Pleistocene. Information on the Mesozoic karstification of the area comes from works on bauxite geology and on the Jurassic/Cretaceous boundary (Rakusz 1937; Dudich and Mindszenty 1984; Császár 2002).

7.1 Alpine-Type (?) Late Paleokarst in the Villány Hills

The hiatus around the Jurassic-Cretaceous boundary in the Villány Hills region and the related terrestrial sediments, though associated with commercial-grade bauxites, are unlike those described from the Transdanubian Range (TR). All the four bauxite horizons of the TR occur at rather large

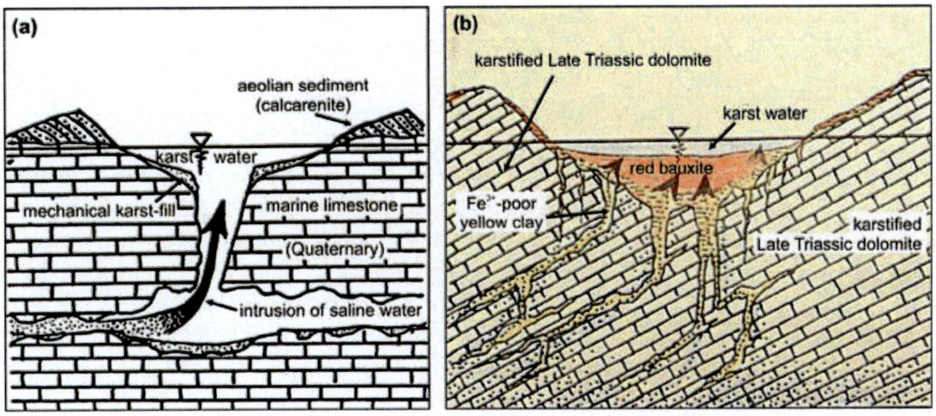

Fig. 10 Cartoons showing the concept of the "blue-hole"-type transgression from below, via the karstic channel ways previously providing the plumbing system of the karst: **a** The Bahamian analogue; **b** Conceptual reconstruction of the Eocene flooding of a bauxite-filled dolina at Gánt

Fig. 11 Paleokarst cavities with infills of various ages in a ~70 m deep quarry in the western Mecsek Mountains (Bükkösd). Baggers and trucks (circled) for scale. Letters and numbers stand for the observed paleokarstic cavities

apparent stratigraphic gaps (cf. with Fig. 2) and between the karstified bedrock, and the cover there is an obvious angular unconformity, suggesting that subaerial exposure was brought about by tectonically controlled uplift *cum* erosion. In the Villány Hills, however, the stratigraphic gap associated with the paleokarst and bauxites is much smaller. Even though the presence of bauxites shows that the duration of the exposure must have reached at least 1 million years, the subaerial period was geologically short, indeed: at places both bedrock and cover are of Berriasian age, and the angular unconformity is either very slight or non-existent. That there must have been at least some tilting associated with the exposure is shown only by the diachronism of the cover and the increase of the apparent gap towards the North (intra-Berriasian in the south, while comprising a lower Kimmeridgian to Mid Aptian time-span in the north). That means that, though also here, the exposure occurs at a major regional unconformity, the tectonic mechanism behind that unconformity must have been different from the one described from the TR. The facts that between bauxite, its bedrock and cover there is an apparent concordance and both bedrock and cover are pure limestones makes these Villány bauxites similar to the karst bauxite deposits of the South-Central Apennines where gentle intraplate deformation (sensu Cloethingh 1986) was invoked to explain the observed situation (D'Argenio and Mindszenty 1995). Unfortunately, the Late Jurassic to Early Cretaceous plate-tectonic history of the Villány segment, belonging to the Bihor-Getic microcontinent which in the Jurassic accommodated also the Villány Unit, is by far too complex to be helpful in the interpretation of the observed phenomena.

Field relationships show that in the Villány various paleokarst phenomena occur in multiple small detached tectonic units of the area. In the southern thrust sheet—supposedly oceanward in Late Jurassic-Early Cretaceous times—Berriasian to Lower Valanginian bauxite is preserved in shallow (<10 m deep) depressions scattered on the surface of Upper Jurassic pelagic limestones (Rakusz 1937; Császár 2002, Fig. 46, chapter "Geology of the Karst Terrains in Hungary"). The bauxite formed under "paralic" conditions, most likely in a coastal position (Dudich and Mindszenty 1984). In the Nagyharsány quarry, at places, in the bauxite, there are intercalations of limestone layers with freshwater biota, and sometimes the karstic cavities are jointly filled with these two rock types (Császár 2002). These point to an oscillating water-table and the gradual flooding of the karstic landscape, with shorter and shorter periods of subaerial conditions and karstification (see also the discussion of the "blue-hole"-type transgression above).

More to the north, in the Szabolcs Valley quarry of the town of Siklós, a vertical paleokarst cavity penetrates the Oxfordian to Early Kimmeridgian pelagic limestones (Fig. 42, chapter "Geology of the Karst Terrains in Hungary"). Its investigation is underway; the first results refer to the temporary emergence of the site to subaerial conditions and then, its renewed flooding still in the Late Jurassic. The basal breccia and calcarenite layers of the cavity-fill with their microfauna indicate that submersion of the cave and sedimentation began in the middle Kimmeridgian, in agitated euhaline water (Sebe et al. in prep). The bulk of the cave-fill is a gray, well-bedded limestone, with red clay (paleosol) clasts, blocks from the host limestone, a low-diversity but abundant gastropod fauna, ostracods and scarce agglutinated foraminifers. It is interpreted to have accumulated after further relative sea-level rise in a low-energy, not euhaline environment, and the cavity can be classified as an anchihaline cave.

The terrestrial gap between Jurassic and Cretaceous rocks widens towards the north within the Villány Hills. The age of the uppermost Jurassic limestone beds varies from Tithonian in the south to Lower Kimmeridgian in the north (Császár 2002). Submergence was diachronous as well, from the Valanginian, perhaps even Berriasian in the south to middle Aptian in the north, while just north of the Villány Hills the lower Cretaceous Urgon-type limestones are completely missing.

7.2 Superimposed Miocene-Quaternary Karstification Phases of the Mecsek Mountains

Up until a couple of years ago, the Mecsek Mountains were hardly studied for paleokarst phenomena. Since 2018, investigations are ongoing with the aim of gaining information on the uplift history of the mountains from the study of paleokarst features. Fieldwork has proved the presence of over 50 karstic cavities with various infills, indicating the repeated activity of the karst system in different geologic time intervals. Homogeneous cave sediments, interpreted to indicate a minimum age for the host cavity, included lower Miocene sand and gravel, lower Miocene rhyolite tuffs, upper Miocene sands, Plio-Pleistocene red clay, and Pleistocene loess and paleosols (Fig. 11). Cave geometries—mostly steep, fault- or joint-controlled fissures and gently sloping, lithologically controlled narrow tunnels following bedding planes—suggest karstification in the vadose zone. Cavities with different infills did not show any altitudinal trends, indicating that carbonate rocks within the same elevation range have been subject to karstification, sedimentary coverage and exhumation during multiple time intervals of the Neogene to Quaternary. In one of the largest paleokarst cavities, huge dolomite blocks occur (Fig. 44, chapter "Geology of the Karst Terrains in Hungary"), which are absent in the modern catchment area of the sinkhole. Their presence can be explained either by the denudation of ~80–100 m of limestone or by—rather improbable—transport across a presently 100 m deep valley, again referring to long-lasting (possibly tens of millions of years old) karstification.

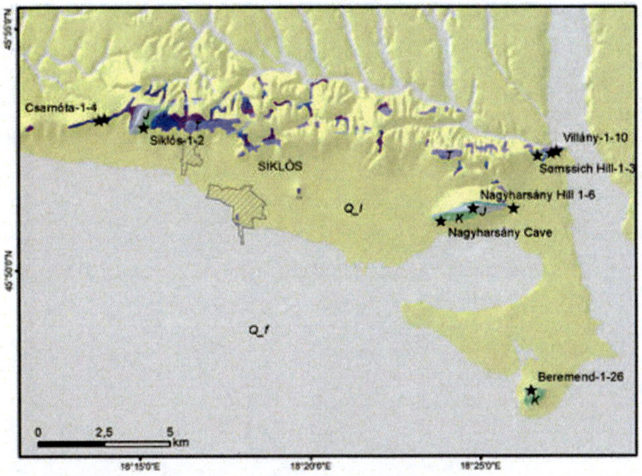

Fig. 12 Location of Plio-Pleistocene paleokarst cavities dated by vertebrate faunas in the Villány Hills. Legend: T: Triassic; J: Jurassic; K: Cretaceous rocks; Q_l: loess and loess-like sediments; Q_f: fluvial and paludal sediments

7.3 Pliocene–Quaternary Paleokarst Cavities of the Villány Hills

The type area of late Neogene-Quaternary vadose-zone karstification in Hungary is the Villány Hills, a gently rolling range of Mesozoic carbonates in the SW of the country. More than 50 fossiliferous and countless more, non-fossil-bearing paleokarstic sites occur in the area (Rakusz and Strausz 1953; Kretzoi 1956; Császár and Kordos 2007) (Fig. 12). The oldest cavities of known age lie at medium elevations within the range and are filled with sands —most probably sediments of the Lake Pannon—deposited in latest Miocene times (Dezső et al. 2007). These indicate that the area of the hills must have been flooded by the lake. The next generation of cavities with well-dated fills are the numerous fissures with an infill of red clay originating from the redeposition of terra rossa-type paleosols. Based on the contained vertebrate faunas, the age of these clays ranges from 3.5 to 1 Ma (Koloszár 2004; Pazonyi et al. 2018). Some of the oldest ones lie at elevations hardly higher than the modern base level, showing that the hills had become exhumed from under the upper Miocene lacustrine sediments practically to their present-day state by the late Pliocene as latest (Sebe et al. 2015). The youngest cavities are filled by reworked loess. Cavities are dominantly vertical, fracture-controlled fissures and shafts, pointing to formation in the vadose zone (Fig. 13).

Fig. 13 Paleokarstic cavity in the Villány Hills filled with Pleistocene fossiliferous red clay (Beremend Quarry)

8 Conclusions

Since carbonate minerals readily dissolve in meteoric water, carbonate sediments/rocks are all exceptionally adapted to karstification. In shallow-water (either marine or continental) depositional environments, the interaction of the sediment with meteoric water begins right after deposition and may continue through all stages of *diagenesis* and—on reexposure—also *telogenesis*, the latter affecting the already lithified rock (cf. early "Bahamian" *versus* late "Alpine"-type karstifiation). Accordingly all carbonate formations in Hungary abound in both early and late paleokarst features often superimposed on each other. *Early (diagenetic) paleokarst* is—as a rule—associated with shallow-water (either marine or freshwater) limestones and/or dolomites. Examples are known from the TR, the Northern Highlands (Bükk, Aggtelek, Uppony) and also from South-Hungary (Villány, Mecsek). In South Hungary also, latest Jurassic to Early Cretaceous limestones while in the Transdanubian Range Late Cretaceous to Eocene and even Pleistocene travertines were also affected by early (diagenetic, "intraformational") karstification. More important are, however, those late, *Alpine-type paleokarst* features which occur at major tectonically controlled subaerial unconformities, and were resulted by the events of the Eoalpine (Cretaceous to Paleogene) and Neoalpine (Miocene to Recent) structural evolution of the Carpatho-Pannonian Region. To separate the effects of the numerous superimposed early and late karst events is often difficult and in addition to thoughtful field-observations requires also sophisticated laboratory methods and the appropriate understanding of the geological evolution of the area concerned.

Acknowledgements The authors are grateful to M. Virág for her invaluable help in grooming both the text and the figures.

References

Allen JR, Matthews RK (1982) Isotope signature associated with early meteoric diagenesis. Sedimentology 29:797–817

Balazs A, Matenco L, Magyar I, Horváth F, Cloetingh S (2016) The link between tectonics and sedimentation in back-arc basins: New genetic coonstraints from the analysis of the Pannonian Basin. Tectonics 35:1526–1559

Balog A, Haas J, Read F, Coruh C (1997) Shallow marine record of orbitally forced cyclicity in a Late Triassic carbonate platform Hungary. J Sed Res 67(4):661–675

Bárdossy G (1982) Karst Bauxites. Elsevier, Amsterdam

Bárdossy G, Kordos L (1989) Paleokarst of Hungary. In: Bosák P (ed) Paleokarst—a systematic and regional review. Academia, Prague & Elsevier, Amsterdam, pp 137–154

Bárdossy G, Mindszenty A (2013) The Iharkút Bauxite. Occasional Papers of Geological and Geophysical Institute of, vol 1

Bosák P (ed) (1989) Paleokarst—a systematic and regional review. academia, Prague & Elsevier, Amsterdam

Bosák P (2002) Karst processes from the beginning to the end. In: Gabrovšek F (ed) Evolution of karst from prekarst to cessation. Postojna-Ljubljana Zalozba ZRC, pp 155–190

Bosák P (2008) Karst processes and time. Geologos 14(1):19–36

Carannante G, D'Argenio B, Ferreri V, Simone L (1987) Cretaceous paleokarst of the Campanian Apennines, from early diagenesis to late filling stage. A case history. Rend Soc Geol It 9:251–256

Carannante G, D'Argenio B, Mindszenty A, Ruberti D, Simone L (1994a) Cretaceous-Miocene shallow water carbonate sequences. Regional unconformities and stacking patterns. Guidebook, Excursion A2, 15th IAS Regional Meeting, Ischia, Italy, pp 29–59

Carannante G, Mindszenty A, Neumann AC, Rasmussen KA, Simone L, Tóth K (1994b) Inland blue-hole type ponds in the Mesozoic-Tertiary karst-filling sequences. Abstracts, 15th IAS Regional Meeting, Ischia, Italy, pp 102–103

Cloetingh S (1986) Intraplate stresses: A new tectonic mechanism for fluctution of relative sea-level. Geology 14:617–621

Combes J-P (1969) Observations et intrerpretations nouvelles sur le bauxites du nord-est de l'Espagne (stratigraphie, paleogeographie, genese) (New observations and interpretations of bauxites in NE Spain) (in French). Ann Inst Geol Publ Hung 54(3):265–280

Combes J-P (1984) Regards sur la geologie des bauxites aspects recentes sur la genese de quelques gisement a substratum carbonate (Remarks ont the geology of karst bauxites on the basis of some recent observations) (in French). Bull Centres Rech Explor Elf-Aquitqine 8(1):251–274

Cvijić J (1918) Hydrographie souterraine et evolution morprhologique du Karst (Underground hydrography and the morphological evolution of Karst) (in French). Rec Trv Inst Geogr Alp 6(4):376–420

Cvijić J (1925) Types morphologiques des terraines calcaires (Morphological types of carbonate terrains) (in French). C.R.Acad.Sci. (Paris), p 180, p 592, p 757, p 1038

Császár G (2002) Urgon formations in Hungary. Geol Hungarica Ser Geol 25:1–209

Császár G, Kordos L (2007) Beremend, kőfejtő. Alsó- és középső-kréta Nagyharsányi Mészkő Formáció, pliocén gerinces lelőhelyek (Beremend, quarry. Lower and Middle Cretaceous Nagyharsány Limestone Fm., Pliocene vertebrate fossil sites). In: Pálfy J, Pazonyi P (eds) Őslénytani kirándulások Magyarországon és Erdélyben (Paleontological excursions in Hungary and Transylvania). Hantken Press, Budapest, pp 172–179 (in Hungarian)

Csoma A (1998) A Környei Mészkő Formáció diagenezis szempontú vitzsgálata (Diagenetic study of the Cretaceous Környe Limestone Formation) (in Hungarian). MSc Theses, Department of Applied Geology, Eötvös Loránd University, Budapest

Csoma AÉ (2003) Calcium-carbonate cementation in the meteoric-mixing zone: processes and criteria for recognition. PhD Theses, University of Kansas, US

Csoma AÉ, Goldstein RH, Mindszenty A, Simone L (2004) Diagenetic salinity cycles and sea-level along a major subaerial unconformity, Monte Campo Sauro Italy. J Sed Res 74(6):889–903

D'Argenio B, Horváth F (1987) Foreword to studies on the Mesozoic and Tertiary geodynamics of the Periadriatic Region, Italy and Hungary. Rend Soc Geol It 9 (1986), Suppl Boll Soc Geol It vol CV, p 147

D'Argenio B, Mindszenty A (1995) Bauxites and related paleokarst: Tectonic and climatic event markers at regional unconformities. Ecl. Geol. Helvetiae 88(3):453–499

D'Argenio B, Amodio S, Ferreri V, Pelosi N (1997) Hierarchy of high-frequency orbital cycles in Cretaceous carbonate plaform strata. Sed Geol 113:169–193

D'Argenio B, Ferreri V, Ardillo F, Buonocunto F (1993) Microstraigrafia e stratigrafia sequentiale. Studi sui depositi di piattaforma carbonatica. Cretacico del Monte Maggiore (Appennino Meridionale). Boll Soc Geol It 112:39–749

Dezső J, Raucsik B, Viczián I (2007) Villányi-hegységi karsztos hasadékkitöltések szemcseösszetételi és ásványtani vizsgálata (Granulometric and mineralogical analysis of karstic fissure filling sediments in the Villány Hills, S Hungary). Acta GGM Debrecina 2:151–180 (in Hungarian)

Dudich E, Komlóssy Gy (1969) Ősföldrajzi és szerkezeti szempontok a magyar bauxit korkérdéséhez (Paleogeographic and tectonic considerations to the problem of the age of Hungarian bauxites). Földtani Közlöny (Bull. Hung. Geol. Soc.) 99(2):155–165 (in Hungarian)

Dudich E, Mindszenty A (1984) Ásványkőzettani-geokémiai adatok a Villányi-hegység és az Erdélyi-középhegység bauxitjainak összehasonlításához (Contribution to the comparative geochemistry and petrology of bauxites in the Villany Mts., SE Transdanubia, Hungary and in the Pădurea Craiului, Bihor Mts. area, W Transylvania, Rumania). Földtani Közlöny 114(1):1–18 (in Hungarian)

Dunkl I (1992) Origin of Eocene-covered karst bauxites of the Transdanubian Central Range (Hungary): evidence for early Eocene volcanism. European Journ.of Mineralogy 4:581–595

Dunkl I, Székely B (2002) Component analysis of detrital FT-ages—with visualization of the fitting. Geotemas 4, Salamanca Intern. Workshop on Fission Track Analysis, Cádiz, 4–7 June 2002, p 63

Dunkl I, Frisch W, Kuhlemann J, Brügel A (2009) Pebble population dating as an additional tool for provenance studies—examples from the Eastern Alps. Geol Soc London Spec Publ 324:125–140

Durn G, Ottner F, Tisljar J, Mindszenty A, Barudzija U (2003) Excursion P8: regional subaerial unconformities in shallow marine carbonate sequences of Istria: Sedimentology, Mineralogy, Geochemistry and Micromorphology of Associated Bauxites, Paleosols and Paleo-Sedimentary Complexes, pp 209–254. In: Vlahovic I, Tisljar J (eds) 22nd IAS Meeting in Sedimentology - Opatija 2003, Field Trip Guidebook: Evolution of Depositional Environments from the Paleozoic to the Quaternary in the Karst Dinarids and the Pannonian Basin. Zagreb, Croatia, Institute of Geology, Zagreb 254 p

Erőss A, Csoma AÉ, Mádl-Szőnyi J (2008) The effects of mixed hydrothermal and meteoric fluids on karst reservoir development, Buda Thermal Karst, Hungary. In: Sasowsky ID, Feazel CT, Mylroie JE, Palmer AN, Palmer MV (eds) Karst from Recent to Reservoirs. Special Publication 14, Karst Waters Institute, Leesburg, Virginia, pp 57–63

Erőss A, Poros Zs, Mádl-Szőnyi J, Mindszenty A, Molnár F, Ronchi P, Csoma AÉ (2011) Role of karstic and basnal fluids in porosity evolution of the Buda Hills, Hungary. AAPG International Conference and Exhibition, Following Da Vinci's Footsteps to Future Energy Resources, Innovations from Outcrops to Assets. AAPG, Paper 1071554

Esteban M, Klappa CF (1983) Subaerial exposure environment. Am Assoc Petrol Geol Mem 33:1–54

Esteban M, Wilson JL (1993) Introduction to karst systems and paleokarst reservoirs. In: Fritz RD, Wilson JL, Yurewicz DA (eds) Paleokarst related hydrocarbon reservoirs. SEPM Core Worskhop 18, New Orleans, 25 April 1993, pp 1–9

Fischer AG (1964) The Lofer cyclothems of the Alpine Triassic. In: Merriam DF (ed) Symposium on Cyclic Sedimentation. Bulletin Kansas Geological Survey, vol 169, pp 107–149

Ford DC, Williams PW (2007) Karst hydrogeology and geomorphology. Wiley, Chichester

Fülöp J (1975) Tatai mezozoós alaphegység-rögök (The Mesozoic basement horst blocks of Tata). Geol Hung Ser Geol 16 (in Hungarian)

Galácz A, Horváth F, Vörös A (1985) Sedimentary and structural evolution of the Bakony Mountains (Transdanubian Central Range, Hungary) paleogeographical implications. Acta Geol Hung 28:85–100

Grund A (1914) Die geographische Zyclus im Karst (the geographic cycle of karstification) (in German). Gesellschaft Für Erdkunde 52:621–640

Grubb PLC (1973) High-level and low-level bauxitization: a criterion for classification. Miner Ser Engin 5, Johannesburg, pp 219–231

Győri O (2014) Paleofluid-migration events recorded in Mesozoic carbonates of the Transdanubian Range, Hungary. PhD theses, Eötvös Loránd University, Budapest (in Hungarian)

Győri O, Orbán R, Mindszenty A, Fodor L, Zs P, Erőss A, Zs B, Molnár F (2014) Red calcite: and indicator of paleo-karst sytems associated with bauxitic unconformities. Geofluids 14(4):459–480

Győri O, Poros Zs, Mindszenty A, Molnár F, Fodor L, Szabó R (2011) Budai-hegységi paleogén karbonátos kőzetek diagenezistörténete (Diagenetic history of the Palaeogene carbonates, Buda Hills, Hungary). Földtani Közlöny (Bull Hung Geol Soc) 141(4):341–361 (in Hungarian)

Haas J (1995) Az Északi Gerecse felső-triász karbonát platform képződményei (Upper Triassic platform carbonates of the northern Gerecse Mts.). Földtani Közlöny (Bull Hung Geol Soc) 125(3–4):259–293 (in Hungarian)

Haas J (2004) Characteristics of peritidal facies and evidences for subaerial exposures in Dachstein type cyclic platform carbonates in the Transdanubian Range, Hungary. Facies 50:263–286

Haas J, Hámor G (2001) Geological garden in the neighborhood of Budapest, Hungary. Episodes 24:257–261

Jakucs L (1977) Morphogenetics of karst regions. Variants of karst evolution. Hilger, Bristol

James NP, Choquette PW (eds) (1988) Paleokarst. Springer, New York

Jennings JN (1971) Karst. The M.I.T Press, Cambridge

Joachimski MM (1994) Subaerial exposure and deposition of shallowing upward sequences: evidence from stable isotopes of Purbeckian peritidal carbonates (basal Cretaceous), Swiss and French Jura mountains. Sedimentology 41:805–824

Juhász E, Korpás L, Balog A (1995) Two hundred million years of karst history. Dachstein limestone hungary. Sedimentology 42: 473–489

Kele S (2009) Édesvízi mészkövek vizsgálata a Kárpát-medencéből: paleoklimatológiai és szedimentológiai elemzések (Study of freshwater limestones from the Carpathian Basin: paleoclimatological and sedimentological implications). PhD theses, Eötvös Loránd University & Hungarian Academy of Sciences, Budapest (in Hungarian)

Kelemen P, Dunkl I, Csillag G, Mindszenty A, von Eynatten H, Józsa S (2017) Tracing multiple resedimentation on an isolated karst plateau: The bauxite-bearing Miocene red clay of the Southern Bakony Mountains Hungary. Sed Geol 358:84–96

Kelemen P, Csillag G, Dunkl I, Mindszenty A,Kovács I, von Eynatten H, Józsa S (2020) Terrestrial kaolin deposits trapped in Miocene karstic sinkholes on planation surface remnants, Transdanubian Range, Pannonan Basin (Hungary). Geol Magazine. https://doi.org/10.1017/S0016756820000515

Knauer J, Gellai MB (1978) A szenon képződmények elrendeződése és kapcsolata az ősdomborzattal a Sümeg-Káptalanfa bauxitkutatási területén (Arrangement of the Senonan formations in the Sümeg-Káptalanfa bauxite-exploration area and their relationship with the paleorelief). Földtani Közlöny (Bull Hung Geol Soc) 108 (4):444–475 (in Hungarian)

Koloszár L (2004) A Tengelici Formáció kifejlődései a DK-Dunántúlon (Developments of the Tengelic Formation in South-Eastern Transdanubia). Földtani Közlöny 134(3):345–368 (in Hungarian)

Komlóssy G (1970) The Iszkaszentgyörgy bauxite (SE Bakony Mts, Hungary). Problems of genesis and mineral formation. Ann Inst Geol Publ Hung 54:347–358

Kormos T (1917) A Villányi-hegység preglaciális képződményei és faunájuk (Preglacial formations of the Villány Hills and their faunas). A Magyar Királyi Földtani Intézet Évi Jelentése 1916-ról (Annual Report of the Hungarian Royal Geological Institute 1916), pp 399–415 (in Hungarian)

Korpás L (1998) Paleokarst studies in Hungary. Occasional papers of the Geological Institute of Hungary, vol 195

Korpás L, E, Juhász (1990) Paleokarszt földtani modellek (Geological models of paleokarsts). Karszt És Barlang II:105–116 (in Hungarian)

Korpás L, Juhász E, Szabó I (1991) Middle and Upper Triassic palaeokarst in the Transdanubian Central Range, Hungary. Abstracts, 13th Reg. Meeting of IAS, Jena (Germany), p 77

Kraus S (1988) Eocén őskarsztos üregek a Mátyás-hegyi-barlangban (Eocene paleokarstic hollows in the Mátyás-hegy Cave). Karszt és Barlang 1988(II):79–80 (in Hungarian)

Kretzoi M (1942) Spelaeus-Fauna aus dem Mecsek-Gebirge ohne Höhlenbären (Spelaeus-Fauna from the Mecsek Mts., without cave-bears). Földtani Közlöny 72(4–12):364 (in German)

Kriván P (1956) Felderítő előzetes festékföld-tűzállóagyag kutatás Sümeg és Cserszegtomaj környékén (Prospecting for fire-clay deposits at Sümeg-Cserszegtomaj). MÁFI (Geological Institute of Hungary) jelentés, Manuscript Report (in Hungarian)

Kriván P (1959) Mezozoós karsztosodási és karsztlefedődési szakaszok, alsóbartoni sziklásparti jelenségek a Budai-hegységben. A szubgresszió fogalma (Periods of Mesozoic karstification and the burial of the resulting karst features in the Buda Hills. The concept of subgression. Földtani Közlöny (Bull Hung Geol Soc) 89 (4):393–401 (in Hungarian)

Lantos Z (2004) Liász neptuni telérek és átülepített medenceüledékek nyomában—Karbonátszedimentológiai esettanulmányok (Liassic neptunian dykes and redepostied basinal sediments—Carbonate-sedimentological case histories). PhD theses, Eötvös Loránd University, Budapest (in Hungarian)

Mindszenty A (1991) Superimposed paleokarst phenomena in the Halimba basin, South Bakomy, Hungary. Abstracts, Field Symp. on Tethyan Cretaceous Formations and Related Mineral Resources, IGCP Project 262, Tirana (Albania), 11–17 October 1991, pp 28–30

Mindszenty A (1992) Diagenezistörténeti vizsgálatok a tatai Kálvária-domb felső triász szelvényében (Diagenetic studies in the Upper Triassic of the Kálvária-Hill, Tata). Manuscript Report, Eötvös Loránd University, Budapest (in Hungarian)

Mindszenty A, Knauer J, Mátéfi-Stefler M (1994) Superimposed paleokarst phenomena in the Halimba basin (South Bakony, Hungary)—The anatomy of a multiple regional unconformity. Abstracts, IAS 15th Reg.Meeting, Ischia (Italy), pp 285–286

Mindszenty A, Deák FJ (1999) Karbonátos paleotalajok a gerecsei felső-triászban (Carbonate paleosols from the Upper Triassic of the Gerecse Mountains, Hungary). Földtani Közlöny (Bull Hung Geol Soc) 129(2):213–248 (in Hungarian)

Mindszenty A, Csoma A, Török, Hips K, Hertelendi E (2000) Flexura jellegű előtéri deformációhoz köthető karsztbauxitszintek a Dunántúli-középhegyégben (Rudistid limestones, paleokarst and geodynamics. The case of the Cretaceous of the Transdaubian Range). Földtani Közlöny (Bull Hung Geol Soc) 131(1–2):107–152 (in Hungarian)

Mindszenty A (2019) Soils or Sediments. The role of R. Langohr's process oriented approach in understanding carbonate-related palaeosols of the stratigraphic record. In: Deák J, Ampe C, Mikkelsen JA (eds) Soils as records of past and present. Proceedings of the Geoarchaeological Meeting, Bruges, 6–8 November 2019, pp 263–269

Müller P (1989) Hydrothermal paleokarst of Hungary. In: Bosák P (ed) Paleokarst—a systematic and regional review. Academia, Prague & Elsevier, Amsterdam, pp 155–164

Nádor A (1991) A Budai-hegység paleokarszt jelenségei és fejlődéstörténetük (Paleokarst evolution of the Buda Hills). PhD theses, Eötvös Loránd University, Budapest (in Hungarian)

Nádor A (1992) Palaeokarstic features in Triassic-Eocene carbonates. Zentralblatt Für Geologie Und Paleontologie 11:1317–1329

Nádor A, Sásdi L (1991) A Budai-hegység paleokarsztjai és fejlődéstörténetük. I. Termális hatást nem tükröző paleokarsztok (Paleokarsts and long-term karst evolution of the Buda Hills. I. Paleokarsts without thermal effect). Karszt és Barlang 1991(I–II):3–10 (in Hungarian)

Nádor A, Korpás L, Juhász E (1993) Palaeokarst controlled by high-frequency sea-level changes, Buda Mountains, Hungary. A MÁFI Évi Jelentése 1991-ről (Annual Report of the Hungarian Geological Institute 1991), pp 111–127

Palmer AN (2007) Cave geology. Cave Books, Dayton, Ohio

Pásztor D (2016) Hipogén karsztosodási indikátorok a Gellért-hegyi Aragonit-barlangban (Indicators of hypogenic karstic phenomena on the mineral paragenesis of the Gellérthegy Aragonit Cave). MSc theses, Department of Physical and Applied Geology, Eötvös Loránd University, Budapest (in Hungarian)
Pazonyi P (2009) A Kárpát-medence felső-pliocén és kvarter emlősfauna közösségeinek paleoökológiai vizsgálata (Palaeoecological investigations of Upper Pliocene and Quaternary mammalian communities in the Carpathian Basin). Földtani Közlöny 139(3):283–304 (in Hungarian)
Pazonyi P, Virág A, Podani J, Pálfy J (2018) Microtus (Microtus) nivaloides from the Somssich Hill 2 site (southern Hungary): an early pleistocene forerunner of modern 'true' Microtus voles revealed by morphometric analyses. Quatern Int 481:61–74
Poros Zs (2011) Fluid migration and porosity evolution in the Buda Hills, Hungary selected examples from Triassic and Paleogene carbonate rocks. PhD theses, Eötvös Loránd University, Budapest
Poros Zs, Mindszenty A, Molnár F, Pironon J, Győri O, Ronchi P, Szekeres Z (2012) Imprints of hydrocarbonbearing basinal fluids on a karst system: mineralogical and fluid inclusion studies from the Buda Hills, Hungary. Int J Earth Sci (Geol Rundsch) 101:429–452
Purdy EG (1974) Reef configurations: cause and effect. In: Laporte LF (ed) Reefs in time and space. SEPM Spec Publ 18:29–76
Rákosi L, Tóth K (1980) Adatok a Déli-Bakony eocén képződményeinek litoés biosztratigráfiájához (Contribution to the litho- and biostratigraphy of the Eocene of the South Bakony area (Hungary). A MÁFI Évi Jelentése 1978-ról (Annual Report of the Hungarian Geological Institute 1978), 239–249 (in Hungarian)
Rakusz Gy, Strausz L (1953) A Villányi-hegység földtana (Geology of the Villány Hills). A Magyar Állami Földtani Intézet Évkönyve (Annals of the Geological Institute of Hungary) 41(2):1–43 (in Hungarian)
Rasmussen KA, Neumann AC (1988) Holocene overprints of Pleistocene paleokarst: bight of Abaco, Bahamas. In: James NP, Choquette PW (eds) Paleokarst. Springer, New York, pp 132–147
Read JF, Horbury AD (1993) Eustatic and tectonic controls on porosity evolution beneath sequence-bounding unconformities and parasequence disconformities on carbonate platforms. In: Read JF, Horbury AD (eds) Diagenesis and Basin development, AAPG Studeis in Geology, vol 36, pp 155–198
Rónaki L (1969) Hévizes barlangüregek és a termálvíz feltörésének nyomai a pécsi Mecsekben (Hot-water cavities and locations of thermal water discharge in the Mecsek around Pécs). Karszt és Barlang 1969(I):25–28 (in Hungarian)
Royden LH, Horváth F (1988) The Pannonian Basin—a study in basin evolution. AAPG Mem. 45, Tulsa, Oklahoma
Schmid SM, Bernoulli D, Fügenschuh L, Matenco S, Schafer R, Schuster R, Tischler M, Ustaszevski K (2008) The Alpine-Carpathian-Dinarid orogenic system—correlation and evolution of tectonic units. Swiss J Geosci 101(1):139–183
Schlager W (1981) The paradox of drowned reefs and carbonate platforms. GSA Bull 92:197–211
Schlager W (1999) Type 3 sequence boundaries. In: Advances in carbonate sequence stratigraphy: applications to reservoirs. outcrops and models. SEPM Spec Publ 63:35–45
Sebe K, Pazonyi P, Gasparik M, Szujó GL (2015) Uplift history of the Villány Hills (SW Hungary) based on paleontological data. Geophysical Research Abstracts 17, EGU2015-13518, EGU General Assembly, 2015
Sloss L (1963) Sequences in the cratonic interior of North America. GSA Bull 74:93–114
Sweeting MM (1973) Karst landforms. MacMillan Publishers Ltd., London
Szabó PZ (1961) A Mecsek és a Villányi-hegység barlangjai (Caves of the Mecsek Mts. and Villány Hills). Karsztés Barlangkutatás 1961 (I):3–20 (in Hungarian)
Szantner F, Knauer J, Mindszenty A (1986) Bauxitprognózis (Bauxite Prognosis). VEAB Kiadvány, Veszprém (in Hungarian)
Tari G (1994) Alpine tectonics of the Pannonian Basin. PhD theses, Rice University, Houston, Texas
Tari G, Linzer H-G (2018) Austrian versus Hungarian bauxites in an Alpine tectonic context – a tribute to Prof. Andrea Mindszenty. Földtani Közlöny (Bull. Hung. Geol. Soc.) 148(1):35–44
Telegdi Roth K (1923) A Dunántúli-középhegység bauxittelepei (The bauxite deposits of Transdanubia). Földtani Szemle I(2):95–103 (in Hungarian)
Telegdi Roth K (1927) Die Bauxitlager des Transdanubischen Mittelgebirges in Ungarn (Bauxite deposits of the transdanubian range, Hungary). Földtani Szemle I(1):33–45 (in German)
Török Á, Szabó-Balog A (1990) Gerecse-hegységi felső triász szelvények paleokarsztja.(Palaeokarst study of Late Triassic carbonate profiles, Gerecse Mountains). Research Report 70, Budapesti Műszaki Egyetem, Ásvány és Földtani Tanszék (Budapet Technical University, Department of Geology and Mineralogy) (in Hungarian)
Török A, Claes H, Brogi A, Liotta D, Tóth Á, Mindszenty A, Kudó I, Kele S, Huntington KW, Shen CC, Swennen R (2019) A multimethodological approach to reconstruct the configuration of a trvertine fissure-ridge system: The case of the Cukor-quarry (Süttő, Gerecse Hills, Hungary). Geomorphology 345:106836
Török Á (2018) The anatomy of a Pleistocene travertine complex—sedimentological, diagenetic and resevoir studies on the Süttő Travertine System (Gerecse Hills, Hungary). PhD, Theses, Eötvös Loránd University, Budapest & KU Leuven
Török Á, Claes H, Brogi A, Liotta D, Tóth Á, Mindszenty A, Kudó I, Kele S, Huntington KW, Shen C-C, Swennen R (2019) A multi-methodological approach to reconstruct the configuration of a travertine fissure ridge system: the case of the Cukor quarry (Süttő, Gerecse Hills, Hungary). Geomorphology 345. https://doi.org/10.1016/j.geomorph.2019.106836
Tóth K (1980) Összefüggések a bauxit előfordulása és a közvetlen fedő eocén rétegek kifejlődése között (Relationship between the occurrence of bauxite and the lithofacies of its imediate cover). Általános Földtani Szemle (General Geol Rev) 14:133–150 (in Hungarian)
Tucker ME, Wright VP (1992) Carbonate sedimentology. Blackwell Scientific Publications, Oxford
Vadász E (1946) A magyar bauxitelőfordulások földtani alkata (Geology of Hungarian bauxites). MÁFI Évk. Ann Geol Inst Hung 37:173–286 (in Hungarian)
Végh-Neubrandt E (1960) A Gerecse-hegység felsőtriászképződményeinek üledékföldtani vizsgálata. Petrologishce Untersuchung der Obertrias-Bildungen des Gerecsgebirges in Ungarn. (Sedimentological study of late Triassic formations of the Gerecse Hills) (in Hungarian and German). Geol Hungarica Ser Geol Tom 12
Vera JA, Ruiz-Ortiz PA, Garcia-Hernandez M, Molina JM (1988) Paleokarst and related pelagic sediments in the Jurassic of the Sub-Betic Zone, Southern Spain. In: James NP, Choquette PW (eds) Paleokarst. Springer, New York, pp 364–384
Vértes L (1952) A Mélyvölgyi-kőfülke és néhány más mecseki barlang kutatásáról (On the exploration of Mélyvölgy rock shelter and some other caves in the Mecsek Mountains). Földtani Közlöny 82(7–9):270–278 (in Hungarian)
Virág M (2018) Hipogén és epigén karsztos folyamatok és hatásuk a budai rózsadombi barlangok fejlődéstörténetében—a Szemlő-hegyi- és a Molnár János-barlang példáján (Hypogene and epigene karstic processes and their influence on the evolution of the Rózsadomb caves in Buda—by the example of Szemlő-hegy Cave and Molnár

János Cave). PhD Theses, Eötvös Loránd University, Budapest (in Hungarian)

Virág M, Mindszenty A, Surányi G, Kele S, Németh T, Hegedűs A, Leél-Őssy Sz (2019) Pleistocene paleohydrological changes recorded by speleothems of the Szemlő-hegy Cave (Buda Thermal Karst, Hungary). In: Blatnik M, Gabrovšek F, Kogovšek B, Mayaud C, Petrič M, Ravbar N (eds) Abstract and guide book of the karst hydrogeology—research trends and applications—27th international Karstological school "Classical Karst", Karst research institute ZRC SAZU, Postojna (Slovenia), 17–21 June 2019, p 134

Waltham T (2008) Fencong, Fenglin, Cone Karst. Cave Karst Sci 35 (3):77–88

Waltham AC, Fookes PG (2003) Engineering classification of karst ground conditions. Q J Eng Geol Hydrogeol 36(2):101–118

Weisse de J-G (1948) Le bauxites de l'Europe centrale. Province dinarique et Hongrie. (Bauxites of Central Europe. Hungary and the Dinaric Province) (in French). Mém de la Soc vaudoise des Sci naturelles 9(1):58

Whitaker FF, Smart PL (1990) Active circulaiton of saline ground waters in carbonate platforms: evidence from the Great Bahama Bank. Geology 18:200–203

Wright VP, Marriott SB (1996) A quantitative approach to soil occurrence in alluvial deposits and its application to the Old Red Sandstone of Britain. J Geol Soc 153:907–913

Wright VP (1982) The recognition and interpretation of paleokarsts: two examples from the Lower Carboniferous of South Wales. J Sed Petr 52:83–94

Wright VP (ed) (1992) Palaeokarsts and palaeokarstic reservoirs. A course book prepared for a University of Reading Short course, 2 May 1991, P.R.I.S. Contribution No.152, Postgraduate Research Institute for Sedimentology, University of Reading

Hydrogeology of the Karst Regions in Hungary

Judit Mádl-Szőnyi, Anita Erőss, Katalin Csondor, Veronika Iván, and Ádám Tóth

Abstract

The Hungarian karst regions with diverse geologic settings and hydrogeologic characteristics are presented and discussed in a basin-scale groundwater flow framework which is the adaptation of groundwater flow systems to carbonate regions and, thus, a novel approach in karst hydrogeology studies. Accordingly, a solid but comprehensive overview of the hydrogeologic environment (topography, climate, and geology), groundwater flow patterns and processes (flow systems and directions, hydrochemistry and temperature conditions), groundwater-related phenomena (springs and caves), and groundwater resources (direct and indirect use) is provided for the Aggtelek Karst, Bükk Mts., SW and NE Transdanubian Mts., Mecsek Mts., and Villány Mts. By the application of this concept, the asymmetric flow patterns, distribution of cold, lukewarm and thermal springs, heat accumulation and heat transport processes, mixing of meteoric and basinal and waters of local, intermediate and regional groundwater flow systems and potential locations of epigene and hypogene speleogenesis can be revealed and explained. The Hungarian karst areas reflect different properties but there are similarities regarding their hydrogeologic environment and flow systems, as well. Therefore, the local-scale karst features and peculiarities fit into a bigger, i.e., basin-scale, picture which can provide a sound base for regional hydrogeologic characterization of karstic carbonate areas.

Keywords

Aquifer • Groundwater flow system • Unconfined and confined carbonate • Spring • Epigene and hypogene karst • Heat utilization

1 Introduction

Karst hydrogeological researches, which refer to the investigation of mineral and thermal springs and understanding the geology of karst regions, have a long-term tradition in the Carpathian Basin, Hungary. The first books in this topic about Transylvanian mineral waters and the water of Buda were published in the eighteenth century (Stocker 1729; Kibédi Mátyus 1787–1793). The first artesian boreholes in karstic rocks were drilled close to springs of Ugod (1825) and Csór (1832, Transdanubian Mts., Dobos 1976). Vilmos Zsigmondy (1871, 1879) revealed the significance of fractures in carbonate rocks during his successful drilling activity in Harkány (1866, Villány Mts.) and Budapest, on the Margaret Island (1867–1870) and in the Városliget (City Park 1868–1878). These deep boreholes provided information about the confined carbonate rocks and underground karstic network. The interpretation of the chemical composition of karst waters was introduced by Than (1890). To understand the connectivity of karstic springs in the Bükk Mountains, first tracer tests were executed in Diósgyőr by Emszt (1913). Underwater mining from the twentieth century required intense hydrogeological exploration to prevent karst water intrusion during the production of coal and bauxite in the Transdanubian Mountains (Vígh 1944; Liebe 2002).

J. Mádl-Szőnyi (✉) · A. Erőss · K. Csondor · Á. Tóth
Department of Geology, ELTE Eötvös Loránd University, Pázmány Péter sétány 1/C, Budapest, 1117, Hungary
e-mail: judit.szonyi@ttk.elte.hu

A. Erőss
e-mail: eross.anita@ttk.elte.hu

K. Csondor
e-mail: katalin.csondor@ttk.elte.hu

Á. Tóth
e-mail: toth.adam@ttk.elte.hu

V. Iván
Department of Geosciences, University of Padova, Via G. Gradenigo 6, 35131 Padova, Italy
e-mail: veronika.ivan@unipd.it

M. Veress and Sz. Leél-Őssy (eds.), *Cave and Karst Systems of Hungary*,
Cave and Karst Systems of the World, https://doi.org/10.1007/978-3-030-92960-2_6

In the remarkably diverse literature of karst hydrogeology of Hungary, the demand for understanding regional interrelationships, flow patterns, and temperature distribution has appeared in the early twentieth century for various karst regions (Schafarzik 1928; Vendel and Kisházi 1964; Kovács and Müller 1980; Papp 1907). The interrelationship between flow and related karstification processes (springs, cave formation, dissolution, precipitation, etc.) sporadically appeared in the literature. These regional flow models remained fully conceptual without scientifically solid pieces of evidence. The regional groundwater flow concept of Tóth (1962, 1963) and the understanding of geological consequences of regional groundwater flow in the upper continental crust (Tóth 1999) opened a new path for scientific evolution in karst studies. Since groundwater is considered as a geologic agent (Tóth 1999), karst phenomena including caves, springs, and the karst system itself can be interpreted as manifestations of flowing groundwater. Therefore, groundwater flow system understanding has a crucial role in karst research and in clarifying the origin and formation of the caves. However, the application of this theory to karst media remained out of scope for some decades because the attention was paid to the revealed high-level heterogeneity of karst, like in other countries (Ford and Williams 2013; Goldscheider and Drew 2014).

The connection between regional groundwater flow and potential dissolution processes in carbonates in the framework of the groundwater flow concept of Tóth (1999) was firstly recognized by Klimchouk (2007). He distinguished between epigene and hypogene karst in the context of basinal groundwater flow (Klimchouk 2007; Klimchouk et al. 2017). He found that epigene karstification can be connected to local flow systems while hypogene cave formation is related to intermediate or regional flow systems. The hydrogeological background of epigene karstification can be understood from this approach, however, based only on the two unconfined basin-halves of Tóth, the location and characteristics (hydrochemistry and temperature) of thermal and lukewarm springs, as well as the near-surface and underground locations of hypogene karstification cannot be properly explained (Mádl-Szőnyi et al. 2017). Goldscheider et al. (2010) published the first paper highlighting the significance of flow system approach in the understanding of the thermal water resources and related geochemical processes. However, the revealed discrepancy was only resolved by the conceptualization of Tóthian-type flow pattern for adjoint confined and unconfined basin-halves based on the joint interpretation of numerical flow and heat transport simulations (Fig. 1, Mádl-Szőnyi and Tóth 2015; Szijártó et al. 2019; Tóth et al. 2020). Based on these studies, we can identify those parts of the system where the carbonate rocks are unconfined or partially covered with siliciclastic sediments (i.e., semi-confined basin-half) and where intense meteoric water recharge takes place. In the other basin-half, where carbonate rocks are fully confined by thick dominantly siliciclastic sediments (with low hydraulic conductivity), the confining layers hold basinal fluids. The interplay of fluids of the two basin-halves and the revealed asymmetric flow pattern can explain the distribution of cold, lukewarm and thermal springs, the heat accumulation and heat transport processes and the mixing of basinal and meteoric waters (of local, intermediate and regional flow systems). Moreover, based on this concept, the potential locations of epigene and hypogene speleogenesis can be revealed. This geometry of two connected basin-halves of unconfined and confined carbonates, and the consequent heat distribution and flow pattern, the occurrence of springs and caves can be handled as a new adaptation of the Tóthian flow concept for carbonate sequences (Fig. 1, Tóth et al. 2020; Mádl-Szőnyi and Tóth 2015; Mádl-Szőnyi 2020).

The hydrogeology of Hungarian karst areas for today can be characterized by flourishing but remarkably diverse literature. However, the knowledge and the depth of understanding are quite different for the various areas. Therefore, during the hydrogeological introduction of the karst areas of Hungary, we put a focus on the understanding of groundwater flow systems (based on the conceptual flow model of Fig. 1) because this conceptual background provides a common basis for the description.

We discuss the karst areas based on their geographic location in the groups of the Northern Hungarian Mountains (Aggtelek Karst, Bükk Mountains), the Transdanubian Mountains (SW Transdanubian Mountains, NE Transdanubian Mountains), the Mecsek Mountains and Baranya Hills (Mecsek Mountains, Villány Mts., Fig. 2). For the karst areas, firstly, the elements of the hydrogeological environment are introduced: topographical conditions which directly influence groundwater flow systems; climate conditions determining the precipitation, evapotranspiration and recharge; and geological conditions. Here, we focus on the Pre-Tertiary structural elements, the position of confined and unconfined (outcropping) carbonates and the main hydrostratigraphic categories (i.e., aquifers, aquitards). As a second step, we characterize the groundwater flow pattern (recharge and discharge areas, flow regimes, groundwater flow systems with their hydrogeochemical character and temperature conditions) and the related manifestations (distribution of springs, epigene and hypogene caves). Due to the hydraulic connection between siliciclastic and carbonate rocks, we did not differentiate between karstic and non-karstic springs. Finally, we discuss the groundwater resources of the areas focusing on drinking water, mineral water, industrial, agricultural, and balneological (i.e., thermal spas) use of groundwater.

The demonstration of the above-mentioned topics for the different karst regions was based on a conceptually uniform

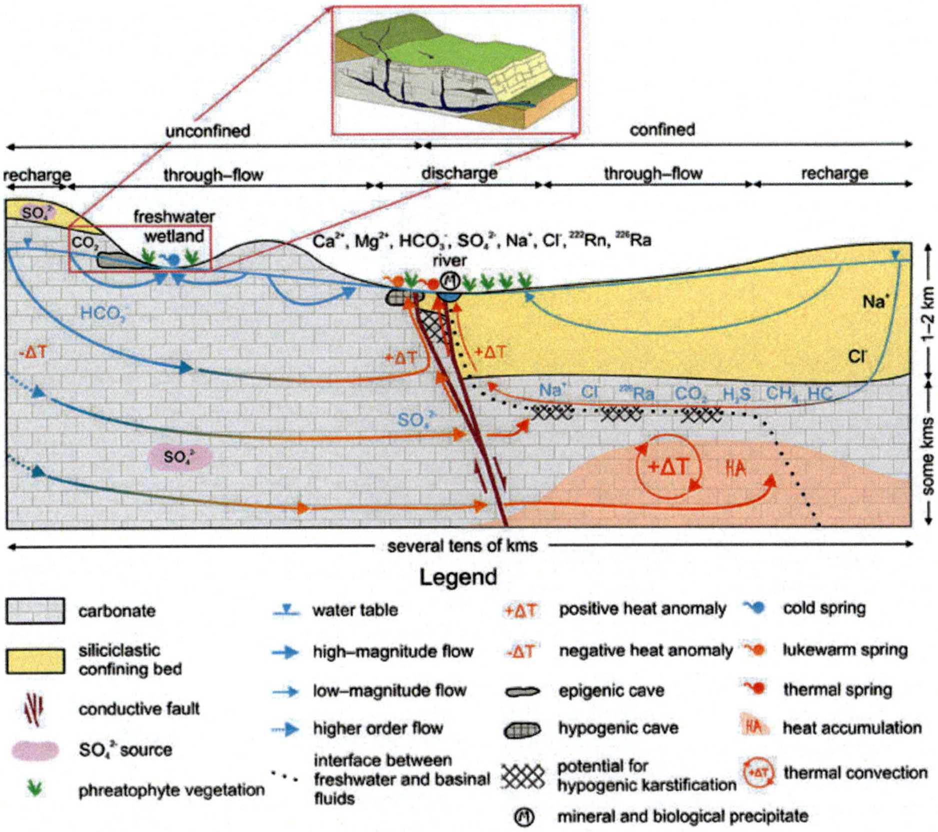

Fig. 1 Conceptual flow pattern for theoretical unconfined and confined basin-halves of carbonate systems. The different orders of flow systems, the fluids with different origin (meteoric and basinal), chemical and heat transport processes are indicated together with the different type of springs (cold, lukewarm, warm) with the potential places for epigene and hypogene karstification (after Mádl-Szőnyi and Tóth 2015). The interplay between different basin-parts, the flow pattern, the areal distribution and chemical character of springs, as well as the temperature field, might change according to different hydrogeological environments

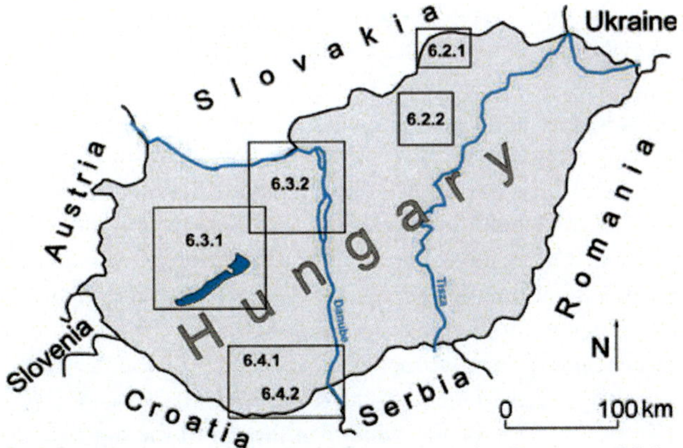

Fig. 2 Index map of displayed karst areas in Hungary: 6.2.1—Aggtelek Karst, 6.2.2—Bükk Mts., 6.3.1—SW Transdanubian Mts., 6.3.2—NE Transdanubian Mts., 6.4.1—Mecsek Mts., 6.4.2—Villány Mts

map series compiled for the goals of the chapter. These are as follows: (i) Hydrostratigraphic map with the indications of the outcrops and structural elements of the Pre-Tertiary basement (adapted from Haas et al. 2010) and the epigene and hypogene caves longer than 50 m (Hungarian Ministry of Agriculture 2020); (ii) Topography with the location of springs (Izápy 2002), main rivers and lakes; (iii) Topography with the indication of direct use of karst groundwater and exploitation of heat from groundwater in carbonate aquifers (Kocsis et al. 2018).

2 North Hungarian Mountains

The North Hungarian Mountains is a mountainous area in the northeastern part of Hungary, where the highest mountains of Hungary are located. Two significant karstic regions, which are rich in caves, can be found here: the Bükk Mountains and the Aggtelek Karst.

2.1 Aggtelek Karst

The Aggtelek Karst (or North-Borsod Karst) is part of the North Hungarian Mountains and located in the northeastern part of Hungary. The karst area extends to the southern Slovakian Slovak Karst. Altogether this transboundary karst system is called Gömör–Torna/Gemer–Turňa Karst, which belongs to the Inner Western Carpathians (Szentpétery and Less 2006). It is characterized by well-developed karst features, transboundary cave systems, and springs. On the Hungarian side of the Aggtelek Karst of according to the structural and geomorphological elements, eight main hydrogeological units can be differentiated: Aggtelek Plateau (the south-eastern part of the Silica Plateau), Alsó Hill, Jósvafő Karst, Bódva Valley, Torna Hill, Galyaság, Szalonna Mts., and Rudabánya Mts. (Fig. 3, Sásdi 1998). It is to note, that from a geographical and hydrogeological standpoint, the units situated on the state border are continuous in the direction of the Slovak Karst.

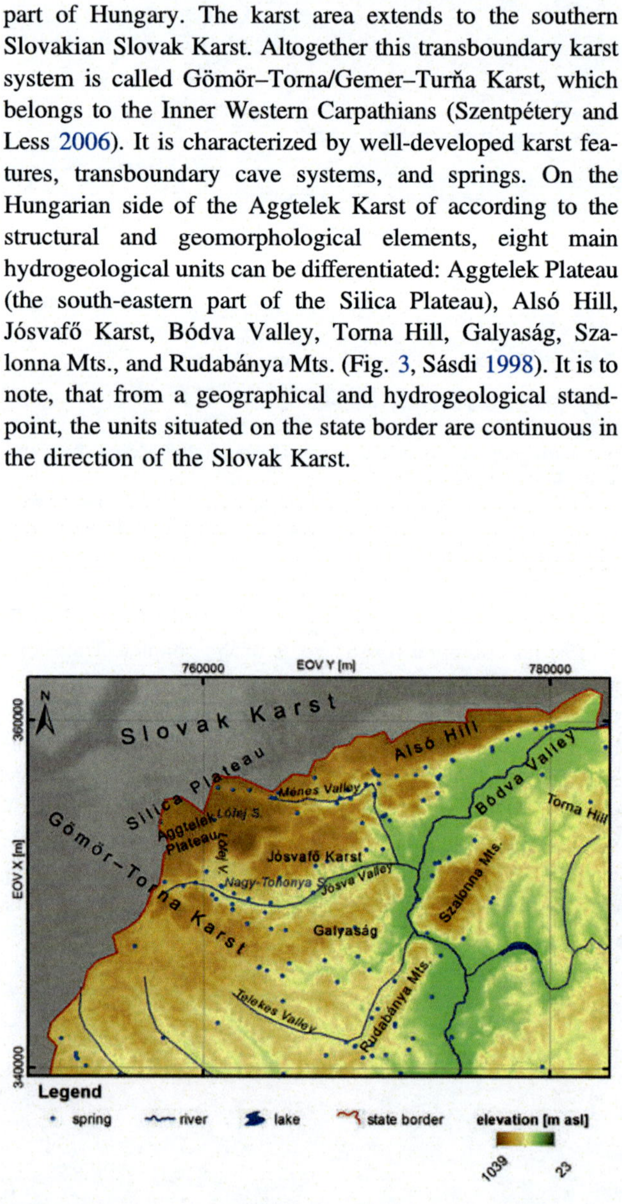

Fig. 3 Topography and location of springs (Izápy 2002) for the Aggtelek Karst

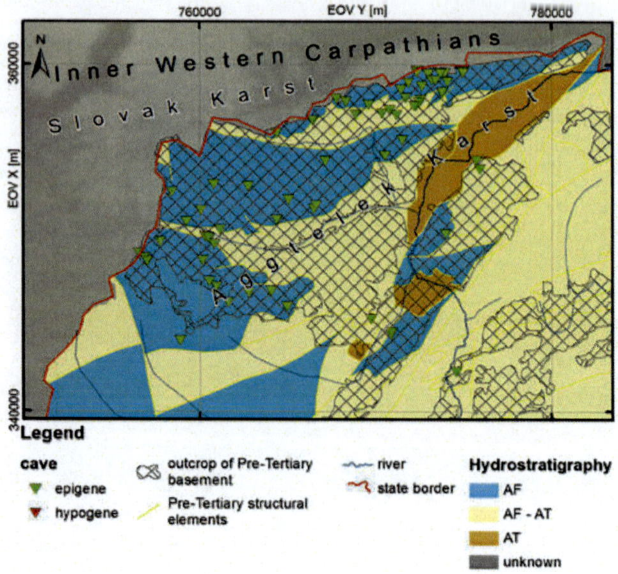

Fig. 4 Hydrostratigraphy, outcrops and structural elements of the Pre-Tertiary basement of the Aggtelek Karst (adapted from Haas et al. 2010)

2.1.1 Hydrogeologic Environment

Topography The Aggtelek Karst is rich in surface and subsurface karst landforms, such as caves, dolines, canyon- and gorge-like valleys, sinkholes and karst springs (Figs. 3 and 4; Gessert 2016; Zámbó 1998). The area can be characterized by mosaic-like topography: the doline-dotted karst plateaus are divided by tectonic–fluvial valleys (Telbisz et al. 2020). The unconfined karst plateaus reach 450–600 m asl elevation. At their edges, the topography drops suddenly. In the Jósva and Bódva Valleys (Fig. 3), the elevation ranges typically between and 140 and 250 m asl.

Climate The climate of the area is humid cold continental, showing orographic differences. The annual mean temperature is 8 °C in the higher regions and 10 °C in the lower ones. The annual precipitation varies between 600 and 750 mm. On average, the evapotranspiration is 71–72% of the precipitation (Gruber et al. 2012; Izápy and Maucha 2004), annually ranging between 440 and 500 mm. The surface runoff usually does not exceed 2% of the precipitation (Izápy and Maucha 2004; Gruber et al. 2012). Consequently, 26–27% of the rainfall infiltrates to the aquifer.

Geology The Aggtelek Karst can be characterized by nappe structures, superposed by folded–imbricated structures. Above these, two horizontal deformation zones in east-western line of the strike are detectable (Fig. 4, Szentpétery and Less 2006). The dominantly Middle (and Upper) Triassic carbonates of the Silica nappe form the main

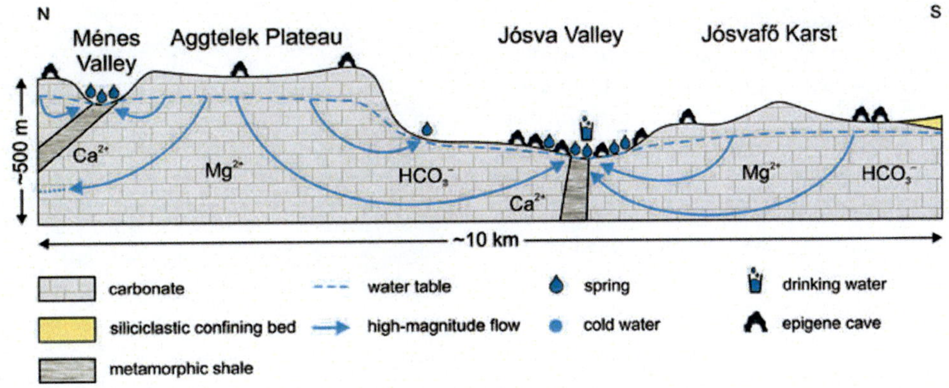

Fig. 5 Generalized hydrogeological cross-section for the plateau of the Aggtelek Karst

karstic aquifers of the area (Szentpétery and Less 2006). Among the Triassic carbonates, a large proportion is highly karstified, constituting an excellent aquifer, while being in complex tectonic relation with the aquitards: Permian evaporites, Triassic sandstones and marls (Szentpétery and Less 2006). The thickness of the karstified carbonate ranges is typically from 400 to 1000 m (Szentpétery and Less 2006). At the edges of the karst plateaus, the Triassic formations become gradually confined by Miocene covering sediments (such as continental, fluvial, and lacustrine sediments) and Quaternary sediments (such as gravels, various fluvial formations, and doline infillings—dominantly red clay—and slope sediments). Fluvial formations are characterized by good hydraulic conductivity, while the doline infillings and slope sediments can be described as aquitards, with a heterogeneous thickness (Szentpétery and Less 2006). The typical karst plateaus, with sinkholes and dolines, are separated by deep fluvial valleys (Zámbó 1998), constituting smaller, connected, but diverse hydrogeological units. The cross-section represents only the karst plateau without confined parts of the system (Fig. 5).

2.1.2 Groundwater Flow Patterns and Processes

The Aggtelek Karst has been in the focus of hydrogeological studies since the early twentieth century. Systematic hydrogeological observations and experiments (e.g., tracer tests) were reported starting from the 1930s (Jakucs 1951; Kessler 1938). Between 1958 and 1993, a complex hydrological monitoring program was conducted at the Ferenc Papp Karst Research Station and a comprehensive hydro-meteorological database was recorded (Maucha 1998). The Aggtelek Karst can be characterized by highly complex structural settings (Hips 2001) and as a mature karst system, by high spatial and temporal heterogeneity. In order to discover the hydrological connections and flow directions of the region, numerous tracer tests were carried out (Gruber et al. 2015; Izápy and Maucha 2004; Sásdi and Szilágyi 1986). The tracer tests revealed the complexity of the flow patterns, e.g., distant points of the karst systems may be in rather rapid hydrological connection even if they are separated by aquitard zones (Gruber et al. 2015).

The uncovered karst plateaus can be characterized by autogenic recharge. In contrast, on the areas south of the Jósva Valley, allogenic recharge reaches the karst springs via sinkholes and sinking streams from the non-karst terrains (Fig. 4, Jakucs 1997).

The springs of the area have typically Ca–HCO_3 water type, or in case of discharging from (partially) dolomite formations, Ca–Mg–HCO_3 facies (Sásdi 1998). The total dissolved solids content varies between 540 and 840 mg/l. The sulfate and TDS content is higher in the springs of the Ménes Valley and the western part of Alsó Hill. This indicates the presence of the Permian evaporite formations in their aquifer (Fig. 3, Sásdi 1998).

The spring water temperatures vary in a relatively narrow range, 80% of the springs are cold, with yearly average temperatures between 8.8 and 12 °C. They show an inverse relation with the discharge elevation (Iván and Mádl-Szőnyi 2018; Maucha 1998). This indicates that the majority of the springs are fed by local flow systems. 20% of the springs can be described as lukewarm with temperatures up to 27 °C (but typically ranging between 12 and 17 °C), which means that they have a flow component from intermediate flow systems as well. This concept is proven for the area through Tritium concentration measurements as well (Izápy and Maucha 2004; Sásdi 1998). The Tritium content of the water is indicative of the time since its infiltration to the aquifer. The evaluation of Tritium concentrations revealed that during baseflow conditions, cold springs are discharging water infiltrated a few months—1–2 years earlier, or in case of dolomite aquifer, 4–6 years earlier. In contrast, the subthermal component of the lukewarm springs can be characterized by significantly longer flow routes and residence time larger than 40 years (Fig. 5, Sásdi 1998).

2.1.3 Groundwater-Related Phenomena

Numerous karst springs are emerging in the area, ranging from the temporary, low discharging ones to the major permanent springs. A spatial–statistical study revealed that spring occurrences are most often connected to low altitudes (145–246 m asl), to the proximity of structural lineaments (≤ 250 m distance) and contact zones between permeable and less permeable formations (≤ 200 m distance, Iván et al. 2020). The springs can be characterized by significant discharge fluctuation: high-flow discharges may exceed baseflow discharge volumes by more than two orders of magnitude (Szentpétery and Less 2006). The springs discharging at lower elevations or from dolomite formations can be characterized by less intense fluctuation and higher baseflow rates (Iván and Mádl-Szőnyi 2018; Sásdi 1998). In the case of two springs (Lófej Spring and Nagy-Tohonya Spring), siphonal eruptions can be observed. This phenomenon occurs periodically and independently from precipitation, causing a drop in discharge in the pre-eruption period, followed by a sudden and intense increase which lasts several hours (Fig. 3, Maucha and Sárváry 1970).

Especially after precipitation or snowmelt, sinkholes transport water from the surface into the karst. According to the water balance equations, on average 25–27% (min. 9%, max. 49%) of the precipitation infiltrates into the karst system through fissures and sinkholes (Szentpétery and Less 2006). The sinkholes of the area are diverse in terms of size, form, infilling, and activity since they evolved in various positions (Zámbó 1998). Based on the latter, three main types of sinkholes can be differentiated (Sásdi 1998): (1) sinkholes developed on the contact zones of karstified aquifers and aquitards (with sinking capacity from 1440 m^3/day up to 14,400 m^3/day, being usually active during snowmelt and intense precipitation events); (2) sinkholes developed on karst plateaus, usually in dolines (generally having small catchment area and not very significant infiltrated volume); and (3) sinking streams situated in the gorge-like karstic valleys (with generally low infiltration capacity but typically active for most of the year, e.g., Lófej Valley, Tohonya Valley, Telekes Valley, Northern parts of Ménes Valley). Along the dry valleys, sinkhole alignments can be found as well (Fig. 3).

Due to the extensive karstification processes around 1200 caves are present in the area (Telbisz et al. 2020), out of which 75 exceed the length of 50 m. From the hydrogeological and genetic standpoint, they are epigene, meaning that the dominant cave development process is connected to the local flow systems of meteoric waters. A wide variety of caves is represented in the area, such as nearly vertical shaft caves, inflow, outflow, and through-flow caves. Their activity (permanent or transient) is dependent on their topographical situation.

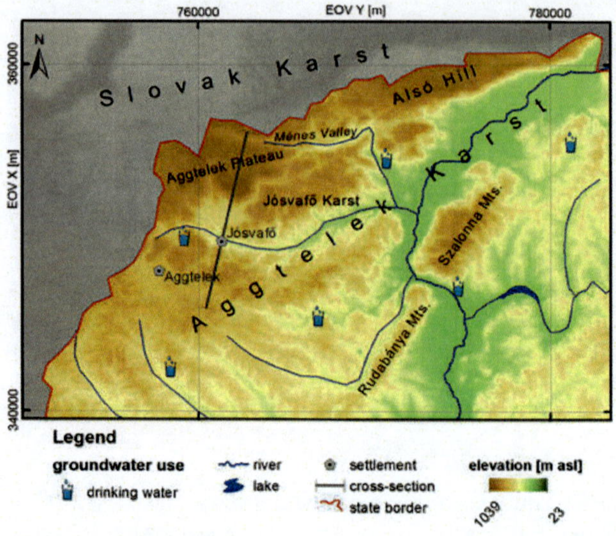

Fig. 6 Topography and location of groundwater use (Kocsis et al. 2018) for the Aggtelek Karst

2.1.4 Groundwater Resources

The Aggtelek Karst lies almost entirely under natural protection, with low industrial activity and very low population density (Dövényi 2010). Thus, the primary use of groundwater resources is to provide drinking water for the inhabitants of the area. In addition, numerous nature conservation values are dependent on groundwater resources (Fig. 6).

2.2 Bükk Mountains

The Bükk Mountains is part of the North Hungarian Mountains, and this is the second-largest karst system of Hungary. The mountain is built up by thick Triassic carbonate rocks which feature well-developed karst with surface karst phenomena, the abundance of karst springs and huge cave systems with the two deepest caves of the country.

2.2.1 Hydrogeologic Environment

Topography The Bükk Mts. is the highest karst mountain range in Hungary. Although the elevation of the most

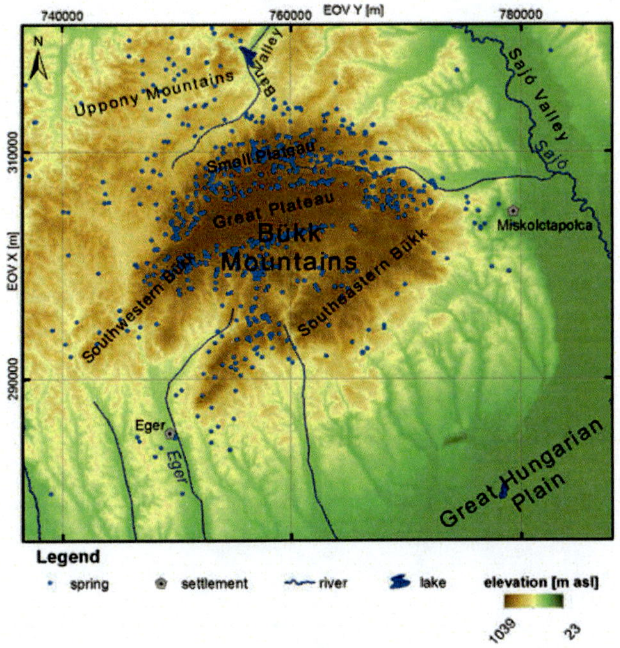

Fig. 7 Topography and location of springs (Izápy 2002) for the Bükk Mts

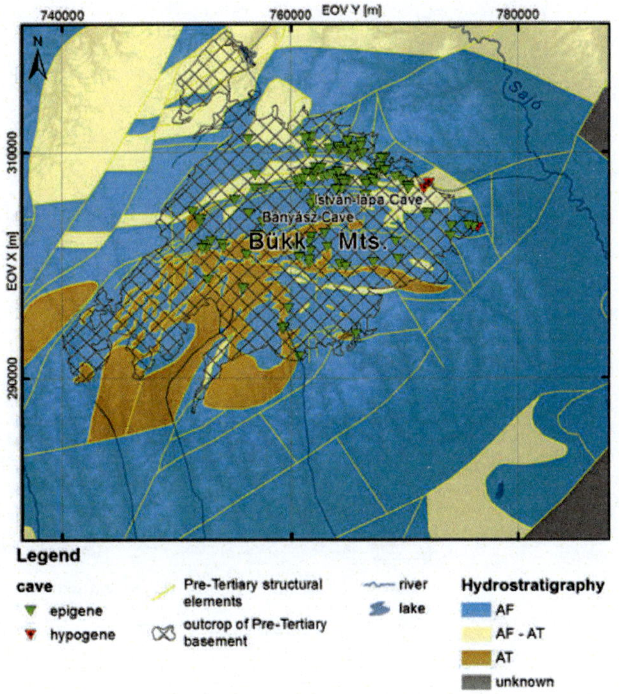

Fig. 8 Hydrostratigraphy, outcrops and structural elements of the Pre-Tertiary basement of the Bükk Mts. (adapted from Haas et al. 2010)

prominent peaks does not reach 1000 m asl, 50 ridges rise above 900 m asl. It is bordered in the east by Sajó Valley, in the north by Bán Valley, the western border is Uppony Mountains (400–450 m asl) and the valley of Eger stream, and the southern border is the Great Hungarian Plain (110–130 m asl). The ~900-m-high plateau is surrounded by a lower degradation surface (300–400 m asl) and wide foothills with valleys. Most of the Bükk Mts. is characterized by outcrops of unconfined karst with abundant surface karst phenomena like different types of karren, dolines and 127 sinkholes (Pelikán et al. 2005). The most abundant surface karst phenomena can be found on the Great Plateau, but favorable rocks for karstification can also be found on the Small Plateau, in the North and South Bükk regions (Fig. 7).

Climate The climate of the Bükk Mts. is determined by the changing altitude and the topographic conditions. The average annual mean temperature is 6–9 °C. Due to the significant continental effect, the winter is cold and the number of snowy days is the highest in the country. The annual mean rainfall is 600–700 mm, but on the Great Plateau is 800 mm (Kocsis et al. 2018). Despite the relatively abundant rainfall, the mountains are poor in rivers and lakes and the infiltrating rainwater appears on the surface in the form of karst springs at the edges of limestone areas (Hevesi 2002). The annual recharge may exceed 36% on the uncovered karst surfaces (Izápy and Maucha 1992).

Geology The Bükk Mts. is widely regarded as a limestone mountain as a whole. However, the geological map reveals that less than half of its surface is built up of carbonate rocks (Pelikán et al. 2005). The majority of rocks were formed from the Carboniferous to the end of the Middle Jurassic period. The middle and upper Triassic platform limestones represent well-karstified and good aquifer units and they are unconfined in most of the Mountains. Besides the thick carbonates, formations were deposited related to volcanism and can be found to a large extent on the surface in the W–SW part of the Mountains (Pelikán et al. 2005). The SW part of the Mountains and the surrounding basin can be considered as confined karst, and the non-carbonate formations are mainly aquitards (Fig. 8). Detailed hydrogeological classification of the formations is provided by Darabos (2017) and Miklós et al. (2020).

2.2.2 Groundwater Flow Patterns and Processes

Based on the geological, mainly folded structures, three different hydrogeological units—the Northern, the Middle and the Southern—are distinguished (Pelikán et al. 2005), which are hydraulically connected on a regional scale (Gondár-Sőregi et al. 2011; Miklós et al. 2020). However, there are several subsystems which are hydraulically separated locally by low-permeability rocks (Kovács et al. 2015).

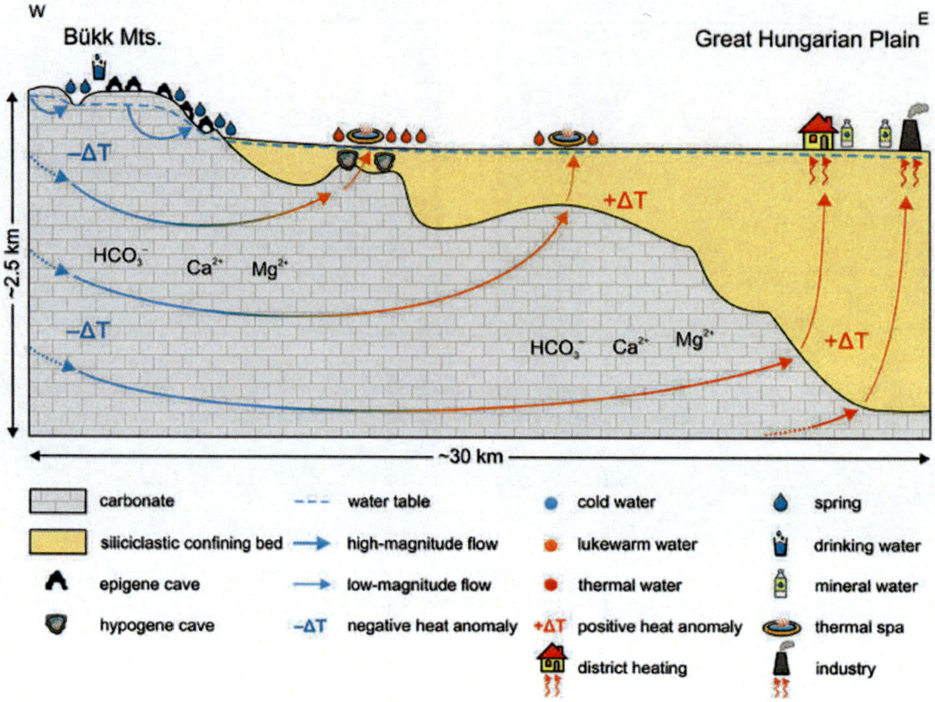

Fig. 9 Generalized hydrogeological cross-section for the Bükk Mts

The karst area is allogenic, the size of the uncovered karst area of the Bükk Mts. is around 207 km^2 and the water catchment area is around 230 km^2 (General Directorate of Water Management 2016). According to the previously established models (Schréter 1923; Aujeszky and Scheuer 1979; Izápy and Sárváry 1991; Lénárt 2011; Lénárt et al. 2014; Lénárt and Lénárt 2008; Gondár-Sőregi et al. 2011), waters infiltrating on uncovered karst feed the local and intermediate flow systems, which generate cold water discharge. Another part of the waters infiltrates and flows deep to the confined part of the system and forms the thermal karst area of the Bükk Mts. in the eastern, western, and southern foothills. There is a well-established monitoring network operated by the University of Miskolc in the area since 1992, which includes monitoring points all over the mountains and in their close vicinity providing water level, temperature and electrical conductivity data (Lénárt et al. 2014).

Detailed geochemical investigation of karst waters in the Southern Unit and the S–SW–SE foreland of the Bükk Mts. showed that—except some thermal wells in the S–SW foreland—most of the karst waters have very similar total dissolved solid content (500–600 mg/l), despite their wide range of temperatures (8–77 °C). The chemical composition of these waters may indicate deep but relatively short flow path and homogeneous aquifer lithology, where a considerable amount of heat is collected but the transit time may not allow acquiring high dissolved solid content. This might be the consequence of the closely located recharge and discharge areas and the considerable elevation difference causing efficient driving force for groundwater flow (500–600 m, Erőss et al. 2014a, b). Based on the chemical characterization of karst waters, the waters show nearly the same hydrochemistry, mainly HCO_3, Cl $_+$ SO_4 anion facies and Ca + Mg, Na + K cation facies are dominant (Fig. 9, Erőss et al. 2014a).

2.2.3 Groundwater-Related Phenomena

There are 720 registered springs in the Bükk Mts. (Fig. 7, Izápy 2002). The most significant karst springs have several thousand litres per minute flow rate (e.g., Garadna spring—7200 m^3/day; Szinva spring—13,000 m^3/day; Tükör spring—3000 m^3/day, Pelikán et al. 2005). The long-term average discharge of all springs is 193,000 m^3/day (Miklós et al. 2020). Most of the karst springs discharge appears at the edge of the mountains. However, in connection with the local erosion base, there are also significant springs within the mountains. The cold springs (6–7 °C) are mainly located in the elevated areas and the big valleys, but there are many lukewarm springs (20–30 °C) located in the foothills. The largest karst springs can be found in Miskolctapolca and Eger (Fig. 10).

In Hungary, most of the caves are located in the Bükk Mts. (1154 according to the Hungarian Cave cadastre, 147

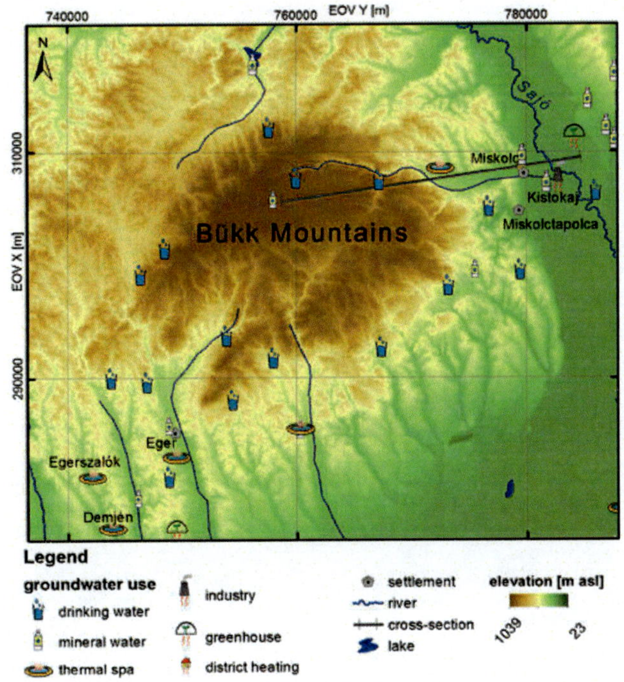

Fig. 10 Topography and location of groundwater use (Kocsis et al. 2018) for the Bükk Mts

longer than 50 m), together with the two deepest ones (Bányász Cave—275 m, István-lápa Cave—254 m deep, Fig. 8). Active sinkhole and stream caves are common in the plateau area, and these are epigene in terms of their genetics. The water of cold springs often passes through caves. Hypogene caves, formed by upwelling thermal water, can be found in the foothills on the eastern part of the mountains.

2.2.4 Groundwater Resources

Cold karst waters of the Bükk Mts. supply the settlements of the area and about 400,000 inhabitants with good quality drinking water. Due to its excellent quality and dissolved solid content, several types of mineral water are bottled in the area (e.g., Lillafüred mineral water, Borsod water). All around the foothills, lukewarm and thermal waters are used for balneological and geothermal heating purposes. Spas of Eger and Miskolctapolca use the naturally discharging thermal karst waters, whereas in Mezőkövesd, Egerszalók and Demjén the thermal water is extracted by wells for the operation of spas. In the foothills region, the thermal karst waters under the thick sedimentary cover provide an excellent opportunity to utilize geothermal energy. The largest geothermal heating plant in the region is located in Kistokaj, providing geothermal energy for district heating of Miskolc. In addition, several smaller investments provide heating of greenhouses and other industrial facilities with geothermal energy (Fig. 10).

3 Transdanubian Mountains

The Transdanubian Mountains is the largest carbonate aquifer system in Hungary with an extent of $\sim$300 km 100 km. It consists of several morphologically and geologically distinct regions, nevertheless, it is a huge interconnected poorly karstified hydrogeologic unit from its southwesternmost part (Keszthely Mts.) to its northeasternmost edge (Buda Hills, Csepregi 2007; Szádeczky-Kardoss 1941). The hydraulic continuity of the system was demonstrated by a long-term (1950–1990) aquifer test induced by mine dewatering and thus the regional hydraulic communication was proved by the propagation of hydraulic head change over the whole reservoir (Alföldi and Kapolyi 2007; Mádl-Szőnyi and Tóth 2015; Tóth 1995). Due to the large extent of the Transdanubian Mts., this system is presented in two parts: the SW and the NE segments which are separated along the Mór trough (Fig. 2).

3.1 SW Transdanubian Mountains

The southwestern part of the Transdanubian Mts. covers the following carbonate–karst regions: Keszthely Mts., Balaton Highland and Bakony Mts. which is subdivided into High

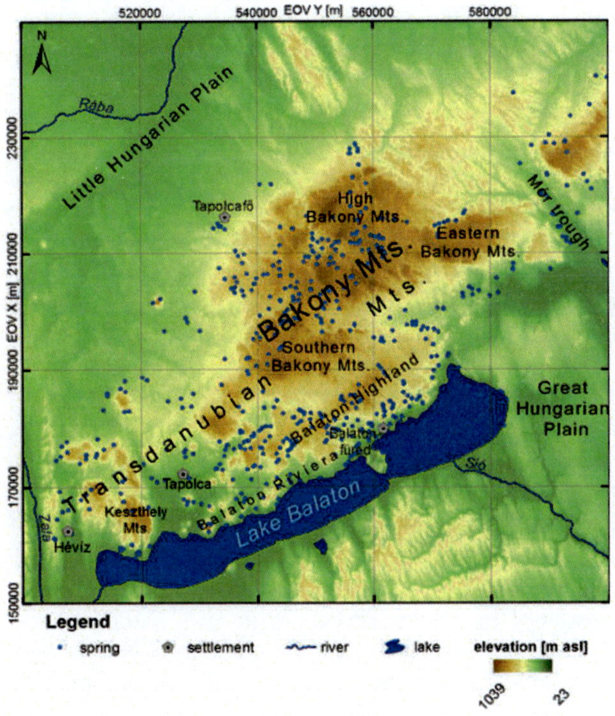

Fig. 11 Topography and location of springs (Izápy 2002) for the SW Transdanubian Mts

Bakony Mts., Eastern Bakony Mts. and Southern Bakony Mts. These subregions are separated by tectonic valleys, troughs and basins and each has a different character (Fig. 11).

3.1.1 Hydrogeologic Environment

Topography The Keszthely Mts. has an average elevation of 300 m asl and its southern part is slightly elevated (300–400 m asl) than the northern segment (200–300 m asl). The Balaton Highland exhibits a gradually decreasing topography from the northwest (250–350 m asl) to the southeast toward the Balaton Riviera (~110 m asl) and Central Europe's largest lake, the Lake Balaton (~105 m asl). The uncovered horsts of the highest region, the High Bakony Mts., reach 600–700 m asl elevation and this part is segmented by tiny intramountainous basins and valleys. The open carbonate plateaus of the Eastern Bakony Mts. can be found at 500–600 m asl. The Southern Bakony Mts. has some buttes with the highest points around 600 m asl, but the average elevation is 300–350 m asl. The SW Transdanubian Mts. is bordered by the Little Hungarian Plain to the W–NW–N and the topography gradually decreases down to ~120 m asl. A sudden drop in topography marks the E–SE border and Transdanubian Mts. adjoins the Great Hungarian Plain (110–130 m asl) here (Fig. 11).

Climate The SW Transdanubian Mts. is characterized by a humid continental climate in general, but local microclimates occur mosaically due to the morphological variations. The annual mean temperature is 8–11 °C for the entire system. The precipitation varies according to the orographic effects: the annual mean rainfall is 500–600 mm in the foothills and it increases up to 800 mm in the High Bakony Mts (Kocsis et al. 2018). The actual evaporation rate is around 500–600 mm/yr and thus only 20–40% of precipitation can infiltrate and supply the groundwater reserves (Stelczer 2000; Felméry et al. 1971).

Geology The Transdanubian Mts. consists mainly of Triassic and Early Jurassic carbonates, and Paleozoic metamorphic and Permian–Cretaceous sedimentary rocks can also be found in this region. The Mesozoic carbonates form the main karst aquifer system of the area which is generally unconfined (or semi-confined), partially confined by thick Neogene sediments at the foothills, dissected by major faults and influenced by folding (Fodor 2010; Haas 2012). From a hydrostratigraphic point of view, the non-carbonate metamorphic and sedimentary units, the confining and basin-fill formations can be considered as aquifer–aquitards or rather aquitards with low hydraulic conductivity (Mádl-Szőnyi 1997; Tóth 2018). In the area of Keszthely Mts. and Bakony Mts., the total thickness of carbonate aquifers can reach

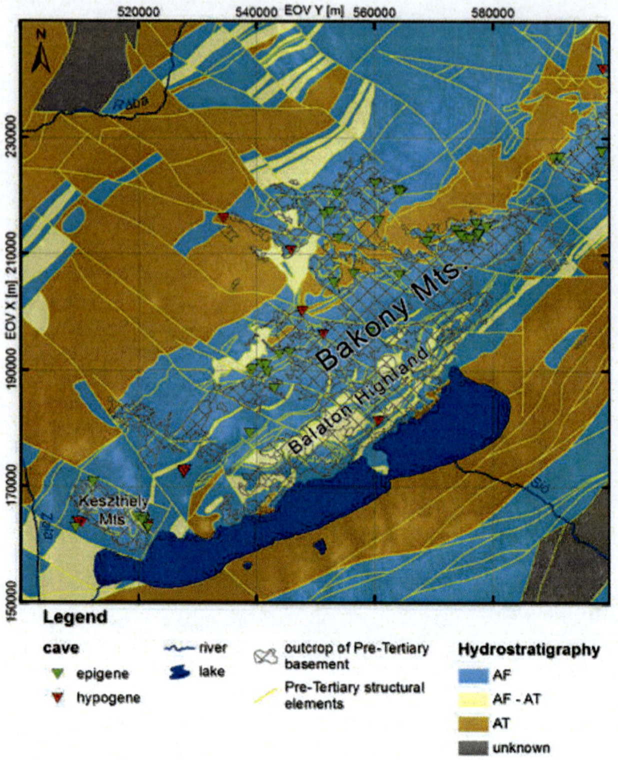

Fig. 12 Hydrostratigraphy, outcrops and structural elements of the Pre-Tertiary basement of the SW Transdanubian Mts. (adapted from Haas et al. 2010)

2000–3000 m, at the same time the groundwater basin is only ~500 m deep at the Balaton Highland where the basement aquitards were thrusted to near-surface and the basin geometry was considerably modified (Figs. 12 and 13).

3.1.2 Groundwater Flow Patterns and Processes

There are hydrostatic pressure conditions within the entire aquifer system which are attributed to the intrinsic properties of carbonate formations (high hydraulic conductivity, low specific storage, high hydraulic diffusivity) in which the pressure changes are equilibrated effectively over large areas (Mádl-Szőnyi and Tóth 2015). Thus, the groundwater table is flattened but it still mimics the basin-scale topographic variations which generate the groundwater flow. Consequently, in elevated areas, such as the Keszthely Mts., High Bakony Mts. and Southern Bakony Mts., regional water table mounds (recharge areas with the intense downward flow) occur and groundwater flows in NW and SE directions toward regions with lower hydraulic heads, i.e., foothills and plains (discharge areas with the upward oriented flow). Between these areas, wide zones of horizontal groundwater flow and corresponding through-flow hydraulic regime dominate. Due to the smoothened undulation of the water table, nested groundwater flow systems with low hierarchy

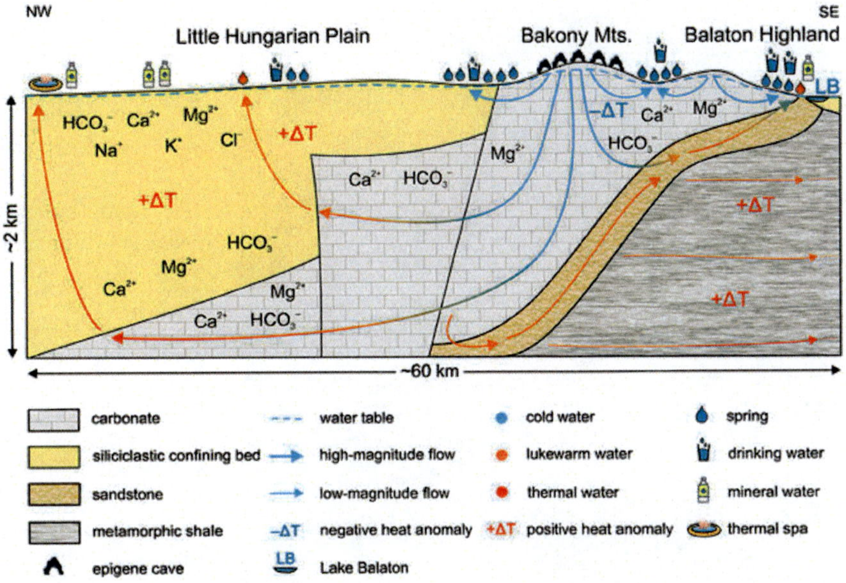

Fig. 13 Generalized hydrogeological cross-section for the SW Transdanubian Mts.

evolve: shorter and shallower local flow systems between local highs and lows are superimposed over the longer and deeper flow systems of higher orders (i.e., intermediate and regional). This can explain the occurrence of bidirectional groundwater flow between the Bakony Mts. and Balaton Highland: >250 m asl the groundwater flows northwards but deeper in the basin, the flow direction is the opposite, i.e., southwards. In this region, the basin geometry restricts the groundwater flow and causes intensified flow toward the area of Balaton Highland and Lake Balaton (Fig. 13; Tóth 2018).

The subsurface temperature field reflects the advective heat transport caused by topography-driven groundwater flow. The recharging cold water can infiltrate down to −3000 m asl ($\sim$20 °C/km of a geothermal gradient), hot water (>30 °C) can be found only in deeper parts of the basin, except some locally discharging lukewarm–thermal springs at the foothills (e.g., Balatonfüred, Hévíz, Tapolca, Tapolcafő with the geothermal gradient up to 100 °C/km). The adjoining confined areas bear higher near-surface temperatures due to the thermal blanket effect of confining sedimentary units and upward moving groundwater (Fig. 13, Lenkey et al. 2002; Tóth et al. 2016; Tóth 2018).

The hydrochemical composition of groundwater is governed by the carbonate rock matrix (limestone and dolomite): Ca^{2+}, Mg^{2+}, and HCO_3^- are the dominant ions even in the confining and basin-fill units as the groundwater is conveyed from carbonate aquifers to these formations. However, groundwater samples from the adjoining sedimentary areas contain also Na^+, K^+, and Cl^- ions in an elevated ratio (Tóth 1990, 2018). In the Balaton Highland area, groundwater is naturally enriched by CO_2 (up to 3000 mg/l) which has mantle origin and transported to the surface by deep regional groundwater flow systems (Fig. 13; Cornides et al. 1986; Gondár-Sőregi and Gondár 1988; Dobos and Scheuer 2018).

3.1.3 Groundwater-Related Phenomena

The carbonate aquifer system is rich in springs: in Keszthely Mts. 41, Bakony Mts. 286 and Balaton Highland 217 springs discharge naturally (Izápy 2002). The physical–chemical parameters of outflowing groundwater can be interpreted regarding the feeding flow systems. Statistical analysis of spring data series revealed a systematic distribution of springs with distinct character in temperature, chloride ion content, discharge rate, and elevation of spring orifice. In the elevated unconfined parts, cold springs are related to local groundwater flow systems, lukewarm and thermal springs around the foothills are fed by either intermediate or regional flow systems (Figs. 11 and 13, Mádl-Szőnyi and Tóth 2015).

Epigene caves ($\sim$60) developed in the unconfined part of the carbonate system and these were formed by dissolution processes caused by rainwater infiltration and near-surface circulation of groundwater. There are only $\sim$10 hypogene caves in the SW Transdanubian Mts. which can be found at the foothills along the border of unconfined and confined carbonates where the cave formation is triggered and controlled by the upward moving lukewarm groundwater. Some inactive hypogene caves occur at higher elevations due to tectonically controlled uplift of the Transdanubian Mts. and the consequent shift of the discharge areas (Fig. 12, Ruszkiczay-Rüdiger et al. 2020).

3.1.4 Groundwater Resources

Naturally discharging springs in the unconfined carbonate parts with low total dissolved solids (TDS, 500–1000 mg/l)

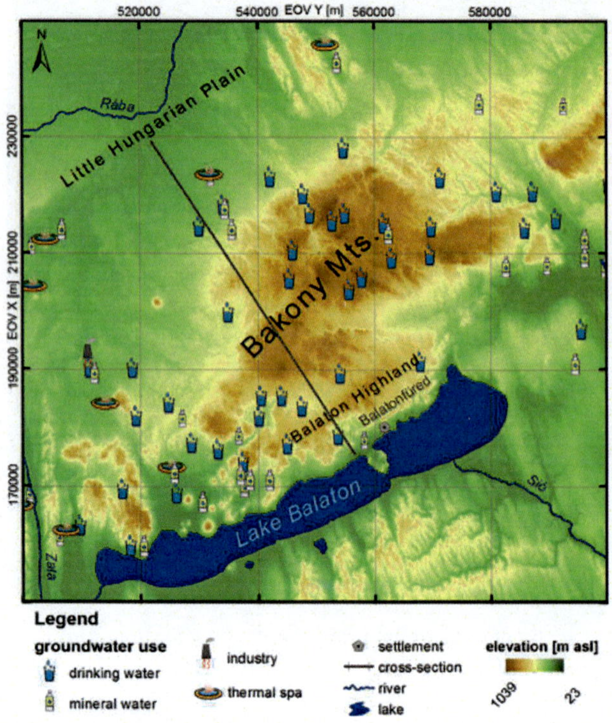

Fig. 14 Topography and location of groundwater use (Kocsis et al. 2018) for the SW Transdanubian Mts

have provided high-quality water as a local drinking water resource in the Bakony Mts.—Balaton Highland area for centuries. Certain areas exhibit special hydrochemical composition of groundwater (e.g., Balaton Highland) which is bottled as mineral water (e.g., Theodora Kékkúti, Theodora Kereki, and Nestlé Aquarel). The city of Balatonfüred is also famous for its medical water for curing purposes (Dobos and Scheuer 2018). Thermal spas in the sedimentary units of adjacent Little Hungarian Plain are tapping groundwater which has been recharged at and transported through the carbonate aquifer system of SW Transdanubian Mts. (Figs. 13 and 14). The area is rich in brown coal and bauxite resources, as well, which were exploited by lowering the water table between 1960 and 1990. At some places, the water table decrease exceeded 100 m and the depression cones propagated through the entire system causing drying up of springs in elevated parts (Alföldi and Kapolyi 2007). After stopping the mine dewatering intervention, water table rebound and related geotechnical problems in urbanized areas are the new challenges (Kovács and Szőcs 2016). This long-term aquifer test showed that there is a strong interrelationship between the surface water and groundwater in the area and these waters form a single and interconnected resource. Therefore, any use of groundwater would affect the surface water and vice versa which is crucial for sustainable water management policy (Tóth 2018).

3.2 NE Transdanubian Mountains

The northeastern part of the Transdanubian Mountains consists of different carbonate–karst regions, such as Vértes Mts., Gerecse Mts., Buda Hills, Pilis Mts., and Danube-E blocks. In the region, we can find the Velence Mts. and Visegrádi Mts. with Paleozoic granite and volcanic origin, respectively. The geographical names are not identical to the hydrogeological terms used for karst regions because the latter hold the confined subregions as well, besides the outcropping carbonates. The best-known part of the region, not only in Hungary but worldwide, is the so-called Buda Thermal Karst, which covers the outcrops of Buda Hills, Pilis Mts., the Danube-E blocks and their confined surroundings in Pest and partially in the Gödöllő Hills (Fig. 15; Mádl-Szőnyi et al. 2019; Mádl-Szőnyi and Tóth 2015).

3.2.1 Hydrogeologic Environment

Topography The Vértes Mts. can be characterized by a carbonate plateau of 400 m asl with some elevated peaks (the highest is 487 m asl) and with its intramountainous basins like the Gánt basin. The Vértes Mts. is bounded by the Mór trough to the southwest and the Tata trough to the northeast. The Gerecse Mts. is a large segment of the region with an average elevation of ~400 m asl (the highest point

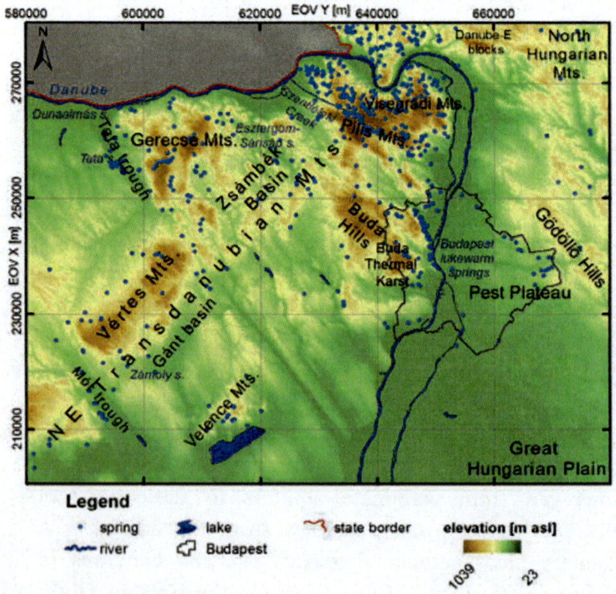

Fig. 15 Topography and location of springs (Izápy 2002) for the NE Transdanubian Mts.

is 634 m asl). Toward the west and east, the elevated carbonates due to tectonic reasons become confined with more than 1000-m-thick siliciclastic cover. The Pilis Mts. is the highest part of this region (up to 765 m asl) and the Szentléleki creek divides it from the Visegrád Mts., which is already part of the North Hungarian Mountains. The Buda Hills can be found to the south of the Pilis Mts. and it is bordered by the Zsámbék basin to the west, the Gerecse to the northwest and the Danube river (Víziváros) to the east. Its highest peak is 559 m asl, but there are also a few peaks with higher than 400 m asl. The highest limestone peak of the Danube-E blocks is 652 m asl. To the south of these blocks, the Gödöllő Hills and the Pest Plateau is part of the Great Hungarian Plain with an elevation of 95–210 m asl (Fig. 15).

Climate The NE Transdanubian Mts. is characterized by a moderately cold and moderately dry climate in the ridge regions and it changes to a warm and wet climate closer to the Danube (Kocsis et al. 2018). The annual mean temperature is 9–11 °C for the whole Transdanubian Mts., but the elevated parts of the region are colder (7–9 °C). The precipitation varies according to the orographic effects. The annual mean rainfall is 500–550 mm in the foothills and it increases up to 650 mm in the highest part of the ranges. In comparison to the SW Transdanubian Mts., however, it is significantly less here. The potential evaporation rate is around 400–600 mm/yr (Kovács 2011). The calculated original infiltration rate through the elevated outcrops of the mountains is 140 m^3/min (Csepregi 2007).

Geology The Transdanubian Mts. is characterized by slightly metamorphosed Variscan formations, overlain by non-metamorphosed Alpine sequences from the Middle and Late Triassic to Early Jurassic (Wein and Konda 1977; Haas et al. 2000). The NE Transdanubian Mts. consists of 2000–3000 m thick Triassic dolomites and limestones, marls and cherty limestones which form the main karst aquifer system of the region. These rocks build up the Vértes Mts., Gerecse Mts., Pilis Mts. and Buda Hills, but one can find Cretaceous sedimentary rocks (grey marl and sandstones), aquifers and aquitards in the Gerecse Mts. From the Late Eocene, terrestrial siliciclastic rocks (Báldi and Báldi-Beke 1985), shallow-water limestone, and deep-water marl deposited behaving as karstified aquifer and aquifer-aquitard, respectively. These formations are overlain by Oligocene anoxic deep marine shales which behave as the main and efficient aquitard. In the Late Oligocene to Early Miocene, the basin was filled up with shallow marine to continental siliciclastics (aquifer–aquitard, Mádl-Szőnyi et al. 2019). The gradual uplift of the Transdanubian Mts. and the elevated heat flux from the Late Miocene (Bada et al. 2007) initiated spring activity with continuous rearrangement of flow systems and

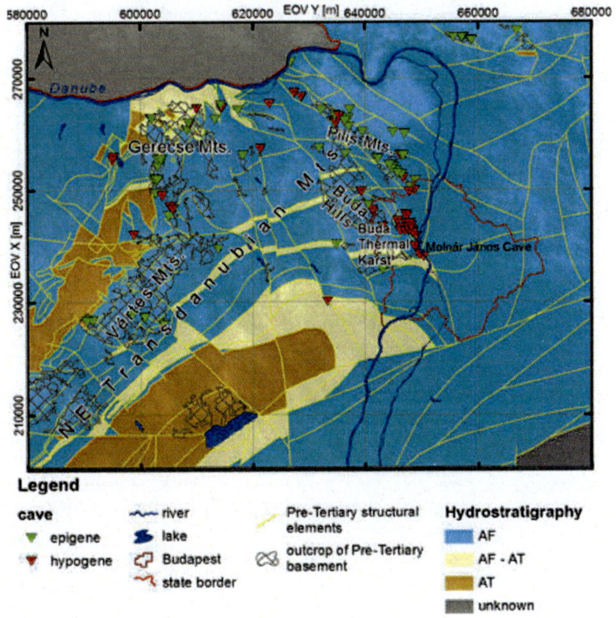

Fig. 16 Hydrostratigraphy, outcrops and structural elements of the Pre-Tertiary basement of the NE Transdanubian Mts. (adapted from Haas et al. 2010)

local base level of erosion (Havril et al. 2016). The former activity of springs is indicated by Pleistocene travertine horizons in the region (Wein and Konda 1977; Ruszkiczay-Rüdiger et al. 2005; Kele et al. 2011) and karstification (Figs. 16 and 17; Leél-Őssy 1995; Erőss 2010).

3.2.2 Groundwater Flow Patterns and Processes

The main driving force of fluid flow can be characterized by the elevation of the water table. It was originally at ~155–160 m asl in the Vértes Mts., ~140–150 m asl in the Gerecse Mts., ~120–125 m asl in the Pilis Mts. And ~120–130 m asl in the Buda Hills (Csepregi 2007; Mádl-Szőnyi 2020). Besides these water table mounds (recharge areas characterized by vertical downward flow), the regional slope of the water table decreases from the NW toward the Danube as the base of erosion in the east and the northeast. We could reveal strong interrelationships between siliciclastic confining layers and carbonate aquifers by the comprehensive basin-hydraulic evaluation of the broader environment of the Buda Thermal Karst (Mádl-Szőnyi 2020; Mádl-Szőnyi et al. 2019; Erhardt et al. 2017). The location of springs stands for discharge areas of the system independently from their source rock due to the complex interrelationships between siliciclastic covering formations (un- and semi-confined areas) and confining layers (in confined regions). Due to the smoothened undulation of the water table compared to siliciclastic basins, nested groundwater flow systems with less hierarchy evolve (Havril et al. 2016). Between recharge and discharge areas in the NE Transdanubian Mts. similarly to

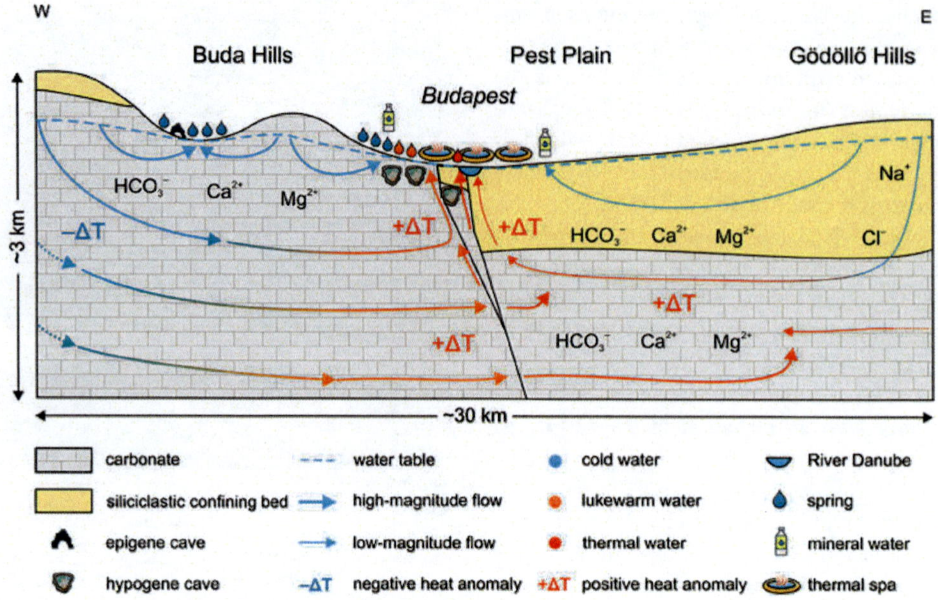

Fig. 17 Generalized hydrogeological cross-section for the NE Transdanubian Mts.

the SW part, wide zones of horizontal groundwater flow, i.e., through-flow hydraulic regime dominate (Fig. 17, Mádl-Szőnyi 2020).

The subsurface temperature field reflects the advective heat transport caused by topography-driven groundwater flow, parallel with the decreasing influence of buoyancy-driven flow from the Late Miocene (Havril et al. 2016; Szijártó et al. 2019). The recharging cold water can infiltrate down to –3000 m asl in the elevated recharge areas. Cold springs connected to local flow systems appear at higher topographic elevations (at about average air temperature) while lukewarm springs (~20 °C) issue around the border of unconfined and confined carbonates (~120 m asl, Mádl-Szőnyi and Tóth 2015). Hot springs (~42 °C) related to regional flow systems can be found only at about 104 m asl in Budapest, close to Danube river. The temperature increase does not follow linear trends and show anomalies close to the main discharge area due to different and superimposing heat transport processes (Figs. 15 and 17; Mádl-Szőnyi 2020; Szijártó et al. 2019).

The governing hydrochemical processes are similar to those which exist in the SW Transdanubian Mts. (Fig. 17, Alföldi 1979; Alföldi et al. 1968; Erőss et al. 2012b; Poyanmehr 2013; Déri-Takács et al. 2015; Kovács and Erőss 2017). The main processes here are also determined by the interaction of infiltrating meteoric water between dolomite and limestone of the main carbonate aquifer. Consequently Ca^{2+}, Mg^{2+}, and HCO_3^- are the dominant ions, however, this basic composition is furtherly modified due to the interaction with the water of siliciclastic cover formations and the existence of evaporites, etc. (Jámbor 1976; Alföldi et al. 1968; Fórizs et al. 2019; Erőss 2010; Mádl-Szőnyi et al. 2018). Besides meteoric fluids, the interplay between adjoining unconfined and confined carbonates, meteoric and basinal fluids modifies the composition of groundwater especially for the regional flow systems at Budapest (Erőss et al. 2012b; Mádl-Szőnyi and Tóth 2015). The NaCl-type water of confining strata on the Pest side and Gödöllő Hills contributes to the regional fluid flow system due to vertical leakage and these covering units are responsible for the elevated chloride content of the discharging thermal water (Alföldi 1979; Erőss 2010; Mádl-Szőnyi and Tóth 2015).

3.2.3 Groundwater-Related Phenomena

The most significant phenomenon in carbonate aquifer systems is the discharge pattern of springs (Fig. 15). The NE Transdanubian Mts. is rich in springs. In the Vértes Mts., there are 22 registered springs, in Gerecse Mts. there are 101, while in Buda Thermal Karst 42 springs can be found. In the remaining parts, 245 springs were issued naturally (Izápy 2002). The springs with the highest volume discharge in the region are characterized by lukewarm temperature (Alföldi et al. 1968; Alföldi and Kapolyi 2007; Mádl-Szőnyi and Tóth 2015) and these are Zámoly (Vértes Mts.), Tata, Dunaalmás (Gerecse Mts.), Esztergom-Sárisáp (Pilis Mts.), Budapest lukewarm springs (Buda Hills). The volume discharge of lukewarm and warm springs of intermediate and regional flow systems in Budapest is stable and does not change due to precipitation events but it correlates rather with the level of the Danube (Bodor et al. 2014, 2015). Due to the mining activity and active dewatering in the NE Transdanubian Mts., the volume discharge of these springs decreased from 1950, and some dried up (Csepregi 2007). When the mining dewatering operation was closed in the 1990s, recovery of the hydraulic head began in the system (Tóth et al. 1999). Consequently, some of the former springs have reactivated (Maller et al. 2018) but reappearing springs cause serious environmental problems as they have been issued in newly built-in areas (Kovács and Szőcs 2016).

We can find many epigene caves (81) in the unconfined part of the NE Transdanubian Mts., where carbonate aquifers outcrop. The location of these caves correlates with local flow systems and cold springs. The hypogene caves (109) appear close to the foothills where unconfined and confined carbonates meet or the carbonates are in a confined situation. A swarm of 50 hypogene caves can be observed along tectonic lines in the Buda Thermal Karst in Buda (Fig. 16). Location of these caves can be connected to intermediate and regional flow systems, i.e., lukewarm and warm water. The hypogene karstification is triggered and controlled here by the upward moving of lukewarm groundwater, mixing corrosion, microbially mediated dissolution, etc. (Kovács and Müller 1980; Erőss 2010; Erőss et al. 2008, 2011, 2012a; Leél-Őssy 2017; Anda et al. 2015, 2016, 2017; Borsodi et al. 2012). Due to tectonically controlled uplift of the Transdanubian Mts. inactive hypogene caves occur at higher elevations (Ruszkiczay-Rüdiger et al. 2020) while active ones can be found at recent discharge areas (Molnár János cave, Fig. 17, Erőss 2010).

3.2.4 Groundwater Resources

Naturally discharging cold, lukewarm, and thermal springs in the NE Transdanubian Mts. have already been used for centuries as drinking water and for healing purposes. Moreover, lukewarm springs of Buda were used for heating even in the Roman Empire (Aquincum). One of the most flourishing periods in thermal water use was the Turkish era and some original thermal spa buildings have remained and been operating since that time in Buda. The first district heating system started to operate in Pest in the first part of the twentieth century and uses the heat content of thermal water exploited by deep wells (Fig. 15, Mádl-Szőnyi 2006).

A dramatic change occurred in the NE Transdanubian Mts. from the 1950s when active dewatering started in the region for coal and bauxite mining (Alföldi and Kapolyi 2007). Before this period, the natural recharge and discharge were balanced for the whole area and the natural changes plus the production did not cause regional influence on the volume discharge of springs and water level. However, from 1950 to 1990, the artificial and natural discharge altogether exceeded the calculated recharge by at least 60% for the

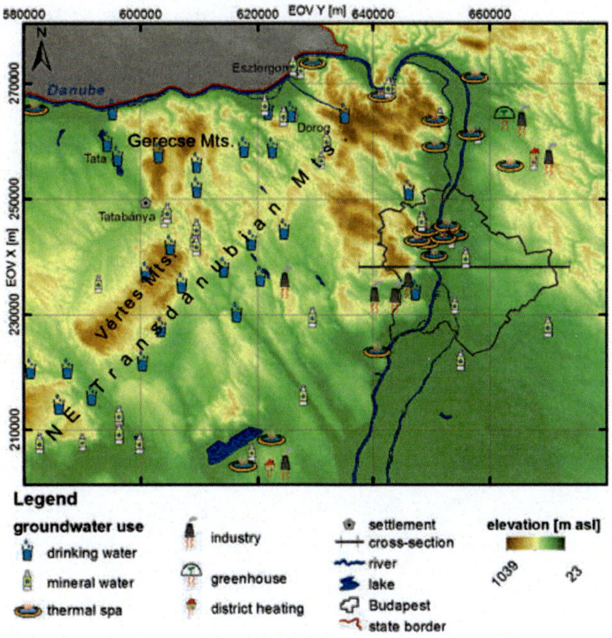

Fig. 18 Topography and location of groundwater use (Kocsis et al. 2018) for the NE Transdanubian Mts.

region (Csepregi 2007). The largest depressions formed in Tatabánya and Dorog coal basins in the Vértes and Gerecse Mts. During this activity, the produced water was partially utilized as drinking water. Consequently, after finishing mining activity, some water producing facilities are still sustained for drinking water purposes, like in the case of Tatabánya water shaft (Csepregi 2007).

Today we actively use the cold karst waters of the region in the drinking water supply in the NE Transdanubian Mts. (Fig. 18). The heat potential of the region is elevated based on the heat content of the springs (Gölz 1982) and deep geothermal potential of Buda Thermal Karst (Mádl-Szőnyi et al. 2015). Tata and Esztergom are representing long-term tradition in spa culture due to lukewarm springs and wells. The utilization of lukewarm and warm waters has been existing in the spas of Budapest since historical times (Budapest Spas Plc.). Some of these spas exist in former natural discharge areas of warm waters, like Gellért, Rudas, and Rác Spas. Others connected to lukewarm and warm waters like Lukács and Király. The Római, Pünkösdfürdő, and Csillaghegy spas use lukewarm water. Parallel with deep-water exploitation, spas appeared related to these artificial discharge points, such as Széchenyi, Paskál, Dandár, Palatinus, and Pesterzsébet. The use of heat content of water in spas is more conscious and enhanced nowadays (Erőss et al. 2015). Some waters of the region stand for traditional mineral water resources. Besides others, one can find the best-known mineral water resources in the region of Buda Thermal Karst. Some of them originated from intermediate flow systems of lukewarm springs like Csillaghegyi or Lukács Római, while others use warm waters of regional flow systems, like Gellérthegyi Kristály, Margitszigeti Kristály, and Dagály.

4 Mecsek Mountains and Baranya Hills

The Mecsek Mountains and the Villány Mountains represent two relatively small karst areas regarding their surface extent in the Mecsek Mts. and Baranya Hills major geomorphological district, SW Hungary. However, the confined carbonates form considerable parts of these systems (Fig. 2 and 20).

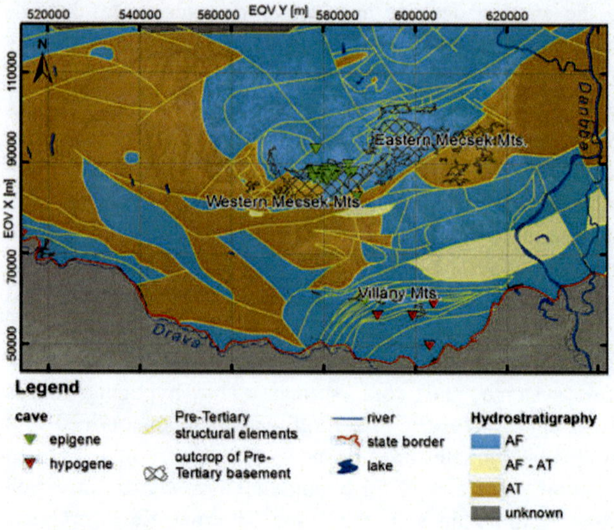

Fig. 20 Hydrostratigraphy, outcrops and structural elements of the Pre-Tertiary basement of the Mecsek Mts. and Villány Mts. (adapted from Haas et al. 2010)

4.1 Mecsek Mountains

The karst area of the Mecsek Mts. is basically restricted to its western part (Rónaki 1973). The hydrogeology of this area was not evaluated regionally in the flow system context as yet, hence the following summary is rather based on literature study.

4.1.1 Hydrogeologic Environment

Topography The mountains cover an area of about 500 km^2 and can be divided morphologically into three parts, the Eastern Mecsek Mts., the Middle Mecsek Mts. and the Western Mecsek Mts., geologically into two parts, the Western and Eastern Mecsek Mts. The highest peak is the Zengő with 682 m asl. The karst area can be found in the northern part of the Western Mecsek Mts. with an extent of 50 km^2, and it is characterized by karstic geomorphological features such as shafts, dolines, dry valleys and sinkholes. This part of the Mecsek Mts. is 350–400 m higher than the surrounding basins. In the Eastern Mecsek Mts., the karstification is less significant (~10 km^2 surface extent) due to the lack of surface carbonates (Figs. 19 and 20, Szabó 1961).

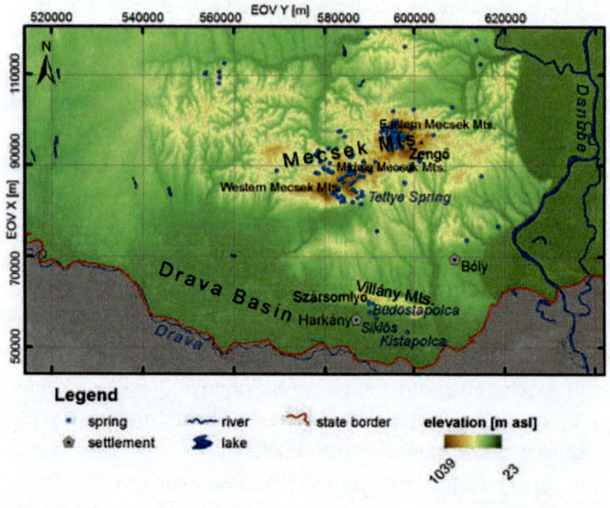

Fig. 19 Topography and location of springs (Izápy 2002) for the Mecsek Mts. and Villány Mts

Climate In the climate of the mountains both Mediterranean and continental influences prevail. There is a significant climatic difference between the northern and southern slopes. The annual mean precipitation is 750 mm in the northern and higher areas, and 600 mm in the other areas. The annual mean temperature is 8–10 °C (Kocsis et al. 2018). In the Western Mecsek Mts., the annual recharge rates vary from 100 mm/year to 200 mm/year (Csurgó and Szulimán 2017).

Geology Focusing on the karst of the Mecsek Mts., the Mesozoic (Middle Triassic) carbonates are dominant which are partly covered by loess. The carbonate rocks (limestones and dolomites) are situated on the northern limb of the anticline structure of the Western Mecsek Mts. and also form the basement of the northern sedimentary basin. The tectonically strongly affected rocks provide favorable conditions for karstification (Fig. 20, Barta and Tarnai 2012).

4.1.2 Groundwater Flow Patterns and Processes

The karst system receives water from non-carbonatic rocks at a topographically higher situation, i.e., it is allogenic karst

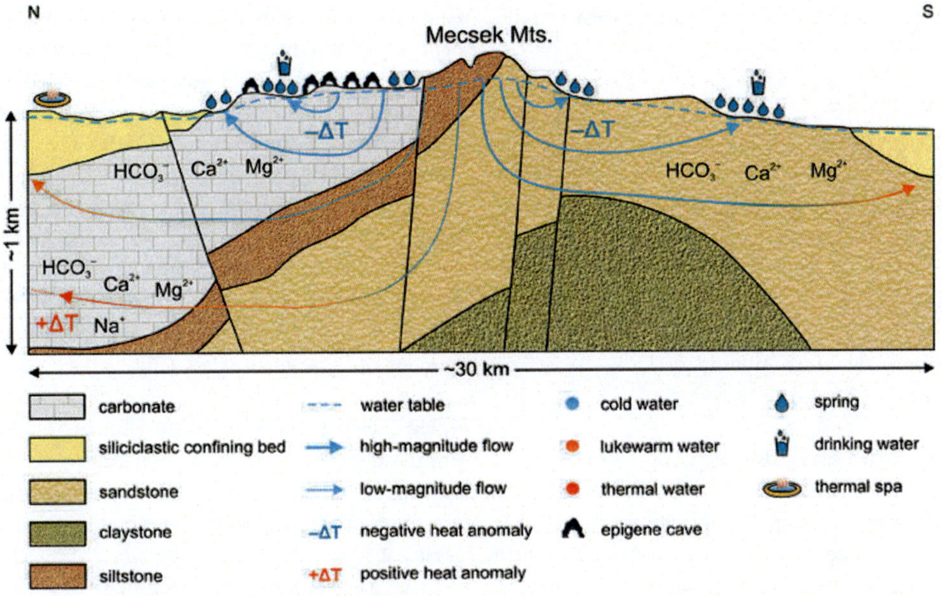

Fig. 21 Generalized hydrogeological cross-section for the Mecsek Mts.

(Jakucs 1971). The hydrogeological research in this area was focused on clarifying local questions such as delineating the recharge area of the karst springs as well as finding connections between swallow holes and the springs by using water tracing techniques (Rónaki 2007). A wide range of flow velocity values (16–1280 m/day) was reported from these experiments. The karst water level of the Western Mecsek Mts. is the highest (260–280 m asl) on the northern slope of the mountains and gradually decreases northward and southward to the discharge level of the major karst springs (220–239 m asl), e.g., in Orfű, Abaliget, Tettye (Fig. 21, Csurgó and Szulimán 2017). The temperature of the springs of this part of the mountains does not exceed 18 °C (Szőke 2012). The spring waters have Ca–Mg–HCO_3 hydrochemical type (Rónaki 1995).

4.1.3 Groundwater-Related Phenomena

The majority of about 200 caves is tectonically controlled shaft and many of them function as a swallow hole. The other group of caves is related to springs and their passages are also strongly influenced by structural features. Considering the origin of the caves from a hydrogeological point of view, they are all related to recharge areas and local flow systems of meteoric water, thus these caves are epigene.

There are 167 registered springs in the whole Mecsek Mts. area (Fig. 19, Izápy 2002), among which there are eight larger karst springs with highly fluctuating discharge rates (150–5200 m^3/day). Most of them can be found mainly in the northern foothills, except the Tettye Spring (4000 m^3/day) which is situated on the southern part of the karst system, and its recharge area is purely autogenic and has a considerable freshwater limestone deposit.

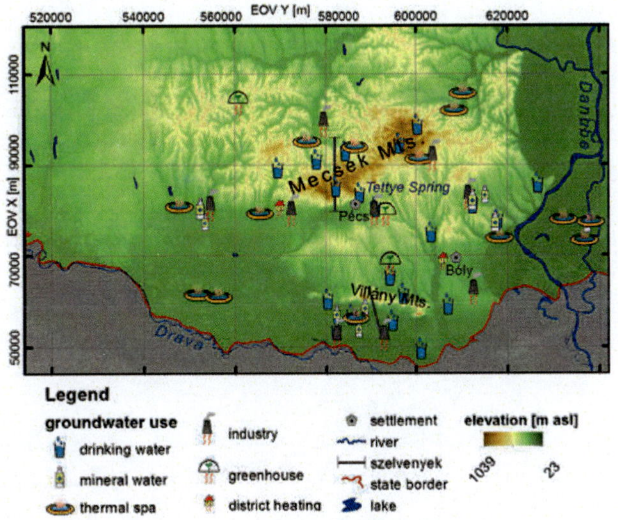

Fig. 22 Topography and location of groundwater use (Kocsis et al. 2018) for the Mecsek Mts. and Villány Mts.

4.1.4 Groundwater Resources

The karst waters are mainly used for the drinking water supply of the area (Fig. 22, Berényi Üveges 2012). The drinking water supply of the fifth largest city of Hungary,

Pécs, mostly relies on the Tettye karst spring with a daily yield of 6000 m^3/day.

4.2 Villány Mountains

The Villány Mountains is the southernmost karstic region of Hungary. There are some smaller carbonate outcrops (Siklós Castle Hill, Kistapolca and Beremend) in the southern foreland of the hills. The area is a highly disturbed karst area, as most of the Villány Mts. have been mined by industrial limestone production, therefore significant karst areas and caves have been destroyed and removed.

4.2.1 Hydrogeologic Environment

Topography The Villány Mts. is a 30 km long hill with flat and narrow ridge surrounded by steep slopes from the north and south sides. The highest peak is Szársomlyó (442 m asl), which is separated from the main mountain range (Fig. 19). The southern side of the hills shows surface karstic forms, however, there are no dolines and sinkholes in the area.

Climate The climate of the Villány Mts. is warm and continental, the average annual temperature is 9–11 °C. On the southern slopes of the hills, a specific Mediterranean microclimate has developed, with temperatures above 0 °C for at least 200 days a year. The mean annual rainfall is 600–750 mm (Kocsis et al. 2018). The average evapotranspiration is 150–300 mm/year and the mean annual recharge amounts from 250 to 350 mm/year (Salem et al. 2019).

Geology The sedimentary rocks building up the hills were deposited from the late Carboniferous to the Cretaceous. The Late Cretaceous nappe formation created seven structural imbrications of the Mesozoic carbonates (Császár 2002; Petrik 2009; Konrád et al. 2010). The various Mesozoic carbonate rock suites form a thick karst reservoir and a good aquifer. The carbonate rocks extend in the foreland basin basement covered by young sediments, which can be considered as aquitard units. Some basement carbonate outcrops can be found in the southern foreland of the Villány Mts., which are also favorable sites for karstification and development of hypogene caves (Fig. 20).

4.2.2 Groundwater Flow Patterns and Processes

For the regional-scale hydraulic assessment of the wider area of the Villány Mts., the methodology of archive pre-production measured data processing and the concept of regional hydraulic continuity were applied (Csondor et al. 2020). The study area included the Drava Basin, the southern foreland of the Mecsek Mts. as well, besides the Villány Mts. (Fig. 19).

In the shallower part of the area (down to about −500 m asl), the flow systems usually reflect the topography, therefore topographically higher areas are characterized by downward flow (recharge) and in the lower areas, there is an upward flow (discharge) related to gravitationally-driven flow systems. On the whole, horizontal flow conditions are dominant, which tends from the mountainous and hilly regions towards the regional base level of the Drava and Danube rivers (Fig. 23). From a hydraulic point of view, the

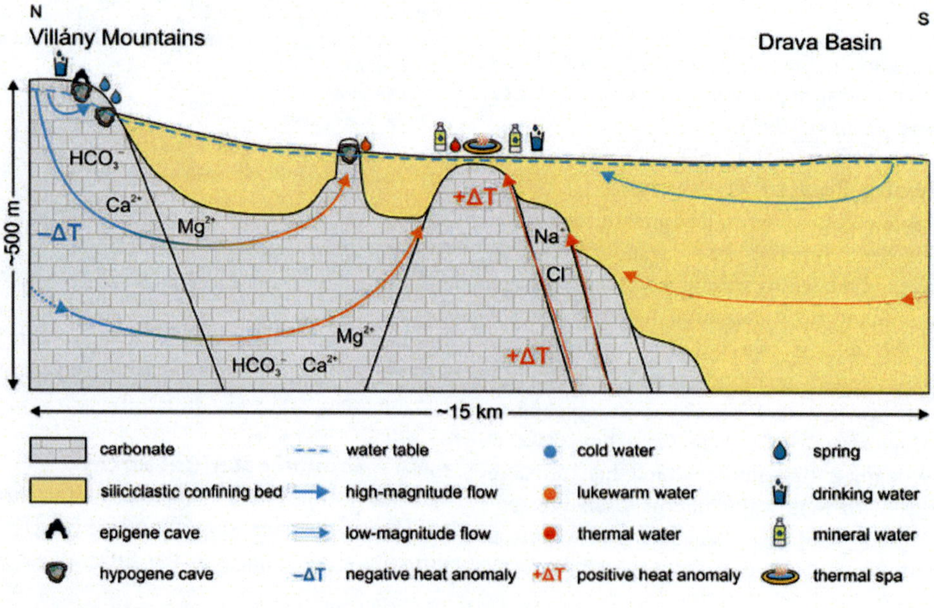

Fig. 23 Generalized hydrogeological cross-section for the Villány Mts.

vertical pressure gradient values close to hydrostatic are typical for most of the area and in the shallow region. Higher gradient values can be observed in some discharge areas, e.g., in Harkány, in the vicinity of the elevated carbonate blocks where the siliciclastic cover causes greater vertical pressure gradient thus inducing upward flow within the underlying aquifer (Csondor et al. 2020). Overpressured flow system was identified in the deeper parts of the Drava Basin (about −1500 to −2000 m asl), but its presence in the southern foreland of the Villány Mts. is also expected.

In terms of chemical composition, the karst waters of the region can be divided into three groups according to their temperatures, i.e., cold, lukewarm, and thermal waters. Cold springs and waters of local flow systems at the foothill area represent Ca–Mg–HCO_3 water type. The lukewarm springs have already higher total dissolved solid content and the SO_4 and Na + K content is also higher. In the thermal waters of Harkány, the effect of the basinal fluids can be seen in the temperature and also in chemical composition. Typically Ca–Na, HCO_3 + Cl–SO_4 facies with higher Cl and Na content is characteristic here, representing a transition between the karst waters of Ca–Mg–HCO_3 facies and the deep basinal fluids of Na–K, Cl–SO_4–HCO_3 facies (Fig. 23, Csondor et al. 2020).

4.2.3 Groundwater-Related Phenomena

The Villány Mts. has a relatively small catchment area (26 km^2), therefore only a few springs can be found (8 springs, Izápy 2002). In the foothills, few cold karst springs issue, nowadays only periodically. However, in the southern foreland of the hills, there are many lukewarm springs with significant (60–700 m^3/day) volume discharge (Fig. 19, in Büdöstapolca, Siklós, and Kistapolca). The spring in Kistapolca is the only naturally outflowing spring of the area, which has thus a crucial role in the flow system understanding. Additionally, it feeds a lake, which can be considered as a groundwater-dependent ecosystem (GDE). In the past decades due to intense water abstraction, the lake dried up several times (Erőss et al. 2020). Thermal water discharge was connected to an uplifted basement block in Harkány, which maintained a swamp. This was drained and captured by the first thermal well of Hungary (Zsigmondy 1873).

There are 23 caves in the Villány Mts. and 5 reach a length of more than 50 m. There are several caves connected to the uplifted basement carbonate blocks in the foreland of the Villány Mts. or the margin of the hills, where the morphology and the minerals decorating the cave walls indicate the effect of thermal waters (Fig. 19, Dezső et al. 2004; Rónaki 2000; Takács-Bolner 2003a, b, 1985; Vigassy et al. 2010).

Regarding their formations, most of the caves are hypogene, but there are epigene pit caves formed by the infiltrating cold water. In some parts of the hills, the combined appearance of epigene and hypogene forms are typical (Rónaki 2000; Dezső et al. 2004). Tectonic preformation played a major role in the formation of each cave. The inactive hypogene caves are located at a higher elevation, only three caves reach the lukewarm karst water table. Classical stream caves are not known in the Villány Mts. (Fig. 20).

4.2.4 Groundwater Resources

Cold and hot karst waters also have been exploited in this area. Cold karst waters are used for drinking water service and industrial purposes in wineries and larger factories. Significant spa culture has been developed on the thermal water. Thermal water is extracted by wells and utilized in several cities as part of balneotherapy and recreation due to the special mineral content of water. In the wider area, in Bóly, a geothermal cascade system harnesses the heat of thermal waters of the confined carbonate reservoir (Fig. 22).

5 Conclusions

In this chapter, the remarkably diverse and rich literature of karst hydrogeology of Hungary was systematically introduced based on the regional groundwater flow concept adapted for adjoining confined and unconfined basin-halves of carbonate regions (Mádl-Szőnyi and Tóth 2015; Mádl-Szőnyi 2020; Tóth et al. 2020). Moving groundwater behaves as a geologic agent (Tóth 1999), therefore the discussed Hungarian karst systems and the consequent karst phenomena (springs, epigene, and hypogene caves) were interpreted as the manifestations of flowing groundwater. We also discussed the diverse use of the karst groundwater as drinking and mineral water and for healing and geothermal purposes. The maps and cross-sections for the different karst areas were constructed based on the same approach and therefore they could depict resemblant components of different karst systems allowing generalization, supporting comparison and thus revealing similarities among Hungarian karst regions.

Based on the comprehensive description of the Hungarian karst areas, we can observe the effects of interplaying flow systems and fluids (meteoric, basinal) of basin-halves characterized by unconfined (or semi-unconfined) and confined carbonates on the hydrogeological sections. The only exception is the Aggtelek Karst (or North-Borsod Karst) because here only shallow geological profiles were available for generalization which, therefore, was limited only for the elevated unconfined karst plateau of Aggtelek. The displayed section stands for the local flow systems, cold springs with low TDS and high level of karstification with only epigene caves. However, the thick Triassic formations (up to 1000 m) here are also confined by Miocene and Quaternary sediments at some locations but there is no information about these confined parts, yet, as the groundwater-related phenomena are more characteristic on the unconfined terrains.

We can characterize the SW Transdanubian Mts. by flow systems heading toward the Little Hungarian Plain (confined part of the system) and the Lake Balaton, where the most significant factor influencing flow systems is the lakeward decrease in the carbonate aquifer thickness. Together with the specific character of the basin geometry, the undulation of the water table and its regional and local trends can explain the flow pattern and the advective heat transport processes and the character of springs. The epigene caves in the elevated Bakony Mts. can be interpreted as a result of the dissolution caused by infiltrating precipitation, while hypogene caves at the foothills were formed due to the upwelling lukewarm water. The NE part of the Transdanubian Mts. shows roughly a similar pattern, but its northeasternmost part, i.e., the Buda Thermal Karst, is quite different. Here we can find two different basin-halves with different geology and hydraulic head distribution, and therefore, the confluence of fluids (meteoric and basinal) from the different basin-halves and a common discharge area around the basin-center can be observed. The thick Triassic carbonate aquifer, the existence and character of thick confining strata can explain the asymmetric flow pattern, heat accumulation, buoyancy-driven flow, interaction of basinal and meteoric fluids and abundance of hypogene caves.

The Bükk Mts. can be represented also by two basin-halves of unconfined and confined carbonates, but the main difference compared to the Buda Thermal Karst (NE Transdanubian Mts.) is the slope of the topography of the confined carbonate basin-half which is flat (or gently sloping onward) in the case of Bükk Mts. This situation and the dominance of local flow systems can explain the well-developed epigene karstification on the plateau and cold springs with low TDS. Moreover, the thermal water occurrence under the confined carbonates due to heat accumulation and position of hypogene caves connected to upwelling warm water can also be attributed to this asymmetric basin geometry. Owing to this hydrogeologic environment, the co-existence of low TDS, absence (or minor ratio) of basinal fluid components in the unconfined basin-half, and thermal water presence under the confined part can be resolved.

The Villány Mts. shows a similar pattern to Bükk Mts. in this sense but on a smaller scale (the thickness of unconfined carbonates and the basin width is less). Moreover, the flow pattern is complemented by an additional basinal fluid component from the confined basin-part (Drava Basin). The Western Mecsek Mts. contains thick carbonates, where we can observe, on one hand, northward flow with well-evolved local flow systems, significant cold springs with low TDS and epigene caves. Under the northern confined part, thermal water can be exploited. On the other hand, in the southern part, where basement rocks are even on the surface, groundwater flows southward across the non-carbonate units.

The displayed karst areas of Hungary show similarities and differences regarding their flow systems and hydrogeologic environment and character, which can be interpreted in the provided basin-scale groundwater flow framework. The local-scale karst features fit into this picture, and we can gain a new perspective to analyze them in a broader context. Moreover, this approach can be adapted to other karst areas worldwide fostering regional characterization and highlighting the role of the boundary between unconfined and confined carbonates, basin geometry and, thus, different groundwater flow systems in determining the character of springs and type of karstification (i.e., epigene or hypogene).

References

Alföldi L (1979) Budapesti hévizek (Thermal waters of Budapest). Budapest, VITUKI. p 102 [in Hungarian]

Alföldi L, Bélteky L, Böcker T, Horváth J, Korim K, Liebe P, Rémi R (1968) Budapest hévizei (Thermal waters of Budapest). Budapest, VITUKI, p 365 [in Hungarian]

Alföldi L, Kapolyi L (2007) Bányászati karsztvízszintsüllyesztés a Dunántúli-középhegységben (Mine dewatering in the Transdanubian Mountains). Budapest, Hungarian Academy of Sciences, p 138 [in Hungarian]

Anda D, Makk J, Krett G, Jurecska L, Márialigeti K, Mádl-Szőnyi J, Borsodi AK (2015) Thermophilic prokaryotic communities inhabiting the biofilm and well water of a thermal karst system located in Budapest (Hungary). Extremophiles 19:787–797

Anda D, Krett G, Makk J, Márialigeti K, Mádl-Szőnyi J, Borsodi AK (2017) Comparison of Bacterial and Archaeal communities from different habitats of the hypogenic Molnár János Cave of the Buda Thermal Karst System (Hungary). J Cave Karst Stud 79:113–121

Anda D, Makk J, Krett G, Márialigeti K, Mádl-Szőnyi J, Borsodi A (2016) A Molnár János-barlang víz és biofilm baktériumközösségei (Bacteria communities in water and biofilm in Molnár János Cave). Hidrológiai Közlöny 96:3–7 [in Hungarian]

Aujeszky G, Scheuer G (1979) A Ny-bükki karsztforrások foglalásainak vízföldtani tapasztalatai (Hydrogeological experience on the capture of karst springs in the Western Bükk Mountains). Hidrológiai Közlöny 59:63–77 [in Hungarian]

Bada G, Horváth F, Dövényi P, Szafián P, Windhoffer G, Cloetingh S (2007) Present-day stress field and tectonic inversion in the Pannonian basin. Glob Planet Chang 58:165–180

Báldi T, Báldi-Beke M (1985) The evolution of the Hungarian Paleogene basins. Acta Geol Hung 28:5–28

Barta K, Tarnai T (2012) A Nyugat-Mecsek karsztja (Karst of the western Mecsek Mountains). Szeged, GeoLitera, p 185 [in Hungarian]

Berényi Üveges I (2012) A Nyugat-Mecsek karsztja (Karst of the western Mecsek Mountains). Szeged, GeoLitera [in Hungarian]

Bodor P, Erőss A, Mádl-Szőnyi J, Czuppon G (2014) A Duna és a felszín alatti vizek kapcsolata a rózsadombi megcsapolódási területen (Interaction between the groundwater and the Danube at the Rózsabomb discharge area). Karsztfejlődés XIX, Szombathely, Hungary, pp 63–75 [in Hungarian]

Bodor P, Erőss A, Mádl-Szőnyi J, Kovács J (2015) A csapadék hatása a rózsadombi források utánpótlódási és megcsapolódási területén (The effect of precipitation at the recharge and discharge area of the springs of Rózsadomb, Buda Thermal Karst, Hungary). Földtani Közlöny 145:385–396 [in Hungarian]

Borsodi AK, Knáb M, Krett G, Makk J, Márialigeti K, Erőss A, Mádl-Szőnyi J (2012) Biofilm bacterial communities inhabiting the cave walls of the Buda Thermal Karst System, Hungary. Geomicrobiol J 29:611–627

Cornides I, Takaoka N, Nagao K, Matsuo S (1986) Contribution of mantle-derived gases to subsurface gases in a tectonically quiescent area, the Carpathian Basin, Hungary revealed by noble gas measurements. Geochem J 20:119–125

Császár G (2002) Urgon formations in Hungary: with special reference to the Eastern Alps, the Western Carpathians and the Apuseni Mountains. Geological Institute of Hungary, Budapest, Hungary

Csepregi A (2007) A karsztvíztermelés hatása a Dunántúli-középhegység vízháztartására. In: Alföldi L, Kapolyi L (eds) Bányászati karsztvízszint-süllyesztés a Dunántúli-középhegységben (Effects of karst water production on the water balance of Transdanubian Mountains). Hungarian Academy of Sciences, Budapest, pp 77–112 [in Hungarian]

Csondor K, Czauner B, Csobaji L, Győri O, Erőss A (2020) Characterization of the regional groundwater flow systems in south Transdanubia (Hungary) to understand karst evolution and development of hydrocarbon and geothermal resources. Hydrogeol J

Csurgó G, Szulimán S (2017) Az uránbányászati bányatérségek szivárgáshidraulikai modellezése. Pécs, Mecsekérc Zrt., p 75 [in Hungarian]

Darabos E (2017) Vízkészlet számítás és idősorok elemzése karsztosodottsági jellemzők meghatározása céljából a Bükki Karsztvízszint Észlelő Rendszer adatai alapján (Water budget calculation and time series analysis of data from the Bükk Karst Water Level Monitoring System for determining karstic features). PhD thesis, University of Miskolc, p 113 [in Hungarian]

Déri-Takács J, Erőss A, Kovács J (2015) The chemical characterization of the thermal waters in Budapest, Hungary by using multivariate exploratory techniques. Environ Earth Sci 74:7475–7486

Dezső J, Sebe K, Horváth G (2004) Villányi-hegység útikalauz (Villány Mountains travel guide). Pécs, Hungary, JPTE Barlangkutató Egyesület, p 159 [in Hungarian]

Dobos I, Scheuer G (2018) Tanulmányok a hazai és külföldi hidrodinamikai rendszerekről (Studies on hydrodynamics systems in Hungary and abroad). Budapest, p 101 [in Hungarian]

Dobos I (1976) A hazai mélységi vízkutatás és feltárás a XIX. Században (The deep water resources research and exploitation in the 19th century in Hungary). Földtani Tudománytörténeti Évkönyv 1975-ről 4:23–36 [in Hungarian]

Dövényi Z (2010) Magyarország kistájainak katasztere (Cadastre of the Hungarian Microregions). Budapest, MTA Földrajztudományi Kutatóintézet. p 876 [in Hungarian]

Emszt K (1913) Bericht über die Tätigkeit des Chemischen Laboratoriums der kgl. ungar. Geologischen Reichsanstalt im Jahre 1911 (Report on the activity of the chemical laboratory of the Royal Hungarian Geological Institute in 1911). Jahresbericht der königlich ungarischen Geologischen Reichsanstalt für 1911:236–243 [in German]

Erhardt I, Ötvös V, Erőss A, Czauner B, Simon S, Mádl-Szőnyi J (2017) Hydraulic evaluation of the hypogenic karst area in Budapest (Hungary). Hydrogeol J 25:1871–1891

Erőss A, Mádl-Szőnyi J, Csoma AÉ (2012a) Hypogenic karst development in a hydrogeological context, Buda Thermal Karst, Budapest, Hungary. In: Maloszewski P, Witczak S, Malina G (eds) Groundwater quality sustainability: IAH selected papers on hydrogeology. CRC Press, Leiden, The Netherlands, pp 119–133

Erőss A, Mádl-Szőnyi J, Surbeck H, Horváth Á, Goldscheider N, Csoma AÉ (2012b) Radionuclides as natural tracers for the characterization of fluids in regional discharge areas, Buda Thermal Karst, Hungary. J Hydrol 426:124–137

Erőss A, Zsemle F, Pulay E (2015) Heat potential evaluation of effluent and used thermal waters in Budapest, Hungary. Central Eur Geol 58:62–71

Erőss A, Csondor K, Czuppon G, Dezső J, Müller I (2020) Groundwater flow system understanding of the lukewarm springs in Kistapolca (South Hungary) and its relevance to hypogene cave formation. Environ Earth Sci 79:132

Erőss A, Csoma É, Mádl-Szőnyi J, Sasowsky I, Feazel C, Mylorie J, Palmer A, Palmer M (2008) The effects of mixed hydrothermal and meteoric fluids on karst reservoir development, Buda Thermal Karst, Hungary. Karst from recent to reservoirs, Karst Waters Institute, Special Publication 14:57–63

Erőss A, Poros Z, Madl-Szőnyi J, Mindszenty A, Molnár F, Ronchi P, Csoma A (2011) Role of karstic and basinal fluids in porosity evolution in the Buda Hills, Hungary. In: AAPG international conference and exhibition 2011: following Da Vinci's footsteps to future energy resources: innovations from outcrops to assets AAPG: Paper 1071554

Erőss A, Csondor K, Surbeck H, Mádlné Szőnyi J, Horváth Á, Lénárt L (2014a) Karsztvizek geokémiai jellemzése különös tekintettel a radionuklidokra a Bükk környezetében (Geochemical characterization of karst waters with special emphasise to radionuclides in the vicinity of Bükk). Karsztfejlődés XIX, Szombathely, Hungary, pp 77–89 [in Hungarian]

Erőss A, Surbeck H, Mádl-Szőnyi J, Csondor K, Horváth Á, Lénárt L (2014b) Characterization of karst waters by radionuclides in the Bükk karst region, Hungary. In: International conference and field seminar "Karst Without boundaries", Trebinje, Bosnia & Herzegovina, p 431

Erőss A (2010) Characterization of fluids and evaluation of their effects on karst development at the Rózsadomb and Gellért Hill, Buda Thermal Karst, Hungary. PhD thesis, Eötvös Loránd University, p 171

Felméry L, Péczely G, Záchné Ruthner M (1971) Tanulmány a Bakony éghajlatáról (Study on the climate of Bakony Mountains). Eötvös Loránd University, Budapest, p 63 [in Hungarian]

Fodor L (2010) Mezozoos-kainozoos feszültségmezők és törésrendszerek a Pannon-medence ÉNy-i részén – módszertan és szerkezeti elemzés (Mesozoic-Cenozoic stress-fields and structural settings in the NW Pannonian Basin: methods and structural analysis). Academic thesis, Hungarian Academy of Sciences, p 167 [in Hungarian]

Ford D, Williams PD (2013) Karst hydrogeology and geomorphology. Wiley, New York, p 562

Fórizs I, Szabó VR, Deák J, Hałas S, Pelc A, Trembaczowski A, Lorberer Á (2019) The origin of dissolved sulphate in the thermal waters of Budapest inferred from stable S and O isotopes. Geosciences 9:433

General Directorate of Water Management (2016) Vízgyűjtőgazdálkodási Terv – 2015 (River basin management of Hungary). General Directorate of Water Management, Budapest, p 698 [in Hungarian]

Gessert A (2016) Geomorphology of the Slovak Karst (Eastern Part). J Maps 12:285–288

Goldscheider N, Mádl-Szőnyi J, Erőss A, Schill E (2010) Thermal water resources in carbonate rock aquifers. Hydrogeol J 18:1303–1318

Goldscheider N, Drew D (2014) Methods in karst hydrogeology. International contribution to hydrogeology, IAH. Taylor and Francis/Balkema, London, p 264

Gölz B (1982) A Dunántúli-középhegység forrásainak természetes hőteljesítménye (Natural effluent heat flux of springs in the Transdanubian Mountains). Földrajzi Értesítő 31:427–447 [in Hungarian]

Gondár-Sőregi K, Gondár K (1988) A Balaton-felvidék karsztvízföldtani vizsgálata (Hydrogeology of karstified formations in the Balaton Highland). Hidrológiai Közlöny 68:348–355 [in Hungarian]

Gondár-Sőregi K, Gondár K, Székvölgyi K, Kun É, Simonffy Z, Ács T, Tahy Á (2011) Climate change and impacts on water supply. Bükk Mountains Test Area. Brussels, CC-WaterS project, European Regional Development Fund (ERDF), p 51

Gruber P, Kereskényi E, Koltai G, Bárány-Kevei I (2012) Data on the hydrogeographical conditions of Baradla cave: seepage and drip waters. Acta Climatologica Et Chorologica 46:125–134

Gruber P, Haviarová D, Balázs I, Mátrahalmi T, Serfőző A, Ambrus M (2015) Víznyomjelzéses vizsgálatok a Haragistya–Szilice–Borzova Karsztterületen (Tracer tests in the Haragistya–Silica–Silická Brezová Karst area). Karsztfejlődés XX, Szombathely, Hungary, pp 63–79 [in Hungarian]

Haas J (2012) Geology of Hungary. Springer Science & Business Media, Heidelberg, p 244

Haas J, Korpás L, Török Á, Dosztály L, Góczán F, Hámor-Vidó M, Oravecz-Scheffer A, Tardi-Filácz E (2000) Felső-triász medence-és lejtőfáciesek a Budai-hegységben a Vérhalom téri fúrás vizsgálatának tükrében. Földtani Közlöny (Upper Triassic basin and slope facies in the Buda Mts. based on study of core drilling Vérhalom tér, Budapest) 130:371–421 [in Hungarian]

Haas J, Budai T, Csontos L, Fodor L, Konrád G (2010) Magyarország pre-kainozoos földtani térképe, 1: 500 000 (1: 500 000 Pre-Tertiary geological map of Hungary). Budapest, Magyar Állami Földtani Intézet [in Hungarian]

Havril T, Molson J, Mádl-Szőnyi J (2016) Evolution of fluid flow and heat distribution over geological time scales at the margin of unconfined and confined carbonate sequences—a numerical investigation based on the Buda Thermal Karst analogue. Mar Pet Geol 78:738–749

Hevesi A (2002) A Bükk hegység földrajzi helyzete, kialakulása, éghajlata. In: Baráz C (ed) A Bükki Nemzeti Park (Geographical situation, formation and climate of the Bükk mountains). Bükk National Park Directorate, Eger, pp 83–108 [in Hungarian]

Hips K (2001) The structural setting of Lower Triassic formations in the Aggtelek-Rudabánya Mountains (northeastern Hungary) as revealed by geological mapping. Geol Carpath 52:287–300

Hungarian Ministry of Agriculture (2020) Országos Barlangnyilvántartás (National Cave Cadastre of Hungary). Budapest, Magyar Állami Természetvédelem [in Hungarian]

Iván V, Mádl-Szőnyi J (2018) Vulnerability assessment and its validation: the Gömör-Torna Karst, Hungary and Slovakia. Geol Soc Lond Spec Publ 466:261–273

Iván V, Stevenazzi S, Pollicino LC, Masetti M, Mádl-Szőnyi J (2020) An enhanced approach to the spatial and statistical analysis of factors influencing spring distribution on a transboundary Karst aquifer. Water 12:2133

Izápy G, Maucha L (1992) A Bükk-hegység vízháztartási vizsgálata (Water management study of the Bükk Mountains). A Bükk karsztja, vizei, barlangjai, Miskolc, Hungary, pp 131–141 [in Hungarian]

Izápy G, Maucha L (2004) Az Aggteleki-hegység karszthidrológiai vizsgálata a Jósvafői kutató állomáson: Karsztvízkutatás Magyarországon (Karst hydrological investigation at the Jósvafő research station, Aggtelek Karst: Karstwater research in Hungary). Budapest, VITUKI. p 40 [in Hungarian]

Izápy G, Sárváry I (1991) Kiemelt jelentőségű karsztos termálvíz-előfordulások vízföldtani vizsgálata, Egri és Eger környéki termálvizek vizsgálata (Hydrogeological investigation of karst thermal water occurrences of special significance, investigation of thermal waters of Eger and in the surroundings of Eger). Budapest, Vízgazdálkodási Tudományos Kutatóintézet, p 29 [in Hungarian]

Izápy G (2002) Magyarország forrásainak katasztere (Natural Spring Cadastre of Hungary). Budapest, VITUKI Hydrological Institute [in Hungarian]

Jakucs L (1951) Vízföldtani megfigyelések a Gömöri-karszton (Hydrogeological observations on the Gömör Karst). Földtani Közlöny:442–445 [in Hungarian]

Jakucs L (1971) A karsztok morfogenetikája. A karsztfejlődés varienciái (Morphogenetics of karst regions. Variants of karst evolution). Budapest, Akadémiai Kiadó, p 310 [in Hungarian]

Jakucs L (1997) Világörökségi barlangjaink kiemelkedő karszttudományi értékei (World heritage caves in Hungary and the karst science). Szeged, p 28 [in Hungarian]

Jámbor Á (1976) Üledékes kéntelep a Zsámbéki-medence szarmata sorozatában (Sedimentary sulfur deposit in the Sarmation formations of the Zsámbék basin). A Magyar Állami Földtani Intézet Évi Jelentése 1974-ről:301–306 [in Hungarian]

Kele S, Scheuer G, Shen C-C, Chang H-W (2011) A Rózsadomb (Budapest) édesvízi mészköveinek U-Th-sorozatos kormeghatározása és stabilizotóp-geokémiai vizsgálata (U/Th dating and stable isotope geochemical investigation of the travertines of the Rózsadomb (Budapest)). Földtani Közlöny 141:293–312 [in Hungarian]

Kessler H (1938) Az aggteleki barlangrendszer hidrográfiája (The hydrology of the Aggtelek Cave System). Földrajzi Közlemények 66:1–30 [in Hungarian]

Kibédi Mátyus I (1787–1793) Ó és Új Diaetetica I-VI (Old and New Diaetetica I-VI). Posony, Füskúti Landerer Mihály. [in Hungarian]

Klimchouk A, Palmer AN, De Waele J, Auler AS, Audra P (2017) Hypogene karst regions and caves of the world. Springer, Cham, Switzerland

Klimchouk A (2007) Hypogene speleogenesis: hydrogeological and morphogenetic perspective. Special Paper no. 1, National Cave and Karst Research Institute Carlsbad, NM, pp 1–106

Kocsis K, Horváth G, Keresztesi Z, Nemerkényi Z (2018) Magyarország nemzeti atlasza 2. kötet: természeti környezet (National Atlas of Hungary Vol. 2: Natural environment). In: Kocsis K, Klinghammer I, Horváth G, Keresztesi Z, Kovács Z, Márton M, Nemerkényi Z & Zentai L (eds.) Magyarország nemzeti atlasza. Budapest, Magyar Tudományos Akadémia, Csillagászati és Földtudományi Kutatóközpont, p 187 [in Hungarian]

Konrád G, Sebe K, Halász A, Halmai Á (2010) A Délkelet-Dunántúl földtani fejlődéstörténete—recens analógiák (The major phases in the geological evolution of South-East Transdanubia and their recent analogues). Földrajzi Közlemények 134:251–266 [in Hungarian]

Kovács J, Erőss A (2017) Statistically optimal grouping using combined cluster and discriminant analysis (CCDA) on a geochemical database of thermal karst waters in Budapest. Appl Geochem 84:76–86

Kovács A, Perrochet P, Darabos E, Lénárt L, Szűcs P (2015) Well hydrograph analysis for the characterisation of flow dynamics and conduit network geometry in a karst aquifer, Bükk Mountains, Hungary. J Hydrol 530:484–499

Kovács J, Müller P (1980) A Budai-hegység hévizes tevékenységének kialakulása és nyomai (Thermal water activity and its imprints in the Buda Hills). Karszt és barlang 2:93–98 [in Hungarian]

Kovács A, Szőcs T (2016) Reactivation of karst springs after regional mine dewatering in the Tata area, Hungary. Karst without Boundaries, pp 337–358

Kovács ÁD (2011) Tó-és területi párolgás becslésének pontosítása és magyarországi alkalmazásai (Lake and spatially distributed areal evaporation estimations and their application in Hungary). PhD thesis, Budapest University of Technology and Economics, p 101 [in Hungarian]

Leél-Őssy S (1995) A budai Rózsadomb és környékének különleges barlangjai (The special caves of Buda Rózsadomb and its surroundings). Földtani Közlöny 125:363–432 [in Hungarian]

Leél-Őssy S (2017) Caves of the Buda Thermal karst. In: Klimchouk A, Palmer AN, De Waele J, Auler AS, Audra P (eds) Hypogene Karst Regions and caves of the world. Springer, Cham, Switzerland, pp 279–297

Lénárt L, Hernádi B, Szegediné Darabos E, Debnár Z, Czesznak L, Tóth M (2014) The importance of Bükk Karst Water Monitoring System (BKWMS) in researching the relations of cold and warm karst waters in the area. GeoSci Eng 3:107–117

Lénárt L, Lénárt EI (2008) A GVOP 3.1.1.-2004-05-0530/3.0 VIMORE projekt keretében végzett bükki barlangi karszthidrológiai mérések összefoglaló értékelése (Assessment of cave karsthydrogeological measurements carried out in the frame of A GVOP 3.1.1.-2004- 05–0530/3.0 VIMORE project). Karsztfejlődés XIII, Szombathely, Hungary, pp 103–116 [in Hungarian]

Lénárt L (2011) Hidegkarsztvíz és melegkarsztvíz kapcsolat rendszere a Bükk térségében (Cold karst water and hot karst water connection system in the Bükk region, research historical remarks on the history of Bükk thermal karst formation). Kutatástörténeti megjegyzések a bükki termálkarszt kialakítás történetéhez. University of Miskolc, Miskolc, p 12 [in Hungarian]

Lenkey L, Dövényi P, Horváth F, Cloetingh S (2002) Geothermics of the Pannonian basin and its bearing on the neotectonics. EGU Stephan Mueller Spec Publ Series 3:29–40

Liebe P (2002) Felszín alatti vizeink (Our groundwaters). Budapest, Hungary, Környezetvédelmi és Vízügyi Minisztérium, VITUKI, p 56 [in Hungarian]

Mádl-Szőnyi J, Tóth Á (2015) Basin-scale conceptual groundwater flow model for an unconfined and confined thick carbonate region. Hydrogeol J 23:1359–1380

Mádl-Szőnyi J, Virág M, Zsemle F (2015) Potential maps for the hydrogeologic prerequisites of the installation of deep geothermal doublets and groundwater source heat pump systems in Budapest, Hungary. Central Eur Geol 58:114–128

Mádl-Szőnyi J, Erőss A, Tóth Á (2017) Fluid flow systems and hypogene Karst of the Transdanubian Range, Hungary—With special emphasis on Buda Thermal Karst. In: Klimchouk A, Palmer A, De Waele J, Auler A, Audra P (eds) Hypogene Karst regions and caves of the world. Springer, Cham, pp 267–278

Mádl-Szőnyi J, Czauner B, Iván V, Tóth Á, Simon S, Erőss A, Bodor P, Havril T, Boncz L, Sőreg V (2019) Confined carbonates—regional scale hydraulic interaction or isolation? Mar Pet Geol 107:591–612

Mádl-Szőnyi J, Erőss A, Havril T, Poros Z, Győri O, Tóth Á, Csoma A, Ronchi P, Mindszenty A (2018) Fluidumok, áramlási rendszerek és ásványtani lenyomataik összefüggései a Budai Termálkarszton (Fluids, flow systems and their mineralogical imprints in the Buda Thermal Karst). Földtani Közlöny 148:75–96 [in Hungarian]

Mádl-Szőnyi J (1997) Víztartó rendszerek sérülékenységi vizsgálata a dunántúli-középhegységi főkarsztvíztároló rendszer (DNy-i rész) példáján (Vulnerability assessment of aquifer systems with the example of main karst reservoir in the SW-Transdanubian Central Range). Földtani Közlöny 127:19–83 [in Hungarian]

Mádl-Szőnyi J (2006) A geotermikus energia: készletek, kutatás, hasznosítás (Geothermal energy: reserves, research and exploitation). Nagykovácsi, Hungary, Grafon, p 144 [in Hungarian]

Mádl-Szőnyi J (2020) Felszínalatti vízáramlások mintázata fedetlen és kapcsolódó fedett karbonátos víztartó rendszerekben a Budai-termálkarszt tágabb környezetének példáján (Groundwater flow pattern in unconfined and adjoining confined carbonate aquifer systems on the example of Buda Thermal Karst). Academic thesis, Hungarian Academy of Sciences, p 131 [in Hungarian]

Maller M, Rehák A, Hajnal G (2018) Water level fluctuation in karst aquifers in the Transdanubian range (Hungary). Pollack Periodica 13:151–162

Maucha L, Sárváry I (1970) Tidal phenomena in the karstic water level. Hydrol Sci J 15:39–45

Maucha L (1998) Az Aggteleki-hegység karszthidrológiai kutatási eredményei és zavartalan hidrológiai adatsorai 1958–1993 (Results and undisturbed data of karst hydrological researches on Aggtelek Hills). Budapest, VITUKI Rt. Hidrológiai kiadványa, p 414 [in Hungarian]

Miklós R, Lénárt L, Darabos E, Kovács A, Pelczéder Á, Szabó NP, Szűcs P (2020) Karst water resources and their complex utilization in the Bükk Mountains, northeast Hungary: an assessment from a regional hydrogeological perspective. Hydrogeol J 28:2159–2172

Papp K (1907) Miskolcz környékének geológiai viszonyai (Geology of Miskolc and its vicinity). Földtani Intézet Évkönyve XVI:89–135 [in Hungarian]

Pelikán P, Less G, Kovács S, Pentelényi L, Sásdi L (2005) Geology of the Bükk mountains. Hungarian Geological Society, Budapest, p 284

Petrik A (2009) A villányi-hegységi mezozoos képződmények mikrotektonikai méréseinek értelmezése (Interpretation of microtectonic measurements of Mesozoic formations in the Villány Mountains). Földtani Közlöny 139:217–236 [in Hungarian]

Poyanmehr Z (2013) Conceptualization and implementation of a groundwater flow model for the Buda thermal-karst system, Hungary. Central Eur Geol 56:359–380

Rónaki L (1973) A mecseki karszt 1:10 000-es méretarányú vízföldtani, morfológiai és speleológiai térképe (1:10000 hydrogeological, morphological and speleological map of the Mecsek karst). Hidrológiai Tájékoztató 78–81 [in Hungarian]

Rónaki L (1995) A mecseki karsztvíz megjelenések kémiai vizsgálatainak áttekintése (Chemical investigation of appearance of karst waters of Mecsek Mountains). Karszt és barlang: 48–50 [in Hungarian]

Rónaki L (2000) A Villányi-hegység és a Beremendi-rög barlangjai Beremend, Hungary, Beremend Nagyközség Önkormányzata (The caves of Villány Mountains and Beremend Hills) [in Hungarian]

Rónaki L (2007) A mecseki karszton történt víznyomjelzések áttekintése (Overview of water tracers of Mecsek karst system). Karsztfejlődés XII, Szombathely, Hungary, pp 91–103 [in Hungarian]

Ruszkiczay-Rüdiger Z, Fodor L, Bada G, Leél-Össy S, Horváth E, Dunai T (2005) Quantification of quaternary vertical movements in the central Pannonian Basin: a review of chronologic data along the Danube River, Hungary. Tectonophysics 410:157–172

Ruszkiczay-Rüdiger Z, Balázs A, Csillag G, Drijkoningen G, Fodor L (2020) Uplift of the Transdanubian Range, Pannonian Basin: how fast and why? Glob Plan Chang 192:103263

Salem A, Dezső J, El-Rawy M (2019) Assessment of groundwater recharge, evaporation, and runoff in the Drava Basin in Hungary with the WetSpass model. Hydrology 6(1):23

Sásdi L, Szilágyi F (1986) Víznyomjelzéses vizsgálatok az Aggteleki-karszton (Drainage tracing investigations on the Aggtelek Karst). Karszt és barlang 1–2:33–38 [in Hungarian]

Sásdi L (1998) Vízföldtan és Vízrajz. In: Baross G (ed) Az Aggteleki Nemzeti Park (Hydrogeology and hydrology). Budapest, Mezőgazda Kiadó, pp 118–159 [in Hungarian]

Schafarzik F (1928) Visszapillantás a budai hévízforrások fejlődéstörténetére (Review on the history of thermal springs in Buda). Hidrológiai Közlöny 1:9–14 [in Hungarian]

Schréter Z (1923) Az egri langyos vizű források (Lukewarm springs of Eger). A magyar királyi Földtani Intézet Évkönyve XXV:1–25 [in Hungarian]

Stelczer K (2000) A vízkészlet-gazdálkodás hidrológiai alapjai: egyetemi és főiskolai tankönyv (Hydrological concepts of water resources management). Budapest, ELTE Eötvös Kiadó, p 411 [in Hungarian]

Stocker L (1729) Thermographia Budensis, seu Scrutinium physico-medicum aquarum mineralium Budae scaturientium (Temperature conditions of Buda, or physical-medical character of discharging mineral water in Buda). Budae, Johannes Georgius Nottenstein [in Latin]

Szabó PZ (1961) A Mecsek és a Villányi-hegység barlangjai (Caves of Mecsek Mountains and Villány Mountains). Karszt és barlang 1:3–20 [in Hungarian]

Szádeczky-Kardoss E (1941) A Dunántúli-Középhegység karsztvizének néhány problémájáról (On the issues of karst water in Transdanubian Mountains). Hidrológiai Közlöny 21:67–92 [in Hungarian]

Szentpétery I, Less G (2006) Az Aggtelek-Rudabányai-hegység földtana. Magyarázó az Aggtelek-Rudabányai-hegység 1988-ban megjelent 1: 25 000 méretarányú fedetlen földtani térképéhez (Geology of the Aggtelek–Rudabánya Hills. Explanatory book to the Pre-Quaternary geological map of the Aggtelek–Rudabánya Hills, 1988, 1: 25 000). Budapest, Magyar Állami Földtani Intézet. p 92 [in Hungarian]

Szijártó M, Galsa A, Tóth Á, Mádl-Szőnyi J (2019) Numerical investigation of the combined effect of forced and free thermal convection in synthetic groundwater basins. J Hydrol 572:364–379

Szőke E (2012) A Nyugat-Mecsek karsztja (Karst of the western Mecsek Mountains). Szeged, Hungary, GeoLitera, p 185 [in Hungarian]

Takács-Bolner K (1985) A Beremendi-kristálybarlang (Beremend crystal cave). Karszt és Barlang I-II:3–12 [in Hungarian]

Takács-Bolner K (2003a) A Beremendi-kristálybarlang (Beremend crystal cave). In: Székely K (ed) Magyarország fokozottan védett barlangjai. Budapest, Mezőgazda Kiadó, pp 399–404 [in Hungarian]

Takács-Bolner K (2003b) Nagyharsányi-kristálybarlang (Nagyharsány crystal cave). In: Székely K (ed) Magyarország fokozottan védett barlangjai. Budapest, Mezőgazda Kiadó, pp 404–407 [in Hungarian]

Telbisz T, Gruber P, Mari L, Kőszegi M, Bottlik Z, Standovár T (2020) Geological heritage, geotourism and local development in Aggtelek National Park (NE Hungary). Geoheritage 12:5

Than K (1890) Az ásványvizeknek chemiai constitutiójáról és összehasonlításáról (On the chemical constitution and comparison of mineral waters). Értekezések a természettudományok köréből 20:1–47 [in Hungarian]

Tóth J (1962) A theory of groundwater motion in small drainage basins in central Alberta, Canada. J Geophys Res 67:4375–4388

Tóth J (1963) A theoretical analysis of groundwater flow in small drainage basins. J Geophys Res 68:4795–4812

Tóth G (1990) Vízföldtan (Hydrogeology). In: Bence G, Bernhardt B, Bihari D, Bálint C, Császár G, Gyalog L, Haas J, Horváh I, Jámbor Á, Kaiser M, Kéri J, Kókay J, Konda J, Lelkesné Felvári G, Majoros G, Peregi Z, Raincsák G, Solti S, Tóth Á, Tóth G (eds) A Bakony hegység földtani képződményei. Magyarázó a Bakony hegység fedetlen földtani térképéhez, 1:50.000. Budapest, Magyar Állami Földtani Intézet, pp 82–83 [in Hungarian]

Tóth J (1995) Hydraulic continuity in large sedimentary basins. Hydrogeol J 3:4–16

Tóth J (1999) Groundwater as a geologic agent: an overview of the causes, processes, and manifestations. Hydrogeol J 7:1–14

Tóth Á (2018) A Balaton-felvidék felszínalatti vizeinek hidraulikai kapcsolata a Bakonnyal és a Balatonnal (Groundwater flow systems and hydraulic connection of the Bakony Mts. - Balaton Highland - Lake Balaton region). PhD thesis, Eötvös Loránd University, p 172 [in Hungarian]

Tóth G, Rman N, Rotár-Szalkai Á, Kerékgyártó T, Szőcs T, Lapanje A, Černák R, Remsík A, Schubert G, Nádor A (2016) Transboundary fresh and thermal groundwater flows in the west part of the Pannonian Basin. Renew Sustain Energy Rev 57:439–454

Tóth M, Dorn F, Fürst Á, Lorberer Á, Sárváry I (1999) A tatai források visszatérésével kapcsolatos vizsgálatok és cselekvési program (Tasks and activities regarding the re-operation of springs at Tata). Tata, Hungary, Hydrosys Ltd, Monumentum Ltd, Equilibrium Ltd. p 33 [in Hungarian]

Tóth Á, Galsa A, Mádl-Szőnyi J (2020) Significance of basin asymmetry and regional groundwater flow conditions in preliminary geothermal potential assessment—implications on extensional geothermal plays. Glob Plan Chang 103344

Vendel M, Kisházi P (1964) Összefüggések melegforrások és karsztvizek között a Dunántúli-középhegységben megfigyelt viszonyok alapján (Relationships between thermal springs and karst waters based on the observations in the Transdanubian Range). MTA Műszaki Tudományos Osztályának Közleményei 97–119 [in Hungarian]

Vigassy T, Leél-Őssy S, Demény A (2010) A Nagyharsányi- és a Beremendi-kristálybarlang és képződményeik genetikai vizsgálata (Genetic study of the Nagyharsányi and Beremend crystal caves and their formations). Földtani Közlöny 140:73–88 [in Hungarian]

Vígh F (1944) Az esztergomi szénmedence hidrológiai viszonyai és a víz elleni védekezés módozatai (The hydrological settings of the coal basin at Esztergom, and methods of dewatering). Bányászati és Kohászati Lapok LXXVII:215–222 [in Hungarian]

Wein G, Konda J (1977) A Budai-hegység tektonikája, Földtani térképek (Tectonics of the Buda Hills, Geological Maps). Budapest, Magyar Állami Földtani Intézet [in Hungarian]

Zámbó L (1998) Az Aggteleki-karszt felszínalaktani jellemzése (Geomorphological characterization of the Aggtelek Karst). Földrajzi Értesítő 47:359–378 [in Hungarian]

Zsigmondy V (1871) Tapasztalataim az artézi szökőkutak fúrása körül (My experiences during drilling artesian wells). Pest, Eggenberger Ferdinánd m. akad. könyvkereskedése [in Hungarian]

Zsigmondy V (1873) Mittheilungen über die Bohrthermen zu Harkány, auf der Margaretheninsel nächst Ofen und zu Lippik, und den Bohrbrunnen zu Alcsúth (Information about the thermal baths at Harkány, on Margaret Island near Buda and Lippik, and the wells at Alcsúth). Pest, Hungary, Kilian [in German]

Zsigmondy V (1879) A városligeti artézi kút Budapesten (The artesian well in Városliget (City Park), Budapest). Földtani Közlöny 3–4:128–131 [in Hungarian]

The Geomorphic Evolution of Karsts and Karstic Surfaces

János Móga and Márton Veress

Abstract

In this chapter the geomorphic evolution of Hungarian karst areas and the history of current karstification are overviewed. The denudation surfaces of the karst areas and their development, as well as their development age and subsequent transformation and the beginning of present karstification are described. Karstification started in the Pleistocene in the Aggtelek-Rudabánya Mountains (maybe in the Upper Pannonian or in the Pliocene) and it began in the Early-Sarmatian in the Bükk Region. In the Western-Mecsek Karst, the landforms of older soil-covered karst is overlain by the landforms of concealed karst and the beginning of concealed karstification dates back to the Würmian or it is older than it. In the Transdanubian Mountains this characteristic feature is also present, but it is less widespread and concealed karstification started in the Würmian or earlier, but its beginning is also different on the blocks belonging to various types.

Keywords

Cover formation • Exhumation • Tropical peneplain • Pediment • Abrasion platform

J. Móga (✉)
Department of Physical Geography, Eötvös Loránd University, Pázmány Péter sétány 1/c, Budapest, 1117, Hungary

M. Veress
Department of Geography, Savaria University Centre, ELTE SEK Eötvös Lóránd University, Károlyi Gáspár tér 4, Szombathely, 9700, Hungary
e-mail: veress.marton@sek.elte.hu

1 Introduction

The karst type of a karst area, its karst features, and the character and intensity of karstification depends on the factors influencing karstification and on its geomorphic evolution too. The beginning of the development of Hungarian karst surfaces and the interpretation of its present image are mostly traced back to the Upper Jurassic and to the end of the Cretaceous. There is no way to reveal surfaces older than this age or if there is any, that surface is in a buried state today thus, it does not belong to the topic of this chapter, but it belongs to chapter "Paleokarst in Hungary".

The development of karst surfaces is determined by several factors which are the following:

- Geological built-up (rock quality, thickness, structure and their age).
- The type and change (degree, velocity) of the climate during development stages.
- Altitude, vertical movements (frequency, degree, velocity).
- Surface forming forces during development stages (their sort, number, intensity, duration).

We are going to see that in Hungarian karsts tropical karstification, fluvial surface formation, abrasion, burial, and exhumation affected mainly the formation of karst surfaces.

- The effect of karstification, the interaction of surface and subsurface karst processes.
- The duration of a stage of geomorphic evolution.

The geology, vertical elevation, and climate of the areas were controlled by movements of plate tectonics. The present surface of a karst area developed as a result of the above-mentioned factors. The difference of the facies of some karst areas is primarily caused by geological structure and the frequency, degree and speed of vertical movements.

M. Veress and Sz. Leél-Őssy (eds.), *Cave and Karst Systems of Hungary*,
Cave and Karst Systems of the World, https://doi.org/10.1007/978-3-030-92960-2_7

Here will be described the role of loess in karstification and landscape evolution since the karstification of the Transdanubian Mountains and the Western Mecsek Karst are significantly influenced by this cover. Loess is denuded easily thus, it thins out in some places, which favours karstification (see chapter "An Interpretation of Karstification and Karst Processes by Hungarian Researchers"). Mixing with other sediments, the denuded loess becomes impermeable. It accumulates in depressions (in epigenetic valleys) where valley floor concealed karstification may happen. Dolines developing on valley floors promote the transportation of valley floor loess into the karst and thus, the valleys can be separated into areic parts. The loess mantle became an important karst genetic factor, it disrupted the karstification of the limestone terrain which on the one hand became covered (concealed) karst, on the other hand—because of the high carbonate content of the typical loess—it makes the waters percolating through it saturated thus, preventing karst corrosion even if the percolating water reaches the limestone beds situated below it. The complete or partial dissolution, the loam formation and soil formation of the carbonate content of the typical loess that deposited onto the karst terrain is important as regards karst development. Experiments prove the quality differences of the water percolating through the soil and the loess which play a crucial role in karstic processes. The waters percolating through the loessy sediment affected by soil formation covering the peneplain blocks and the plateau may have a limestone dissolving capacity of various degree depending on the soil quality, the thickness, and porosity of the subsoil loess bed.

Where the cover is thin, the loess can be completely affected by leaching and soil formation in its total depth even to the bedrock therefore, significantly increasing the dissolution capacity (aggressivity) of the meteoric water percolating through it which reaches the carbonate rocks. If there is loessy sediment below the soil cover, the water percolating through this loses its limestone dissolving capacity to a significant extent by dissolving the calcium carbonate content of the loess (Kiss et al. 2007). After all, it depends on the absolute amount of the carbonate content of the loess and on its change in the loess whether dissolution takes place in the bedrock. Self-generating processes may have an effect on loessy terrains. Surface waters flowing into the developing depressions increase decalcification, which increases surface runoff and thus, the development of the depression (Veress 1982, 2000b). The passages of depressions become plugged, intermittent lakes develop and the slowly seeping water of lakes increases the duration of dissolution (Veress 2000b).

Depending on its clay content, the loess mantle may have different behaviour regarding waters originating from precipitation: they drain some of the waters enabling them to infiltrate into deeper beds, while the other amount of precipitation increases surface runoff.

The grain size of loess influences doline development. In case of small grain size, the infiltrating water moves laterally, while in case of a larger grain size, it moves vertically (Deák et al. 2015; Veress et al.) and gets into the karst. The relation between the loess and the epikarst affects the development of subsidence dolines. Where the epikarst is well-developed due to former karstification, the loess may be washed-in into its passages without their plugging. On the loessy surface a lot of dolines are formed therefore, during its denudation the loess accumulates in the dolines. The loessy surface only becomes dissected to a small degree. Such concealed karst is the Western Mecsek Karst. Where the epikarst is underdeveloped, the loess may be washed-in into the passages of the epikarst to a small degree or it may not either. The development of the epikarst will be local, few shafts develop, and thus, the number of dolines (situated above the dolines) will also be low. Therefore, the catchment area of the dolines will be relatively large, and by this the waters flowing towards them create creeks contributing to the dissection of the loess mantle. Such concealed karst terrains occur in the Transdanubian Mountains.

2 Karst Areas

2.1 Aggtelek-Rudabánya Mountains (Aggtelek-Rudabánya Hills)

The Aggtelek Karst is plateau karst according to its dissected character, renewed allogenic covered karst from evolutionary point of view, mixed allogenic-autogenic karst hydrologically and morphologically and mostly soil-covered karst according to coveredness (Jakucs 1956, 1977; Hevesi 1991a, b; Móga 2002a, b; Veress 2016, see chapter "A General Description of Karsts in Hungary"). Valleys with doline rows, dolines (drawdown dolines), uvalas, ponors with blind valleys (with karst marginal ponors and ponors inside the karst), depressions of superficial deposit, subsoil karren (Zámbó 1998a, b; Móga 2002a, b; Veress 2010, 2017, see chapter "The Surface Morphology of Karsts in Hungary") are characteristic karst features in its area.

According to Láng (1955a, b), the Aggtelek Karst (and its plateaus) are built up of peneplanated surface parts which developed by the end of Miocene and during the Pliocene. Sásdi (1998) states that in the former terrestrial part of the karst, the tropical-subtropical peneplain developed at the boundary of the Oligocene and the Miocene. The opinion dating back its tropical karstification to the Cretaceous and Late Cretaceous is more predominant (Jakucs 1977; Gyurica and Sásdi 2009). The mounds of the plateaus are probably but not provably the remnant features of tropical karstification. The solution dolines of the karst, except if they are not on valley floor, are mostly situated on such mounds (Móga

2002a). According to Jakucs (1977), the surface was affected by continuous karstification from the Cretaceous to our present days since he thinks that the terra rossa (or any weathered material of such character) being widespread on the karst is the final product of tropical karstification. However, the theory of continuous karstification is hardly valid since the karst was covered by sediment many times (see below). According to Gyurica and Sásdi (2009), the karst became covered with marine sediment in the Oligocene at least partially, which is indicated by the distribution border of the abrasion conglomerate of the Bretka Gravel Formation: its northern border can be marked along the line between the settlements of Aggtelek and Égerszög (Fig. 1).

During the Miocene, rhyolite tuff accumulated originating from volcanic activity taking place in several stages (Gyurica and Sásdi 2009), from which a terra rossa like mantle of chemical weathering material developed (Bidló and Maucha 1964). However, according to other opinions, the terra rossa which is still widespread on the karst, mainly in the depressions of the karst and has a significant role in karstification processes (see chapter "The Surface Morphology of Karsts in Hungary") redeposited (Andrusov et al. 1958), and developed during karstification (Jakucs 1964; Stefanovits 1976; Less 1998).

Jakucs (1977) states that the Medve-szikla (Bear Cliff) being exposed on the slope of Lake Vörös is a tropical karst remnant feature. A notch being similar to the features of the South-Chinese stone forest can be seen on the cliff with terra rossa environment, and these morphological characteristics refer to a development under a climate warmer than today's maybe Tertiary, subtropical one. The quartz crystalline transformations of the ferricrete fills of the karren here indicate that the fill is several million years old (Beck and Borger 1999). According to the above-mentioned authors,

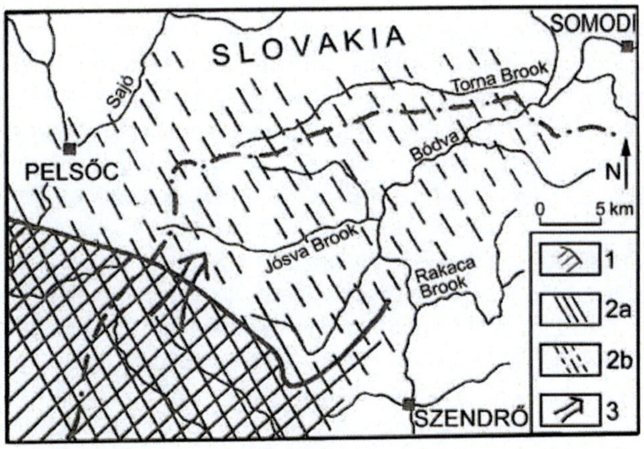

Fig. 1 Former distribution of Oligocene-Miocene sediments (Sásdi 1990). Legend: 1. Bretka Formation (limestone and conglomerate), 2. Putnok (=Szécsényi Slir) Formation (a = certain, b = supposed), 3. direction of transgression

the host material is of crystalline origin, which reached its present place by fluvial transportation and was transformed by chemical weathering. They think that this latter process took place on a former tropical peneplain of Cretaceous/Early Eocene age.

The flow of karstwater was of NS direction both in the Oligocene (disregarding transgression) and the Miocene (Gyurica and Sásdi 2009). This also refers to the presence of karstification although Sásdi (1998) dates the karstification of this time to the end of the Miocene when the superficial deposit denuded.

In the Pannonian age karstification related to karstwater took place: there are Pannonian cover cavity fills around the settlements of Alsótelekes and Felsőtelekes (Sásdi 1998). The freshwater limestones occurring in the karst marginal Pannonian beds (Teresztenye Plateau) also refer to karstification (Sásdi 1998; Gyurica and Sásdi 2009). However, according to Sásdi (2005), the Pannonian freshwater limestones are not related to the karstification of the mountains. It has to be mentioned that according to Zámbó's calculations (1986), there are dolines on the karst which started to develop in the Upper Pannonian.

In the Early Pannonian age a newer transgression took place which spread to the present terrains with an elevation of 300 m and only the more elevated parts of the plateaus protruded from this (Sásdi 1998, Fig. 2). As already mentioned, the mountains became covered with rhyolite tuff in the Middle Miocene (Jakucs and Móga 1997). In the Late Pannonian-Pliocene the alluvial cone of Ős-Sajó (Ancient Sajó) developed on the currently covered karst part, but on the plateaus as well and the latter became covered with gravel thinly (Sásdi 1998, Fig. 3). According to Zámbó (1998a), in the Pannonian the mountains were transformed into pediment and then later it became dissected and getting into various elevations it constituted the plateaus though as it will be seen below, other authors date pediment development later.

From the beginning of Pliocene, vertical and horizontal movements of significant degree began (Gyurica and Sásdi 2009) which continued in the Pleistocene: the degree of elevations was 50–100 m (Mezősi 1998), it may even have reached 300 m in the northern part of the karst (Zámbó 1998a). Láng (1955a) states that the final (today also existing) morphological separation of the currently covered karst and buried karst took place at the end of the Pliocene, and then in the Pleistocene, the plateaus became elevated to such extent that surface karstification could have begun and karst hydrography could have developed. If shaft caves developed from ponors situated at former rock boundary (Dénes 1972, see chapter "An Interpretation of Karstification and Karst Processes by Hungarian Researchers"), during the denudation of the cover, the plateau of Alsó Hill may also have lost its cover at this time since there are several karst features of

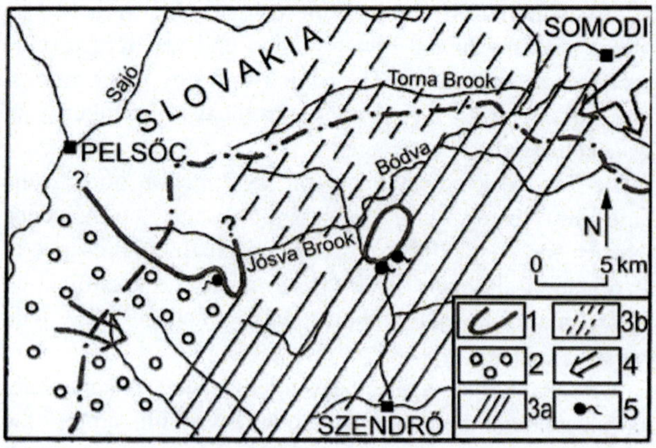

Fig. 2 Former distribution of Pannonian rocks (Sásdi 1990). Legend: 1. uncovered karst, 2. clayey gravel beds, 3. clayey, sandy groups of beds with lignite (a = certain, b = supposed), 4. direction of material transport, 5. traces of spring activity

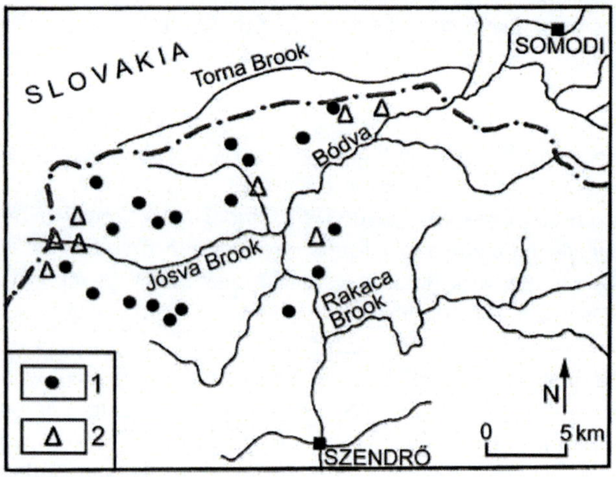

Fig. 3 Sporadic remnants gravel occurrences on the karst (Sásdi 1990). Legend: 1. quartz gravels on the surface, 2. quartz gravels in the cave

such character. However, the degree of elevations is indicated by the fact that the Late Pliocene freshwater limestones are at an altitude of 250–280 m, while those of Pleistocene age are at an altitude of 145 m (Alföldi et al. 1975). The inheritance of the valleys of the karst onto the limestone began at the end of the Pliocene (Sásdi 1998).

Bella et al. (2016) think that the Jósvafő paleopolje continued its development in the Late Pliocene by horizontal dissolution taking place on the surface of former pediment. The pediment (the authors do not give a development stage) probably developed during the formation of the Borsod Gravel Formation on the karst. The development of the pediment also means that the already existing karst features (primarily depressions) may have been destroyed as a result of the significant planation taking place in its area.

According to Jakucs (1956), in the Pannonian, the valleys with NS direction being situated on the northwards inclining surface of the area of the Aggtelek Karst (he meant Aggtelek Plateau by this) and currently having doline rows (for example Baradla Valley, Mész Valley, Hideg Valley) inherited (the author calls this Stage I), the cover became destroyed in these valleys and in their environs. The rock boundary with ponors of blind valleys developed at the interface of the soil-covered karst and covered karst by this. He thinks that this process resulted in the further erosional development of the large caves of the karst (Baradla-Domica, Béke Cave, etc.). However, on the contrary, the karst tilted at the Pliocene-Pleistocene boundary (Sásdi 1990), or in the Pontian stage (Gaál 2014) and its surface became tilted in SSE direction (Sásdi 1990). As a result of this, gravel material (Borsod Gravel Formation) was transported from NW and W (Gaál 2014), from the Slovak Ore Mountains (Móga 2002a) to the uncovered terrains and to the Pannonian terrains. This gravel cover refers to the Pliocene development of the pediment similarly to the case of other Hungarian mountains (Pécsi 1980). The age of the accumulation of the Borsod Gravel Formation is Pliocene according to Jaskó (1932), while Gaál (2014) dates it to the Pontian stage. On the floor of epigenetic valleys (the already mentioned Baradla Valley, Mész Valley, Hideg Valley) with a southern orientation developing on this surface, the ancestors of today existing ponors were formed contributing to the development of the Baradla-Domica cave system (Veress 2010, 2012, Fig. 4). Subsequently, the Aggtelek Plateau tilted oppositely during the separation of the karst into blocks (plateaus) thus, its surface began to dip in northern direction, but partly also the plateau part eastwards, the western part of Galyaság and also the covered karst situated south from them (Veress 2010, 2012). As a result of this tilting, surface waters flowed and flow towards the Aggtelek Plateau from the southern covered part of the karst (Fig. 4). This resulted in the development of newer ponors at the rock boundary and the further erosional development of the Baradla-Domica cave system (Veress 2010). At the ponors of the capture line along the Baradla-Domica cave system, basin-like depressions (depressions of superficial deposit) are arranged (Láng 1971; Veress 2017; Fig. 15 in chapter "The Surface Morphology of Karsts in Hungary"). Below them, closed depressions can be found on the bedrock (chapter "The Surface Morphology of Karsts in Hungary"). The development of basin-like depressions took place by the partial destruction of the fill of the depressions. The material of the fills was transported into the cave since the fill is the material of the Borsod Gravel Formation there (Piros and Gyurica 1986). According to

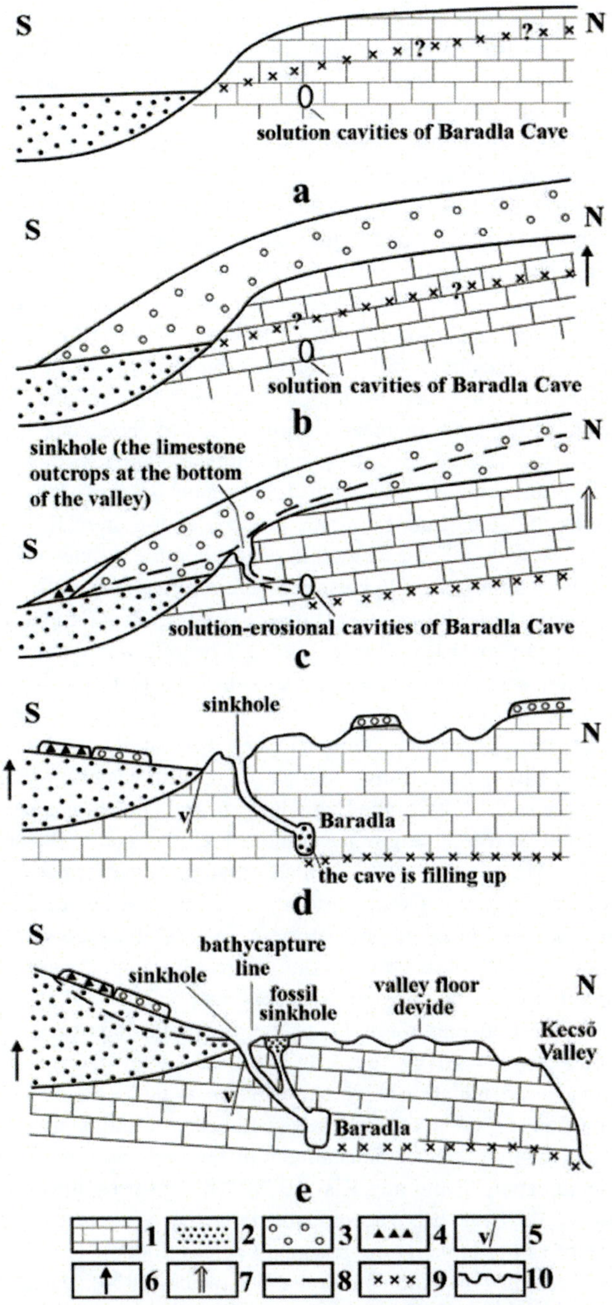

Fig. 4 Surface development and valley development on the Aggtelek Plateau (Veress 2012). Legend: a. uncovered features are formed on the Aggtelek Plateau, features of the Edelény Formation develop southwards from the plateau, b. Pliocene-Pleistocene: the plateau tilted, its SSE inclined surface inclined, but the surface of the covered karst which was covered by the material of the Borsod Gravel Formation also inclined, c. Pleistocene (?): the southern margin of the plateau and the areas in a southern direction from this (the present covered karst) lifted, the plateau had an almost horizontal position, but its valleys had inherited to the limestone in their total by this time, solution dolines were formed on valley floors, e. end of Pleistocene and beginning of Holocene: tilting continued, the plateau is inclined in northern direction, the second generation of ponors developed at the margin of the plateau, 1. limestone, 2. gravel, sand, clay (Edelény Formation), 3. gravel (Borsod Formation), 4. reworked gravel, 5. fault, 6. tilting, 7. uplift, 8. valley floor, 9. karstwater level, 10. dolines and uvalas

Veress (2017), cave development and the exposure of paleokarstic depressions were connected through the ponors and took place affecting each other (Fig. 16 in chapter "The Surface Morphology of Karsts in Hungary"). The gravel material was transported from the depressions into the Baradla-Domica cave system thus, maintaining erosion there. When the denudation reached fine-grained beds in the fill, accumulation increased in the cave system. The cave growth reacted upon the deepening of the depression since the cave system was able to receive the denuded sediment. The transformation of the catchments area of the ponor(s) into a basin also increased the amount of water getting into the ponors (Veress 2017).

Focusing on Aggtelek Plateau and Jósvafő Plateau, Jakucs (1964) distinguished the following stages of karstification related to landscape evolution.

- Pre-pannonian stage: the karst is separated into plateaus.
- Pannonian stage: the karst (or some of its parts) become covered.
- Post-pannonian stage: Valley development on Aggtelek Plateau and Jósvafő Plateau from the end of the Pannonian. The latter is transformed into a polje.
- Pleistocene stage: Downcutting of the River Jósva, the ponors of the capture line develop.
- Early Holocene stage: The Jósvafő polje opens. The lower levels of the large caves of the karst develop. Tributary valleys develop which divide up the floor of the Jósvafő polje. Doline development begins on the floor at this time.

2.2 The Bükk Region

There are diverse karst types in the area of the Bükk Region thus, soil-covered karst, mixed allogenic-autogenic karst (Bükk Plateau, Southeastern Bükk). Concealed karst is not significant, some larger and smaller patches can be mentioned from the floor of karst depressions. Its karst features are diverse (Fig. 5). Solution dolines with various morphology and morphological environment can primarily be mentioned, but some collapse dolines (Fig. 6) and ponors with blind valley occur too.

According to Láng (1954), the Bükk Region is a peneplain remnant where the parts of former peneplain got into different elevations. In the Late Jurassic and Early Cretaceous, the mountains were situated in great depth and their structure was transformed (Hevesi 2002). Sásdi (1997) thinks that it has been a terrestrial area since the middle of the Eocene and then sediment formation began in the Oligocene. In the Eocene the area of the mountains had hot climate (Andreánszky and Kovács 1955) therefore, tropical peneplanation happened (Bulla 1968); Pinczés 1968, 1980; Láng et al. 1970; Hevesi 1978, 1980, 1990). According to Tóth (1984, 2001), Fejes and Tóth (1984), the area of the

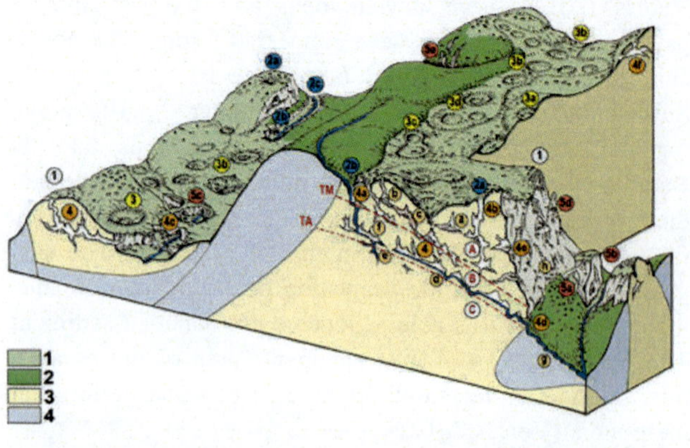

Fig. 5 Karst landform of Bükk Mountains after Baráz Cs. Legend: A—Vadose zone, B—Epiphreatic zone, C—Phreatic zone, TA—Low karstwater level, TM—High karstwater level, Surface landform: 1—karren (lapies), 2—swallow, 2a—opening of a shaft on the summit, 2b—sinkhole with active cave, 2c—blind valley with a resurgent stream, 3—doline, 3a—lonely (hanging) doline, 3b—series of dolines, 3c—uvala, 3d—polje, Underground landform: 4—cave, 4a—sinkhole cave (active), 4b—shaft (inactive vertical cave), 4c—active stream cave, 4d—uplifted collapsing cave, 4e—ruin cave, 5—Surface destroyed karst landforms: 5a—cave, rock gate, 5b—karst gorge, roofless cave, 5c—collapsed doline, 5d—stone pillars, 5e—cuesta, Surface and subsurface Karst relicts and traces: a—collapsed chamber with avens, b—aven with collapsed rockfield, c—syphon, d—inner cave swallow hole passages, e—inactive tetaratas, f—travertine deposit, g—rockfall and debris flow

Central Bükk is the remnant of several peneplanation lasting from the Cretaceous to the Middle Eocene and on this peneplain an older Pliocene denudation level develops at an altitude of 850–870 m and then a younger Pliocene surface is formed at an elevation of 650–700 m (Fig. 23 in chapter "The Surface Morphology of Karsts in Hungary").

A general view exists that the mountains were peneplanated into a tropical peneplain under hot climate (Bulla 1964; Pinczés 1968, 1980; Láng et al. 1970; Hevesi 1978, 1980, 1990). Therefore, the unevenly denuded surface of the Triassic became overlain by the Early Tertiary rocks (Bércziné Makk 1980). According to Hevesi (2002), the Bükk Plateau is the remnant of this peneplain and qualifying the development age of the peneplain he considers it to be Early-Middle Eocene. He believes that the peneplain continued its formation by terrestrial denudation (karstification), the features preceding tropical karstification were destroyed because of this, between the Lower Eocene and Middle Oligocene. Thus, karstification affected almost the same surface. He thinks that this view supports Jakucs's (1977) opinion that on karsts of Aggtelek character, present surface karst features are well-developed and of great density because karstification took place on their karst surfaces without interruption for a long time. However, as it can be seen below, the mountains became covered subsequently thus, the Lower Eocene-Middle Oligocene karst surface is not the direct ancestor of the current karst surface. Hevesi (2002) thinks that in the Oligocene, the middle parts of the mountains protruded from the sea that flooded their margins. According to him, the gravel beds of the marginal parts of the mountains and the sand-gravel beds around Noszvaj also refer to this (Báldi 1983). However, on the Kis Plateau (Csókás) there are Upper Oligocene sediments. Sásdi (1997) thinks that this refers to the fact that the mantle of the Kis Plateau thrust over the Palaeozoic rocks of North-eastern Bükk after the Oligocene, but according to Hevesi (2002), it refers to transgression in the mountains at the end of the Oligocene. However, with a newer uplift the mountains became dry land again and during the Eggenburgian age debris was formed under less warm climate, which promoted valley development and cave development (Hevesi 2002). Along the Répáshuta limestone stripe (South-eastern Bükk) red clay with debris accumulated and filled the dolines here and buried the inselberg karst in their environ (Hevesi 2002).

In the Miocene, rhyolite tuff covering took place many times (Jámbor 1959; Balogh 1964). The tuff was denuded within a short time and accumulated in karst features (cave, fissure) (Sásdi 1997).

In the Ottnangian-Carpathian era, the mountains were covered by the sea transgrading from a north-eastern direction (Hevesi 2002). In the Eggenburgian and Ottnangian age, tuffs of volcanic origin accumulated in the south (Pantó 1961; Hámori et al. 1978). In the Carpathian stage, sea cover became complete (Nagymarosy 1981), then pediment formation took place in the northern part of the mountains (Hevesi 2002) after the marine regression under warmer Mediterranean climate (Andreánszky and Kovács 1955)

In the Badenian the uplift of the mountains happened by upwarping (Moldvay 1972). In the inner area of the mountains, the denudation of the substratum is indicated by the limestone gravels of the Badenian beds of the marginal areas, while the calcareous sinters of the beds refer to their karstification (Bohn and Kiss 1978). The uplift by upwarping could have been accompanied by the rejuvenation of karstification towards the margins (Veress 2000a).

According to Andreánszky and Kovács (1955), the climate was Mediterranean (with an amount of precipitation of 850 mm, then 1500 mm, the average temperature in July was 24.5 °C) in the Sarmatian-Early Pannonian age. Hevesi (2002) thinks that the remaining superficial deposit denuded from the karst and tuff accumulated as a result of volcanic activity (Varga 1981).

Sásdi (1997) states that as a result of the significant uplift in the Pannonian, buried mounds were exposed on Nagy Plateau. However, he believes that Miocene sediments occurred on the plateaus even in the Pleistocene.

Fig. 6 Collapsed doline in the Bükk Mountains (Udvar-kő cave) (Photo Baráz Cs.)

By the end of the Miocene, the Bükk Region or some of it was transformed into mountains with significant elevation which separated from its environment and was dissected by valleys (Balogh 1964; Pinczés 1968, 1980; Hevesi 1978, 1990), and already had an individual drainage network at this time (Kerekes 1936; Pinczés 1957, 1980; Hevesi 1978, 1990).

Although according to Dunkl et al. (1994), at the end of the Miocene there was still thick, non-karstic cover in the mountains, several researchers (Kleb 1976; Pinczés 1957) reported on terrestrial limestone gravels from the margins due to drillings. Based on the latter data, Hevesi (2002) thinks that the karstification of the mountains lasting to present day began in the Sarmatian. They could have been transformed from buried karst into mixed allogenic-autogenic karst.

The doline fills and the freshwater limestones refer to their partial exposed character and karstification. Based on Pannonian flora elements, Pannonian fill was found in one of the dolines of the South-eastern Bükk (Vitális 1977; Láng et al. 1970). Sásdi (1997) described Miocene and Late Oligocene doline fill from the mountains, particularly from the South-eastern Bükk. The Lower Miocene karstification of the South-western Bükk is also proved by the freshwater limestone ruins of sediments of such age (Sásdi 1993).

According to Hevesi (2002), the surface karst features of Late Oligocene-Early Miocene were completely destroyed and only those cave parts and features could have survived which had developed below the karstwater level. However, since we think that the mounds of Nagy Plateau could have developed by karstification, these relict landforms partially survived. Hevesi (2002) thinks that the exposure of the mountains continued in the Sarmatian (valley development), but he also assumes pediment development, though it was accompanied by dissection since the climate became wetter.

At the beginning of the Pannonian, the mountains became elevated significantly (Sásdi 1997), but according to other opinions, sinking took place and the uplift only happened at the end of the Early Pannonian (Hevesi 2002). Sásdi (1997) states that as a result of the uplift, the mounds of the Nagy Plateau became exposed, while Hevesi (2002) believes that sinking was accompanied by the retardation of denudation. The average temperature in January was 6–8 °C, in July it was 22–24 °C, and the amount of precipitation was 1100–1200 mm in the mountains (Andreánszky and Kovács 1955).

Surface waters mainly flowed in northern and north-eastern direction (Hevesi 2002), which fact coincides with the opinion that the entrances of the spring caves in the environs of Nagymező (Nagy Plateau) have a northern direction (Sásdi 1997). The northern orientation of streams is proved by the fact that the sediment transportation from the mountains took place in northern direction. Since there are no gravels originating from the substratum in the Southern Bükk until the Pliocene (Pinczés 1968). Another evidence for this is that the streams of the valleys with doline rows of Nagy Plateau also left the karst in the above-mentioned direction (Hevesi 1980). According to Hevesi (2002), the beheadings of the streams of the plateau happened at the end of the Early Pannonian.

At the time of the Late Pannonian-Pliocene, the climate became drier (Pécsi 1980). The development of the

Fig. 7 One of the Bükk rocks, Őr-kő (Photo Baráz Cs.)

pediments of Bükkalja and Bükkhát started (Pinczés 1957, 1977; Hevesi 1986, 1990). However, at the end of the Pliocene, the climate became wet (Kretzoi and Pécsi 1972), and because of the sinking of the Great Hungarian Plain (Sümeghy 1944) and the uplift of the mountains, the erosional dissection of pediments began (Hevesi 2002). The karstification being significant from the Late Sarmatian to the Early Pannonian, and then slowing down, became more intensive in the Late Pliocene (Hevesi 2002). The valleys with doline rows of the Nagy Plateau were formed by the end of the Pliocene (Hevesi 1980, 1990). Hevesi (2002) states that the pattern of the current valley network developed by the end of the Pliocene which was accompanied by the development of streams with a southern orientation in the Southern Bükk. This is proved by the fact that the entrances of younger spring caves already look to the south (Sásdi 1997). By the end of the Pliocene, the mounds of the southern margin of the plateau developed (the "Rock", Fig. 7), and its regions (micro regions) became separated.

The inactive spring caves refer to the Pleistocene state of the mountains. Jánossy (1979) thinks that the archaeological findings (fossils) of the fill of the karstic fissure of Köves--Várad which are situated below and near Pungor-lyuk cave are of Günz-Mindel interglacial age. Therefore, the Pungor-lyuk was already above the karst water level at that time and according to Hevesi (2002), it is a representative of the older spring caves generation of the mountains (Hevesi believes that it developed at the end of the Pliocene.) According to the archaeological findings, the fill of the younger spring cave generation of the mountains (for example, Istállóskő Cave, Subalyuk) developed in the Early and Middle Würmian (Jánossy 1979). Thus, the development of these caves must be older than this. Sásdi's opinion (1997) is that there are fills referring to ponor activity in several inactive spring caves, but he did not give their age. This refers to an activity of katavothron character and to the fact that the karstwater level was close to the surface. According to Pinczés (1977) and Hevesi (1978, 1980), the pediments of Late Pliocene widened in the glacials. The spring caves became destroyed, the dolines functioned as snow traps and gorge development became more intensive (Hevesi 2002).

The older spring cave generation developed in the Günz-Mindel interglacial and then the younger generation was formed in the Mindel-Riss interglacial and the current, active spring caves appeared in the Riss-Würmian interglacial (Hevesi 2002). Newer gorges were formed in the Würmian by collapse of caves (Kerekes 1936).

2.3 The Transdanubian Mountains

As a result of the different elevation, structure, and extent of exposure of horsts, various karst types occur in the mountains (Veress and Vetési-Foith 2019a, b). Thus, on horsts covered with loess there is concealed karst (Tés Plateau, Hárskút Basin, Gerecse, etc.) with subsidence dolines, on horsts with impermeable cover there is cryptokarst (Kab Mountain, some parts of Pilis Mountain, Naszály?) with

Fig. 8 One of the largest basalt depressions, Macska-lyuk (Cat Hole) ponor on the Kab Mount

caprock dolines, on horsts partly covered with impermeable beds, mixed allogenic-autogenic karst occurs with ponors with blind valleys (Kab Mountain, Figs. 8 and 9), on horsts of lower elevation if they are uncovered, soil-covered karst is present (Tapolca Karst, some dolomite terrains) with solution dolines (Fig. 10), in case of cover (the karst in the environs of Devecser) there is soil-covered karst with partly filled solution dolines or conglomerate karst (among Magyarpolány, Városlőd and Farkasgyepű), with depressions that developed in superficial deposit.

Fig. 9 Regression valleys joined to basalt ponor on Kab Mount

Fig. 10 Tapolca karst (with solution dolines, basalt buttes in the background)

Láng (1952) and Bulla (1958, 1964) interpreted the horsts of the Transdanubian Mountains with different age (and elevation) as peneplain levels since they think that tropical peneplanation took place from the Upper Cretaceous to the Middle Miocene continuously. However, in other earlier works peneplanation and lowering was dated later. According to Leél-Őssy (1958), it took place during the Miocene-Pliocene in the Pilis Mountains and according to Láng (1956) in the Miocene in the Gerecse Mountains.

However, Pécsi (1980) thinks that patches of gravel cover also occurring on the roof levels of horsts are not tropical chemical weathering products, but they indicate pedimentation taking place under warm, sub-humid and semi-arid climate (the gravel material originated from the adjacent mountains of crystalline material that subsided by now). According to Pécsi (1980), the surface of the horsts of the Transdanubian Mountains developed through Mesozoic peneplanation during tropical karstification as a result of which surfaces of cone karst (inselberg karst) developed. He thinks that peneplanation took place from the Upper Jurassic to the Lower and Middle Cretaceous. During planation, the surface denuded almost to the base level of erosion and some areas of the mountains got below transgression. The inselberg karst did probably not develop uniformly and varieties of various maturity were formed (Veress 2000b). The exposure of surface parts with dolines by manganese and bauxite mining at Úrkút (Szabó 1956), Nyirád and Iharkút (Pataki 1983) refer to this.

According to Pécsi (1980), the peneplain started to dissect tectonically at the end of the Cretaceous. The separating horsts could have risen, sunk, or oscillated. Pécsi (1987a) thinks that the peneplain remnants could be covered with Eocene limestone (Bakony Region), with Oligocene sandstone (Buda Mountains, Pilis Mountains), with Miocene (Middle Oligocene—Lower Miocene) gravel completely (Bakony Region, the area among Magyarpolány, Városlőd, and Farkasgyepű and some of the basins) or partially (Eastern Gerecse).

The mounds of inselberg karst could only have survived at places where the caprock protected them against denudation, but it is also probable that they were also destroyed to a great extent as a result of permanent karstification. The more elevated a horst (at an absolute or relative altitude), the greater the chance of the denudation of its karstic landforms at its surface and thus, of the truncation of various degrees of the mounds was. At present, surface features remaining from tropical karstification are probable only at some sites. These are the following:

- On Kab Mountain, where outcrops of limestone patch occur at the surface with basalt cover (Németh 2005; Móga and Németh 2005; Veress and Unger 2015, see chapter "The Surface Morphology of Karsts in Hungary"). These are the karstic mounds of the dissected bedrock whose outcrop is very young if they became covered. However, if they were not covered, they could not have been destroyed significantly either since its degree is determined by the denudation of the basalt surface. Though, the extent of the denudation of the latter is insignificant because of the little elevation difference of the basalt surface.
- On horsts covered with Middle Cretaceous limestone in the Bakony Region where the exhumation of the mounds

takes place from below the superficial deposits (see chapter "The Surface Morphology of Karsts in Hungary").

– The surfaces built up of Triassic Main Dolomite such as the Veszprém-Devecser Graben or the environs of Nyirád and Ódörögd are dissected by mounds and groups of mounds. Csillag and Sebe (2015) interpreted their development by Early Miocene tropical planation. However, some parts of the dolomite surface (Veszprém Plateau) were transformed by Pliocene pedimentation (Pécsi 1987b). The survival of the landscape with mounds can be explained by two factors: since they are not of mountain front position, pedimentation affected the above-mentioned areas less or it did not affect them at all and/or because of their low position (at least at present), denudation was less intensive.

Between the Middle Oligocene and Lower Miocene, terrestrial sediment formation took place (Csatka Gravel Formation) on the low surface being more or less separated into horsts and disrupted by archipelago parts (Middle Eocene limestones were formed and at the north-eastern part of the mountains, Oligocene sandstones were formed). In the area of terrestrial sediment formation, the gravel was accumulated by rivers arriving from the south and the southeast (Korpás 1981) since the mountains were a low piedmont zone at that time (Láng 1958). Gravel formation gradually ceased until the beginning of the Pannonian (Pécsi 1980). The widespread distribution of this gravel in the Transdanubian Mountains (it occurs in the Bakony Region and in Gerecse too) indicates that the mountains were a uniform, hardly dissected area at that time. Pécsi (1980) regards this gravel accumulation as the development age and development of the older pediment of the Transdanubian Mountains. He thinks that its age is of Middle Oligocene and Middle Miocene. The patches of gravel cover are at different altitudes now (Láng 1955b, 1958) disregarding the reworked occurrences. This refers to the fact that since the development of the gravel cover the peneplain has become tectonically dissected definitively and the peneplain parts got into various altitudes. The Late Pannonian calcareous sinter occurrence at an altitude of 340–350 m below the Pisznice Cave (situated at an altitude of 460 m) in the Central Gerecse also refers to this (Scheuer and Schweitzer 1983). Thus, the cave was already inactive at this time and this also marks the elevated position of the bearing block (Nagy Pisznice) and also the fact that karstification took place in some parts of Gerecse.

According to Pécsi (1980), the gravel developed under warm, sub-humid, and semi-arid climate during pedimentation and pediplanation. Pécsi (1963, 1970) states that the horsts of the Transdanubian Mountains were alternating the sedimentary basins and pediments of the surrounding mountains that subsided into the depth by now. He also establishes that the recurring gravel covers refer to recurrent pedimentation. The surface features of older karstification may have been destroyed particularly during this latter process.

In the Pannonian, abrasion levels (Lóczy 1913; Prinz 1926) and Pannonian limestones were formed in the mountain front areas and in the basins inside the mountains such as in the Zirc Basin and Bicske Basin as well as in their grabens for example in the area of the Veszprém-Nagyvázsony Graben (Pécsi 1980).

In the Pliocene, pedimentation occurred again (Pécsi 1980) due to the streams leaving the mountains (Pécsi 1963). As a result of the uplifts, the pediments can be situated at various altitudes and may also form two levels because of their transformation in the Pleistocene (Pécsi 1987a). The pediment formation happened on the Pannonian sediments, but it also spread onto older surfaces (Pécsi 1980) or it developed on the Pontian abrasion platform (Pécsi 1987a) and penetrated into some valleys too (Pécsi 1963). The piedmont half-planes were overlain by Upper Pliocene calcareous sinters and the older terraces of the River Danube and its tributary rivers (Pécsi 1980, Fig. 11). Pliocene

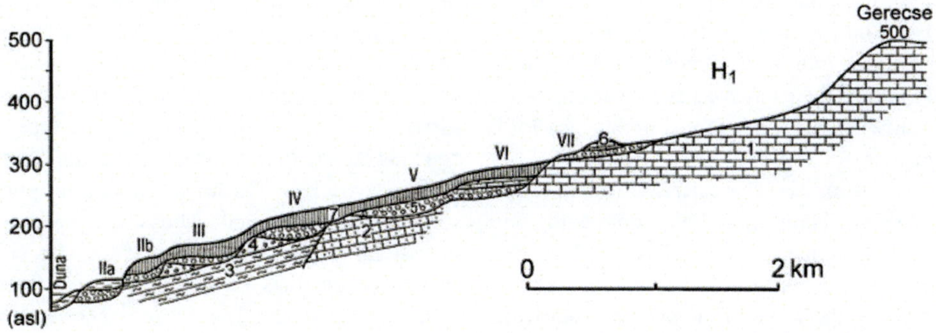

Fig. 11 Danube terraces and Upper Pliocene pediments at Lábatlan at the margin of the Gerecse Mountains (Pécsi 1980). Legend: H_1 Upper Pliocene pediment, IIa-VII Danube terraces, VII Late Upper Pliocene terrace (?) covered with calcareous sinter, VI. Pre-günz (Danube terrace with freshwater limestone), V. Günzian terrace, IV Mindel glacial terrace, III. Riss glacial terrace, IIa-IIb. Würmian glacial terraces, 1. Mesozoic rocks in general, 2. Cretaceous sandstone, 3. Eocene marl, 4. Oligocene gravel, conglomerate, 5. terrace gravel, 6. freshwater limestone, 7. slope loess

freshwater limestones for example in Gerecse occur at various altitudes as a result of the upheaval of different degree of some horsts (Láng 1956).

In the south-western part of the Transdanubian Mountains, the basalt lava covered Pannonian surface (Tapolca Basin), Pliocene pediments and probably Cretaceous peneplains (Kab Mountain) (Pécsi 1980) during the Late Pliocene basalt volcanism (Lóczy 1913; Jugovics 1954), whose absolute age is 2.7 million years (Balogh et al. 1987). The peneplain of the Kab Mountain was probably transformed by that time. According to Láng (1958), the basalt lava covered a surface dissected by valleys and mounds.

It is probable that the development of the older epigenetic valleys of the mountains and the dissection of denudation levels already started in the Pliocene. In the Pleistocene, valley development became more intensive particularly in the interglacials. The gravelly terrains and Pannonian surfaces (especially in the basins) were dissected by a valley network. In the Pleistocene, the Pliocene pediments were dissected by valleys in the interglacials and preceding it, a newer pediment may have developed on the already existing pediments in the glacials (Pécsi 1963). Loess was formed in the mountains which dates back to the Würmian in the Bakony Region. Pécsi (1982) states that the loess is younger than 30 thousand years in the Bakony Region (but it may be older than this on other mountain parts). Therefore, particularly in the Bakony Region, the development age of syngenetic subsidence dolines that were formed on non-reworked loess is younger than 30 thousand years (Veress and Vetési 2019a, b). The development age of postgenetic dolines may also be older than this.

According to Veress and Vetési (2019b), the relative age of the development of subsidence dolines is different. In case of dolines with a valley floor position, this age is younger than the sediment lining the valley floor, on the terrains between exhumed mounds it is younger than the development of the surrounding valley (superficial deposit was transported into them from the mounds and their environs), and it is also younger on the floor of solution dolines than the accumulation of floor sediments. Since valley development on the horsts and in their environs is of various time, the beginning of concealed karstification of horsts also took place at a different time.

Postgenetic karstification is characteristic of the Tés Plateau (Bakony Region). Here, mixed allogenic-autogenic karst developed at the patches of the Oligocene Miocene gravel cover (at present, its material exists in the form of sporadic gravel mostly in the caves of the plateau) following the dissection of the cover. After the denudation of gravel cover patches and the development of loess, former, filled ponors redeveloped as subsidence dolines (Veress 2000b).

Taking into consideration the geomorphic evolution of horsts, Pécsi (1980, 1991) distinguished the following horst types in the Transdanubian Mountains (Fig. 46 in chapter "The Surface Morphology of Karsts in Hungary"):

- Cryptopeneplain: the altitude of its surface is below 300 m. The cryptopeneplain is covered with Pliocene gravel and Middle Miocene limestone and gravel. The peneplain (block) may occur at the mountain front or inside the mountains. In the latter case, the block (or block group) is a basin or graben morphologically.
- Low threshold surface: the altitude of its surface is below 300 m (400 m) and it is built up of old (mainly Triassic) carbonate rocks. The blocks of this type lost their superficial deposits by pedimentation.
- Horst elevated to summit position: the altitude of its surface may be between 400 m (maybe 300 m) and 550 m. Several varieties of this type can be distinguished. Thus, in the area of the covered horsts elevated to summit position, Middle Eocene limestone and Oligocene sandstone covers the Mesozoic (Triassic, Jurassic) substratum which often has a surface of cone karst. (We think that the horsts where the cover is Cretaceous limestone can also be put into this type.) In the area of half exhumed horsts elevated to summit position, the Tertiary cover only survived in patches, while the exhumed horsts elevated to summit position are terrains where Tertiary sediments were completely destroyed.
- Horst in summit position: the altitude of its surface is 600–700 m and it is built up of Triassic and Jurassic carbonate rocks.
- The above classification can be supplemented by the block type with basalt cover (Kab Mountain) which occurs in the southern part of the Bakony Region.

The blocks belonging to the above types will have a different degree of coveredness and thus, various karstification as well (see chapter "An Interpretation of Karstification and Karst Processes by Hungarian Researchers"). The surfaces of horsts are being karstified now if the circumstances are suitable there particularly the denudation (or accumulation surfaces) of various horst types. The type of karstification depends on whether there is non-karstic cover and if there is, of what type it is. On horsts in summit position, peneplain surfaces are karstified where the patches of permeable superficial deposit (loess) survived. Its condition is only present at few sites for example on the floor of old depressions (see chapter "The Surface Morphology of Karsts in Hungary"). On horsts elevated to summit position, concealed karstification is more widespread. Here, there are karstifying patches where the superficial deposit survived between the exhuming karstic mounds (horst covered with Middle Cretaceous limestone) in a thinned out state or at that place where covered mounds occur under the loose superficial deposit (Tés Plateau) or below consolidated cover (Kab Mountain) (Veress 2000b, Veress and Vetési-Foith 2019a, b). It has to be highlighted that the lower parts of this horst type being stripped of gravel cover belong to the older pediment. Karstification may also happen here mainly on the lined floors of epigenetic valleys (Hárskút Basin, maybe some parts of Tés Plateau, the environs of Márvány Valley, see chapter "The Surface Morphology of Karsts in Hungary").

On lower threshold surfaces, the peneplain (dolomite terrain), the older pediment, but the younger one may also be karstified. Tapolca Karst (which is mostly uncovered soil-covered karst) or the karst being covered with reworked superficial deposit in the environs of Devecser may be mentioned as examples for the latter.

The horsts with cryptopeneplains of the mountains are buried karsts. Either conglomerate karst developed (during the dissolution of the limestone material of the cover) on these or waters flowing from surfaces with gravel cover contribute the development of the gorges of the mountains and the opening-up of karstic cavities there (Veress 2000b; Veress and Vetési-Foith 2019a, b).

2.4 The Western Mecsek Karst

The Western Mecsek Karst is situated at the northern margin of the Mecsek Mountains which are surrounded by badly karstifying carbonate rocks and non-karstic rocks from the south (Barta and Tarnai 1997, Fig. 63 in chapter "The Surface Morphology of Karsts in Hungary"). The whole Mecsek is a system of peneplain surface parts (Pécsi 1963; Lovász 1981). Pécsi (1963) thinks that peneplanation took place at the end of the Cretaceous, but according to Lovász (1981), Eocene–Oligocene tropical peneplain remnants also occur (Fig. 12, Jakab Mountain, Misina-tető). The peneplain parts were later elevated into different altitudes by tectonic movements. Particularly the peneplain parts of marginal areas and within these the parts built up of karstic rocks were transformed by outer forces.

During Cretaceous-Early Tertiary karstification, polje development happened and the polje was transformed by the Orfű Valley (Szabó 1956). In the Carpathian stage, the area of the karst was transformed into pediment (its remnant is at an altitude of 450–470 m), which or a part of which developed into abrasion platform during transgression in the Middle Miocene (Lovász 1981). During the uplift of the mountains, the northern margin of this platform was also transformed into an abrasion platform due to a newer

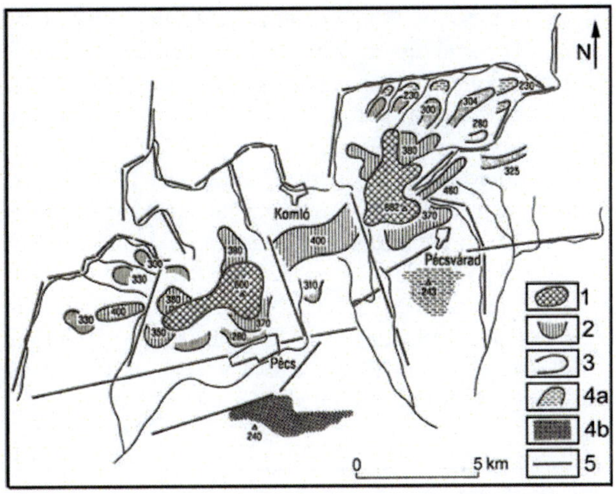

Fig. 12 Main structural lines and denudation surfaces of the Mecsek Mountains (Lovász 1981). Legend: 1. Eocene–Oligocene surface, 2. Carpathian abrasion surface, 3. Lower Pannonian abrasion surface, 4a. piedmont half-planes, 4b. buried alluvial cones, 5. main structural line

transgression in the Pannonian (Lower Pannonian). The two levels are separated from each other by a steeper slope (Fig. 62 in chapter "The Surface Morphology of Karsts in Hungary").

Valleys with a SN orientation go transversely to the area of the karst, the valleys primarily received and still receive their streams from the surrounding non-karstic terrain. These valleys either inherited from superficial deposits being denuded by now to the karstifying rock or they developed because the karstwater level situated close to the surface and they could deepen following the lowering of this level (Hevesi 2001). The beginning of their development is younger than the Pannonian, but older than the Pleistocene (Lovász 1977; Hevesi 2001). According to Hevesi (1991a, b), the older abrasion level became covered in the Middle Miocene.

Loess, but clay and sand occurring together with loess are widespread in the Mecsek Mountains and thus, in the karst area too (Veress 2011). The latter probably originates from the sandstone of Jakab Mountain.

The karstification represented by the development of solution dolines may have begun after the formation of the valleys, which is proved by the fact that the distribution of dolines is determined by valleys to a great degree (for example doline rows occur in the continuation of valley heads). However, there are no solution dolines on valley floors which, as already mentioned, may refer to the fact that during their development the valley floors were situated at the karstwater level. According to Hevesi (1991b), karstification is continuous on the upper abrasion level from the Sarmatian and on the lower abrasion level from the Pliocene to these days. Szabó (1968) dates back the development of solution dolines to the Early Tertiary. However, as there are no solution dolines on the upper level (Lovász 1971, 1977), they were destroyed and at most the surviving shafts of former dolines can represent this.

In the Pleistocene, loess was formed which developed uniformly and since it does not contain limestone thus, it may have been dissolved for a longer time and it is probably older than the Würmian. The subsidence dolines of the karst developed in the loess. Thus, concealed karstification is young (Szabó 1968; Lippmann et al. 2008), it is younger than the development of the loess (Veress 2011) and its beginning is probably Würmian and younger than this.

3 Conclusions

It is a common character in the history of Hungarian karst areas that denudation terrains with low inclination (peneplains, pediments, and abrasion platforms) developed during the geomorphic evolution of karst areas. These surfaces are primarily the tropical karst peneplains whose geomorphic evolution was very different. Therefore, the degree of exposure and denudation as well as the elevation of peneplain parts also differs. However, no tropical peneplain can be found on the Western Mecsek Karst. Fluvial surface formation always plays a role in the past of karst areas. On the one hand, it denuded the superficial deposit, on the other hand, the developing valleys, being inherited to the karstic bedrock had a significant influence on karstification.

The duration and frequency of periods with terrestrial areas, the degree and frequency of coveredness, the type, thickness, and present distribution of the covering non-karstic rocks are different on various karst areas.

The present surface of karst areas developed during extremely different landscape evolution. The transformation of former denudation surfaces is very different. Karstification also played a role in geomorphic evolution. The time from which karstification is continuous until the present is uncertain to determine. The present surface karstification seems to be continuous from the Pleistocene (maybe from the Upper Pannonian or the Pliocene) on the Aggtelek-Rudabánya Mountains, while in the Bükk Mountains it may be continuous, being present from the Late Sarmatian or from the Pannonian though not with the same intensity. The beginning of the concealed karstification of the Transdanubian Mountains and the Western Mecsek is young. In case of the Transdanubian Mountains, it is Würmian or younger, and it began at a different time on blocks of various blocks, while in case of the Western Mecsek, it may also be older than the Würmian. On the latter karst a change of karst type happened since the older soil-covered karst was overlain by concealed karst. This overlying character occurs less frequently in case of the karsts of the Transdanubian Mountains. Such karst terrains can be mentioned from the Bakony Region (Tés Plateau, Kab Mountain).

References

Alföldi L, Balogh K, Rónai A (1975) Magyarázó Magyarország 200000 földtani térképsorozatához (Explanatory book to the geological map series 200000 of Hungary): M-34-XXXIII. Miskoc-MÁFI, Budapest, 156 pp (in Hungarian)

Andreánszky G, Kovács SÉ (1955) A hazai fiatalabb harmadidőszaki flórák tagolódása és ökológiája (The dissection and ecology of younger Tertiary flora of Hungary). MÁFI Évkönyve XLIV(1):326 (in Hungarian)

Andrusov D, Borza K, Martiny E, Pospisil A (1958) O povode a dobe vzniku tzv. terra rossy juzného a stredného. Slovenska. Geol. Zbornik SAV 9:1. Bratislava

Báldi T (1983) Magyarországi oligocén és alsó-miocén formációk (Hungarian Oligocene and Lower Miocene formations). Akadémia Kiadó, Budapest, 293 pp (in Hungarian)

Balogh K (1964) A Bükk-hegység földtani képződményei (Geological features of the Bükk Mountains). Magyar Állami Földtani Intézet Évkönyve 48(2):719 (in Hungarian)

Balogh K, Árva-Sós E, Pécskoy Z, Ravasz-Baranyai I (1987) K/Ar dating of post-Sarmation alkali basaltic rocks in Hungary. Acta Mineral Petrogr, Szeged, 28 pp

Baráz Cs (2010) The magic karst. Protection of the natural and cultural heritage in the Bükk National Park. Carline booklets (Bábakalács füzetek) – 12. Eger, 40 pp

Barta K, Tarnai T (1997) Karsztkutatás az orfűi Vízfő-forrás vízgyűjtő területén (Karst research on the catchment of the Vízfő spring, Orfű). Karszt és Barlang I–II:12–19 (in Hungarian)

Beck RK, Borger H (1999) Soils and relief of the Aggtelek karst (NE Hungary): a record of the ecological impact of paleoweathering effects and human activity. In: Bárány Kevei I, Gunn J (eds) Essays in the ecology and conservation of karst. Szeged-Budapest-Miskolc, Hungary, pp 13–30

Bella P, Veselsky M, Gaál L, Miár J (2016) Jósvafő paleo-polje: morphology and relation to the landform evolution of Aggtelek karst and Jósva River valley, Hungary. Z Geomorphol 60(3):219–235

Bérciné MA (1980) Eocén-oligocén határképződmények a Bükkalján (Eocene-Oligocene boundary features in Bükkalja). Őslénytani Viták 25:123–141 (in Hungarian)

Bidló G, Maucha L (1964) A jósvafő környéki karsztüledék vizsgálata (Investigation of karst sediment in the environs of Jósvafő). Az Építőipari És Közlekedési Műszaki Egyetem Tudományos Közleményei 10:71–89 (in Hungarian)

Bohn P, Kiss K (1978) Magyarország mélyfúrási alapadatai (Basic data on deep drilling in Hungary). MÁFI Évi Jelentés 1976-ról, Budapest, pp 599–671 (in Hungarian)

Bulla B (1958) Néhány megjegyzés a tönkfelszínek kialakulásának kérdéséhez (Some remarks on the question of the development of peneplains). Földr Ért VII(3):257–274 (in Hungarian)

Bulla B (1964) Magyarország természeti földrajza (Physical geography of Hungary). Tankönyvkiadó, Budapest, 424 pp (in Hungarian)

Bulla B (1968) Válogatott természeti földrajzi tanulmányok (Selected studies of physical geography). Akadémia Kiadó, Budapest, 143 pp

Csillag G, Sebe K (2015) Long-term geomorphological evolution. In: Lóczy D (ed) Landscapes and landforms of Hungary. Springer, Cham Heidelberg, New York, Dordrecht, London, pp 29–38. https://doi.org/10.1007/978-3-319-08997-3

Deák G, Szemes M, Veress M (2015) A gipsz fedőjének vízmozgásai analóg modelleken (Water movements of the plaster cover on physical analogue models). Karsztfejlődés XX:215–229. https://doi.org/10.17701/15.215-229

Dénes Gy (1972) A fokozatosan lepusztuló vízzáró takaró szerepe az exhumálódó karszt morfológiai fejlődésében (The role of the gradually denuding impermeable cover in the morphological development of exhuming karst). Karszt és Barlang I:5–8 (in Hungarian)

Dunkl I, Árkai P, Balogh K, Csontos L, Nagy G (1994) A hőtörténet modellezése fission track adatok felhasználásával – a Bükk-hegység kiemelkedés története (Thermal modelling based on apatite fissure track dating: the uplift history of the Bükk Mts.). Földtani Közlöny 124(1):1–24 (in Hungarian)

Gaál L (2014) A földtani fejlődés áttekintése (An overview of geological evolution). In: Gruber P, Gaál L (szerk) A Baradla-Domica-barlangrendszer – Aggteleki Nemzeti Park, Jósvafő, pp 19–30 (in Hungarian)

Gyuricza Gy, Sásdi L (2009) A Baradla-barlangrendszer kialakulásának kérdései a tágabb környezet földtani fejlődésének tükrében (Theory of the evolution of the Baradla Cave as mirrored in the geological evolution of its neighbourhood). Földt Köz 139(1):83–92 (in Hungarian)

Hámori G, Ravaszné Baranyai I, Balogh K, Árváné Soós É (1980) A magyarországi riolittufa szintek kora (The age of Hungarian rhyolite tuff levels). MÁFI Évi Jelentés az 1978. Évről:65–73 (in Hungarian)

Hevesi A (1978) A Bükk szerkezet- és felszínfejlődésének vázlata (A sketch of the structure and landform development of the Bükk Mountains). Föld Ért XXVII(2):169–203 (in Hungarian)

Hevesi A (1980) Adatok a Bükk hegység negyedidőszaki ősföldrajzi képéhez (Data to the quaternary paleogeographical features of the Bükk Mountains). Földtani Közlöny 110(3–4):540–550 (in Hungarian)

Hevesi A (1990) A Bükk felszínfejlődése különös tekintettel a karsztosodására (The geomorphic evolution of the Bükk Mountains with special regard to its karstification). Magyar Földrajzi Társaság, Budapest, 137 pp (in Hungarian)

Hevesi A (1991a) Magyarország karsztvidékeinek kialakulása és formakincse, I. rész (Development and landforms of karst regions in Hungary I). Földrajzi Közlemények XV(1–2):25–35 (in Hungarian)

Hevesi A (1991b) Magyarország karsztvidékeinek kialakulása és formakincse, II. rész (Development and landforms of karst regions in Hungary II). Földrajzi Közlemények CXV(3–4):99–120 (in Hungarian)

Hevesi A (2001) A Nyugat-Mecsek felszíni karsztosodásának kérdései (Questions of the surface karstification of the Western Mecsek). Karsztfejlődés VI:103–111 (in Hungarian)

Hevesi A (2002) Fejlődéstörténet II. Felszínfejlődés (Geological evolution II. Geomorphic evolution). In: Baráz Cs (ed) A Bükki Nemzeti Park, Bükki Nemzeti Park Igazgatóság, Eger, pp 81–108 (in Hungarian)

Jakucs L (1956) Adatok az Aggteleki-hegység és barlangjainak morfogenetikájához (Some data to the morphogenetics of the mountains and caves of Aggtelek). Földrajzi Közlemények IV:25–38 (in Hungarian)

Jakucs L (1964) Geomorfológiai problémák az Észak-Borsodi-karsztvidéken (Geomorphological problems in the Northern Borsod karst region). Borsodi Földrajzi Évkönyv 5:12–23 (in Hungarian)

Jakucs L (1977) A magyarországi karsztok fejlődéstörténeti típusai (Evolutionary types of Hungarian karsts). Karszt és Barlang I–II:1–16 (in Hungarian)

Jakucs L, Móga J (1997) A Gömör-Tornai-karszt (Gömör-Torna Karst). In: Karátson D, Száraz M Gy (eds) Magyarország földje. Kitekintéssel a Kárpát-medence egészére. Magyar Könyvklub, pp 378–384 (in Hungarian)

Jámbor Á (1959) A Bükk-fennsík pleisztocén „vályog" képződményei (The Pleistocene „loam' features of the Bükk Plateau). Földtani Közlöny LXXXIX(2):181–184 (in Hungarian)

Jánossy D (1979) A magyarországi pleisztocén tagolása gerinces faunák alapján (Hungarian Pleistocene dissection based on vertebrate fauna). Akadémia Kiadó, Budapest, 205 pp (in Hungarian)

Jaskó S (1932) A Baradla új részéről (On the new part of Baradla). Földrajzi Közlemények 60(9):157–159 (in Hungarian)

Jugovics L (1954) A Déli-Bakony és a Balaton-felvidék bazaltterületei (Basalt terrains of Southern Bakony and of Balaton Upland). Földtani Intézet Évi Jelentése 1953-ról, pp 65–88, Budapest (in Hungarian)

Kiss K, Zámbó L, Fehér K, Móga J (2007) A lösztakaró karsztosodásban játszott szerepének vizsgálata a Tési-fennsíkon (An investigation of the role of the loess mantle in karstification on the Tési plateau). Karsztfejlődés XII:193–205 (in Hungarian)

Kerekes J (1936) A Tárkányi öböl morfológiája (Morphology of the Tárkány bay). Földrajzi Közlemények LXIV(6–7):80–87 (in Hungarian)

Kleb B (1976) Észlelési magyarázó Eger 1:10000-es építésföldtani térképsorozatához. (Explanatory notes to the engineering geological map series 1:10000 of Eger) Eger-Felnémet. Eger-Lajosváros, Budapest, Közlekedési Dokumentációs Vállalat, 298 pp (in Hungarian)

Korpás L (1981) A Dunántúli-középhegység oligocén-alsó-miocén képződményei (Oligocene-Lower Miocene features of the Transdanubian Mountains). MÁFI Évkönyve 64. 142 pp (in Hungarian)

Kretzoi M, Pécs M (1972) A Pannóniai-medence pliocén és pleisztocén időszakának tagolása (Pliocene and Pleistocene dissection of the Pannonian Basin). Földrajzi Közlemények XXXI(4):300–326 (in Hungarian)

Láng S (1952) Geomorfológiai-karsztmorfológiai kérdések (Geomorphological-karstmorphological questions). Földr Ért 1:120–126 (in Hungarian)

Láng S (1954) Hidrológiai és morfológiai tanulmányok a Bükkben (Hydrological and morphological studies in the Bükk Mountains). Hidrológiai Közlöny XXXIV:103 (in Hungarian)

Láng S (1955a) Geomorfológiai tanulmányok az Aggteleki-karszton (Geomorphological studies in Aggtelek Karst). Földr Ért 4(1):1–17 (in Hungarian)

Láng S (1955b) A Gerecse peremhegységi részének geomorfológiája (Geomorphology of the offset part of Gerecse). Földr Ért IV (2):157–194 (in Hungarian)

Láng S (1956) A Központi Gerecse geomorfológiája (Geomorphology of Central Gerecse). Földr Ért V(3):265–281 (in Hungarian)

Láng S (1958) A Bakony geomorfológiai képe (The geomorphology of the Bakony Mountains). Földrajzi Közlemények 6:325–343 (in Hungarian)

Láng S, Mihâltz I-né M, Gy V (1970) A miskolctapolcai Nagykőmázsa dolináinak morfológiai és földtani vizsgálata (Morphological and geological investigations of the dolines of Nagykőmázsa of Miskolctapolca). Földr Ért XIXI:77–85 (in Hungarian)

Láng S (1971) A hazai karsztok és környékük lepusztulásának egyes kérdései (Some questions on the denudation of Hungarian karsts and their environs). Karszt és Barlang I:1–4 (in Hungarian)

Leél-Őssy S (1958) A Kevély-hegycsoport karsztmorfológiája és barlangjai (Karstmorphology and caves of Kevély Mountain Group). Földr Ért VII:17–52 (in Hungarian)

Less Gy (1998) Földtani felépítés (Geological characterization). In: Boross G (ed) Az Aggteleki Nemzeti Park. Mezőgazda, Budapest, pp 26–66 (in Hungarian)

Lippmann L, Kiss K, Móga J (2008) Az Abaliget-Orfűi karszt karsztos felszínformáinak vizsgálata térinformatikai módszerekkel (Investigation of the karstic phenomenon near Orfű and Abaliget by GIS methods). Karsztfejlődés XIII:151–166 (in Hungarian)

Lóczy L (1913) A Balaton környékének geológiai képződményei és ezeknek vidékek szerinti telepedése (The geological features of the surroundings of Lake Balaton and their regional distribution). – A Balaton tudományos tanulmányozásának eredményei, I. K. Budapest, 617 pp (in Hungarian)

Lovász Gy (1971) Adatok az Abaligeti-karszt geomorfológiai és hidrológiai jellemzéséhez (Data on the geomorphological and hydrological description of Abaliget Karst). Földr Ért XX(3):283–296 (in Hungarian)

Lovász Gy (1977) Geomorfológia: Kőzetek (Geomorphology: rocks). In: Lovász Gy (ed) Baranya megye természeti földrajza, Baranya Megyei Levéltár, Pécs, pp 62–68 (in Hungarian)

Lovász Gy (1981) A Baranyai-dombság, a Mecsek és a Villányi-hegység (The Baranya Hills, the Mecsek Mountains and the Villány Mountains). In: Pécsi M (szerk) A Dunántúli-dombság (Dél-Dunántúl). Akadémia Kiadó, Budapest, pp 124–136 (in Hungarian)

Mezősi G (1998) A Borsodi-dombság tájföldrajzi jellemzése (Landscape geographical description of the Borsod Hills). Földr Ért 79 (1):395–408 (in Hungarian)

Móga J (2002a) Felszínalaktani vizsgálatok a Galyaság területén (Geomorphological investigations in the area of Galyaság). Karsztfejlődés VII:173–186 (in Hungarian)

Móga J (2002b) A tornai Alsó-hegy (The Alsó Hill of Torna). Karszt és Barlang I–II:95–104 (in Hungarian)

Móga J, Németh R (2005) The morphological research of the basalt and loess covered plateaus in the Bakony Mts. (Transdanubian middle mts.-Hungary). Acta Carsologica 34(2):397–414

Moldvay L (1972) A neotektonikus felszínalakulás jelenségei a magyarországi középhegységben (Phenomena of neotectonic surface development in Hungarian Mountains). A Magyar Állami Földtani Intézet Jelentése 1970-ről, Budapest, pp 587–637 (in Hungarian)

Nagymarosy A (1981) Chrono- and biostratigraphy of the Pannonian Basin: a review based mainly of data from Hungary. Earth Evol Sci 3–4:183–194

Németh R (2005) A Kab-hegyi bazalttakaró depresszióinak vizsgálata (Investigation of the depressions of the basalt cover of Kab Hill). Karszt és Barlang 2000–2001 évf.:33–41 (in Hungarian)

Pantó G (1961) Az ignimbrit-kérdés alakulása és magyarországi vetülete (Ignimbrite question and its Hungarian relation). MTA Műszaki Tudományos Osztályának Közleményei XXIX(1–4):295–332 (in Hungarian)

Pataki A (1983) Karsztmorfológiai megfigyelések a nyirádi és az iharkúti bauxit-előfordulás területén (Karstmorphological Observations in the Bauxite Occurrence Area of Nyirád and Iharkút), MÁFI Évi jelentés az 1983 évről: 131–133 (in Hungarian)

Pécsi M (1963) Hegylábi (pediment) felszínek a magyarországi középhegységekben (Pediment surfaces in Hungarian Central Mountains). Földrajzi Közlemények 11(3):195–212 (in Hungarian)

Pécsi M (1970) Surfaces planation in the Hungarian Mountains and their relevance pedimentation (Studies in geography in Hungary 8). Akadémia Kiadó, Budapest, pp 29–40

Pécsi M (1980) A Pannóniái-medence morfogenetikája (The Morphogenesis of the Pannonian Basin). Földr Ért 29:105–127 (in Hungarian)

Pécsi M (1982) Quaternary studies in Hungary. Elmélet-Módszer-Gyakorlat. Akadémia Kiadó, Budapest, 313 pp

Pécsi M (1987a) Domborzatfejlődés és a geomorfológiai szintek korrelációja (Relief evolution and correlation of geomorphological levels). In: Pécsi M (szerk) A Dunántúli-középhegység A. Akadémia Kiadó, Budapest, pp 131–139 (in Hungarian)

Pécsi M (1987b) A domborzat jellegzetesebb exogén formái (Characteristic exogene features of the relief). In: Pécsi M (szerk) A Dunántúli-középhegység A). Akadémia Kiadó, Budapest, pp 140–186 (in Hungarian)

Pécsi M (1991) Geomorfológia és domborzatminősítés (Geomorphology and topography classification). Magyar Tudományos Akadémia Földrajztudományi Kutató Intézet, Budapest, 296 pp

Pinczés Z (1957) Az Eger-völgy problémái (Problems of Eger Valley). Földr Ért VI:29–43 (in Hungarian)

Pinczés Z (1968) Herausbildung der Tertiären Oberflächen des Bükk-Gebirges. Acta Geographica Debrecina 189–200

Pinczés Z (1977) Hazai középhegységek periglaciális planációs felszínei és üledékei (Periglacial planation surfaces and sediments in the Hungarian Mountains). Földrajzi Közlemények XXV(CI)(1–3):41–45 (in Hungarian)

Pinczés Z (1980) Production of planation surfaces and their types as ilustrated on the examples of a tertiary volcanic and a Mesozoic Mountains. Acta Geographica, Geologica Et Meteorologica Debrecina XIV–XV(1975–1976):5–29

Piros O, Gyuricza Gy (1986) A Baradla-barlang eróziós genetikai vizsgálata (Erosional Genetic Investigation of Baradla Cave) – NME Közleményei, I. sorozat, Bányászat 33(1–4):47–55 (in Hungarian)

Prinz Gy (1926) Magyarország földrajza (Geography of Hungary). Tudományos Gyűjtemény 16, Danubia, Pécs, 202 pp (in Hungarian)

Sásdi L (1990) Az Aggtelek-Rudabányai-hegység karsztjának földtani fejlődéstörténete (Geological Evolution of the Karst of the Aggtelek-Rudabánya Mountains). Karszt és Barlang I:3–8 (in Hungarian)

Sásdi L (1993) A Bükk-hegység paleokarsztjára vonatkozó megfigyelések (Observations on the Paleokarst of the Bükk Mountains). A bükki barlangok kutatásának védelmének és hasznosításának legújabb eredményei konferencia kiadványa, Miskolc:35–44 (in Hungarian)

Sásdi L (1997) Újabb földtani adatok a Bükk-hegység karsztjának fejlődéstörténetéhez (Newer geological data on the karst of the Bükk Mountains). Karszt és Barlang I–II:20–25 (in Hungarian)

Sásdi L (1998) Vízföldtan és vízrajz (Water geology and hydrography). In: Baross G (szerk) Az Aggteleki Nemzeti Park. Mezőgazda Kiadó, Budapest, pp 118–159 (in Hungarian)

Sásdi L (2005) Az Aggteleki-Rudabányai-hegység édesvízi mészkő előfordulásai (The calcareous sinter occurrences of the Aggtelek-Rudabánya Mountains). Karsztfejlődés X:137–151 (in Hungarian)

Scheuer Gy, Schweitzer F (1983) A Gerecse és a Budai-hegység édesvízi mészkőösszletei és képződésüknek geomorfológiai sajátosságai (Calcareous sinter beds of Gerecse and the Buda Hills and the geomorphological characteristics of their development). Kandidátusi értekezés, Budapest, Kézirat, 184 pp (in Hungarian)

Stefanovits P (1976) Talajtan (Soil science). Mezőgazdasági Kiadó, Budapest, 380 pp (in Hungarian)

Sümeghy J (1944) A Tiszántúl (The Trans-Tisza Region). Magyar Tájak földtani leírása VI. Földtani Intézet, Budapest, 208 pp (in Hungarian)

Szabó P (1956) Magyarországi karsztformák klímatörténeti vonatkozásai (Relations between climate and karst features in Hungary). Földrajzi Közlemények 80:183–189 (in Hungarian)

Szabó PZ (1968) A magyarországi karsztosodás fejlődéstörténeti vázlata (An Evolutionary Outline of Hungarian karstification). Dunántúli Tud. Gyűjtemény 13–25 (in Hungarian)

Tóth G (1984) A Központi-Bükk és geomorfológiai körzetei (The Central Bükk and its geomorphological districts). Földr Ért XXXIII (4):333–345 (in Hungarian)

Tóth G (2001) A karsztos résrendszer fejlődésének tényezői a Bükkben (Factors of the development of the karstic joint system in the Bükk Mountains). Karsztfejlődés VI:113–125 (in Hungarian)

Tóth G, Fejes P (1984) Idősebb pliocén lepusztulási szint kimutatása a Bükkben (Report on an older Pliocene denudation level in the Bükk Mountains). Karszt és Barlang II:65–72 (in Hungarian)

Varga Gy (1981) Újabb adatok az összesült tufatelepek és ignimbritek ismeretéhez (New data on welded tuffs and ignimbrites). MÁFI Évi Jelentése az 1978 évről:499–509 (in Hungarian)

Veress M (1982) Adatok a Hárskúti-fennsík karsztmorfogenetikájához (Data on the karst morphogenetics of the Hárskút Plateau). Karszt és Barlang II:71–82 (in Hungarian)

Veress M (2000a) Középhegységi karsztok néhány típusa (Some types of middle mountain karsts). Földrajzi Közlemények CXXIV(1–4):1–28 (in Hungarian)

Veress M (2000b) Covered karst evolution Northern Bakony mountains, W-Hungary. A Bakony Természettud. Kut. Eredményei 23, Bakonyi Természettudományi Múzeum, Zirc, 167 pp

Veress M (2010) Adalékok az Aggteleki-fennsík völgyeinek fejlődéséhez (Some data to the valley evolution of the Aggtelek Plateau). Karszt és Barlang, 2008 évf. (I–II):3–12 (in Hungarian)

Veress M (2011) Adatok a Mecsek-hegység fedett karsztosodásához a Cigány földi mintaterületről vett példák felhasználásával (Some data to the covered karstification of the Mecsek Mountains using examples taken from the Cigány föld research area). Karszt és Barlang, 2010 évf. I–II:9–30 (in Hungarian)

Veress M (2012) New data on development of the Baradla Cave. Acta Carsologica 42(2–3):193–204

Veress M (2016) Covered karst. Springer, Berlin, Heidelberg, New York, 536 pp. https://doi.org/10.1007/978-94-017-7518-2

Veress M (2017) A depresszió-exhumálás és a barlangképződés kapcsolata az Aggteleki karszton (The relationship between depresszion exhumation and cave development on the Aggtelek karst). Földrajzi Közlemények 141(2):97–111 (in Hungarian)

Veress M, Unger Z (2015) Kab Mountain: karst under a Basalt Cap. In: Lóczy D (eds) Landscapes and landforms of Hungary. Springer, Heidelberg, New York, Dordrecht, London, pp 55–62. https://doi.org/10.1007/978-3-319-08997-3

Veress M, Vetési-Foith Sz (2019a) The Distribution of Surface Karst Features in the Bakony Region (Transdanubian Mountains Hungary). J Geol Res 1(1):21–25

Veress M, Vetési-Foith Sz (2019b) A Bakonyvidék felszíni karsztja (The surface karst of the Bakony Region in Hungary). Földrajzi Közlemények 43(2):167–186 (in Hungarian)

Vitális Gy (1977) Földtani és vízföldtani megfigyelések a miskolc-tapolcai Nagykőmázsán (Geological and hydrogeological observations in Nagykőmázsa of Miskolc-Tapolca). Hidrológiai Közlöny 50(2):49–55 (in Hungarian)

Zámbó L (1998a) Az Aggteleki-karszt felszínalaktani jellemzése (Surface morphological characterisation of Aggtelek karst). Földr Ért XLVII(3):359–378 (in Hungarian)

Zámbó L (1998b) Felszínalaktan (Surface morphology). In: Baross G (ed) Az Aggteleki Nemzeti Park. Mezőgazda Kiadó, Budapest, pp 70–96 (in Hungarian)

The Surface Morphology of Karsts in Hungary

Márton Veress

Abstract

This chapter presents the surface karst of Hungarian karst areas which are as follows: Aggtelek-Rudabánya Mountains, the Bükk Region, the Bakony Region, the Vértes Plateau, the Gerecse Region, the Buda Hills, the Pilis Mountains, and the Western Mecsek Karst. Their description involves their research history, the general description of their karstification, the morphometric study of the karst features belonging to their area, the morphology, morphological environment, genetics, and development of karst features.

Keywords

Karst area • Karst feature • Research history • Morphometry • Genetics of a karst feature

1 Introduction

Karst features are the following: karren, dolines, ponors, karst lakes, complex depressions, poljes, karst gorges, karst remnant features, and concretion features.

Karren are small features of karsts which develop in diverse environments and they can be classified in several ways (Bögli 1960; White 1988; Ford and Williams 2007; Veress 2010a, 2016, 2019). According to their environment, karren can develop on covered surfaces, on partially covered surfaces, and on bare surfaces (Bögli 1960). Among Hungarian researchers, Leél-Őssy (1952a) distinguished karren field and karren slope. He thinks that the former develops on surfaces of low inclination, while the latter are formed on surfaces of high inclination. According to their development, karren can develop by percolating water or flowing water (Veress 2010a, 2019). The former can develop both under the soil (and the cover) and on bare surfaces, while those of flow origin develop on bare surfaces. Some karren types such as root karren owe their origin to the action of organic acids produced by plant roots (Jakucs 1956a, b, c; Veress 2010a). Karren features under cover deposits are grikes, kamenitzas, root karren, pitkarren, bed head karren (bedding grike), and on partially covered terrains or at the margin of the fill of larger karren features there are karren notches and Schichtfugenkarren. On bare surfaces, in addition to kamenitzas, grikes, bed head karren and pitkarren, rillenkarren, rinnenkarren, wandkarren, meanderkarren, trittkarren, roundkarren (they partly develop under the soil), scallops, Schichtfugenkarren, and karren caves occur. Karren do not only develop on carbonate rocks (thus to the action of waters of low pH), but also on evaporites and even on sandstones in case of high pH as well.

Dolines are arheic closed depressions which can develop on soil-covered karst and bare karst and on covered karst. The dolines of soil-covered karst and bare karst are solution dolines or collapse dolines (Sweeting 1973; Waltham and Fookes 2003; Williams 2004; Waltham et al. 2005). Among solution dolines, drawdown dolines, point recharge dolines, and inception dolines are distinguished (Sauro 2012). Drawdown dolines emerge if the passages of epikarst develop during water transport through karren. Hungarian researchers distinguish further varieties of solution (drawdown) dolines according to their position. Thus, row dolines of lower position on the valley-floor (Jakucs 1968, 1977a) and hanging dolines of higher position on mounds (Hevesi 1980). Point recharge dolines develop on the floor of inheriting valleys, where the deepening of the depression is caused by the material transport of the water that flows into the subsurface passages (Williams 1985, 2004). According to Hungarian researchers (Jakucs 1968, 1977a), the depressions developing on valley floor are ponors originated by subsurface capture and later transformed into dolines, into

M. Veress (✉)
Department of Geography, Eötvös Lóránd University, Károlyi Gáspár tér 4, Szombathely, 9700, Hungary
e-mail: veress.marton@sek.elte.hu

M. Veress and Sz. Leél-Őssy (eds.), *Cave and Karst Systems of Hungary*,
Cave and Karst Systems of the World, https://doi.org/10.1007/978-3-030-92960-2_8

swallow dolines (Hevesi 1980). However, according to Jakucs (1977a), solution dolines can also develop on the floor of valleys inherited to the limestone they are drawdown dolines according to more up-to-date classification.

Covered karst features include karren, caprock dolines, subsidence dolines, depressions of superficial deposit, and covered karst ponors. Caprock dolines develop on cryptokarst (Veress 2016) if the cover is consolidated rock and the void of the bedrock is balanced by collapse (Williams 2004; Waltham et al. 2005). Subsidence dolines are formed on concealed karst, their varieties (Waltham and Fookes 2003; Williams 2004; Waltham et al. 2005) are dropout dolines (the cover is cohesive and the material lack of the bedrock is balanced by collapse), suffosion dolines (the cover is non-cohesive, the material void is balanced by suffosion), and compaction dolines (the material of the infilled doline is compacted). Depressions of superficial deposit (DSDs) are depressions of the superficial deposit with a larger diameter, they develop because subsidence dolines or ponors that are formed in their area transport the superficial deposit into the karst through their intermittent watercourses (Veress 2009, 2016). Covered karst ponors (Veress 2016), similar to the ponors of the mixed allogenic-autogenic karst, are situated at rock boundaries and at the end of blind valleys. However, in the case of such features, the karstic rock does not outcrop beyond the blind valley, but there a permeable superficial deposit characteristic of covered karst is present.

Although some researchers usually consider large-sized depressions of a complex morphology as uvalas, here they are called complex depressions. On their floor, various karst features may occur such as solution dolines, subsidence dolines, blind valleys, and ponors. Ponors develop on mixed allogenic-autogenic karst, along rock boundary, at the termination of the blind valleys of non-karstic terrain. A rock boundary can develop at wedged-in non-karstic rock or at superimposed non-karstic rock. Thus, ponors can occur on valley floor, in depressions of karstic or non-karstic origin, at the foot of a karstic mound.

The valleys of the karst form gorges or gorge-like section partially or completely. In karst areas, the conditions of valley development are given as follows:

- There has to be an impermeable cover on the karst. If the valley floor reaches the karstic rock during its deepening, valley development can even continue there if the stream of the valley receives water from the non-karstic terrain of the catchment area. If the cover is removed, the valley can only receive water from the catchment area, if the valley head is surrounded by non-karstic terrain or it begins at a karst spring (pocket valley). In lack of these conditions, the valley turns into a dry valley, or its stream continues to incise if the karstwater table is situated at the valley floor. Dry valleys develop in a karstic way since, according to Jakucs (1977a), solution (drawdown) dolines emerge on their floor.
- In karsts dismembered into blocks, complex valley evolution is common. Epigenetic-regressional valleys and epigenetic-antecedent valleys occur. In the former case, inheritance occurs during the retreat of the valley head. In the latter case, inheritance takes place in the karstic rock of a block that is covered but above its environs.
- The valley deepens into the wedged-in rock or into the surrounding non-karstic rock of karst areas (blind valley).
- The ceiling of the cave collapses.
- On the cover of the karst with a small thickness and local expansion, gullies and creeks develop which may merge to form karst depressions.

Karst mounds are remnant features. The smaller features are karst hills and the larger are karst mountains. During the widening of adjacent dolines, narrow dividing walls are left behind from the original terrain.

Calcareous sinter is the material that is precipitated from karstwaters (freshwater limestone). Calcareous sinter can constitute travertine mound and in the channels of streams it can create rimstone dams with or without pools and escarpments. Spring cones are mounds which occur at the emergence sites of karstwater.

Hungarian karst areas are found in the North Hungarian Mountains, the Transdanubian Mountains and the Western Mecsek. The karst areas of the North Hungarian Mountains are the Aggtelek-Rudabánya Mountains, the Bükk Region, and the Nézsa-Csővár Hills. The karst features of Csővár blocks are described here in short. Karren fields, ponors, and caprock dolines are mentioned from these blocks (Ozoray and Láng 1957; Leél-Őssy 1952a, b; Gánti 1957). They involve the Bakony Region, the Vértes-Velence Mountains, and the Dunazug Mountains (the micro-region groups of the latter are the Gerecse Region, the Pilis Mountains, and the Buda Mountains).

2 The Aggtelek-Rudabánya Mountains

2.1 General Characteristics

Mainly the Aggtelek Karst section of the mountains is rich in karst phenomena. The Rudabánya Mountains and Szalonna Mountains are two plateaus separated from the Aggtelek Karst, surface karst phenomena are mentioned from the Szalonna Mountains (Esztramos) (Leél-Őssy 1953a). Hungarian and Slovakian karst research has paid special attention to the Aggtelek Karst because of its diversity and to the

Baradla-Domica Cave System due to its size. Therefore, among Hungarian karst areas this is the area that is studied in the most detailed and diverse way.

The description and classification of its karst was done by Jakucs (1956a, 1977b), Hevesi (1991a, b), and Veress (2014, 2016). Zámbó (1970, 1986a, b), Zámbó and Telbisz (2000) dealt with its karstification characteristics thus with factors influencing dissolution and its intensity. The morphometry of karst features was studied by Mezősi (1984), Mezősi et al. (1978), Telbisz (2001), Kevei Bárány et al. (2015), and Látos et al. (2015). Jaskó (1933), Balázs (1960), Láng (1955a, 1971), Jakucs (1956a, 1964, 1977a), Dénes (1971), Zámbó (1998a, b), Móga (2002a, b), Veress (2010b), and Sümegi Jr. et al. (2012) investigated the morphology of its surface and the origin of these features. The hydrological relations of the karst were described by Jakucs (1956b, 1977a), Sásdi (1998), Samu and Kevei Bárány (2008), Hazslinszky (1965, 1974), Gruber et al. (2014), and Gruber et al. (2015). The development of surface features and karst features was analysed by Láng (1971), Sásdi (1990), Zámbó (1998a, b), Gyurica and Sásdi (2009), Veress (2010b), and Gaál (2014). The relation between surface karstification and cave formation was analysed by Jakucs (1956a), Gyurica and Sásdi (2009), and Veress (2017). Sásdi (2005) described its freshwater limestones.

The main characteristics of karst are as follows:

Although it is a karst with small extension and low elevation even at a Hungarian scale (see chapter "A General Description of Karsts in Hungar"), it has a rich and diverse karst feature assemblage.

Its karstification is of cold-water origin, however there are lukewarm water springs on its marginal plateau (Szalonna Plateau, Sásdi 1998) and here cavity formation may have been influenced by warm-water effect (Takács Bolner 1998).

The karst is separated into plateaus mainly by water drainage valleys which developed along non-karstic rock zones or tectonic lines (Fig. 1.4).

For the classification of karst, Jakucs (1977b) uses the term karst of Aggtelek type, while Hevesi (1991a, b) uses the expression 'of Aggtelek nature' instead of type. Veress (2016) calls it renewed allogenic karst since it has been separated from its sediment sources during the development of the Torna Valley (Fig. 1a, b), which is separated into karst zones (Veress 2014, Fig. 2, see chapter "A General Description of Karsts in Hungary"). However, some of its

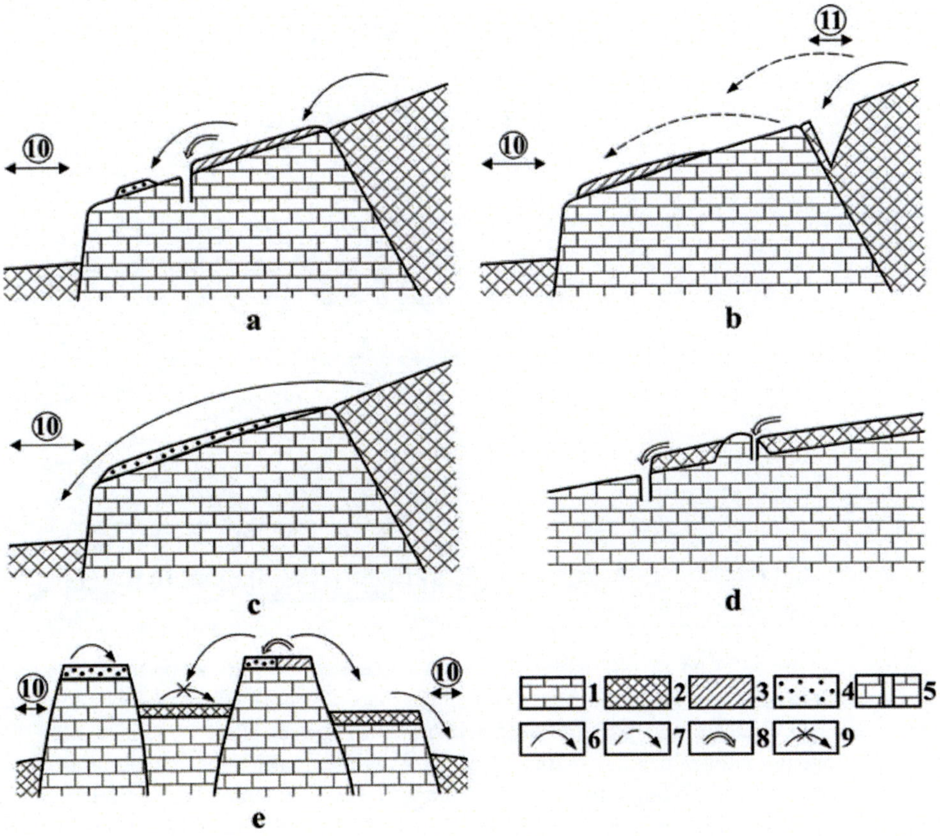

Fig. 1 Theoretical profiles of medium-mountain-covered karst types (Veress 2016) (1) limestone; (2) consolidated non-karstic rock; (3) unconsolidated, impermeable superficial deposit; (4) permeable superficial deposit, (5) chimney, shaft; (6) sediment transport; (7) former sediment transport; (8) transport of superficial deposits into the karst hollows; (9) no transport of superficial deposit from the block; (10) mountain foreland; (11) valley **a** recent allogenic covered karst, **b** renewed allogenic covered karst, **c** semi-allogenic covered karst, **d** mantled allogenic karst, **e** horst covered karst

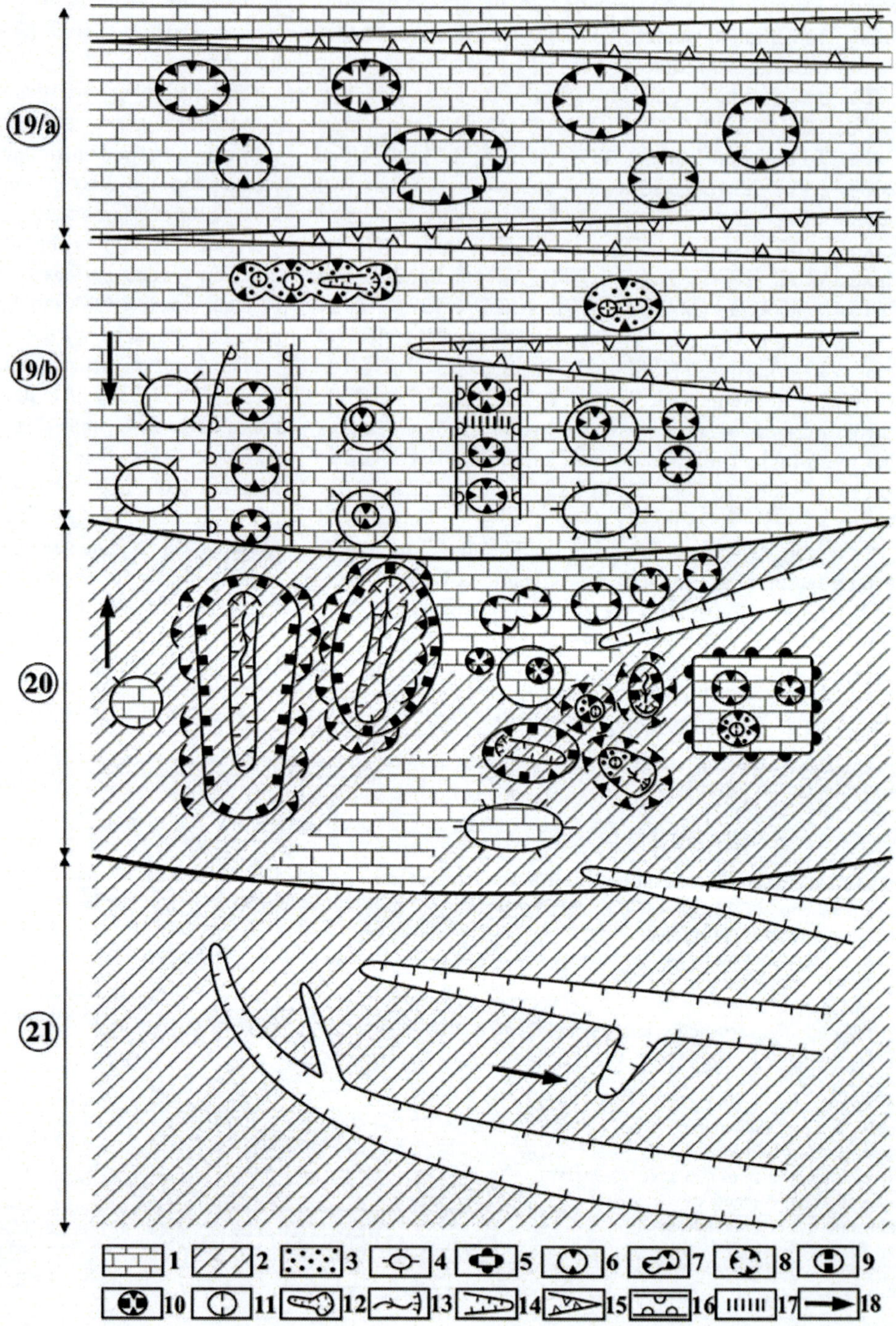

Fig. 2 Karst zones of renewed allogenic covered karst (not to scale, example of the Aggtelek Karst) (Veress 2016) (1) limestone; (2) impermeable cover sediment of marine or fluvial origin; (3) cover sediment of weathering residue origin; (4) karstic mound; (5) block; (6) solution doline; (7) uvala; (8) buried doline or uvala; (9) depression of superficial deposit; (10) partially exhumed paleodoline; (11) suffosion doline; (12) suffosion doline with gully; (13) ponor,; (14) valley in superficial deposit; (15) valley formed along anticline; (16) epigenetic valley; (17) valley divide; (18) dip direction of surface; (19a) uncovered karst zone, (19b) patches formed in the uvalas and dolines of the uncovered karst zone through the accumulation of weathering residue (concealed karst and cryptokarst), (20) mixed composite covered karst zone with mixed allogenic-autogenic parts, (21) buried karst zone

sections can be referred to different karst types since its dissection into plateaus. Thus, the Aggtelek Plateau, part of the Alsó-hegy Plateau, and part of the Galyaság form a mixed allogenic-autogenic karst (Móga 2002a, b), while the Haragistya, the Alsó-hegy Plateau, the Szinpetri Plateau, and the Nagyoldal are autogenic karsts (Móga 2001, 2002b).

Various parts of the karst experienced vertical movements of different degree and direction (Zámbó 1998a). Mainly the areas of synclines rose (Alsó-hegy Plateau, Haragistya), while the anticlines of the Ménesvölgy, the Jósva Valley, and the Telekes relatively subsided (Zámbó 1998a). These are mainly karstification sites and the latter are the zones of valley formation.

The karst has various geomorphological surfaces (Zámbó 1988a). Thus, horst-like plateaus of summit position the elevation of which can be various, but ranges between 400 and 600 m(Aggtelek Plateau, Szinpetri-Szögliget Plateau, Nagyoldal, Alsó-hegy Plateau). The exhumed horsts of summit position with a small extension are dissected by valleys. The blocks of Szalonna Karst and Égerszög Karst (Galyaság) belong to here. Erosion-derasion basin terrains covered by Neogene sediment subsided plateaus in basin position with karst features and structural valleys (Jósvafő Plateau, Szelce Valley), and erosion structural valleys (Bódva Valley, Ménes Valley, Kecső Valley, Jósva Valley, Tóth Valley, Telekes Valley) also occur.

On the majority of the plateaus there are mounds of various elevation, extension, and origin (for example, the mound of Mount Pitics is a horst according to Móga 2002a). It may happen that the summits are separated from their environs by epigenetic valleys (Aggtelek Plateau, Szinpetri Plateau), but it is also possible that the mounds rise above flat plateaus. The occurrence of valleys is rare or they may be absent on plateaus belonging to the latter group.

Genetically, at least four valley types of the karst can be distinguished.

- The already mentioned valleys that developed at structural or non-karstic rock zone mainly of W to E alignment. These mostly transport karst waters out of the karst.
- Epigenetic valleys that became inherited from the former cover to the karstic rock. They developed by corrosional-erosional deepening at karstwater level situated close to the surface. Such valleys occur mainly on Aggtelek Plateau (Zámbó 1998a; Veress 2010b), and on Szinpetri Plateau and in the area of the Galyaság. Dolines, uvalas, and blind valleys are aligned on the floor of these valleys (valley with doline rows).
- Valleys related to karstification can be corrosional valleys and blind valleys according to Zámbó (1998a, b). Corrosional valleys completely or predominantly develop by dissolution (their erosional evolution is restricted or insignificant). These are mainly tributary valleys or valleys connected to uvalas. Their development took place at that site and at that time when the karstwater table was at the surface of the karst. There are no dolines on the floor of such valleys. Zámbó (1998a) distinguished two varieties of blind valleys: valleys on the karst margin and valleys inside the karst. In case of the former, the valleys of the non-karstic terrain terminate at the rock boundary on the margin of the karst. According to Zámbó (1998a, b), the latter involves uvalas that are constituted by covered part dolines which coalesced by dissolution (with a flat floor) and there is water drainage into the karst inside them. Zámbó calls this uvala variety 'lápa'.
- Erosion valleys, erosion-derasion valleys, creeks, gullies, and erosion channels of covered karst areas.

According to Zámbó (1998a, b), some valleys of the karst are dry valleys. Such are the valleys with doline rows and the corrosional valleys of the soil-covered karst on the plateaus or the shallow valley-like features which are dissected or constituted by solution dolines.

On karst and in the dolines red clays occur in various thickness (Zámbó 1970) including terra rossa and fossil rendzina (Jakucs 1964; Zámbó 1970), however, in doline fills not only red, but also dark-brown-black clayey beds occur (Zámbó 1970). The clayey sediment (mostly red clay) is 0.1–0.2 m thick at some places, in depressions its thickness is variable, it can reach 20 m too (Jakucs 1964; Zámbó 1998a; Fig. 3). It is primarily widespread in the centre of the karst (Aggtelek Plateau, Szinpetri Plateau, Haragistya Plateau, in the western and middle parts of Galyaság, Móga 2002a). At some places, it occurs on terrains between the dolines, at some sites only in dolines, but not in all of them.

According to the data of VES measurements, in the fill of many depressions, mainly in the southern part of the karst

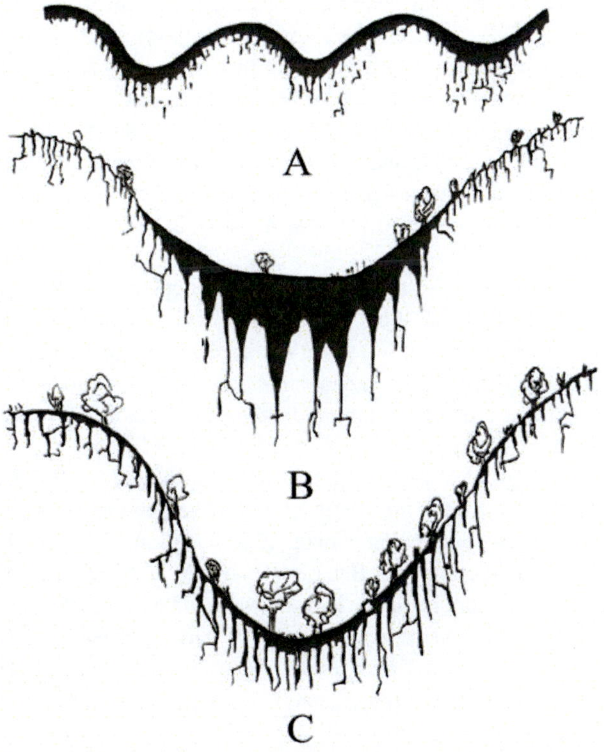

Fig. 3 Doline types according to Jakucs (1964). Legend: **a** dolines of low-elevated terrain with terra fusca, **b** medium-elevated terrain with terra rossa and pinnacles, **c** high-elevated terrain with rendzina soil, **a** lower level (270–280 m), **b** middle level (310–350 m), **c** upper level (above 500 m). Notice: Their distribution can be seen in Fig. 5

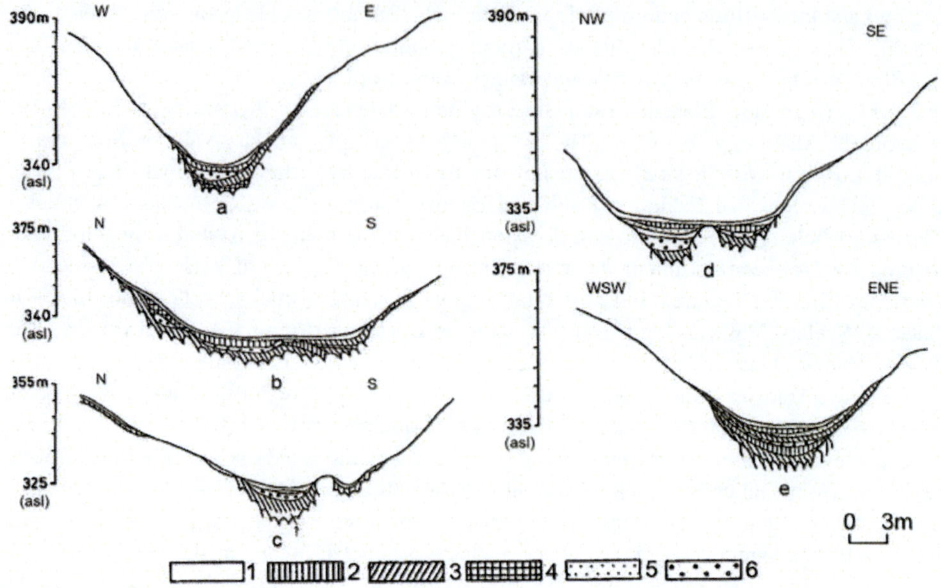

Fig. 4 Doline and uvala sediments from the Aggtelek Plateau (Zámbó 1998b). Legend: 1. Terrae calcis with humus, recent soils, 2. Terra fusca, or light terra rossa, 3. Clayey terra rossa, 4. Vermillion red variety of terra rossa, 5. Terra fusca with soil, 6. Fossil terra rossa, level 'A'. Notice: Elevation distortion of the profiles is threefold

(Teresztenye Plateau) a mixture of sand, gravel, loess, and their clayey variety with limestone debris also occurs (Veress 2010b, 2017).

Clayey fills have various effects on karstification. The clayey veneer protects the rock against dissolution and forces the water to percolate laterally thus promoting horizontal dissolution (Zámbó 1970, 1998a, b) which results in uvalas of various sizes and complexities (Zámbó 1998a, b). During widening, the dividing walls between the dolines become narrower and narrower (Fig. 4c), lower and lower, they become covered and uvalas with flat floor (Fig. 4b, d) are formed.

According to Veress (2017), some depressions have a Pannonian fill and, thus, they date before the Pannonian age. Thus, the Dász doline is filled by Pannonian sediments (Móga 2002a), but a Pannonian cover can be found in the catchment area of the ponors feeding the Baradla Cave too (Jakucs 1956a).

Taking rates of dissolution into consideration, Zámbó (1986b) regarded the onset of the development of some dolines as Upper Pannonian (5.35 million years ago). However, according to Bella et al. (2016), in the area of the Jósvafő paleopolje, dolines and uvalas date to the Quaternary. Therefore, the development of surface karst features began at various times. Older features could have been preserved in their original states or they became filled and subsequently covered.

2.2 Distribution and Morphometry of Karst Features

According to Jakucs (1964), dolines of various size (Fig. 3) occur at different elevations on the karst, their distribution is described in Fig. 5. Thus, at an altitude of 270–280 m small-sized dolines with diameters of 5–30 m and depths of 2–8 m can be found. Here, the (yellow–brown) terra fusca-like cover is of continuous development and of the same thickness in the dolines and between them. The bedrock underwent karren formation (grikes). Such an area is the Jósvafő Plateau. At altitudes of 310–350 m, dolines have diameters of 50–200 m and depths of 14–40 m. Dolines merge into uvalas and are aligned into valley floor row dolines. Dark red terra rossa (red clay) is present on the doline floor, but it is absent on slopes. The doline floor was affected by karren predominantly by solution spike formation (Fig. 3). Such areas are the Aggtelek Plateau, the Szinpetri Plateau, and the western part of the Galyaság. Dolines at and above an altitude of about 500 m have diameters of 50–200 m and their depth is 20–50 m. Terra rossa is missing from their floor and the bedrock did not undergo karren formation. The Szilas Plateau (the eastern part of the Alsó-hegy Plateau), the Haragistya Plateau, and Nagyoldal are such areas.

The morphometry of dolines was investigated by Mezősi et al. (1978), Mezősi (1984), and with the digitalization of

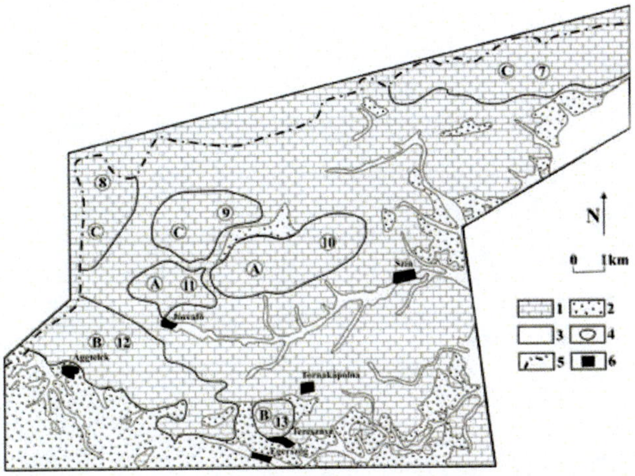

Fig. 5 Dolines occurring at various elevations according to Jakucs (1964) (modified). Legend: 1. Carbonate rocks, 2. Non-karstic rocks, 3. Alluvium, 4. Studied areas, 5. Country border, 6. Settlement, 7. Szilas Plateau, 8. Haragistya, 9. Nagyoldal, 10. Szinpetri Plateau, 11. Jósvafő Plateau, 12. Aggtelek Plateau with the western part of Galyaság, 13. Teresztenye Plateau, A. low-elevated terrain with small dolines without terra rossa, B. medium-elevated terrain with large dolines, with terra rossa and tropical pinnacles, C. high-elevated terrain with large dolines, without terra rossa and pinnacles

doline data by Telbisz (2001), Kevei Bárány et al. (2015), and Látos et al. (2015). In a study of Kevei Bárány et al. (2015), various morphometric parameters of the dolines of Aggtelek, Bükk, and Mecsek Mountains were studied and compared. Telbisz (2001) scanned and digitalized maps of dolines (Fig. 6, Table 1). Látos et al. (2015) studied doline morphometry, in the total area of the karst and per area with TOPO (based on maps at 1:10,000 scale, they chose the doline by the digitalization of the outmost, closed contour line), KRIG (with the digitalization of the contour lines and elevation points, then they created a 10 m digital terrain model of horizontal resolution made by kriging interpolation) and LIDAR (the choice of surface points was done after taking photos of the terrain, then they made a digital terrain model of a resolution of 2.5 m/px) methods.

According to Kevei Bárány et al. (2015), the dimensions (length, width, depth) of the dolines of Aggtelek Karst are the largest and those of the Mecsek Karst are the smallest (Table 2). According to Telbisz (2001), the lognormal distribution of the groundplans of the doline is the most regular on Alsó-hegy Plateau and on the Haragistya: larger dolines are present in ever lower numbers.

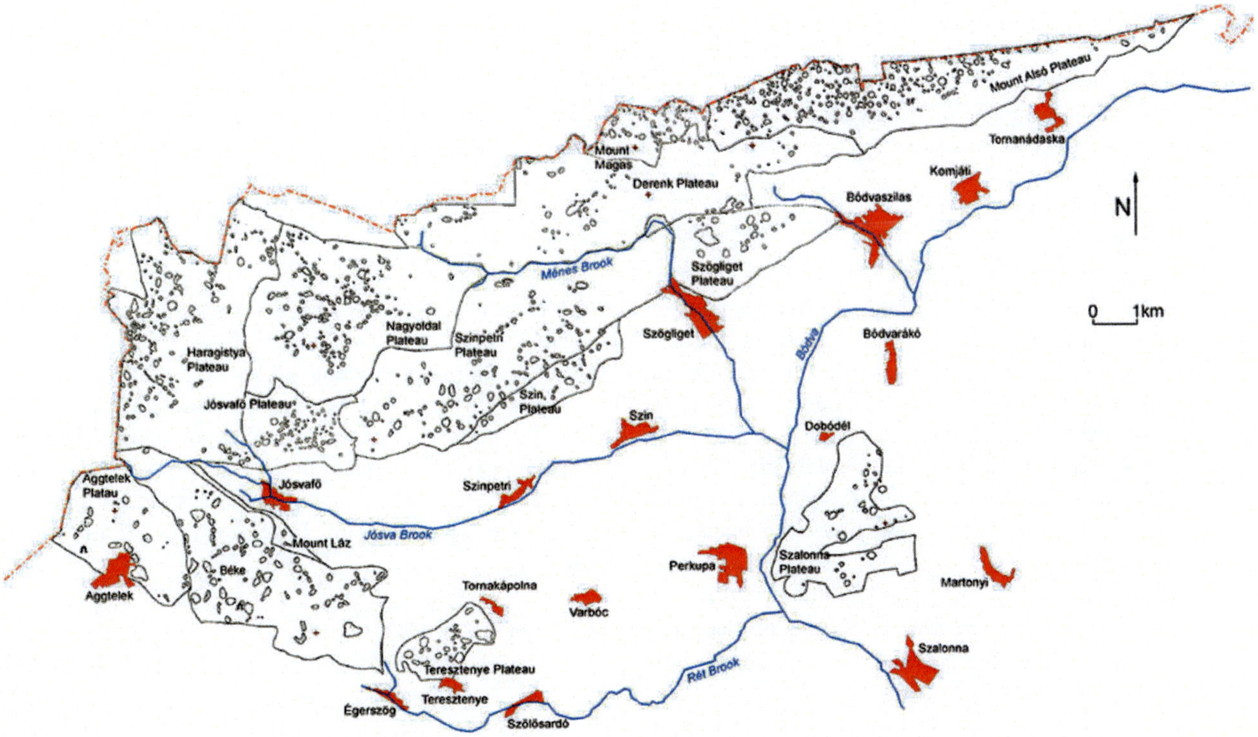

Fig. 6 Dolines on Aggtelek Karst (Telbisz 2001): parameters based on their data are described in Tables 1 and 7

Table 1 The areas of Aggtelek Karst with dolines (Telbisz 2001)

Name	Rock	Area (km^2)	Doline number
Aggtelek Plateau	Wm	5.77	23
Alsó Mount (Szilasi Plateau)	Wm	12.41	244
Béke	Srr	12.35	91
Derenk Plateau	Wm, Sm, Gm, Hm, Pm, Dm, Bh	14.42	77
Haragistya Plateau	Wm, Wd	14.75	166
Mount Láz	Gm	0.58	3
Mount Magas	Wt	2.35	58
Nagyoldal Plateau	Wm, Wd	12.95	140
Szalonna Plateau	Sm, Gm, Dtm, Szm	6.04	28
Szín Plateau	Sm, (Gm)	2.50	26
Szinpetri Plateau	Wm, Wd	16.43	107
Szögliget Plateau	Wm	4.63	22
Teresztenye Plateau	Sm, (Gm, Et)	2.34	28
Jósvafői Plateau	Wm, Wd	3.41	64
Total		110.95	1077

Legend:
Karstic rocks
Wm—Wetterstein limestone, Wd—Wetterstein dolomite, Gm—Gutenstein limestone, Sm—Steinalm limestone, Hm—Hallstatt limestone, Pm—Pötschen limestone, Dm—Derenk limestone, Dtm—Dunnatető limestone, Stm—Szinpetri limestone
Non-karstic rocks:
Bh—Bódvaszilas sandstone, Et—Edelény variegated clay
Notice: See the site of areas in Fig. 6

Table 2 Summary statistic data of average doline length, width, and depth (Kevei Bárány et al. 2015)

Site	Length (m)	Width (m)	Depth (m)
Aggtelek	116.38	78.02	15.99
Bükk	70.65	48.21	9.69
Mecsek	35.36	29.72	11.13

The total area of the dolines belonging to various parts of the karst was compared. Mezősi (1984) stated that the proportion of the total area of row dolines is smaller than that of plateau dolines as compared to the bearing area (Table 3). However, the average area of row dolines (uvalas) is larger than that of plateau dolines, especially on dolomite terrains. As far as the share the share of doline formation is considered. Látos et al. (2015) claim that the highest value can be found on the Páska-Bükk (Dusa) Plateau, but it is also high in the eastern part of Alsó-hegy Plateau and on Aggtelek Plateau (Table 4).

A relationship can be identified between the inclination of the bearing slope and the occurrence of dolines. According to Zámbó (1998a), no dolines occur on slopes with an inclination above 20°. According to the terrain model made by the KRIG method, 96% of the dolines were formed on surfaces with an inclination lower than 10° (Látos et al. 2015).

Table 3 The results of the morphometric investigations of dolines and doline types (Mezősi 1984)

Name	I. Row doline type	II. Plateau doline type	
		On dolomite	On limestone
Doline density doline/km^2	11–13	32–36	7–9
Total area of dolines in % of karst surface	23	32	31
Average area of dolines/km^2	0.01	0.002[a]	0.016

Notice: [a]Uvalas

Table 4 Doline area ratio and median of the doline base-area data compared to the TOPO values. The percentages are calculated by using TOPO values as reference data (Látos et al. 2015)

Plateau	Ratio of doline formation (total area of dolines/expansion of the area bearing dolines)					Median of the doline base area (m^2)				
	TOPO	*KRIG*	%	*LiDAR*	%	*TOPO*	*KRIG*	%	*LiDAR*	%
Aggtelek	0.17	0.18	106.4	0.17	100.5	6297	6367	101.1	5716	90.8
Szinpetri	0.11	0.12	108.3	0.13	120.5	4392	4779	108.8	4899	111.5
Nagyoldal	0.13	0.14	104.2	0.14	108.3	3317	3586	108.1	2984	90.0
Haragistya	0.10	0.10	92.8	0.12	113.1	2659	2632	99.0	2426	91.2
Western Alsó-hegy	0.07	0.05	79.8	0.08	117.5	1864	2192	117.6	2697	144.7
Páska-Bükk	0.30	0.23	78.4	0.29	98.5	6248	3912	62.6	7010	112.2
Eastern Alsó-hegy	0.15	0.16	106.6	0.17	117.8	2568	2864	111.6	2703	105.3
Teresztenye	0.14	0.11	83.6	0.12	89.3	6366	5653	88.8	6377	100.0
Jósvafő	0.10	0.11	112.3	0.12	120.6	3104	2654	85.5	3074	99.0
Total/average	0.13	0.13	96.9	0.14	103.5	4091	3849	98.1	4071	103.0

According to Mezősi et al. (1978), the relief proportion (depth/average diameter) of row dolines is smaller than that of plateau dolines. Kevei Bárány (2015) states that the relief proportion is smaller in the Aggtelek Karst and in the Bükk Mountains than in the Mecsek Mountains thus, while 10% of the dolines of Aggtelek and Bükk have a 0.6 relief proportion (this value is lower for the others), 60% of those situated in the Mecsek Mountains shows such value. Consequently, the side slopes of the dolines of the Mecsek Mountains are steeper (20–25°) than in Aggtelek Karst (10–15°). In addition, the inner asymmetry of the dolines was also studied (Kevei et al. 2015, Table 5), which can be expressed by the length proportion (marked with R_l, it is the quotient of the two longitudinal axis parts calculated from the deepest point), by the width proportion (marked with R_w, it is the quotient of the two width axis parts calculated from the deepest point), and by the elongation proportion (marked with R_{lw}, it is the quotient of total length and total width). This value is the highest in Aggtelek Karst (the largest width is the farthest from the deepest point here) and the lowest in Mecsek Karst. (In the Bükk Mountains, the values are similar to those in Aggtelek Karst, but they are somewhat lower).

The orientation of dolines has been studied by several researchers. Thus, according to Mezősi et al. (1978), the longer axis of the studied 64 dolines has a NNE-SSW and EW tectonic directions. Studying orientation frequency, Telbisz (2001) stated that longitudinal axes gather in a SSW-NNE direction and perpendicular to this (NNW-SSE, Fig. 7). According to Telbisz (2001), a third direction is WNW-ESE and perpendicular to it, which, the author believes, follows the direction of valleys that developed during the covered karst state. Regarding the orientation of the closest adjacent dolines there is regularity in the distribution with the exception of the Alsó-hegy Plateau. At sites with regular distribution, the directions of longitudinal axes interchange (Fig. 7a, b). Telbisz (2001) explains this by the fact that at the intersection of fractures, the dolines (parent doline) rather develop in the direction of the more pronounced fracture and are more elongated, while their adjacent doline develops perpendicular to the main fracture direction. According to Telbisz (2001), elongation averages are larger at sites where the proportion of row dolines is larger (Tables 6 and 7). Thus, the valley affects doline elongation.

Studying the orientation of dolines, Látos et al. (2015) received a 345°–165° direction to longitudinal axes as dominant directions by all three procedures (TOPO, KRIG, LIDAR). According to Kevei Bárány et al. (2015), the orientation of dolines is NS and WSW-ENE. In contrast to result of Telbisz (2001), she found that the closest neighbour orientations show random distribution. A significant characteristic is that the alignment of the dry valleys does not affect the direction of the longest axes of the dolines (Kevei Bárány et al. 2015).

Table 5 Average elongation ratios of dolines (Kevei Bárány et al. 2015)

Measurement site	Length ratio (R_l)	Width ratio (R_w)	Elongation ratio (R_{lw})
Aggtelek	1.93	1.18	1.63
Bükk	1.56	1.11	1.53
Mecsek	1.12	1.33	1.22

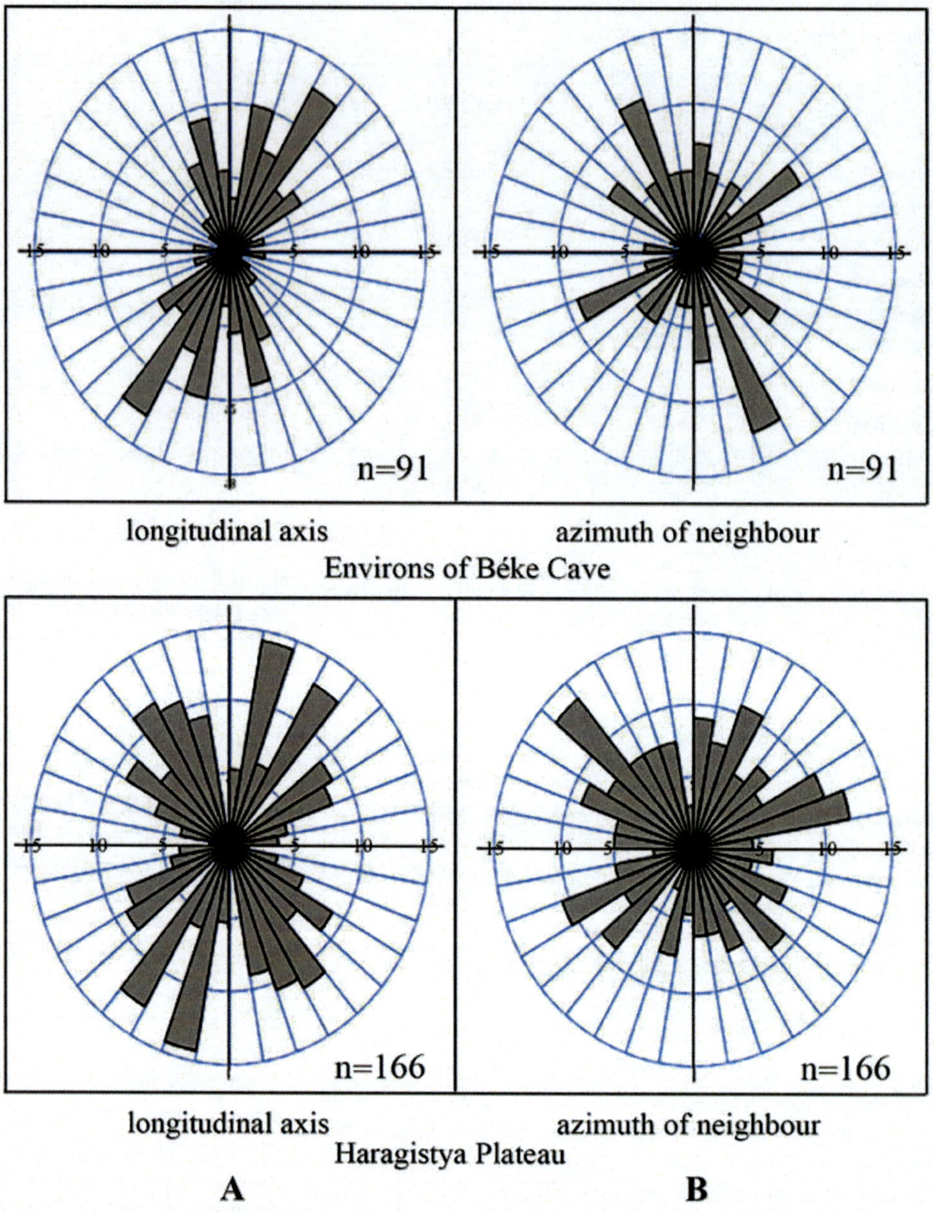

Fig. 7 **a** The distribution of the longitudinal axes of the dolines of Béke Cave and Haragistya Plateau (A) and of the neighbours of these dolines (B) (Telbisz 2001). **b** The distribution of the longitudinal axes of the plateau dolines of Szinpetri Plateau and Alsó-hegy Plateau (A) and of the neighbours of these dolines (B) (Telbisz 2001)

Investigating doline density, Mezősi (1984) stated that the density of row dolines is higher than that of plateau dolines (Table 3). In addition, based on his further studies, plateau dolines have a lower occurrence density on limestone than on dolomite terrains (the latter area is the Jósvafő Plateau). According to Telbisz (2001), regarding rock development, the doline density of the whole karst is 9.71 dolines/km^2, highest on Wetterstein limestone, while regarding the areas this value is the greatest on Alsó-hegy (Látos et al. 2015, Table 7). However, as it has already been mentioned, doline density is also affected by doline size. Telbisz claims that the reason for this is that larger dolines may have developed by coalescence which reduces doline number. In the southern part of the karst, their groundplan is larger than that could be concluded from the doline density (Fig. 8). According to Telbisz (2001), these dolines are close to non-karst terrains. As a consequence, they may have received more water by which they could have grown faster than those farther positioned from covered karst terrains.

According to Telbisz (2001), the nearest neighbour index of dolines is below 1, thus doline pattern is not considered random (Table 6). He explains this by the fact that the presence of row dolines is decisive on the karst. Kevei Bárány et al. (2015) also received a similar result.

Fig. 7 (continued)

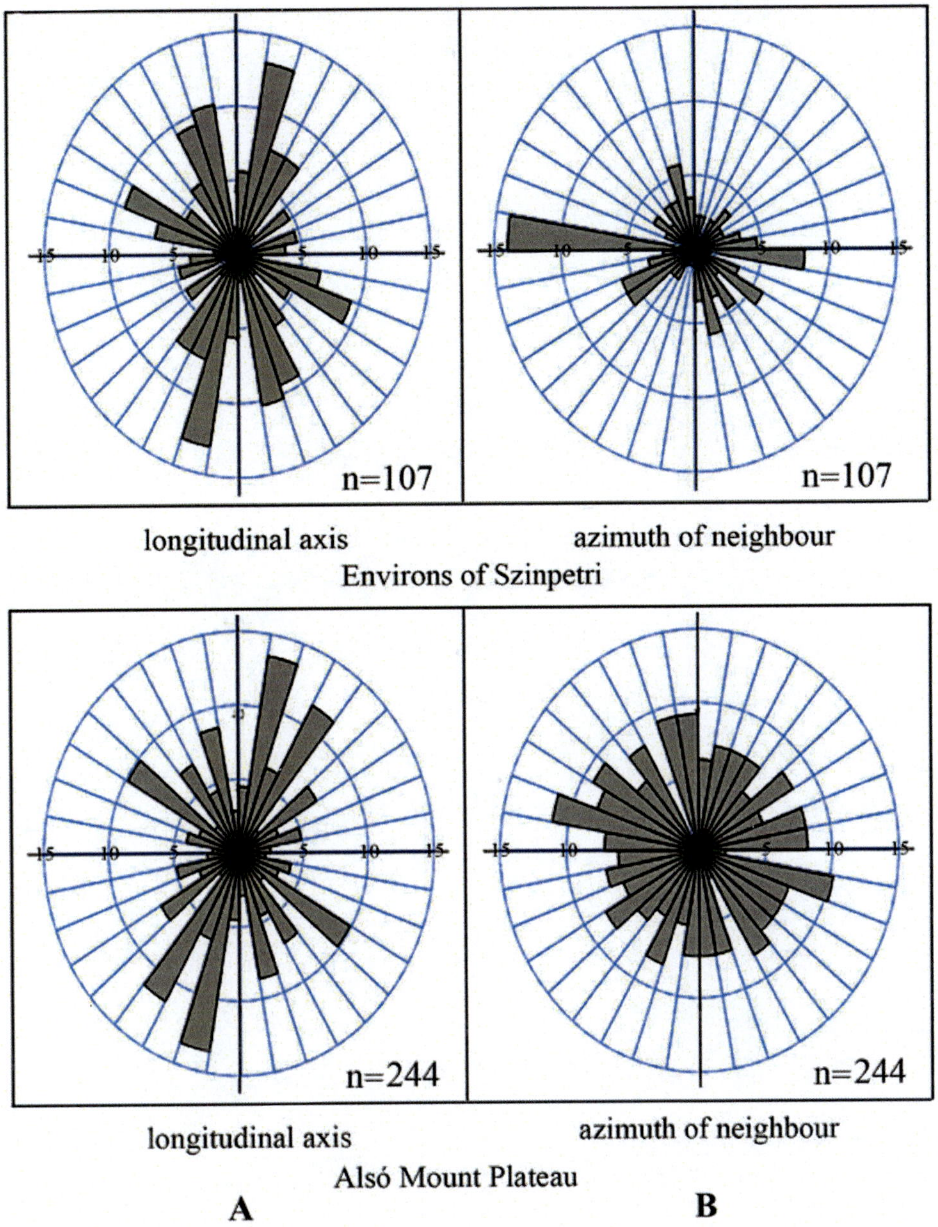

2.3 Karst Features

Their karst features are karren, solution dolines, uvalas, subsidence dolines, complex depressions, depressions of superficial deposit (DSDs), ponors, karst lakes, gorges, and concretions.

Leél-Őssy (1952a) mentions several karrenfields from the area of Aggtelek Karst thus from the slopes around Lake Aggtelek and from the area of the Alsó-hegy Plateau. The karrenfield around Lake Aggtelek ('Ördögszántás') is due to subsoil karren formation from which the soil denuded by deforestations and crop management. In this karren field, in addition to root karren solution mainly took place along the bedding planes of large-inclined beds and, thus, a chain of the remnants of bed heads constitutes shorter and longer mounds (bed head karren). Leél-Őssy (1953a) mentions large-sized kamenitzas from the Szalonna Karst (with a diameter and depth of 30–40 cm). Jakucs (1964) emphasizes karren formation on doline floors.

Table 6 Morphometric data by subareas. The percentages are calculated by using TOPO values as reference data (Látos et al. 2015)

Plateau	Area (km^2)	Doline area (km^2)	Doline number (piece)					Doline density (doline/km^2)		
			TOPO	*KRIG*	%	*LiDAR*	%	*TOPO*	*KRIG*	*LiDAR*
Aggtelek	12.31	7.20	113	119	105.3	125	110.6	15.68	16.52	17.35
Szinpetri	12.99	7.73	135	132	97.8	154	114.1	17.46	17.08	19.92
Nagyoldal	8.24	5.17	151	155	102.6	177	117.2	29.23	30.01	34.27
Haragistya	9.98	7.43	152	147	96.7	176	115.8	20.45	19.78	23.68
Western Alsó-hegy	7.25	3.88	63	56	88.9	65	103.2	16.25	14.45	16.77
Páska-Bükk	2.46	1.26	26	29	111.5	27	103.8	20.59	22.97	21.39
Eastern Alsó-hegy	12.40	8.44	311	308	99.0	357	114.8	36.87	36.51	42.32
Teresztenye	3.68	3.07	33	34	103.0	36	109.1	10.76	11.09	11.74
Jósvafő	3.51	3.08	69	72	104.3	79	114.5	22.37	23.35	25.62
Total/average	72.85	47.26	1053	1052	99.9	1196	113.5	22.28	22.26	25.31

Table 7 Morphometric data of the dolines of the Aggtelek Karst (Telbisz 2001)

Name of the area	Doline formation ratio (%)	Average doline area (m^2)	Degree of elongation	Nearest neighbour index (R)
Aggtelek Plateau	4.56	11,439	1.86	0.41
Mount Alsó (Szilas Plateau)	6.56	3335	1.44	0.63
Béke	7.09	9637	1.87	0.62
Derenk Plateau	2.20	4118	1.67	0.44
Haragistya Plateau	4.38	3890	1.66	0.57
Mount Láz	2.29	4426	1.65	Little data
Mount Magas	7.13	2889	1.44	0.59
Nagyoldal Plateau	4.13	3818	1.63	0.62
Szalonna Plateau	2.84	6128	1.48	0.35
Szin Plateau	4.11	3955	1.78	0.41
Szinpetri Plateau	3.65[a] (4.8[a])	5599[a](4769[a])	1.70	0.57
Szögliget Plateau	5.69[a] (11.0[a])	11,980	1.49	0.40
Teresztenye Plateau	12.17	10,187	1.48	0.51
Jósvafő Plateau	5.55	2957	1.72	0.57
Average	5.85	5710	1.60	0.51

[a]Value calculated only for the limestone area
See the location of the areas in Fig. 6
See the explanation for the abbreviation of rocks in Table 1

Karren

Zámbó (1998a, b) distinguishes karren feature groups on the karst. Thus, he identifies microkarren, karren of bare surfaces, and subsoil (covered) karren. According to him, bare karren occurs on steep slopes of cliffs, towers, valleys, ponors, springs, and cave entrances, as well as on the slopes of low-inclination mounds and of the dividing walls between dolines. Features developing at these sites are rillenkarren, rinnenkarren, wider, and narrower varieties of grikes. Subsoil karren may directly develop by subsoil solution or by solution under various terra rosa varieties. According to Zámbó (1998a), the intensity of subsoil corrosion fivefold exceeds the karren formation of bare surfaces on the average.

Zámbó (1998a, b) emphasizes that subsoil karren grows by widening and karren feature becomes rounded. Zámbó (1998a) mentions grikes, bed head karren, rinnenkarren, pitkarren, pinnacles, and rootkarren among subsoil karren.

Zámbó (1998a, b) lists karren slopes from steep slopes, from the boundary slopes of Aggtelek Plateau, and from the slopes of valleys (Ménes Valley, Jósva Valley, Baradla

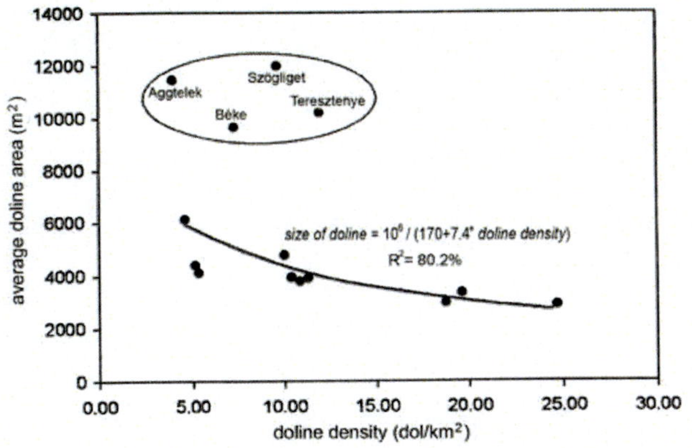

Fig. 8 The average doline area as a function of doline density (Telbisz 2001)

Valley); karren fields from the Alsó-hegy Plateau. These surface parts became exhumed, and thus karren of bare surfaces and karren of covered origin often occur together and the development of the latter was transformed as a result of cover removal. In the southern part of the karst (Aggtelek Plateau, the western part of the Galyaság) semi-exhumed karren features, especially bed head karren and pinnacles are common. Bare karren features (rillenkarren, small kamenitzas) are in a developing stage on exposing, exposed figures. Jakucs (1977a) regarded one of the exposed karren types (Medvesziklák) as a result of the tropical karst formation, which continues to the present (complex paleokarstic feature). 'Karrennasen' is widespread in the cover. They are constituted of rock debris rounded by dissolution (Zámbó 1998a).

Dolines

The density and size of solution dolines, which mostly occur together with uvalas (Fig. 3a, chapter "GIS Research of Karst Terrains in Hungary"), are large though because of the coalescence of dolines by dissolution. The density is lower than expected. According to Zámbó (1998a), dolines being closer to covered karst are of allogenic character, while those being farther away are rather of autogenic character. The floor of these features is frequently filled with sediment of significant thickness. The dolines have wide and expanded floors both in the cover and in the bedrock. Mainly the surface of the cover is flat, the side slopes are relatively gentle (the dolines of the Szilas Plateau are exceptions where the inclination of side slopes is high). The reason for this is that their widening exceeds their deepening because of the dissolution of the side slopes (Zámbó 1970, 1986a). It is common that the depressions are just apparently dolines (no dividing walls are visible on their floor), they are actually uvalas (Fig. 4b, d).

The dolines are elongated (mainly in tectonic directions) and frequently constitute rows (alternating with uvalas), mainly those situated on valley floors. The side slopes are often partially bare where the soil is patchy. Mainly the doline floors have been affected by karren formation. In karst literature, according to morphological environment, dolines of valley floor and plateau dolines are distinguished. We think that their morphological environments are the following: the floor of dry valleys, and possibly their side slope (Figs. 9 and 10), the flat parts of plateaus (Fig. 9), the mounds of plateaus, the top or side slopes of mounds (Fig. 9), and the floor of larger depressions (possibly uvalas) with small depth (Szilas Plateau).

According to Hevesi (1991a, b), the row dolines of Aggtelek Karst developed from ponors. Zámbó (1998a, b) andMóga (2002a) share the same opinion. However, Zámbó (1998a) believes that if the karstwater table was at the floor of dry valleys with doline rows at the time of their inheritance, no ponors developed. Although Jakucs (1977a) refers to the retreat of the sinkhole site, he regards the valley floor dolines of Aggtelek Karst as the solution dolines of the valley floors that became inherited over to the limestone. According to Veress (2010b, 2012), for example, neither the dolines or the uvalas of the valleys (Baradla Valley, Mész Valley, and Hideg Valley) of the Aggtelek Plateau are former ponors (Figs. 4, chapter "A General Description of Karsts in Hungary" and 10). This is proved by the fact that

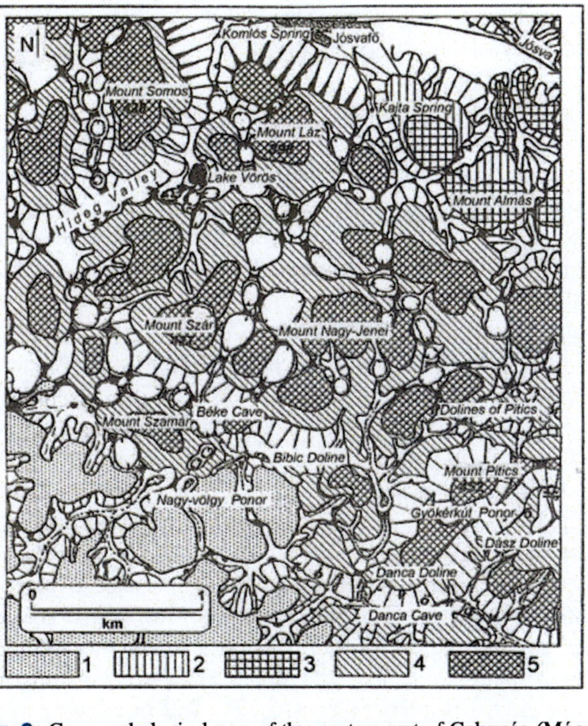

Fig. 9 Geomorphological map of the western part of Galyaság (Móga 2002a). Legend: 1. Pannonian-Pontian sediments, 2. interfluve that developed on non-karstifying rocks, 3. The roof of the interfluve that developed on non-karstifying rocks, 4. surface of karst plateau and interfluve that developed on karstifying rocks, 5. More elevated roof of the karst plateau and the roof of the interfluve that developed on karstifying rocks

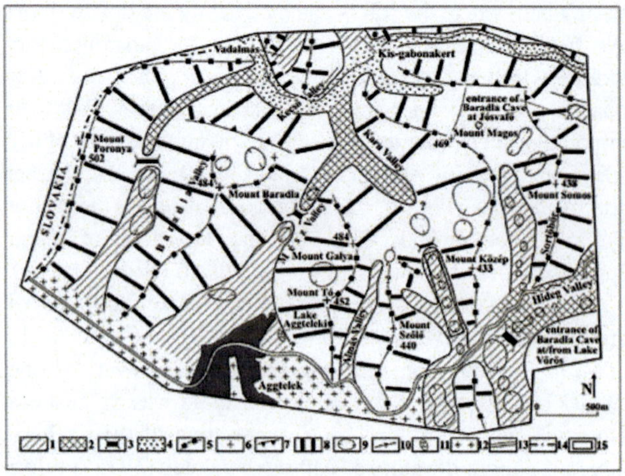

Fig. 10 The valleys of the Aggtelek Plateau (Veress 2012 modified). Legend: 1. Bottom of older valley and the bottoms of tributaries of the older valley, 2. Bottom of the tributary valley which developed on the floor of the older valley by the headward erosion of the Kecső and Jósva, 3. Valley-floor divide (saddle), 4. Bottom of the Kecső and its tributary valley (which did not develop on older valley bottom), 5. Ridge, 6. Mount, 7. The margin of the Kecső Valley, 8. Valley side, 9. Doline, 10. Stream, 11. Lake, 12. Buried karst, 13. Road, 14. Frontier, 15. Uvalas described in Fig. 11

VES measurements indicate no fluvial sediment in the fill of the uvala (Fig. 11) of one of the valleys, between Szőlő Hill and Közép Hill.

Complex karst depressions

Complex karst depressions are referred into three groups here: based on their above-mentioned morphological characteristics into uvalas (1), into larger depressions having no uvala characteristics from a morphological point of view (2), and poljes (3). According to Zámbó (1998a), the coalescence of dolines results in the development of blind valleys if the floor of blind valleys filled with red earth is flat, they constitute a depression called in Hungarian 'lápa'. However, larger depressions may also develop if closed, arheic valley sections are formed through doline and uvala development (Fig. 11).

Uvalas (1) may also occur in valley-floor position (Fig. 11) or in plateau position. Dolines in valley floor position are aligned in a row, while those in plateau position are not necessarily so arranged (Szilas Plateau). Uvalas in valley-floor position are often built up of two partial dolines and rarely of three or more dolines. Among valley-floor uvalas, there are uvalas with slopes not separated from the slope of the bearing valley (Fig. 11) and there are uvalas where the slope is separated. The elevation and size valley floor uvalas does not change regularly in any direction. Plateau uvalas may also indicate such a character, but there may also occur uvalas where the change of these parameters shows some regularity. In the latter case, older and larger main part dolines (parent dolines) are connected by ever smaller 'child dolines'. Their floor is ever higher moving away from parent doline (Sümeghy Jr. et al. 2012).

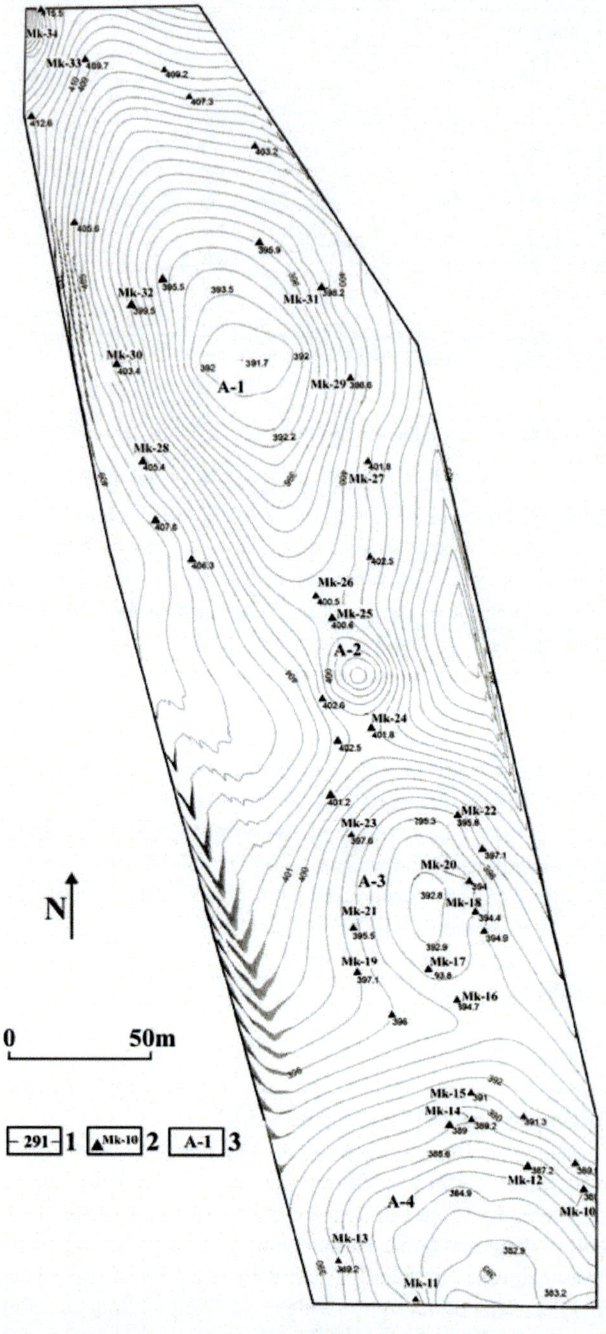

Fig. 11 The uvalas of the valley between Közép Hill and Szőlő Hill (Veress 2010b). Legend: 1. Contour line, 2. Rock outcrop, 3. Identification mark of karstic depression

In the eastern part of the Alsó-hegy Plateau (Szilas Plateau) dolines coalesced in rows, but in diverse directions create uvalas. These uvalas (or some of them) are situated in larger (of a diameter of 1–2 km), shallow depressions, but doline rows are not on valley floors here. Between partial dolines there are narrow, ruined dividing walls, locally with smaller mounds. Large karst depressions (2) do not show an uvala character, but probably depressions also occur among them which were formed by the dissolutional coalescence of dolines. Their properties are as follows:

- Their number is limited, probably not more than some dozens (Fig. 12).
- They probably developed during old (pre-Pannonian) karstification.
- They are of large size (the expansion of one depression in the Bába Valley is ca. 500 m along its longitudinal axis), but they do not exceed the size of the largest dolines.
- Sometimes they are polje-like, for example, they have a spring, a ponor,, and their floor may be flat. Their side slope is not steep, their margin is often straight, but the margin of those with no straight margin may be dissected (for example, the margin of the lápa of Lake Keserű) or arched (Dász doline).
- They are partly filled with superficial deposit, their floor is flat, the bedrock may be dissected (Fig. 13), but it may be (partially) flat (Fig. 14). Thus, they may be filled uvalas, but depressions of other origin too (where the bedrock is flat).

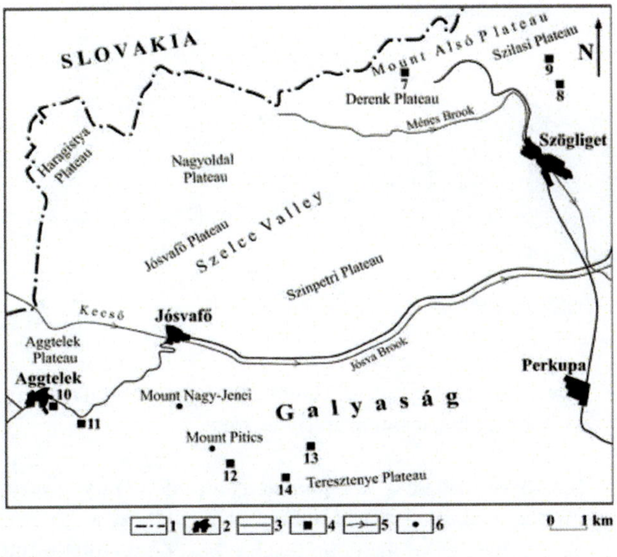

Fig. 12 The site of all complex karst depressions on Aggtelek Karst that were mentioned in the chapter (Veress 2014 modified). Legend: 1. State border, 2. Settlement, 3. Road, 4. Complex depression, 5. Stream, 6. Hill, complex depressions: 7. Derenk depression, 8. The depression of Acskó-rét, 9. Depression row of Bába Valley, 10. Ponor depression of Zombor-lyuk, 11. The depression of Kis-Ravasz ponor and Nagy-Ravasz ponor, 12. Dász doline, 13. Vizetes, 14. Keserű-tó-lápa

- The bedrock is not always limestone, for example, in the case of the depression of the Bába Valley it is partly or completely sandstone (Sásdi 1998).
- They have a karstic drainage.
- The depressions may create a row, the surface catchment area of the members of the row is separated, but the karstic drainage system of those belonging to a row may constitute a hydrologically uniform system.
- Their morphology is diverse, there may be blind valleys (the depression of Zombor-lyuk, the depressions of Nagy-Ravaszlyuk and Kis-Ravaszlyuk ponors), channels with blind termination (Vizetes), creeks, gullies, ponors, subsidence dolines (the 'ponor' of Acskó-rét) on their floor. Hazslinszky (1965) mentions subsidence dolines from one of the depression systems of the Bába Valley.
- They can also constitute a depression system (the Lake Keserű lápa and the neighbouring depression).
- They are mostly, but not always in summit position.
- They often develop at a rock boundary (Derenk depression, the Lake Keserű lápa, the depression of Zsombor-lyuk ponor). In this case, one of the side slopes is built of limestone. The floor valleys of depressions are directed towards the rock boundary, where or near it there are ponors (see below).
- They are real DSDs, which means that the material filling the depression of the limestone bedrock is transported into the karst. Thus, a depression develops on the superficial deposit then the depression of the bedrock becomes gradually exhumed too. Depressions formed on the superficial deposit can be found at the ponors of Baradla Cave (for example, the sinkhole depression of Zombor-lyuk), which are called sinkhole basins by Láng (1971). The denudation of the superficial deposit is caused by channel erosion then by the denudation of the side slope of the valley due to pluvial erosion. Depression growth is favoured by the fact that there are depressions on the bedrock too. This can be seen for the depression bearing the Zombor-lyuk ponor (Fig. 15). Here, limestone outcrops along the margins of the depression. The existence of the depression on the bedrock is also proved by VES measurements. The bedrock is separated into partial depressions ('B-B' profile of Fig. 16), and minor depressions deepened into the bedrock floor ('A-A' profile of Fig. 16). The fill of the depression in the bedrock has been transported into Baradla Cave through Zombor-lyuk ponor. The rate of the denudation of the superficial deposit is 10–12 m/1 million year (Láng 1971).
- They may also occur along the karst margin or inside the karst.
- Zámbó (1998a) regards these features uvalas inside the karst and karstic blind valleys, while according to Móga

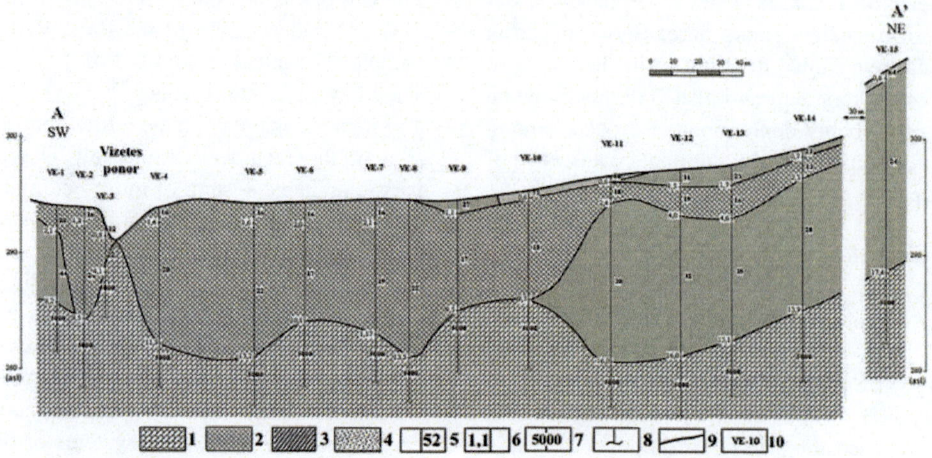

Fig. 13 The geoelectric-geological profile of the environs of 'Vizetes'ponor (Veress 2017). Legend: 1. Limestone; 2. Clay (with limestone debris and sand); 3. Clayey sand, gravel; 4. Sand gravel (with limestone debris); 5. Geoelectric resistivity of the series (Ohmm); 6. Base depth of the geoelectric series (m); 7. Geoelectric resistivity of bedrock (Ohmm); 8. Approximate depth of penetration of VES measurement; 9. Boundary of the geoelectric beds; 10. Site and identification code of VES measurement

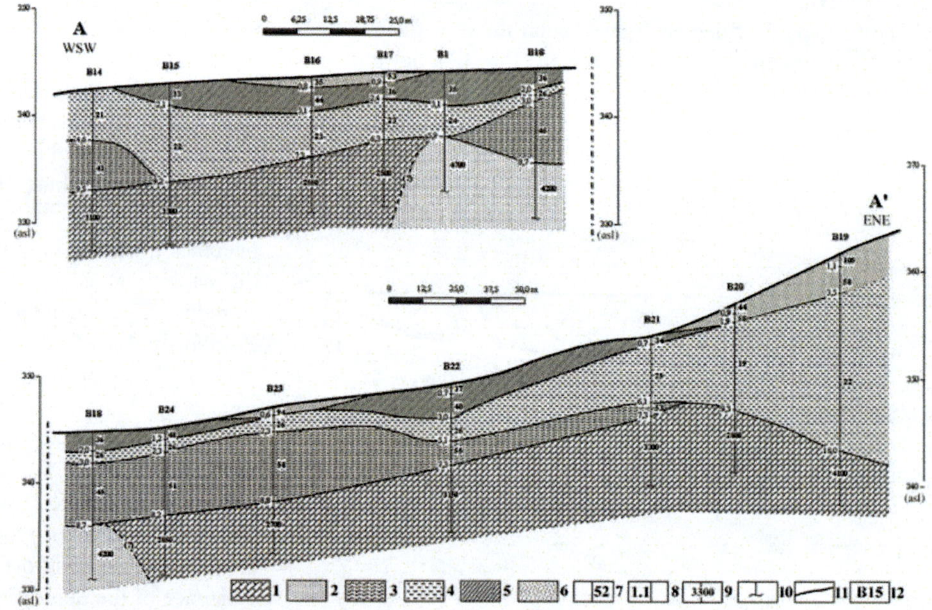

Fig. 14 The geoelectrical-geological profile of one depression of Bába Valley (Veress 2016) 1. Limestone, 2. Sandstone, 3. Limestone debris (clayey), 4. Clay, 5. Clayey sand and gravel (loess), 6. Sand and gravel or loess (with limestone debris), 7. Geoelectric resistivity of the series (Ohmm), 8. Base depth of geoelectric series, 9. Geoelectric resistivity of bedrock (Ohmm), 10. Approximate penetration of measurement, 11. Geoelectric series boundary, 12. Site and identification code of VES measurement. Notice: Track of profile can be seen in Fig. 20

(1999), some of them, for example, the Derenk Basin (Móga 2002b), are subsurface tapped basins.

- Major depressions and large caves of the karst (for example, Baradla Cave and the depressions bearing its ponors) mutually affect their development (Veress 2017) since the filling material of the bedrock depression could not have been transported away at the surface, but only into the cavity system of the karst. (The larger the cave, the greater extent is the exhumation.) The origin and large size of caves are due to the filling of depressions particularly if this material was gravel (erosional development). However, cave growth is enhanced by the chance of sediment transportation into the karst.

The Jósvafő Plateau is regarded as a polje (3) by several researchers (Jakucs 1964; Zámbó 1998a; Bella et al. 2016). However, the area of the polje reaches beyond the plateau in SW direction and involves the Lófej and Tohonya Valleys and their environs (Fig. 17). The polje has been transformed into a paleopolje (Bella et al. 2016). According to Lehmann's classification (1959), Jósvafő polje is a valley polje. A valley polje can be open both downstream and upstream

Fig. 15 Zombor-lyuk ponor and its depression

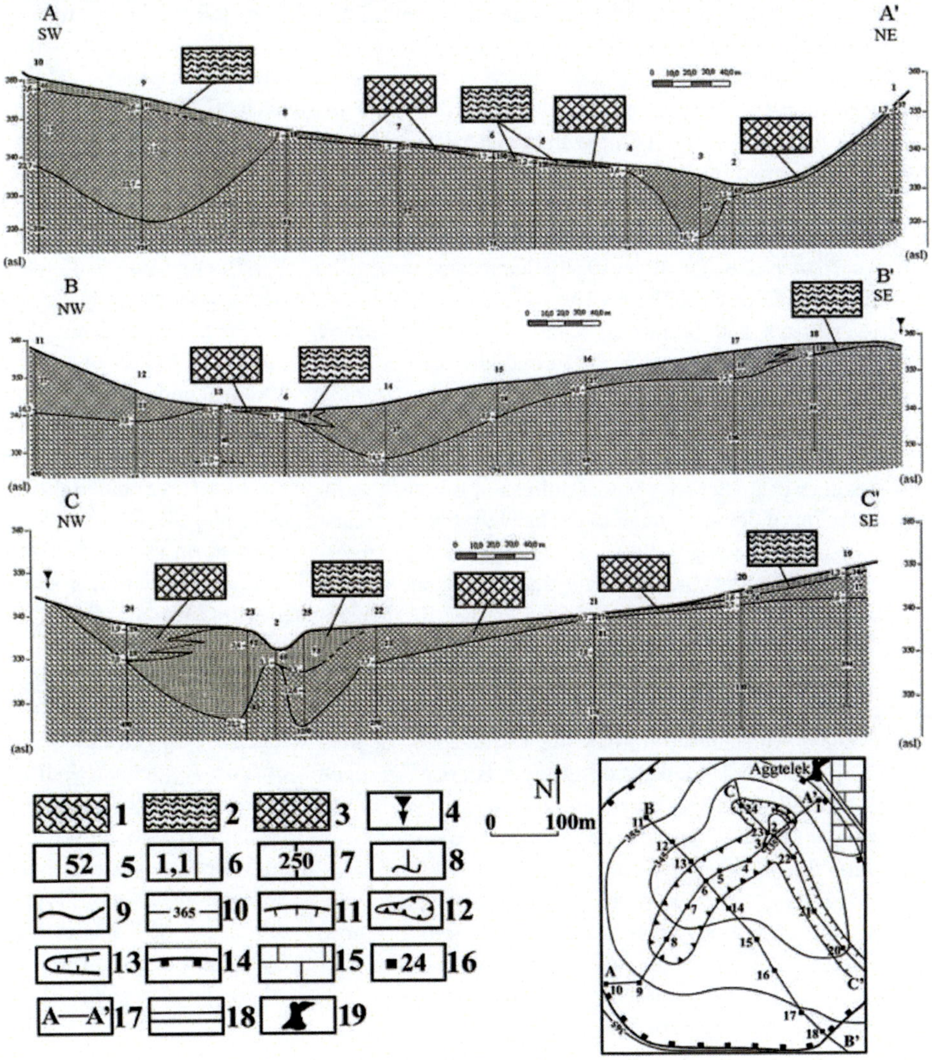

Fig. 16 Geoelectric-geological profile of the depression of superficial deposit bearing the Zombor-lyuk ponor (Veress 2017). Legend: 1. Limestone debris, fragmented limestone, 2. Limestone debris (clayey), 3. Clay (with limestone debris and sand), 4. Limestone outcrop, 5. Geoelectric resistivity of the series (Ohmm), 6. Base depth of the geoelectric series (m), 7. Geoelectric resistivity of the bedrock (Ohmm), 8. Approximate depth of penetration of VES measurement, 9. Geoelectric series boundary, 10. Contour line, 11. Ponor, 12. Blind valley, 13. Ravine, 14. Depression of superficial deposit, 15. Limestone, 16. Site and identification code of VES measurement, 17. Profile location, 18. Aggtelek-Jósvafő road, 19. Built-up area

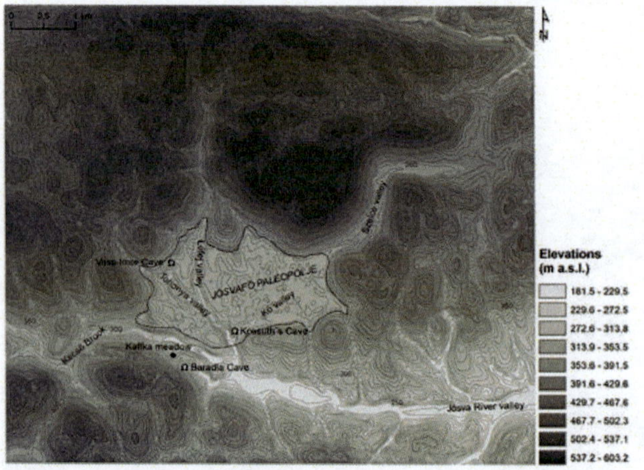

Fig. 17 Map of the Jósvafő paleopolje and its environs (Bella et al. 2016)

(Bonacci 2004). The inlet valley of Jósvafő polje is the Szelce Valley, and its outlet valley is the Tohonya Valley. According to Bella et al. (2016), the development of the polje started in the Late Miocene by tectonic differentiation (subsidence). Its development took place in two phases (Bella et al. 2016). During the first phase, in the late Pliocene, the flat floor of the polje with an elevation of 275–300 m was formed by horizontal solution in the area of the former pediment. Horizontal dissolution was enabled by flooding since the karstwater level was close to the surface. In the second phase (Quaternary), the karstwater level subsided in the area of the polje because of the downcutting of the River Jósva and horizontal solution was replaced by vertical solution (doline and uvala formations).

Subsidence dolines

The number and size of the subsidence dolines of the karst is low. Another characteristics is that dolines of this type (with the exception of Jósvafő Plateau) exclusively occur in karst depressions (drawdown dolines, uvalas, large-sized depression, possibly ponors). It may occur that subsidence dolines are not separated morphologically from the bearing depression. An example for this is the Nagyvizes doline-ponor of Alsó-hegy described by Hazslinszky (1965). No subsidence doline is found either on terrains between karst depressions or in valleys. Subsidence dolines are restricted to depressions if the superficial deposit of the floor is coarse-grained gravel or sand as in the 'lápa' of Lake Keserű or on the Teresztenye Plateau (Veress 2008), but they do not occur in the doline fills with terra rossa.

Among them there may be subsidence dolines with channel or creek (they have a ponor function) or those without these features. Ponor function may be more permanent when the channel has a groundwater spring.

Ponors

In Aggtelek Karst, Zámbó (1998a, b) identifies open ponors (with a passage), and concealed ponor (water is drained through the fill of the depression) as well as outer ponors and inner ponors, while Sásdi (1998) also distinguishes channel ponors (drainage point on the channel floor) from the valleys of the karst, on the floor of which the water of the stream percolates partially or completely. Zámbó (1998a, b) thinks that outer ponors develop at the boundary of uncovered and covered karst, while inner ponors are formed in surface depressions (doline, valley) inside the karst. Specific outer ponors developed at the rock boundary marking the termination of the terrain covered by Pannonian sediments (Sásdi 1998) and bordering the karst from the south as a capture line (Jakucs 1956a; Zámbó 1998b; Móga 2002a) where karst terrain ends in a fault scarp (Láng 1955a). The emergence of ponors of the capture line was probably enabled by Pleistocene tilting that turned the inclination of the surface of Aggtelek Plateau and of the covered karst surrounding it from the south to a northern direction. As a result, streams flowed from the covered karst to the north creating ponors with blind valleys. Along this line, the following ponors occur from the west to the east: Ördög-lyuk, Domica-lyuk, ponors of Demik-lápa (on the Slovakian side), the plugged ponor of Lake Cserna, Kis-Baradla ponor, Acheron ponor, Zombor-lyuk ponor, Kis-Ravaszlyuk ponor, and Nagy-Ravaszlyuk ponor (Fig. 18). However, the Kis-Baradla ponor is not at rock boundary, but it is in the depression of a doline (Gruber et al. 2014). Thus, it is probably a subsidence doline. In the area of Galyaság, the ponors of Béke Cave (the ponors of Szomor Hill, the Nagy Valley,, and of Bibic doline), and the Danca ponor also belong to the ponor row. These ponors and the water flowing inward through them are responsible for the development of Baradla Cave and Béke Cave. Another outer ponor is, for example, the Ördöglyuk ponor near Imola Village which is situated on the floor of the Tóberke Valley. It has a significant catchment area, which is built up of Pannonian clayey gravel (Sásdi and Szilágyi 1986), according to Sásdi (1998), its expansion is 1,5 km^2.

The site and position of ponors changed because of the shift of the capture site (Veress 2010b). Thus, the presently inactive plugged sink of Zombor-lyuk ponor was in the place of Lake Aggtelek (Sásdi 1998). Therefore, it is probable that rock boundary as a whole or in parts shifted (Veress 2010b). Because of the significant inclination of the karst surface, and because some ponors developed in the side of karst depressions (Zombor-lyuk ponor), vertical shifting surpassed the extent of horizontal shifting. However, there are karst marginal ponors on the Alsó-hegy Plateau (Szilas Plateau) where ponors occur along the margin of two wedged-in

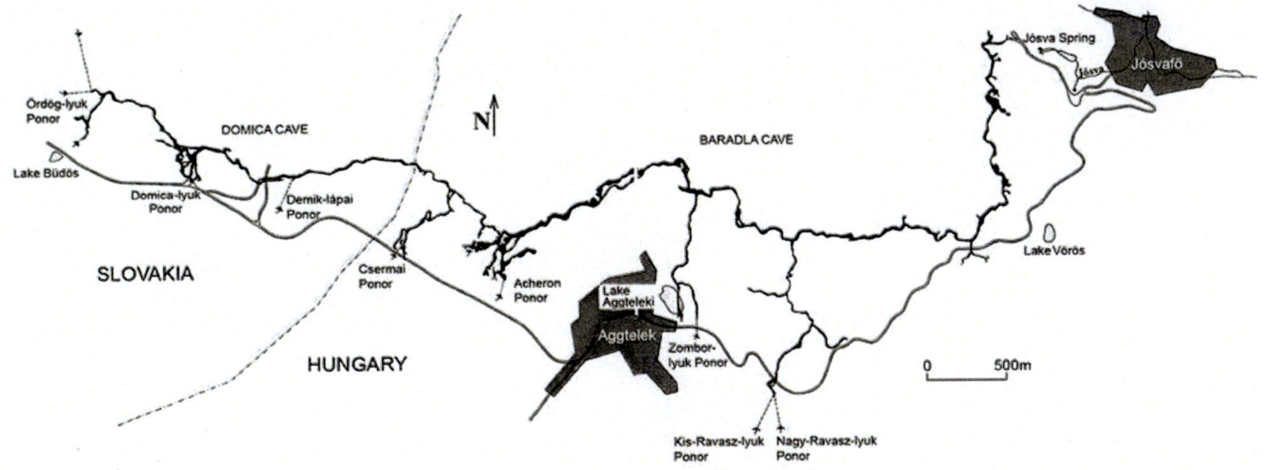

Fig. 18 The Baradla-Domica Cave System and its ponors (Kordos 1984)

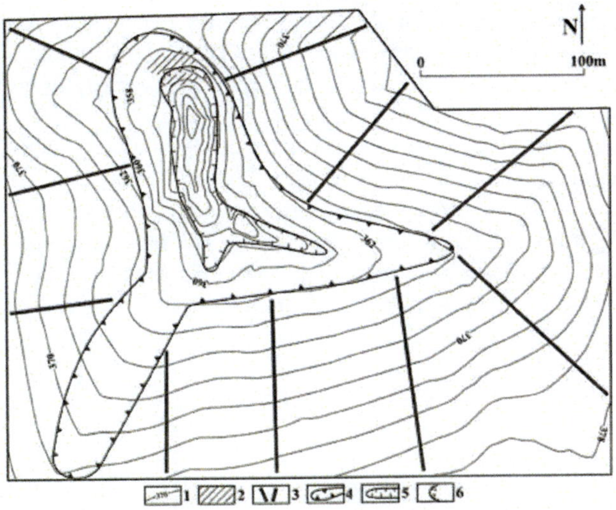

Fig. 19 Morphological map of the Dász doline (Aggtelek Karst) (Veress 2016) 1. Contour line, 2. Limestone outcrop, 3. Slope of paleodoline, 4. Blind valley, 5. Ravine (internal valley), 6. Ponor

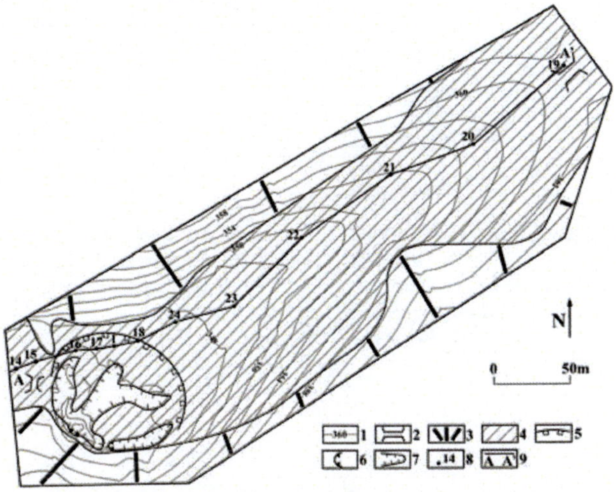

Fig. 20 A map of a part of Bába Valley (environs of the ponor Nr. 2) (Aggtelek Karst) (Veress 2016) 1. Contour line, 2. Col, 3. Uncovered side slope of depression of superficial deposit, 4. Floor of depression of superficial deposit, 5. Internal depression of depression of superficial deposit, 6. Ponor, 7. Ravine of ponor, 8. Site and identification code of VES measurement, 9. Alignment of geoelectric-geological profile

sandstone stripes (Móga 2002b). Some of these ponors are also part of a karst depression (the ponors of the depressions of Bába Valley). Ponors inside of the karst are primarily connected to karst depressions, they are less widespread on valley floors. The number of ponors inside the karst which are in karst depressions (uvalas, dolines, Figs. 19 and 20) surpasses the number of karst marginal ponors, most of them have a small catchment area (this is the area of the bearing depression) and major caves are rare in their continuation.

Ponors can conduct water onto the karst reservoir as in the case of the ponors in the depressions of Bába Valley. These ponors are not fed by permanent streams; however, according to water tracing, their water appears in the water of the permanent Köszörű Spring fed by karstwater (Sásdi 1998). But they often constitute a separated hydrographic system. Thus, for example, the water of the ponors of Baradla Cave appears in Jósva Spring through the cave, or the water of the ponors of Béke Cave appears in Komlós Spring. The drainage passages of the adjacent ponor may or may not be connected. In the former case, several ponors are linked to a spring (see above). A good example for a spring fed by a single ponor is the ponor of Bükk-lápa the water of which appears in Szénhely Spring (Szilas Plateau, Hazslinszky 1965).

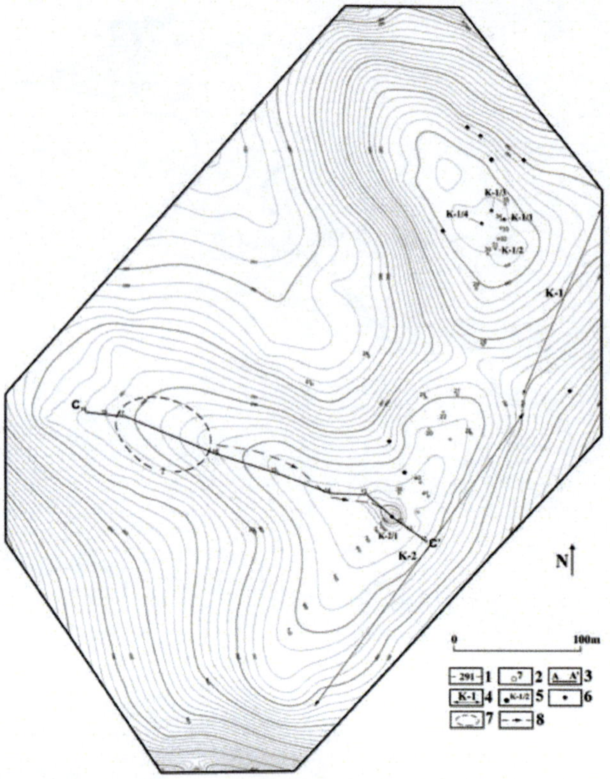

Fig. 21 Keserű-tó lápa and a depression of superficial deposit northeast from it (Veress 2017). Legend: 1. Contour line, 2. Site of VES measurement, 3. Track of profile, 4. Partly exhumed depression, 5. Karstic depression (covered karstic ponor and subsidence doline), 6. Rock outcrop, 7. Buried ponor, 8. Channel

The ponors of depressions may be situated at the end of blind valleys (Dász doline), or a channel or gully leads to the ponor (the ponor of the lápa of Lake Keserű, Figs. 21 and 22). They are rarely of permanent nature (the ponor of Vizetes) more often intermittent. The fill is not always completely impermeable. In this case, a subsidence doline also develops on the superficial deposit (lápa of Lake Keserű).

Sásdi (1998) also distinguishes channel ponors (drainage point on the channel floor) from the valleys of the karst, on the floor of which the water of the stream seeps away partially or completely.

Lakes

Karst lakes can be of intermittent (during rainy seasons, impoundment takes place at the karst drainage) or of permanent nature. Intermittent lakes can survive for several days. Such a lake may develop at the ponor of Zombor-lyuk (Sásdi 1998). Intermittent lakes can reach beyond the inner depression that drains the water and they may flood some parts of the bearing depression. Móga (2002a), for example, mentions an intermittent lake in Vizetes ponor partly filling the bearing depression (Vizetes).

Lakes of permanent water may occur in ponors or in dolines (Samu and Kevei Bárány 2008). Doline lakes are Lake Vörös and the Lake of the Teresztenye Plateau. Lake Aggtelek and Lake Cserna are sinkhole lakes that developed in former karst margin ponors. Presently neither of them exists, at most a marsh or hydrophilic vegetation can be

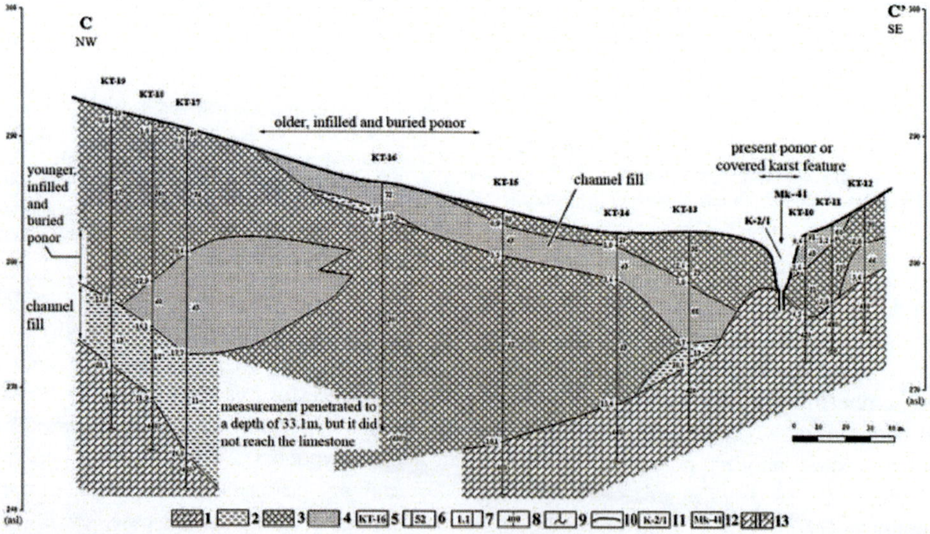

Fig. 22 Geoelectric-geological profile 'C–C' from the area of Keserű-tó lápa (K-1) (Veress 2017). Legend: 1. Limestone, 2. Clay, 3. Clay (with limestone debris and sand), 4. Sand-gravel-loess (with limestone debris), 5. VES measurement site with identification number, 6. Geoelectric resistivity of the series (Ohmm), 7. Base depth of the geoelectric series (m), 8. Geoelectric resistivity of bedrock (Ohmm), 9. Approximate depth of penetration of VES measurement, 10. Geoelectric series boundary, 11. Karstic feature and its identification mark, 12. Rock outcrop with identification number, 13. Chimney, shaftNotice: The track of profile can be seen in Fig. 21

found in their place. Lake Derenk and Lake of Acskó-rét developed by the plugging of ponors inside the karst or of subsidence dolines (Sásdi 1998).

Gorges

The gorges of the mountains are not significant which can probably be traced back to the fact that streams of greater importance are not forced to antecedence and the cave ceilings do not show collapses yet. Zámbó (1998a) thinks that some sections of the erosion valleys (Telekes Valley, Jósva Valley, Ménes Valley), and of their tributary valleys that developed along karstic stripes are gorges (the valleys of the Kecső, Tohonya, and Mocsolyás Streams), but he also regards the part of the Bódva Valley between Telekes side and Csipkés Hill as sections with gorges. Hevesi (1991b) regards the Vár Valley section of the Ménes Valley as a gorge of probably antecedent origin. Sásdi (1998) mentions valley sections of gorge character from the Sárogkert, Mocsolyás, and Bába Valleys (tributary valleys of Ménes Valley), and on the Szalonna Karst he regards the Mész and Bik Valleys as gorges. Opinions on the origin of the gorge of the Telekes Valley (between the eastern part of Galyaság and the Rudabánya Mountains) vary: according to Leél-Őssy (1953a), it is antecedent (its lower section) while according to Hevesi (1991b) it is of opening-up origin.

Concretions

As compared to its karstification, the concretions of the karst are not significant. Their description and classification was done by Sásdi (2005). According to age, he distinguished Early Pannonian, Plio-Pleistocene, Middle-Pleistocene, Late Pleistocene, Late Pleistocene-Holocene, and contemporary calcareous sinters. Among them, the Early Pannonian freshwater limestones do not belong to calcareous sinters accompanying the karstification of the mountains. The youngest is the travertine mound in front of Jósva Spring originating from 1957. According to development environment, Sásdi distinguished calcareous sinter in the foreground of springs and those in the channel of streams (Sásdi 2005). Morphologically, he distinguished veneers (Kecső Valley, Tohonya Valley, Lófej Valley, Mocsolyás Valley), escarpments (Jósva Brook, Kecső Valley, Sárogkert Valley, Patkós Valley), travertine mounds (Jósva Spring, Csurgó-kút of Bába Valley), and those of braided character (Mocsolyás Valley, Bába Valley). The Sárogkert, Mocsolyás, Patkós, and Bába Valleys are the tributaries of the Ménes Valley.

3 The Bükk Region

3.1 General Characteristics

The classification of its karst was done by Jakucs (1977b), Hevesi (1986a, b, 1991a, b), and Veress (2000a). Juhász (1976), Farsang and M. Tóth (1992), Zboray and Kevei Bárány (2004), Kevei et al. (2015) carried out morphometric studies on its surface karst. Factors influencing karstification were described by Juhász (1976), Bárány and Mezősi (1978), Hevesi (1978, 1980), while the morphology of its surface was described by Leél-Őssy (1954, 1959a, b, c), Láng (1954, 1964), Horváth (1962), Láng et al. (1970), Hevesi (1980, 1984, 1986a, b, c, 2000, 2002), Tóth (1984), Kevei Bárány (1992), Veress (1992), Hevesi and Záhorszky (2000), Záhorszky (2001), Veress and Zentai (2009), and Kozma and Holló (2010). The fill of its depressions was investigated by Jámbor (1959), Láng (1970), Pelikán (1992), Veress and Zentai (2009). Láng (1954), Hevesi (2007), Czesznak et al. (2014), and Szegediné Darabos et al. (2014) dealt with karst hydrology. Freshwater limestones were described by Hevesi (1972), Scheuer (1975), Tóth (1983), and Bódi et al. (2016). The vegetation of the dolines of the Nagy Plateau was analysed in detail (Bárány-Kevei 1991; Kevei Bárány 1992). The main characteristic features of its karst are as follows:

Its karst areas of continuous development are the most significant in Hungary. In spite of this, it is not a uniformly developed karst. Its karst terrains are interrupted by non-karstic rocks and non-karstic superficial deposits as well as by valleys and by escarpments developed at the boundaries of karstic and non-karstic rocks. In the mountains, karstification is present in three areas, somewhat separated from one another (elevation, morphology): Bükk Plateau (Nagy Plateau and Kis Plateau), the Southern Bükk (with the Southwestern Bükk and the Southeastern Bükk), and the Northern Bükk (Kemesnye Hill). Karst is the most uniform and the most expanded on the Bükk Plateau (mainly on the Nagy Plateau).

According to Jakucs (1977a), its karst is of Aggtelek type, Hevesi (1991a, b) claims that it is of Aggtelek character, while Veress (2000a) states that it is an upwarping karst. The former classifications focus on the doline row character of the karst surface, while the latter concentrates on the fact that since its inner part is more elevated and thus, more denuded, the character and extent (perhaps the age) of karstification here is different from the margins. Soil-covered karst, mixed allogenic-autogenic karst, and allogenic karst are widespread in the area. A uniform concealed karst is missing, concealed karst features occur on the sediments of the depressions or the dolines of doline rows become activated as subsidence dolines (see below).

Hydrothermal effect is also present mainly in the southern Bükk Region, where the temperature of karstwater is ever higher towards the margin of the mountains in southern direction. This can be traced back to the fact that waters arriving into greater depths. Thus, they became increasingly upheated approaching the surface.

The mountains are dismembered by a series of folds with a southern dip, creating an imbricate structure (boundaries of thrust sheet determined surface morphology, for example, the site and direction of valleys), with boundaries of WE direction (Balogh 1964). The anticlines of non-karstic rocks are denuded to a large extent (replaced by valleys) by non-karstic processes. Karst developed in the synclines of the folds.

The mountains form a peneplain elevated to different altitudes (Láng 1954). In some parts of the area of the Central Bükk, younger (Pliocene) denudation surfaces occur (Fejes and Tóth 1984, 2001; Fig. 23).

The surface of the Bükk Plateau is dissected since there are elevation differences of 100–150 m in its area. According to elevation, two surfaces are identified: mounds and flat parts separating them. The lower plateau surface is not completely flat either, but subdivided into flat sections (1), the floor level of epigenetic valleys incised into this surface (2) and surface of larger karst systems (3). According to Hevesi (2002), mounds can be situated inside and along the margin of the plateau. The relative elevation of inner mounds is 150 m, the diameter of their floors is 400–800 m, and the inclination of their slopes is 8–20°. The length of the side slopes of marginal mounds is 400–800 m and their relative elevation ranges between 100 and 300 m here. Slopes are steeper than 30° and may be vertical as well. The 'Rocks' (Fig. 7.7) ('kő or kövek' in Hungarian) represent one variety of outer mounds which developed at sites where the limestone is in contact with schist. The steeper, vertical cliffs of the 'Rocks' are due to intensive denudation. The 'Rocks' constitute three rows along the margin of the plateau: at the southwestern margin of Nagy Plateau (for example, Oltár-kő, Tar-kő, Három-kő, etc.), at the southwestern, southern margin of Kis Plateau (Köpüs-kő, Sólyom-kő, etc.), and at the northwestern, northern margin of Kis Plateau (Látó-kövek, Örvény-kő). The mounds have both karstic and non-karstic features (see below). According to Tóth and Fejes (1984), at least some of the mounds are the transformed remnants of the elevated sections of the Upper Cretaceous-Middle Eocene peneplain. According to Hevesi (2002), the mounds are peneplain remnants, but they obtained their mound character during epigenetic valley formation as remnants of interfluves.

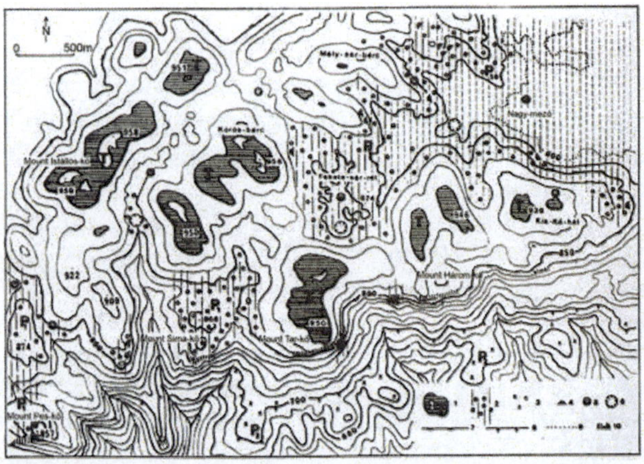

Fig. 23 Denudation levels on the Bükk Plateau (Fejes and Tóth 1984). Legend: 1. Areas of peneplain more elevated than 925 m, 2. Remnants of older Pliocene planation surface (P_1), 3. Margin of the younger Pliocene planation surface with an elevation of 650–700 m (P_2), 4. Inactive spring caves, 5. Old inactive or intermittent ponors, 6. Specific large dolines, 7. Contour line of 850 m, 8. Contour line of 700 m, 9. Contour line of 800 m, 10. Cliff wall

In the Bükk Mountains, there are several ruined caves (Hevesi 1978, 1986c, 2002), which also contribute to the diversity of landforms. Depression fills (these were mainly studied in dolines) are diverse. According to Jámbor (1959), the dolines in row contain red clayey limestone debris, while in other dolines which are not aligned in rows (on the mounds of Istállóskő, Hutabérc, Nagy István erőse) carbonate-free loess (mountain loess), then below it clayey loess and still below red clays can be found, a reworked, transformed variety of Miocene rhyolite tuff. Láng et al. (1970) studied the doline fill of the Nagykőmázsa Hill near Miskolctapolca. In one of the dolines, the fill is 9 m thick and consists of silt, silty mud, and muddy silt. Based on pollen analysis, the upper part of the fill originates from the Pleistocene and it is reworked, while its lower part is silty, clayey mud with Tertiary (Pannonian) flora elements. Exploring the fill of one of the dolines of the Zsidó-rét (Fekete-Sár) Tóth (1986) described a carbonate-free falling dust formation which is separated into two parts by the debris of the bedrock. During the remapping of the Bükk Mountains, Pelikán (1992) bored the fill of several depressions and revealed the following sediments: silt of loess character, red clay (in the grikes of the bedrock of the doline floor) reddish-brown clay with schist debris, and rhyolite tuff. Based on the data of Veress and Zentai's (2009) VES measurements, the sediments described from the dolines of Nagy mező, Zsidó rét (Fekete-sár), and Kis Plateau were clayey-muddy loess, clay with limestone debris, and loess with limestone debris and clay.

Specific, common features of the Bükk Region are valleys. Valleys with doline rows, valleys without dolines, and blind valleys can be distinguished. These are mainly dry valleys, but valleys with permanent streams also occur.

- Valleys with doline rows (Fig. 24) are epigenetic valleys inherited over to the karstic rocks of the Bükk Mountains, on the floor of which dolines are aligned (Hevesi 1978, 2000, 2002). According to Hevesi (2002), these valleys are characterized by low gradient (sections with a gradient of 0.38% occur), small depth, large width (70–300 m), and gentle side slope (5–8°, rarely 8–20°). Valleys are often several km long and may constitute valley systems (main valleys are connected by tributary valleys). Tributary valleys join main valleys by steeper sections (hanging valleys). Particularly, the floor of longer main valleys can be of various dip directions and these are separated from each other by valley-floor divides. Between valley-floor divides or between a valley head and valley-floor divides there may be arheic valley sections of significant length (up to several km) (Hevesi 1980). Taking into consideration the dolines of valley floors, since they were ponors formerly, Hevesi (1980) described streams and their valleys that developed on the former superficial deposits of the Nagy (Fig. 25) and Kis Plateaus (Fig. 26) and their direction too. Valley retreat to dolines is a common phenomenon which results in open dolines (Leél-Őssy 1954). Valleys with doline rows are particularly widespread on the Nagy and Kis Plateaus, but they also occur in the Southern and Northern Bükk. Valleys with doline rows are of WE direction on the Nagy and Kis Plateaus, but S to N and N to S directed valleys also occur on the Nagy Plateau.

Fig. 24 Morphological map of Káposztáskert Valley and its environs (Hevesi 2000). Legend: 1. Roof of main interfluve, 2. Interfluve, 3. Saddle, 4. Saddle on interfluve, 5. Slope between interfluve and valley floor, 6. Erosion valley, 7. Derasion valley, 8. Plate-shaped hanging doline, 9. Valley-floor doline, 10. Valley-floor twin doline

- Valleys without dolines have a V shape. Specific valleys without doline rows occur in the side of the mounds of the mountains, which are either wedged out or they are connected to the main valleys. In the area of the Southwestern and Southeastern Bükk, there are often valleys which start from a non-karstic terrain (Tatár Valley), and cutting through the limestone zone they continue on a non-karstic terrain. Sections on the limestone zone are epigenetic and of gorge character (Hevesi 2002). However, such an epigenetic valley with doline rows also occurs which continues on non-karstic terrain and terminates as a blind valley. A reversed situation is also possible. In this case, the blind valley can be found on the upper section of the valley with doline rows. For example, the section of the valley with doline rows above the valley head of Felső-Sebes Stream which falls on a shaley terrain, the blind valley of Jávorkút ponor (Hevesi 2002).
- If the main valley of the non-karstic terrain has tributary valleys on karstic terrain, the latter are transformed into an arheic valley with doline rows which turn into hanging valleys since their deepening lags behind the deepening of the main valley. Such a valley is, for example, one of the tributaries of the Vadalmás Valley on the Répáshuta limestone zone in the Southeastern Bükk (Hevesi 2002).
- The karstwater table may be located at the floor of valleys with permanent streams (if they are cut into karstic rock). Valleys with permanent streams start at plateau margins with a karst spring in their valley head (pocket valley), e.g. the Szalajka Valley. It also occurs that the erosion valley started in the zone of the non-karstic rock (Garadna). However, another valley (Hór Valley) can also be mentioned from the Bükk Mountains which was formed at the interface of limestone and non-karstic rock (Láng 1964).
- There are blind valleys at the margin of the wedged-in, non-karstic rocks (the blind valley of Jávorkút ponor), but they may also occur on the floor of partially inherited epigenetic valleys at valley-floor rock boundary. Thus, on the Kis Plateau, blind valleys occur at the valleys of the nappe outlier enclosing Andó-Bikk, Galya-ér, Hárs-tető (Hevesi 1980). The former valley rock boundary on the floor of epigenetic valleys and the termination of blind

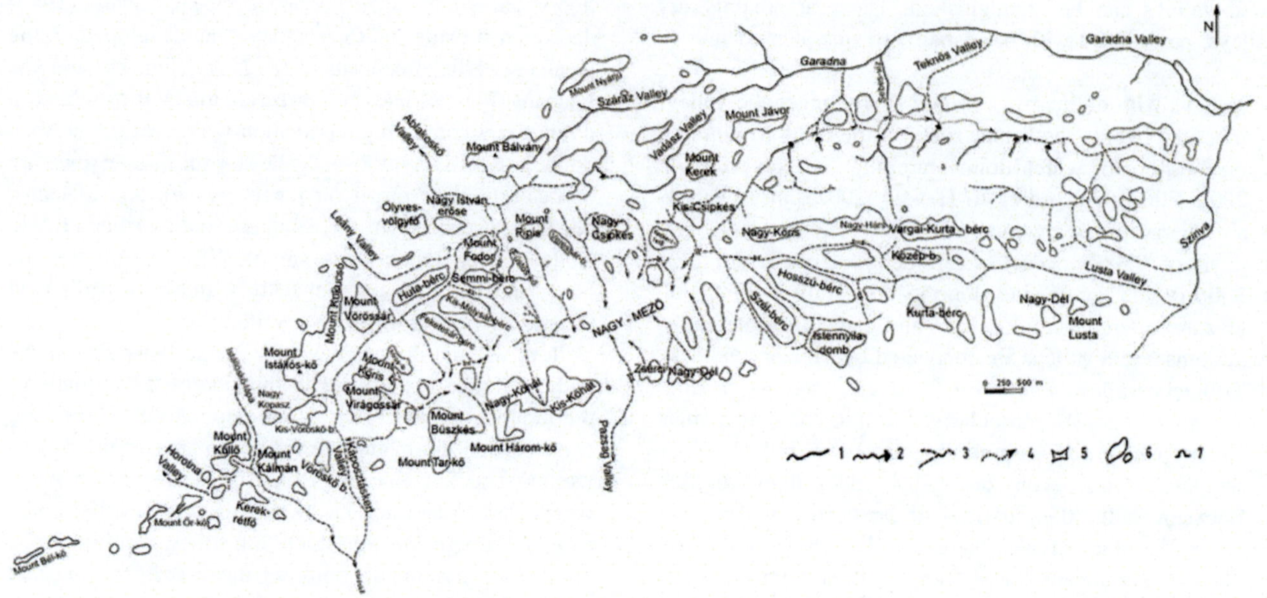

Fig. 25 The epigenetic valleys of Nagy Plateau (Hevesi 1980). Legend: 1. Permanent stream, 2. Active ponor with permanent stream, 3. Probable former stream based on valley doline rows, 4. Less identifiable former stream, 5. Valley floor, saddle, 6. Mound, 7. Cave

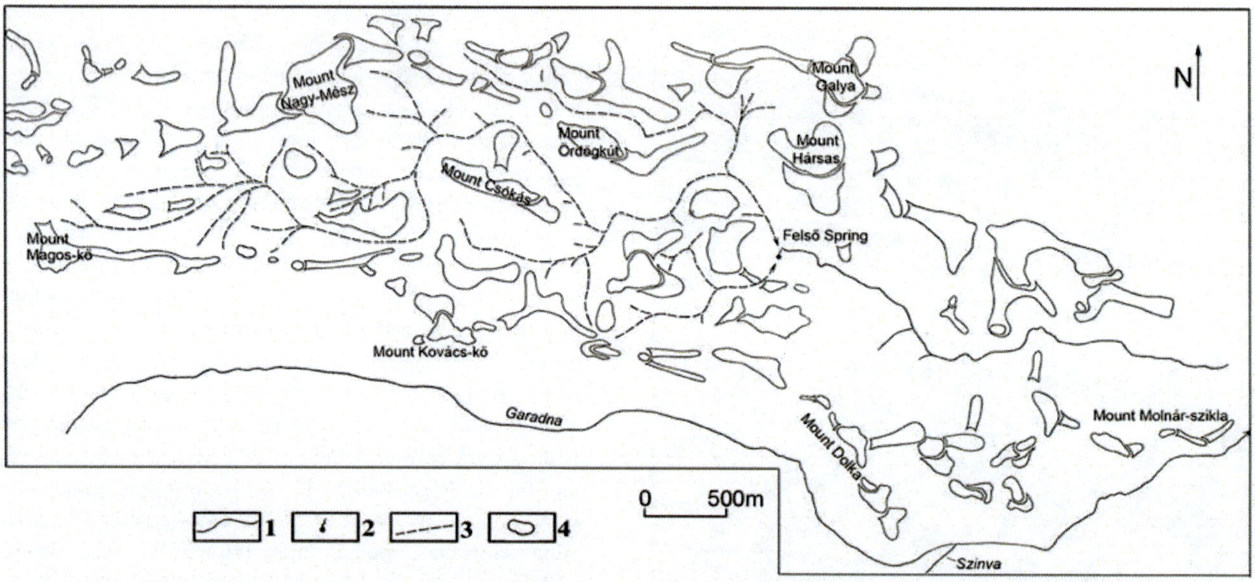

Fig. 26 The epigenetic valleys of Kis Plateau (Hevesi 1980). Legend: 1. Permanent stream, 2. Spring, 3. Former stream that can be marked by valley doline rows, 4. Mound

valleys is marked by the boundary of counterslope sections of the valley floor. Valleys (blind valleys) can develop in larger depressions (separated epigenetic valley sections too) such as the blind valley of Kaszásrét ponor, or Dorongós (in the area of the Répáshuta limestone ridge) in the Southeastern Bükk. However, a large blind valley can also be mentioned from the Nagy-mező depression that has been transformed by now (Fig. 27) with subsidence dolines inside (Veress and Zentai 2009).

The escarpment surrounding the Bükk Plateau at an elevation of 200 m between the Hór and Eger Streams surrounds the Bükk Plateau which is particularly developed at

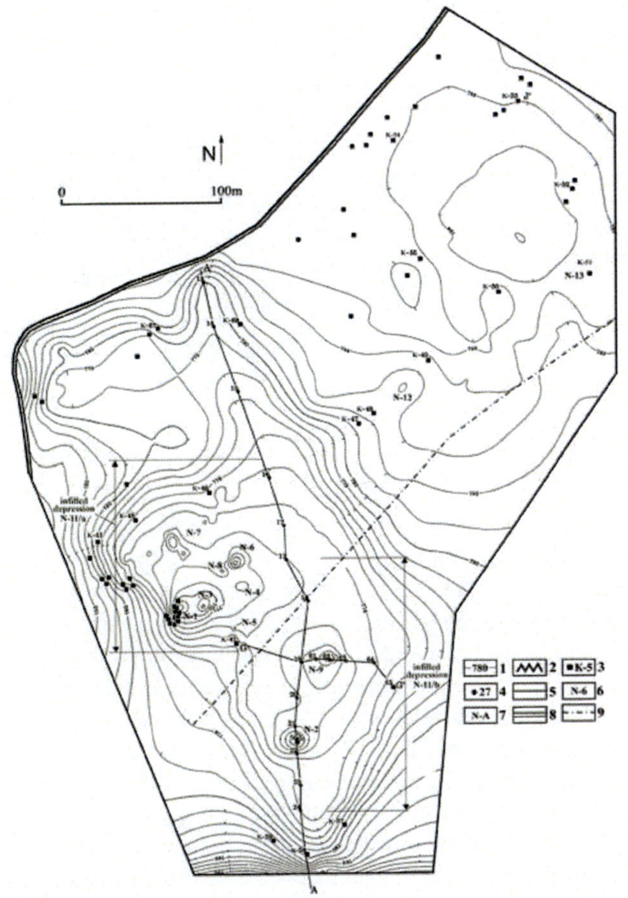

Fig. 27 The relief draught of a part of Nagy-mező (Veress and Zentai 2009). Legend: 1. Contour line, 2. Cliff wall, 3. Rock outcrop with identification number, 4. VES measurement site with identification number, 5. Track of profile, 6. Identification mark of karst depression, 7. Identification mark of buried karst depression, 8. Road, 9. Fence

the southern margin of the Nagy Plateau, a cliff of various gradient (vertical in some places) dissected by block streams of up to several 100 m length. Its elevation is diverse, at the 'Rocks' (Fig. 7.7) more pronounced, interrupted by erosional (Garadna Valley) and epigenetic valleys (Káposztáskert Valley). Tóth (1983) holds that it was created by pedimentation. He states that the escarpment is the front of the pediment that developed through younger Pliocene planation. The front and the planation surface was shaped by differential weathering: the escarpment was formed at the interface of clay shale (Southern Bükk) and limestone (Bükk Plateau). Structure also played a role in its development. Some sections are along thrust faults, while others developed along strike-slip faults (Hevesi 1978, 2002).

3.2 Morphometry

Juhász (1976) studied the density of karst features (mainly dolines) on various rocks of the mountains. On Anisian dolomite terrains he found weak karstification where the density of karst features is below 5 features/km^2. On Ladinian cherty limestone terrains, there is medium karstification where the density is 5–10 features/km^2. Ladinian white plateau limestone terrains are heavily karstified (10–20 features/km^2, while Anisian limestones show excellent conditions for karstification (above 20 features/km^2).

Based on the data from aerial photographs, Zboray and Kevei Bárány (2004) analysed the gradient of doline slopes with various exposure in an area of 1.5 km^2 on the Bükk Plateau. They established that the studied doline slopes with western (SW, W, NW) exposure are gentle, while slopes with eastern (SE, E, NE) exposure have a smaller and higher inclination.

The morphometric investigations by Kevei Bárány et al. (2015) (performed on the dolines of study areas both on the Kis and Nagy Plateaus) led to the following results.

In the Bükk Mountains, the average diameter of dolines is 70.65 m, their average width is 48.21 m, and their average depth is 9.69 m (Table 2). Here, dolines are smaller than on the Aggtelek Karst and larger than on the Mecsek Karst.

Elongated row dolines are grouped by their longitudinal axes into NS and WNW-ESE directed classes (Fig. 28). According to the authors, this is controlled by the direction of the bearing valleys. They also studied the orientation of the longitudinal axis of dolines with various diameters. In the Bükk Mountains, dolines with a longitudinal axis shorter than 70 m are mostly of NS direction, but an ENE to WSW orientation also occurs subordinately. In case of dolines with a longitudinal axis shorter than 70 m, the main direction was WNW to ESE and N–S was subordinate. In the Aggtelek Karst, there was no direction difference due to diameter size, but similarly to the Bükk Mountains it was provable in the Mecsek. (The orientation of dolines with a diameter larger than 30 m was N–S, while that of those with a lower diameter a WNW to ESE orientation was demonstrated.) Although the authors do not mention, in the case of the Bükk Region, the dolines more than 70 m across are probably in elevated position.

The nearest neighbour orientation in the Bükk Mountains does not show regularity (Fig. 28), and Kevei Bárány et al. (2015) claim that neither in the Aggtelek Karst either, while Telbisz (2001) points out that irregular distribution is only characteristic of the Alsó-hegy Plateau in the latter (Fig. 7b).

The relief ratio of the dolines of the Bükk Mountains is similar to that of the Aggtelek Karst, but it is larger than in the Mecsek. For the dolines of the Bükk Mountains and

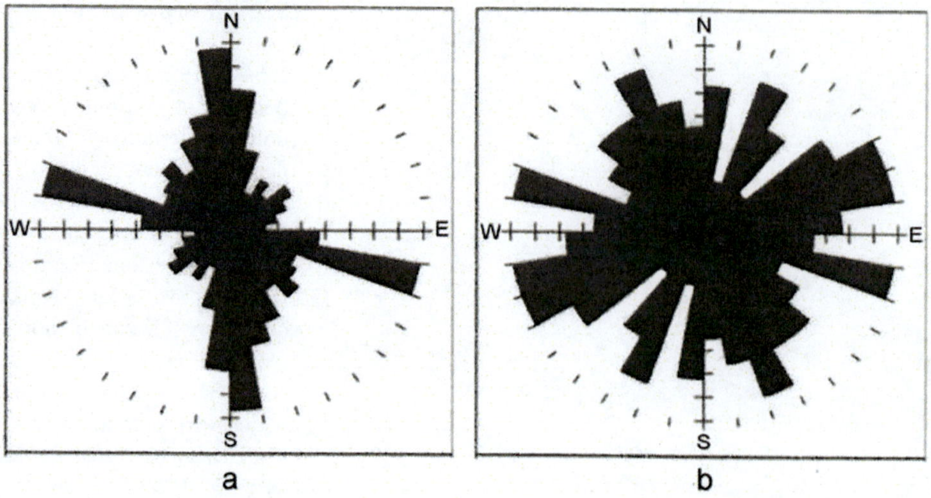

Fig. 28 The long axis (**a**) and the nearest neighbour (**b**) orientation of dolines in Bükk Mountains (Hungary) (bi-directional rose diagram, total data points 266 and 176, Keveiné Bárány 2015)

Aggtelek Karst, this can be traced back to the lower inclination of the side slopes which can be attributed to horizontal solution.

The average inner asymmetry of dolines (measured by average elongation ratio) is similar to that in the Aggtelek Karst and although lower there, it is still higher than for dolines of the Mecsek Karst (Table 5). The difference is due to the fact that the dolines of the Bükk Mountains and of Aggtelek are more asymmetric (the inclination of the doline slopes shows a larger difference for the same doline than at those of the Mecsek Mountains).

The nearest neighbour index of the dolines of a study area of 1 km^2 is 0.66, which indicates transition from a cluster to a random pattern.

3.3 Karst Features

The karst features of the Bükk Region are karren, solution dolines and uvalas, collapse dolines, complex karst depressions, subsidence dolines, ponors, gorges, concretions, and karstic relict landforms (Fig. 5, chapter "The Geomorphic Evolution of Karsts and Karstic Surfaces").

Karren

In the Bükk Region, karren formation (mainly subsoil) is widespread, but primarily on the mounds of the Bükk Plateau. According to Horváth (1962), covered, subsoil karren are recognizable in the mountains especially at sites where the soil is denuded. He mentions exhumed karren from the area of Tar-kő, Kiskőhát, Füstös-kő, and Varga-Kurta horst. These karren are karren knife edges (former bed heads probably dissolved and rounded) and between them solution grikes along bedding planes, here the beds are of low inclination, kamenitzas are also characteristic. He claims that such debris is formed by the denudation of karren knife edges that became rounded by dissolution ('Karrennasen').

Tóth (1983) describes karren fields from the mounds of the plateau (Tar-kő, Sima-kő, Cserepes-kő) too. In his opinion, karren develop at sites where the limestone is pure and the beds are vertical. Here, bare karren can also develop. He mentions bare karren (bed head karren) from Veres-sár Hill, Fekete-Sár Hill, Zsérci-Nagy-Dél, Kálmán Hill, and from the top of the 'Rocks' of the plateau margin. According to him, exhumed paleokarren occur on the top of the mounds too.

Hevesi (2002) includes bed head karren (where grikes between bedding planes can be 2–6 m wide) among the karren features of the mountains and mentioned rock corridors that developed along bedding planes near Bél-kő. He also describes similar features from the upper part of the Istállós-kő lápa.

Solution dolines, uvalas

The karst landscape of the Bükk Region is dominated by solution dolines (Figs. 5, chapter "The Geomorphic Evolution of Karsts and Karstic Surfaces" and 3b, chapter "GIS Research of Karst Terrains in Hungary"). According to morphological environment, Hevesi (1978, 1980, 2000, 2002) distinguishes valley floor row dolines (1) and (hanging) dolines in elevated position (2). However, dolines may occur on flat (3) and in larger karst depressions (4), e.g. in the area of Zsidó-rét and Nagy-mező. In cross section, dolines have a conical shape (Juhász 1976), while valley-floor dolines show a funnel shape (Hevesi 2002). Jakucs (1977a) states that the dolines of the Nagy Plateau are asymmetric in cross section. Asymmetry is also characteristic of row dolines and uvalas (see below).

Valley-floor dolines (1) are row dolines (Figs. 29, 30 and 31). Dolines with a depth of 2–5 m have a 5–10 m diameter,

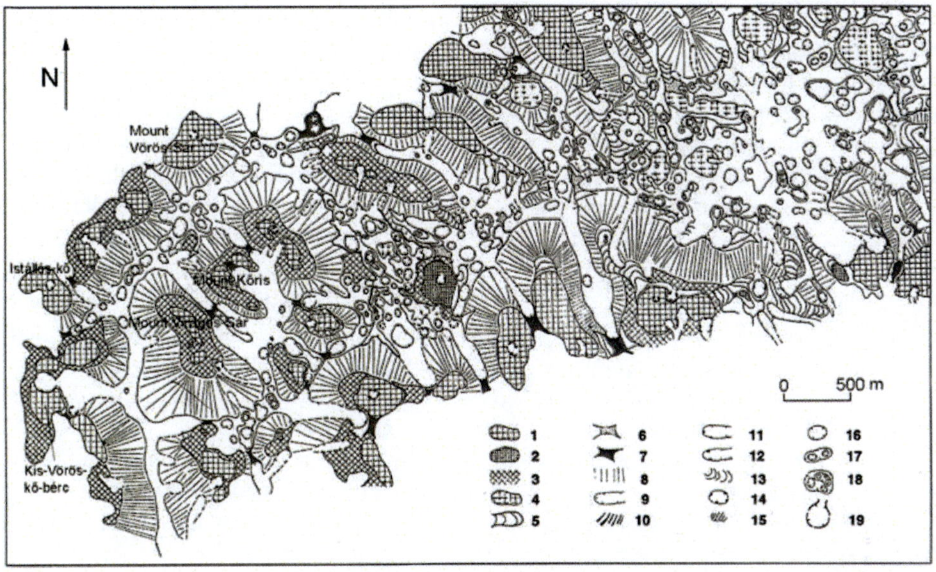

Fig. 29 The morphological map of the part of Nagy Plateau between Káposztáskert Valley and Nagy mező (Hevesi 2002). Legend: 1. Roof of main interfluve; 2. Roof of tributary interfluve, lower elevated roof of main interfluve; 3. Interfluve; 4. Low, flat shelf-like island separated from the lowering part of the interfluve; 5. Landing (its lines on the main and assistant contour lines), 6. Saddle at valley-floor divide; 7. Saddle on interfluve; 8. Slope between interfluve and valley floor; 9. Erosion valley, valley floor; 10. Valley escarpment; 11. Erosion and derasion (solifluctional) valley; 12. Erosion and derasion valley that developed by landslide; 13. Tongue; 14. Plate-shaped hanging doline; 15. Cliff wall; 16. Valley-floor doline; 17. Twin doline; 18. Smaller karst vápa (polje); 19. Large karst vápa

Fig. 30 The morphological map of Kis Plateau (Hevesi 1980). Legend: 1. Roof of interfluve; 2. Interfluve; 3. Landing of interfluve; 4. Saddle; 5. Boundary of Kis Plateau; 6. Erosion valley; 7. Derasion valley; 8. Slide scar; 9. Erosion and derasion valley; 10. Karstic blind valley; 11. Tongue; 12. Slope being wavy because of slide; 13. Creek, 14. Alluvial cone; 15. Spring; 16. Permanent stream; 17. Intermittent stream, 18. Terrace; 19. Valley floor doline; 20. Doline in karst vápa; 21. Stalagmite fragment occurrence site on interfluve, 22. Cliff wall, D. Dolka-tető, Cs. Csókás-tető, G. Galya-tető, H. Hármas-tető, K. Kovács-kő, M. Magos-kő, NM. Nagy Mész-tető, Ö. Ördögkút-tető

those with a depth of 5–10 m have a diameter of 50–100 m, and 10–25-m-deep dolines are about 100 m across (Hevesi 2002). Valley-floor row dolines mostly coalesce into uvalas and they rarely occur individually. In uvalas, partial dolines occur by twos, threes, and fours and they constitute larger karst units of valley floors. On valley floors, uvalas with various numbers of partial dolines may alternate. Partial dolines and uvalas are separated from each other by saddles of various elevations. Valley floor dolines occur in all karst areas of the Bükk Mountains. They are the most specific to the Nagy and Kis Plateaus, but they are also widespread in some parts of the Southeastern Bükk. (For example, on the Kisgyőr-Tapolca limestone ridge and in the area of the Répáshuta limestone zone.) According to Hevesi (2002), in the area of the Répáshuta limestone zone, doline rows consisting of two or three dolines only occur in the tributary valleys of main valleys. However, on the Kisgyőr-Tapolca limestone ridge, the floor of the valleys inherited over to the limestone, spreading across to the non-karstic terrain, is rich in doline rows. Among valley-floor dolines, drawdown dolines, point recharge dolines, and dolines that were developed (transformed) from ponors probably also occur. The following facts prove that some row dolines have a non-ponor origin.

- No sediment of fluvial origin could have been demonstrated on the Bükk Plateau (Pelikán 1992).
- At some sites such as at the valleys of Zsidó-rét, Nagy and Kis Kőmázsa, (Figs. 31 and 32) dolines occur in double or triple rows (Tóth 1983; Veress 1992; Záhorszky 2001). At a certain site of the valley floor, there may be only one former ponor and there cannot be several ponors along the direction perpendicular to the valley. Thus, at sites where there are more dolines, they are not former ponors.
- Among the dolines of Zsidó-rét and Nagy-mező, there are dolines whose bedrock floor is flat (Figs. 33 and 34). Such dolines cannot have been either ponors or point recharge dolines either formerly.
- In valley floor position, there are depressions which were formed in the superficial deposit (Fig. 35, Veress and Zentai 2009). This points to the fact that some valley-floor dolines are subsidence compaction dolines.

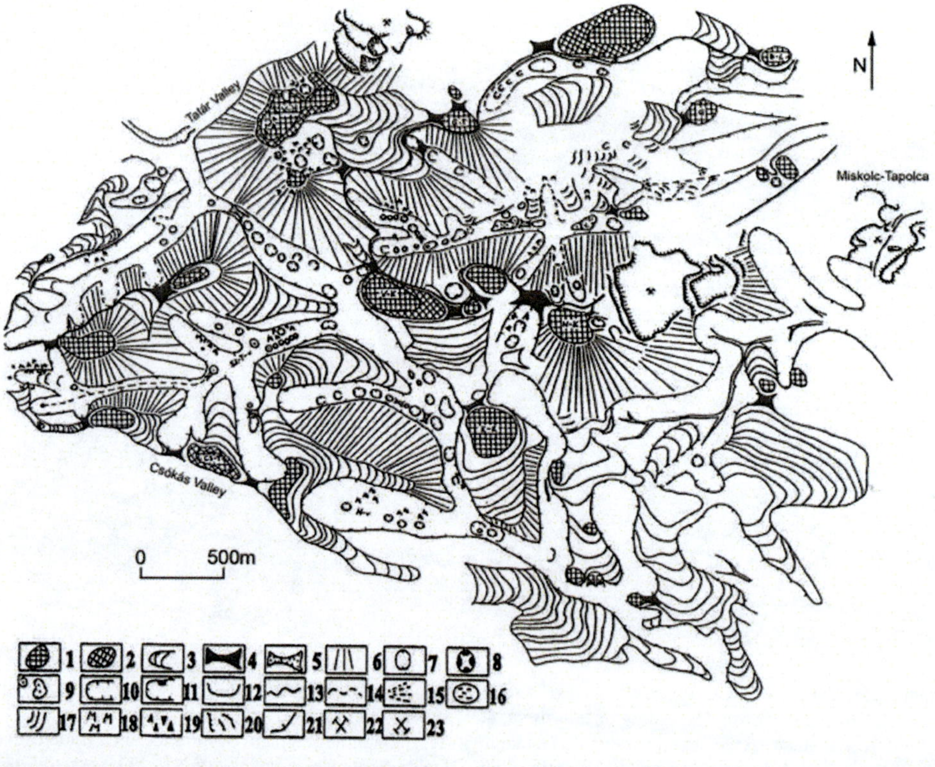

Fig. 31 Morphological map of the environs of Nagy-Kőmázsa (Záhorszky 2001). Legend: 1. Mound, roof of interfluve; 2. Interfluve; 3. Landing; 4. Saddle close to the roof; 5. Saddle between valleys; 6. Valley side; 7. Valley-floor doline; 8. hanging doline; 9. Twin doline with subsidence doline; 10. erosion valley; 11. Erosion-derasion valley; 12. Valley escarpment; 13. Stream; 14. Intermittent stream; 15. Alluvial cone;16. Lake; 17. Mass movement; 18. Karren slope; 19. Rockfall; 20. Quarry, dump; 21. Boundary of main valley; 22. Operating quarry; 23. Disused quarry, S-B Sűrű-bérc, CS-T Csókás-tető, V-11 Vásár-hely, V-k Kis-kőmázsa, G-T Galya-tető, G Galya, B- Bán-les, SZ-T-v Száraz Valley, M-v Mocsolyás Valley

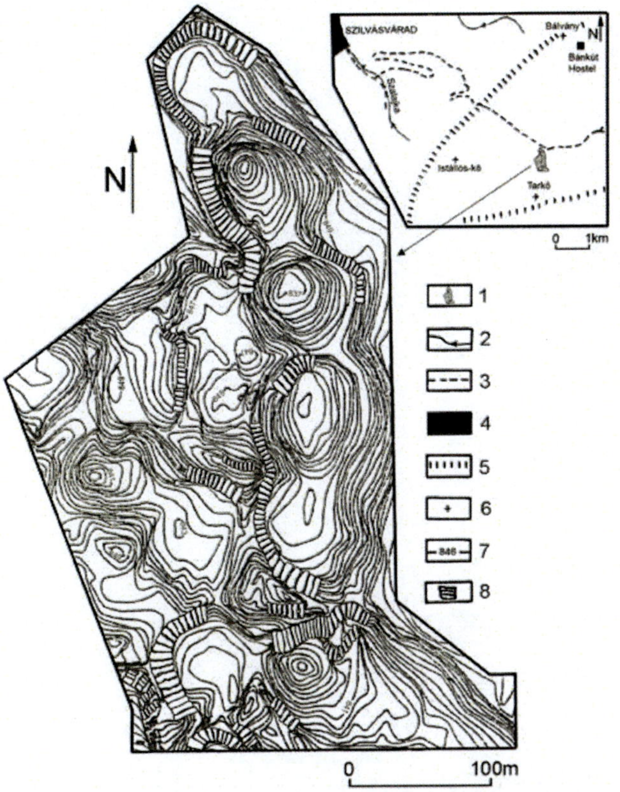

Fig. 32 Topographic map of Zsidó-rét (Veress 1992). Legend: 1. Mapped area; 2. Stream; 3. Road; 4. Settlement; 5. Margin of High-Bükk; 6. Mound; 7. Contour line; 8. Steep, barren rocky slope

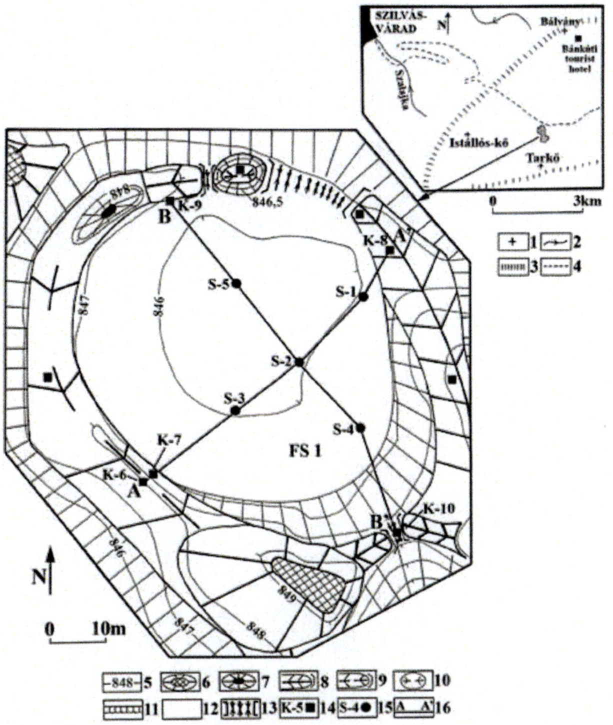

Fig. 33 A plate-shaped doline of Zsidó-rét (Veress and Zentai 2009). Legend: 1. Mountain, 2. Stream, 3. Plateau margin, 4. Road, 5. Contour line, 6. Karst hill, 7. Karst mound on a dividing wall, 8. Half-dividing wall, 9. Truncated half-dividing wall, 10. Relict form of a dividing wall, 11. Side slope of the doline, 12. Doline floor, 13. Karst saddle, 14. Rock outcrop and its mark, 15. Site and identification code of VES measurement, 16. Track of a geoelectric-geological profile

- At some sites, row dolines are uvalas. There is only a narrow ridge between them at the surface or on the bedrock (Figs. 24, 29, 32 and 35). It is not probable that formerly ponors were developed in such a density.
- For the majority of valley-floor dolines, there is no counterslope floor section on the valley floor (Fig. 35). A former blind valley can occur at the doline only on counterslope sections. Inner valleys leading to former ponors are absent on the floors of epigenetic valleys.

The hanging dolines of the mounds (2) can be situated at the roof level or in the side of the mounds, on saddles between partial mounds. Uvala formation is less common here. If a coalescence occurs, uvalas are constituted by two dolines. Hevesi (2002) considers hanging dolines as former ponors which developed when there had been superficial deposit cover in the environs of mounds. Hanging dolines are older and of larger diameter than valley floor dolines whose development took place on floors through downcutting and inheriting over to the bedrock later while the formation of hanging dolines probably began on mounds standing out from the cover during the time when terrains in lower elevation were covered (Hevesi 1978, 2002). The diameter of hanging dolines is 90–200 m, but they are shallow, only some metres deep. According to Hevesi (2002), these depressions are former ponors whose floor was shaped by accumulation and their side slope is partially missing since the water leaving them destroyed their downstream margin. Hanging dolines are often located near retreating valleys. If the valley reaches the doline during its retreat, it is only separated from the valley by a saddle. If the valley retreat continues, the saddle is destroyed and the doline opens up. Valleys retreat from two directions to the dolines (or uvalas) situated on the saddles of the mounds.

Among the depressions of flat terrains, dolines are common (3) and uvalas are rarer. If there is an uvala, it is constituted by two partial dolines. The dolines of larger karst depressions (4) are not necessarily of row pattern (if they are, they may be constituted of several rows), but they may be of grouped pattern too. In the latter case, the uvala that develops as a result of the coalescence of the dolines with various directions will be of patchy distribution. Such an uvala can be seen at the bottom of Fig. 32.

A specific variety of hanging dolines is the valley margin (valley side) doline (Fig. 32), or the doline at the margin of

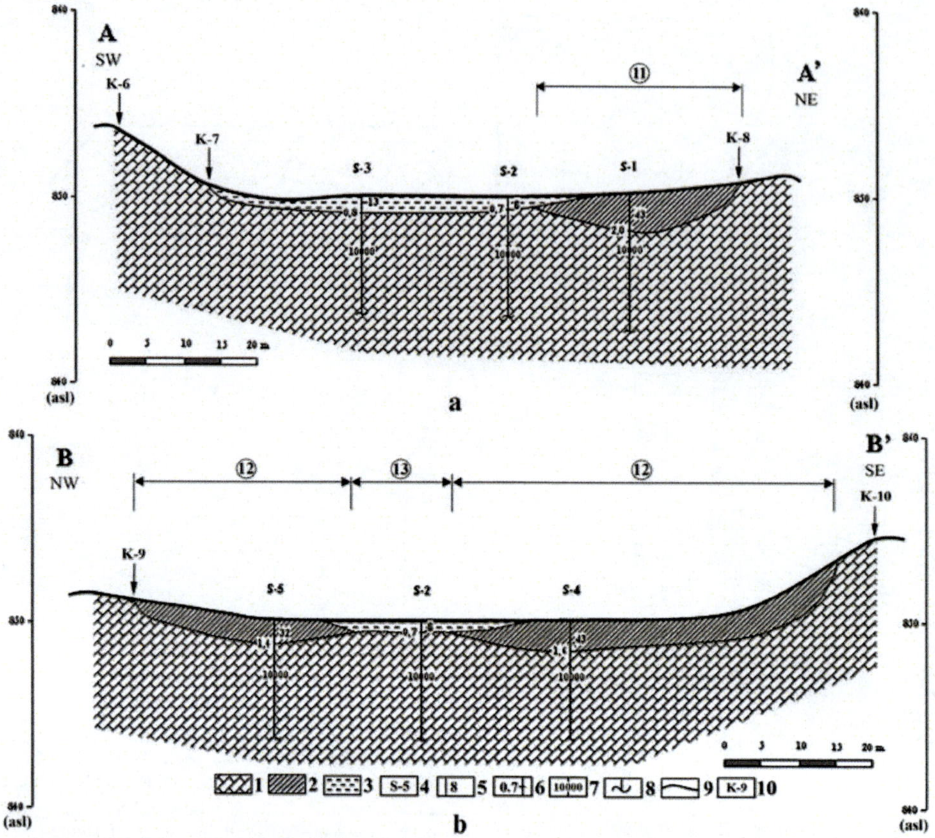

Fig. 34 Geoelectric-geological profiles of the plate-shaped doline shown in Fig. 8.33 (Veress and Zentai 2009). Legend: 1. Limestone, 2. Loess (clayey-muddy) or clay with limestone debris, 3. Clay, 4. Site and identification code of VES measurement, 5. Geoelectric resistance of series (Ohmm), 6. Base depth of geoelectric series (m), 7. Geoelectric resistance of bedrock (Ohmm), 8. Approximate depth of penetration of VES measurement, 9. Geoelectric series boundary, 10. Identification code of rock outcrop, 11. Bedrock depression, 12. Partial doline, 13. Covered dividing wall

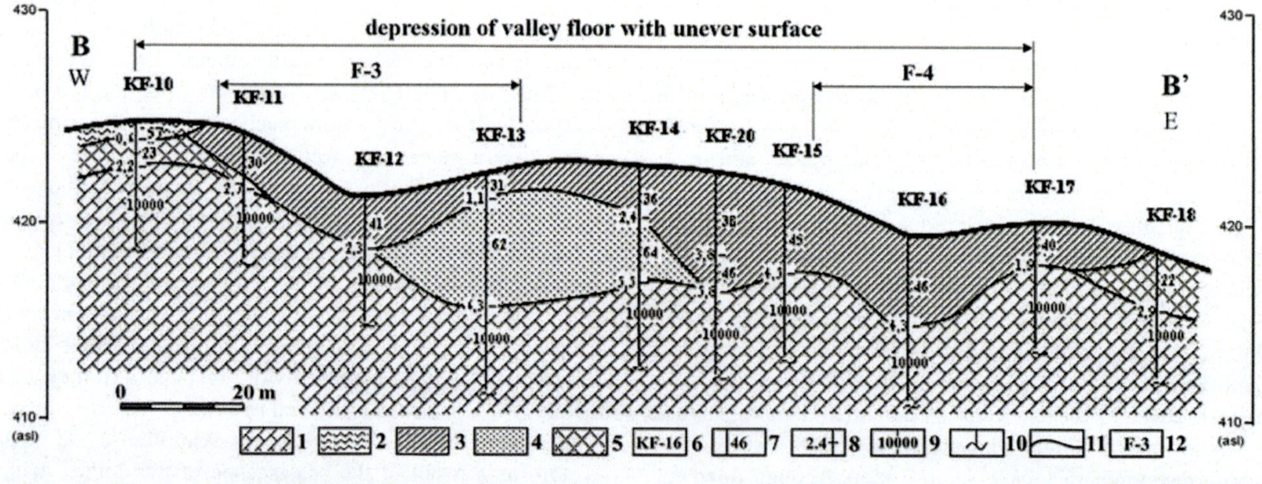

Fig. 35 Geoelectric-geological profiles of one of the valley parts with doline rows of Kis Plateau (Veress and Zentai 2009). Legend: 1. Limestone, 2. Limestone debris (clayey), 3. Loess (clayey-muddy) or clay with limestone debris, 4. Loess (sandy or with limestone debris), 5. Clay (with loess and limestone debris), 6. Identification code of VES measurement, 7. Geoelectric resistance of series (Ohmm), 8. Base depth of geoelectric series (m), 9. Geoelectric resistance of bedrock (Ohmm), 10. Approximate depth of penetration of VES measurement, 11. Geoelectric series boundary, 12. Karst depression

karst depression (Fig. 27). These dolines are separated from the valley floor or from the uvala by saddle. The saddle may be absent. In this case, the hanging doline is open and not closed (Figs. 27 and 32). Hanging dolines are mainly common on the Nagy and Kis Plateaus, but they also occur in the Southern Bükk, for example, in the eastern part of the Kisgyőr-Tapolca limestone ridge (Hevesi 1986a, b).

A characteristic feature noticeable of uvalas (Zsidó-rét) is that the eastern and western sides are less gentle and the inclination of these slopes is nearly similar. However, the gradients of northern and southern slopes show differences (Fig. 32). In Zsidó-rét, we can also observe that the floor of one of the part dolines of the smaller units of uvalas consisting of two part dolines has a flat floor and this partial doline is in hanging position over the floor of the other partial doline. The northern side slope of the partial doline without a flat floor is always steeper (Fig. 32).

A variety of solution dolines is the flat-floored plate-shaped doline which is mostly, but not always a hanging doline (see below), 80–200 m across, with side slopes often incomplete. They cannot be former ponors if their floor is flat. Their development is due to horizontal solution caused by the water moving laterally in the fine-grained superficial deposit (Veress 2020). According to their environment, they can occur on mounds, at the margin of uvalas, on flat terrains, individually or with other depressions together.

As it has already been mentioned, plate-shaped dolines do not only occur on mounds, but on plain terrains in larger depressions (Figs. 27 and 32) too. The presence of these latter indicates that plate-shaped dolines are not necessarily older than the dolines of the valley floors. Their plain floor, incomplete side slopes, and their coalescence with dolines in their environs prove that they are developed by horizontal dissolution (Veress 2020). According to their position, their morphology may be given as follows:

Plate-shaped dolines of the slopes of the mounds are situated near valley heads or in the valley head. In the latter case, the doline does not have a slope towards the direction of the valley or its side slope is incomplete.

In case of a plate-shaped doline at the margin of an uvala, the doline floor may be closed (there is a smaller depression on the floor, and there is a saddle at the coalescence) or open (there is no saddle at the coalescence), or there may be smaller dividing wall remnants (Fig. 32).

In case of a group pattern, adjacent dolines can be non-plate-shaped and plate-shaped dolines too. Uvala formation is significant in both cases. If the neighbours are also plate-shaped dolines, karst saddles, dividing walls, and karst hills occur among them and they constitute uvalas built up of three or four dolines (Figs. 36 and 37). From the saddles to the karst hills, dolines are separated by dividing walls. The dividing walls branch out from karst hills. Karst hills, which are the most elevated parts of the uvala, represent the surface preceding doline development. Karst hills are truncated cone-like features with a height of some metres, being elongated and having an almost flat top.

According to Leél-Őssy (1954), the collapse dolines of the Bükk Mountains are Udvarkő depression (Fig. 6, chapter "The Geomorphic Evolution of Karsts and Karstic Surfaces"), Füstös-kő depression, and the depression of Csipkés-kút ponor. Hevesi (2002) only considers Udvar-kő depression as a collapse doline. Collapse dolines are caved-in cavities of ponor entrances.

Fig. 36 Plate-shaped doline (Zsidó-rét). Legend: 1. Plain floor, 2. Inner floor depression (compaction dolines), 3. Karst hill, 4. Karst saddle

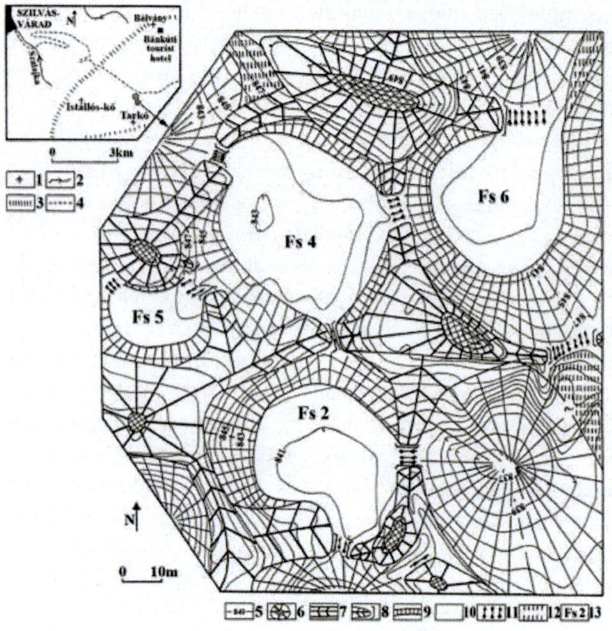

Fig. 37 Doline group of Zsidó-rét (Veress and Zentai 2009). Legend: 1. Mountain, 2. Stream, 3. Plateau margin, 4. Road, 5. Contour line, 6. Karst hill, 7. Dividing wall, 8. Half-dividing wall, 9. Doline side slope, 10. Doline floor, 11. Karst saddle, 12. Slope of karstic origin, 13. Identification mark of a plate-shaped doline

Complex depressions

Leél-Őssy (1954) distinguished 'lápa' and polje. 'Lápa' is originally a folk term which was used as a geographical common noun both in the Bükk Mountains and in the Aggtelek Karst mainly at valleys, arheic valley sections, and major karst depressions. (However, it is to be noted that the original folk term is not consistent. The term 'lápa' is not always used in the above sense.) According to Leél-Őssy (1954), lápák (the plural of 'lápa') are long (even longer than 6 km) valley-like depressions of tectonic origin in the Bükk Mountains. He does not differentiate 'lápa' from polje genetically. One or more uvalas may also occur in lápák (Figs. 29, 30 and 31) and they became arheic during karstification. Hevesi (2002) calls larger and complex depressions of the Bükk Mountains 'vápa', situated on the floor of epigenetic valleys, and he regards these features as valley-floor poljes. Among 'vápák' (the plural of 'vápa') or poljes, he distinguishes smaller and larger ones. However, according to the above author, entire areas of the Nagy and Kis Plateaus surrounded by mounds can also be regarded as poljes.

Leél-Őssy (1954) mentions two poljes from the Nagy Plateau from the Bükk Mountains: Zsidó-rét (with an area of 1.5 km^2) and Nagy-mező (with an area of 3 km^2). He considers both of them as of tectonic origin. We think that these two, undoubtedly arheic karst systems are not poljes because they do not show any polje characteristics such as steep, straight side slopes (in fact the side slope is not uniform and in some places, it is given by the slopes of depressions there) and if there is a cover on their floor, it is not alluvium. Their arheic character is due to the development of dolines and uvalas. Similarly, arheic valley sections cannot be considered as poljes either.

In our opinion, according to their morphology, at least three varieties of complex karst depressions can be distinguished. Arheic valley sections (1), which developed during the karstification of the floor of epigenetic valleys possibly in several phases. First, drawdown dolines, point recharge dolines,, and ponors developed. (Some sections of epigenetic valleys of various length may already have become arheic at this stage.) Then, the length of arheic sections increased partially by the coalescence of the above features and maybe by the local solutional deepening of valley-floor parts. (The valley section between the KF-12 and KF-14 VES measurement sites may represent such valley-floor dissolution because of its small depth which can be seen in Fig. 35.) Finally, arheic valley sections could have continued their deepening as a result of sediment transportation into the developing subsidence dolines (activating former ponors) and into their drainages. During this latter process, these valley sections could have turned into DSDs.

Non-valley depressions may be depressions with dolines and uvalas (2) and depressions with ponors (3). A depression with dolines (2) is Zsidó-rét (with plate-shaped dolines, hanging plate-shaped dolines and uvalas), and Nagy-mező (with plate-shaped doline, with hanging plate-shaped doline, with aggraded blind valley and subsidence dolines).

In depressions with ponors (3) there are ponors with blind valleys (Kaszásrét ponor and Örvénykő ponor on the Kis Plateau, Dorongós ponor in the Southeastern Bükk). Among them, the valley of Örvénykő depression is a terraced valley (Leél-Őssy 1954).

Subsidence dolines

Among subsidence dolines, there are suffosion dolines, dropout dolines, and compaction dolines as well. The subsidence dolines of the Bükk Mountains have not been studied in detail. The denomination of these features is not unambiguous thus, for example. They are mentioned in literature as 'breakings' (Lénárt 1977). Their genetic relationship with other features (row dolines) has not been studied much either. Pelikán (1992) distinguishes real dolines and pseudo-dolines. The former are associated with limestone, while the latter with superficial deposit (their occurrence is not mentioned) since the superficial deposit is washed into the cracks of the limestone. Thus, pseudo-dolines can be regarded as suffosion dolines.

Another doline types can be found in the stream loads of slopes in the valleys near Nagy-Kőmázsa and Kis-Kőmázsa (Hevesi and Záhorszky 2000), and they open up occasionally. Such openings are also mentioned from saddles between row dolines here (Hevesi and Záhorszky 2000; Záhorszky 2001). Záhorszky's (2001) map marks subsidence dolines too (Fig. 31) since some depressions are small and they occur in groups (not in rows). In the mountains, there are no subsidence dolines in the majority of depressions, similarly to the Aggtelek Karst, because of fine-grained doline fills. In the fill of some dolines, the material with a grain diameter smaller than 0.05 makes up 90% of the fill (Jámbor 1959).

Subsidence dolines can also develop in partially filled depressions thus, in drawdown dolines (which can be valley-floor doline or doline of flat terrain), in blind valleys, in former valley-floor ponor, or on valley floor having a stream. If the valley-floor doline is a former ponor, the clearing of its chimney may result in the development of subsidence doline in the fill. Hevesi (1980) describes a doline near Sólyom-kő on the Kis Plateau which was formed in doline fill. The activation of former ponors is enabled by the fact that the degree of coveredness as well as the thickness and composition of the cover can be different in the environment (catchment area) of valley floor dolines (Hevesi 1980). If the cover of the environment ensures a larger amount of water supply, the clearing of the chimney may trigger suffosion or collapse in the superficial deposit. In this way, former ponors are transformed into valley-floor dolines and then into subsidence dolines. A subsidence doline developed on the floor of one of the valleys of the Kis Plateau too (Veress and Zentai 2009). Here, the valley floor was aggraded to a significant degree. There is only a partial similarity between the morphologies of the bedrock dissected by karstification (which causes the arheic character of the valley floor section) and of the depressions of the cover. It may occur that there is a depression both on the cover and on the bedrock, but there may be a depression only on the bedrock or only on the cover (Fig. 35). At this latter site (between the VES observation sites of KF-11 and KF-13), a subsidence doline (marked F-3) originated by compaction.

One of the dropout dolines of Berva Valley (Southwestern Bükk) developed on active valley floor in 2009 (Kozma and Holló 2010) which drained off the water of the stream. Its development was primarily caused by the fact that the karstwater level subsided below the fill (clay, gravel) of the valley floor because of water drainage which resulted in the collapse of the cover (Kozma and Holló 2010). Other young dropout dolines are also reported from the environs (Kozma and Holló 2010).

A specific, well-studied subsidence doline group is located in the former, partially alluviated blind valley of the Nagy-mező (Veress and Zentai 2009). Here, the group of the dolines marked N-3, N-6, N-4, N-1, and N-3 developed in the floor fill of a hanging drawdown doline (Fig. 27). The doline marked N-2 was formed on the dividing wall between the two parts of the depression (ponor?) of the former blind valley (Fig. 38), while the doline marked N-9 developed in a

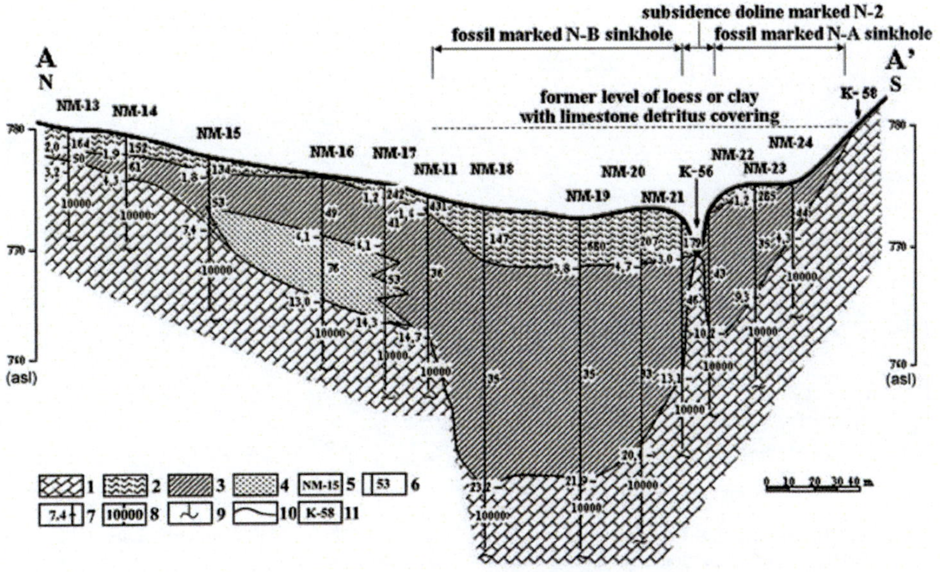

Fig. 38 Geoelectric-geological profile 'A-A' of the infilled blind valley of the Nagy-mező (Bükk Mountains, Hungary, see Fig. 27, Veress 2016) 1. Limestone, 2. (Clayey) limestone debris, 3. (Clayey-silty) loess or clay with limestone debris, 4. Loess (with sand or limestone debris), 5. Identification code of VES measurement, 6. Geoelectric resistivity of the series (Ohmm), 7. Base depth of the geoelectric series (m), 8. Geoelectric resistivity of bedrock (Ohmm), 9. Approximate depth of penetration of VES measurement, 10. Boundary of geoelectric series, 11. Number of limestone outcrop

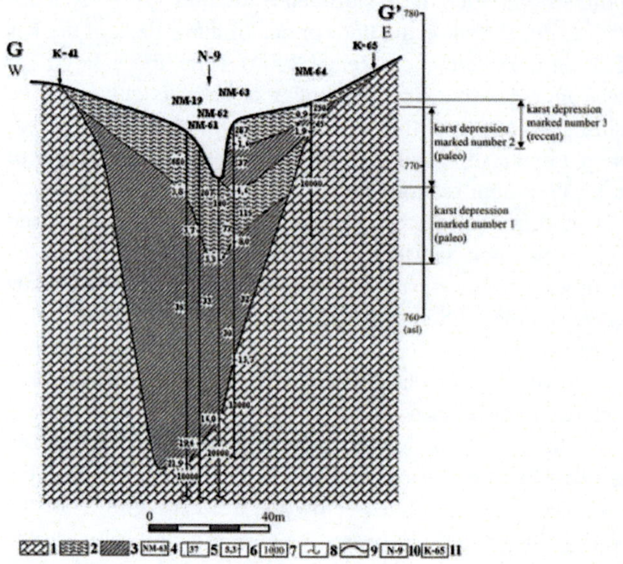

Fig. 39 The geoelectric-geological profile of the subsidence doline marked N-9 of Nagy-mező (Fig. 27) (Veress and Zentai 2009). Legend: 1. Limestone, 2. Limestone debris (clayey), 3. Loess (clayey-muddy), or clay with limestone debris, 4. Number of VES measurement, 5. Geoelectric resistivity of the series (Ohmm), 6. Base depth of the geoelectric series (m), 7. Geoelectric resistivity of bedrock (Ohmm), 8. Approximate depth of penetration of VES measurement, 9. Boundary of geoelectric series, 10. Identification code of karst depression, 11. Number of limestone outcrop

part depression of the depression (Fig. 39). At this latter, there are two or three filled former depressions in the fill below the present depression. Older depressions refer to multiple doline development and indicate that the present doline is postgenetic.

Examples for subsidence dolines that developed in dolines on flat terrain are the depressions of Mohos doline (Zsidó-rét).

Ponors

The ponors of the Bükk Mountains primarily occur on the Nagy Plateau and in the Southeastern Bükk. Ponors can be outer and inner ponors (Leél-Ősssy 1959a). The outer (-marginal) ponors of the Nagy Plateau are aligned western and eastern rows. (These sites bear the mixed allogenic-autogenic karst areas of Nagy Plateau.) The two rows of ponors with blind valleys developed at the margin of two porphyry zones being wedged in into the limestone (Fig. 40). Members of the western ponor row are the ponors of Bánkút and Csipkéskút, while the eastern row includes the ponors of Jávorkút, Bolhás, Disznóspatak, Létrás I, and Létrás II. The Örvény-kő ponor of the Kis Plateau is also a karst marginal ponor (Leél-Őssy 1959a).

Karst marginal ponors also occur in the Southeastern Bükk. Here, the ponor row with blind valleys developed at the boundary of shaly beds and limestone (Hevesi 1986a,

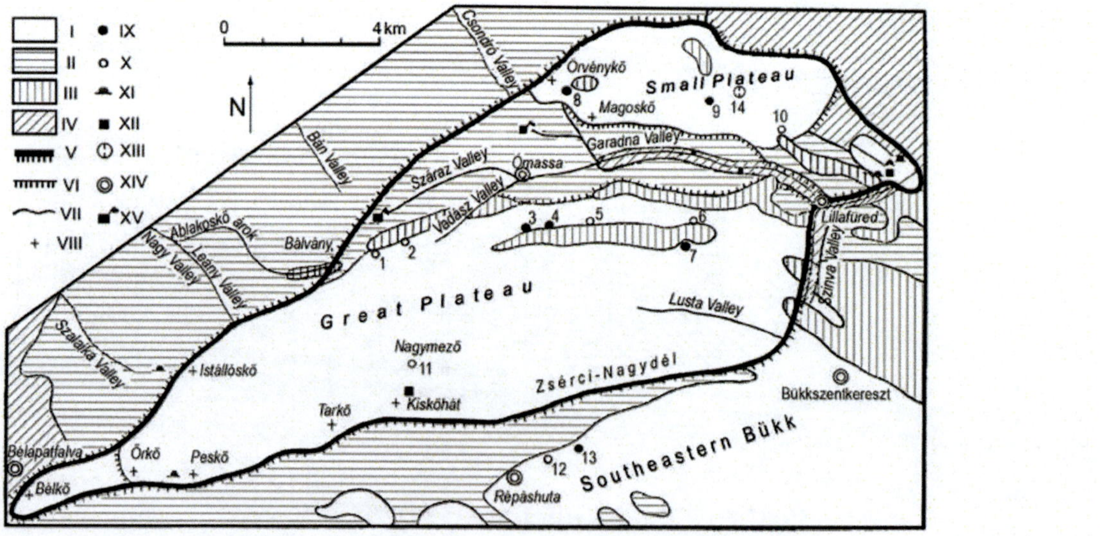

Fig. 40 Geomorphological map outline of Bükk Plateau (Leél-Őssy 1959a, b, c). Legend: I. Middle Triassic limestone, II. Carboniferous Permian–Triassic shale, III. Porphyyrite stripes, IV. Tertiary-Quaternary cover layers, V. Boundary of High-Bükk, VI. Boundary of Kis Plateau and Nagy Plateau, VII. Stream and its valley, VIII. Mound, IX. Open ponor with cave, X. Choked ponor, XI. Spring cave, XII. Shaft, XIII. Collapse doline, XIV. Village, XV. Tourist hostel, 1. Bánkút ponor, 2. Csipkéskút ponor, 3. Jávorkút ponor, 4. Bolhás ponor, 5. Disznóspatak ponor, 6. Létrás I ponor, 7. Létrás II ponor, 8. Örvénykő ponor, 9. Csókás Cave, 10. Kaszásrét ponor, 11. Nagymező ponor, 12. Dióspatak ponor, 13. Pénzpatak ponor, 14. Udvarkő collapse doline

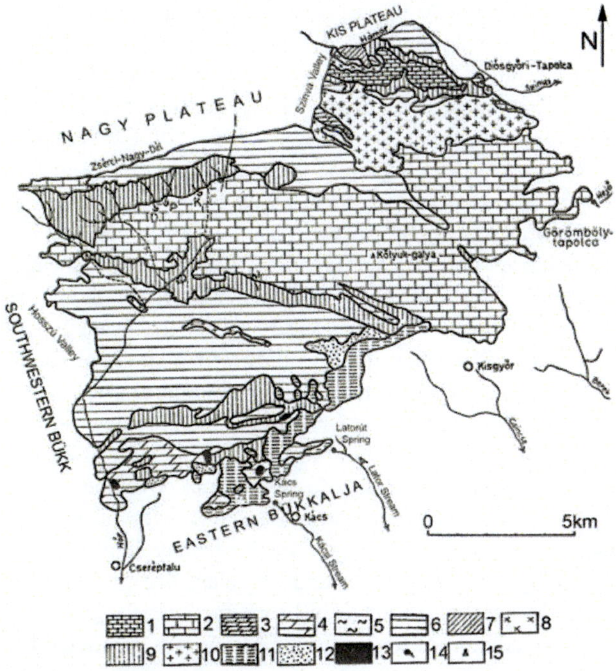

Fig. 41 The ponors of Southeastern Bükk (Hevesi 1986a). Legend: 1. Anisian white limestone; 2. Ladinian 'Répáshuta limestone'; 3. Ladinian 'plateau limestone'; 4. Karnian Berva limestone; 5. Anisian grey dolomite; 6. Ladinian cherty grey limestone; 7. Lower Triassic sandstone, shale; 8. Anisian porphyrite, diabase, and tuffs; 9. Ladinian shale and cherty limestone; 10. Ladinian diabase, porphyrite, quartz porphyry; 11. Upper Eocene limestone calcareous marl; 12. Eocene continental variegated clay, sand, gravel; 13. Middle-Miocene rhyolite tuff; 14. Karst spring; 15. Mound, D. Diós-kút ponor, I. Ilona-kút ponor, K. ponor of Kajla Stream, P. ponor of Pénz Stream

Fig. 41) and include the ponors of Bella Valley, Diós Stream, Pénz Stream, Kajla Stream, Dél-sár, Hollós-tető, Rókafarm,, and Ilona-kút. There are several intermittent water receiving chimneys aligned on the valley floor below the ponor of Diós Stream (Hevesi 1986a).

Ponors inside the karst can occur on the floor of epigenetic valleys and in major karst depressions. However, Hevesi (2002) also mentions a so-called channel ponor where the water of the stream seeps away in the channel of limestone. Seepage sites can be found, for example, in the Southeastern Bükk, in Pazsag Valley, in Balla Valley, and in Szarvaskúti-Csúnya Valley. Csókás ponor is a valley floor ponor near Sólyom-kút where ponors are situated on the floor of valleys incised into the clay nappe enclosed by Andó-bikk, Galya-tető, and Hársas-tető (Hevesi 1980). The ponor of Gyertyán Valley is also a valley-floor ponor in the Southeastern Bükk. The latter is a katavothron since it also functions as an intermittent spring (Hevesi 1986a). Nagy-Kőmázsa ponor, Csengős ponor, and Osztrigás ponor are also valley-floor ponors situated near Tapolca in Southeastern Bükk (Ferenczy 2002). The ponor with blind valleys in the valley head of Nyerges-kő lápa can be mentioned from the Southwestern Bükk (Hevesi 1986b). On the floor of epigenetic valleys, a number of dolines which function as ponors intermittently since they get water from the surface with superficial deposit of their environment. However, the waters of arheic terrains (Nagy-mező) are also conducted into the karst through subsidence dolines.

The ponor of Kaszás-rét (Kis Plateauis located in a karst depression fed by an about 400-m-long stream. According to Hevesi (2002), Dorongós (Répáshuta limestone ridge) in the Southeastern Bükk is a former valley which was transformed into a closed depression following its inheritance and ponor development. The depression bearing the ponor is 1050 m long.

Ponors can be fed by permanent streams (the ponors of Létrás, Disznós Stream, Bolhás, Csipkés-kút, Kaszás-rét, and Pénz Stream) or by intermittent streams (the ponors of Bánkút, Csókás, and Örvényes).

The shaft caves of the Bükk Mountains (for example, the shafts of Kis-kőhát, Nagy-Hárs, Lyukas ridge, etc.) owe their origin to the truncation of former ponors (Tóth 1983; Hevesi 2002).

Gorges

Karst gorges primarily occur in the Southern Bükk, particularly in the Southwestern Bükk (Fig. 42). In the Bükk Mountains, shorter and longer sections of epigenetic valleys constitute gorges. Valleys may have only one or several gorges (Lök Valley). According to Hevesi (2002), they may be longer (some 100 m or 1–2 km) or short (some 10 m) which he calls gorge gate. Gorges developed at sites where allogenic karst is transformed into mixed allogenic-autogenic karst during its exposition. Such conditions were present in the Southern Bükk where the stream valleys were inherited over to limestone zones enclosed by non-karstic rock.

Hevesi (1978, 2019) claims that former caves of gorges could have developed by one time or multiple beheading. In the former case, the cave or caves open up in one level, while in the latter, they open up in several levels (Fig. 43). The Hámori Gorge of the Szinva Stream (Puskaporos Gorge), the gorges of Lök Valley (Kő-köz, Mészvölgy Gorge), the lower section of Forrás Valley, and the Esztázkő Gorge in Gyetra Valley developed by the opening of only one trough cave. Tatár Valley in Southeastern Bükk developed by the opening of several trough caves.Several gorges were formed at the plateau margins (margin gorge): the gorges of Ablakos-kő Valley, Vörös-kő Valley, Savós Valley, Tekenes Valley, and Sebes Valley on Nagy Plateau, and the gorges of Csondró Valley, Háromkürtő Valley, Szentlélek Valley, and Felső-Forrás Valley on Kis Plateau. Leél-

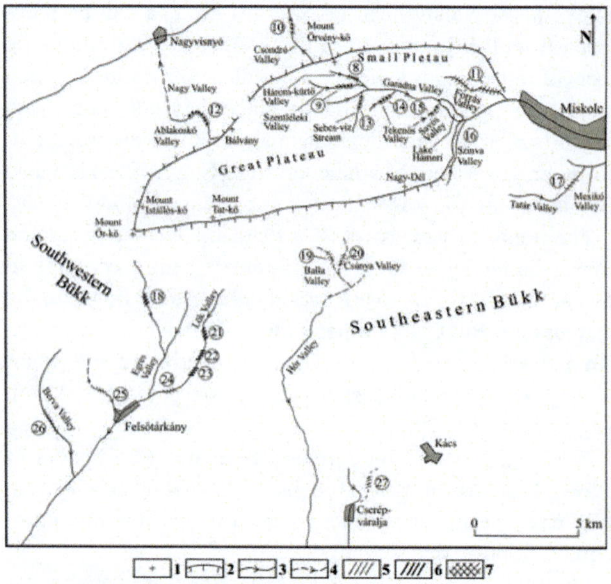

Fig. 42 The gorges of the Bükk Region. Legend: 1. Mound; 2. Plateau margin; 3. Permanent stream; 4. Intermittent stream; 5. Gorge that developed by cave opening; 6. Gorge gate; 7. Plateau marginal gorge; 8. Háromkürtő Gorge; 9. Szentlélek Gorge; 10. Csondró Gorge; 11. Gorges of Forrás Valley; 12. Ablakos-kő Valley; 13. Gorge of Sebes Stream; 14. Tekenes Valley; 15. Savós Valley 16. Puskaporos Gorge; 17. Gorge of Tatár Valley, 18. Esztáz-kő Gorge; 19. Gorge of Balla Valley; 20. Gorge of Csúnya Valley; 21. Vaskapu; 22. Bujdosó-kő; 23. Macska-kő; 24. Kő-köz; 25. Mészvölgy Gorge; 26. Gorge of Berva Valley; 27. Gorge of Felső-Szoros

Őssy (1954) states that plateau margin gorges are of regression-epigenetic origin.

Concretions

In the Bükk Region, freshwater limestone occurs at several places. Hevesi (1972) identified 26 sites and Scheuer (1975) 35 occurrence sites (Fig. 44). The travertine hill of Egerszalók originated in 1961 from the concretion of the thermal water of 65.5 °C gushing out during the drilling of a hydrocarbon well, its volume reached 2700–3000 m^3 by 2011 (Benkhard 2015).

According to measurements (Hevesi 1972), the rate of calcareous sinter development in the mountains reaches 2–3 mm/year (Sebesvíz) and 4 mm/year (Felső-Forrás Valley), but in Szalajka Valley this value may be 5 mm/year too. In the case of moss occurrence, this rate may also be 20 mm/year (Szalajka Valley). At some sites, calcareous sinter already develops at emergence sites (spring), while at other places it appears some 100 m or some 1000 m(Szalajka Valley) away from the spring. Its deposition is often interrupted. The site of the deposition is affected by the gradient and morphology of the valley (the valley will have a wide floor). The erosion of the stream decreases at these sites.

Hevesi (1972) distinguished freshwater limestone occurrences at the mountain front and inside the mountains. Those of the mountain front are either lukewarm water calcareous sinters (the springs of Eger, Diósgyőr, and Latorút) or cold-water calcareous sinters (the springs of Bélapátfalva and Mályinka). The largest is the occurrence of Monosbél with an estimated area of 5 km^2. It is probably a travertine mound with basins and rimstone bars.

The most significant occurrences inside the mountains are at Dobrica-kút, Pacsirta Spring, Sebesvíz, Lillafüred, and in Szalajka Valley (Hevesi 1972). The occurrence of Dobrica-kút is in valley-floor position with a length of about 30 m and with a basin and dam development. The occurrence of Pacsirta Spring is in slope position, conical on the top, and stepped at the bottom. In the valley of Sebesvíz calcareous sinter of valley-floor position developed in a length of about 700 m. About 60 dams with basins are superimposed on the mound. The freshwater limestone of Lillafüred is a travertine mound of valley-floor position at the confluence of the Szinva and Garadna Streams (Fig. 45). The accumulation of freshwater limestone was promoted by the different elevations of the channels of the two streams.

The calcareous sinter of Szalajka Valley is in valley-floor position and its deposition begins about 500 m away from the spring. It consists of a travertine hill (travertine mound) with a length of 245 m and closer to the spring and a concretion section below on the valley floor dissected by rimstone bars. The most spectacular freshwater limestone dam system of Hungary developed on this travertine hill with 16 dams and basins. The dams are of various height (the highest is 205 cm), their width is 4–5 m. In addition to the active, developing dams of the channel there are also dry, abandoned dams nearby as a result of the shift of the stream.

Relict landforms

The relict karst landforms of the mountains are the above-described mounds, karst hills, dividing walls between dolines, and opened dolines. Further relict landforms are the doline rows and 'lápák' that opened by retrogressive valleys. Leél-Őssy (1954) states that an opened doline row is, for example, the upper section of the Vörös-kő Valley. A variety of mounds that has not been mentioned so far is a smaller cone-shaped mound group (with a height of 5–10 m) described by Hevesi (2002). These features occur in large number and density in the area of the Répáshuta limestone ridge which are the remnants of Middle-Miocene inselberg karst according to Hevesi (2002).

Further relict landforms can be mentioned from plateau margin mounds, but these features can also develop at destroying caves. During their destruction, marginal mounds, particularly the 'Rocks' are separated into ribs between pits and into towers between pits (Hevesi 1978, 2002). Frost weathering, mass movements (collapses), and karstification play a role in the process. Here, karren

Fig. 43 Development of limestone gorge in case of multiple beheading from through cave (Hevesi 2019)

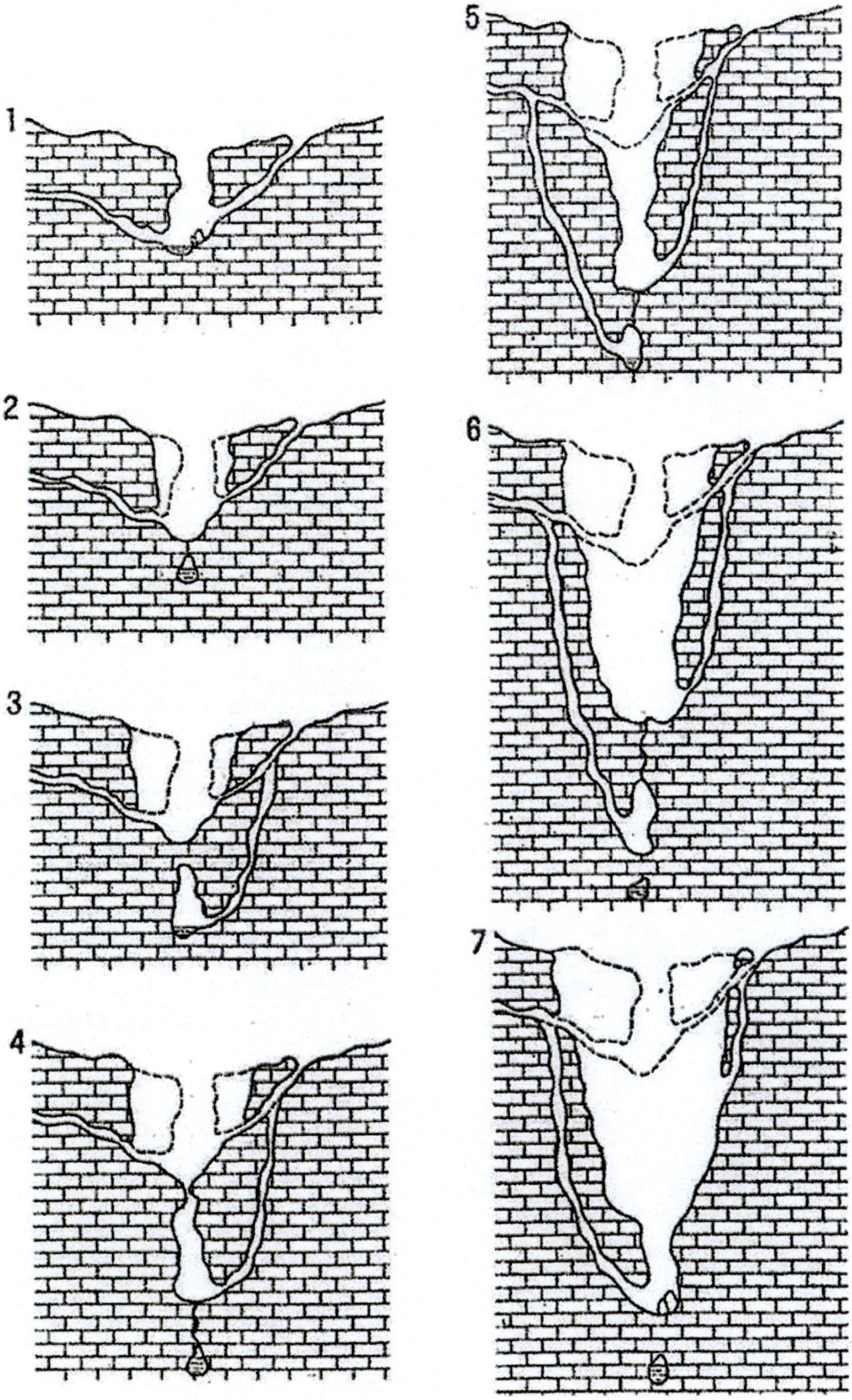

formation also contributes to the denudation of slopes. Slope denudation causes the exposure and separation of karst cavities (former ponor pits). The opening karstic chimneys are transformed into rock corridors, block stream and alluvial cones develop at their termination which unite into debris slopes (Hevesi 2002).

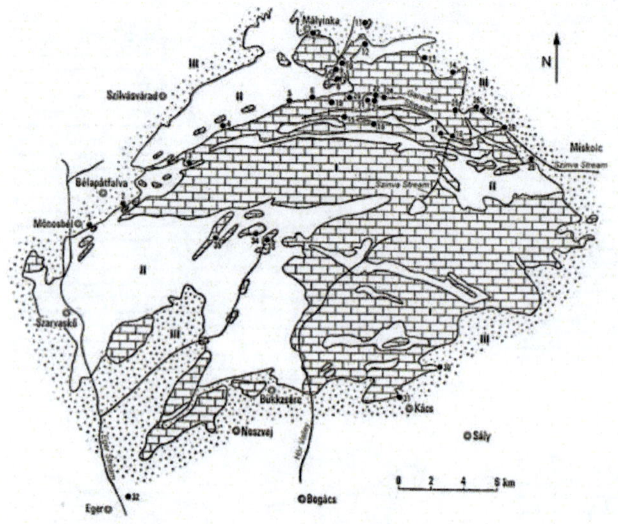

Fig. 44 Freshwater limestones in the Bükk Region (Scheuer 1975). Legend: 1. Mónosbél; 2. Mária Spring and Salátás Spring of Bélapátfalva; 3. Szalajka; 4. Leány Spring; 5. Bán Spring; 6. Springs of Moldva Valley; 7. Mályinka; 8–9. Springs of Kerekhegy Valley and Kemesnye Valley (Csondró Valley); 10. Spring of Nagyszállás Valley; 11. Hársas Spring at Tardonána; 12. Harica Springs; 13. Dobrica-kút; 14. Galya Spring; 15. Spring of Garadna; 16. Sebesvíz; 17. Eszperantó Spring; 18. Lillafüred; 19. Meteor Spring; 20. Springs of Csikorgó Valley; 21. Pacsirta Spring; 22. Lower spring of Szentlélek Valley; 23. Köpüs-kút; 24. Springs of Háromkút Valley; 25. Felső Spring; 26. Flóra Spring; 27. Középső Spring; 28. Puskaporos pass of Szinva Valley; 29. Springs of Diósgyőr; 30. Sály Spring; 31. Springs of Kács; 32. Eger; 33. Fekete-len Spring; 34. Imó; 35. Vöröskő Spring, I. Karst rocks, II. Impermeable and poorly permeable Paleozoic and Mesozoic rocks, III. Tertiary sediments

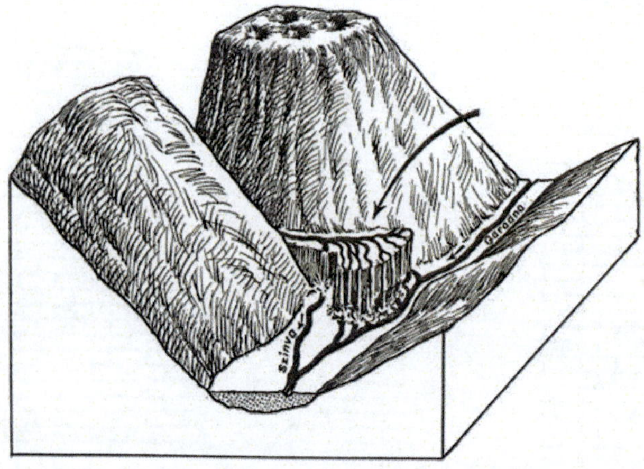

Fig. 45 Travertine mound of Lillafüred (marked with arrow, Jakucs 1977b)

Near-surface caves (spring caves or inflow caves) constitute various landforms during their denudation. Gorges or sections with gorges are formed from completely denuded through caves, while rock corridors develop from other, shorter, non-trough caves and cavities. Pits, chimneys (short caves of some metres), and arches survive from partially destroyed caves and between arches, windows develop (Tóth 1983; Hevesi 1986c, 2002). Such remnants (relict caves) are on Körös-bérc, Magas-kő, Tamás-kő, Tar-kő, etc. (Hevesi 1986c).

4 The Bakony Region

The karst types of the Bakony Region can be identified on the basis of Pécsi's (1980, 1991) block classification. The block types are as follows (see chapter "The Geomorphic Evolution of Karsts and Karstic Surfaces"): cryptopeneplain, low threshold surface, horst elevated to summit position, and horst of summit position. In the area of cryptopeneplain, buried karst is typical, in the area of the horst elevated to summit position, there is concealed karst and soil-covered karst. On horsts of threshold type and on blocks of horsts in summit position, soil-covered karst may occur, but at sites where the patches of permeable superficial deposit survived, concealed karsts may also occur. Concealed karst patches also occur on cryptopeneplains where the impermeable cover was destroyed, but the loess cover survived (Fig. 46). The coveredness and thus the karstification of the blocks belonging to the above-mentioned block types will be different (see chapter "An Interpretation of Karstification and Karst Processes by Hungarian Researchers") and, therefore, the Bakony Region and other parts of the Transdanubian Mountains are also horst-type karsts (Veress 2000b, 2016).

4.1 General Characteristic Features

The general properties of its karstification were studied by Láng (1948, 1958, 1962), Leél-Őssy (1959b), Jakucs (1977b), Hevesi (1991a, b), Veress (2000b, 2016), Veress and Vetési-Foith (2019a, b). Surface karst was investigated and described by Gergely (1938), Révész (1947), Láng (1948, 1958, 1962), Veress (1982, 2000b, 2009), Hevesi (1991b), Szabó (2000), Veress and Szabó (2000), Rybár (2009), Veress and Tóth (2015), Veress and Unger (2015) and Veress et al. (2015) Veress and Vetési-Foith (2019a, b). Jaskó (1961), Bratán et al. (1967), and Schmidt Elegius et al. (1962) investigated the hydrology of the karst of the mountains.

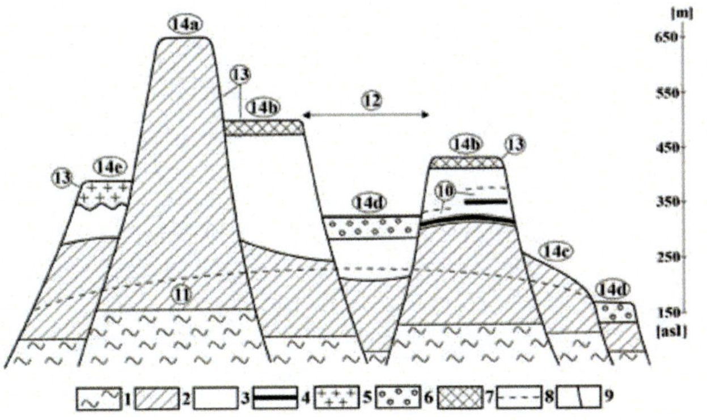

Fig. 46 Block types in the Transdanubian Mountains (Veress and Vetési-Foith 2019a). Legend: 1. Marl; 2. Main dolomite; 3. Dachstein, Jurassic, Cretaceous, and Eocene limestones; 4. Impermeable intercalation, clay, marly limestone, silica etc.; 5. Basalt; 6. Gravel; 7. Loess; 8. Karstwater table; 9. Fault; 10. Karst water storey; 11. Main karst water; 12. Basin, graben; 13. Mount; 14a. Horst in summit position; 14b. Horst elevated to summit position; 14c. Threshold surface; 14d. Cryptopeneplain; 14e. Block covered with basalt

The Bakony Region is the largest partially karstic meso-region of Hungary. It is the most specific representative of the karst of Transdanubian type (Jakucs 1977b; Hevesi 1991a, b). The general characteristics of its karst are as follows:

- Its karst areas are relatively far from each other separated by non-karstic and buried karst areas.
- Its karstification has been in part interdependent from fluvial geomorphic evolution.
- Its karst types are diverse: soil-covered karst, concealed karst, cryptokarst, mixed allogenic-autogenic karst, allogenic karst, and buried karst are found next to each other (Veress and Vetési 2019a, b). Mixed allogenic-autogenic karst is on Kab Mountain (at the termination of the basalt cap with ponors with blind valleys), cryptokarst is found here and on Fekete Mountain (with caprock dolines) too, and buried karst is in the basins of the mountains and on horsts with a carbonate bedrock covered with basalt.
- Paleokarstification can be detected at several sites (Vadász 1946, 1951, Szabó 1956, Bulla 1968, Bárdossy et al. 1983, Csillag and Sebe 2015, see chapter "Paleokarst in Hungary").
- The denudation levels of horsts of various age become karstified, including the Mesozoic peneplain, older pediments (developed between the Middle Oligocene and Middle Miocene), and younger (Pliocene, maybe Pleistocene) pediments.
- In addition to cold-water karstification, subordinate warm-water cavity formation also occurs at the southern margin (Leél-Őssy 1953b, 1958a; Láng 1958; Plózer 1977).
- Their adjacent blocks are of different evolution, and as a result of their various exposure, they have different geomorphology (Pécsi 1980) and karstification (Veress 2000b, 2016) as well. Such karst is karst with horsts.
- As a result of near-surface cavity formation (due to close-to-surface impermeable beds), cavity destruction is significant which is also manifested in landscape evolution (Veress 2000b).

4.2 Karst Features

The karst features include karren (mainly subsoil karren), various dolines like subsidence dolines (mainly suffosion dolines), caprock dolines, solution (drawdown) dolines, collapse dolines (or dolines of such character) depressions of superficial deposit, ponors, gorges, concretions, and karstic relict landforms. (Katavothra as a unique feature are mentioned from the Tapolca Karst.) On dolomite terrains, particularly at lower elevations a several-metre-thick rubble bed developed. Since rubble beds are created by dissolution (Jakucs 1977a, b), in our opinion it was formed during karstification (see chapter "An Interpretation of Karstification and Karst Processes by Hungarian Researchers"). The karst terrains of the mountains are described in Figs. 47 and 48.

Karren

Disregarding some exceptions, the karren of the Bakony Region are of subsoil type. Both subsoil karren and bare karren are small sized. Bare karren features those developed on dolomite, but mainly on sandstone and on calcareous sinter. Bare karren on limestone are mentioned by Láng (1958) from Középső-Hajag (bed head karren) and from Som Hill (karren slope) by Bertalan (1963).

Subsoil karren are rootkarren (they occur sporadically on exposed rock blocks) and grikes. Exposed grikes occur near Dörgicse (Balaton Uplands) and Likaskő of Hódos Stream (Szent László-erdő Hill). Other karren features are also recognizable near Dörgicse (Veress and Vetési-Foith 2019a, b), thus lattice-like karren, kamenitzas, and pitkarren. Rinnenkarren can only be found at a single site of the mountains, probably developed on bare surface.

The side slopes of the thermal spring cones of the Tihany Peninsula are exempt from vegetation at several places. On these surface small-sized karren are visible in an expansion of some m^2. The Aranyház thermal spring cone has the richest and most diverse features (pitkarren, thimble karren, rillenkarren, and grikes).

The boulders of the Kál Basin are much richer in karren (Veress and Tóth 2015), primarily the block field of Szentbékkálla, but that of Salföld too (most of the block field of

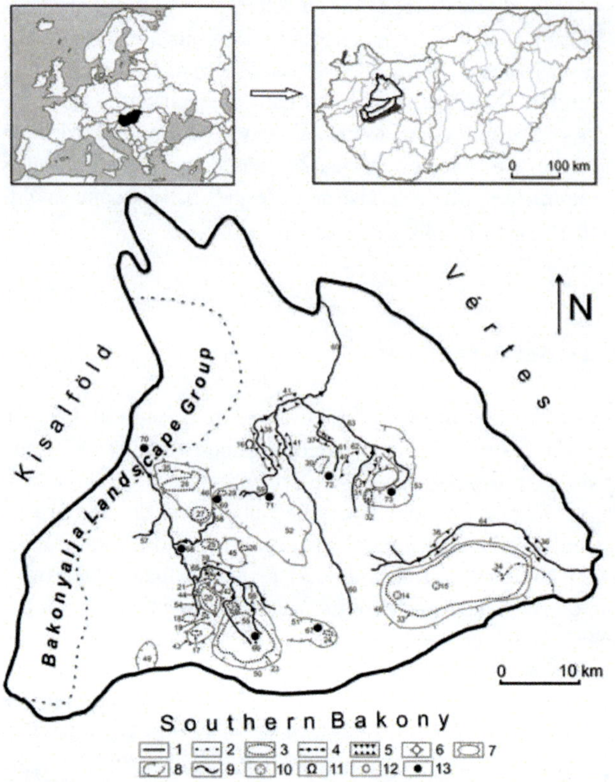

Fig. 47 The karstic features of the northern part of the Bakony Region (Veress and Vetési-Foith 2019a). Legend:1. Boundary of the Northern Bakony; 2. Boundary of micro-region group; 3. Concealed karst; 4. Dry valley; 5. Gorge; 6. Block roof; 7. Plateau; 8. Basin; 9. Stream and its valley; 10. Karst depression; 11. Cave opening; 12. Spring; 13. Settlement; 14. Depression of Csengő shaft (marked I–51); 15. Depression of Háromkürtő shaft (marked I–12); 16. Cave of Likas-kő of Hódos Stream. Concealed karst areas: 17. On the Középső-Hajag; 18. On the Felső-Hajag; 19. Near Fehér-kő Valley; 20. On the Mester-Hajag; 21. Between the Iharos block and Judit Spring; 22. On the Égett block; 23. In the Hárskút Basin; 24. In the Lókút Basin; 25. Between the Som Mountain and Száraz-Gerence Stream; 26. Doline group at Ősbükkös; 27. Doline group by Eleven-förtés; 28. By the Márvány Valley; 29. Near Újszépalmapuszta; 30. Between the Cuha Stream and Gézaházapuszta; 31. On the Sűrű Mountain; 32. Near Dudar; 33. On the Tés Plateau; 34. Tábla Valley; 35. Márvány Valley; 36. Gaja Gorge; 37. Kőmosó Gorge; 38. Gorge of Hódos Stream; 39. Kerteskő Gorge; 40. Gorges of Ördög Valley; 41. Cuha Gorges; 42. Iharos block; 43. Középső-Hajag block; 44. Mester-Hajag block; 45. Som block; 46. Kőris block; 47. Magos block; 48. Tés Plateau; 49. Csehbánya Basin; 50. Hárskút Basin; 51. Lókút Basin; 52. Porva Basin; 53. Dudar Basin; 54. Stream of Fehér-kő Valley; 55. Öreg Stream; 56. Gerence Stream; 57. Vörös János Stream; 58. Száraz-Gerence Stream; 59. Hódos Stream; 60. Cuha Stream; 61. Stream of Kő Valley; 62. Stream of Ördög Valley; 63. Dudar Stream; 64. Gaja Stream; 65. Judit Spring; 66. Hárskút; 67. Lókút; 68. Bakonybél; 69. Újszépalmapuszta; 70. Bakonykoppány; 71. Porva; 72. Gézaházapuszta; 73. Dudar

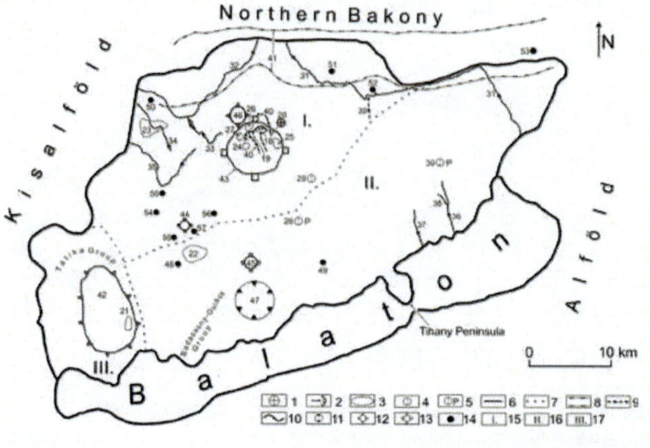

Fig. 48 The karstic features of the southern part of the Bakony Region (Veress and Vetési-Foith 2019a modified). 1. Ponor; 2. Blind valley; 3. Concealed karst; 4. Complex karst depression; 5. Paleokarst depression; 6. Boundary of the southern part of the Bakony Region; 7. Boundary of micro-region group; 8. Tectonic graben; 9. Dry valley, 10. Stream and its valley; 11. Basin; 12. Plateau, block roof; 13. Block with basalt cap; 14. Settlement; 15. Southern Bakony micro-region group; 16. Balaton Uplands micro-region group; 17. Keszthely Mountains micro-region group; 18. Öreg-köves ponor; 19. Macska-lik ponor; 20. Tönkölyös ponor. Concealed karst areas: 21. On the Balatonederics Plateau; 22. On the Tapolca Karst; 23. Nearby Devecser; 24–26. On the Kab Mountain; 27. Depression system marked M-2; 28. Vörös-tó depression; 29. Köves-gyűr depression; 30. Meggyes-puszta depression; 31. Séd Stream; 32. Torna Stream; 33. Padrag Stream; 34. Egres Stream; 35. Kígyós Stream; 36. Stream of Malom Valley; 37. Koloska Stream; 38. Királykút Stream; 39. Tekeres Valley; 40. Dry valleys on Kab Mountain; 41. Veszprém-Devecser Graben; 42. Keszthely Mountains; 43. Kab Mountain; 44. Agár block; 45. Fekete Mountain; 46. Bocskor Mountain; 47. Kál Basin; 48. Tapolca; 49. Dörgicse; 50. Devecser; 51. Márkó; 52. Veszprém; 53. Öskü; 54. Ódörögd; 55. Nyirád; 56. Sáska; 57. Diszel; 58. Zalahaláp

Kővágóörs has been destroyed by now). The largest and densest karren features of Hungary that developed on bare surfaces can be found on the boulders of these block fields of sandstone.

The most common features of the rocks of the Szentbékkálla block field are kamenitzas with various size and morphology (their diameter can be several metres and they are often dissected by floor kamenitzas) and on the side slopes of blocks or on their slightly inclined upper surfaces there are short rinnenkarren (morphologically they are more like rillenkarren, but regarding their size, they are similar to rinnenkarren, Fig. 49). Notches of various size (their height and depth can be several metres) are common and widespread. Pits, stone polygons, and grikes also occur (Veress 2003a). On the Salföld block field primarily kamenitzas and pits are characteristic. Relict landforms (karren buttes with a height of some tens of centimetres) also occur on the Szentbékkálla and Kővágóörs block fields produced by surface solution on sandstone blocks. The development of karren features is enabled by the dissolution of amorphous silica (Veress 2003a; Veress and Tóth 2015).

Mostly small-sized and diverse bare karren features or those below soil patches are on the dolomite surfaces near Veszprém. Karren primarily occur on blocks of various size

Fig. 49 Sandstone karren on one of the blocks of the Szentbékkálla block field. Legend: 1. Sandstone polygon, 2. Kamenitza, 3. Rinnenkarren

and in patches of valley sides. Veress (2003b) distinguished kamenitzas, grikes, thimble karren, notches, rinnenkarren, and pits. Particularly the morphology of kamenitzas is diverse. Some kamenitzas and thimble karren are still developing below vegetation and soil patches. Szabó (2000) described another karren feature, and thus micro-cracks, splitkarren, and cave-like passages. He classified dolomite surfaces according to the maturity of their karren and distinguished karren terrain in embryonal state, immature karren slope, and surface of mature karren. Surfaces of embryonal karren are characterized by a network of micro-cracks, small-sized kamenitzas and small thimble karren. (In the area of a mine yard near Litér.) The immature karren slope is characterized by thimble karren and grikes of various size and shape. On slopes of mature karren, notches, thimble karren, and bag karren, splitkarren and grikes occur at different elevations.

Rybar (2009) studied the karren features (these are mainly Schichtfugenkarren) of the bare slopes of Tekeres Valley developed in dolomite and then he classified the shape and evolution of dolomite slopes. In case of horizontal bedding, Schichtfugenkarren of horizontal position develop along bedding planes. Schichtfugenkarren are younger and younger downwards because of soil and debris denudation. With the collapse of bed parts situated above upper Schichtfugenkarren, stepped slopes (scarp front is vertical) are formed. At thicker beds, the collapse is stopped and an overhanging slope is created by soil subsidence (since dissolution is the most intensive at soil level). If the beds are inclined opposite the valley slope, counterslope Schichtfugenkarren occur. Trock only collapses above the initial section of Schichtfugenkarren (where it is thin enough) and then the process is interrupted. In this case, the scarp fronts are inclined towards the axis of the valley. The developing slope is covered by the accumulating debris (on which soil forms) since scarp floors dipping opposite the valley slope act as traps.

Subsidence dolines

The predominant karst type of the mountains is concealed karst. Concealed karsts develop at sites where the superficial deposit is partly or completely permeable, it is mostly loess and its varieties and not too thick (Veress 2000b, 2006; Veress and Vetési-Foith 2019a, b). As a result of the patchy occurrence of superficial deposits, the concealed karst is of discontinuous development, separated into patches of various expansion.

The distribution of concealed karst patches is described in Figs. 47 and 48. The doline number of the patches and the block type on which they occur are summarized in Table 8. There are about 700 subsidence dolines on 22 concealed karst patches of the mountains (Probably, their real number is higher since there are several concealed karst patches which have not been studied properly). In various manuscripts, subsidence dolines are mentioned, for example, from the environs of a Vörös János-Séd (Northern Bakony), and of Szentgál. (Southern Bakony) and from the area of the saddle between Tűzköves Hill and Üsti Hill.)

Concealed karst patches primarily occur on horsts elevated to summit position (Veress 2000b; Veress and Vetési-Foith 2019a). Among them, there are covered horsts elevated to summit position, half-exhumed horsts, and exhumed horsts. This block type bears 16 out of all

Table 8 The occurrence of subsidence doline groups in the mountains (Veress and Vetési-Foith 2019a)

Place of occurrence	Elevation (m)	Number (piece)	Morphological environment	Type of the bearing rock
Tés Plateau	400–500	137	Plain terrain, valley floor, depression of superficial deposit	Horst elevated to summit position
Environs of Márvány Valley (northern part of Kőris Block)	360–500	120	On interfluves, in valleys	Horst elevated to summit position
Mester-Hajag	440–500	85	On terrains between mounds and in their depressions of superficial deposit	Horst elevated to summit position
Area between Som Block and Száraz-Gerence	400–500	76	On plain terrain, on valley floor	Horst elevated to summit position
Hárskút Basin	420–500	60	Plain terrain, valley floor, valley side	Horst elevated to summit position
Égett Block	440–470	25	On terrains between mounds and in their depressions of superficial deposit	Horst elevated to summit position
Kab Mountain	400–450	20[1]	On valley floor, in caprock doline, in ponor, on plain terrain	Horst with basalt cap
Area between Iharos Block and Judit Spring (Kertes-kő)	390–410	21	On plain terrain	Horst elevated to summit position
Felső-Hajag	440–500	20[1]	On terrain between mounds	Horst elevated to summit position
Area between Cuha and Gézaháza-puszta	450–470	14	On plain terrain, in depression of superficial deposit	Horst elevated to summit position
Balatonederics Plateau (Keszthely Mountains)	400–420	20	On plain terrain	Horst elevated to summit position
near Devecser	160–230	10[1]	On plain terrain	Threshold surface
near Tapolca	120–180	10[1]	On plain terrain	Threshold surface
Between Kőris Block and Parajos Block (Eleven Förtés Doline Group)	670–680	9	On plain terrain, on valley floor	Horst in summit position
Near Dudar (marginal part of Dudar Basin)	350–400	7	On plain terrain	Marginal part of cryptopeneplain
Középső-Hajag	580–600	6	On plain terrain	Horst in summit position
Near Tündér-major (Kő Mountain)	350–400	5	On plain terrain	horst elevated to summit position or marginal part of cryptopeneplain
Near Szépalma-puszta (at the marginal part of Porva Basin)	400–450	4	On valley floor	Horst elevated to summit position
at the margin of Fehérkő Valley	340–360	4	in depression of superficial deposit between mounds	Horst elevated to summit position
Near Lókút (Lókút Basin)	400–450	3	On valley floor (?)	Horst elevated to summit position
Ősbükkös doline group	400–450	3	On valley floor	Horst elevated to summit position
Sűrű Mountain	400–440	2	On plain terrain	Horst elevated to summit position

Notice: [1]Estimated

concealed karst areas. Both on low threshold surfaces and on horsts in summit position there are two concealed karst patches, while on cryptopeneplain and horst with basalt cover there is one concealed karst patch. On the latter horst types, doline number is also low. Thus, on blocks of horsts in summit position there are only 28 dolines. In the Northern Bakony (which bears 19 out of 22 concealed karst patches), 30% of horsts elevated to summit position bear concealed karst, while only 9.1% of horsts in summit position and 12.5% of cryptopeneplains show them. Concealed karst features are subsidence dolines, covered karst ponors and DSDs.

On blocks of horsts elevated to summit position, concealed karst is more widespread because

- Loess development could still have happened on these blocks.
- The chance of survival of the cover was greater on the blocks of this type since the inclination of their surface is low and the expansion of some blocks is large. On blocks with smaller expansion, the mounds of the bedrock (Mester-Hajag) could hinder the denudation of superficial deposit.
- The blocks of this type are often surrounded by blocks because of their position inside the mountains, which might also have limited the denudation of the superficial deposit.

The density of subsidence dolines is low (for example, in the Hárskút Basin it is 0.2 doline/100 m^2), but their sizes are diverse. Depressions with a diameter and depth of some metres are more widespread (there are exclusively such depressions, for example, in the patch in the environs of the Márvány Valley), but on some concealed karst patches (Tés Plateau, Hárskút Basin) there are depressions with a larger diameter even those exceeding 10 m.

There is a relation between the shape of dolines and bedrock stratification. If the beds are of low dip (lower than 10°) the doline has a symmetric cross section, if they have a larger dip (about 11°–42°) their cross section is asymmetric (Veress 1982, 2016). In the latter case, the bed heads of the bedrock outcrop on the steeper slope of the doline.

Dolines rarely coalesce and there are few uvala-like depressions, but very often even those of non-valley floor position are aligned in rows. The floor of many is largely alluviated. In this case, there may be inner, small-sized mainly dropout subsidence dolines on doline floors. Mass movements (collapses, slides) are common on their slopes. They are often significantly filled (this can be also marked by the fact that only the crowns of trees stand out above the cover). Inactive, denuding or destroyed dolines are also common. This may happen by the complete filling of the depression (a shallow, marshy spot survives in the place of the doline), or following the denudation of the cover the depression is also destroyed and the shaft below the floor is exposed. Some dolines are connected by gullies and creeks.

The concealed karst patches of the mountains occur in diverse morphological environments. Thus, they can develop on plain terrains, on valley floor, in valley sides, on interfluves, on the terrains between the exhumed mounds of the uneven bedrock, in DSDs, and in older (paleokarstic) depressions (Veress and Vetési-Foith 2019a).

Bedrock and surface morphology affect the denudation of the superficial deposit, and thus cover thickness and doline development too. Therefore, doline development environments can be the following (Veress 2000b, 2008, 2009, 2016; Veress and Vetési-Foith 2019a, b).

On blocks with a small area and with dissected bedrock, the majority of the superficial deposit is transported into the surrounding valleys. As a consequence, the mounds of the bedrock are exhumed. The bedrock, though thinning out, may survive on the terrains between the exhuming mounds where dolines may develop. This doline development is characteristic of Cretaceous limestone blocks. The pattern of subsidence dolines depends on the degree of block tilting and the relation of the dip direction of the block relative to the direction of the surrounding valleys. If the block has a low inclination, dolines constitute an irregular group (Fig. 50a). If the block dips in valley direction and to a larger degree, exhumed mound rows and doline rows being parallel to the valley develop since terrains with superficial deposit are elongated in a similar direction (Figs. 50b and 51a). If the block dips perpendicularly to the direction of the valley, the row of exhumed mounds, the elongatedness of terrains of superficial deposit, and thus doline rows are also perpendicular to the valley (Fig. 50c). Mounds are occasionally exhumed because the valley reaches their area during its retreat (Fig. 50d). In this case, subsidence dolines may develop at the branches defining the beginning of the valley (creeks) and tributaries of the valleys as well.

In case of high valley density, the proximity of valleys to each other favours the thinning out of the superficial deposit, and thus doline development on interfluves too (the environs of Márvány Valley Fig. 51).

The thickness of the accumulating superficial deposit is controlled by bedrock morphology. The superficial deposit is thinner on the bedrock, and thus the chance of doline emergence increases where there is a covered mound on the bedrock (Figs. 52 and 53).

Superficial deposit may be accumulated on the floor of epigenetic valleys. If the superficial deposit is of moderate thickness, subsidence dolines may develop on the valley floor. If smaller valleys are retreating back from the epigenetic valley to the surrounding terrain, the superficial deposit thins out during the process and, therefore, their floors may also become karstified (the environs of the Márvány Valley).

Subsidence dolines also occur at the margin of epigenetic valleys in low numbers (Kertes-kő Gorge). These are rather dropout dolines. Their development can be ascribed to the fact that the superficial deposit thins out near valley margins as a result of denudation. The superficial deposit thins out on the floor of gullies and creeks.

In the area of basins, valley evolution is slower if the valley reaches the bedrock at the basin margin (Hárskút Basin). In the basin, the gradient of the valley decreases, which enhances infiltration. At sites where the gradient is locally larger, as a result of its reworking, the superficial deposit thins out on the valley floor where dolines may develop (Veress 1982).

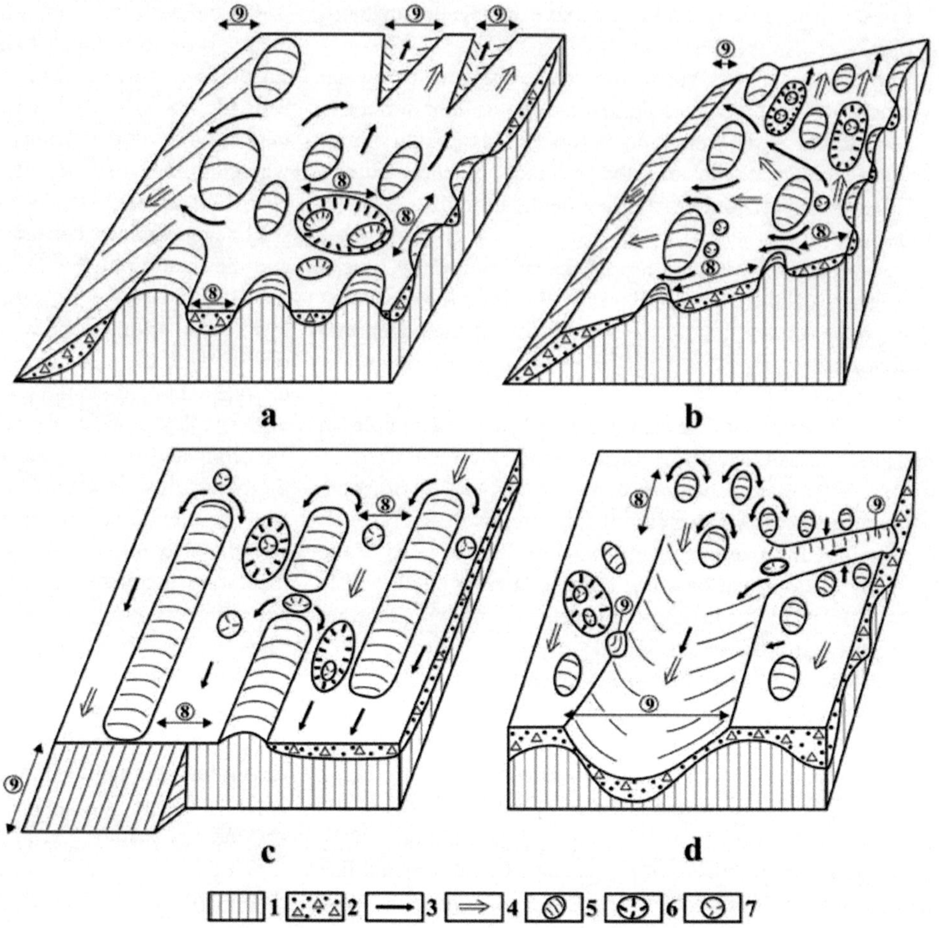

Fig. 50 Karstification of blocks dissected by limestone mounds. Legend: a. The block and its surface have no significant inclination; b. The block and its surface are inclined in two directions, one of the directions is parallel with the surrounding and sediment receiving valley; c. The block and its surface are inclined in the same direction which is perpendicular with the surrounding and sediment receiving valley; d. The erosion valley retreats into the spaces between the cones; 1. Limestone; 2. Superficial deposit and limestone debris; 3. Denudation of the superficial deposit by pluvial erosion and its direction; 4. Dip direction of surface; 5. Exhumed limestone mound; 6. Depression of superficial deposit; 7. Subsidence doline; 8. Lower, flat (or low-inclined) surface with superficial deposit between cones; 9. Valley, valley side

Dolines may also be formed in the fill of paleokarst depressions. Meteoric water accumulating in the depression may contribute to this process (Eleven-förtés doline group).

DSDs and enclosed subsidence dolines develop during processes mutually affecting each other. Developing subsidence dolines receive the superficial deposits of their environs, and thus they contribute to DSD development. The developing DSD promotes the development of further subsidence dolines. Partially, because the meteoric water arriving into the DSD cannot flow down on the surface and partly because the superficial deposit thins out on the floor of these features as well. In addition to this, waters leaving the dolines (which are on the floor of DSDs) filled with impermeable sediment increase infiltration, which favours the development of newer dolines.

In the mountains, rock zones of various age and composition often alternate with one another. Thus, for example, at the marginal part of the Hárskút Basin, the Cretaceous Requienia limestone is confined by Muineria clay, where the latter dips towards the former, and in the contact zone meteoric water arrives from Muineria clay to the loess covering the limestone and the chance of doline development increases there (and nearby). A doline row (consisting of 15 dolines) formed along the rock boundary (of several kilometres) in the above-mentioned area. A similar doline development may also occur (Hárskút Basin) at places where the waters of the gravel cover get to the surrounding terrain covered by loess. Such a rock boundary with gravel cover occurs in valley sides cut into gravel cover.

Doline development is also influenced by human activity. On Bocskor Hill, situated northwest from Kab Mountain (southeast from the town of Ajka), coal mining of Cretaceous limestone was practised. During the collapse above the cave galleries and the shifts of the blocks depressions (mainly grikes) developed on the limestone surface. At sites where the limestone is covered, the void caused by mining inherited over too. Thus, rounded (or elongated) depressions with steep sides (subsidence pseudokarst depressions) were formed. Such depressions may develop by cover collapse or suffosion.

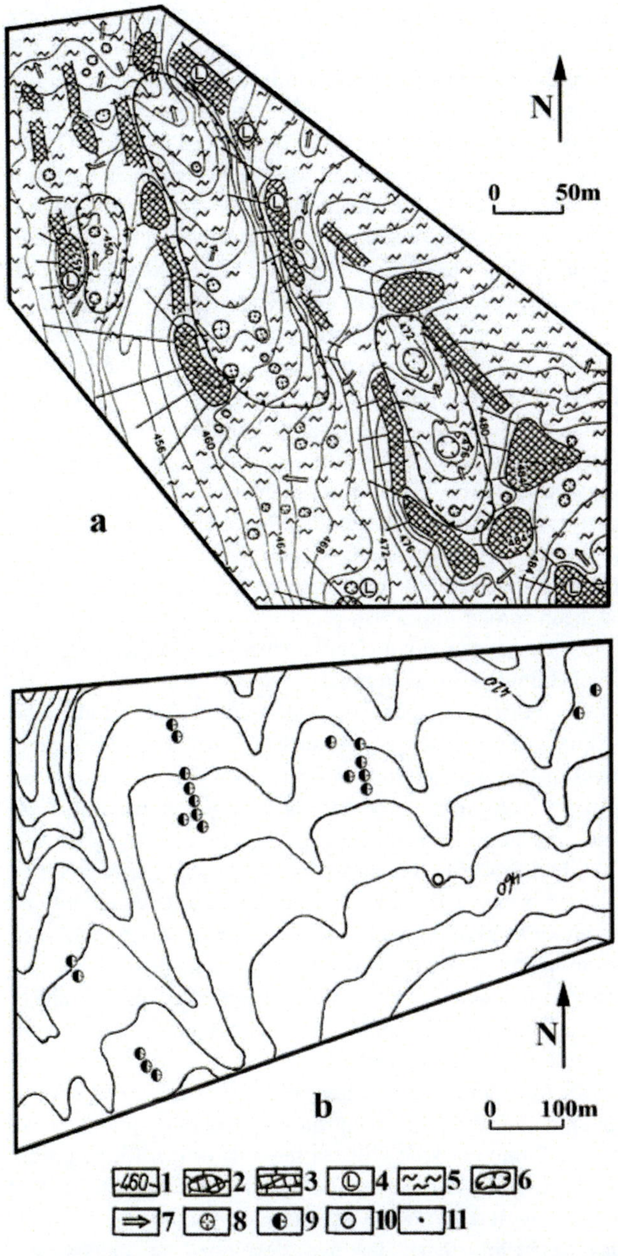

Fig. 51 Concealed karst: **a** between exhumed limestone mounds from Mester-Hajag, **b** from interfluves from the environs of Márvány Valley (Veress and Vetési-Foith 2019a)Legend: 1. Contour line, 2. Exhumed mound, 3. Semi-exhumed feature (it is ridge-like, but it does not constitute a mound), 4. Limestone outcrop on the mound, 5. Karstifying terrain with superficial deposit between mounds that developed by exhumation, 6. Depression of superficial deposit, 7. Material redeposition, 8. Suffosion doline between exhumed limestone mounds, 9. Suffosion doline on interfluves, 10. Suffosion doline on valley floor, 11. Water drainage conduit in a karstic depression

Some covered karst ponors (Hárskút Basin and Tés Plateau) also occur in the mountains. The ponor of Hárskút Basin has about 1-km-long valley that developed in gravel cover.

Depressions of superficial deposit (DSDs) are specific features of the concealed karst of the mountains. DSDs are pseudodepressions, real depressions only occur in the area of the Eleven-förtés doline group (Kőris Hill). Depressions of superficial deposit occur on the Tés Plateau (on plain terrain, on valley floors) and on blocks of Cretaceous limestone (on terrains between exhumed mounds, Fig. 51a). The depression of Három-kürtő shaft developed on the plain surface part of the Tés Plateau (Figs. 52 and 53), while the DSDs of Mester-Hajag were formed on terrains between exhumed mounds (Fig. 51a). Their development is favoured by limited surface material transport which can be caused by the low inclination of the surface or by exhumed mounds. The morphology of the DSDs of Middle Cretaceous areas is simple: mostly there are some suffosion dolines inside them. The landforms of the DSDs of the Tés Plateau are more diverse. They may be dropout dolines and suffosion dolines of floor and marginal position, with scars of mass movements in their sides and creeks may be connected to them (or situated inside them). Well-developed shafts may have been formed in their dolines.

Solution dolines

Subordinately, solution dolines (drawdown dolines) occur in some areas of the mountains in moderate numbers. The development of solution dolines was probably not favoured by the fact that the loess washed-in from the surface and enabled the development of the epikarst only locally. It may also have been contributed by the fact that there are only some uncovered horst roofs of low inclination in the mountains, but in the Transdanubian Mountains too. The areas of solution dolines occur at the following sites: karst near Tapolca (Fig. 10, chapter "The Geomorphic Evolution of Karsts and Karstic Surfaces"), the karst in the environs of Devecser, and some dolomite terrains. We mention here that Győrffy (1957) considers one of the lakes of the Kál Basin as a doline lake. Solution dolines occur on threshold surfaces since threshold surfaces are uncovered and the inclination of their surface is low. On the Sarmatian limestone of Tapolca Karst (Tinnye Formation), dolines, and uvalas also developed. On the floor of some dolines, subsidence dolines occur too. From the floor of one of these depressions (Nyelőke) Jaskó (1957) describes a permanent water cover which can be explained by its closeness to the karstwater level. There is no permanent lake on its floor presently. The plant waste that accumulated at its margin refers to an intermittent lake. On its floor there is a large dropout doline with vertical walls. Now, it functions as a water drainage (a valley is connected to it). However, the presence of the former lake may prove the appearance of the karstwater, and thus it can probably be regarded as katavothron.

At Devecser dozens of depressions (partially infilled) even with a diameter exceeding 100 m occur which are covered by superficial deposit (some of them may contain

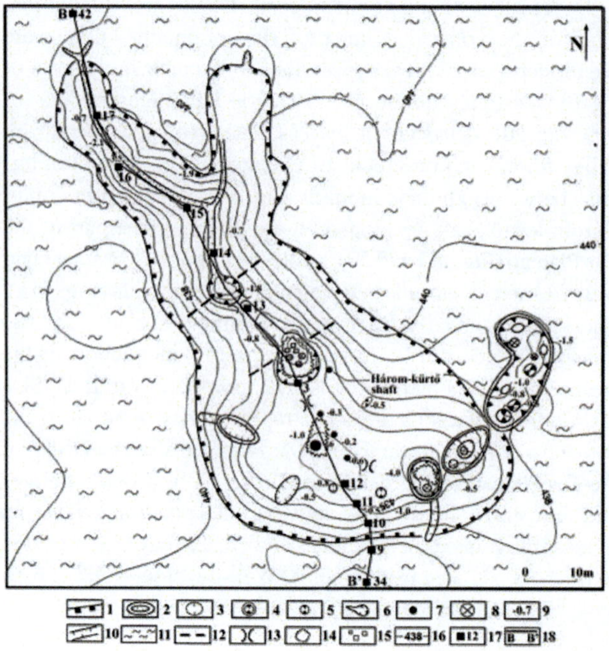

Fig. 52 The morphological map of the depression of superficial deposit (marked I-12) bearing Háromkürtő-shaft (Veress 2016 modified). Legend: 1. Margin of the depression of superficial deposit; 2. Suffosion doline (diameter is larger than 2 m); 3. Suffosion doline (diameter is smaller than 2 m); 4. Dropout doline (diameter is larger than 2 m); 5. Dropout doline (diameter smaller than 2 m); 6. Ponor; 7. Chimney, shaft; 8. Conduit in the cover; 9. Depth of the feature; 10. Gully, channel; 11. Surrounding terrain; 12. Rock boundary; 13. Col; 14. Outcrop; 15. Rock debris; 16. Contour line; 17. Site of VES measurement; 18. Line of VES measurement profile

small, young subsidence dolines) in the Middle-Miocene Limestone (Lajta Limestone Formation). West from the valley of the Egres Stream, where the altitude of the surface is below 190 m, the topographical map (this reflects a state from the 1960s) indicates an intermittent water cover on their floor (eastwards from the valley, where the elevation of the surface is above 210 m a water cover is less characteristic or it is not specific of the depressions at all) which can be explained by the rise of karstwater table to the effect of precipitation.

Solution dolines can also be found on dolomite terrains. Two types of closed depressions occur, their common characteristic is that their diameter is large as compared to their depth. The depressions of one variety have a smaller diameter (some 10 m) (1) and they may occur in various levels between and on the dolomite mounds (Fig. 54) in the depressions of the other variety or on the floor of valley-like features. The depressions of the other variety have a larger diameter (several hundred metres) (2) and are surrounded by mounds.

Depressions belonging to the latter variety (2) occur in that part of the Veszprém-Devecser graben which is between the settlements of Márkó and Öskü, and in Southern Bakony and in Balaton Uplands on the dolomite terrain between Veszprém and Nagyvázsony. Dolines of the former variety (1) are characteristic in the area of the Veszprém-Devecser graben which is between Márkó and Öskü. Láng (1948, 1958) mentions three dolines from the large ones (2) from Southern Bakony: the depressions of Köves-gyűr-puszta, Vörös-tó, and Meggyes-puszta. Láng (1948) considers them as of polje character. Trájer et al. (2020) mention Meggyes-puszta depression as paleodoline. The diameter of the depression of Kövesgyűr-puszta exceeds 1 km, the bedrock is uneven and in some places it is covered. A ponor (Láng 1958), and ponors (Jaskó 1961) are mentioned from its inner part (probably a subsidence doline or dolines). The depression of Vöröstó is situated between the villages of Nagyvázsony and Vöröstó. Its diameter is 2–3 km. Several valleys with permanent streams are connected to it. On its floor there is an inner depression which probably had a significant water drainage formerly. It has been filled and it has been turned into a fish pond.

On the floor of small-sized depressions (1) there is gravel which could have accumulated inside them subsequently. These features are recent solution dolines (Veress and Szabó 2000), which developed on surfaces that lost their gravel cover or that were not covered with gravel at all. As a result of their large size, those having a larger diameter (2) are older and non-active (though inside they can be karstified). They had probably developed preceding the gravel cover of their environs and they did not become covered by the material of Csatka Gravel Formation subsequently either, but according to the data of the geological map, they became lined with loess, fluvial sand, and reworked gravel subsequently like, for example, the depression of Köves-gyűr puszta (Császár et al. 1981).

Shallow depressions often with a diameter of several hundred metres mainly occur on Kab Mountain (but they can also be found on Apát Hill, on the hills of the Tátika group, on the Kovács Hill), these depressions have permanent or intermittent lakes (Leél-Őssy 1959c). According to Leél-Őssy (1959c), these features developed at paleokarstic depressions (solution dolines?) if these depressions were only partially filled with basalt lava.

Caprock dolines and complex karst depression

Caprock dolines develop on the basalt-covered blocks of the Bakony Region where the basalt is thin because there are paleokarst mounds on the bedrock (Veress 2016). Karstification takes place inside the basalt cover at these places (karst windows) (Németh 2005; Móga and Németh 2005; Veress and Unger 2015). The basalt continually collapses (Németh 2005) above the shafts of the bedrock (Veress 2016). Caprock dolines do not have steep sides, but they

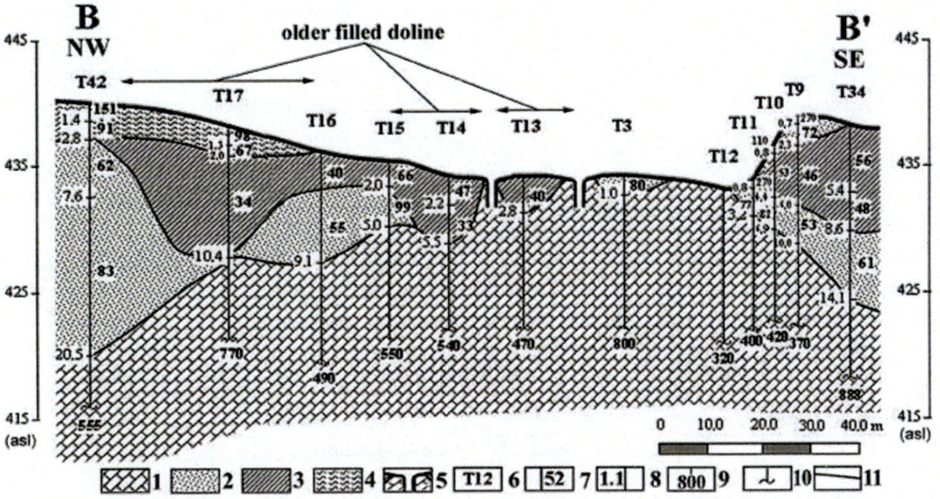

Fig. 53 The geoelectric-geological profile of the depression of superficial deposit presented in Fig. 52 (Veress 2016 modified). Legend: 1. Limestone, 2. (Sandy) loess (with limestone debris), 3. Loess (clayey or silty) or clay with limestone debris, 4. (Clayey) limestone debris, 5. Chimney, 6. Code of VES measurement, 7. Geoelectric resistivity of series (Ohmm), 8. Base depth of geoelectric series (m), 9. Geoelectric resistivity of bedrock (Ohmm), 10. Approximate penetration depth of VES measurement, 11. Boundary of geoelectric series

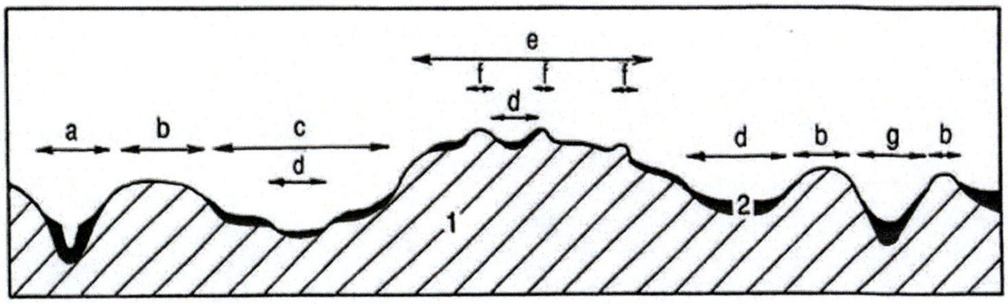

Fig. 54 Principal profile of a dolomite surface (Veress and Szabó 2000). Legend 1. Rock, 2. Rubble (active zone), **a** complex karstic valley, **b** Ridge, **c** Polje-like karstic valley or doline, **d** doline, **e** residual hill, **f** cone, **g** partially karstic valley

may constitute coalescing uvalas with smaller caprock dolines or subsidence dolines which developed on their floor in pairs (Veress 2016). Caprock dolines often have strikingly plain floors. Creeks lead to them and because of this, they function as ponors. Limestone outcrops during the continuous collapse of the basalt, and thus they can be transformed into ponors with blind valleys after a while.

Caprock dolines can also be found on Fekete Hill southeastwards from Kab Mountain (Győrffy 1957). Here, basalt collapses are explained by the dissolution of the calcareous basalt tuff (Győrffy 1957) or by the material lack that developed by the transportation of the Pannonian sand from the bedrock (Cholnoky 1932).

Complex depressions occur at karst windows on Kab Mountain (Fig. 55). The complex depressions of the Kab Mountain are of an expansion of several 100 m. The depressions developed by the karstic development of epigenetic valleys. This is evidenced by their elongated character, their meandering ground plan and while there is limestone inside them, the limestone is covered by basalt in their environment. Complexity is characteristic of these features both morphologically and genetically. On their floor, ponors (which may be inactive and active), caprock dolines, and subsidence (mainly suffosion) dolines occur (Veress and Unger 2015).

Collapse-like dolines

The collapse dolines or collapse-like depressions of the mountains occur on Tapolca Karst, in the environs of Likas-kő of Hódos Stream (Dörgő Hill, Szent László forest), and on the area between Gézaháza-puszta and Ördög Valley (Sűrű Hill group). The collapse dolines of Tapolca Karst develop in Sarmatian limestone, Kincses-gödör is the largest among them which is 102 m long, 12 m deep, and 25 m wide and during heavy rainfalls it can be filled with water (Móga et al. 2011). The reason for their development is the Tapolca Cave System below them. The relative large size

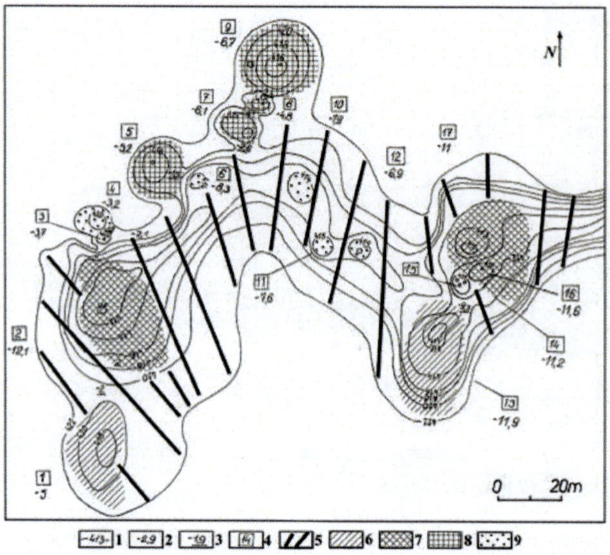

Fig. 55 Geomorphological map of the M-2 system (Kab Mountain, Hungary, Veress and Unger 2015) 1. Contour lines, 2. Deepest point of a karstic depression (as compared to an elevation of 422 m), 3. Depth of sill between subdepressions (as compared to an elevation of 422 m), 4. Code of depression, 5. Epigenetic valley, 6. Passive ponor (doline), 7. Intermittently active ponor, 8. Caprock doline, 9. Subsidence doline

and thin ceiling of this system favoured collapses. Depressions of the other two areas are depression with a round, mostly elongated ground plan and with a small depth (1–2 m) that developed in Eocene limestone. In the environs of Hódos Stream, the Middle Eocene limestone is thin (its thickness ranges from some tens of centimetres to 1–2 m). Its bedrock is karstified main dolomite which is exposed to the surface in patches at several places. The depressions occur close to these outcrops (Veress 2000b) since the Eocene limestone is thin at the karst mounds of the dolomite. These depressions developed by the collapse of the ceiling of the cavities which had formed in the thin Eocene limestone (thus, the ceiling of the cavities was also thin).

Ponors

Ponors that developed at rock boundary can be found on Kab Mountain (Figs. 8 and 9, chapter "The Geomorphic Evolution of Karsts and Karstic Surfaces"). Here, the ponors occur in the following morphological environment (Veress and Unger 2015).

At the termination of the basalt cover (for example, Macska-lik), where well-developed valleys with a length of several 100 m lead to the ponors.

At karst windows (Figs. 1d and 56) where they developed at the basalt margin which surrounds the outcropping limestone patch. There are karst windows where the basalt is not absent, but it only occurs in small thickness. At these sites caprock dolines developed which are led by valleys, and thus their transformation into ponors is taking place now.

On the floor of the valley deepening into the basalt where it exposes the limestone bedrock (for example, Tönkölyös ponor).

Gorges

The gorges of the Bakony Region develop at those sections of epigenetic valleys where the block bearing the valley section is elevated which is evidenced by the fact that gorges are often formed on the floor of incised meandering valleys (meandering took place preceding the elevation of the block, on its low surface). As a result of this, gorges are of epigenetic-antecedent origin (Veress 2000b). Gorges begin at the termination of the impermeable cover. Thus, gorges and their relating features and phenomena develop in the environment of the buried within this, isolatedly buried (according to block-type cryptopeneplain) karst terrains of the mountains. However, epigenetic-regression valleys may also have sections of gorge character mostly at their upper part. Epigenetic-antecedent gorges are the Kőmosó Gorge (the gorge of Kő Valley Stream), Kerteskő Gorge (the gorge of Gerence Stream), a part of the valley of Hódos-Stream, and Kő Gorge (the gorge of Öreg-Séd). It also occurs that not only one, but several blocks are elevated along the valley between which blocks of lower position are wedged in. In

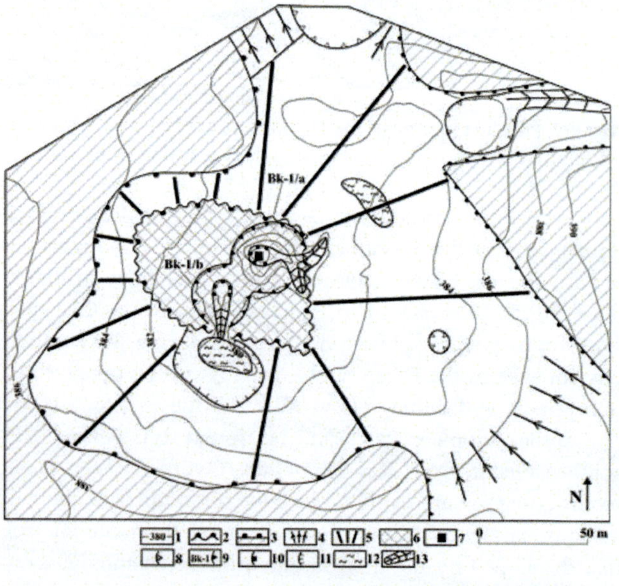

Fig. 56 Morphological map of a karst window and its environs (Kab Mountain, Veress and Unger 2015) 1. Contour line; 2. Boundary of distribution of basalt blocks; 3. Scarp, rampart created from basalt blocks; 4. Slightly sloping surface without scarps; 5. Gently sloping terrain between scarp and karst window; 6. Karst window; 7. Limestone outcrop; 8. Caprock doline; 9. Ponor and its code; 10. Dropout doline; 11. Fossil doline; 12. Lake; 13. Gully, ravine

this case, the valley may have several sections with gorges. Ördög Valley, Cuha Valley, Gaja Gorge, and Gerence Valley (if we consider the rocky part with steep valley sides of the latter situated north from Bakonybél as a gorge too) have several gorges. The catchment areas of the streams of the gorges are the basins or impermeable terrains (cryptopeneplains) of the mountains. The catchment areas are as follows: Lókút Basin and Hárskút Basin (Kertes-kő Gorge), the clayey terrain between Sűrű hill group and Csesznek (Kőmosó Gorge), Porva Basin (the part of Hódos Stream with gorges), Hárskút Basin (Kő-szoros), the gravel-covered surface between Zirc and Eplény (the gorges of Cuha), and Dudar Basin (the gorges of Ördög Valley and Gaja). However, the catchment area of their streams is significant. For example, the catchment area of Gerence is 95 km^2, while that of the Cuha Stream is 27 km^2 (Schmidt Elegius et al. 1962). In the sides of gorges, larger and smaller cavities of opening up genetics are common which are of hanging position and may have spherical cavities (Veress 2000b; Veress and Vetési 2019c).

Water flowing both in epigenetic-regression valleys and in epigenetic-antecedent valley sections (gorges) gets into the karst partially or completely (Jaskó 1961). Water drainage can be seepage along the total length of the gorge or point-like as well. Thus, in one of the crevices of the channel of Vörös János-Séd drainage was 50 l/min in 1959-ben (Jaskó 1961). In Kerteskő Gorge at a discharge of 600 l/min, 75% of the discharge of Gerence Stream seeped away (Bratán et al. 1967). Gorges function as ponors of the karst in the Bakony Region. In seven gorges or in valleys with gorges of Northern Bakony, 8 million m^3 water is discharged into the karst annually (Schmidt Elegius et al. 1962).

The majority of gorges developed in Cretaceous and Eocene limestones in which there are often impermeable intercalations, local karstwater storeys developed above the main karstwater of the mountains (high karstwater level). The seeping water of the stream of the developing gorge significantly contributed to the development of the karstwater storey. The cavities of the karstwater storey promoted the deepening of the gorge.

Concretions

The Bakony Region is not rich in recent calcareous sinter (tufa). Rimstone dams can be mentioned from the Northern Bakony, from the channel of the stream of Judit Spring (which drains off the water of the karstwater storey of Cretaceous limestone) (Fig. 57) and from the channels of the streams of the Balaton Uplands (for example, the streams of Koloska Valley, Malom Valley, and Királykút Valley) (Veress 2002). The thermal spring cones of the Tihany Peninsula (about 75 are recorded) developed during postvolcanic activity. The karstwater of the Bakony Region flew into magma reservoir, and thus it contributed to phreatomagmatic volcanism (Németh et al. 2001), and then to postvolcanic activity (Veress et al. 2015). Therefore, the material of the thermal spring cones partly originated from the karstwater.

Relict landforms

Karstic relict landforms are mounds and destroyed caves that lost their ceiling. Mounds occur sporadically and the expansion of terrains with mounds is not significant. They are exposed or spared covering. In the lack of superficial deposit, their development age cannot be given or it is

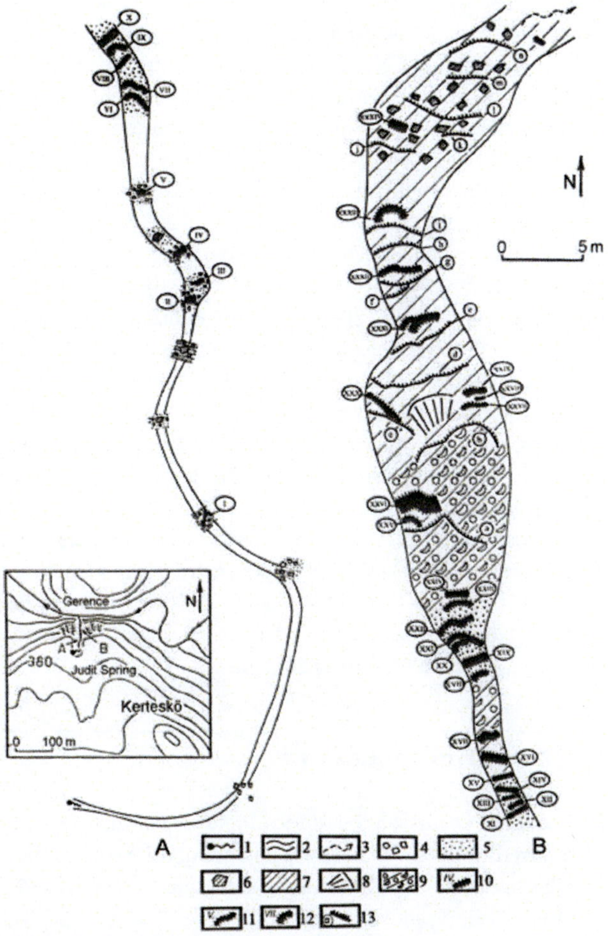

Fig. 57 Calcareous sinters of the stream of Judit Spring in ground plan (Veress 2002). Legend: 1. Spring, 2. Channel, 3. Percolation site, 4. Rock debris, 5. Mud, 6. Rock debris and block with calcareous veneer, 7. Calcareous sinter veneer on channel floor, 8. Calcareous mud, 9. Calcareous sinter veneer on channel floor with micro-dams and micro-basins, 10. Rimstone dam of straight ground plan, 11. Rimstone dam of irregular ground plan, 12. Arched rimstone dam, 13. Calcareous sinter escarpment (identification codes of rimstone dams with Roman numerals, those of escarpments with letters)

uncertain to be determined. Mounds occur on surfaces built up of main dolomite and Cretaceous limestone (Figs. 50 and 51a). Cretaceous conical blocks are as follows: Mester-Hajag, Égett Hill, Középső-Hajag, the area between Fehérkő Valley and Szekrényes-kő-Valley (the environs of Augusztin farm), and the area between Kerteskő Gorge and Iharos Hill. These areas constitute a block group interrupted by valleys between Gerence Valley and Hárskút Basin. The exposure of the mounds is still happening. The superficial deposit is reworked, fine-grained material with autochthonous limestone debris in some places. Exposure takes place by pluvial erosion or channel erosion.

Although dolomite terrains are widespread in the mountains, dolomite terrains dissected by mounds only occur at two places on the surface: the part of the graben between Veszprém and Devecser that is between Márkó and Öskü, and in the area among the villages of Zalahaláp, Diszel, Sáska, Ódörögd, and Nyirád. On the terrain between Márkó and Öskü the density and size of the mounds, and thus the dissected character by mounds increases in eastern direction. According to their morphology, four types can be distinguished (Szabó 1998). The mounds can constitute mound groups which may also have a common base. The mounds are asymmetric, the length of their opposite slopes is different (Szabó 1968, Veress and Vetési-Foith 2019b). According Móga et al. (2011), they are exhumed paleokarst features. They are probably the relict landforms of Miocene tropical planation (Csillag and Sebe 2015).

Rock corridors are the remnants of phreatic cavities that lost their ceilings, they often occur in groups in gorges. However, they can also be found farther from gorges in mounds where phreatic cavity formation took place (Veress and Vetési-Foith 2019c). They are narrow features with vertical walls and a length of some metres, but sometimes they can be wider as well (Veress 2000b). The former developed by the destruction of cave passages, while the latter by the destruction of chambers.

5 The Vértes Mountains

Surface karst occurs on the Vértes Plateau. Its karst features are referred to by Láng's (1948) and Sásdi's (2016) studies. A collapse doline and subsidence (suffosion) dolines can be mentioned among its karst features.

The collapse doline (Várgesztes abyss) developed in Eocene limestone which is a feature with a diameter of 15–20 m and a depth of 5–7 m. The number of subsidence dolines is small, their size and maturity are not significant, and they are mostly simple depressions which are rarely connected by channels or creeks. The dolines were formed in two patches on the plateau, but their density is also small within the plateau. The two patches are near Várgesztes in

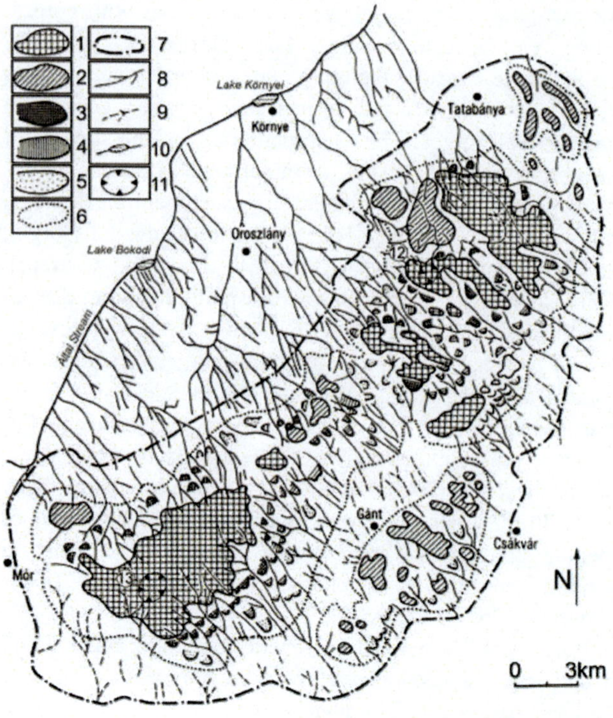

Fig. 58 Morphological map of the Vértes Mountains (Ádám and Pécsi 1988). Legend: 1. Roof levels of Vértes Plateau (elevation of 400–500 m), 2. Plateaus of medium and low horsts (380–420 m), 3. Higher levels of interfluves (350–380 m), 4. Medium levels of interfluves (220–300 m), 5. Lower levels of interfluves (150–200 m), 6. Foot of horsts and the lower, 220–300 m boundary of the basin margin, 7. Boundary of piedmont surface at 180–220 m, 8. Erosion valley, 9. Erosion-derasion valley, 10. Lake, 11. concealed karst, 12. Concealed karst in the environs of Lófő Valley, 13. Concealed karst in the environs of Csóka-kő

Lófő Valley and in its environs (11 dolines), and NE from Csóka Hill (10 dolines) (Fig. 58).

According to morphological environment, the depressions occur on valley floor (four depressions), in valley side (one depression), on plain terrain (nine depression), on mound (two depressions), and in exhumed paleokarst depressions (five depressions). The dolines occur at an elevation of 420–450 m, and thus they are on horsts elevated to summit position.

6 The Gerecse Region

First Láng (1948, 1955b, 1956) described its karst features and thus he mentions dolines and ponors from Gerecse-tető from the area of the Pisznice group, from Lukas-kő, and from Nagy-Gete. Leél-Őssy (1959b) classified its karst, and then based on this, Hevesi (1991b) outlined its karst features. Sásdi (2004) deals with the evolution of its karst. Cave exploration groups of Tatabánya studied its surface karst in order to explore caves. Our description of the karst features is based on their experience and observations.

Its karst is similar to the karst of the Bakony Region. It is a karst with horsts where concealed karst is characteristic, but cryptokarst also occurs. The distribution and frequency of the karst features is not significant. The most common features of its surface karst are subsidence dolines, but caprock dolines and exposed subsoil karren patches also occur. One of the karren patches developed on freshwater limestone (Babály ridge). Láng (1955b) mentions karren, barren limestone cliffs (surface) from the side slopes of Nagy-Gete and Hegyes-kő (Eastern Gerecse). In addition to the above-mentioned places, exposed karren also occur on Öreg-Kovács Hill, on Pes-kő, and on Nagy-Gerecse in the mountains.

The subsidence dolines of the Gerecse Region are suffosion dolines. They mainly occur in Central Gerecse, but they can also be found in Eastern Gerecse (Fig. 59). The superficial deposit, in which these features developed, is loess (Láng 1956) or loess-like permeable sediment. Its thickness is variable, but mainly small, according to cave explorers' data it is thinner than 1 m in some places. In some cases, there are no surface depressions above the shafts because of thin superficial deposit either. Dropout dolines or depressions of superficial deposits are not known in the mountains. Suffosion dolines occur in several patches in the mountains, their diameter may range from some 100 m to 1–2 km. The patches are primarily at surfaces with an elevation of 400–550 m, but dolines also occur at elevations of 250 m and 600 m (Table 9). The doline density of the patches is low (it lags behind the doline density of the Bakony Region). The diameter of the dolines is mostly some metres, but those with a diameter of some 10 m also occur. The dolines are predominantly on flat terrains and in summit position, only some dolines are on old, alluviated valley floor (Table 9). Their depth is small, some metres.

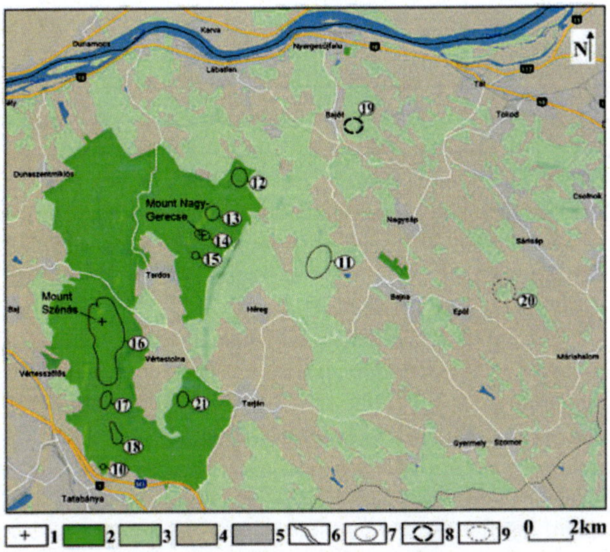

Fig. 59 Concealed karst areas of Gerecse. Legend: 1. Summit, 2. Forest, 3. Meadow, 4. Arable land, 5. Settlement, 6. Road, 7. Concealed karst, 8. Cryptokarst, 9. Karren, 10. Kő Hill, 11. Lukas-kő, 12. Domoszló, 13. Kis-Gerecse, 14. Nagy-Gerecse, I. 15. Nagy-Gerecse II, 16. Öreg-Kovács Hill, 17. Halyagos, 18. Environs of the hunting lodge of Kis-Rét, 19. Domonkos Hill, 20. Babál Hill, 21. Pes-kő

Table 9 Subsidence (suffosion) dolines in the Gerecse Region

Site	Their mark in Fig. 59	Elevation	Piece	Approximate expansion of doline patch (m)	Morphological environment	Type of the bearing block
Kő Mountain	10	250–300	7	300	On plain surface, in summit position	Threshold surface
Lukas-kő	11	320–350	5	800–1000	On plain surface, in summit position	Threshold surface
Domoszló	12	400–440	10	500	On plain surface, in summit position	Horst elevated to summit position
Kis-Gerecse	13	450–490	6 (?)	300	On plain surface, in summit position	Horst elevated to summit position
Nagy-Gerecse I	14	600–630	9	250	On plain surface, in summit position	Horst in summit position
Nagy-Gerecse II	15	580–600	7	300	On plain surface, in summit position	Horst in summit position
Öreg-Kovács Hill	16	500–550	55	2000	On plain surface, in summit position	Horst elevated to summit position
Halyagos	17	400–420	13	500	On valley floor	Horst elevated to summit position
The environs of the hunting lodge of Kis-rét	18	350–400	10	600	In valley side	Horst elevated to summit position

No valleys or creeks are connected to the dolines. Their ground plan is mostly round, rarely elongated. Their morphology is simple. They have a symmetric cross section and gentle slopes. On their floor, there are no inner, smaller (dropout dolines or dolines of such character) with steep sides (except a doline) and the scars of mass movements are absent on their side slopes. However, in case of some dolines it is noticeable that one or more small depressions (in this case they have a wreath-like arrangement) with a diameter of 1–2 m also developed at their margins.

The alluviation of the depressions is not significant (this phenomenon is only observable in case of a doline where the funnel-like floor was filled up to a plain during some decades). Because of weak alluviation there are only some inactive subsidence dolines.

Patches of suffosion dolines occur on horsts elevated to summit position (five patches), on horsts in summit position (two patches), and on threshold surfaces (two patches) (Table 9). Thus, similar to the Bakony Region, mainly the area of the horsts elevated to summit position is karstifying though the proportion of surface karstification of the blocks of the other two block types is larger as compared to the Bakony Mountains. The patches of suffosion dolines occur in the following areas: Kő Hill, Lukas-kő, Domoszló, Kis-Gerecse, Nagy-Gerecse, Öreg-Kovács Hill, Halyagos, and the environs of the hunting lodge of Kis-rét (Fig. 59, Table 9).

Caprock dolines (estimated five dolines) can be found on Domonkos Hill. Here, the depressions developed on Eocene sandstone. The limestone is 12–15 m deep as compared to the surface (above this there is a 2–3-m-thick marl and a 10–12-m-thick sandstone). The depressions are not large, at one of them, the shaft upwarping from the bedrock has a diameter of some metres.

A stone bridge survived above the cave by the collapse of the gallery of Jankovich Cave. A depression similar to a collapse doline developed during the collapse.

7 The Buda Hills (Buda Mountains)

Their karst is soil-covered karst and concealed karst. Leél-Őssy (1952a), Marosi (1959), and Jakucs (1994) studied their surface karst features, while Dénes (2001), Korpás et al. (2003), and Leél-Őssy (2012) investigated their concretions.

Leél-Őssy (1952a) mentions karren slopes from Remete Hill, from Fazekas Hill, Hosszú-erdő Hill, and from Fekete Hill. He also mentions karren (karren slopes) from dolomite, thus from Tündérhegy klippe, from Apáthy cliff, from the areas of Nagy-Ördög-torony and Kis-Ördög-torony, from the Danubian slope of Gellért Hill, and from the slopes of Csiki Hills. Marosi (1959) mentions small-sized dolines from Nagy-Kopasz Hill, from Nagy-Szénás Hill, and from Remete Hill. These are probably subsidence dolines.

According to Leél-Őssy (1952a) and Jakucs (1994), in the Buda Hills, rubble (in some places dolomite powder) beds are widespread on dolomite and on limestone at the surface that developed to hydrothermal effect.

The marginal fault system of the hills is the Buda thermal line along which hydrothermal springs (Buda thermal springs) were and are fed by the rising hydrothermal water. Freshwater limestones developed from the rising thermal waters in a significant amount, their absolute age was given by Kele et al. (2009, 2011, 2012, see chapter "Age, Depositional Environment and Geochemistry of Freshwater Carbonates (Travertine, Tufa) from Hungary"). Thus, the freshwater limestone nappe of the Buda Castle Hill can be mentioned which was described by Korpás et al. (2003) and Leél-Őssy (2012). Its development age is Pleistocene according to Jakucs (1994), while Korpás et al. (2003) consider it 3 million years. It was formed from lakes which had been fed by thermal springs (Korpás et al. 2003). Calcareous sinter was also described from the Gellért Hill by Dénes (2001).

8 The Pilis Mountains

The surface karst of the Pilis Mountains is soil-covered karst and concealed karst, but because of the presence of the nearby volcanic rocks and sandstone patches, cryptokarst and mixed allogenic-autogenic karst also occur in some places.

Their surface karst was studied and investigated by Leél-Őssy (1952a, c, 1958b), Láng (1953), Sásdi (2000), Veress et al. (2005), and Deák et al. (2009).

Leél-Őssy (1952a) mentions karren fields from the Pilis Hill, the Nagy-Kevély Hill, the Kis-Kevély Hill and the Kétágú Hill and karren slopes from the Kevély Hill, the Kétágú Hill, the Pilis Hill, and the Strázsa Hill. Deák et al. (2009) mention weak karren formation from the area of Csévi klippen. Veress et al. (2005) described bare karren formation from the Strázsa Hill near Dorog. Here, the karren (rinnenkarren) that developed on Triassic limestone were filled with Eocene sandstone. Subsequently, diaclases developed in these rocks which became filled with calcite. Small-sized recent karren (rillenkarren) developed in the calcite fills (Figs. 60 and 61).

Their concealed karst is represented by some subsidence dolines which occur on the plain terrains of the blocks or on valley floors. Altogether 10 depressions can be identified that developed in superficial deposit (loess). The dolines occur at an elevation of 400–550 m, and thus they are on horsts elevated to summit position.

The origin of the small depressions of the Pomáz-Budakalász freshwater limestone terrain is uncertain. Here, more

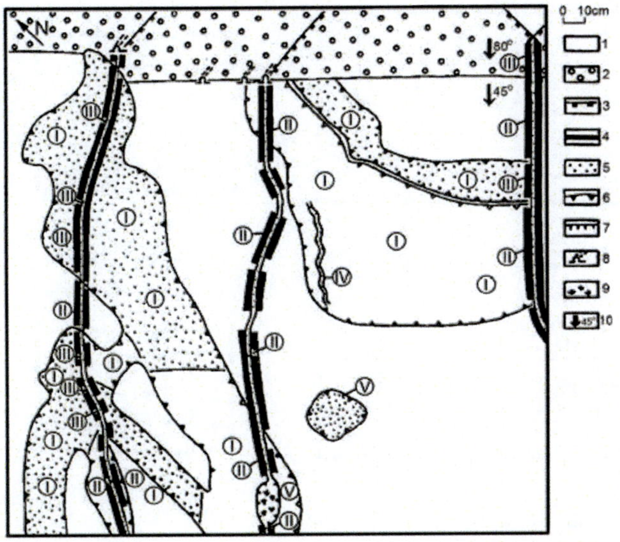

Fig. 60 Karren features of a part of Strázsa Hill (Veress et al. 2005 modified). Legend: 1. Slope part of Triassic limestone, 2. Slope part of Eocene sandstone, 3. Joint, 4. Calcite fill in diaclase, 5. Paleokarren feature filled with Eocene sandstone, 6. Exposed paleokarren feature, 7. Recent karren feature, 8. Karren cavity, 9. Vegetation in paleokarren feature, 10. Direction of slope inclination, I. Paleo rinnenkarren, II. Recent rillenkarren in the fill of Triassic limestone diaclase, III. Recent rillenkarren in the fill of Eocene sandstone diaclase, IV. Recent rillenkarren on Triassic limestone, V. Paleo kamenitza

than 100 depressions with a diameter of some metres can be found on the terrain covered by loess (Leél-Őssy 1958b). Some of them can be of natural origin while the others are mining scars. Today their origin cannot be determined because the area is built up and artificial transformations took place at the surface.

However, ponors also occur in the area of the Pilis Mountains. At least three ponors are known and all of them have special characteristics. The most interesting is the ponor of Macska Cave which is situated on a valley floor that developed on Eocene sandstone. It is a caprock doline and the upper part of its passage is also in sandstone. The depression without shaft that developed in the gorge of the Dera Stream with epigenetic origin (Cholnoky 1940) is also particular (Pilisszentkereszt Gorge). This is an active water recipient nowadays too. (Sásdi 2000). Above the level of the channel there are two further non-active shafts (probably former water drainage passages). The development of the shaft was probably enabled by the fact that there are lots of non-karstic (volcanic) rock debris in the channel which functions as a kind of rock boundary where it thins out. Ördög-lyuk can also be mentioned which might have been a ponor that developed at rock boundary. By now, the volcanic rock has been denuded nearby (from the former catchment area), by which its depression was destroyed and is not active any more. Therefore, in the area of the Pilis Mountains, there is no mixed allogenic-autogenic karst but it might have been present subordinate formerly. In literature, further ponors are also mentioned, thus the ponors of Üröm, Harapovács, Pomáz, and near Aranylyuk (Leél-Őssy 1952c, 1958b; Hevesi 1991b) which were destroyed by human activity and were built up. The depression of Üröm ponor was formed in Eocene marl and only its drainage passage is in Eocene limestone (Leél-Őssy 1952c), and thus it is a caprock doline. The other above-listed depressions were probably subsidence dolines based on the descriptions (Leél-Őssy 1958c).

Destroyed caves (rock corridors) and a depression that developed by the collapse of a cave are also described from the area of Csévi klippen (Deák et al. 2009).

9 The Western Mecsek Karst

9.1 General Characteristic Features

The general characterization of the karstification of the Western Mecsek Karst was dealt with by Szabó (1968), Jakucs (1977b), Barta and Tarnai (1997), Hevesi (1991b, 2001), and Veress (2011, 2016). Its surface karst features were studied by Hevesi (1991b, 2001), Barta and Tarnai (1991), Veress (2011), and Vetési-Foith et al. (2017); the distribution of the features was investigated by Lovász (1971, 1977); its pattern was studied by Lovász (1977), Czigány and Lovász (2006); and its morphometry by Hoyk (2002), Lippmann et al. (2008), and Kevei Bárány et al. (2015). Data on its calcareous sinter was given by Kraft et al. (1986) and Koltai et al. (2012). Several rearches have dealt with the vegetation of the dolines of the karst (Kevey and Csiky 2012; Bátori et al. 2012).

The surface of the karst area (with an expansion of 30 km^2) is separated into two levels with different elevations, into a more elevated, Middle-Miocene level (450–470 m), and into a lower (250–380 m) Pannonian abrasion platform (Fig. 62, Lovász 1971, 1977), which is divided into the karst areas (parts) of Abaliget, Orfű, and Misina Tubes (Lovász 1977).

The karst is mixed allogenic-autogenic karst, since from the south it is bordered by sandstone and badly karstifying limestone and by dolomite stripes (Fig. 63, Barta and Tarnai 1997), but according to Czigány and Lovász (2006), its eastern part (Misina Tubes Karst) is autogenic karst because of its elevated position. Since the autogenic part of the whole Mecsek Karst is covered by loess, it is predominantly concealed karst (Hevesi 2001). Its surface dipping northwards is dissected by through valleys which begin from the non-karstic terrain, then going through the karst they leave the area at its northern margin (Hevesi 2001) and join the

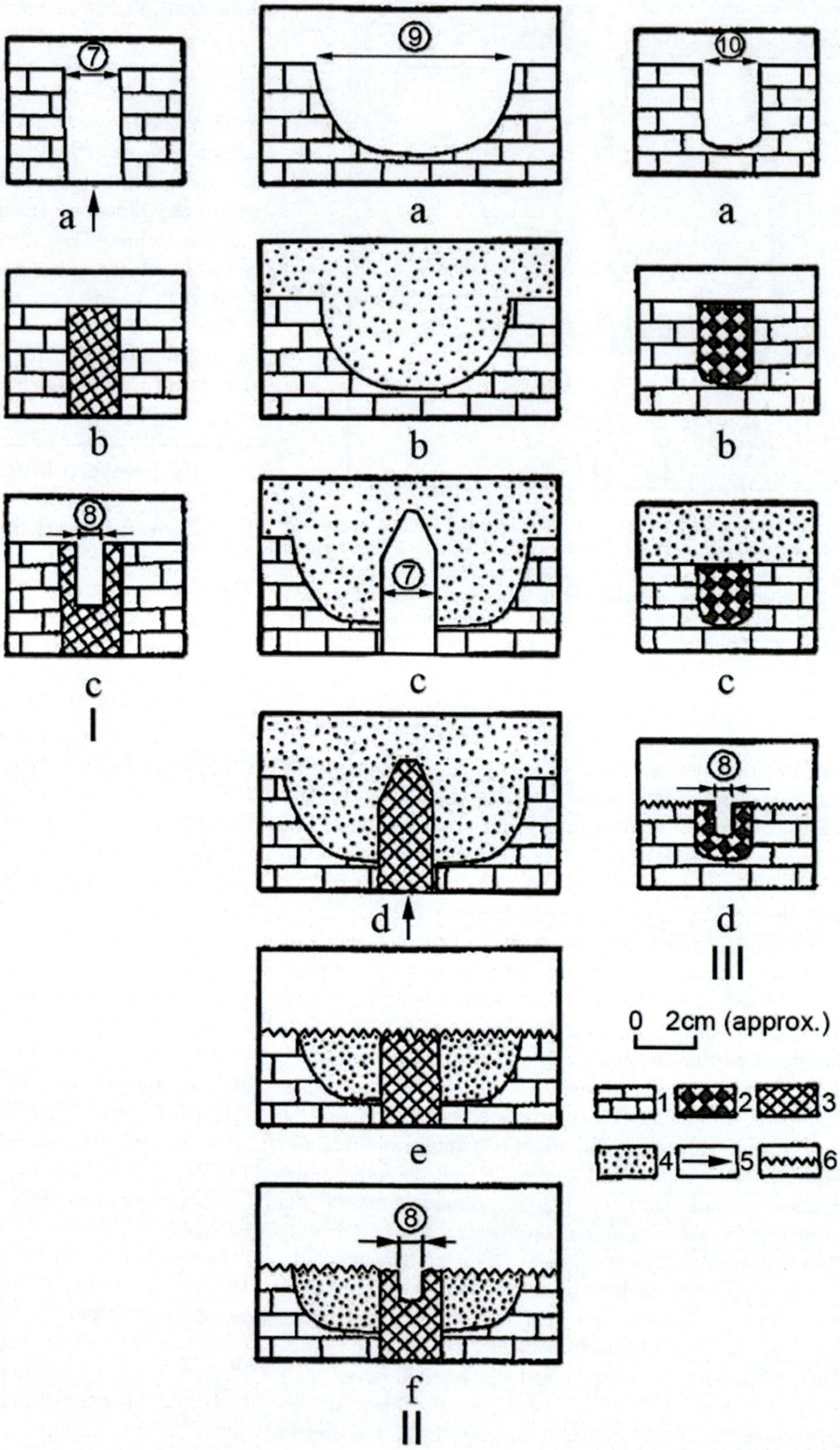

Fig. 61 Development of karren feature generations on Strázsa Hill (Veress et al. 2005, modified). Legend: 1. Triassic limestone, 2. Old calcite, 3. Young calcite, 4. Eocene sandstone, 5. Hot water, 6. Denudation surface, 7. Diaclase, 8. Rillenkarren, 9. Rinnenkarren, 10. grike, I. Diaclase develops (a), in its young calcite (b) rillenkarren develops (c), II. Rinnenkarren develops (a), which is filled and covered by Eocene sandstone (b), diaclase develops (c), which is filled by young calcite (d), the Eocene sandstone is denuded, (e), rillenkarren develops on the exposed calcite (f), III. grike develops (a), the grike is filled with old calcite (b), the filled grike is covered by sandstone (c), rillenkarren develops on the exposed calcite after the denudation of sandstone (d)

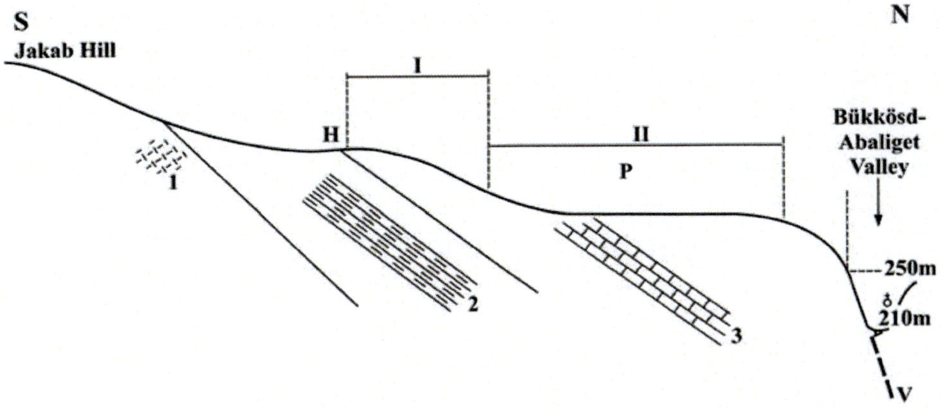

Fig. 62 Geological and geomorphological section across the Mecsek karst (Lovász 1971). H. Middle-Miocene abrasion platform; P. Pliocene abrasion platform; I. zone of large dolines of low density; II. zone of large dolines of high density; 1. Permian sandstone and Triassic dolomite; 2. Lower Triassic limestone; 3. Middle Triassic limestone; V: fault

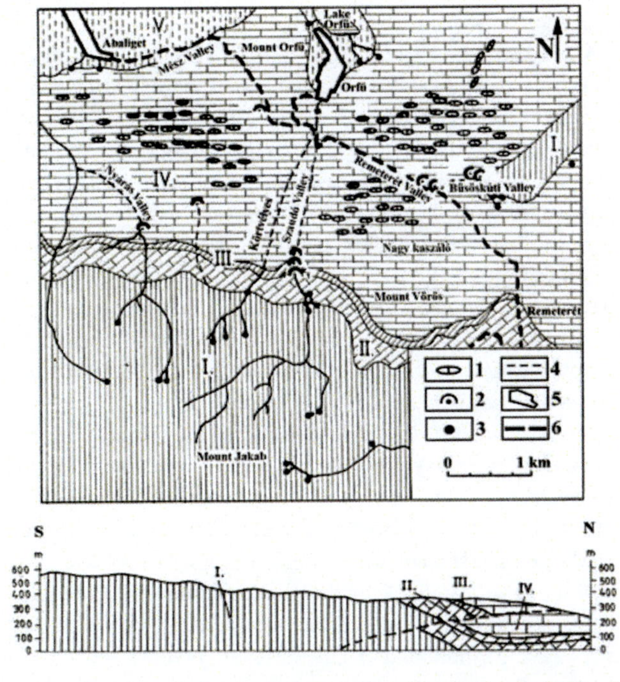

Fig. 63 Karst of the Mecsek Mountains (Barta–Tarnai 1997). 1. Doline field; 2. Ponor; 3. Spring; 4. Intermittent and permanent watercourses; 5. Built-up area; 6. Road; I. Non-karstic rocks (sandstones, conglomerates, siltstones); II. Poorly karstifying limestones; III. Dolomite; IV. Well-karstifying rocks below loess; V. Tertiary and Quaternary deposits

Abaliget Valley. Therefore, Veress (2016) calls the Mecsek Karst semi-allogenic covered karst (Fig. 1c).

The cover of the karst area is primarily Würmian (Hevesi 2001) or older loess, but according to the data of VES measurements, sand and clay also occur mainly intercalated into loess or mixed sandy loess with limestone debris (Veress 2011). Loess covers the karst in an almost uniform development with a thickness of 1–2 m. Thickening in a northern direction, its thickness may also reach 3–4 m in some places (Lippmann et al. 2008), but it may significantly thicken out (it can reach 10 m) in some solution dolines (Veress 2011). The karstic and non-karstic landscapes of the karst are demonstrated by the geomorphological map of a terrain part near Abaliget (Fig. 64).

Its karst features are as follows: solution (drawdown) dolines, uvalas, subsidence dolines, ponors (point recharge dolines), and concretions. Various authors do not differentiate the two doline types genetically, but they consider subsidence dolines younger only based on their smaller size and considering different morphometric characteristics (Kevei Bárány et al. 2015). They often investigate the two doline types together, thus mostly common data can be given for their different characteristics.

9.2 Distribution and Morphometry of Karst Features

The frequency and size of karst depressions increases from the south to the north and from the east to the west (usually in the whole karst), but especially on Abaliget Karst (Lovász 1971). Karst depressions are more widespread on the lower, younger abrasion surface (Lovász 1977; Lippmann et al. 2008). However, both doline types (particularly on Orfű Karst) are present on the slope separating the two terraces, at the upper level, there are only subsidence dolines (Lovász 1971, 1977). Lovász (1971) explained the poverty (or lack) of the upper level in dolines by the small thickness of the limestone here. Small thickness can be attributed to the fact that the non-karstic rocks bordering the karst from the south and dipping in northern direction are still close to the surface here (Fig. 62).

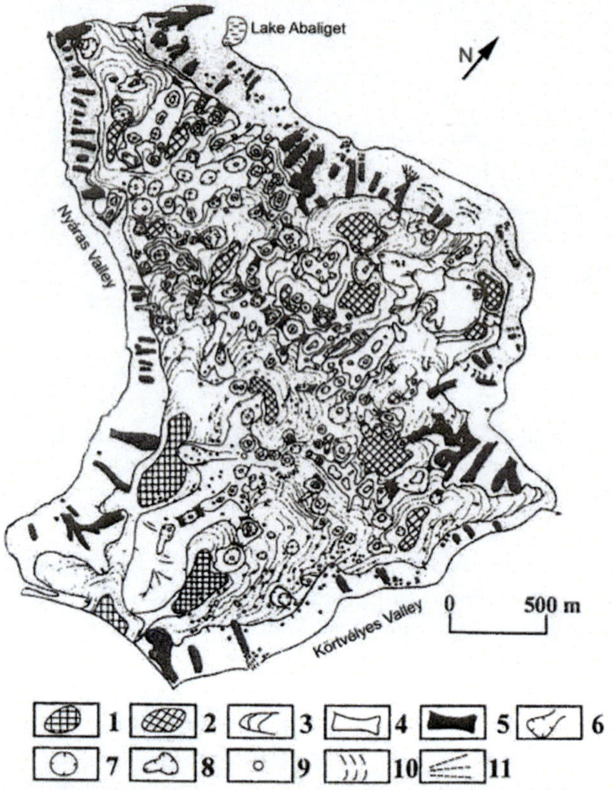

Fig. 64 Geomorphological sketch map of Mt. Bodó and Domokos (Abaliget) (Hevesi 2001). Legend: 1. Top of interfluve (asl.: 400–420 m), 2. Top of interfluve (asl.: 310–325 m), 3. Resting mild slope of interfluve (contour lines at every 5 m), 4. Col, 5. Col between doline, 6. Valley, valley-like depression, 7. Doline, 8. Complex karst depression, 9. Initiative of doline (subsidence doline), 10. Tongue of mass movement, 11. Alluvial cone

Dolines occur at an elevation of 260–552 m, their majority is at a surface with an elevation of 260–380 m, but if we only consider the dolines of the lower abrasion platform, the dolines of this terrain are grouped at an elevation of 280–310 m (Lippmann et al. 2008).

In the western part of the karst, majority of the valleys do not have tributary valleys. Hevesi (2001) sees a relation between the existence (and lack) of tributary valleys and doline density. According to him, doline density is high on interfluves where the valley bordering the ridges does not have any tributary valleys and doline density is low where there are tributary valleys.

The main characteristic of Mecsek Karst is that doline density is the largest here among Hungarian karst areas. While the doline density is 9,7 doline/km^2 on Aggtelek Karst (Telbisz 2001) according to Hevesi (2001), it is 73,4 doline/km^2 on Abaliget Karst (near Zsidó Valley). According to Lippmann et al. (2008), most of the dolines, 1227 doline can be found in an area of the karst with an expansion of 22,7 km^2. Lippmann et al. (2008) state that doline density reaches 80 doline/km^2 at several places on Abaliget Karst (Fig. 65), but in an area of Orfű Karst with an expansion of 2 km^2 there are 164 doline/km^2. Hoyk (2002) gives data on doline number different from the above. In an area of 14 km^2 studied by him, there are 1540 dolines and the doline density reaches 380 doline/km^2. The reason for the difference is that the data were obtained from maps (M1:10,000) in the former study (where the smallest were not marked, and thus not taken into consideration), while in the latter smaller dolines were also taken into consideration in the field research. These high doline density values can only be explained by the fact that majority of dolines is given by small dolines, thus subsidence dolines (see below).

Regarding the inclination of the bearing slope, Lippmann et al. (2008) demonstrated that considering their areas, they occur on slopes with an inclination of 2–7° in the most significant proportion, but the proportion of doline formation is the largest on slopes with an inclination of 2–4° (dolines are absent on slopes steeper than 13°).

As regards geological data, almost 75% of the dolines developed on Middle Triassic fine-nodular limestone (Lippmann et al. 2008).

In case of 30% of the dolines of Mecsek Karst (data originate from two sample areas) the relief proportion is 0.4, in case of 60% of the dolines this value is 0.6 and larger than at the dolines of Aggtelek Karst or the Bükk Mountains (Kevei Bárány et al. 2015). This means steeper side slopes in case of the dolines of the Mecsek Karst, which can be explained here by the lower degree of horizontal dissolution. However, this can probably explained by the fact too that subsidence dolines were also included in data processing among which there are dolines with steep side slopes as well. Regarding inner asymmetry this value is the smallest on Mecsek Karst (Table 5). According to Kevei Bárány et al. (2015), this can be attributed to the fact that the dolines of the Mecsek Karst were and are less widening as compared to the dolines of Aggtelek or the Bükk Mountains. Probably, subsidence dolines modified the data partly in this case too since their formation and development favours the development of circular features and their survival during their growth.

On Abaliget Karst, the longitudinal axis of the dolines coincides with fault lines (Czigány and Lovász 2006; Lippmann et al. 2008), while on Orfű Karst it coincides with the valleys (Lippmann et al. 2008). On Orfű Karst, the longitudinal axis of subsidence dolines is identical with the dip direction of the surface (Lippmann et al. 2008). However, according to the investigations of Kevei Bárány et al. (2015), the relation between the longitudinal axes of the dolines and dry valleys is the strongest here among Hungarian karsts. According to the studies of Kevei Bárány et al. (2015), the orientation of the dolines is NS, while small dolines (with a diameter smaller than 30 m) show a

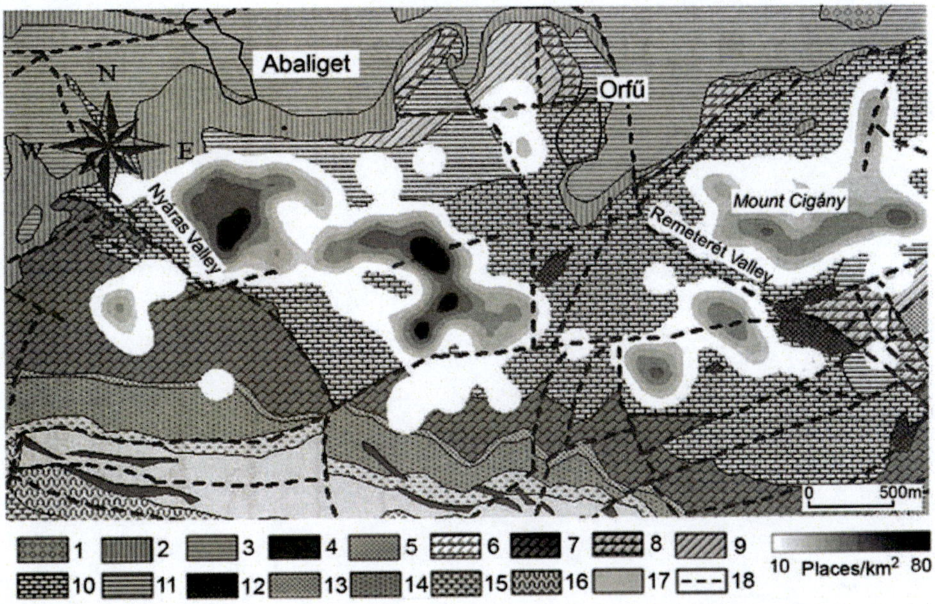

Fig. 65 Doline densities (Lippmann et al. 2008). Legend: 1. Gravel, sand, sandstone; 2. Conglomerate; 3. Sandstone, clay marl; 4. Alcalic-diabase product; 5. Limestone with brachipods; 6. Dark grey argittaceous limestone; 7. Thick-bedded, fine-grained limestone; 8. Accharoidal dolomite; 9. Dolomite marl, marl; 10. Arenose limestone; 11. Limestone by mud movement; 12. Biogen limestone; 13. marl, claystone; 14. Calcareous marl lined limestone; 15. Dolomite marl, claystone; 16. Red and green sandstone, aleurit; 17. Dolomite marl, marl; 18. Fault

maximum value in WNW-ESE direction. The value of the nearest neighbour index indicates that there is no regularity in the doline pattern (Kevei Bárány et al. 2015).

9.3 Karst Features

Solution (drawdown) dolines

Majority of solution dolines (Fig. 3c, chapter "GIS Research of Karst Terrains in Hungary") is of plateau position (Fig. 66b). Plateau position indicates that they cannot be point recharge dolines since these are of valley floor position. They can be differentiated from subsidence dolines because of the following.

Since solution dolines are larger (Figs. 67 and 88) regarding both their marginal diameter and depth, their diameter is larger than some 10 m (it can reach 70–80 m) and their depth can be 20 m (Veress 2011). The diameter of the largest is 240 m and its depth is 25–30 m (Lovász 1971). The diameter of subsidence dolines is smaller than 10 m (their depth is at most some metres), but mostly only 1–3 m (Figs. 67 and 68).

As solution dolines did not develop in the superficial deposit (loess), they are only lined with superficial deposit (Fig. 69) or they are filled with superficial deposit to various degree (Figs. 70 and 71), while subsidence dolines were formed in superficial deposit. The latter can occur on the superficial deposit that is intercalated into drawdown dolines (Fig. 69), or in the fill of the infilled dolines (Figs. 70 and

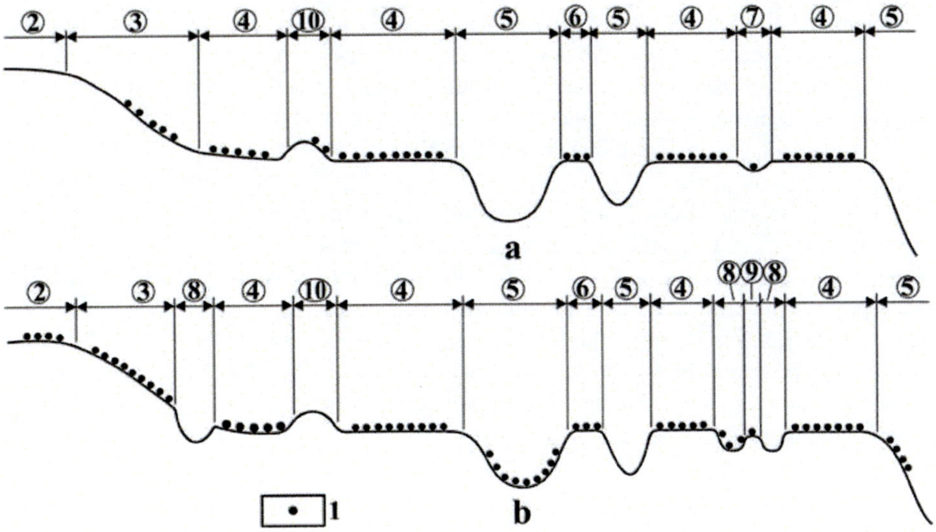

Fig. 66 Drawdown dolines and subsidence dolines on the lower abrasion level in various morphological environments. Legend: 1. Subsidence doline, 2. Upper abrasion platform, 3. Slope between the two platforms, 4. Plain terrain (plateau), 5. Valley, 6. Interfluve, 7. Valley head of valley, 8. Drawdown doline, 9. Dividing wall between dolines, 10. Mound on abrasion platform

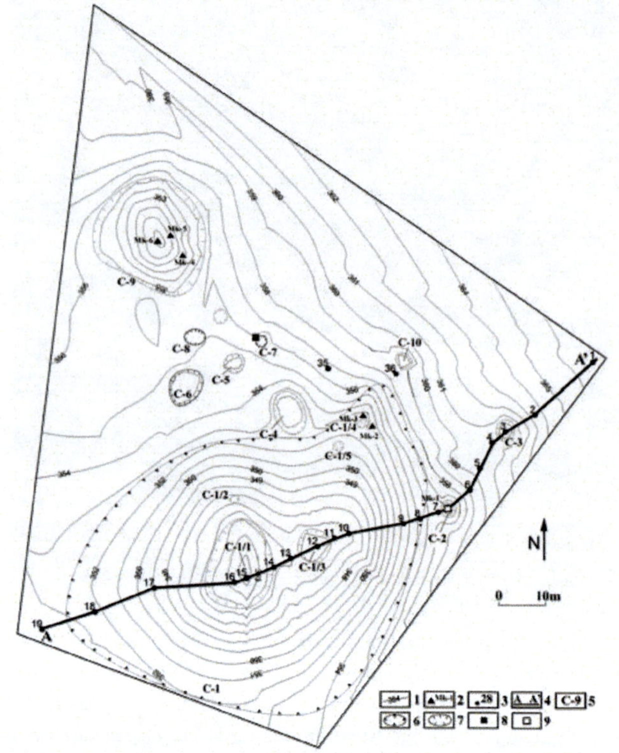

Fig. 67 Solution dolines and subsidence dolines from the western part of Orfű Karst (Orfű Karst, Cigány Hill, Veress 2011). Legend: 1. Contour line, 2. Rock outcrop and its identification code, 3. VES observation site with identification number, 4. Track of profile, 5. Identification mark of karst depression, 6. Solution doline, 7. subsidence doline, 8. Conduit, 9. Chimney, shaft

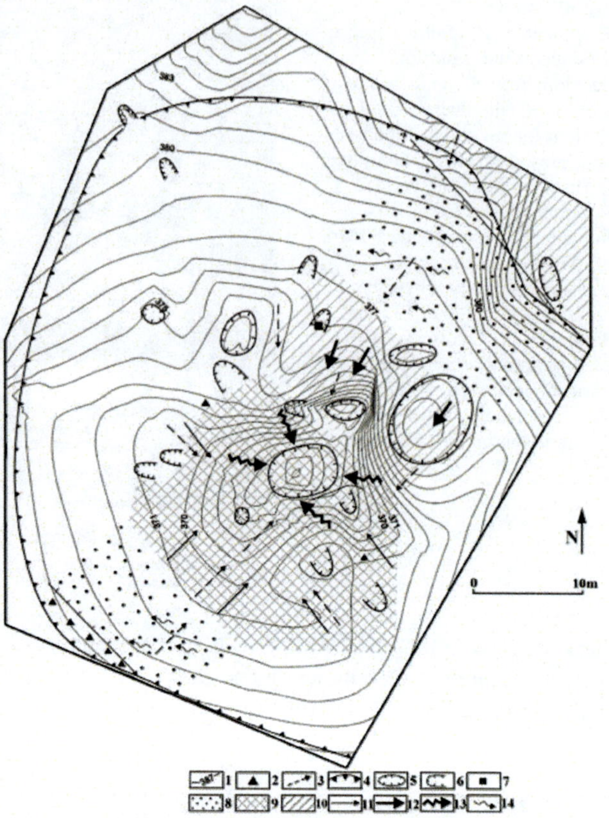

Fig. 68 Supposed percolations of the cover of a solution doline and the distribution of subsidence dolines of the doline floor on Orfű Karst (Orfű Karst, Cigány Hill, Veress 2011, modified). Legend: 1. Contour line, 2. Rock outcrop, 3. Inclination of the surface of limestone bedrock, 4. Lined solution doline, 5. Subsidence doline, 6. Open depression on the cover, 7. Conduit in the cover, 8. sand-loess (with limestone debris), 9. Clay (with limestone debris, sandy) on the bedrock, 10. Clay (with limestone debris, sandy) in superficial deposit, 11. Percolation above the clay, 12. Percolation below the clay, 13. Percolation through the clay, 14. Percolation through the sand-loess beds

71), or on the superficial deposit of a bedrock without dolines (Fig. 72).

Since the occurrence of subsidence dolines especially on Orfű Karst developed in solution dolines (Figs. 67, 68 and 69).

Smaller dolines are undoubtedly of shorter life span than larger ones as stated by several authors (Hevesi 2001; Lippmann et al. 2008). However, as they developed in the superficial deposit, they have different genetics (of subsidence origin), and they were formed as a result of younger karstification because, as it has already been mentioned, a lot of them developed in the superficial deposit lining, filling, or covering solution dolines.

Since drawdown dolines are lined with superficial deposit, their formation and development preceded loess development and the development of subsidence dolines in the superficial deposit. According to Szabó P (1968), their age is Tertiary, while Hevesi (2001) thinks that they developed in the Pliocene. In our opinion, the development of drawdown dolines was retarded and modified by the superficial deposit inside them.

Drawdown dolines are often elongated and have an asymmetric cross section along their longitudinal axis. Their dip slope is shorter and steeper. Their longitudinal axis coincides with fractures (Czigány and Lovász 2006), and valley directions (Lovász 1971) and with the direction of the doline rows (see below).

According to their morphological environment, they can be situated close to the summit (mound), on valley heads or on interfluves (Hevesi 2001). They are common on plain (plateau) terrains, they more rarely occur in smaller valleys of the karst (Vetési-Foith et al. 2017).

The floor of drawdown dolines can be slightly flat (probably because of their partial accumulation) while the floor of others is narrowing funnel-like. Inside them, the

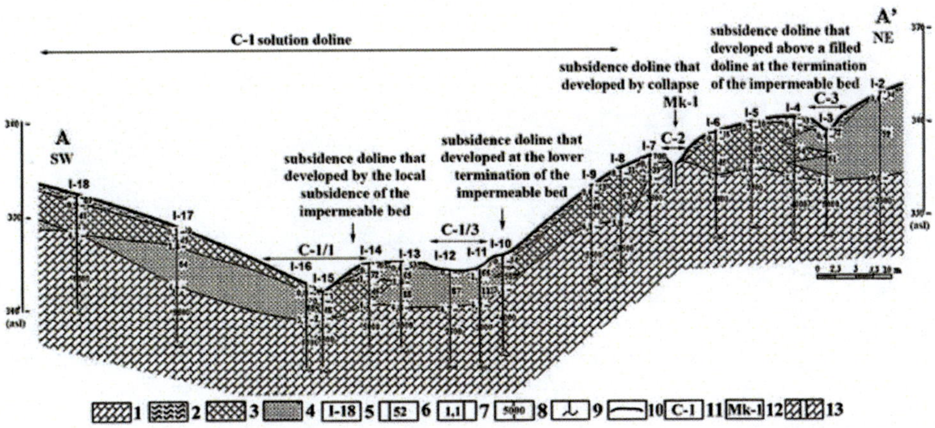

Fig. 69 Geoelectric-geological profile 'A-A' of the solution doline that can be seen in Fig. 67 (Orfű Karst, Czigány Hill, Veress 2011). Legend: 1. Limestone; 2. Soil, sand, sand flour; 3. Clay (with limestone debris, sandy); 4. Sand-loess (with limestone debris); 5. VES measurement site with identification number; 6. Geoelectric resistivity of the series (Ohmm); 7. Base depth of the geoelectric series (m); 8. Geoelectric resistivity of bedrock (Ohmm); 9. Approximate depth of penetration of VES measurement; 10. Geoelectric series boundary; 11. Identification mark of karst depression; 12. Identification number of rock outcrop; 13. Chimney

bedrock outcrops rarely or only in small patches from below the cover and the soil.

Dolines may be solitary (maybe twins), complex (Hevesi 2001), and grouped. On the floor of complex dolines, both intermittent drainage network and drainage site (subsidence doline) can occur as well. Grouped dolines constitute rows

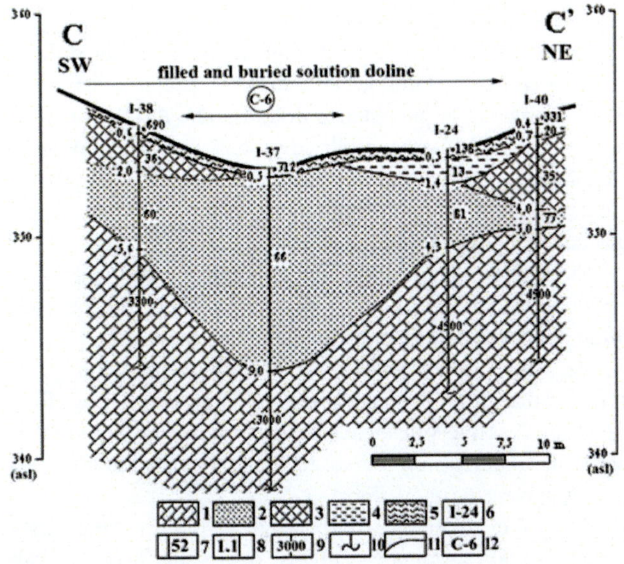

Fig. 70 Filled solution doline (Veress 2011). Legend: 1. Limestone; 2. Sand-loess (with limestone debris); 3. Clay (with limestone debris, sandy); 4. Clay; 5. Soil, sand, sand flour; 6. VES measurement site with identification number; 7. Geoelectric resistivity of the series (Ohmm); 8. Base depth of the geoelectric series (m); 9. Geoelectric resistivity of bedrock (Ohmm); 10. Approximate depth of penetration of VES measurement; 11. Geoelectric series boundary; 12. Identification code of karst depression

(Lovász 1971, 1977; Czigány and Lovász 2006) which may be separated into uvalas built up of constituting dolines of various number. Doline rows are arched and the doline number is variable in the rows. Doline rows primarily occur at the valley heads of valleys and in their continuation on plain terrains. A difference can be made between the direction of the doline row (the position of the row is given by their cardinal points) and their orientation (the cardinal point in the direction of which doline size increases). Orientation is identical with the dip direction of the bearing terrain. On Abaliget Karst, dolines are arranged in three doline row systems and into some doline rows. The rows and systems are on terrains with an elevation below 350 m and have a direction of N-S or NW–SE and an orientation of N, NW (Lovász 1971, 1977). Its longest expansion is close to 3 km. In the systems, the main doline rows are connected by tributary doline rows (in some cases they also have tributary rows) (Lovász 1971, Fig. 73). The doline rows occur on the lower platform exclusively, but the subsidence doline rows constituting their continuation are also present on the slope separating the two levels (Lovász 1971).

On Orfű Karst, where the dolines are grouped in four rows (Fig. 74) the doline rows are shorter than on Abaliget Karst, they are less complex, they are situated in more diverse directions, and they are also present on the slope between the two levels (Lovász 1977). Solitary drawdown dolines are common here. They follow structural directions to a smaller extent (Lovász 1977).

Few doline rows can be detected in the area of Misina Tubes (Fig. 74). The rows are less complex even than the doline rows of Orfű Karst. The rows are partly constituted by subsidence dolines (Lovász 1977). It can be experienced (Czigány and Lovász 2006) that subsidence dolines occur

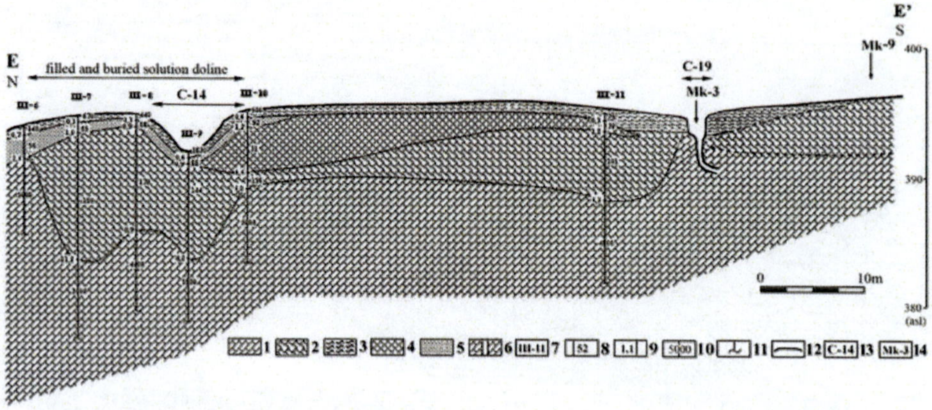

Fig. 71 Geoelectric-geological profile of subsidence dolines that developed above dissected bedrock on Orfű Karst (Veress 2011). Legend: 1. Limestone; 2. Limestone debris (sand?); 3. Soil, sand, sand flour; 4. Clay (with limestone debris, sandy); 5. Sand, loess (with limestone debris); 6. Conduit in the cover and in the bedrock; 7. VES measurement site with identification number; 8. Geoelectric resistivity of series (Ohmm); 9. Base depth of the geoelectric series (m); 10. Geoelectric resistivity of bedrock (Ohmm); 11. Approximate depth of penetration of VES measurement; 12. Geoelectric series boundary; 13. Identification code of karst depression; 14. Rock outcrop with identification code

next to solution doline rows, thus increasing the complexity and density of the rows. On karst area sections dissected by valleys, doline rows occur on interfluves and in the valley heads of large valleys (Lovász 1977).

Taking into consideration the morphological characteristics of dolines and the development characteristics of the rows, Czigány and Lovász (2006) distinguished the following varieties of doline rows:

- Short doline row (of some 100 m long) which is an uvala or uvala-like since the elevation of the dividing wall between the dolines is smaller than the elevation of its environment. The doline row is on plateau.
- The doline row is short and it is situated on the slope dipping towards the valley head of the valley. The dolines of the row grow towards the valley head.
- The doline row is longer (consists of more dolines) than in the above cases. Their upper (more elevated) end is on plateau or on interfluve, but the lower end of the row stretches into the valley and terminates above its floor. The doline row has a tributary branch, the dolines of the main branch are the parts of uvalas. They mostly have an orientation of N and NW.
- The doline row is several km long and it has tributary doline rows of significant length. They are of northern orientation, their upper end is on plateau, and their lower end terminates at valley margin. The dolines of lower and lower elevation of the doline row are larger and larger and form parts of uvalas to a larger degree.

Subsidence dolines

The depressions of these doline type give the majority of the dolines of the karst. Their characteristics are as follows:

Subsidence dolines are present both on lower elevated terraces and higher elevated terraces, while drawdown dolines can only be found on lower elevated terraces. Their size, mainly their diameter is small, mostly some metres (their depth is even smaller).

Suffosion dolines are predominant, but dropout dolines are also common. Dropout dolines are especially characteristic among minor subsidence dolines. The development of dolines probably begins with cover collapse. The slopes of small-sized dropout dolines (soil-covered dropout dolines, Veress 2016) of such origin become gentle subsequently and the doline is turning into a suffosion doline.

Their floor is not filled up (the pile of the collapsed, subsided material can be locally noticed, but the bedrock often outcrops). This can be explained by the low inclination of the bearing terrain and by the lack (and immaturity) of features that drain the water into the depression. Therefore, filled-up, inactive subsidence dolines (normally common on concealed karsts) may be absent here. The swallow hole is mostly absent on their floor.

According to their morphological environment, they occur on plain terrains (on plateaus), on interfluves, in valleys, in (the rows of) drawdown dolines, and in their environs (Fig. 66). Valley dolines can be found in valley sides, on valley floors, and in their channels too. In smaller, partly filled valleys they may be uvala-like. They may be present

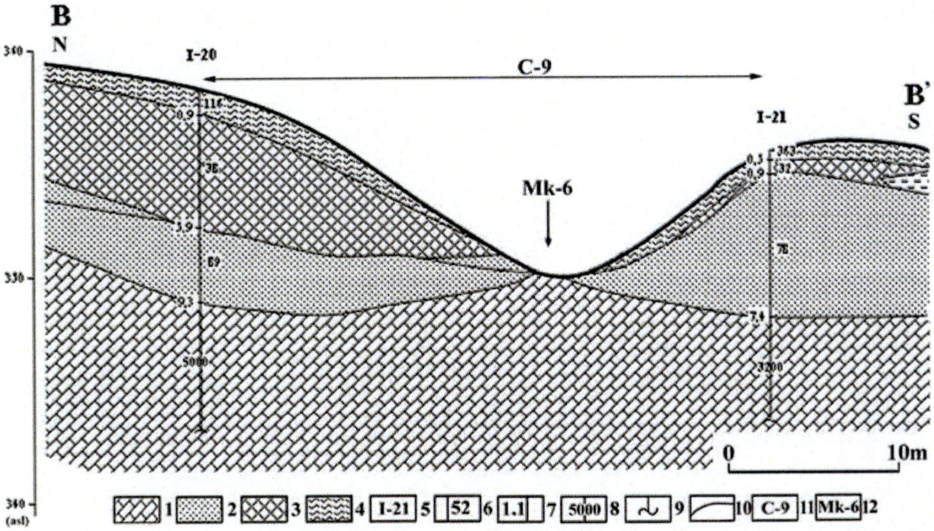

Fig. 72 VES profile of subsidence doline that developed on plain bedrock (Orfű Karst, Cigány Hill, Veress 2011). Legend: 1. Limestone; 2. Sand-loess (with limestone debris); 3. Clay (with limestone debris, sandy); 4. Soil, sand, sand flour; 5. VES measurement site with identification number; 6. Geoelectric resistivity of series (Ohmm); 7. Base depth of the geoelectric series (m); 8. Geoelectric resistivity of bedrock (Ohmm); 9. Approximate depth of penetration of VES measurement; 10. Geoelectric series boundary; 11. Identification code of karst depression; 12. Rock outcrop with identification code

on the doline floors and slopes in drawdown dolines, but they are also observed at doline margins and on the dividing walls between adjacent drawdown dolines and on the dividing walls of the constituting dolines of uvalas. In the steep valley side of the northern margin of Abaliget Karst, dropout dolines alternate with features of mass movements.

According to their pattern, they occur individually or in groups. The latter form irregular groups (piles) or are arranged in rows. In drawdown dolines where they also occur in groups they form a row or rows.

– On steeper slopes (for example, in valley side position), they are elongated in downslope direction (Lipmann et al. 2008). The larger the inclination of the bearing slope, the longer the dolines (Vetési-Foith 2017). Elongated dolines are asymmetric: their lower slope is short and steeper, the opposite slope is long and less steep.

According to VES measurements, they are situated in three positions as compared to the morphology of the bedrock (Veress 2011). Thus, they can develop above a plain bedrock (Fig. 72), above filled-up drawdown doline (Figs. 70 and 71) and above the mound of the bedrock (right part of Fig. 71). The side slopes of the dolines above the plain bedrock are constituted by wedging-out cover beds (Fig. 70). The fill of drawdown dolines can also be significant below subsidence dolines, it can reach 10 m (Fig. 70). Where the upper beds are curved, the incurvation is caused by the suffosional material loss from the lower ones (the left part of Fig. 71), and where they are not curved, the material loss happened by collapse (right part of Fig. 71). It is more probable that epikarst passages are responsible for the material loss of the dolines above drawdown dolines, while in case of those situated above the plain bedrock, the epikarst passages may have developed later as well. The older age of the bedrock passages of the subsidence dolines above drawdown dolines (thus the dolines are postgenetic) is due to the fact that the doline fill below them is limestone debris (Fig. 71) or contains limestone debris too (Fig. 70). Since the thickness of the superficial deposit approaches (and it also exceeds) 10 m, the infiltrating water probably arrives at the bedrock in a saturated state, and thus dissolution on the bedrock took place preceding the development of the superficial deposit and of subsidence dolines.

The development site of dolines is associated with clay and clayey beds (Veress 2011). Subsidence dolines primarily develop at places where the clay intercalation wedges out (Figs. 69, 70 and 71). Here, suffosion, and thus doline development is promoted by the water reaching the clay bed and percolating above it. The water gathering at the wedging-out of the clay bed, and thus having a larger quantity may infiltrate downwards, strengthening suffosion. It may happen that the clay beds bearing the subsidence doline are also curved (below the subsidence doline situated at the lowest point of the doline in Fig. 69). In this case, the material was transported from the superficial deposit below

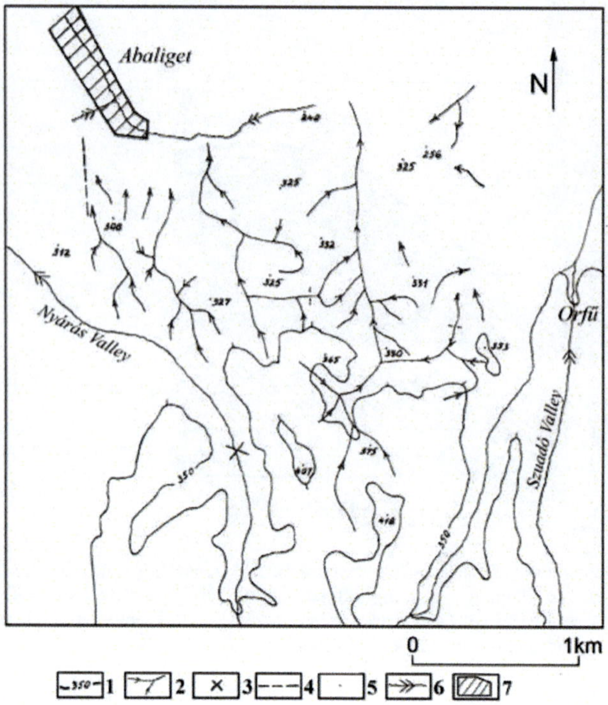

Fig. 73 Doline rows on Abaliget Karst (Lovász 1977). Legend: 1. Contour line, 2. Doline row and its orientation, 3. Ponor, 4. Abaliget Cave, 5. Mound, 6. Valley and its stream, 7. Settlement

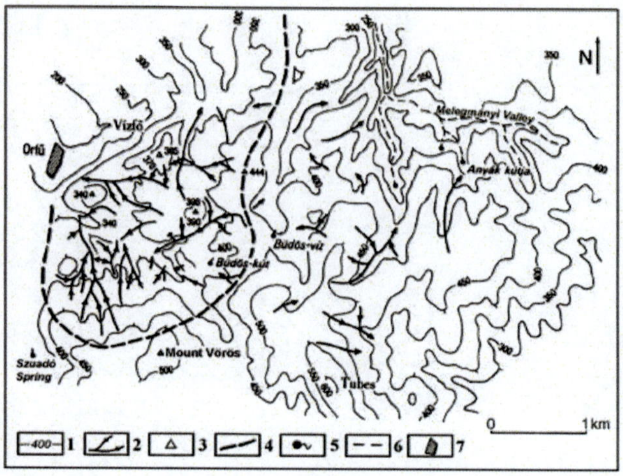

Fig. 74 Doline rows on Orfű Karst and on Misina Tubes Karst (Lovász 1977). Legend: 1. Contour line, 2. Doline row and its orientation, 3. Mound, 4. Watershed, 5. Spring, 6. Stream, 7. Settlement

the clay bed into the karst. This is possible in a way that the meteoric water infiltrating below the clay (which reaches the deepest point of the drawdown doline) causes suffosion.

The fact that the number of subsidence dolines is high on the karst can probably be explained by several reasons.

- The karst, especially because of the elevation of the lower northern part, favoured loess accumulation.
- The exemption of loess from calcareous material since it does not decrease the solution capacity of the carbonated water arriving from above (though according to VES measurements, the superficial deposit contains limestone debris at several places).
- The small thickness of the superficial deposit (although the thickness of the superficial deposit below subsidence dolines in drawdown dolines often exceeds the values given in literature for the terrains between depressions, it does not exceed or slightly exceeds 10 m).
- The already mentioned clayey, but wedging-out intercalations which favour concentrated water drainage into the karst.
- The low extent of dissection and inclination of the bearing surface, which favour infiltration, the survival of loess, and the uniform development of loess. The infiltration of small degree is proved by the fact that on the karst of Misina Tubes where there are several valleys, and thus a larger quantity of water can leave the karst and flow on the surface, there are fewer dolines. Another evidence is that on interfluves without tributary valleys, thus on terrains being more suitable for infiltration, doline density is higher.
- The presence of drawdown dolines which function not only as sediment traps, but also as meteoric water traps.
- Epikarst dated to older karstification.

Ponors

Ponors are aligned at the southern margin of the Abaliget and Orfű part of the karst, where well-karstifying rocks are in contact with non-karstic and poorly karstifying rocks (Fig. 63), mostly on the floors of through valleys. Permanent streams (of the Szuadó and Nyárás Valleys) and intermittent streams (Büdöskút Valley) lead to the depressions (Barta and Tarnai 1997).

Although as it has been mentioned above, in Hungarian karst literature these features are regarded as ponors, it is more probable that they are point recharge dolines. This is proved by the following observations:

- Some features are not situated at rock boundary, but they are inside the karst (the ponors of Nyárás Valley and Büdöskút Valley), others are at rock boundary, but at the interface of dolomite and limestone.
- They constitute a row on the same valley floor (the ponors of Szuadó and Büdöskút Valleys).
- There is no counterslope section and, thus, a blind valley at the depressions on the valley floor in all cases. In its

lack, in case of floods, the water flows through the depressions of upper position into the depressions of lower position as it was observed at the depressions of the Szuadó Valley (Barta and Tarnai 1997).

The ponors of the karst or depressions functioning as ponors are in the following valleys (Barta and Tarnai 1997, Fig. 63). Two depressions from the catchment area of Orfű Spring in the Büdöskút Valley and two other depressions on its section joining the Remeterét Valley. In the Szuadó Valley, there are three (Szuadó, Gilisztás, and Trió ponors) depressions functioning as ponors. (In the Körtvélyes Valley, there is only percolation.) In the catchment area of Abaliget Cave, the ponors of the Nyárás and Vigánvár Valleys are worth mentioning. These depressions are located on the floor of valleys extending over the karst.

Concretions

Kraft et al. (1986) described 13 calcareous sinter occurrences in the Mecsek Mountains (only 9 originates from karstwater), their development age is young (Holocene), most intensively formed 2500–5000 years ago (Kraft et al. 1986). The most significant is the freshwater limestone of Tettye Spring where sinter accumulated in a length of 300 m. According to the above authors, the calcareous sinter is older here, it probably began to accumulate in the Early Pleistocene. In their studies, they distinguished calcareous sinter developed on valley floor (in channel) and that formed in valley side (on slope). Among the occurrences, six are of valley-floor position and seven of slope position (however, the freshwater limestone of Tettye Spring also has a part of slope environment). Three occurrences have travertine mounds and ten have dams with basins, but in two cases micro-dams also occur (Figs. 73 and 74).

10 Conclusions

The surface karst of the Aggtelek-Rudabánya Mountains, the Bükk Region, the Bakony Region, the Vértes Mountains, the Gerecse Region, the Buda Mountains, the Pilis Mountains, and the Western Mecsek was described. Among them, the Aggtelek-Rudabánya Mountains and the Bükk Region have specific, diverse large karst landforms. A common characteristic of Hungarian karsts is the lack of bare karren and poljes.

On the plateaus of the Aggtelek-Rudabánya Mountains, solution dolines are predominant on valley floors, on mounds, and on plain terrains. The horizontal growth of dolines leads to uvala formation. Some depressions are exhuming paleokarst features, whose karstic development stopped when they were sediment traps. Their exposure is associated with a change of their function (ponors) and this promoted the formation of major caves. Ponors can be karst marginal ponors (some of them contributed to the development of significant cave systems) and ponors inside the karst (the latter mainly in larger karst depressions). Subsidence dolines occur in karst depressions.

In the Bükk Region, valley floor dolines are predominant, during their development uvalas and valley sections turning into arheic karst depressions of various sizes were formed. The dolines of the mounds are specific, among which flat-floored plate-shaped dolines are common (these also occur on plain terrains). Karst marginal ponors are characteristic on the Nagy Plateau and in the Southeastern Bükk in the Bükk Region. Ponors inside the karst are located in larger karst depressions and on the floor of epigenetic valleys. The widespread character of through caves favoured the development of gorges.

In the mountain parts of the Transdanubian Mountains, karstification takes place on blocks and is determined by the presence, thickness, and quality of the superficial deposit. The permeable superficial deposit (mainly loess) on some block types (horst elevated to summit position) resulted in the development of concealed karst with subsidence dolines. However, smaller patches of the mixed allogenic-autogenic karst with karst marginal ponors (Kab Mountain) and patches of cryptokarst with caprock dolines (Gerecse Region, Pilis Mountains) also occur. The gorges of the Bakony Region developed at epigenetic-antecedent valley sections.

The karst of the Western Mecsek was formed on abrasion platforms, its landforms include solution dolines and subsidence dolines. Doline density is high (the highest among Hungarian karsts). Solution dolines constitute doline rows and doline systems which are built up of main and tributary doline rows. Their orientation coincides with the slope direction of the surface. The doline rows occur in the continuation of valleys, on terrains surrounding valleys and sloping towards them. Subsidence dolines often occur on the floor of solution dolines.

References

Ádám L, Pécsi M (1988) A Vértes-Velencei-hegyvidék (The Vértes-Velence Mountains). In: Pécsi M (ed) A Dunántúli-középhegység B, Akadémia Kiadó Budapest, pp 190–311 (in Hungarian)

Balázs D (1960) A Tereszteny ei barlangrendszer víznyelőinek és forrásainak összefüggéseire vonatkozó kísérletek (Experiments on the relations between the ponors and springs of Teresztenye cave system). Karszt- és Barlangkutatási Tájékoztató 227–229 (in Hungarian)

Balogh K (1964) A Bükk-hegység földtani képződményei (Geological features of the Bükk Mountains). Magyar Állami Földtani Intézet Évkönyve 48(2):719 (in Hungarian)

Bárdossy Gy.-Pataki A.-Nándori Gy (1983) Bányaföldtani térképsorozat módszertani kidolgozása és gyakorlati alkalmazása az iharkúti külfejtéses bauxitbányászatban (Methodological elaboration and practical application of a minegeological map series in the open cut bauxite mining of Iharkút). Földtani Kutatás 26:3–10 (in Hungarian)

Barta K, Tarnai T (1997) Karsztkutatás az orfűi Vízfő-forrás vízgyűjtő területén (Karst research on the catchment of the Vízfő spring, Orfű). Karszt és Barlang I-II:12–19 (in Hungarian)

Bárány I, Mezősi G (1978) Adatok a karsztos dolinák talajökológiai viszonyaihoz (Data on the soil-ecological relations of karst dolines). Földrajzi Értesítő XXVI I(1):65–73 (in Hungarian)

Bátori Z, Körmöczi L, Erdős L, Zalatnai M, Csiky J (2012) Importance of karst sinkholes in preserving relict, mountain and wet woodland plant species under sub-Mediterranean climate: a case study from southern Hungary. J Cave Karst Stud 74:127–144

Bella P, Veselsky M, Gaál L, Miár J (2016) Jósvafő paleo-polje: morphology and relation to the landform evolution of Aggtelek karst and Jósva River valley, Hungary. Z Geomorphol 60(3):219–235

Benkhard B (2015) Travertine Mound of Egerszalók: The Hungarian Pamukkale. In: Lóczy D (ed) Landscapes and landforms of Hungary. Springer, Heidelberg, New York, Dordrecht, London, pp 201–207. https://doi.org/10.1007/978-3-319-08997-3

Bertalan K (1963) A Bakonybéli Somhegy barlangjainak kutatástörténete (Research history of the caves of the Som Hill of Bakonybél). Karszt És Barlang II:75–78 (in Hungarian)

Bonacci O (2004) Poljes. In: Gunn J (ed) Encyclopedia of caves and Karst Science. Fitzory Derborn, New York, pp 599–600

Bódai B, Móga J, Barth A, Holló S, Kele S (2016) A szilvásváradi Szalajka-patak mésztufáinak szezonális geokémiai vizsgálata. (Seasonal geochemical study of freshwater tufa of Szalajka-creek (Bükk Mts, Hungary). Karsztfejlődés XXI. 45–64. https://doi.org/10.17701/16.45-64:45-64 (in Hungarian)

Bögli A (1960) Kalklösung und Karrenbilding. Zeit. F. Geomorph. n. e. Suppl. 2:4–21

Bratán M, Mohos P, Zsuffa I (1967) A Gerence-patak hidrológiai tanulmánya (The hydrological study of Gerence stream). Hidrológiai Közlöny 47:284–296 (in Hungarian)

Bulla B (1968) A Magyar föld domborzata fejlődésének ritmusai, az újharmadkor óta a korszerű geomorfológiai szemlélet megvilágításában (The phases of the Terrain Development of Hungary since the Lower Tertiary in perspective of the modern geomorphological view). Válogatott természeti földrajzi tanulmányok, Akadémia Kiadó, Budapest, pp 90–104 (in Hungarian)

Cholnoky J (1932) A mészkőhegységek földrajzi jellemvonása (Geographical characteristics of limestone mountains). Földgömb III 5:194–201 (in Hungarian)

Cholnoky J (1940) A csillagoktól a tengerfenékig III (From stars to the sea floor). Franklin Társulat, Budapest 496 p (in Hungarian)

Császár G, Csereklei E, Gyalog L (1981) A Bakony-hegység földtani térképe (szerk.) (Geological map of the Bakony Mountains) Magyar Állami Földtani Intézet, Budapest (in Hungarian)

Csillag G, Sebe K (2015) Long-term geomorphological evolution. In: Lóczy D (ed) Landscapes and landforms of Hungary. Springer, Cham Heidelberg, New York, Dordrecht, London, pp 29–38. https://doi.org/10.1007/978-3-319-08997-3

Czigány Sz, Lovász Gy (2006) A mecseki karszt térképezésének újabb eredményei (Additional data on the mapping of the Western Mecsek karst area). Természetföldrajzi közlemények a Pécsi Tudományegyetem Földrajzi Intézetéből, Pécsi Egyetem, Pécs 28:3–14 (in Hungarian)

Czesznak L, Szegediné Darabos E, Tóth M, Lénárt L, Hernádi B (2014) Új típusú vízkészlet meghatározás a Bükkben (New method of karstwater resources calculation in Bükk mountains). Karsztfejlődés XIX:125–136 (in Hungarian)

Deák I, Leél-Őssy Sz, Kövér Sz, Surányi G (2009) A Csévi-szirtek barlangjai (Caves of the Csév klippen). Karszt és Barlang 2007 (1-II):17–36 (in Hungarian)

Dénes Gy (1971a) A fokozatosan lepusztuló vízzáró takaró szerepe az exhumálódó karszt morfológiai fejlődésében (The role of the gradually denuding impermeable cover in the morphological development of exhuming karst). Karszt és Barlang I.:5–8 (in Hungarian)

Dénes Gy (2001) A Gellért-hegy karsztja és barlangjai (The karst and caves of Gellért Hill). Karsztfejlődés VI:133–141 (in Hungarian)

Farsang A, M Tóth T (1992) A Bükki dolinák morfometriai vizsgálata (A morphometric investigation of dolines in the Bükk Mountains). In: A Bükk karsztja, vizei, barlangjai II:39–50 (in Hungarian)

Ferenczy G (2002) A hegységben található barlangok általános jellemzése (A general description of the caves in the mountains). In: Baráz Cs (szerk): A Bükki Nemzeti Park, Bükki Nemzeti Park Igazgatóság, Eger 167–176 (in Hungarian)

Ford DC, Williams PW (2007) Karst hydrogeology and geomorphology. Wiley, Chichester, p 562

Gaál L (2014) A földtani fejlődés áttekintése (An overview of geological development). In: Gruber P.-Gaál L. (szerk.) A Baradla-Domica-barlangrendszer – Aggteleki Nemzeti Park, Jósvafő 19–30 (in Hungarian)

Gánti T (1957) A naszályi karszt (Naszály karst). Hidrológiai Közlöny 37(4):380–384 (in Hungarian)

Gergely F (1938) Geomorfológiai megfigyelések az Északi-Bakony területén (Geomorphological observations in the area of Northern Bakony). Bölcsészdoktori Értekezés, Kézirat (in Hungarian)

Gruber P, Keresknyéi E, Koltai G, Kevei-Bárány I (2014) Adatok a Baradla-barlang hidrogeográfiai viszonyaihoz, különös tekintettel a befolyó és csepegő vizekre (Contributions to the hydrogeography of Cave Baradla with special regards on inflowing and dripping waters). APN füzetek. XI:9–26 (in Hungarian)

Gruber P, Dagmar H, Balázs I, Mátrahalmi T, Serfőző A, Ambrus M (2015) Víznyomjelzési vizsgálatok a Haragistya-Szilice-Boszova karsztterületen (Trace test in the Haragistya-Silica-Silická, Berzova karst area) Karsztfejlődés XX:63–79 (in Hungarian)

Győrffy D (1957) Geomorfológiai tanulmányok a Káli-medencében (Geomorphological studies in the Kál Basin). Földrajzi Értesítő VI. 3:265–299 (in Hungarian)

Gyuricza Gy.–Sásdi L (2009) A Baradla-barlangrendszer kialakulásának kérdései a tágabb környezet földtani fejlődésének tükrében (Theory of the evolution of the Baradla Cave as mirrored in the geological evolution of its neighbourhood). Földt. Köz. 139(1):83–92 (in Hungarian)

Hazslinszky T (1965) Az Északborsodi Alsóhegy karsztjának néhány hidrográfiai kérdése (Some hydrographic questions on the karst of Alsó-hegy of Northern Borsod). Hidrológiai Közlöny 6:259–266 (in Hungarian)

Hazslinszky T (1974) Az Alsó-hegy vázlatos vízháztartási mérlege (A sketchy balance of water balance of Alsó-hegy). Karszt és Barlang I:11–17 (in Hungarian)

Hevesi A (1972) Forrásmészkő képződés a Bükkben (Freshwater limestone development in the Bükk Mountains). Földrajzi Értesítő XX I(2–3):187–205 (in Hungarian)

Hevesi A (1978) A Bükk szerkezet- és felszínfejlődésének vázlata (A Sketch of the Structure and Landscape Evolution of the Bükk Mountains). Földrajzi Értesítő, XXVII 2:169–203 (in Hungarian)

Hevesi A (1980) Adatok a Bükk hegység negyedidőszaki ősföldrajzi képéhez (Data to the Quaternary Paleogeographical Features of the Bükk Mountains) Földtani Közlöny 110(3–4):540–550 (in Hungarian)

Hevesi A (1984) Karsztformák kormeghatározásáról és mészkőhegységeink újharmadidőszak végi – jégkori arculatának megrajzolásában játszott szerepükről, a Bükk hegység példáján (On the dating of karst features and their role in the Late Tertiary – Pleistocene evolution: example of the Bükk Mountains). Földrajzi Értesítő XXXIII 1–2:25–36 (in Hungarian)

Hevesi A (1986a) A Déli-Bükk karsztja II. rész: Délnyugati-Bükk (The karst of Southern Bükk Part II: South-western Bükk). Karszt És Barlang II:87–94 (in Hungarian)

Hevesi A (1986b) A Déli-Bükk karsztja I. rész: Délkeleti-Bükk (The karst of Southern Bükk Part I: South-eastern Bükk). Karszt és Barlang I.:3–14 (in Hungarian)

Hevesi A (1986c) Rombarlangok a Bükkben (Collapsed caves in the Bükk Mountains). Nehézipari Műszaki Egyetem Közleményei 33 (1–4):167–179 (in Hungarian)

Hevesi A (1991a) Magyarország karsztvidékeinek kialakulása és formakincse, I. rész (Development and landforms of karst regions in Hungary I). Földrajzi Közlemények XV(1–2):25–35 (in Hungarian)

Hevesi A (1991b) Magyarország karsztvidékeinek kialakulása és formakincse, II. rész (Development and landforms of karst regions in Hungary II). Földrajzi Közlemények CXV 3–4:99–120 (in Hungarian)

Hevesi A (2000) Karsztos hegységeink arculata (The aspect of our karst mountains). In: Karátson D (ed) Pannon Enciklopédia – Kertek, Budapest, pp 288–293 (in Hungarian)

Hevesi A (2001) A Nyugat-Mecsek felszíni karsztosodásának kérdései (Questions of the surface karstification of the Western Mecsek). Karsztfejlődés VI:103–111 (in Hungarian)

Hevesi A (2002) Felszínalaktani jellemzés, karsztformakincs (Geomorphological characterisation, karst landform). In: Baráz Cs (ed) A Bükki Nemzeti Park, Bükki Nemzeti Park Igazgatóság, Eger, pp 109–148 (in Hungarian)

Hevesi A (2007) A legutóbbi három tél (2004/05, 2005/06, 2006/07) hóviszonyainak hatása a Bükk-fennsík töbörfejlődéséhez (The effect of the snowfall relations of the last three winters - 2004/05, 2005/06, 2006/07 - on the doline formation of the Bükk Plateau). Karsztfejlődés XII:165–170 (in Hungarian)

Hevesi A (2019) Valamit a mészkőbarlangok és a mészkőszurdokok származási kapcsolatáról (About the relation of the origin of limestone caves and limestone gorges). Karszt- és Barlang, Különszám 23–30 (in Hungarian)

Hevesi A, Záhorszky A (2000) A miskolc-tapolcai kőmázsák és környékük 1:10000-es felszínalaktani térképezésének eddigi eredményei (Results of 1:10000 scale geomorphological mapping of Miskolc-Tapolca kőmázsák and their environs). Karsztfejlődés V:41–45 (in Hungarian)

Horváth S (1962) Fedett karrok a Bükkben (Covered karren in the Bükk Mountains). Karszt És Barlang I:25–26 (in Hungarian)

Hoyk E (2002) A Nyugat-Mecsek karszt dolináinak morfometriai vizsgálata (A morphometric investigation of the dolines of the Western Mecsek karst). Karsztfejlődés VII:161–171 (in Hungarian)

Jakucs L (1956a) Adatok az Aggteleki-hegység és barlangjainak morfogenetikájához (Some data to the morphogenetics of the mountains and caves of Aggtelek). Földrajzi Közlemények IV:25–38 (in Hungarian)

Jakucs L (1956b) A barlangi árvizekről (On cave floods). Földrajzi Közlemények IV:281–400 (in Hungarian)

Jakucs P (1956c) Karrosodás és növényzet (Karren formation and vegetation). Földrajzi Közlemények 3:241–249 (in Hungarian)

Jakucs L (1964) Geomorfológiai problémák az Észak-Borsodi-karsztvidéken (Geomorphological problems in the Northern Borsod karst region). Borsodi Földrajzi Évkönyv 5:12–23 (in Hungarian)

Jakucs L (1968) Szempontok a karsztos tájak denudációs folyamatainak és morfológiájának értékeléséhez (Viewpoints to the evaluation of the denudation processes and morphogenesis of karst landscapes). Földrajzi Értesítő 17:17–46 (in Hungarian)

Jakucs L (1977a) Morphogenetics of Karst Regions. Adam Hilgar, Bristol, 284 pp

Jakucs L (1977b) A magyarországi karsztok fejlődéstörténeti típusai (Evolutionary types of Hungarian karsts). Karszt és Barlang I-II:1–16 (in Hungarian)

Jakucs L (1994) A Budai-hegység hidrotermális karsztja (The hydrothermal karst of the Buda Hills). Földrajzi Értesítő XLII I (3–4):235–245 (in Hungarian)

Jaskó S (1933) Morfológiai megfigyelések és problémák a Gömör-Tornai karsztvidék délkeleti részében (Morphological observations and problems in the south-eastern part of the Gömör-Torna karst region). Földrajzi Közlemények LX I(9–10):125–144 (in Hungarian)

Jaskó S (1957) A földtani felépítés és a karsztvíz elterjedésének kapcsolata a Dunántúli-középhegységben (The relation between geological structure and the distribution of karstwater in the Transdanubian Mountains). Kézirat, Magyar Állami Földtani Intézet Adattár, Budapest (in Hungarian)

Jaskó S (1961) A balatonfelvidéki és északbakonyi patakok vízhozamának kapcsolata a földtani felépítéssel (The relation of discharge of the streams of Balaton Highlands and Northern Bakony with geological structure). Hidrológiai Közlöny 41(1):75–84 (in Hungarian)

Jámbor Á (1959) A Bükk-fennsík pleisztocén „vályog" képződményei (The Pleistocene "adobe" features of the Bükk Plateau). Földtani Közlöny LXXXIX(2):181–184 (in Hungarian)

Juhász A (1976) A földtani viszonyok és karsztosodás összefüggése a Bükk-hegységben (The relation of geological conditions and karstification in the Bükk Mountains). Karszt és Barlang I-II:1–8 (in Hungarian)

Kevei Bárány I (1992) Gondolatok a karsztfelszíni formák rendszerezéséhez különös tekintettel a bükki dolinákra A Bükk karsztja, vizei, barlangjai II:269–277 (Thoughts to the classification of karst surface features with special regard to the dolines of the Bükk Mountains. The karst, streams and caves of the Bükk Mountains II:269–277) (in Hungarian)

Kevei Bárány I, Kiss M, Simon N (2015) Néhány további adat a hazai karszt dolinák aszimmetriájának kialakulásához (Some additional data on asymmetry in the formation of Hungarian karst dolines). Karsztfejlődés XX:125–144 (in Hungary). https://doi.org/10.17701/15.125-144

Kevey B, Csiky J (2012) A Nyugat-Mecsek növényvilága (Vegetation of the western Mecsek Mountains). In: Barta K, Tarnai T (eds) A Nyugat-Mecsek karsztja (Karst areas of the western Mecsek Mountains). GeoLitera, Szeged, pp 127–141 (in Hungarian)

Kele S, Scheuer Gy, Demény A, Shen C-C, Chiang H-W (2009) U-series dating and isotope geochemical study of the Gellért Hill (Budapest) travertine. Centr Eur Geol 52/3–4:199–224

Kele S, Scheuer Gy, Demény A, Shen C-C, Hong-Wei C (2011) A Rózsadomb (Budapest) édesvízi mészköveinek U/Th sorozatos kormeghatározása és stabilizotóp-geokémiai vizsgálata (Uranium-series dating and geochemical study of the travertines located on the Rózsadomb Hill (Budapest). Földt Közl 141/2:445–468 (in Hungarian)

Kele S, Scheuer Gy (2012) Az Ördögárok-völgyéhez kapcsolódó Kondor úti és Nyéki úti édesvízi mészkő előfordulások U-soros kormeghatározása és képződési környezetük rekonstrukciója (Uranium-series dating and paleoenvironmental reconstruction of the freshwater limestones of the Kondor and Nyéki streets, Ördögárok valley). Hidr Táj 1:1, 53–55 (in Hungarian)

Koltai G, Kele S, Kármán K, Kevei Bárány I. (2012) Édesvízi mészkövet lerakó források geokémiai vizsgálata a Mecsekben (A geochemical study of springs depositing freshwater limestone in the Mecsek Mountains). Karsztfejlődés XVII:35–46 (in Hungarian)

Korpás L, Kovács-Pálffy P, Lantos M, Földvári M, Kordos L, Krolopp E., Stüben D, Berner Zs (2003) A budai Vár-hegy édesvízi mészköveinek szedimentológiája, geokémiája, kronológiája és paleokarszt fejlődése (The sedimentology, geochemistry, chronology of the freshwater limestones and the paleokarst development of the Buda Vár Hill). Karsztfejlődés VIII:81–105 (in Hungarian)

Kozma K, Holló S (2010) A Berva-pataki víznyelő kialakulása és pusztulása az Egri Berva-bérc lábánál (The development and denudation of the ponor of Berva Stream at the foot of the Berva horst of Eger). Karsztfejlődés XV:103–112 (in Hungarian)

Kraft J, Scheuer Gy, Schweitzer F (1986) A Pécs-környéki forráslerakódások vizsgálata (A study of spring depositions in the environs of Pécs). Pécsi Műszaki Szemle XXXI(3):13–18 (in Hungarian)

Láng S (1948) Karszttanulmányok a Dunántúli-középhegységben (Karst studies in the Transdanubian Mountains) Hidrológiai Közlöny 28:49–52 (in Hungarian)

Láng S (1953) A Pilis geomorfológiája (The geomorphology of the Pilis Mountains). Földrajzi Értesítő 36–363 (in Hungarian)

Láng S (1954) Hidrológiai és morfológiai tanulmányok a Bükkben (Hydrological and morphological studies in the Bükk Mountains). Hidrológiai Közlöny XXXIV:103 (in Hungarian)

Láng S (1955a) Geomorfológiai tanulmányok az Aggteleki-karszton (Geomorphological studies in Aggtelek Karst). Földr Ért 4(1):1–17 (in Hungarian)

Láng S (1955b) A Gerecse peremhegységi részének geomorfológiája (Geomorphology of the offset part of Gerecse). Földrajzi Értesítő IV (2):157–194 (in Hungarian)

Láng S (1956) A Központi Gerecse geomorfológiája (Geomorphology of Central Gerecse) Földrajzi Értesítő V(3):265–281 (in Hungarian)

Láng S (1958) A Bakony geomorfológiai képe (The geomorphology of the Bakony Mountains). Földrajzi Közlemények 6:325–343 (in Hungarian)

Láng S (1962) A Bakony geomorfológiai vázlata (A geomorphological outline of the Bakony Mountains). Karszt És Barlangkut Táj 7:86–91 (in Hungarian)

Láng S (1964) A Bükk geomorfológiai vázlata (A geomorphological outline of the Bükk Mountains). Karszt És Barlangkutatási Tájékoztató 5–6:83–88 (in Hungarian)

Láng S (1971) A hazai karsztok és környékük lepusztulásának egyes kérdései (Some questions on the denudation of Hungarian karsts and their environs). Karszt és Barlang I.:1–4 (in Hungarian)

Láng S, Miháltz I-né, Vitális Gy (1970) A miskolctapolcai Nagykőmázsa dolináinak morfológiai és földtani vizsgálata (Morphological and geological investigations of the dolines of Nagykőmázsa of Miskolctapolca). Földr Ért XIX(I):77–85 (in Hungarian)

Látos T, Telbisz T, Deák M, Székely B, Kozma Zs, Standovár T, (2015) Lidar- és topográfiai térkép alapú digitális terepmodellekből levezetett, illetve kézzel digitalizált töbör-körvonalak morfometriai összenasonlítása az Aggteleki karszt példáján (Comparison of doline contours derived from lidar and topographic map-based DTMS with doline contours created by manual digitisation, the case example of Aggteleki karst) Karsztfejlődés XX:145–165 (in Hungarian)

Leél-Őssy S (1952a) Karrosodás és karros formák (Karren formation and karren features). Hidrológiai Közlöny 32(7–8):298–303 (in Hungarian)

Leél-Őssy S (1952b) Morfológiai vizsgálatok a Váckörnyéki triászrögökön (Morphological investigations on Triassic blocks in the environs of Vác). Földrajzi Értesítő I 2:363 (in Hungarian)

Leél-Őssy S (1952c) Az ürömi víznyelő barlang (The inflow cave of Üröm). Hidrológiai Közlöny XXXII 3–4:122–128 (in Hungarian)

Leél-Őssy S (1953a) Karszt és barlangkutatás a Szalonnai-karszton (Karst and cave research on Szalonna Karst). Hidrológiai Közlöny 33(1–2):67–70 (in Hungarian)

Leél-Őssy S (1953b) A Cserszegtomaji-kútbarlang (The well cave of Cserszegtomaj). Hidrológiai Közlöny 33:309–313 (in Hungarian)

Leél-Őssy S (1954) A Magas Bükk geomorfológiája (The geomorphology of High Bükk). Földrajzi Értesítő III(2):323–336 (in Hungarian)

Leél-Őssy S (1958a) Karsztmorfológiai vizsgálatok a balatonfüredi Lóczy-barlangban és környékén (Karstmorphological investigations in the Lóczy cave in Balatonfüred and in its environs). Földrajzi Értesítő VII 3:379–381 (in Hungarian)

Leél-Őssy S (1958b) A Kevély-hegycsoport karsztmorfológiája és barlangjai (The karst morphology and caves of Kevély Mountain Group). Földrajzi Értesítő VII:17–32 (in Hungarian)

Leél-Őssy S (1959a) A Bükk víznyelőinek és víznyelőbarlangjainak tanulmányozása (A study of the ponors and inflow caves of the Bükk Mountains). Földrajzi Értesítő VIII:179–190 (in Hungarian)

Leél-Őssy S (1959b) Magyarország karsztvidékei (Karst regions of Hungary). Karszt És Barlangkutatás I:79–88 (in Hungarian)

Leél-Őssy S (1959c) Adatok a dunántúli bazaltfelszíneken kialakult lefolyástalan mélyedések (kőtálak) eredetéhez (Data on the origin of arheic depressions (stone bowls) that developed on the Transdanubian basalt surfaces). Földrajzi Értesítő 2:271–273 (in Hungarian)

Leél-Őssy Sz. (2012) A Budai Vár-hegy geológiai viszonyai (Geological relations of the Buda Vár Hill). Karsztfejlődés XVII.5–22 (in Hungarian)

Lehmann H (1959) Studien über Poljen in den Venezianischen Voralpen und im Hochapennin. Erdkunde 13:258–289

Lénárt L (1977) Hidrogeológiai kirándulások a Bükkben (Hydrogeological excursions in the Bükk Mountains). Tankönyvkiadó, Budapest 292 pp (in Hungarian)

Lippmann L, Kiss K, Móga J (2008) Az Abaliget-Orfűi karszt karsztos felszínformáinak vizsgálata térinformatikai módszerekkel (Investigation of the karstic phenomenon near Orfű and Abaliget by GIS methods). Karsztfejlődés XIII:151–166 (in Hungarian)

Lovász Gy (1971) Adatok az Abaligeti-karszt geomorfológiai és hidrológiai jellemzéséhez (Data on the geomorphological and hydrological description of the Abaliget Karst). Földrajzi Értesítő XX(3):283–296 (in Hungarian)

Lovász Gy (1977) Geomorfológia: Kőzetek (Geomorphology: Rocks). In: Lovász Gy (ed) Baranya megye természeti földrajza, Baranya Megyei Levéltár, Pécs, pp 62–68 (in Hungarian)

Marosi S (1959) A Budai-hegység barlangjai és felszíni karsztos formái (The caves and surface karst features of the Buda Hills). In: Pécsi M (ed) Budapest természeti földrajza, Akadémia Kiadó, Budapest, pp 129–152 (in Hungarian)

Mezősi G (1984) A Sajó-Bódva köz felszínfejlődése (The landscape evolution of Sajó-Bódva Köz). Földrajzi Értesítő XXXII I(3):181–206 (in Hungarian)

Mezősi G, Bárány I, Tóth L (1978) Karstmorphometrische Untersuchungen im Gebirge Aggtelek (Nordungarn). Acta Geogr XVIII:131–140

Móga J (1999) Vannak-e poljék a Gömér-Tornai karszton? (Are there any poljes on Gömör-Torna Karst?) Karsztfejlődés III:125–138 (in Hungarian)

Móga J (2001) A szerkezet és kőzetfelépítés szerepe a Szilicei-fennsík felszínformáinak a kialakulásában (The role of structure and lithology in the formation of karstic landforms of the Szilice Plateau). Karsztfejlődés VI:143–159 (in Hungarian)

Móga J (2002a) Felszínalaktani vizsgálatok a Galyaság területén (Geomorphological investigations in the area of Galyaság). Karsztfejlődés VII:173–186 (in Hungarian)

Móga J (2002b) A tornai Alsó-hegy (The Alsó-hegy of Torna). Karszt és Barlang I-II.:95–104 (in Hungarian)

Móga J, Németh R, (2005) The morphological research of the basalt and loess covered plateaus in the Bakony Mts (Transdanubian Middle Mts.-Hungary). Acta carsologica 34(2):397–414

Móga J, Kiss K, Szabó M, Kériné Borsodi A, Kéri A, Mari L, Knáb M, Iván V (2011) Természeti és antropogén hatások vizsgálata a Tapolcai-karszt epikarsztos rendszerében (The study of natural and anthropogenic effects in the epikarstic system of Tapolca karst) Karsztfejlődés XVI:185–201 (in Hungarian)

Németh K, Martin U, Harangi Sz (2001) Miocene phreatomagmatic volcanism at Tihany (Pannonian Basin, Hungary). J Volcanol Geoth Res 111(1-4):111–135

Németh R (2005) A Kab-hegyi bazalttakaró depresszióinak vizsgálata (Investigation of the depressions of the basalt cover of Kab Hill). Karszt és Barlang 2000–2001 évf.:33–41 (in Hungarian)

Ozoray Gy, Láng G (1957) Karsztjelenségek a csővári rögcsoportban (Karst phenomena in the block groups of Csővár). Karszt és Barlangkutató Tájékoztató 5:36–37 (in Hungarian)

Pelikán P (1992) Adatok a Bükk-hegység felszíni karsztfejlődéséhez (Data to the surface karst development of the Bükk Mountains). A Bükk Karsztja, Vizei, Barlangjai II:259–268 (in Hungarian)

Pécsi M (1980) A Pannóniái-medence morfogenetikája (The Morphogenesis of the Pannonian basin) Földrajzi Értesítő 29:105–127 (in Hungarian)

Pécsi M (1991) Geomorfológia és domborzatminősítés (Geomorphology and relief classification). Földrajz Tudományi Kutató Intézet, Budapest 296 pp (in Hungarian)

Plózer I (1977) A Hévízi-tó forrásbarlangjának feltárása (Exploration of the spring caves of Lake Hévíz). Karszt és Barlang I-II:65–66 (in Hungarian)

Révész T (1947) Adatok az Északi-Bakony karsztosodásának ismeretéhez (Data on the knowledge of the karstification of Northern Bakony). Bölcsészdoktori értekezés, Kézirat (in Hungarian)

Rybár O (2009) Dolomittérszínek karros felszínfejlődésének néhány típusa (Some types of karren surface formation of dolomite terrains). Karsztfejlődés XIV.:161–174 (in Hungarian)

Samu A, Kevei Bárány I (2008) Karsztos tavak története és állapotváltozásai az Aggteleki és Szlovák karsztokon (The history and changes in the state of the karstic lakes located on Aggtelek Karst and Slovakian Karst). Karsztfejlődés XIII:117–134 (in Hungarian)

Sásdi L (1990) Az Aggtelek-Rudabányai-hegység karsztjának földtani fejlődéstörténete (Geological Evolution of the Karst of the Aggtelek-Rudabánya Mountains). Karszt És Barlang I:3–8 (in Hungarian)

Sásdi L (1998) Vízföldtan és vízrajz (Water geology and hydrography). In: Baross G (szerk): Az Aggteleki Nemzeti Park, Mezőgazda Kiadó, Budapest, pp 118–159 (in Hungarian)

Sásdi L (2000) A Pilis-hegység karsztjának fejlődéstörténete (Karst evolution of the Pilis Mountains). Karsztfejlődés V:77–83 (in Hungarian)

Sásdi L (2004) A Gerecse karsztjának fejlődéstörténete (Karst evolution of Gerecse) Karsztfejlődés IX:215–228 (in Hungarian)

Sásdi L (2005) Az Aggteleki-Rudabányai-hegység édesvízi mészkő előfordulásai (The calcareous sinter occurrences of the Aggtelek-Rudabánya Mountains). Karsztfejlődés X:137–151 (in Hungarian)

Sásdi L (2016) A Vértes karsztjának fejlődéstörténete (Karst evolution of the Vértes Mountains). Karszt És Barlang 2013–2014:47–56 (in Hungarian)

Sásdi L, Szilágyi F (1986) Víznyomjelzéses vizsgálatok az Aggteleki-karszton (Water tracing investigations on Aggtelek Karst), Karszt és Barlang I.:33–38 (in Hungarian)

Sauro U (2012) Closed depressions in karst areas. In: White WB, Culver DC (szerk) Encyclopedia of caves. Elsevier, Amsterdam, pp 140–155

Scheuer Gy (1975) Kiegészítő adatok a Bükk hegységi édesvízi mészkövek előfordulásához (Complimentary data to the calcareous sinter occurrences of the Bükk Mountains). Földrajzi Értesítő XXIV (1):75–78 (in Hungarian)

Schmidt Elegius R, Láng S, Ozoray Gy (1962) Adatok egyes középhegységeink vízháztartásához (Data on the water balance of some middle mountains of Hungary). Vázlatok és tanulmányok Magyarország vízföldtani atlaszához, Műszaki Könyvkiadó, Budapest, pp 48–56 (in Hungarian)

Sümegi Gy ifj, Sümegi Gy id, Varga B (2012) Töbörcsoportok sajátos vonásai a Jósvafői fennsíkon (Specific characteristics of doline groups on Jósvafő Plateau). Karsztfejlődés XVII:165–178 (in Hungarian)

Sweeting MM (1973) Karst landforms. Columbia University Press, New York, p 362

Szabó L (1998) Előzetes morfológiai megfigyelések a Veszprém-Várpalotai-fennsíkon (Preliminary morphological observations on the Veszprém-Várpalota Plateau. A szombathelyi Berzsenyi Dániel Tanárképző Főiskola Tudományos Közleményei, Természettudományok 6 (Procedings of Berzsenyi College). Nat Sci 115–128 (in Hungarian)

Szabó L (2000) Karros formák vizsgálata a Veszprémi-fennsík dolomitterületein (Investigation of karren features in the dolomite areas of the Veszprém Plateau). Földrajzi Értesítő XLIX(1–2):27–41 (in Hungarian)

Szabó PZ (1956) Magyarországi karsztformák klímatörténeti vonatkozásai (Relations between climate and karst features in Hungary). Dunántúli Tudományos Gyűjtemény 80:183–189 (in Hungarian)

Szabó PZ (1968) A magyarországi karsztosodás fejlődéstörténeti vázlata (An evolutionary outline of Hungarian karstification) Dunántúli Tud. Gyűjtemény 80:13–25 (in Hungarian)

Szegediné Darabos E, Lénárt L, Tóth K, Hernádi B, Kovács P (2014) A bükki karsztvízszint észlelő rendszer keretében gyűjtött hidrometeorológiai adatok elemzése (Analysis of hydrometeorological data of Bükk karstwater level monitoring system) Karsztfejlődés XIX:137–146 (in Hungarian)

Takács Bolner K (1998) Kristálybarlangok az Esztramos gyomrában (Crystal caves in Esztramos Hill). In: Baross G (szerk): Az Aggteleki Nemzeti Park, Mezőgazda Kiadó, Budapest, pp 194–200 (in Hungarian)

Telbisz T (2001) Új megközelítések a töbör-morfológiában az Aggteleki-karszt példáján (New approaches in doline morphology based on the example of Aggtelek Karst). Földrajzi Közlemények CXXV(1–2):85–108 (in Hungarian)

Tóth G. (1983) A Bükk felszínfejlődése és mai formakincse (Geomorphic evolution and present landforms of the Bükk Mountains). In: Sándor A (ed) Bükki Nemzeti Park, Mezőgazda Kiadó, Budapest, pp 62–106 (in Hungarian)

Tóth G (1984) A Központi-Bükk és geomorfológiai körzetei (The Central Bükk and its geomorphological districts). Földrajzi Értesítő XXXIII(4):333–345 (in Hungarian)

Tóth G (1986) Adatok a Magas-Bükk nem karsztosodó üledéktakaróinak ismeretéhez (Data on the non-karstic sediment layers of the High-Bükk). Nehézipari Műszaki Egyetem Közleményei I Sorozat, Bányászat 33(1–4):197–202 (in Hungarian)

Tóth G (2001) A karsztos résrendszer fejlődésének tényezői a Bükkben (Factors of the development of the karstic joint system in the Bükk Mountains). Karsztfejlődés VI:113–125 (in Hungarian)

Tóth G, Fejes P (1984) Idősebb pliocén lepusztulási szint kimutatása a Bükkben (Report on an older Pliocene denudation level in the Bükk Mountains). Karszt és Barlang II:65–72 (in Hungarian)

Trájer AJ, Mlinárik L, Hammer T, Földényi R, Somlai J, Bede-Fazekas Á (2020) Investigation of the vulnerability of a partly covered karst feature in Veszprém, Hungary. Environ Sci Pollut Res. https://doi.org/10.1007/s11356-020-08407-x

Vadász E (1946) A magyar bauxit – előfordulások földtani alkata (Geological characteristics of Hungarian bauxite occurrences). MÁFI Évkönyve, Magyar Állami Földtani Intézet, Budapest 37 (2):171–234 (in Hungarian)

Vadász E (1951) Bauxitföldtan (Bauxite geology) – Akadémia Kiadó, Budapest, 130 pp (in Hungarian)

Veress M (1982) Adatok a Hárskúti-fennsík morfogenetikájához (Data on the karst morphogenetics of the Hárskút Plateau). Karszt És Barlang II:71–82 (in Hungarian)

Veress M (1992) Adatok a Fekete-Sár-rét karsztmorfogenetikájához (Data on the morphogenetics of Fekete-Sár-rét). A Bükk Karsztja, Vizei, Barlangjai I:5–18 (in Hungarian)

Veress M (2000a) Középhegységi karsztok néhány típusa (Some types of middle mountain karsts). Földrajzi Közlemények CXXIV(1–4):1–28 (in Hungarian)

Veress M (2000b) Covered karst evolution Northern Bakony Mountains, W-Hungary. A Bakony Természettud. Kut. Eredményei 23, Bakonyi Természettudományi Múzeum, Zirc, 167 pp

Veress M (2002) Néhány bakonyi patakmeder mésztufa kiválásainak morfogenetikai típusai (Morphogenetic types of the calcareous sinter concretions of some stream channels in the Bakony Mountains). Hidrológiai Közlöny 82(1):15–22 (in Hungarian)

Veress M (2003a) Adalékok a homokkő anyagú kőtengerek (Káli-medence) pszeudokarrjainak morfogenetikájához (Data on the morphogenetics of the pseudokarren of sandstone stone fields (Kál Basin). Folia Musei Historica. Naturalis Bakonyiensis, 2001–2003a Zirc:7–46 (in Hungarian)

Veress M (2003b) A karrok (Karren) – Akadémiai doktori értekezés Kézirat, 365 pp (in Hungarian)

Veress M (2006) Adatok a Tési-fennsík két térszínrészletének fedett karsztosodásához (Data on the covered karstification of two terrain parts of Tés Plateau). Karsztfejlődés XI:171–184 (in Hungarian)

Veress M (2008) A mészkőfekü morfológiájának hatása a fedett karsztosodásra az Északi-Bakonyban (The effect of the morphology of the limestone bedrock on covered karstification in the Northern Bakony Mountains). Karszt és Barlang, 2004–2005. évf.:33–54 (in Hungarian)

Veress M (2009) Investigation of covered karst form development using geophysical measurements. Zeits f Geomorph 53(4):469–486

Veress M (2010a) Karst environments—Karren formation in high mountains. Springer, Dordrecht Heidelberg London New York, p 230

Veress M (2010b) Adalékok az Aggteleki-fennsík völgyeinek fejlődéséhez (Some data to the valley evolution of the Aggtelek Plateau). Karszt és Barlang, 2008 évf. (I-II):3–12 (in Hungarian)

Veress M (2011) Adatok a Mecsek-hegység fedett karsztosodásához a Cigány földi mintaterületről vett példák felhasználásával (Some data to the covered karstification of the Mecsek Mountains using examples taken from the Cigány föld research area). Karszt és Barlang, 2010 évf. I-II:9–30 (in Hungarian)

Veress M (2012) New data on development of the Baradla Cave. Acta Carsologica 42(2–3):193–204

Veress M (2014) Az Aggteleki karszt karsztos zónái (Karstic zones of Aggtelek Karst). ANP füzetek XI:27–43 (in Hungarian)

Veress M (2016) Covered Karst. Springer, Berlin, Heidelberg, New York, 536 pp. https://doi.org/10.1007/978-94-017-7518-2

Veress M (2017) A depresszió-exhumálás és a barlangképződés kapcsolata az Aggteleki karszton (The relationship between depression exhumation and cave development on the Aggtelek karst). Földrajzi Közlemények 141(2):97–111 (in Hungarian)

Veress M (2019) The karren formation of bare slopes. Earth-Sci Rev 188:272–290

Veress M (2020) The evolution and development of solution dolines with horizontal growth and the processes of their floors: a case study on the plate-shaped dolines of the Bükk Mountains, Aggtelek Karst and Pádis Plateau. Earth 1:49–74. https://doi.org/10.3390/earth1010005

Veress M, Szabó L (2000) Adalékok a dolomit térszínek formáinak morfogenetikájához (Data on the morphogenetics of the features of dolomite terrains). Földrajzi Értesítő XLIX 1–2:13–26 (in Hungarian)

Veress M, Zentai Z, Bauer N (2005) Paleokarrok a dorogi Strázsa-hegyen (Paleokarren on the Strázsa Hill of Dorog). Karszt És Barlang 2000–2001:51–62 (in Hungarian)

Veress M, Zentai Z (2009) Karsztjelenségek minősítése a Bükk-hegység néhány mintaterületén a mészkőfekü morfológiájának és a fedőüledékek szerkezetének értékelésével (Classification of karst phenomena on a few sample areas of the Bükk Mountains by evaluating the morphology of the limestone bedrock and the structure of superficial deposits). Karszt és Barlang 2007 I-II:37–54 (in Hungarian)

Veress M, Tóth G (2015) Kál-Basin: Pseudokarren on Sandstone. In: Lóczy D (ed) Landscapes and landforms of Hungary. Springer, Heidelberg, New York, Dordrecht, London pp 79–87. https://doi.org/10.1007/978-3-319-08997-3

Veress M, Unger Z (2015) Kab Mountain: Karst under a Basalt Cap. In: Lóczy D (ed) Landscapes and landforms of Hungary. Springer, Heidelberg, New York, Dordrecht, London, pp 55–62. https://doi.org/10.1007/978-3-319-08997-3

Veress M, Gadányi P, Tóth G (2015) Thermal spring cones of Tihany Peninsula. In: Lóczy D (ed) Landscapes and landforms of Hungary. Springer, Heidelberg/New York/Dordrecht/London, pp 71–78

Veress M, Vetési-Foith Sz (2019a) The distribution of surface karst features in the Bakony region (Transdanubian Mountains Hungary). J Geol Res 1(1):21–25

Veress M, Vetési-Foith Sz (2019b) A Bakonyvidék felszíni karsztja (Surface karst of the Bakony Region in Hungary). Földrajzi Közlemények 43(2):167–186 (in Hungarian)

Veress M, Vetési-Foith Sz (2019c) The cave opening types of the Bakony region (Transdanubion Mountains, Hungary). Int J Curr Res. 11605/pp.3546–3562. https://doi.org/10.24941/ijer.34579.05.2019c

Vetési-Foith Sz, Kiss Z, Gárdonyi I (2017) A Mecseki karszt oldódásos- és utánsüllyedéses töbreinek morfometriai elemzése (A morphometric analysis of the solution dolines and subsidence dolines of the Mecsek Karst). Karsztfejődés XXII:119–138 (in Hungarian)

Waltham AC, Fookes PG (2003) Engineering classification of karst ground conditions. Q J Eng Geol Hydrogeol 36:101–118

Waltham T, Bell F, Culshaw M (2005) Sinkholes and subsidence. Springer, Berlin Heidelberg, p 382

White WB (1988) Geomorphology and hydrology of karst terrains. Oxford University Press, New York, Oxford

Williams PW (1985) Subcutaneous hydrology and the development of doline and cockpit karst. Zeits f Geomorphol 29(4):463–482

Williams PW (2004) Dolines. In: Gunn J (ed) Encyclopedia of caves and karst science. Fitzroy Dearborn, New York, London, pp 304–310

Záhorszky A (2001) A Délkeleti Bükk DK-i részének peremkarsztja (Marginal karst of the south-eastern part of South-eastern Bükk). Karsztfejlődés VI:127–132 (in Hungarian)

Zámbó L (1970) A vörösagyagok és a felszíni karsztosodás kapcsolata az Aggteleki-karszt délnyugati részén (The relation of red clays and surface karstification in the south-western part of the Aggtelek Mountains). Földrajzi Közlemények 94(4):281–293 (in Hungarian)

Zámbó L (1986a) Karsztvörösagyagok CO_2 termelés és a karsztkorrózió összefüggése (The connection between the CO_2 production of karst red clays and karst corrosion). A Nehézipari Műszaki Egyetem Közleményei I. sorozat Bányászat 33/1–4:125–138 (in Hungarian)
Zámbó L (1986b) Paleomorfológiai rekonstrukció a karsztos oldódás intenzitásának mértéke alapján (Paleomorphological reconstruction based on the degree of karstic dissolution intensity). Nehézipari Műszaki Egyetem Közleményei I. sorozat Bányászat 33(1–4):137–165 (in Hungarian)
Zámbó L (1998a) Az Aggteleki-karszt felszínalaktani jellemzése (Geomorphological characterisation of Aggtelek Karst). Földrajzi Értesítő XLVII(3):359–378 (in Hungarian)
Zámbó L (1998b) Felszínalaktan (Geomorphology). In: Baross G. (ed) Az Aggteleki Nemzeti Park, Mezőgazda Kiadó, Budapest, pp 70–96 (in Hungarian)
Zámbó L, Telbisz T (2000) A karsztkorróziós talajhatás érvényesülése a karsztfejlődésben (The effectiveness of karst corrosion soil impact in karst development). Karsztfejlődés V:103–114 (in Hungarian)
Zboray Z, Kevei Bárány I (2004) Domborzatértékelés a Bükk-fennsíkon légifelvételek felhasználásával (Terrain assessment based on aerial images on the Bükk Plateau) Karsztfejlődés IX:207–213 (in Hungarian)

GIS Research of Karst Terrains in Hungary

Tamás Telbisz, László Mari, and Gáspár Albert

Abstract

Since the 1990s, GIS tools have provided several new possibilities in the study of karst terrains. Classical maps were gradually replaced by GIS databases and visualizations, but the analytical capabilities of GIS were also increasingly applied in karst research. Data collection has been revolutionized by continuously evolving remote sensing methods (e.g. LiDAR). GIS became crucial in doline morphometry, thus in this chapter, we present a detailed statistical analysis of sinkholes focussing on Aggtelek, Bükk and Mecsek mountains, because these are the karst terrains of Hungary, where dolines are the most abundant. The large-scale relief characteristics of the karst mountains can be studied by digital terrain analysis and these results are also presented in this chapter. Landform classification is the "new form" of geomorphological mapping, thus, some case studies using this approach are also provided here. During the last centuries, Hungarian karst terrains have undergone significant landscape changes. Based on earlier series of military maps, satellite images and other databases (e.g. CORINE), these landscape changes have been studied by several authors using GIS methodology. More recently, special GIS methods, 3D modelling software and terrestrial laser scanning were introduced into cave mapping as well. Besides these main topics, we also briefly mention in this chapter some further GIS-related research on karst, namely karst vulnerability mapping, karst geodiversity mapping and the study of human–environment relations on karst.

Keywords

Morphometry • Land cover • LiDAR • DTM • Landform classification • Cave models

1 Introduction

In Hungary, the first geographical applications of GIS appeared in the 1990s, but their real spreading occurred in the 2000s, and this was also the period, when karst researchers started to use varied GIS tools in their studies. The quick development of GIS software as well as the increasing availability of large datasets made it possible to use GIS in more and more karst studies. Classical maps (of topography, geology, geomorphology, etc.) were gradually replaced by GIS databases and visualizations, but the analytical capabilities of GIS were also increasingly applied in karst research. Among the many global or European scale datasets, we outline here the SRTM digital terrain model (Rabus et al. 2003) and the CORINE land cover dataset (Büttner et al. 2004), which were probably the most frequently used datasets in karst studies in Hungary. The old and recent topographic maps meant also an important data source, especially, the military maps, which were created in several periods since the end of the eighteenth century, and the 1:10,000 scale EOTR (Unified National Map System) topographic maps, which were published between 1975 and 2000. In addition, individual surveys also had some significance in karst research. Aerial images and GPS measurements can be mentioned first, but more recently, airborne LiDAR datasets and UAV-based photos were also applied in karst research. Laser distometers and terrestrial laser scanning (TLS) are used mostly in cave environments as GIS-related mapping tools (Albert 2017). We briefly note here that while access to global datasets is simple, fast and free in many cases, the national GIS datasets in Hungary are often difficult to access or they are available only in paid

T. Telbisz (✉) · L. Mari
Department of Physical Geography, Eötvös Loránd University, Pázmány Péter sétány 1/C, Budapest, 1117, Hungary
e-mail: tamas.telbisz@ttk.elte.hu

L. Mari
e-mail: laszlo.mari@ttk.elte.hu

G. Albert
Department of Cartography and Geoinformatics, Eötvös Loránd University, Pázmány Péter sétány 1/A, Budapest, 1117, Hungary
e-mail: albert@ludens.elte.hu

M. Veress and Sz. Leél-Őssy (eds.), *Cave and Karst Systems of Hungary*,
Cave and Karst Systems of the World, https://doi.org/10.1007/978-3-030-92960-2_9

services. Further on, the GIS-based research results often remain isolated. Thus, the GIS integration and management of karst-related GIS data should be an important near-future task in Hungarian karst research.

In this chapter, we present the karst-related research issues where GIS plays a significant role. Among these fields, the two most frequently studied topics, i.e. the morphometry and the land cover changes on karst terrains are exposed with more details in this chapter, whereas other topics are only briefly introduced here. These topics include the vulnerability and hydrogeology of karst terrains, the 3D modelling of caves and the human–environment relations on karst terrains.

2 Morphometric Analysis

Morphometric analysis means the quantitative investigation of landform shapes. In the study of microforms (i.e. karren), GIS is generally not relevant, though digital elevation models of karren features have been created by Telbisz (2004). The number of large-scale karst forms (plateaus, poljes, gorges) is relatively limited in Hungary, thus there are only few studies about the GIS analysis of these forms so far. On the other hand, dolines are abundant on some Hungarian karst terrains, notably on the so-called "Aggtelek type karsts", which include Aggtelek–Rudabánya Mts, Bükk Mts and Mecsek Mts. Therefore, most of the karst morphometric studies focus on these mountains.

2.1 Doline Morphometry

The general morphological features of dolines are presented in chapter "The Surface Morphology of Karsts in Hungary" of this book, thus we summarize here only the directly morphometrical studies in relation with dolines. Even before the advent of GIS, some Hungarian karst terrains were already studied by morphometrical approach, and the publications of Mezősi (1984), and Bárány-Kevei and Mezősi (1994) focussing on Aggtelek Karst can be considered as the pioneering works in this issue. These authors analysed the morphometrical parameters of ca. 80 dolines, and they concluded that the orientation of dolines, i.e. the azimuth of the doline long axis basically coincides with the predominant NNE–SSW and E–W joint and fissure directions. Further on, they also observed that the asymmetrical shape of certain dolines reflects the aspect-related differences in the soil and microbial activity, which are influenced by microclimatic conditions. They found that the slopes with western and southern exposures are gentler than the opposite sides. In addition, they discriminated row dolines found in chains at the bottom of valleys and plateau dolines, which are located on the relatively flat plateau surfaces. They also observed that row dolines are more strongly influenced by tectonic lines than plateau dolines. The work of Farsang and Tóth (1993) provided certain methodological development in doline morphometry. They thoroughly surveyed 27 dolines in Bükk Mts and created contour lines at 1 m intervals. Based on their detailed map, they found that the asymmetry was varied at different levels turning from N–S to W–E direction. They explained this complex pattern partly by structural elements and partly by the microclimatic effect.

The first studies targeting the complete doline morphometric analysis of whole mountains were carried out in the 2000s. Based on 1:10,000 scale topographic maps, Dutkó (2000) and Varga (2007) studied Bükk dolines. Dutkó (2000) identified 889 dolines in the Bükk Mts. He described that four geological formations bear most of the dolines in that mountains: the Bükk-fennsík Limestone, which is a thick-bedded carbonate platform facies limestone of several hundred meters thickness, the Kis-fennsík Limestone, the Fehér-kő Limestone and finally, the Répáshuta Limestone, which has thin layers, contains chert nodules and whose thickness is less than 100 m. These formations are dominant at distinct parts of the mountains and the typical morphometric parameters of dolines are also different for these distinct areas. Dutkó (2000) investigated doline asymmetry as well, and he observed that the western (including southwestern and northwestern) sides of dolines are the steepest in the Nagy-fennsík area. He explained this fact by the general tilt of layers at this area, but he noted that in some cases this explanation does not hold true, therefore microclimatic effects may also have an impact on doline asymmetry.

Telbisz (2001) also used 1:10,000 scale topographic maps and identified 1077 dolines in Aggtelek Karst. His statistical analysis revealed that the frequency distribution of doline planform area follows a lognormal distribution. Moreover, he observed that the structural directions are important factors in the elongation of doline planforms. According to his theory, row dolines often coalesce along the main structural direction, whereas in a lateral direction (along the secondary fissures) smaller but separate dolines are frequently found in the close neighbourhood of larger dolines. He calculated doline density for the whole area and concluded that the Wetterstein Limestone, which is a carbonate platform facies with almost 1000 m thickness is the most suitable for the formation of dolines, but lithology is not the only influencer of density, because landform evolution and relief characteristics also strongly influence the size and density of dolines. Based on the relationship of doline area and density, he could discriminate areas, where allogenic streams significantly contributed to the formation and growth of dolines during certain phases of landform evolution.

The Western Mecsek Karst is somewhat special in the group of Aggtelek type karsts, because the well karstifiable limestone is covered by loess deposits here. The loess overlay is thinner in the south but reaches 4–6 m thickness in the north. Hevesi (2001) pointed out that unlike in the Aggtelek and Bükk Mts, the proportion of row dolines are significantly lower in the Mecsek Mts and the differences in the morphometrical properties of row versus plateau dolines are less remarkable. A further special phenomenon in the Mecsek Mts is that doline density has extreme values. Hevesi (2001) calculated 73.4 dolines per km^2 for Abaliget plateau. He also emphasized that this area is rich in small size, suffosion dolines. Hoyk (2002) identified 1540 dolines in the Western Mecsek Karst based on topographic maps, and she selected 64 out of them for a more detailed morphometric analysis. She outlined that the morphometrical parameters of Mecsek dolines are significantly different from the parameters of dolines in Aggtelek Karst or Bükk Mts. For a smaller study area, she calculated 380 dolines per km^2 as doline density. She observed that larger depressions often contain smaller dolines, which suggests that the dissection of the already existing dolines is a significant process, but in some other cases, she acknowledged that the coalescence of dolines is also plausible.

The most thorough doline morphometrical study about Western Mecsek Karst was elaborated so far by Lippmann et al. (2008). Besides topographic maps, they used orienteering maps and their own GPS field survey, too. They identified 1277 dolines in the 1:10,000 scale topographic maps, out of which 468 are outlined by contours, whereas the remaining forms are marked only by symbols. In addition, they detected 266 further dolines in the field that were not marked at all on the maps. They calculated a doline density of 164 dolines per km^2 for the Orfű plateau. However, they pointed out that doline area ratio, i.e. the proportion of surface covered by dolines is very small (3.1%) due to the small size of dolines. They found no direct relationship between doline long axes and structural directions. Vetési-Foith et al. (2017) examined doline morphometry for a part of the Western Mecsek Karst by a detailed field survey, but they did not use GIS in their analysis. They investigated 49 dolines and outlined the differences between the large size solution dolines and the small size suffosion dolines. They noted that doline long axes coincide with valley directions at many but not all locations. In addition, they observed that valley directions are determined by structural lines. For their small study area, they calculated an extreme high doline density of 2,440 dolines per km^2 that is the consequence of the fact that they considered even the smallest depressions to be a doline.

The Transdanubian Mts are of a different karst type, where exokarst landforms, especially dolines are relatively rare phenomena, thus doline morphometric studies are hardly available about these mountains. Vetési-Foith (2018) created a survey about a small study area of Mester-Hajag Hills found in the Bakony Mts. He identified 55 small size suffosion dolines, whose long axes were mostly parallel with either the slope direction or the structural lines. In a larger scale study, Veress and Vetési-Foith (2019) characterized the spatial distribution of suffosion dolines in the Bakony Mts. They counted 691 suffosion dolines in the area. Most of these forms are very small in size, and generally they are located in larger size depressions, which are covered by a varied thickness of deposit (loess or other material). Eighty-nine per cent of these small size dolines are found on elevated horsts in cryptokarst settings.

The above-mentioned studies used mostly topographic maps and field surveys. However, other data sources have also been applied in doline morphometry. Based on aerial images, Zboray and Keveiné Bárány (2004) created a digital terrain model (DTM) about a 15 km^2 sample area on Nagy-fennsík in the Bükk Mts using stereo-photogrammetric methodology. Their DTM had 2 m horizontal resolution and 275 dolines were delineated based on this dataset. Slope histograms were also created, and by interpreting these histograms, the authors emphasized again that most of the dolines are asymmetrical, i.e. the western facing sides of dolines are gentler than the eastern facing sides.

So far, LiDAR data about Hungarian karst territories is very limited, and out of the doline karst terrains, only Aggtelek Karst was surveyed by airborne LiDAR. This LiDAR dataset was used for doline morphometry by Látos et al. (2015) and Telbisz et al. (2016). These studies compared the morphometrical parameters derived from the LiDAR dataset to the results drawn from the topographic map-based analysis. Further on, they tested the efficacity of DTM-based automatic doline delineation methods using either the topographic map-based DTM created by Kriging interpolation or the LiDAR-based DTM. They found that the statistical distributions of the morphometrical properties are similar for all datasets (lognormal distribution for doline area and gamma distribution for doline depth), but the DTM-based methodology resulted somewhat larger dolines than the classical, outermost closed contour line method. The planform area (and related characteristics) showed very high correlations between the datasets. Depth values were less correlated and the lowest (moderately strong) correlations were observed between the circularity values of the different datasets. The LiDAR dataset resulted 5.4% higher doline density value than the 1:10,000 scale topographic map.

As a conclusion, we can state that if the scale and accuracy of the topographic map is detailed enough to show dolines, then the classical closed contour method leads to acceptable results in doline morphometry. In case of Aggtelek Karst and Bükk, the 1:10,000 scale Hungarian topographic maps are appropriate for doline morphometry (though the LiDAR dataset provides certain improvement even in that case), but in case of Mecsek, this scale is not perfect, and a future LiDAR survey would be necessary to improve the morphometrical analysis of dolines in the latter mountains.

The creation of DTMs based on UAV photos using Structure from Motion (SfM) technology could be a relatively cheap alternative in the survey and morphometrical analysis of doline terrains (Látos and Telbisz 2018). A small-scale study including 17 dolines at Jósvafő Plateau (Aggtelek Karst) demonstrated that UAV-based morphometrical parameters were very similar to LiDAR-based values. However, as most of the Hungarian doline terrains are mostly covered by forests, the potential use of UAV surveys is limited.

Doline morphometry is an important issue from the viewpoint of vegetation, too. Closed depressions modify the microclimate, including insolation, temperature and humidity, thus the living conditions of both the microorganisms and the macro-vegetation are influenced (Bárány-Kevei and Mezősi 1995; Bátori et al. 2014a, b). Dolines have a higher number of species per unit area, and the same is valid if only protected species are taken into account (Hoyk and Keveiné Bárány 2000). The topics in relation with vegetation are more thoroughly explicated in chapters "The Vegetation of Karsts in Hungary" and "Geoecology of Hungarian Karsts", but some relationships, which are directly related to morphometry, are emphasized here. Székely et al. (2015) calculated vegetation height based on the LiDAR survey performed for the area of Aggtelek National Park. Vegetation height can be calculated as the difference between the digital surface model (DSM), which represents vegetation tops and the digital terrain model (DTM), which represents the land surface. In addition, they took into consideration the results of a very detailed, grid-based vegetation mapping. As a conclusion, they demonstrated that dolines significantly influence vegetation height, namely that trees grow higher in the central parts of dolines, thus the DSM surface is much smoother in terms of elevation than the DTM, i.e. the topography. Further on, they found that vegetation type is also influenced by the microtopography of dolines. For instance, the proportion of beech is significantly lower within the dolines than outside the dolines. Moreover, slope aspect and gradient also modify vegetation height and composition. According to Bátori et al. (2014a, b), dolines serve as island-like habitats and karst dolines of Hungary can be considered as potential refugia for many vascular plant species under future global warming, therefore they are particularly important from a conservation point of view.

In the following, we provide some statistics of the most basic doline morphometric parameters for the Hungarian Aggtelek-type karsts using a unified methodology. As a common data source, we have the 1:10,000 scale topographic maps, so it is used for this data summary. However, we underline that due to the generally small size of dolines in the Western Mecsek Karst, a significant part of dolines are represented in the maps only as symbols and a further proportion is absolutely missing from the maps, thus the data for the Mecsek dolines is distorted to some extent.

First, the spatial distribution of karstifiable rocks (limestones and dolomites) was determined for each mountain based on geological maps. Second, the doline-dotted terrains (subunits) were outlined within the karstifiable rock areas taking into consideration the slope map and the real presence of dolines (Fig. 1). Altogether, ten subunits were delineated in the Aggtelek–Rudabánya Mts, and four–four subunits in the Bükk and the Western Mecsek Mts. These subunits incorporate most of the dolines within these mountains, but sporadic dolines also occur in each area (Table 1). It is an interesting fact that only a portion of the carbonate terrains are doline-dotted. For Aggtelek-Rudabánya Mts this portion is 40%, for Bükk it is 29%, and for W-Mecsek, it is 57%. One reason for this relatively low portion is that dolines are absent or rarely found on terrains, where the general slope is higher than 15°. Another reason comes from the lithological properties, i.e. certain limestones and dolostones are not pure, consequently they are less suitable for karstification.

Considering the subunits, one can observe remarkable differences in doline density (Fig. 2). In Aggtelek Karst and Bükk, doline density values of subunits vary from 5.8 to 25.8 dolines per km^2, while in the Western Mecsek, significantly higher values occur from 12.3 to 38.5 dolines per km^2, even if only contour marked dolines are taken into account. However, if the symbol marked dolines are also included, then we get extreme high values like 72.2 dolines per km^2 for Abaliget Plateau and 123.3 dolines per km^2 for Orfű Plateau. The precise spatial pattern of doline density can be represented by doline density maps. In the present study, we used Kernel Density calculation method and 1 km^2 circular area (i.e. 564 m search radius) and 30 m cell size to create these maps for all three mountains (Fig. 3). The main reasons for these large differences are the thick loess cover and the partly deviating landform evolution of the Western Mecsek area.

Doline sizes are also varied among subunits. The frequency distribution of doline planform area can be approached by lognormal distribution in all mountains. It means that the frequency curve of the logarithm of area roughly follows a Gaussian bell shape (Fig. 4). This figure also demonstrates that the mean doline planform area is the largest in Aggtelek Karst and the smallest in the Western Mecsek even if the symbol marked dolines are not taken into account. If the subunits are compared (Fig. 5), we can observe that there is a large internal variation even among the Aggtelek subunits. The largest dolines are found in Páska-bükk, Teresztenye and Aggtelek subunits either the mean or the maximal values are considered. The largest closed depression belongs to Vizetes sinkpoint next to

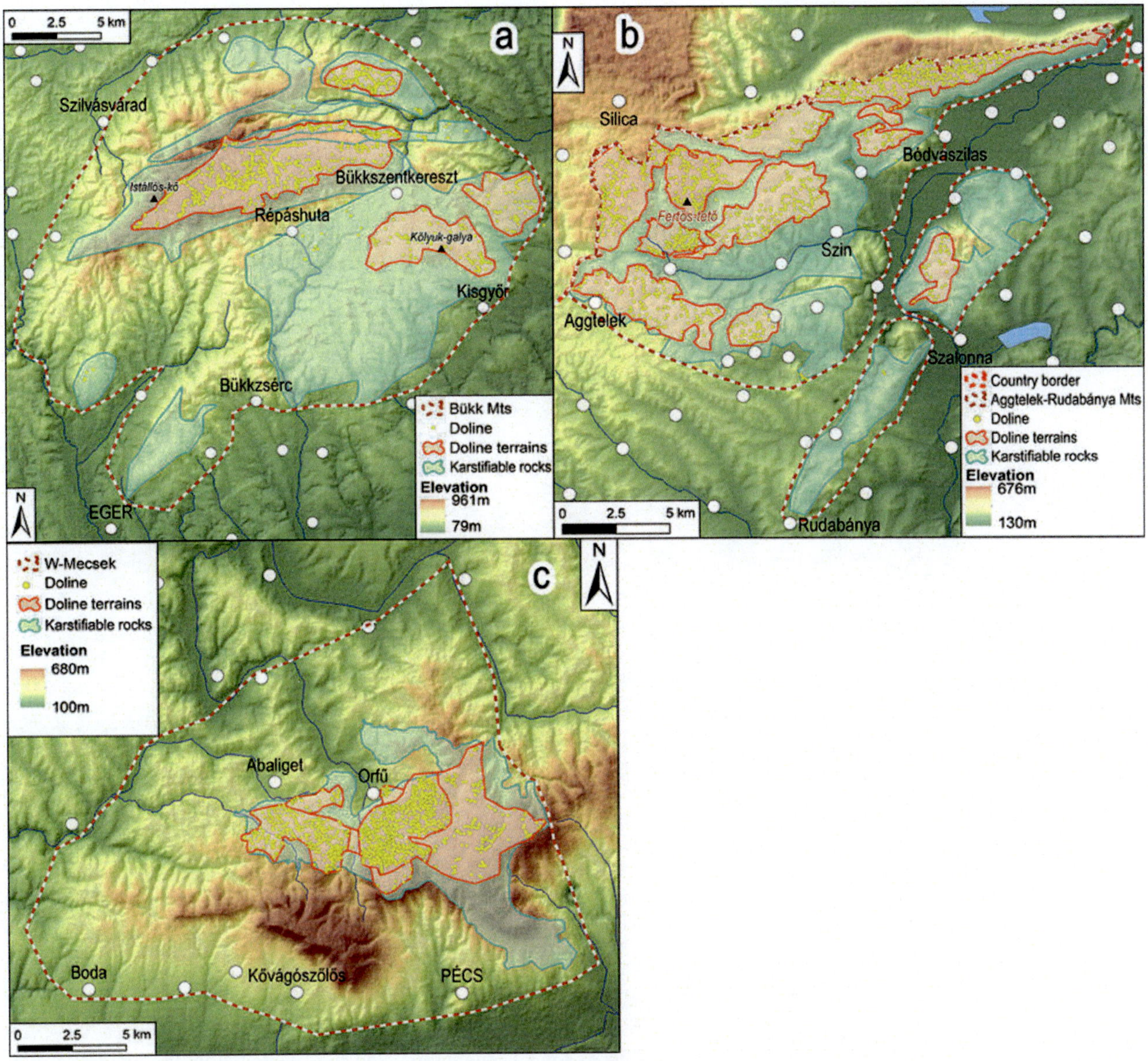

Fig. 1 Distribution of karstifiable rocks in Aggtelek–Rudabánya Mts (**a**), Bükk Mts (**b**) and Western Mecsek Mts (**c**)

Teresztenye Plateau, and it has an area of 0.2 km^2. Naturally, it is not a simple doline but a compound form, but it is included in the doline dataset as a closed depression. Bükk subunits are less variable in terms of doline size, whereas Mecsek subunits are again more varied, but here, the differences are between small and smaller. As for doline depth, Aggtelek dolines are also the deepest with 7.1 m as a mean value (and 35 m as the deepest form), whereas Bükk dolines are surprisingly shallow (only 4.4 m as a mean), and Mecsek dolines are in-between with 5.3 m mean depth. In this latter value, only the contour marked dolines are included. The depth can be related to size by calculating the width-to-diameter ratio. Considering this ratio, the contour marked dolines of Mecsek are the relatively deepest forms with the mean value of 0.14, whereas Aggtelek and Bükk dolines have a slightly smaller mean value of 0.09. As for internal variations, the relatively deepest dolines are found on the eastern part of Alsó-hegy (ALE) within Aggtelek Karst, on the Kis-fennsík Plateau within Bükk Mts and on the Orfű Plateau within the Western Mecsek.

Doline shapes can be characterized by elongation (i.e. length-to-width ratio), circularity (i.e. 4 * π * area/perimeter) and long axis direction. As elongation increases, circularity decreases for geometrical reasons (Fig. 6). However, circularity may decrease also if the outline of the doline is more dissected even if the length and width of the envelope

Table 1 Comparison of doline morphometric parameters in the subunits of Aggtelek-Rudabánya, Bükk and Western Mecsek Mts (N_d: number of dolines; A_d: mean doline planform area; D_d: mean doline depth; E_d: mean doline elongation; C_d: mean doline circularity)

Mountains	Subunit	Area	N_d	Density	A_d	D_d	E_d	C_d
Aggtelek-Rudabánya	Aggtelek (AGG)	12.0	114	9.5	10,530	7.1	1.63	0.85
	Alsó-hegy East (ALE)	12.5	322	25.8	4083	9.0	1.35	0.93
	Alsó-hegy West (ALW)	7.3	64	8.8	4401	4.4	1.55	0.89
	Haragistya (HAR)	10.0	154	15.4	5053	5.5	1.57	0.89
	Jósvafő (JOS)	3.5	69	19.7	4285	4.0	1.65	0.87
	Nagyoldal (NAG)	8.2	151	18.3	4510	6.7	1.50	0.89
	Páska-bükk (PAS)	2.5	26	10.6	14,343	10.0	1.45	0.90
	Szalonna (SZA)	3.9	30	7.7	7896	8.4	1.39	0.94
	Szin (SZI)	12.7	135	10.6	6152	6.6	1.57	0.89
	Teresztenye (TER)	3.4	35	10.3	18,190	8.4	1.57	0.86
	Total	**76.0**	**1100**	**14.5**	**5982**	**7.1**	**1.50**	**0.90**
Bükk	Kis-fennsík (KIS)	5.5	129	23.3	1906	4.8	1.52	0.90
	Kőlyuk-galya (KGA)	16.3	96	5.9	2362	4.5	1.55	0.90
	Nagy-fennsík (NAG)	41.0	555	13.5	2716	4.4	1.52	0.89
	Nagy-Kőmázsa (NKM)	9.5	55	5.8	3054	4.8	1.81	0.85
	Total	**72.3**	**835**	**11.5**	**2541**	**4.4**	**1.56**	**0.89**
Mecsek	Abaliget (ABA)	4.3	150 (308)[a]	35.2 (72.2)[a]	2757	7.4	1.34	0.92
	Lipóc-hegy (LIP)	0.9	11 (25)[a]	12.3 (27.9)[a]	3823	5.3	1.59	0.85
	Melegmány (MEL)	8.2	126 (175)[a]	15.4 (21.3)[a]	698	3.2	1.42	0.89
	Orfű (ORF)	5.9	228 (731)[a]	38.5 (123.3)[a]	1041	5.1	1.38	0.91
	Total	**19.3**	**515 (1239)[a]**	**26.7 (64.2)[a]**	**1512**	**5.3**	**1.39**	**0.91**

[a] including dolines represented by symbols only. In all mountains, there are some sporadic dolines outside these subunits whose data are not included in this table (Aggtelek–Rudabánya: 15; Bükk: 44; Mecsek: 5 sporadic dolines)

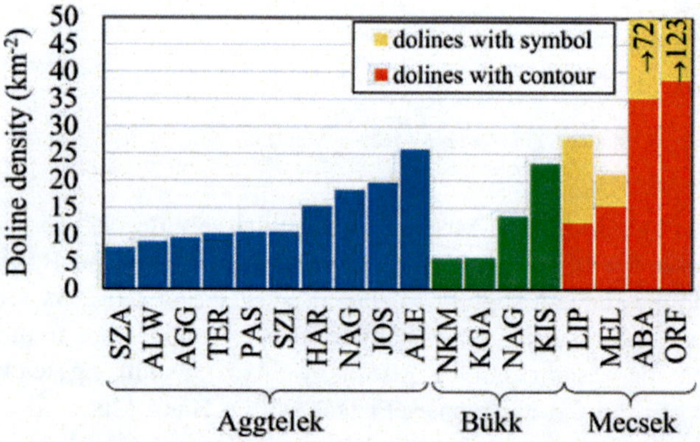

Fig. 2 Mean doline density of subunits in Aggtelek-Rudabánya, Bükk and Mecsek Mts (for the abbreviations of subunits see Table 1)

polygon is unchanged. Thus, elongation and circularity are moderately strongly correlated (with R^2 values ca. 0.7–0.75 for all mountains), and the points below the regression line in Fig. 6 typically mark the dolines, which have a more zigzaggy outline. In general, zigzaggy outlines are characteristic to larger dolines, which were created by coalescence, or which are connected to small dry valleys. Anyway, the shape parameters do not show large variations among the mountains. The mean circularity is around 0.9 for all mountains, and even the subunits are relatively similar. Lower values that is more dissected doline outlines are typical for the Aggtelek (AGG) and Teresztenye (TER) subunits in Aggtelek-Rudabánya Mts, for the Nagy-kőmázsa (NKM) subunit in Bükk and for the Lipóc

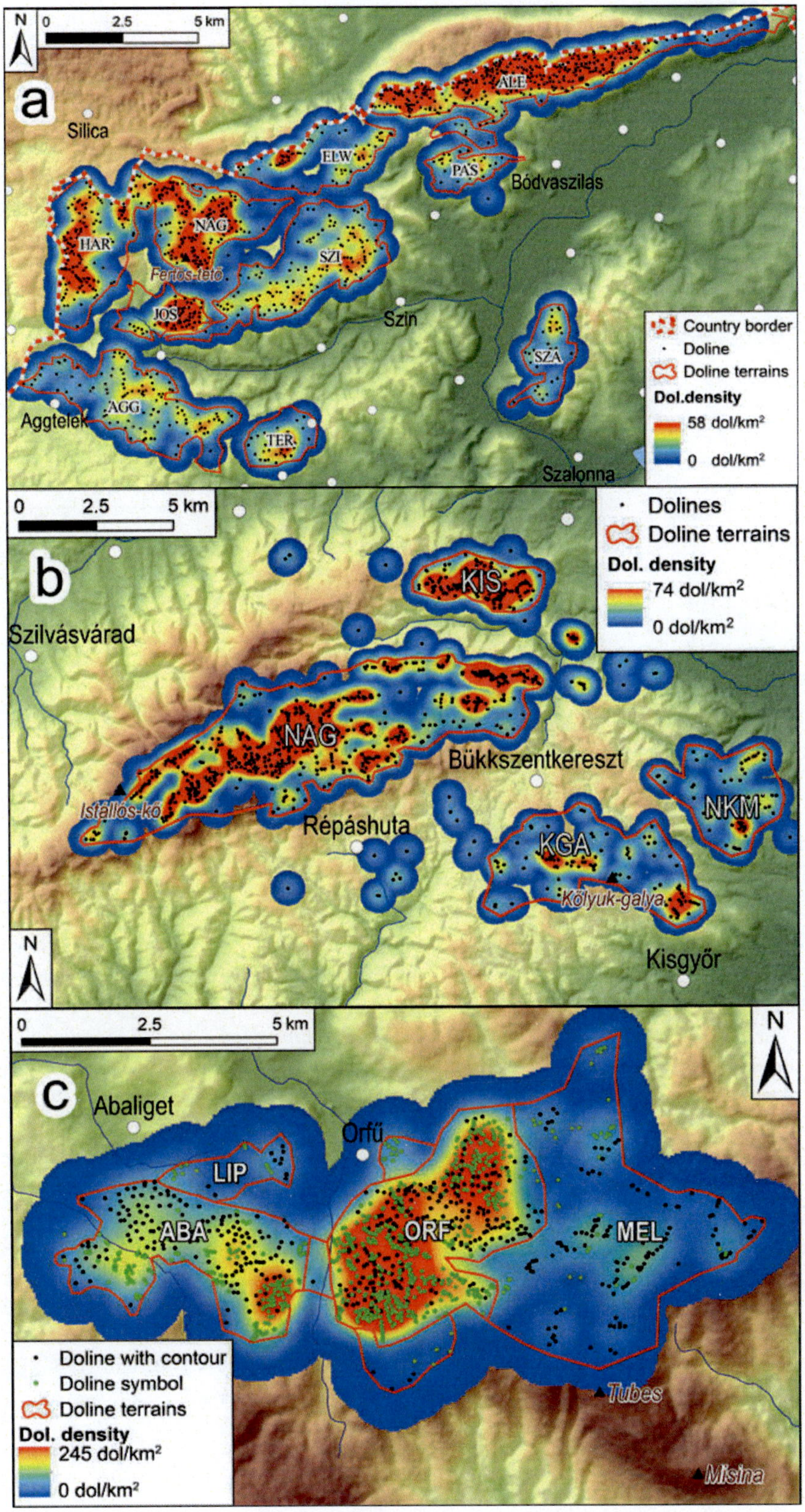

Fig. 3 Doline density maps of Aggtelek–Rudabánya (**a**), Bükk (**b**) and Mecsek (**c**) Mts (for the abbreviations of subunits see Table 1)

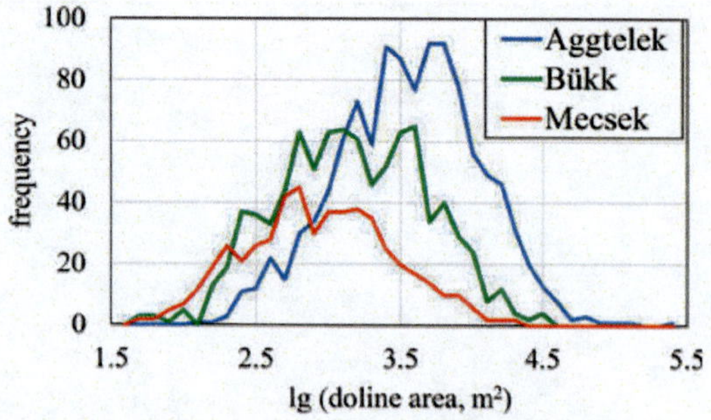

Fig. 4 Frequency distributions of doline planform area

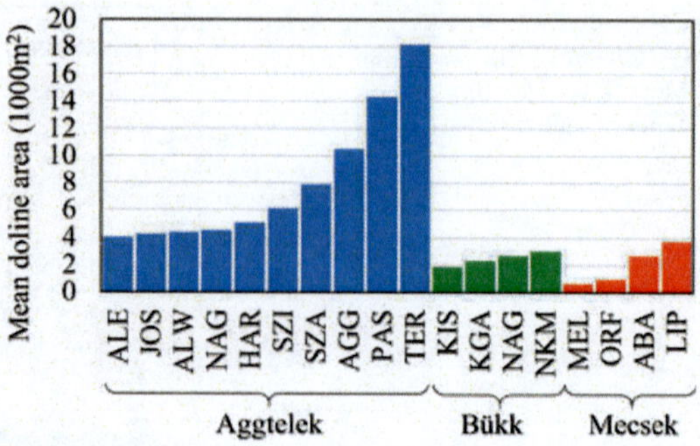

Fig. 5 Mean values of doline planform area of subunits in Aggtelek–Rudabánya, Bükk and Mecsek Mts (for the abbreviations of subunits see Table 1)

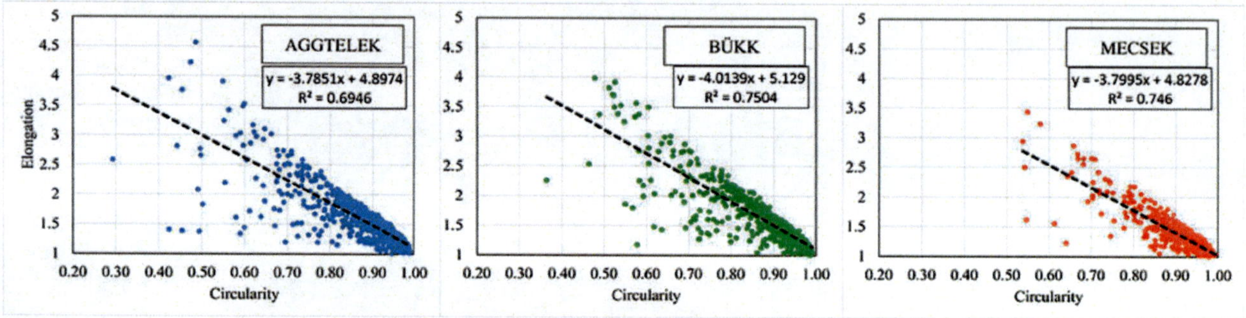

Fig. 6 Doline circularity vs elongation for Aggtelek–Rudabánya, Bükk and Mecsek Mts

Hill (LIP) in Mecsek. The most circular dolines are dominant on the eastern part of Alsó-hegy Plateau (ALE), on the Szalonna Karst (SZA) and on the Abaliget Plateau (ABA). The elongation values show similar variations, i.e. all subunits have relatively similar mean values from 1.35 to 1.65 and the extreme values are typical for the same subunits.

In general, the long axis direction of a doline is influenced more strongly by the structural directions. However, the mean aspect of the topographic surface, the tilt of rock layers as well as the present or the former valley directions may also have an impact on this parameter. As for Aggtelek Karst, a more detailed analysis of doline orientation can be read in chapter "The Surface Morphology of Karsts in Hungary". Here, we present only the summarized rose diagrams for the three mountains (Fig. 7). As these diagrams include each subunit within one area, the individual directions are mixed and smoothed to some extent. In order to filter out the almost random data of (sub)circular dolines, we took into account in the creation of long axis rose diagrams only those dolines, which have an elongation value higher than 1.25. Similarly, for the nearest neighbour rose diagrams, we took into consideration only those dolines, which have a nearest neighbour distance less than 160 m. A general conclusion is that the rose diagrams of the nearest neighbours show larger circular variations than the rose diagrams of long axes. Moreover, the predominant direction of nearest neighbours, if it is recognizable at all, is almost never parallel with the primary long axis direction, but in several cases, it is close to the secondary long axis direction. The primary directions are the following: in Aggtelek karst, the NNW–SSE long axis direction is relatively unambiguous, while the nearest neighbours do not show any special direction. In Bükk Mts, the primary long axis direction is WNW-ESE, but a secondary N–S direction is also recognizable. The nearest neighbours have an outstanding but narrow peak at NNE–SSW, but further direction maxima can be observed at W–E, NW–SE and N–S. In case of the Western Mecsek, the long axis directions show a clear N–S orientation, whereas the primary nearest neighbour direction slightly deviates from this orientation and shows a maximum at NNE–SSW, and further local maxima in nearest neighbour directions can be observed at W-E and NW–SE.

2.2 Mountain Scale Relief

Among the large-scale karst landforms, we can mention primarily the karst plateaus and poljes, and in general all morphological elements of karst terrains that have several

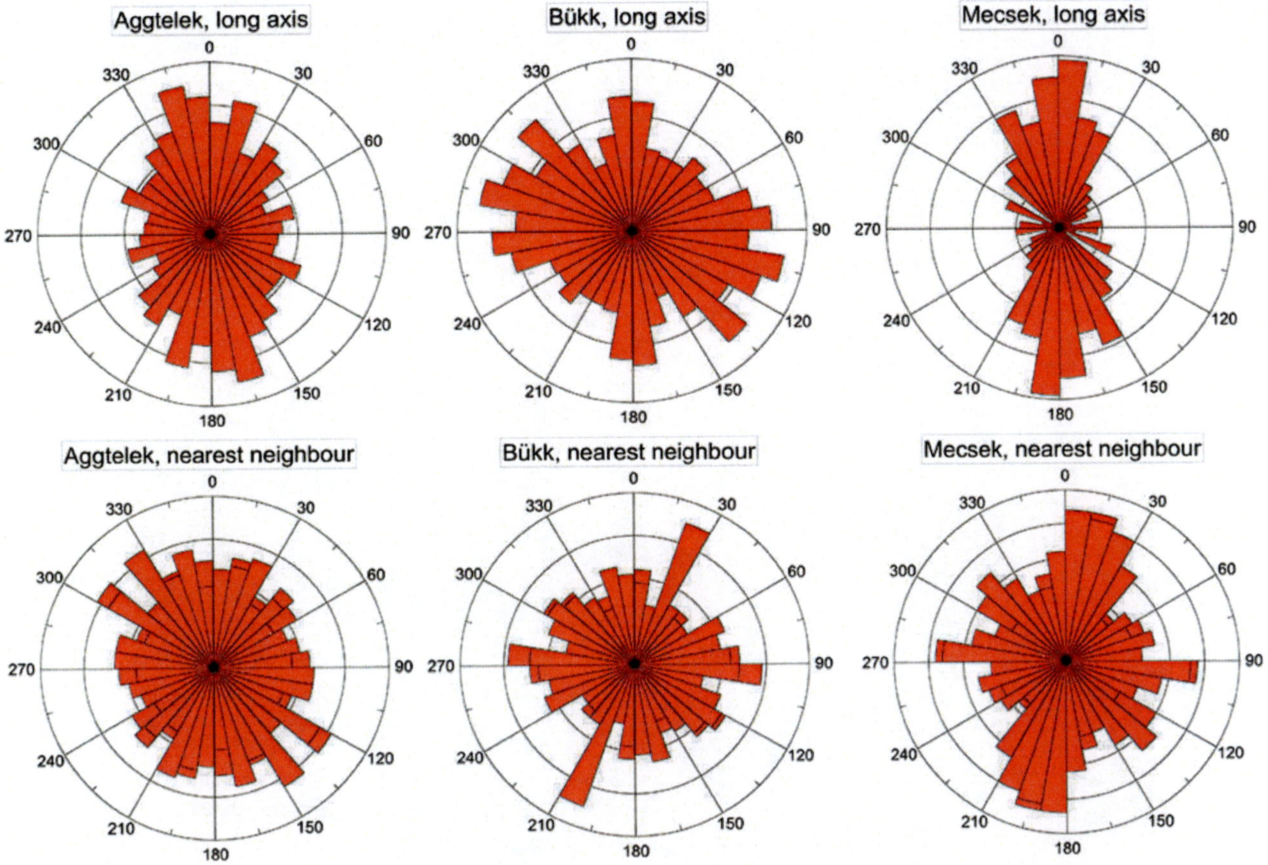

Fig. 7 Rose diagrams of doline long axes and nearest neighbour directions for Aggtelek–Rudabánya, Bükk and Mecsek Mts

km^2 surface extension. These large landforms can be homogeneous or dissected, and the latter is characteristic if they were affected by significant tectonic or erosional impacts after their primary formation. The extension, the elevation and slope characteristics of these landforms can be studied based on DTMs in a GIS framework. As these landforms are large, these examinations generally do not require high resolution DTMs, therefore the SRTM or similar global or local datasets are suitable for this purpose. The morphometric characterization of these landforms may help to answer questions in relation with their formation, too. Karst plateaus are numerous in Hungary, but in fact, true poljes do not exist in the country, only one paleo-polje is present, the so-called Jósvafő Plateau, which is found in Aggtelek Karst.

Aggtelek Karst is the southern part of the Gömör–Torna (Gemer-Turňa) Karst, which extends to both Hungary and Slovakia. This landscape was formed by large-scale pediplanation during the Late Miocene and Pliocene, and this process has been studied by several authors (for a list of these studies see Telbisz 2011). This process resulted a planation surface, which was later dissected by tectonic movements and fluvial erosion. Long topographic swath profiles (Telbisz et al. 2013) are suitable to measure and visualize the large-scale slope of the surface. The whole karst area from the Rudohorie Mts (Szlovák Érchegység) to the Putnok Hills has an N–S direction, gentle, ca. 1°, general slope (Fig. 8). The mean aspect of the karst plateaus shows a "fan-like" spatial arrangement of topography especially near the margins of the karst area (at Koniar and Jasov plateaus in Slovakia). The differential uplift of the tectonic blocks is recognizable in escarpments between blocks and at plateau edges as well as in elevation histograms. The formerly predominant N–S fluvial valley direction is preserved at some locations as "wind gaps". These landforms are uplifted and dried out N–S valley sections. Examples are the Jablonovské sedlo in Slovakia and the valley near Derenk (Telbisz 2011).

Veselský et al. (2015) worked out a DTM-based automatic procedure to recognize flat surfaces. They used 5 m resolution DTMs interpolated from 1:10,000 scale topographic maps. Their procedure used slope and curvature thresholds multi-scale analysis and landform classification principles to identify flat surfaces. They applied this

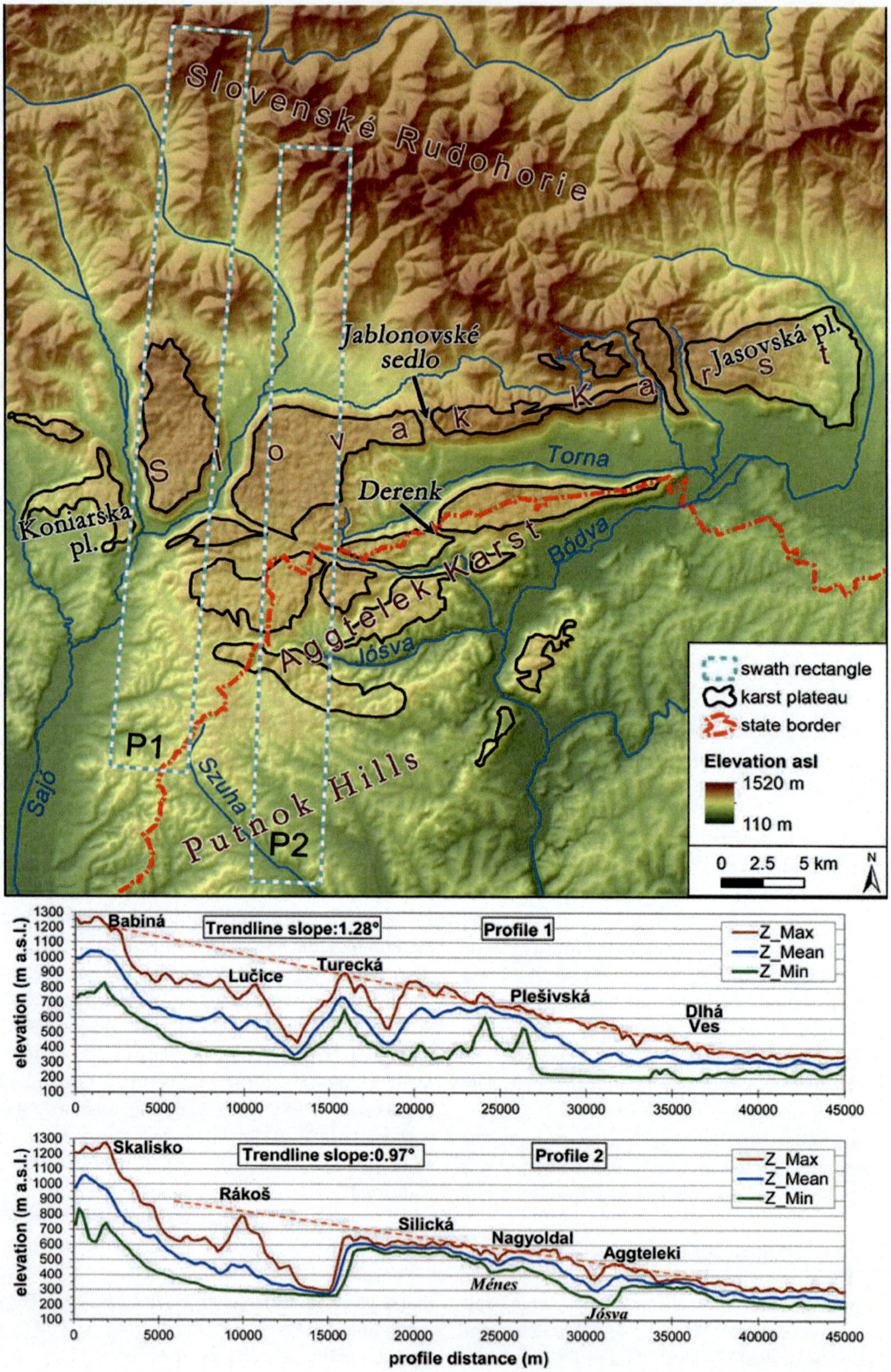

Fig. 8 N–S topographic swath profiles across Gömör–Torna Karst (consisting of Slovak Karst and Aggtelek Karst)

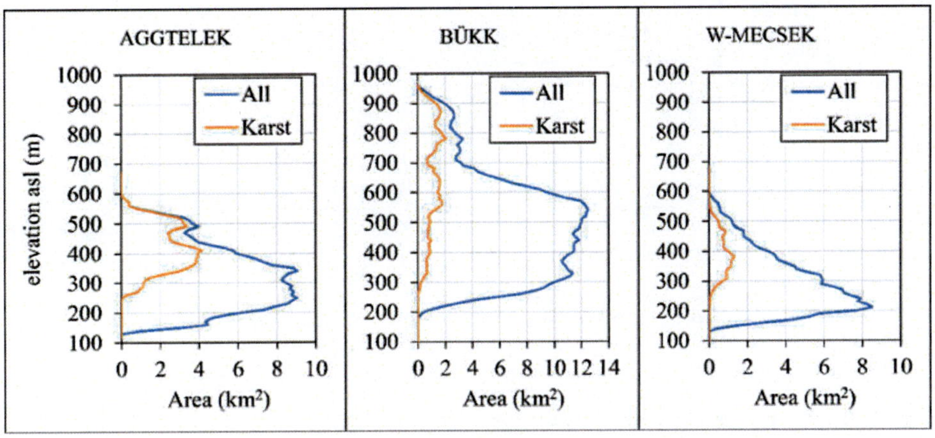

Fig. 9 Elevation distributions for the whole areas (ALL) and for the doline-dotted karst areas (KARST) in the Aggtelek–Rudabánya, Bükk and Mecsek Mts

procedure to the lower, small extension flat surfaces of the Jósva valley. They managed to discriminate several terrace surfaces along Jósva river at different altitudes (at 130–180 m asl; 190–275 m asl and 320–380 m asl as the main levels). These levels can be linked to incision periods of the river, and finally to the formation of the levels in the Baradla–Domica Cave system.

Bella et al. (2016) thoroughly examined the Jósvafő plateau by using morphometrical parameters derived from the 5 m resolution DTM including hillshade, slope, mountain front sinuosity, doline and uvala area and depth. Further on, they also took into account the fault line direction diagrams, which were created based on the neighbouring cave sections. By synthesizing the above data, they concluded that Jósvafő plateau is a piezometric level paleo-polje formed by lateral enlargement of the paleo-valley floor (for more details see chapter "The Surface Morphology of Karsts in Hungary").

The large-scale landforms of the Western Mecsek have been studied by Sebe (2006) and Kovács et al. (2007) using DTM analysis, although none of them focussed, especially on the karst features of the mountains.

In Fig. 9, we compare the elevation histograms of the three Hungarian "Aggtelek type" karsts (Aggtelek-Rudabánya, Bükk and Western Mecsek) by using the 1″ resolution SRTM DTMs. The elevation distributions have been calculated for the whole areas of the mountains, and also for the somewhat smaller doline karst terrains. The whole areas were delineated based on the topography (see Fig. 1), while the doline karst terrains were delineated using geological and geomorphological criteria as well. Among the mountains, the lowermost elevation frequency peak is present in the Western Mecsek at around 210 m asl. The frequency of the elevation zones linearly decreases upwards. This type of elevation histogram is typical for the fluvially formed mountains (Telbisz et al. 2011). It means that in the Western Mecsek, the larger scale karst landforms are less remarkable in topography even if there exists some well-karstified surfaces here. If the elevation histogram of the doline terrains is examined, then a narrower range from ca. 300 to 500 m asl with a maximum at 380 m asl can be observed. The elevation distribution of Aggtelek–Rudabánya Mts is more step-like. It suggests that karst plateaus are influential features in the relief of these mountains, and this fact also indicates that the terrain is more complex with several smaller tectonic blocks. The elevation distribution shows a wide maximum frequency zone between 250 and 350 m asl, and a secondary maximum at 490 m asl. The elevation distribution of karst terrains is bimodal, there is an upper maximum near the 490 m asl maximum of the whole surface and a lower maximum at 410 m asl. The lowermost parts of Aggtelek–Rudabánya Mts are free from dolines like in case of the Mecsek. The Bükk shows a slightly different picture. The elevation distribution is similarly step-liked, but the whole range is much larger. There is a wide frequency maximum in the elevation histogram from 300 to 600 m asl, while the well-known Nagy-fennsík Plateau generates a small, secondary maximum from 730 to 860 m asl. The elevation range of the doline-dotted surfaces has a much wider range than in the two other mountains. The local maxima at (and above) 560 m asl can be linked to the Kőlyuk–Galya area in the southeastern Bükk, while the Nagy-fennsík results maxima at 780 m asl and 870 m asl. The relatively low-lying (below 500 m asl) doline covered surfaces, which are relatively homogenously distributed in the elevation histogram are found in the Nagy–Kőmázsa area and in the Kis-fennsík.

The slope histograms (Fig. 10) are similar for all three mountains. There is a relatively outstanding frequency maximum at a relatively low slope angle and a decreasing and flattening curve towards the higher angles. However, the values are different for the three mountains. In case of Aggtelek, the maximum frequency is at 8° for the whole mountains and for the doline terrains alike. Doline terrains

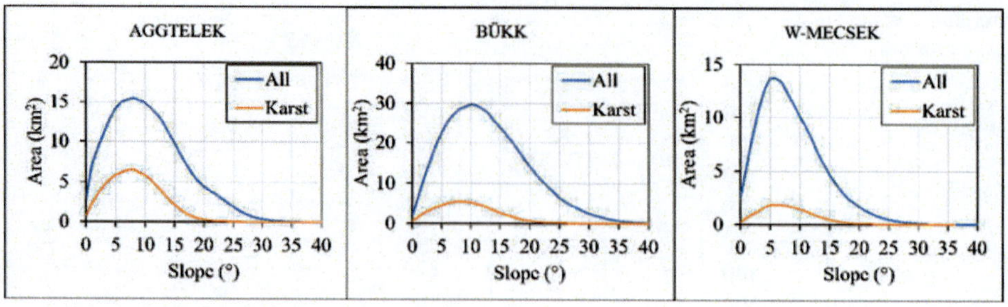

Fig. 10 Slope distributions for the whole areas (ALL) and for the doline-dotted karst areas (KARST) in the Aggtelek–Rudabánya, Bükk and Mecsek Mts

practically do not have slopes higher than 20°, and the proportion of slopes steeper than 30° is negligible for the whole mountains. The Bükk is somewhat richer in steeper slopes, thus the peak frequency is at 10° slope for the whole mountains and even the 30–40° slope range is not negligible. However, the modus of the doline terrains is at 8° in that case, too. The Western Mecsek has the smoothest relief of the three study areas that is reflected in the fact that the modus of the whole mountain histogram is at 5°, and that of the doline terrains is similar, it is at 6° slope.

The above-mentioned differences in the elevation and slope distributions of the "Aggtelek type" karst mountains can be explained by tectonic and landform evolution reasons, which are presented in more details in the chapters dealing one by one with these mountains.

Karst gorges are linearly long but narrow features, which could be also the targets of DTM-based morphometric analysis. However, in Hungary, only relatively small gorges are found on karst terrains. Their analysis would require very high resolution DTMs, but at present, there are no available data sources, thus DTM-based morphometric analysis has not yet been published about Hungarian karst gorges.

2.3 Geomorphologic Maps and Landform Classification

Although each country has a particular approach of geomorphological mapping, in general, we can state that geomorphological maps provide the synthesis of an area from a geomorphological viewpoint by presenting the landforms and processes as well as the age and lithology of a given area. Due to the complex nature of classical geomorphological maps, they are often overladen and difficult to digest, thus simplified or thematic geomorphological maps are more frequently presented in the literature. Nowadays, hand-edited geomorphological maps are less typical, instead, landform classification maps are more widely distributed and favoured (Dikau et al. 1995; van Asselen and Seijmonsbergen 2006; Gökgöz and Baker 2015). Landform classification is based on measurable morphometric parameters and is carried out by GIS tools. The fundamental data source for landform classification is the DTM, because it is suitable to calculate morphometric parameters (slope, aspect, curvature, relative relief, focal statistics such as local maximum and minimum, mean, standard deviation, etc.) and hydrological parameters (flow direction and accumulation, watershed, sinkpoint). The landform classes can be defined by a combination of these parameters, using varied thresholds and approaches.

In the followings, we present two Hungarian sample areas, for which we created DTM-based landform classification maps, and we compare these maps to the classical geomorphological maps, which were already available about these areas (Telbisz 2010). The basic data source for these samples is the 1:10,000 scale topographic map, which was digitized and interpolated by Kriging method to create the DTMs. The first sample area is the Jósvafő plateau in Aggtelek Karst, of which Móga (1998) created a classical geomorphological map. In this case, we defined the following landform classes: summit, lower summit, valley bottom, steep slope, moderate slope, nearly flat surface, doline, saddle. The threshold values were adjusted to make the landform classification results as similar to the classical geomorphological map as possible. The morphometric definitions of the landforms are presented in Table 2, whereas the derived map is shown in Fig. 11. One of the most difficult tasks is the correct automatic delineation of saddles based on the DTM. In fact, we do not know any well-tested methodology in the literature for saddle outlining. We applied the following procedure: first, the flow directions were calculated. Second, we selected the pixels, for which all 8 possible flow direction values occur in the 50 m circular neighbourhood. These pixels can be saddles, peaks or pits. Pixels, which were classified previously as part of a summit or valley or doline were deleted, and the remaining selected pixels were classified as saddles.

The second sample area is the western part of Nagy-fennsík Plateau in the Bükk, of which Hevesi (2002) created a classical geomorphological map. Since the legend

Table 2 Morphometric definitions of landform classes. Z: elevation of the pixel, Z_{max}: maximum elevation in the r radius neighbourhood, Z_{mean} mean elevation in the r radius neighbourhood; RR: elevation range in the r radius neighbourhood

Name of landform class	Morphometrical definition
Summit	$(Z_{max} - Z) < 20$ m (if r = 500 m) or $(Z_{max} - Z) < 10$ m (if r = 250 m) – multiscale criteria lower summit: below 320 m asl
Valley bottom	$(Z_{mean} - Z) > 25$ m (if r = 500 m) or $(Z_{mean} - Z) > 5$ m (if r = 100 m)
Steep slope	RR > 50 m (if r = 100 m)
Moderate slope	20 m < RR < 50 m (if r = 100 m)
Nearly flat surface	RR < 20 m (if r = 100 m)
Doline	Closed depression delineated by hydrologic modelling tools (for a detailed methodology see Telbisz et al. 2016)
Saddle	Flow directions get all 8 possible values within the 50 m radius neighbourhood and the pixel is not part of a summit, valley or doline

of the classical map of Hevesi (2002) is somewhat different from the map of Móga (1998) in the previous study, we refined some landform classes and adjusted thresholds values to make the derived map more comparable to the classical map. Thus, in the second sample, the summit is defined as a pixel, whose elevation is closer to the 250 m radius maximum elevation than 7 m. A pixel is defined as a ridge, if its elevation difference from the 500 m radius maximum elevation is less than 30 m. A pixel is a slope, if the 100 m radius relative relief is larger than 20 m, and a pixel is a valley, if it is at least 15 m lower than the mean elevation of its 300 m radius neighbourhood. Those forms, which are identified by hydrologic modelling as one compound depression are marked as twin (coalesced) dolines. Dolines were distributed into two categories: hanging dolines, which are at the level of summits and valley dolines, which are at the valley bottoms. The results are shown in Fig. 12. It is unambigous that the landform classification map and the classical geomorphological map show good agreement, in both cases.

3 Land Cover Changes

For most part of Hungary, the natural vegetation has been completely transformed due to the human activities during the recent centuries. So is the case with our karst mountains, where mostly transformed plant communities are found (see details in chapter "The Vegetation of Karsts in Hungary", in which vegetation is presented). Land cover changes are apparent manifestations of vegetation transformations and of social transitions, too. Land use and land cover changes are generally studied by GIS methods, and these methods were used on karst areas as well (e.g. Keveiné Bárány and Zboray 2001; Mari and Fehér 1999; Mari and Mattányi 2005; Móga et al. 2013, 2014, 2017). In this section, we briefly present the methodology and results of land cover change studies using two historical dates and the three Aggtelek-type karst mountains (Aggtelek Karst, Bükk and Mecsek) as case examples. The land cover changes of Aggtelek Karst have been thoroughly studied by several authors (Mari 2003; Gruber 2003; Tanács et al. 2007; Nagy 2014).

The starting point of our comparison is the second part of the eighteenth century, because the first detailed military survey of the whole country took place that time, mainly between 1780 and 1784. The maps of this survey are plan view maps based on detailed field survey and they show basic land use and land cover categories (Klinghammer and Győrffy 1997). These maps do not have a strictly defined projection, therefore it is cumbersome and not fully precise to georeference them and to make precise area calculations. However, they provide a good base to observe long-term trends in land cover changes. The georeferencing of these maps can be done by "image-to-image"-type transformation methods, and generally the EOV (Hungarian National Projection) is applied as a common projection system.

The endpoint of the land cover comparison is the twenty-first century, when a number of databases about land use and land cover are available. A freely available European database is the CORINE Land Cover Dataset (Heymann 1993; Mari and Mattányi 2002), which is based on high-resolution satellite images.

The interpretation of satellite images basically includes two different type methods. The first type of methods means a reclassification of pixels using different algorithms based on the similarities or differences of pixel spectral properties. The reclassification can be partly or fully automatic. The unsupervised classification uses the pixel histograms of the different bands. In that method, the number of classes can be set by the user. However, this method often leads to misinterpretation, i.e. different land cover-type pixels are classified to the same class if their spectral properties are similar. Or the opposite, certain land cover types incorporate spectrally different pixels, which are consequently classified to

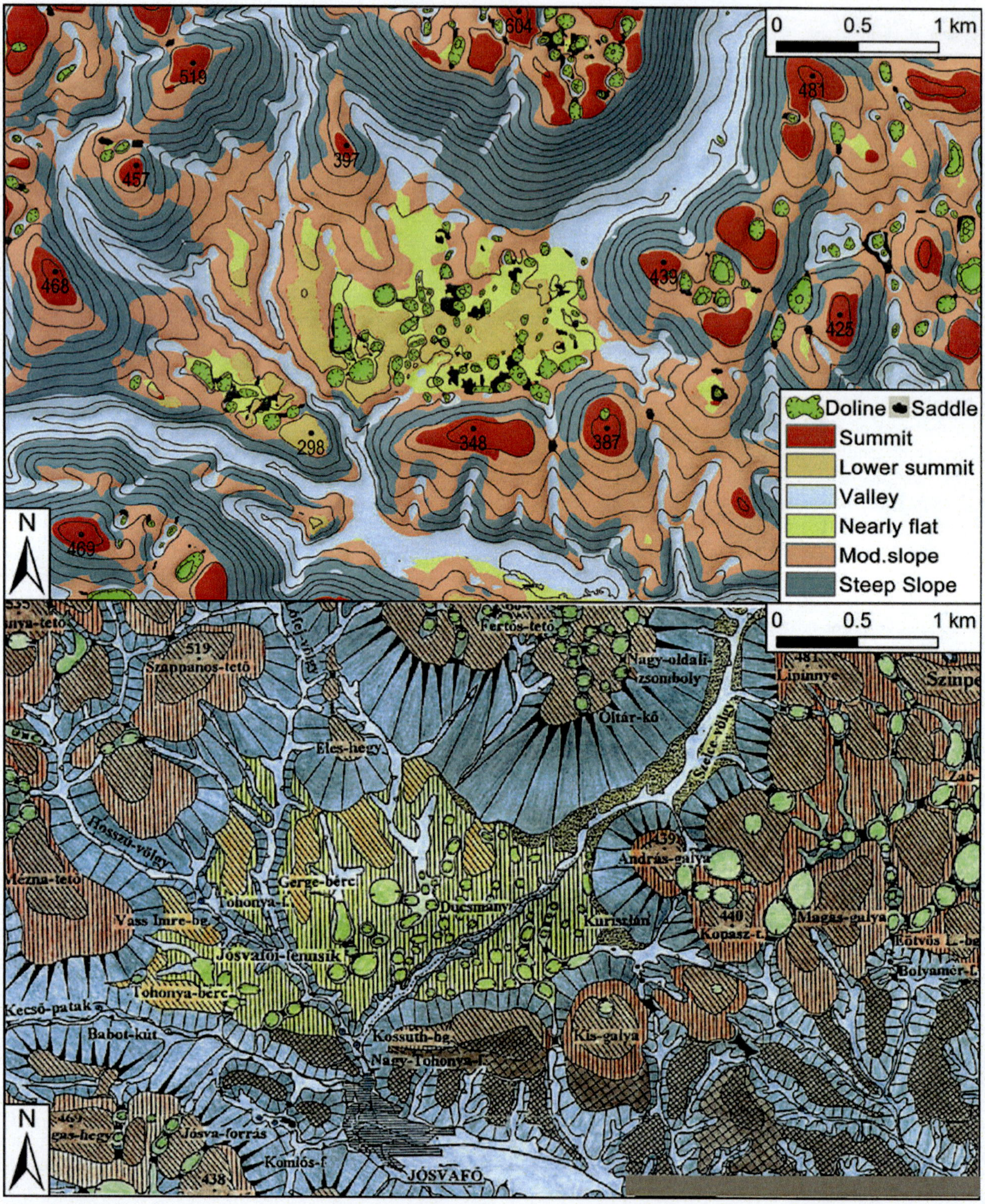

Fig. 11 DTM-based landform classification map (above) and the classical geomorphological map (bottom, after Móga 1998) of Jósvafő plateau. Contour interval is 10 m. The legend of the classical geomorphological map can be seen in the cited work

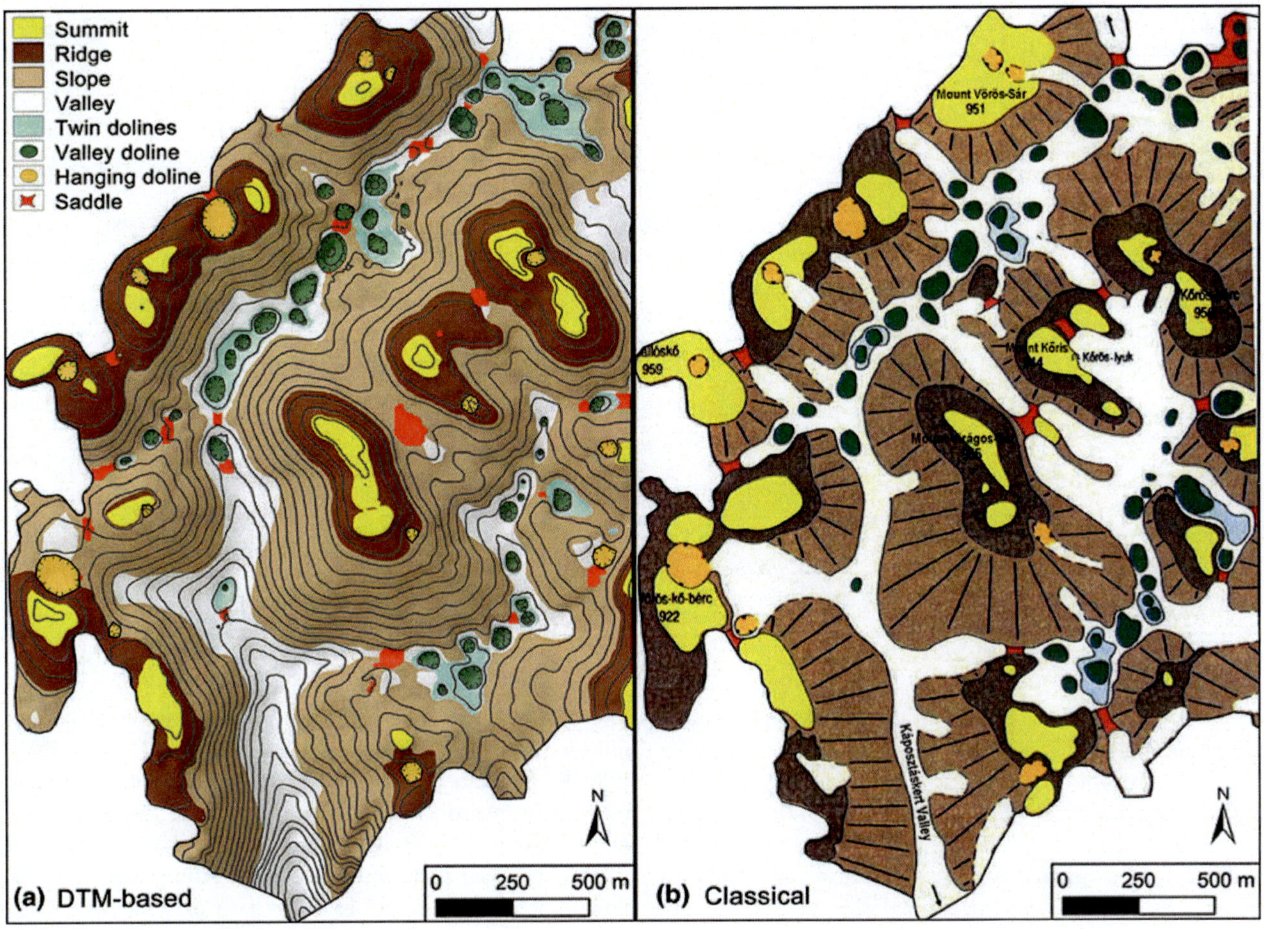

Fig. 12 DTM-based landform classification map (above) and the classical geomorphological map (bottom, after Hevesi 2002) of the western part of Nagy-fennsík (Bükk Mts). Contour interval is 20 m

different classes. Therefore, the interpretation of the classes requires some field knowledge even in case of automatic reclassification. Nonetheless, in order to improve the results, supervised classification can be applied, which means that training areas (representative samples) are defined first and the pixel classifying algorithms use the extra information given by the spectral properties of the training areas. A completely different method is the computer-aided but mostly manually performed photo interpretation. In that case, the polygons of land cover classes are delineated by human digitization (vectorization) using the satellite images. The interpretation can be helped and checked by topographic maps and field surveys (Mari and Mattányi 2002).

The European CORINE Land Cover (COoRdination of INformation on the Environment) project started in 1985 and it includes five main categories and 44 land cover classes (Heymann 1993). The original scale of the CORINE dataset is 1:100,000, the minimum size of area elements is 25 ha and the minimum width of linear elements is 100 m (Mari and Mattányi 2002; Büttner et al. 2004).

When old military maps and present land cover datasets are compared, we have to create common categories and transform all data to these categories to make the comparison meaningful and quantifiable (Mari 2003; Nagy 2003). In certain cases, the old maps do not provide enough information to discriminate CORINE land cover types, which are similar to each other (e.g. pastures vs natural grasslands). In these cases, we took into account physical geographical settings and other data sources and selected the category, which was more likely in the questionable cases. In the first military maps, we identified nine categories (settlements, arable lands, vineyards, pastures/natural grasslands, gardens with mixed agriculture; forests; scrubs; marshes; lakes). In the CORINE 2018 dataset, we omitted categories, which are not present at the study areas and merged categories, which are rare and similar to each other. Finally, we got 14 categories. Five of the CORINE categories, which are present in the study areas are absolutely missing in the first military maps. Three of them did not even exist that time (industrial units; mines/dumps/constructions; sport and leisure

Table 3 Proportions of land cover categories (in per cent) at the end of the eighteenth century and in 2018. AK1: Aggtelek Karst, eighteenth century; AK2: Aggtelek Karst 2018; B1: Bükk, eighteenth century; B2: Bükk 2018; M1: Mecsek, eighteenth century; M2: Mecsek 2018

CORINE code	Land cover type	AK1	AK2	M1	M2	B1	B2
112	Discontinuous urban fabric	0.93	1.61	1.13	4.60	0.10	0.94
121	Industrial units	0.00	0.00	0.00	0.23	0.00	0.07
131, 132, 133	Mines/Dumps/Constructions	0.00	0.49	0.00	0.73	0.00	0.54
142	Sport and leisure facilities	0.00	0.00	0.00	0.52	0.00	0.21
211	Arable land	22.08	5.70	26.24	11.98	3.82	0.63
221	Vineyards	5.08	0.00	5.76	0.23	2.18	0.96
222	Fruit trees	0.00	0.00	0.00	0.12	0.00	0.05
231	Pastures, grassland	10.80	10.42	11.54	4.52	4.55	1.73
242	Complex cultivation patterns	0.15	1.17	1.16	6.25	0.07	0.50
243	Principally agriculture with significant natural vegetation	0.00	4.51	0.00	3.95	0.00	1.12
311	Forests	52.84	67.20	51.89	62.16	87.89	85.37
324	Transitional woodland-scrub	7.70	8.90	2.28	4.45	1.38	7.89
411	Inland marshes	0.42	0.00	0.00	0.05	0.00	0.00
512	Water bodies	0.02	0.00	0.00	0.20	0.00	0.00

facilities), whereas two of them (fruit trees; principally agriculture with significant natural vegetation) could not be discriminated from other types in the first military maps.

The numerical results are presented in Table 3 and Fig. 13. For comparison, the maps of both dates are also provided here (Fig. 14). Almost two and a half centuries elapsed between the two dates, and this period brought profound changes in the socio-economic conditions of the whole country and of the study areas as well. As a result, the land use also significantly changed during that period, though not as drastically as in urban agglomerations. A remarkable change can be observed in the proportion of arable lands, which significantly decreased in all study areas (especially in Aggtelek Karst and Mecsek, where arable lands occupied about one quarter of the surface two and a half centuries ago). Vineyards, which had proportions over 5% at the end of the eighteenth century, almost completely disappeared. It is partly the result of the phylloxera disaster at the end of the nineteenth century, and partly the result of the loss of former vine markets due to the state border changes after the First World War (Szuhay 1998). Inland marshes also disappeared as a result of river management operations. However, the proportion of marshes was low even at the beginning of the study period since karst mountains are generally characterized by water scarcity. On the other hand, the area of forests significantly increased during the study period in Aggtelek Karst and Mecsek. In case of Bükk, we can see an increase in the proportion of the transitional wood and scrub category, which includes the forest clearings as well. The increase is due to the fact that forest management has become more conscious since the nineteenth century and cuttings were regulated (Járási 1998). In addition, the need for charcoal, which was very important previously, has drastically decreased in the twentieth century. Further on, a significant part of these mountains became national park, thus they are strictly protected areas now. Another clear trend is the growth of settlement areas even though they still occupy relatively little proportion of the land. The proportion of pastures and grasslands remained relatively stable in Aggtelek Karst, while it significantly decreased in the Mecsek and Bükk mountains. As a conclusion, it is obvious from the analysis that the present land cover is closer to a semi-natural state than two and a half centuries ago, which is in connection with the less active land use of the decreasing rural population and the increasing importance of nature protection goals.

4 Karst Vulnerability and Hydrogeology

The issues concerning karst vulnerability and hydrogeology are often studied by GIS methods; therefore, we only briefly review these issues here, and more detailed introductions to these topics are written in Chapters 6 (hydrogeology) and 21 (geoecology).

The protection of water resources is one of the most basic issues in karst vulnerability. Karst vulnerability maps are important data visualization techniques in this topic. These maps are generally created by GIS tools as these maps require the integration of several type spatial information.

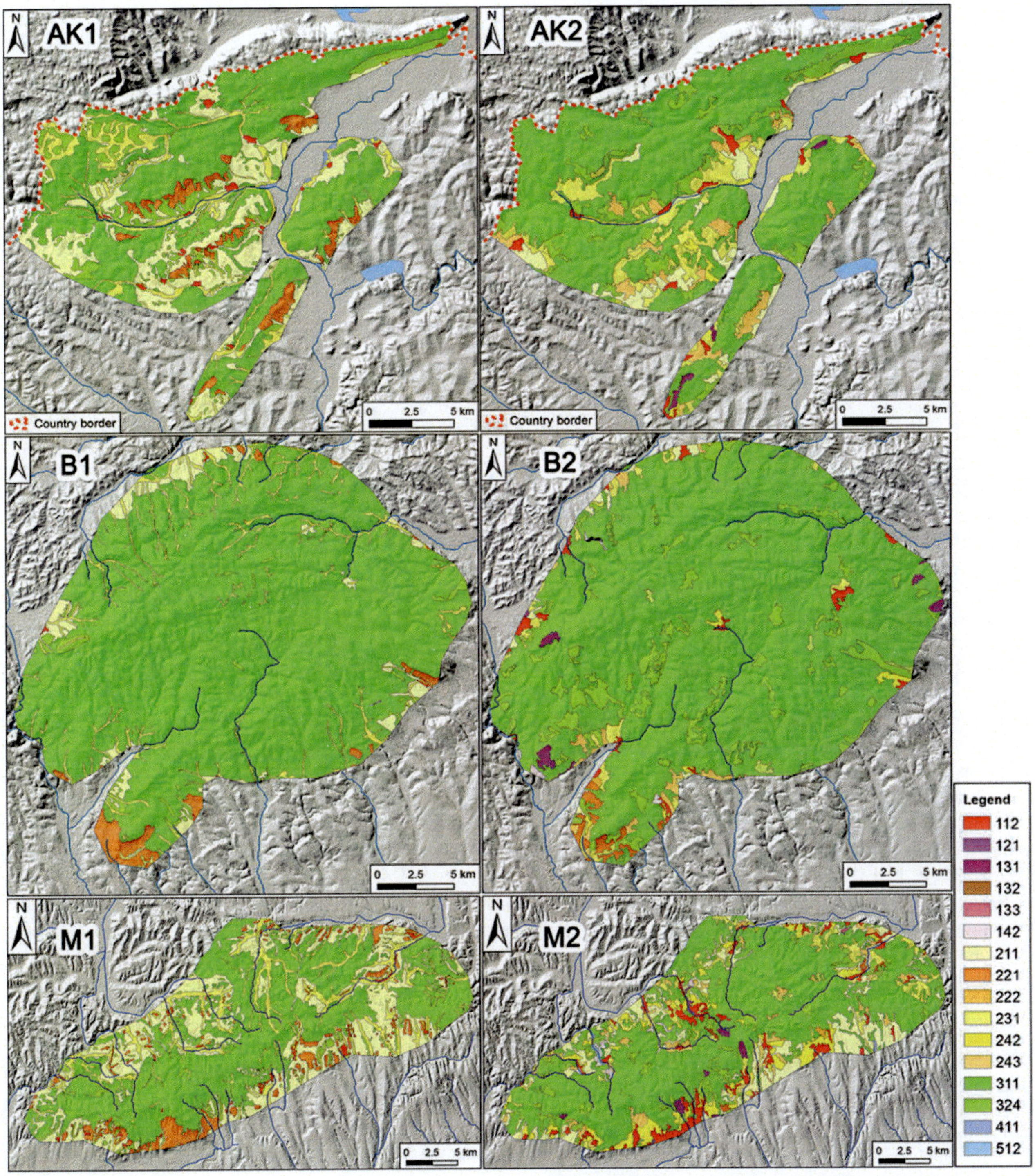

Fig. 13 Comparison of land cover maps of the studied karst areas based on the first military survey (1782–1785) and the CORINE 2018 databases. AK1: Aggtelek Karst, 1st military; AK2: Aggtelek Karst 2018; B1: Bükk, 1st military; B2: Bükk 2018; M1: Mecsek, 1st military; M2: Mecsek 2018. For the land cover codes, see Table 3

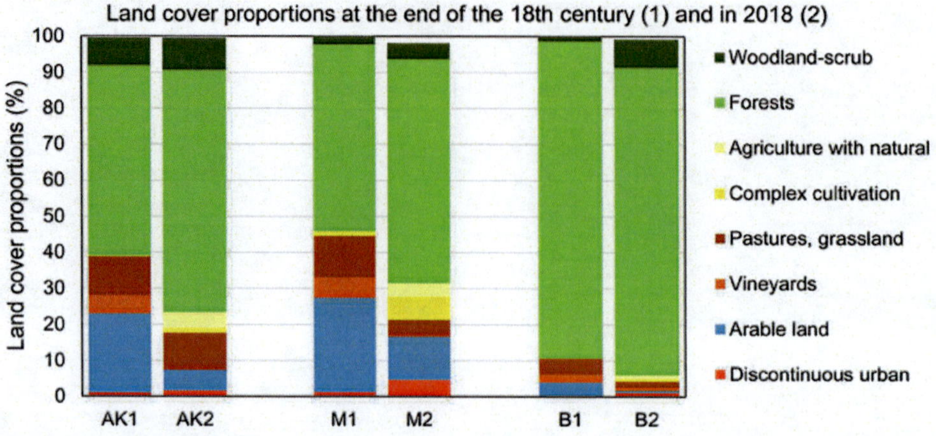

Fig. 14 Diagrams showing the changes in the proportions of land cover categories from the end of the eighteenth century to 2018. Only categories with larger than 1% at least in one cell are drawn in the figure. AK1: Aggtelek Karst, Eighteenth century; AK2: Aggtelek Karst 2018; B1: Bükk, eighteenth century; B2: Bükk 2018; M1: Mecsek, eighteenth century; M2: Mecsek 2018

Karst vulnerability maps of different Hungarian karst areas have been created by Iván et al. (2011), Móga et al. (2017) and Iván and Mádl-Szőnyi (2018). The karst hydrological data of Bükk Mts are stored in an integrated GIS system containing monitoring data of 700 caves, 1138 sinks and of several water tracer tests (Hernádi et al. 2012, 2013, Szegediné Darabos et al. 2015).

The vulnerability of karsts can be regarded not only from a hydrological viewpoint, but in geological or geomorphological terms, too. The notion of geodiversity was coined at the end of the 1990s, obviously using the term of biodiversity as a model, and it denotes the diversity of rocks, minerals, fossils, forms, processes, soils and waters. The necessity to investigate the geodiversity of karst terrains in Hungary has been proposed for the first time by Keveiné Bárány (2007), and Őrsi (2009) worked out a geodiversity evaluation for a sample area in the western part of the Bükk Mts using GIS methodology. Further on, the vulnerability of karsts incorporates the vegetation issues, and in an even broader meaning, the whole landscape as well. The vegetation of the Western Mecsek Karst has been mapped by Hoyk and Keveiné Bárány (2000) using GIS methods. These investigations can serve as a basis for nature protection management.

5 Subsurface Karst Objects in GIS

GIS has traditionally been concerned with integrating spatial data from ground objects. As with dolines, the position of cave entrances can be expressed with coordinates, and using GIS tools, the frequency of cave occurrences in an area can be analysed (Shofner et al. 2001). The National Cave Register of Hungary stores the coordinates, parameters and descriptions of 4146 caves of the country (Ministry of Rural Development 2020). Although this cadastre contains information about the length and vertical extension of caves, a thorough spatial analysis can be done only if the exact shape of the passages is known (Tarsoly 2014). The integration of caves into 3D GIS is a natural need in the case of a karst analysis, and this has become an increasingly viable option as 3D capabilities and measuring tools have improved in 3D programs (Mucsi et al. 2004).

5.1 Data of Cave Surveys

The cave survey is traditionally based on distance and compass direction (dip and azimuth) measurements in cave passages. This is complemented by the passage cross-sectional drawings at key points, also based on measurements (Jeannin and Häuselmann 2014). The data of the survey can be used to construct the cave polygon skeleton, and subsequently, a cave map (Fig. 15). The results of the surveys, producing either analogue or digital data, can be used to construct the spatial model of the cave (Albert 2010). Early software enabled the 3D polygon skeleton to be displayed, but still used a local coordinate system. Later, the cave data were also adapted to the cartographic projections (in Hungary, typically the EOV—Hungarian National Projection), which made it possible to represent the surface data and the cave passages in a common reference system.

An important step in the development of measuring instruments was the replacement of measure tape and clinometric compass with laser telemetry, which has taken place at the early years of the first decade of the twenty-first century (Albert 2010). The most widespread tool for this type of surveying was the combination of DistoX laser

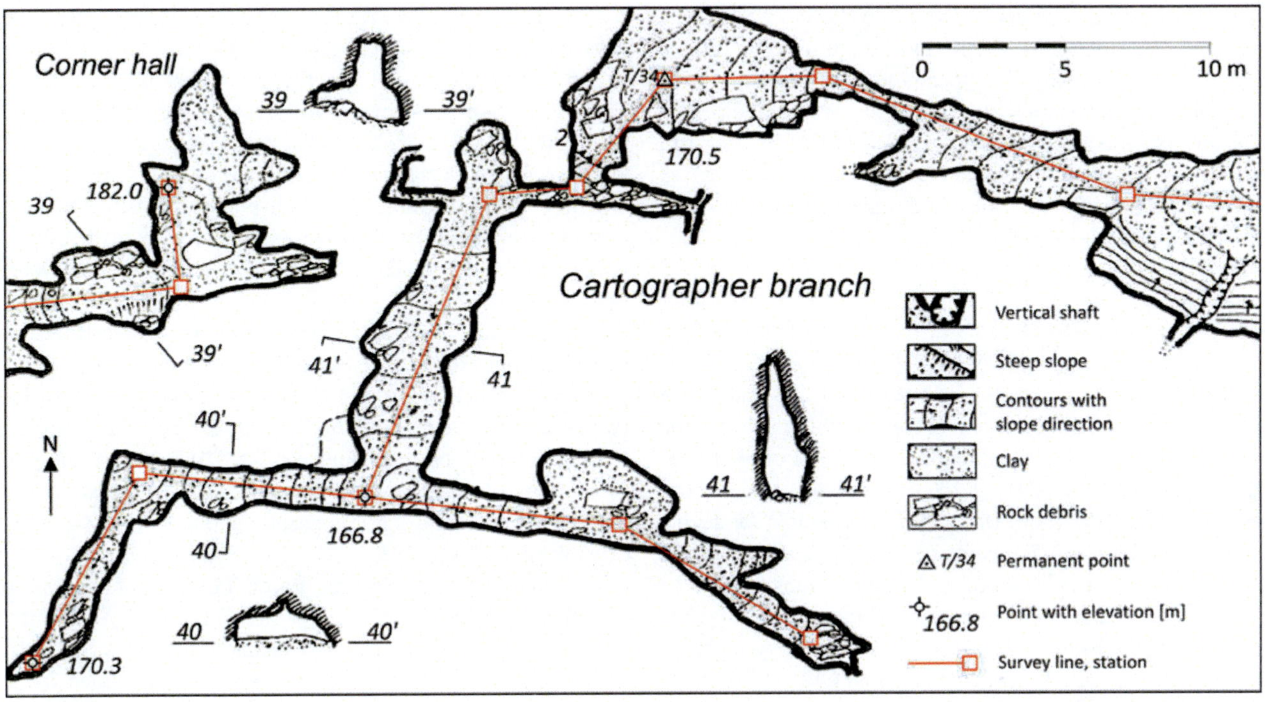

Fig. 15 The relation of a polygonal surveying and the compiled cave map showing three cross-sectional profiles in a small part of the Pál-völgy Cave (modified after Kárpát 1983)

telemeter with a handheld computer, allowing the cave surveyor to build the database and compile the map on the spot (Hegedűs and Szabó 2014). However, the really big step was the introduction of the terrestrial laser scanning (TLS), which has undergone significant technical progress over the last decade (Albert 2017). The first TLS cave surveys were done in large, easily passable caves, because of the size of the instrument.

The laser scanner produces millions of point-like data in a virtual 3D environment around the measure station gathering information about the cave wall surface in sub-centimetre resolution. To create a realistic 3D model of a cave, several measure station should be established throughout the passage system and powerful computers are needed to process the hundreds of millions of individual points into complex 3D surfaces. Such point cloud processing packages are getting more efficient in terms of handling large data volume and reduced processing time with advanced and more powerful functionalities to visualize and generate different products (Oludare Idrees and Pradhan 2016). However, due to the complexity of the post-processing and the larger size of the surveying equipment, the use of the TLS in caves is limited so far, and in small, narrow passages, the use of DistoX is still more viable for the time being.

5.2 The Role of Cave Surveys

The surveying of caves is important not only for orientation, aids in further exploration, finding connections to nearby passages, but also important as part of the morphological characterization of the cave. Cave morphology helps us to understand the hydrological processes, and to trace back the geomorphic history of the cave (Palmer 1989). The porosity structure of a karstic rock body includes micro-, meso- and macroporosity, and only the latter (in the form of cave passages) can be considered as potential part of a GIS (Albert et al. 2015). Knowledge of the relationship between the surface and the cave can be also important for exploring additional access points, for creating artificial entrances and for understanding the environmental effects.

5.3 Complexity Levels of Cave GIS

The complexity of GIS, created from cave surveys, can greatly vary. Obviously, the level of complexity, correlating primarily with the number of measured points, depends mainly on the used surveying tools and thus, on the time when the survey was done. In many cases, several different types of source data (paper notes, records, laser telemetry and even TLS data) should be handled if a cave GIS is established. Albert (2017) emphasized that managing such an inhomogeneous source, one has to be very cautious when building the GIS to avoid loss of data, redundancy and to ensure quality control. Basically, three types of data are distinguished: entrance (point-like objects); passage (3D polygons); realistic model (3D surfaces).

5.4 Entrance Database

If only the entrance coordinates are stored and each cave entrance is represented as one record in the database, the functionalities of the system require the use of ordinary 2D GIS tools (such as *ArcGIS*, *QGIS*, or a webmap server). The National Cave Register of Hungary is an example for such a database. Although it is maintained continuously from the data of reported expeditions of speleologist teams (registered cavers of the Hungarian Speleological Society), it does not give the chance to understand the 3D morphology of caves through queries and 3D visualizations in a GIS environment.

Cave cadastres serve as sources of data not only for morphologists, hydrologists and cavers but also play important roles in determining the natural heritage sites and planning tourist destinations. In order to build a 3D GIS of the surveyed data, tools that are capable to visualize and query complex 3D structures are needed.

5.5 Passage Pattern in 3D

When cave passages are modelled either physically or virtually, one has to invert the phenomenon, representing the void as an object (Albert 2018). The directional and length measurements are good sources to create a 3D "skeleton" of the cave. The earliest cave GIS tools (like the *Polygon* software, which was the most widespread in Hungary) provided access to both the survey records and the interactive visualization of the 3D passage system. However, even if the profile sectioning of the passages provides valuable information about the volume of the cave, only a few of the documenting GIS tools (such as *Therion* and *Compass*) were capable to represent the approximate shape of each passage.

5.6 Realistic Models

For scientific (e.g. morphological) analyses, the TLS cave models are invaluable data sources. However, due to the surveying methods, the model may lack important details of cave passages, where the morphology of the cave is highly complex, and inaccessible wall-sections are shaded by other cave structures (e.g. stalactites, curtains, etc.) from the surveying instrument. In Hungary, the first attempts to scan caves with TLS occurred in the first years of the 2010s. These exploratory surveys concentrated on larger caves, that are easily passable with the instruments, and even these wider sections were only partly surveyed. The results, though the data is still available, have been published only as a written report and a web-viewer in the case of the Pál--völgy Cave in the Buda Hills (Gede et al. 2013), and were used to create a traditional, large-scale 2D map in the case of the main branch of Baradla Cave in Aggtelek Karst (Borzsák and Szunyogh 2015). The data of a TLS survey of another cave in Aggtelek Karst, the Béke Cave, were also used to create a cave map (Restás-Göndör 2015). The second longest dolomite cave in Hungary, the 420 m long Szentgáli Kő-lik Cave, was also surveyed using TLS, and a common map of the cave and the topography above it was compiled (Tarsoly and Bekk 2019).

The integration of high-resolution TLS cave models into a GIS would require large storage and memory capacity from the aspect of information technology. However, it is not yet clear, how such a system could be maintained for a long use, especially if more than one cave is involved. In Hungary, the browser-based visualization of TLS scan images of the Szemlő-hegy Cave, combined with a morphological analysis tool, were the only attempt to publish the data of such a survey in a publicly available format (Gede et al. 2015).

5.7 Possible Use of the Data

For traditional 2D GIS analyses, the entrance database of caves, combined with the vertical extent and length attributes, has been created and maintained in Hungary in the form of the National Cave Register (Ministry of Rural Development 2020). For safety reasons, visitors of the website do not have access to the coordinates of the entrances, only the attributes and descriptions are displayed. However, for scientific purposes, one can require the GIS-compatible version of the data.

If one attempts to use the data of a cave survey in Hungary, the usual way is to require the traditional polygon records and profile sections from the speleological group, which is responsible for the monitoring of the given cave. Nowadays, these data are at least partly digitized and usually the 3D polygonal model of the cave is also accessible.

In urban areas, depicting the cave polygons together with the surface topography can be informative in itself. Cave systems under the city of Budapest was depicted in this way, for example, by Virág (in Mindszenty 2013) in the case of the Buda Thermal Karst System, or by Móga et al. (2017) in the case of Tapolca.

The polygonal data can provide useful information for the analysis if it is integrated into a GIS. Using polygonal data of Pál-völgy Cave as well as archive cave maps and profiles, a volumetric model was created, through which the estimated size of the cave was also calculated (Albert 2010). The then unknown length of the cave was estimated to be two times the then known 12.6 km length of the passages. After a decade of further exploration, in 2020, the known length of the Pál-völgy Cave System is 32.05 km. In the mentioned study of Albert (2010), not only the size but also the proportion of macroporosity was estimated (1.46%), which is an important input parameter in hydrological modelling. Similar modelling—based on polygonal data and archive maps—was carried out for other caves of the Buda Thermal Karst System (Albert et al. 2015), namely for the Szemlő-hegy Cave and the Molnár János Cave.

Comparing polygon-based volumetric models with realistic TLS models is one way to get a better understanding of the capabilities of a 3D cave GIS (Fig. 16). Merk (2014) attempted to carry out this task for a very small part of the Pál-völgy Cave (for some tens of metres of passage length), and she found that the volume of the 3D model created from cave polygons and cross sections was only two-thirds of the realistic TLS-based 3D model.

In larger caves, which are of special importance due to their natural or cultural heritage, TLS can certainly play an important role. Survey data not only records the current state of the cave but also can be used for map making, and finally it can be used to show cave details that are inaccessible to the public for some reason. Attempts to establish a database of 3D models (including rock surfaces) for geological investigation has been made several times (Bemis et al. 2014). Even online repositories such as Sketchfab could be used to store and disseminate the information gained by modelling. However, the size of these models (in terms of terabytes) is limited due to the finite server capacity. Albert (2017) summarized the possible technical trajectories of cave surveying in the near future. As the DistoX was once popular, and the TLS became the leading edge during the last decade, the emerging technologies like LiDAR-based mobile mapping systems, and/or close-range photogrammetry will probably produce the bulk of cave survey data in the future. What is sure that caves will remain places where both the surveying instruments and the surveyors face with challenges.

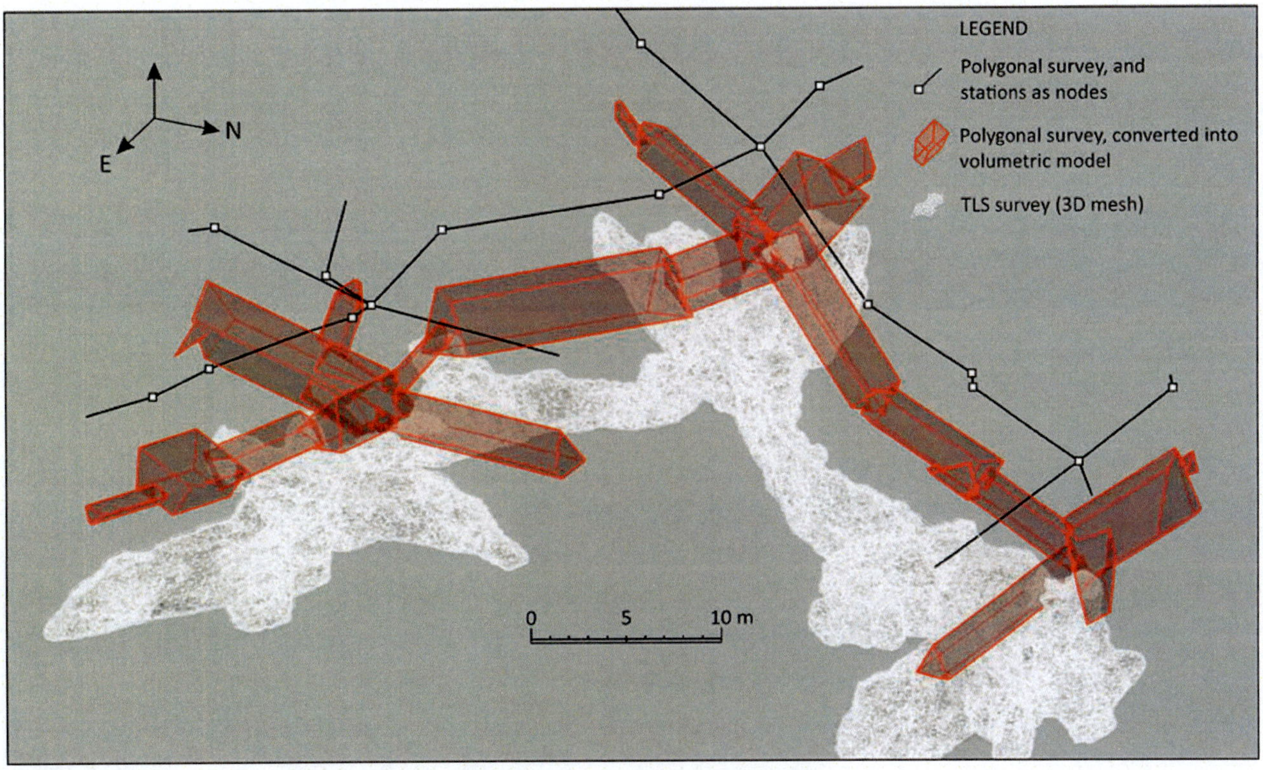

Fig. 16 Representation of three types of cave model: the polygonal "skeleton"; the volumetric and the realistic model. Each require a different data structure in a cave GIS. The models show a small part of the Pál-völgy Cave system

6 Human–Environment Relations on Karst

This book primarily presents the natural characteristics of Hungarian karst terrains. However, as human–environment relations have some specific features on karst terrains, certain social aspects are also briefly mentioned here. The relationship of natural settings and social characteristics is one of the main issues in geography, and it can be studied in the theoretical framework of geographic possibilism (Kőszegi et al. 2015). From the human viewpoint, it is argued that due to the special hydrologic, pedologic, relief and geological settings, karst terrains are less favourable for settling and traditional agriculture. Some selected social characteristics of Gömör–Torna (Gemer-Turňa) Karst (GTK) were compared to the neighbouring non-karst areas using GIS methodology by Telbisz et al. (2014, 2015). The GTK and its neighbourhood were distributed into 8 units in that research (Fig. 17). It was demonstrated that the karst area has lower population density than the neighbouring non-karst territories, and demographic decline and ageing are also more typical in the karst area than in the other units. In addition, the spatial distribution of population is uneven even within the karst area, since the plateaus are practically uninhabited, whereas the foothill area, where karst springs are present, hosts a chain of settlements.

In order to find quantitative human–environment relationships, some numerical parameters were selected. Mean elevation, mean slope and mean distance to the closest river were determined for each landscape units using the SRTM DTM and drainage network data. Further on, population density, settlement density, mean settlement population, road density and population change from 1828 to 2011 were also calculated for each landscape unit based on social databases. Thereafter, statistical regression analysis was used to infer functional relationships among these factors (Table 4).

It was found that settlement density did not correlate with any of the studied factors. In turn, population density, mean settlement population and population change were in moderately strong relationship with river distance, whereas the road density was more strongly correlated with the mean slope of the landscape unit. In almost all cases, the social parameters of GTK (the red diamond in Fig. 18) are below the regression curve. It suggests that the karst area is less developed than the value estimated simply from these regressions, thus the karst really means some kind of drawback from the viewpoint of socio-economic

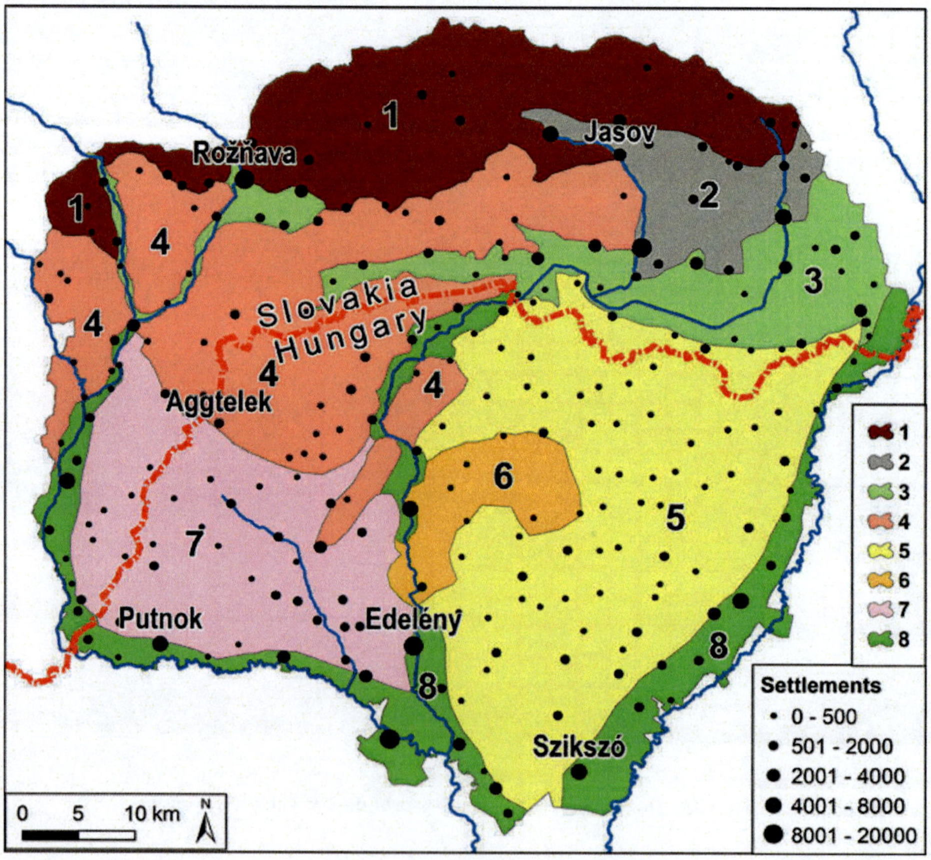

Fig. 17 Landscape units of Gömör-Torna Karst and its neighbourhood. 1: Rudohorie Mts; 2: Rudohorie Foothills; 3: Northern Valleys and Basins; 4: Gömör-Torna Karst; 5: Cserehát Hills; 6: Szendrő Hills; 7: Putnok Hills; 8: Southern Valleys and Basins

Table 4 Best-fit function types and determination coefficients (R^2) of the studied relationships (bold: highest R^2)

Dependent variable	Best-fit function	Independent variable			
		Elevation	Slope	Height	River distance
Settlement density	Linear	0.0180	0.00003	0.0011	**0.1409**
Mean settlement population	Logarithmic	0.2024	0.6494	0.5132	**0.7657**
Population density	Logarithmic	0.3232	0.7958	0.6663	**0.8119**
Road density	Exponential	0.4698	**0.7713**	0.6473	0.4388
Population change (1828–2011)	Logarithmic	0.2910	0.6594	0.5461	**0.8628**

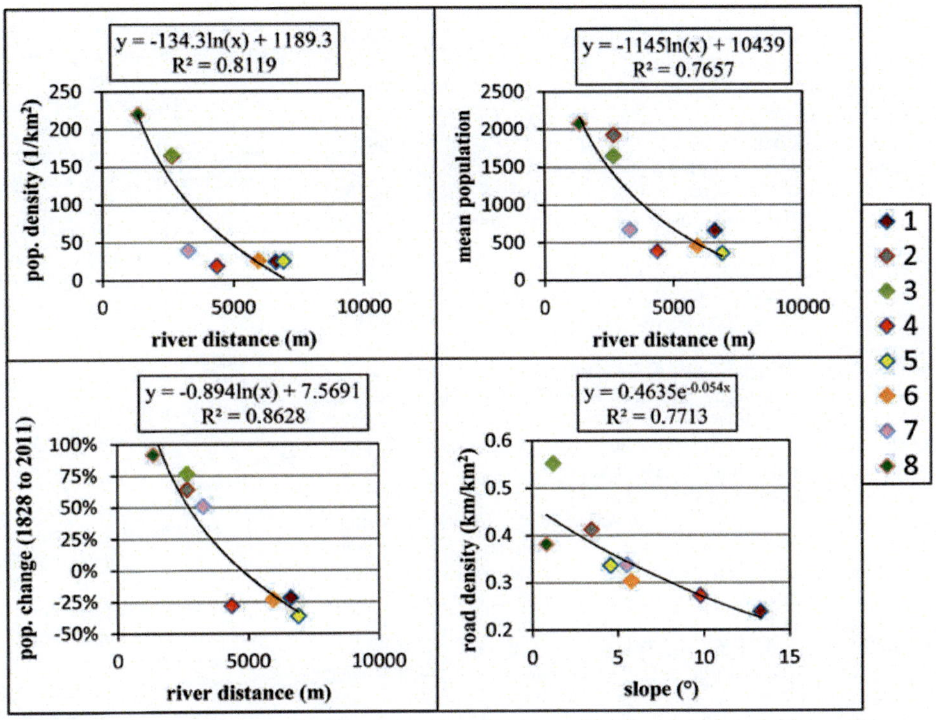

Fig. 18 Best-fit functional relationships between physical parameters and social parameters (Gömör-Torna Karst is red in each diagram). 1: Rudohorie Mts; 2: Rudohorie Foothills; 3: Northern Valleys and Basins; 4: Gömör–Torna Karst; 5: Cserehát Hills; 6: Szendrő Hills; 7: Putnok Hills; 8: Southern Valleys and Basins

development. However, tourism, which is connected to caves, can generate a significant positive impact, but it is limited to few settlements only. Thus, the aim of spatial planning in this area should be the preservation of the positive impacts and the extension of these advantages to as many settlements as possible.

7 Conclusions

In this chapter, we presented a statistical analysis of the most important doline morphometrical parameters of the Hungarian Aggtelek-type karsts using a unified methodology. We presented that differences in doline density of subunits varies from 5.8 to 123.3 dolines per km^2. However, we underlined that in case of karst terrains, where dolines are typically very small, namely in Mecsek or on the covered karst terrains of Bakony, further studies are recommended based on LiDAR data, when it becomes available. The elevation and slope characteristics of the above mountains have been presented using digital terrain analysis. Landform classification maps were also created using DTMs, and it was demonstrated that classical geomorphological maps and these semi-automatically derived maps are in good agreement. The analysis of land cover changes of Aggtelek, Bükk and Mecsek Mountains proved that the present land cover is closer to a semi-natural state than two and a half centuries ago, which is in connection with the less active land use of the decreasing rural population and the increasing importance of nature protection goals. Different types of modern cave survey methods (including terrestrial laser scanning, TLS) and their integration into GIS systems have been also presented in this chapter. At present, TLS surveys were carried out in Pál-völgy, Baradla, Béke, Szentgáli-kőlik and Szemlő-hegy Caves. The GIS analysis of human–environment relations on karst were demonstrated using the Gömör–Torna Karst as an example, and it was concluded that the karst really means certain drawback from the viewpoint of socio-economic development. However, tourism, which is connected to caves, can generate a significant positive impact, but it is limited to few settlements only.

References

Albert G (2010) A budapesti Pál-völgyi-barlang üreg- és pórustérfogatának modellezése (Modelling the cavity and porosity volume of Pál-völgy Cave, Budapest). Földtani Közlöny 140:263–280 (in Hungarian)

Albert G (2017) Aspects of cave data use in a GIS. In: Karabulut S, Cinku MC (eds) Cave Investigations. IntechOpen, pp 25–44

Albert G (2018) 3D geológiai modellek fejlesztése térinformatikai szemlélettel (Development of 3D geological models using GIS). GlobeEdit, Beau Bassin (in Hungarian)

Albert G, Virág M, Erőss A (2015) Karst porosity estimations from archive cave surveys—studies in the Buda Thermal Karst System (Hungary). Int J Speleol 44:151–165. https://doi.org/10.5038/1827-806X.44.2.5

Bárány-Kevei I, Mezősi G (1994) New morphometrical parameters for explanation of karst development. Acta Geographica Szegediensis 31:27–33

Bárány-Kevei I, Mezősi G (1995) Ecological characteristics of doline types in Aggtelek hills, North-Hungary. Acta Universitatis Szegediensis: Acta Geographica 30:135–146

Bátori Z, Csiky J, Farkas T, Vojtkó AE, Erdős L, Kovács D, Wirth T, Körmöczi L, Vojtkó A (2014a) The conservation value of karst dolines for vascular plants in woodland habitats of Hungary: refugia and climate change. Int J Speleol 43:15–26. https://doi.org/10.5038/1827-806X.43.1.2

Bátori Z, Farkas T, Erdős L, Tölgyesi C, Körmöczi L, Vojtkó A (2014b) A comparison of the vegetation of forested and non-forested solution dolines in Hungary: a preliminary study. Biologia 69:1339–1348. https://doi.org/10.2478/s11756-014-0430-4

Bella P, Veselsky M, Gaal L, Minar J (2016) Josvafo paleo-polje: morphology and relation to the landform evolution of Aggtelek Karst and Josva River valley, Hungary. Z Geomorphol 60:219–235. https://doi.org/10.1127/zfg/2016/0212

Bemis SP, Micklethwaite S, Turner D, James MR, Akciz S, Thiele ST, Bangash HA (2014) Ground-based and UAV-Based photogrammetry: a multi-scale, high-resolution mapping tool for structural geology and paleoseismology. J Struct Geol 69:163–178. https://doi.org/10.1016/j.jsg.2014.10.007

Borzsák S, Szunyogh G (2015) A Baradla-barlang lézerszkenneralapú, nagyléptékű térképezése (High scale mapping of Baradla Cave based on laser scanning survey). Zöld Horizont 10:3 (in Hungarian)

Büttner G, Feranec J, Jaffrain G, Mari L, Maucha G, Soukup T (2004) The CORINE land cover 2000 project. EARSeL eProceedings 3:331–346

Dikau R, Brabb EE, Mark RK, Pike RJ (1995) Morphometric landform analysis of New Mexico. Z Geomorphol Suppl 101:109–126

Dutkó A (2000) A Bükk dolináinak statisztikai elemzése (Statistical analysis of Bükk dolines). M.Sc. thesis, Department of Physical Geography, Eötvös Loránd University (in Hungarian)

Farsang A, Tóth T (1993) Morphometric investigation of dolines in Bükk mountains. Acta Geographica Szegediensis 31:53–60

Gede M, Petters C, Nagy G, Nagy A, Mészáros J, Kovács B, Egri C (2013) Laser Scanning Survey in the Pál-völgy Cave, Budapest. In: Buchroithner MF (ed) Proceedings of the 26th international cartographic conference. Dresden, Germany, p 905

Gede M, Ungvári Z, Kiss K, Nagy G (2015) Open-source web-based viewer application for TLS surveys in caves. In: Gartner G, Haosheng H (eds) Proceedings of the 1st ICA European symposium on cartography. International Cartographic Association, Vienna, pp 321–328

Gökgöz T, Baker MKM (2015) Large scale landform mapping using lidar DEM. ISPRS Int J Geo Inf 4:1336–1345. https://doi.org/10.3390/ijgi4031336

Gruber P (2003) Tájtörténeti kutatások a Baradla-barlang vízgyűjtőjén (Landscape historical research of Baradla Cave drainage basin). Karsztfejlődés 8:243–251 (in Hungarian)

Hegedűs A, Szabó Z (2014) A barlangok felmérése. Tanfolyami jegyzet (Cave survey. Course textbook). MKBT Oktatási Szakosztály, Budapest (in Hungarian)

Hernádi B, Lénárt L, Horányiné Csiszár G, Tóth K (2012) A bükki nyílt karszt vertikális karsztosodottsága (Vertical karstification of the open karst of Bükk). Karsztfejlődés 17:63–78 (in Hungarian)

Hernádi B, Balla B, Czesznak L, Horányiné Csiszár G, Sűrű P, Tóth K (2013) A vízvédelem érdekében végzett barlang és töbörvizsgálatok eredményeinek beintegrálása a Bükki Karsztvíz Észlelő Rendszer (BKÉR) vizsgálati rendszerébe (Integration of cave and doline water research data collected for water resources protection into Bükk Karst Water Monitoring System). Karsztfejlődés 18:9–21 (in Hungarian)

Hevesi A (2001) A Nyugati-Mecsek felszíni karsztosodásának kérdései (Issues of surface karstification in Western Mecsek). Karsztfejlődés 6:103–111 (in Hungarian)

Hevesi A (2002) A magyarországi karsztok fejlődéstörténet és formakincs szerinti csoportosítása; az aggteleki jellegű karsztok felszíni formakincse (The classification of Hungarian karst terrains based on landform evolution and geomorphology; the surface landforms of Aggtelek type karsts). Ph.D. thesis, Miskolci Egyetem (in Hungarian)

Heymann Y (1993) Land cover technical guide—EUR 12585. Office for Official Publications of the European Communities, Luxembourg

Hoyk E (2002) A nyugat-mecseki karszt dolináinak morfometriai vizsgálata (Morphometrical analysis of Western Mecsek Karst). Karsztfejlődés 7:161–171 (in Hungarian)

Hoyk E, Keveiné Bárány I (2000) Tájökológiai szempontú vegetációelemzés a nyugat-mecseki karszton (Analysis of vegetation of Western Mecsek karst from the viewpoint of landscape ecology). Karsztfejlődés 5:47–55 (in Hungarian)

Iván V, Mádl-Szőnyi J (2018) Vulnerability assessment and its validation: the Gömör-Torna Karst, Hungary and Slovakia. Geol Soc London, Special Publications 466:261–273. https://doi.org/10.1144/SP466.15

Iván V, Móga J, Fehér K, Miklós B (2011) Karsztsérülékenységi vizsgálatok a bükki Kis-fennsík karsztján (Karst vulnerability research on the Kis-fennsík plateau of Bükk). Karsztfejlődés 16:167–183 (in Hungarian)

Járási L (1998) Az erdőgazdálkodás múltja (The history of forest management). In: Baross G (ed) Az Aggteleki Nemzeti Park (Aggtelek National Park). Mezőgazda Kiadó, Budapest, pp 395–410 (in Hungarian)

Jeannin P-Y, Häuselmann P (2014) Speleological investigations. In: Goldscheider N, Drew D (eds) Methods in Karst hydrogeology. Taylor & Francis Books, London, pp 25–44

Kárpát J (1983) A Pál-völgyi-barlang (Map of the Pál-völgy Cave). Scale 1:250 (in Hungarian)

Keveiné Bárány I (2007) Geodiverzitás a karsztokon (Geodiversity on karst). Karsztfejlődés 12:215–223 (in Hungarian)

Keveiné Bárány I, Zboray Z (2001) Karszttájak változásainak vizsgálata térinformatikai módszerekkel (Analysis of karst landscape changes using GIS methodology). Karsztfejlődés 6:45–59 (in Hungarian)

Klinghammer I, Győrffy J (1997) A katonai térképezés (Military mapping). In: Pannon Enciklopédia, Magyarország földje (The land of Hungary). Kertek 2000 Kiadó, Budapest, pp 26–29 (in Hungarian)

Kőszegi M, Bottlik Z, Telbisz T, Mari L (2015) Human-environment relationships in modern and postmodern geography. Hungarian Geographical Bulletin 64:87–99. https://doi.org/10.15201/hungeobull.64.2.1

Kovács IP, Lampért K, Bugya T, Lovász G, Varga G (2007) Planation surfaces of the southern part of Western Mecsek. Studia Geomorphologica Carpatho-Balcanica 41:45–60

Látos T, Telbisz T (2018) LiDAR és UAV alapú digitális domborzatmodellek összevetése töbör-morfometria szempontjából a Jósvafői-fennsík példáján (The comparison of LiDAR and UAV based digital elevation models for the purpose of doline morphometry using Jósvafő Plateau as an example). Karsztfejlődés 23:19–30. https://doi.org/10.17701/18.19-30 (in Hungarian)

Látos T, Telbisz T, Deák M, Székely B, Koma Z, Standovár T (2015) Lidar és topográfiai térkép alapú digitális terepmodellekből levezetett, illetve kézzel digitalizált töbör-körvonalak morfometriai összehasonlítása az Aggteleki-karszt példáján (Morphometrical comparison of dolines derived from LiDAR data and manually digitized from topographical maps using Aggtelek Karst as an example). Karsztfejlődés 20:145–165. https://doi.org/10.17701/15.145-165 (in Hungarian)

Lippmann L, Kiss K, Móga J (2008) Az Abaliget-Orfűi Karszt karsztos felszínformáinak vizsgálata térinformatikai módszerekkel (Analysis of karstic landforms on Abaliget-Orfű Karst using GIS methodology). Karsztfejlődés 13:151–166 (in Hungarian)

Mari L (2003) Felszínborítás-változás vizsgálata térinformatikai módszerekkel az Aggteleki Nemzeti Park területén (Analysis of landscape changes on Aggtelek Karst using GIS methodology). Karsztfejlődés 8:231–242 (in Hungarian)

Mari L, Fehér K (1999) The impacts of land use change on the Budapest hydrothermal-karst: a study of Szemlő-hegy cave. Acta Geographica Szegediensis 36:104–111

Mari L, Mattányi Z (2002) Egységes európai felszínborítási adatbázis a CORINE Land Cover program (An integrated European land cover dataset, the CORINE Land Cover program). Földrajzi Közlemények 126:31–38 (in Hungarian)

Mari L, Mattányi Z (2005) Földrajzi információs rendszerrel támogatott tájföldrajzi vizsgálatok az Ipoly-völgy példáján (Landscape geographical analysis of Ipoly valley using GIS methodology). Földrajzi Értesítő 54:265–284 (in Hungarian)

Merk Z (2014) A Pál-völgyi-barlang 3D modelljének elkészítése és összevetése korábbi kutatási eredményekkel (Creation of the 3D model of Pál-völgy Cave and its comparison to earlier research results). M.Sc. thesis, Department of Cartography and GIS, Eötvös Loránd University (in Hungarian)

Mezősi G (1984) A Sajó-Bódva köze felszínfejlődése (Landform evolution of Sajó-Bódva-köz). Földrajzi Értesítő 33:181–205 (in Hungarian)

Mindszenty A (ed) (2013) Budapest. Földtani értékek és az ember. Városgeológiai tanulmányok (Budapest. Geological values and humans. Urban geological studies). ELTE Eötvös Kiadó, Budapest (in Hungarian)

Ministry of Rural Development, Hungary (2020) National Cave Register. http://www.termeszetvedelem.hu. Accessed 28 Mar 2020

Móga J (1998) Felszínalaktani megfigyelések a Gömör-Tornai-karszton (Geomorphological observations on Gömör-Torna Karst). Ph.D. thesis, Department of Physical Geography, Eötvös Loránd University (in Hungarian)

Móga J, Kiss K, Szabó M, Kériné Borsodi A, Kéri A, Mari L, Knáb M, Iván V (2013) Hazards and landscape changes (degradations) on Hungarian Karst mountains due to natural and human effects. J Mt Sci 10:16–28. https://doi.org/10.1007/s11629-013-2400-7

Móga J, Strat D, Iván V, Mari L, Kiss K, Szabó M, Borsodi A, Csüllög G (2017) Changes of the karst landscape and epikarst system in the area of the Tapolca karst terrains, North-West Balaton Highlands, Hungary. Forum Geografic 16:12–25. https://doi.org/10.5775/fg.2017.074.i

Móga J, Szabó M, Mari L, Borsodi A, Kéri A, Knáb M, Kiss K, Iván V (2014) Természeti és antropogén hatásokra végbemenő tájváltozások vizsgálata a Bakonyban (Analysis of landscape changes in the Bakony Mountains due to natural and anthropogenic impacts). Földrajzi Közlemények 138:89–106 (in Hungarian)

Mucsi L, Balog A, Juhász G (2004) Interaktív, 3D-s barlangi információs rendszer a világhálón (Interactive, 3D cave information system on world wide web). Geodézia és Kartográfia 56:3–8 (in Hungarian)

Nagy D (2003) Tájtörténeti kutatások a Gömör-Tornai-karszton I. A történelmi táj rekonstrukciója az ANP környezetében az I-III Katonai Felmérések alapján (Landscape historical research on Gömör-Torna Karst I. The reconstruction of the historical landscape in and around ANP using the I-III military survey data). In: ANP Füzetek II., ANP. ANP, Jósvafő, pp 107–143 (in Hungarian)

Oludare Idrees M, Pradhan B (2016) A decade of modern cave surveying with terrestrial laser scanning: a review of sensors, method and application development. Int J Speleol 45:71–88. https://doi.org/10.5038/1827-806X.45.1.1923

Őrsi A (2009) Geodiverzitás-vizsgálat egy nyugat-bükki mintaterületen (Geodiversity analysis on a Western Bükk study area). M.Sc. thesis, Department of Physical Geography, Eötvös Loránd University (in Hungarian)

Palmer AN (1989) Geomorphic history of the Mammoth cave system. In: White WB, White EL (eds) Karst hydrology: concepts from the Mammoth cave area. Springer, US, Boston, MA, pp 317–337

Rabus B, Eineder M, Roth A, Bamler R (2003) The shuttle radar topography mission—a new class of digital elevation models acquired by spaceborne radar. ISPRS J Photogramm Remote Sens 57:241–262. https://doi.org/10.1016/S0924-2716(02)00124-7

Restás-Göndör A (2015) Barlangtérkép szerkesztése TLS felmérés adatainak felhasználásával (Creation of cave maps using TLS survey data). M.Sc. thesis, Department of Cartography and GIS, Eötvös Loránd University (in Hungarian)

Sebe K (2006) Domborzatmodell alkalmazhatósága a geomorfológiai elemzésben a Nyugat-Mecsek példáján (The applicability of digital elevation models in geomorphological analysis using Western Mecsek as an example). Hungarian Geograph Bull 55:5–23 (in Hungarian)

Shofner GA, Mills HH, Duke JE (2001) A simple map index of karstification and its relationship to sinkhole and cave distribution in Tennessee. J Cave Karst Stud 63:67–75

Szegediné Darabos E, Miklós R, Tóth M, Lénárt L (2015) A 2014/2015-ös év fontosabb kutatási irányai és eredményei a Bükkben (Some important research directions and results in Bükk Mountains during the year 2014/2015). Karsztfejlődés 20:29–47. https://doi.org/10.17701/15.29-47 (in Hungarian)

Székely B, Telbisz T, Koma Z, Kelemen K, Szmorad F, Deák M, Látos T, Standovár T (2015) Geomorphometric correlations of vegetation cover properties and topographic karst features based on high-resolution LiDAR DTM of Aggtelek Karst, NE Hungary. In: EGU General Assembly conference abstracts. Vienna

Szuhay P (1998) Az Aggteleki-karszt falvainak szociográfiája (Sociography of Aggtelek Karst villages). In: Baross G (ed) Az Aggteleki Nemzeti Park (Aggtelek National Park). Mezőgazda Kiadó, Budapest, pp 378–394

Tanács E, Szmorad F, Bárány-Kevei I (2007) Review of the forest management history and present state of the Haragistya karst plateau (Aggtelek Karst, Hungary). Acta Carsologica 36:441–451

Tarsoly P (2014) Lézerszkenner alkalmazása ingókövek és gyapjúzsákok felmérésében a Velencei-hegységben (The application of laser scanner in the survey of balanced rocks and boulders). Karsztfejlődés 19:261–272 (in Hungarian)

Tarsoly P, Bekk T (2019) A Szentgáli-kőlik felmérése lézerszkenneléssel (Laser scan survey of Szentgáli-kőlik Cave). 71:18–24. https://doi.org/10.30921/GK.71.2019.1.4 (in Hungarian)

Telbisz T (2001) Új megközelítések a töbör-morfológiában az Aggteleki-karszt példáján (New perspectives in doline morphology using Aggtelek Karst as an example). Földrajzi Közlemények 125:95–108 (in Hungarian)

Telbisz T (2004) Digitális domborzatmodellek használata a karsztkutatásban (The use of digital elevation models in karst studies). Karsztfejlődés 9:21–33 (in Hungarian)

Telbisz T (2010) Digitális domborzatelemzési módszerek alkalmazása karsztos területek geomorfológiai térképezésében (The application of digital terrain analysis in the geomorphological mapping of karst terrains). In: Hegedűs A (ed) Geoinformatika és domborzatmodellezés 2009: A HunDEM 2009 és a GeoInfo 2009 konferencia és kerekasztal válogatott tanulmányai. Miskolci Egyetem, Földrajz Intézet, Miskolc, pp 1–13 (in Hungarian)

Telbisz T (2011) Large-scale relief of the Slovak Karst and Aggtelek Karst (Gömör–Torna/Gemer–Turňa Karst)—a DEM-based study. Hungarian Geograph Bull 60:379–396

Telbisz T, Bottlik Z, Mari L, Petrvalská A (2015) Exploring relationships between Karst terrains and social features by the example of Gömör-Torna Karst (Hungary-Slovakia). Acta Carsologica 44:121–137. https://doi.org/10.3986/ac.v44i1.1739

Telbisz T, Bottlik Z, Mari L, Petrvalská A, Kőszegi M, Szalkai G (2014) Természeti tényezők hatása a népesség területi eloszlására a Gömör-Tornai-karszt és környezete példáján (The impact of natural factors on the spatial distribution of population—the case study of Gömör-Torna Karst and its surroundings). Földrajzi Közlemények 138:277–292 (in Hungarian)

Telbisz T, Kovács G, Székely B, Szabó J (2013) Topographic swath profile analysis: a generalization and sensitivity evaluation of a digital terrain analysis tool. Z Geomorphol 57:485–513. https://doi.org/10.1127/0372-8854/2013/0110

Telbisz T, Látos T, Deák M, Székely B, Koma Z, Standovár T (2016) The advantage of lidar digital terrain models in doline morphometry compared to topographic map based datasets—Aggtelek karst (Hungary) as an example. Acta Carsologica 45:5–18. https://doi.org/10.3986/ac.v45i1.4138

Telbisz T, Mari L, Szabó L (2011) Geomorphological characteristics of the Italian side of Canin Massif (Julian Alps) using digital terrain analysis and field observations. Acta Carsologica 40:255–266. https://doi.org/10.3986/ac.v40i2.10

van Asselen S, Seijmonsbergen AC (2006) Expert-driven semi-automated geomorphological mapping for a mountainous area using a laser DTM. Geomorphology 78:309–320. https://doi.org/10.1016/j.geomorph.2006.01.037

Varga Á (2007) Az Aggteleki-karszt és a Bükk hegység összehasonlító töbör-morfometriai elemzése térinformatikai módszerek segítségével (Comparative doline morphometrical analysis of Aggtelek Karst and Bükk Mountains using GIS methodology). M.Sc. thesis, Department of Physical Geography, Eötvös Loránd University (in Hungarian)

Veress M, Vetési-Foith S (2019) A Bakonyvidék felszíni karsztja (The surface karst of Bakony Mountains). Földrajzi Közlemények 143:167–186. https://doi.org/10.32643/fk.143.2.6 (in Hungarian)

Veselský M, Bandura P, Burian L, Harciníková T, Bella P (2015) Semi-automated recognition of planation surfaces and other flat landforms: a case study from the Aggtelek Karst, Hungary. Open Geosci 7:799–811. https://doi.org/10.1515/geo-2015-0063

Vetési-Foith S (2018) A Mester-Hajag fedett karsztos formáinak morfometriai vizsgálata (Morphometrical analysis of covered karst landforms on Mester Hajag Mountain). Karsztfejlődés 23:95–111. https://doi.org/10.17701/18.95-111 (in Hungarian)

Vetési-Foith S, Kiss Z, Gárdonyi I (2017) A mecseki karszt oldódásos- és utánsüllyedéses töbreinek morfometriai elemzése (Morphometrical analysis of dissolution and suffusion dolines of Mecsek Karst). Karsztfejlődés 22:119–138. https://doi.org/10.17701/17.119-138 (in Hungarian)

Zboray Z, Keveiné Bárány I (2004) Domborzatértékelés a Bükk-fennsíkon légifelvételek felhasználásával (Analysis of relief of Bükk Plateau using aerial imagery). Karsztfejlődés 9:207–213 (in Hungarian)

The Caves of Aggtelek Karst, Szalonna Karst and Rudabánya Mountains

Péter Gruber, Gábor Szunyogh, and Tamás Telbisz

Abstract

The Aggtelek Karst lies along the Hungarian–Slovakian border and it is a part of the Western Carpathian mountain range. It represents a typical example of well-developed plateau karsts in the humid temperate climate zone. Numerous surface and subsurface landforms (285 caves) were created by the intense karstification of Triassic limestones. The compound multi-phase evolution of karst from the Late Cretaceous to the present is recorded in the morphology and sedimentary deposits of caves. The formation types of caves are highly variable that resulted a compound morphology including speleothems and erosional features. The rare cave fauna and archaeological findings are also of special value in this trans-boundary karst area. Many abysses formed at high densities on the top of karst plateaus (e.g. on Alsó-hegy Plateau, there are up to 40 shaft caves per 1 km^2). The 30 km long Baradla–Domica cave system is an excellent example of a fluvially modelled 'ideal watertable cave'. Several multi-level caves were formed by allogenic sinking streams at the edge of karst plateaus or along traverse valleys. The Baradla–Domica cave system hosts more than 500 animal species and played an important role in regional and global biospeleology. Many archaeological artefacts created by cave dwelling humans were preserved from the Palaeolithic, Mesolithic and mostly the Neolithic period. However, traces of younger cave settlements are also documented by rich artefacts. In the eastern part of the karst, in Esztramos Hill, thermal caves are also found and they form an interesting hybrid system with artificial former mining passages.

Keywords

Aggtelek Karst • Plateau Karst • Cave • Diversity • World Heritage

1 Introduction

The Aggtelek Karst is not a separate geographical unit, but the southern part of the Gömör–Torna Karst. It forms both geologically and hydrographically an organic unit with the Slovak Karst on the other side of the border. The caves and formations of the area are of outstanding significance due to their extraordinary richness of form, complexity, relative integrity and concentration in a small area. Karstic processes have created a diversity of forms and habitats that are significant biologically, geologically and paleontologically. In terms of topography, the Aggtelek Karst is currently characterized by medium mountains and temperate continental climatic conditions. However, throughout the geological history, it has undergone significant changes in both topography and climate, of which accumulated sediments and preserved surface forms provide abundant information. Based on these, it can be concluded that tropical-subtropical karstification took place here during the Late Cretaceous and Paleogene, while periglacial geomorphological processes prevailed during the glacial periods. Such diversity of karstification processes in a relatively small area is rare worldwide, and these phenomena have been documented extremely thoroughly here. Taking all these values into account, in 1995 the caves of the Aggtelek and Slovak Karst were jointly inscribed on the UNESCO World Heritage List (Fig. 1).

P. Gruber (✉)
Aggtelek National Park Directorate, Tengerszem oldal 1., 3758 Jósvafő, Hungary

G. Szunyogh
University of Óbuda, Budapest, Hungary
e-mail: szunyogh.gabor@bgk.uni-obuda.hu

T. Telbisz
Department of Physical Geography, Eötvös Loránd University, Pázmány Péter sétány 1/C, 1117 Budapest, Hungary
e-mail: tamas.telbisz@ttk.elte.hu

M. Veress and Sz. Leél-Őssy (eds.), *Cave and Karst Systems of Hungary*,
Cave and Karst Systems of the World, https://doi.org/10.1007/978-3-030-92960-2_10

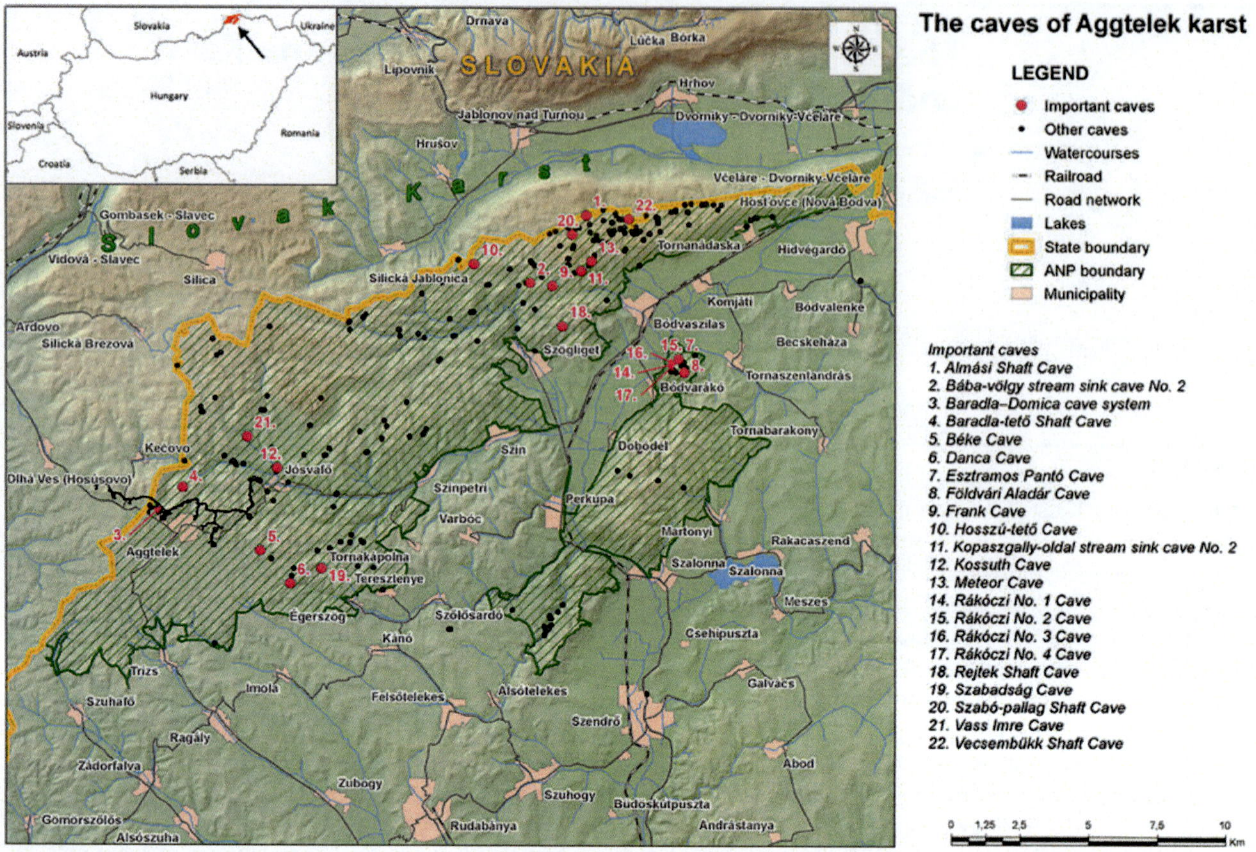

Fig. 1 Location map of the caves of Aggtelek Karst presented in this chapter

The most significant cave in the area is the *Baradla–Domica cave system*, richly decorated with stalactites, stalagmites and dripstone columns. This cave system crosses the border of Hungary and Slovakia. As one of the most important active stream caves in the temperate zone, it has also been included in the Ramsar Convention on the Protection of Wetlands.

More than 99% of the caves in the Aggtelek Karst are still in their fully original natural state, the preservation of which is guaranteed by adequate protection. The remaining 1%, on the other hand, is made up of show caves that allow human use, in which significant alterations have also taken place. The annual number of visitors exceeds 100,000 per year.

Under Hungarian law, all caves are state-owned and their protection includes their related surface area as well. As karst systems are particularly sensitive to environmental changes (agricultural pollution, deforestation, soil erosion), maintaining the entirety of geological, hydrological and biological processes requires the integral management of the whole drainage basin.

2 General Review

In the area of the Gömör–Torna Karst, caves with different genetics and morphology occur at a relatively high density (Fig. 2). A significant part of them were largely shaped by cave streams. Some of the caves still have an almost continuously active watercourse, but others have water flow only intermittently. Both the dissolution and abrasion effects of subsurface streams have contributed to the development of these caves. The majority of them consist of almost horizontal, often multi-level passages. Their typical shapes are tunnel-like, the corridors are often winding with notches or scallops on the sides and ceiling channels at the top. The second type of caves are active, or periodically active, stream sink caves formed at lithological boundaries. For both types, it is characteristic that collapses along the tectonic fractures formed large rooms within them.

Quite different types of caves were formed where corrosion played an almost exclusive role, primarily in the

Fig. 2 Genetic types of the caves of Aggtelek Karst. **a** Intermittently active stream cave (excerpt from the Discovery branch of Béke Cave, National Cave Register); **b** main stream branch of Baradla–Domica cave system (photo Egri); **c** Intermittently active stream sink cave (Pócsakő Stream Sink Cave, National Cave Register); **d** Entrance to Bába-völgy No. 2. stream sink cave; **e, g** Szabó-pallag Shaft Cave of corrosion origin (**e** National Cave Register, **g** photo Egri); **f, h** Vecsembükk Shaft Cave of corrosion origin (**f** National Cave Register, **h** photo Egri); **I** Rákóczi No. 3. Cave of phreatic origin (National Cave Register); **j** Phreatic passages in Rákóczi No. 4. Cave (photo Kovács); **k** Typical popcorns from a cave of Esztramos Hill (photo Kovács)

Table 1 Thedeepest shaftsand vertically extended caves inthe Aggtelek Karst(status of August 31, 2020)

	Name	Karst area	Depth/vertical extent (m)
1	Vecsembükk Shaft Cave	Alsó-hegy Plateau	236
2	Szabó-pallag Shaft Cave	Alsó-hegy Plateau	151
3	Meteor Cave	Alsó-hegy Plateau	127
4	Stream Sink Cave No. 2 in Kopasz-gally Hillside	Alsó-hegy Plateau	118
5	Baradla–Domica cave system	Aggtelek Plateau	112
6	Almási Shaft Cave	Alsó-hegy Plateau	100
7	Béke Cave	Aggtelek Plateau	97
8	Baradla-tető Shaft Cave	Aggtelek Plateau	87
9	Tektonik Shaft Cave	Alsó-hegy Plateau	80
10	Rákóczi No. 1 cave	Esztramos Hill	79

tectonically weakened parts of the limestone. This is how the shaft caves (locally called "*zsomboly*"), which are particularly characteristic of the Gömör–Torna Karst, were created. The walls of these caves are often sculptured by vertical, solution runnels. Cave depth often exceeds 100 m (Table 1) and in several cases these caves consist of a series of interconnected underground shafts. On the other hand, the planform area of these caves is relatively small: in some cases, only an area the size of a tennis court covers caves of significant total length, such as the *Szabó-pallag-Shaft Cave* (length: 1035 m) or the *Vecsembükk-Shaft Cave* (length: 900 m). The shaft caves have an extraordinary density in some places, namely in some parts of the *Alsó-hegy Plateau* there are 40 shaft caves per one km^2.

Another type is caves created by mixing corrosion in a predominantly phreatic environment, which are characterized by solution cupolas. Caves of this type can be found inside *Esztramos Hill*. Due to the lack of natural entrances, these caves were only discovered through mining.

The special feature of Aggtelek Karst is thus the co-occurrence of epigenetic and hypogenetic caves, the multitude of independent, large river systems and the unique density of shaft caves. Moreover, long-term (tens of millions of years) karst development has resulted in a very heterogeneous morphology, which, in addition to the present continental climate, also reflects traces of evolution under tropical and periglacial climatic conditions.

As for the mineral deposits, stalactites and stalagmites, draperies, and flowstones are the most common forms in the caves of Aggtelek Karst (Fig. 3). Besides, other rarer types also occur, such as cave pearls, speleothem pools, limestone tuff and most of the formation types associated with infiltrating or stagnant water can also be found here, such as aragonite frostwork, popcorns, helictites, shelfstones, calcite rafts, cave shields and overgrown calcite crystals also occur. In addition to calcite deposits, aragonite, gypsum and temporary ice formations can also be found in the area. *Rákóczi No. 1 Cave* popcorns are among the most characteristic type forms according to the international literature (Hill and Forti 1997).

Although the 30 km long *Baradla–Domica cave system* (Table 2) is not among the longest caves in the world, in the temperate zone there is not any active stream cave longer than that, which is similarly rich in speleothems. Formation-rich caves longer than Baradla either represent a completely different cave type or have developed in wet (sub)tropical karst areas.

The underground world of the Aggtelek Karst provides habitat for more than 500 troglobionts (cave dwellers) and troglophiles (cave-loving) invertebrate species, 38 of which were described first here (Fig. 4). Among the endemic troglobionts, the Hungarian blind ground beetle *Duvalius hungaricus* occurs only here, and the *Helodrilus mozsaryorum* worm is known only from the *Baradla Short Lower Cave*. Due to its large population, the blind cave amphipod *Niphargus aggtelekiensis* and the woodlice *Mesoniscus graniger* are also outstanding.

Bats are the most important trogloxene species that use the cave for only part of their life. Of the 28 European bat species, 21 species are present in the area. Of European importance are the pond bat (*Myotis dasycneme*) and the greater mouse-eared bat (*Myotis myotis*), which are registered as vulnerable species on the 1994 IUCN Red List. Also significant are the wintering colonies of the Mediterranean horseshoe bat (*Rhinolophus euryale*) whose number may reach 5,000 during their winter stay in the *Baradla–Domica cave system*.

The caves of Aggtelek Karst are also significant from an archaeological point of view. The presence of prehistoric people has been detected in more than 30 caves of the Gömör–Torna Karst so far. The findings demonstrate the occurrence of prehistoric cultures spanning at least 35,000 years (Fig. 5). The most significant archaeological site is the *Baradla-Domica cave system* with its foreground,

Fig. 3 Various types of carbonate speleothems. **a** Rimstone dams in Béke Cave (photo Egri); **b** stalagmite with a height of 18 m in Baradla–Domica cave system (photo Egri); **c** popcorns in Rákóczi No. 1 Cave (photo Egri); **d** helictites in Meteor Cave (photo Egri); **e** draperies in Meteor Cave (photo Egri)

Table 2 The longest caves in the Aggtelek Karst (status of August 31, 2020)

	Name	Karst area	Length (m)
1	Baradla–Domica cave system	Aggtelek Plateau	30,100
2	Béke Cave	Aggtelek Plateau	7,183
3	Szabadság Cave	Galyaság	3,030
4	Vass Imre Cave	Aggtelek Plateau	2,185
5	Kossuth Cave	Jósvafő Plateau	1,610
6	Danca Cave	Galyaság	1,390
7	Baradla Short Lower Cave	Aggtelek Plateau	1,000
8	Szabó-pallag Shaft Cave	Alsó-hegy Plateau	950
9	Vecsembükk Shaft Cave	Alsó-hegy Plateau	900
10	Rákóczi No. 1 cave	Esztramos Hill	750

Fig. 4 Cave fauna. **a** *Eschatocephalus* (*Ixodes*) *vespertilionis*, a tick parasitizing on bats (photo Ľuptáčik); **b** *Typhloiulus* sp. the biggest troglobiont (2.6 cm) of the area (photo Kováč and Mock); **c** *Heteromurus nitidus* a troglophil springtail common in Aggtelek Karst caves (photo Rahmé); **d** *Rhinolophus hipposideros* one of the most common wintering species in Baradla Cave (photo Boldogh)

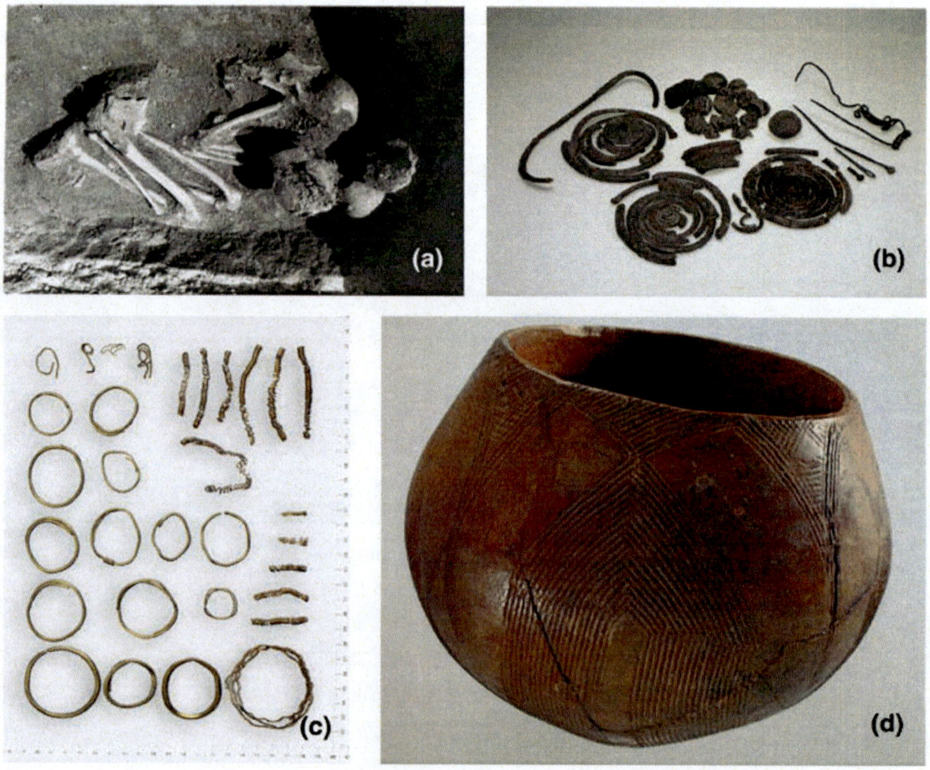

Fig. 5 Archaeological findings: **a** grave from the Baradla–Domica cave system (photo Korek); **b** bronze treasure from the Baradla–Domica cave system (photo Kardos); **c** gold treasure from the Baradla–Domica cave system (photo Dabasi); **d** typical Neolithic pot from the Baradla–Domica cave system (photo Kardos)

where settlements belonging to the Bükk Culture and charcoal drawings unique in Central Europe can be found. Also noteworthy is the *Hosszú-tető Cave*, which was periodically used by the people of the Kyatice Culture in the Bronze Age. The series of prehistoric finds is complemented by cave inscriptions dating from the thirteenth century.

The caves of the area are also of outstanding significance from a paleontological point of view. The 7 km long main corridor of the *Baradla–Domica cave system* explores the Triassic Wetterstein, Steinalm and Gutenstein Limestone Formations. In these layers, characteristic alga, gastropod, ammonite, brachiopod and crinoid remnants belonging to seven different reef and lagoon facies can be observed. For this reason, these profiles have been declared a geological type section (stratotype) and these are the only cave type section in the world.

Sediments from smaller caves, fissures and palaeokarst cavities are also rich in vertebrate remains ranging from the Middle Pliocene to the Upper Pleistocene. Of particular significance are the finds from the Esztramos No. 23 Site, which reflect the characteristic species of the Middle and Upper Pliocene as well as the Lower Pleistocene of the Northern Carpathians. On this basis, and taking into account other cave sites in Hungary, a unique microstratigraphic sequence could be established. The internationally recognized Middle Pliocene '*Estramontinum*' vertebrate bio-stratigraphic stage acquired its name from this location. The excavation site No. 7 is an international reference site indicating the boundary between the Pliocene and the Pleistocene.

Naturally, the origin of caves and the deposits found in them are closely related to surface karst forms and processes. Karst forms of Aggtelek Karst include all characteristic elements of temperate medium mountain karstification, thus providing an excellent opportunity to study the relationships between surface and subsurface karst development. Surface karst phenomena in the area include doline-dotted karst plateaus, karst valleys with doline series, uvalas (in a meaning of coalesced dolines), sinkhole lakes, karren fields, narrow and deep limestone gorges of several km in length and blind valleys with allogenic streams (see also chapters "The Geomorphic Evolution of Karsts and Karstic Surfaces" and "The Surface Morphology of Karsts in Hungary"). Groundwater is poured to the surface by 230 permanent and temporary karst springs.

The caves of Aggtelek Karst have had a special place in the history of cave research as well (cf. chapter "History of Karst and Cave Research in Hungary"). The most significant cave in the region, Baradla, appeared in various scientific publications as early as the sixteenth century. György Buholcz, Mátyás Bél, Robert Townson, József Sartory and Imre Vass are the most important authors from the early centuries who described the cave (Fig. 6). Such early documents of scientific research in caves are rare worldwide. The *Baradla Cave* is also one of the first tourist caves in the world, where organized visits have taken place since the beginning of the eighteenth century.

Scientific research in the area has yielded significant results in the fields of geology, palaeontology, archaeology, biospeleology, cave climatology, cave therapy, mineralogy, karst morphology and karst hydrology.

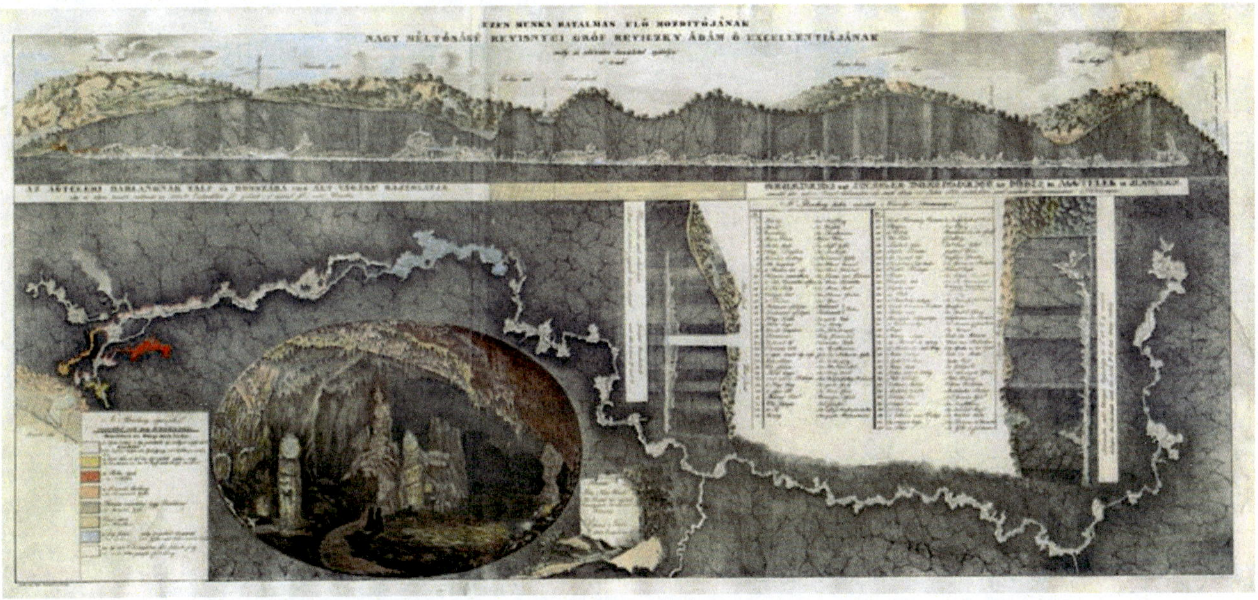

Fig. 6 The historical map of Baradla Cave drawn by Imre Vass, 1829

3 Presentation of the Significant Caves in Alphabetical Order

3.1 Almási Shaft Cave

The *Almási Shaft Cave* is located on the *Alsó-hegy Plateau* above Bódvaszilas, ca. 500 m north of the Szabó-pallag hunting lodge, and it opens right next to the Hungary-Slovakia state border. Its entrance is located near the upper edge of a large-sized doline, typical of the landscape, among rock formations densely decorated by solution microforms (karren). The estimated length of the shaft cave is 358 m and its vertical extent is 100 m (Nyerges 2003a). The first explorations of the cave, which had been open since ancient times, reached a depth of 43 m (Scholtz 1911) in 1911 and 93 m (Kessler 1927) in 1927. Caving is only possible for people skilled in cave rope technique.

The almost completely vertical passage system with significant shafts was formed in the Middle Triassic Wetterstein Limestone (Piros 2006) along an approximately north–south-oriented fracture (Less 1998). The formation of the cave is mainly due to the tectonic settings and the dissolving effect of the seeping waters (Sárváry 1970; Kósa 1992; Zámbó 1998). In order to clarify the hydrological connection between the cave and the springs at the foot of *Alsó-hegy Plateau*, a water tracer test was carried out in 1969. The tracer appeared in the *Tapolca Spring* 2 km north of the shaft cave and in the *Andród-kert Spring* (Sárváry 1971), too.

The lowermost point of the cave can be reached by descending in two paths, in fact at the two ends of the same fissure, through several shafts of significant dimensions (Nyerges 1997). The shaft walls are decorated with extremely rich speleothem formations, and some of them follows the path of the descent for several 10 m. The most famous and perhaps the most beautiful formation, a flowstone extends from the bottom of the cave's third and most impressive shaft, to a barely visible height. From the shaft bottom we can reach the lowermost point of the cave at a depth of −100 m among stones cemented by dripstones and through slightly narrower passages (Nyerges 2001).

In general, the shafts and speleothems of the cave are extremely spectacular, its walls are richly covered with formations. Its passages are decorated with speleothems of various shapes and colours, stalactites and stalagmites, draperies, helictites and travertine pools. The shaft walls are largely decorated with cave popcorns of subaerial origin.

Descending at the entrance of the *Almási Shaft Cave*, after ca. 5 m, the previously tubular shaft expands in a northerly direction. In the large-sized shaft, we can first descend 11 m and then descend another 29 m to the depths. The wall of the elongated fissure with an average width of 1–2 m, and a length of 15–20 m, is decorated with speleothems from a depth of 10 m. The so-called *Új-rész (New Section)*, which can be reached by exposed climbing above the fissure is particularly spectacular and rich in formations. 60 m from here, from the bottom of the 29th shaft, one can reach the most spectacular part of the cave, the shaft base decorated with flowstones (Fig. 7), with a further 35 m of descent. There are now dry tetarata pools, and spectacular coloured bands on the walls that testifies the former water level. Moving from here towards the end point, however, some tetarata pools are still filled with water, proving the current hydrological activity of the cave.

The red clay typical of the area is also significant in the deposits of the shaft cave, and it dominates the bottom of the shafts from a depth of 10 m. In many places, this material mudded the formations, partly in a natural way and partly due to the trampling of cavers. The infiltrating water holds the walls wet and has dissolved some of the formations, but at certain points speleothems are still developing.

Fig. 7 The largest flowstone of Aggtelek Karst in Almási Shaft Cave (photo Egri)

The cave is a known bat living habitat, and precise observational data are available about greater horseshoe bat (*Rhinolophus ferrumequinum*) lesser horseshoe bat (*Rhinolophus hipposideros*), lesser mouse-eared bat (*Myotis blythii*), greater mouse-eared bat (*Myotis myotis*), Geoffroy's bat (*Myotis emarginatus*), and western barbastelle (*Barbastella barbastellus*; Boldogh 2007).

3.2 Bába-völgy Stream Sink Cave No. 2

The entrance of the *Bába-völgy Stream Sink Cave* No. 2 is about 2 km north of the village of Szögliget, and it opens in the second depression of the series of sinkholes in the *Bába Valley*, at the foot of a rocky outcrop, in a wooded environment. The measured length of its passages is 70 m, whereas its vertical extent is 22 m (Székely 2003).

The *Bába Valley* with its series of sinkholes run on the western side of the *Nagy-Kopasz-gally* area of the *Alsó-hegy Plateau*, with an orientation of ENE–WSW. Members of this sinkhole line periodically drain rainfall water into the deep. Part of the catchment area of the karstic depressions is a non-karstic rock belt, which are built up of the sandstone and siltstone members of the Lower Triassic Bódvaszilas Sandstone formation. The karstic part of the catchment area is built up of the Middle and Upper Triassic Wetterstein, Derenk and Hallstatt Limestone formations (Szentes 1963; Piros 2006). The tectonic structure of this area of the *Alsó-hegy Plateau* is quite complex. The plateau includes multiple overthrusts and is further divided by faults. This tectonically fragmented rock structure and the non-karstic rock band together created a good basis for the formation of sinkhole caves (Szentes 1963). Runoff from the non-karst area is forced into the deep along the bedding planes and fractures, reaching the margin of the limestone area.

Following the discovery of the cave in 1957, in 1959 fluorescent tracer tests were applied for the water sunk in the cave. This study clearly confirmed the hydrological relationship with the nearby *Borz Spring* (Dénes 1963).

The orientation of the cave passages is determined by an east–west fracture. The first section is mainly characterized by debris, then after nearly 10 m, we can advance in a slightly larger erosion channel, which is still not high enough to walk upright. At the end of this section, the passage bifurcates, and we can reach the western low point through a vertical fissure dominated mainly by debris blocks. Returning to the fork, we can reach the end point of the eastern, slightly sloping branch in a passage richly decorated with speleothems.

The formation of the cave can be traced back to the collapses along tectonically directed fissures and the erosion–corrosion effects of water infiltrated or directly sunk from the catchment area.

The original colour and structure of the enclosing rock of the cave are revealed from the thin, almost ubiquitous clay film or other clay formations in places actively washed by the watercourse. After the entrance section, from a depth of ca. 7 m, a red, reddish-white, breccia-textured rock decorated with calcite veins is revealed. It belongs to the Hallstatt Limestone formation, and it represents a special value of the cave (Székely 2003).

The cave is one of the known habitats of the *Niphargus aggtelekiensis* a blind amphipod and a temporary habitat for various bat species, as evidenced by the scattered guano at the bottom of the cave at several points.

3.3 Baradla–Domica Cave System

The *Baradla–Domica system* is the most significant cave in the Aggtelek Karst and in all Hungary. Its total length is more than 30 km. Most of it is located in Hungary as Baradla, but a smaller section called *Domica Cave* (8 km) is located in Slovakia. The most famous of the 11 entrances known today is the natural main entrance at the bottom of a 50 m high rock wall near Aggtelek. This entrance has been open since ancient times.

One of the outstanding values of the *Baradla–Domica cave system* is that its main branch crosses the almost complete sequence of the local Triassic carbonate sediments (Fig. 8). In the Mesozoic Era, the Triassic limestones uplifting from the Tethys Sea in the Cretaceous, about 90 million years ago, slid down as large slabs towards the lower parts, creating parts of the present-day Silica Nappe system with layers tilting towards southwest at a dip angle of 25–30°. The east–west main branch of Baradla reveals the structure of this sequence. In the *Medvekarmos Room*, close to the Jósvafő entrance, one can see the oldest rock of the cave system, the ca. 240 million-year-old, brownish-yellow, marly, laminated limestone, which belongs to the Lower Triassic Szin Marl Formation (Szentes 1965). Going further towards Aggtelek, all the way to the *Óriások terme (Giants' Hall)*, the cave passage cuts through the dark grey or black, well-bedded limestone of the Gutenstein Formation. These layers also include subordinately grey, bedded dolomites, which formed in the early part of the Middle Triassic about 230–240 million years ago (Kovács et al. 1988; Piros 1993, 2002). From the *Óriások terme* to the *Denevértemető* (*Bat Cemetery*) there is a completely different rock type, the so-called Steinalm Limestone (Piros et al. 1989a, b, c, d). Unlike the previous one, it is a white or light brown, partly dolomitized, limestone of 30–200 cm thick beds formed in well-ventilated waters of the open shallow sea, on the one hand from the vases of reef-forming organisms and on the other hand from algae mud deposited in the lagoons behind them. Near the *Sárkányfej* (*Dragon's Head*), the cave

Fig. 8 Geological layout of Baradla–Domica cave system. **a** geological overview map of the cave (Balciar); **b** Szin Marl Formation (photo Egri); **c** Crinoidea fossils in Steinalm Limestone (photo Egri); **d** Ammonites fossils in Wetterstein Limestone (photo Egri); **e** "Dobos cake" fossil in Gutenstein Limestone (photo Egri); **f** Gastropoda fossils in Wetterstein Limestone (photo Gaál); **g** boundary of covered (left) and uncovered (right) karst (photo Rózsa)

passage crosses the few metres thick, reddish, fine-grained layers of the so-called Jenei Limestone, in which remnants of Ammonites, Brachiopods, and Crinoids occur (Velledits et al. 2011). The lagoon facies limestone of the Wetterstein Formation incorporates the most significant part of the *Baradla–Domica cave system*. The entire *Domica cave* as well as the main branch of the Baradla is situated in this limestone up to 1800 m from the Aggtelek entrance (approx. to the *Vaskapu/Iron Gate*; Piros and Kovács 1993). In the bed of the above formation there is a reef facies limestone up to the *Denevértemető*, which forms a transition between the Steinalm and Wetterstein Limestones (Piros et al. 1989e).

The thick limestone beds of 20–30° dip are clearly visible in many places, and in certain places the collapse of the thick-bedded ceiling can also be observed (*Libanon hegye, Morea-hegy*).

The main branch of the cave is slightly inclined from west to east from the initial section of the Domica (*Ördög-lyuk/ Devil's Hole* at 375 m asl), to the Jósvafő exit (261 m asl). Meanwhile, it changes direction several times, caused by structural fissures of different orientation. The main branch is connected to the south by a number of narrower tributaries, such as the *Csernai-ág*, the *Acheron-ág*, the *Törökmecset (Turkish Mosque)-ág* and the *Retek (Radish)-ág*. These tributaries drain water accumulated during heavy rains and snowmelt into the main branch from a number of sink points. As a result, the morphology of the cave is characterized by a duality: the main branch is the most developed, consisting of spacious and, in some places, monumental passages (*Nádor utcája, Budai-Nagy-alagút*). In contrast, the tributaries are mostly narrow and sometimes have meanders that are very difficult to pass (*Törökmecset-ág, Retek-ág, Vörös-ág, Kúszó-ág*).

The cave corridors have been formed by allogenic streams sinking at the foot of karst plateaus, where the water arrives from gravel covered terrains. The gravel formations were deposited on the subsided tectonic blocks of Triassic carbonates near the margin of karst plateaus. The cave system was created mostly in the shallow phreatic zone or along the piezometric surface of the underground karst water. The evolution of the cave took place gradually and several levels have been created in accordance with the phases of valley incision (Kessler 1938; Droppa 1961, 1972; Jakucs 1971, 1975; Gyuricza and Sásdi 2009; Veress 2012; Veress and Unger 2015).

The slightly inclined corridors leading from the *Domica Cave* through the state boundary and continuing along the main branch of the *Baradla Cave* to the Jósvafő demonstrate that this system is an excellent example of an 'ideal water-table cave' (Ford and Ewers 1978; Ford 2000). The main corridors have been formed in relation to the formation of large pediments in the *Jósva River valley*. Based on ^{10}Be and ^{26}Al cosmogenic nuclide dating, allochthonous fluvial sediments (mostly quartz gravels) deposited in the upper level of the cave system in the Middle and Late Pliocene, whereas in the main level deposition took place mostly in the Early Pleistocene (Gyuricza and Gaál 2014; Bella et al. 2019). In the *Baradla Cave*, the formation of the phreatic loops was controlled by the bedding planes of Wetterstein limestones with a dip angle of 20–30° towards southwest (Gaál 2014). Phreatic loops were filled by allochthonous pebbles and completely remodelled by ceiling erosion mostly in the Early Pleistocene (Bella et al. 2019). Where the piezometric surface is located above the phreatic loops of the cave and allochthonous sediments are fluvially transported into the loop sections, the irregular and long passages are gradually aligned by paragenetic ceiling erosion along the downward parts of the phreatic loops as the sediment accumulation forces the water to flow along the ceiling, and even *bypass tubes* may develop above loops, which are completely filled with sediments. As a result of continued alluviation, i.e. the ever-growing sedimentary fill, the paragenetic development of the ceiling parts of loop passages progresses upwards until the piezometric surface is reached (Pasini 1967, 1975; Ford 1965, 2000; Ford and Ewers 1978; Farrant and Smart 2011).

After the partial removal of fluvial sediments (during the post-paragenetic erosion phase), the large corridor with ceiling channel, wall channels and an aggraded streambed slightly inclined downstream towards the former *Jósva Spring* at Jósvafő village was created as a result of this long-term cave evolution. Multiple undulating wall channels and meander notches occur in many places as proofs of the repeated streambed aggradations and lateral wall erosion. The youngest passages, the *Rövid-Alsó (Short Lower) Cave* and the *Hosszú-Alsó (Long Lower) Cave*, developed about 40–50 m below the main branch of the cave. The vast majority of their passages are constantly under water (Fig. 9). They drain the water of the Styx stream in the main branch after heavy rains or snowmelt.

On the walls of the tunnel-like and meandering passages, the previous lateral incisions of the stream, the notches, can be clearly seen. The slope of the notches is equal to the gradient of the watercourse (Lauritzen and Lundberg 2000). During the cave development, a huge amount of gravel was washed into the cave through the sinks several times, which often filled the cave up to the ceiling. The dissolving and abrasive work of the cave stream, which was uplifted by the sediments to the ceiling formed the ceiling channels (Roth 1940; Bella and Móga 2014), and the ceiling hollows formed as a result of the swirling movement of the water. At several points in the Baradla, one can observe the so-called pendants (Fig. 10), which hang from the ceiling into the passage, and are dissolved around (Bella and Móga 2014). Of particular interest are the scallops formed on the cave wall, whose asymmetrical shape indicates the direction of water flow (Bögli 1984). In some cases, animals also contribute to the enrichment of the morphology. Infiltrating water becomes acidic due to bat feces, and guano hollows are formed by increased dissolution in the bedrock or deposits below the infiltration pathways (Bella and Móga 2014). Pits formed in this way are common on speleothem crusts, flowstones and wider stalagmites.

The former and present drainage network of the cave, as well as the direction of most passages, are primarily determined by structural fissures. In addition to the typical NE–SW and NW–SE joints, Domica has a large number of WNW–ESE faults, while Baradla is dominated by NNE–

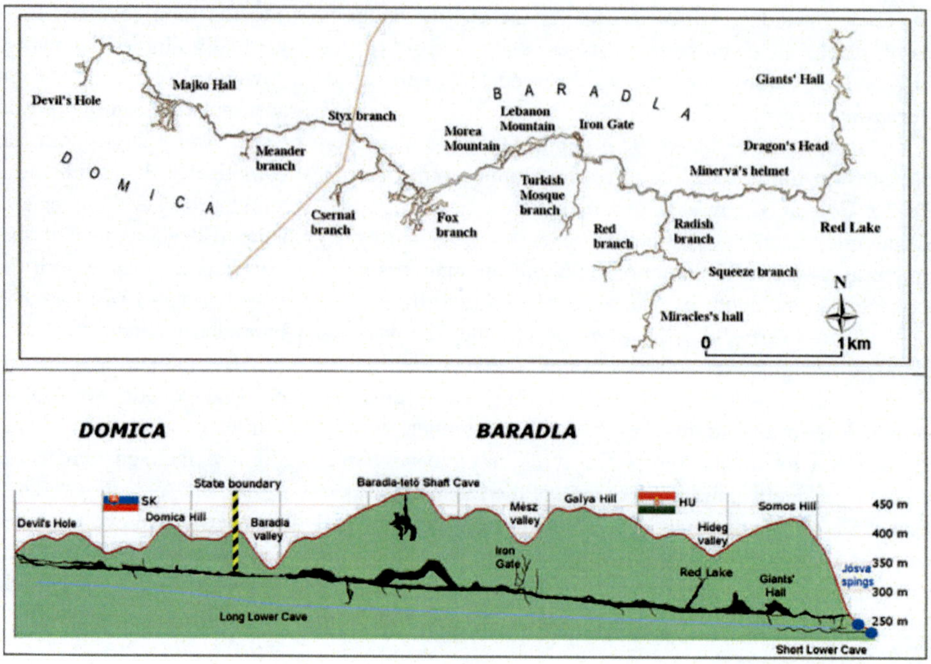

Fig. 9 Location and longitudinal profile of Baradla–Domica cave system

SSW and N–S oriented faults (Gaál 2014). Where several structural lines intersect, the seeping water has dissolved high halls that have further increased their size by upward growing collapses (*Majko-dóm, Oszlopok csarnoka, Mórea-hegy, Libanon hegye, Óriások terme*). The largest of them is the *Óriások terme*, which is 130 m long, 40 m wide and 27 m high. The debris material of the collapses has already been transported away by floods in some places, while in others the highly accumulated debris blocks clearly proves the collapse origin of these sections.

From the main branch of the Baradla, steeply sloping passages lead down to the lower level in some places. In the Aggtelek section, the 40 m deep shaft of the *Dancza-nyelő* leads to the *Hosszú-Alsó-barlang*, and in the Jósvafő part, the huge sink of the *Óriások terme* drains water to the *Rövid-Alsó-barlang* (Gruber et al. 2012). Its waters come to the surface through the artificial tunnel of the *Jósva Spring*.

The *Baradla–Domica cave system* is extremely rich in depositional formations (Fig. 11). Of the 38 calcium carbonate formations known worldwide (Hill and Forti 1997), 17 have been identified here so far (Gaál et al. 2014). The vast majority of these consist of calcite, but in some cases aragonite formations also occur. The deposits include various forms of stalactites, which sometimes appear as icicles and elsewhere as radishes (bulbous forms). Helictites are less common, but also occur. Among the stalagmites, pagoda, palm, or wand-like forms are extremely spectacular. The cave system is particularly characterized by beautiful dripstone columns, but various speleothem coatings, flowstones and draperies also occur. Limestone tuff formations, often with spherical stones or coralloid popcorns in their pools, offer a stunning view. Rarer are cave pearls, stalagmite cups, cave rafts floating on the surface of the water, or muddy vermiculites adhering to the walls like worms. Aerosol-derived popcorn coatings cover the walls at many points in the cave system. The colour of the various formations is predominantly light or yellowish brown, caused by slight iron oxide contamination. To a lesser extent, clear white formations also occur, where pure calcium carbonate is present. The black colour basically means a manganese coating, but in many of Baradla's passages the black colour marks the trace of torched visitors from historical and prehistoric times. Guano mounds are also valuable from a mineral point of view. Infiltrating water causes important chemical changes in bat feces that lead to the formation of special minerals such as apatite, gypsum or brushite (Sztrókay 1959). The most significant deposits of the latter are in the *Róka (Fox)-ág* and on *Libanon hegye*.

From the point of view of cave biology, the *Baradla–Domica cave system* is known as the most significant cave in the Gömör–Torna Karst and in the entire Western Carpathians. So far, more than 500 animal species have been identified from where (Fig. 12). Biologists published notes about Baradla as early as the middle of the nineteenth century. János Petényi Salamon studied the distribution of bats in 1844 (Petényi 1844). In 1865, János Frivaldszky found the snow-white woodlice, which was introduced into science as *Mesoniscus graniger* (Frivaldszky 1865). It is still one of the best known representatives of our real cave-dwelling animals. At the beginning of the twentieth century, Ernő

Fig. 10 Erosional features. **a** Meandering ceiling channel (photo Bella); **b** pendants (photo Bella); **c** river passage (photo Egri)

Csiki, who described a blind ground beetle, the so-called *Duvalius hungaricus* (Csiki 1903), had another great success. This discovery drew the attention of several renowned researchers to Baradla, which eventually led to the very thorough and extensive biological work of Endre Dudich. Dudich's work was already characterized by a modern ecological approach, that is, he evaluated cave creatures with their nutritional relationships, and their habitats with their environmental conditions. His monograph was published in 1932, in which he mentions 262 species and subspecies from Baradla, of which 18 are real cave-dwelling types. Dudich later established a cave biology laboratory in the cave. One of the most significant species he described is a cave amphipod, the *Niphargus aggtelekiensis*. This species prefers clear waters, so it is also a reliable indicator of the quality of the cave environment. Several other species were first recorded scientifically in Baradla. These include the spider species *Porrhomma profundum*, the springtails *Pseudosinella aggtelekiensis* and *Pygmarrhopalites aggtelekiensis*, the rotifer *Proales baradlana*, and the flies *Lycoria (Sciara) ofencaulis* and *Lycoria baradlana* (Dudich 1932). Blue-green algae have also been studied, of which a new species called *Baradlana speluncaecola* has been discovered, and a new microscopic fungal species named *Trichosporon aggtelekiense* has been found in the cave sediment of Domica (Dudich 1932). The worm species *Helodrilus mozsaryorum* is known only from the *Rövid-Alsó-barlang* below Baradla. Endemic species in the area include the millipedes *Typhloiulus sp.*, which is the largest invertebrate cave dweller in the karst. Its body length is almost 3 cm and the number of legs is 147 pairs (Salamon et al. 2014).

There are species that use the cave for only part of their life, the so-called troglophils. Their most significant representatives are bats. To date, 16 bat species have been described by experts from here (Boldogh 2006). The pond

Fig. 11 Varied types of speleothems. **a** Schematic figure of speleothems in Baradla–Domica cave system (Balciar) 1. stalactite, 2. stalagmite, 3. helictite, 4. stalagnate, 5. coating, 6. flowstone, 7. drapery, 8. shield, 9. rimstone dams, 10. shelfstone, 11. corallites, 12. pisolites, 13. montmilch, 14. boxwork, 15. cave cup, 16. cave raft, 17. vermiculite; **b** shield (photo Staník); **c** "Mother-in-Law's Tongue"—a stalactite–drapery combination (photo Egri); **d** "József palatine" memorial columne (photo Egri); **e** "Lebanon cedar" stalagmite (photo Egri); **f** "Snowy mountains" rimstone dams (photo Egri); **g** "Dripstone forest" in the Retek branch (photo Egri); **h** "Helmet of Minerva" compound stalagmite (photo Egri)

bat (*Myotis dasycneme*) and the greater mouse-eared bat (*Myotis myotis*) are registered as vulnerable species. Winter colonies of the Mediterranean horseshoe bat (*Rhinolophus euryale*) are of European importance. During winter months, as many as 4,500 specimens may winter here each year (Uhrin et al. 2012).

The natural entrance to the Baradla has been open since ancient times, so people from many cultures have found refuge in it. Most of the thousands of finds unearthed here during the excavations are from the Neolithic Era. The thin-walled well-shaped vessels of people living here 6,000–7,000 years ago were made by hand, without a wheel, and decorated with parallel lines.

Archaeological research at Baradla began in 1876. The 2-year excavations were led by Jenő Nyáry (Nyáry 1881). He first dug in the *Csontház* (*Bone House*) and then continued his work in the *Pitvar* (*Lobby*) and *Denevér-ág* (*Bat Branch*). Later, in the 1910s and 1920s, Ottokár Kadić, Lajos Márton, Miklós Kretzoi and others also dug in the cave, thanks to which Baradla became one of the most famous Neolithic sites in Hungary (Rezi Kató 2014). From the cultural layers of the cave, mainly the artistically worked pottery and utensils of the Bükk culture were found, but it was proved that its passages were already inhabited even earlier. Items mainly from the Middle Neolithic period belonging to the Great Plains Linear Pottery Culture, were

Fig. 12 Cave fauna. **a** *Duvalius hungaricus* (photo Balázs and Dányi); **b** *Trechoblemus micros* (photo Rahmé); **c** *Porrhomma profundum* (photo Kováč and Ľuptáčik); **d** *Egaenus convexus* (photo Balázs) a spider merging into its environment

found, but some findings also from the Paleolithic Era: spearheads, arrowheads, chisels, scrapers, amulets, stone spears and stone axes were found. They also found human skeletons, probably from the Neolithic Era, but some of them may be Bronze Age or Iron Age. The 13 whole human skeletons unearthed reveal that the dead people were buried with knees pulled below the abdomen, face towards the ground, flat stone on their napes. Alongside them were placed wheat, barley, millet, lentils, and similar crops. Excavations have revealed the remains of several animal species, especially cave bears. In addition, a number of finds have become known which belong to the Late Bronze Age Kyete Culture. The gold collection consisting of 27 gold wire hoops and coils hidden between two rocks is presumably of Iron Age. During the construction of the *Róka-ág* sanatorium, Middle Neolithic objects and human bones were discovered, as well as late Bronze Age finds. A set of 59 objects, mostly round, decorated bronze artifacts and swallowtail-shaped pendants, gained world fame, and archaeologists said they may have been accessories for a ceremonial dress.

The first written mention of the cave dates from 1549, its first survey was carried out in 1794 by József Sartory. Until 1825, it was known only 1.8 km long. This is depicted in a map of Keresztély Raisz made in 1802, which is also the first printed drawing of the cave. In 1825, Imre Vass penetrated the *Vaskapu* (*Iron Gate*), which was the then end point, and explored the continuation of the main branch of the cave for about 5 km. He created an accurate map and description of the cave and the surface above it. His work was published in print in Hungarian and German in 1831 (Székely 2014).

Hubert Kessler was the person who succeeded in proving the direct relationship between Baradla and Domica by penetrating the siphons of the Styx stream in 1932 (Kessler 1932). After finding the connection, the two caves were separated by an underground border lattice, which was abolished after Hungary and Slovakia joined the Schengen Agreement.

The temperature of the cave is almost constantly around 10 °C, the humidity is 95–98%, so it gives the visitor a damp, cold feeling. The carbon dioxide content is very low in the main branch, the instrumental tests detected only 0.1%. The highest carbon content, more than 4%, was measured at the *Vaskapu-víznyelő*. The average radon activity of the air is 2 kBq/m^3 and the maximum is 13 kBq/m^3, which is below the health limit. Continuous climatological studies from the 1950s onwards have shown that, due to the special climate of the cave, the hygienic-microbiological condition of the air, it has a healing effect on people suffering from respiratory diseases.

Together with the *Domica Cave*, the entire cave system and the catchment area (Fig. 13) were brought under the 2001 Ramsar Convention, which sets out the protection of wetlands of international importance.

Present-day scientific investigations in *Baradla–Domica cave system* focus mainly on cave genesis theories (Gyuricza and Sásdi 2009; Veress and Unger 2015; Bosák et al. 2004; Bella et al. 2019), hydrogeologic studies (Borbás et al. 2011; Gruber et al. 2012), exploration for new passages at the lower, inundated levels, speleobiology, speleothem dating (Zámbó et al. 2002; Galbács et al. 2011; Czuppon et al. 2018) and the analysis of palaeoclimate signals in cave deposits (Siklósy et al. 2011; Demény et al. 2017).

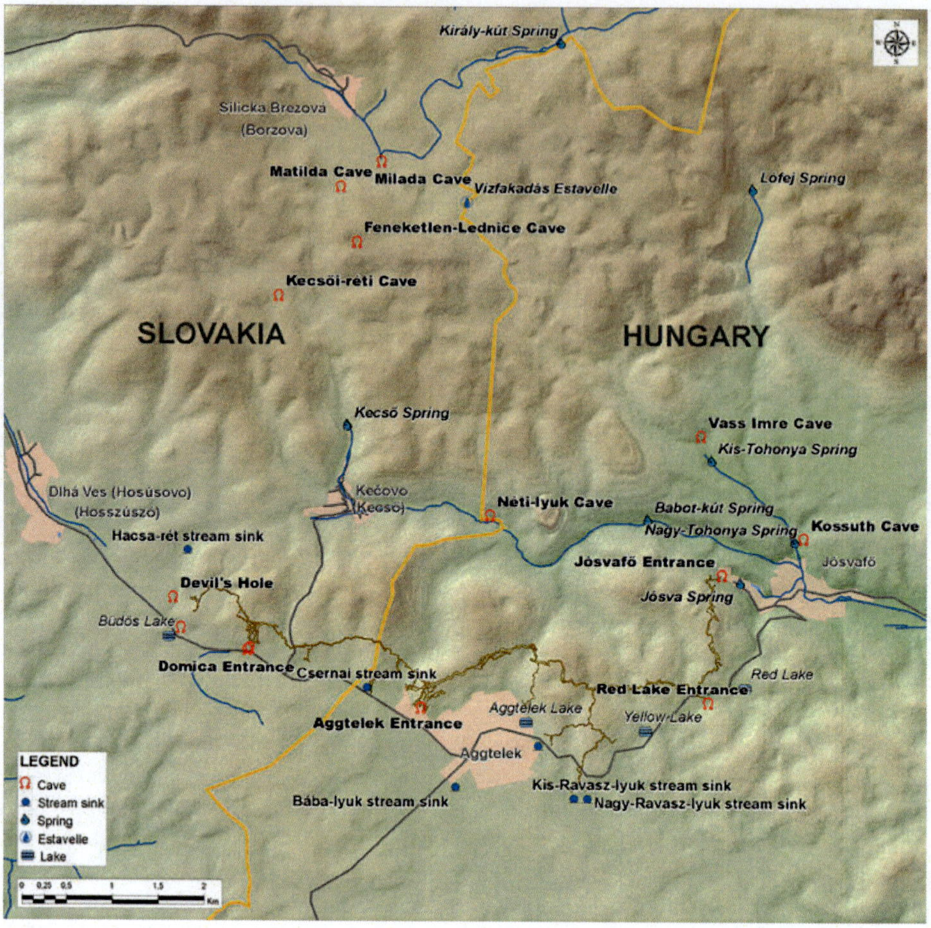

Fig. 13 Hydrological objects in the catchment area of Baradla–Domica cave system

3.4 Baradla-tető Shaft Cave

The *Baradla-tető Shaft Cave* ("*zsomboly*" in Hungarian) is located in the *Baradla-tető* area of the *Aggtelek Plateau*, 200 m north of the summit rising to 477 m above sea level. Its entrance opens at the upper edge of a large doline typical of the landscape, among rocks heavily shaped by karren. The estimated length of the shaft cave is 387 m and its vertical extent is 87 m. Caving is only possible for people skilled in caving rope technique.

The shaft system formed in the Middle Triassic Wetterstein Limestone (Gruber 2003) has several steps of significant size, and it follows NE–SW-oriented joint fractures. In addition to tectonics, the dissolving effect of the seeping water and the collapse of large cavities played a major role in its formation. At a depth of −45 m from the entrance, the cave bifurcates. The passage towards southwest leads to a room richly decorated with stalactites of considerable size, and in the debris-laden fissure shaft formed below we can reach a depth of −84 m. If we continue our journey descending below the apparent lowermost point formed at a depth of −45 m, we can reach the true lowermost point of the cave at a depth of −87 m through a series of stepped shafts richly decorated with stalactites. The cave passages are generally rich in stalactite formations. They contain flowstones of various shapes and colours, stalagmites and stalactites, as well as travertine pools.

The shaft is horizontally 600 m from the *Libanon hegye* formation found in the *Baradla–Domica cave system* at 50 m deeper than the shaft endpoint. The formation of the cave can be explained in two ways. The presence of shafts testifies classical dissolution genesis, however, both the morphology of the shafts and the presence of large, collapse chambers reveal significant tectonic activity. If we explain the formation of the cave by the classical dissolution genesis, then any passages that may continue in the deep need not be connected to the passages of the *Baradla Cave*. However, if the currently known passages were created primarily by the collapse of a lower, as yet unknown cave section—and only secondarily by the dissolving effect of the seeping waters—it can rightly be assumed that there is not only a genetic but possibly a hydrological link to the *Baradla Cave*.

The first section of the cave, which is without collapse forms to a depth of almost −45 m, can be reached through a shaft, initially 1.5 m in diametre. At a depth of ca. 18 m, this shaft expands significantly to the northeast. From here to the bottom of the shaft on the opposite walls very nice, untouched flowstones can be studied. After a narrow squeeze on the shaft bottom, heading southwest, we can reach the *Rom-terem (Ruin Chamber)* on gradually expanding passages, which is the largest contiguous room in the cave (Fig. 14). Its length is 30 m, its width is 6–10 m. In this section of the passage—and already at the base of the *Bejárati-akna (Entrance Shaft)*—the red clay filling colours the walls.

The *Rom-terem* and the corridor leading to it is extremely rich in formations (Fekete 1987). Travertine pools and stalagmites grew on the huge collapse blocks at its bottom. Almost all types of stalactites, straw stalactites, helictites and draperies can be found here. The walls are only partially coloured by red clay, so some of the formations are snow-white, almost transparent. However, the passages formed in the debris under the chamber are relatively poor in formations, and the passage walls are muddy. In these passages, the walls and the edges of the collapse blocks are covered on a large surface by coralloid formations.

Descending below the apparent lowermost point of the *Bejárati-akna*, getting deeper and deeper through the narrow but deep shafts, we can pass along walls decorated by stalactites, stalagmites and flowstones. This section of the passage is characterized by white and red clay-coated formations covering the walls of the narrow fissure. Going deeper and deeper through the smaller shafts and narrow fissures, we reach the ca. 20 m deep final shaft. In this section of the cave, the formations become rarer. One can descend to the end point on an extremely corroded shaft decorated with sharp stone ribs (Nyerges 2001). In this zone, the corrosive effect of water seeping along the walls prevails, but the tectonic orientation is also striking. The base of the fissure shaft is covered with debris which is so loose in several places that it is possible to see deeper through the fissures and cracks.

The cave is a temporary habitat for various bat species, as evidenced by the scattered guano on the base at several points.

Fig. 14 The Ruin Chamber in the Baradla-tető Shaft Cave (photo Polacsek)

3.5 Béke Cave

The *Béke Cave* is the second longest subsurface system in the Aggtelek Karst. Its enclosing rocks are basically Wetterstein, subordinately Gutenstein and Campil Limestones. Although the direction of its passages was determined by the tectonic fissures in the host rock, the lateral erosion of its still active underground stream has removed traces of marked tectonic effects. The plan view of the cave reflects the branching system of an "ideal" underground water network. Its backbone is formed by the 4350 m long *Fő-ág (Main Branch)*, to which its first-order tributaries (the 438 m long *Kígyós-ág/SnakeBranch* on the left and the 684 m long *Felfedező-ág/Discovery Branch* on the right) and their even shorter and narrower inferior tributaries join (Szunyogh and Kisbán 2004). The currently known length of the cave is 7183 m, and its vertical extent is 97 m.

The formation of the *Béke Cave* took place in a similar way to that of the neighboring *Baradla Cave*, i.e. the carbonated water seeping into the limestone dissolved the initial cavities at the end of the Pliocene or around the Pliocene–Pleistocene boundary. However, they probably did not form a connected, high throughput capacity system because (in the absence of adequate hydraulic fall) the karst water stored in the limestone could only change slowly, so its aggressiveness remained low (Gyuricza and Gaál 2014). A significant change occurred when the receding valley of the *Jósva Creek* reached the limestone area and drained the karst water reservoir. It was then that the previously stagnant karst water began flowing towards the valley of the *Jósva Creek*, and the initial cavities merged into passages suitable for draining a large amount of water. It is assumed that during this period the *Baradla Cave* and the *Béke Cave* formed a common groundwater system. However, later on, the further headward erosion of the surface separated *Jósva Spring*, which pours the water of *Baradla Cave* to the surface and *Komlós Spring*, which is the outlet point of the *Béke Cave* stream. As a result, they form now two completely independent systems.

The development of the cavities was accompanied by the expansion of sinkholes, which allowed a large amount of quartz gravel to enter the cave from the non-karst area southwest of the sinkholes. The large size of the *Béke Cave* is partly due to the intense erosion of these quartz gravels (Jakucs 1977). Like in *Baradla Cave*, in certain periods, the alluvial deposits transported into the cave filled the previously formed cavities to a significant proportion, thus the water flowing in the cave exerted its cavity-forming effect above the filling, that is in the epiphreatic zone of the karst (Veress 2012). In some places, the deposits almost completely fill the passages. According to boreholes deepened into these deposits, the thickness exceeds 5 metres at some locations (Berényi Üveges et al. 2004).

The cave was discovered in 1952 under the leadership of László Jakucs by digging the *Bíbic Sinkhole* (Jakucs 1953). The first explorers getting to the cave were László Gera and Sándor Papp. By the end of the first half of the 1950s, almost the entire passage network of the cave had become known. A further significant progress was made only in 1998–2000. It took place in the period when Márton Kucsera and Attila Nyerges, under the leadership of Attila Nyerges, climbed a number of unknown chimneys driving upwards from the ceiling of the *Fő-ág*. These new discoveries required special rock-climbing techniques.

The *Béke Cave* is an active stream system (Fig. 15). The trunk of its water network is the *Komlós Stream* with an average gradient of 0.75%. Its underground upstream area includes a relatively narrow and low passage of about 60 m length at the upper end of the Fő-ág. It is joined on all sides by long, narrow and impassable passages, which collect the infiltrating and inflowing water from the sinkholes of Nagy valley. The discharge of the *Komlós Stream* is increased by its two periodic tributaries and the ten cave springs found in the middle and lower segments. The continuous hydrological activity of the cave is due to the fact that more than four

Fig. 15 Béke Cave. **a** Part of the main branch (photo Egri); **b** water-filled pool behind a rimstone dam (photo Egri); **c** another part of the main branch formed by Komlós stream (photo Egri); **d** typical image of the main branch (photo Egri); **e** the "Five-tuff" rimstone dam (photo Egri)

hundred continuously developing travertine dams have been formed in its main branch, delaying and equalizing the flow of water in the *Komlós Stream*. These dams created a chain of 123 small lake basins.

Recently, there has been a significant change in the location of water disappearance near the end of the cave. Previously, the stream disappeared in a high-capacity cave sink 250 m upstream the end of the cave, then it flowed to the surface at the *Komlós Spring*, ca. 500 m away from this sinking point. However, in the early 2010s, the yield of the *Komlós Spring* gradually decreased, and the water leaving the cave appeared in the higher parts of the hillside above the spring. The change in the path of the water in the cave has also become apparent: the former sinkhole has become completely clogged, and since then the *Komlós Stream* has sunk through a new sinkhole formed near the end of the cave. The protrusions of the passage section in front of the new sink dammed the stream and increased the water level that resulted in the formation of several siphons in the cave without an air gap.

The appearance of these siphons brought a fundamental change in the composition of the air in the cave (Stieber and Leél-Őssy 2015). Previously, at the end points of the cave (partly through the natural and partly through the artificial entrances), air could flow freely into the cave, ensuring the continuous ventilation of the cave system. During this period, the carbon dioxide content of the air was 0.1–0.3% (with a constant temperature of 10.4 °C and humidity of 99.9%). From 2013 onwards, however, the aforementioned siphons developed and prevented the movement of air. As a result, the carbon dioxide content of the air has risen to 1–3% and may even have reached 4% in some places. This had serious consequences: the air in the cave became dangerous to humans, the number of bat deaths increased, and stalactite formation slowed. At present, there are ongoing works to remove the plug in the former cave sinkhole and improve air ventilation in the cave.

The *Fő-ág* of the cave can be divided into three sections. The uppermost part is located between the *Tizenkétágú cseppkő terme (Twelve-Branch Stalactite Hall)* and the artificial main entrance. This part is strongly winding, moderately rich in stalactites, its side walls are decorated with well-developed scallops, and its bottom are covered with clay-free, coarse-grained, loose gravel. It is dry in most part of the year, and it carries water only during rainy periods. From the artificial main entrance to the *Kötélhágcsós (Rope Ladder) Siphon*, there is the middle part of the *Fő-ág*, which is extremely rich in stalactite formations, where the *Komlós Stream* is already generally active, though it gradually dries out after rainless periods. Due to the cave lakes found downstream, the corridor of the *Fő-ág* is almost constantly covered with water from the *Kötélhágcsós Siphon* to the Jósvafő exit. The depth of the water varies from 0.5 to 1 m from basin to basin, but it also reaches 2 m at *Lake Tarajos*. This continuous, large-surface water lends a unique appearance to the lower section of the cave: its stalactites and rock walls are always gleaming due to the condensing mist; the orange, red or white travertines are shining and clear; the roar of its waterfalls, the rumble of waves, are constantly echoed by its halls.

The size of the *Fő-ág* varies from section to section. Its average width is 4 m in the upper section, 5.5 m in the middle and 3.7 m in the lower section. The average heights of these sections are 7 m, 10 m and 4 m, respectively. The largest cavity of the cave is the *Óriás-terem (Giant Hall)*, which is 60 m long, 20 m wide and 18 m high. Its most significant tributary is the *Felfedező-ág*, which departs from the *Bíbic Sinkhole*. It has an average width of 1.6 m and a height of 7 m. From its main branch connection to the *Kavicsos-ág (Gravel Branch)*, it has a meandering course.

The passages of the *Béke Cave* are relatively intact, and the collapses played only a subordinate role in shaping their image. Notable are the so-called *Recsegő-omlás (Crackling Collapse)*, which was still in a moving state in the times after the discovery, as well as the *Kis (Small)* and *Nagy (Large) Collapses*, which are stable, but made it difficult to reach the lower section of the *Fő-ág*. Rock falls occasionally occur at the end of the *Felfedező-ág* beyond the *Disznófürdő (Pig Bath)*. Meandering sections of hundreds of metres are typical especially in the *Felfedező-ág* and in the upper section of the *Fő-ág* (Szunyogh 1994). The process of meander development can be well traced, because the constant subsidence of the *Komlós stream* bed helped to preserve the meanders formed at different stages of development. These former meanders can be observed as overpasses, siphon bypasses, or possibly as asymmetrical notches. The most obvious consequence of meander development is the fact that the meanders at different levels are also horizontally offset from each other (Fig. 16).

The rock surfaces of the *Béke Cave* are characterized by "positive" formations, which consist of the same bedrock, but emerge from the enclosing rock with a variety of shapes. These features include 30–60 cm high rock mushrooms protruding from the stream bed; ribs protruding from the sides of the passages, often resembling a sharp knife blade; eared rocks connected to the wall at several points, pendants; and bedrock columns often several metres high, left out of the process of cavity formation (Szunyogh 1997). Also noteworthy are the rock blocks (balanced rocks or pedestal rocks), which fell into the creek bed and obstructed the flow of water. In some places, the *Komlós Creek* has not yet formed a new bed to avoid these obstructions, but simply found a way below the rock. At present, these rock blocks no longer touch with their entire bottom the base of the passage, but they rest on a few points only (the pedestal), which were left out by erosion (Fig. 17).

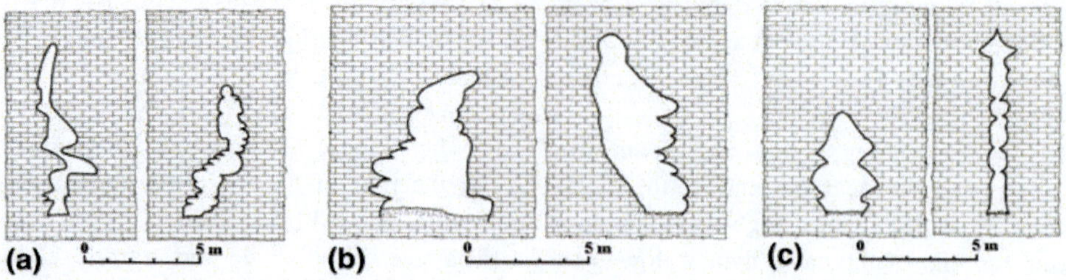

Fig. 16 Typical cross sections in Béke Cave. **a** assymmetrical notches; **b** one-side notches; **c** symmetrical notches

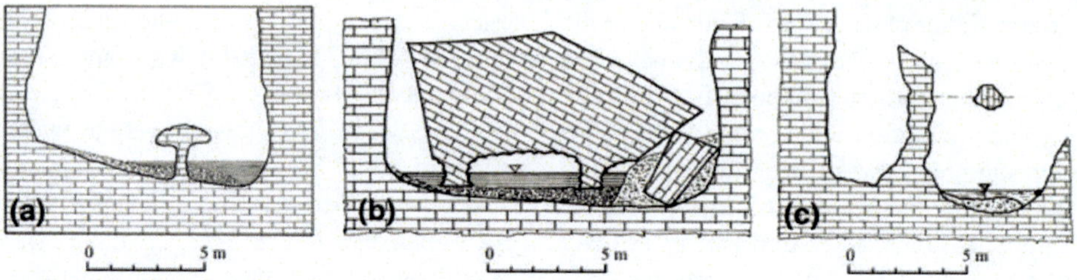

Fig. 17 Rock features from Béke Cave. **a** Schematic drawing of rock mushrooms; **b** Draft of a pedestal rock; **c** sketch of bedrock columns

The system is one of the richest caves in Hungary from the viewpoint of stalactites. Its stalactites are huge in size, usually "living" meaning that they are in a continuous stage of development. Their colours and shapes are very diverse, providing examples of almost all types presented in the literature. Their condition is good overall, with traces of devastation seen in relatively few places. The development of stalactites is particularly affected by the fact that the water level of the *Komlós Stream* (due to the growth of travertine dams) occasionally rose to high levels. When the water reaches the stalactites their growth stops, but their sides are covered with rich mineral precipitates (acicular calcites, stalactite collars). After the lowering of water level, they dry up again and decorate the cave like bush-shaped stalactites.

Also very spectacular are the travertine formations (the so-called *Nagy-tufa* or the travertine "sofa" at the entrance of the *Kígyós-ág*), which are often hanging over the creek, and include hundreds of small, 1–2 cm^2 area, water-filled pools, settling on top of each other as large "pillows" (Fig. 18). Their formation is due to water supersaturated in calcium carbonate flowing from tributaries or cave springs. It is worth mentioning the loose-structured 2–4 cm diametre freshwater limestone spheres, the "cave eggs", which were precipitated from the water of the *Komlós Stream*, covering in large number the bottom of the *Fő-ág* in certain sections. The cave popcorns (or acicular calcites) are generally found below the mean level of the stream water. Cave coralloids originating from the aerosol content of the air, are frequently

Fig. 18 Rimstone dam types in Béke Cave. **a** Balcony type rimstone dams; **b** profile of barrel and chalice-like rimstone dams; **c** omega and spur like rimstone dams; **d** parabola and S shaped rimstone dams

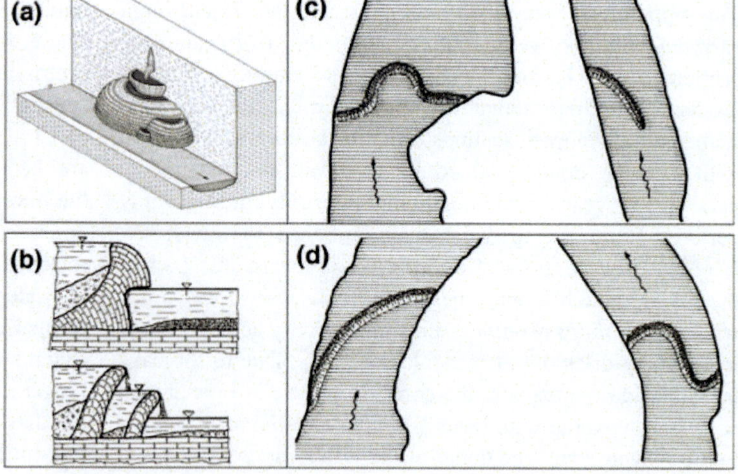

located on the side of stalactites, and helicites are also present in large numbers in some places. The upper passages of the cave including the higher parts of the *Fő-ág*, are extremely clayey, diversified by clay pyramids, furrows and wells, as well as by popcorns formed on the clay.

The environment of the *Béke Cave* is suitable for investigating the role of soil effect in karst corrosion. Therefore, László Zámbó built a measuring system that made it possible to study the chemical composition of karst water in the entire vertical section from the surface through the epikarst to the cave's ceiling (Zámbó and Ford 1997). Near the artificial main entrance, measuring stations were built within the cave, as well as in the sinkhole above this cave passage, at different levels in the soil and on the surface. The calcium and magnesium carbonate content of the infiltrating water was measured monthly for two years, thus obtaining numerical data on the spatial and temporal changes of the dissolution capacity of karst water (Telbisz et al. 1999).

Soon after the discovery of the cave, it became clear that its air was suitable for the treatment of chronic bronchitis and asthma, so in 1959, following the investigations of Márton Kirchknopf (1966), an underground sanatorium was established in the rooms between the artificial main entrance and the travertine dam of *Nagy-tufa*. Encouraged by the initial successes, in 1964 the Borsod Coal Mining Trust opened the Jósvafő exit by creating an artificial horizontal passage, allowing asthmatic patients to enter the cave without much fatigue. Soon the new sanatorium building near the *Komlós Spring* was completed, and after the Ministry of Health declared the *Béke Cave* a healing cave in 1969, the cures could become regular. Its healing power can be attributed, among other things, to the high calcium ion content in its air that has an anti-inflammatory and antispasmodic effect, as well as to the high carbon dioxide level, which stimulates breathing. At present, the healing activities are paused, but there are initiatives to restart this service.

The cave is a bat habitat with observation data about Mediterranean horseshoe bat (*Rhinolophus euryale*), greater horseshoe bat (*Rhinolophus ferrumequinum*), lesser horseshoe bat (*Rhinolophus hipposideros*), common bent-wing bat (*Miniopterus schreibersii*), Bechstein's bat (*Myotis bechsteinii*), lesser mouse-eared bat (*Myotis blythii*), greater mouse-eared bat (*Myotis myotis*), Daubenton's bat (*Myotis daubentonii*), Geoffroy's bat (*Myotis emarginatus*), Natterer's bat (*Myotis nattereri*), noctule bat (*Nyctalus noctula*), grey long-eared bat (*Plecotus austriacus*), brown long-eared bat (*Plecotus auritus*) and Brandt's bat (*Myotis brandtii*; Boldogh 2006).

3.6 Danca Cave

The *Danca Cave* is located south of Jósvafő, on the northern side of the *Tót Valley*, west of Égerszög, at 320 m above sea level, with a natural entrance. It is a periodically active spring cave, which collects waters through the *Névtelen* and *Gyökérkúti* sinks, located at the western foot of the nearby *Pitics Mountain*. The cave is 1390 m long, its horizontal extent is 570 m, whereas its vertical extent is 30 m.

The enclosing rock of the *Danca Cave* is the Triassic Wetterstein Limestone and, in some sections, dolomite. The evolution of the cave has been strongly influenced by tectonic settings, and these factors also determine the current image of the passages. The passage structure of the cave is simple: it consists of a tortuous main branch and one major tributary. There are several relatively large halls in the cave: the largest one is the *Nagy-omladékterem* (*Great Debris Hall*), which is 17 m long, 6 m wide, and 10 m high.

Hydrologically, the cave stream is intermittently active: the stream bed is usually dry, and flowing water is present only during very heavy rains. The maximum discharge measured so far is 5 m^3/min. Nevertheless, smaller lakes with permanent water formed in some sections (*Fürdőszoba*, *Szauna*, *Szifon-terem*).

The relatively spacious, long-known entrance section of the cave terminates in an extensive collapse zone. The "new part" that follows, begins with a chaotic and narrow labyrinth among the collapse blocks. Beyond the ruined section (from *Viktória-terem* to the *Szifon*), the cave is actually a narrow and low (sometimes only 30 cm high) underground creek with a small gradient. This passage is characterized by sharp turns and straight sections in between, chimneys extending from the ceiling to a height of 10 m, collapse halls as well as debris blocks damming the streambed. The rock ceiling of the *Szifon* drops 2.4 m below the stationary water level, and to make it passable under average hydrological conditions approx. 30 m^3 of water must be pumped out. The crossing is hard even after dewatering, because its bottom is covered with 1.3 m deep dilute mud.

Beyond the siphon, the image of the cave fundamentally changes. Its height increases to 4–10 m and its width to 1–2 m. Its route still consists of a series of straight sections and sharp turns, although from the *Elágazás-terem* onwards, the incising meander bends (Fig. 19), on average 0.2–0.5 m wide, become gradually more characteristic in the morphology of the passage bottom.

At a distance of 500 m from the siphon, the cave widens and flattens: its height decreases to 1.5–3.5 m. Its previous fracture-dominated character is lost, instead a tunnel profile with rounded ceiling becomes dominant. The corridor is

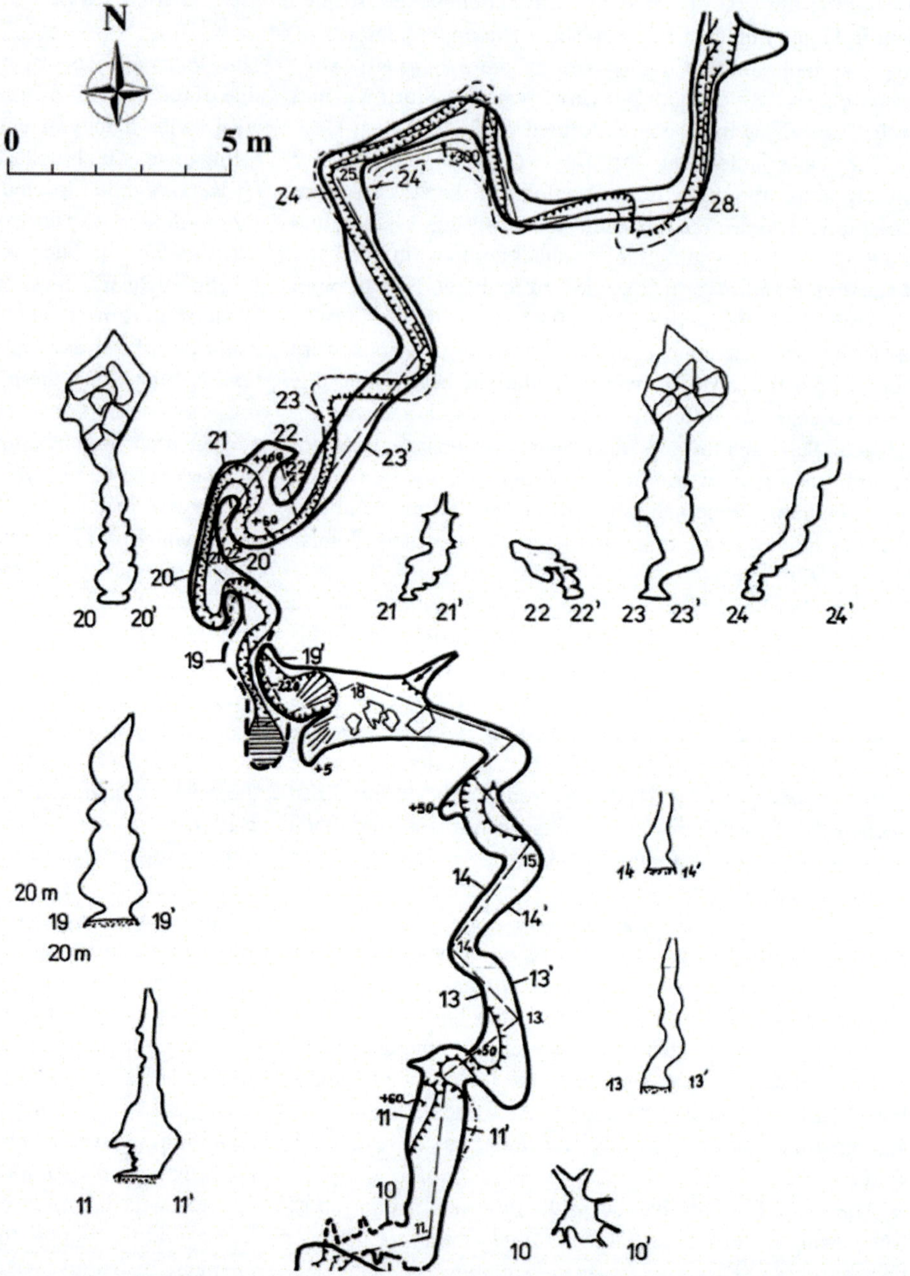

Fig. 19 Meanders in Danca Cave

filled with several metres thick, cemented gravel and clay layers, into which the current stream bed cuts 1.5 m deep, thus the profile of the passage has a characteristic keyhole shape. On the surface of the terrace-like gravel deposits, stalagmites of various shapes and sizes form a real forest. The clay accumulations are covered by 1.5–2 cm thick flowstones, which, in turn, bear 13–15 cm high dripstone columns. The ceiling is decorated with stalactites, soda straws and bulbous stalactites. The end point of the cave is a speleothem covered pile of debris, below which it is possible to advance 10–12 m by crawling in the creek bed, however, due to the unstable ceiling, the crossing is very dangerous.

Although the *Danca Cave* is rich in speleothems, its most important speleological values are related to the morphology formed by the cave stream. In several places, the overdeveloped (self-intersecting) meanders created "meander tables" that emerge as islands from their surroundings. Short side channels returning to the main branch are rarely seen in other stream caves, however, they are common here, as a result of the deep incision of the side channels. The bedrock columns enclosed by such channels are, in fact, conserved remnants of the islands of the stream formed very long ago. Although the length and depth of these tributaries are several metres, their width is usually only 10–20 cm. Unlike the nearby stream caves (*Baradla*, *Béke*, *Szabadság*), the stream bed of the *Danca Cave* is not filled up of gravel or sand, but it is made up of brightly polished limestone. Fantastically shaped ribs extend from the side of the passage. Unstable debris can be found in several places in the cave. Among the precipitations of the cave are noteworthy the large and completely intact dripstones and clay stalagmites in the sections beyond the siphon, as well as the helictites seen in the *Tündér-ág (Fairy branch)*.

The exploration of the cave was started in the period 1956–1959 by the staff of the Kinizsi Cave Research League, by breaking up the collapse that closed the entrance section and by widening the mentioned sinks. Although progress seemed very promising, work was stopped due to a series of collapses. A major discovery was made only about twenty years later, in 1981: the researchers of the Ferencváros Friends of Nature Association, led by Zoltánné Vidics, bypassed the sloping collapse blocks, and advanced 408 m in the cave until reaching a water-filled siphon. The water was pumped out in 1983, opening another 700 m long section. Finally, on January, 28, 1984, with the exploration of the *Tündér-ág*, the *Danca Cave* reached its current length of 1390 m. Its end point is below the *Gyökérkúti* sink.

The *Danca Cave* is highly protected based on its hydrological and morphological values (1988). The cave can be visited with the permission of the Aggtelek National Park Directorate. Its entrance is open, but the "new" section discovered in 1980 is closed at the *Szilvamag*. The "old" part is relatively easy to walk through, but the "new" part can only be accessed through a dangerous collapse zone and the extremely narrow maze of debris that follows. From there, one has to walk all the way to the siphon on all fours or in many places on your belly, which is especially difficult as the bottom of the passage is covered with sharp rock debris in many places. The siphon is constantly closed, thus it is only possible to proceed further after dewatering. However, the passage is not only difficult but also risky: on the other side of the siphon, one has to climb over a clay dam behind which a significant amount of water remains even after pumping. In case this dam is accidentally broken through, the siphon can be recharged, trapping the cavers, who are inside.

The cave is a bat habitat with observation data about greater horseshoe bat (*Rhinolophus ferrumequinum*), Bechstein's bat (*Myotis bechsteinii*), greater mouse-eared bat (*Myotis myotis*), and brown long-eared bat (*Plecotus auratus;* Boldogh 2006).

3.7 Esztramos Pantó Cave

The *Pantó Cave* in Esztramos is found between the mining levels V and VI of *Esztramos Hill* (Esztramos Mountain, Fig. 20). The tunnel leading to the cave entrance opens about 1.5 km northeast of Bódvarákó, in the middle of the NW side slope of *Esztramos Hill* facing the *Bódva Valley*. The length of the cave is 470 m and its vertical extent is 29 m.

The cave formed in the Middle Triassic light grey, crystalline metamorphic limestone of the Steinalm Formation under phreatic conditions (Piros 2006). A characteristic

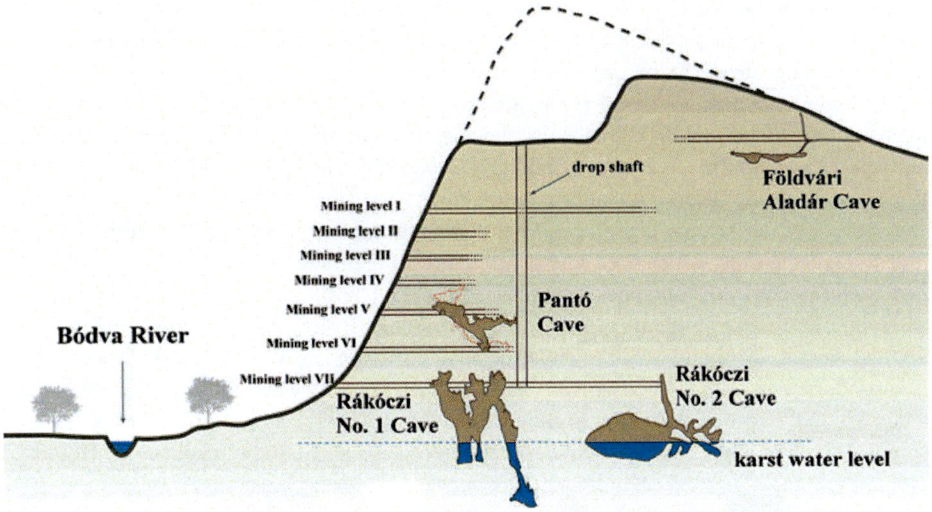

Fig. 20 Longitudinal section of Esztramos Hill with the significant mining levels and caves

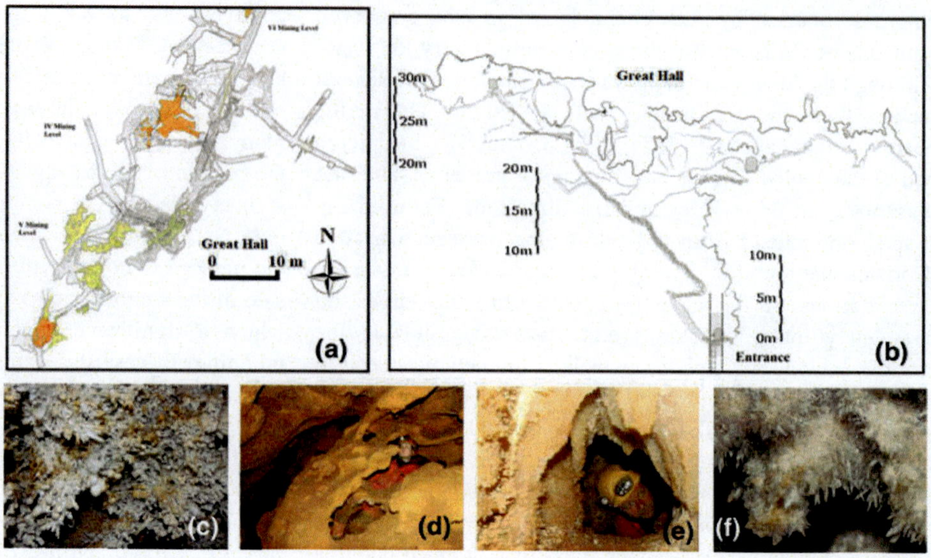

Fig. 21 Pantó Cave in Esztramos Hill. **a** Plan view of the cave (National Cave Register); **b** longitudinal section of the cave (National Cave Register); **c**–**f** calcite formations (photo Kovács); **d** passage formed in the phreatic zone (photo Kovács)

feature of the tectonically preformed cave is that the natural cavities, fissures and the artificial tunnel sections and excavation chambers follow or intersect each other in close neighbourhood. Nearly 1,000 m^3 of mining debris has been removed from the cave, however, at many points the passages are still partially or completely blocked by debris generated by mining (Fig. 21).

The morphology of the cave is characterized by limonite–calcite-filled fissures formed along the limonite veins (Szakáll and Kovács 1997), narrow bypasses connecting larger rooms and fissures, and impassable dissolved niches. The most beautiful corrosion forms can be observed in the roundabout starting from the *Elágazó* passage, as the depositional formations do not hide here the heavily rounded walls, which are divided/connected by corrosion hollows, stone bridges and small windows. Between the formation-free cavities, in a smaller oval room, inactive, strongly redissolved dripstones, differentially eroded limonite layers, popcorn covered redissolved flowstones, and massive calcite fibres can be observed.

The cave is characterized by the overlapping of various deposits (e.g. popcorns or occasionally acicular crystals or helictites on limonite or calcite coatings, or possibly on flowstones). The most spectacular formations are the water-clear, scalenohedral crystals and helictites that appear in the smaller niches, which are more protected from natural and anthropogenic disturbances. Next to the *Nagy-terem*, there is a helictite-rich small niche, which terminates in a vertical fissure. In this room, the superimposed formations richly adorn the originally dissolved walls: draperies, tiny tetaratas, and finger-thick helictites grew on the active flowstones. From the latter, further small curved speleothems and spherical popcorns or needle crystals branch out. Stalactites, stalagmites and columns are common in this room.

3.8 Földvári Aladár Cave

The *Földvári Aladár Cave* is located in the abandoned quarry of *Esztramos Hill*, at the upper mining level of 312 m asl, in a protective pillar created in order to preserve the cave. Its entrance opens on the eastern side of the pillar. The cave essentially consists of three spacious chambers and a smaller side room. It has a maximum horizontal extent of 84 m and a vertical extent of just over 7 m.

The cave, named after the former professor of geology at the University of Miskolc, was opened in 1964 during mining (Dénes 1964). Its fate had long been disputed, but the designation of a 50 × 150 m protective pillar, allowed it to survive even though the mining was continued (Koleszár 2000). The enclosing rock of the 410 m long cave was formed in the light grey, crystalline, metamorphic limestone of the Middle Triassic Steinalm Formation (Piros 2006), which is the raw material of the quarry. This rock is almost only visible on the ceiling, all other surfaces are covered with mineral deposits. The most striking element of the cave is the nearly horizontal ceiling of the chambers, which are uniformly at 306–307 m above sea level, and from which, at the time of excavation, huge mushroom-shaped stone tongues were hanging connected to the ceiling only by a narrow remnant rock. Most of these forms were torn off as a result of the mining blasts, but above their residues lying on the substrate, their former connections are still clearly visible today (Takácsné Bolner 2003).

The cave essentially consists of a huge room divided by a row of speleothem columns, from which smaller or larger side passages branch off (Fig. 22). The entrance side of the cave hall was once covered by a 1.5 m deep lake, whose shelfstones deposited on the stalactites. Inside, the speleothem columns are covered with cave popcorns. The only cave shield of the system can be seen here, too. On the ceiling, partly dissolved stalactites (Kraus 1997) and the truncated remnants of the broken stone mushrooms can be seen.

This system is one of the oldest caves in Hungary (Kordos 1973). It has been dissolved below the karst water level more than 2 million years ago in the upper Pliocene (or early Pleistocene), and the plane of the ceiling indicates the former karst water level. Corrosion due to the mixing of the karst water and the underground water flowing in the Pannonian sediments may have played a significant role in the dissolution of the cave (Kordos 1974).

Cave carbonate precipitates are divided into two main groups based on their origin: subaqueous and subaerial precipitates. The most spectacular of those formed below the water level are the well-developed calcite crystals. They form mostly 1–5 mm long crystals, but some specimens are as long as 10 and 2.5 cm thick. Crystals are characterized by rhombohedral cleavage and are often intergrown. The tiny transparent or opalescent white calcite needles cover huge wall surfaces in several places, and one of the small rooms is decorated with a cluster of intergrown, orange, 3–5 cm long calcite needles.

The most varied morphological elements of the cave were created by the water infiltrating through the ceiling and flowing along the side walls. Stalactites and stalagmites, columns, draperies and flowstones of various shapes and sizes are characteristic. Also typical are the coralloid shapes, which sometimes form 5–20 cm thick crusts, covering the walls or the older formations that appear in patches under

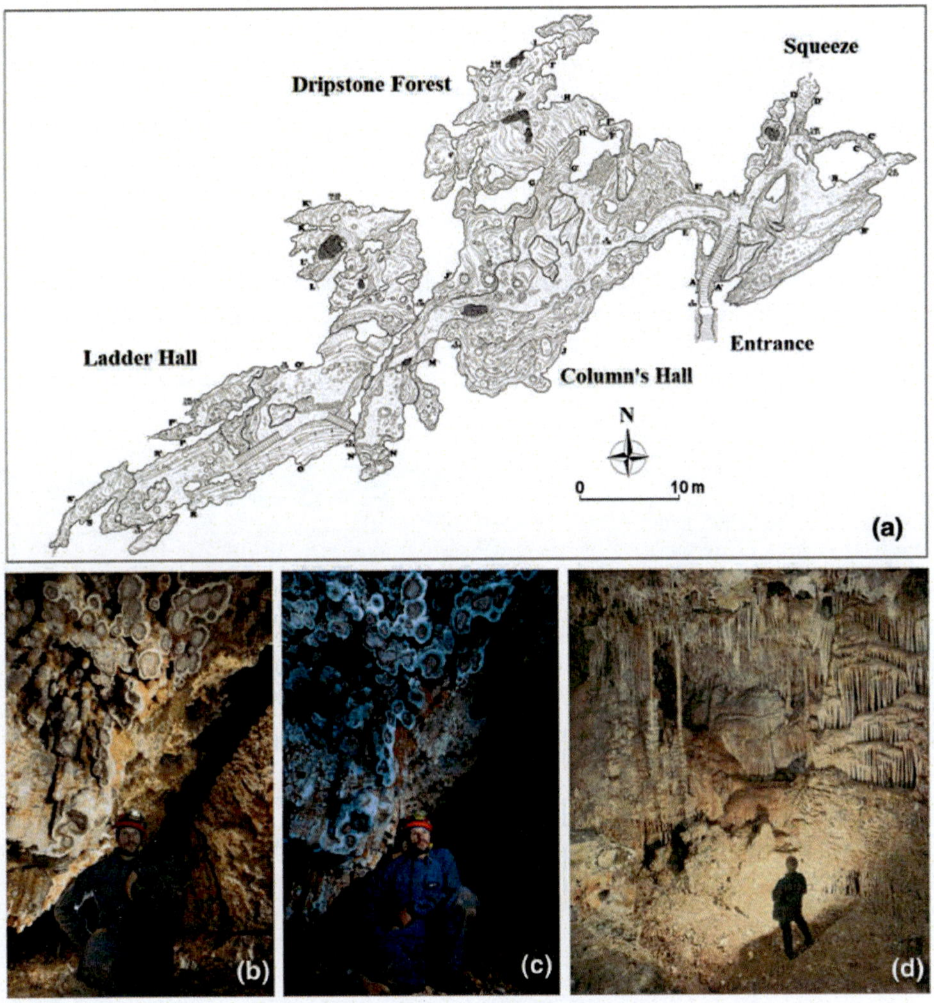

Fig. 22 Földvári Aladár Cave. **a** Plan view of the cave (National Cave Register); **b**, **c** calcite formations in normal and UV light (photo Berentés); **d** typical room in the cave (photo Borzsák)

them in some places. The popcorns cover the speleothems in some places, or they are located on the shelfstones, and sometimes they appear on the calcite needles. Helictites, i.e. deflected stalactites are an eye-catching sight, too. In some places, twisted specimens have been formed, in other places, clear straight crystal fibres reaching 10–15 cm also occur. Dissolved stalactites can be observed in several places. On some speleothems, moonmilk skirts of 2–5 cm wide can be seen. Large areas are also covered by reddish-brown, various sized coralloids, which include both spherical, mushroom-like and coral-like features. The corrosion forms of the cave include mostly spheroidal pockets ranging in diameter from a few centimetres to a few decimetres. In some places, steep, cylindrical chimneys were also formed.

The cave is a bat habitat with observation data about greater horseshoe bat (*Rhinolophus ferrumequinum*), lesser horseshoe bat (*Rhinolophus hipposideros*), lesser mouse-eared bat (*Myotis blythii*) and Natterer's bat (*Myotis nattereri*; Boldogh 2006).

After the cessation of mining, a lighting system, stairs and handrails were built in the cave for small visitor groups, and the cave could be visited for a few years. Today, however, the lighting cannot be used any more, and the visits has ceased.

3.9 Frank Cave

The *Frank Cave* is located on the southwest side of *Alsó-hegy Plateau*. Its entrance opens about 2.5 km north of Szögliget, on the hillside above the *Acskó Spring*, at the bottom of a wooded suffosion depression. The cave is currently known to have a length of 182 m and a vertical extent of 46.7 m (Nyerges 2003c).

The passages of the cave, excavated in 1971, formed in the Middle Triassic Gutenstein Limestone (Plózer 1974; Piros 2006). Morphologically, the *Frank Cave* can be divided into two parts. The first part consists of a passage that gradually expands, diverges in two, and then reunites after a narrow entrance. Thereafter a 60 m long sloping passage is followed by a stepping-down section, which ends in a large, 16 m deep shaft. Well-developed meanders formed in the western passage of the entrance section. This part of the cave is poor in depositional formations, with only a few smaller stalactites.

The second part of the *Frank Cave* consists of a large stream passage with a constantly active watercourse. The entrance shaft leads to its largest hall. From the room, in upstream direction, heading NW, the *Forrás siphon* can be reached after a short, 35 m long, meandering passage. The *Nyelő siphon* can be reached from the room in downstream direction, passing 25 m, at the end of a sudden detour, 12 m long and gradually decreasing. The passage of the main branch has a gravel bed, 2–4 m wide. Continuing from the deviation in the main branch, the passage reaches a high fissure, which approaches the surface. The fissure closes 25 m above the passage bottom. The passages are poor in speleothem formations, but clay pyramids have formed on the clay filling of the stream branch (Fig. 23).

Based on water tracings carried out in 1969 and 1980, the *Acskó Spring*, located directly below the *Frank Cave*, gets its water from the *Hangyás sinkhole* (Hazslinszky 1965), which is located 450 m from the spring as the crow flies, as well as from the *Bába-völgy Stream Sink Cave No. 3* (Dénes 1969),

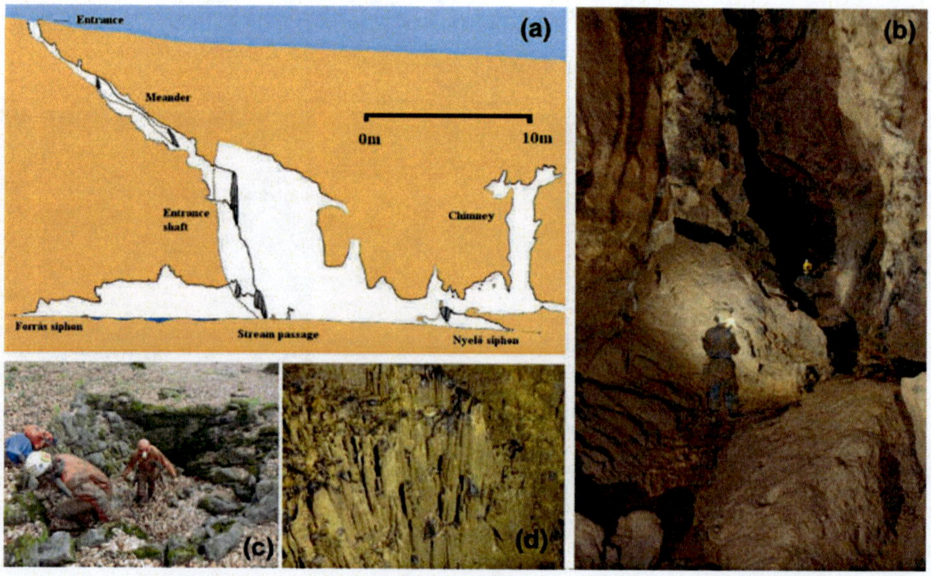

Fig. 23 Frank Cave. **a** Plan view of the cave (National Cave Register); **b** stream branch in the cave (photo Egri); **c** cave entrance (photo Szabó); **d** clay pyramids (photo Sági)

which is at 650 horizontal m from the spring. The stream that appears in the Frank Cave is part of this water system. The water that disappeared in the *Nyelő siphon* appeared after 16 h in the *Acskó Spring* at a horizontal distance of 200 m, then sank again in the flat, swampy area of the *Acskó meadow*, and reappeared in the *Csörgő Spring* in Szögliget after 44 h. In 1976, the members of the Debrecen Scuba Diving Club tried to swim across the *Forrás siphon*, from where they got into a small subaerial room after a 3 m underwater passage. From here, they managed to swim a further 10 m long and 5 m deep section, but clay accumulation at the end of this passage prevented their progression (Boros 1976).

The cave is known as bat habitat with observation data greater horseshoe bat (*Rhinolophus ferrumequinum*), greater mouse-eared bat (*Myotis myotis*) and Daubenton's bat (*Myotis daubentonii*; Boldogh 2006).

3.10 Hosszú-tető Cave

The *Hosszú-tető Cave* is located 500 m north of *Vidomáj-puszta* near the village of Szögliget, on a hillside just below *Hosszú-tető*, at an altitude of 487 m above sea level. Its entrance opens in a semicircular rock courtyard at the foot of a 6–8 m high rock wall. The cave is 54 m long, has a maximum horizontal extent of 20 m and a vertical extent of 20 m (Gruber 2003).

The cave is a simple form and it can be easily got around with basic equipment. It was formed in the Middle Triassic Wetterstein Limestone, and it could once function as a spring cave (Jakucs 1951; Gruber 2003). In the cave we can see mostly clean, whitish grey, nicely dissolved wall surfaces, which are decorated with moonmilk in some places and small and large speleothem formations in most places. Although the cave is largely dry, most of the formations are periodically active, developing. The entrance hall and the interior chamber were created by the expansion of a single common fissure. The only side branch of the cave opens from the interior of the fossil, almost horizontal fissure passage. In the interior chamber, stalactites, soda straws, draperies and flowstones developed. At several points in the cave we can find solution niches of different size. The popcorns formed in the cave are mainly crystalline popcorns originally precipitated from the aerosol, but they are partly redissolved.

The first 15 m of the cave have long been known (Jakucs 1951). However, its true value was only revealed in 2000, when during the exploratory research of the cave, charcoal fragments, grains reminiscent of bright reddish-brown burnt tiles, and unglazed pottery fragments belonging to the Bükk Culture were found in the filling. Later, during instrumental examinations carried out by the experts of the Hungarian National Museum, golden circles belonging to the late Bronze Age Kyjatice Culture (Rezi Kató 2003) were revealed from many points of the cave. During the archaeological excavations, pottery of the Bükk Culture and the late Bronze Age Kyjatice Culture (Rezi Kató 2004), notched arrowheads, iron buckles, fireplace (Rezi Kató 2005) from the fifteenth to sixteenth centuries, pottery from the eighteenth century, and cartridge cases were found. During archaeological excavations, the bottom sediment was removed to a depth of several metres in some places, and the original filling level is indicated by a line cut into the side wall, also marked with red paint in the first part of the cave. In order to provide effective protection, the Cultural Heritage Directorate closed the cave and in 2001 the cave was declared specially protected by the Ministry of the Environment.

3.11 Stream Sink Cave No. 2 in Kopasz-gally Hillside

The entrance to the sinkhole cave opens about 2 km NW of Bódvaszilas, in the lower part of the southeastern slope of the *Kopasz-gally Hill*. The entrance is on the south side of a large-diameter, asymmetrical sinkhole, at the foot of rock outcrops in a wooded setting. The north–south-oriented cave, with a total length of 450 m, has a horizontal extent of about 130 m and a vertical extent of 116 m (Nyerges 2003d).

Looking at the southern area of *Kopasz-gally–Nagy-Kopasz-gally* on the *Alsó-hegy Plateau*, we can see a series of sinkholes running from WSW to ENE direction along a curving line, and almost every sinkhole member hides a cave entrance (Fig. 24). These are, in fact, intermittent sinks, which have a small surface catchment area. Their position is not accidental either, because they follow the southern edge of the non-karstic Bódvaszilas Sandstone rock band, which arcuately stretches here in the karstifiable rock mass (Móga 2002).

The simple, downward step-like passage system formed in the Triassic Hallstatt Limestone follows roughly north–south-oriented fractures (Nyerges 2003d). Later on, it follows NE–SW and NW–SE fractures. The cave passages developed along these remarkable fractures by collapse and the erosive–corrosive effect of water flowing and infiltrating from the catchment area. In the section in front of *Molnár-utca*, the greenish-gray, silky, wedged layers of the Bódvaszilas Sandstone Formation are revealed.

The discovery of the cave dates back to 1969, when it became known to a depth of about 60 m. Exploratory research was revived in 1976 and continued in the 1980s, based on the hydrogeological connection of the cave to the *Vecsem Spring*, as evidenced by water tracing (Nyerges 2003d).

At the bottom of the intermittently active sink funnel, the waters flowing into the cave follow a path through the collapse maze, which is impassable to humans. Thereafter, the

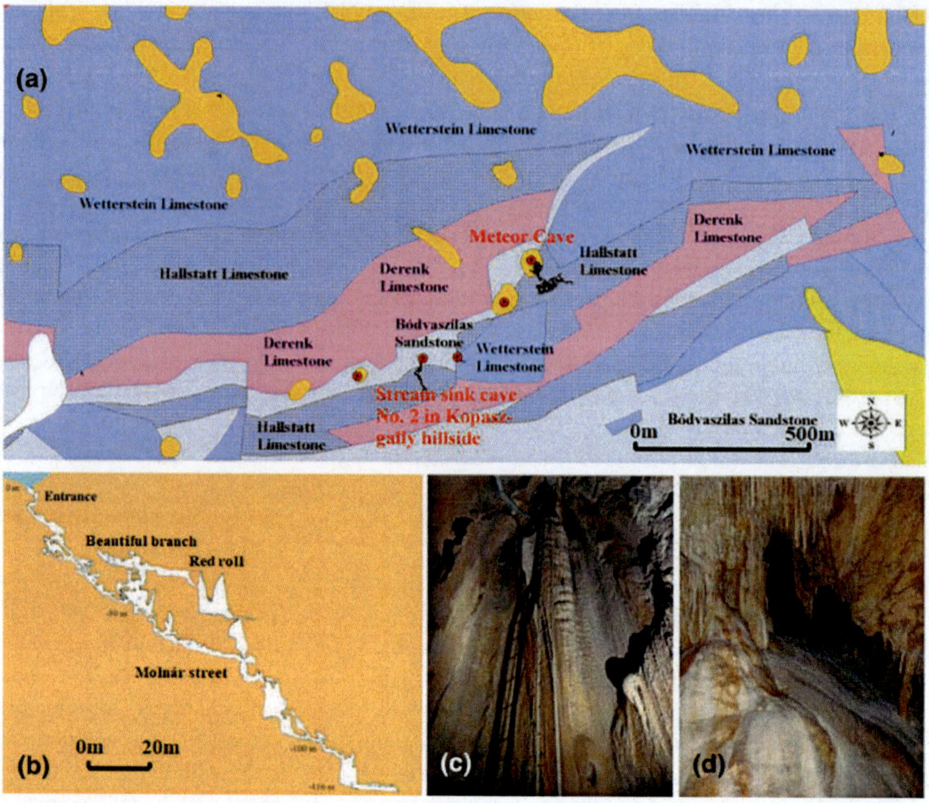

Fig. 24 Stream sink cave No. 2 in Kopasz-gally hillside. **a** Location of Alsó-hegy Plateau stream sink caves in the geological map; **b** longitudinal profile of the cave (National Cave Register); **c**, **d** typical speleothems from the cave (photo Egri)

different water paths unite at a depth of about 30 m and then the waters continue their journey in a definite channel to the end point at a depth of 118 m.

The middle section of the cave is divided into two levels. At the lower, active level, the stream gets deeper and deeper through smaller steps in the passages of a few metres wide, mainly decorated with stalactites and flowstones. The upper, "fossil" level richly covered with speleothem formations is the most spectacular and vulnerable part of the cave. This branch is characterized by stalactites, flowstones of varied size and colour as well as small and large travertine basins. The formations are snow-shite in some places and even the slightest contamination is noticeable on them. The other, longer passage of the upper level, which is poor in spectacular formations, connects back to the main branch through two shafts after a very narrow meander.

In the lower part of the main branch, the density of the formations decreases. In addition to limestone debris, more rounded purple-brown pebbles from the non-karst catchment also occur on the bed, typically with a black manganese coating. The scallops and the notch-lines of the passage walls not affected by speleothem formation are clear signs of stream corrosion activity.

The slightly sloping, narrow passages are connected by four corrosion shafts with a depth of 7–14 m. On the water-washed wall surfaces the grey and reddish colour of the enclosing rock is beautifully revealed. Particularly spectacular is the 14 m break of the spacious shaft number 2, which is adorned with a flowstone extending parallel to the built-in ladder.

The cave is a bat habitat with observation data about greater horseshoe bat (*Rhinolophus ferrumequinum*), lesser horseshoe bat (*Rhinolophus hipposideros*), and greater mouse-eared bat (*Myotis myotis*; Boldogh 2006).

3.12 Kossuth Cave

The *Kossuth Cave* is located on the edge of Jósvafő village, in the *Tohonya Valley*, behind the eponymous *Nagy-Tohonya Spring*, as a spring cave. Its currently known length is 1610 m and its vertical extent is 57 m.

In order to explore the system behind the spring, on behalf of Hubert Kessler, under the direction of Attila Chambre and Ödön Rádai, the blasting master György Verbály of Aggtelek and his three companions began to create a tunnel on September 1, 1953. After three years of enthusiastic and persistent work, on February 7, 1956, at the end of the tunnel about 40 m long, they entered the main branch of the cave.

Kessler and Rádai tried to swim through the siphon found at the end point several times, but only managed to determine the direction of the passage, as they did not venture into the narrow fissure with their simple equipment. They wanted to break through the narrowing by blasting, but the work was unsuccessful, and even the blast debris partially blocked the siphon. Since the late 1950s, several people have tried to swim across the *Reménytelen (Hopeless) siphon* with more or less success. In 1988, the divers reached a distance of 50 m from the water surface, where they found that the passage widened and flattened, however the rock pendants hanging from the ceiling hampered them to get through. During the research camp organized in 1997, the divers of the István Plózer Underwater Cave Research Group used specially designed equipment to swim through the narrowing that had proved invincible for several decades. On July 8, they managed to overcome the narrowing, which was only 30 cm high. The subsequent section, explored to a length of nearly 80 m, begins with a spacious corridor and ends into a transverse fissure at a depth of 32 m. From the end of the guide rope fixed at 130 m from the entrance to the siphon the upward passage was already visible, when the exploration was interrupted by the fatal accident of one of the divers. From the 2000s, members of the Amphora Diving Club began researching the siphon, and thanks to their years of hard work, they finally managed to swim across the siphon on November 13, 2009. Thereafter, they explored the *Apáink (Our Fathers) Room*, which is a continuation of the cave, and found another siphon at its end (Fig. 25).

The cave was formed mostly in Middle Triassic Wetterstein Limestone, but in some places in dolomite. Based on water tracings, its catchment area is large: it has an area of 24.4 km^2 (Sásdi and Szilágyi 1986). The passages of the cave typically extend in an ENE direction from the spring and run parallel to (and below) a rock-walled, currently dry but fluvially formed lateral valley. The genetic relationship between the cave and the valley is still unclear.

Following the entrance tunnel, the main sink within the cave can be reached after a 180 m long, mostly dry passage. This sink leads the cave stream to the spring at a lower level. However, in the event of periodic floods or siphon-type outbursts, this sink cannot completely drain the water of the stream, thus the cave stream also feeds the spring through other, intermittent sinks. During large-scale floods, much of the remarkably elevated water also leaves the cave through the much higher, artificial entrance tunnel. At such times, certain sections of the cave, such as the *Három Királyok Branch*, are completely flooded.

In the upper parts, the stream swells into a 50 m long, narrow lake and continues into the *Fő (Main) Branch*. This part is determined by the narrow, tilted fissures characteristic of the whole cave and is richly decorated with solution forms. Above the *Fő Branch*, there is a 300 m long upper level passage, which is rich in depositional formations, but is heavily filled with clay. This senile level is connected to the lower passage at several points. The parts of the upper level

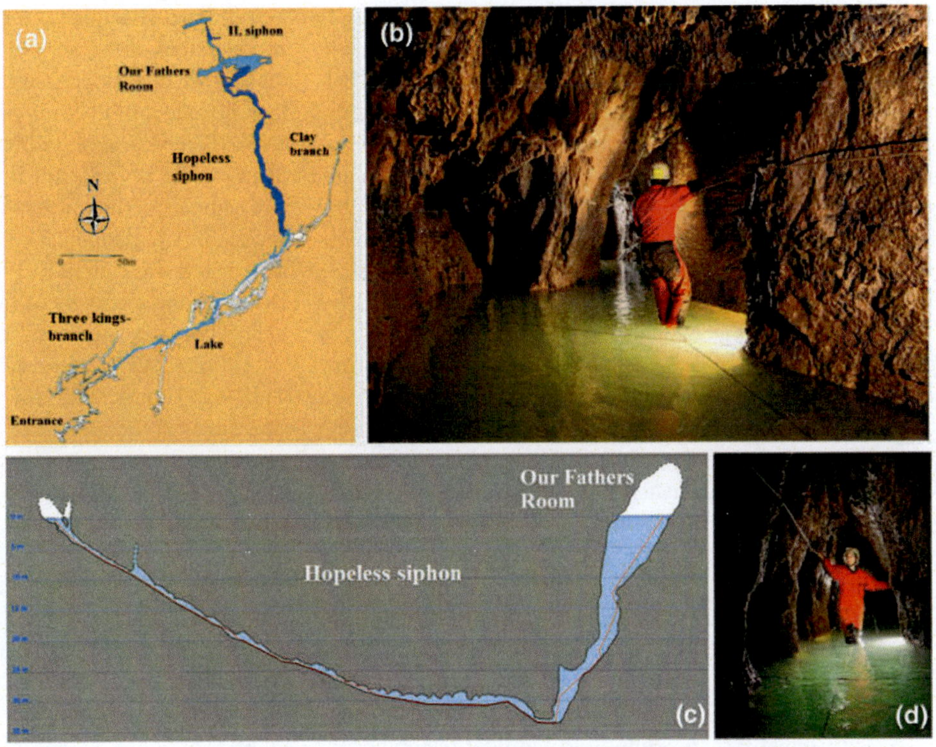

Fig. 25 Kossuth Cave. **a** Plan view of the cave (National Cave Register); **b** part of the cave, where the active stream is dammed to a lake (photo Egri); **c** longitudinal profile of Reménytelen siphon (Balázs); **d** fixed wire ropes help the crossing (photo Egri)

beyond the *Agyagos siphon* consist of short and clayey creeping passages, as well as larger sections with surprisingly rich formations. These parts are closed by bifurcating, flattening and claying endpoints.

The *Nagy-Tohonya Spring* is the second highest discharge spring in the area after the *Jósva Spring* group. At the Hydrological Research Station in Jósvafő, careful hydrological measurements were carried out during 30 years (Gádoros 1971; Maucha 1998), and as a result, it was established that the discharge of *Nagy-Tohonya Spring* varies between 1000 and 68,400 l/min, while its average discharge is 6,607 l/min. Changes in spring flow are influenced not only by precipitation but also by many other factors. The spring has confirmed hydrological connection with six surface sink points (*Lófej-forrás sinks, Mogyorós-kúti sink, Hazug well, Kuriszlán-völgyi sink, Gergés-lápai sink, Mogyoróstöbri sink*). The spring attracted the attention of researchers not only with its abundance of water, but also with its two extremely interesting properties. The spring water is warmer in winter, reaching a temperature of 16 °C during both high and low water conditions, while in summer it is cooler, 11 °C. The explanation for this phenomenon is that part of its water comes from the rising waters of the deep karst, which are warmer than the water of the surrounding karst springs. As the proportion of water infiltrating or sinking from the surface is smaller in winter, the spring water will be warmer in that season.

Another interesting feature of the spring is that in certain periods, its discharge increases many times its original value at regular intervals and then decreases in a few hours. Floods shorter than a day are usually siphon type outbursts that can occur even without precipitation. According to model calculations, if a single cavity is hypothesized behind the siphon, then its volume must be 3800 m^3 to explain the outbursts of the *Nagy-Tohonya Spring*, and this cavity can be filled in an average of four days. The siphon-cavity of the spring must be on a tributary of the main branch, as measurements show that the spring discharge does not decrease before the outbursts of the spring. It has been proved by statistical analysis that the pressure waves of the outbursts of the *Lófej Spring* located upstream in the catchment also contribute to the discharge changes of the *Nagy-Tohonya Spring*, as the waters of the *Lófej Spring* flow into the *Nagy-Tohonya Spring* after sinking. It has been also shown that air temperature fluctuations create discharge fluctuations during the period of snowmelt. A time-series study of the siphon type working of the spring not only helped to identify direct and indirect siphon type outbursts, but also pointed out that the tidal changes of cracks in the solid crust also influence the spring discharge, because it was observed that in the cave streams, water flow fluctuates for a period of about six hours in the middle stage of each flood event. It has also been shown that, in addition to the tidal phenomenon, earthquakes and larger air pressure changes also cause changes in the fissure volume through rock deformation, thus leading to measurable discharge changes.

Based on general bat observations and using bat nets near the spring, it can be stated that the *Kossuth Cave* is a spawning, breeding and a wintering place for bats. The species detected here are (Boldogh 2006): *Rhinolophus euryale, Rhinolophus ferrumequinum, Rhinolophus hipposideros, Myotis myotis, Myotis mystacinus, Eptesicus serotinus* and *Plecotus auritus*. Our knowledge of the invertebrate fauna of the cave is incomplete. Among the insects, species belonging to the order of beetles (*Coleoptera*) and flies (*Diptera*) and subclass of mites (*Acari*) have been observed. Blind freshwater amphipods (*Niphargus aggtelekiensis*) can be observed in the underwater parts. As a curiosity, the walls are covered with a living bacterial coating of *Lepthothrix* in many places.

3.13 Meteor Cave

The *Meteor Cave* is located on the southwest side of the *Alsó-hegy Plateau.* Its entrance opens about 2 km NW of Bódvaszilas, in a wooded environment, at the bottom of a large periodic stream sink called *Kis-vizes sinkhole*, on the eastern side of the sinkhole at the bottom of rock outcrops. The cave, with a total length of 1672 m (Nyerges 2003e), is oriented predominantly NW–SE, although the *Titánok Csarnoka (Hall of Titans)*, famous for its imposing size, is almost perpendicular to this direction. The maximum horizontal extent of the cave of about 170 m can be measured in the former direction. Due to its size, it is worth mentioning that the extent of the *Titánok Csarnoka* is about 100 m in the SW–NE direction and about 35 m perpendicular to it. The depth of the cave is 127 m (Nyerges 2003e).

Originally, the cave did not have an open entrance accessible to people. The exploration of the periodically active sink was started in 1961 under the direction of György Dénes by the members of the Vörös Meteor Speleoclub, and their work was successful even in that year (Dénes 1961).

Although the *Meteor Cave* periodically acts as a stream sink, it is significantly different in nature from other stream sink caves in the area, and from typical stream sink caves in general. Instead of a step-like deepening including of shafts, the cave is characterized by an extensive maze of debris and a huge room richly decorated with speleothem formations. The collapse maze is typical of the first part of the cave near the sinking zone, and it is connected to the spacious lower part created mainly by collapses only by a narrow "neck". The stream sink cave character is most advanced in the "neck" part. Here, through several smaller corrosion shafts, the water gets deeper and deeper, eventually disappearing under the collapse blocks of the *Titánok Csarnoka* (Fig. 26).

Fig. 26 Meteor Cave. **a** Plan view of the cave (National Cave Register); **b** One of the Titans—a large stalagnate (photo Egri); **c** drapery from the Hall of Titans (photo Egri); **d** Hall of Titans (photo Egri)

The *Kis-vizes sinkhole* is a member of the series of stream sinks located along an arcuate line in roughly SW–NE direction on the hillside of the *Kopasz-gally–Nagy-Kopasz-gally* area. Here, the stream sinks follow the arcuate southern edge of the non-karstic rock band (Bódvaszilas Sandstone Formation). In the first section of the cave, the red Hallstatt Limestone with white calcite veins can be observed, the internal parts typically formed in the grey or yellowish-white Wetterstein Limestone, which has reddish patches or veins. Szentes (1963) described the multicoloured Seizi Slate as being tectonically wedged between the layers of the Wetterstein Limestone Formation. At the beginning of the NE side branch of the *Titánok Csarnoka*, as well as in a small patch on the eastern side of the *Omladékos-terem*, the layers of the Perkupa Evaporite Formation can also be observed (Nyerges 2003e).

The cave can be divided into four main parts, which are markedly different in nature. The section closest to the entrance consists of a chain of wide but flat, predominantly collapse chambers. The bedrock, which is mostly covered with debris blocks, roughly follows the southeasterly tilt of the ceiling, so there are no significant heights here. In this section, it is only possible to move on by getting around the rock blocks, crawling and sliding on bedrock strata, though in some places, there is enough room for standing up. The western half of this section is practically a rugged sinking zone, with several recharge points, without a dominant watercourse. On the smaller and larger rock blocks, traces of the corrosive effect of the infiltrating and flowing water appear in the form of scallops and runnels of varied size. Although the image of the first part is dominated by debris, on the ceiling surfaces of certain rooms, beautiful groups of speleothem formations (stalactites, draperies, soda straws and flowstones) formed in places. Less common, but in some places larger sized stalagmites and flowstones also formed (in the *Majmos-terem* and *Nagy-terem*). The stalactite baldachin of *Nagy-terem* and the gravel material cemented to the ceiling are clear signs that the cave is functioning as a sink of varying intensity.

The second section of the cave, i.e. the middle part, is the incised bed of the stream that has so far merged the smaller watercourses. This section is typically a high but narrow passage with 2–10 m deep breaks and shafts. The erosion–corrosion forms of the narrow bedrock channel intersect not only the enclosing bedrock, but also the flowstones.

The most significant and impressive section of the cave is the *Titánok Csarnoka* and the connected *Lapos-terem (Flat

Room). Horizontally, these two rooms are not separated from each other, the significant difference is reflected in their vertical extent. While the height of the former in many places exceeds 10–15 m, the latter often does not even reach a height of 1.5–2 m. The collapses played a very significant role in the formation of both parts of the room. These events were determined by the marked tectonic shear zone (Szentes 1964), which determines the west–east strike of the *Titánok Csarnoka*. The giant rock block, which forms the base of the *Lapos-terem* fell off the ceiling of the southern part of the ancient *Titánok Csarnoka*, practically in one single block. On the other hand, the collapses in the area of the present *Titánok Csarnoka* resulted in smaller blocks, among which, however, several m^3 in size are still present. The possibility of collapses was due to the tectonically shattered rock mass, but presumably the cavity-forming and sediment-transporting effect of the formerly more intensive water flows also contributed to this process. On the side walls of the present room, erosion–corrosion forms related to streams are not visible, but rather flat surfaces of collapse or tectonic origin are characteristic (Szentes 1964).

In addition to its large size, the *Titánok Csarnoka* is also famous for its speleothem formations. Stalagmites, dripstone columns, flowstones and smaller or larger speleothem pools have evolved on the rock debris in extraordinary variety and quantity, in some places to such an extent that the debris is completely covered. The formations have the highest density in the central part of the room, around the *Titans* (giant stalagmites), near the entrance to the room, and around the lakes. In the *Lapos-terem*, the dripstone formations (stalactites and stalagmites, columns, draperies) typically occur in smaller, discrete patches. The bedrock is covered by varying amounts of debris. Interesting and vulnerable formations of this section are the helictites, which appear on the southwestern part of the ceiling on a surface of several tens of square metres.

The lower levels of the cave are the *Patakos-ág* and the *Agyagos-ág*. The former drains the watercourse of the central part of the cave under the blocks of the *Titánok Csarnoka* and the *Lapos-terem*. The walls of these passages are covered with clayey sediment, they are typically without formations, and the vegetal remnants adhered to the ceiling demonstrate the intermittent swelling of water in this section. In the stream bed, similar to the upper, active sections, purple-brown, rounded pebbles can be observed, which are often covered by black (manganese) coating. The water seeps into the sediment of the endpoint and reappears in a room called *Agyagos*, which is the deepest point of the cave.

As the cave acts as an intermittent stream sink, organic nutrients are easily transported into the cave from the surface. Due to this fact, both the number of animal species and the number of individual specimens are significant. In an article published in 1969 (Bajomi 1969), Dániel Bajomi described 90 animal species, which at that time raised the Meteor Cave to be the fourth cave in Hungary from the viewpoint of animal species. Of these, he determined 13 species as troglobiont species.

In addition to the insects and spiders commonly observed near the cave entrance, blind cave amphipods (*Niphargus sp.*) can be relatively easily observed by the naked eye. Guano accumulations in the cave suggest the presence of bats outside the resting period, too.

Observational data show that the wintering population of the Mediterranean horseshoe bat (*Rhinolophus Euryale*) is significant. Other bat species such as greater horseshoe bat (*Rhinolophus ferrumequinum*), lesser horseshoe bat (*Rhinolophus hipposideros*) and Geoffroy's bat (*Myotis emarginatus*; Boldogh 2007) have also been observed in the cave.

3.14 Rákóczi No. 1 Cave

Rákóczi No. 1 Cave is located at mining level VII of *Esztramos Hill* belonging to the area of Szalonna Karst. The entrance to the cave opens about 1.5 km northeast of Bódvarákó, in the lower third of the NW slope of *Esztramos Hill* facing the *Bódva Valley*, in an artificial environment. According to current surveys, the branched, two-level cave has a horizontal extent of about 70 m, a total length of 750 m, and a total vertical extent of 79 m (Fig. 27).

The cave was formed in the Middle Triassic light grey, crystalline, metamorphic limestone of Steinalm Formation (Piros 2006) under phreatic conditions. The cave consists of two main fissures with vertical walls following NE–SW orientation, which are connected to each other for a short distance. The large and irregularly shaped, spherical niches are interconnected along the tectonically preformed fissure passages, and they continue in a vertical fissure system below the karst water level.

The special feature of the cave is the variety of different formations and their building on each other. Its walls are mostly covered up to the ceiling by large-scale and varied mineral deposits: large stalactites and stalagmites, curtains, draperies, typical spherical popcorns, coralloids and cauliflowers, "Christmas trees" from calcite rafts, limonite–hematite calcite crusts, flowstones of various colours, soda straws and helictites (Kraus 2003).

The lower part can also be visited by tourists. In the upper level of the fissure, which is not accessible to tourists, in the zone of the "stalactite roots" we find the most spectacular and vulnerable part of the cave, where even the bottom is grown by popcorns 1–2 cm in diametre. In addition to the coralloid cover and large stalactites, spectacular curtains and helictites can also be found in this section of the cave.

In addition to depositional formations, especially in the upper zone of the side wall of the second main fissure,

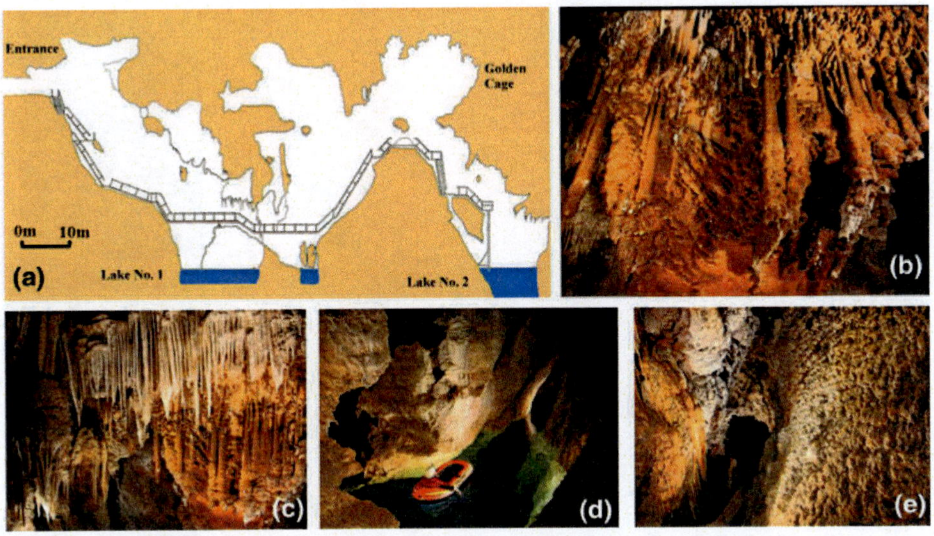

Fig. 27 Rákóczi No. 1 Cave. **a** Longitudinal profile of the cave (National Cave Register); **b** popcorn covered, downward thickening stalactites of the Golden Cage (photo Egri); **c** the Golden Cage (photo Egri); **d** part of Lake No. 2 (photo Egri); **e** fissure with richly decorated popcorn-covered walls (photo Egri)

corroded walls and redissolved formations are also common. There is a lake at the bottom of both main fissures of the cave. The passages continue here in the (phreatic) zone below the karst water level. Corroded walls are characteristic in the zone just above the lakes, close to the karst water level, where the walls are mostly free of depositional formations, thus here the spherical hollows formed by dissolution are more striking.

The first section and the first lake of the cave were known to the miners, who worked in the tunnel crossing the cave, and a sketch map of 1958 is available about this section (Kordos 1984). In 1964, the explorers of the cave, breaking through the narrowing on the upper level, came across another lake and found that there were further passages under the water (Szilvássy 1965). Later, during diving, additional underwater passages and partly subaerial rooms, halls and siphon openings were found. *Rákóczi No. 1 cave* is associated with the largest cave rescue operation in Hungary that occurred in the 2000s. About two hundred people took part in the 5-day rescue, with divers making a total of 286 dives. The lost diver, due to his weakened condition, was eventually brought up through an 11 m long, exploded tunnel into an already known part of the cave (Adamkó and Hegedűs 2002). The save operation completed successfully.

The wildlife of the cave includes greater horseshoe bats (*Rhinolophus ferrumequinum*) lesser horseshoe bats (*Rhinolophus hipposideros*), grey long-eared bats (*Plecotus austriacus*) and serotine bats (*Eptesicus serotinus*; Boldogh 2006). During the recent years, divers observed blind cave amphipod traces in the deposits, and in 2013 they even managed to find a living *Niphargus aggtelekiensis* specimen.

3.15 Rákóczi No. 2 Cave

The *Rákóczi No. 2 Cave* (or *Surrantós Cave*) is located at mining level VII of *Esztramos Hill* belonging to the area of Szalonna Karst. The tunnel leading to the entrance of the cave opens about 1.5 km northeast of Bódvarákó, in the lower third of the NW hillslope of *Esztramos Hill* facing the *Bódva Valley*. The cave was formed along the tectonically determined NE–SW (and subordinately NNW–SSE) fissure system. It is currently known to have a length of 533 m and a total vertical extent of 53 m (Nyerges 2003i).

Like the other caves on *Esztramos Hill*, its exploration took place due to mining. The extraction of iron ore began in 1928, and in the following period several ore exploration tunnels were created, which crossed caverns at level VII. During the creation of a mining railway station within level VII, as the SE directed tunnel was continued, the miners reached a larger cavity (Koleszár 2000). As long as the limestone mining was active, *Rákóczi No. 2 Cave* had a strong inward air flow, thus, the cave was used for ventilation. As a result, the dust whipped by stones thrown into a nearby 125 m deep drop-down shaft completely covered and polluted the parts near the entrance.

The cave is a parallel system of high but narrow fissures extending below the karst water level (Fig. 28). In the passages of the tectonically determined cave formed in the metamorphic limestone of the Triassic Steinalm Formation (Piros 2006), the water surface representing the karst water level can be seen at 14 points (e.g. at *Pond Chimney*, at the lakes near the *Pass*, at the *Round Lake*, etc.). The largest lake in the SW ending of the cave is 42 m long and 9 m deep (Nyerges 2003i).

Fig. 28 Rákóczi No. 2, 3, 4 Caves. **a** Plan view of Rákóczi No. 2 Cave (National Cave Register); **b** lake of Rákóczi No. 2 Cave (photo Egri); **c** acicular calcite from Rákóczi No. 3 Cave (photo Kovács); **d** calcite formations from Rákóczi No. 4 Cave (photo Kovács)

The cave morphology is characterized by the narrow, 2–3 m wide and 8–10 m high NE–SW fissures (main fissure direction) and the secondary, NNW–SSE fissures at an angle of ~60°. At the intersections, narrow passages in labyrinth structure were formed (e.g. *Penalty Circles*, *Witch*). The side walls of the fissures are not vertical but slightly sloping. The passages are nowhere expanded to more spacious rooms. Their upper ends are usually tapered, and sometimes solution forms can be seen at their tops. Their lower ends are filled with clayey–silty lake sediments.

The spatial distribution and types of the formations are very diverse. In the upper part of the fissures, boxwork patterns or dissolved, bare wall surfaces are common. In the lower parts, different variants of popcorns and speleothems occur en masse, but crystallized carbonate minerals (calcite scalenoeders and rhombohedrons, aragonite variants) and sulphate minerals (gypsum, barite) can also be observed (Nyerges 2003i). The corrosional development of the cave below the karst water level is still going on, as evidenced by the predominance of bare, formation-free wall surfaces near the lakes. Above the water level, masses of snow-white popcorns cover the walls.

Like the other caves in *Esztramos Hill*, it is also characterized in some places by dusty, dirty and damaged formations due to blast debris and mining traffic. Primarily in the areas near the entrance, the inwardly ventilated dust adheres to the walls due to moisture and masks the formations. The wildlife of the cave includes only a few small and large horseshoe bats.

3.16 Rákóczi No. 3 Cave

Rákóczi No. 3 Cave (formerly known as the *Omlás No. 2 Cave* as mentioned by Majoros 1986; Lénárt 1997 and Koleszár 2000) is part of the *Esztramos Hill* mining level VII. The tunnel leading to its entrance opens about 1.5 km northeast of Bódvarákó, in the lower third of the NW slope of *Esztramos Hill* facing the *Bódva Valley*. The multi-level shaft cave was formed along a vertical fissure. It has a horizontal extent of 150 m, a total vertical extent of 34 m and a depth of 27 m relative to the entrance.

The tectonically preformed cave was formed in the metamorphic limestone of the Triassic Steinalm Formation by collapse. It is a stepped, largely vertical fissure. Due to its slope conditions, rope technology is required for safe visiting.

The cave is practically formed along a single fissure (with 15–20° to 195–200° orientation), from which only a few smaller passages or subordinate fissures differ. The multi-level shaft cave consisting of several smaller chimneys and shafts, extends below the karst water level. Going down the cave, the frequency of mineral precipitation decreases, and near the lakes, the spherical niches formed by phreatic dissolution are the predominant morphological elements.

Based on the morphology of the cave and the varying type and quantity of the fillings, it can be divided into several levels. In the upper section, in addition to the collapse blocks, explosive debris falling into the fissure during the excavation of the tunnel is also common, and in several places, the ceiling of passages are formed in the mining debris. A boxwork structure is recognizable on the softened side walls. Going one level down, the sidewalls are richly covered in different colour and shape formations: speleothem crusts, acicular calcites of varying sizes, different types of popcorns, and aragonite crystals. Interestingly, the hard, black crust found below the various formations has separated from the wall in certain places. In the further, downward part of the fissure, depositional formations are fewer, whereas dissolution forms become predominant. The only significant side branch of the cave is the *Nyaktiló-ág*, which consists of two parallel fissures. Its orientation is the same as that of the main fissure, and it opens from the upper side of the shaft leading down to the lowermost point. Its first section is poor in speleothem formations, and its walls are covered with limonite crust in several places. On the other hand, its inner, downward fissure is extremely rich in formations. Its walls are covered with a continuous layer of popcorns. The lake No. 4 is found at the bottom of this passage.

The bright green, crystal-clear four cave lakes appearing at the karst water level are presumably connected to each other, their current water level shows a correlation with the water level of *Bódva River*. The underwater sections of the lakes await further exploration by diving methods.

3.17 Rákóczi No. 4 Cave

The *Rákóczi No. 4 Cave* (formerly known as *Reprendszer* as mentioned by Majoros 1986; Lénárt 1997 and Koleszár 2000) is located at the mining level VII of *Esztramos Hill*. The tunnel leading to the entrance to the cave opens about 1.5 km northeast of Bódvarákó, in the lower third of the NW slope of *Esztramos Hill* facing the *Bódva Valley*. The length of the cave is 150 m and its vertical extent is 50.3 m.

The tectonically preformed cave was formed in the Triassic metamorphic limestone of the Steinalm Formation, and it consists of four levels. The levels are connected by vertical shafts or fissures spanning occasionally several levels.

The lowest level of the cave is a fissure about 1 m wide, strongly redissolved, cut with spherical hollows. The walls are in some places covered with small, redissolved calcite needles. The most conspicuous formation is a piece of bedrock hanging from the ceiling dissolved in the shape of a tongue. On the second level, even more spectacular dissolution forms developed: small pockets, bedrock crests, windows, stone bridges, and smaller spherical niches. Among the depositional forms, coralloids appear, and beneath them thick acicular calcite covers the redissolved walls. Remains of limonite crusts and weathered calcite crusts also occur in patches. Calcite rafts remained on the bedrock closing the fissure are special. The third level is actually a long, planar fissure, tapering downward and its walls and bottom are decorated by popcorns. The *Reprendszer* is quite poor in mineral depositions compared to most other *Esztramos Hill* caves. Level four is an exception, where a number of spectacular formations can be seen in the more hidden areas, such as acicular calcites grown on the sidewalls with needle-shaped or spherical popcorns, popcorn Christmas trees (logomites), popcorns formed along calcite veins, and coralloids. In a smaller narrowing side passage of level four, coralloids reminiscent of huge oyster mushrooms grew on needle calcites, and acicular calcites formed in groups on the calcite rafts of the pierced side walls.

3.18 Rejtek Shaft Cave

The cave is located on the *Dusa Plateau*, which is south of *Alsó-hegy Plateau* and surrounded by the *Bódva River* and the *Ménes Stream*. The cave is 1300 m NE of Szögliget, on a flattened ridge, at 412 m above sea level, in a wooded environment. It is very difficult to find the entrance, which opens in a characterless place, without a rock outcrop or a depression. Its passages have a total length of 450 m and a maximum vertical extent of 66.5 m (Nyerges 2003f). Its formations are protected by ladders built into the old part. Caving the new parts requires knowledge of cave roping technique.

The *Rejtek Shaft Cave* is the only significant cavity on the *Dusa Plateau*, formed in Middle Triassic Wetterstein Dolomite and Limestone (Szentes 1963; Kósa 1976; Piros 2006). It can be divided into two vertical sections, which are connected by a horizontal section in the upper third part of the cave. The horizontal extent of the cavity is almost 70 m, extending towards the slopes of the nearby hillside along a definite ESE–WNW tectonic zone (Nyerges 2003f). The special value of the cave is the high number and variety of popcorns despite being a shaft cave (Fig. 29). These formations occur typically between −22 and −50 m depth, and in many places can only be approached by damaging them.

During World War II, the inhabitants of the area wanted to establish a hiding place in the cave, hence its name ("*Rejtek*" means "hiding place" in Hungarian). They used primitive wooden torches, but they could not go deeper than the first chamber. In 1957, the cave was explored first to −22 m and then in 1958 to −63 m (Kósa 1963). Finally, in the autumn of 1974, explorers discovered the *New Part* on the SE side of the cave (Kordos 1984).

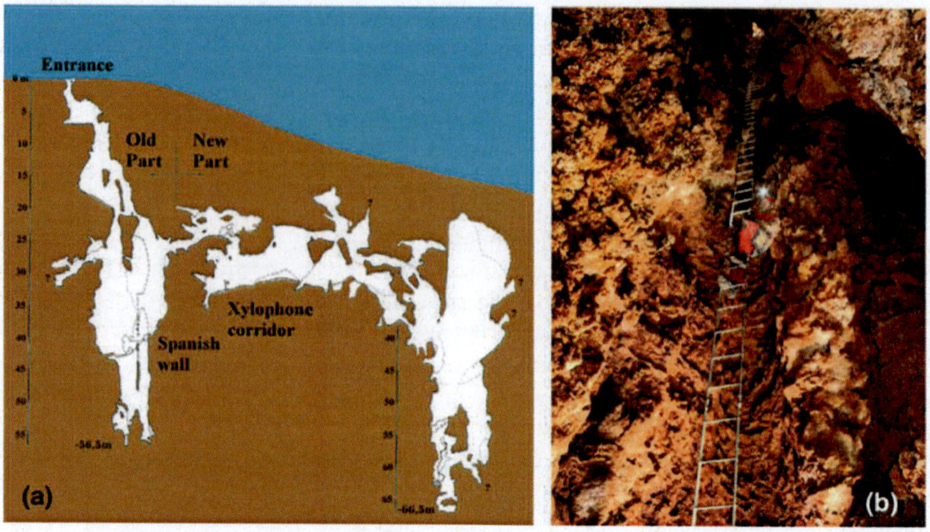

Fig. 29 Rejtek shaft cave. **a** Longitudinal profile of the cave (National Cave Register); **b** popcorn decorated shaft (photo Egri)

The formations in the cave can be divided into three zones (Kósa 1963). In the uppermost passages, down to −15–20 m, popcorns or flowstones can be observed only in small patches. Below this level there is a zone, where speleothems are predominant, and further downwards, it passes continuously into the popcorn dominated zone. In the upper part, the popcorns are settled on the well-developed speleothems, while in the deeper sections this reverses and the popcorns are covered by smaller undeveloped speleothems. The walls are covered in extreme richness by coralloids of various shapes and origin. We can find a multitude of flat, mushroom-like or laterally branching coral-like precipitates. In some places, these formations do not settle on the bedrock, but on the clay layer covering the walls. The coralloid formations include tiny yellow calcite needles and a variety of popcorns and corals in all shades from white to dark brown. In the end-shaft on the left, large, bulbous, cauliflower-like formations can be seen, interwoven in places with white layers, near which traces of dissolution and calcite shelves can be observed. The formations can be divided into four groups (Kósa 1963): hard popcorns (coral and mushroom-like); speleothems; soft popcorns (coral and mushroom-like); and crystal needles on bedrock. The hard version is mostly ocher yellow, reddish-brown in colour, while the crumbly one is pale yellow. The latter is present in a small amount in the −25 m zone. The size of the formations varies greatly, from 2–3 mm to 6–7 cm all versions can be found. The harder version is suitable to make thin sections. Microscope examinations show that there is an inclusion in the central part, around which a radial fibrous structure can be seen (up to a maximum of 1.5 mm), while the outer parts are crystallized in an annular structure (Kósa 1963). Based on the X-ray examinations of the soft version, it can be stated that it consists of very fine-grained calcite and 5% aragonite (Kósa 1963).

The spherical formations of various sizes and spatial distribution were presumably formed under periodically changing or stagnant water level conditions. Aside from a few metres depth near the surface, which developed as a collapse, the main agent of the cave formation was the corrosion of water infiltrating from the surface along tectonic fractures (Kósa 1963). However, based on the morphology of the cave and the small amount of filling deposits, it seems to be closer to cavities of phreatic origin, although the cross sections of its shafts show many similarities with the typical *Alsó-hegy Plateau* shaft caves.

Large quantities of bat guano, and some bat corpses suggest that bats use the cave as a dwelling place, but observational data are available only about greater horseshoe bat (*Rhinolophus ferrumequinum*) and lesser horseshoe bat (*Rhinolophus hipposideros*; Boldogh 2006). Occasionally, smaller animals fall into the top shaft of the cave.

3.19 Szabadság Cave

The third longest cave in the Aggtelek Karst is the *Szabadság (Freedom) Cave*, which is located northeast of the village of Égerszög, at the bottom of the *Dász doline*, 322 m above sea level (Szunyogh 2003). The currently known length of the cave is 3030 m, its vertical extent is 48 m. The cave is nearly horizontal, its average slope is only 1.3%.

The 900 m long section near the entrance, called the *Megvalósult álmok (Realized Dreams)* branch was formed in

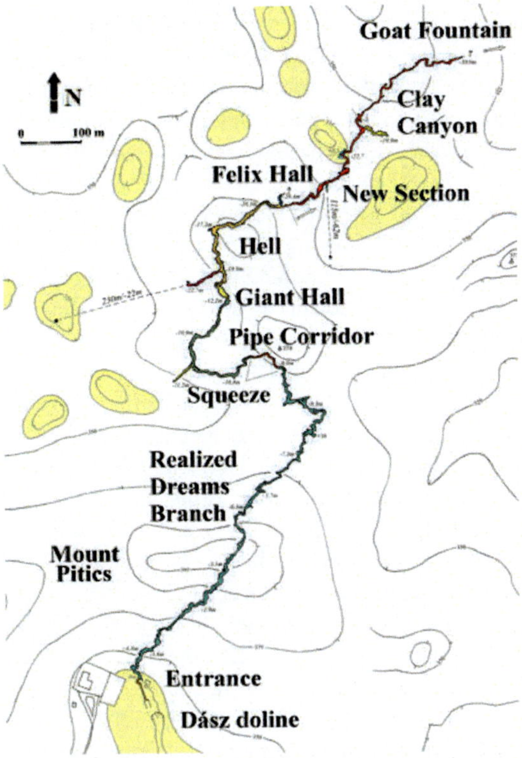

Fig. 30 Surface map of Szabadság Cave (National Cave Register)

the well-karstifiable, thick-bedded Wetterstein Limestone. Further sections are located in rocks less suitable for cave formation. For example (Fig. 30), the *Kuszoda (Squeeze)* was formed in the dark Gutenstein Dolomite, the *Csőfolyosó (Pipe Corridor)* and the *Óriás-terem (Giant Hall)* were formed in Lower Triassic laminated limestone and slate, while the *Pokol (Hell)* was formed in dark, shaly Gutenstein Limestone, which is densely woven by calcite veins (Balázs 1961). Between the current end point of the cave and the *Kecske-kút* (*Goat Fountain*), which is the source of the hydrogeological system of the *Szabadság Cave*, there are Lower Triassic (Campili) laminated slates and limestones.

The rock enclosing the *Szabadság Cave* became open karst at the end of the Pliocene due to the tectonic uplift of the area (Balázs 1961). It was at this time that the retreat of the *Jósva River valley* and the formation of the underground hydrological network connected to the *Jósva valley* began. In the area of *Pitics Mountain*, located above the *Szabadság Cave*, the formation of dolines, stream sinks and cavities began due to karst corrosion. Later, with the development of stream sinks, Pannonian gravel was transported into the cave from the covered karst area south of the open karst, contributing to the erosional work of the underground watercourses. As a result, a duality appeared in the image of the cave: the erosionally formed *Megvalósult álmok branch* became relatively spacious, while the *Gyöngy-folyó (Pearl River)*, which gained water from a larger but gravel-free, mostly open karst catchment area, remained a narrow, difficult-to-pass side branch. In the further stages of development, the *Szabadság Cave*, along with the enclosing rock mass, slightly tilted to the north, and its passages to the north were lowered to a somewhat deeper altitude than their original position. This movement is demonstrated by the gradual lowering of the former waterline notch compared to the current bottom level of the passage along the *Megvalósult álmok branch*, and by the parallel thickening of the sediments deposited at the passage bottom.

The exploration of the cave was started in 1952 by a brigade recruited from the inhabitants of *Égerszög* village. They digged out a plugged stream sink funnel at the deepest point of the *Dász doline*. The work was continued by the team of László Jakucs, but it was soon stopped due to the risk of collapse (Balázs 1961). The works were started again by the Kinizsi Cave Research Group of the Ministry of Food, led by Dénes Balázs. The cave was finally entered on November 15, 1954, and 2717 m length was explored that time. Subsequently, no significant new section was explored until on October 27, 1991, when the members of the cave research group of the Ferencváros Nature Friends Sport Club (led by József Frecska and Zoltánné Vidics) discovered a 561 m long passage by excavating the clogged end point of the cave (Vidics 1993).

Outstanding among the scientific research on the cave are the biological studies of Imre Loksa (Loksa 1959, 1961). He discovered two previously unknown cave-dwelling beetle species (*Thalassophüus Longicornis* and *Athaeta Spelaea*) in this cave.

The water entering the *Szabadság Cave* gets to the surface in a karst spring called *Kecske-kút* (at an altitude of 250 m asl) 1650 m from its stream sink (Balázs 1961). The catchment area is 2.3 km^2. Fluctuation of water discharge is small compared to other karst springs in the area: $Q_{max}/Q_{min} = 20$ (for comparison, in the case of the *Béke Cave*, $Q_{max}/Q_{min} = 400$). This small discharge fluctuation suggests that the underground passages are likely to be narrow, impeding and decelerating the flow of water. According to chemical analyses, the hardness of the water appearing in the spring is 22.5 dGH, and the dissolved calcium–magnesium ratio is 4. The average water temperature is 9.8 °C. Although the spring is permanent with an average discharge of 0.19 m^3/h and a maximum of 5.3 m^3/h, the stream in the *Szabadság Cave* flows only intermittently, for 30–60 days a year.

The structure of the system is simple, it consists of a single, winding main branch multi-level in some sections and a few small side branches connected to it. The cave can be divided into several sections based on its morphology. The first, 875 m long section is the *Megvalósult álmok branch* (Balázs 1961). Near the entrance the passage is meandering. Its remarkable features are chimneys extending to a height of 15–20 m, smaller and larger collapse chambers, siphons and siphon bypasses. Its sidewalls are sculpted by well-developed, sharply corroded waterline notches. On a section of about 400 m, above the relatively narrow, variable-height passage that includes the current stream bed, there is an upper floor consisting of a series of spacious, collapse chambers. After about 500 m there are some (now clogged) sinks. Subsequently, the morphology of the cave changes: it does not separate into floors, but it consists of a chain of straight fissures (canyons), which are 50–80 cm wide, 5–6 m high, long, and sometimes oblique due to tectonic settings.

The next morphologically distinct part of the cave is the 95 m long *Kuszoda*, which is a 60 m long, extremely low (on average 50 cm, but sometimes only 30 cm high), flat, mostly 120 cm wide (Fig. 31), elliptical section formed in Gutenstein Dolomite. Its ceiling is stalactite-free, and its bottom is covered with thick, hard-cemented gravel. During floods, this passage is filled with water up to a height of 10–20 cm.

The *Kuszoda* leads to the slightly wider, oval cross section, but also stalactite-free passage, the 300 m long *Cső-folyosó*, which is joined from the west by a periodically active side branch called the *Gyöngy-folyó*, which drains the water of the nearby sinkholes.

The *Gyöngy-folyó* ends at the *Óriás-terem*, 21 m long, 12 m wide and high. This room, which is orders of magnitude larger than the other rooms of the cave, developed in a laminated, low-strength rock, therefore, its morphology is primarily determined by the accumulation of collapse debris. This is where the all-encompassing, washy clay filling, typical of the further sections, appears.

After the *Óriás-terem* follows the 330 m long *Pokol*, where the cave becomes multi-level again. Its upper level is a spacious, oval cross-sectional passage. Into its bottom a meandering canyon (on average 5 m deep and 20–30 cm wide) with a constant water flow deepens, thus a keyhole cross section is formed. The bottom of this canyon is a bit wider, so it also forms a passable though extremely difficult level for humans. In some parts, even a third floor can be separated. The 80 m long *Víznyelős-oldalág (Sink Branch)* joins the *Pokol*. The *Pokol* ends with a sink, in which the water (periodically) flowing in the cave leaves the known passages to appear later at the *Kecske-kút Spring*.

Above the mentioned sink is the very narrow entrance of the *Új-rész (New Section)* discovered in 1991 (Nyerges 2003b). The *Új-rész* itself is a 561 m long inactive overpass filled with clay. Morphologically, it can be divided into three parts: the first rises slightly and continues to a relatively large (20 × 5 × 6 m) cavity called the *Félix-terem (Felix Hall)*. Beyond that, the *Új-rész* forms an evenly sloping, spacious passage with speleothems and then ends with a narrow, muddy corridor called *Agyagos-kanyon (Clay Canyon)*. Two smaller side branches (13 and 51 m long) are connected to it. From a morphological point of view, the *Sötétszürke-kürtő (Dark Grey Chimney)* should be highlighted, which approaches the surface upwards and forms a shaft extending downwards to a depth of 15 m.

The *Megvalósult álmok branch* is extremely rich in speleothems. Its stalactites represent almost all sizes and shapes, from straw stalactites to thicker "icicle" forms, but stalagmites 1–2 m high and 15–20 cm in diameter also occur (Kraus 1995). Radish-like bulbous stalactites, draperies, "stalactite lamps" and baldacchino canopies also occur. Many speleothems are alive, evolving, but the majority are highly shiny (i.e. recrystallized), and therefore no longer evolving specimens. The value of the speleothems of the cave is increased by the fact that they are in a natural state, practically untouched.

Helictites and tetarata formations, which appear frequently in some places, are also very spectacular. In the

Fig. 31 Szabadság Cave. **a** Longitudinal profile of Realized Dreams branch (National Cave Register); **b** Squeeze (photo Berentés); **c** Speleothems from the Realized Dreams branch (photo Berentés)

150 m section close to the entrance, popcorns also occur, richly covering the walls, the fallen boulders and the stalactites hanging from the ceiling. Their shape is similar to bunches of grapes and cauliflowers. The "booted dripstones" of the upper passages, where the popcorns grew on stalagmites, are also famous. They are mostly directed, patched and protrude from the surface towards the entrance. Commonly, popcorns cover only one side of the stalactites. In the curved passages, their position corresponds to the movement of the inward flowing air. The most significant limestone tuff formation in the *Szabadság Cave* is located in the *Csőfolyosó*, at the confluence of the *Gyöngy-folyó*.

Due to its geological, morphological and mineralogical values, it has been declared a highly protected cave since 1982 (including a surface protection zone of 210 ha).

The cave is a known bat habitat with observational data about greater horseshoe bat (*Rhinolophus ferrumequinum*), lesser horseshoe bat (*Rhinolophus hipposideros*), greater mouse-eared bat (*Myotis myotis*), whiskered bat (*Myotis mystacinus*), Daubenton's bat (*Myotis daubentonii*), and western barbastelle (*Barbastella barbastellus*; Boldogh 2006).

3.20 Szabó-pallag Shaft Cave

The *Szabó-pallag Shaft Cave*, also known as the *Owl Abyss*, on the plateau of the *Alsó-hegy Plateau* is located to the west from the *Szabó-pallag* hunting lodge. It opens ca. 300 m to the right of the road leading directly to the plateau. Its spectacular entrance opens at the side of a large doline, typical of the area, among characteristic rock formations (Fig. 32). The surroundings were once covered with pine forest, but they are now clear cut.

The length of the shaft cave is 950 m, it has a horizontal extent of only 51 m, and a vertical extent of 151 m (Nyerges 2003g). Caving is only possible for people skilled in cave roping technique. Its large, spectacular entrance has been open since ancient times. The first explorers reached a depth of -63 m in 1927 (Kessler 1927). This remarkable shaft cave reached its current length and depth by 1992 (Elekes et al. 1992).

The predominantly vertical passage system of large-sized shafts was formed in Middle Triassic Wetterstein Limestone (Piros 2006) at the intersection of several significant fractures.

We know two independent lowermost points of the cave. At the bottom of the entrance shaft, which is 65 m deep, we can slide through a squeeze to reach the former endpoint of the cave at a depth of −76 m. However, through the *Ablak (Window)* explored in 1978 (Kordos 1984) and some smaller shafts, we can get to the large, collapse block of *Mozdony (Locomotive)* decorated with speleothems. From here, there are two routes, the *Sáros (Muddy) Shaft* and the *Drúz Shaft*, through which we can get to the "common" base of the shafts, to a depth of about −121 m. After another descent of 30 m, there is the lowermost point of the shaft at −151 m depth. However, due to the rupture of the safety scaffolding, this is currently not available. Between the *Mozdony* and the lowermost point, there are several huge chimneys.

Fig. 32 Entrance to Szabó-pallag Shaft Cave (photo Egri)

In general, the speleothems of the cave passages are marvellous. The walls of the shafts are ornamented with flowstones of various shapes and colours, stalactites, stalagmites, and massive popcorn formations. The colour of the formations is predominantly red, mainly due to the red clay flowing on it and already incorporated during their growths. Because the image of the entire cave is mostly dominated by the red clay (Fig. 33), the trampling pollution caused by cavers is less conspicuous than in the other caves of the *Alsó-hegy Plateau*.

The formation of the cave is mainly due to the tectonics and the dissolving effect of the seeping waters. However, the almost vertical tilting of the rock layers can not be neglected either (Sárváry 1970; Kósa 1992; Zámbó 1998). Based on its size and complexity, the *Szabó-pallag Shaft Cave* is one of the most developed shaft caves in the region. This complex cave is the mesh of at least a dozen shafts. The vertical shafts are usually connected to each other by smaller or larger windows. Subsequent collapses also played a significant role in the formation of the cave. The enlarged fissures formed mainly along north–south tectonic strike fused in several places to create characteristic L and E shaped cross sections. In some places, the chimneys and shafts are separated by only a few decimetres of rock wall. From a genetical viewpoint, the zone of the *Szent (Holy) Shaft*, which joins the bottom of the *Drúz* and *Sáros Shafts* is especially interesting. The picture of this section is quite different from the rest of the cave. In this passage, the morphology, i.e. the

Fig. 33 Descent in a characteristic shaft of Szabó-pallag Shaft Cave (photo Egri)

scallops observed on the walls suggests rather a steeply sloping, stream cave section (Nyerges 2001). Presumably, this section was formed independently of the shaft, and their "connection" was random.

The cave is registered as a significant wintering and breeding location for bats. Observational data show the occurrence of a wide variety of bat species including lesser (*Rhinolophus hipposideros*) and greater horseshoe bats (*Rhinolophus ferrumequinum*), as well as greater mouse-eared bats (*Myotis myotis*), lesser mouse-eared bats (*Myotis blythii*), Bechstein's bats (*Myotis bechsteinii*), Natterer's bats (*Myotis nattereri*), Daubenton's bats (*Myotis daubentonii*), Geoffroy's bats (*Myotis emarginatus*), Alcathoe bats (*Myotis alcathoe*), brown long-eared bats (*Plecotus auritus*) and western barbastelles (*Barbastella barbastellus*; Boldogh 2006, 2007). In the entrance shaft of the cave, a pair of owls had previously nestled for decades.

3.21 Vass Imre Cave

The *Vass Imre Cave* is located N–NW from the village of Jósvafő, at the north end of the *Tohonya Valley*, which is a typical blind valley. The entrance opens in a sparsely undergrown, hornbeam-oak forest, 8–10 m from the edge of the forest, on rocky terrain. The cave is east–west oriented, 2185 m long, with 424 m horizontal extent and 56 m vertical extent.

The *Vass Imre Cave* was formed in Middle Triassic Wetterstein limestone and dolomite rocks, along NW–SE fractures. The cave has no natural entrance. The members of the Cave Research Group of the Budapest University of Technology explored the cave in 1954–55 by digging the flood spring funnel of the *Kis-Tohonya Spring*. It is a typical periodic spring cave, the passages of which are predominantly horizontal. Due to the long-term cave evolution, several levels formed in it as the erosion base level gradually subsided.

The cave belongs to the hydrological system of the *Kis-Tohonya Spring*, which is located at an altitude of 258 m asl, which collects the waters of the *Haragistya Plateau*. The 3.6 km^2 catchment area of the cave extends in the direction of the *Hosszú*. It is bordered by the *Nagy-Tohonya* catchment area in the north, the *Kecső Spring* catchment area in the west, and its southern boundary is probably along the ridge of the *Holt-kút Summit* and the *Mezna Summit*. The connection of the *Agancsos-rét Sink*, the *Papverme Sink* and the *Mustang Cave* with the *Kis-Tohonya Spring* was confirmed by water tracing studies.

The most frequently visited middle level of the cave can be characterized as a fossil, stalactite-rich former stream bed. Occasionally, during major floods, water may still appear here. Traces of the fossil stream bed can still be recognized in the upper levels (e.g. *Eldorádó*). The lowest level, the *Hádész, Tetarata corridor, Vizes-ág*, has the typical morphology of active stream beds (Fig. 34). Near the end point of the cave (*Eldorádó, Halál-torka, Hádész*), the levels were collapsed to each other, thus levels can be traversed vertically here to a significant extent. The parts of the cave with the highest vertical extent are the tributaries called *Tündérkert* and *Walhalla*, which end in upward developing chimneys. The present end point of the cave is a huge debris pile formed in the *Ciklopszok Csarnoka (Hall of Cyclopes)* along a NW–SE fault plane. From the debris, a stream with more or less constant water flows out.

Due to the diversity and beauty of the cave's formations, it stands out among the dripstone caves of Hungary. At the main passage level, the area around the *Narancs-zuhatag (Orange waterfall)* and the *Baradlai-cseppkőoszlop (Baradla dripstone column)* is the richest part from the viewpoint of speleothems, but unfortunately there are also many damaged, incomplete formations in this section that can also be visited by tourists. The more hidden, hard-to-reach, lesser known and visited upper passages (such as the *Vörös-tavi-ág, Sajt, Eldorádó*) still contain mostly untouched parts, and they are also the most beautiful parts of the cave. In addition

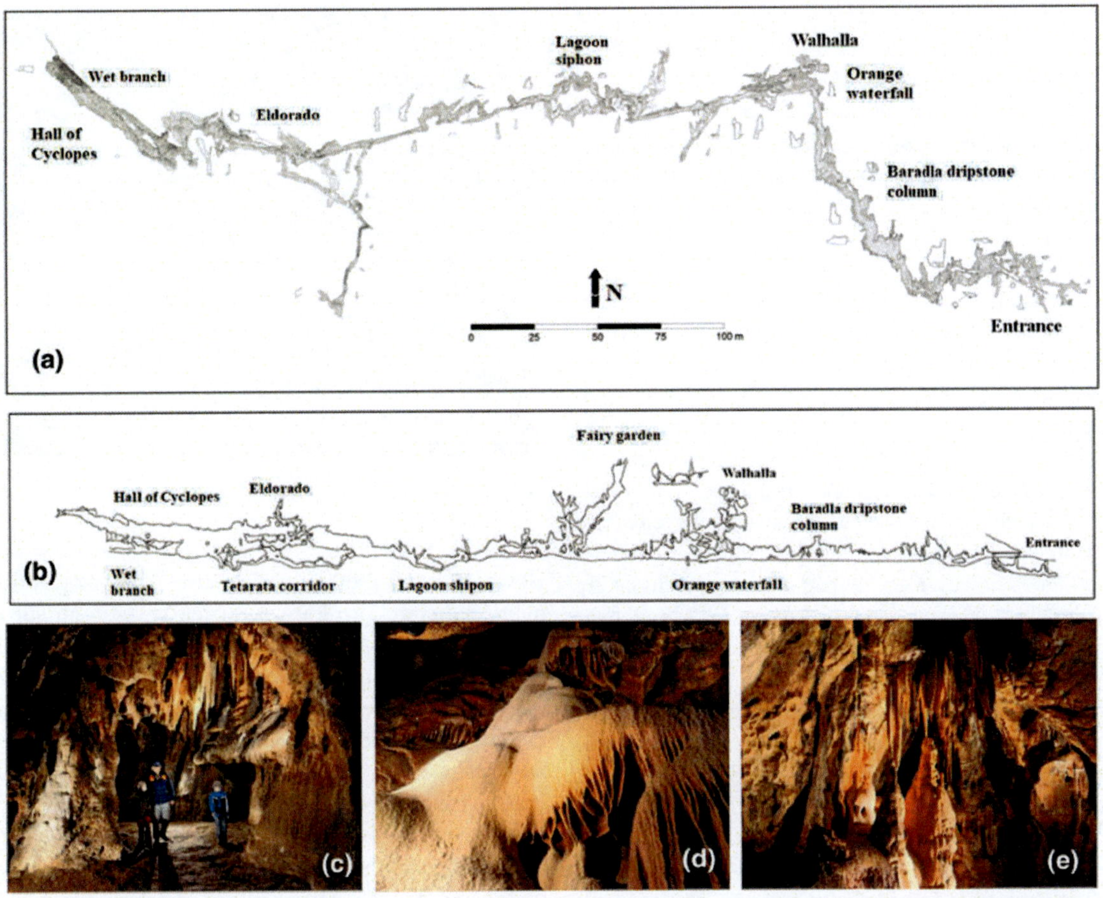

Fig. 34 Vass Imre Cave. **a** Plan view of the cave (National Cave Register); **b** longitudinal profile of the cave (National Cave Register); **c** in the main branch (photo Egri); **d** Orange Waterfall (photo Egri); **E** speleothem formations (photo Egri)

to the most common stalactites and stalagmites, dripstone columns, soda straws up to a metre in size, flowstones, curtains, draperies and canopies also appear. Of special value are the rich helictite coatings, which are white in colour, branched in shape and can reach lengths of up to 5–10 cm. Shelfstones are also valuable, as well as the collar speleothems developing in tetarata pools.

Despite its beauty, the cave has unfortunately been damaged by unprofesssional exploration, careless hiking, inappropriate attitude of certain hikers and explorers, and thus many formations and rare mineral deposits are now known only from earlier descriptions. Based on the geological environment of the cave and the results of the water tracings, it is likely that significant, unexplored passages may still be hidden behind the current endpoint, so the cave is worth further research.

Thanks to the studies carried out here, the cave is perhaps one of the most researched caves in Hungary from a scientific point of view. For decades after its discovery, significant and systematic research was carried out here in the framework of the activities of the Papp Ferenc Karst Research Station. The research station was founded in 1957 by the Budapest Technical University of Construction and Transport. Its aim was the climatological, hydrological, mineralogical and petrological research of the Aggtelek Karst, as well as the instrumental study of the interaction of karst processes. At various points in the cave, automatic measuring instruments were placed, the data of which (via a properly constructed cable system) could be continuously recorded at the research station. This complex cave laboratory was unique in its time, and a number of new scientific results have been produced here, which have been presented in a total of 95 publications (Maucha 1995). Measurements and interpretations included the rate of speleothem formation, the increase of helictites (Maucha 1968), lunisolar and seismic fissure size fluctuations (Maucha 1973; Gádoros 1969), the relationship between karst spring discharge fluctuations and crustal tidal phenomena (Maucha 1977), the relationship between spring discharge and Ca–Mg ratio of the water, the relationship between precipitation in the area and the yield and chemical composition of dripping water in the cave (Gádoros 1966a), as well as the penetration distance and time delay of changes in surface climate elements (Gádoros 1966b). Hungary's first radon emission studies were also carried out in the *Vass Imre Cave*.

From the faunistic point of view, the cave is one of the most significant caves in the Aggtelek and Slovak Karst World Heritage. Of the invertebrates, the most valuable are the *Eukoenenia spelaea* palpigrades, *Plusiocampa sp.* diplurans and *Typhloiulus sp.* millipedes. The cave is a documented bat habitat with observational data about western barbastelle (*Barbastella barbastellus*), greater horseshoe bat (*Rhinolophus ferrumequinum*), lesser horseshoe bat (*Rhinolophus hipposideros*), Bechstein's bat (*Myotis bechsteinii*), greater mouse-eared bat (*Myotis myotis*), whiskered bat (*Myotis mystacinus*), Daubenton's bat (*Myotis daubentonii*), Natterer's bat (*Myotis nattereri*), Brandt's bat (*Myotis brandtii*) and serotine bat (*Eptesicus serotinus*; Boldogh 2006). The inner parts of the cave are also likely to be used as wintering place by a significant number of bats, however these parts can be completely blocked from the exit by the *Lagúnás siphon* for even longer periods during major floods. As a result, a large number of bat corpses and skeletons can be found in the inner parts, which probably died because they could not reach the surface at the time of such closures. The 260 m long section of the cave can be visited in small groups with electric lighting, but without built-in concrete sidewalks.

3.22 Vecsembükk Shaft Cave

The *Vecsembükk Shaft Cave* on the *Alsó-hegy Plateau* is located ca. 2 km east–northeast from the *Szabó-pallag* hunting lodge, right next to the Hungary–Slovakia state border. Its entrance opens between characteristic rock formations in the side of a significant doline typical of the landscape. The surveyed length of the shaft cave is 900 m, its horizontal extent is only 28 m, while its vertical extent is 236 m (Nyerges 2003h). Caving is only possible for people skilled in cave roping technique.

Its large entrance with significant landscape value has been open since ancient times (Fig. 35). Its first explorers tried to descend into the deep as early as 1911 (Scholtz 1911), but only later, in 1927, did they reach a depth of 91 m (Kessler 1927), and then in 1969 to a depth of 178 m (Szenthe 1969). This extremely spectacular shaft cave is known today for a depth of 236 m, which was reached in 1970 by explorers (Szenthe 1970).

The predominantly vertical passage system of large-sized shafts was formed in Middle Triassic Wetterstein limestone (Piros 2006) at the intersection of several significant fractures. The lowermost point of the cave can be reached by a series of deep shafts, which are extremely spectacular in a Hungarian context and have a unique morphology.

Fig. 35 Large entrance to Vecsembükk Shaft Cave (photo Egri)

However, several other vertical cave sections parallel to the main shafts are also known, which are also unique in their size and attractions. At the lowest few metres, the passages cross crinoidal limestone beds (Müller 1971). Based on red clay samples taken at several points in the cave, significant quantities of material were found that could not be counted as limestone dissolution residues, such as dihexagonal quartz grains, metamorphic rock grains, quartz gravels, and limonite grains (Szenthe 1971). The dihexagonal quartz grains clearly indicate their origin from rhyolite tuff layers, whereas the quartz gravels suggest a former debris cover of the plateau (Szenthe 1971).

The spacious entrance shaft of the cave is 67 m deep, with almost daylight at its bottom. An additional 16 m deep shaft reaches the *shaft No. 90*, which is extremely rich in formations. From the bottom, or through the so-called *Új-rész (New Section)*, which opens at 30 m height from its side, we can reach the lowermost point of the Vecsem-bükk Shaft Cave at a depth of −236 m through several shafts. In several places, smaller or larger chimneys also join the main shafts, the largest of which is the *Hideg álom (Cold Dream)* section.

In general, the speleothems of the cave passages are really remarkable. From the entrance to the end, the walls of the shafts are ornamented with flowstones of various shapes and colours, stalactites, stalagmites, and massive popcorn formations (Fig. 36). Rare coralloid precipitate types can be studied from the lower zone of *shaft No. 90* to the deeper parts. The colour of the formations is red in many places, mainly due to the red clay flowing on it and already incorporated during their growths.

The formation of the cave is mainly due to the tectonics and the dissolving effect of the seeping waters. However, the almost vertical tilting of the rock layers can not be neglected either (Sárváry 1970; Kósa 1992; Zámbó 1998). Based on its size and complexity, the *Vecsembükk Shaft Cave* is one of

Fig. 36 In a large, formation-rich part of Vecsembükk Shaft Cave (photo Berentés)

the most developed shaft caves in the region. The large-depth vertical shafts are usually connected to each other by smaller or larger windows. In some places, the chimneys and shafts are separated by only a few decimetres of rock wall (Nyerges 2001). Subsequent collapses also played a significant role in the formation of the cave, and the debris material sometimes fills significant parts of certain section.

In order to clarify the possible hydrological connection between the cave and the springs at the foot of *Alsó-hegy Plateau*, water tracer tests were carried out in 1969 and in 1970. As a result, it was demonstrated that the tracer material was found in the *Zsámány well* 2 km north of the shaft cave and not far from it, the tracer appeared in the *Kör well*, too (Sárváry 1971).

The cave is a bat habitat with observational data about Bechstein's bat (*Myotis bechsteinii*), lesser mouse-eared bat (*Myotis blythii*), greater mouse-eared bat (*Myotis myotis*), Daubenton's bat (*Myotis daubentonii*), Geoffroy's bat (*Myotis emarginatus*), pond bat (*Myotis dasycneme*), Natterer's bat (*Myotis nattereri*), common pipistrelle (*Pipistrellus pipistrellus*), serotine bat (*Eptesicus serotinus*), western barbastelle (*Barbastella barbastellus*) and brown long-eared bat (*Plecotus auritus*; Boldogh 2006). The cave is also an important bat wintering location for several species (Boldogh 2006), thus, there is a restriction on visits during winter months.

4 Conclusions

Aggtelek Karst is one of the most prominent and complex representatives of the temperate zone, medium mountain karst terrains, where surface and underground forms are similarly diverse. Throughout the geological history, the area has undergone significant changes in both topography and climate. As a result, caves with different genetics and morphology occur at a high density, thus in 1995, the caves of the Aggtelek and Slovak Karst were jointly inscribed on the UNESCO World Heritage. The most significant part of the caves were created by meteoric water and shaped by cave streams. The *Baradla–Domica cave system* is the longest dripstone cave with an active stream in the temperate zone. Shaft caves represent another characteristic type (mainly on *Alsó-hegy Plateau*), but hypogenic caves also occur (mostly in *Esztramos Hill*). Cave deposits have an outstanding diversity in Aggtelek Karst. In addition to the most common stalactites, stalagmites, draperies and flowstones, rarer types are also present such as cave pearls, aragonite frostwork, popcorns, shelfstones, cave shields or helictites. The caves of Aggtelek Karst have an extraordinarily rich fauna with several endemic species. Besides, some caves (especially Baradla and *Hosszú-tető Cave*) have an archaeological significance, too. One of the geological curiosities of the area is that the only geological type section, which is located in a cave is found just here, in *Baradla Cave*. There are also several palaeontological sites of European importance, and the internationally recognized Middle Pliocene 'Estramontinum' vertebrate bio-stratigraphic stage acquired its name from *Esztramos Hill*. The caves of the area (especially Baradla) also played a decisive role in the history of cave research in Hungary, and are still in the focus of scientific research in several fields (geology, archaeology, biospeleology, karst morphology and hydrology, speleothem dating, climatology and cave therapy).

References

Adamkó P, Hegedűs G (2002) 118 óra a barlang fogságában (118 hours trapped in the cave). Budapest 111 p (in Hungarian)

Balázs D (1961) A Szabadság-barlang (The Freedom Cave) - Karszt és Barlang 2:61–75 (in Hungarian)

Bella P, Móga J (2014) A barlangrendszer alaktana (Morphology of the cave). In: Gruber P, Gaál L (eds) The Baradla-Domica cave system. Aggteleki Nemzeti Park Igazgatóság, Jósvafő, pp 73–92 (in Hungarian)

Bella P, Bosák P, Braucher R, Pruner P, Hercman H, Minár J, Veselský M, Holec J, Léanni L (2019) Multi-level Domica-Baradla cave system (Slovakia, Hungary): middle Pliocene-Pleistocene evolution and implications for the denudation chronology of the Western Carpathians. Geomorphology 327:62–79

Berényi Üveges J, Berényi Üveges I, Lovas Gy, Vid G (2004) Vizsgálatok a Baradla- és Béke-barlang kitöltéseiben (Investigation of sedimentary deposits in Baradla and Béke caves) Karsztfejlődés 9:311–320 (in Hungarian)

Bajomi D (1969) A Meteor-barlang faunisztikai vizsgálata(Research on the fauna of Meteor Cave). Karszt és Barlang II. pp 61–64 (in Hungarian)

Boldogh S (2006) The bat fauna of the Aggtelek National Park and its surroundings. Vespertilio 9–10:33–56

Boldogh S (2007) Földalatti denevérszállások az Aggteleki-karszt és a Galyaság területén (Underground bat habitats in Aggtelek Karszt and Galyaság. ANP Füzetek 3:164–185 (in Hungarian)

Borbás E, Kovács J, Fehér K, Vid G, Hatvani IG (2011) Water chemistry analysis in the sediment of Baradla Cave, Hungary. Cent Eur Geol 54:367–380

Boros L (1976) A VMTE Tektonik (időközben Lakatos László) Barlangkutató Csoport 1976. évi jelentése az évi munkájáról (1976 Annual Report of VMTE Tektonik Cave Research Group). MKBT Beszámoló 236 p (in Hungarian)

Bosák P, Hercman H, Kadlec J, Móga J, Pruner P (2004) Palaeomagnetic and U-series dating of cave sediments in Baradla Cave, Hungary. Acta Carsologica 33:219–238

Bögli A (1984) Flieβfazetten – ein karsthydrographisch wichtiges Merkzeichen des Flieβverhaltens von Karstwasser. Die Höhle 35(3–4):119–126

Csiki E (1903) Magyarországi új bogarak (Coleoptera nova ex Hungaria). Annales hist.-nat. Mus Nat Hung 1:441–446 (in Hungarian)

Czuppon G, Demény A, Leél-Őssy S, Óvari M, Molnár M, Stieber J, Kiss K, Kármán K, Surányi G, Haszpra L (2018) Cave monitoring in the Béke and Baradla caves (Northeastern Hungary): implications for the conditions for the formation cave carbonates. Int J Speleol 47:13–28

Demény A, Németh A, Kern Z, Czuppon G, Molnár M, Sz L-Ő, Óvári M, Stieber J (2017) Recently forming stalagmites from the Baradla Cave and their suitability assessment for climate - proxy relationships. Central Eur Geol 60:1–34

Denes G (1961) A Meteor-barlang feltárása (Exploration of Meteor Cave). Karszt- És Barlang 1961(2):83–85 (in Hungarian)

Denes G (1963) Az Alsó-hegy karsztjának hidrográfiai viszonyairól (Hydrography of Alsó-hegy Plateau)). Karszt- És Barlangkutatási Tájékoztató 1963(9):165 (in Hungarian)

Denes G (1964) Esztramosi-barlang (Esztramos Caves). Karszt- És Barlangkutatási Tájékoztató (4):79 (in Hungarian)

Dénes Gy (1961) A Meteor-barlang feltárása (Exploration of Meteor Cave). Karszt- És Barlang 1961(2):83–85 (in Hungarian)

Dénes Gy (1969) Újabb eredményes víznyelő-festés az Alsóhegyen (Successful tracing test on the Alsó-hegy). Karszt- és Barlangkutatási Tájékoztató 1969(2):3 (in Hungarian)

Droppa A (1961) Domica—Baradla, jaskyne predhistorického človeka. Šport, Bratislava, 1–151

Droppa A (1972) Príspevok k vývoju jaskyne Domica. Československý Kras 22:65–72

Dudich E (1932) Biologie der Aggteleker Tropfsteinhöhle Baradla in Ungarn. Wien, Speleolog Monographien, 12. XII+246

Elekes B, Nyerges A, Nyerges M, Rose G (1992) A Szabó-pallagi-zsomboly (Baglyok Szakadéka) kutatásának újabb eredményei (New results in the research of Szabó-Pallag Shaft Cave). Karszt És Barlang 1–2:3–8 (in Hungarian)

Farrant AR, Smart PL (2011) Role of sediment in speleogenesis; sedimentation and paragenesis. Geomorphology 134(1–2):79–93

Fekete L (1987) A Baradla-tetői-zsomboly (The Baradla-tető Shaft Cave). Föld és Ég (1987)1:2–4 (in Hungarian)

Ford DC (1965) The origin of limestone caverns: a model from the central Mendips Hills, England. Bull Nat Speleol Soc 27(4):109–132

Ford DC (2000) Caves Branch, Belize, and the Baradla-Domica System, Hungary and Slovakia. In: Klimchouk AB, Ford DC, Palmer AN, Dreybrodt W (eds) Speleogenesis. National Speleological Society, Huntsville, Evolution of Karst Aquifers, pp 391–396

Ford DC, Ewers RO (1978) The development of limestone cave systems in the dimensions of length and depth. Can J Earth Sci 15:1783–1798

Frivaldszky I (1865) Adatok a magyarhoni barlangok faunájához (Data for the fauna of Hungarian Caves). Mathematikai És Természettudományi Közlemények 3(1):17–53 (in Hungarian)

Gaál L (2014) Szerkezeti viszonyok (Tectonic conditions). In: Gruber P, Gaál L (eds) The Baradla-Domica Cave system. Aggteleki Nemzeti Park Igazgatóság, Jósvafő, pp 45–60 (in Hungarian)

Gaál L, Gruber P, Móga J (2014) A barlangrendszer szilárd halmazállapotú kitöltése (Solid deposits in the cave system). In: Gruber P, Gaál L (eds) The Baradla-Domica Cave system. Aggteleki Nemzeti Park Igazgatóság, Jósvafő, pp 117–158 (in Hungarian)

Gádoros M (1966a) Measuring instruments for observation of drop-water in caves. Int Kong für speleologie. Abhandlungen, Part. II:115–128

Gádoros M (1966b) Electrical equipment for the transmission of climathological and hydrological data from Vass Imre Cave. Hung Nat Speleol Bull 28(2):80–89

Gádoros M (1969) Registrirung red Litoklasenbewegung unter der lunasolaren Einflüssen. Int. Kong. für speleologie. Abhandlungen, Bd. 3. Stuttgart, p. S. 42/1–4

Gádoros M (1971) A Complex investigation of the Nagy-Tohonya spring of Jósvafő. Karszt És Barlangkutatás 6:79–102

Galbács G, Kevei-Bárány I, Szőke E, Jedlinszki N, Gornushkin IB, Galbács MZ (2011) A study of stalagmite samples from Baradla Cave (Hungary) by laser induced plasma spectrometry with automatic signal correction. Microchem J 99:406–414

Gruber P (2003) Hosszú-tetői-barlang (Hosszú-tető Cave). In: Székely K (ed) Magyarország fokozottan védett barlangjai (Highly protected caves in Hungary) Mezőgazda Kiadó, pp 63–64 (in Hungarian)

Gruber P, Kereskényi E, Koltai G, Bárany-Kevei I (2012) Data on the hydrogeographical conditions of Baradla Cave: seepage and drip waters. Acta Climatologica Et Chorologica 46:125–134

Gyuricza G, Gaál L (2014) A barlangrendszer keletkezése és fejlődése (The formation and development of the cave system). In: Gruber P, Gaál L (eds) The Baradla-Domica Cave System. Aggteleki Nemzeti Park Igazgatóság, Jósvafő, pp 93–116

Gyuricza G, Sásdi L (2009) A Baradla-barlangrendszer kialakulásának kérdései a tágabb környezet földtani fejlődésének tükrében (Questions related to the formation of Baradla Cave system in the light of geologic evolution of the larger surroundings). Földtani Közlöny 139:83–92

Hazslinszky T (1965) Az északborsodi Alsóhegy karsztjának néhány hidrográfiai kérdése (Some hydrographic questions of the Alsó-hegy Karst). Hidrológiai Közlöny 6:259–266 (in Hungarian)

Hill CA, Forti P (1997) Cave minerals of the World. Nat Speleol Soc, Michigan, 463 p

Jakucs L (1951) Vízföldtani megfigyelések a Gömöri-karszton (Hydrogeologic observations on Gömör Karst). Földtani Közlöny 81(10–12):445 (in Hungarian)

Jakucs L (1953) A Béke-barlang felfedezése (Exploration of the Béke Cave). Művelt nép Kiadó. Budapest. 94 p (in Hungarian)

Jakucs L (1971) A karsztok morfogenetikája (Morphogenetics of Karst Regions). Akadémiai kiadó, Budapest, 310 p (in Hungarian)

Jakucs L (1975) Aggteleki karsztvidék. Útikalauz (Aggtelek Karst. Guide). Sport Kiadó, 407 p (in Hungarian)

Jakucs L (1977) Morphogenetics of karst regions, variants of karst evolution. Akadémiai kiadó. Budapest, 283 p

Kessler H (1927) A Vecsembükki és az Almási zsombolyok első bemászása (First descents to Vecsembükk and Almási shaft caves). Turistaság És Alpinizmus 17(7–8):123–127 (in Hungarian)

Kessler H (1932) Aggteleki felfedezések (Explorations at Aggtelek). Turistaság És Alpinizmus 22(10):236–243 (in Hungarian)

Kessler H (1938) Az aggteleki barlangrendszer hidrográfiája (Hydrography of Aggtelek cave system). Földr Közl 66(1–3):1–30 (in Hungarian)

Kirchknopf M (1966) Légúti betegségek klimatikus kezelése a Béke-barlangban (Climatic treatment of respiratory diseases in the Béke Cave). Karszt És Barlang 1966(1):9–10

Koleszár K (2000) A bányászat során föltárult esztramosi barlangok megismerésének története a kezdetektől 1965-ig (Cave exploration history of Esztramos caves explored during mining from the beginnings to 1965). Karsztfejlődés 5:293–308 (in Hungarian)

Kordos L (1973) Adatok az Esztramos barlangjainak ismeretéhez (Some data for Esztramos caves). Karszt És Barlang 1973(7–8):11–12 (in Hungarian)

Kordos L (1974) Az Esztramos barlanggenetikai, hegységszerkezeti és üledékföldtani vizsgálata (Genetical, Tectonic and Sedimentary Research of Esztramos Hill). Karszt És Barlang 1974(1):22–23 (in Hungarian)

Kordos L (1984) Magyarország barlangjai (Caves of Hungary). Gondolat Könyvkiadó, Budapest, 253–255 (in Hungarian)

Kovács S, Less Gy, Piros O, Róth L (1988) Az Aggtelek–Rudabányai-hegység triász formációi (Triassic formations of Aggtelek-Rudabánya Mountains). A Magyar Állami Földtani Intézet Évi Jel. 1986-ról, Budapest, 19–43 (in Hungarian)

Kósa A (1963) A szögligeti Rejtek-zsomboly (Rejtek Shaft Cave in Szögliget). Karszt És Barlang 1963(2):66–70 (in Hungarian)

Kósa A (1976) A Rejtek-zsomboly (Rejtek Shaft Cave). Élet és Tudomány (in Hungarian)

Kósa A (1992) Nyolcvan év az Alsó-hegyen. Még egy szó a zsombolyokról (80 years on Alsó-hegy Plateau. One more word about shaft caves). Karszt és Barlang 1992(1–2):9–14 (in Hungarian)

Kraus S (1995) A Szabadság-barlang karbonátos kiválásai (Carbonate precipitates of the Freedom Cave) - Karszt és Barlangkutatás. MKBT, Budapest 10:161–164 (in Hungarian)

Kraus S (1997) A legnagyobb esztramosi barlangok kalcium-karbonát kiválásairól (On the calcium-carbonate precipitations of the largest Esztramos caves). In: Szakáll S, Papp G (eds) Az Esztramos-hegy ásványai. Topographia Mineralogica Hungariae, 1997(5):75–76 (in Hungarian)

Kraus S (2003) Rákóczi 1. sz. barlang. In: Székely K (ed) Magyarország fokozottan védett barlangjai. (Highly protected caves in Hungary) Mezőgazda Kiadó, pp 93–96

Lauritzen SE, Lundberg J (2000) Solutional and erosional morphology. In: Klimchouk AB, Ford DC, Palmer AN, Dreybrodt W (eds) Speleogenesis. Nat Speleol Soc, Huntsville, Alabama, USA, Evolution of Karst Aquifers, pp 408–426

Less G (1998) Földtani felépítés (Geology of the area). In: Baross G (ed) Az Aggteleki Nemzeti Park, Mezőgazda Kiadó, pp 26–41 (in Hungarian)

Lénárt L (1997) Az Esztramos-hegy hidrogeológiai és szpeleológiai kutatásának eredményei. (Results of the hydrogeological and speleological research of Esztramos Hill). In: Szakáll S, Papp G (eds) Az Esztramos-hegy ásványai. Topographia Mineralogica Hungariae 1997(5):55–74 (in Hungarian)

Loksa I (1959) Das Vorkommen einer neuen Höhlencollembola (Folsomia antricola n. sp.) und von Folsomia multiseta Stach in Ungarn. Opuscula Zoologica 3(1):37–42

Loksa I (1961) Ökologisch-faunistische Untersuchungen in der Freiheitshöhle bei Égerszög. Acta Zoologica Academiae Scientiarum Hungaricae 191(7):219–230

Majoros Z (1986) Esztramosi barlangok (Esztramos caves). A Nehézipari Műszaki Egyetem Közleményei, I. sorozat, Bányászat 33(1–4):189–195 (in Hungarian)

Maucha L (1968) Contribution on the origin of extremic concretions. Karszt És Barlangkutatás 5:97–100

Maucha L (1973) Instrumental measurements of crust deformations causing tidal phenomena of karst waters. Stud Mater Energy Flows Earth. Akadémiai kiadó. Budapest 203–206

Maucha L (1977) Study of movements of karst waters and karstic rocks. Ann Geophys 33:151–156

Maucha L (1995) A Vass Imre-barlangban végzett tudományos kutatások eredményeinek összefoglalása (Summary of scientific research carried out in Vass Imre Cave). Karszt És Barlangkutatás 10:35–52 (in Hungarian)

Maucha L (1998) Az Aggteleki-hegység karszthidrológiai kutatási eredményei és zavartalan hidrológiai adatsorai (Results of the karst hydrological research on Aggtelek Karst and undisturbed hydrological time series). VITUKI, Budapest, 417 p (in Hungarian)

Móga J (2002) A tornai Alsó-hegy felszínalaktani vizsgálatának új eredményei (New results of the geomorphological research of Alsó-hegy Plateau). Karszt És Barlang 1998–1999:101–102 (in Hungarian)

Müller P (1971) Nagy Megalodontidae-maradvány az Alsó-hegyről (A large Megalodontidae find from Alsó-hegy Plateau). Karszt-És Barlangkutatási Tájékoztató 1971(4/16):18 (in Hungarian)

Nyáry J (1881) Az aggteleki barlang mint őskori temető (Aggtelek Cave as a prehistoric cemetery), MTA, Könyvkiadó hivatala, Budapest, 179 p (in Hungarian)

Nyerges A (1997) Az Alsó-hegy zsombolyainak felszín alatti morfológiája (Underground morphology of Alsó-hegy shaft caves). MSc Thesis, Manuscript (in Hungarian)

Nyerges A (2001) Az Aggteleki-karszt zsombolyai (Shaft caves of Aggtelek Karst). Karsztfejlődés 6:267–268 (in Hungarian)

Nyerges A (2003a) Almási-zsomboly (Almási Shaft Cave). In: Székely K (ed) Magyarország fokozottan védett barlangjai (Highly protected caves in Hungary) Mezőgazda Kiadó, pp 58–60 (in Hungarian)

Nyerges A (2003b) A Szabadság-barlang felmérése (Survey of the Freedom Cave) - KvVM TvH BTO, Manuscript (in Hungarian)

Nyerges A (2003c) Frank-barlang. In: Székely K (ed) Magyarország fokozottan védett barlangjai. (Highly protected caves in Hungary) Mezőgazda Kiadó, pp 61–62 (in Hungarian)

Nyerges A (2003d) Kopaszgaly-oldali 2. sz. víznyelőbarlang. In: Székely K (ed) Magyarország fokozottan védett barlangjai. (Highly protected caves in Hungary) Mezőgazda Kiadó, pp 65–67. old (in Hungarian)

Nyerges A (2003e) Meteor-barlang. In: Székely K (ed) Magyarország fokozottan védett barlangjai. (Highly protected caves in Hungary) Mezőgazda Kiadó, pp 70–73 (in Hungarian)

Nyerges A (2003f) Rejtek-zsomboly. In: Székely K (ed) Magyarország fokozottan védett barlangjai. (Highly protected caves in Hungary) Mezőgazda Kiadó, pp 74–76 (in Hungarian)

Nyerges A (2003g) Szabó-pallagi-zsomboly. In: Székely K (ed) Magyarország fokozottan védett barlangjai. (Highly protected caves in Hungary) Mezőgazda Kiadó, pp 76–80 (in Hungarian)

Nyerges A (2003h) Vecsembükki-zsomboly. In: Székely K (ed) Magyarország fokozottan védett barlangjai. (Highly protected caves in Hungary) Mezőgazda Kiadó, pp 82–80 (in Hungarian)

Nyerges A (2003i) Rákóczi 2. sz. barlang. In: Székely K (ed) Magyarország fokozottan védett barlangjai. (Highly protected caves in Hungary) Mezőgazda Kiadó, pp 96–98 (in Hungarian)

Pasini G (1967) Nota preliminare sul ruolo speleogenetico dell'erosione "antigravitativa." Le Grotte D'italia 4(1):75–90

Pasini G (1975) Sull'importanza speleogenetica dell' "Erosione antigravitativa". Atti del Seminario Internazionale di Speleogenesi, Varenna (Como, Italy), 1972. Le Grotte d'Italia, 4(4):297–322

Petényi SJ (1844) Pár szó az emlősökről általában és a magyar honiakról különösen (Some words about mammals in general and especially about Hungarian types). Magyar Orvosok és Természetvizsgálók IV. (pécsi) Nagygyűlésének munkálatai, Pest (in Hungarian)

Piros O, Szilágyi F, Borka Zs (1989a) Aggtelek–Rudabányai-hegység, Baradla–barlang, Vetődéses terem, Gutensteini Mészkő Formáció (Aggtelek-Rudabánya Mountains, Baradla Cave, Fault Hall,

Gutenstein Limestone Formation). Magyarország geológiai alapszelvényei 112. Magyar Állami Földtani Intézet, Budapest (in Hungarian)

Piros O, Szilágyi F, Borka Zs (1989b) Aggtelek–Rudabányai-hegység, Jósvafő, Baradla-barlang, Színpad, Gutensteini Mészkő Formáció (Aggtelek-Rudabánya Mountains, Baradla Cave, Theater, Gutenstein Limestone Formation). Magyarország geológiai alapszelvényei 118. Magyar Állami Földtani Intézet, Budapest (in Hungarian)

Piros O, Szilágyi F, Borka Z (1989c) Aggtelek–Rudabányai-hegység, Jósvafő, Baradla-barlang, Sárkányfej, Steinalmi Mészkő Formáció (Aggtelek-Rudabánya Mountains, Baradla Cave, Dragon's Head, Steinalm Limestone Formation). Magyarország geológiai alapszelvényei 117. Magyar Állami Földtani Intézet, Budapest (in Hungarian)

Piros O, Szilágyi F, Borka Z (1989d) Aggtelek–Rudabányai-hegység, Baradla-barlang, Óriástermi víznyelő, Steinalmi Mészkő Formáció (Aggtelek-Rudabánya Mountains, Baradla Cave, Giant's Hall sink, Steinalm Limestone Formation). Magyarország geológiai alapszelvényei 111. Magyar Állami Földtani Intézet, Budapest (in Hungarian)

Piros O, Szilágyi F, Borka Z (1989e) Aggtelek–Rudabányai-hegység, Jósvafő, Baradla-barlang, főág 5700 m (Aggtelek-Rudabánya Mountains, Baradla Cave, 5700 m). Magyarország geológiai alapszelvényei 124. Magyar Állami Földtani Intézet, Budapest (in Hungarian)

Piros O (1993) Gutensteini Formáció (Gutenstein Formation). In: Haas J (ed) Magyarország litosztratigráfiai alapegységei. MÁFI, Budapest, pp 168–169 (in Hungarian)

Piros O, Kovács S (1993) Wettersteini Mészkő Formáció (Wetterstein Limestone Formation). In: Haas J (ed) Magyarország litosztratigráfiai alapegységei. Budapest, pp 172–175 (in Hungarian)

Piros O (2002) Anisian to Carnian carbonate platform facies and dasycladacean biostratigraphy of the Aggtelek Mts, Northeastern Hungary. Acta Geol Hung 45:119–151

Piros O (2006) Földtani természeti értékek az Aggtelek–Rudabányai--hegység területén (Geological natural values in the area of Aggtelek-Rudabány Mountains). In: Szentpétery I, Less Gy (eds) Az Aggtelek–Rudabányai-hegység földtana. Magyarázó az Aggtelek–Rudabányai-hegység 1988-ban megjelent 1:25 000 méretarányú fedetlen földtani térképéhez. Magyar Állami Földtani Intézet, pp 73–77 (in Hungarian)

Plózer I (1974) A magyarországi búvár-barlangkutatás története és bibliográfiája (1908–1973) (History and bibliography of Hungarian cave diving research). Karszt És Barlang 1974(2):59–60 (in Hungarian)

Rezi Kató G (2003) Szögliget, Hosszú-tető, Julcsa-barlang. In: Régészeti Kutatások Magyarországon 275. (Archaeological Investigations in Hungary), KÖH, Budapest, p 227 (in Hungarian)

Rezi Kató G (2004) Szögliget, Hosszú-tető. In: Régészeti Kutatások Magyarországon 292. (Archaeological Investigations in Hungary), KÖH, Budapest, p 280 (in Hungarian)

Rezi Kató G (2005) Szögliget, Hosszútetői-barlang. In: Régészeti Kutatások Magyarországon 325. (Archaeological Investigations in Hungary), KÖH, Budapest, p 280 (in Hungarian)

Rezi Kató G (2014) Baradla a történelem előtti időkben (Baradla in prehistoric times). In: Gruber P, Gaál L (eds) The Baradla-Domica Cave System. Aggteleki Nemzeti Park Igazgatóság, Jósvafő, 327–354 (in Hungarian)

Roth Z (1940) Vývojový vztah jeskyně Baradel k jeskyni Domica v Jihoslovenském Krase. Věstník Královské české společnosti nauk, Praha, 1–9

Salamon G, Dányi L, Angyal D, Balázs G, Forró L (2014) A Baradla gerinctelen faunája (Invertebrate fauna of Baradla). In: Gruber P, Gaál L (eds) The Baradla-Domica Cave System. Aggteleki Nemzeti Park Igazgatóság, Jósvafő, pp 279–306 (in Hungarian)

Sárváry I (1970) A zsombolygenetika kérdéseiről (Some questions of shaft cave genetics). Karszt És Barlang 1970(1):5–14 (in Hungarian)

Sárváry I (1971) Víznyomjelzés az Alsó-hegy zsombolyaiban (Water tracing in the shaft caves of Alsó-hegy Plateau). Karszt És Barlang 1971(1):25–29 (in Hungarian)

Sásdi and Szilágyi (1986) Víznyomjelzéses vizsgálatok az Aggteleki-karszton (Water tracing on the Aggtelek Karst). Karszt és Barlang, I. pp 33–38 (in Hungarian)

Scholtz PK (1911) Beszámoló az 1911. évi június hó 10–13 közötti tett expedícióról a szilasi fennsíkon. Napló (Report about the expedition on Szila Plateau on 10–13, June, 1911. Diary), Manuscript (in Hungarian)

Siklósy Z, Demény A, Sz L-Ő, Szenthe I, Lauritzen SE, Shen CC (2011) A cseppkövek kormeghatározása és azok paleoklimatológiai jelentősége (The dating of stalagmites and their palaeoclimatological significance). Földtani Közlöny 141(1):73–88 (in Hungarian)

Stieber J, Leél-Őssy Sz (2015) Megváltozott vagy csak visszaváltozott a Béke barlang klímája? (Had the climate of Béke cave changed to a new state or returned to a previous state?) Karsztfejlődés 20:263–282 (in Hungarian)

Szakáll S, Kovács Á (1997) Az esztramos-hegyi vasérctelep ásványai. (Minerals of the Esztramos Hill iron ore deposit) In: Szakáll S, Papp G (eds) Az Esztramos-hegy ásványai. Topographia Mineralogica Hungariae 1997(5):131–144 (in Hungarian)

Szentes G (1963) A bódvaszilasi Meteor-barlang környékének kőzetföldtani viszonyai (Geology of Meteor Cave near Bódvaszilas). Karszt És Barlang 1963(2):62–64 (in Hungarian)

Szentes G (1964) A Bódvaszilas környéki karszterület genetikai kérdései (Genetic questions of the karst area around Bódvaszilas). Karszt És Barlang 1964(1):21–24 (in Hungarian)

Szentes G (1965) Földtani, tektonikai és genetikai vizsgálatok a Baradla-és a Béke-barlangban (Geological, Tectonic and Genetical Investigations in Baradla and Béke Caves). Karszt És Barlang 1965 (2):71–78 (in Hungarian)

Szenthe I (1969) 178 m mély a Vecsembükki-zsomboly (The Vecsembükk Shaft Cave is 178 deep). Karszt- És Barlangkutatási Tájékoztató 1969(7):3 (in Hungarian)

Szenthe I (1970) Sikeres feltáró eredmények a Vecsembükki-zsombolyban (Succesful explorations in Vecsembükk Shaft Cave). Karszt És Barlang 1970(1):15–16 (in Hungarian)

Szenthe I (1971) Vízföldtani vizsgálatok a Vecsembükki-zsombolyban (Hydrogeological investigations in Vecsembükk Shaft Cave). Karszt És Barlang 1971(2):57–60 (in Hungarian)

Székely K (2003) Bába-völgyi 2. sz. víznyelő barlangja. In: Székely K (ed) Magyarország fokozottan védett barlangjai (Highly protected caves in Hungary) Mezőgazda Kiadó, pp 60–61 (in Hungarian)

Székely K (2014) A Baradla története (History of Baradla). In: Gruber P, Gaál L (eds) The Baradla-Domica Cave System. Aggteleki Nemzeti Park Igazgatóság, Jósvafő, pp 367–420 (in Hungarian)

Szilvássy G (1965) A Vámőrségi Barlangkutató Csoport 1964 évi beszámolója (1964 Annual Report of Customs Guards' Cave Research Group). Karszt- És Barlangkutatási Tájékoztató 1965(1–2):16–18 (in Hungarian)

Sztrókay K (1959) Mineralogical observations from Aggtelek Cave. Földtani Közlöny 89:280–285 (in Hungarian)

Szunyogh G (1994) A comparative analysis of the meanders in caves in the vicinity of Aggtelek. In: Conference on the Karst and Cave Research Activities of Educational and Research Intitutions in Hungary, 1993. Budapest-Szombathely, pp 55–58

Szunyogh G (1997) A review and evaluation of the speleological values of the Béke Cave Aiming at a Comprehensive Scientific Research Program. In: Proceedings of the research, conservation, management conference. Aggtelek National Park Directorate 2:241–247

Szunyogh G (2003) Szabadság-barlang (The Freedom Cave) In: Székely K (ed) Magyarország fokozottan védett barlangjai (Highly protected caves in Hungary). Mezőgazda Kiadó, Budapest, pp 51–53 (in Hungarian)

Szunyogh G, Kisbán J (2004) Béke-barlang. Komlós patak földalatti útja. (Béke Cave. The underground path of the Komlós stream). Aggteleki Nemzeti Park Igazgatóság, Jósvafő, 140 p (in Hungarian)

Takácsné Bolner K (2003) Esztramosi Földvári Aladár-barlang. In: Székely K (ed) Magyarország fokozottan védett barlangjai. Mezőgazda Kiadó, pp 89–92 (in Hungarian)

Telbisz T, Vigassy T, Zámbó L (1999) Variances of karst corrosion on the basis of differences in the solution of Ca-and Mg-carbonates. In: IGU international symposium on nature conservation and sustainable development on Karst Terrains, Budapest-Miskolc, 5–9 September, pp 193–200

Uhrin M, Boldogh S, Bücs S, Paunović M, Miková E, Juhász M, Csősz I, Estók P, Fulín M, Gombkötő P, Jére C, Barti L, Karapandža B, Matis Š, Nagy ZL, Szodoray-Paradi F, Benda P (2012) Revision of the occurrence of *Rhinolophus euryale* in the Carpathian region, Central Europe. Vespertilio, 16, Praha – Revúca, 289–328

Velledits F, Cs P, Blau J, Senowbari-Daryan B, Kovács S, Piros O, Pocsai T, Szügyi Simon H, Dumitrica P, Pálfy J (2011) The oldest Triassic Platform margin reef from the Alpine-Carpathian region (Aggtelek, NE Hungary): Platform evolution, reefal biota and biostratigraphic framework. Riv Ital Paleontol Stratigr 11762:221–268

Veress M (2012) New data on the development of the Baradla Cave (Hungary, Aggtelek Karst) Acta Carsol 40(2–3):193–204

Veress M, Unger Z (2015) Baradla-domica: large cave system on the Hungarian-Slovak border. In: Lóczy D (ed) Landscapes and Landforms of Hungary, World Geomorphological Landscapes. Springer International Publishing, Cham, pp 167–175

Vidics Z (1993) Új feltárás a Szabadság-barlangban (New exploration in the Freedom Cave). Karszt És Barlang 1993(1):59–60 (in Hungarian)

Zámbó L (1998) Az Aggteleki-karszt felszínalaktani jellemzése (Geomorphological description of Aggtelek Karst). In: Baross G (ed) Az Aggteleki Nemzeti Park, Mezőgazda Kiadó, pp 70–95 (in Hungarian)

Zámbó L, Ford DC (1997) Limestone dissolution processes in Béke doline, Aggtelek National Park, Hungary. Earth Surf Proc Land 22:531–543

Zámbó L, Ford D, Telbisz T (2002) Baradla-barlangi cseppkőkoradatok a késő-negyedidőszaki klímaingadozások tükrében (Speleothem age data from Baradla Cave compared to Late Quaternary climate changes). Földtani Közlöny 132:231–238 (in Hungarian)

The Caves of the Bükk Mountains

Szilárd Vetési-Foith, László Lénárt, Szabolcs Leél-Őssy, and Magdolna Virág

Abstract

In this chapter, we present the most significant caves in the Bükk Mountains, which are part of the mountain range Északi-középhegység, without the need for completeness (since the Bükk Mountains is the richest in caves in Hungary). By the selection, we aimed to include the most significant caves based on the most representative cave types and sizes, speleothem or archaeological values of the mountains. We review the genetic types of the caves in the mountains and then characterize the selected caves in alphabetical order. We discuss their geographical location, enclosing rock, their genetics, their speleothem and, where necessary and important, the hydrology of cave systems, as well as the archaeological values of these caves.

Keywords

Bükk Mountains • Karstic cave • Ponor cave • Cave system • Cave genetics • Cave morphology

S. Vetési-Foith (✉)
University of Pécs, Pécs, Hungary

L. Lénárt
Department of Hydrogeology and Engineering Geology, University of Miskolc, Miskolc-Egyetemváros, Miskolc, 3515, Hungary
e-mail: hgll@uni-miskolc.hu

S. Leél-Őssy
Department of Physical and Applied Geology, Eötvös Loránd University, Pázmány Péter sétány 1/C, Budapest, 1117, Hungary
e-mail: losz@caesar.elte.hu

M. Virág
Department of Geology, Eötvös Loránd University, Pázmány P. stny. 1/C, Budapest, 1117, Hungary

1 Introduction

The Bükk Mountains are the richest caves in Hungary. According to their size, apart from some large caves in the Aggtelek Karst and the Buda Hills, not only the longest caves are found here, but Hungary's deepest cave as well. According to the Hungarian cave register, 1157 of 4177 caves in the country are situated here, and 52 of the highly protected caves (145) are in this area. Of the 104 archaeologically protected caves, 59 are in the Bükk Mountains. In terms of dimensions, there are 11 caves longer than 1000 m and also 11 which are deeper than 100 m. The total length of the caves is more than 66 km. In terms of their size, the length of the István Lápai Cave (8700 m) and the depth of the Bányász Cave (300 m) should be emphasized. Among the caves of these mountains, we find ones that are used for tourism, spa and water extraction. The genetic cave types of the mountains, in the order of their frequency of occurrence are the following:

- Active (periodically active) and inactive spring caves (Tóth 1983, 1984; Tóth and Fejes 1984), such as the Kecske-lyuk Cave and the Istállós-kői Cave.
- Active and inactive ponor caves (Hevesi 2002), such as the Bolhás Ponor Cave. These are often complex systems. The nearly vertical entrance sections are crossed by almost horizontal passages, some of which are of phreatic origin. This phenomenon occurs only in the Bükk Mountains of the karsts of Hungary.
- Shaft caves that may be the continuation of the former ponors (Dénes 1971; Hevesi 1980, 2002), or, according to the analogies of Aggtelek (Sárváry 1970), epikarstic passages, as well as, fractures and vadose shafts formed along the bed plane, such as the Kőrislyuk Cave, based on the analogies of the Bakony Mountains (Veress 2019).
- Cave remnants (collapsed caves) can be former ponors or remnants of former spring caves, according to Hevesi (1986a), such as the Három-kút Cave.

M. Veress and Sz. Leél-Őssy (eds.), *Cave and Karst Systems of Hungary*,
Cave and Karst Systems of the World, https://doi.org/10.1007/978-3-030-92960-2_11

- Catavothrons that can be inactive (along previous karst water levels, Sásdi 1997) or can be active (along current karst water levels), such as the Gyertán-völgy Ponor Cave (Hevesi (1986b).
- Caves exposed to hot water (Lénárt 2011), such as the Tavas Cave in Miskolctapolca.
- Caves discovered by mining, such as the caves of the Alsó-kőbánya quarry in Miskolctapolca (Lénárt 1984b).
- Calcareous sinter caves such as the Anna Cave (Lénárt 1981b).
- Destroyed caves, gorges, such as the estuary gorge section of the Szarvaskút-Csúnya Valley (Hevesi 1986b, 2019).
- Through-caves can be mentioned also in the Bükk Mountains such as the Szivárvány-Sebes Cave System (Sűrű and Piri 2014).

Due to its intermittent uplift, the spring caves of the Bükk Mountains form spring cave levels (two inactive and one active below them, Jánossy 1979, Hevesi 2002, see chapter "The Geomorphic Evolution of Karsts and Karstic Surfaces"). This is unique in terms of the karsts in Hungary.

Inactive caves at inactive cave levels, due to their early formation, are rich in paleontological and anthropological (Regős 2002) finds (see chapter "Sedimentologic, Taphonomic Processes and Paleontological Values of Hungarian Caves"). The former population of the caves can be explained by several factors: the large size of the entrances (width, height), the fact that they are not deep, due to the characteristics of spring caves, and also because there are often active spring caves or their watercourses situated below them, which were once water extraction sites. The large size of the cave entrances was due to the former abundant karst springs (the large volume of the reservoir rock favoured this) as well as the permanent resting position of the mountains (this allowed karst water to keep exiting in the same place for a long time).

There are also inactive caves, often inactive ponors, which in many cases are in roof position. (The roof position of some of the caves can be traced back to the fact that the surface of the mountains was previously located near the erosion base of the karst.) Caves in roof position have therefore a thin vault, which promotes their destruction. This was also facilitated by the old age of some of the caves. During their intensive destruction, their vaults were damaged (windows have formed), then during their further destruction they separated into parts, or shortened into passages, and then formed into rock gates and arches (Hevesi 1986a, c). With the partial destruction of the former through-caves, cave remnants were also formed in the enclosing non-karstic rocks and on the sides of the gorges formed by collapsing, and the former side passages also transformed into cave remnants (Kerekes 1936; Hevesi 1978). Especially during multiple beheadings, the opening side passages or phreatic branches were also separated into parts (Hevesi 2019).

Ponor caves are common in the Nagy-fennsík plateau and the South-Eastern Bükk. They were formed at the inclusion of non-karstic rocks (mixed autogenous-allogenic karst) in these areas. At the end of the settled non-karstic rocks, ponor caves occur on the Kis-fennsík plateau. Ponor caves are often several kilometres long. The significant length can be explained by the large drainage basin of the ponors and the long-term karstification. Some of the sinkhole caves were connected to epiphreatic or phreatic passages.

In the nineteenth century, there was still no evidence that prehistoric man lived in the area which is called Hungary today. In the last decade of the century, flint tools were found during the foundation diggings of a house in Miskolc, at the foothills of the Bükk Mountains, and in the floodplain of the Szinva stream coming from the mountains. Although the authenticity of these was widely disputed, excavations started in 1906 by the initiative of polyhistor Ottó Herman under the leadership of geologist Ottokár Kadić. The large number of spring caves with spacious entrances at different levels in the mountains provided excellent locations.

They started the work on the Kis-fennsík plateau, in the Kecske-lyuk and the Büdös-pest caves. At first, they were unsuccessful, but after the research site was moved to the Szeleta Cave, tools and pottery of the Ice Age man were found from the Paleolithic to the Iron Age. Then they returned to the Büdös-pest Cave and the Kecske-lyuk Cave, and in this second attempt, rich finds were discovered there as well. The oldest find in the Szeleta Cave was found to be 130,000 years old, but Neolithic, Bronze Age and Iron Age finds were also found here. The concept of the internationally recognized "Szeleta culture" (circa 45,000–22,000 B.C.) was introduced, but finds from several other cultural levels (e.g. solutréeni, taubachi, moustéri, aurignaci cultures) were also identified. The most famous finds from this location are the polished stone axes and spearheads. (The name "Bükk Culture" may be misleading since it is a Neolithic culture less than 6,000 years old, and its finds were mainly found in the Baradla Cave (see chapter "Sedimentologic, Taphonomic Processes and Paleontological Values of Hungarian Caves")).

Significant finds of a similar age were also discovered in the Hillebrand Jenő Cave of the Kis-fennsík plateau (wooden post holes and scratches made by Paleolithic man were also found here), in the Pes-kő Cave on the Nagy-fennsík Plateau, in the Herman Ottó Cave and the Suba-lyuk Cave, in the Southern Bükk, and in Istállós-kő Cave in the Northern Bükk; the remains of the famous "Prehistoric fire pit" and several musical instruments come from here.

Archaeological excavations were carried out in the caves of the Bükk Mountains during the twentieth century, and these were renewed in several stages with the lead of Ottokár Kadić, Jenő Hillebrand, Andor Saád, van Giffen, János Nemeskéri, János Dancza, László Vértes, and most recently Árpád Ringer (without the need for completeness).

In addition to archaeological finds, very significant Pleistocene prehistoric remains have been found in these spring caves and several other similar sites (e.g. the Tar-kői-kőfülke Cave, see chapter "Sedimentologic, Taphonomic Processes and Paleontological Values of Hungarian Caves").

2 Presentation of the Caves

There are three caves in the Bükk Mountains belonging to the categories of touris, medicinal-tourism and partly water-yielding caves. These are the Anna Cave, the Szent István Cave and the Miskolctapolcai-Tavasbarlang ("Barlangfürdő"—Cave Spa) (see below). Each year, the first and the second are visited by 60,000–70,000 people, and many hundred thousand bath in the latter (Lénárt 1979). Postal stamps were also issued about the Anna Cave, the Tavas Cave and the Szeleta Cave, which are internationally recognized from a palaeoarchaeological perspective. These caves are strictly protected.

2.1 Anna Cave

The syngenetic sinter cave (Table 1) is situated in Miskolc, in the Szinva Valley (Figs. 1 and 2). It has two entrances, the so-called "waterworks" entrance opens at an altitude of 264 m, and the "tourist" entrance at 272 m, on the brim of the Szinva sinter cone (Fig. 3a). The grade difference in the cave is 14 m, its length is 568 m, and a significant part of it consists of the tunnels created by drifting connecting the 34 larger sinter caverns that were closed off by natural arching on the top. A greater part of it (380 m), including 32 caverns are part of the tourist section (Kessler and Megay 1963).

Table 1 Data of the Anna Cave

General information	
Cadastral reference	5391-1
Region	Southern Bükk
County	Borsod-Abaúj-Zemplén
Municipality	Miskolc
Genetics	Syngenetic sinter cave
Length	568 m
Vertical extent	13 m
Depth	4 m
Protection	Highly protected

The sinter cave was found in 1833 when drifting a 93 m long tunnel prospecting water. The cave is unique in Hungary due to its size. At first, it was used for tourism in its primitive form, and between 1951 and 1954 the waterworks of the town developed it for water production. There are brown, yellow and grey fossilized pine branches, root remnants, moss, the impressions of decayed tree trunks and various detrital sinter beds in the cave. Sometimes these have some covering of growing cave popcorn (Fig. 3c), active, small stalactites (Fig. 3b), senile thick dripstone crust and calcite crystal group.

The air temperature is seriously affected by the visitors and lighting fittings; its maximum can reach 14 °C (Lénárt 1984a). The green growth on the cave surfaces due to the lamps was significant between 1960 and 1989 (Rajczy et al. 1986), but today it is minimal. In the cave 28 Articulata species were described.

For its water supply, one spring was tapped in sinter, and two in dolomite. The spring in the sinter has a yield of 1000 m^3/day even today, and the water temperature is 6–14 °C.

2.2 Balla Cave

The entrance (Fig. 4a, b) to the Balla Cave opens on the eastern, rocky slope of Mount Balla in Southern Bükk (Fig. 1 and Table 2). The existence of the cave was already known at the beginning of the last century, thanks to the prehistoric bone and artefact finds found in it (Hillebrand 1911a, b, 1912). Its huge 8 m high and 10 m wide entrance can be seen from afar.

The enclosing rock of the cave is Triassic Bükkfennsík limestone (Bükkfennsík Limestone Formation, Regős and Ringer 2003a). At the entrance, layers with a 30° inclination angle and a 75° strike direction can be observed, which also affect the formation of the cave. The enclosing rock consists of thick and finely layered limestone in some places, but in other places red clay cemented limestone breccia characteristic (Regős and Ringer 2003a).

The cave actually consists of a single 6 m high, 8–10 m wide hall. There are eight hollows of various sizes in this hall. One of these opens to the surface. The speleothems of the cave can be divided into two groups. As a representative of the karstic speleothems lacy-dissolved wall surfaces and wall surfaces with solution pockets formed by the mixing corrosion in the flowing water, as well as in some places dissolved dripstone runoff, dripstone flags and botryoid can be mentioned. Fragmented wall surfaces are the representatives of the frost-fragmented phenomena (Regős and Ringer 2003a). Interestingly, there are impressive ice formations near the entrance in winter. Their location is indicated by dripping wells in summer (Székely 2002).

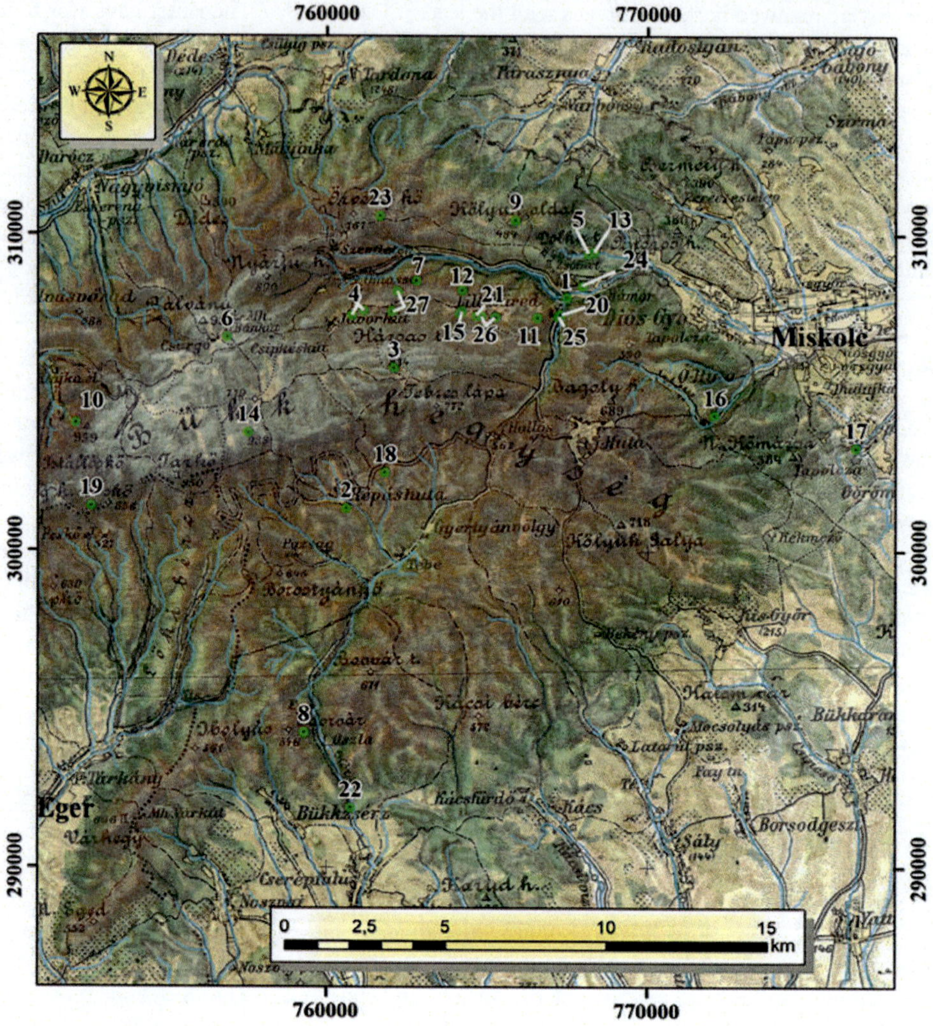

Fig. 1 Locations of the caves mentioned in this chapter (The map was created by Hernádi 2020 by combining the 1932 Erdey-Hubay-Vigyázó tourist map of the Bükk Mountains and the coloured relief map of the Bükk Mountains). Legend: 1. Anna Cave, 2. Balla Cave, 3. Bányász Cave, 4. Bolhás-Jávorkút Cave System, 5. Büdös-pest Cave, 6. Diabáz Cave, 7. Fekete Cave 8. Hajnóczy Cave, 9. Hillebrand Jenő Cave, 10. Istállóskő Cave, 11. István-lápa Cave, 12. Jáspis Cave 13. Kecske-lyuk Cave, 14. Kis-kőhát Ponor Cav 15. Létrás Vizes Cave, 16. Mexikó-völgy Ponor Cave, 17. Miskolctapolca-tavas Cave, 18. Pénz-patak Ponor Cave, 19. Pes-kő Cave, 20. Soltész-kert Sinter Cave, 21. Speiz Cave, 22. Suba-lyuk Cave, 23. Szamentu Cave, 24. Szeleta Cave, 25. Szent István Cave, 26. Szepes-Láner Cave System, 27. Szivárvány-Sebes Cave System

The Balla Cave has been under archaeological protection since 1951 due to the finds found in it, and since 1982 it has been declared as a highly protected cave due to its landscape value.

2.3 Bányász Cave

The elliptical entrance (1 m wide and 2 m high) of the shaft cave is located in a meadow dissected by solution dolines, south from the Mount Nagy Hársas which is one of the peaks of the Nagy-fennsík plateau (Fig. 1), in the central part of the Bükk-fennsík plateau (Table 3). The demolition of its first shaft began in 1964 by the members of the Miskolc Mining Cave Research Group. They entered the open passages by removing 100 m^3 of debris. According to contemporary descriptions, the lower section of the cave, which reached a depth of 130 m, collapsed in 1966 (Nyerges 2003a).

Exploration of the cave continued in 2010. From a depth of 89 m (the former endpoint) new mines were constantly revealed. The last shaft to date, which is 16 m deep (Tető-felújítás Shaft) was found in 2020. This leads to a depth of 300 m compared to the entrance. With this, the Bányász Cave is not only the deepest cave of the Bükk Mountains, but also the deepest cave in Hungary.

Nyerges (2003a) describes it as an inactive ponor cave; however, we classify it as an active shaft cave. Its zone below 200 m is abundant in small streams. The enclosing rock of the cave is (Middle) Triassic Bükkfennsík Limestone (Bükkfennsík Limestone Formation, Pelikán 2005). The cave was formed along the NW–SE nearly 80° inclined structural planes (faults) of this limestone (Pelikán 2005), which can be traced throughout the cave. Like shaft caves in general, it consists mainly of shafts with vertical development and a cascading system of high-drop, narrow passages connecting them.

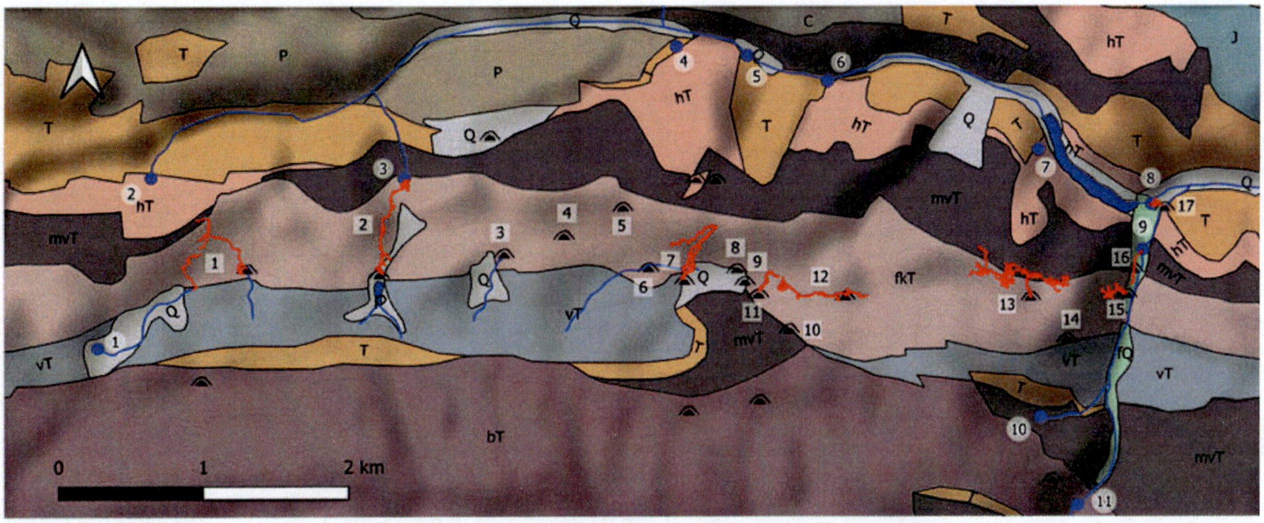

Fig. 2 Geological map with of the eastern part of the Nagy-fennsík plateau (Less 2002, M = 1:50 000 geological map, using data from National Cave Registry and OpenStreetMap contributors Fekete 2020). Legend: Karstic rocks: fQ (Pleistocene-Holocene calcareous sinter); bT (Bükkfennsík Limestone Formation); fkT (Fehérkő Limestone Formation); hT (Hámor Dolomite Formation); T (Triassic). Non karstic or nor kartsified rocks: Q (Pleistocene-Holocene alluvium); J (Jurassic Vaskapu Sandstone Formation); T (Ablakosvölgy Formation, Savós Marls); mvT (Metavolcanic phenomena); vT (Vesszős Formation, shale, marl); P (Permian, Nagyvisnyó Limestone Formation); C (Carbon Mályinka Formation). Springs ●: ① Jávor-kút Spring; ② Garadna Spring; ③ Huba Spring ④ Margit Spring ⑤ Wekerle Spring ⑥ Levente Spring ⑦ Eszperantó Spring ⑧ Anna Spring ⑨ Soltész-kert Spring ⑩ Vesszős Spring ①① Szinva spring. Caves (cave entrance sign): 1. Bolhás-Jávorkút Cave System; 2. Szivárvány-Sebes Cave System; 3. Fenyves-rét Ponor Caves; 4. Útmenti Ponor Cave; 5. Cubákos Ponor Cave; 6. Sinkhole of the Létrás-Vizes Cave; 7. Létrás-Vizes Cave; 8. Kismogyorós Ponor Cave; 9. Ikertebri Ponor Cave; 10. Bükkös Ponor Cave; 11. Speiz Cave; 12. Szepes-Láner Cave System; 13. István-lápa Cave; 14. Király-Lajos Ponor; 15. Szent István Cave; 16. Soltész-kert calcareous sinter cave; 17. Anna Cave

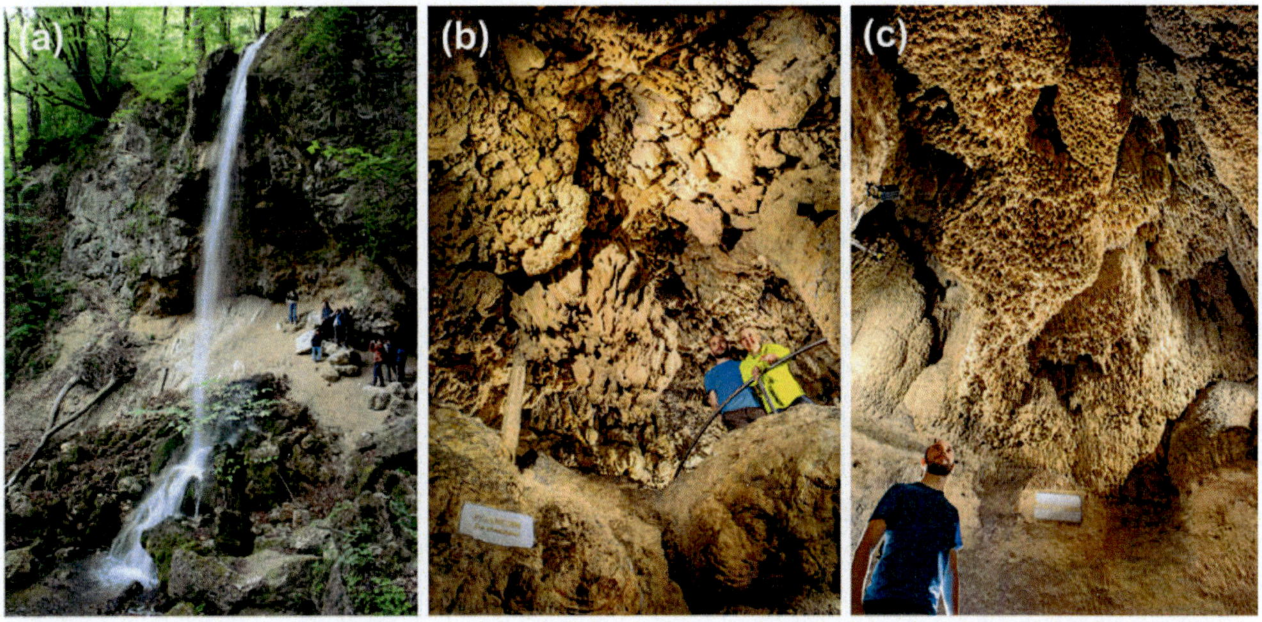

Fig. 3 Anna Cave. **a** Famous waterfall of the Szinva stream Next to the entrance of the Anna Cave (photo by Virág); **b** The Szomorúfűz, the most known phenomenon of the Anna Cave (photo by Egri); **c** Plants crusted by freshwater limestone are often recognized by visitors in the Északi-fény Chamber (photo by Egri)

Fig. 4 Balla Cave and Büdös-pest Cave. **a** ext to Répáshuta opens the Balla Cave with prehistoric finds 50 m above the Balla Valley (photo by Egri); **b** The entrance to the 50 m cave chamber is 10 m wide and 8 m high (photo by Egri); **c** The 6 × 5 m entrance of Büdös-pest Cave opens 50 m above the Forrás Valley on the Kis-fennsík plateau (photo by Egri); **d** Diggings in the Bükk Mountains, led by Ottokár Kadić, began in 1906 in the nearly 40-m-long cave hall, next to the Kecske-lyuk Cave (photo by Egri)

Table 2 Data of the Balla Cave

General information	
Cadastral number	5381-1
Region	Southern Bükk
County	Borsod-Abaúj-Zemplén
Municipality	Répáshuta
Genetics	Inactive spring cave
Length	128 m
Vertical extent	9 m
Depth	1 m
Protection	Highly protected

Table 3 Data of the Bányász Cave

General information	
Cadastral number	5372-31
Region	Bükk Plateau
County	Borsod-Abaúj-Zemplén
Municipality	Miskolc
Genetics	Active shaft cave
Length	931 m
Vertical extent	300 m
Depth	300 m
Protection	Highly protected

The cave was dissolved by the waters coming here and leaking towards the Szinva spring (Rántó and Sűrű 2014). The shaft cave actually consists of 11 vertical, mostly dissolved-walled shafts and the narrow passages connecting them (Fig. 5a–c). Half tubes can also be observed in some places, from which, according to Veress (2019)—taking into account analogies from the Bakony Mountains—it can be concluded that during heavy rains the cave is not able to drain large amounts of water and is filled to a certain depth. The end of the highest half tube is in a depth of 28 m from the entrance.

Many forms of dripstone formations can be found in the spacious, 2–3 m in diameter, 24 m deep shaft which is the entrance to the cave (Rántó 2019, oral communication). In the lower parts of the cave, dripstone formations (dripstone crusts and dripstone runoffs) can be found in some places, but in a negligible number and size compared to the Hungarian karst caves. Limestone ribs dissolved in a whimsical shape can also be seen in several places (Fig. 5d).

It has been a highly protected cave since 1993 (Székely 2002) due to its pitch-like shafts and its depth, as well as the bat population here (Rántó 2019).

2.4 Bolhás-Jávorkút Cave System

The Bolhás-Jávorkút Cave System is a hydrologically related system of two caves (Bolhás Ponor Cave, Jávorkút Ponor Cave) that open barely 400 m apart on the Bolhás meadow (Figs. 1 and 2 and Table 4). To prove this-mit proveolsz, nem egyertelmu-water dye tracing tests were conducted in 1954 and then in 1997 (Börcsök et al. 1997). The system drains the

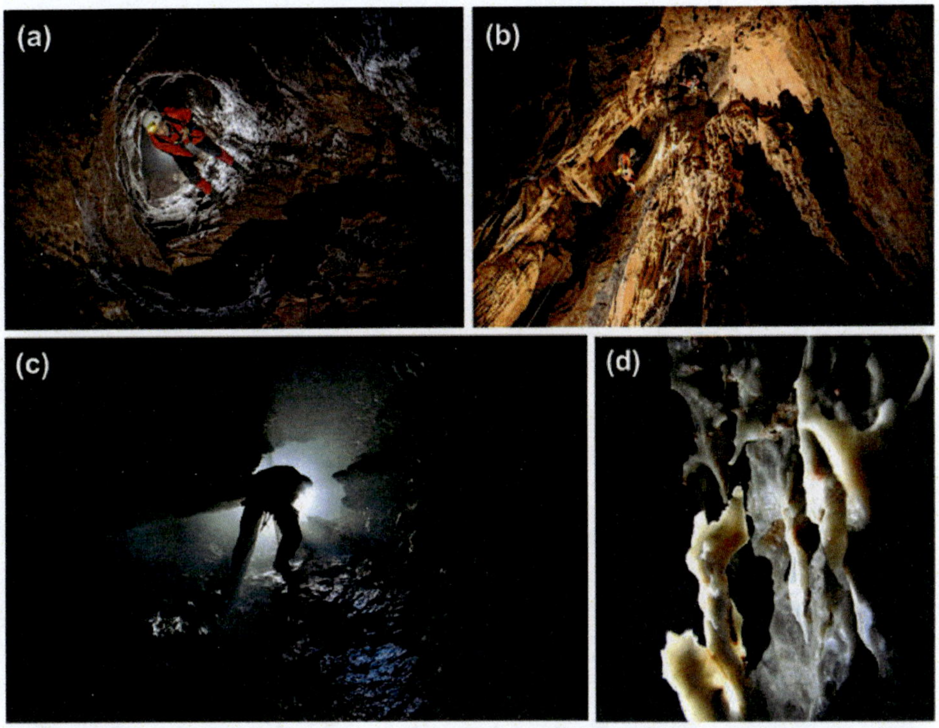

Fig. 5 Bányász Cave. **a** One of the shafts of the Bányász Cave near the entrance (photo by Berentés); **b** The spacious Törpilla Shaft look from below (photo by Berentés); **c** The Anna Shaft in a depth of about 250 m (Berentés); **d** Limestone residue dissolved in a whimsical shape on the side-wall (photo by Berentés)

Table 4 Data of the Bolhás Jávorkút Cave System

General information	
Cadastral number	5372-4
Region	Bükk Plateau
County	Borsod-Abaúj-Zemplén
Municipality	Miskolc
Genetics	Active ponor
Length	5100 m
Vertical extent	131 m
Depth	125 m
Protection	Highly protected

waters of the Bolhás Stream. The two caves were dug and explored independently by the members of the Magyar Hidrológiai Társaság Zsombolykutató Munkabizottság (Ponor Research Committee of the Hungarian Hydrological Society) in 1953. The first major achievement of the committee was the digging of the Bolhás Ponor Cave in the spring of 1953 (Nyerges 2003b). In the summer of this year, the Jávorkút Ponor Cave in the vicinity of the former was also explored. The Bolhás Ponor Cave and the Jávorkút Ponor Cave are most likely part of the Bánkút-Bolhás-Jávorkút-Garadna hydrological system, of which not much is known to this day (Sásdi 1992). Taking the geological conditions of the region into account, according to Hegedűs (2004), the assumed length of this system may exceed 15 km.

The enclosing rock of the cave system is typically a strongly tilted, well-stratified Triassic (Ladini) period white limestone stretching in a narrow band in an east-westerly direction. This limestone is surrounded on two sides by volcanic and various slates (Kárpát 1984). To the northwest of the entrance to the Bolhás Ponor Cave, the impermeable volcanic layer is replaced by Triassic (Anisus) dolomite. The presence of this dolomite allows deep-drawn waters to make their way to the Garadna spring (Nyerges 2003b).

The cave system with several entrances, with an extensive, step-deepening sinking zone, has a very complex structure. It can be divided into three parts based on its morphology. The first part is the near-surface swallowing system. Several inactive branches of this approach the level of the entrance, but their former sink funnels are now largely destroyed. In these parts, the branches are characterized by staggering. A maze-like network of shafts, chimneys (Fig. 6b, e) is connected to the larger halls of the cave system. Both serrated and dripstone-rich zones occur (Fig. 6a, d). The second characteristic section is formed by the horizontally developed passages (Fig. 6c). These are drainage passages formed along high but narrow tectonic lines. The forms of these branches were created by erosion and corrosion together. These can be divided into main and secondary branches. The main branches drain the surface waters entering the given sinkhole. The secondary branches are in part generally parallel to these and periodically active. The third main section is the section of the active creek main

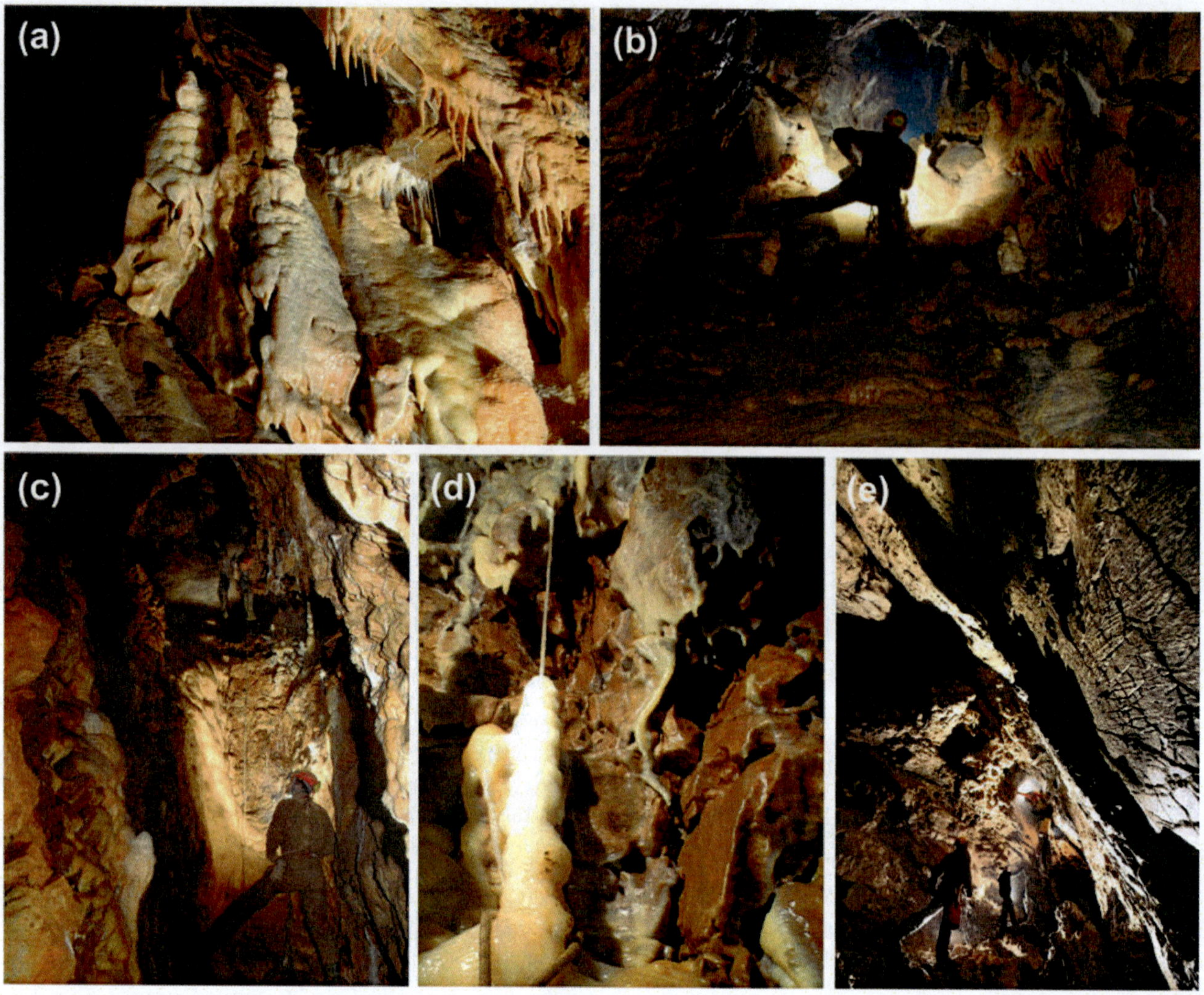

Fig. 6 Bolhás-Jávorkút Cave System. **a** In the upper section of the Jávorkút Ponor Cave, in the Luna-termi Branch, we can find several completely untouched dripstone formations (photo by Hegedűs); **b** The vertical chimney of the Hősök-ága Branch in the Jávorkút Ponor Cave (photo by Hegedűs); **c** The chimney systems of the Jávorkút Ponor Cave are independent no upper level has been formed. The Hősök-ága Branch can also be assessed using rope technology (photo by Hegedűs); **d** A location in the Luna-termi Branch of the Jávorkút Ponor Cave which is rich in dripstones (photo by Hegedűs); **e** In many places in the Bolhás Ponor Cave, steep bedding planes form the side wall of the passages (photo by Berentés)

branch, which can only be found in the case of the Jávorkút Ponor Cave. Its known length is about 100 m, its width is 1–3 m, its height is 5–12 m, its average water flow is 100–300 l/min. The end of this section is a series of siphons. The stream that disappears there does not appear in the passages of the Bolhás Ponor Cave. The reason for this is that the main branch of the Bolhás Ponor Cave is only one inactive floor passage of the whole system (Nyerges 2003b).

The filling and the formations of the cave system are extremely varied, and can also be divided into three parts based on the previous morphological classification. In the swallow zone close to the entrance, there are karrentified or smooth-polished wall surfaces on the active parts. Some near-surface inactive passages are rich in dripstone formations. Accumulation of sediment washed from the surface in the horizontal passages of the cave is typical. Filling them with gravelly, fine-grained stone debris continues today. Scallops, dissolved surfaces and solution pockets are characteristic on the walls. Dripstone formations of different sizes and colours (dripstone runoff, dripstone crust, stalactite) can be found almost everywhere in these parts. Stalagmites and helictites also occur in inactive floor levels. The active main branch with a cave creek is poor in speleothem (Nyerges 2003b).

The cave system is waiting for further exploration. It would be expedient to carry out this exploratory research in the direction of the Garadna spring. Mainly based on their hydrological and genetic significance, the ponor caves were separately declared highly protected in 1982 (Székely 2002).

2.5 Büdös-Pest Cave

The cave opens at a height of 307 m above sea level on the Kis-fennsík plateau (Fig. 1 and Table 5) on the north-facing ridge of the Dolka-tető hill at a relative height of 47 m above the stream of the Forrás Valley. Its 6.5 m high and 5.2 m wide entrance provides an impressive view (Fig. 4c, d). Excavations were carried out in the area of the cave at the beginning of the twentieth century, during which the experts separated the glacial and contemporary remains (Székely 2002).

The Büdös-pest Cave was formed in Triassic limestone (Fehérkő Limestone Formation). This limestone has a non-homogeneous composition, and in places syngenetic breccia and thin layers of flint indicate that the deposition of the limestone did not occur undisturbed. Its large entrance continues in a 37 m long 3–5 m wide and about 3 m high passage. This passage is first horizontal and then gradually becomes steeper until it opens to the surface with a chimney that is 5.6 m wide, 4 m in diameter and 7 m high. In this area, the cave is dangerous because of collapsing, as the vertical side-wall of this chimney is held together only by the roots of the plants (Regős and Ringer 2003b).

The cave is relatively poor in speleothems. Only in some places we can observe dripstone flag, botryoid and straw stalactite which are covering the cave wall in spots. It was placed under archaeological protection in 1951, because it has the most complete Upper Pleistocene stratigraphy in the Bükk Mountains in addition to the Suba-lyuk Cave. Its anthropological material belongs to the Middle Paleolithic (Kordos 1984).

Table 5 Data of the Büdös-pest Cave

General information	
Cadastral number	5363-3
Region	Bükk Plateau
County	Borsod-Abaúj-Zemplén
Municipality	Miskolc
Genetics	Inactive spring cave
Length	37 m
Vertical extent	13 m
Depth	0 m
Protection	Highly protected

2.6 Diabáz Cave

The entrance to the Diabáz Cave, formerly known as the Bánkúti-visszafolyó Cave, opens to the west from Bánkút, next to a forestry road (Fig. 1). The cave was explored in 1975 when speleologists reached their lowest point. It was named because of an unfortunate mistake. In the previous literature (Kordos 1983, 1984) it is mentioned that the cave was formed on the border of limestone and diabase. In contrast, laboratory studies have shown that this finding is incorrect (see below).

The periodically active ponor cave (Fig. 7a–c and Table 6) was formed in Middle Triassic limestone (Fehérkő Limestone Formation) and on the border of Ladin-era Szentistvánhegy Metaandesite (porphyrite) (Székely 2002; Nyerges 2003a). Not all parts of the cave are active (Fig. 8). Its inactive parts contain spectacular and variable-shaped dripstone formations (stalagmites and stalactites, dripstone runoffs and dripstone flags). Its most beautiful part is the Szép Branch (Figs. 7d, e and 8). This section of the cave cannot be visited in order to protect the formations. In addition, helictites can be seen in some other locations as well.

The ponor at the entrance to the cave is only periodically active. A spring with a constant flow of water at a depth of 25 m continuously forms the main passage of the cave, the activity of which is assisted from time to time by waters that occasionally flow from the surface. A watercourse from another surface area is also involved in the formation of the cave. This transports volcanic rock material from the surface to the cave, which is similar to the sediment found in other passages but has a smaller grain size (Nyerges 2003a).

Several people studied the possible surfacing locations of the water leaving the cave. According to Jakucs (1951), it comes to the surface in the Garadna Spring, but it was previously thought that it was drained by the Szalajka Spring. Jakucs's statements were confirmed in 1991 by the MÁFI Cave Research Group.

The cave was intended to be used as a reservoir in the 1980s. For this purpose, the endpoint of one of its branches (Bell Shaft) was constrained by a wall, but the experiment did not end successfully. Ladders were installed in the upper section of the cave (Hajóköteles Shaft, Körte Chimney, Fig. 7c) to Waterfall Shaft (Fig. 7a) with its imposing size. The cave has been protected since 1988 due to its hydrological significance and vulnerability.

2.7 Fekete Cave

The entrance to the sixth-longest and the second-deepest cave in the Bükk Mountains opens at the northeastern edge of the Nagy-fennsík plateau (Table 7) and at the southern end of the Tekenősi Valley (Fig. 1). The entrance to the

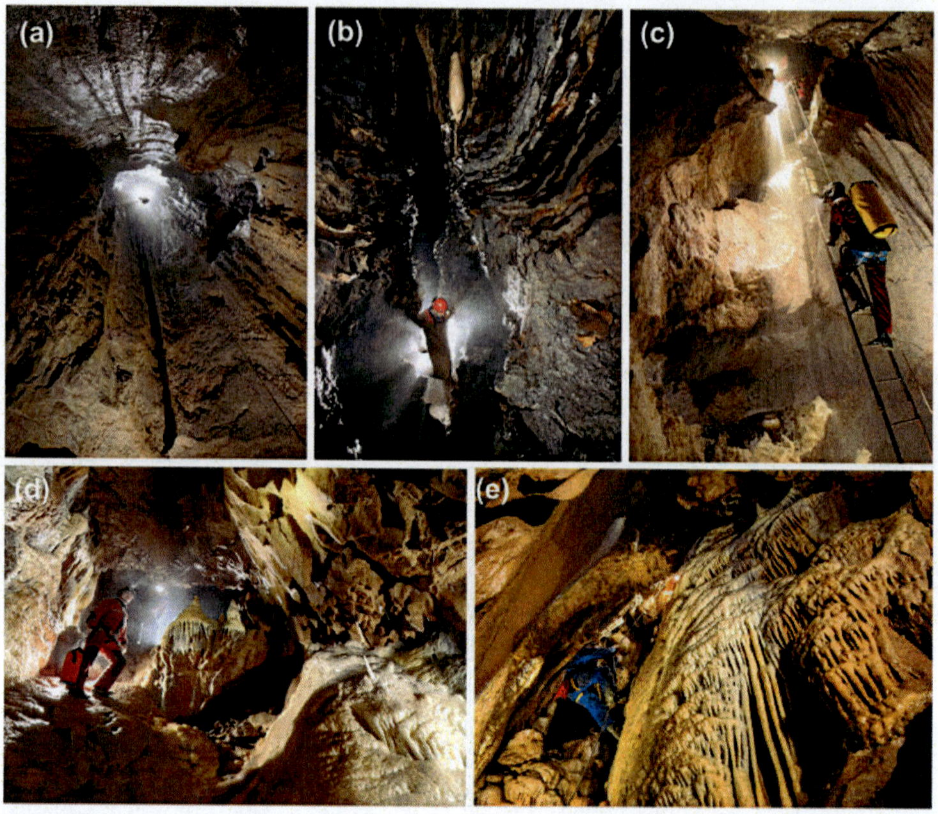

Fig. 7 Diabáz Cave. **a** The 30 m deep Vízesés Shaft is very spectacular in dry weather (photo by Berentés); **b** The Elefánt-lépcső on the Fehérkő Limestone beds (photo by Berentés); **c** The ladder of the Körte-akna Shaft is crossed by an almost constantly active waterfall (photo by Berentés); **d** A part from the diverse dripstone formations of the Szép-ág Branch (photo by Berentés); **e** The large stalactite runoff of the Szép-ág Branch (photo by Egri)

Table 6 Data of the Diabáz Cave

General information	
Cadastral number	5331-9
Region	Bükk Plateau
County	Borsod-Abaúj-Zemplén
Municipality	Miskolc
Genetics	Periodically active ponor cave
Length	1000 m
Vertical extent	161 m
Depth	161 m
Protection	Highly protected

periodically active ponor cave was discovered in the summer of 1975. Its drainage basin is barely 1 km^2, which is mainly limited to the area of the Tekenősi meadow and the surrounding hillsides. In the case of sudden heavy rainfalls, the amount of water absorbed in the ponor can exceed up to 500 l/min. Otherwise, the stream channel leading to the ponor is usually dry (Nyerges 2003b).

The Fekete Cave, is a sinkhole cave, and it was formed on a rock boundary. It is interesting that in addition to the Triassic limestone, its passages cross not only Triassic dolomite but also porphyrite and Permian limestone of the same age (Székely 2002, Fig. 9a). It is named after the dark grey calcareous Anisus dolomite in which the entrance zone developed (Kordos 1984). Their enclosing rocks change from chamber to chamber. Dolomite is replaced by light grey thin-bedded limestone of the Lower Triassic. This is followed by light grey, calcareous crystalline limestone, but the cave also contains porphyrite in several places formed as a result of the Ladinian volcanism, typical of the area (Nyerges 2003b).

The cave was formed in a tectonically heavily stressed area (Simon 1980; Kordos 1984). A significant part of the branches is covered by the above-mentioned dark grey, brittle Anisus dolomite (Fig. 9b). Extensive fracture systems have been formed in this rock due to tectonic stress (Fig. 10). The passages of the fracture system were either not or only partially covered by filling of secondary origin, and due to the dolomite nature of the enclosing rock, dripstone formation is not significant in these sections either (Nyerges 2003b). Various dripstone formations make the middle section of the cave special. Here the enclosing rock of the branches is limestone (Székely 2002).

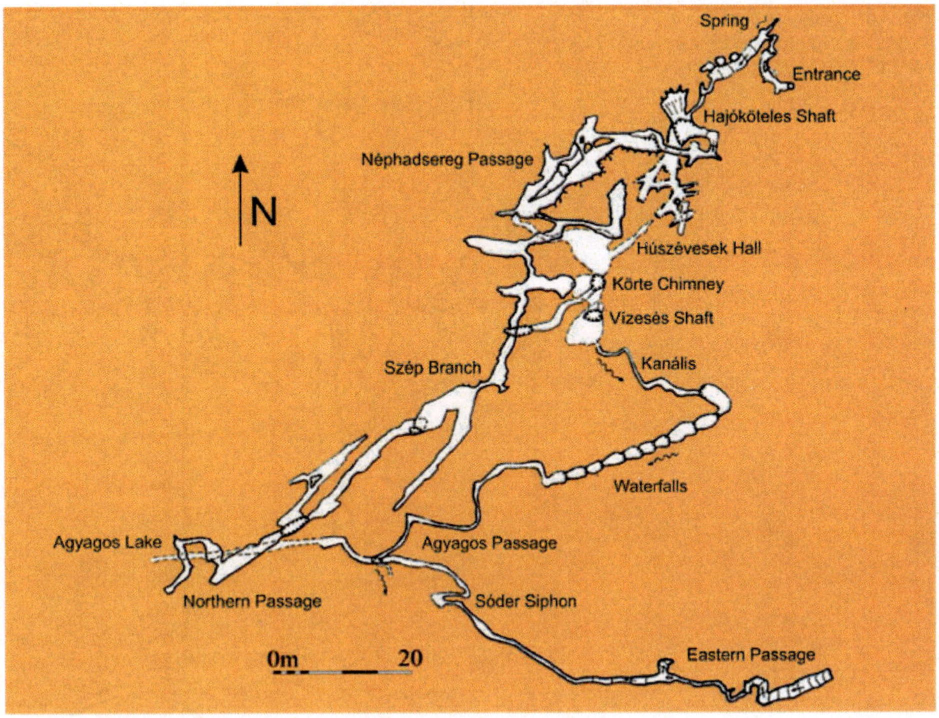

Fig. 8 Floor plan of the Diabáz Cave. National Cave Register

Table 7 Data of the Fekete Cave

General information	
Cadastral number	5372-8
Region	Bükk Plateau
County	Borsod-Abaúj-Zemplén
Municipality	Miskolc
Genetics	Periodically active ponor cave
Length	3000 m
Vertical extent	174 m
Depth	174 m
Protection	Highly protected

The Fekete Cave has been highly protected since 1982. It is due to the variety of its enclosing rock, the integrity of its formations, and last but not least, its size.

2.8 Hajnóczy Cave

The entrance to the Hajnóczy Cave opens on Ódor Hill in Southern Bükk (Table 8), on the border of the village Cserépfalu (Fig. 1). The cave was discovered in 1971 by high school students. Its professional exploration began in that year and the length of its branches discovered increased year by year.

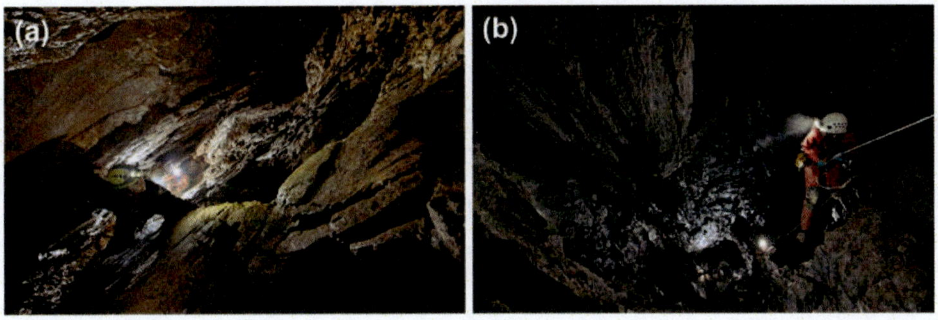

Fig. 9 Fekete Cave. **a** Shaley rock before the siphon row (photo by Berentés); **b** View of the Nagy-akna Shaft in the eponymous, dark-coloured dolomite rock (photo by Berentés)

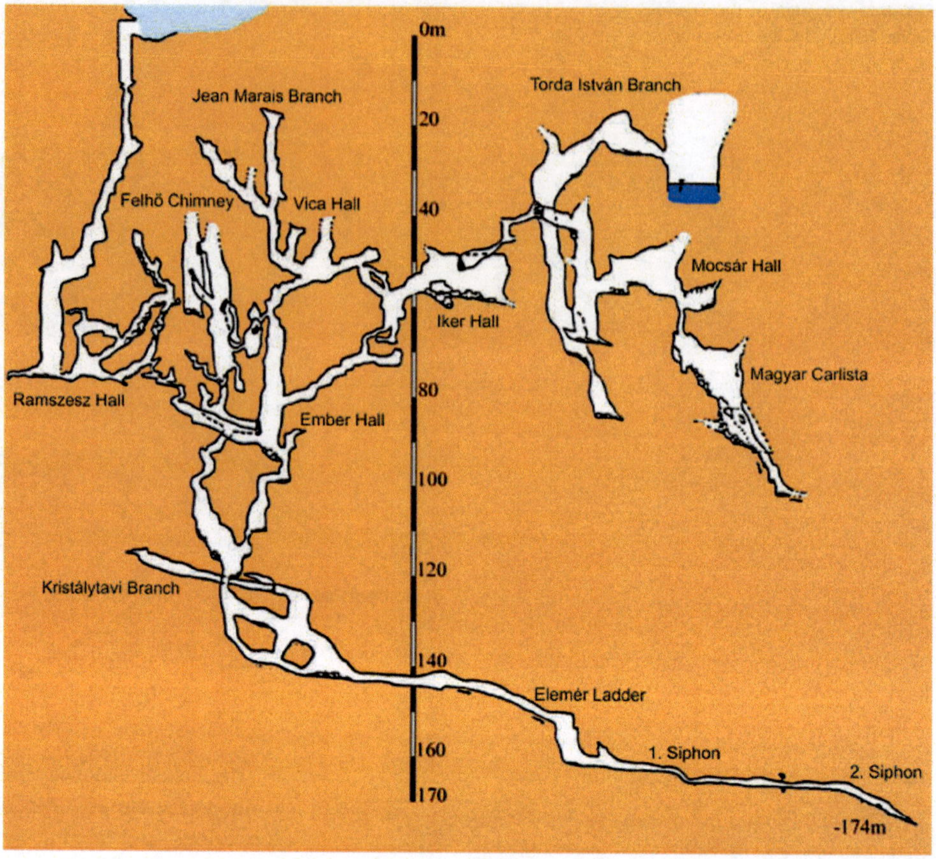

Fig. 10 Profile of the Fekete Cave. National Cave Register, MAFC, Németh and Nyerges 1998

Table 8 Data of the Hajnóczy Cave

General information	
Cadastral number	5382-2
Region	Southern Bükk
County	Borsod-Abaúj-Zemplén
Municipality	Cserépfalu
Genetics	Inactive shaft cave
Length	4250 m
Vertical extent	124 m
Depth	81 m
Protection	Highly protected

The enclosing rock of the cave is Triassic (Ladini-karni) flint and flint-free limestone (Balogh 1964). Its coverage was previously higher, as evidenced by the fact that there are significant slate cones in parts of the Hajnóczy Cave above which there is now no slate on the surface. The direction of the cave's passages is NE-SW which is typical in the area and the lithoclase system perpendicular to this direction (Varga 1978).

Morphologically, the cave consists of two well-separable parts. The parts closer to the surface were mostly formed by bedding plane-side dissolution and then the subsequent stage was no longer formed by bedding plane-side corrosion processes. A characteristic feature of the formation along the bedding plane is the maze-like structure (Varga 1978). These meandering passages are narrow and rich in constrictions. The passages of the other well-separable, deeper part are characterized by large formations, as well as the largest halls of the cave system (Giant Hall, Grand Canyon, Tsitsogó). These chambers are connected by narrow passages, in which the pathway is hampered by piles of debris made of limestone and shale carried from the surface (Varga 2003).

The cave is rich in dripstone formations (Fig. 11a, c, d, e). The most beautiful and largest are located in the Csodák-terme Hall and the Galéria Hall. The latter is about 70 m long, 1–8 m wide and 4–13 m high. The cause of richness in dripstones is the fact that the vault of the chamber is made up of vertically standing bedding planes, so more water could seep through there. The dripstones here have a diameter of 6–15 cm and a length of 180–310 cm (Varga 2003). Stalactites, huge dripstone runoffs, cave pearls and helictites of varying size, shape and colour are also present in several places (Székely 2002).

The scientific research of the cave began in the mid-1970s. Its paleontological finds were studied by Kordos

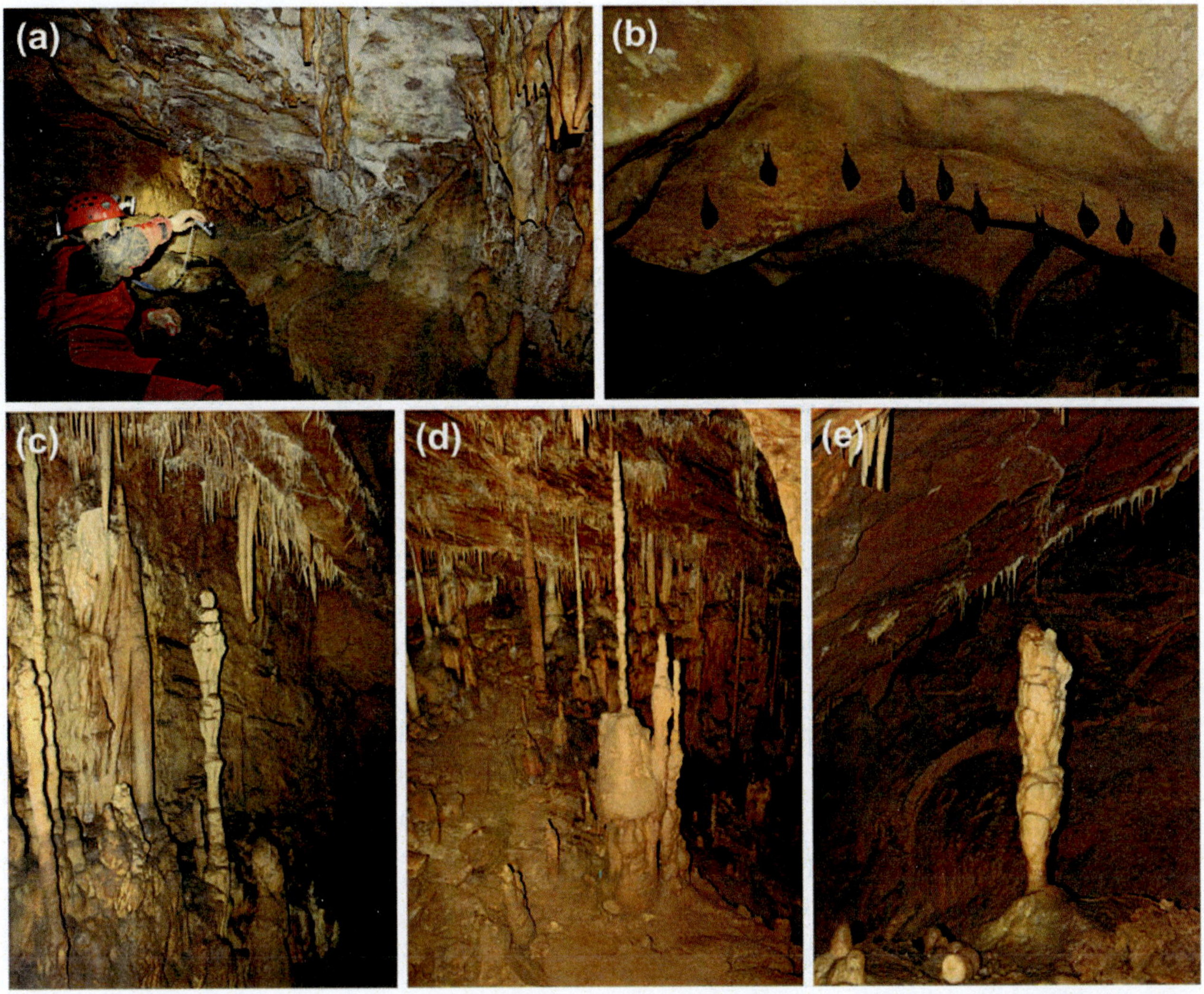

Fig. 11 Hajnóczy Cave. **a** A part of the Szőlőskert section in the Nagy-terem Chamber; **b** Wintering bats in the Óriás-terem Hall; **c** A dripstone called Botond botja (Botond's stick) in the Gallery Hall; **d** The richest part of the cave in dripstones is the Gallery Hall; **f** The dripstone named Nyárfa (Aspen) in the Gallery Hall (photos by Timkó)

(1974) and Hír (1985). One of the Pleistocene-extinct bear species and several other important age-indicator vole species emerged from the sediments. Similarly, old cave filling has so far only been found in three caves in the Bükk Mountains. The climate of the cave was studied by Miklós (1978). He found that the average air temperature of the cave fluctuated between relatively low values (8.5–10.5 °C), allowing bats to live here (Fig. 11b).

Based on its genetic and morphological values as well as because of its rich dripstone concretions, it was declared highly protected in 1982.

2.9 Hillebrand Jenő Cave

The entrance to the Hillebrand Jenő Cave opens on the southeast slope of the Hill Kőlyuk-galya (Fig. 1 and Table 9), a few meters from the Kő-lyuk Cave, so it was formerly known as Kő-lyuk2 Cave. The narrow entrance to the cave, explored in 1947, leads to a cave rich in speleothems.

As in some caves of the Bükk Mountains syngenetic breccia arose in the Hillebrand Jenő Cave as well. The enclosing rock is Upper Triassic limestone (Kisfennsík

Table 9 Data of the Hillebrand Jenő Cave

General information	
Cadastral number	5363-5
Region	Bükk Plateau
County	Borsod-Abaúj-Zemplén
Municipality	Parasznya
Genetics	Phreatic cave
Length	240 m
Vertical extent	23
Depth	16
Protection	Highly protected

Limestone Formation, Székely 2002). The cave is a typical example of phreatic branch formation. After the tectonic preparation of the enclosing rock, passage formation below the karst water level took place (Regős and Ringer 2003c). As the height of the karst water level decreased, a new active passage developed while the one above it became inactive. The approximate age of the decrease in the height of the karst water level can be determined by the findings found in the finely layered sediments deposited from the surface water.

The cave is characterized by a variety of speleothems. Channels can be observed on the vault, suggesting it was filled with water previously. There are solution pockets on the walls as a result of mixing corrosion. The waters flowing from the surface developed a meandering stream channel. Dripstone formations include stalagmites and stalagtites, dripstone runoffs, dripstone flags, botryoids and hieroglyph-shaped dripstone formations (Regős and Ringer 2003c).

The peculiarity of the cave is that a layer can be well distinguished, from which skeletons, fragments of pottery and other objects were found (Székely 2002). These were left behind by the man of the Neolithic Bükk culture, along with the holes of wooden plots. One of the halls of the cave got its name from these holes (Stieber 1956). The Paleolithic scratches in the Cölöplyukas Hall—"prehistoric drawings"—are the only scratches of this age in Hungary, and at the same time the easternmost occurrence of such scratches in Europe.

Due to the uniqueness of the finds found in it, the cave was placed under archaeological protection in 1951 and then declared highly protected in 1982.

2.10 Istállós-kő Cave

The Istállós-kő Cave opens at the upper end of the Szalajka Valley (Fig. 1) at an altitude of 546 m above sea level. It is one of the highest positioned prehistoric cave sites. It became the focus of prehistoric research in the early twentieth century (Hillebrand 1914, 1917, 1919). Its huge (10 m wide, 7.5 m high) triangular entrance facing southwest draws attention from afar (Fig. 12a).

The now inactive spring cave (Table 10) was formed in Triassic limestone (Bükkfennsík Limestone Formation), where the folded, plate-like structure of the rock can be well observed. In addition to the signs of formation along the bed plate, forms resulting from both tectonic stress and dissolution can be found. The entrance to the cave was a water outlet for the former spring water, and it gradually dried up as the spring's water level sank. This was the period when the sediment accumulated. The sediment preserves the memory of the Aurignac culture (Fig. 12b), which was present in Hungary only here and in the Pes-kő Cave (Kadić 1934; Hillebrand 1935). Based on radiocarbon measurements, the researchers were able to determine the age of the beds; according to this data, the top bed is about 40,000 years old (Ringer 1989).

The Istállós-kő Cave, together with the one-hectare surface area around it, was declared protected in 1944. It was also granted archaeological protection in 1951, and since 1982 it has been classified as a highly protected cave.

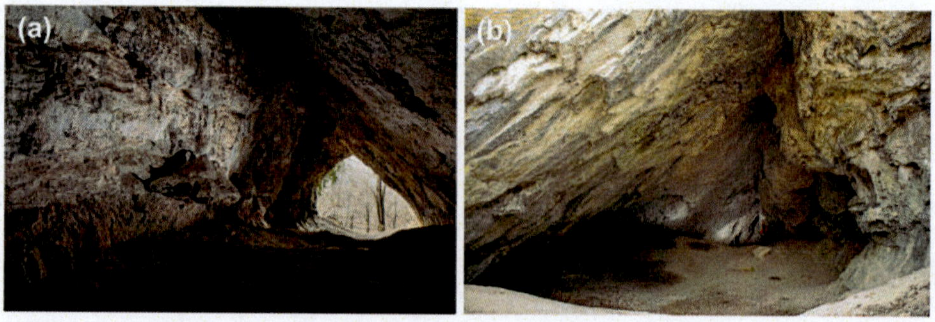

Fig. 12 Istállós-kő Cave. **a** Above the upper end of the Szalajka Valley, on the side of the highest peak of Bükk Mountains, next to the hiking trail, opens the Istállóskő Cave (photo by Egri); **b** The cave, which consists of nearly 60 m long hall, is famous for the prehistoric fire pit found here in 1912 (photo by Egri)

Table 10 Data of the Istállós-kő Cave

General informations	
Cadastral number	5341-1
Region	Eastern Bükk
County	Heves
Municipality	Szilvásvárad
Genetics	Inactive spring cave
Length	57 m
Vertical extent	16 m
Depth	0 m
Protection	Highly protected

2.11 István-Lápa Cave

The cave is located on the edge of the Nagy-fennsík plateau on the Bükk Plateau (Figs. 1 and 2). Here, in a solution doline opens the entrance to the István-Lápai Cave. The speleologists entered the cave in 1964 first (Nyerges 2003c), although its digging had begun earlier (Székely 2002). It has a maximum depth of 254 m and a total length of more than 8.7 km. With these parameters, it is considered to be the second deepest cave in the Bükk Mountains and Hungary, the longest cave in the Bükk Mountains and the fifth-longest cave in Hungary (Table 11).

The cave is located in an arheic valley called Istvánlápa, surrounded on two sides by impermeable rocks (Sárváry 1969). The caves formed here have a considerable drainage basin. The ponor cave formed in Middle Triassic limestone (Fehérkői Limestone Formation) is also enclosed on the one hand by Middle Triassic volcanics, and on the other hand by Upper Triassic shale (Nyerges 2003c), hornstoney limestone and diabase (Kordos 1984). The former authors believe that the caves in the Istvánlápa Valley (Létrási-vizes Cave, Szepesi Cave, István-lápai-Cave, Szent István Cave) form a unified system, between which there is a hitherto unexplored connection. According to Nyerges (2003c), the length of this cave system might exceed 20 km.

Table 11 Data of the István-lápa Cave

General information	
Cadastral number	5372-6
Region	Bükk Plateau
County	Borsod-Abaúj-Zemplén
Municipality	Miskolc
Genetics	Active ponor cave
Length	8700 m
Vertical extent	254 m
Depth	254 m
Protection	Highly protected

The entrance section of the cave is a passage of a former ponor formed in the plane of a north–south direction fault line (Kordos 1984). The system of nearly 20 cascading, almost vertical shafts (maximum depth 17 m; Fig. 13a, d, e) is followed by a nearly horizontally developed (Fig. 13b, c) branch with a periodically active stream, the Fő-ág (Main Branch). The Main Branch is divided into two branches from the connection of the shaft system, the longer Nyugati-ág (Western Branch) and the shorter Keleti-ág (Eastern Branch) (Nyerges 2003c). Both branches developed on more levels. Of these two, the Keleti-ág Branch is an inactive passage; only its deepest part is filled sometimes with karst water when the water level is high. On the other hand, the Nyugati-ág Branch has a very active watercourse, which can have a water flow of up to 700–800 l/min. The water of the Szepesi-Láner Cave System enters the cave through its so-called siphon nr. 4 (Fig. 13c) to its endpoint. Large solution pockets are common which are developed by the effect of turbulent water flow (Fig. 13b, c; Kordos 1984). The Nyugati-ág Branch is the most spectacular active, creeky cave in Hungary. In the event of a prolonged drought, the cave creek sometimes desiccates completely, making the largely gravelly filling of the channel visible. There are three gravel-filled siphons, often filled with water, at the end of the Nyugati-ág Branch. These siphons, if they are filled with water, can be bypassed via other branches. To facilitate ease of access, a horizontal wire rope more than 20 m long was also installed in the Bivak Corridor.

The pore deposit and the speleothems of the branches are very varied. There are bare limestone surfaces covered with grey and brown clay in the shaft system of the entrance section; dripstone formation is rare. At the active level of the horizontal branch of the cave system, the grey limestone is covered with a black, manganese-iron crust, which is probably of bacterial origin. In the horizontal branch, the relative relief is significant, exceeding 10 m and not only downwards. The hydrological system of the István-lápa Cave is very complex and not yet fully clarified. In the upper, inactive, dry zones, instead of the black manganese crust, white, spherical dissolved, porous wall surfaces are typical. The separate, huge shaft systems of the upper level, approaching the surface at 55 m, were only discovered in the eighties of the last century, thanks to the development of single-rope technique-related technology (Sárkányfal Hall, Bea Branch, Meteor Branch; Fig. 13a, d, e). Forms of erosion-corrosion origin (stone pockets, pedestal rocks, stone ridges), scallops, and karrentified wall

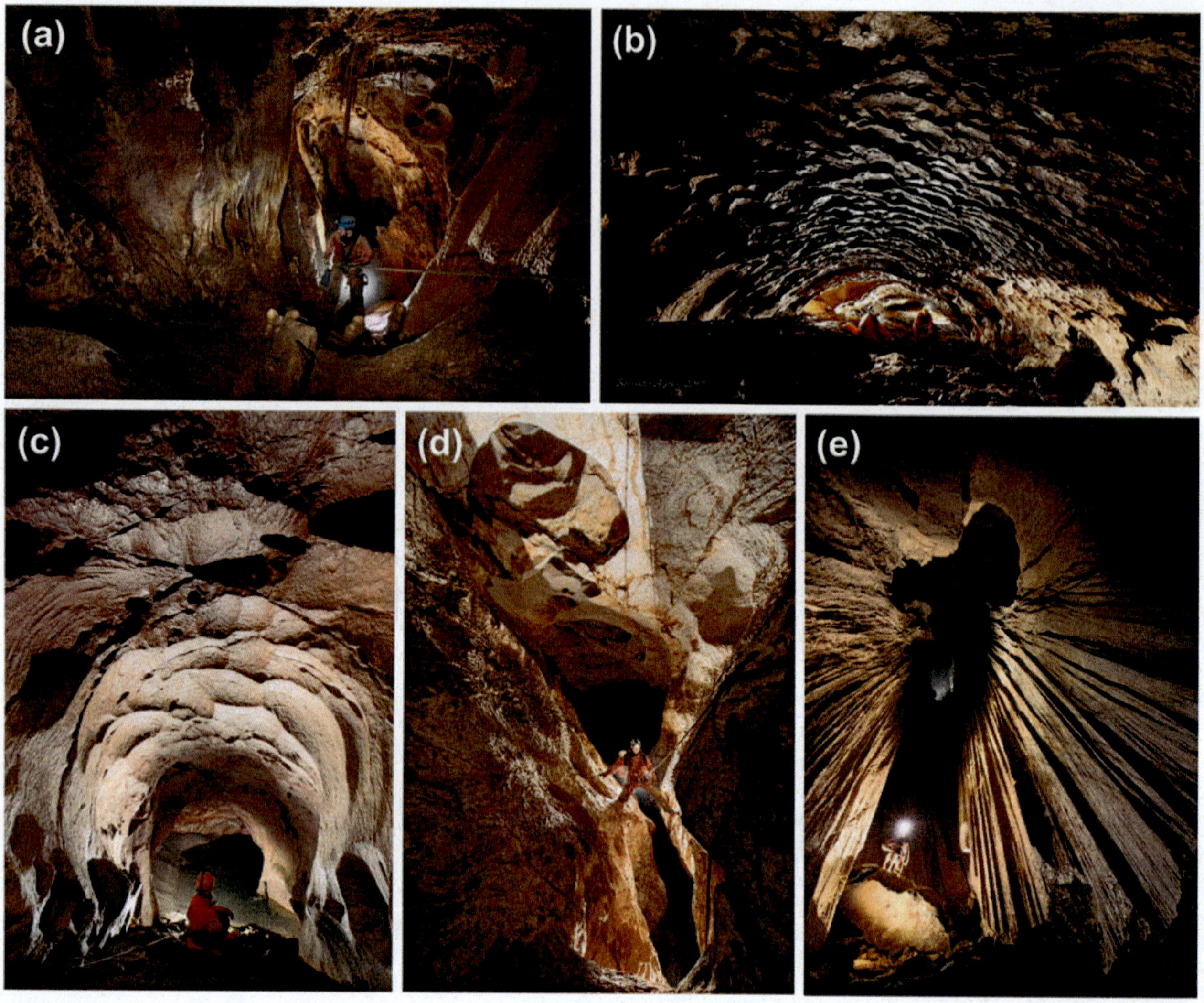

Fig. 13 István-lápa Cave. **a** In the drainage shaft system of the 250 m deep cave there are ladders, but the half tubes starting from the horizontal branch, which also approaches the surface, can only be accessed with the help of rope technology. The picture shows the Sárkányfal (Dragon Wall) Shaft (photo by Berentés); **b** In front of Siphon 3 of the streamy Nyugati-ág Branch, the vault is decorated with spectacular solution forms (photo by Berentés); **c** The Siphon 4 of the Nyugati-ág Branch is the endpoint of the cave (photo by Berentés); **d** The upper section of the Keleti-ág Branch is the Föld szíve (Earth's heart) section (photo by Berentés); **e** The Meteor-kürtő Chimney developed above the Keleti-ág Branch (photo by Berentés)

surfaces (Fig. 13b) are common in the corridors connecting the shafts, and in the case of siphons, smoothly dissolved pocket-like forms are typical. Various dripstone formations (Fig. 13a), such as dripstone runoffs, dripstone crusts, stalagmites, stalactites, and helictites can be found throughout the cave (Nyerges 2003c).

Due to its prestigious extent, varied form and hydrological significance, it has been declared as a highly protected cave since 1982.

2.12 Jáspis Cave

The extremely narrow entrance of the Jáspis Cave was discovered in 1993 on the northern slope of Szárdóka Hill on the Nagy-fennsík plateau (Figs. 1 and 2) with the excavation of the periodically active Szilfás-nyelő Ponor. It got its name from the reddish jasper stones that washed into the cave from the drainage basin of the ponor. It is the second deepest cave in the Bükk Mountains (Table 12).

Table 12 Data of the Jáspis Cave

General information	
Cadastral number	5372-100
Region	Bükk Plateau
County	Borsod-Abaúj-Zemplén
Municipality	Miskolc
Genetics	Periodically active ponor cave
Length	1051 m
Vertical extent	189 m
Depth	189 m
Protection	Highly protected

The passages of the Jáspis Cave were formed on the boundary of the Middle Triassic limestone (Nyavalyás Limestone Section) and dolomite (Hámor Dolomite Formation), mainly along bedding plane (Székely 2002). Thus, both dissolution and erosion were involved in the formation of the current morphology of the cave. Accordingly, in some sections of the passages the erosive forms are typical, in others the soluble forms dominate. In some places, limestone alternates with dolomite. In these places, the 5–20 cm thick dolomite beds which are more impervious to dissolution are made of less impervious limestone (Ferenczy 2003a). Due to the intensive erosion and dissolution, the passages of the cave, except for the narrow sections of the entrance, are characterized by a large cross-section (Székely 2002).

Regarding the hydrology of the cave, it is worth mentioning that in rainy periods, four 40–100 m side branches drain water. In this case, the water flow of the main branch can exceed 300 l/min (Ferenczy 1993). The water absorbed in the siphon, which can be considered as the endpoint, together with the water of the Balekina Cave, comes to the surface in the Wekerle Spring in the Garadna Valley (Ferenczy 2003a).

On several sections concretions are characteristic. In the Helictite Hall, there are several batches made of aragonite-aragonite-calcite material (Fig. 14b, d, e) developed from one point. What is interesting about them is that aragonite as a concretion formed in warm water caves. The Jáspis Cave, on the other hand, is of cold water origin. Concretions of aragonite from cold water are very rare, occurring only in the presence of certain metals (strontium) acting as catalysts. Based on this feature, it can be said to be unique in Hungary, as no similar phenomenon has been described so far. At the bottom of the Heliktites Hall, there is an imposing dripstone column about 4 m high (Ferenczy 1993, 2003a). Cave pearls can also be found in some parts of the cave (Fig. 14a). Limestone solutions are also interesting (Fig. 14c).

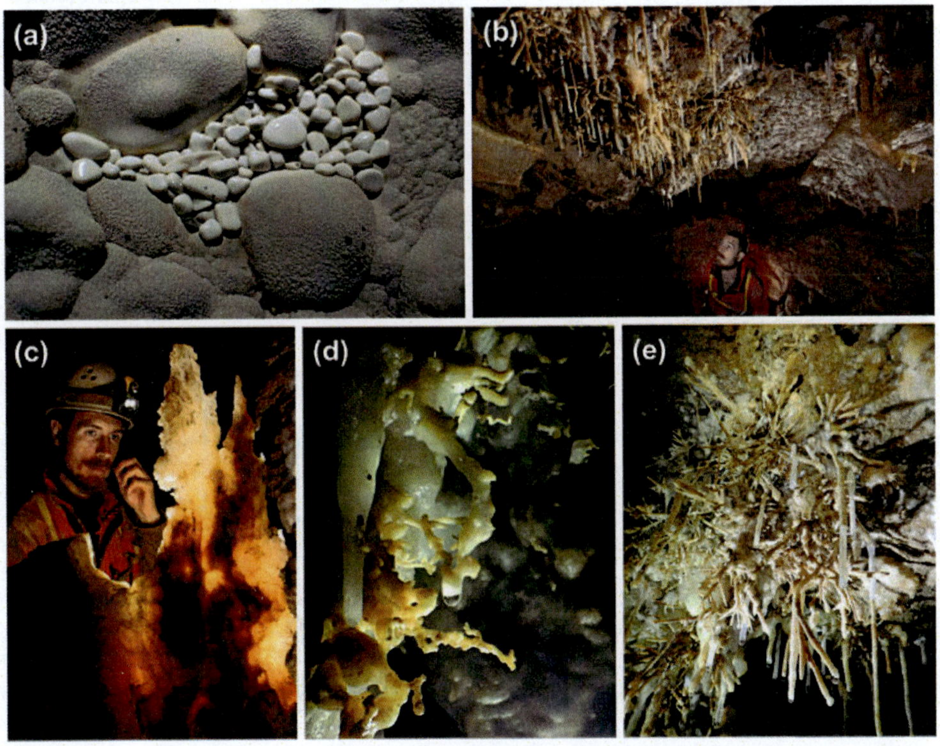

Fig. 14 Jáspis Cave. **a** Carbonate-rich dripping water collected in the depressions turns the debris into "cave pearls" (photo by Berentés); **b** The most beautiful part of the cave is the Heliktites-terem Hall (photo by Berentés); **c** Limestone residue dissolved in a whimsical form (photo by Berentés); **d** The material of the helictites is aragonite (photo by Berentés); **e** Diverse formations reminded the explorers of the Lechuguilla Cave in New Mexico (photo by Berentés)

The Jáspis Cave has been under high protection since 1996, not only because of its genetics and mineral values but also because of its considerable depth.

2.13 Kecske-lyuk Cave

The entrance to the Kecske-lyuk Cave opens on the northern side of the Forrás Valley, west of Miskolc (Table 13; Figs. 1 and 2), at an altitude of 256 m above sea level. Since it is located almost on the outskirts of the city, it is easily accessible. Its imposing entrance visible from afar is 6 m high and 6 m wide (Fig. 15a). More than 100 m can be visited wearing street clothes (Fig. 15b), making it a popular destination for tourists.

The periodically active spring cave was formed in Triassic limestone (Székely 2002) along the N-S direction fissure. In its initial and final stages it takes on a southwestern direction, but it develops mostly perpendicular to this direction (Ferenczy 2003b). In the cave there is a meandering stream channel, which is especially characterized by perpendicular changes of direction. The cave system does not conduct water for most of the year. At that time, pretty much the whole cave can be easily explored, from the entrance to the doline located southwards under the ridge of the Hársas Hill, which is the catchment of the system. During heavy rains, only the entrance section is walkable. The cave stream then drains the amount of rainfall absorbed into the 3.5-ha doline mentioned above. Its channel is dissected by smaller stairs and rimstone bars.

It is filled with mainly Eocene-Miocene clay and quartz gravel washed in from its drainage basin. The latter played a major role in the erosion development of the cave system. Highly developed abrasion platforms of erosion origin located on several levels on the outer part of the meanders can be observed (Ferenczy 2003b). The position of these abrasion platforms can be explained by the fluctuation of the incoming water volume.

Table 13 Data of the Kecske-lyuk Cave

General information	
Cadastral number	5363-2
Region	Bükk Plateau
County	Borsod-Abaúj-Zemplén
Municipality	Miskolc
Genetics	Periodically active spring cave
Length	480 m
Vertical extent	23 m
Depth	0 m
Protection	Highly protected

After the completely barren entrance section, the cave is characterized by dripstones (Fig. 15b, c, d). The most common are white as well as yellowish-brown dripstone runoffs, micro-rimstone runoffs and dripstone bumps. Unfortunately, apart from these dripstone formations, no other types of dripstones can be found nowadays, as visitors have broken them off. On the vault of the 20 m long and 8 m high entrance hall, there is an upward extending chimney, which approaches the surface but does not reach it. Due to the once flowing and turbulent water, there are solution pockets and dissolved channels on the vault (Ferenczy 2003b). The latter probably developed during paragenesis.

In the case of the Kecske-lyuk Cave, faunistic studies must be mentioned. The rich fauna of the cave has been known for a long time. In the 1940s, 45 animal and plant species were counted. The cave is particularly rich in arachnids, but the bat population is also significant.

The Kecske-lyuk Cave was placed under archaeological protection in 1951 as one of the first excavation points of the Bükk Prehistoric Research Program. It was later placed in a highly protected category due to its geological curiosity and extremely rich fauna.

2.14 Kis-Kőhát Shaft Cave

On the northeastern side of Kis-Kőhát Hill (Figs. 1 and 2), which is located on the southern edge of Nagymező, the entrance of the Kis-Kőhát Shaft surrounded by pine trees opens high above the meadow. Its large entrance is located in a 30 × 15 mdepression (Fig. 16a; Székely 2002). Ottokár Kádić first started researching the cave, which has long been known to the locals, in 1929, looking for traces of the prehistoric man of Bükk Mountains. The current low point of the cave was reached in the 1990s, but the trying circumstances made further explorations impossible.

The passages of the shaft cave (Table 14) were formed in Middle-Upper Triassic limestone (Bükkfennsík Limestone Formation). Based on the forms and the sediment filling of the cave passages, both dissolution and erosion played a role in their formation (Leél-Őssy 1954). The tectonic preformation of the cave passages can be observed here as well as in most of the caves in the Bükk Mountains (Fig. 17). The walls of the Óriás Hall show several fracture surfaces with different inclinations. The most characteristic fracture surface is parallel to the longitudinal axis of the chamber, and the other fractures are perpendicular to it (Kovács 2003). The chamber is formed at the intersection of the fractures. Corrosion development is the main characteristic along these fractures. Debris once carried by water flowing at the mouth of the cave played a leading role in shaping the further passages of the shaft cave. According to Kovács (2003), the development of the passages can be attributed mainly to the Pleistocene,

Fig. 15 Kecske-lyuk Cave. **a** The characteristic triangular entrance to the Kecske-lyuk Cave is 6 m wide and 6 m high and opens just 5 m above the Forrás Valley at the eastern end of the Kis-fennsík plateau (photo by Egri); **b** The 1–2 m wide, almost horizontal cave passage meanders (photo by Virág); **c** The most beautiful dripstone formation in the overall barren cave (photo by Virág); **d** Although there are some nice dripstone runoffs in the cave (photo by Virág)

when, due to the prominence of the mountain range, the waters accumulated in the thinning Miocene debris cover were deeply absorbed. When the cover sediment was denuded, the drainage role of the cave ceased, and consequently the development of the passages caused by erosion has stopped as well. From that point in time the branches developed due to the dissolving effect of the leaking waters.

The 117 m deep cave consists of three larger and some half tubes developed in parallel, as well as several smaller shafts and the Óriás Hall. There is a horizontal section of passage intersecting the main shaft at a depth of 40–50 m. The half tubes on the vault of this section suggest that the cave is filled with water to a depth of 25–30 m during heavy rains. Perpendicular, horizontal and vertical passages were formed in the deeper parts of the inactive shaft cave. The stress of enclosing rock was probably higher here, which may explain the zig-zag development.

The dripstone formations of the cave are varied (Fig. 16 c). Typically, ribbed stalagmites, dripstone columns, dripstone crusts, dripstone runoffs, small dripstone pools and botryoids occur. Under the Óriás Hall, a rope left by former speleologists is covered by dripstone now (Fig. 16b). In some places, these formations started to erode naturally (Kovács 2003).

The paleontological study of the cave was carried out by Kordos (1976). Based on the remains found in the filling of the Óriás Hall, the age of the sediment filling could be recorded. In addition, it was possible to determine a new biolayer unit, the Kőhát section. Due to the relatively cold (4–6 °C) average temperature, the cave provides excellent winter accommodation for thousands of bat colonies (Székely 2002).

Based on its geological value and significant bat population, it was declared highly protected in 1982.

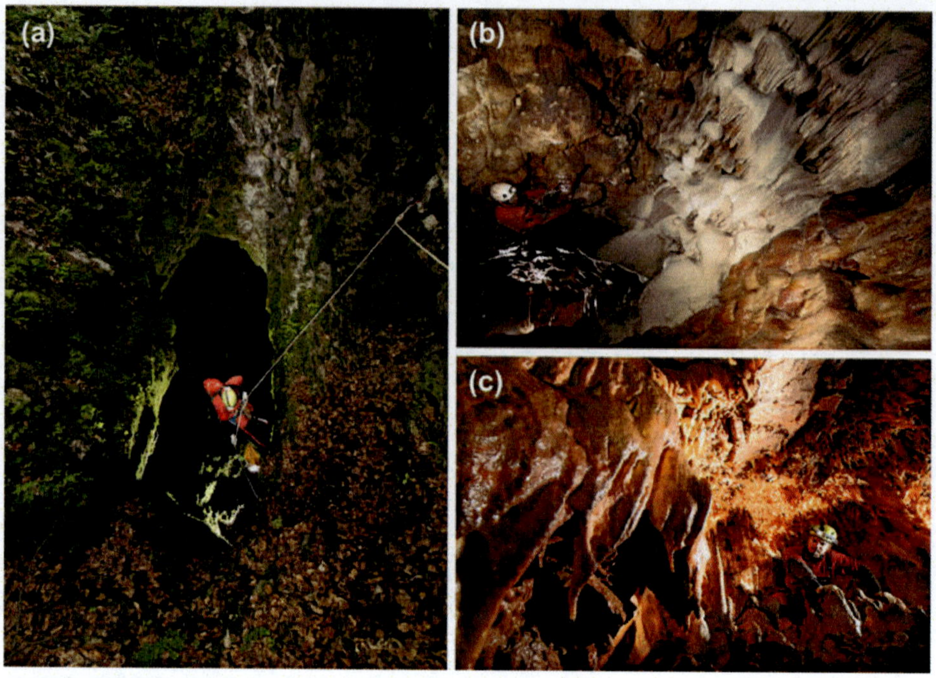

Fig. 16 Kis-kőhát Shaft. **a** The Kis-kőhát Shaft starts with a 35 m deep shaft (photo by Berentés); **b** An old rope covered with dripstone can also be seen in a dripstone runoff in the shaft under the 40 deep, 30 × 50 m Óriás-terem Hall (photo by Berentés); **c** Helictites are also found at the bottom of the dripstone flags in the lowest section of the cave (photo by Berentés)

Table 14 Data of the Kis-Kőhát Shaft Cave

General information	
Cadastral number	5343-3
Region	Bükk Plateau
County	Borsod-Abaúj-Zemplén
Municipality	Nagyvisnyó
Genetics	Inactive shaft cave
Length	480 m
Vertical extent	117 m
Depth	117 m
Protection	Highly protected

2.15 Létrás-Vizes Cave

It is an active ponor cave (Table 15), situated on the plateau Nagy-fennsík (Figs. 1 and 2). Its natural entrance (Entrance I) opens at an altitude of 565 m, on the boundary of karstic and non-karstic rocks. A seasonal creek from the non-karstic area enters the cave through this entrance. The cave has an artificial entrance as well (Entrance IV), situated at an altitude of an additional 5 m, artificially dug and connecting into a chimney of the cave. A permanent stream also enters the cave through the Tavi-nyelő ponor; this is clogged up with debris and is not suitable for human entry. The surface (morphological) catchment area of the cave is 1.8 km^2; approximately half of this area is well-karstified limestone, the other half is non-karstic rocks (clay shale, volcanics, tuffs and sandstone covered by the detritus of present-day creek bed—see on Fig. 18a, b, d; Lénárt 1983).

It was mentioned the first time in an article by László Jakucs in 1951. Its exploration took place in many stages between 1963 and 1982 (Kordos 1984; Kovács 2003; Lénárt 2003).

To get an insight into the processes ongoing in the cave, a lengthy diverse measurement series has been carried out (Lénárt 1977a, 1986a, 2005). Therefore the cave has become the most well-researched cave of Hungary and the Bükk Mountains from the perspective of cave processes.

The first map of the cave was drawn by the MHT Karsztvízkutató Szakosztálya team in the winter of 1963 and 1964, using a miner's compass. Since then, other partial and complemented maps have been prepared, but the Tavi-ág Branch still only has the polygon line recorded (National Cave Registry). The length of the cave is 3757 m (Fig. 2), and its grade difference is 74 m (Fig. 19).

The cave was formed in the Fehérkő Limestone Formation and is a multi-level sinkhole cave with erosion type passages mostly. The passages were shaped by corrosion in the first stages of cave formation, next, there was an erosion stage, and later corrosion played a role again. The effect of gravitation was also present to a small extent in the laminated limestone. Five levels can be distinguished in the cave, of which the lowest level (Level V) is only accessible for humans at certain places and only for a few meters at those

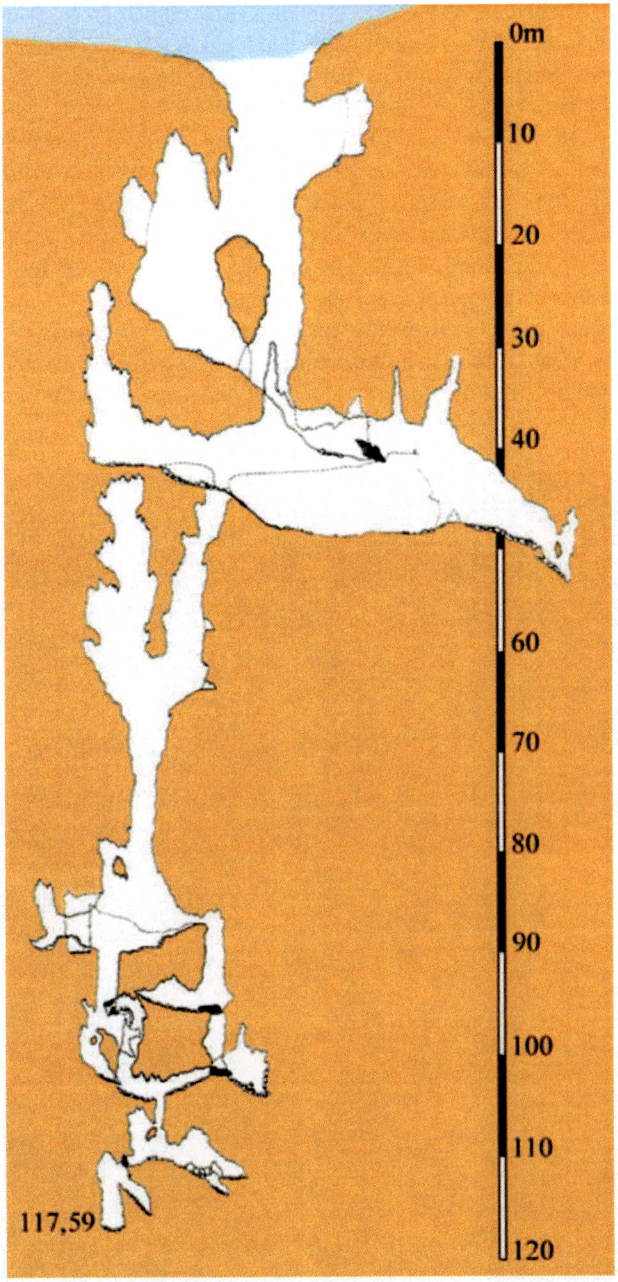

Fig. 17 The profile of the Kis-kőhát Shaft. National Cave Register, BEAC, Elekes, Huber 2000

locations, but most of the water flow takes place here. The cave creek appears on the fourth level after this level has filled up with water.

The natural entrance of the cave opens at the boundary of clay shale and limestone (Lénárt 1977b). The various limestone beds can be determined based on the exploration of the cave. The lowermost laminated limestone, about 20 m thick, can be considered "boundary limestone". This is followed by an approximately 30 m thick red and banded ("firestone") thinly laminated limestone, easily distinguishable from the main mass of the formation, a 350–400 m thick grey, thick-bedded limestone. (Most of the passages of the cave are situated in this stage.) The calcium carbonate content of the rock is 70% in the case of the laminated limestone, and 93–94% in the case of the other two types. Considering trace elements, the amount of Sr is significant in the area of Tó (Lake).

In wintertime, the ice formations at the natural entrance resemble dripstones. (In cold winters, the ice tongue can stretch in as deep as 20 m.) The angle of dip of the strata is 18–22°, gradually steepening, and from approximately the middle of the cave, it is a permanent 52–55°. (The direction of dip is a permanent north, north-northeast.) About 60% of the passages have a north-northeast–south-southwest line of strike. The main direction of the cave was shaped by the strata's direction of dip, and the faults' line of strike (Majoros and Lénárt 1975, Lénárt 1976).

Between 1970 and 1995, measurements of the air, water and soil temperature were taken using 24 mercury thermometers. The temperature inside a cave is determined by the relationship between the creek water and the inflowing cold air. The air temperature is 2–8 °C, depending on the season and the distance from the entrance; in the innermost parts of the cave it is a permanent 8 °C. Water temperature is between 3–6 °C (Lénárt 1975, 1983, 1987).

The average radon content of the cave is 2 kBq/m^3 (with a maximum of 9 kBq/m^3); in wintertime, it is almost 0 and has a background value, based on the decade-long measurement series of water, air and clay cave fill (Somogyi and Lénárt 1986; Hakl et al. 1997).

The streams on the fourth level and the level of the Tó (Lake) mark the maximum of the local karst water level (Fig. 18c). The creek is only active at Entrance I in springtime and in case of more intense precipitation; its yield is 5–10 l/min on average. From the side of the Tavi-nyelő sinkhole the creek is active almost all year long; its yield is an estimated 300–400 l/min. Since 2004, measurements are being taken continually with the frequency of sampling once every 15 min to record the water level, temperature and conductivity of the Tó (Lake). The water level raised 5.5 m in 2010, in the biggest karst flood of the last 60 years. The temperature of the water was 9.3 °C, and the fluctuation was between 6 and 11.5 °C. Water conductivity was 0.2 miliSi/cm, and the interval of change was 0.15–0.5 miliSi/cm (Hernádi et al. 2014; Lénárt et al. 2014; Lénárt 2018).

The Tó (Lake) usually has a surface of 10 m^2 and a depth of 2.5 m, but sometimes it can dry out. Based on water tracing, most of the water leaving the Lake appears in the springs of Szinva Valley, and some surfaces in the springs of Garadna Valley (Sárváry 1969; Kovács 2003).

Table 15 Data of the Létrás-Vizes Cave

General information	
Cadastral number	5372-2
Region	Bükk Plateau
County	Borsod-Abaúj-Zemplén
Municipality	Miskolc
Genetics	Active ponor cave
Length	3757 m
Vertical extent	74 m
Depth	59 m
Protection	Highly protected

The waters of the cave contain hydrogen-carbonate, with a hardness of 9–17 nk°, and unusually high sulphate content. The softest waters are the dripping waters in the cave (Lénárt, 1980).

On one occasion, CO_2 content measurements were carried out in the entire length of the cave. The highest value was 0.45 tf% at Lake, which is the deepest point and the endpoint of the cave as well (Ek et al. 1986).

On the fourth level, 90% of the detritus of the fresh sediment originates from the current surface, a non-karstic catchment area; it mostly consists of shale (well-worn and flat), and 10% is local limestone, lime formation debris and coating. The number and amount of trace elements are

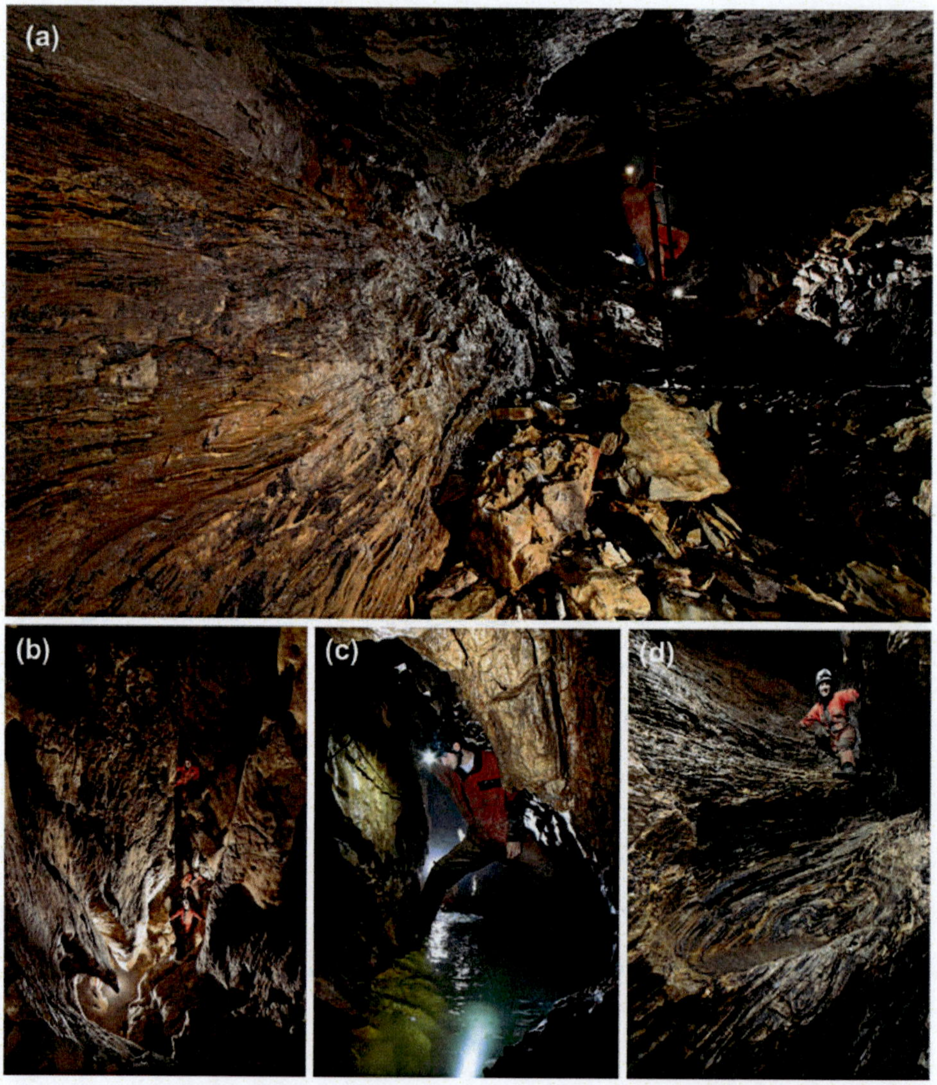

Fig. 18 Létrási-vizes Cave. **a** In the Hágcsós-terem Hall of the Létrás-vizes Cave plate limestone forms the side-wall (photo by Berentés); **b** The labyrinth after the Sóhajok hídja (photo by Berentés); **c** The siphon of the Patakos-ág Branch (photo by Berentés); **d** Foldings in the slightly metamorphized Triassic period limestone in the Zebra-folyosó Corridor (photo by Berentés)

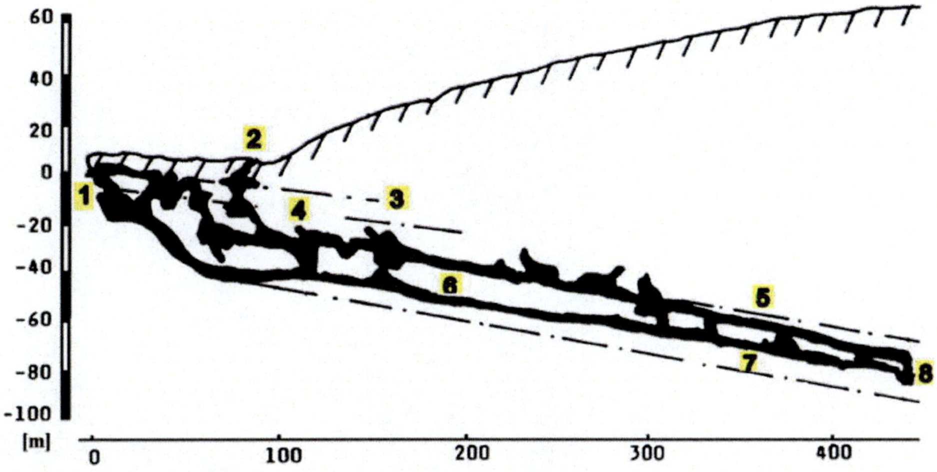

Fig. 19 A simplified vertical section of Létrási-vizes Cave (Lénárt 1983). Legend: 1. I. (naturally) entrance; 2. IV. (artificial) entrance; 3. Level I.; 4. Level II.; 5. Level III.; 6. Level IV.; 7. Level V.; 8. Lake

higher than the usual in case of limestones. Al, Si, S pollution was detected in the cave popcorn and stalactite formations, and the Ba content was also high. The paleontological vertebrate findings in the sediment haven't indicated a fauna from earlier than the Upper Pleistocene. The spore and pollen found are younger than the Pliocene. In the inner parts of the cave, the white, pumice-like rock debris, round quartz gravel and mica considered to be the remnants of the Miocene sediment bed are present in a significant number, and the average amount of heavy minerals is 5–6% (Kovács 1991).

Based on over 20 years of bat observation, mainly Myotis and Rhinolophus species are present in the cave, partly in greater colonies and partly scattered (Lénárt and Balla 1989, 1991). The greatest number of bats in the cave was recorded in the winter of 1989–1990; it consisted of 245 specimens.

The growth of the lime formations is taking place in stages. Precipitation and dissolution are taking place sequentially. Using the coating on a drip yield vessel, we detected 0.08 mm/year dripstone growth rate. The growth with the greatest intensity originates from water with low sulphate content (Lénárt 1986b).

2.16 Mexikó-völgy Ponor Cave

The entrance to the Mexikó-völgy Ponor Cave is located in a rock wall a few meters above the channel of the Mexikó Creek in the Mexikó Valley (Fig. 1) on the border of the village Bükkszentlászló, which administratively belongs to Miskolc (Fig. 20b). The cave was discovered by cave speleologists from Miskolc in the spring of 1969 and explored with nearly two decades of digging work.

Contrary to its name, the cave is not a ponor cave, as it was not formed on a rock boundary, it merely drained and occasionally drains the water of the Mexikó Creek. The enclosing rock of the cave passages is Triassic limestone (Bükkfennsík Limestone Formation, Székely 2002). As the drainer of the surface waters, its passages were shaped by both karstic corrosion and erosion. The cavity system is characterized by a complex space form. The cave was formed on several levels and has a zig-zag development. The system with a branch length of 414 m was developed on a total area of 54 × 45 m (Szabó and Verespusztai 2003). The cave is a typical shaft cave that is dissected with gradually deepening shafts (Fig. 20e; Table 16).

It can be divided into two parts by its morphology (Fig. 21). The main branch of the cave slopes slightly to a depth of about 35 m. The slightly sloping sections are connected by a few meters deep, steep-walled shafts. Half tubes formed on the vault are more common in these parts. These passages of the system are inactive (Szabó and Verespusztai 2003). The passages in the deeper parts of the cave (deeper than 35 m) have a steep slope. These steeply sloping sections are connected by deep (more than 15 m) near-vertical shaft sections. These sections are characterized by laterally developed half tubes. This part of the cave is periodically active.

It can also be divided into two parts by its speleothems. In the inactive passages close to the surface, the most characteristic forms are the various concretions. The dripstones, with their size and richness of colour, give this part of the cave a unique look. Stalactites, draperies, dripstone flags, helictites and various coloured crusts are typical (Szabó and Verespusztai 2003). The image of the deeper parts is different. Here, periodically active watercourses prevent the formation of secondary formations or damage them (Székely 2002). These parts are characterized by a variety of solution forms, solution pockets formed by mixing corrosion, and incisions caused by other erosion.

The water circulation of the cave is very much dependent on the weather. It only gets water from the Mexikó Creek

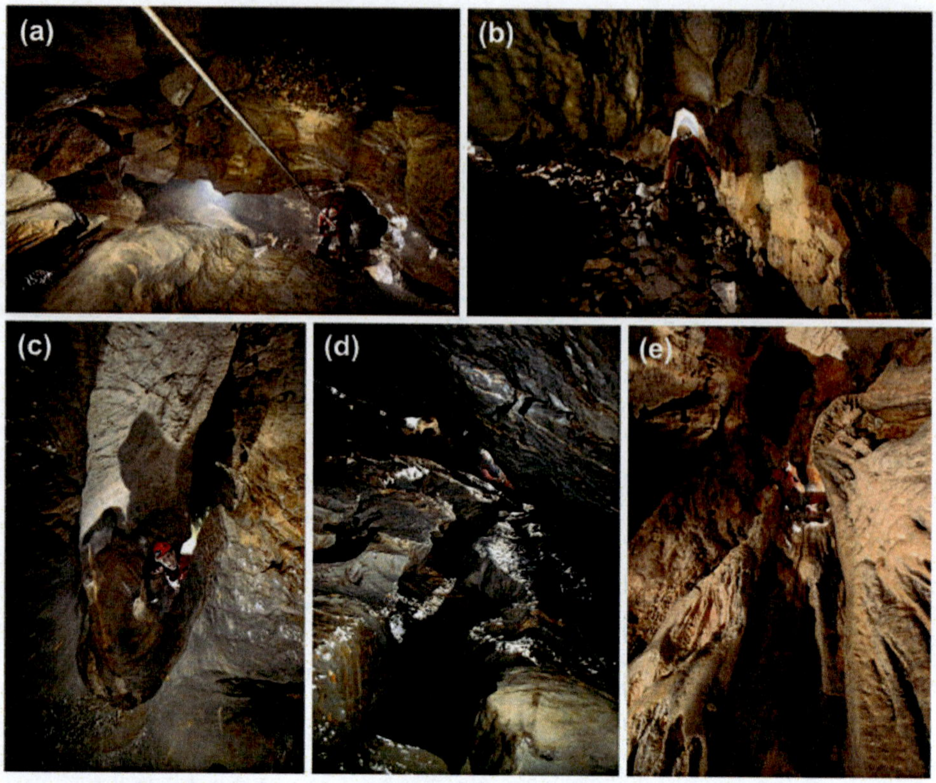

Fig. 20 Mexikó-völgy Sinkhole Cave and Pénz-patak Sinkhole Cave. **a** The Jutka-kürtő Shaft in the Pénz-patak Ponor Cave (photo by Berentés); **b** Clastic entrance sextion of the Mexiokó-völgy Ponor Cave (photo by Berentés); **c** Part from the creeky branch of the Pénz-patak Ponor Cave (photo by Berentés); **d** The famous Nagy-fal (Great Wall) of the Pénz-patak Ponor Cave (photo by Berentés); **e** Entrance shaft of the Mexikó-völgy Ponor Cave (photo by Berentés)

Table 16 Data of the Mexikó-völgy Ponor Cave

General information	
Cadastral number	5392-37
Region	Southern Bükk
County	Borsod-Abaúj-Zemplén
Municipality	Miskolc
Genetics	Periodically active shaft cave
Length	414 m
Vertical extent	79 m
Depth	79 m
Protection	Highly protected

during heavy rains and snowmelt. It has only three periodically active locations with concentrated water entry. Presumably, in these three sections, the water of the creek seeps through the rock cracks of the channel of the Mexikó Creek, reaching the deeper parts along the strata.

Due to its morphological and hydrological significance, it has been highly protected since 1988. In the event of sudden water intrusions, the cave can also be life-threatening, so it has been closed.

2.17 Miskolctapolca-Tavas Cave

The Tavas Cave (Cave with lake, Fig. 1) was already known for the prehistoric man; the cave was developed into a cave spa in 1959, following a long preparatory period. Pea gravel was spread on the bottom of the cave, and electric light was installed (Kessler and Megay1963). In 1959, Kessler considered both the air of the cave (containing radium emanation) and the thermal karst water originating both in the cave and coming from other springs to have medicinal properties. The springs were covered by concrete in 1969 and 1970, but they sometimes still find ways to surface.

Beginning from the end of the 1970s, a new tunnel was drifted west of the cave. Some of the caverns thus discovered were connected to the Cave spa. It has been planned for a long time to start a warm air cave therapy in one of the new large caverns.

The only warm spring of Miskolctapolca is the Termál Spring, and it has an underwater cavern at the bottom. Its water contains Ca–Mg–HCO_3 (Fig. 22), and its temperature is 30–31 C. Its daily yield is 3700 m^3, two-thirds of which provides the water supply of the spa complex. (The springs once situated in the cave and its area disappeared after

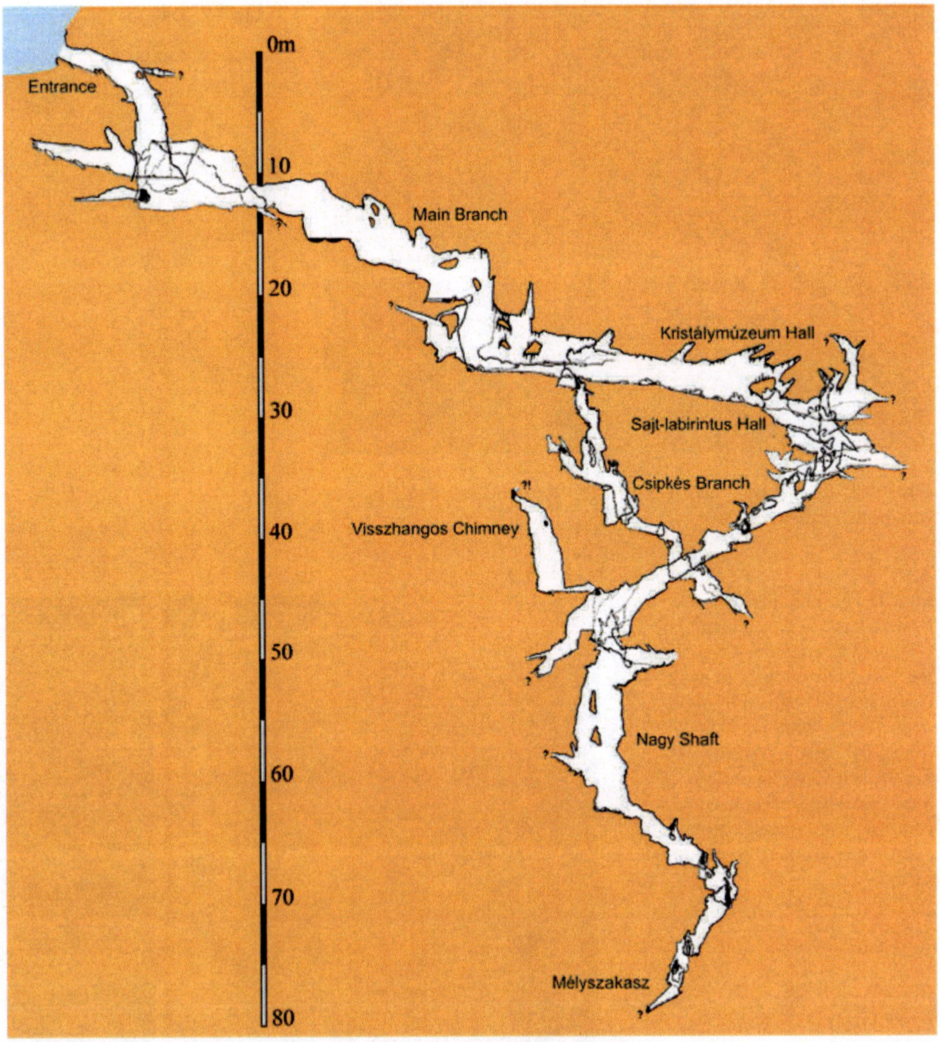

Fig. 21 Longitudinal profile of the Mexikó-völgy Ponor Cave. National Cave Register, Pizolit Barlangkutató Sportegyesület (Pizolit Cave Research Sport Association), Szabó 1998

tapping the spring.) The typical formations in the Tavas Cave are the chimneys developing upwards from the bottom, containing niches or rows of niches, and also the very fast lime precipitation, often with rimstone (Lénárt and Stuhán 1999). A 30 °C isotherm, which runs in about 1300 m depth in the Bükk Mountains, is on the surface here.

Warm waters (Table 17) upwelling from the depths created many caverns (including the Tavas Cave) with corrosion in the Vár Hill. Mining in the area between 1891 and 1936 destroyed many of these caves or opened up cave ruins; for instance, the strictly protected Fecske-lyuk Cave. The Miskolctapolca-Tavas Cave is one of the most visited caves in the Bükk Mountains. Its touristical role is significant. Since it is easy to visit them, the spring caves of the mountains have the same attribute; for example, the Suba-lyuk Cave (Fig. 23a).

2.18 Pénz-patak Ponor Cave

The entrance to the Pénz-pataki Ponor Cave, which drains the water of the Pénz Creek, opens in the Csúnya Valley between Kerek Mountain and Kis-kerek Mountain. The roughly triangular entrance between the large stone blocks continues in a vertical downward passage. The exploration of the system took place in 1953; at this time the speleologist team led by László Jakucs reached the siphon at a depth of 140 m.

The ponor cave (Table 18) formed in the Triassic limestone (Bükkfensíki Limestone Formation) is the most developed member of the series of caves located between Répáshuta and Hollóstető. The caves formed in this area drain the water of streams flowing through the Middle-Upper Jurassic shale on the surface (Csepreghy 2003). As these

Fig. 22 Limestone concretions of the Dögönyöző in the Miskolctapolca-tavas Cave

Table 17 Data of the Miskolctapolca-Tavas Cave

General information	
Cadastral number	5392-1
Region	Bükkalja
County	Borsod-Abaúj-Zemplén
Municipality	Miskolc
Genetics	Hydrothermal cave
Length	561 m
Vertical extent	38 m
Depth	20 m
Protection	Highly protected

Fig. 23 Suba-lyuk Cave and Szeleta Cave **a** North from the village of Cserépfalu, 45 m above the Hór Valley, the Suba-lyuk Cave, rich in prehistoric finds, opens in the Southern Bükk (photo by Egri); **b** The entrance to the Szeleta Cave (photo by Virág); **c** The Szeleta Cave is comfortably walkable almost along its entire length (photo by Egri)

streams transport significant amounts of sediment from non-karstic areas, in addition to tectonic factors and corrosion, erosion also plays a significant role in the formation of cavities in these caves (e.g. Jutka Chimney, Fig. 20a). Marks of this can be seen in the narrow, 2–3 m high, stepped erosion corridor embellished by a few dripstones following the entrance debris zone (Székely 2002). A crack wall of about 60 m, the Nagy Fal (Great Wall, Fig. 20d), then leads to the siphon at the endpoint. Based on the morphology of its passages, the cave is divided into two parts. Up to a depth of 60 m from the entrance, the passages are zig-zagged, stepped sections. There are a significant number and size of half tubes in this zone. Dripstone formations also occur in places (the largest stalagmite is 1 m high). The walls (covered with scallops) of 1–5 m cascades and waterfalls offer a particularly impressive sight (Csepreghy 2003). From a depth of 60 m to the endpoint of the cave, the passages have a large slope (the branch called the Great Wall is almost vertical) and the maze-like structure is not typical.

The water of the Pénz Creek flows from the entrance of the cave to the siphon forming the endpoint (Fig. 20c). According to the results of the water dye tracing carried out in 1960, the waters absorbed in the depths of the cave appeared in the Lator-út-, Szinva- and Margit Springs. Later,

Table 18 Data of the Pénz Patak Ponor Cave

General information	
Cadastral number	5381-2
Region	Southern Bükk
County	Borsod-Abaúj-Zemplén
Municipality	Bükkszentkereszt
Genetics	Active ponor cave
Length	1989 m
Vertical extent	155 m
Depth	150 m
Protection	Highly protected

only the relationship with the Szinva Spring could be proved (Csepreghy 2003).

Based on its hydrological significance, it was declared a highly protected cave in 1982.

2.19 Pes-kő Cave

The huge, 8 m high and 2 m wide entrance of the Pes-kő Cave opens to the southwest of the Pes-kő Hill on the southern edge of the Nagy-fennsík plateau. This cave has been known for a very long time; Elek Fényes mentioned it as early as 1851 (Fényes 1851).

The Pes-kő Cave was formed in the Middle Triassic limestone (Bükkfennsík Limestone Formation). Its formation was determined by two tectonic fault lines (Hír and Regős 2003). The 62 m long (Table 19) cave consists of an average 12 m high hall (Székely 2002). The fillings of the spring cave, which has now become inactive, include materials brought from the chemical weathering of the rock and water, loess supplied by the wind and debris of frost weather origin. Several excavations were carried out by specialists in the twentieth century to examine these beds. By studying the sediments, conclusions were drawn about the prehistoric climate (Vértes 1956, 1965). It was found that there may have been a steppe-like environment, with a gradual cooling indicated by the appearance of some alpine and arctic species. The real peculiarity of the archaeological find found here is that such Aurignacian culture material as in the Pes-kő Cave has so far been known in Hungary only from the Istállós-kő Cave (Ringer 1989; Regős and Ringer 2003d).

The cave has been under increased protection since 1982 due to its paleontological and archaeological significance.

Table 19 Data of the Pes-kő Cave

General information	
Cadastral number	5343-1
Region	Bükk Plateau
County	Heves
Municipality	Felsőtárkány
Genetics	Inactive spring cave
Length	62 m
Vertical extent	10 m
Depth	0 m
Protection	Highly protected

2.20 Szamentu Cave

The entrance to the Szamentu Cave, known as the longest system in the Kis-fennsík plateau, opens from a permanently active ponor (Barátságkerti-visszafolyó Ponor) located southeast of the Örvény-kő Peak. Its first 50 m were already explored in the middle of the twentieth century, but the progress was blocked by a clay plug. Thus, further exploration of the cave took place in 1967, new halls and passages were discovered even in 1998 (Kárpát 2003).

The cave was formed in Middle Triassic (Ladini) limestone. The 945 m long cave (Table 20; according to Székely 2002, its length is 975 m) drains the waters of the Barátság-kerti Spring. Therefore, a small watercourse almost always trickles in the cave, and it sometimes swells to a significant size during rainy periods (Kárpát 2003). The initial section of the cave is a narrow corridor about 120 m long, dissected by different solution forms. Most of the cave can be visited only by crawling, and even that is significantly hampered by the few centimetres of water that is constantly present in the stream channel. The passage first narrows and then widens more and more, all the way to the Teenager Hall (Fig. 24). The Teenager Hall is one of the largest cave halls in the Bükk Mountains. The steeply sloping, 10–15 m high room, with a maximum width of 25 m and a length of 35 m,

Table 20 Data of the Szamentu Cave

General information	
Cadastral number	5361–43
Region	Bükk Plateau
County	Borsod-Abaúj-Zemplén
Municipality	Varbó
Genetics	Active ponor cave
Length	945 m
Vertical extent	41 m
Depth	36 m
Protection	Highly protected

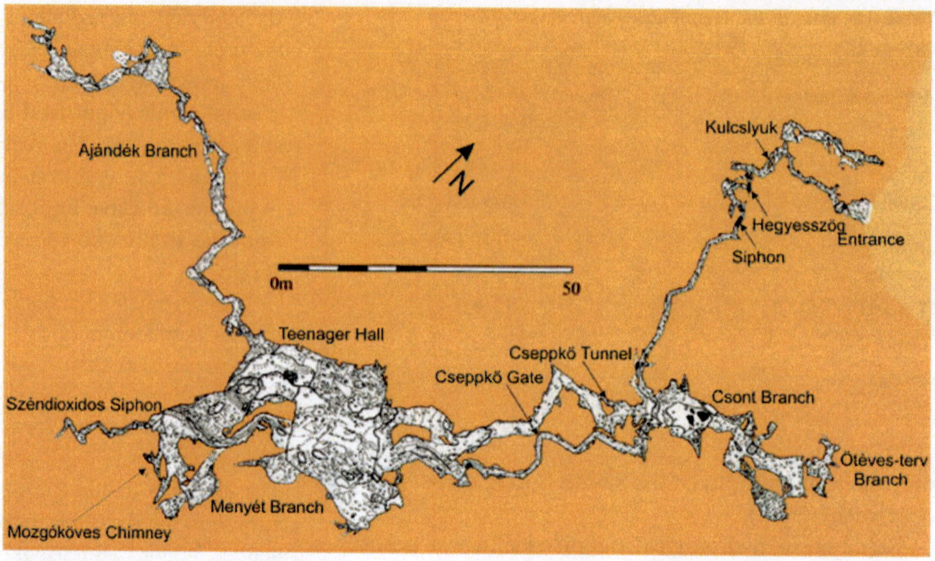

Fig. 24 The floor plan of the Szamentu Cave. National Cave Register, Szabó 2000

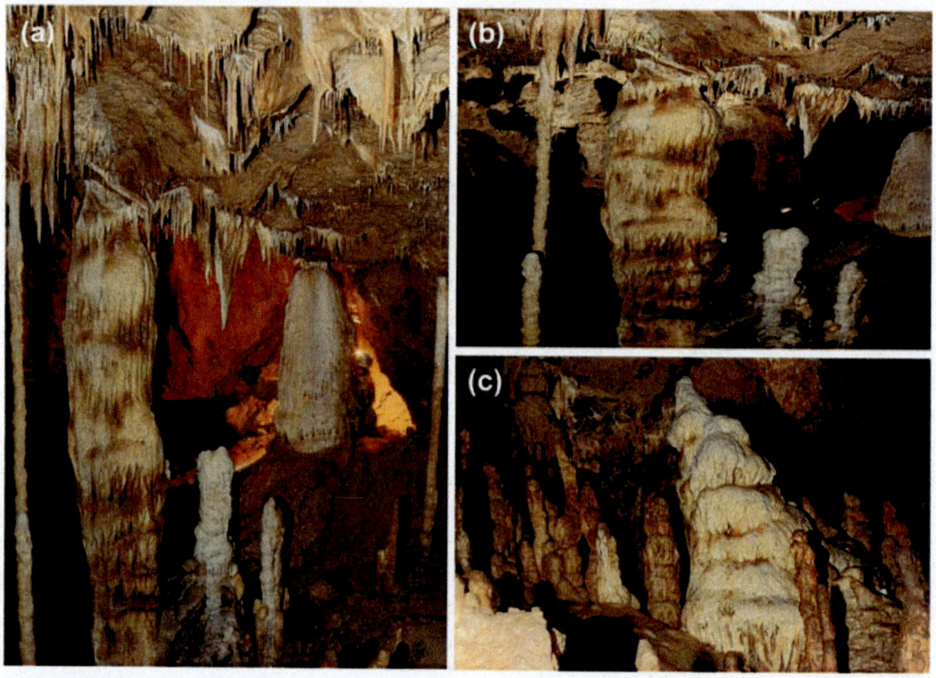

Fig. 25 Dripstone formations of the Szamentu Cave

is extremely rich in dripstone formations (Székely 2002; Fig. 25). Waist-thick, 3–5 m high stalagmites, dripstone candles, dripstone columns, dripstone flags and rimstones decorate the hall. Thin, a few centimetres or decimetres long stalactites hang from the vault of the hall. The vault of the room is perhaps not even 20 m deep below the surface, so it has previously been suggested that it could be made accessible by an artificial shaft. The Teenager Hall is connected to the Erzsébet Hall by a short passage. This chamber of untouched beauty was found in 1998. This chamber is smaller in size and poorer in formations (Kárpát 2003).

Certain parts of the cave cannot be explored due to the occasional accumulation of carbon dioxide (2–3%). The cave was declared highly protected in 1982. This was mainly due to its unique richness of formations.

2.21 Szeleta Cave

The most significant prehistoric find in Hungary comes from the Szeleta Cave (Table 21), which opens on the slope of the Szeleta Peak on the rocky hillside above the Puskaporos

Table 21 Data of the Szeleta Cave

General information	
Cadastral number	5363-1
Region	Bükk Plateau
Count	Borsod-Abaúj-Zemplén
Municipality	Hámor
Genetics	Inactive spring cave
Length	115 m
Vertical extent	16 m
Depth	3 m
Protection	Highly protected

Table 22 Data of the Szent István Cave

General information	
Cadastral number	5372-1
Region	Bükk Plateau
Count	Borsod-Abaúj-Zemplén
Municipality	Miskolc
Genetics	Inactive spring cave
Length	1470 m
Vertical extent	101 m
Depth	24 m
Protection	Highly protected

Valley (Szinva Valley; Fig. 23b). Its historical significance lies in the fact that in the first years of the twentieth century, the regular and professional archaeological excavations of Hungarian caves and the processing of the found finds began here (Hillebrand 1911c). The most important archaeological finds are discussed in Chap. 17, so they are not mentioned here.

The enclosing rock of the Szeleta Cave is a Middle Triassic (Ladini) limestone (Fehérkő Limestone Formation, Székely 2002). The pinkish, yellowish, light red breccia characteristic of the formation can be observed at several points in the cave (Regős and Ringer 2003e). This breccia formation coincided with the deposition of the limestone, so it is syngenetic. In places where breccia settlements are absent, karstic forms are characteristic. These are mainly curved, rounded shapes, but there are also shapes in the Cseppköves Hall that are like hieroglyphs and wasp nests. In the more hidden parts of the cave there are also Karren forms (finger pads, clint pools). In some places, fragmented wall portions caused by frost weathering are typical (Regős and Ringer 2003e). There is a clearly visible black stripe on the wall of the hall, marking the level of the former filling of the cave (Lénárt 1979).

The spring system once led the waters of the plateau above it to the Szinva Stream, which was still close to it at this level. The only monumental hall of the now inactive spring cave (Fig. 23c), like most of the prehistoric caves, is relatively poor in speleothems, concretes, and the existing ones (dripstones) are in a state of extinction. As a result of natural redissolution, they appear as layered, inner-shelled, sometimes radially structured formations (Regős and Ringer 2003e). During winter, very spectacular ice-stalagmites with a height of 1 m, with a thickness of an arm, develop in the near-entrance part. The vault of the hall is dissected by a few chimneys. Sometimes bats can be observed in these. Two gradually narrowing side branches start from the hall. The left is narrower and shorter, the right is wider. At the end of the latter is the Cseppköves Hall mentioned above.

The cave was placed under archaeological protection in 1951, due to the fact that the bay leaf-shaped stone tools found here are the eponym of the world-famous prehistoric Szeleta Culture (Kordos 1984). In 1982, it was declared highly protected.

Table 23 Data of the Szepes-Láner Cave System

General information	
Cadastral number	5372-3
Region	Bükk Plateau
Count	Borsod-Abaúj-Zemplén
Municipality	Miskolc
Genetics	Periodically active shaft cave
Length	2500 m
Vertical extent	159 m
Depth	159 m
Protection	Highly protected

2.22 Szent István Cave

The artificial entrance of the cave (Figs. 1 and 2) opens in the Szinva Valley at an altitude of 318 m. Its natural entrance is a spring chimney (the so-called "Kutya-lyuk"), situated 18 m higher. The "Kutya-cave" was published in 1921 by Kadić. Pávai Vajna (1931) considers some parts of the cave to be of hydrothermal origin (Table 22). The deepest point of the cave is the Pokol (Hell), filled up with mining debris.

The exploration of the cave sections possible to visit today started in 1927, the same time when the building of the Palotaszálló (Hotel Palota) began, and the Anna Cave was re-opened. In 1928, work started to drift a 55 m long artificial tunnel and to install lights in the cave, and it was opened to the public on 20 August 1931 (Fig. 26a, c).

Owing to the excessive lighting in the cave, algae, moss and fern growth appeared on the surfaces (Lénárt and Bereczki 1987); in 1984, 60 algae, 13 moss and 5 fern species were documented. As a consequence, the so-called

Fig. 26 Szent István Cave. **a** The most visited cave in the Bükk Mountains is the Szent István Cave, which is rich in driptone formations (photo by Egri); **b** To bypass the difference in level there are stairs in the cave (photo by Egri); **c** The endpoint of the built section is spectacular as well (photo by Egri); **d** The Oszlopok Csarnoka Hall is the most famous part of the Cave (photo by Egri); **e** The Bástya built in 1931 (photo by Egri)

"biodripstones" were formed (Lénárt 1985), meaning "biologically activated dripstone development". (Today the green growth is minimal, thanks to the radical diminishing of lighted time.)

In 1990, speleotherapy sessions were started in the Fekete Hall; patients with respiratory problems had the chance to spend 3 h in the Hall each day for 3 weeks with the supervision of hospital staff. Following the measurements published by Szabó in 1963, radon, drip yields, air temperature, humidity and draught measurements took place between 1988 and 2003 (Juhász 1962; Juhász et al. 2019), and later water-level observation as well. In the chamber where the therapy sessions take place, the air temperature is a constant 10.1 °C, and it might be raised by the lighting fittings and the patients by a further 1.5 °C maximum. (The complete regeneration of the temperature takes place at night and at weekends.) The hardness of dripping water is 16–21 nk°, the Rn content of the air is 860 in summertime and 250 Bq/m^3 in the winter, and the humidity is 8.8 g/m^3 (95–100%, Lénárt et al. 1992).

The total length of the passages is 1470 m, and its vertical expansion is 101 m. Tourists can visit the main passage in 150 m length (Fig. 26b, d, e). The section where the cave therapy takes place is 60 m, outside of the show route.

2.23 Szepes-Láner Cave System

The Szepesi-Láner Cave System (Table 23) was formed in Triassic Fehérkő Limestone (Figs. 1 and 2). The entrance of

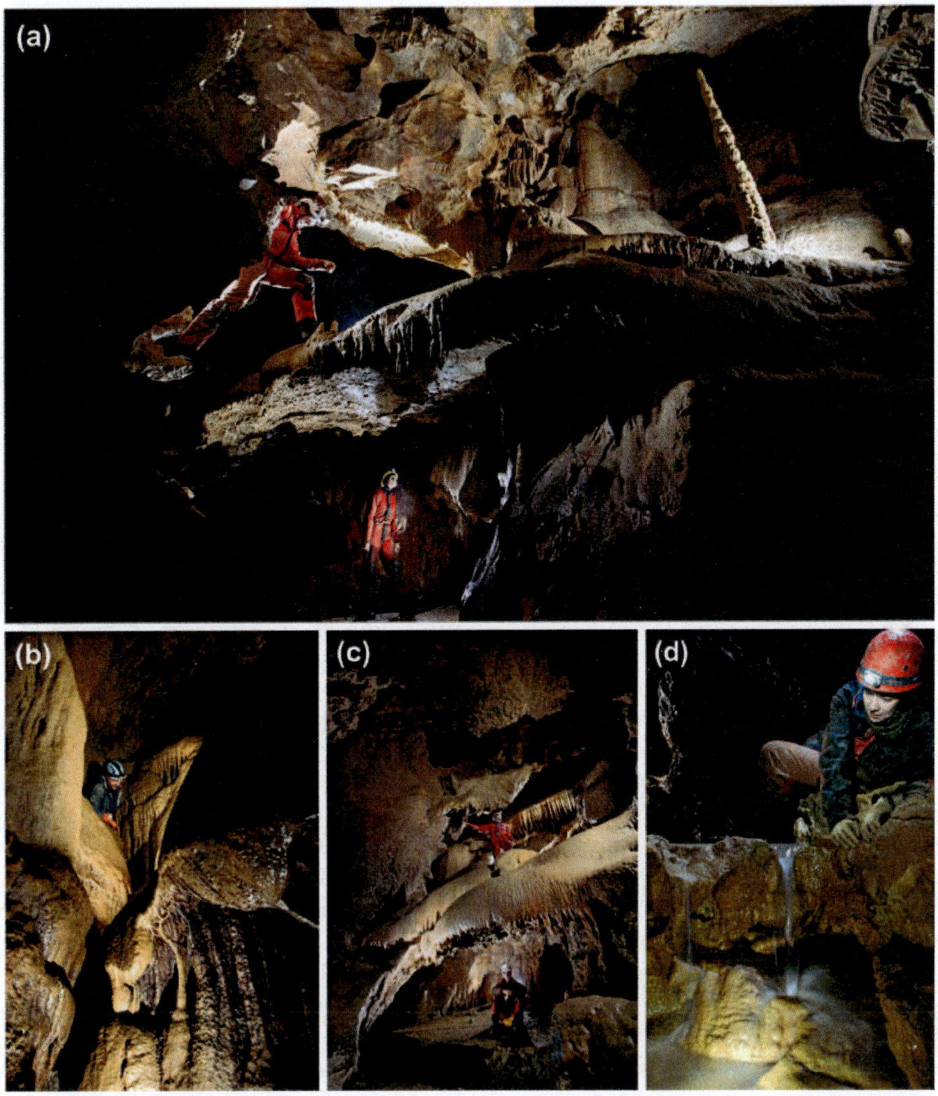

Fig. 27 Szepesi-Láner Cave System. **a** Dripstone formation named Kis-baldachin in the Patakos-ág Branch (photo by Berentés); **b** A part of the Nyugati-ág Branch dripstone-ful section (photo by Berentés); **c** Formation named Nagy-baldachin in the Patakos-ág Branch (photo by Berentés); **d** Characteristic rimstone bars of the Három aranyásó Hall. (photo by Berentés)

the Láner Branch of the cave is located at an altitude of 548 m and opens at the bottom of a steep rock wall in a doline. When the cave was first found and explored in 1952, it was named Spirál Cave. Following the discovery of significant new cave passages in 1985, the cave was re-named after speleologist Olivér Láner. The other entrance of the cave, leading to the Szepesi Branch, is more widely known, and it is situated about 600 m to the East in a hillside insignificant from a karstology perspective, at an altitude of 552. (Both entrances of the cave system are secured with a locked steel door each.) The full exploration of the cave has taken many years of work by two cave exploration teams, the MHT Zsombolykutató Szakosztálya and Miskolci Bányász Barlangkutató Csoport. The horizontal branch of the cave was reached in 1962 through the shaft system by István Bödör, Tamás Kutas and Kálmán Molnár (Kovács 1992).

The cave system is a strictly protected, seasonal sinkhole cave, with shaft systems leading into the active water conduit passage. The Szepesi Entrance is muddy and, at places, fairly tight. Iron ladders were installed in the 130 mshaft system to help with ascending and descending. The formations in the "Szintes ág" (horizontal passage) are beautifully rich in shapes and dripstones. The flowstones in the Nyugati-ág passage cover hundreds of square meters, and the baldachin canopies about 10–12 m^2 (Fig. 27b). At certain locations, there is only one path to go on: high above the level of the stream, between the stalagnats. At the same place, a series of sinter barriers form storeys both horizontally and "vertically". The pools behind the sinter barriers are rarely empty of water (Lénárt 1978a).

This is one of the most beautiful caves in the Bükk Mountains (Fig. 27a–d). The end of the Nyugati-ág passage

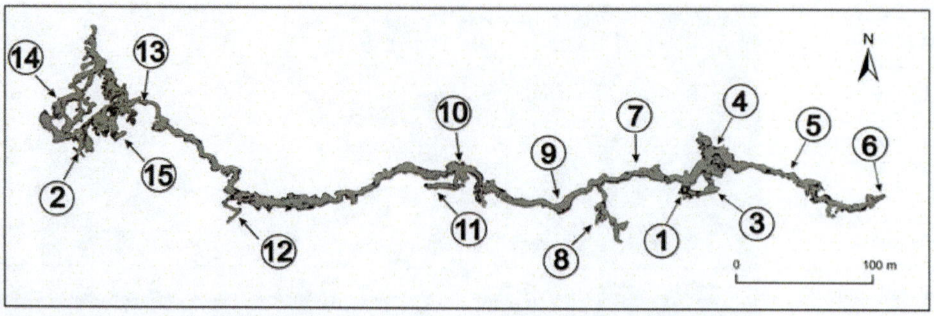

Fig. 28 Simplified floor plan of the Szepes Láner Cave System (Assessed by the Marcel Loubens Cave Research Group 1984–2004, edited by Kovács 2005). Legend: 1. Entrance of the Szepes Láner Cave System from the Szepes-ág Branch; 2. Entrance of the Szepes Láner Cave System from the Láner-ág Branch; 3. Shaft system; 4. Három aranyásó Hall; 5. Keleti-ág Branch; 6. Lake; 7. Start of the Nyugati-ág Branch; 8. Borsó-ág Branch; 9. Kis Baldachin; 10. Nagy Baldachin; 11. Kristály-tavi Branch; 12. Tündér-tavi Branch; 13. Sand siphon, the connection point of the Szepes and the Láner caves; 14. Fofó-ág Branch; 15. Láner-ág Branch

is a series of siphons filled up with water and sand. These were dug through after long and persistent work by the members of the Marcel Loubens Barlangkutató Egyesület speleologist team. Thus they connected the Szepesi Cave with the Láner Olivér Cave; this latter had also been known already all the way to the siphons filled up with sand. There are side flues and short and tight side passages near the siphons. The connecting Láner Entrance passage consists of dry shafts, some of which are parallel, and there is some really spectacular dripstone formations.

The lower part of the Keleti-ág passage is still active and rich in dripstones. It ends in a siphon lake with a surface of about 15–18 m^2, but until now, cave divers were unable to conquer it. The greatest water level fluctuation detected took place in 2010, at the time of the karst flood in the Bükk Mountains; it reached 6 m. (It was recorded using a measurement tool installed in the cave, recording with a frequency of one measurement per 15 min; the water was increasing for 9 h, the peak flow lasted for 15 h, and it took 24 h for the water to decrease and reach the original level.) Until 2006, the water level fluctuations of the lake never exceeded more than a few centimetres—measured with a gauge.

In 1984, the temperature at the upper part of the shaft system was between 4.8 and 6.2 °C and at the middle and lower part it was between 7.5 and 8.0 °C; however, at the endpoint (deepest point—at the Lake), the temperature was only a steady 7.2 °C. The radon level is a stable low of 0.6 kB/m^3 in the shaft system (Hakl et al. 1993).

The entire length of the cave system is about 2500 m (Fig. 28), and its greatest depth is 159 m. Due to its size and beauty, the cave is strictly protected. This is the first cave system in the Bükk Mountains which was created by connecting two caves.

2.24 Szivárvány-Sebes Cave System

The Szivárvány-Sebes Cave System was created by connecting two caves, the Szivárvány and Sebes Caves. The entrance to the Szivárvány Cave opens from a doline on the northern edge of the Nagy-fennsík plateau (Figs. 1, and 2; Table 24). The entrance to the Sebes Cave is located not far to the north (Fig. 1). Its exploration began in the 1950s. At that time, its length was estimated at 4–5 km (Jakucs 1951). The caves belonging to the Sebes Spring's drainage area (including the Szivárvány and Sebes caves) were explored by members of the Bányász Kutató Csoport (Miner Cave Research Group) in the 1960s and 1970s. The connection between the two caves was discovered in 2009. It is the only typical through-cave in the Bükk Mountains.

The cave system's enclosing rock is the excellent karstifiable Fehérkői limestone (Fehérkői Limestone Formation). This limestone is bordered on the north by a non-karstic rock (metandesite). Thus, a local water system was created in the well-karstified rock (Sűrű and Piri 2014), establishing the ponor character of the cave system (Fig. 29a, b).

Table 24 Data of the Szivárvány-Sebes Cave System

General information	
Cadastral number	5372-23
Region	Bükk Pateau
County	Borsod-Abaúj-Zemplén
Municipality	Miskolc
Genetics	Through cave
Length	3500 m
Vertical extent	106 m
Depth	106 m
Protection	Highly protected

Fig. 29 Szivárvány-Sebes Cave System. **a** The steeply sloping bedding planes of the Szalámis Hall in the Szivárvány Sebes Cave System connected a few years ago (photo by Berentés); **b** The side-wall and the vault of the Szalámis Hall is made by bedding planes (photo by Berentés)

The cave system has two levels. Morphological research has revealed that active springs and sinkholes have been relocated several times to a lesser extent. This is evidenced by the scallops of passages and the cave floors that head in the direction of each sinkhole. The relocation of these springs and sinkholes was caused by the valley incision and the re-swelling effect of the debris generated from the surface and the debris of the cave (Sűrű and Piri 2014).

The Szivárvány-Sebes Cave System is extremely rich in dripstone formations. This is due to the fact that although its deepest point is at a depth of 106 m, the distance of most branches from the surface is less than 50 m. The inactive upper sections in particular are rich in formations. Particularly rare formations, dripstone drums and radish-shaped stalactites can also be found in the cave system.

The cave system has been highly protected since 2005. This is due to its great length, richness in formations and unique character.

3 Conclusions

Of all the Hungarian mountains, the Bükk Mountains are the richest in caves. According to the cave registration system, there are 1157 caves in the area on about 60 × 60 km. Thus, the cave density typical of the area is 0.32 caves/km^2, which is considered high.

The caves of the Bükk Mountains are characterized by large dimensions in terms of both length and depth. Caves of outstanding length are an interconnected system of several caves at a short distance from each other. Such system is the Bolhás-Jávorkút Cave System, the assumed but unproven total length of which can exceed 15 km. The formation of such systems is made possible by the well-karstifiable Triassic limestone that makes up these mountains. This rock is tectonically heavily stressed, full of fractures and faults. This feature also advances the development of long branches. However, there are also unique caves that do not connect to any cave system, and still have a significant branch length (István-lápa-Cave, Hajnóczy Cave).

In some places, limestone is surrounded or bordered by non-karstic rock. In these locations, ponor caves are formed, which drain the surface waters, providing a water entry point into the karstic rock full of fractures and faults. Sometimes these ponor caves are arranged in rows, which favours the formation of the above-mentioned cave systems.

Shaft caves were formed in the karstic rock in the inner parts of the mountains. These can be active or inactive. They were formed mainly by solution along the bedding plane, like the deepest cave in Hungary: the Bányász Cave, with a depth of 300 m.

Spring caves have been developed at the rim of the mountains. Most of these have become inactive by now, after the surface of the drainage area has been deepened due to denudation. Inactive spring caves are usually rich in archaeological sites; for instance, the Pes-kő Cave or the Szeleta Cave.

References

Balogh K (1964) A Bükkhegység földtani képződményei (Geological formations of the Bükk Mountains). MÁFI Almanac 48(2):245–553 (in Hungarian)

Börcsök P, Elekes B, Németh T, Nyerges A (1997) A BEAC Barlangkutató Csoport 1997. évi jelentése (Annual report of the BEAC Cave Research Group in 1997). Budapest, p 11 (in Hungarian)

Csepreghy F (2003) Pénz-pataki-víznyelőbarlang (Pénz-patak Ponor Cave). In: Székely K (ed) Magyarország fokozottan védett barlangjai (Highly protected caves in Hungary). Mezőgazda Kiadó, Budapest, pp 214–216 (in Hungarian)

Dénes Gy (1971) A fokozatosan lepusztuló vízzáró takaró szerepe az exhumálódó karszt morfológiai fejlődésében (The role of the gradually denuding impermeable cover in the morphological development of exhuming karst). Karszt és Barlang I:5–8 (in Hungarian)

Ek C, Gábris Gy, Hevesi A, Lénárt L (1986) Teneur en CO_2 de l'air de quatre grottes de hongrie. In: Proceedings of the 9th international congress of speleology I. 89–93 (Barcelona)

Fekete Zs (2020) Földtani térkép a Nagyfennsík K-i részéről (Geological map of eastern part of the Nagy-fennsík plateau) (Less 2002, M=1:50 000 geological map, using data from National Cave Registry and OpenStreetMap contributors) (in Hungarian)

Ferenczy G (1993) A Jáspis-barlang feltárása, kutatási eredményei (Exploration and research results of the Jáspis Cave). Karszt és Barlang I–II:16–20 (in Hungarian)

Ferenczy G (2003a) Jáspis-barlang (Jáspis Cave). In: Székely K (ed) Magyarország fokozottan védett barlangjai (Highly protected caves in Hungary). Mezőgazda Kiadó, Budapest, pp 130–132 (in Hungarian)

Ferenczy G (2003b) Kecske-lyuk (Kecske-lyuk Cave). In: Székely K (ed) Magyarország fokozottan védett barlangjai (Highly protected caves in Hungary). Mezőgazda Kiadó, Budapest, pp 167–170 (in Hungarian)

Hakl J, Lénárt L, Hunyadi I (1993) Radon measurements in Caves and Springs located in the area of the Bükk National Park. Newest results in research, protection and use of caves of Bükk Mountains. Miskolci Egyetem, 11–13. November 1993. I:5–13. Miskolc

Hakl J, Hunyadi I, Csige I, Géczy G, Lénárt L, Várhegyi A (1997) Radon transport phenomena studies caves-international experiences on radon levels and exposures. Radiat Meas 28(1–6):675–684

Hegedűs A (2004) A Bolhási-Jávorkúti-víznyelőbarlang új részeinek feltáró kutatása és dokumentálása (Exploratory research and documentation of the new parts of the Bolhás-Jávorkút Ponor Cave). Karsztfejlődés IX: Szombathely, 359–369 (in Hungarian)

Hernádi B, Czesznak L, Juhász B, Kovács P, Lénárt L, Tóth M, Tóth K (2014) Karsztárvizek keletkezése a Keleti Bükkben (Formations of the karstfloods in the Eastern Bükk). Karsztfejlődés XIX:105–124, Szombathely (in Hungarian)

Hevesi A (1978) A Bükk szerkezet- és felszínfejlődésének vázlata (A Sketch of the Structure and Landscape Evolution of the Bükk Mountains). Földrajzi Értesítő XXVII(2):169–203 (in Hungarian)

Hevesi A (1980) Adatok a Bükk hegység negyedidőszaki ősföldrajzi képéhez (Data to the Quaternary Paleogeographical Features of the Bükk Mountains) Földtani Közlöny 110(3–4):540–550 (in Hungarian)

Hevesi A (1986a) A Déli-Bükk karsztja II. rész: Délnyugati Bükk (The karst of the Southern Bükk Part I: Southwestern Bükk) Karszt és Barlang II:87–95 (in Hungarian)

Hevesi A (1986b) A Déli-Bükk karsztja I. rész: Délkeleti-Bükk (The karst of Southern Bükk Part I: South-eastern Bükk). Karszt és Barlang I:3–14 (in Hungarian)

Hevesi A (1986c) Rombarlangok a Bükkben (Collapsed caves in the Bükk Mountains). Nehézipari Műszaki Egyetem Közleményei 33 (1–4):167–179 (in Hungarian)

Hevesi A (2002) Felszínalaktani jellemzés, karsztformakincs (Geomorphological characterisation, karst landform). In: Baráz Cs (ed) A Bükki Nemzeti Park, Bükki Nemzeti Park Igazgatóság, Eger, pp 109–148 (in Hungarian)

Hevesi A (2019) Valamit a mészkőbarlangok és a mészkőszurdokok származási kapcsolatáról (About the relation of the origin of limestone caves and limestone gorges). Karszt és Barlang, Különszám, 23–30 (in Hungarian)

Hillebrand J (1911a) A Répáshutai Ballabarlangban talált diluviláis gyermekcsontok maradványai (Remains of the diluvial pediatric bones found in the Balla Cave by Répáshuta). Földtani Közlöny 41:452–464 (in Hungarian)

Hillebrand J (1911b) Az ősember csontjai a Bükkhegység Balla-barlangjában (Bones of the pre-historic man in the Bükk Mountain's Balla Cave). Földtani Közlöny 41:92–95 (in Hungarian)

Hillebrand J (1911c) A Szeleta-barlangot kitöltő rétegek geológiai koráról (Something about the strata that fill the Szeleta Cave). Földtani Közlöny 11–12:788–794 (in Hungarian)

Hillebrand J (1912) A Balla-barlangban 1911. évben végzett ásatások eredményei (Results of the diggings in the Balla Cave in 1911). Földtani Közlöny 42:763 (in Hungarian)

Hillebrand J (1914) Az 1913. évi barlangkutatásaim eredményei (Results of my cave researches in 1913). Barlangkutatás 2 (3):117–123 (in Hungarian)

Hillebrand J (1917) Az 1916. évi barlangkutatásaim eredményéről (Results of my cave researches in 1916). Barlangkutatás 5(2):106–108 (in Hungarian)

Hillebrand J (1919) Az 1917. évben végzett ásatásaim eredményei (Results of my caves researches in 1917) Barlangkutatás 7(1–4):10–13 (in Hungarian)

Hillebrand J (1935) Die altere Steinzeit in Ungarn. Arch. Hung. 17

Hír J (1985) A Hajnóczy-barlang üledékei (Sediments of the Hajnóczy Cave). Karszt és Barlang 1–2:17–24 (in Hungarian)

Hír J, Regős J (2003) Pes-kő-barlang (Pes-kő Cave). In: Székely K (ed) Magyarország fokozottan védett barlangjai (Highly protected caves in Hungary). Mezőgazda Kiadó, Budapest, pp 141–143 (in Hungarian)

Jakucs L (1951) A Bükk-hegység még feltáratlan, ismeretlen barlangrendszerei (The yet unexplored, unknown cave systems of the Bükk Mountains). Földtani Közlöny 81(3):200–201 (in Hungarian)

Jánossy D (1979) A magyarországi pleisztocén tagolása gerinces faunák alapján (Classification of the Pleistocene in Hungary on the basis of vertebrate fauna). Akadémiai Kiadó, Budapest, 205 pp (in Hungarian)

Juhász A (1962) A Létrástetői barlang (Létrástető Cave). Karszt és Barlang, 1962 II:45–50 (in Hungarian)

Juhász E, Munkácsi A, Lénárt L, Papp Á, Kiss-Tóth E, Lukács A (2019) Medical assessment of patients relating to cave therapy in Lillafüred in the Bükk. Geosci Eng 10:94–104, Miskolci Egyetem, Miskolc

Kadić O (1921) A Hámori-barlangok Borsod megyében (Hámori Caves in County Borsod). Természet 17:56–58 (in Hungarian)

Kadić O (1934) Der Mensch zur Eiszeit in Ungarn. Földt. Int. Évk. 30

Kárpát J (1984) Jávorkúti-víznyelőbarlang (Jávorkút Ponor Cave). Magyarország barlangtérképei 6, MKBT, Budapest, 1986 (in Hungarian)

Kárpát J (2003) Szamentu-barlang (Szamentu Cave). In: Székely K (ed) Magyarország fokozottan védett barlangjai (Highly protected caves in Hungary). Mezőgazda Kiadó, Budapest, pp 176–178 (in Hungarian)

Kerekes J (1936) A Tárkányi öböl morfológiája (Morphology of the Tárkány bay). Földrajzi Közlemények LXIV(6–7):80–87 (in Hungarian)

Kessler H, Megay G (1963) Lillafüred barlangjai (Caves of Lillafüred). Harmadik, átdolg. Kiadás, p 67. Miskolc (in Hungarian)

Kordos L (1974) Jelentés a Hajnóczy-barlangban végzett őslénytani és üledékföldtani vizsgálatokról (Report of the paleontological and sedimentary geological studies in the Hajnóczy Cave). Karszt és Barlangkutatási Táj 5–6:20–25 (in Hungarian)

Kordos L (1976) A Kis-kőháti-zsomboly szubfosszilis denevér populációjának vizsgálata (Examination of the sub-fossil bat population of the Kis-kőhát Shaft). A Miskolci Hermann Ottó Múzeum Évkönyve XIII–XIV:567–585 (in Hungarian)

Kordos L (1983) A bükki barlangok világa (World of the Bükk's caves). In: Sándor A (ed) Bükki Nemzeti Park – Kilátás a Kövekről (Bükk National Park—View from the stones). Mezőgazdasági Kiadó, Budapest, pp 139–171 (in Hungarian)

Kordos L (1984) Magyarország barlangjai (Caves in Hungary). Gondolat Könyvkiadó, Budapest (in Hungarian)

Kovács A (2003) 50 év barlangkutatás. Személyes, események, eredmények a Marcel Loubens Barlangkutató Egyesület és elődei 1952–2002 (50 years of cave research. Personal events, results, the Marcel Loubens Cave Research Group and its predecessors). Marcel Loubens Barlangkutató Egyesület, Miskolc, p 365 (in Hungarian)

Kovács Zs (1991) Törmelékes üledék rétegsor vizsgálata a Létrási-Vizes-barlang Y ágában (Investigation of debris sediment strata in the Y branch of the Létrás-vizes Cave). Karszt és Barlang I–II:25–34 (in Hungarian)

Kovács Zs (1992) A létrástetői Szepessy- és Láner Olivér-barlangok kutatásának új eredményei (New results of the research of the Szepessy and Láner Olivér Caves on Létrástető). A Bükk karsztjai, vizei, barlangjai c. konf. előadásai I:189–197 (in Hungarian)

Leél-Őssy S (1954) A Magas-Bükk geomorfológiája (Geomorphology of the Magas-Bükk Mountains). Földrajzi Értesítő III(3):594–603 (in Hungarian)

Lénárt L (1975) Klimatológiai mérések a Bükk-fennsíkon lévő Létrási-Vizes-barlangban (Climatological measurements in the Létrás-Vizes Cave on the Bükk-fennsík plateau). Baradla 150 nemzetközi konferencia, Budapest-Aggtelek, augusztus 26–29, MKBT kiadvány, 217–246 (in Hungarian)

Lénárt L (1976) Tektonikai vizsgálatok a Létrási-Vizes-barlangban és környezetében (Tectonical measurements in the Létrás-Vizes Cave and its surroundings). Karszt és Barlang, 9–14 (in Hungarian)

Lénárt L (1977a) A Zsombolyosoktól a Marcel Loubens Csoportig. Nehézipari Műszaki Egyetem, Miskolc, p 92

Lénárt L. (1977b) Hidrogeológiai kirándulások a Bükkben (Hydrogeological trips in the Bükk Mountains). Egyetemi jegyzet, Tankönyvkiadó, p 29 (in Hungarian)

Lénárt L (1978) A Szepesi-barlang bejárási útmutatója (Tour guide of the Szepesi Cave). Bejárási útmutató a Karszt és Barlangkutatók I. Országos Tudományos Diákköri Találkozója tanulmányútjaihoz. NME-MKBT kiadvány, 4–7 +map (in Hungarian)

Lénárt L (1979) Barlangok a Bükkben (Caves in Bükk Mountains). Miskolc, BAZ Megyei Idegenforgalmi Hivatal, 41 pp (in Hungarian)

Lénárt L (1980) A Létrási-Vizesbarlang vízkémiai vizsgálata (Hydrochemical study of the Létrás-Vizes Cave). Karszt és Barlang II:57–64 (in Hungarian)

Lénárt L (1981) Adalékok a lillafüredi mésztufabarlangok kutatásához (Additives for the research of limestone tuff caves in Lillafüred). Karszt és Barlang 1–8 (in Hungarian)

Lénárt L (1983) A Létrási-Vizes-barlang komplex barlangtani vizsgálata (Complex cavernological study of the Létrás-Vizes Cave), NME egyetemi doktori disszertáció, Miskolc (in Hungarian)

Lénárt L (1984a) A látogatók hatása a lillafüredi Anna-mésztufabarlang klímájára (The impact of the visitors on the climate of the Anna sinter cave in Lillafüred). VII. Nemzetközi Speleoterápiai Szimpózium, Keszthely-Tapolca, 1982. November 2–6, MKBT-TBCTT kiadvány, 138–152 (in Hungarian)

Lénárt L (1984b) A Miskolc-tapolcai alsó (Várhegyi, felhagyott) kőbánya barlangjai (Caves of the quarry by Miskolctapolca). BMTE Év-könyv 1983:48–55 (in Hungarian)

Lénárt L (1985) Adalékok az idegenforgalmi barlangok növényzetesedése miatt kialakuló biocseppkövek kérdésköréhez (Additives to the issue of bio dripstones formed due to the vegetation of tourist caves). Nemzetközi Lámpaflóra Kollokvium. 1984. október. 10–13. MKBT kiadvány, 117–128 (in Hungarian)

Lénárt L (1986a) A Létrási-Vizes-barlang komplex barlangtani vizsgálatának főbb eredményei (The main results of the complex cavernological study of the Létrási-Vizes Cave). NME Közleményei, I. sorozat, Bányászat 33(1–4):33–46 (in Hungarian)

Lénárt L (1986b) Karszthidrológiai vizsgálatok a Létrási-Vizes-barlangban (Karsthydrological measurements in the Létrási-Vizes Cave). A Bányamérnöki Kar kutatási eredményei 1983–1986b. június 4–5. NME-OMBKE kiadvány, 385–398 (in Hungarian)

Lénárt L (1987) A Létrási-Vizes-barlang klimatológiai vázlata (Climatologic sketch of the Létrási-Vizes Cave). Oktatási intézmények karszt-és barlangkutató tevékenységének II. országos tudományos konferenciája, 55–64 (in Hungarian)

Lénárt L (2003) Létrási-vizesbarlang (Létrási-vizes Cave). In: Székely K (ed) Magyarország fokozottan védett barlangjai (Highly protected caves in Hungary) Mezőgazda, Budapest, pp 138–141 (in Hungarian)

Lénárt L (2005) Some aspects of the "3E's" (Economics-Environment-Ethics) model for sustainable water usage in the transboundary Slovakian and Aggtelek karst region based on some examples from the Bükk Mountains. Ph.D. thesis work, Kassa/Kosice

Lénárt L (2011) A Bükk környéki termálkarszt lehatárolásának tudománytörténeti áttekintése 1970-től napjainkig (History of delimitation of the thermal karst of the Bükk region from 1970 to our days). Karsztfejlődés XVI:9–20 (in Hungarian)

Lénárt L (2018) A 25 éves Bükki Karsztvízszint Észlelő Rendszer (BKÉR) története, adatrendszere, felhasználása, további céljaink (History, data system, use of the 25-year-old Bükk Karst Water Level Detection System (BKÉR), our further goals). Műszaki Földtudományi Közlemények. A Miskolci Egyetem Közleményei 86(3):18–26 (2017) (in Hungarian)

Lénárt L, Bereczki L (1987) Fényerősség, hőmérséklet és vízkémiai mérések a lillafüredi kiépített barlangokban (Brightness, temperature and hydrochemical measurements in the built-up caves in Lillafüred). Borsodi Műszaki-Gazdasági Élet 32(1):24–26 (in Hungarian)

Lénárt L, Balla B (1989) On the fluctuation of the number of the bats living in the Letrasi-Vizes cave (Bükk Mountains, Hungary) during the years 1975–1989. In: Proceedings of International Congress of Speleology, Budapest, 13–20. August 1989. I, 316–319

Lénárt L, Balla B (1991) Denevérmegfigyelések a Létrási-Vizes-barlangban (Bat monitorings in the Létrás Vizes Cave). Borsodi Műszaki-Gazdasági Élet 36(3):45–48 (in Hungarian)

Lénárt L, Stuhán G (1999) Miskolc-Tapolca Barlangfürdő (598) p 16; Miskolc-Tapolca Höhlenbad (598A), p 16; Miskolc-Tapolca Cave Bath (598B), p 16; Miskolc-Tapolca Jaskynné Kúpele (598C), p 16. Tájak-Korok-Múzeumok Kiskönyvtára, Budapest

Lénárt L, Hakl J, Hunyadi I, Balla B (1992) Barlangterápiai célú klimatológiai mérések a lillafüredi Szent István-barlangban (Climatological measurements for cave therapy in the Szent István Cave in Lillafüred). Kossuth-emléknapok, Aggtelek-Jósvafő, szeptember 18–20:47–53 (in Hungarian)

Lénárt L, Hernádi B, Szegediné Darabos E, Debnár Zs, Czesznak L, Tóth M (2014) The importance of Bükk Karst water monitoring system in researching the relations of cold and warm Karst waters in the area. Geosci Eng 3(3):107–118

Less Gy (2002) Geological map of Bükk Mountains, M = 1:50 000. Geological Institute of Hungary, Budapest

Liebe P (2002) Felszín alatti vizeink (Our groundwaters). Környezetvédelmi Minisztérium. VITUKI Kiadvány, Budapest (in Hungarian)

Majoros Zs, Lénárt L (1975) A tektonika, a rétegdőlés és egyéb tényezők hatása a barlangjáratok kialakulásában (Influence of tectonics, stratification and other factors on the formation of caverns). V. Ifjúsági Napok, Győr, November 19–21, MHT kiadvány II:17–23 (in Hungarian)

Miklós G (1978) A Hajnóczy-barlang mikroklímája (Microclimate of the Hajnóczy Cave). Karszt és Barlang 1–2:11–18 (in Hungarian)

Nyerges A (2003a) Bányász-barlang (Bányász Cave). In: Székely K (ed) Magyarország fokozottan védett barlangjai (Highly protected caves in Hungary). Mezőgazda Kiadó, Budapest, pp 109–110 (in Hungarian)

Nyerges A (2003b) Bolhási-Jávorkúti-barlangrendszer (Bolhás-Jávorkút Cave System). In: Székely K (ed) Magyarország fokozottan védett barlangjai (Highly protected caves in Hungary). Mezőgazda Kiadó, Budapest, pp 111–116 (in Hungarian)

Nyerges A (2003c) István-lápai-barlang (István-lápa Cave). In: Székely K (ed) Magyarország fokozottan védett barlangjai (Highly protected caves in Hungary). Mezőgazda Kiadó, Budapest, pp 125–129 (in Hungarian)

OpenStreetMap contributors. https://www.openstreetmap.org/copyright/hu

Pávai Vajna F (1931) A forró oldatok és gőzök-gázok szerepe a barlangképződésnél (The role of hot solutions and vapors gases in cave formation). Hidrológiai Közlöny 10:115–122 (in Hungarian)

Pelikán P (2005) A Bükk-hegység földtana (Geology of the Bükk Mountains). MÁFI, Budapest 284 pp (in Hungarian)

Rajczy M, Padisák JP, Komáromi Zs, Lénárt L (1986) A lillafüredi Anna-barlang lámpaflórája (Lantern flora of the Anna Cave in Lillafüred). NME Közleményei, 1. sor. Bányászat 33(1–4):257–264 (in Hungarian)

Rántó A (2019) A Bányász-barlang kutatása 2018-ban (Research of the Bányász Cave in 2018), 12 pp (in Hungarian)

Rántó A, Sűrű P (2014) Még egy lépés a dobogón (A Bányász-barlang kutatása az elmúlt években) (One more step on the podium (Research of the Bányász Cave in the in the last years). MKBT Tájékoztató, március-április, 6 pp (in Hungarian)

Regős J (2002) Régészeti szempontból jelentős barlangok (Archaeologically significant caves) In: Baráz Cs (ed) A Bükki Nemzeti Park. Hegyek, erdők, emberek (National Park of Bükk. Mountains, Forests, People). BNPI kiadás, pp 315–329 (in Hungarian)

Regős J, Ringer Á (2003a) Balla-barlang (Balla Cave). In: Székely K (ed) Magyarország fokozottan védett barlangjai (Highly protected caves in Hungary). Mezőgazda Kiadó, Budapest, pp 197–199 (in Hungarian)

Regős J, Ringer Á (2003b) Büdös-pest (Büdös-pest Cave). In: Székely K (ed) Magyarország fokozottan védett barlangjai (Highly protected caves in Hungary). Mezőgazda Kiadó, Budapest, pp 158–159 (in Hungarian)

Regős J, Ringer Á (2003c) Hillebrand Jenő-barlang (Hillebrand Jenő Cave). In: Székely K (ed) Magyarország fokozottan védett barlangjai (Highly protected caves in Hungary). Mezőgazda Kiadó, Budapest, pp 165–167 (in Hungarian)

Regős J, Ringer Á (2003d) Istállós-kői-barlang (Istállós-kő Cave). In: Székely K (ed) Magyarország fokozottan védett barlangjai Highly protected caves in Hungary). Mezőgazda Kiadó, Budapest, pp 224–227 (in Hungarian)

Regős J, Ringer Á (2003e) Szeleta-barlang (Szeleta Cave). In: Székely K (ed) Magyarország fokozottan védett barlangjai (Highly protected caves in Hungary). Mezőgazda Kiadó, Budapest, pp 178–181 (in Hungarian)

Ringer Á (1989) The role of cave sites and their chronostratigraphy in the research of the Paleolithic of Hungary. Karszt és Barlang, 1989. Special Issue 51–58

Sárváry I (1969) A Létrás-Istvánlápai-barlangrendszer (The Létrás-Istvánlápa Cave System). Karszt és Barlang II:53–56 (in Hungarian)

Sárváry I (1970) A zsombolygenetika kérdéseiről (On the issues of shaft genetics). Karszt és Barlang I:5–14 (in Hungarian)

Sásdi L (1992) Víznyomjelzés a Diabáz-barlangban (Water dying in the Diabáz Cave)—MÁFI Csoportjelentés (in Hungarian)

Sásdi L (1997) Újabb földtani adatok a Bükk-hegység karsztjának fejlődéstörténetéhez (Newer Geological Data on the Karst of the Bükk Mountains). Karszt és Barlang I-II:20–25 (in Hungarian)

Simon E (1980) A Tekenősi Fekete-barlang geológiai felépítése és genetikája (Geological structure and genetics of the Tekenős Fekete Cave). Nehézipari Műszaki Egyetem Közleményei, I. Bányászat 28 (3–4):139–146 (in Hungarian)

Somogyi Gy, Lénár L (1986) Bükki barlangok radioaktivitásának vizsgálata (Investigation of the radioactivity of Bükk caves). NME Közleményei 1. sorozat, Bányászat 33(1–4):221–234 (in Hungarian)

Stieber J (1956) A Hillebrand Jenő-barlang újkőkori cölöpmaradványai (Neolithic wooden post holes in the Hillebrand Jenő Cave). Botanikai Közlemények 287–290 (in Hungarian)

Sűrű P, Piri A (2014) Szivárvány-Sebes-barlangrendszer (Szivárvány--Sebes Cave System). In: Slíz Gy (ed) Felfedezések a föld alatt. Magyarország új barlangjai 2003–2013. Discoveries under the ground. The new caves at Hungary, pp 192–196 (in Hungarian)

Szabó Gy (1963) Angaben zum Mikroklima der Höhlen bei Lillafüred. Acta Climatologica 2–3(1–4):13–31

Szabó RZ, Verespusztai G (2003) Mexikó-völgyi-víznyelőbarlang (Mexilkó-völgy Ponor Cave). In: Székely K (ed) Magyarország fokozottan védett barlangjai (Highly protected caves in Hungary). Mezőgazda Kiadó, Budapest, pp 206–208 (in Hungary)

Székely K (2002) Fokozottan védett barlangok (Highly protected caves). In: Baráz Cs (ed) A Bükki Nemzeti Park. Hegyek, erdők, emberek (National Park of Bükk. Mountains, Forests, People). Bükki Nemzeti Parki Igazgatóság, Eger, pp 177–195 (in Hungarian)

Tóth G (1983) A Bükk felszínfejlődése és mai formakincse (Geomorphic evolution and present landforms of the Bükk Mountains). In: Sándor A (ed) Bükki Nemzeti Park, Mezőgazda Kiadó, Budapest, pp 62–106 (in Hungarian)

Tóth G (1984) A Központi-Bükk és geomorfológiai körzetei (The Central Bükk and its geomorphological districts). Földrajzi Értesítő XXXIII(4):333–345 (in Hungarian)

Tóth G, Fejes P (1984) Idősebb pliocén lepusztulási szint kimutatása a Bükkben (Report on an older Pliocene denudation level in the Bükk Mountains). Karszt és Barlang II:65–72 (in Hungarian)

Varga Cs (2003) Hajnóczy-barlang (Hajnóczy Cave). In: Székely K (ed) Magyarország fokozottan védett barlangjai (Highly protected caves in Hungary). Mezőgazda Kiadó, Budapest, pp 200–204 (in Hungarian)

Veress M (2019) Shaft lengths and shaft development types in the vadose zone of the Bakony Region (Transdanubian Mountains, Hungary). J Soil Water Sci 3(1):54–74

Vértes L (1956) Ausgrabungen in der Petényi- und Peskő-Höhle. Folia Archaeologica, 9, pp 3–23

Vértes L (1965) Az őskőkor és az átmeneti kőkor emlékei Magyarországon. Akadémiai Kiadó, Budapest

The Caves of the Northeastern Part of the Transdanubian Mountains

Szabolcs Leél-Őssy and Magdolna Virág

Abstract

The thermal waters of famous springs and spas of Budapest have dissolved more than 240 hypogene caves in the Buda Hills, mostly in Eocone limestone and marl, during the Middle and Late Pleistocene. Recently we have 60 km of known cave passages mapped, chiefly in the Rózsadomb (Rose Hill) part of the city. Every cave was discovered by accident (stone quarrying, digging basements, trenching for gas and water pipes, canalization, etc.). These caves are characterized by speleothems of thermal-water origin, mostly calcite, aragonite (cave coralloids, mammillary crusts, cave rafts, frostworks) and some types of gypsum crystals, as well as shapely solution pockets, especially cupolas. Three caves, Pál--völgy Cave, Szemlő-hegy Cave and Buda Castle Cave are open to the public as show caves. The tourists can also visit Mátyás-hegy Cave for an adventure tour, and cave divers can explore warm thermal water passages in Molnár János Cave. József-hegy Cave, Citadella Crystal Cave and Királylaki Cave are accessible for research projects only. There are fewer, but very interesting, caves in the Triassic limestone of the Pilis Hills. The intensively investigated Ariadne Cave System is now nearly 20 km long. Sátorkőpuszta Cave, with many superb cupolas and gypsum precipitates, is the world type example of one of the principal hypogene cave morphologies. The Gerecse Hills are not rich in caves, but it is well worth viewing Pisznice, Lengyel and Keselő caves, which are also in the Triassic limestone.

Keywords

Hypogene caves • Speleothems of thermal-water origin • Solution pockets • Cupolas • Rózsadomb • Buda Thermal Karst

1 Introduction

Three of the four mountain blocks that compose the Transdanubian Range (~Transdanubian Mountains), the Buda Hills, Pilis Hills and Gerecse Hills are rich in caves (Fig. 1). There are none in the Visegrád Hills, which are nearest to the Danube Bend, because they are composed of Miocene volcanic rocks.

The Buda Hills are the richest in caves. The Buda Thermal Karst is famous for its hypogene caves which were created by dissolution mainly during the Pleistocene, by mixing corrosion between ascending thermal waters rich in CO_2 (and H_2S), and the infiltrating meteoric waters (e.g. Kovács and Müller 1980; Leél-Őssy Sz. 1997; Leél-Őssy Sz. and Surányi 2003; Erőss 2010; Erőss et al. 2012; Mádl-Szőnyi and Tóth 2015; Virág 2016, 2018; Leél-Őssy Sz. 2017; Leél-Őssy Sz. and Virág 2018; etc.).

2 Caves of the Buda Hills (Buda Mountains)

Altogether 240 caves are known in the Buda Hills at the present time. The aggregate length of these caves is approximately 70 km.

The genetics of Buda caves are most interesting. The difference between hypogene and epigene cave types is significant (cf. Klimchouk 2007; Klimchouk et al. 2017). Almost all caves in this area are of thermal water origin.

The Buda Thermal Karst hypogene caves are hosted mainly by Eocene limestones or marls, plus a few caves in Triassic limestones or dolomites (Fig. 2). They are thought

Sz. Leél-Őssy (✉) · M. Virág
Department of Geology, Eötvös Loránd University, Pázmány P. stny. 1/C, Budapest, 1117, Hungary
e-mail: szabolcs.leel-Ossy@ttk.elte.hu

M. Veress and Sz. Leél-Őssy (eds.), *Cave and Karst Systems of Hungary*,
Cave and Karst Systems of the World, https://doi.org/10.1007/978-3-030-92960-2_12

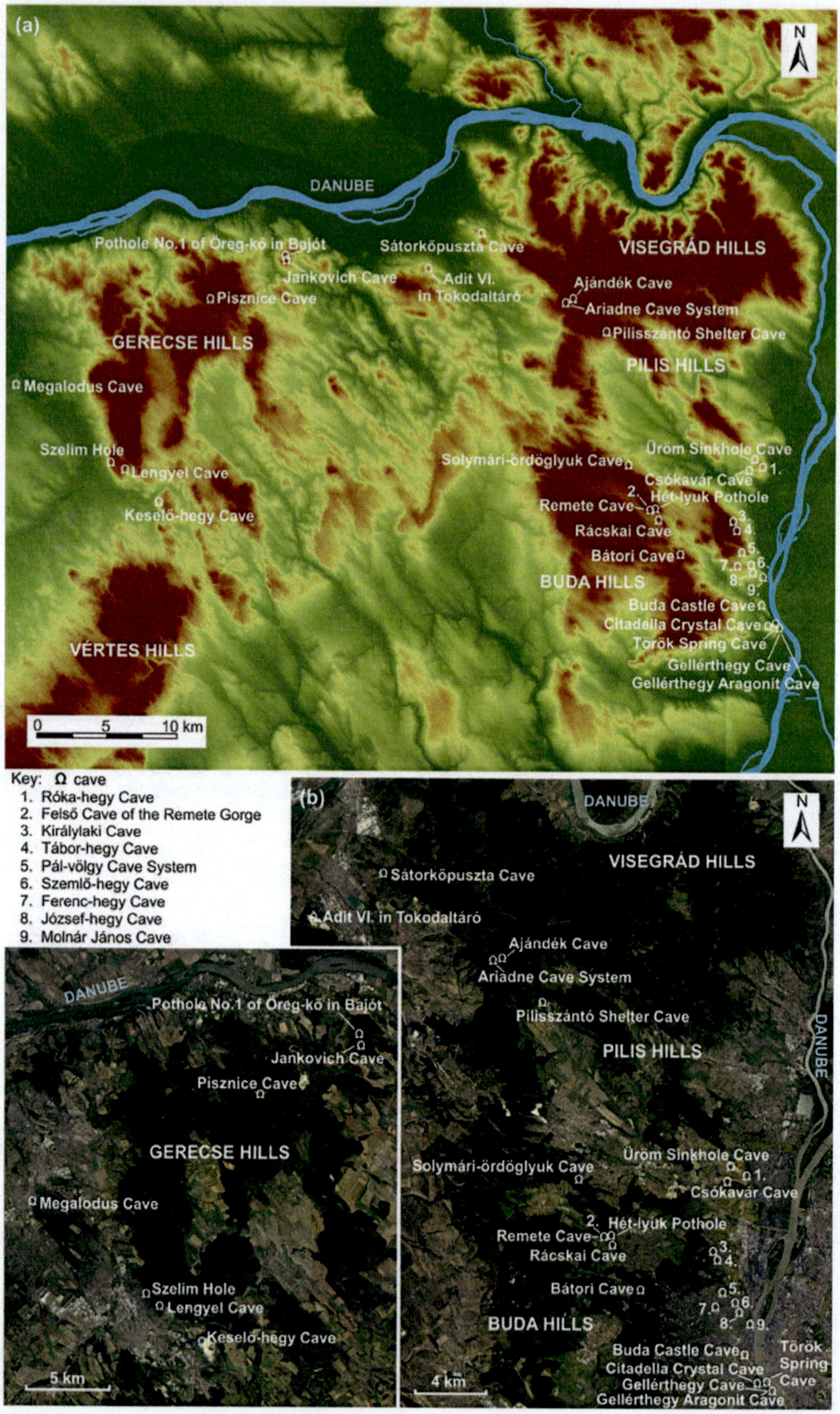

Fig. 1 Locality of the caves of the Buda Hills, Pilis Hills and Gerecse Hills mentioned in this chapter. They are marked by Ω (base maps: **a** SRTM, **b** satellite image by Google Earth IKONOS)

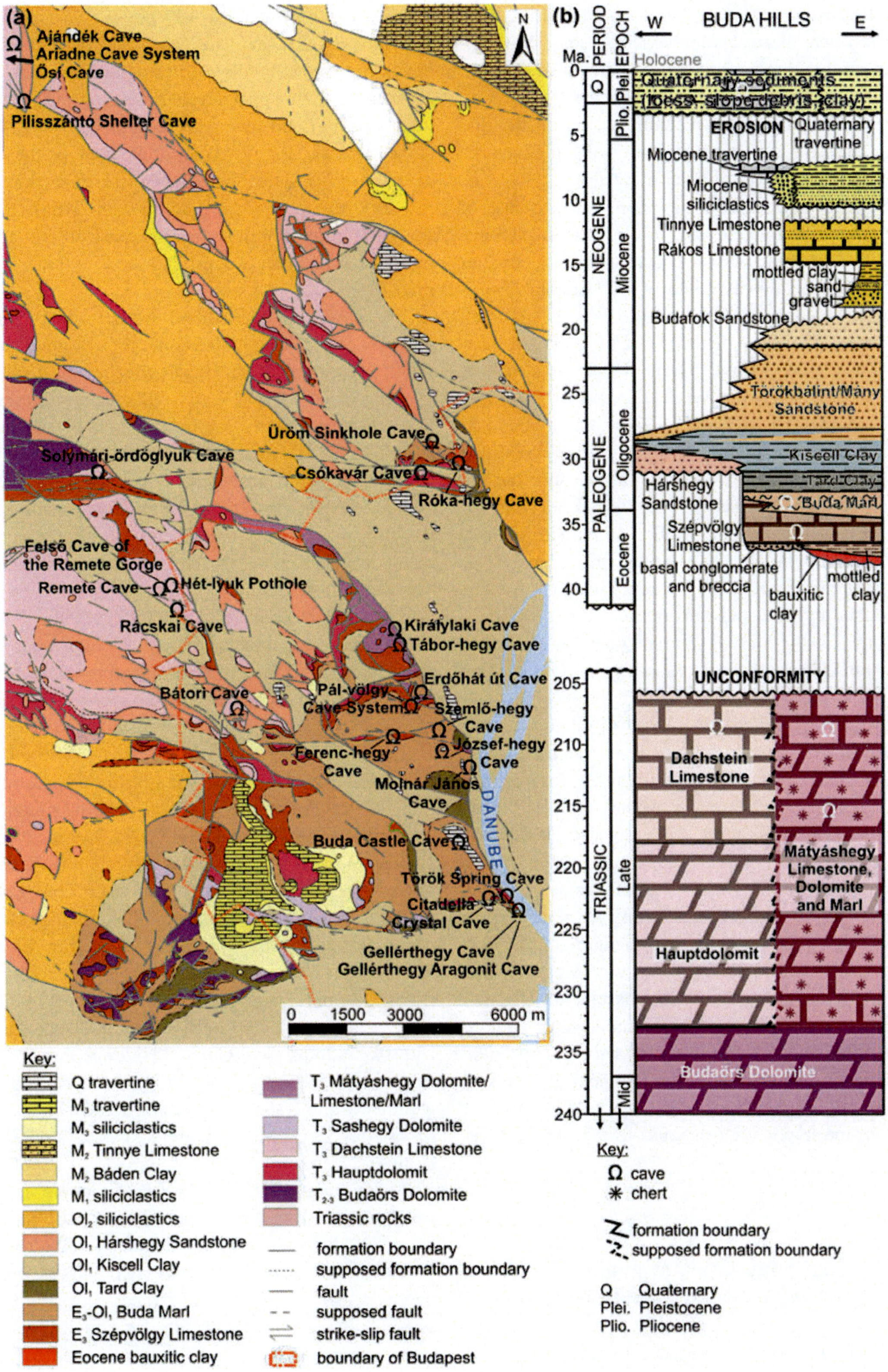

Fig. 2 Geological map of the Pilis and Buda Hills (compiled by M. Virág, modified after the geological map of the Buda Hills without Quaternary sediments (L. Fodor in Mindszenty et al. 2000). Travertines from Wein (1977) and Kele (2009). The caves mentioned in this chapter are marked by Ω (**a**). Stratigraphic column of the Buda Hills and adjacent area (compiled by M. Virág in Mindszenty 2013: based on Wein 1977; Nagymarosy and Báldi-Beke 1988; Fodor et al. 1994; Korpás 1998; Haas (ed) 2001; Haas et al. 2000; Müller and Magyar 2008; Scheuer and Schweitzer 1988; Kele 2009; Esteban et al. 2009). Note, that time-scales for the Mesozoic and Cenozoic formations are not the same! The caves are marked by Ω (**b**)

to have developed in Pleistocene times by interaction with CO_2 (and H_2S)-rich thermal waters rising along Miocene fractures and discharging at the local base-level of erosion. In most of the caves, their corrosivity was apparently enhanced by mixing with local meteoric waters in the discharge zone. Due to tectonic uplift, many of the caves are now dry and far above the actual base level.

There are scattered features showing that cavities of various size were also formed prior to the Pleistocene cave development in the Buda Hills. These generally smaller paleokarst cavities are mostly filled with clastic or chemically precipitated deposits but in a few cases they are till open and unchanged. Many of them can be found in Rózsadomb, in Mátyás-hegy Cave (Kraus 1988), József-hegy Cave (Leél-Őssy Sz. and Surányi 2003), and in Pál-völgy Cave (Takács-Bolner 1980). Such paleokarstic cavities had already been recognised in the northern part of the Buda Hills some decades ago: e.g. Róka-hegy Cave and Rácskai Cave, developed in the Triassic Dachstein Limestone of Hosszúerdő Hill (Kriván 1959; Müller 1989; Korpás and Juhász 1990; Nádor 1991, 1992; Nádor and Sásdi 1991; Nádor et al. 1993).Győri et al. (2011) investigated sediment-filled paleokarstic cavities and mineral veins in the Eocene Szépvölgy Limestone at Fenyőgyöngye Quarry as well as in the Eocene and Oligocene Buda Marl in the southeastern Mátyás-hegy Quarry. Mainly based on findings from these quarries, they described a complex story of diagenesis and paleokarstification. Other paleokarstic cavities can be found in numerous smaller caves in Triassic dolomite, for example in Zsindely Cave, the cavities of the Apáthy Cliff and those close to Tündér Cliff (Zugliget); for more details of paleokarst features see Chap. "Paleokarst in Hungary".

Hypogene caves of the Buda Thermal Karst that now located high above the actual thermal water level developed in Pleistocene times as a result of dissolution of the carbonate rocks by thermal waters rising along pre-existent fractures and discharging at the former base-level. Because of subsequent tectonic uplift, most caves are now dry and far above the thermal-water level of today. Based on research on the speleothems and dissolution morphological elements, it has become possible to reconstruct the story of the shift of the caves from the hypogene (phreatic) to the epigene (vadose) zone.

2.1 Caves of Rózsadomb (E.G. Ferenc-hegy Cave, József-hegy Cave, Molnár János Cave, Pál-völgy Cave System, Szemlő-hegy Cave)

Most of caves (100 caves) are situated around the Rózsadomb area in Budapest, west of the Danube. Here, more than one hundred caves with an aggregate length of more than 60 km are known today (Fig. 3a). The total true length of these caves may exceed 100 km because there are many unexplored passages. The caves of Rózsadomb are mostly developed in Szépvölgy Limestone (Eocone), with some sections of their upper levels extending into the Buda Marl (also Eocene) and lower levels into the Mátyáshegy Limestone (Triassic; Fig. 2). West of the Danube the karstic rocks are exposed at the surface. East of the river, they are at 1000 to 2000 m in depth due to the North–South faults down-stepping abruptly to the east: there are younger, non-karstic sediments overlying the limestone, dolomite and marl beds.

The caves of Rózsadomb were formed mainly by the process of "mixing corrosion" (cf. Bögli 1965; Runnels 1969; Plummer 1975; Bakalowicz et al. 1987; Ford and Williams 1989, 2007; Ford 1995). Here, dissolution results from different waters saturated in $CaCO_3$ at different temperatures and/or ion concentrations mixing and becoming aggressive again, able to dissolve limestone or other carbonates. Dissolution continues until the new, mixed solution itself reaches equilibrium. The dissimilar ion concentrations result from the role that the partial pressure of CO_2 plays. The main contributors in this process were and still are the ascending thermal waters (Cser and Szenthe 1986). Several Hungarian researchers had already recognized the hydrothermal origin of the caves before the theory of mixing corrosion was worked out (Pávai-Vajna 1930; Jakucs 1948; Leél-Őssy S. 1957; Balázs 1966).

The thermal water causing dissolution in the caves of Rózsadomb is not of juvenile origin. It is part of a gravitationally driven large hierarchical regional groundwater flow system (cf. Tóth 1963, 1999, 2009). The local base-level of erosion in the Buda Thermal Karst is the Danube River. Not only the regional system (warm waters) but also the intermediate (lukewarm waters) and local (cold waters) groundwater flow systems are discharging into the river bed. According to the classification of Klimchouk (2007), the Buda Thermal Karst is a typical hypogene karstic area. For more details of the hydrogeology of the Buda Thermal Karst and Rózsadomb see Chap. "Hydrogeology of the Karst Regions in Hungary". (e.g. Fig. 6.17 in chapter "Hydrogeology of the Karst Regions in Hungary", plus Erőss 2010; Erőss et al. 2012; Mádl-Szőnyi and Tóth 2015; Mádl-Szőnyi et al. 2017; etc.).

A small proportion of the precipitation falling on the Buda Hills and adjacent areas where karstic rocks are at the surface infiltrates into the rocks mainly via tectonic fractures. This meteoric water flows downward and eastward, reaching depths of 1000–1500 m and is heated up along its route of several tens of kms in length. Due to the variety of host rocks encountered, the solute mineral content differs from place to place, while the heating depends on the depths attained. The geothermal gradient in this area based on the geothermal database of Hungary is 47.3 °C/km (e.g.

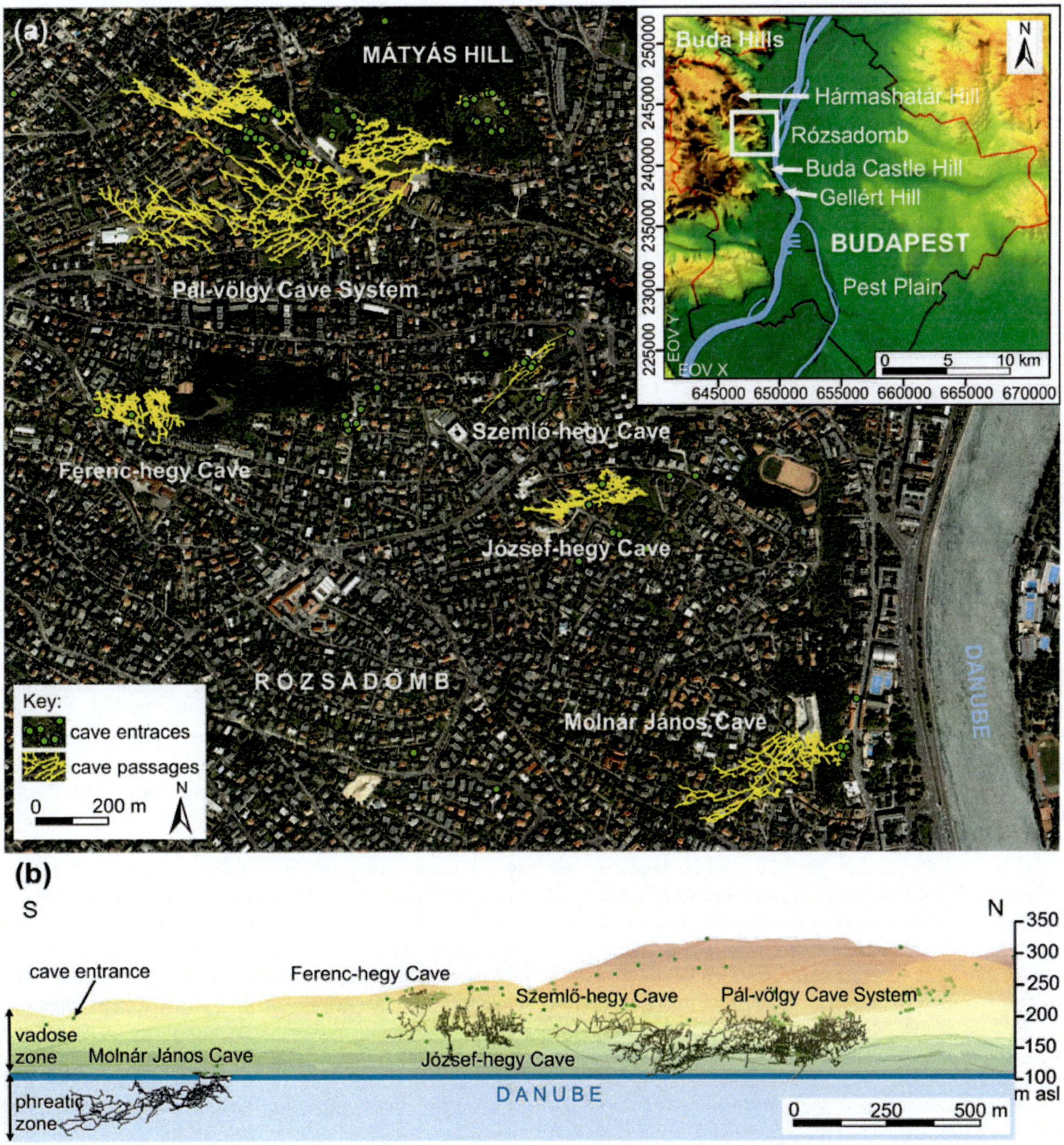

Fig. 3 Location of caves longer than 1 km (shown by cave-polygons) and entrances of all caves in Rózsadomb (base maps: SRTM—terrain map of Budapest; satellite image by Google Earth IKONOS) (**a**); Cross-section of caves in Rózsadomb (base map: SRTM; after Albert et al. 2015) (**b**). Speleological data from the National Cave Register

Mádl-Szőnyi et al. 2015), and the regional geothermal gradient in the Pannonian Basin is 42–55 °C/km (e.g. Völgyi 1979; Kontra and Magyar 2019). According to isotope geochemical studies, the age of groundwater travelling these distances is as great as 5 to 10 ky or even 30 to 40 ky in extreme cases (Deák 1978; Maucha 1989; Izápy and Maucha 1993). Having passed under the Danube, groundwater flow reaches the Pest Plain and encounters high hydrostatic pressure because of the several hundred metres thick dense "impervious" clays overlying the carbonate rocks. These warm waters (subject to convection under conditions of high pressure and temperature) may also mix with other groundwaters coming from the Gödöllő Hills to the east (Mádl-Szőnyi et al. 2015, 2017; Mádl-Szőnyi and Tóth 2015). As a result, they are diverted, ascend and upwell at the only possible place, along the Buda Thermal Fault. Largely on the Buda side of the river, different components of the flow-systems (e.g. with different temperature, TDS, dissolved CO_2 and redox potential) can mix with each other, resulting in mixing corrosion. Not only the mixing corrosion but also the cooling of the upwelling warm water, and the admixture of CO_2 of deep-origin and sulfuric acid released either from oxidation of pyrite and/or derived from maturation of organic matter (cf. H_2S) and from microbial activity also cause dissolution and hypogene karstification (cf. Goldscheider et al. 2010).

Finally, close to the surface, thermal waters may then mix with the infiltrating cold meteoric water discharging along the Danube River at the foot of the Buda Hills. Although the local infiltration water component is negligible, it endangers the purity of the active medicinal springs in many cases owing to anthropogenic pollution (e.g. leaking pipelines of the city).

Cave passages are dissolved by these hypogene processes along the fractures in the carbonate host rocks. The waters coming from the depth reach the surface along the western river banks. Budapest is famous for its spas (e.g. Rudas-, Gellért-, Császár-, Király- and St. Lukács Spa) and the related thermal-water springs. Travertine deposits are often found related to the Pliocene and Pleistocene predecessors of the modern springs (Scheuer and Schweitzer 1980, 1988; Müller and Magyar 2008; Kele 2009, 2011; Kele et al. 2009, 2011; Kele and Scheuer 2012; for more details of the travertines, see Chap. "Age, depositional environment and geochemistry of freshwater carbonates (travertine, tufa) from Hungary").

The springs of the spas of our days occur at 100–110 m asl but, based on the assumed age of the Rózsadomb caves, the first discharge points occurred 500 ky to 1.5 million years earlier at sites that are now at 150–250 m asl. These caves were filled up with lukewarm or warm (20–50 °C) thermal water at that time. Molnár János Cave, discharging at the level of the Danube (104 m asl) is the only known and accessible active thermal-water cave in Budapest. The entrances of the dry and now inactive caves in Rózsadomb are between 160 and 260 masl. (e.g. Virág et al. 2013a; Fig. 3b).

The mixing of different waters is continuous from the upper few meters down to tens of meters within the reservoir. The level of the Danube seems to be the approximate contact zone between the meteoric and the thermal waters. In the active Molnár János Cave, however, wide passages are known more than 90 m below the entrance, i.e. far below the present level of the Danube. Thus, mixing dissolution must have been sustained for a long time and large, 10 m high, chambers could form in the main levels in these caves.

The rate of uplift of the mountains hosting the caves (0.1 to 0.2 mm a year—Leél-Őssy Sz. 1997; Ruszkiczay et al. 2005; Szanyi et al. 2009, 2012) and the fall of the base-level of erosion has resulted in the relocation of the mixing zones and caves. This, and the periodicity of the uplift rates (Fodor et al. 1991, 1994), resulted in the considerable width of the cave levels and the multi-storey nature of the cave system that has been discovered.

This area has been quite intensely dissected by tectonic movements, providing good opportunities for the migration of different waters and, as a result, for the formation of the Rózsadomb caves. Numerous separate thermal spring outlets made the development of the different caves possible.

The places where cave passages were to be dissolved were determined in advance by the patterns of fissures and fractures in the rock, i.e. by dilatational tectonics and lateral displacement (Fig. 4). This evident tectonic determinism was recognized some decades ago (e.g. Jakucs 1948; Leél-Őssy S. 1957; Kraus 1982). More exhaustive tectonic surveys may also help determine the ages of the different fracturing orientations and recognize some distinct peculiarities such as dilational jogs (between overlapping fault segments, partially or completely filled with calcite minerals; Fodor et al. 1991; Benkovics et al. 1995, 1999).

The mixing zone was mainly in the relatively thin (40–60 m thickness) Szépvölgy Limestone (Eocene) during the Pleistocene. The majority of the cave passages in Rózsadomb therefore developed in these carbonate rocks at this time. As a result, large numbers of Eocene fossils (e.g. Nummulites, Discocyclina, Lithothamnium, Chlamys, Echinoidea, Ostrea, Crinoidea) can be seen in the cavities (Fig. 5). Due to its higher clay content, this formation is less able to karstify than the underlying Mátyáshegy Limestone (Triassic), which also has distinct marly horizons. The dimensions of the passages depend not only on the purity of rocks, however, but also on the solutional aggressivity of the waters and the length of time that the cave was forming.

The caves of Rózsadomb are thus rather young in geological terms. The radiometric ages of some of the mineral precipitates (speleothems) in József-hegy Cave, Pál-völgy Cave System, Ferenc-hegy Cave and Szemlő-hegy Cave have been determined by U-series dating (Ford and Takács-Bolner 1991; Leél-Őssy Sz. 1997, 2010; Leél-Őssy et al. 2011; Szanyi et al. 2012; Leél-Őssy Sz. and Virág 2018; Virág 2018; Virág et al. 2019). Based on the analysis of calcite speleothems of the same age as the cave development, such as the mammillary crusts or cave rafts that formed in the thermal water lakes, it can be concluded that the passages of the dry caves of Rózsadomb were dissolved about 0.5–1 million or 1.5 million years ago.

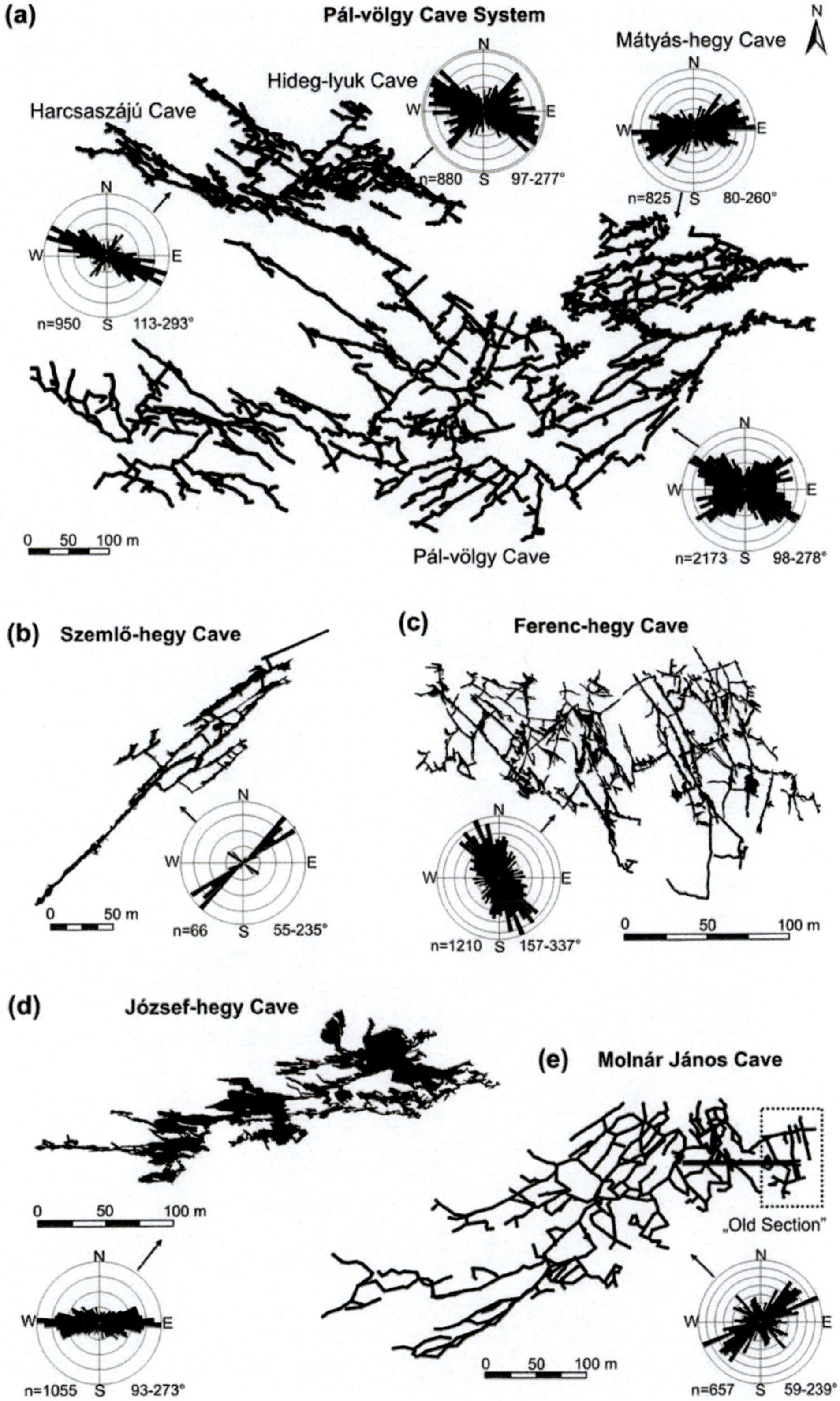

Fig. 4 Cave-polygon maps of Pál-völgy Cave System, Szemlő-hegy Cave, Ferenc-hegy Cave, József-hegy Cave and Molnár János Cave (speleological data from the National Cave Register). The orientation of the cave passages is illustrated by GEOrient

Fig. 5 Fossils in the host rock of caves in Rózsadomb. **a** Echinoidea (Pál-völgy Cave, photo by A. Kiss); **b** Echinoidea, Nummulites and Discocyclina in the phreatic zone (Molnár János Cave, photo by A. Sári); **c** Echinoidea (Pál-völgy Cave, photo by Cs. Egri); **d** Echinoidea, Nummulites and Discocyclina (Szemlő-hegy Cave); **e** Crinoidea (Pál-völgy Cave, photo by A. Kiss); **f** Lithothamnium (Szemlő-hegy Cave); **g**, **h** Chlamys (Pál-völgy Cave)

The appearance of these caves changed considerably after drying up. On the one hand, mineral precipitation (now vadose speleothems) continued, while the bedrock phreatic morphology began to decay, their typical features including a variety of rock collapses and the blocking of passages by clay accumulation (Jakucs 1994).

The most characteristic morphological elements in these caves are the spherical cupolas and the hemispherical solution pockets. According to Müller (1974) and Müller and Sárváry (1977), the cupolas originated above the warm karstic waters, where condensing vapour caused dissolution. On the walls of fissures above the thermal pools, there is a temperature gradient: CO_2 and vapour rise in convection cells to condense on the much colder rock above and dissolve it. Szunyogh (1982, 1984, 1989) verified this theory with thermodynamic calculations. Audra et al. (2007) and Vattano et al. (2017), describe the cupolas as the product of condensation-corrosion processes in saturated air. Jakucs (1948) and Dublyansky (1987, 1995) contend that many cupolas can be created below the water table, too, as a consequence of aqueous convection currents. Osborne (2004) has discussed the question of the origin of cupolas in detail.

In the Rózsadomb area, the host rocks of the caves are Triassic limestones/dolomites and Eocene limestones and Eocene/Oligocene marls. The Eocene carbonate sediments also contain finely dispersed (scattered) pyrite. In Miocene times some of the tectonic fractures became filled with calcite, barite and pyrite from hydrothermal fluids, well before the formation of the hypogene caves. In the ceilings of many of the thermal karstic cavities it is possible to see these old mineralized veins, indicating that the dissolution in Pleistocene times followed—at least partly—the old mineralized fractures (Poros 2011; Poros et al. 2012; Fig. 6).

Siliceous altered zones occur in various places in most caves, principally in the Pál-völgy Cave System, but also in Molnár János-, Szemlő-hegy-, Ferenc-hegy- and József-hegy caves (Fig. 7). In older literature (after Schréter 1912) every researcher associated the siliceous zones with post-volcanic activity in the Eocene or Miocene (Takács-Bolner and Kraus 1989). However, Sásdi (1993) suggested that it is due to preferential solution of the carbonate content in these clayey carbonate rocks.

Győri et al. (2011) analysed the siliceous zones in southeastern Mátyás-hegy Quarry. They assumed that the highly porous zone contains clay-minerals and small amounts of alunite. They suggested that the mineral association probably resulted from sulfuric acid dissolution of the marl, probably in Pliocene times, and proposed that the source of sulfur was either the oxidation of pyrite or H_2S in the thermal water.

According to Polyak and Provencio (2001), Polyak et al. (2006) and Palmer (2007) the interaction of clay (illite) and K-feldspar-rich marl with fluids rich in sulfuric acid normally results in precipitation of alunite, jarosite and silica. In oxidized environments, due to the reaction of the sulfuric acid with the carbonate in the marl, gypsum precipitates in the pores.

According to Vörös et al. (2013) and Virág et al. (2017), oxidation of the pyrite in the hydrothermal mineral veins gives rise to sulfuric acid dissolution in the marl. Mainly based on detailed mapping in passages in Pál-völgy Cave System, a colorful and highly porous, often 0.5 to 2.0 m thick, altered siliceous "halo" zone can be observed around the originally pyrite vein fill in the fracture-controlled corridors in the marl. Quartz/silica precipitation replaces the dissolved carbonate shell of a Pecten, and kaolinite and quartz/silica mixtures are precipitated in the white band of the altered zone. Liesegang band-type "haloes" can be observed along large, oxidized goethite (originally pyrite) tubers in the marl. Moreover, partly degraded barite is found in the middle of the siliceous altered zones around the original mineral veins in Ferenc-hegy Cave and Pál-völgy Cave System. Detailed mineralogical study of the individual bands of these altered zones shows that they consist of quartz, kaolinite, goethite, and minor amounts of alunite and jarosite, all arranged parallel to the vein. This mineral paragenesis suggests that the altered zones are the results of in situoxidation of the iron-sulfide minerals in the veins. This unusual mineral association is probably the product of the interaction of the sulfuric acid with the clay and carbonate minerals of the host rock. The sulfuric acid may derive from the oxidation of pyrite but, as noted, an alternative source of the sulfur may be the oxidation of H_2S in the thermal waters. These altered zones are assumed to be evidence for sulfuric acid speleogenesis in the marl ever since the Pleistocene began. This dissolution and chemical reaction occurred in the oxidizing environment either in pores in the upper phreatic zone just below the thermal water table, or in the vadose zone due to the capillary rise of water from the vadose/phreatic boundary. It is apparent that not only carbonic acid dissolution but also sulfuric acid dissolution and precipitation phenomena occurred in the caves of Rózsadomb. (For more details of (sub)recent siliceous altered zones in Molnár János Cave see Sect. 12.2.1.3).

There are not a great many dripstones in the Rózsadomb caves because the beds above them (mainly the Buda Marl) are rich in clay. Therefore, the infiltrating water is mostly poor in carbonate and does not contain enough HCO_3^- to create abundant stalactites, stalagmites, etc. Nevertheless, there are some beautiful dripstones in Pál-völgy Cave itself and in the Harcsaszájú and Hideg-lyuk cave sections of the greater Pál-völgy Cave System, and also in József-hegy Cave (Fig. 8). However, the caves in this area are richer in calcite, aragonite, gypsum, barite and other crystals of mainly thermal-water origin (Figs. 9 and 10). They are

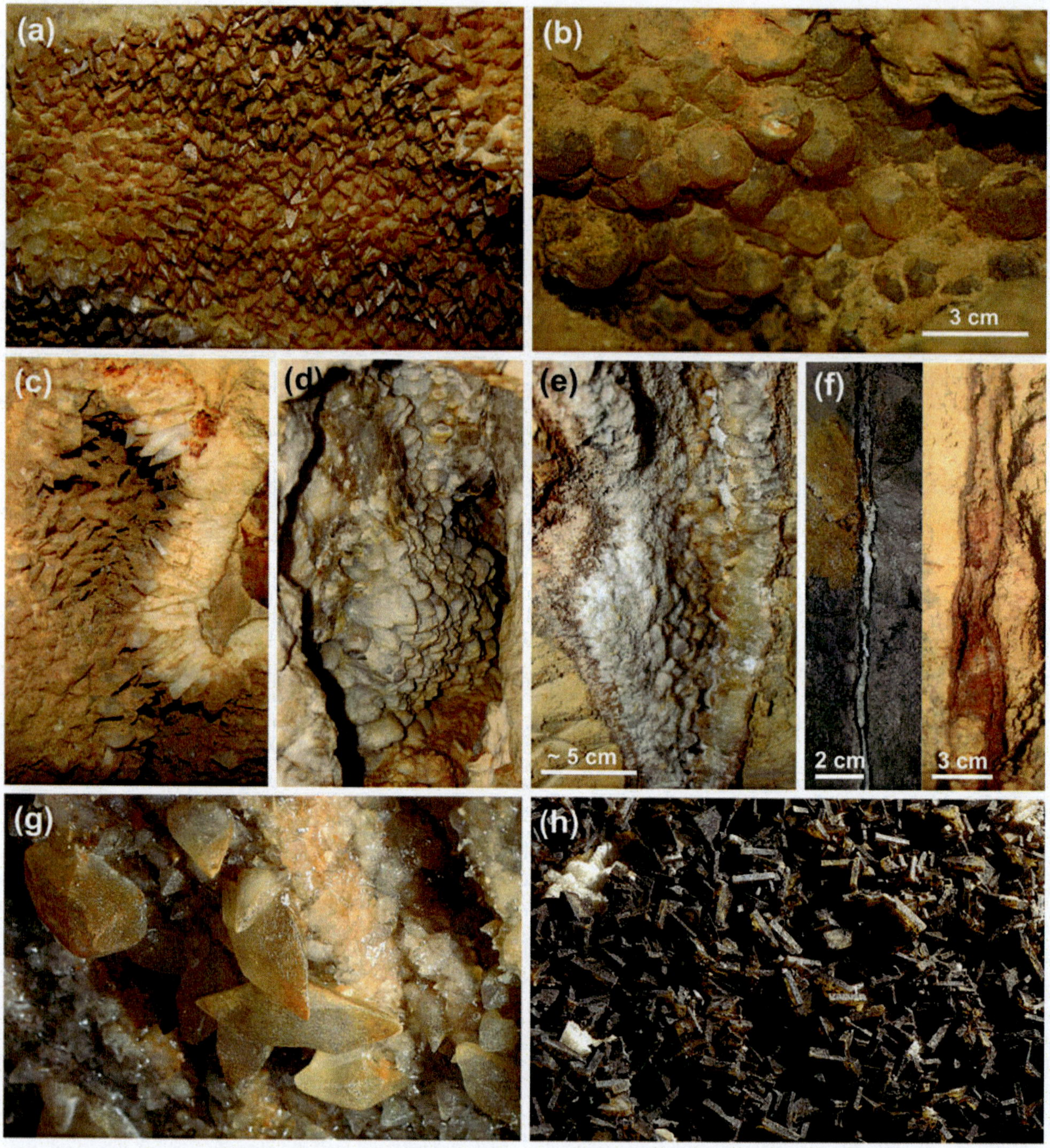

Fig. 6 Cave minerals precipitated before the main phase of speleogenesis (supposedly in Miocene and/or Pliocene times). **a** Dogtooth spar calcite with scalenohedral crystals (József-hegy Cave, photo by Cs. Egri); **b** cavity-filling nailhead sparry calcite with rombohedral crystals (Mátyás-hegy Cave); **c** cavity- and vein-filling dogtooth spar calcite with scalenohedral crystals (Pál-völgy Cave); **d** cavity-filling dogtooth spar calcite crystals and calcite crust (Szemlő-hegy Cave); **e** cavity-filling dogtooth spar calcite crystals and finely-crystalline gypsum crust (Molnár János Cave); **f** pyrite- and calcite-filled vein cross-cutting the marl (tunnel to the Molnár János Cave), and goethite- and hematite-containing oxidized pyrite-vein cross-cutting the marl (Molnár János Cave); **g** nailhead sparry calcite with rombohedral crystals (Pál-völgy Cave, photo by A. Kiss); **h** yellow tabular barite crystals with black Mn-oxide coating (József-hegy Cave, photo by Cs. Egri)

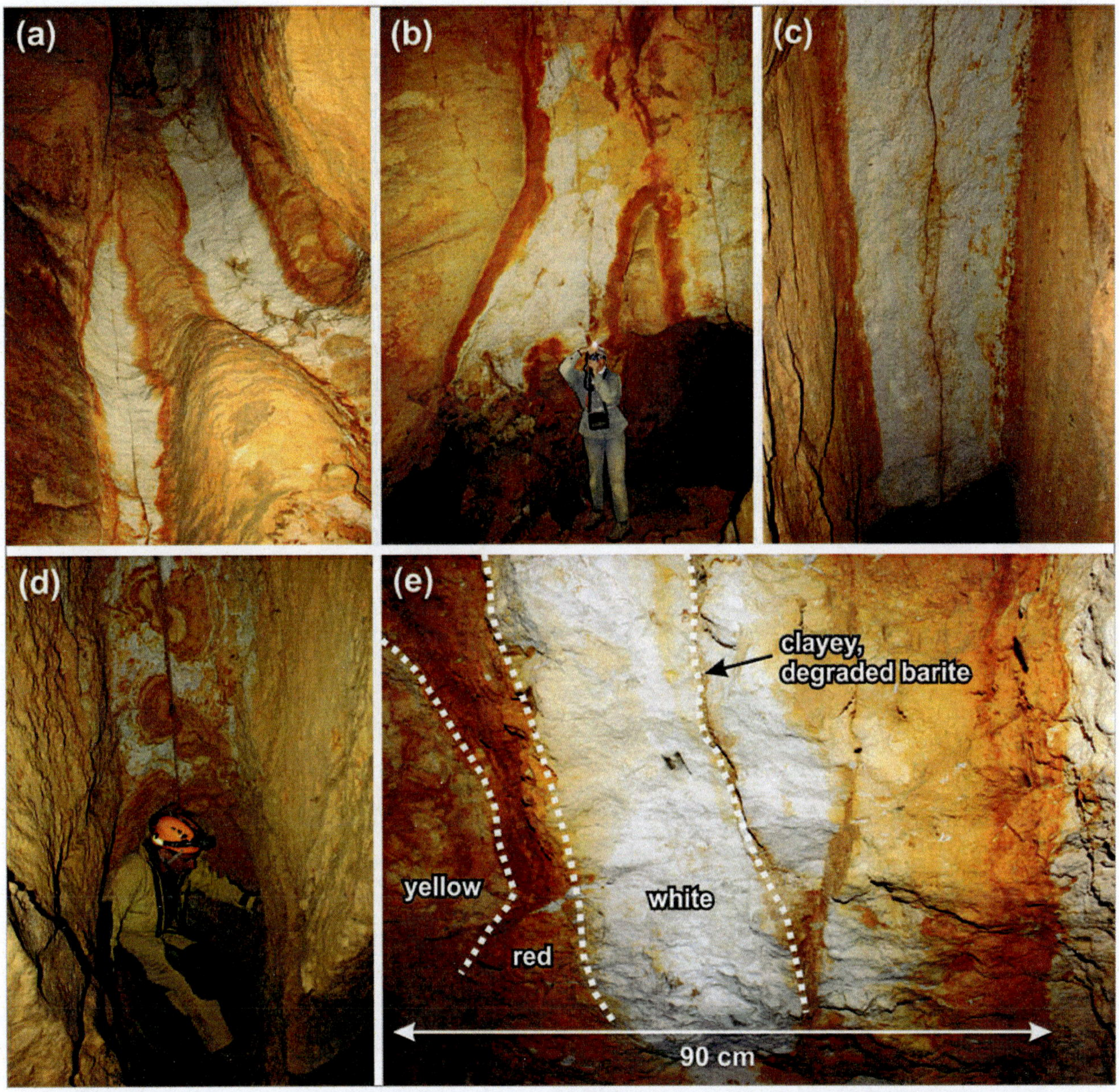

Fig. 7 Siliceous altered zones in the Pál-völgy Cave System. **a** Altered zones in the Mátyás-hegy Cave; (b-c) altered zones in the Pál-völgy Cave; **d** Liesegang band-type 'haloes' of red-coloured goethite in the Mátyás-hegy Cave; **e** photo, showing the anatomy and details of typical siliceous altered zone in the Pál-völgy Cave

sometimes older than the cave, sometimes precipitated from the thermal water itself and thus have nearly the same age as the caves, and sometimes precipitated from thermal water vapour condensed from the cave air. More than thirty minerals have been identified in the explored caves (Leél-Őssy Sz. and Virág 2018), only around ten of which can be detected without the use of lab apparatus (e.g. XRD, SEM, TEM, DTA, CL, optical and UV-fluorescence microscopes—Table 12.1). The most common minerals are calcite, aragonite, barite, gypsum, hydromagnesite, huntite, dolomite, goethite, kaolinite, quartz and romanechite. Most interestingly, some of these minerals appear in many of the different habits or morphologies (cf. speleothems, that Hill and Forti (1997) defined as secondary mineral deposits formed in caves). For example, seventeen different forms of calcite and eight of gypsum (Fig. 10) are known (see later).

The caves have no natural entrances because they were formed by ascending thermal water. The caves of

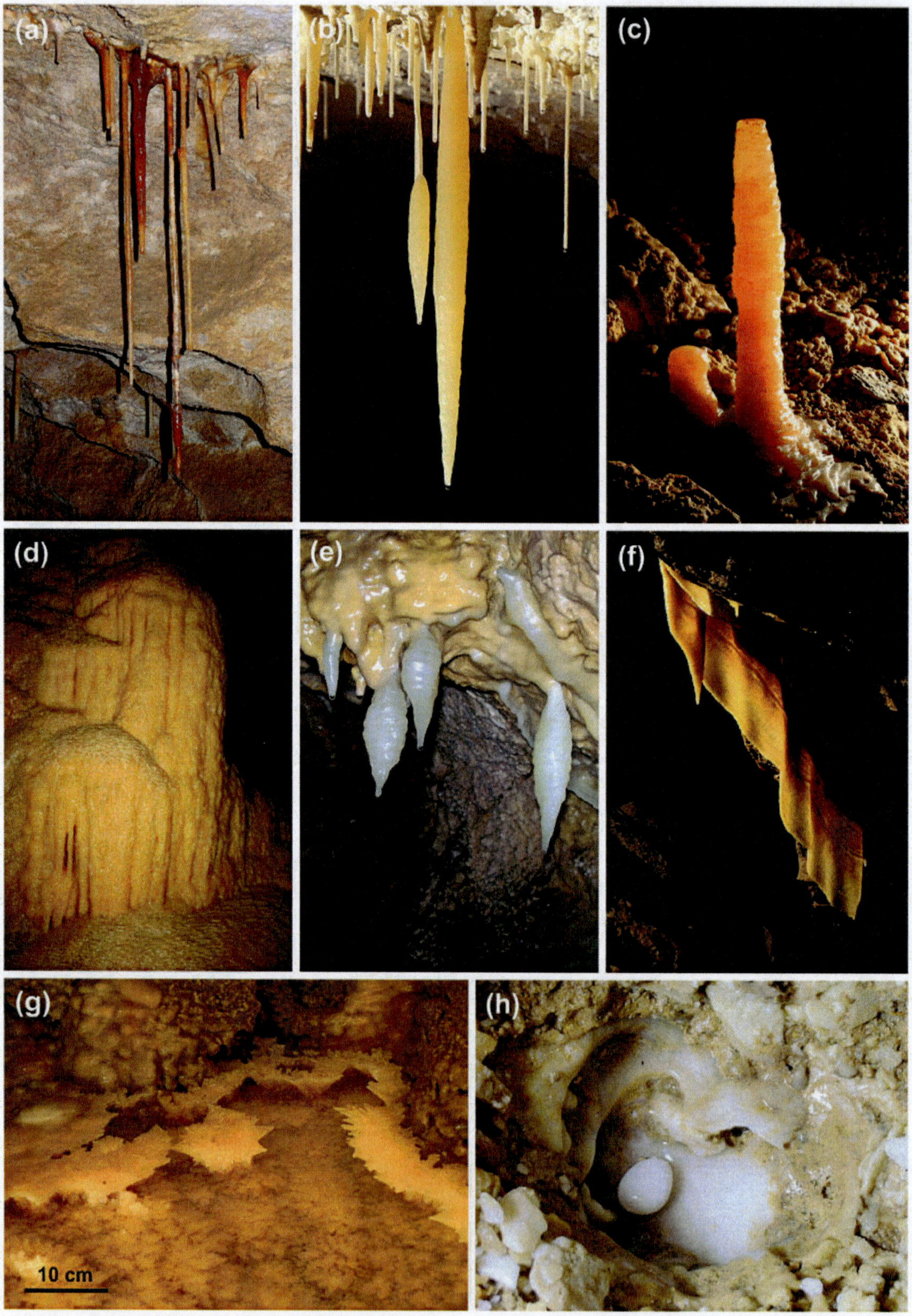

Fig. 8 Speleothems related to infiltrating meteoric water. **a** Red-coloured soda-straw stalactites (Harcsaszájú Cave, photo by R. Kovács); **b** Stalactites (Pál-völgy Cave, photo by Cs. Egri); **c** Stalagmites (Pál-völgy Cave, photo by A. Kiss); **d** the Búboskemence flowstone (Pál-völgy Cave); **e** white stalactites and draperies (Hideg-lyuk Cave, photo by Á. Polyák); **f** dripstone drapery (Pál-völgy Cave, photo by Cs. Egri); **g** pool spars, cave cups and shelfstones in the cave pool (Pál-völgy Cave, photo by A. Kiss); **h** cave pearl in the cave pool (infiltrating meteoric water dripping intensively to the pool, Szemlő-hegy Cave, photo by A. Hegedűs)

Fig. 9 Speleothems of thermal-water origin. **a** Mammillary crust (Szemlő-hegy Cave, photo by Cs. Egri); **b** folia (Pál-völgy Cave, photo by Cs. Egri); **c** floating, paper-thin cave rafts recently precipitating on the surface of the small thermal-water pool (St. Lukács Passage, Molnár János Cave); **d** cemented cave rafts with phreatic overgrowth (Szemlő-hegy Cave); **e** cave cone (~heap of broken cave rafts, Pál-völgy Cave, photo by A. Kiss); **f** cave cone (Szemlő-hegy Cave, photo by Cs. Egri); **g** yellow cave coralloids (Szemlő-hegy Cave); **h** white acicular (needle-like) aragonite frostworks (József-hegy Cave, photo by Cs. Egri); (i) pool fingers (József-hegy Cave, photo by R. Kovács)

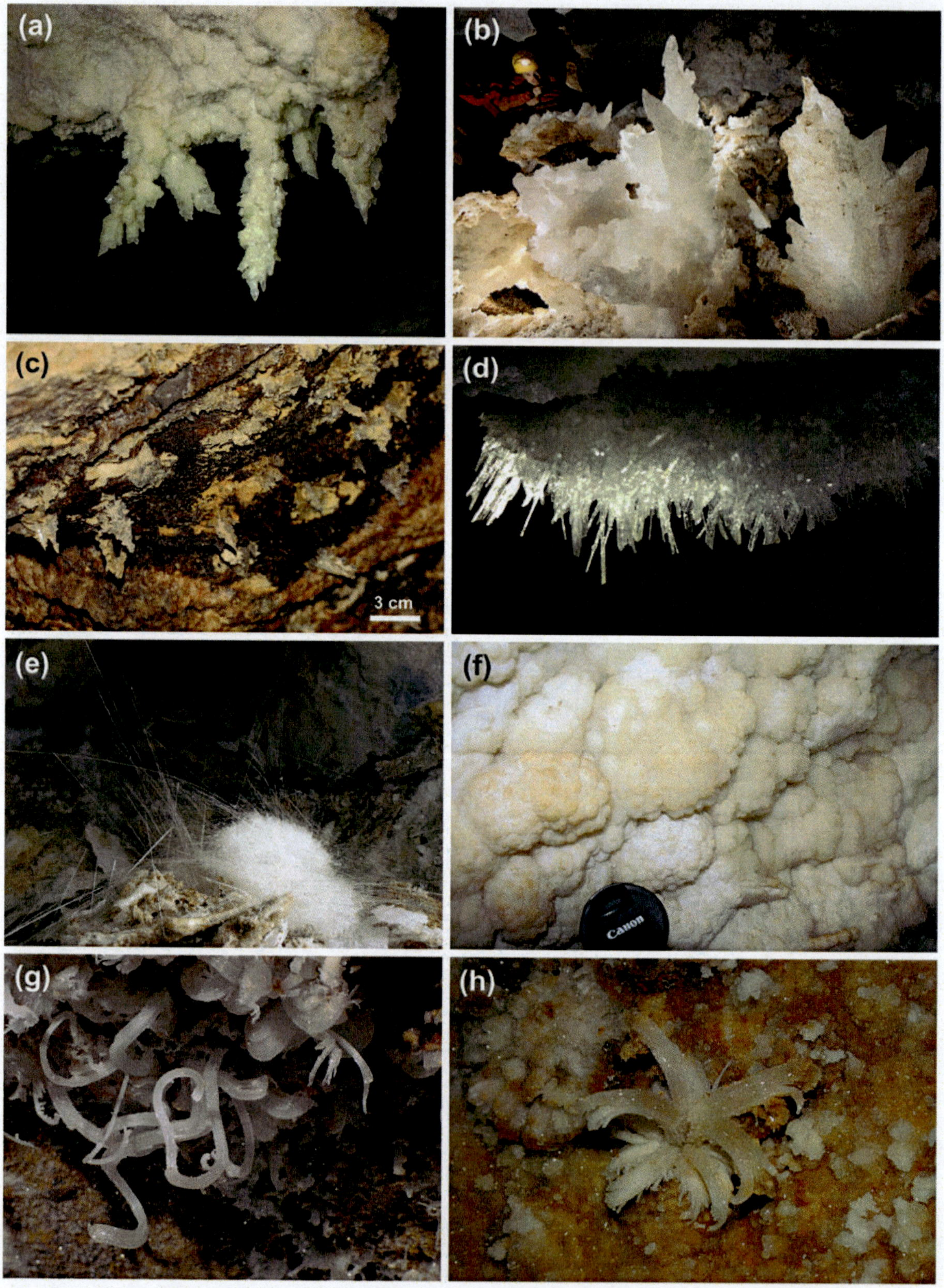

Fig. 10 Gypsum speleothems. **a** Coarse-crystalline, translucent and white gypsum chandelier and gypsum crust (József-hegy Cave); **b** translucent and white coarse-crystalline gypsum crystals detached and fell-down to the bottom of the cave (József-hegy Cave, photo by Á. Berentés); **c** translucent coarse-crystalline gypsum: spine- or thorn-like selenite (Molnár János Cave); **d** small translucent gypsum crystals of the crust (József-hegy Cave, photo by R. Kovács); **e** fibrous cotton-like string and "cave hair" or thread of gypsum (József-hegy Cave, photo by A. Hegedűs); **f** mammillary crusts encrusted by gypsum (Szemlő-hegy Cave); **g** gypsum snakes (József-hegy Cave, photo by A. Hegedűs); **h** gypsum flower (József-hegy Cave, photo by P. Adamkó jr.)

Table 1 Cave minerals and speleothems in the caves of the Buda Thermal Karst (compiled by Magdolna Virág based on the followings: Leél-Őssy Sz. and Virág 2018; Virág 2018; moreover Kraus 1990, 1993, 2006; Leél-Őssy Sz. 1997, 2017; Leél-Őssy and Surányi 2003; Leél-Őssy et al. 2011; Takács-Bolner and Kraus 1989; Takács-Bolner 2005; Győri et al. 2011; Pásztor 2016; and others). Terminology according to Hill (1973, 1987), Hill and Forti (1997) and Palmer (2007)

Cave mineral	Chemical composition	Speleothem type	Precipitation environment
Speleothems (postdating the main stage of cave-formation/speleogenesis); * before the main phase of speleogenesis; (*) siliceous altered zones:			
CARBONATE MINERALS			
		spar, vein*/ dogtooth spar calcite, nailhead sparry calcite (knobby growth)	deep-phreatic zone (thermal water/hydrothermal fluids)
		boxwork*	phreatic zone
		mammillary crust	phreatic zone (at the water table of thermal-lake), pool deposit
		cave raft (heap of cave rafts: cave cone ~ raft cone)	on water table of thermal water lake, pool deposit
		folia	on water table of thermal water lake, pool deposit
		pool fingers (~biothems)	on water table of thermal water lake, pool deposit
		cave coralloid (~popcorn)	vadose zone (evaporation)
		logomite, logotite, cave coral	vadose zone (evaporation)
calcite	$CaCO_3$	tray, rim	vadose zone (evaporation)
		cave pearl	meteoric water dripping to the pool (infiltrating water)
		dripstone: (soda) straw stalactite, stalactite, stalagmite, column	meteoric dripping water (infiltrating water), vadose zone
		flowstone, drapery, crust	meteoric seepage water (infiltrating water), vadose zone
		helictite	capillary of infiltrating meteoric water, vadose zone
		rimstone dam (~gour), rimstone pool	flowing meteoric water
		shelfstone, pool spar, hollow triangular crystal	cave pool (meteoric water), pool deposit
		coral pipe, cave cup, conulite	meteoric dripping water (infiltrating water), vadose zone
		septaria boxwork	meteoric infiltrating water, vadose zone
		moonmilk, powder	evaporation (infiltrating meteoric water), vadose zone
aragonite	$CaCO_3$	frostwork, needle, (anthodite?)	vadose zone (evaporation)
hydromagnesite	$Mg_5(CO_3)_4(OH)_2 \times 4H_2O$	moonmilk, powder	vadose zone (evaporation)
huntite	$CaMg_3(CO_3)_4$	moonmilk, powder	vadose zone (evaporation)
dolomite	$CaMg(CO_3)_2$	moonmilk, powder	vadose zone (microbially mediated)
magnesite	$MgCO_3$	cave coralloid (~popcorn), powder	vadose zone (evaporation)
OXIDES and HYDROXIDES			
quartz, silica(*)	SiO_2	micro-spar filling in pores (*), cave sediment	siliceous altered zone/cave sediment
goethite (*)	$FeO(OH)$	coating, vein*, tuber* (by oxidation)/ snottite (microbially mediated ~biothems), "stalactite", cave sediment	above water table of (thermal water) lake/oxidation/ siliceous altered zone (vadose or phreatic zone)
hematite	Fe_2O_3	coating (microbially mediated?), vein * (by oxidation), cave sediment	above water table of (thermal water) lake/oxidation (vadose or phreatic zone)
romanechite	$(Ba,H_2O)_2(Mn^{+4},Mn^{+3})_5O_{10}$	coating (microbially mediated? ~ biothems), cave sediment	above water table of (thermal water) lake/oxidation (vadose or phreatic zone)

(continued)

Table 1 (continued)

Cave mineral	Chemical composition	Speleothem type	Precipitation environment
OXIDES and HYDROXIDES			
hollandite	$Ba(Mn^{4+}{}_6Mn^{3+}{}_2)O_{16}$		
Cave mineral	**Chemical composition**	**Speleothem type**	**Precipitation environment**
SULFATE MINERALS			
barite	$BaSO_4$	sparry, vein-filling*/ micro-spar, coating (microbially mediated?), cave hood	deep-phreatic zone (thermal water/hydrothermal fluids), vadose zone?
		chandelier (translucent coarse crystalline gypsum)	phreatic/vadose? zone (thermal water)
		coarse crystalline gypsum crust, microcrystalline coating	vadose zone (evaporation)
		gypsum flower, gypsum snake, cave spider	vadose zone (evaporation)
gypsum	$CaSO_4 \times 2H_2O$	fibrous "cave hair", cotton-like string, thread, needle	vadose zone (evaporation)
		gypsum "stalactite", gypsum tower, gypsum column	vadose zone (evaporation of seepage meteoric water)
		selenite (translucent coarse crystalline gypsum), spine or thorn	vadose zone (evaporation)
		gypsum star (microbially mediated?)	vadose zone (evaporation)
celestite	$SrSO_4$	micro-spar	vadose zone (contact zone of carbonate and gypsum)
thenardite	Na_2SO_4	fibrous "cave hair", cotton-like string	vadose zone (evaporation)
pentahydrite	$MgSO_4 \times 5H_2O$	fibrous, powder	vadose zone (evaporation)
hexahydrite	$MgSO_4 \times 6H_2O$	fibrous, powder	vadose zone (evaporation)
epsomite	$MgSO_4 \times 7H_2O$	flower, snake	vadose zone (evaporation)
alunite (*)	$KAl_3(SO_4)_2(OH)_6$	micro-spar filling in pores	siliceous altered zone
jarosite (*)	$KFe_3(SO_4)_2(OH)_6$	micro-spar filling in pores	siliceous altered zone
OTHERS (minerals and group of minerals not closely related to speleogenesis):			
SULFIDE MINERALS			
pyrite*	FeS_2	vein-filling, tuber (oxidation to goethite), scattered (in host rock)	deep-phreatic zone (hydrothermal fluids), in sedimentary rocks
marcasite*	FeS_2	tuber (oxidation to goethite)	deep-phreatic zone (hydrothermal fluids), in sedimentary rocks
cinnabar	HgS	spar, vein-filling, scattered*	deep-phreatic zone (hydrothermal fluids)
metacinnabar	HgS	spar, vein-filling, scattered*	deep-phreatic zone (hydrothermal fluids)
HALOGENIDES			
fluorite*	CaF_2	vein-filling spar	deep-phreatic zone (hydrothermal fluids)
"Clay minerals" (SILICATES)			
kaolinite (*)	$Al_4[Si_4O_{10}(OH)_8]$	micro-spar filling in pores (*), powder, cave sediment	siliceous altered zone/cave sediment
halloysite	$Al_4[Si_4O_{10}(OH)_8](H_2O)_4$	micro-spar filling in pores	siliceous altered zone
illite*	$(K,H_3O)(Al,Mg,Fe)_2(Si,Al)_4O_{10}[(OH)_2,(H_2O)]$	cave sediment, (siliceous altered zone)	cave sediment/ siliceous altered zone
smectite group	$(Na,Ca)_{0.33}(Al,Mg)_2(Si_4O_{10})(OH)_2 \times nH_2O$	cave sediment, (siliceous altered zone)	cave sediment/ siliceous altered zone
zircon	$ZrSiO_4$	in cave sediment	cave sediment

Rózsadomb are all tectonic-controlled, concealed caverns. The form and location of the passages are entirely independent of the surface geomorphology. No fluviatile sediments occur in these caves because no water was able to percolate from the surface in sufficient quantity.

This explains why all these caves were discovered in the last 120 years as consequences of human activities. The discovery of the Rózsadomb caves was due to stone mining, excavation of construction pits, foundation works, canalization and the laying of gas and water pipelines, well-digging and geological boring, tunnel and canal ruptures, road construction and ground levelling.

Only the first sections of the thermal-water cave behind the Lukács Spa (a small part of Molnár János Cave) and Erdőhát út Cave on Mátyás Hill were known more than 160 years ago. Both were shorter than 20 m.

The view of the Danube bank is on the list of UNESCO World Heritage sites. There are many medicinal spas now along the River. The most famous are the Gellért, Rudas and Lukács Thermal spas. At present, artificial bores feed the Gellért and Rudas spas: their natural springs are still discharging directly into the Danube.

In short, these dry-cave passages are old solutional outlets of the springs. In Molnár János Cave, the only active cave today, 95% of the explored passages are below the thermal water surface in the phreatic zone. It is possible that there are similar active caves yet unknown here, for example behind Török Spring. Who knows?

Five of the one hundred known caves in the Rózsadomb area are longer than one km. The Pál-völgy Cave System has a total mapped length of more than 32 km, the longest cave in Hungary. This was the first large cave discovered in Budapest, when a limestone quarry breached it in 1904. Szemlő-hegy Cave was second: the entrance of the cave was found during quarrying and/or construction works in 1930. Three years later as a new underground pipe was being laid two km from Szemlő-hegy Cave the workers intercepted the entrance to Ferenc-hegy Cave six m below the surface. In the 1970s, divers discovered the first part of Molnár János Cave behind the artificial Malom (Mill) Lake. The last major discovery in this area was the discovery of József-hegy Cave under the foundations of a planned housing estate in 1984.

When we find the entrance of a new cave in the Rózsadomb area, we often have to dig down to 10–30 m to reach open passages; this was the case, for example, at József-hegy Cave. Usually, not only the entrance zone will be obstructed but also infilled sections and very narrow passages must be cleared. There are many collapses and lots of clay accumulation in these caves. They have been extended in 15–30 successive digging or other clearing operations.

Almost all cave passages are under the houses and streets in the Rózsadomb area. In certain cases, the caves can mean risk for the buildings or, on the other hand, the new buildings endanger conditions in the caves. Recently, there are rigorous rules to regulate building activity above the caves.

Caves and the construction work above them interact. This tends to have mostly negative effects, especially in the long run. The one positive aspect is the fact that construction exposes indications of caves in the first place, making their exploration possible!

Air currents (strong drafts) from fractures, etc. are important evidences because the underground air is fresh and suitable for breathing in all caves discovered in Rózsadomb. Cave air quality is not influenced by the partial housing coverage of the area. In this leafy quarter of Budapest at most only 20–30% of the area is built-up.

However, the conditions for the infiltration of water are different. The active, currently developing forms of cave concretions—such as dripstones, "glassball"-like cave coralloids (cave popcorns) or gypsum flowers—may dry up, cease developing and die where the surface is covered by roadways, pavements, buildings, etc. Consequently, building coverage of less than 15% is required by the environmental protection agency in areas where caves are likely to occur, and the area of additional impervious cover cannot exceed 5%.

The other danger caused by the building on the area is leakage or diversion of water.

A few examples:

- Twice in recent decades the water level of the lake in Mátyás-hegy Cave (probably it is part of the regional karst watertable?) was several meters higher for a couple of months because of a hidden breakage in a pipe. After the fault had been fixed, the drastic inflow stopped and the original water level was restored.
- In 1986 water was leaking for about two months from a 200 mm water main in front of No. 26, Józsefhegyi Street. It eventually washed out a cavity the size of a truck under the road. The 40 m deep József-hegy Cave No. 2. was discovered by excavating this cavity. A few months later, another breakage in the area caused the collapse of the excavated second shaft and made the entrance to the new shaft perilous.
- Earlier, water had been leaking for months in the garden of the same house. It entered a research shaft in Áfonya Cave that had been dug out during nearly two decades of work. The water saturated the detrital fills in the caves, threatening the area with collapse.
- Since caves have, technically, unlimited water absorbing capacity, large underground cavities are readily formed by the washing out of infilled natural caves. This leads to damage to buildings. Most serious damage was caused to a house at No. 50, Rómer Flóris Street, where 40% of the

material under the foundations vanished due to wash-out. Cracks of several centimeters were formed, and the building was imperilled.

Among the factors endangering the caves of Rózsadomb, the quarrying has resulted in the destruction or blockage of some of the natural passages. In areas like this, the advance of urbanization on the surface may necessarily endanger known or unknown caves underneath. Leaks and breaks of drainage pipes (e.g. carrying sulfates, nitrates), seepage of household chemicals into the depths (e.g. phosphates, nitrates), plus contaminants from old dehydrators or the anti-skid salt used in winter (NaCl, $MgCl_2$) may all infiltrate into the cave passages, contaminating the dripping water and the speleothems (dripstones and various other mineral precipitates). With the increase of the area of paved surfaces, water can seep into the soil at only a few places, which changes the location and intensity of dripping in the caves beneath. Vibration caused by vehicles or the construction of new buildings may also trigger rock falls in passages close to the surface (e.g. Szunyogh 2018). Analyses of drip waters in some caves have been undertaken since the 1980s (cf. Takács-Bolner et al. 1989; Takács-Bolner 1995; Mari and Fehér 1999; Fehér et al. 2009, 2016; Virág et al. 2009a,b; Virág and Mádl-Szőnyi 2013; Bergmann et al. 2011; Kiss et al. 2013; Borbás and Fehér 2013; Rehák and Szieberth 2018; Szombathy et al. 2019a; etc.). For more details of the findings of recent cave monitoring, see Chap. "Monitoring and geochemical investigations of caves in Hungary: implications for climatological, hydrological, and speleothem formation processes".

2.1.1 Ferenc-hegy Cave

Among the large caves of the Rózsadomb area, Ferenc-hegy Cave is closest to the surface and is at the highest elevation (more than 100 m above the Danube). The total length of the known cave is now 6700 m. Most of the passages are in an upper 40 m zone (mainly the "Old Section" and the "New Section"), but some lower ones ("Deep Section") extend to 80 m below the entrance (Fig. 11).

When its entrance was opened during trench digging in 1933, Géza Miklóssy and Hubert Kessler discovered the first 500 m. After 30 years of further discoveries, the length of the "Old Section" was extended to around 2 km, mainly due to the efforts of Gyula and Andor Szilvássy. They also discovered the "New Section" of the cave in 1963, making it longer than four kms (Sásdi 2003a). The Guru Speleo Group (led by Sándor Nagy) found a "Deep Section" and some new passages in the middle level in 2005.

The most fatal tragedy in the Rózsadomb caves happened here when four young men took an unauthorized tour in the cave in 1975. They got lost and nobody knew where they were. They were found dead after a month of searching.

Since 1997 special visits are made available for children who suffer from asthma. The clean, humid air of the caves is very good for them, particularly when they are exercising ("speleotherapy").

Originally the cave had several entrances but nowadays only one is kept open, in the middle of the cave. A ditch was dug from a new house to the main sewer under Törökvész Street close to the discovery entrance in the summer of 2014. An unknown cupola with a diameter of about 3 m was found three meters under the road. The bottom was covered by detrital material. Cavers of Kinizsi SG removed this and found a new passage that led into the known part of Ferenc-hegy Cave. They also found the concrete-stopper under the original discovery entrance of 1933. They closed the new entrance safely and returned the important Törökvész Street for traffic. The new part can be reached from the public entrance.

Almost all passages of the Ferenc-hegy Cave are dissolved in the Eocene Szépvölgy Limestone. The cave does not extend down into the Triassic carbonates, but the Eocene basal breccia can be found in the "Deep Section".

The typical labyrinth-like system is characterized by very narrow passages. There are six to eight parallel main fissures in the "Old" and "New" sections. Their widths are between 0.5 m and 3 m. The tectonic control is most striking here. The main passages of the "Old Section" are oriented NW–SE, while the "New Section" trends N-S and NW–SE, a difference of 25° and 30° (Fig. 4c). In general, the passages follow original fractures filled with earlier calcite and barite crystals. Cave coralloids ~300 ky old at 230 m asl (Ford and Takács-Bolner 1991) cover the cave passages everywhere in the middle level. According to Szanyi et al. (2009), ~450 ky old paper-thin cave rafts can be found in the "Deep Section" (180 m asl). There is lot of snow-white moonmilk (calcite) in the upper parts. Among the dissolution forms, huge spherical cupolas (their diameter is several meters) characterize the uppermost sections of the cave, where they are the upward-termination forms of this system. Thermal outlet channels or CO_2-upwelling tubes ("gun barrels") are among the unusual forms in this cave, as well as solution scallops. There are cemented quartz pebbles on the barite vein in Törekvés Passage, which is very significant because these pebbles came from the surface of a Pliocene sandstone formation that is now totally eroded away at the surface (Sásdi 2003a).

2.1.2 József-hegy Cave

József-hegy Cave is the richest in white crystals not only in the Buda Hills but in all of Hungary (Figs. 12 and 13). It was

Fig. 11 Ferenc-hegy Cave. **a** Huge spherical cupola; **b** thermal outlet channel or "gun barrel" (photo by P. Vörös); **c** thermal outlet channel or CO_2-upwelling-tube; **d** cave coralloids and tiny stalactites; **e** dissolved surface of cave coralloids; **f** spongy-like calcite speleothems; **g** spongy-like calcite speleothems and cave coralloids

Fig. 12 József-hegy Cave. **a** Coarse-crystalline, translucent and white gypsum chandelier and gypsum crust in the huge Kinizsi Chamber (photo by Cs. Egri); **b** mammillary crusts and cracked-bottom of the cave section (photo by Á. Berentés); **c** spherical cupolas (photo by Cs. Egri); **d** spherical cupolas covered by gypsum crust (photo by R. Kovács); **e** characteristic high and narrow, fissure-like cave passage with black Mn-oxide (romanechite) coatings covered by white cave coralloids (photo by Cs. Egri)

Fig. 13 József-hegy Cave. **a** Red–black-white coloured Fe–Mn-oxidic and carbonatic mineral-coatings on the host rock (photos by G. Ambrus); **b** huge cupola in the marl (photo by Á. Berentés); **c** white cave cones (~heaps of broken cave rafts, "Christmas-trees", photo by Cs. Egri); **d** white cave coralloids and aragonite frostworks (photo by R. Kovács); **e** aragonite frostwork and gypsum crystals (photo by A. Hegedűs); **f** gypsum snakes (photo by Cs. Egri); **g** fibrous gypsum "cave hair" (József-hegy Cave, photo by A. Hegedűs); **h** gypsum crust detached and fell-down to the bottom of the cavity (photo by P. Adamkó jr.); **i** pink-coloured "ice cream"-like stalagmite (photo by I. Czájlik)

discovered after 65 days of hard work by the Kinizsi Speleo Group of Rózsadomb (Kinizsi SG), under the leadership of Szabolcs Leél-Őssy and Péter Adamkó. A company wanted to build an extensive housing estate with more than thirty small blocks (altogether 135 flats) on the clay-rich Buda Marl on the top of József Hill, 100–120 m above the Danube and offering beautiful views of the city. The workers dug several deep basement pits, one of which broke into the ceiling of one-meter diameter cave cupola. It was at the end of January 1984 and a large vapour column rose immediately from the cupola because the air temperature on the surface was -10 °C at the time. The workers thought they had broken into a water canal of the plumbing system. This happened nine meters below the surface, in a garden, away from other homes and the street.

After cavers heard what had happened, many came to dig to find a new cave. Kinizsi SG were successful after a week, digging a 50 m long, 29 m deep artificial passage—without any open cavities—by just following the cave draft. They worked in winter, when the cold outside air favours flow out of the topmost "entrance". In summer, the motion of the airflow is reversed.

The discovery passage is in a large, 20 m wide E-W direction fault zone in a very hard cave collapse-breccia, so the Kinizsi SG could work without scaffolding. They found some white cave coralloid and a few small gypsum crystals during digging. József-hegy Cave is a hypogene cave like the others in Rózsadomb, i.e. it was dissolved upwards from below and therefore has no entrance from the surface above. The open passages of the caves in Rózsadomb are usually 20 m under the upper cupola. Between the upper cupola and the principal open passages there is no passable way, only small fissures.

The Kinizsi SG discovered the first part of the cave, narrow passages with small rooms, at 29 m depth after two months of work. This upper part of the cave is dissolved in carbonate-rich Buda Marl (Bryozoan marl). There were lots of clay and stone fragments on the floors of the passages. The Kinizsi SG opened an entrance into a larger cave at 51 m depth after a week's further digging (Adamkó and Leél-Őssy 1984; Leél-Őssy Sz. et al. 2011). The popcorns or cave coralloids, aragonite crystals (frostworks) and gypsum crusts covering the rocks there made the walls snow-white everywhere.

The upper parts of József-hegy Cave, developed in Eocene and Oligocene Buda Marl, have only a few crystals but are very rich in beautiful large cupolas. The diameter of the greatest is six meters. The most beautiful solution forms in the marl are seen in the Kísérletes (Experimental) Passage. The insoluble components of the marl (mainly kaolinite) have accumulated on the floor beneath.

In the first part of the middle section, the Kinizsi SG discovered the entrance to the huge Kinizsi Chamber (Kinizsi Railway Station). This is 70 m in length, 20 to 25 m wide, and has a height of between 8 and 15 m.

The middle section is dissolved in the Eocene Szépvölgy Limestone, which is extremely rich in Nummulites and Discocyclina fossils, mentioned above, but sometimes the uppermost portion of it extends into the Buda Marl. There is a distinct boundary between the two formations. The Szépvölgy Limestone passages are larger, with widths of two–three m, heights of three-five m, and there are many larger rooms between five m and 20 m. There are large crystals on the walls, and fills of one-three m on the floors.

The lower section of the known cave is in several separate parts. Four of them have been named—Természet temploma (Church of Nature), Üvegpalota (Glass Palace), Szolárium (Solarium) Shaft and Csapda (Trap) Shaft. The latter two are in Szépvölgy Limestone, and the others in the underlying cherty Mátyáshegy Limestone (Triassic). In places it also has high clay content, so that it can almost be considered "marl". The floors are clean without fill. The walls are adorned with aragonite and calcite crystals. There are no large rooms, only high (sometimes more than 10 m) and narrow passages here.

The major passages in the cave follow an E-W direction quite precisely (Fig. 4d). Interesting tectonic features, such as dilational jogs (between overlapping fault segments, partially or completely filled with calcite minerals), fault-striation, fault breccias, etc. to be seen in the walls (Fodor et al. 1991).

The known depth of the cave is 105 m, and the length is about 6.5 km, but the Kinizsi SG are working on the new discovery every week: probably, they know only a half (the northern) part of the cave. Maybe the other half of the cave is behind the large fault zone, so the group may find further new passages on the other side of it and at the known ends of other passages.

The temperature in József-hegy Cave (around 14 °C) is higher than in the other caves of Rózsadomb because it is closer to the thermal water springs. There is air-communication between József-hegy Cave and Molnár János Cave (or another, yet unknown, spring cave). The Kinizsi SG are aware of points in the cave where the temperature (15.5 °C) and the CO_2 content (9500 ppm) are higher than in other places: for more data, see Table 12.2.

Percolation from rainfall reaches the cave rapidly. Although the main passages are 50 to 60 m under the surface, most of it arrives along the large fissures within a few hours. In 1992 the Kinizsi SG conducted an interesting experiment to determine the exact rates of infiltration in the marl. A ditch was dug above the Kísérletes-ág (Experimental Passage) and 24 kg NaCl was placed on its 267 m^2 floor as trace indicator. 50 m^3 water (simulating 200 mm of natural precipitation) was then poured into the ditch. If the drip yield grew in the cave, the question was: was this the direct discharge of the artificial precipitation, or a pressure pulse

Table 12.2 Cave-climate parameters of the caves in Rózsadomb (analysed by József Stieber between 1995 and 2020, air- circulation in winter and summer)

Analysed parameters	Szemlő-hegy Cave	József-hegy Cave	Molnár János Cave	Mátyás-hegy Cave	Ferenc-hegy Cave	Pál-völgy Cave
Temperature (°C)	12.6 to 14.2	12.5 to 14.4	26.8 to 27.2	10.0 to 12.3	11.3 to 12.5	9.9 to 10.2
Relative humidity (rel.%)	98 to 100	98 to 100	99.5 to 100	95 to 100	95 to 100	97 to 100
CO_2 (ppm)	2200 to 4200	2300 to 4500	4500 to 5500	1200 to 3000	1600 to 3200	1200 to 2800
Concentration of radon (Bq/m^3)	3200 to 8600	4500 to 9200	6200 to 12,500	900 to 2500	no data	850 to 2200
Chemical reactivity of the aerosol (pH)	5.06 to 5.82	no data	5.81 to 6.25	no data	no data	6.4

(kinematic wave) expelling older precipitation already stored in the fissures? Four drip points, 25–45 m under the surface, were mapped in the cave earlier. The intensity of dripping started to increase after 33–41 h, suggesting 1.0–1.3 m/h infiltration velocities. The Cl^- appeared in the dripping water only eight hours after the yield started to grow. Its content grew from 20–60 mg/l to 500–600 mg/l. The drip rate decreased from 200–300 ml/h to 0.5–1.0 ml/h. The yield grew only in those drip points that were directly under the ditch and was unchanged at 4–5 ml/h at other drips that were not under it. This indicated that the vertical fissures played the major role in the infiltration, while the 20–24° stratal dip had no role (Leél-Őssy Sz. 1997).

József-hegy Cave is the richest in non-dripstone crystals in Hungary. Naturally, the most frequently occurring mineral is calcite. Visitors can find calcite veins, scalenohedron crystals, several types of cave coralloids, cave rafts, mammillary crusts, folia, pool fingers, tetaratas (rimstone dams), cold-water rhombohedrons, and there are some dripstones, too.

The snow-white gypsum crystals are the most striking formations. Chandeliers are the largest form, sometimes being almost one meter in length. Gypsum crusts are very frequent in the middle and lower levels, with thicknesses ranging 0.5–5.0 cm. Often, 1–2 cm long and barely 1–3 mm wide crystalline needles stand out from their surfaces.

One can see fibrous speleothems in a few places, sometimes 90 cm long gypsum hair (~"needle-grass"), "cave cotton". Gypsum flowers are wide-spread. The diameters of spirally twisted crystals are 1–2 cm, and their length is 10–15 cm (anthodite-like speleothems, cf. Ghergari and Onac 1995; Onac et al. 1995). Balloon-like "cave blisters" and hollow gypsum stalagmites are also found.

There are many aragonite frostworks, and sometimes of dolomite, in the middle and lower levels. Unfortunately, aragonite crystals easily re-dissolve in the infiltrating waters.

One can find barite in places in the cave, always within or connected to the tectonic fractures and thus are older than the cave. These were precipitated from 60 to 75 °C thermal waters (Dublyansky 1995). Edge lengths of the barite crystals ranges 1–3 cm. These crystals are often covered by manganese coating. Manganese-oxide coatings (e.g. romanechite) appears elsewhere on the walls also. Their genesis is aided by bacteria (e.g. Leél-Őssy Sz. 2010; cf. Molnár János Cave in Borsodi et al. 2012).

Other macroscopic minerals noted in the cave include hydromagnesite, huntite, kaolinite and montmorillonite.

The Kinizsi SG got interesting results from the U-series dating of the speleothems in the cave (Leél-Őssy Sz. 1997; Leél-Őssy Sz. et al. 2011; Szanyi et al. 2009, 2012). The cave rafts that had been precipitated on the surfaces of past thermal-water lakes were ~ 400–500 ky old in the main level (50-60 m above the level of Danube River). The mammillary crusts which were precipitated below the waterline in the lakes had the same ages. The cave coralloids and aragonite needles or frostworks in the main level were significantly younger, ~100–250 ky.

The speleothems in the lower level are much younger, precipitated directly from the thermal water ~65–105 ky ago, the age of aragonite crystals (on the top of the cave coralloids) being only ~30 ky. An interesting form, the "glassball"-like cave coralloid (cave popcorns), can be found both in middle and lower levels in the cave. It is precipitated from meteoric water infiltrating from the surface. The age of these forms is between ~9 ky and ~17 ky.

It is very difficult to make a show cave from the most spectacular József-hegy Cave because its pure white crystals are very fragile. The Kinizsi SG recommended (Leél-Őssy and Adamkó 2006) construction of a 50 m deep elevator near the Kinizsi Chamber first, then footpaths made of stainless steel under plexiglass covers. After the Kinizsi Chamber, from the Kessler Room across the Ice-cream Corridor there should be a 3 m artificial tunnel into the Castle Room, and after the Láng Sándor Chamber, a similar exit corridor out to the open hillside at lower elevation. The white crystals can be protected by these means.

2.1.3 Molnár János Cave

Molnár János Cave currently is mainly in the phreatic zone at and below the local watertable, 7 to 137 m asl. A large proportion of its passages are filled by thermal water (Fig. 14). Therefore, study of Molnár János Cave is considered a good analogue which may help to improve our understanding of the thermal-water related paleokarst phenomena in the now-dry caves higher up the Rózsadomb. Recently the total length of cave corridors has been extended to six km and its explored maximum depth is 100 m below the level of Danube River (in the phreatic zone). The uppermost parts of the cave are 30 m above the watertable (in the vadose zone). The lukewarm water flowing through the cave is pumped via pipes to St. Lukács Spa.

The Malom (Mill) Lake is situated at the foot of the Rózsadomb (48 Frankel Leó Street). The presence of a small inactive cavity above the lake has been long known, linked to passages descending the thermal water level. Water flows from the Alagút (Tunnel) Spring (31 to 33 °C) and Boltív (Arch) Spring (25 to 27 °C) into Malom Lake. The small

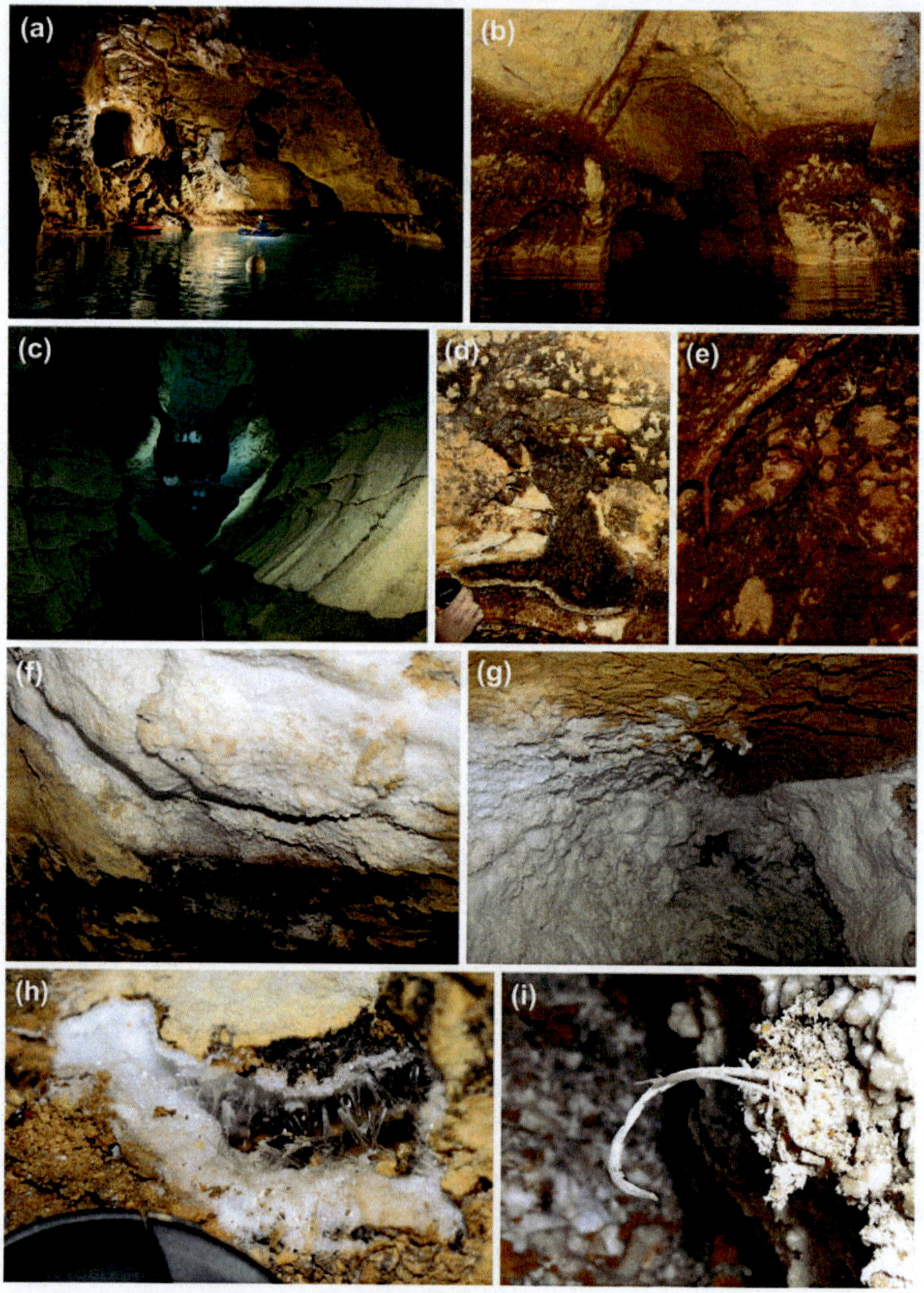

Fig. 14 Molnár János Cave. **a** The lake of the Kessler Chamber: the boundary between the vadose and phreatic zone in the active thermal cave (one of the largest underground lake-surfaces in Hungary, photo by Cs. Egri); **b** spherical cupolas above the thermal-water level; **c** characteristic high and narrow fissure-like cave passage and dissolution forms in the phreatic zone (photo by A. Sári); **d** dogtooth spar-type scalenohedral calcite crystals covered by Mn-Fe-oxide coatings in the middle and lower parts of the cave wall; **e** Fe–Mn oxide coatings and snottites as biothems; **f** Mn-Fe oxide coatings above small lake in vadose zone on the lower part, and gypsum crust alternates with calcite-aragonite-huntite-dolomite-celestite as "evaporative" frostwork-and-powder on the upper part of the cave wall (St. Lukács Passage; **g** ~6500 to ~7000-ky-old calcite-aragonite evaporative frostwork covered by gypsum crust grown on the marl (St. Lukács Passage; **h** finely-crystalline powder-like and needle-like gypsum crystals in small vugs of clayey clastic sediment (St. Lukács Passage); **i** gypsum "snake" and gypsum crust grown on the marl (St. Lukács Passage, photo by R. Kovács)

cavity was named after János Molnár, who described it first in 1858 (Molnár 1869). He only explored the dry 12 m upper cavity, the bottom of which ends at the active thermal-water cave.

János Molnár also made the first chemical analyses of the thermal water. From his descriptions (Molnár 1869) it was found that the nearby springs, Iszaptó (Sludge) Springs, have a high content of CO_2 and H_2S and similar sulfur-containing water and precipitates are also present at the Császár (Imperial) Spa and Király (King) Spa. The discharge from the thermal wells in the park between the Danube and the spa buildings is collected in a sewer today. In the 1500 s the lukewarm waters were retained by dams to create Malom Lake and used to drive mills.

Exploration of the underwater sections of the cave began in 1953 following the development of improved diving technology. In 1960, MHS BEKSZ divers entered the cave from Alagút Spring. In 1972, the FTSK Delfin Diver Club started regular exploration with a scuba system, with the support of Hubert Kessler (Plózer 1972, 1974). The explorations continued with the organized collaboration of divers Sándor Kalinovits, Attila K. Kollár, László Maróthy and István Plózer. After their furthest dives, a total length of 420 m of passages (the "Old Section") had been explored (Kalinovits 1984).

In 2002, when the pipe conveying lukewarm water to the Lukács Spa was being replaced, a small hole opened in a fallen mud wall (the "Black Wall"—barite crystals covered by black Mn-oxide coatings), exposing a continuation upstream into further parts of the cave. At the inflow of warm water at the "Black Wall", Zsolt Gyurka further opened the narrow hole in 2002. As a result of which the previously unknown, further section of the cave was opened (Kalinovits 2003, 2006). The new sections were explored and mapped mainly by Sándor Kalinovits, Zsolt Gyurka, László Müllner, Zsolt Szilágyi, Zoltán Szabó, Attila Hosszú, Szabolcs Storozinsky, Péter Zsoldos and József Spanyol. Water and sediment samples for hydrodynamical, hydrogeochemical and mineralogical studies have been regularly collected by Dénes Szieberth (e.g. by Dénes Szieberth in Szombathy et al. 2019b) in the cave. Guest dives to the underwater sections are now regularly organized (https://mjcave.hu/en/).

In 1976 and 1977, the József-hegy Tunnel was built inside the Rózsadomb behind a concrete "Turkish bath". In 2008, after preliminary geophysical surveys, a 15-m-long corridor was dug to connect the tunnel to the Kessler Chamber (Surányi et al. 2010). As a result, this 10 m high, 20 m diameter room was also accessible by dry foot. However, 90% of the 22,000 m^3 full volume of the chamber (2/3 of its vertical extent) is below the waterline. The air in the room was unsuitable for inhalation at the time (6.5% carbon dioxide) but is now replaced by outside air. The St. Lukács Passage, an extension of Molnár János Cave, was discovered in 2007 from the József-hegy Tunnel.

The cave passages have strong tectonic control. The longest corridors are dissolved mainly along NE-SW faults. There are lesser NW–SE and N-S passages, creating and irregular grid pattern (Fig. 4e). Usually the passages follow the general dip (30°) of the host Eocene Limestone and Eocene–Oligocene Marl (S-SE). Kessler Chamber and the St. Lukács Passage also extend upwards into the Buda Marl.

Kessler Chamber straddles the vadose and phreatic zones today. Siliceous altered zones can be investigated within dissolutional cavities and also in the marl. A pyrite-calcite-filled vein can be observed in unaltered Buda Marl in the dug tunnel. Based on the detailed investigations, the following theory of formation can be outlined. Due to the oxidation of pyrite, sulfuric acid was released along the mineral vein and, in this highly acidic environment, the carbonate content of the marl was dissolved and replaced with kaolinite, quartz/silica precipitates. Various forms of gypsum and minor amounts of jarosite were also precipitated, plus minor ferrihydrite (to goethite) in the pores.

Waters with two different temperatures and chemistry mix in the cave today, thus mixing corrosion is mostly responsible for the modern dissolution. Not only carbonic acid but also sulfuric acid may have played a role in the genesis of the cave passages, as noted. The mixing and the dissolution by the 18–20 °C and 27 °C components of the water explains why only Mn-Fe-oxide coatings and coarse sparitic calcite, plus barite crystal veneers (as "cave hoods" on the calcite crystals) formed before the development of the main passages can be seen in the phreatic zone. The re-dissolution of calcite and the oxidation of pyrite can be observed. The clastic sediments come partly from the siliceous altered zone material and partly from the dissolution residues of the host rock—quartz, varying amounts of kaolinite depending on the type of rock, and a small amount of illite. Depending on the redox conditions, pyrite can also be seen in certain sections. There are NE-SW and NW–SE oriented fractures and small cavities—exposed by the recent cave passages—partially filled by calcite, barite and pyrite that clearly pre-date the cave formation.

The underwater phreatic sections are characterized by solutional scallops and dissolution pockets. Spherical cupolas are found above the thermal water table. Recent growth of spherical cupolas resulting from condensation-corrosion can be also observed in the cave (Müller 1974).

Dissolution experiments by Erőss (2010) and Erőss et al. (2011) have shown that dissolution, mainly due to mixing corrosion, is currently active. Based on radionuclide analyses and NaCl tracer tests of the thermal water in the cave and at the Alagút and Boltív springs, Csondor et al. (2016, 2017) suggest that the water flows via an unknown cave passage from István Chamber to Boltív Spring.

From investigations of the biofilm microbial communities (Borsodi et al. 2012; Makk et al. 2016; Anda et al. 2017; Enyedi et al. 2019), it is proven that sulfuric acid dissolution and oxidation-reduction process-related microbial-activity are also taking place in the cave. The chemolithoautotrophic microbial communities ($\sim$microbial mats) are oxidizing the iron and manganese in the water, deriving their energy from these processes. The same microbes also oxidize the sulfur in the water and other members of the community reduce it, respectively (cf. sulfide oxidation and sulfate reduction).

In the vadose zone of the cave, the carbonate, sulfate and other mineral groups are equally represented. Among the speleothems above the thermal-water lake in the Kessler Chamber, Fe–Mn-oxide speleothems are present as vertical coatings, and the sulfates can be observed as translucent, coarse-grained gypsum "thorns" (spicules) in the clayey rotted walls, and subordinately as gypsum "snakes", "flowers" and white fine-crystal gypsum crusts. It is supposed that the formation of sulfate minerals is mainly the result of the evaporation of the thermal water and precipitation associated with the local oxidation of the pyrite or pyrite-containing veins in the marl. In the St. Lukács Passage, which is largely in the vadose zone, there are also carbonate speleothems: calcite and aragonite in the form of evaporative frostwork, huntite forming common paragenesis with gypsum, dolomite formed by microbial influences, and celestite. Recent paper-thin cave rafts are precipitated vigorously on the surfaces of the shallower thermal-water lakes. On the basis of the U/Th age determination of the tiny white cave coralloids and frostworks developing in the vadose zone, the highest endpoint of the side branch of St. Lukács Passage may already have been above the surface of the water $\sim$6500–7000 years ago. The role of the infiltrating water from the surface is subordinate.

2.1.4 Pál-völgy Cave System

At the present time, 32 km of the passages in this hypogene cave system are known, making it the longest explored cave in Hungary. Originally it comprised several independent but neighbouring caves. Pál-völgy Cave was the first large cave to be discovered in Budapest. Its entrance was found during stone quarrying in 1904, as were the other parts of this cave system—Harcsaszájú (Catfish Mouth) Cave (1902), which originally was small; Hideg-lyuk (Cold Hole) Cave, Bekey Cave, Kis-hideglyuk (Small Cold Hole) Cave, and Gábor Áron Cave, found at almost at the same time (e.g. Fritz and Tóth 2016; Leél-Őssy Sz. 2013).

Opposite the Pál-völgy quarry a second stone quarry was opened on western side of Mátyás Hill, the northeastern side of the valley, before the First World War. This quarry revealed the entrance to Mátyás-hegy Cave in the 1930s.

After the exposure of *Pál-völgy Cave* (Fig. 15) in 1904, Imre Gábor Bekey, János Bagyura, Pál Kornél Scholtz and Károly Jordán together mapped approximately one km of passages in the cave in the next few years. Guided tours started early in the years immediately following its discovery. The cave was built up and opened for tourist visitors in 1919. This was the first Hungarian show cave that had electric lighting, in 1927. Some further passages and a new exit into the lower part of the quarry were made accessible for tourists in 1989 when Budapest hosted the quadrennial Congress of the International Speleological Union. Today a 500 msection of Pál-völgy Cave is open for public (https://www.dunaipoly.hu/en/places/interpretation-sites/pal-volgyi-cave). In 1980, Attila Kiss and József Kurucz followed a draught from the Theatre Room, and discovered a new section of the cave. From this time, the Bekey Imre Speleo Group led by Attila Kiss, Katalin Takács-Bolner and now Attila Tóth discovered more and more new passages every year, increasing the length of Pál-völgy Cave to more than 18 km today—this is the longest section of the Pál-völgy Cave System (Takács-Bolner and Sásdi 2003).

The first part of *Mátyás-hegy Cave* was used as an air-raid shelter during the Second World War. Now there is an earthquake observatory installed in it. One can go into the Fireman's Cave from here: supposedly, firemen rescued children who had lost their way there. In 1948, Béla Mohos broke through the end of the cave to discover the Centenáris (Centenary) Section—celebrating the 100 years anniversary of the 1848 Hungarian revoluted against the Habsburg Empire. All further discoveries in the Mátyás-hegy Cave started from this section. Most important was the discovery of a connection between Pál-völgy Cave and Mátyás-hegy Cave, by the Acheron and Bekey Speleo Group in 2001.

Cavers started to work in two other, originally small, caves in the old Pál-völgy quarry after 2000. The Szabó József Speleo Group (led by András Nagy) was here from 2005. They extended the 170 m long *Hideg-lyuk Cave* to greater than 4 km in 2009. Since the entrance into the new part of the cave was very dangerous, they also opened a second entrance from Gábor Áron Cave. The Barit Speleo Group (led by Gergely Domonkos Nagy, Lénárd Szabó and Attila Nyerges) searched for new passages in *Harcsaszájú Cave* from 2006. They found a rear section of the cave in 2008 that is now longer than 4 km. The cavers connected Hideg-lyuk Cave and Harcsaszájú Cave together in 2010, and then linked them to Pál-völgy Cave in 2011. The Pál--völgy Cave System came into being in this way.

There are lots of unusual dripstones in the Buda Hills and you can find large and beautiful examples in this cave system. Chiefly Harcsaszájú Cave and Pál-völgy Cave, but also Hideglyuk Cave, have flowstones. Although Mátyás-hegy Cave has no spectacular speleothems, it is very rich in dissolution forms. The oldest dripstone column of the Buda Hills (and until recently, in Hungary) is found in Pál-völgy Cave: the "Giant" dripstone is older than $\sim$350 ky

Fig. 15 Pál-völgy Cave, Pál-völgy Cave System. **a** The 0.5 million-year-old Giant Dripstone (recently the oldest dated dripstone column and speleothem in Hungary, photo by A. Kiss); **b** typical stalactites, stalagmites and dripstone columns (photo by A. Kiss); **c** characteristic high and narrow, fissure-like cave passage (photo by Á. Berentés); **d**, **e** hemispherical solution pockets (photos by Cs. Egri); **f** cave cones (~heaps of broken cave rafts) partly covered by cave coralloids (photo by Cs. Egri); **g** folia-rich cave wall (photo by Cs. Egri); **h** cemented cave rafts (photo by A. Kiss); **i** Millipede within stalagmite (photo by Cs. Egri); **j** various stalactites, soda-straw stalactites and draperies (photo by Cs. Egri)

(Takács-Bolner 1998). According to newer age determinations in Virág et al. (2013b), this column is older than ~500 ky. Other age-determined dripstones proved much younger (e.g. ~5–1 ky in Demény et al. 2013; ~222 ky, ~54 ky and ~12 ky to Recent in Virág et al. 2011, 2013c). You can also find small gypsum stars, many cave rafts (this cave has the largest quantity of this significant form of calcite), and folia. The cave cones (heaps of sunken cave rafts) in the Jubileum Section are largest and most beautiful found in the Buda Hills. Researchers obtained valuable age data from these cave rafts (Ford and Takács-Bolner 1991; Szanyi et al. 2012). The deepest sample was situated at 148 m asl, while the highest one at 175 m asl. The youngest dated cave raft from the Pál-völgy Cave was formed ~269 ky, but some samples are older than ~450 ky. Interestingly, there are lesser amounts of cave coralloids and no aragonite, but there are beautiful sparite calcite crystals (e.g. pool spars) of cold-water lake origin. One can find fossils almost everywhere on the walls. The main level of the caves is developed in Eocene Szépvölgy Limestone, but the uppermost parts extend into the overlying Buda Marl, where there are conspicuous siliceous altered zones. Hemispherical solution pockets as well as spherical cupolas very often decorate the passages in the limestone. Most of the tectonic-controlled passages follow NE-SW oriented fractures but NW–SE directions are characteristic, too (Fig. 4a). The passages also follow the 30° S-SE dip of limestone and marl beds.

Mátyás-hegy Cave (Fig. 16) is also developed in all three formations: the upper parts in the Buda Marl, the middle in Szépvölgy Limestone, and the lowest—in the Agyagos ("Muddy") stream channel—is in the cherty Triassic Mátyáshegy Limestone. As in the lower part of József-hegy Cave, there is no clay on the floor. The karst(?) water table is reached at the deepest point in the cave (91 m under the entrance), where there is a pool 5–6 m^2 in area and 3–4 m deep. This is the only cave in this area that collects percolating waters. Some fluviatile debris can be found in several passages (e.g. Leél-Őssy Sz. 2015). This is a relatively "bare" cave, without any crystals, but it is the richest in hemispherical solution pockets. However, in the eastern part of Mikulás-ág (Santa Claus Section) ~500 ky old mammillary crusts and ~339 ky old paper-thin cave rafts as well as ~402 ky old thick cave rafts can be found (Virág et al. 2014; Virág 2018). There are also small quantities of gypsum flowers, thin crusts and needles. Due to the easy access, Mátyás-hegy Cave has become a training cave for Hungarian cavers and a popular target for exploration. Every Hungarian caver with a cave-tour guide or research leader certificate can ask permission to take a tour in the cave. This cave is also open for "adventure caving" tours for small groups led by qualified English language cave guides. This cave is open for tourists throughout the year, but pre-registration is requested (Caving under Budapest: www.caving.hu).

Every crystal and the larger part of the floors are intact in *Hideg-lyuk Cave* and in *Harcsaszájú Cave*—these are "new" caves (Fig. 17). One can enter them only with special permission. The rear of Harcsaszájú Cave has an extremely thick clay accumulation on the floor. There is much clay-rich Buda Marl above this section of the cave, so it is suggested that the clays are mainly the insoluble residues (and some from the Szépvölgy Limestone).

Every winter there are about 500 hibernating bats in the Pál-völgy Cave System, most of them in Pál-völgy Cave.

2.1.5 Szemlő-hegy Cave

Szemlő-hegy Cave lies under the eastern part of Ferenc Hill, a highly urbanized and built-up area. The original discoverers' entrance was at 10 Barlang Street. There is now an artificial entrance behind a tourist reception building at 35 Pusztaszeri Road. The total length of cave is 2230 m and the vertical extent is 50.4 m (Fig. 18).

Szemlő-hegy Cave was discovered during quarrying and/or construction works in 1930. From the narrow "Discoverers" "Entrance", on September 20 that year Hubert Kessler and András Futó followed the "Vortex Corridor" to the thermal-water origin speleothem-rich Giant's Corridor 30 m below, and then the Long Corridor. This spectacular cave was described as "the Underground Flower Garden of Budapest" because of the unique richness of cave coralloids, mammillary crusts, cave rafts and gypsum crusts. Due to the large amounts of these types of speleothems, speleologists first recognized here that this cave has a thermal-water origin (Kessler 1931, 1936).

In 1958 cave researchers of the Kinizsi Speleo Group, leaded by János Palánkai, discovered the 280 mlong Kadić Section. In 1973 members of the Ferencváros Speleo Group discovered the Muddy Section and its North-Eastern continuation. The Ferencváros Chamber is the largest room in the cave. Jenő Kiss and SZIKKTI Sport and Speleo Group have been exploring the cave since 1976 (Kraus and Hazslinszky 2003).

In 1961 and 1962, a 1:100 scale map of the cave was completed under the direction of János Horváth (1965). The total length at that time was 1962 m. This was the first (and for a long time, the only) cave map in Hungary with such detail and accuracy. On the basis of this map, János Horváth and Gábor Szunyogh prepared a 3D plaster model of the cave for the 10th Congress of the International Speleological Union in Budapest in 1989.

Based on Horváth's 1:100 scale map, Albert et al. (2015) modeled the volumes of the cave passages. Volume modeling of the proportion of the bulk porosity contributed by caves in Rózsadomb (Szemlő-hegy Cave and a significant extent of Pál-völgy Cave) revealed similar macroporosities (cavities more than 0.5 m in diameter). They are chiefly in the Szépvölgy Limestone, lesser in the Buda Marl (Szemlő-hegy Cave: 1.66% and Pál-völgy Cave: 1.46%). In contrast, the macroporosity of the Buda Marl (the host rock of the approximately 450 m long "Old Section" in active hypogene Molnár János Cave) is only 0.92%.

The spectacular speleothems of thermal-water origin, the dissolution forms (such as cupolas) and the high and narrow passages invited entrepreneurs to develop the cave further

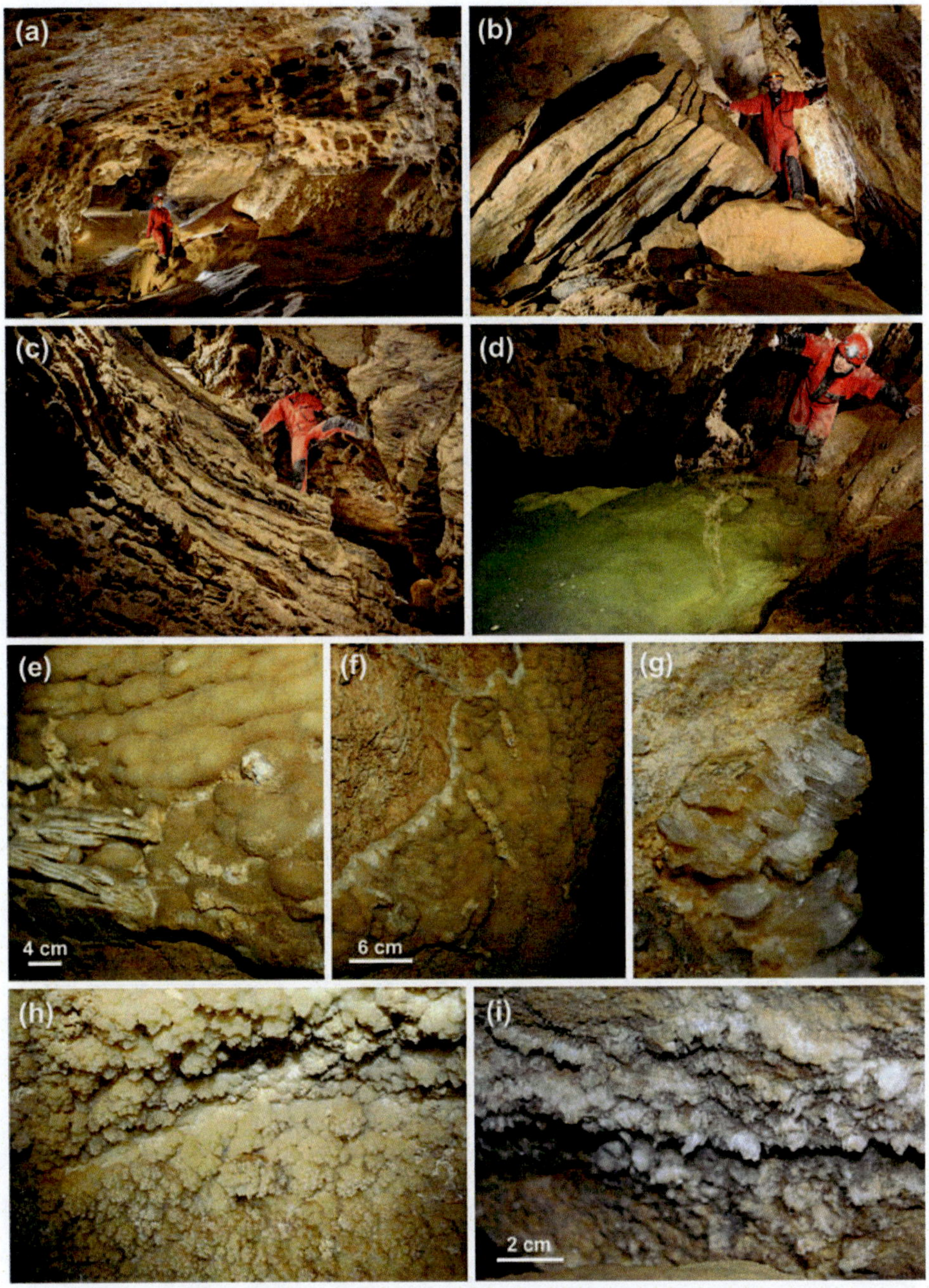

Fig. 16 Mátyás-hegy Cave, Pál-völgy Cave System. **a** The largest characteristic cave chamber in the Mátyás-hegy Cave rich in hemispherical solution pockets (Theatre Chamber, photo by Cs. Egri); **b** huge fell-down blocks of Eocene limestone (photo by Cs. Egri); **c** cave passage dissolved in Triassic cherty limestone (photo by Cs. Egri); **d** the "Lake" at a depth of 91 m below the cave entrance (photo by Cs. Egri); **e** cemented cave rafts and mammillary crusts; **f** folia-like speleothems and mammillary crusts on the host rock limestone; **g** translucent gypsum flowers; **h** pool spar-like speleothems; (i) gypsum flowers and thin gypsum crusts

and open it up to the public shortly after it was discovered. Construction began in 1974 and the cave was opened to the public in 1986. The construction included mining an entrance passage from Pusztaszeri Road, deepening and widening the cave floors, fitting them with concrete pavements and electric lights. Initially a 300 m long section of Szemlő-hegy Cave was open for public, but the total length of the cave tour is now 530 m (https://www.dunaipoly.hu/en/places/interpretation-sites/szemlo-hegyi-cave).

The cave was declared a therapeutic cave in 1991, organized by St. John's Hospital. The temperature of the cave air is 12 to 13 °C all the year round, and the relative humidity is nearly 100%. As a consequence, the ionized cave air does not contain dust or pollen. Its beneficial medical effects have been recognized and are used for people suffering from respiratory diseases such as bronchitis or asthma.

The thermal-water speleothems are located mainly in the lower passages of Szemlő-hegy Cave that are open to the

Fig. 17 Harcsaszájú Cave and Hideg-lyuk Cave, Pál-völgy Cave System. **a** Characteristic high and narrow fissure-like cave passage with solution pockets (Harcsaszájú Cave, photo by G.D. Nagy); **b** high and narrow cave corridor with karren-like dissolution forms on the cave wall (Harcsaszájú Cave, photo by A. Hegedűs); **c** ring-like break-up of the cave roof (Hideg-lyuk Cave, photo by Zs. Kalotai); **d** coral pipe-like drip-forms in the muddy-clayey cave sediment (Harcsaszájú Cave, photo by E. Szabó); **e** cave pearls in the cave pool (infiltrating meteoric water dripping intensively to the pool, Hideg-lyuk Cave, photo by L. Tisza); **f** small carbonate crystals party filling a vug (Hideg-lyuk Cave, photo by Zs. Kalotai); **g** yellowish-brownish banded draperies (Hideg-lyuk Cave, photo by Zs. Kalotai)

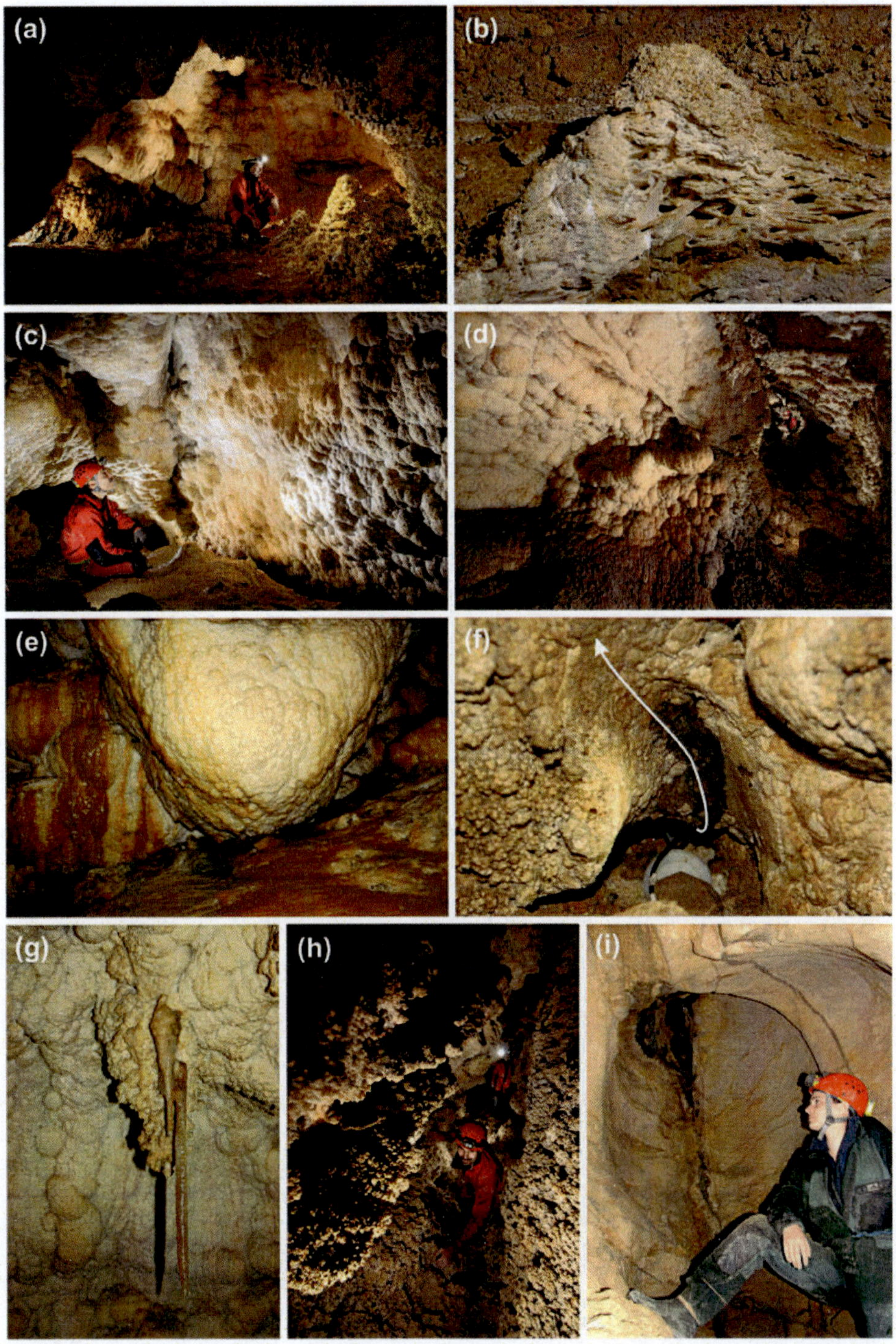

Fig. 18 Szemlő-hegy Cave. **a** Typical hypogene cave section with speleothems of thermal-water origin (photo by Á. Berentés); **b** cave cone (~heap of broken cave rafts) covered by cave coralloids (photo by A. Hegedűs); **c** cave chamber covered by mammillary crusts (photo by Cs. Egri); **d** cave section covered by mammillary crusts (photo by Á. Berentés); **e** red-coloured flowstone partly covering the mammillary crusts and cave rafts (Hearth of Earth; **f** according to meteorological measurements, along NW–SE fractures, there is still a definite upward flow of CO_2 dissolving the surface of older speleothems and even the carbonate host rock (marked by white arrow); **g** reddish and dark brown-coloured stalactites and cave coralloids on the mammillary crusts; **h** cave coralloid-rich high and narrow cave passage (photo by Á. Berentés); (i) spherical cupola (photo by A. Hegedűs)

public. The upper, non-tourist parts are almost free of such speleothems but characterized by thermal solution morphology and spherical cupolas of various sizes.

The genesis and precipitates of the cave and the paleo thermal-water level fluctuations were first studied morphologically by Sándor Kraus (1982, 1990, 1993, 2006).

The major proportion of the Szemlő-hegy Cave passages are in the Eocene Szépvölgy Limestone, developed mostly along NE-SW fractures and some minor crossing fissures (Fig. 4b). These fracture directions are associated with a major dextral strike-slip zone and its Riedel fractures (cf. Benkovics et al. 1999). Diameters of passages increase to the

NW. The main passages are found at 170 m asl but the Discoverers' entrance is at 206 m asl. The tops of passages almost reach the surface in the NE and central sectors. Large (several m diameter) spherical cupolas are the common upward terminations in the limestone, dissolved by condensation-corrosion in CO_2-rich cave air above the thermal-water lakes (cf. Audra et al. 2007). While the upper parts of the cave are speleothem-free, the lower parts are richly decorated with thermal-water origin speleothems. In addition, the SW area (Kadić Section) contains striking dripstones. The upper parts of this section extend up into the dipping (30° S-SE) Buda Marl.

The sediments in the cave are mostly insoluble residues of the host rocks (e.g. quartz, kaolinite, illite, Fe-oxides). Abundant quartz and significant enrichment of kaolinite are observed, which may be due to redeposition of a clayey "(volcanic) tuff layer" (Virág 2018).

This typical hypogene cave is now in the vadose zone (165 to 215 m asl). It is presumed to have formed under phreatic conditions, its present position being the result of tectonic uplift in Pleistocene times (cf. "recent thermal karstic paleokarst"). Szemlő-hegy Cave as a thermal-water cave and its speleothems nicely reveal the history of its passages from the hypogene (phreatic) through to the epigene (vadose) phases (Virág et al. 2015, 2016).

It is suggested that dissolution was the predominant process of cave formation about one million years ago in the phreatic zone. In Mid-Pleistocene times, as uplift moved the cave gradually up to the vadose zone, isolated thermal-water lakes a few meters deep survived in the NE-SW corridors. Due to evaporation and CO_2 degassing from the lukewarm waters, there was condensation-corrosion in the ceilings, resulting in the formation of cupolas. At the same time, apparently, in the lower reaches of the cave, thermal-water origin speleothems with variegated morphology were precipitated at and around the phreatic/vadose interface. Below the water-table, there were isopachous calcite crusts 3–10 cm thick, with their crystals oriented perpendicular to the cave walls. Cave rafts formed on the water surface and subsequently sank to the bottom. Above the surface of the thermal-water lakes, evaporation resulted in precipitation of cave coralloids and aragonite frostworks. In the Kadić Section and Giant's Corridor, 0.5–3.0 cm thick gypsum crusts grew on the cave walls, often covering earlier-precipitated calcite speleothems. They are believed to have precipitated due to sulfuric acid, from oxidation of either H_2S (above the water-table of the lakes) and/or pyrite, interacting with the carbonate host rock.

In Szemlő-hegy Cave carbonate speleothems are dominant and the other minerals are subordinate. As described above, among the mainly calcite speleothems, mammillary crusts are considered to be phreatic precipitates, cave rafts are water-table indicators, and the cave coralloids and dripstones are vadose features. Lesser quantities of gypsum as well as bare aragonite frostwork, Mn-oxides, Fe-oxides have been identified, and in the pores of the carbonate rocks there are small barite and dolomite crystals, plus celestite at the contact between the gypsum crust and carbonate rock or speleothems. The NE-SW and NW–SE fractures are partially filled by calcite, minor barite (e.g. as "cave hoods" or caps on calcite crystals) and pyrite that obviously pre-date the cave formation. As the cave is thought to have developed in Pleistocene times, this fracture-filling paragenesis must be of pre-Pleistocene age.

The speleothems facilitate reconstruction of the changes of the thermal water-table (Virág 2018; Virág et al. 2019). U/Th data reveal that thermal-water speleothems were precipitated in Mid Pleistocene times, from approximately 500 to 160 ky. At that time the carbonate-supersaturated thermal water was in the lower sections of the passages. Cave rafts now found at 179 m asl, were precipitated ~500 ky ago on the surface of larger or smaller thermal-water lakes separated by dams in NE-SW passages. Cave rafts that sank to the bottoms of the lakes accreted mammillary crust overgrowths dated ~390–410 ky. ~290–300 ky ago the thermal water table dropped by about 8 m (the related rafts and mammillary crusts are found now at 172 m asl) and most passages became partially air-filled (vadose zone). Some of the earlier calcite rafts became overgrown by vadose speleothems (e.g. cave coralloids), then the thermal water-table probably ascended around 280 ky, this time by about 5–6 m because the related speleothems are now at 174–175 m asl. At about 186 ky, waterline speleothems began growing at 168 m asl. The last of them are dated at ~160 ky and the cave has been dry ever since.

There are also typical vadose precipitates (dripstones) in the cave, as noted. They formed in Pleistocene to Holocene times and belong to the epigene phase of the cave. The age obtained on a brownish stalactite suggests that meteoric water infiltration from the surface was already occurring more than ~227 ky ago.

Along the dominating NW–SE fractures there is still a definite upward flow of CO_2 (Weidinger et al. 2017; Stieber 2016 and Table 12.2). It is dissolving surfaces of older speleothems and even the carbonate rock. According to meteorological measurements, this CO_2-enriched "warmer" air is rising from deeper, still unknown, sections of the cave. This indicates that the cave today is still in dynamic contact with deeper, perhaps still active, phreatic passages, i.e. active hypogene condensation-corrosion dissolution is still taking place locally below, and even within the known cave.

2.2 Caves of Buda Castle Hill

Budapest has caves not only in the Rózsadomb area. The "Capital City of Caves" (cf. Kadić 1931, 1936) has also famous caves on Vár-hegy (Castle Hill), one km to the south under the old city of Budapest. These caves are also hypogene but differ considerably from those Rózsadomb, their age being only ~150–200 ky. They are dissolved in the base of the Pleistocene travertine "cap" on Buda Castle Hill, a 6–10 m thick layer of travertine aged between ~350 and ~190 ky (according to Krolopp et al. 1976) or ~350–160 ky (Scheuer 1986). U series ages of ~350–450 ky (Kele 2009) have been obtained on the travertine deposits themselves, where they rest on underlying fluvial sediments and the Buda Marl (Kele 2009; Leél-Őssy Sz. 2012). These caves were discovered when the old city was built approximately 700 years ago and the habitants dug the cellars and wells. The aggregate length of these 50 or 60 caves is about four km, including some partly artificial passages. Many of the small caves were interconnected by tunnels before and during the Second World War (and also in earlier centuries), and the length of the longest section is around 3 km. The traces and tools of primitive man (Buda Culture) were found there (Takács-Bolner 2003a). They lived at the springs that were depositing the travertine. The caves are mainly without any speleothems, but some small soda-straw stalactites and stalagmites grew on concrete and slabs of Dachstein Limestone used to reinforce the ceiling during the Second World War. There are some support pillars from the Turkish (Ottoman) times, too. Three parts of the largest cave are open for public. Organized and managed by Danube-Ipoly National Park Directorate (https://www.dunaipoly.hu/en/places/interpretation-sites/buda-castle-cave), visitors taking the 1.5 km walking tours in *Buda Castle Cave* take pleasure in the shapely solution pockets in travertine ceilings of the 3300 mlabyrinth of caves hidden beneath the streets of the Castle District and discover the history and the unique geology of Buda Castle Hill (Fig. 19). Buda Castle Cave is open for tourists throughout the year but pre-registration is requested. Anybody can enter the *Labyrinth* from Úri Street and the *Rock Hospital* from Lovas Street outside the Castle walls. The hospital was built in natural cave rooms connected by artificial corridors during the Second World War.

2.3 Caves of Gellért Hill

Gellért Hill, two km the south of Rózsadomb, is also an interesting cave area. There is a several-meter high and wide entrance arch, *Gellérthegy Cave* or Iván Cave or "Sziklatemplom" (Rock Church), which has been known for millenia. Probably the second element (Pest) of the name of the Hungarian capital originates here: "pest" means cave, cavity, or furnace in Slavic languages. The host rock is Upper Eocene breccia. A small (one-room) hypogene cave, *Gellérthegy Aragonit Cave,* lies six m below Gellerthegy Cave, discovered by a geological borehole in 1961 (Fig. 20). It was dissolved chiefly at the contact of the Eocene chert-breccia with underlying Triassic dolomite. The lower part of chamber has been artificially dug deeper into the dolomite. The natural part of the cavity is rich in mainly thermal-water origin speleothems: ~130 ky old thick mammillary crust, ~55 ky old relatively thin calcite crust ("septaria-boxwork") filling desiccation cracks in layered goethite sediments, aragonite frostworks, hexahydrite, small cave coralloids (cf. in Virág et al. 2013d; Leél-Őssy Sz. and Virág 2018). There is also hydromagnesite on aragonite frostworks, thin gypsum crusts, epsomite flowers and snakes or coils (cf. Pásztor 2016). The 60 m long and 18 m deep *Citadella Crystal Cave,* dissolved in Eocene Buda Marl and discovered during house building in 2007 (Leél-Őssy Sz. et al. 2007) is another hypogene nearby (Fig. 21). This cave is rich in white calcite, aragonite, gypsum, and small amounts of huntite, halloysite, dolomite minerals (Fig. 22). The age of the cave rafts is ~195 ky. The air temperature is 18 °C at the deepest point because modern thermal springs are very close. Gellért Hill is built chiefly of Triassic dolomite: there are further two or three dozens small (a few-meters long) caves in this rock. There are similar but still active spring caves at the foot of Gellért Hill. One of them, *Török Spring Cave,* is recently formed by ca. 40 °C thermal water that partly filled it (Fig. 20). According to Erőss (2010), Erőss et al. (2012), the dissolution is not only by CO_2 and cooling of thermal water but also sulfuric acid dissolution and related microbial-activity. Investigations of the biofilm microbial communities (~microbial mats) have been published in Borsodi et al. (2012, 2018); Anda et al. (2014); Makk et al. (2016); Kovács-Bodor et al. (2018) and Enyedi et al. (2019). Mainly due to the hypogene processes, there are ferrihydrite (to goethite), gypsum crusts, folia and cave raft precipitates. For more details of the hydrogeology, see Chap. "Hydrogeology of the Karst Regions in Hungary".

2.4 Caves of Hármashatár Hill and Remete Gorge

On Hármashatár Hill, still in Budapest, there are two interesting caves developed in the Triassic cherty dolomite. *Tábor-hegy Cave* is longer than 150 m, deeper than 20 m. *Királylaki Cave,* 380 m long and 34 m deep was discovered by Richárd Kovács, Pál Borka and István Szenthe only a few years ago (Szabó and Leél-Őssy Sz. 2013). Királylaki Cave is very rich in thermal-water origin speleothems, including cave coralloids, cave rafts, calcite, aragonite and hydromagnesite frostworks: see Leél-Őssy and Virág (2018) and

Fig. 19 Buda Castle Cave. **a** Characteristic cave section of the labyrinth system with solution pockets in the roof of the passage, reflecting the results of long-time natural and anthropogenic influence (photo by Cs. Egri); **b** imprint of a Mammoth's tooth in the travertine; **c** the largest chamber in the travertine (photo by Cs. Egri); **d** well from Medieval times; **e** Solution pockets in the cave roof (in travertine); **f** soda-straw stalactites grown on concrete ceiling; **g** marl overlain by Pleistocene travertine; (h) "fried-egg" stalagmite

Fig. 20 Gellérthegy Aragonit Cave and Török Spring Cave. **a** Characteristic cave section with mammillary crust, goethite sediments and calcite-filled desiccation cracks ("septaria-boxworks") on the Eocene chert-breccia (Gellérthegy Aragonit Cave); **b** details of ~130 ky old thick mammillary crust, ~55 ky old relatively thin calcite crust ("septaria-boxworks"), filling the desiccation cracks of goethite sediments (Gellérthegy Aragonit Cave); **c** aragonite frostwork on mammillary crust (Gellérthegy Aragonit Cave); **d** hydromagnesite moonmilk on the needle-like aragonite frostwork (Gellérthegy Aragonit Cave); **e** mammillary crust (Gellérthegy Aragonit Cave); **f** epsomite-flower/snake with thin gypsum crust (Gellérthegy Aragonit Cave); **g** thermal-water lake in the Török Spring Cave with recently precipitating paper-thin cave rafts floating on the water table, calcite and ferrihydrite precipitates at the edge of the lake on the dolomite (Török Spring Cave); **h** Gypsum crust on the dolomite (Török Spring Cave)

Fig. 23). The age of the cave rafts is ~550 ky (G. Surányi, pers.comm. 2013).

In the Remete Gorge on the edge of the city there are a dozen normal karst spring caves. All of them are found in the few tens of meters thick Triassic Dachstein Limestone. The largest is *Remete (Hermit) Cave*, which is 46 m long with a vertical extension of 15 m. A monk lived here as a hermit at the end of 18th Century, after King Joseph II banned the monastic orders in Hungary in 1782. The most interesting cave is *Hét-lyuk (Seven Hole) Pothole* (Leél-Őssy S. 1950), which is 70 m long and 33 m deep. Valuable archaeological findings from the Bronze and Palaeolithic Ages (tools, and the teeth of *Homo neanderthalensis*) were found in *Remete Cave* and in *Felső (Upper) Cave of the Remete Gorge* (Borzsák and Székely 2003). South of the Remete Gorge, there is a typical small hypogene cave, *Rácskai Cave,* in Rácskai Quarry (Kraus 2003a). It is a 176 m long and 25 m deep cave rich in cave coralloids and small stalactites, logomites, cave rafts as well as folia (Fig. 24g).

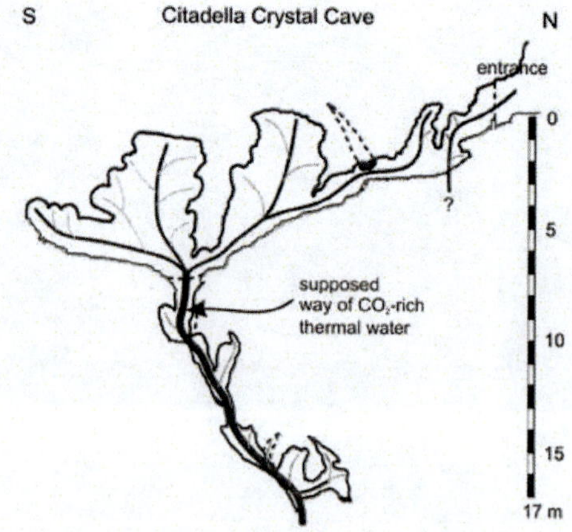

Fig. 21 Cross section of Citadella Crystal Cave (speleological data from the National Cave Register; after Leél-Őssy Cs. et al. 2007)

2.5 Bátori Cave

A few kms from the other caves but still Budapest, Bátori Cave is in the Triassic Dachstein Limestone near the top of Hárs Hill. Almost everywhere else on the Hill, the Hárshegy Sandstone (Oligocene) covers the limestone. This cave has been known since prehistoric times. The name is from a famous monk, László Bátori, who lived in it between 1437 and 1457 and wrote tracts about Bible.

The cave is rich in hematite of hydrothermal origin and in limonite. This iron ore was mined in the XVth century. In the 1960s many traces of the mining and relict smelting-furnace were found here (Vajna 1973), plus older archeological finds from the Neolithic, Copper, Bronze and Iron ages (Székely 2003).

The 360 m long, 50 m deep Bátori Cave is famous for its beautiful, typical huge cupolas (like the Sátorkőpuszta Cave, see Sect. 12.3.3) in the upper parts. These cupolas include distinctive forms like clusters of grapes. The lower parts are rich in cave coralloids (Fig. 24c–f).

2.6 Solymári-ördöglyuk Cave (or Solymár Devil's Hole)

The last hypogene cave known in the region, also in Dachstein Limestone, is the 5.5 km long 78 m deep Solymári-ördöglyuk (Devil's Hole) Cave, just outside of Budapest (Fig. 24a, b).

This is a very popular cave, with a natural entrance. The surfaces of the limestone are smoothed by wear from visitors. There are also speleotherapy sessions for children with asthma, as in Ferenc-hegy Cave. Wooden steps were built at the entrance in 1929 to give easy access to the cave. They were changed to concrete in 1936 and remain in use today. The cave has a lot of beautiful cupolas and a few cave coralloids and dripstones. Unfortunately, the passages are very narrow between the large rooms in this 5.5 km long cave, and one has to climb up 3 to 5 m at several places; therefore, the caving is difficult and some ladders and ramps have been put in place in the past (Hazslinszky 2003).

There were rich palaeontological discoveries of old mammals here. Bones of rodents, *Ursus spelaeus*, cave lion, rhinoceros, deer, etc. were found in the entrance zone. Many bats live in the cave now.

3 Caves of the Pilis Hills (Pilis Mountains)

The Pilis Hills, built of Triassic limestones (mainly Dachstein Limestone) and dolomites are a single NW–SE range between major faults. There are many cave passages, altogether almost 25 km, in the 500–700 m high horst blocks.

Here one can take a visit to three particularly interesting caves. *Sátorkőpuszta Cave* was discovered in a small limestone quarry by István Venkovits and László Jakucs after the Second World War. This 325 m long and 45 m deep cave is a prototype of hypogene cave morphology. The *Ariadne Cave System* includes the long-known Legény (Boy) Cave and Leány (Girl) Cave, plus caves only discovered in the last two decades—Ariadne Cave, Ősi Cave, Rejtekút Cave, Vacska (Wild Cat) Cave and Indikációs Cave. *Ajándék (Gift) Cave* is nearby and there is a chance of finding connections with the Ariadne Cave System. The vertical extension of the system would then be more than 300 m, a new record for Hungary. Like the Ariadne Cave System, Ajándék Cave was discovered by the Ariadne Speleo Group, led by Gergely Surányi and Richárd Kovács.

Unfortunately, one cannot enter two other interesting and beautiful caves: Róka-hegy (Fox Hill) Cave on the edge of Budapest which is very rich in cave coralloids, and the nearby Csókavár Cave. The entrances to *Róka-hegy Cave* are in the bottom of an old limestone quarry that was filled with building rubble forty years ago. Probably not only building rubble, because the temperature is very high (about 30 °C) in a 12 mshaft left open between the surface and the cave! Recently there were poisonous gases inside (Kraus 2003b).

Csókavár Cave is found in the bottom of another limestone quarry. This quarry is circular in form with a diameter of about 100 m at the surface and 40 m deep. It was filled with the clayey-organics brown coal waste from gasworks for eleven years after 1965. The waste contained very strong poisons (e.g. cyanides, and various Cd-, Pb-, and As-compounds). The thermal springs of Csillag-hegy (Csillag Hill) are very close to this quarry and there was danger that infiltrating water would pollute the springs. As a consequence, the waste was removed from the quarry in 2011 and 2012: 58 000 tons were cleared by machines, exposing

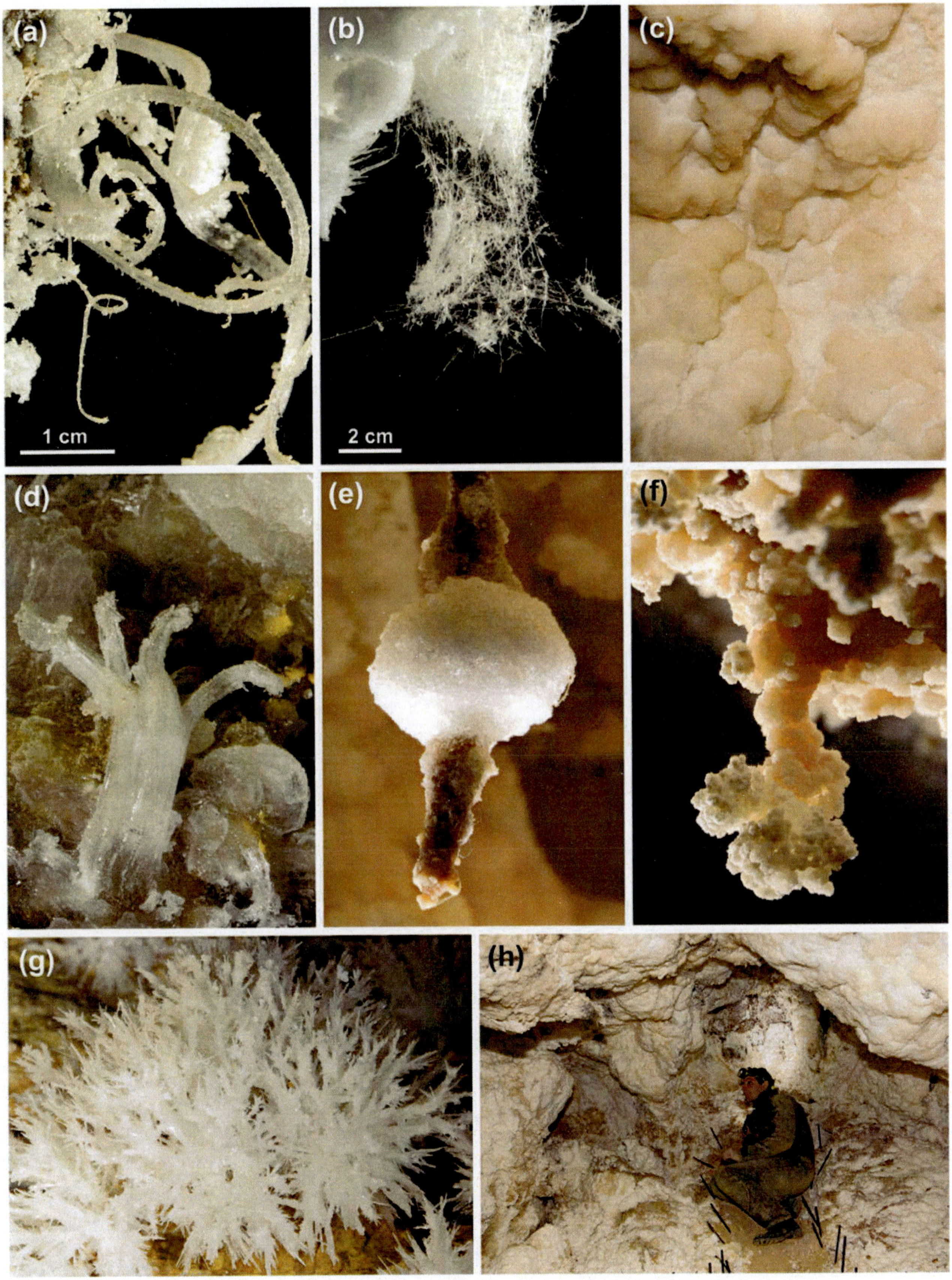

Fig. 22 Citadella Crystal Cave. **a** Gypsum snakes; **b** fibrous gypsum, "cave hairs" and cotton-like strings; **c** gypsum crust; **d** gypsum flower; **e** fibrous, cotton-like strings by gypsum; **f** cave coralloid; **g** aragonite frostwork; **h** small cave chamber covered by gypsum and aragonite speleothems. Photos by R. Kovács

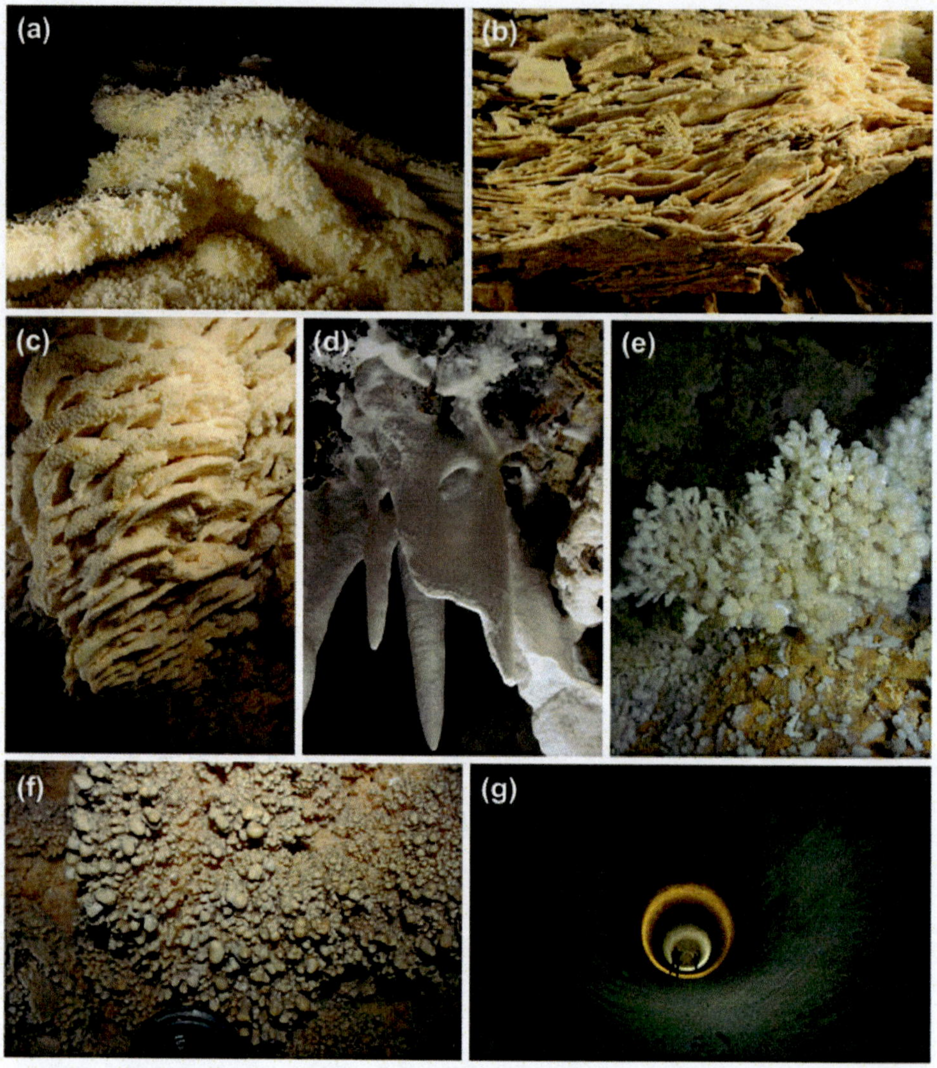

Fig. 23 Királylaki Cave. **a** Cemented cave rafts covered by sparry-calcite crust as phreatic overgrowths (photo by R. Kovács); **b** paper-thin cave rafts (photo by R. Kovács); **c** folia-like speleothems (photo by R. Kovács); **d** white drapery and stalactites. Note their partially dissolved surface! (photos by R. Kovács); **e** white cave coralloids and frostworks (photo by R. Kovács); **f** cave coralloids; **g** the cave open from the giant tunnel in the Hármashatár Hill (photo by A. Hegedűs)

some small caves in the quarry walls. A 20 m long cave at the bottom had been known since 1958. The wastes flowed into the caves for 40 years. It was impossible to clear the constricted cave passages with machines, so cavers wore chemical defensive clothes and gas masks to remove 220 000 kgof debris (Leél-Őssy Sz. and Szabó 2013). The work was dangerous indeed. They opened a ~300 m long cave which descends to the karst watertable. Because removal of the waste was not 100% successful, the air in the cave is still dangerous (e.g. sulphuric acid in the aerosol drops, cyanides, etc.) and entering without a gas mask is prohibited. There are magnificent calcite scalenohedrons, cave coralloids and needle-like aragonite crystals (frostworks) in the cave (Fig. 25).

The only active sinking stream cave around Budapest lies between Róka-hegy Cave and Csókavár Cave in the village of Üröm. It is the 250 m long, 40 m deep *Üröm Sinkhole Cave* (Leél-Őssy S.1952).

Although very small, a cave known as *Pilisszántó Shelter Cave* is also of interest. The valuable findings discovered here include 61 pieces of Bronze Age tools, the fossils of 60 species of mammals and 83 species of birds (Szablyár 2003).

3.1 Ajándék Cave

The highest point of the Pilis Hills is Mount Pilis (757 m asl). There is an almost 200 m high rocky face on the southwestern side—Csévi Cliffs. The entrances of Ajándék Cave and the Ariadne Cave System are these cliffs, *Ajándék Cave* near the top and the *Ariadne Cave System* close to the bottom.

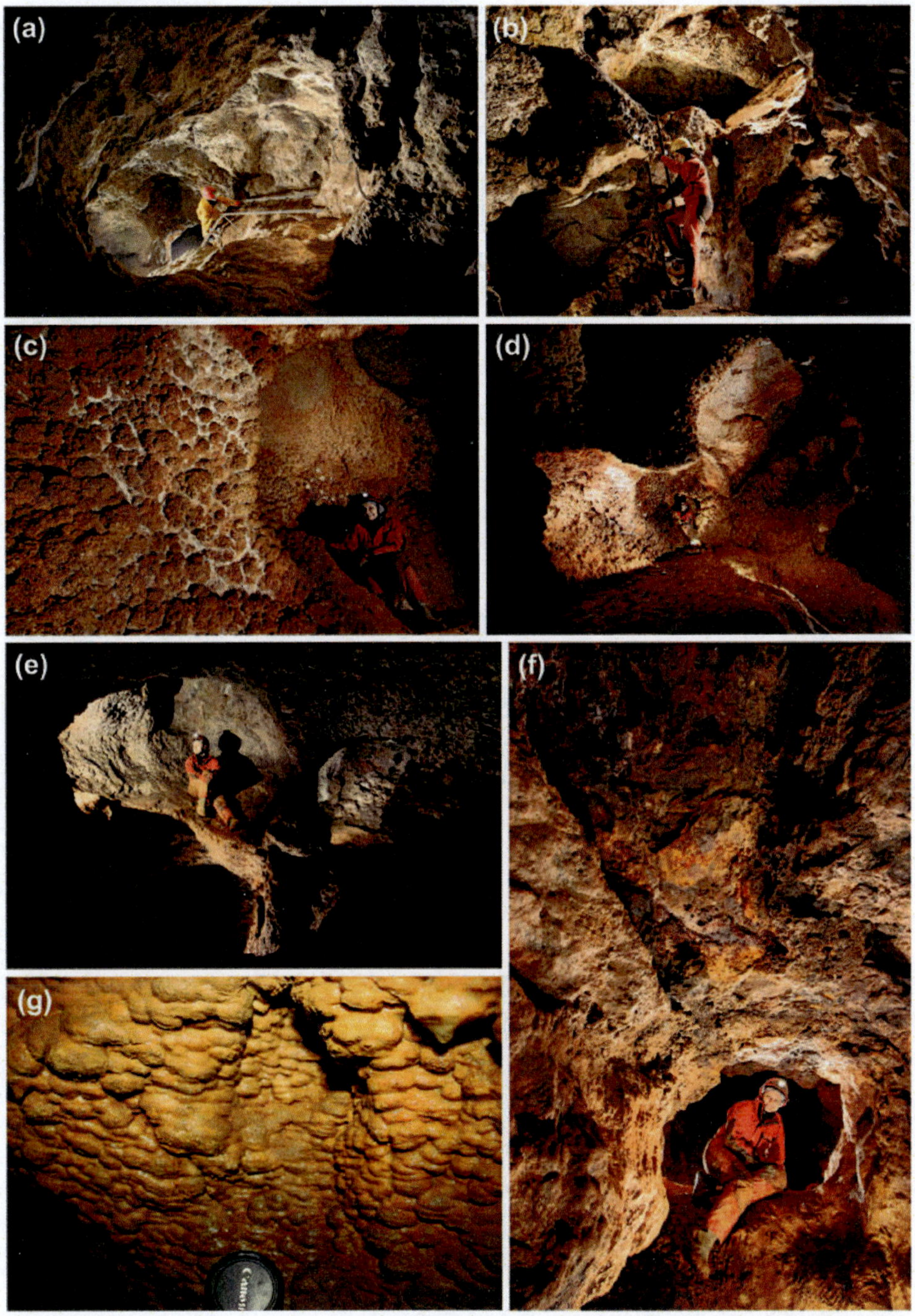

Fig. 24 Solymári-ördöglyuk Cave, Bátori Cave and Rácskai Cave. **a**, **b** Characteristic sections of the Solymári-ördöglyuk Cave in Dachstein Limestone (photos by Cs. Egri); **c** cave coralloids in the Bátori Cave (photo by Á. Berentés); **d** cave coralloids and characteristic dissolution forms in the Bátori Cave (photo by Á. Berentés); **e** huge cupolas (Bátori Cave, photo by D. Marosffy); **f** hematite and limonite on the Fe-oxide-rich roof of the Bátori Cave (photo by Á. Berentés); **g** folia-like speleothems ("donuts") in the Rácskai Cave

Ajándék Cave is 2860 m in length and 180 m deep. The cave entrance is a beautiful cupola, 3–4 m in diameter. It was hidden with wood during the Second World War (it may have been a secret storeroom). The wood decayed and the entrance opened after 2000. Members of the Ariadne Speleo Group started digging at the back of the cupola and discovered the cave in 2001. The discovery passage, formed in cave breakdown, collapsed shortly after the discovery but, fortunately, there was now a precise map of the cave, so explode an 8 m long artificial adit was blasted by Gergely Surányi into the first room of the cave.

There is a beautiful lake at the deepest point of the 180 m deep cave (Fig. 26). The water level of the lake is very variable, a 15 m difference being observed between the highest and the lowest positions. In the summer of 2020, the lake was in 14 m below its maximum level, which is marked

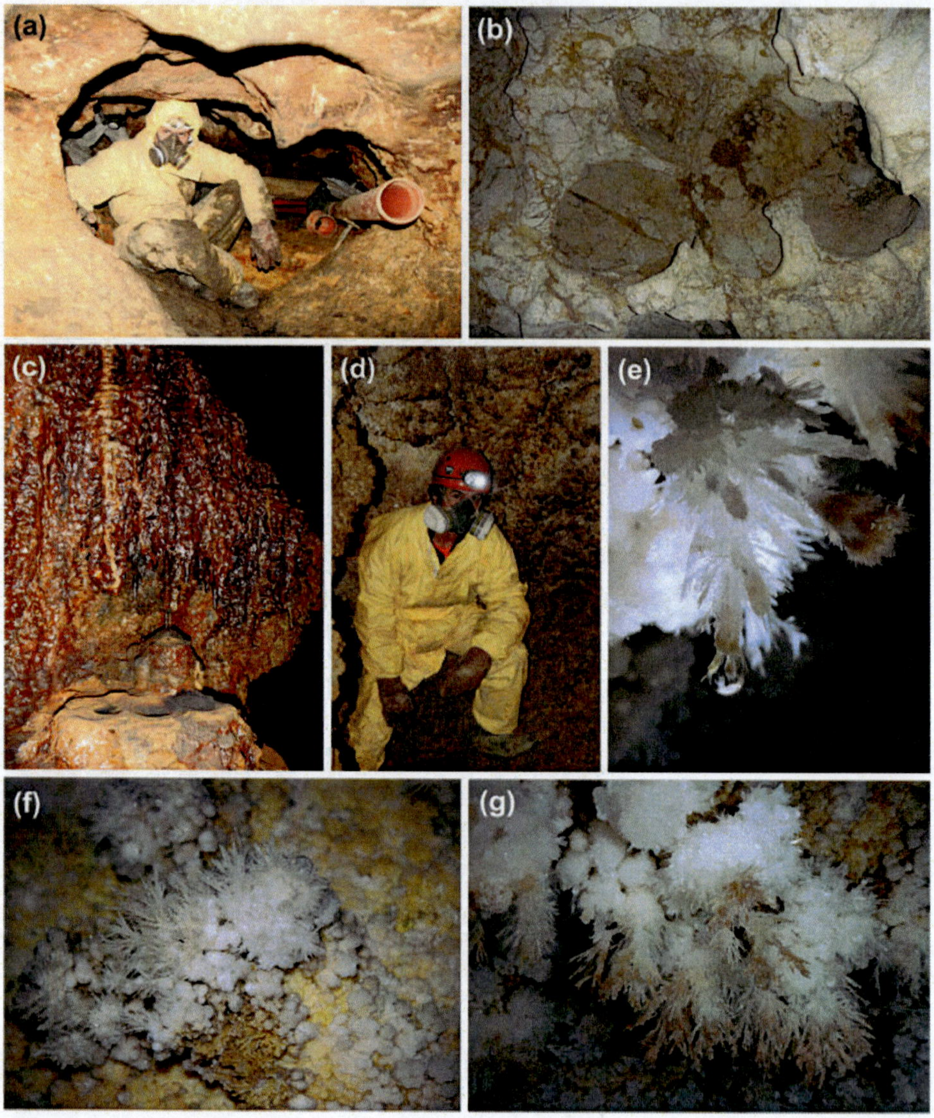

Fig. 25 Csókavár Cave. **a** Cupola-like dissolution forms (photo by Z. Szabó); **b** Cupolas of the cave roof (photo by R. Kovács); **c** Fe-oxide minerals in the flowstone and small stalactites (photo by R. Kovács); **d** cave coralloid-rich cave section (photo by Sz. Leél-Őssy); **e** aragonite frostwork (photo by R. Kovács); **f** aragonite frostworks on the cave coralloids; **g** aragonite frostworks on the white cave coralloids (photos by R. Kovács)

by a deep solution groove. The depth of the lake in the summer of 2020 was 14.5 m. This lake is about 80 m higher than the deepest point of the Ariadne Cave System, so it is certain that it is not part of the karst watertable. There are very thick clay deposits in the lower part of the cave. The upper parts are very well decorated with stalactites (Fig. 27).

3.2 Ariadne Cave System

The *Legény Cave* and *Leány Cave* entrances to this system were always open. The prominent triangular entrance of the former is 5–6 m high and about 10 m wide, and that of Leány Cave is somewhat smaller. Behind them much smaller passages lead into the caves. István Venkovits discovered the first few-hundred of meters beyond them after 1930. Legény Cave was about 350 m long, and Leány Cave about 200 m long at that time (Leél-Őssy S. 1954a). The Ariadne Group found a connection between the two caves in 1997 and the combined length was increased to 3500 m (Sásdi 2003b). The group has worked in all caves on the Csévi Cliffs every weekend for the last 25 years, led by Gergely Surányi and Richárd Kovács. They first discovered some new separate caves *(Indikációs Cave, Rejtekút Cave, Ariadne Cave, Vacska Cave* and *Ősi Cave)* and later connected them together step by step. Only Ősi Cave remains to be joined to the greater Ariadne Cave System. Most recent was the addition of Indikációs Cave in 2019. The largest discovery was Vacska Cave in 2010: this cave is now longer than 6 km.

The caves that were known before the Second World War were poor in dripstones, but the newer parts of Legény Cave,

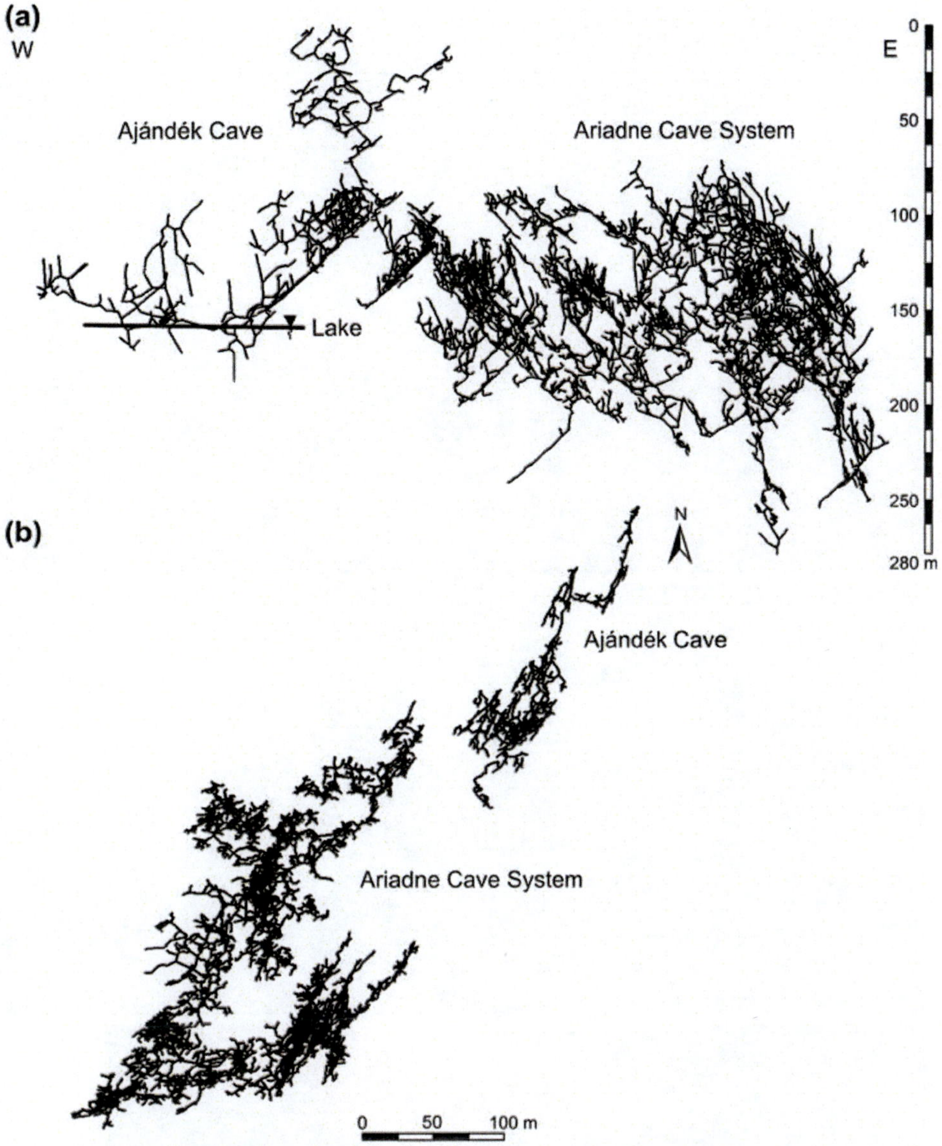

Fig. 26 Cross sections (**a**) and cave-polygon maps (**b**) of Ariadne Cave System and Ajándék Cave (speleological data from the National Cave Register). Compiled by Richárd Kovács

Leány Cave and, particularly, Vacska Cave are very well decorated with them. There are also several large (20–30 m wide) chambers in Vacska Cave. A small siphon 140 m below its entrance level is the deepest point currently known in the entire system. No large passages have yet been found beyond the siphon.

Indikációs Cave is notorious for its narrow passages, lots of mud and lack of speleothems: not a popular target!

The most beautiful part of Legény Cave is the Bat Section, found a few years ago. Here there are beautiful white rimstone flows (tetaratas), although some parts are blackened by bat guano and others are red. The flowstone is 30 to 40 m long and carries a small stream during intensive rains outside (Fig. 27).

Legény Cave is famous for its very small and narrow passages ("squeezes"), such as Pénztár (Cash-desk) behind the entrance, and Betyár (Rogue) Hole. The Ariadne Speleo Group has now excavated short bypasses to these tight spots (Deák et al. 2007).

The original Ariadne Cave is short (a little bit more than 200 m long), but the walls and the dripstones are spectacularly white.

The Ariadne Speleo Group built some ladders in the vertical parts of Vacska Cave and Legény Cave to make cave tours easier, and enlarged other narrowest passages (e.g. the first section of Vacska Cave, or the Csúzlik in Leány Cave).

Excavations in the entrances of Legény Cave and Leány Cave made some archaeological findings in 1913 but they

Fig. 27 Ajándék Cave and Ariadne Cave System. **a** Stalactites, soda-straw stalactites and stalagmites in the Ajándék Cave; **b** brownish-red flowstone on the calcite crust; **c** flowstone; **d** stalagmites; **e** cave pool with rimstone dams (~ gour, tetarata) and pool spars; **f** flowstones and stalagmites; **g** small horseshoe bat hanging on the serrate edge of a drapery; **h** lake at the deepest point of the 180 m deep Ajándék Cave. Photos by R. Kovács

are of little importance (Bella 1913). Bones were found in Ősi Cave a few years ago. Many bats live in Legény Cave and Leány Cave.

The known length of the Ariadne Cave System is now more than 19 km, the third longest cave in Hungary (after the Pál-völgy Cave System in Budapest and Baradla Cave in Aggtelek and Jósvafő). The total explored depth in the cave system is more than 200 m (Fig. 26).

Demény et al. (2013) published H isotope data for inclusion-hosted water and the carbonate C and O isotope data for stalagmites from Leány Cave (and Pál-völgy Cave) that covered the period, ~9–5.4 ky BP. Comparisons of the results with other regional speleothem records indicated that variations in moisture sources from the Atlantic and Mediterranean regions depended on the strength of the North Atlantic Oscillation.

3.3 Sátorkőpuszta Cave

The geologists, László Jakucs and István Venkovits, recognized the exceptional significance of this cave in 1946. In it an upward sequence of cupolas opens from each other like a string of pearls. The diameter of individual cupolas ranges between 1 and 4 m. Sometimes there are junctions from which new "strings of pearls" depart, like a bunch of grapes (Jakucs 1948, 1977; Fig. 28). There is a large (35 × 15 m) chamber at the bottom of the cave. This got its original name, Kővirágok terme (Chamber of Stone Flowers) from the beautiful snow-white gypsum and aragonite crystals. It is

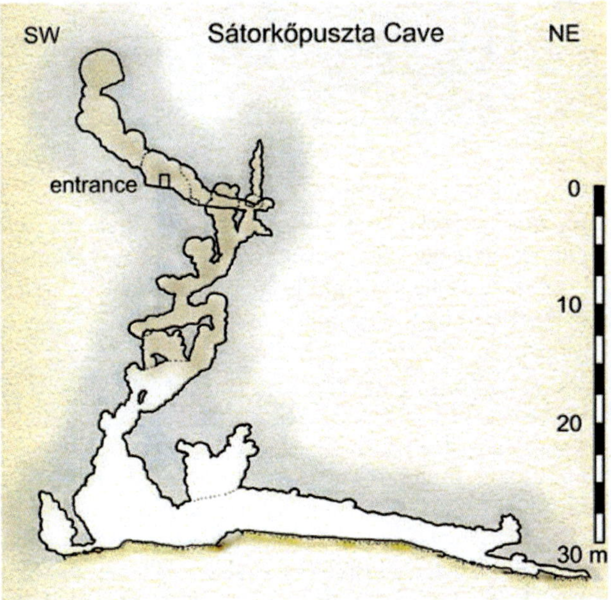

Fig. 28 Cross section of Sátorkőpuszta Cave (speleological data from the National Cave Register). Compiled by S. Borzsák

now known as the Benedek Endre Chamber, after the first leader of the spelo group working in the cave. Crystal needles a few cm long combine to create 10–20 cm diameter aragonite "bushes" (frostwork). There are boxworks of sparry calcite crystals. Many gypsum flowers stand out a few cm from the thick gypsum crusts. The flowers are as thick as a pencil, and their length is around 10 to 20 cm. The largest gypsum crystal is a column two m high and one m wide (Fig. 29). The most famous gypsum crystal is the 70 cm long, "fencing foil"-like, chandelier (Takács-Bolner 2003b).

For 30 years there was a drilling ground for Soviet soldiers around the cave entrance. The cave suffered a loss of around 90% of its decorations, most of its finest crystals being destroyed (Lieber 2010).

There are some old coal mines a few kms away from the cave in the Dorog Basin. There are abundant pyrite and marcasite crystals around the coal bands, thus sulfuric acid solution may have formed the cave, as it has in many other caves in the United States of America and in Europe (Egermeier 1981; Hill 1987, 1990; Audra et al. 2007). These sulphides may explain the genesis of the gypsum in Sátorkőpuszta Cave.

Research has begun again here, by the Benedek Endre Speleo Group (Tamás Lieber leads the group). It is trying to discover new parts to this outstanding cave.

4 Caves of the Gerecse Hills (Gerecse Mountains)

The horst ranges of the Gerecse Hills are little bit lower than the Pilis Hills. They are 500 to 600 m high (asl), and consist not only of Triassic limestone and dolomite but also Jurassic and Cretaceous sediments (limestone, marl and sandstone) and Pliocene/Pleistocene travertines (freshwater limestones). The Gerecse Hills are poorer in caves than the Pilis or Buda Hills. They are to be found chiefly in the western part of the Gerecse, in well-bedded Triassic Dachstein Limestone. The longest caves there are *Pisznice Cave* and *Lengyel (Polish) Cave* (both around 600 m in length). Jankovich Cave and Szelim Hole are very interesting and spectacular. Each has a large circular breach in the ceiling (10 m to 20 m in diameter) that is open to the sky, and a broad entrance below (Fig. 30). They are ancient erosional relict features. There were rich archaeological findings from the Neolithic and the Bronze ages at the bottom of *Jankovich Cave,* also palaeontological findings—bones of *Ursus spelaeus* and reindeer (Juhász 2003a). Palaeolithic stone tools and traces of Neanderthal man were found at the bottom of *Szelim Hole,* and also *Ursus spelaeus*, cave hyena, etc. (Juhász 2003b).

The 80 m long, 35 m deep *Pothole No. 1. of Öreg-kő (Old Rock) in Bajót* had extremely large barite crystals (10–

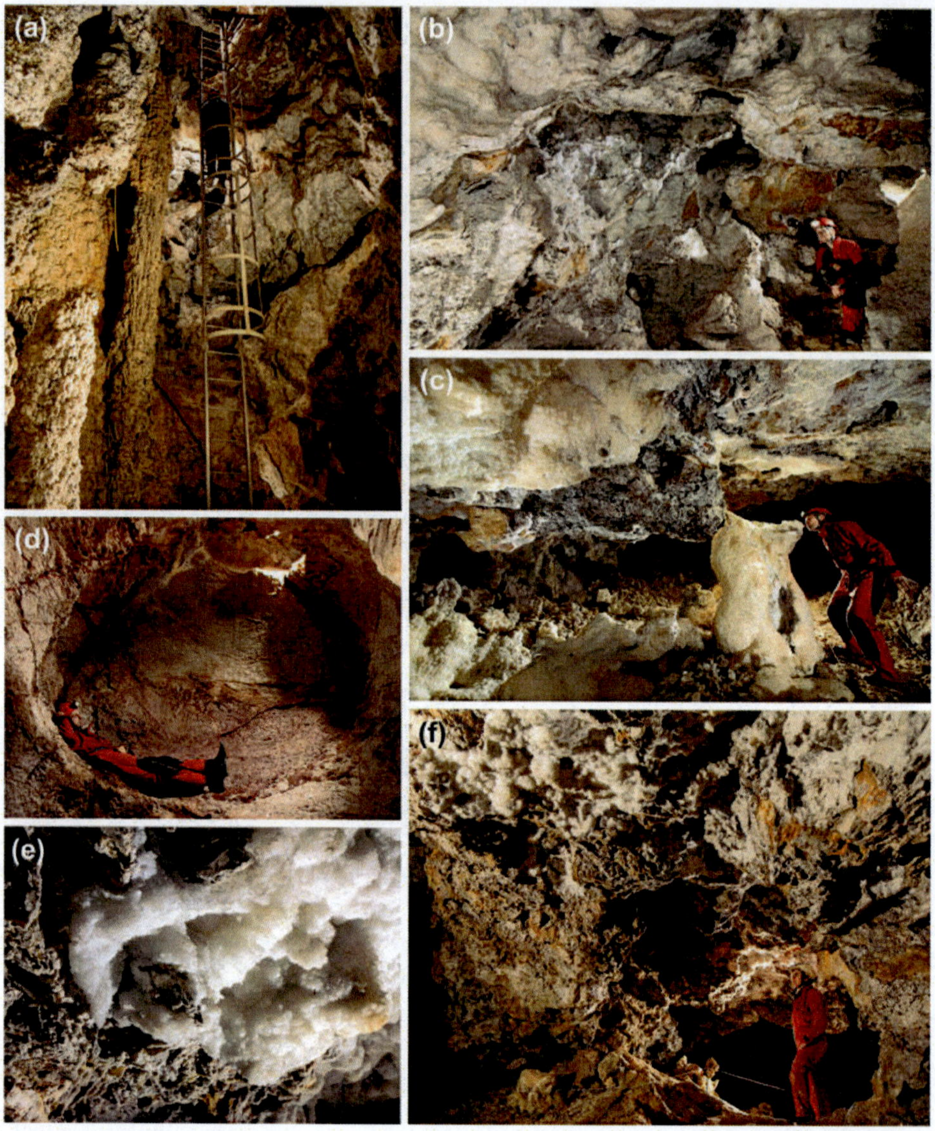

Fig. 29 Sátorkőpuszta Cave. **a** Ladder to the large chamber at the deepest part of the cave (photo by Cs. Egri); **b** "Chamber of the Stone Flowers": a characteristic hypogene cave chamber rich in gypsum and other speleothems of thermal-water origin, e.g. cave coralloids and frostworks (photo by Á. Berentés); **c** large hollow gypsum "column" (or gypsum tower), detached and fell-down thick gypsum crusts on the bottom as well as on the ceiling (photo by Á. Berentés); **d** huge cupola in the upper section of the cave (photo by Á. Berentés); **e** snow white gypsum crust (photo by Cs. Egri); **f** thick gypsum crusts on the cave walls (photo by Cs. Egri)

12 cm) and gypsum flowers (Leél-Őssy S. 1954b). In the summer the CO_2 content is very high in the cave, making it dangerous to descend in this season (Juhász 2003c). Many bats sleep in the cave in winter.

One of the largest Hungarian active limestone quarries is on Keselő Hill. Dachstein Limestone is mined, plus Triassic dolomite in a small part of the quarry. More than two dozen caves have been breached by quarry faces in recent decades. One of them is the very interesting, 115 m deep and 400 m long, *Keselő-hegy Cave*. This cave is dissolved in a large, 1–5 m wide fault. There are wonderful re-dissolved surfaces of calcite veins, cave coralloids and aragonite crystals (1–2 cm long needles) exposed the walls (Nyerges 2003).

Several coal mines were worked in the Dorog Basin between the Pilis and Gerecse Hills in the twentieth Century. A few natural *cave passages* were broken into *Adit VI.* in *Tokodaltáró*. There were beautiful cave coralloids and needle-like aragonite frostworks in these passages—perhaps the largest in Hungary. Unfortunately, most of the crystals were destroyed by the mining.

4.1 Lengyel Cave

The 0.5 × 0.8 m entrance to this 600 m long, 70 m deep cave at the top of Kő Hill near Tatabánya is found in a small rocky pit. Gyula Vígh and István Gaál were the first visitors in the cave, after 1930. On a map of the cave the tectonic control is striking. Mixing corrosion dissolution forms are visible in the upper level of the cave and the walls are adorned with recent

Fig. 30 Caves in Gerecse Hills. **a** Szelim Hole and its open parts formed in Triassic Dachstein Limestone (photo by Cs. Egri); **b** large circular breach in the ceiling of the Szelim Hole; **c** Jankovich Cave and its large circular breach in the ceiling (photo by Cs. Egri); **d** huge cupola in the uppermost section of the Pisznice Cave; **e** cave coralloids and stalagmite (Pisznice Cave); **f** characteristic cave passage with dissolution forms in the Pisznice Cave (photo by Á. Kocsis); **g** stalactites and cave coralloids (Pisznice Cave)

dissolutional microforms in the lower level. Since the cave functioned as a swallet in later stages, there are lots of sediments on the floors of the passages. Beautiful dripstones are also seen. Megalodus fossils stand out from the walls in some places. There are lots of bats in the cave.

Lengyel Cave is famous for its bad air. The CO_2 content in the lower part of the cave rises to 3–4% in summer and is least at the end of the winter (Juhász 2003d).

There were some archaeological and palaeontological findings, now in the Hungarian Natural History Museum: bones of Megaloceras, giant deer, cave lions, cave wolf, etc. as well as tools made from these bones by the Mousterian culture. They are marked as originating from Lengyel Cave, but this labelling may be erroneous.

4.2 Megalodus Cave

Megalodus Cave is a more interesting small (280 m long) cave at the foot of the Gerecse Hills. The cave is famous for its Megalodus fossils. They are easily observable on the roof of the recently active cave passage, protruding from the Triassic Dachstein Limestone because of differential weathering. Many cavity-filling isopachous radial calcite crystals were precipitated under phreatic conditions. These crystals cover the host rock ($\sim$thick mammillary crust; Fig. 31). The lower, horizontal passages of the cave are now almost always under the watertable.

The cave is situated at the deepest point of the "Geological Garden" in Tata, in an old limestone quarry. The geological sequence from the Upper Triassic to the Upper Cretaceous can be seen in the quarry, where the beds are very rich in fossils (e.g. bedding surfaces are full of Ammonites). Lots of large rock samples from everywhere else in Hungary are on display on the floor of the quarry, to show the geological sequence and variety of rocks found in this country.

4.3 Pisznice Cave

The 545 m Mount Pisznice is in the middle of the Gerecse Hills. The cave is entered in its rocky, steep southern side. This 550 m long cave is characterized by mainly horizontal passages (Fig. 32). Its vertical extension is 20 m. People lived at the entrance of the cave and in the surroundings of the cave from the Bronze Age until the Turkish (Ottoman) times. Many bones of domestic animals (horse, pig, cow, sheep) have been found here. The deep guano of several thousand bats was mined out here at the end of the 19th Century. Only a few dozen bats live in the cave now (Egri and Juhász 2003a, b, c, d).

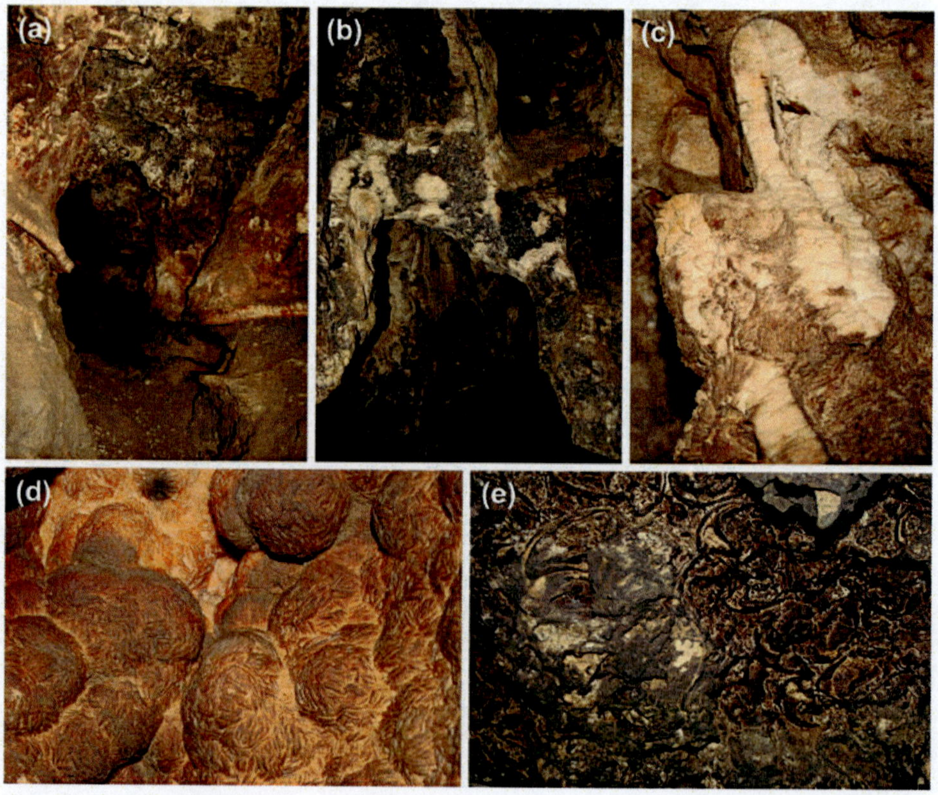

Fig. 31 Megalodus Cave. **a** Characteristic horizontal cave passage; **b** cave passage with dissolution forms, cavity-filling and vein-filling calcite crystals; **c**, **d** details of the isopachous radial calcite crystals ($\sim$mammillary crusts); **e** Megalodus fossils protruding from out from the Triassic Dachstein Limestone because of differential weathering. They are easily observable on the roof of the recently active horizontal cave passage

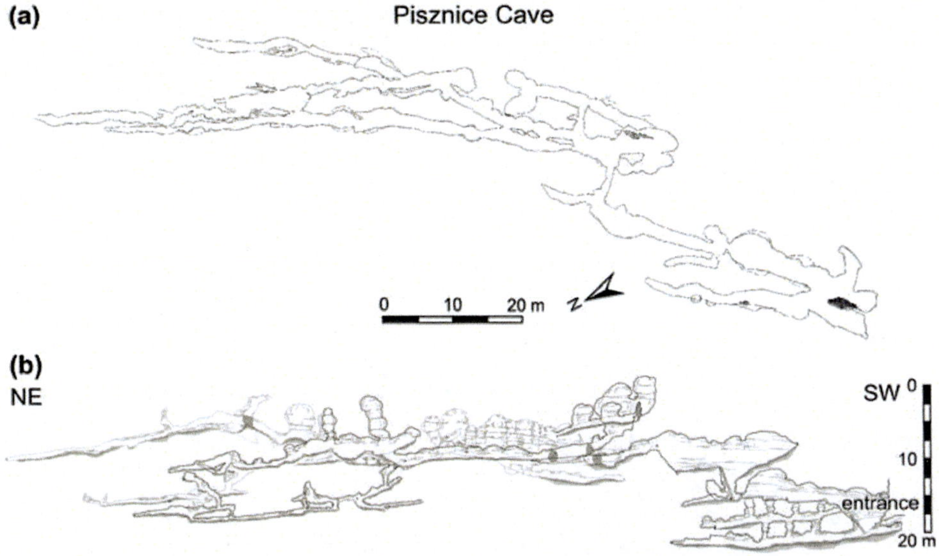

Fig. 32 Map (**a**) and cross-section (**b**) of Pisznice Cave (speleological data from the National Cave Register). Compiled by Zs. Polacsek and J. Ba

The length of the main corridor is 110 m, their width is a few meters, and the height is 2.5 m; large cupolas in the upper main passages rise a further 6–8 m. Cave coralloids combined with dripstones and helictites can be seen in the cave (Fig. 30).

5 Conclusions

There are opportunities to discover new caves or new extensions to known caves around Rózsadomb in Budapest, and in Ariadne Cave System in the Pilis Hills. There is an active stone quarry on Keselő Hill in the Gerecse Hills which may also expose previously unknown caves during the work.

The known hypogene caves in Budapest are dry caves by now, except for the active Molnár János Cave, as all caves have thermal water origin. The Buda Hills have risen more than 100 m as a consequence of tectonic movements in the last few hundred thousand years. These caves are very rich in speleothems of thermal-water origin (chiefly József-hegy Cave, Szemlő-hegy Cave, Citadella Crystal Cave, Királylaki Cave and the newer parts of Pál-völgy Cave). Since the white crystals are very sensitive, visiting the "unbuilt" (non-tourist) and highly protected caves is only possible with the permission of the Danube-Ipoly National Park.

Research on the speleothems and the cave development is continuing.

Acknowledgements The authors are grateful to Derek Ford for his invaluable corrections and comments to an earlier version of the manuscript, and to Andrea Mindszenty for her help in correcting the tables and figure captions. Thanks are due to Csaba Egri and Richárd Kovács for cave polygons, maps and speleological data from the National Cave Register, furthermore to Ágnes Berentés, Csaba Egri, András Hegedűs, Attila Kiss and Richárd Kovács, as well as to Péter Adamkó jr., Gergely Ambrus, Zsófia Kalotai, Ákos Kocsis, Gergely Domonkos Nagy, Ádám Polyák, Attila Sári, Emőke Szabó, Zoltán Szabó, Levente Tisza and Péter Vörös for their cave photographs.

References

Adamkó P and Leél-Őssy Sz (1984) Budapest új csodája: a József-hegyi-barlang (The New Wonder of Budapest: the József-hegy Cave). Karszt És Barlang I:1–8 (in Hungarian)

Albert G, Virág M, Erőss A (2015) Karst porosity estimations from archive cave surveys—studies in the Buda Thermal Karst System (Hungary). Int J Speleol 44(2):151–165

Anda D, Büki G, Krett G, Makk J, Márialigeti K, Erőss A, Mádl-Szőnyi J, Borsodi AK (2014) Diversity and morphological structure of bacterial communities inhabiting the Diana-Hygieia Thermal Spring (Budapest, Hungary). Acta Microbiol Immunol Hung 61(3):329–346

Anda D, Krett G, Makk J, Márialigeti K, Mádl-Szőnyi J, Borsodi AK (2017) Comparison of bacterial and archaeal communities from different habitats of the hypogenic Molnár János Cave of the Buda Thermal Karst System (Hungary). J Cave Karst Stud 79(2):113–121

Audra P, Hoblea F, Bigot J-Y, Nobecourt J-C (2007) The role of condensation-corrosion in thermal spelogenesis: study of a hypogenic sulfidic cave in Aix-Les-Bains France. Acta Carsol 36 (2):185–194

Bakalowicz MJ, Ford DC, Miller TE, Palmer AN (1987) Thermal genesis of solution caves in the Black Hills, South Dakota. Geol Soc Am Bull 99:729–738

Balázs D (1966) A keveredési korrózió szerepe a karsztosodásban (The role of mixing corrosion in karstification). Hidrológiai Közlöny 4:179–185 (in Hungarian)

Bella L (1913) A Legény-és Leány-barlang ásatásának régészeti méltatása (Evaluation of the archeological excavations in the Legény and Leány Caves). Barlangkutatás I:170–211 (in Hungarian)

Benkovics L, Török Á, Nádor A (1995) A Ferenc-hegyi vonulat barlangjainak geológiája (Geology of caves of the Ferenc Hill). Karszt-és Barlangkutatás X:193–210 (in Hungarian)

Benkovics L, Obert D, Bergerat F, Mansy JL, Dubois M (1999) Brittle tectonics and major dextral strike-slip zone in the Buda karst (Budapest, Hungary). Geodin Acta 12(3–4):201–211

Bergmann C, Leél-Őssy Sz, Fehér K, Fórizs I (2011) Vízvizsgálatok a Molnár János-barlangban (Water-chemistry of the Molnár János Cave). Karsztfejlődés XVI:223–245 (in Hungarian)

Borbás E, Fehér K (2013) Monitoring vizsgálatok a Pál-völgyi-barlangban (Long-term monitoring in the Pál-völgy Cave). Karsztfejlődés XVIII:23–36 (in Hungarian)

Borsodi AK, Knáb M, Krett G, Makk J, Márialigeti K, Erőss A, Mádl-Szőnyi J (2012) Biofilm bacterial communities inhabiting the cave walls of the Buda thermal karst system Hungary. Geomicrobiol J 29(7):611–627

Borsodi AK, Anda D, Makk J, Krett G, Dobosy P, Büki G, Erőss A, Mádl-Szőnyi J (2018) Biofilm forming bacteria and archaea in thermal karst springs of Gellért Hill discharge area (Hungary). Geomicrobiology J Basic Microbiol 58(11):928–937

Borzsák S, Székely K (2003) Remete-völgyi Felső-barlang (Felső Cave of the Remete Gorge). In: Székely K (ed) Magyarország fokozottan védett barlangjai (Strictly protected caves of Hungary). Mezőgazda Kiadó, Budapest, pp 280–281 (in Hungarian)

Bögli A (1965) The role of corrosium by mixed water in cave forming. In: Stekl O (ed) Problems of the speleological research, Czechoslovakia. Academy of Science, Prague, pp 125–131

Cser F, Szenthe I (1986) The way of cave formation by mixing corrosion. In: Proceedings of the 9th international congress of speleology, Barcelona, 1–7 August 1986, vol 1, pp 227–280

Csondor K, Erőss A, Horváth Á, Szieberth D (2016) Radon as a natural tracer for underwater cave exploration. J Environ Radioact 173:51–57

Csondor K, Erőss A, Horváth Á, Szieberth D (2017) Természetes nyomjelzők alkalmazása a Molnár János-barlangban (The application of natural tracers in the Molnár János Cave). Karsztfejlődés XXII:35–45 (in Hungarian)

Deák I, Leél-Őssy Sz, Kövér Sz, Surányi G (2007) A Csévi-szirtek barlangjai (Caves of the Csévi Cliffs). Karszt és Barlang 2007(I–II):17–36 (in Hungarian)

Deák J (1978) Environmental isotopes and water chemical studies for groundwater research in Hungary. Isotope Hydrol 1:221–249

Demény A, Gy C, Siklósy Z, Leél-Őssy Sz, Lin K, Shen C-C, Gulyás K (2013) Mid-Holocene climate conditions and moisture source variations based on stable H, C and O isotope compositions of speleothems in Hungary. Quatern Int 293:150–156

Dublyansky JV (1987) Teoretyicseszkoje mogyelirovanije gyinamiki formirovanyija gidrotermokarsztovüh polosztyej (The theoretical modelling of the dynamics of the formation of hydrothermal karst). Metodü izucsenyija geologicseszkih javlenyij (The methods of studying geological phenomena). Novoszibirszk, pp 97–111 (in Russian)

Dublyansky JV (1995) Speleogenetic history of the Hungarian hydrothermal karst. Environ Geol 25:24–35

Egemeier SJ (1981) Cavern development by thermal waters. Nat Speleol Soc Bull 43(2):31–51

Egri Cs, Juhász M (2003) Pisznice-barlang (Pisznice Cave). In: Székely K (ed) Magyarország fokozottan védett barlangjai (Strictly protected caves of Hungary). Mezőgazda Kiadó, Budapest, pp 322–324 (in Hungarian)

Enyedi NT, Anda D, Borsodi AK, Szabó A, Pál SE, Óvári M, Márialigeti K, Kovács-Bodor P, Mádl-Szőnyi J, Makk J (2019) Radioactive environment adapted bacterial communities constituting the biofilms of hydrothermal spring caves (Budapest, Hungary). J Environ Radioact 203:8–17

Erőss A (2010) Characterization of fluids and evaluation of their effects on karst development at the Rózsadomb and Gellért Hill, Buda Thermal Karst, Hungary. PhD Theses, Eötvös Loránd University, Budapest

Erőss A, Mádl-Szőnyi J, Borsodi A, Knáb M, Csoma AÉ, Mindszenty A (2011) Results of in situ dissolution experiment to understand hypogenic karstification processes, Buda Thermal Karst, Hungary. In: Bertrand C, Carry N, Mudry J, Pronk M, Zwahlen F (eds) Proceedings H2Karst, 9th conference on limestone hydrogeology, Besançon (France). UMR 6249 Chrono-Environnement, pp 161–164

Erőss A, Mádl-Szőnyi J, Surbeck H, Horváth Á, Goldscheider N, Csoma AÉ (2012) Radionuclides as natural tracers for the characterization of fluids in regional discharge areas, Buda Thermal Karst, Hungary. J Hydrol 426–427:124–137

Esteban M, Budai T, Juhász E, Lapointe O (2009) Alteration of Triassic carbonates in the Buda Mountains—a hydrothermal model. Central European Geol 52(1):1–29

Fehér K, Kiss K, Kovács J, Kiss A (2009) Beszivárgás-vizsgálatok a rózsadombi termálkarszton (Infiltration studies on the Rózsadomb Thermal Karst). Karsztfejlődés XIV:45–55 (in Hungarian)

Fehér K, Kovács J, Márkus L, Borbás E, Tanos P, Hatvani IG (2016) Analysis of drip water in an urban karst cave beneath the Hungarian Capital (Budapest). Acta Carsologica 45(3):213–231

Fodor L, Leél-Őssy Sz, Tari G (1991) En-echelon fractures in a dextral shear zone—tectonic heritage for a hydrothermal cave (Budapest, Hungary). Terra Nova 4:165–170

Fodor L, Magyari Á, Fogarasi A, Palotás K (1994) Tercier szerkezetfejlődés és késő paleogén üledékképződés a Budai-hegységben. A Budai-vonal új értelmezése (Tertiary tectonics and Late Paleogene sedimentation in the Buda Hills, Hungary. A new interpretation of the Buda Line). Földtani Közlöny (Bull. Hung. Geol. Soc.) 124(2):129–305 (in Hungarian)

Ford DC (1995) Some thoughts on hydrothermal caves. Cave Karst Sci 22(3):107–118

Ford DC, Takács-Bolner K (1991) Abszolút kormeghatározás és stabil izotóp vizsgálatok budai barlangi kalcitmintákon (U series dating and stable isotope analyses on calcite precipitates from Buda caves). Karszt és Barlang 1991(I–II):11–18 (in Hungarian)

Ford DC, Williams PW (1989) Karst geomorphology and hydrology. Unwin Hyman, London

Ford DC, Williams PW (2007) Karst hydrogeology and geomorphology. John Wiley & Sons Ltd., Chichester

Zs F, Tóth A (2016) A Pál-völgyi-barlangrendszer születése (The birth of the Pál-völgy Cave System). Karszt És Barlang 2012–2014:93–96 (in Hungarian)

Ghergari L, Onac BP (1995) The crystallogenesis of gypsum flowers. Caves Karst Sci 22:119–122

Goldscheider N, Mádl-Szőnyi J, Erőss A, Schill E (2010) Review: thermal water resources in carbonate rock aquifers. Hydrogeol J 18 (6):1303–1318

Győri O, Poros Zs, Mindszenty A, Molnár F, Fodor L, Szabó R (2011) Budai-hegységi paleogén karbonátos kőzetek diagenezistörténete (Diagenetic history of the Palaeogene carbonates, Buda Hills, Hungary). Földtani Közlöny (Bull. Hung. Geol. Soc.) 141(4):341–361 (in Hungarian)

Haas J (ed) (2001) Geology of Hungary. Eötvös University Press, Budapest

Haas J, Korpás L, Török Á, Dosztály I, Góczán F, Hámor-Vidó M, Oravecz-Scheffer A, Tardi-Filácz E (2000) Felső-triász medence- és lejtőfáciesek a Budai-hegységben—a Vérhalom téri fúrás vizsgálatának tükrében (Upper Triassic basin and slope facies in the Buda Mts. —based on study of core drilling Vérhalom tér, Budapest). Földtani Közlöny 130(3):371–421 (in Hungarian)

Hazslinszky T (2003) Solymári-ördöglyuk (Solymári-ördöglyuk Cave). In: Székely K (ed) Magyarország fokozottan védett barlangjai (Strictly protected caves of Hungary). Mezőgazda Kiadó, Budapest, pp 282–285 (in Hungarian)

Hill CA (1973) Cave minerals. National Speleological Society, Huntsville

Hill CA (1987) Geology of Carlsbad Cavern and other caves in the Guadalupe Mountains, New Mexico and Texas. New Mexico Bureau of Mines and Mineral Resources Bulletin 117

Hill CA (1990) Sulfuric acid speleogenesis of Carlsbad Cavern and its relationship to hydrocarbons, Delaware Basin New Mexico and Texas. AAPG Bull 74(1):1685–1694

Hill CA, Forti P (1997) Cave minerals of the World. National Speleological Society, Huntsville

Horváth J (1965) A Szemlőhegyi-barlang 1961–62. évi felmérése (Geodetic Survey of the Szemlő-hegy Cave, 1961–1962). Karszt és Barlang 1965(I):21–30 (in Hungarian)

Izápy G, Maucha L (1993) The discharge-quality relationship interpreted for karst springs by a new karst model. Bull De La Société Géographique De Liége 29:53–60

Jakucs L (1948) A hévforrásos barlangkeletkezés földtani és fizikai tényezői (Geological and physical factors of thermal cave genesis). Hidrológiai Közlöny XXVII I(1–4):53–58 (in Hungarian)

Jakucs L (1977) Morphogenetics of karst regions. Variants of karst evolution. Hilger, Bristol

Jakucs L (1994) A Budai-hegység hidrotermális karsztja (The hydrothermal karst of the Buda Hills). Földrajzi Értesítő XLII I (3–4):235–246 (in Hungarian)

Juhász M (2003a) Jankovich-barlang (Jankovich Cave). In: Székely K (ed) Magyarország fokozottan védett barlangjai (Strictly protected caves of Hungary). Mezőgazda Kiadó, Budapest, pp 313–314 (in Hungarian)

Juhász M (2003b) Szelim-lyuk (Szelim Hole). In: Székely K (ed) Magyarország fokozottan védett barlangjai (Strictly protected caves of Hungary). Mezőgazda Kiadó, Budapest, pp 325–327 (in Hungarian)

Juhász M (2003c) Öreg-kői 1. sz. zsomboly (Pothole No. 1. of Öreg-kő). In: Székely K (ed) Magyarország fokozottan védett barlangjai (Strictly protected caves of Hungary). Mezőgazda Kiadó, Budapest, pp 320–321 (in Hungarian)

Juhász M (2003d) Lengyel-barlang (Lengyel Cave). In: Székely K (ed) Magyarország fokozottan védett barlangjai (Strictly protected caves of Hungary). Mezőgazda Kiadó, Budapest, pp 317–319 (in Hungarian)

Kadić O (1931) Budapest—barlangváros (Budapest—the Capital of Caves). Turisták Lapja XLIII:249–250 (in Hungarian)

Kadić O (1936) Budapest a barlangok városa (Budapest, the Capital of Caves). Különlenyomat a Földtani Értesítő Új folyam I/4 és következő számaiból (II/4-ig) (Special Edition). Magyar Barlangkutató Társaság (Hungarian Cave Research Society), Budapest (in Hungarian)

Kalinovits S (1984) Molnár János-barlang (Molnár János Cave). Magyarország barlangtérképei 5 (Map of caves in Hungary 5). Magyar Karszt-és Barlangkutató Társulat (Hungarian Speleological Society), Budapest (in Hungarian)

Kalinovits S (2003) Molnár János-barlang (Molnár János Cave). In: Székely K (ed) Magyarország fokozottan védett barlangjai (Strictly protected caves of Hungary). Mezőgazda Kiadó, Budapest, pp 260–263 (in Hungarian)

Kalinovits S (2006) A Molnár János-barlang új szakaszainak felfedezése (The exploration of new passages of Molnár János Cave). Karszt És Barlang 2002–2003:3–10 (in Hungarian)

Kele S (2009) Édesvízi mészkövek vizsgálata a Kárpát-medencéből: paleoklimatológiai és szedimentológiai elemzések (Study of freshwater limestones from the Carpathian Basin: paleoclimatological and sedimentological implications). PhD theses, Eötvös Loránd University & Hungarian Academy of Sciences, Budapest (in Hungarian)

Kele S, Gy S, Demény A, Shen C-C, Chiang H-W (2009) U-series dating and isotope geochemical study of the Gellért Hill (Budapest) travertine. Central European Geol 52(3–4):199–224

Kele S, Scheuer Gy, Demény A, Shen C-C, Chiang H-W (2011) A Rózsadomb (Budapest) édesvízi mészköveinek U/Th sorozatos kormeghatározása és stabilizotóp-geokémiai vizsgálata (Uranium-series dating and geochemical study of the travertines located on the Rózsadomb, Budapest). Földtani Közlöny (Bull Hung Geol Soc) 141(3):293–312 (in Hungarian)

Kele S (2011) A máriaremetei édesvízi mészkő U-soros kormeghatározása és paleokörnyezeti rekonstrukciója (Uranium-series dating and paleoenvironmental reconstruction of the freshwater limestone of Máriaremete). Hidrológiai Tájékoztató 50(1):87–90 (in Hungarian)

Kele S, Scheuer G (2012) Az Ördögárok-völgyéhez kapcsolódó Kondor úti és Nyéki úti édesvízi mészkő előfordulások U-soros kormeghatározása és képződési környezetük rekonstrukciója (Uranium-series dating and paleoenvironmental reconstruction of the freshwater limestones of the Kondor and Nyéki streets, Ördögárok Valley). Hidrológiai Tájékoztató 51(1):53–55 (in Hungarian)

Kessler H (1931) A Szemlőhegyi cseppkőbarlang (The Szemlő-hegy Cave). Turisták Lapja XLIII:250–252 (in Hungarian)

Kessler H (1936) Barlangok mélyén (In the depth of caves). Franklin Társulat, Budapest (in Hungarian)

Kiss K, Szalai Z, Szabó E (2013) A Harcsaszájú-barlang beszivárgó vizeinek vizsgálata (Analysis of infiltrating waters of the Harcsaszájú Cave). In: Mindszenty A (ed) Budapest: földtani értékek és az ember. Városgeológiai tanulmányok (Budapest: Geoheritage and Mankind. Urban geological studies—"In urbe et pro urbe"). ELTE Eötvös Kiadó (Eötvös University Press), Budapest, pp 237–239 (in Hungarian)

Klimchouk AB (2007) Hypogene speleogenesis: hydrogeological and morphogenetic perspective. Special Paper No.1, National Cave and Karst Research Institute, Carlsbad, New Mexico

Klimchouk A, Palmer AN, Waele JD, Aulker A, Audra P (2017) Hypogene Karst regions and caves of the world. Springer, Switzerland

Kontra J, Magyar Z (2019) Using geothermal water resources in Hungary. In: Palocz-Andresen M, Szalay D, Gosztom A, Sípos L, Taligás T (eds) International Climate Protection. Springer, Cham

Korpás L (1998) Paleokarst studies in Hungary. Occasional papers of the geological institute of Hungary, Budapest, vol 195

Korpás L, Juhász E (1990) Paleokarszt földtani modellek (Geological models of paleokarsts). Karszt és Barlang 1990(II):105–116 (in Hungarian)

Kovács J, Müller P (1980) A Budai-hegyek hévizes tevékenységének kialakulása és nyomai (Origin and traces of hydrothermal activities in the Buda Range). Karszt és Barlang 1980(II):93–98 (in Hungarian)

Kovács-Bodor P, Anda D, Jurecska L, Óvári M, Horváth Á, Makk J, Post V, Müller I, Mádl-Szőnyi J (2018) Integration of in situ experiments and numerical simulations to reveal the physicochemical circumstances of organic and inorganic precipitation at a thermal spring. Aquat Geochem 24:231–255

Kraus S (1982) A Budai-hegység hévizes barlangjainak fejlődéstörténete (Evolution of the hydrothermal caves in the Buda Mountains). Karszt és Barlang 1982(I):29–34 (in Hungarian)

Kraus S (1988) Eocén őskarsztos üregek a Mátyás-hegyi-barlangban (Eocene paleokarstic hollows in the Mátyás-hegy Cave). Karszt és Barlang 1988(II):79–80 (in Hungarian)

Kraus S (1990) A budai barlangok hévizes karbonátkiválásai (Carbonate deposition in the thermal-water caves of Buda Mountains). Karszt és Barlang 1990(II):91–96 (in Hungarian)

Kraus S (1993) A Szemlő-hegyi-barlang vízszintváltozásai (Water table fluctuation in the Szemlő-hegy Cave). Karszt és Barlang 1993(I–II):47–53 (in Hungarian)

Kraus S (2003a) Rácskai-barlang (Rácskai Cave). In: Székely K (ed) Magyarország fokozottan védett barlangjai (Strictly protected caves of Hungary). Mezőgazda Kiadó, Budapest, pp 276–278 (in Hungarian)

Kraus S (2003b) Róka-hegyi-barlang (Róka-hegy Cave). In: Székely K (ed) Magyarország fokozottan védett barlangjai (Strictly protected caves of Hungary). Mezőgazda Kiadó, Budapest, pp 302–303 (in Hungarian)

Kraus S (2006) Újabban megismert barlangi kiválások (Newly discovered speleothems). Karszt És Barlang 1995–1996:9–12 (in Hungarian)

Kraus S, Hazslinszky T (2003) Szemlő-hegyi-barlang (Szemlő-hegy Cave). In: Székely K (ed) Magyarország fokozottan védett barlangjai (Strictly protected caves of Hungary). Mezőgazda Kiadó, Budapest, pp 286–290 (in Hungarian)

Kriván P (1959) Mezozóos karsztosodási és karsztlefedési szakaszok, alsóbartoni sziklásparti jelenségek a Budai-hegységben. A szubgresszió fogalma (Periods of Mesozoic karstification and the burial of the resulted karst features in Buda Hills. The concept of subgression). Földtani Közlöny (Bull. Hung. Geol. Soc.) 89(4):393–401 (in Hungarian)

Krolopp E, Schweitzer F, Scheuer Gy, Dénes Gy, Kordos L, Skoflek I, Jánossy D (1976) A budai Várhegy negyedkori képződményei (Quaternary formations of the Castle Hill of Buda). Földtani Közlöny (Bull Hung Geol Soc) 106:193–228 (in Hungarian)

Leél-Őssy Cs, Leél-Őssy Sz, Adamkó P (2007) A Citadella-kristálybarlang (The Citadella Crystal Cave). Karszt és Barlang 2007(I–II):67–78 (in Hungarian)

Leél-Őssy S (1950) Geomorfológiai és hidrológiai vizsgálatok a Máriaremetei-szorosban (Geomorphological and hydrological investigations in the Máriaremete Gorge). Hidrológiai Közlöny 30 (11–12):473–476 (in Hungarian)

Leél-Őssy S (1952) Az Ürömi víznyelőbarlang (The Üröm Sinkhole Cave). Hidrológiai Közlöny 32(3–4):122–128 (in Hungarian)

Leél-Őssy Sz (1954a) A pilisi Legény-és Leány-barlangok (Legény and Leány Caves in the Pilis Hills). Földrajzi Értesítő III(3):594–603 (in Hungarian)

Leél-Őssy Sz (1954b) A bajóti Öregkő és barlangjai a Gerecsében (The Öreg-kő of Bajót and its caves—Gerecse Hills). Földrajzi Értesítő III(1):60–69 (in Hungarian)

Leél-Őssy S (1957) A Budai-hegység barlangjai (The caves of Buda Hills). Földrajzi Értesítő V I(2):155–167 (in Hungarian)

Leél-Őssy Sz (1997) A József-hegyi-barlang (Budapest) termálkarsztos keletkezése és fejlődése (Thermal karstic origin and evolution of the József-hegy Cave, Budapest). PhD theses, Hungarian Academy of Sciences, Budapest (in Hungarian)

Leél-Őssy Sz (2010) Mineralogy and speleology of the Szemlőhegy and Mátyáshegy Caves with an introduction to the geology and speleology of the Rózsadomb area, Budapest, Hungary. Acta Mineralogica-Petrographica, Field guide series for IMA 2010, vol 10

Leél-Őssy Sz (2012) A Budai Vár-hegy geológiai viszonyai (Geology of the Buda Castle Hill). Karsztfejlődés XVII:5–22 (in Hungarian)

Leél-Őssy Sz (2013) The longest cave in Hungary. In: Filippi M, Bosák P (eds) Proceedings of the 16th International Congress of Speleology, Brno, 21–28 July 2013, vol 2, 116–118

Leél-Őssy Sz (2015) Caves in the Buda mountains. In: Lóczy D (ed) Landscapes and Landforms of Hungary. Springer, Heidelberg

Leél-Őssy Sz (2017) Caves of Buda thermal karst. In: Klimchouk AN, Palmer A, De Waele J, Auler AS, Audra P (eds) Hypogene Karst regions and caves of the world. Springer, Cham, Switzerland, pp 279–298

Leél-Őssy Sz, Surányi G (2003) The peculiar hydrothermal caves in Budapest (Hungary). Acta Geol Hung 46(4):407–436

Leél-Őssy Sz, Adamkó P (2006) Természetvédelmi problémák és a kiépítési lehetőségek a József-hegyi-barlangban (Environmental protection problems and building development possibilities in the József-hegy Cave). Karszt És Barlang 1995–1996:35–38 (in Hungarian)

Leél-Őssy Sz, Szanyi G, Surányi G (2011) Minerals and speleothems of the József-hegy Cave (Budapest, Hungary). Int J Speleol 40(2):191–203

Leél-Őssy Sz, Szabó Z (2013) A Csókavári-kőfejtő barlangjai, szennyezése és kármentesítése (Pollution and clean-up in the caves of the Csókavár Quarry). In: Mindszenty A (ed) Budapest: földtani értékek és az ember. Városgeológiai tanulmányok (Budapest: Geoheritage and Mankind. Urban geological studies—"In urbe et pro urbe"). ELTE Eötvös Kiadó (Eötvös University Press), Budapest, pp 267–270 (in Hungarian)

Leél-Őssy Sz, Virág M (2018) Az utóbbi 20 év barlangkutatási eredményei a Budai-hegységben—különös tekintettel a Rózsadomb környékére (Developments in the speleology of the Buda Hills in the last two decades—with a special focus on the Rózsadomb and its surroundings). Földtani Közlöny (Bull Hung Geol Soc) 148 (1):45–74 (in Hungarian)

Lieber T (ed) (2010) A Sátorkőpusztai-barlang monográfiája (Monography of the Sátorkőpuszta Cave). BEBTE, Budapest (in Hungarian)

Mádl-Szőnyi J, Pulay E, Tóth Á, Bodor P (2015) Regional underpressure: a factor of uncertainty in the geothermal exploration of deep carbonates, Gödöllő Region Hungary. Environ Earth Sci 74 (12):7523–7538

Mádl-Szőnyi J, Tóth Á (2015) Basin-scale conceptual groundwater flow model for an unconfined and confined thick carbonate region. Hydrogeol J 23(7):1359–1380

Mádl-Szőnyi J, Erőss A, Tóth Á (2017) Fluid flow systems and hypogene karst of the transdanubian range, Hungary—with special emphasis on Buda thermal karst. In: Klimchouk AN, Palmer A, De Waele J, Auler AS, Audra P (eds) Hypogene karst regions and caves of the world. Springer, Cham, Switzerland, pp 267–278

Makk J, Tóth EM, Anda D, Pál SE, Schumann P, Kovács AL, Mádl-Szőnyi J, Márialigeti K, Borsodi AK (2016) Deinococcus budaensis sp.nov., a mesophilic species isolated from a biofilm sample of a hydrothermal spring cave. Int J Syst Evol Microbiol 66:5345–5351

Mari L, Fehér K (1999) The impacts of land use change on the Buda termal karst: a study of Szemlő-hegy cave. In: Bárány-Kevei I, Gunn J (eds) Essays in the ecology and conservation of karst, IGU comission sustainable development and management of Karst Terrains, Acta Geographica Szegediensis, vol 36, pp 104–111

Maucha L (1989) Karst water resources research in Hungary and its significance. Karszt és Barlang (Special Issue):39–50

Mindszenty A (ed) (2013) Budapest: földtani értékek és az ember. Városgeológiai tanulmányok (Budapest: Geoheritage and Mankind. Urban geological studies—"In urbe et pro urbe"). ELTE Eötvös Kiadó (Eötvös University Press), Budapest (in Hungarian)

Mindszenty A, Mádl-Szőnyi J, Pethő S, Kovács J, Müller I, Fodor L, Kádár M, Angelus B, Erőss A, Nyúl K, Poyanmehr Z, Varga R (2000) A Rózsadombi Termálkarszt Monitoring Optimalizálása. Zárójelentés a 2000. évben végzett munkáról (Final report on the works of the year 2000 about the optimization of Rózsadomb Thermal Karst monitoring). Manuscript, Department of Applied Geology, Eötvös Loránd University, Budapest (in Hungarian)

Molnár J (1869) A hévvizek Buda környékén (Thermal Waters in the surroundings of Budapest). Math. és Természettudományi Közlemények VII:163–244 (in Hungarian)

Müller P (1974) A melegforrás-barlangok és gömbfülkék keletkezéséről (On the origin of thermal caves and spherical niches). Karszt és Barlang 1974(I):7–10 (in Hungarian)

Müller P (1989) Hydrothermal paleokarst of Hungary. In: Bosák P (ed) Paleokarst—a systematic and regional review. Academia, Prague & Elsevier, Amsterdam, pp 155–164

Müller P, Magyar I (2008) A budai pannóniai képződmények (The Pannonian deposits of the Buda mountains). Földtani Közlöny 138 (4):345–356 (in Hungarian)

Müller P, Sárváry I (1977) Some aspects of developments in Hungarian speleology theories during the last 10 years. Karszt és Barlang (Special Issue):53–60

Nádor A (1991) A Budai-hegység paleokarszt jelenségei és fejlődéstörténetük (Paleokarst evolution of the Buda Hills). PhD theses, Eötvös Loránd University, Budapest (in Hungarian)

Nádor A (1992) Palaeokarstic features in Triassic-Eocene carbonates. Zentralblatt Für Geologie Und Paleontologie 11:1317–1329

Nádor A, Korpás L, Juhász E (1993) Palaeokarst controlled by high-frequency sea-level changes, Buda Mountains, Hungary. A MÁFI Évi Jelentése 1991-ről (Annual Report of the Hungarian Geological Institute 1991):111–127

Nádor A, Sásdi L (1991) A Budai-hegység paleokarsztjai és fejlődéstörténetük. I. Termális hatást nem tükröző paleokarsztok (Paleokarsts and long-term karst evolution of the Buda Hills. I. Paleokarsts without thermal effect). Karszt és Barlang 1991(I–II):3–10 (in Hungarian)

Nagymarosy A, Báldi-Beke M (1988) The position of the Paleogene formations of Hungary in the standard Nannoplankton zonation. Annal Univ Sci Bud de Rolando Eötvös Nominatae Sect Geolog 28:3–25

Nyerges A (2003) Keselő-hegyi-barlang (Keselő-hegy Cave). In: Székely K (ed) Magyarország fokozottan védett barlangjai (Strictly protected caves of Hungary). Mezőgazda Kiadó, Budapest, pp 315–316 (in Hungarian)

Onac BP, Ghergari L, Gàl A (1995) Crytallographical studies on gypsum crystals in Ponoraş Cave (Pădurea Craiului Mountains, Romania). Theoretical and Applied Karstology 8:63–67

Osborne RAL (2004) The troubles with cupolas. Acta Carsologica 33 (2):9–36

Palmer AN (2007) Cave geology. Cave Books, Dayton, Ohio

Pásztor D (2016) Hipogén karsztosodási indikátorok a Gellért-hegyi Aragonit-barlangban (Indicators of hypogenic karstic phenomena on the mineral paragenesis of the Gellérthegy Aragonit Cave). MSc theses, Department of Physical and Applied Geology, Eötvös Loránd University, Budapest (in Hungarian)

Pávai-Vajna F (1930) A forró oldatok, gőzök és gázok szerepe a barlangképződésnél (The role of the hot solutions and vapours and gases in the speleogenesis). Hidrológiai Közlöny 10:115–122 (in Hungarian)

Plózer I (1972) A malom-tavi Molnár János-barlang vízalatti járatainak kutatása (Exploration of underwater passages in the Molnár János Cave at Malom Lake). Karszt és Barlang 1972(I–II):13–17 (in Hungarian)

Plózer I (1974) A magyarországi búvár-barlangkutatás története és bibliográfiája, 1908–1973 (History and bibliography of cave explorations by divers in Hungary, 1908–1973). Karszt és Barlang 1974(II):55–64 (in Hungarian)

Plummer LN (1975) Mixing of seawater with calcium carbonate groundwater. Geol Soc Amer Memoirs 142:219–236

Polyak VJ, Provencio P (2001) By-product materials related to H_2SO_4 influenced speleogenesis of Carlsbad, Lechuguilla, and aother caves of Guadalupe Mountains, New Mexico. J Cave Karst Stud 63 (1):23–32

Polyak VJ, Mcintosh WC, Provencio P, Guven N (2006) Alunite and natroalunite tell a strory—the age and origin of Carlsbad Cavern, Lechuguilla Cave, and other sulfuric-acid type caves of the Guadalupe Mountains. In: New Mexico Geological Society Guidebook, 57th Field Conference, Caves and Karst of Southeastern New Mexico, pp 203–210

Poros Zs (2011) Fluid migration and porosity evolution in the Buda Hills, Hungary selected examples from Triassic and Paleogene carbonate rocks. PhD theses, Eötvös Loránd University, Budapest

Poros Z, Mindszenty A, Molnár F, Pironon J, Győri O, Ronchi P, Szekeres Z (2012) Imprints of hydrocarbonbearing basinal fluids on a karst system: mineralogical and fluid inclusion studies from the Buda Hills, Hungary. Int J Earth Sci (Geol Rundsch) 101:429–452

Rehák A, Szieberth D (2018) Csepegővizek vizsgálata a Molnár János-barlangban (Drip-water analysis in the Molnár János Cave). Karsztfejlődés XXIII:237–248 (in Hungarian)

Runnels DD (1969) Diagenesis, chemical sediments and the mixing of natural waters. J Sed Petr 39:1188–1201

Ruszkiczay-Rüdiger Z, Fodor L, Bada G, Leél-Őssy Sz, Horváth E, Dunai T (2005) Quantification of quaternary vertical movements in the central Pannonian Basin. A review of chronologic data along the Danube River Hungary. Tectonophysics 410:157–172

Sásdi L (1993) A Budai-hegység paleokarsztjai és fejlődéstörténetük. II. Termális hatást tükröző paleokarsztok (Paleokarsts and long-term karst evolution of the Buda Hills. II. Paleokarsts with thermal effect). Karszt és Barlang 1993(I–II):39–46 (in Hungarian)

Sásdi L (2003a) Ferenc-hegyi-barlang (Ferenc-hegy Cave). In: Székely K (ed) Magyarország fokozottan védett barlangjai (Strictly protected caves of Hungary). Mezőgazda Kiadó, Budapest, pp 249–251 (in Hungarian)

Sásdi L (2003b) Leány-Legény-barlangrendszer (Leány-Legény Cave System). In: Székely K (ed) Magyarország fokozottan védett barlangjai (Strictly protected caves of Hungary). Mezőgazda Kiadó, Budapest, pp 294–296 (in Hungarian)

Scheuer G (1986) A budai Vár-barlang geológiai vizsgálata, geológiai állapotfelvétel (Geological studies in the Buda Castle Cave in order to assess its conditions). Manuscript, Budapest (in Hungarian)

Scheuer G, Schweitzer F (1980) A budai hévforrások fejlődéstörténete a felsőpannontól napjainkig (Evolution of thermal springs in the Buda Hills from Upper Pannonian to recent). Hidrológiai Közlöny 60(11):492–501 (in Hungarian)

Scheuer G, Schweitzer F (1988) A Gerecse és a Budai-hegység édesvízi mészkőösszletei (Freshwater limestones of the Gerecse and Buda Hills). Földrajzi Tanulmányok 20, Akadémiai Kiadó, Budapest (in Hungarian)

Schréter Z (1912) Harmadkori és pleisztocén hévforrások tevékenységének nyomai a Budai hegyekben (Traces of Tertiary and Pleistocene thermal spring activity in the Buda Hills). Magyar Királyi Állami Földtani Intézet Évkönyve XIX(5):181–231 (in Hungarian)

Stieber J (2016) A Szemlő-hegyi-barlang aeroszol-monitoring programjának eredményei (Results of the aerosol-monitoring program in the Szemlő-hegy Cave). Karszt És Barlang 2012–2014:71–78 (in Hungarian)

Surányi G, Dombrádi E, Leél-Őssy Sz (2010) Contributions of geophysical techniques to the exploration of the Molnár János Cave (Budapest, Hungary). Acta Carsologica 39(3):565–576

Szablyár P (2003) Piliszántói-kőfülke (Pilisszántó Shelter Cave). In: Székely K (ed) Magyarország fokozottan védett barlangjai (Strictly protected caves of Hungary). Mezőgazda Kiadó, Budapest, pp 301–302 (in Hungarian)

Szabó Z, Leél-Őssy Sz (2013) Az óbudai Királylaki-barlang—a budai barlangok legújabb "ékszerdoboza" (Királylaki Cave, Óbuda—New "Jewel-box" of Buda). In: Mindszenty A (ed) Budapest: földtani értékek és az ember. Városgeológiai tanulmányok (Budapest: Geoheritage and Mankind. Urban geological studies—"In urbe et pro urbe"). ELTE Eötvös Kiadó (Eötvös University Press), Budapest, pp 232–234 (in Hungarian)

Szanyi Gy, Bada G, Surányi G, Leél-Őssy Sz, Varga Z (2009) A Budai-hegység pleisztocén kiemelkedéstörténete barlangi lemezes kalcitkiválások uránsoros kormeghatározása alapján (Pleistocene uplift history of the Buda Hills (Hungary), using uranium-series dating of cave rafts). Földtani Közlöny (Bull Hung Geol Soc) 139 (4):353–366 (in Hungarian)

Szanyi G, Surányi G, Leél-Őssy Sz (2012) Cave development and quaternary uplift history in the Central Pannonian basin derived from speletothem ages. Quat Geochronol 14:18–25

Székely K (2003) Bátori-barlang (Bátori Cave). In: Székely K (ed) Magyarország fokozottan védett barlangjai (Strictly protected caves of Hungary). Mezőgazda Kiadó, Budapest, pp 241–244 (in Hungarian)

Szombathy P, Rehák A, Szieberth D (2019a) Antropogenic pollution in the Molnár János Cave. In: Blatnik M, Gabrovšek F, Kogovšek B, Mayaud C, Petrič M, Ravbar N (eds) Abstract and guide book of the Karst hydrogeology—research trends and applications—27th international karstological school "Classical Karst", Karst Research Institute ZRC SAZU, Postojna (Slovenia), 17–21 June 2019a, p 128

Szombathy P, Mátyási J, Nagy J, Hajnal G, Szieberth D (2019b) Investigation of the flow system of the Molnár János Cave. In: Blatnik M, Gabrovšek F, Kogovšek B, Mayaud C, Petrič M, Ravbar N (eds) Abstract and guide book of the Karst hydrogeology—research trends and applications—27th international karstological school "Classical Karst", Karst Research Institute ZRC SAZU, Postojna (Slovenia), 17–21 June 2019b, pp 127–128

Szunyogh G (1982) A hévizes eredetű gömbfülkék kioldódásának elméleti vizsgálata (Theory of the dissolving of spherical cavities formed by thermal water). Karszt és Barlang 1982(II):83–88 (in Hungarian)

Szunyogh G (1984) A gömbfülkék kondenzvíz-korróziós kialakulásának elméleti fizikai leírása (The theoretical investigation of the origin of spherical caverns of thermal origin—second approach). Karszt és Barlang 1984(I):19–24 (in Hungarian)

Szunyogh G (1989) Theoretical investigation of the development of spheroidal niches of thermal water origin, second approximation. In: Hazslinszky T, Takács-Bolner K (eds) Proceedings of the 10th international congress of speleology, Budapest, 13–20 August 1989, vol 3, 766–768

Szunyogh G (2018) Rock-mechanics stability and safety of Ferenc-hegy Cave, Budapest, Hungary. Karsztfejlődés XXIII:219–236

Takács-Bolner K (1980) Új feltárások a Pál-völgyi-barlangban (New excavations in the Pál-völgy Cave). Karszt és Barlang 1980(II):87–92 (in Hungarian)

Takács-Bolner K (1995) Karst water protection problems indicated by dripping water analyses in Buda Thermal Karst area. Acta Carsologica XXIV:523–534

Takács-Bolner K, Tardy J, Némedi L (1989) Evaluation of the environmental impacts in Budapest's caves on the basis of the study of the quality of dripping waters. In: Kósa A (ed) Proceedings of the 10th international congress of speleology, Budapest, 13–20 August 1989, vol 2, 634–639

Takács-Bolner K (1998) Külső szakértők által végzett speciális vizsgálatok a Pál-völgyi-barlangban (Special investigations carried out by external experts in the Pál-völgy Cave). In: Kiss A, Takács-Bolner K (eds) Jelentés a Bekey Imre Gábor Barlangkutató Csoport 1997. évi munkájáról. Manuscript, Magyar Karszt-és Barlangkutató Társulat (Hungarian Speleological Society), pp 48–51 (in Hungarian)

Takács-Bolner K, Kraus S (1989) The results of research into caves of thermal water origin. Karszt és Barlang (Special Issue):31–38

Takács-Bolner K (2003a) Budai Vár-barlang (Buda Castle Cave). In: Székely K (ed) Magyarország fokozottan védett barlangjai (Strictly protected caves of Hungary). Mezőgazda Kiadó, Budapest, pp 244–248 (in Hungarian)

Takács-Bolner K (2003b) Sátorkőpusztai-barlang (Sátorkőpuszta Cave). In: Székely K (ed) Magyarország fokozottan védett barlangjai (Strictly protected caves of Hungary). Mezőgazda Kiadó, Budapest, pp 304–306 (in Hungarian)

Takács-Bolner K, Sásdi L (2003) Pálvölgyi-Mátyáshegyi-barlangrendszer (Pálvölgy-Mátyáshegy Cave System). In: Székely K (ed) Magyarország fokozottan védett barlangjai (Strictly protected caves of Hungary). Mezőgazda Kiadó, Budapest, pp 264–275 (in Hungarian)

Takács-Bolner K (2005) Comparative analysis on the morphology and mineralogy of Rózsadomb caves. In: Hazslinszky T (ed) Genesis and formations of hydrothermal caves. Proceedings of the international conference on the occasion of the 100th anniversary of the discovery of Pál-völgy Cave, Budapest, 21–24 June 2004, Magyar Karszt-és Barlangkutató Társulat (Hungarian Speleological Society), pp 125–131

Tóth J (1963) A theoretical analysis of groundwater flow in small drainage basins. J Geophys Res 68(16):4795–4812

Tóth J (1999) Groundwater as a geologic agent: an overview of the cases, processes and manifestations. Hydrogeol J 7(1):1–14

Tóth J (2009) Gravitational Systems of Groundwater Flow: Theory, Evaluation. Cambridge University Press, Cambridge, Utilization

Vajna G (1973) A rejtélyes Bátori-barlang (The Mysterious Bátori Cave). Gondolat Kiadó, Budapest (in Hungarian)

Vattano M, Madonia G, Audra P, D'Angeli IM, Galli E, Bigot JY, Nobécourt JD, Waele JD (2017) An overview of the hypogene caves of Sicily. In: Klimchouk AN, Palmer A, De Waele J, Auler AS, Audra P (eds) Hypogene Karst regions and caves of the world. Springer, Cham, Switzerland, pp 199–210

Virág M (2016) Történeti áttekintés a budai barlangok keletkezéséről—a klasszikus karsztfejlődéstől a hipogén barlangképződésig (Speleogenesis in the Buda Thermal Karst—a historical review from the "classical" epigenic to the hypogenic theories). Karszt És Barlang 2012–2014:24–46 (in Hungarian)

Virág M (2018) Hipogén és epigén karsztos folyamatok és hatásuk a budai rózsadombi barlangok fejlődéstörténetében—a Szemlő-hegyi- és a Molnár János-barlang példáján (Hypogene and epigene karstic processes and their influence on the evolution of the Rózsadomb caves in Buda—by the example of Szemlő-hegy Cave and Molnár János Cave). PhD theses, Eötvös Loránd University, Budapest (in Hungarian)

Virág M, Mádl-Szőnyi J, Leél-Őssy Sz, Erőss A, Zih-Perényi K, Mindszenty A (2009a) The effects of urbanization on the underlying thermal karst as reflected by dripwaters in a cave (Buda Hills, Hungary). In: Barbieri F (ed) Abstracts of the 13th congress of regional comittee on mediterranean neogene stratigraphy (RCMNS), Naples (Italy), 2–6 September 2009a, Acta Naturalia de L'Ateneo Parmense", vol 45(1/4), pp 255–256

Virág M, Mádl-Szőnyi J, Mindszenty A, Zih-Perényi K, Leél-Őssy Sz, Erőss A, Siklósy Z (2009b) Az urbanizáció hatása a budai barlangok csepegő vizeire a Szemlő-hegyi-barlang példáján (The effect of urbanization on the caves of the Buda Thermal Karst as reflected by dripwaters in the Szemlő-hegy Cave). Karsztfejlődés XIV:57–81 (in Hungarian)

Virág M, Mindszenty A, Surányi G, Leél-Őssy Sz (2011) Infiltration history of the last 50 thousand years as reflected by the anatomy of a flowstone type speleothem precipitated from dripwaters of the Pálvölgy-Mátyáshegy Cave System (Buda Thermal Karst, Budapest, Hungary). In: Capezzuoli E, Brogi A, Ricci M, Bertini A (eds) Proceedings of the International School on Travertine and Tufa (ISTT), Abbadia San Salvatore, Siena (Italy), 5–9 September 2011, Rendiconti Online Societa Geologica Italiana, vol 16, 39–41

Virág M, Mádl-Szőnyi J (2013) A Szemlő-hegyi-barlang csepegővizeinek vizsgálata a rózsadombi törmeléktakaró beszivárgásmódosító hatásának értékelése céljából (Investigation of dripwaters in the Szemlő-hegy Cave in order to evaluate the infiltration-modifying effect of the debris cover of Rózsadomb). In: Mindszenty A (ed) Budapest: földtani értékek és az ember. Városgeológiai tanulmányok (Budapest: Geoheritage and Mankind. Urban geological studies—"In urbe et pro urbe"). ELTE Eötvös Kiadó (Eötvös University Press), Budapest, pp 239–243 (in Hungarian)

Virág M, Kálmánfi-Ast H, Mindszenty A (2013a) Barlangszintek, barlangi kiválások és travertínók térbeli jellemzése (Spatial characterization of cave levels, cave minerals and travertines). In: Mindszenty A (ed) Budapest: földtani értékek és az ember. Városgeológiai tanulmányok (Budapest: Geoheritage and Mankind. Urban geological studies—"In urbe et pro urbe"). ELTE Eötvös Kiadó (Eötvös University Press), Budapest, pp 259–263 (in Hungarian)

Virág M, Takács-Bolner K, Surányi G (2013b) A Budai-hegység legidősebb vizsgált cseppköve: az Óriás-cseppkő (Giant dripstone —The oldest dripstone of Buda Hills). In: Mindszenty A (ed) Budapest: földtani értékek és az ember. Városgeológiai tanulmányok (Budapest: Geoheritage and Mankind. Urban geological studies—"In urbe et pro urbe"). ELTE Eötvös Kiadó (Eötvös University Press), Budapest, pp 243–245 (in Hungarian)

Virág M, Mindszenty A, Surányi G, Molnár M, Leél-Őssy Sz (2013c) A Búboskemence cseppkőlefolyás (The Búboskemence flowstone). In: Mindszenty A (ed) Budapest: földtani értékek és az ember. Városgeológiai tanulmányok (Budapest: Geoheritage and Mankind. Urban geological studies—"In urbe et pro urbe"). ELTE Eötvös Kiadó (Eötvös University Press), Budapest, pp 245–248 (in Hungarian)

Virág M, Mindszenty A, Leél-Őssy Sz, Surányi G (2013d) Gellérthegyi-aragonitbarlang (Gellérthegy Aragonit Cave). In: Mindszenty A (ed) Budapest: földtani értékek és az ember. Városgeológiai tanulmányok (Budapest: Geoheritage and Mankind. Urban geological studies—"In urbe et pro urbe"). ELTE Eötvös Kiadó (Eötvös University Press), Budapest, pp 181–186 (in Hungarian)

Virág M, Hegedűs A, Surányi G (2014) Szpeleotéma-térképezéssel az egykori termálvizek nyomában (Mapping of speleothems to understand the behaviour of paleo-thermal waters). In: Veress M, Zentai Z (eds) Abstracts of the 17th conference of Karsztfejlődés, Bük–Szombathely, 30 May to 1 June 2014, p 13 (in Hungarian)

Virág M, Mindszenty A, Kiss K, Kele S, Czuppon Gy, Palcsu L, Surányi G, Braun M, Weidinger T, Hegedűs A, Leél-Őssy Sz (2015) Superimposition of hypogenic and epigenic karst processes documented by speleothems of the Szemlő-hegy Cave (Budapest, Hungary). In: Abstract Book of the 31st IAS meeting of sedimentology, international association of sedimentologists, polish geological society, Krakow (Poland), 22–25 June 2015, p 564

Virág M, Mindszenty A, Kele S, Czuppon Gy, Surányi G, Braun M, Palcsu L, Futó I, Hegedűs A, Kiss K, Szieberth D, Leél-Őssy Sz (2016) Study of speleothems and other thermal-karst features in the caves of Rózsadomb (Budapest, Hungary). In: Otonicar B, Gostinčar P (eds) Abstract and Guide Book of the Paleokarst—24th International Karstological School "Classical Karst", Karst Research Institute ZRC SAZU, Postojna (Slovenia), 13–17 June 2016, p 34

Virág M, Vörös P, Mindszenty A, Németh T, Bendő Zs, Leél-Őssy Sz (2017) Altered zones as evidences for an unusual sulphuric acid related carbonate dissolution in the Buda Thermal Karst (Hungary). In: Gostinčar P (ed) Abstract and Guide Book of the Milestones and Challenges in Karstology—25th International Karstological School "Classical Karst", Karst Research Institute ZRC SAZU, Postojna (Slovenia), 19–23 June 2017, p 57

Virág M, Mindszenty A, Surányi G, Kele S, Németh T, Hegedűs A, Leél-Őssy Sz (2019) Pleistocene paleohydrological changes recorded by speleothems of the Szemlő-hegy Cave (Buda Thermal Karst, Hungary). In: Blatnik M, Gabrovšek F, Kogovšek B, Mayaud C, Petrič M, Ravbar N (eds) Abstract and Guide Book of the Karst Hydrogeology—Research Trends and Applications—27th International Karstological School "Classical Karst", Karst Research Institute ZRC SAZU, Postojna (Slovenia), 17–21 June 2019, p 134

Völgyi L (1979) Geothermal inhomogenity in the Hungarian Great Plain (Pannonian Basin). Acta Mineralogica-Petrographica XXIV (1):137–147

Vörös P, Virág M, Mindszenty A, Bendő Zs, Németh T, Leél-Őssy Sz (2013) Eocén kőzeteket átszelő repedéskitöltésekhez kapcsolódó elváltozott zónák a budai barlangokban (Altered zones associated with vein-filling mineral parageneses occurring in the Eocene host rocks of the caves of the Buda Thermal Karst). In: Mindszenty A (ed) Budapest: földtani értékek és az ember. Városgeológiai tanulmányok (Budapest: Geoheritage and Mankind. Urban geological studies—"In urbe et pro urbe"). ELTE Eötvös Kiadó (Eötvös University Press), Budapest, pp 34–39 (in Hungarian)

Weidinger T, Virág M, Tordai Á, Lukács D, Leél-Őssy Sz, Mindszenty A (2017) Barlangklimatológiai mérések a Budai-termálkarsztban (Cave-climatological measurements in the Buda Thermal Karst). In: Cserny T, Alpek BL (eds) Proceedings of the Conference of HUNGEO 2017 "Földtudományok és környezet —harmóniában", Pécs, August 2017, Magyarhoni Földtani Társulat (Hungarian Geological Society), pp 129–134 (in Hungarian)

Wein G (1977) A Budai-hegység tektonikája (Tectonics of Buda Hills). MÁFI Alkalmi Kiadvány (Occasional Papers of the Geological Institute of Hungary), Budapest (in Hungarian)

The Caves of the Southwestern Part of the Transdanubian Mountains (Transdanubian Range)

Szilárd Vetési-Foith

Abstract

In this chapter, we describe the caves of the southwestern part of the Transdanubian Mountains, the Bakony Region and the Vértes Mountains, which we consider the most significant. We present the general genetic grouping of the caves and then characterize the major caves of the two mountains according to a system of criteria. Such as their geographical location, their host rock, the genetics and the speleothem of the caves.

Keywords

Bakony region • Vértes mountains • Karstic cave • Pseudokarstic cave • Cave genetics • Cave morphology

1 Introduction

The southwestern part of the Transdanubian Mountains can be divided into the Bakony Region and the Vértes Mountains. The Bakony Region is richer in caves. The Bakony Region as a mesoregion is made up of five micro-region groups (Keszthely Mountains, Balaton Uplands, Southern Bakony, Northern Bakony, Bakonyalja). Different karst phenomena, including caves, occur in each of the above-mentioned micro-region groups, but at different densities. In 2020, 676 caves are known from the Bakony Region, 116 from the Vértes Mountains (http://www.termeszetvedelem.hu/index.php?pg=caves).

The earliest written mention of the caves of the Bakony Mountains, of scientific value, can be found in the work of Mátyás Bél (Bél 1737), "Notitia". The beginning of the substantive research of the caves of the Bakony Mountains dates back to the first third of the twentieth century. At first the general description of the caves (Bertalan 1938, 1955, 1958) was in focus, which was followed by the morphological and genetic (Eszterhás 1976, 1978, 1980, 1981, 2013; Veress 1980a, b, 2000; Veress and Vetési-Foith 2019a) described later. The caves of the Keszthely Mountains and the Balaton Basin were researched by Leél-Őssy (1953, 1958) and Láng (1958).

In recent decades, cave research groups have carried out significant excavations in the area of the Bakony Region and the Vértes Mountains. Such as the Alba Regia Cave Research Group (Tés Plateau), the Labyrinth Karst- and Cave Research Sports Association, the Acheron Cave Research Group, and the Styx Cave Research Group (the Keszthely Mountains, the Balaton Uplands), and the Gerecse Cave Research Group Association (the Vértes Mountains) can be mentioned.

The main constituting rock of the Bakony Region is Triassic Main Dolomite (Main Dolomite Formation). The bedrock of the main dolomite is Veszprém marl (Veszprém Marl Formation) in the Northern Bakony (Haas 1993), Sándorhegy limestone (Sándorhegy Limestone Formation) in the Balaton Uplands (Gyalog 2005), and Ederics limestone (Ederics Limestone Formation) in the Keszthely Mountains (Haas 1993). The main dolomite is overlain by Triassic Dachstein limestone (Dachstein Limestone Formation), and younger Jurassic, Cretaceous and Eocene limestones. The Bakony Region can be divided into horsts or horst groups with different elevations and morphogenetics (Pécsi 1980, 1991).

The Balaton Basin structurally belongs to the Transdanubian Hills, it is one of its mesoregions. The bedrock of this basin-like, sometimes bare karstic, sometimes covered karstic area is the Main Dolomite Formation and the Sédvölgy Dolomite Formation (Móga et al. 2011). The dolomite bedrock is overlain here mostly by Sarmatian limestone. The basin character is interrupted here and there by basalt blocks and buttes, in which the pseudokarstic caves of the studied area were formed.

S. Vetési-Foith (✉)
University of Pécs, Pécs, Hungary

M. Veress and Sz. Leél-Őssy (eds.), *Cave and Karst Systems of Hungary*,
Cave and Karst Systems of the World, https://doi.org/10.1007/978-3-030-92960-2_13

The main constituting rock of the Vértes Mountains is the main dolomite. The main dolomite is overlain on the NW by Dachstein limestone with a transitional sequence. The surface extent of the Jurassic and Cretaceous formations is relatively small and is present mainly in the SW and NE parts of the Vértes Mountains. The Vértes Mountains can be divided into different morphogenetic horsts as well (Pécsi 1970, 1980).

2 Genetic Classification of the Caves in the Bakony Region

The cold water origin caves of the Bakony Region can develop in the vadose and the phreatic zone (Veress 2000; Veress and Vetési-Foith 2019a). The caves of the vadose zone can be shafts and pits. The caves of the phreatic zone may have formed by the outlets of the karst water (effluent cave), while in the interior of the mountains they are relic caves. In some places on the margin of the mountains, a hydrothermal effect also played a role in the formation of the caves. Due to the widespread distribution of volcanic rocks (basalt) in the Southern Bakony, non-karst caves also occur.

Depending on their formation environment, shafts may occur below subsidence dolines, sinkholes, or former sinkholes. If they develop under subsidence dolines, their formation environment is concealed karst, where the karst is covered by water-permeable non-karstic rock (decalcified loess).

Pit caves develop on the mixed allogenic-autogenic karst of the mountains. In addition to dissolution, the development caused by erosion also plays a role by these caves. These are at the margin of the Kab Mountain's basalt cover as well as inside the basalt cover. There are also shafts on the Tés Plateau that reflect an erosional effect (Eszterhás 1983; Veress 2000; Veress and Vetési-Foith 2019a). These caves are shafts formed at the edge of the former gravel cover. These are the shafts of the subsidence dolines of the loess covered areas which are evolving by solution nowadays. Veress (2018) distinguished different types of shafts according to their pattern. Thus, a simple shaft, complex shaft, a shaft system, a branching shaft, a tiered shaft and a shaft with a phreatic branch.

The caves in the phreatic zone are nearly horizontal in development and formed by mixing corrosion. The caves formed in the phreatic environment in the inner parts of the mountains were formed mainly on the side of epigenetic valleys, by opening up. Most of the phreatic cavities in the Bakony Region were formed by the water of the karst water storeys (Fig. 1). Karst water storeys developed locally above the level of the karst water (which is called main karst water) in the mountains by impermeable intercalations (Fig. 1). Where Triassic rocks are on the surface (possibly if the younger limestones settled on them do not have an impermeable intercalation), phreatic cavities formed in the level of the main karst water (Fig. 1). Morphological evidence of their open up origin is that the cavities branch at and towards their entrance and that several caves of the Bakony Mountains are through caves (or crossover caves), i.e. cavities with two entrances going transversely through the bearing rock mass. Their positional and altitudinal distribution also proves their open up origin, as they never occur in a valley head, but in the valley sides, in groups, at different altitudes.

The distribution of caves of hydrothermal origin is subordinate in the Bakony Region. Cave formation of thermal water origin caused by volcanic and post-volcanic activity is limited locally to the margin of the Keszthely Mountains, the Balaton Uplands and the Balaton Basin (Leél-Őssy 1953). These are mostly former cold water origin caves transformed by hydrothermal effect, but there are caves of exclusively hydrothermal origin as well (Plózer 1977; Kárpát 1982, 1983a, b).

Pseudokarstic caves are common in basalt or basalt tuff of basalt-covered blocks. Some of the caves were formed at the same time as the lava rock solidified, i.e. syngenetically, or postgenetically after the lava rock solidified. The former have a rounded spatial shape, the latter are more fissure-like formations (Eszterhás 2013). The caves in this chapter are shown in Fig. 2.

3 Presentation of the Caves

3.1 Alba Regia Cave

On the Tés Plateau is one of the most significant caves in the Northern Bakony, the Alba Regia Cave, which is the water intake of several subsidence dolines (some of which may be former sinkholes). The cave was discovered in 1975 by members of the Alba Regia Cave Research Group. According to the newest data, its total length exceeds 3.5 km and its depth is 200 m (Table 1).

The active shaft cave begins on the border of the Upper Triassic and Lower Jurassic limestone. The lower, sloping branch system of the cave was formed in Upper Triassic Dachstein limestone and in marly limestone strata surrounded by it (Eszterhás 1984; Szolga 2003a). According to Veress's shaft classification (2018, 2019), the cave is a tiered shaft (a cave formed along bedding planes), its main branch has several floors (Veress and Vetési-Foith 2019a).

The cross section of the cave's branches is predominantly "clover" shaped. The vertically elongated cave widens by the marl intercalation (these are called flattenings), which according to Eszterhás (1983), is of erosive origin. There is a small (a few dm wide) drainage channel at the bottom of the main branch, there are chimneys and solution channels on the

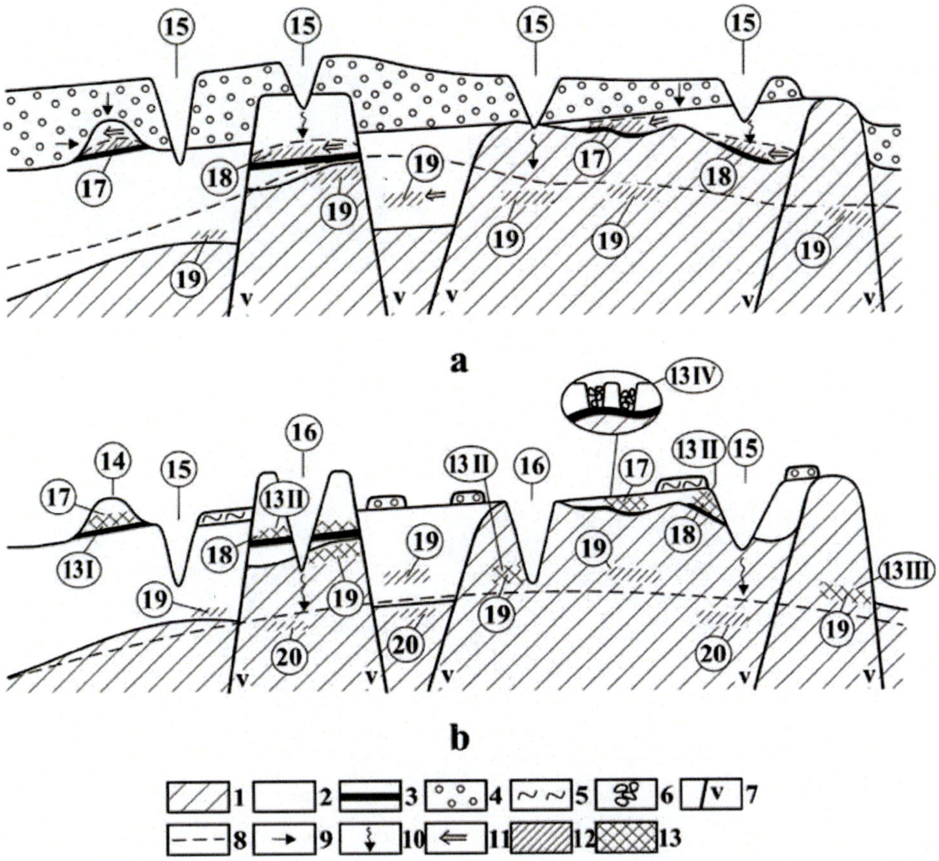

Fig. 1 The phreatic caves of the Bakony Mountains according to development environment and the way of opening up. Legend: a. former state, b. present state, 1. main dolomite, 2. Eocene and/or Cretaceous limestone, 3. impermeable beds or partially impermeable intercalation, 4. Oligocene-Miocene gravel, 5. loess, 6. collapse blocks of the ceiling, 7. fault, 8. karst water table, 9. water flow in the cover, 10. seeping water of a stream, 11. horizontal flow of karst water, 12. closed, non-opened up cavities, 13. opened-up cavities: 13.I. cavity exposure during the denudation of the cover, 13.II. opening up during valley development, 13.III. opening up during the denudation of the side slope of the block, 13. IV. opening up by the collapse of the ceiling, 14. limestone mound (hill), 15. epigenetic valley, 16. epigenetic-antecedent gorge, 17. karst water storey that developed from the water of the cover and cavity formation, 18. karst water storey that developed from the seeping water of a stream and cavity formation, 19. former cavity formation that can be associated with main karst water, 20. present cavity formation primarily in the main dolomite (supposed) (Veress and Vetési-Foith 2019a)

vault, and scallops on the walls (Eszterhás 1983). Surface waters formed the branches of the cave along fractures, but the water moving along the bedding plane also played a role in the formation of them. This is proved by the fact that the position of the cave branches is the same as the slope of the bedding planes and that the branches were formed on several floors. Erosion may be indicated by the widening formed in the marl intercalation, which may have developed when the depression of the cave received gravel from the area (sinkhole state). Vault channels and blind shafts indicate that paragenetic development has also taken place in the cave. Scallops indicate water filling and strong water flow, while the channel of the bottom indicates low and sustained water flow.

Through the shafts of the covered karstic depressions, the tributaries of the main branch drain water continuously, so the deeper levels of the cave can develop further. The combined catchment area of the covered karstic depressions leading water to the cave is 30–40 thousand m^2. Approximately 2000 m^3 of water enters the Alba Regia Cave from this area yearly (Eszterhás 1983). Concerning ventilation, the cave can be divided into two zones. The parts close to the surface are dynamically aerated, well ventilated, presumably due to the small depth. The deeper levels are characterized by slow air flow (Eszterhás 1983; Szolga 2003a). The air temperature in the Alba Regia Cave varies between 8.1 and 8.5 °C (Kordos 1984).

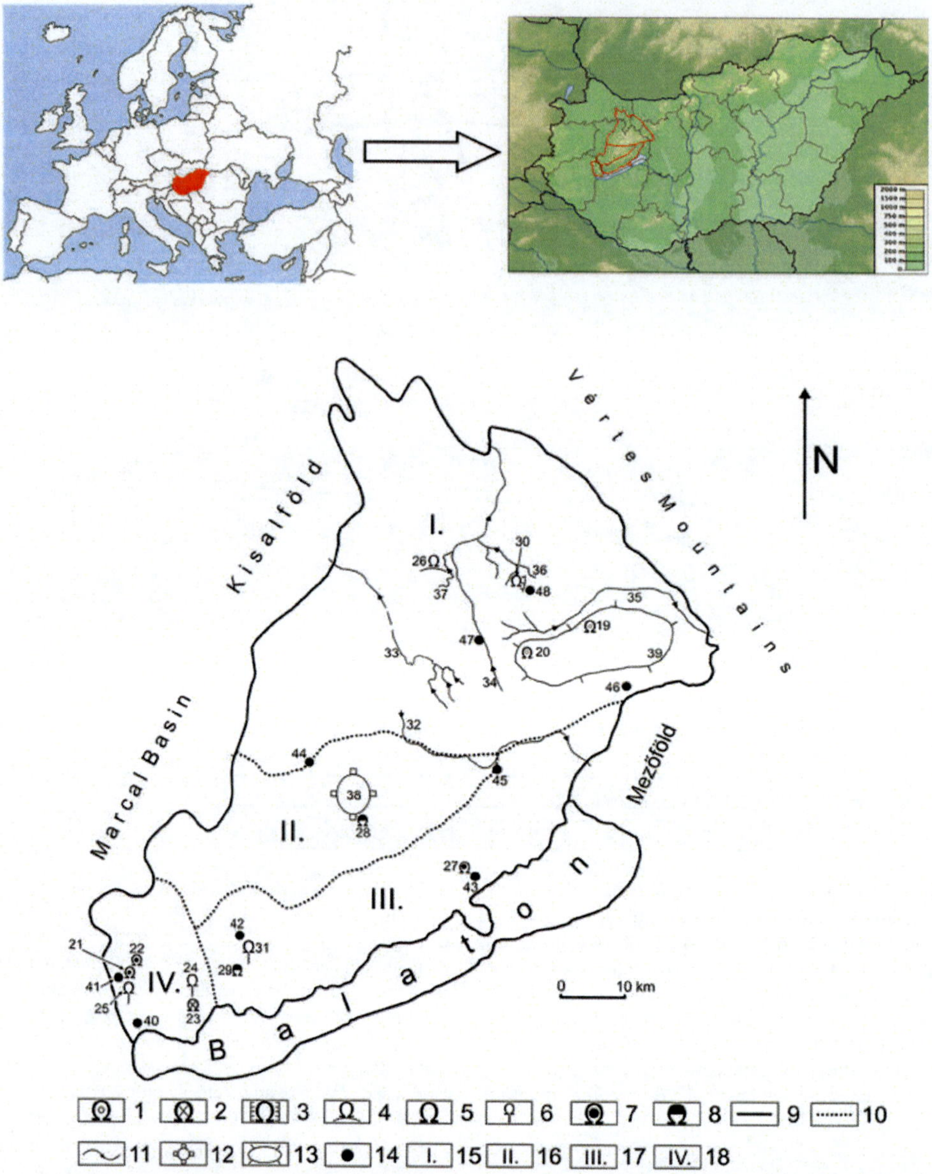

Fig. 2 The occurrence of the caves in the Bakony Region mentioned in this chapter. Legend: 1. inactive shaft cave, 2. tectonic cave, 3. cave with opened up origin in valleyside, 4. cave with opened up origin on block roof, 5. effluent cave, 6. spring cave affected by hydrothermal water, 7. hydrothermal cave, 8. non-karstic cave, 9. boundary of the mountain, 10. boundary of micro-region group, 11. stream and its valley, 12. block with basalt cap, 13. plateau, 14. settlement, 15. Northern Bakony micro-region group, 16. Southern Bakony micro-region group, 17. Balaton Uplands micro-region group, 18. Keszthely Mountains micro-region group, 19. Alba Regia Cave, 20. Csengő Shaft, 21. Cserszegtomaj Well Cave, 22. Acheron Well Cave, 23. Csodabogyós Cave, 24. Döme Cave, 25. Hévíz Spring Cave, 26. Likas-kő Cave of Hódos Stream, 27. Lóczy Cave, 28. Pula Basalt Cave, 29. Nagy Sárkány Cave on the Szent György Hill, 30. Ördög-lik Cave of the Sűrű Mountain, 31. Tapolca Cave System, 32. Séd Stream, 33. Gerence Stream, 34. Cuha Stream, 35. Gaja Stream, 36. Dudar Stream, 37. Hódos-ér Stream, 38. Kab Mountain, 39. Tés Plateau, 40. Keszthely, 41. Hévíz, 42. Tapolca, 43. Balatonfüred, 44. Ajka, 45. Veszprém, 46. Várpalota, 47. Zirc, 48. Dudar (Veress and Vetési-Foith 2019a modified)

Its speleothems are crust stones, draperies, calcite grains and calcite tubers. In the deeper parts of the cave system, where air circulation is minimal, the carbon dioxide in the air (3.5–4.0 vol%) is enriched, resulting in the redissolution of dripstone formations. As a result, they are transformed into calcite needle-like crust residues or pine cone-like formations.

In the case of the Alba Regia Cave, various faunistic studies are worth mentioning (Eszterhás 1984; Paulovics

Table 1 Data of the Alba Regia Cave

General information	
Cadastral reference	4422-1
Region	Northern Bakony
County	Fejér
Municipality	Isztimér
Genetics	Active shaft cave
Length	3600 m
Vertical extent	200.2 m
Depth	200 m
Protection	Highly protected

1998). In the cave 216 animal species were isolated thus, having the second most diverse fauna of Hungarian caves.

The Alba Regia Cave has been highly protected since 1982. Nevertheless, researches (Hegyi and Koch 1983; Zentai 1989) prove that the parts of the cave below the agriculturally cultivated area are subject to chemical loading by the water seeping through the soil.

3.2 Csengő Shaft

The entrance to the Csengő shaft is located next to the long-known (Bél 1737) cave, referred to as the Ördög-lyuk cave, on the NW side of Csengő Hill, in a collapse origin depression. The two caves belong to a cave system, the Ördög-lyuk cave is a clogged entrance to the shaft (Kordos 1984). The cave was excavated in 1979 by the Alba Regia Cave Research Group. Nowadays, its maximum depth is 134 m (Table 2), making it the second deepest cave in the Tés Plateau with similar genetics.

At the entrance to the shaft, there are Upper Triassic Dachstein-type limestones and Lower Jurassic limestones, but in the deeper parts, the host rock of the cave is only Upper Triassic Dachstein limestone (Szolga 2003b). Concerning its genetics, it can be classified as a complex shaft cave, as its branches have a zigzag pattern (Veress 2018; Veress and Vetési-Foith 2019a).

The cave has 4 larger "rooms". They are all surrounded by vertical or nearly vertical walls and widen downwards. The largest room, the Óriás Mine, has a vertical extent of 25 m (Kökény et al. 1987). The blind shafts in the shaft cave suggest that the cave is sometimes filled with water to a depth of 15 m from the surface, and the blind shafts develop paragenetically. The vertical rooms are connected by narrow sloping, narrow cross-sectional branches.

Similarly to other shaft caves, it is poor in dripstone formations. However, a small number of stalactites, 20–30 cm long draperies, light coloured crust stones can be mentioned. Due to its shaft cave nature, it ventilates difficultly, so the carbon dioxide level in the lower part reaches 2 vol% (Szolga 2003b).

Based on the bat fauna living here, and the size, the speleothem and the geological values of the cave, it was declared highly protected in 1988.

3.3 Cserszegtomaj Well Cave

One of the most interesting caves in Hungary is hidden by a rim-free well which is about 60 m deep (Table 3) in the corner of one of the cemeteries in the Village of Cserszegtomaj on the SW edge of the Keszthely Mountains. The Cserszegtomaj well cave was accidentally discovered while digging a well in 1930 at a depth of 54 m. It is now more than 3 km long and has 4 relatively larger rooms (Kárpát 2003b).

The horizontally developed cave was formed at rock boundary. Its bedrock consists of Triassic main dolomite which is covered by Lower Pannonian siliceous sandstone (Leél-Őssy 1953; Kárpát 2003b; Sásdi 2020) on which Pannonian clay and sandstone with pyrite were settled (Tóth 1965). The bedrock of the area was terrigenous from the end of the Triassic to the beginning of the Pannonian, so it was exposed to surface karstic denudation. The sandstone cover

Table 2 Data of the Csengő Shaft

General information	
Cadastral reference	4422-4
Region	Northern Bakony
County	Veszprém
Municipality	Olaszfalu
Genetics	Active shaft cave
Length	230 m
Vertical extent	133 m
Depth	133 m
Protection	Highly protected

Table 3 Data of the Cserszegtomaj Well Cave

General information	
Cadastral reference	4440-2
Region	Balaton Basin/Keszthely Riviera
County	Zala
Municipality	Cserszegtomaj
Geenetics	Thermal water origin
Length	3320 m
Vertical extent	12 m
Depth	10 m
Protection	Highly protected

settled on a dissected karstic surface in the Pannonian, filling the karstic depressions of the dolomite (Kárpát 2003b). The depressions of the dolomite, which were filled with sandstone, may have formed at karren features (Kárpát 2003b), spring funnels (Leél-Őssy 1953), solution dolines (Szentes 1946), or the collapse of dolomite cavities (Kárpát 1982). These are symmetrical, funnel-shaped, closed 1–3 m deep depressions with a 3–8 m diameter. Therefore, the cave is a cavity formed at rock boundary, which is separated into rooms and branches by sandstone pillars of karstic depressions. Thermal water flowing through the cracks of the dolomite played a role in the formation of the cave, at the end of the Pliocene, as a result of volcanic and post-volcanic activity (Leél-Őssy 1953). This is proved by the concretions of hydrothermal origin, the strong dusting and physical weathering of the dolomite, or even the complex map of the cave. As the sandstone settled on the dolomite is impermeable, the inflowing thermal water, mixed with cold karst water on the border of the dolomite and the sandstone, started its cave-forming activity due to mixing corrosion (Kárpát 1982).

The spatial position of the Cserszegtomaj well cave's branch system is determined by the spatial position of the beds. The width and height of the branches vary greatly. The central and northern zones of the cave are characterized by narrow branches, while in the southern parts, we can find a chain of halls which are 4–18 m in diameter and 2–3 m high.

The Labyrinth Karst and Cave Research Sports Association, together with the Pagony Cave Research Group, collected water samples at two points in the cave and performed quality tests on them. According to them, the infiltrating water reaches the cave in a very long time because the total hardness of the water samples was high. The sulphate ion concentration measured in the water samples (which is the result of the redissolution of the pyrite in the cover) also proves this (Fehér 2015).

The concentration of carbon dioxide in the cave is high. It has a seasonal fluctuation, but on average it is around 1–1.5 vol%, however in the furthermost zones of the southern zone it has already been measured 4 vol% (Kárpát 1982). The enrichment of carbon dioxide is caused, among other things, by the fact that the cave is ventilated only through the outlet well, so it is only influenced by the outdoor air pressure, generating a barometric draft. In addition, the high carbon dioxide value may be due to the pyrite dissolution processes of the cover rock, its air impermeable characteristics and its thickness. The temperature of the cave is high (12.5–12.8 °C) compared to the caves in Hungary. Despite the fact that the branches are completely dry, the relative humidity is 100 vol% (Kárpát 2003b).

It's speleothems are gypsum depositions formed during the oxidation of the pyrite-containing cover, which appear in fibrous, twisted crystal groups. In the deeper sections, in addition to gypsum, there are also coloured hematite and manganese, which form blue and red crusts on the cave wall (Kárpát 2003b).

The Cserszegtomaj well cave is of great value, because it is unique concerning its formation. The Acheron well cave nearby has an analogy with the Cserszegtomaj well cave in its genetics thus, in its morphology and in its development (Kárpát 2003a).

3.4 Csodabogyós Cave

The Csodabogyós Cave (Table 4) is located in the Keszthely Mountains on the eastern edge of the Ederics Plateau. The cave was already known in the '70s. Its maximum known length is 6400 m and its depth is 147 m (Polacsek 2014).

Almost the entire cave system was formed in Upper Triassic Ederics limestone (Ederics Limestone Formation). In the cave, there are only a few sections whose host rock is that Upper Triassic main dolomite which is characteristic in the whole Keszthely Mountains (Kárpát 2003c). According to Palmer's cave pattern classification (2004), it is a spongework cave system, i.e. it is made up of several floors, has an irregular map, and made up mainly of rooms. The rooms enclose irregular remains of the host rock. However, it can be observed on its length profile (Acheron-Labirint 1992) that the excavated branches are parallel to each other, so a kind of tectonic preformation (Sásdi 2020) is also characteristic of the cave system (Fig. 3).

The cave system consists of several large and a few small rooms and the branches connecting them. The Colosseum Hall, the 52-es Csapdája Hall, and the Feneketlen Hasadék Hall are the most notable of its rooms. Two of these three, the Colosseum Hall and the 52-es Csapdája Hall, also form a separate shaft system within the cave (Kárpát 2003c).

Among the caves of the Bakony Region, some sections of the Csodabogyós Cave are the richest in various dripstone formations (Fig. 4). In addition to the numerous stalactites

Table 4 Data of the Csodabogyós Cave

General information	
Cadastral reference	4440-4
Region	Keszthely Mountains
County	Veszprém
Municipality	Balatonederics
Genetics	Tectonic cave
Length	6400 m
Vertical extent	147 m
Depth	147 m
Protection	Highly protected

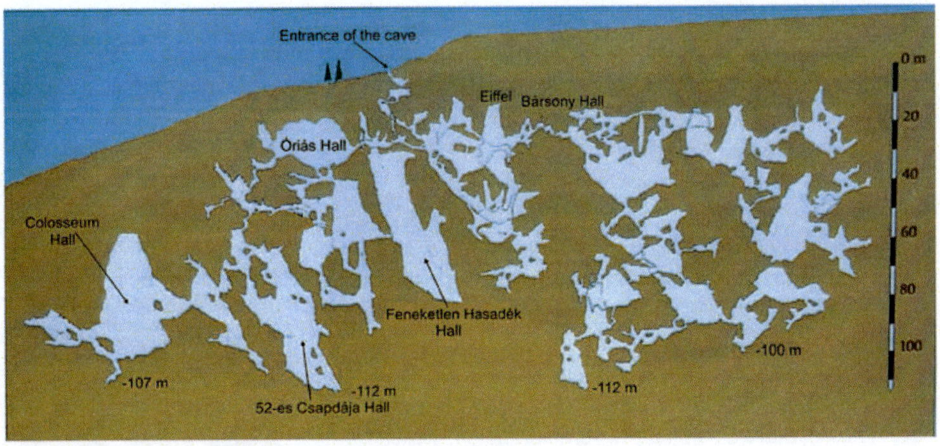

Fig. 3 The map of the Csodabogyós Cave (Kárpát 2003c modified)

Fig. 4 Dripstones in the Csodabogyós Cave (photo by Csaba Egri)

and huge stalagmites, there are dripstone columns, drip curtains, helictites and balcony stalagmites on these dripstone sections, which are still active nowadays (Kárpát 1991).

The atmospheric circulation of the cave system is greatly influenced by the fact that it forms an air flow system with the Szél-lik cave, which opens at a distance of 100 m and 40 m deeper from it. Because of this, the cold winter air flows into the lower Szél-lik cave and leaves at the exit of the Csodabogyós Cave as a clearly visible column of steam. In summer, the atmospheric circulation is reversed, in which case the air flows inwards at the entrance of the Csodabogyós Cave (Kárpát 2003c).

Due to the winter-summer atmospheric circulation, the carbon dioxide level of the rooms and branches of the cave system never rises above 0.3 vol%. The temperature of the sections farther from the surface is constant, fluctuating around 9.5 °C (Kárpát 2003c), in the temperature of the near-surface parts can be felt the influencing factor of the current season's temperature. The relative humidity is 100 vol%.

3.5 Döme Cave

The Döme Cave opens on the side of Szabad Hill in the eastern part of the Keszthely Mountains. The entrance to the shaft, which was filled with soil-stone debris before, was unearthed and excavated in 1998 by the Styx Cave Research Group. The largest nowadays known length of the cave is 353 m and its depth is 107 m (Table 5). Its significance is due to its unique speleothems in Hungary.

The cave was formed in Ederics limestone, the bedrock of which is the Upper Triassic main dolomite which is the characteristic rock of the entire Keszthely Mountains (Nyerges 2003). The morphology of the cave's rooms and branches suggests that the thermal water played a significant role in its formation. Due to this characteristics, and since it is tectonically not preformed, it stands out from other caves in the eastern part of the Keszthely Mountains. Its further development was also supported secondarily by the cold water solution (Eszterhás 2000a).

The cave can be divided into three parts morphologically. The first section runs to the depth of 15 m near the surface. At this section, solution pockets on the walls of the spacious

Table 5 Data of the Döme Cave

General information	
Cadastral reference	4440-5
Region	Keszthely Mountains
County	Veszprém
Municipality	Balatonederics
Genetics	Spring cave influenced by thermal water effect
Length	353 m
Vertical extent	107 m
Depth	107 m
Protection	Highly protected

cavities and a network of roots extending from the surface are characteristic. The aggregation of roots, the so-called root stalagmites which made the cave known nationally, but worldwide as well can be found here. Such a formation was first described from a limestone cave here, as well as one which's base species is downy oak. Its characteristics is that their structure is leaner and special “dog shaped”. Such forms are called “Ederics type root stalagmite” as suggested by Eszterhás (2000a). The entrance to the cave may have been opened earlier. This is proved by the Neolithic human bones and tools found at this section, as well as by other fossil large mammalian bones (Eszterhás 2000a). The second section, which runs from 15 m to the endpoint of the cave, is essentially a completely vertical branch. Its walls are rich in karstic forms due to the constantly active dripping, dripstones can also be found in some places. The third section (Pannonia branch) is a branch formed at a depth of approximately 80 m (Bán 2001) which is perpendicular to the second section. Here, the walls are covered with brown clay and beneath it there is a fine sand-like filling. From the presence of this crust and the re-dissolved dripstones and the solution forms of the vault found here (Nyerges 2003), it can be concluded that sometimes, during heavy rains, the cave receives a significant amount of water, which is not able to leak out quickly, so this part of the cave becomes filled with water. While the first and second sections are pretty much a vertical branch, here in the Pannonia branch there are rooms developed garland-like (Nyerges 2003).

The temperature of the cave is constant. Its value fluctuates between 10 and 10.7 °C (Bán et al. 1999). Due to poor ventilation, the carbon dioxide is enriched in the lower parts, it can reach up to 4–6 vol%. Thanks to the root stalagmite formations and archeological finds here, the cave has been highly protected since 2001.

3.6 Gánt Cave

The entrance to the Gánt Cave opens in a cliff called the Angyallépcső by the locals on the border of the village of Gánt in the Vértes Mountains (Fig. 5). The first description of the cave comes from Kocsis (1975). Its total length is 149 m (Table 6), making it the second longest cave in the Vértes Mountains.

The Gánt Cave was formed in the Upper Triassic, stratified dolomite (Kocsis 1975; Polacsek and Ba 2018b), although according to Gönczöl (1986) and Sivó (2005), dolomite is characteristic only in its near-surface parts, the host rock of its endpoint is limestone. Based on its mountain front position, low altitude (255 m), host rock, speleothem,

Fig. 5 The entrance to the Gánt Cave (photo by Attila Gazda)

Table 6 Data of the Gánt Cave

General information	
Cadastral reference	4521-16
Region	Vértes Mountains
County	Fejér
Municipality	Gánt
Genetics	Effluent cave influenced by thermal water effect
Length	149 m
Vertical extent	14 m
Depth	14 m
Protection	Distinctly protected

and phreatic nature, the Gánt cave is an effluent cave, sustained by hydrothermal effect. The cave was formed on two levels (Gönczöl 1986). On the walls and the vault of the cave there are solution pockets in different sizes, formed by mixing corrosion. The size of these forms increases towards the endpoint of the cave (Polacsek and Ba 2018b). On the upper level, they can be the size of a fist, deeper, they can be up to 3 m in diameter by the merging of several solution pockets (Sivó 2005). Its morphological curiosity is the so-called Graboplast wall, which is located in the Végponti Room, a fragment of the wall laced with 1–2 mm dissolved holes (Gönczöl 1986).

The cave consists of six bigger and smaller halls and the sometimes very narrow corridors connecting them. Its temperature is relatively constant, 10.1° C, regardless of the year or season. Its air is dry, the value of humidity fluctuates around 80 vol% (Sivó 2005).

As a form of speleothems, small drip-stones can be mentioned which were formed in the deeper parts of the cave. Its archaeological significance is due to the more than 4,000-year-old archeological heritage containing various objects for personal use in it (Sivó 2005).

3.7 Hévíz Spring Cave

The Hévíz spring cave (Table 7) is located on the SW edge of the Bakony Region, west from the Keszthely Mountains in the area of the town of Hévíz. The lake itself was already known in Roman times for its warm water and healing effect. Its use by the Romans is proved by Roman coins found in the water (Kovács 2003). The entrance to the spring cave of Lake Hévíz was discovered by divers (István Plózer and György Kovács) in 1972 (Kordos 1984) approximately at a depth of 40 m.

Lake Hévíz is actually the spring funnel of the Hévíz spring cave. According to Plózer (1974), the spring funnel was formed by thermal water upwelled from the Triassic main dolomite in such a way, that the upwelling thermal water made the sandstone collapse below the peat layer forming the spring funnel with a diameter of 70 × 90 m, according to Sásdi (2020). At the end of this spring funnel, approximately at a depth of 38 m, there is the corridor of the spring cave, as well as the spherical cavity itself. The bottom of the Amphora spring room which is 17 m in diameter and 14 m high, is covered with thick, clayey mud, and the vault is made of Pannonian sandstone (the bedrock is probably Triassic dolomite) (Plózer 1977). The spherical cavity is divided into two parts by a mud saddle in the middle. The depth and water temperature of the two parts are different. Cold (17.2 °C) water temperatures were measured in the eastern (shallower depth) sub-basin and warm (40 °C) water temperatures were measured in the western (deeper depth) sub-basin. As the amount of cold water is only a negligible part of the yield, the average water temperature typical of the whole spherical cavity is 38–39 °C (Plózer 1977; Kovács 2003).

The only significant (5–6 kg) speleothem of the spring cave is the marcasite blocks at the bottom of the western (warmer) part of the spherical cavity. Its fauna is diverse: bacteria, blue-green algae, sponges. Presumably, the presence of these species contributes to the healing effect of Lake Hévíz (Plózer 1977; Kordos 1984; Kovács 2003).

Due to its hydrological and geological significance, the cave has been highly protected since 1982. Its a highly frequented place by tourists.

Table 7 Data of the Hévíz Spring Cave

General information	
Cadastral reference	4440-1
Region	Keszthely Mountains
County	Zala
Municipality	Hévíz
Genetics	Spring cave influenced by thermal water effect
Length	27 m
Vertical extent	17 m
Depth	12 m
Protection	Highly protected

3.8 Likas-kő Cave of Hódos Stream

The Likas-kő Cave of Hódos Stream (through cave of Hódos Stream) is located in the Northern Bakony, in the northern part of Dörgő Hill, between the Hódos River and the Hálóeresztő Trench, at an altitude of 300–320 m, at a limestone height (cliff). Its earliest mention in literature dates from 1864 (Pesty 1864). The Likas-kő Cave of Hódos Stream is not worth mentioning because of its size, but because of the uniqueness of its development (Table 8).

The cave is tubular, 15 m long, approximately 2 m wide and 1.5 m high corridor with a NE-S direction which goes

Table 8 Data of the Likas-kő Cave of Hódos Stream

General information	
Cadastral number	4413-8
Region	Northern Bakony
County	Győr-Moson-Sopron
Municipality	Bakonyszentlászló
Genetics	Opening up origin in the inner part of the mountain
Length	36 m
Vertical extent	2 m
Depth	0 m
Protection	Protected

Fig. 6 The entrances to the Likas-kő Cave of Hódos Stream

through the cliff and has several entrances at both ends (Fig. 6). There are rooms that have lost their vaults above the cave. Solution pockets and chimneys formed by mixing corrosion are on the walls of the cave and on the walls of the cliff hosting the cave.

The main mass of Dörgő Hill is made up of Norian Dachstein limestone and dolomite. Several small Eocene limestone cliffs emerge from this horst. The Likas-kő Cave of Hódos Stream was also formed in such a gravel-covered cliff in a phreatic environment since karst water level was formed above the marly settlement of the Eocene limestone (Fig. 1). Its water dissolved the cavities, each of which, or parts of it, can be seen today (Veress and Futó 1987). The cliff lost its gravel cover (only its fragments remained), so the previously covered cavities were exhumated and started to perish (Veress and Futó 1987).

Concerning its genetics, the Likas-kő Cave of Hódos Stream is a cave with an opening up origin in the inner part of the mountain. The morphological evidence of its opening up origin is that the cave branches out at its entrances and in the direction of its entrances (Veress and Futó 1987; Veress and Vetési-Foith 2019a). During the denudation of the cliff rock, some cavities lost their cavity nature by losing their vaults, while the cave, ruined by partial denudation, forms a remnant, as do several open caves in the Bakony Region (Veress 1980a, b, 2000; Veress and Vetési-Foith 2019b). Above the cave, there are several hall-like or branching rock-corridors, which before the loss of their vaults were cave halls and cave corridors.

3.9 Lóczy Cave

The Lóczy Cave is located on the western side of Tamás Hill, which is on the northern border of the town of Balatonfüred. The cave was accidentally discovered in October 1882 by quarrying in the area (Kordos 1984). The greatest length of the cave known today is 154 m (Table 9).

Table 9 Data of the Lóczy Cave

General information	
Cadastral number	4462-1
Region	Balaton Basin/Balaton Riviera
County	Veszprém
Municipality	Balatonfüred
Genetics	Hydrothermal origin
Length	154 m
Vertical extent	21 m
Depth	15 m
Protection	Highly protected

The upper part of the cave is on the boundary of Middle Triassic, bluish-grey Füred Limestone and Veszprém marl, the lower part was formed in Middle Triassic marl (Leél-Őssy 1958). The Lóczy Cave is a tectonically preformed hydrothermal cave (Láng 1958; Leél-Őssy 1958).

It can be divided into two parts based on the plan (Szabó 1998) and the morphology of the cave. The first part runs from the entrance of the cave to the beginning of the Lejtős-folyosó branch. There are wider branches in this part, as well as the three halls of the cave. Solution pockets are typical here (Fig. 7), the walls are covered with black, crystalline manganese-containing calcite crust and anthodites. Its hydrothermal origin has been explained by these speleothems, the labyrinthine plan of the cave and the presence of solution pockets (Leél-Őssy 1958). The second part of the cave is the Lejtős-folyosó branch, which is an approximately 40-m-long narrow branch with a slope angle of 20–30°, where aragonite crystals also occur (Hazslinszky 2003). At its end point, karst water sometimes appears according to fluctuations in its level. Due to poor ventilation, carbon dioxide gas accumulates here.

The Lóczy Cave is highly protected since 1982 due to its unique genetics and morphological values. Currently, it is a show cave with equipped electric lighting, and can be visited with a guide.

3.10 Pula Basalt Cave

The Pula Basalt Cave is located in the southwestern foreground of Kab Hill in the Southern Bakony. The 151 m long and 22 m deep cave formed in basalt (Table 10) is not only the Bakony Region's largest basalt cave, but it is also the largest basalt cave in Hungary. The cave was discovered by László Markó in 1955, the first literary data also came from him (Markó 1960).

Fig. 7 Solution pockets in the Lóczy Cave (photo by Csaba Egri)

Table 10 Data of the Pula Basalt Cave

General information	
Cadastral number	4430-11
Region	Southern Bakony
County	Veszpém
Municipality	Pula
Genetics	Non-karstic cave
Length	150 m
Vertical extent	21 m
Depth	21 m
Protection	Distinctly protected

Material loss was formed during karstic dissolution in the limestone beds under the Kab Mountain's basalt. This material loss was incised to the basalt above it (Eszterhás 2013).

The cave consists of a large room and several passages branching out from it. This 15 m-long, 8 m-wide room is divided into three levels by two human-sized stairs. On the vault, there is a plate-separated fibrous basalt, in some places in the form of rather large plate scales. Smaller pieces of these scales can be found at the bottom of the three levels of the hall (Eszterhás 1986).

The cave is poor in concretions. In its upper zone, 1–2 mm calcite crystals crust the wall, and one of the walls of the only larger room in the cave is covered with a snow-white cotton-like coating, which according to studies is a collection of thin aragonite yarns and fungal yarns (Eszterhás 1985, 1986).

3.11 Nagy Sárkány Cave on the Szent György Hill

The Nagy Sárkány Cave on the Szent György Hill (Table 11) is called Dragon cave in the vernacular because according to folklore, it was once inhabited by a dragon, which St. George killed himself.

The Nagy-Sárkány Cave was formed during mass movement, i.e. in terms of its genetics, it is a talus cave of accumulation origin (Eszterhás 2000b, 2013). Talus caves form in the debris slope below the basalt edges. The most significant talus caves in the Bakony Region were formed in the basalt of the Szent György Hill.

The Nagy Sárkány Cave was mentioned as early as the 18th century, and the excavation and research of the cave was carried out in the 1990s (Eszterhás 1994, 1998; John and Sikos 1998).

The Nagy Sárkány Cave on the Szent György Hill is unique due to its climate. The cave air in the summer months is just over 0 °C. Cold air flows in through the gaps between the rock blocks in winter, causing the wall of the rock that forms the cave to cool.

Condensed water freezes on the cold surface of the rock, from which ice formations form. These formations do not melt in summer either, as the cooling effect of the cold air trapped in the cave in winter still prevails (Eszterhás 2000b). Based on climatological measurements, the Nagy Sárkány Cave on the Szent György Hill is a dynamic ice cave (Eszterhás 2000b, 2002), in which ice formations also appear in the summer period (ice crust on the wall).

Table 11 Data of the Nagy Sárkány Cave

General information	
Cadastral numbers	4450-7
Region	Balaton Basin/Balaton Riviera
County	Veszprém
Municipality	Kisapáti
Genetics	Non-karstic cave
Length	32 m
Vertical extent	10 m
Depth	10 m
Protection	Protected

3.12 Ördög-lik Cave of the Sűrű Mountain

The Ördög-lik Cave of the Sűrű Mountain (Table 12) is in the Ördög Trench in the Northern Bakony. The cave, which opens at high relative heights, has long been mentioned in the literature (Rómer 1860). Like the other caves of the Ördög Trench (Veress 1980b), the Ördög-lik Cave of the Sűrű Mountain was formed by opening up (Veress 1981) from one of the phreatic cavities of the karst water level formed in the Eocene limestone (Fig. 1).

The cave has two levels (Kassai 1963): the upper level developed on the border of the Middle Eocene nummulite limestone (Szőc Formation) and the Dachstein-type limestone, while the lower level developed in the dolomite-limestone formation of the Fenyőfő section. Due to the presence of the rock boundary, its upper level dissolved along the rock boundary while its lower level rather along fractures (Veress et al. 1992).

Both levels are separated by halls and corridors connecting them. Both levels' characteristics is that the formed branches were aligned with the nearly perpendicular fractures, so tectonic preformation played a major role in the formation of the cave. The upper level, with its passages wider and poorer in debris, is richer in solution pockets and various limestone concretions. According to Veress et al. (1992), a part of the cave's upper level was formed by the concrescence of solution pockets.

Concretions can be found only rarely in the cave, in some places only drip-stones with rudimentary development are typical (Kassai 1963).

3.13 Tapolca Cave System

The Tapolca Cave System includes three caves, the Kórház Cave, the Károly Berger Cave and the Tapolca Lake Cave (Table 13). Due to its importance for tourism, the latter is perhaps the best known, with 150,000 visitors a year (Szilaj and Leél-Őssy 2011a, b).

Table 12 Data of the Ördög-lik Cave of the Sűrű Mountain

General information	
Cadastral number	4423-74
Region	Northern Bakony
County	Veszprém
Municipality	Dudar
Genetics	Opening up origin
Length	150 m
Vertical extent	18 m
Depth	11 m
Protection	Distinctly protected

Table 13 Data of the Tapolca Cave System

General information	
Cadastral number	4450-1
Region	Balaton Basin/Tapolca Basin
County	Veszprém
Municipality	Tapolca
Genetics	Effluent cave
Length	14,780 m
Vertical extent	34 m
Depth	34 m
Protection	Highly protected

As part of the Tapolca Karst, the cave system is located below the town of Tapolca at a depth of 12–15 m. The Tapolca karst is one of the micro-regions of the Balaton Basin. The basin-like micro-region, consisting mainly of Sarmatian limestone (Tinnye Formation), is located between the Keszthely Mountains and the Balaton Uplands.

The host rock of the Tapolca Cave System is Sarmatian limestone. The cave system developed in the phreatic zone on the margin of the mountain, with a hydrothermal influence, thus, the cave system is a spring cave that has been affected by thermal water. This is also proved by the fact that its branches developed almost horizontally.

Between 3 and 8 m below the surface, there is a bed of marl, which has a water-barrier property, so the entry of the precipitation into the cave system is prevented. Because of this, precipitation has no or subordinate importance in the development of the caves (Kolláth 2003a, b), but the karst water in the mountains has, which partly occurs at the spring of the nearby Malom Lake, with a water flow of 22 thousand l/min.

The Tapolca Tavas Cave was discovered in January 1902 during the excavation of a well at a depth of 14 m.

In 1925, during the construction of the lung medicine pavilion belonging to the hospital, another cave was found. It was named Kórház Cave, which at that time was not yet connected to the Tavas Cave. In 1937–38, the cave was completely excavated and turned into a shelter. The Kórház Cave soon became known for the healing effect of its air. This is the only therapeutic cave in Hungary that can be operated with a direct hospital background. Its healing effect is due to the constantly pleasant temperature, the relative humidity of almost 100 vol%, the absence of dust, pollen and spores and the special ionic composition of the cave air (Kolláth 1991, 1995). In 1981, it was declared a healing cave by the National Directorate of Spas.

The third part of the Tapolca cave system, the Berger Károly Cave, was found in 2002 during the preparation for a hotel construction. Prior to 2010, it was known to be about 900 m long, but these branches forming only the upper level of the cave. In 2010, the cavers of Tapolca reached the lower level of the cave which is at least 4 km long. The Berger Károly Cave was connected to the two caves mentioned above on 19 December 2012, thus forming the Tapolca Cave System.

The formation of the cave system's branches was influenced by both fractures and stratification. The direction of its branches matches the direction of the fractures of the host area (NE-SW and perpendicular to it). The position and shape of the branches were also influenced by the beds that included the cave system. Where there were well-crystallized beds, there was the most significant effect of mixing corrosion. Almost exclusively dissolved, tubular branches with a diameter of 1–1.5 m can be observed in these places. Where these passages came close enough to each other, the vault collapsed along the slip plane formed along the clay strips. During such collapses, vaulted rooms were formed, which were further formed by dissolution (Kolláth 2003a, b).

Among the caves of the cave system, the Berger Károly Cave developed in two, the Tavas Cave and the Kórház Cave in three levels. Due to its higher position, in the Kórház Cave and in the Berger Károly Cave we can only observe water on the lower level, which in some places accumulates in underground lakes. Here are the two most beautiful natural cave lakes of Hungary with the largest diameter (Fig. 8). In the Tavas Cave, only the upper level is located above the karst water level, some parts of the second and third levels are under water.

The temperature of the air in the caves and its fluctuations vary from level to level. In the dry sections on the upper levels, the air temperature is the lowest and the most variable (10.4–17.6 °C). The deeper sections (thus, containing water) show the balancing and heating effect of the lukewarm water formed from the mixture of cold and warm water. The air temperature in these places is between 17.6 and 20.2 °C (Kordos 1984). Where there is a water surface, the relative humidity of the system reaches 100 vol%.

The members of the Tapolca Cave System (although thermal water is also participated in their formation) are relatively poor in hot water concretions. This is probably due to the fact that the thermal water is involved in the formation was mixed with cold karst water (Sásdi 2020) and not alone formed the cavities. There are thermal water concretions here and there, which can be of two types in the case of the Tavas Cave. On the one hand, there are calcite crystals on the walls, and on the other hand, aragonite concretions on the edges of the chimneys and fissures. Dripstones are not typical in this cave. However, they can be mentioned from the Kórház Cave, together with calcified root nodules and limonite concretions (Kolláth 2003a, b). In the Berger Károly Cave, a columnar botryoid and logomite are characteristic.

The Tapolca Karst is one of the most sensitive karst areas in Hungary. Anthropogenic activities and the effects that result from them are manifold and concentrated, as a town lives its daily life above the cave system. The main reasons for its vulnerability are the lack of regulations or their

Fig. 8 Underground lake in the Berger Károly Cave (photo by Csaba Egri)

circumvention, as well as the growing tourism (Móga et al. 2011).

3.14 Tihany Spring Cave

The Tihany Forrás Cave (Table 14) is located on the Tihany Peninsula, in the interior of the village of Tihany, opposite the main entrance of the Tihany Abbey.

The Tihany Spring Cave is unique in its formation and by its enclosing rock. The caves in the area of the peninsula were formed in sinter cones. Their formation can be related to the thermal spring activity caused by the post-volcanic operation following the late Pliocene basalt volcanism. According to the latest studies (Veress et al. 2015), the material of the sinter cones concreted from the thermal karst water in the magma chamber of a Pliocene basalt volcano. From the spouting thermal springs, layered, steep sinter cones of different sizes, were formed. These consist of lime and hydroquartzite (Hoffer 1943) assemblages. These sinter cones include cavities and caves. These were previously thought to be the channels of former geysers (Lóczy 1913). Pávai Vajna (1931) and Hoffer (1943) thought that due to the solution forms on the walls of the caves and the top-closing nature of the cavities these cavities were subsequently dissolved in their previously deposited sediments by hot waters erupting from the depths.

The cavities in the geyser cones of the Tihany Peninsula were already known at the beginning of the twentieth century, but this cave was discovered only in 1951. The interior of the cave bears the marks of artificial enlargements. For this reason, it is conceivable that the cave was already under utilization in the Middle Ages (Takácsné Bolner 2003a).

The Tihany Spring Cave consists of two interconnected small halls. The first room is 5 m long and 4 m wide, the second room is also 5 m long, but only 2.5 m wide. The rooms are both 2 m high. These two larger rooms are naturally dissolved cavities that were later artificially interconnected. There are opal concretes on the walls, and small, rounded solution forms along the fractures on the cave's vault.

Due to its unique genetics, it was declared a highly protected cave in 1982 as the most significant representative of the sinter cone cavities.

Table 14 Data of the Tihany Spring Cave

General information	
Cadastral number	4463-1
Region	Balaton Basin/Tapolca Basin
County	Veszprém
Municipality	Tihany
Genetics	Hydrothermal origin
Length	14 m
Vertical extent	4.5 m
Depth	1 m
Protection	Highly protected

3.15 Vértessomló Cave

The Vértessomló Cave is located next to the village of Vértessomló, on the side of the Nagy-Somlyó Hill, which is almost insularly separated from the main mass of the Vértes Mountains. The cave was discovered at the end of 2000, and after only a few months it was declared a highly protected cave due to its mineral specialty. With a length of 180 m, it is the longest cave in the Vértes Mountains (Table 15).

The Vértessomló Cave was formed in Dachstein limestone (Takácsné Bolner 2003b; Kocsis 2006; Polacsek and Ba 2018a). In terms of its genetics, it is a spring cave which was formed with hydrothermal influence. The evidence of its hot water influenced spring cave origin is its low altitude, the fact that it is at the margin of the mountains and that it is formed in the phreatic zone, and on the walls there are solution pockets which were formed by mixing corrosion. On the walls of the cave, there are barite crystals here and there in large size. The latter proves the former presence of thermal water. According to Kocsis (2006), these barite concretions are post-genetic, as they do not occur as lodes, as in the case of the typically thermal water caves of the Buda Hills. Tectonics also played a role in the formation of the cave (Polacsek and Ba 2018a).

The cave consists of two large halls and the narrow branches connecting them (Kocsis 2006). Based on its morphological features, the cave can be divided into two parts, a near-surface part and a deeper part. The upper, smaller hall is poorer in dissolution and concretions due to subsequent processes. Small dripstones and perpendicular (boxwork structure) prepared calciters adorn the walls

Table 15 Data of the Vértessomló Cave

General information	
Cadastral number	4522-1
Region	Vértes Mountains
County	Komárom-Esztergom
Municipality	Vértessomló
Genetics	Effluent cave influenced by thermal water effect
Length	180 m
Vertical extent	36 m
Depth	36 m
Protection	Highly protected

(Takácsné Bolner 2003b; Polacsek and Ba 2018a). Forms of dissolution and concretions suggestive of thermal karst genetics will become increasingly common in the lower room. The walls are characterized by solution pockets, bag-like side passages, and on the vault there is a towering chimney (Takácsné Bolner 2003b). Its concretions are botryoids and aragonite formations, as well as large barite deposits, to which the cave owes its reputation. Large amount of limonite was also formed in the deepest part of the cave. Although dripstones are less frequent, some dripstone runoff and stalactites can be found in the lower hall (Polacsek and Ba 2018a).

Near the entrance to the cave and the chimney mentioned above, the air movement is significant, but in the rest of the cave not. The carbon dioxide content of the air gradually increases towards the lower room, reaching 4 vol% at the lowest point (Polacsek and Ba 2018a). In addition to its mineralogical and genetic curiosities, the Vértessomló Cave is a significant bat habitat.

4 Conclusions

The southwestern part of the Transdanubian Mountains, the Bakony Region and the Vértes Mountains are rich in caves. The density of caves is higher in the Vértes Mountains (0.36 caves/km^2, while in the Bakony Region this value is 0.17 caves/km^2). The extent and genetic diversity is greater by the caves of the Bakony Region.

The caves of the Vértes Mountains are mostly small, the longest of them is the VértessomlóCave, but it is only 180 m long. The reason for this is that the main mass of the mountains consists of Triassic, hardly soluble main dolomite, on which younger, more easily karstifiable limestones are settled only in small thickness and locally. The caves of the Vértes Mountains were mostly formed in the main dolomite, at the margin of the mountain, and in the phreatic zone. In addition to cold karst water, thermal water also played a role in their development.

The main constituting rock of the Bakony Region is also the Triassic Main Dolomite, but Dachstein and younger well-soluble limestones are settled on it. Due to the basaltic volcanism in the Pliocene, the Southern Bakony and the Balaton Uplands are characterized by basalt caps and pseudokarstic caves developed in them. In this chapter, we presented the Nagy Sárkány Cave on the Szent György Hill and the Pula Basalt Cave. The karstic caves of the Bakony Region can be cold and hot water caves. The latter occur subordinately, mainly in the micro-region group of the Keszthely Mountains, in the Balaton Basin and the Balaton Uplands, but one cave of the Vértes Mountains is also mentioned. Such are the Cserszegtomaj Well Cave, the Döme Cave, the Gánt Cave, the Hévíz Spring Cave, the Lóczy Cave, the Tapolca Cave System, the Tihany Spring Cave, and the Vértessomló Cave. Cold water caves developed in vadose and phreatic environments. The latter may have formed in a marginal (effluent cave) or in an interior position. The most significant of the effluent caves in the Bakony Region is the Tapolca Cave System consisting of three caves, with a total length of almost 15 km. The interior caves of the phreatic zone were formed in the main dolomite in the level of the main karst water or in the former karst water levels formed in the Jurassic, Cretaceous and Eocene limestones settled on the main dolomite. Leaking water from watercourses of antecedent valleys or sections of valleys caused cavities where there was an impermeable intercalation below the valley or the section of the valley. Due to the continuous erosion of the watercourse, the valley sooner or later crossed the phreatic cavities, so that they opened up. Therefore, this type is called opening up cave. The caves presented are the Likas-kő Cave of Hódos Stream and the Ördög-lik Cave of the Sűrű Mountain. The caves (shafts) in the vadose zone were formed in places where there is water-permeable superficial deposit on the carbonate rock. Covered karstic depressions formed on the superficial deposit concentrate the surface runoff, thus promoting the formation of the shaft. Among those presented in the chapter are the Alba Regia Cave and the Csengő Shaft.

References

Acheron és Labirint Barlangkutató Csoport (1992) A Csodabogyós-barlang hossz-szelvénye (Longitudinal profile of the Csodabogyós Cave). Annual report of the Acheron Barlangkutató Szakosztály, manuscript, MKBT 14 pp (in Hungarian)

Bán B (2001) A Döme-barlang hossz-szelvénye (Longitudinal profile of the Döme Cave). Almanac of the Styx Barlangkutató Csoport, manuscript 59 pp (in Hungarian)

Bán B, Bán T, John Sz, Kocsis Zs, Sikos Z (1999) Döme-barlang (Döme Cave). Almanac of the Styx Barlangkutató Csoport, manuscript 5–20 (in Hungarian)

Bél M (1737) Notitia geographico-historica Comitates Vesprimiensis, Simeghiensis et Szalaiensis. manuscript in the OSZK, Budapest

Bertalan K (1938) A Bakony-hegység barlangjai (Caves of the Bakony Mountains). Turisták Lapja 50:154–208 (in Hungarian)

Bertalan K (1955) Kiegészítés a bakonyi barlangok ismeretéhez (Amendment to the knowledge of the Bakony Mountains' caves). Földrajzi Értesítő 4:55–62 (in Hungarian)

Bertalan K (1958) Magyarország nem karsztos eredetű barlangjai (Non-karstic caves of Hungary). Karszt és Barlangkutatási Tájékoztató 6:3–21 (in Hungarian)

Eszterhás I (1976) Faunisztikai vizsgálatok az Alba Regia-barlangban (Faunistic studies in the Alba Regia Cave). Almanac of the Alba Regia Barlangkutató Csoport, manuscript 99–113 (in Hungarian)

Eszterhás I (1978) Az Alba Regia-barlang állatvilágának vizsgálata (Studying the fauna of the Alba Regia Cave). Almanac of the Alba Regia Barlangkutató Csoport, manuscript 52–60 (in Hungarian)

Eszterhás I (1980) Troglofaunisztikai vizsgálatok az Alba Regia barlangban (Troglofaunistic studies in the Alba Regia Cave). Almanac of the Alba Regia Barlangkutató Csoport, manuscript 66–71 (in Hungarian)

Eszterhás I (1981) A Burok-völgy karsztmonográfiája (Karstmonography of the Burok Valley). A Veszprém megyei Múzeumok Közleményei 16:15–30 (in Hungarian)
Eszterhás I (1983) Az Alba-Regia barlang, a Bakony legnagyobb ismert balrangja (The Alba Regia Cave, the biggest known cave in the Bakony Mountains). A Bakonyi Természettudományi Múzeumok Közleményei 2:7–28 (in Hungarian)
Eszterhás I (1984) Az Alba Regia-barlang állatvilága (Fauna of the Alba Regia Cave). Karszt és Barlang I:25–30 (in Hungarian)
Eszterhás I (1985) A Pulai-bazaltbarlangban és a környékén végzett kutatások (Studying the Pula Basalt Cave and its surroundings). Almanac of the Alba Regia Barlangkutató Csoport, manuscript, 45–55 (in Hungarian)
Eszterhás I (1986) A Pulai-bazaltbarlang és környéke (The Pula Basalt Cave and its surroundings). Karszt és Barlang I:23–32 (in Hungarian)
Eszterhás I (1994) Magyarország jégbarlangjai (Ice caves in Hungary). Lychnis 36–42 (in Hungarian)
Eszterhás I (1998) Sárkány-barlang 2. sz. szelelőlyuk (Sárkány Cave, air hole No. 2). Manuscript in the Almanac of the Vulkánszpeleológiai Kollektíva, MKBT (in Hungarian)
Eszterhás I (2000a) Gyökérsztalagmitek egy magyarországi mészkőbarlangban (Stalagmiform rootes in a limestone cave in Hungary). Karsztfejlődés 5:257–272 (in Hungarian)
Eszterhás I (2000b) Szent György-hegyi barlangkutatás (Speleology on the Szent György Mountain) Manuscript in the Almanac of the Vulkánszpeleológiai Kollektíva, MKBT, 29–73 (in Hungarian)
Eszterhás I (2002) A mérsékelt öv jégbarlangjai bazaltban (Ice caves of the temperate zone in basalt). Karsztfejlődés 7:259–267 (in Hungarian)
Eszterhás I (2013) Bazaltbarlangok a Bakonyban (Basalt caves in the Bakony Mountains Hungary). Karsztfejlődés 18:183–204 (in Hungarian)
Fehér K (2015) Vízminőségi vizsgálatok a Cserszegtomai-kútbarlangban (Studying the water quality in the Cserszegtomaj Well Cave). Report of the Labirint Karszt- és Barlangkutató Sport Egyesület, 5–8 (in Hungarian)
Gönczöl I (1986) A Vértes-hegység barlangkataszteri felmérése (Cave cadastral survey of the Vértes Mountains). Karszt és Barlang 2:115–118 (in Hungarian)
Gyalog L (2005) Magyarázó Magyarország fedett földtani térképéhez (Explanatory for the covered geological map of Hungary) Magyar Állami Földtani Intézet Budapest, 189 pp (in Hungarian)
Haas J (1993) A Budaörsi Dolomit Formáció, Fődolomit Formáció, Kösseni Formáció. In: Haas J (ed) Magyarország litosztratigráfiai alapegységei. Triász (Litostratigraphycal base units of Hungary, Triassic) Földtani Intézet kiadványa, pp 49–52; pp 77–81; pp 91–94 (in Hungarian)
Hazslinszky T (2003) Lóczy-barlang (Lóczy Cave). In: Székely K (ed) Magyarország fokozottan védett barlangjai (Highly protected caves in Hungary). Mezőgazda Kiadó, Budapest, pp 365–367 (in Hungarian)
Hegyi E, Koch Z (1983) A mezőgazdasági kemizálás és hatása a Tési-fennsíkon (Agricultural chemistry and its impact on the Tés Plateau) Almanac of the Alba Regia Barlangkutató Csoport, manuscript 112–120 (in Hungarian)
Hoffer A (1943) A Tihanyi-félsziget vulkáni képződményei (Volcanic phenomena of the Tihany Peninsula). Földtani Közlöny 73:375–429 (in Hungarian)
John Sz, Síkos Z (1998) A Szent György-hegyi Sárkány-barlang (jégbarlang) (The Sárkány Cave (ice cave) on the Szent György Mountain). MKBT Tájékoztató 3 pp (in Hungarian)
Kárpát J (1982) A Cserszegtomaji-kútbarlang (The Cserszegtomaj Well Cave). Karszt és Barlang I:35–40 (in Hungarian)
Kárpát J (1983a) Az Acheron-kútbarlang morfogenetikai jellemzői (Morphogenetic characteristics of the Acheron Well Cave). Manuscript in the Almanac of the Acheron csoport 32–37 (in Hungarian)
Kárpát J (1983b) Az Acheron-kútbarlang (The Acheron Well Cave). Karszt és Barlang I-II:25–28 (in Hungarian)
Kárpát J (1991) Cseppkőbarlang a fennsíkon (Dripstone cave on the plateau) – Búvár 6:32–33 (in Hungarian)
Kárpát J (2003a) Acheron-kútbarlang (Acheron Well Cave). In: Székely K (ed) Magyarország fokozottan védett barlangjai (Highly protected caves in Hungary). Mezőgazda Kiadó, Budapest, pp 361–362 (in Hungarian)
Kárpát J (2003b) Cserszegtomaji-kútbarlang (Cserszegtomaj Well Cave). In: Székely K (ed) Magyarország fokozottan védett barlangjai (Highly protected caves in Hungary). Mezőgazda Kiadó, Budapest, pp 363–365 (in Hungarian)
Kárpát J (2003c) Csodabogyós barlang (Csodabogyós Cave). In: Székely K (ed) Magyarország fokozottan védett barlangjai (Highly protected caves in Hungary). Mezőgazda Kiadó, Budapest, pp 375–378 (in Hungarian)
Kassai M (1963) A sűrűhegyi Ördög-lik új felmérése (New survey of the Ördög-lik Cave of the Sűrű Mountain). Karszt és Barlang 1:21–26 (in Hungarian)
Kocsis A (1975) A Vértes-hegység barlangjai (Caves in the Vértes Mountains). Magánkiadás 16 pp (in Hungarian)
Kocsis Á (2006) A legújabban megismert karsztobjektumok, barlangok a Vértesben (The most recently discovered karst objects and caves in the Vértes Mountains). Karszt és Barlang 1–2:57–61 (in Hungarian)
Kolláth J (1991) Tapolcai Tavasbarlang (Tapolca Lake Cave). Tájak – Korok – Múzeumok Könyvtára 401, 16 pp (in Hungarian)
Kolláth J (1995) A tapolcai Kórház-barlang klimatikus sajátosságai (Climatic features of the Kórház Cave in Tapolca). Környezeti Ártalmak és Légzőrendszer, V. Hévíz (in Hungarian)
Kolláth J (2003a) Tapolcai Kórház-barlang (Tapolca Hospital Cave). In: Székely K (ed) Magyarország fokozottan védett barlangjai (Highly protected caves in Hungary). Mezőgazda Kiadó, Budapest, pp 368–370 (in Hungarian)
Kolláth J (2003b) Tapolcai-tavasbarlang (Tapolca Lake Cave). In: Székely K (ed) Magyarország fokozottan védett barlangjai (Highly protected caves in Hungary). Mezőgazda Kiadó, Budapest, pp 370–373 (in Hungarian)
Kordos L (1984) Magyarország barlangjai (Caves in Hungary). Gondolat Könyvkiadó, Budapest (in Hungarian)
Kovács Gy (2003) Hévízi-forrásbarlang (Hévíz Spring Cave). In: Székely K (eds) Magyarország fokozottan védett barlangjai (Highly protected caves in Hungary). Mezőgazda Kiadó, Budapest, pp 380–382 (in Hungarian)
Kökény K, Vaskor J, Gyebnár J (1987) A Csengő-zsomboly hossz-szelvénye (Longitudinal profile of the Csengő Shaft). Almanac of the Alba Regia Barlangkutató Csoport, manuscript, p 21 (in Hungarian)
Láng S (1958) A Bakony geomorfológiai képe (Geomorphological view of the Bakony Mountains). Földrajzi Közlemények 6:325–343 (in Hungarian)
Leél-Őssy S (1953) A Cserszegtomaji-kútbarlang (The Cserszegtomaj Well Cave). Hidrológiai Közlöny 33:309–313 (in Hungarian)
Leél-Őssy S (1958) Karsztmorfológiai vizsgálatok a balatonfüredi Lóczy-barlangban és környékén (Karstmorphological investigations in the Lóczy Cave and its surroundings in Balatonfüred). Földrajzi Értesítő 7(3):379–381 (in Hungarian)

Lóczy L (1913) A Balaton környékének geológiai képződményei és ezeknek vidékek szerinti telepedése (The geological features of the surroundings of Lake Balaton and their regional distribution). A Balaton tudományos tanulmányozásának eredményei Budapest, p 617 (in Hungarian)

Markó L (1960) Beszámoló a veszprémi barlangkutató csoport 1954–59. évi munkájáról (Report of the Veszprém Cave Research Group on its work between 1954–59). Karszt- és Barlangkutatási Tájékoztató 323–325 (in Hungarian)

Móga J, Kiss K, Szabó M, Kériné Borsodi A, Kéri A, Mari L, Knáb M, Iván V (2011) Természeti és antropogén hatások vizsgálata a Tapolcai-karszt epikarsztos rendszerében (Investigation of natural and anthropogenic effects in the epikarstic system of the Tapolca karst). Karsztfejlődés 16:185–201 (in Hungarian)

National cave register database: http://www.termeszetvedelem.hu/index.php?pg=caves. Date of access 19 May 2020

Nyerges A (2003) Döme-barlang (Döme Cave). In: Székely K (ed) Magyarország fokozottan védett barlangjai (Highly protected caves in Hungary). Mezőgazda Kiadó, Budapest, pp 378–379 (in Hungarian)

Palmer A (2004) Patterns of caves. In: Gunn J (ed) Encyclopedia of cave and Karst science. Fitzroy Dearborn, New York, pp 573–575

Paulovics P (1998) Denevérfaunisztikai megfigyelések a Tési-fennsík barlangjainál 1993–1998 években (Bat faunistic observations in the caves of the Tés Plateau in 1993–1998). Almanac of the Alba Regia Barlangkutató Csoport, Országos Barlangnyilvántartás, manuscript 53–60 (in Hungarian)

Pávai Vajna F (1931) A forró oldatok és gőzök-gázok szerepe a barlangképződésnél (The role of hot solutions and vapors gases in cave formation). Hidrológiai Közlöny 10:115–122 (in Hungarian)

Pécsi M (1970) Problems of relief planation. Akadémiai Kiadó, Budapest, p 151

Pécsi M (1980) A Pannóniai-medence morfogenetikája (Morphogenetics of the Pannonian Basin). Földrajzi Értesítő 29:105–127 (in Hungarian)

Pécsi M (1991) Geomorfológia és domborzatminősítés (Geomorphology and topography classification) Földrajz Tudományi Kutató Intézet, Budapest, p 296 (in Hungarian)

Pesty F (1864) Magyarország helyneveinek összeírása (Census of Hungarian place names). Országos Széchenyi Könyvtár, manuscript (in Hungarian)

Plózer I (1974) A Hévízi-tó forráskráterének barlangjai (Caves of the spring crater of Lake Hévíz). Karszt és Barlang 2:73–78 (in Hungarian)

Plózer I (1977) A Hévízi-tó forrásbarlangjának feltárása (Exploration of the spring cave of Lake Hévíz). Karszt és Barlang 1–2:65–66 (in Hungarian)

Polacsek Zs (2014) Kutatási jelentés a Csodabogyós-barlang 2013. évi kutatásáról (Research report of the Csodabogyós Cave's research in 2013). A Magyar Állami Természetvédelem Hivatalos Honlapja, 3 pp (in Hungarian)

Polacsek Zs, Ba J (2018a) Tatabányai barlangkutatás (Cave research in Tatabánya). Barlangi kutatásvezetők válogatott kutatási jelentései, manuscript, 70–74 (in Hungarian)

Polacsek Zs, Ba J (2018b) Tatabányai barlangkutatás (Cave research in Tatabánya). Barlangi kutatásvezetők válogatott kutatási jelentései, manuscript, 101–103 (in Hungarian)

Rómer F (1860) A Bakony: természetrajzi és régészeti vázlat (The Bakony Mountains: natural history and archelogic scheme). Győr, p 216 (in Hungarian)

Sásdi L (2020) A Keszthelyi-hegység karsztja és fejlődéstörténete (The karst and development of the Keszthely Mountains). Karszt és Barlang 2015–16:33–52 (in Hungarian)

Sivó Zs (2005) A Gánti-barlang kutatásának 25 éve (25 years of the research of the Gánt Cave). Almanac of the Alba Regia Barlangkutató Csoport, manuscript 178–182 (in Hungarian)

Szabó Z (1998) A Lóczy-barlang alaprajza (Layout of the Lóczy Cave). Országos Barlangnyilvántartás (in Hungarian)

Szentes F (1946) A kénkovand előfordulásai a Keszthelyi-hegységben (Occurrences of pyrite in the Keszthely Mountains). Beszámoló a jövedéki mélykutatásokról 51–103 (in Hungarian)

Szilaj R, Sz L-Ő (2011a) Tapolca barlangjai és a Berger Károly-barlang (The caves of Tapolca and the Berger Károly Cave). Élet és Tudomány 66(3):80–82 (in Hunngarian)

Szilaj R, Sz L-Ő (2011b) Tapolca barlangjai és a Berger Károly-barlang (The caves of Tapolca and the Berger Károly Cave). Élet és Tudomány 66(4):112–115 (in Hungarian)

Szolga F (2003a) Alba Regia-barlang (Alba Regia Cave). In: Székely K (ed) Magyarország fokozottan védett barlangjai (Highly protected caves in Hungary). Mezőgazda Kiadó, Budapest, pp 344–348 (in Hungarian)

Szolga F (2003b) Csengő-zsomboly (Csengő Shaft). In: Székely K (ed) Magyarország fokozottan védett barlangjai (Highly protected caves in Hungary). Mezőgazda Kiadó, Budapest, pp 351–353 (in Hungarian)

Takácsné Bolner K (2003a) Tihanyi Forrás-barlang (Tihany Spring Cave) In: Székely K (ed) Magyarország fokozottan védett barlangjai (Highly protected caves in Hungary). Mezőgazda Kiadó, Budapest, pp 374–375 (in Hungarian)

Takácsné Bolner K (2003b) Vértessomlói-barlang (Vértessomló Cave). In: Székely K (ed) Magyarország fokozottan védett barlangjai (Highly protected caves in Hungary). Mezőgazda Kiadó, Budapest, pp 336–338 (in Hungarian)

Tóth L (1965) Újabb feltárások és megfigyelések a Cserszegtomaji-kútbarlangban és környékén (Recent excavations and observations in and around the Cserszegtomaj Well Cave). Karszt és Barlang 2:47–50 (in Hungarian)

Veress M (1980a) A Csesznek környéki völgyoldalak barlangtorzóinak vizsgálata (Investigation of cave torsos in the valley sides around Csesznek). Karszt és Barlang 2:65–70 (in Hungarian)

Veress M (1980b) Adatok a dudari Ördög-árok barlangjainak morfogenetikájához (Data on the morphogenetics of the caves in the Ördög Valley near Dudar)—A Veszprém Megyei Múzeumok Közleményei. Természettudomány 15:49–60 (in Hungarian)

Veress M (1981) A Csesznek környéki barlangok genetikájának vizsgálata (Genetical observation of the caves near Csesznek). A Bakony Természettudományi Vizsgálatának Eredményei, Zirc 14:63 (in Hungarian)

Veress M (2000) Covered karst evolution Northern Bakony Mountains, W-Hungary. A Bakony Természettud. Kut. Eredményei 23, Bakonyi Természettudományi Múzeum, Zirc, p 167

Veress M (2018) A Bakonyvidék genetikai aknatípusai (The genetic shaft types of the Bakony Region). Karsztfejlődés 23:55–84 (in Hungarian)

Veress M (2019) Shaft lengths and shaft development types in the Vadose zone of the Bakony Region (Transdanubian Mountains, Hungary). J Soil Water Sci 3(1):54–74

Veress M, Futó J (1987) Adatok a Hódos-éri Likas-kő morfogenetikájához (Data on the morphogenetics of the Likas-kő Cave of Hódos Stream). Karszt és Barlang 1–2:9–16 (in Hungarian)

Veress M, Gadányi P, Tóth G (2015) Thermal spring cones of Tihany Peninsula. In: Lóczy D (ed) Landscapes and landforms of Hungary. Springer, Heidelberg/New York/Dordrecht/London, pp 71–78

Veress M, Péntek K, Horváth ET (1992) Evolution of corrosion caves: Ördög-lik Cave, Bakony, Hungary. Cave Sci 19(2):41–50

Veress M, Vetési-Foith Sz (2019a) A Bakonyvidék barlangjainak genetikai típusai (The genetic shaft types of the caves of the Bakony Region). Karszt és Barlang under publication (in Hungarian)

Veress M, Sz V-F (2019b) A Bakonyvidék felszíni karsztja (The surface karst of the Bakony Region). Földrajzi Közlemények 143 (2):167–186 (in Hungarian)

Zentai F (1989) Védelmet kérünk az Alba Regia-barlangnak (We want protection for the Alba Regia Cave). Almanac of the Alba Regia Barlangkutató Csoport, manuscript 112–114 (in Hungarian)

Caves in the Mecsek Mountains

Károly Barta and Csaba Egri

Abstract

There are two karst regions in Southwest Hungary: the Mecsek and the Villány Mountains. The Mecsek is the bigger one with a strongly karstified limestone area. This karst area is similar to the Aggtelek karst and Bükk Mountains (see earlier), basically it is formed by infiltrating cold water and allogenic streams. Besides the rich surface karst forms there are several swallow holes, spring caves and chimneys up to 70 m. This chapter shows the most important caves in the Western Mecsek karst.

Keywords

Mecsek • Allogenic karst • Chimney • Swallow hole • Spring cave • Water supply

1 Introduction

In the Mecsek 220–230 caves have been registered (National Cave Registry, Hungary), more than half of which are 5–50 m deep chimneys. Their origin is due to infiltrating cold waters and most of them are obviously tectonically preformed, i.e. some of their details are aligned along tectonic cracks or their brecciated rocks point to tectonic stress. The number of accessible open chimneys changes dynamically since, following humid spells, through the reworking of the several metre deep loess cover by water and mass movements new chimneys open up and old ones are plugged. The process is intensified by the location of most of the chimneys at the bottom of doline floors, where the sedimentary cover is considerably thicker. The shaft diameters are around 1–2 m, larger sizes are rare (e.g. Büdöskút Chimney). In addition to solution residues and loess and clay washed in from cover deposits, local accumulations of red clay or the much more widespread red sandstone gravels (e.g. in the Achilles Swallow Cave) are worth mentioning as fills. Pliocene age is determined for the formation of red clays trapped in the sheltered cracks and corners of chimneys and larger caves (Sebe et al. 2008; Dezső et al. 2012). The presence of sandstone gravels in chimneys clearly evidences that this is a semi-allogenic covered karst, as mentioned above, and from the southern part of the area built up of sandstone large amounts of deposit had been previously transported onto the limestone surface only mantled with some metres of loess. This evolution is also attested, in addition to the sediment trapped in the cave, by sporadic remnants of sediment and doline fill. Along with the sandstone gravels found in the caves, the morphology of numerous caves also supports the previously allogenic character: now these caves are deepened into terrain without features pointing to linear surface watercourses opening in summit position or their remnants (e.g. doline rows), but still they are swallow holes with steep terraced slopes and erosional features (e.g. Pietró Cave, Vásárosút Swallow Cave, Fig. 1).

The deepest chimneys (Remény Chimney, Achilles Swallow Cave, Jószerencsét Shaft Cave) were explored as early as in the 1960s and 1970s. Each of them was deeper than 50 m, but to avoid collapse, narrowing and deposition

Translated by Dénes Lóczy

K. Barta (✉)
Department of Geoinformatics, Physical and Environmental Geography, University of Szeged, Egyetem u. 2., 6720 Szeged, Hungary
e-mail: barta@geo.u-szeged.hu

C. Egri
Department of National Parks and Landscape Protection, Deputy State Secretariat for Nature Conservation, Ministry of Agriculture, Apáczai Csere János u. 9., 1052 Budapest, Hungary
e-mail: csaba.egri@fm.gov.hu

M. Veress and Sz. Leél-Őssy (eds.), *Cave and Karst Systems of Hungary*,
Cave and Karst Systems of the World, https://doi.org/10.1007/978-3-030-92960-2_14

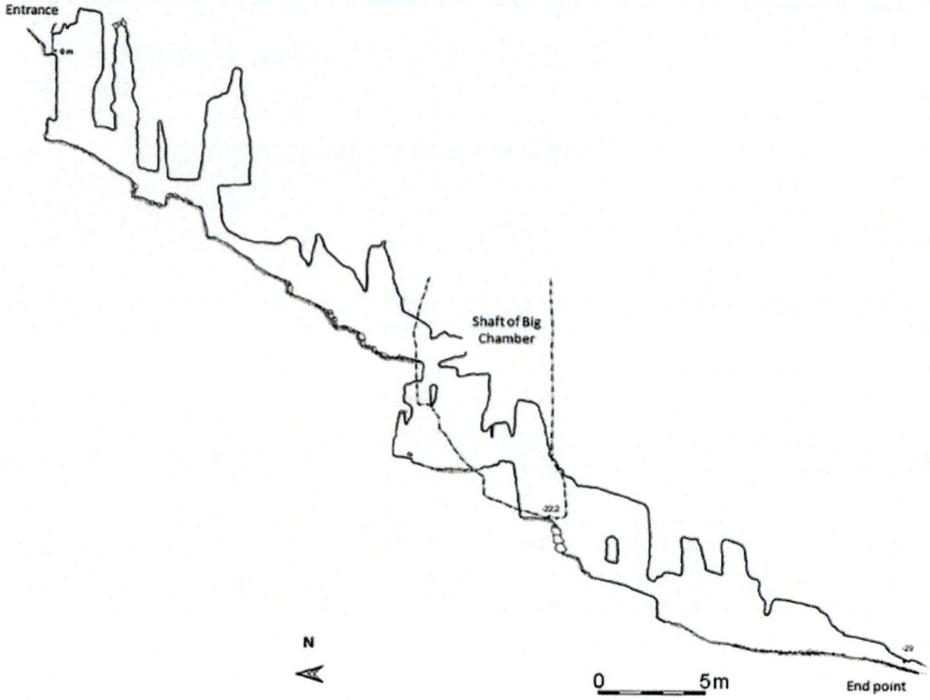

Fig. 1 Longitudinal section of the Vásárosút Swallow Hole (*source* National Cave Registry)

hazards their exploration was interrupted (Rónaki 1970; Miklós 1980).

Two other major groups of caves are associated with the presently active karst hydrological system: even today from the sandstone surface of the Jakab Hill in the south several permanent or intermittent watercourses arrive upon the northward sloping limestone surface, where they reach underground forming swallow caves and, re-emerging at the northern foothills, spring caves (Barta and Tarnai 1997, Fig. 2).

The joint number of swallow and spring caves is an order of magnitude lower than that of the chimneys, but because of their dimensions, richness of forms and utilization (particularly the spring caves) the former are of much greater importance.

The underdeveloped nature of the karst region and the swallow holes is attested by the general lack of counterslope steps beyond the points of bathycapture of streams, the position of swallow holes on valley floors and the further flow of streams in the valleys following higher discharges and floods (e.g. Büdöskút, Szuadó, Körtvélyes, Nyárás Valleys). There are 3–4 swallow holes in a row in the Szuadó and Büdöskút Valleys, while only two are known from the Körvélyes (Rónaki 1962b). An interesting exception is the Sózó Swallow Hole, which, in spite of its only minimal non-karstic catchment (0.15 km^2), is found in a blind valley which continues in a doline row beyond a 10–15-m step. Among the swallow caves of the Western Mecsek Karst even the larger ones begin with a narrow crawl at the entrance and follow with a (series of) shaft(s) and a wider corridor and, finally, narrow down to become impassable at the karst water level. A through cave passable from an active swallow hole to the spring has not been discovered yet in the Mecsek.

As it has been mentioned above, the major spring caves of the mountains are found in the northern foothills of the karst region, near Abaliget, Orfű and Mánfa. An exception is the Tettye Spring (to be described below). The longest cave of the mountains is the Abaliget Cave, a show cave of ca. 2,000 m length. All the other spring caves are passable at maximum some hundreds metres of length and all end in a siphon.

The caves of the Western Mecsek Karst are formed in Triassic limestones and dolomites, with exceptions of some minor sandstone hollows in Jakab Hill and some caves in the northern section of the mountains, which deepen into Miocene limestone or conglomerate (e.g. Törökpince, Füstös-lik) and the caves in tufa of Tettye Hill in Pécs (see below).

With regard to their significance, the introduction of the Mecsek caves starts with spring caves and then the major swallow holes will be described.

2 Abaliget Cave

The longest cave of the Mecsek winds underground for ca. 2,000 m length, including the main branch with stream and the side branches. For its origin it is a typical allogenic

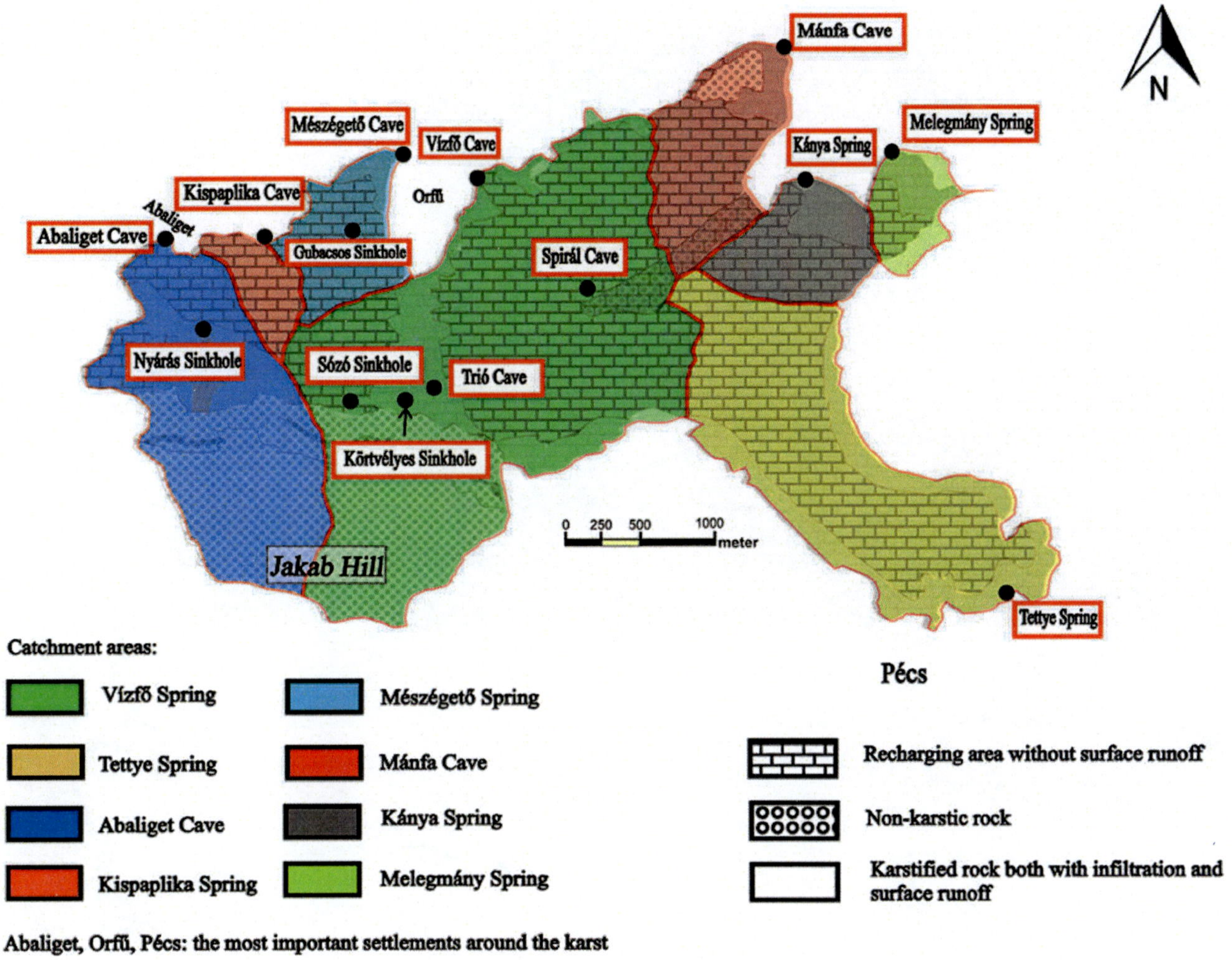

Fig. 2 Karst hydrological system of the Western Mecsek showing the major swallow holes and springs (modified after Barta et al. 2012)

stream cave (Jakucs 1977) with a catchment of 6.37 km^2 area (Dezső et al. 2012) the more elevated southern part of which is composed of non-karstic rocks, mostly sandstone, in ca. 2.4 km^2 area (Barta and Tarnai 1999). The high-discharge stream draining this terrain disappears in the Nyárás Valley sinkhole (Fig. 2). On the cave catchment no more active swallow holes are known, but there are some caves which belong to the hydrological system of the Abaliget Cave and used to drain undoubtedly larger areas, such as the Törökpince and the Akácos Swallow Hole, which have been detached and become inactive (Havasi et al. 2003). In addition, it is also interesting that, in contrast to the caves of Triassic limestone in the Western Mecsek Karst, both caves are formed in Miocene conglomerate or sandstone. The Akácos Swallow Hole is connected with the largest side branch of the Abaliget Cave, the branch West-2, while the hydrological links of the Törökpince are only confirmed by dye tracing.

Similarly to other caves in the Mecsek, the Abaliget Cave originated in Triassic (Anisian) limestone. It is remarkably performed tectonically and the main branch is aligned along a N-S faultline (Havasi et al. 2003). The typical cross section of the main passage is 2-m wide and 3-m high. The landforms are governed by an active allogenic cave: in addition to the meandering cave channel and erosional features, large amounts of collapsed material accumulated along some sections. The main branch is poor in dripstones (Fig. 3). More dripstones are found in the higher parts (e.g. in the Big Chamber) and in side branch West-2. The oldest speleothems were dated at 300,000–400,000 years (Dezső et al. 2012).

The entrance of the spring cave has been known since ancient times. Its first scientific description derives from Vince Kölesi from 1820. He published a map of the cave which suggests that he explored almost the whole length of the Main Branch of today. He describes wonderful

Fig. 3 Sketch map of the Abaliget Cave and detail of a typical passage (*source of map* National Cave Registry)

dripstones and traces of former human constructions which have since then completely disappeared. Although after Kölesi's activities numerous visits, documents and publications occurred but detailed explorations only happened in the twentieth century. Side branches were explored one after the other and, with the terminal collapsed material climbed, in 1954 cavers discovered the Big Chamber, which is the end point today (Havasi et al. 2003). Exploration in the 1960s led by Béla Vass concentrated on the study of further side branches and passing the siphon under the Big Chamber. Among the side branches, the best results were achieved in West-2: the former length of 70 m was multiplied tenfold and in 1969 the end point of the passage was reached from the direction of the Akácos Swallow Hole. The cavers managed to drain the siphon of the main branch and swim through it, but they could not proceed further because of the accumulation of large amounts of mud (Vass 2012).

As early as the second half of the nineteenth century, guided tours were organized and tickets were on sale for cave visits, but until the 1950s there also have been periods of total neglect at irregular intervals. Following infrastructural development, it became a show cave again in 1957. In 1996 the management of the cave was taken over by the Danube-Drava National Park Directorate. By the end of 2001 the infrastructure serving tourism, including lighting, had been modernized at the technological level of the age (web1). At present, the cave is visited by ca. 70,000 people annually. In the 1960s the beneficial balneological effects of cave air were recognized and its utilization for curing pulmonary diseases began. In 2000 the Abaliget Cave was officially declared a medicinal cave. In the 1960s and 70s regular micro climatological measurements also took place in the cave (Havasi et al. 2003).

Along with the extraordinarily diverse morphological, hydrological, archaeological and sedimentological research in the cave (e.g. Lovász 1971; Dezső et al. 2012), zoological investigations are also outstanding. The most eminent scientists, Lajos Méhely, Elemér Bokor, Endre Dudich and

Antal Gebhardt identified 190 animal species in the cave. The biggest scientific sensation was caused by the discovery of *Niphargus foreli gebhardti* and *Stenasellus hungaricus*. In the 1920s, together with the Baradla Cave, the Abaliget Cave was regarded the most explored cave in Hungary (Kordos 1984). Bats were observed by Vince Kölesi in the nineteenth century, but regular bat monitoring only started in the 1990s and revealed that the cave is frequented by 19 bat species for reproduction and wintering. It is the most important mating cave in South-Transdanubia (Kéki and Havasi 2010). In the last ten years Dorottya Angyal conducted faunistic research in the cave (Angyal and Korsós 2013; Angyal 2015).

Above the cave, on the cave catchment, the almost exclusive land use is forestry. Although formerly considerable uranium ore deep mining had been practised on Jakab Hill before this activity stopped in the early 2000s. Due to the closure of mines there are no severe threats to the cave, the sole natural hazard derives from the natural hydrological conditions of the cave: precipitation-induced floods or flash floods caused by rapid thunderstorms often make utilization for tourism impossible. A good example is provided by the extremely humid year of 2010 and in winter and spring 2018 the cave had to be closed for days on several occasions because of destructive floods. Then the long-term average discharge of 100–150 l/min can rise 100-fold, to 10–15 m^3/min, and pass along the main branch (Fig. 4).

Since 1982 the cave is strictly protected and since 2009 its entire catchment is part of the Western Mecsek Protected Landscape.

3 Orfű Vízfő Cave

The Vízfő Spring at the northern foot of the Mecsek is the most abundant karst spring in the mountains and outlet of a spring cave of ca. 400 m total length. The hydrological conditions of the spring are similar to those of the Abaliget Cave, it is fed by a typical allogenic stream. Cave catchment area is almost 16 km^2, out of which ca. 10.5 km^2 is built up of karstic rocks and 5.5 km^2 of non-karstic rocks, primarily sandstone. Thus, the heavily fluctuating discharge of the spring partly derives from stream and flood water carrying large amounts of gravelly sediments which disappear in the swallow holes on the northern slopes of Jakab Hill and partly from corroding seepage water. The maximum yield of the spring measured to date exceeded 2.5 m^3/s (Berényi Üveges 2012) and annually can even amount to several millions of cubic metres. The movement of both karstwater types and, thus, the pattern of cave passages are basically controlled by the junction of tectonic lines. The main strikes of cave passages are N to S and E to W. In spite of the allogenic catchment of remarkable size non-karstic sediment is only present as sand and silt fractions in the cave (Rónaki 1962a; Bencsik et al. 2015). This indicates that in the internal sections of the cave not yet explored extensive settling basins, siphons, chambers may exist. The sediment transported and then deposited by the stream is mostly reworked eolian loess and quartz-based fine sand.

The remotest swallow hole which feeds the spring lies at more than 3 km distance from the cave (Fig. 2), and it is assumed that the length of the cave system beyond the spring

Fig. 4 The cave entrance at high water stage and the cave passages under a flash flood in 2010 (*source of photo* National Cave Registry)

may reach 5–6 km (Barta and Tarnai 1997). At the same time, based on tectonic investigations, in the area between swallows and the spring there is a zone where no well-developed water-conduits are expected (Bauer and Tóth 2011, 2015).

The cave itself took shape in Middle Triassic (Anisian) dolomite (Rónaki 1962a), but, according to other sources (Parrag 2003a, b), limestone also occurs in the inner parts. The natural landform assemblage of the cave is characterized by marked tectonic cracks, active passages with streams and siphons (Fig. 5).

Certain more elevated sections (e.g. the Cseppkőerdő) are particularly rich in dripstones (Fig. 6). A special feature of the cave is the iron-manganese crusting of up to 1 cm thickness which mantles the dripstones and the walls, widely spread on the Western Mecsek Karst but most typical here. Laboratory analyses confirmed the biogenic origin of the precipitations (Rónaki 1962a).

Since 1982 the cave is strictly protected and since 2009 its catchment is part of the Western Mecsek Protected Landscape (established by the ministerial decree 4/2009 (IV.10.)), the northern boundary of which runs close to the cave entrance.

The high discharge of the Orfű Vízfő Cave Spring has had century-long social and economic impacts. It seems very probable that the spring and the cave lying beyond it may have had an important role to play in the foundation of the Árpád-age monastery built close to the spring. In later times spring water was used both for industrial purposes (in the paper mill, grain mill and water retting of hemp) and as potable water (Parrag 1998). It should be remembered that the main touristic attraction of Orfű, the system of artificial lakes established in the 1960s, also owes its origin and good water quality to the abundant karst spring and, in the same way, many other touristic destinations (such as the Mill Museum) could not operate without karstwater from this spring.

Deeper exploration of the cave, which begins with the spring siphon, commenced under the leadership of Hubert Kessler in 1952. He intended to access the cave through

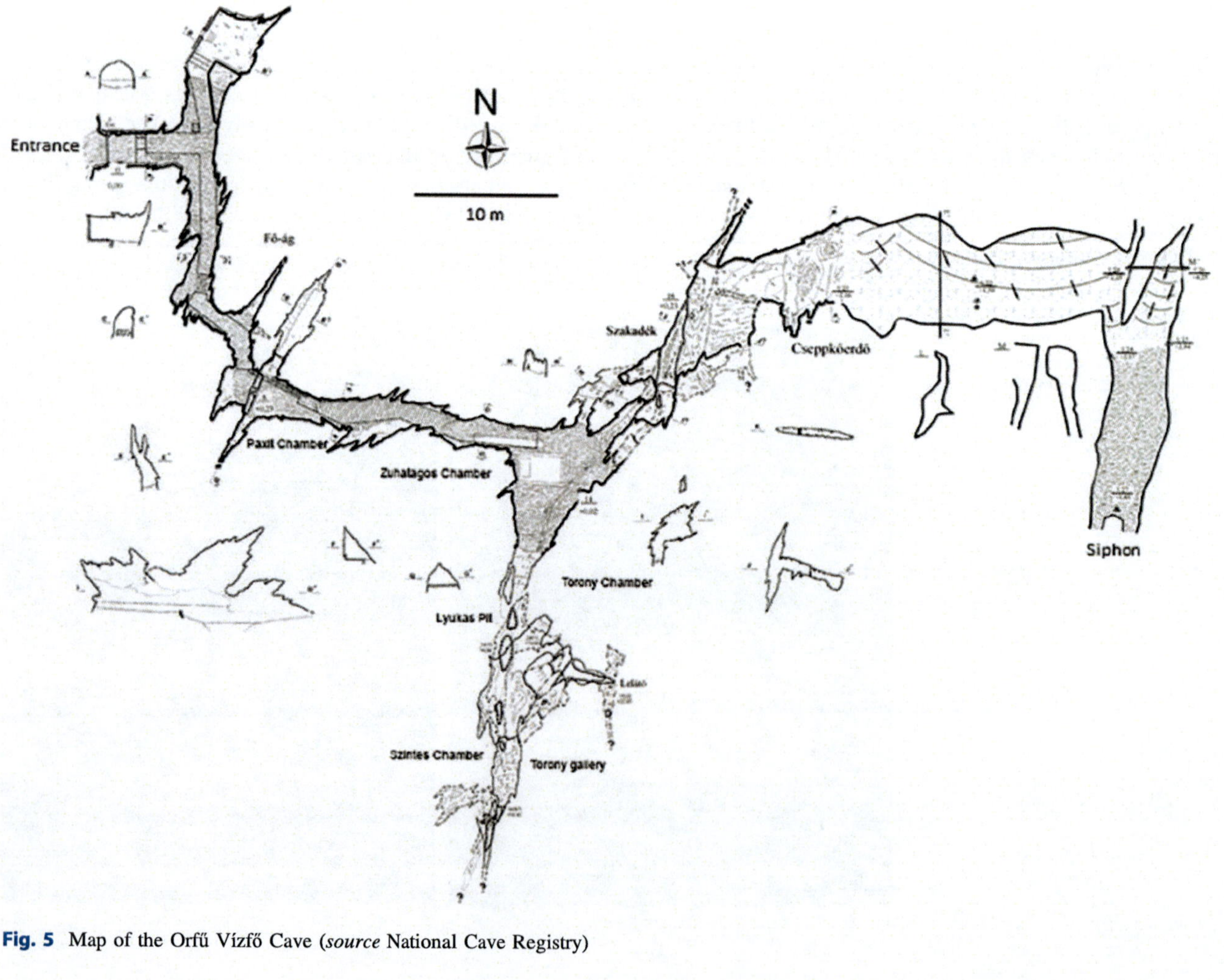

Fig. 5 Map of the Orfű Vízfő Cave (*source* National Cave Registry)

Fig. 6 Dripstones in the inner sections of the cave

lowering the water level of the entrance siphon. The high discharge and the technological limitations of the age, however, prevented him to realize his plans. In the late 1950s a great impetus was given to the exploration of the cave by the acute water shortage of Komló, a nearby coal-mining centre, and the plans to solve the drinking water supply of that town from the Vízfő Spring of Orfű. Swimming through the spring siphon was accomplished in October 1958 (Parrag 2003a, b; Rónaki 1970) and, thanks primarily to Béla Vass and László Rónaki the cave system as we know it today was opened by 1960. The first siphon was eliminated through a gallery above the water surface (Fig. 7) and the second was cut through by exploding 5 kg of Paxit. The third siphon marks the end point of the cave to this day. Swimming through the third siphon was attempted on several occasions, the most memorable trial being by Béla Vass and László Rónaki in 1966, when, using an AGA compressed air device, borrowed from Switzerland, they penetrated at 42 m distance and to 26 m depth. At that point, however, the passage narrowed down and the explorers could not reach further (Szabó 1961; Rónaki 1970; Vass 2012). The next attempt was made only in 2016, but it also failed.

The spring water was conducted to Komló in the 1970s, applying a technological solution unique at that time: instead of a pipeline to Komló, located in 9-km distance, an only 2.3-km long pressure pipe was built, designed by Béla Vass, which conducted water from the Vízfő, with the help of high-capacity pumps, into a doline (called Vízműnyelő, i.e. Waterworks swallow hole) on the catchment of the Mánfa Kőlyuk cave which was lined with concrete for this special purpose. From there the water flow followed a natural system of passages until the Kőlyuk, where, together with waters from the cave spring and surface waters in the surroundings, the water of the Vízfő Spring of Orfű was joined into the water supply of Komló. As a related investment, the water intake structure in front of the Orfű Vízfő Cave, the so-called Mushroom, now an industrial monument, was built and the cave itself was turned into waterworks (Fig. 8). The latter development included the closure of the cave, the construction of a concrete pavement, providing electricity and water pipe and the installation of other structures. With the great fluctuation of discharge, water extraction values also showed a wide range. The daily water transfer to the Kőlyuk of Mánfa amounted to 2,000 m^3 on the average (Havasi 2003), but its maximum reached 8,000 m^3/day. In order to make the utilizable water reserves more equable, the choice of impoundment above the spring level was also implemented at the Vízfő. In essence, in the internal part of spring cave foreseen for utilization, in an appropriate section, a concrete wall was raised to impound and retain the arriving discharge in the cavity system beyond. The wall is supplied with a water conduit and tap, which regulates water extraction. Thus, the surplus water from the wet spring and early summer period can be stored in the joint system of the hill for the drier summer. In the Mecsek, at the Tettye Spring and the Kőlyuk of Mánfa such impoundments are still active today but it failed in the case of the Orfű Vízfő Cave: during

Fig. 7 The cave entrance at low discharge (2008) and during flood in 2010

Fig. 8 The former waterworks, the Mushroom, now an industrial monument

test operation overpressure resulted in water outburst on two occasions and threatened the lower-lying parts of Orfű village. On the second occasion the outburst occurred at 2-bar pressure, which equals a 20-m high impounded water level (Vass 2012).

The waterworks were heavily criticized because of large-volume transformations: they were accused of the destruction of the natural appearance of the cave and its industry-like operation. Since in the 1980s water transfer to Komló stopped and by the turn of the millennia the spring water did not contribute to the water supply of Komló, the removable artificial elements (metal structures, reinforced concrete dam, pipes) were demolished in the cave in 2012–2013, the original siphon, which was subsequently plugged, was cleaned and a new cave closure, passable by bats, was established. Naturally, near the entrance the traces of passage broadening and explosions are still visible, but the internal parts were restored to close-to-natural conditions.

The cave equally provides habitat for aquatic and terrestrial organisms. Demeter (1994) documented a range of molluscs in its water, such as *Pupilla muscorum*, *Trichia hispida*, *Pisidium casertanum*. Unfortunately, according to the 2010–2013 survey by Dorottya Angyal covering 14 caves of the Mecsek, the Vízfő contains one of the invertebrate fauna poorest in species, among molluscs solely *Oxychilus draparnaudi* was described. Moreover, in her opinion, this species may be an indicator of negative human impacts (Angyal 2015). Bats regularly visit the cave, but, relative to its size, in small numbers. This means not more than 10–30 wintering specimens.

It has to be mentioned that the spring of the cave is a principal source of water for the Orfű lake system. Therefore, the protection of its water quality is of outstanding importance for both nature conservation and regional tourism. The cave catchment of almost 16 km^2 area is almost completely forested. A major threat is the Pécs-Abaliget public road, the route with the heaviest traffic in the karst region and the highest potential contamination risk. Within a very short time pollution of transport origin reaches the Vízfő Spring Cave from the road through dolines, chimneys and swallow holes which are only in some metres' distance from the road. Closing these caves and raising their entrance levels above the level of the road is important for the preservation of cave conditions and for the reduction of contamination of karstwater as well.

4 The Kőlyuk of Mánfa

The Kőlyuk of Mánfa (Mánfa Cave) is the easternmost of the major karst springs at the northern foot of the Western Mecsek Karst (Fig. 2). The natural entrance, once a spring, is located at the junction of two valleys, at the base of a steep cliff, opens at 225 m above sea level and has a length of 180 m. Since 1982 the cave is strictly protected and its catchment has also been under nature conservation for decades, since 2009 it is part of the Western Mecsek Protected Landscape (by the ministerial decree of 4/2009 (IV.10.)). Among the Mecsek caves, the Kőlyuk has the broadest entrance, ca. 6 m wide and 6.5 m high (Fig. 9).

The imposing entrance has been known since ancient times, but the first written mention is associated with the name of Miksa Hölbling, who visited the cave in 1844. The first serious interventions began in 1928, led by Gyula Ozanich, when both the lower passage leading to the Gyula Spring and the upper passage were broadened by explosions and large amounts (more than 100 m^3) rubble and sediment were removed from them (Havasi 2003). Because of the siphon near the entrance, the true exploration of the cave could only start in the 1950s. The first two siphons of the cave spring were first passed through in 1955. The water of the cave outlet, the Gyula Spring, was led into the system of the Town Waterworks of Komló as early as 1957 (Nagy 1975) and in the second half of the 1960s, when, to remedy the acute water shortage of the dynamically developing Komló, the cave was fundamentally transformed (Vass 2012; Berényi Üveges 2012).

Fig. 9 The entrance of the Kőlyuk of Mánfa

In its original state, it is a two-storey spring cave with a stream branch dissected by siphons at the upper level. Moving outwards, beyond siphon 1, the water disappeared through an impassable swallow hole to the lower passage. The section close to the entrance of the upper storey is of the nature of a broad and already inactive spring cave and the water burst out below, through the narrow lower passage with an active stream to the surface in the Gyula Spring. Today out of the mentioned sections only the outer inactive section of the upper branch was preserved in a close-to-natural state, while the other sections suffered transformations to the extent that the character of a natural cave was lost for good (Fig. 10). Because of utilization as waterworks from the east, from the direction of the Nagy-mély Valley, an artificial gallery was cut joining the main branch in the second half of the 1960s. The gallery reaches the main branch about halfway between the former siphon 1 and the entrance. The corridor in the continuation of the gallery, where it was necessary, was broadened to a size conveniently passable and the cave siphons were eliminated. As a consequence of utilization as waterworks, this section is rather similar to an artificial gallery than a natural cave passage. This character is supported by the concrete pavement and water pipeline and the abutment along some parts.

Today the principal function of the cave is the drinking water supply of Komló and the neighbouring settlements. It is to be emphasized that also here the impoundment above the spring level makes the widely fluctuating underground discharge more equable (Fig. 11). It is only of value for industrial history that in the 1970s and 1980s thousands of m^3 daily supplementary drinking water amount flowed into the cave from the Vízfő Spring through a 2.3-km-long pressure pipeline over 178 m relief into the „Vízműnyelő" (Waterworks swallow hole), from where it continued through an underground passage system to the Kőlyuk.

The rock enclosing the cave is Middle Triassic (Anisian) limestone (web2). Its catchment extends to 2.8 km^2, about a quarter of which is non-karstic terrain (Fig. 2). This non-karstic sandstone imbrication creates a particular

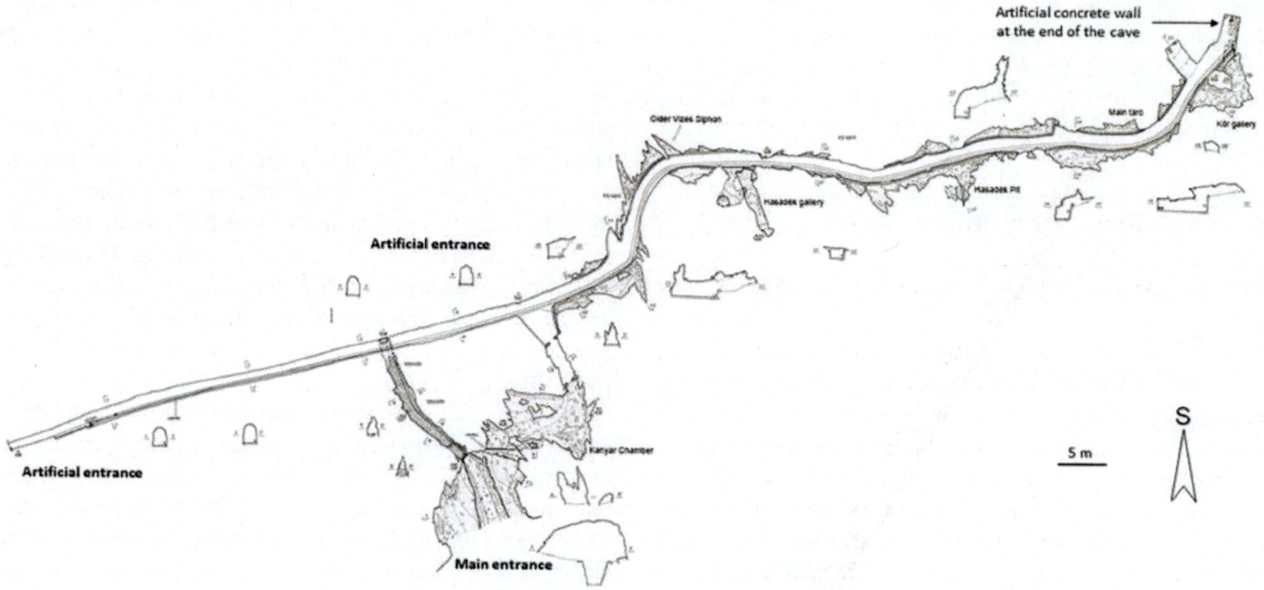

Fig. 10 Map of the Kőlyuk of Mánfa today (*source* National Cave Registry)

Fig. 11 The impounding reinforced concrete wall at the end of the cave and the access gallery

hydrological situation: because of the impounding effect of the imbrication the precipitation fallen on the limestones of the southernmost margin infiltrate and issues in springs. The thus generated intermittent watercourses leave the sandstone terrain, join with one another and disappear underground in the Zsidó Valley swallow hole to re-emerge, together with water infiltrating from the limestone surface, in the Kőlyuk on the surface. Consequently, this is an active spring cave originated by descending cold-water karstification, typical of the Western Mecsek Karst. The annual discharge fluctuates around 100,000 m^3.

The inventory of the cave fauna was completed by Elemér Bokor and Antal Gebhardt in the 1920s and 1930s. They identified almost 160 species, the following described as true cave-dwellers: *Lartetia gebhardti*, *Stenasellus hungaricus robustus*, *Niphargus molnári*, *Dendrocoelides pannonicus* and *Pilycelis tóthi* (Havasi 2003). With the utilization of the cave in the decades after World War II, faunistic research was sporadic and a new upswing of studies only started in the 1990s and 2000s, when, under the guidance of Miklós Szatyor, within the Pro Natura Karst and Cave Exploration Association regular bat monitoring was launched in the cave (web3). It resulted in the description of a rich bat community of ca. 20 species, which rivals with the water resources as the foremost value of the cave. From the last decade the investigations of invertebrate cave fauna by Dorottya Angyal stand out and underpin the observation that, even under intensive human use (as waterworks), numerous rare species can live or survive in the cave. Among such species, only *Paladilhia hungarica* is mentioned here (Angyal 2012, 2015).

5 Cave of the Mészégető Springs

The morphological and karst hydrological positions of this cave is similar to those of the Abaliget and the Vízfő caves: it opens at the base of the steep cliff bordering the Western Mecsek Karst in the north and its catchment lies to the south on a much more elevated terrain (Fig. 2). Its origin is also mainly due to descending cold karstwaters. A remarkable difference compared to the above-described caves, however, is that the present catchment does not expose non-karstic bedrocks, the Jakab Hill Sandstone only occurs reworked in the thick alluvium of the Mész Valley, which also belongs to the catchment. Another essential difference is the highly diverse land use of the catchment (arable land, meadows, riding schools, resort), which is also a source of potential contamination (see later).

The cave opens in the built-up area of Orfű, at the base of the wall of an abandoned stone quarry at 191 m above sea level. The enclosing rock is Triassic (Anisian) limestone. It is a ca. 360-m long spring cave, which consists of a sub-horizontal active stream passage and a vertical spring delta of three or four branches (Fig. 12). The lower, permanently active spring cave, which used to function as waterworks, is not yet explored. The middle spring branch, which is intermittently active during floods, is known along 15–20 m distance from the spring, but has no link passable by humans with the horizontal passage. The uppermost'overflow', the continuation of the internal horizontal branch, encompasses the cave entrance and opens to the surface through a closed siphon. Another branch is assumed to exist the lower section of which became submerged when Lake Orfű was dammed. The annual yield of springs changes around 150,000 m^3. The catchment area is 1.8 km^2 (Parrag 2003a, b). Previously the catchment also included the lowermost one-third of the present catchment of the Vízfő Spring, with a sandstone terrain of several km^2 area, but by now, due to the uplift of the mountains and intensive valley incision induced by the rapid subsidence of the northern foreland, these areas were lost (Barta et al. 2012). Today the principal swallow hole is the Gubacsos in the Mész Valley, which receives large amounts of sandstone gravels from the valley alluvium.

Typical of the tectonically preformed general appearance of the cave are the corridors of NW to SE or N to S strike and the minor chambers developed at their junctions. The cave is easily passable due to its relatively large average dimensions (1.5–3 m height and roughly the same width of corridors). In cave morphology cracks and meanders predominate, while among the microfeatures solutional forms are abundant: heads of strata, stone blades, hourglass features, cave karren and, at many sites, benches (Fig. 13). The cave is relatively poor in dripstones, only smaller stalactites, flowstones, crustings and small brimstone basins are observable, mostly in the inner stream sections. All along the cave walls are characteristically dark, which is partly due to the dark grey colour of the enclosing limestone and partly to widespread black manganese crusts. Among cave sediments, in addition to in situ or inwashed fine materials, the large amounts of sandstone gravels are certainly worth mentioning.

The cavities of the stone quarry where the cave is located were mentioned by László Vértes (Vértes 1952), who probably described the still passable entrance section of the middle branch in the spring delta. In 1955 László Kevi and Béla Vass employed water dyeing to find a link between the Gubacsos Swallow Hole and the Mészégető springs: the fluorescein fed into the Gubacsos appeared in the spring after 166 h (Szabó 1961; Vass 2012). The cave itself was explored under the guidance of Béla Zalán on 17 October 1986, when it was accessed along 200 m length down to Siphon 2. At the end of the same month the Caving Group of the Baranya County Touristic Office was successful in draining the siphon and the cavers reached Siphon 3, which is located ca. 20 m further. On 16 January 1988 members of the Amphora Diver SC swam through Siphon 3 and in the continuation of the passage found Siphon 4. The diver who crossed Siphon 4 found a spacious and high passage beyond Siphon 4, but did not proceed further. Between 14 and 23 July 1989, organized by the Pro Natura Caver Group and the Bauxite Caver Group, the water level of the siphons was lowered by pumping and after two days of work cavers managed to cross Siphon 4 and reach the inundated Siphon 5. In lack of time and energy the draining of Siphon 5 was not attempted. The last major trial to get further took place in 2001, when, in the frame of the joint camp of the Pro Natura Karst and Cave Exploration Association and the Szeged Karst and Cave Exploration Association, Siphon 5 was crossed and another 52 m of the cave was explored. The actual end point is Siphon 6, the section beyond Siphon 5, however, could not be passed through again after the summer of 2001. Beyond Siphon 2 CO_2 concentration is increasing and becoming dangerous beyond Siphon 5.

In addition to exploration, the cave was surveyed for zoology, radon emanation and water quality (Czigány et al. 1999; Parrag 2003a, b; Illés and Feketéné 2011). The latter are worth emphasizing since in the 1990s the Mészégető Spring was one of the most contaminated karst caves of the Mecsek Mountains and, unfortunately, the most recent microbiological measurements available (from 2010) do not indicate any improvement in water quality. Most of the investigated parameters (e.g. Coliform, *E. coli*) invariably surpass the drinking water limits.

In spite of the strict protection of the cave since 1993 and the inclusion of the overwhelming part of the catchment into the Western Mecsek Protected Landscape in 2009, the

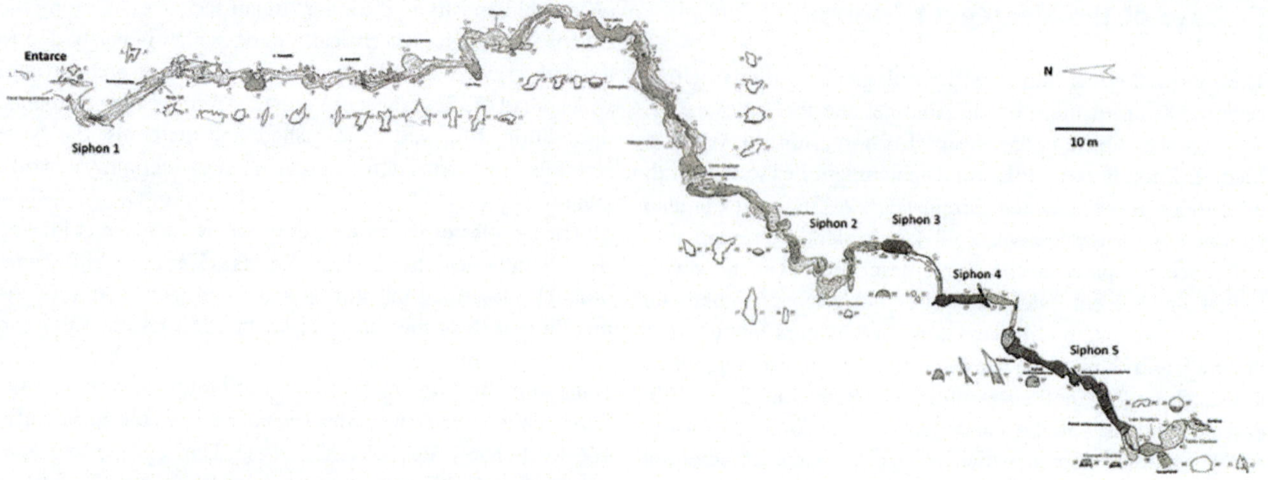

Fig. 12 Map of the Mészégető Cave (*source* National Cave Registry)

Fig. 13 Details of the active stream corridors of the Mészégető Cave

restrictions aimed at preserving water quality could not be fully enforced. The main hazard to the state of the cave is land use on the catchment. The following threats are worth mentioning:

1. The proportion of rear estates without canalization in the residential and resort area which lies above the cave entrance south of the spring still amounts to 20–25%. These are serious sources of pollution on the karst terrain which is only covered by a weathered loess mantle of maximum several metres' thickness.
2. Besides the waters infiltrating into the cave, another source of contamination is the activities practised on the catchment of the Gubacsos Swallow Hole, which exert a direct impact on the water of the stream which flows into the cave. Among them the pension and the riding school are of primary importance. From the more than 10-hectare area liquid manure, communal sewage and waste can reach the karst system. Also the Gubacsos can swallow untreated sewage and communal waste from the nearby holiday chalet.
3. Another hazard is represented by the Pécs-Abaliget public road aligned along the axis of the catchment, which has a heavy traffic, particularly in summer. In addition to normal operation (emission of heavy metals from the wearing parts of vehicles, deposition of exhaust gases, dripping oil), an accident could cause the contamination of the cave and the water of the cave stream not only through the Gubacsos Swallow Hole but also through the nearby Sziklás Swallow Hole. Along a ca. 1.5-km long road section the road and the channel of the intermittent stream running parallel with it conduct water into both swallow holes. In winter materials strewn on

road surfaces to prevent slipperiness may also reach the cave.
4. There are meadows and arable fields between the road and the marginal houses of the village Orfű. One of the arable fields is located right on the edge of a swallow hole which receives huge amounts of runoff during rainfall events. Moreover, at 20 m distance from the swallow hole manure is regularly deposed.

A solution could be negotiations with land users in order to agree on an integrated catchment utilization plan or regulation which would be acceptable for all stakeholders.

6 The Tettye Spring Cave and the Tettye Tufa Cave

Second in discharge to the Vízfő Spring of Orfű in the Mecsek Mountains is the Tettye Spring of Pécs. For its main characteristics it is almost completely different from all other springs and caves in the Mecsek:

(a) it is the only major karst spring which issues on the southern side of the Western Mecsek Karst;
(b) it is an autogenic karst spring, i.e. its catchment is exclusively composed of carbonate rocks;
(c) it does not open on a valley floor but occurs in hanging position;
(d) although its water is used in an institutionalized form from the oldest times, its cave had not been explored before the 1980s and 1990s.

The spring is located on the southern slope of the mountains, in the NE part of the county seat, Pécs. Its particular morphology is due to two factors: on the one hand, the mountains are uplifted along the marked tectonic line (the so-called „Mecsekalja" line) bordering the Western Mecsek in the south and, on the other hand, the Middle Triassic limestone aquifer (which stores the spring water) is underlain by an aquiclude of Lower Triassic (Werfen) marls which has a considerably higher topographic position than the base level of erosion (Aujeszky and Scheuer 1972; Haas 2004). The impervious series prevents that, parallel to mountain uplift, the karst spring should issue at a relatively lower level. Therefore, the spring is found at 233 m above sea level, i.e. much higher than the southern mountain foreland of 100–150 m mean elevation (Szabó 1953).

The spring has been a crucial factor of settlement for people inhabiting the region since ancient times. In the Roman age an important provincial capital, Sopianae, was here and archaeological finds reveal that the well-developed aqueduct system of the town was supported by the Tettye Spring situated high above the town. In the vicissitudes of history the town and its pipeline system was destroyed several times. In the 16-seventeenth centuries, however, both archaeological and written evidence attests to the well-developed water supply of the town, with baths built during the Ottoman Occupation (1543–1686). From this period there is unambiguous evidence that it was not only the water of the Tettye on which water supply was based, but also numerous smaller springs on the slopes of the Mecsek (Rőth 1997). The modern development of the Tettye spring for water supply began in 1892 and continues to this day (Fig. 14). It has been feeding in water to the pipeline system of the town uninterrupted ever since. As for any karst spring, a key problem of its use for water supply is also the balancing of extreme yields and the exploitation of the annual yield as fully as possible. Today, similarly to the Kőlyuk of Mánfa, the Tettye operates with storage above the spring threshold. In the twentieth century, however, there were numerous attempts to exploit the water reserves below the spring level, too. To this end, a gallery system of more than 800 m total length was constructed in the environs of the spring. The attempts were unsuccessful and today these galleries function as underground water storage reservoirs (Vass 1997). Thanks to the increased storage capacity and impoundment above the spring threshold the previous 30–50% exploitation ratio of the spring yield could be raised to more than 94% (Vass 2012).

The cave itself was only explored on the turn of the 1980s and 1990s. Since the main objective of exploration was to reach a section in the bedrock suitable for impoundment above the spring threshold, after passing through the debris zone of the entrance and the siphon beyond it, classic exploration activities were stopped and instead a gallery parallel with the cave passage was excavated to access the section beyond the siphon. The total length of known passages is only 40 m and, in fact, the cave is nothing more than a 16-m deep siphon (Fig. 14). The explored sections are stream passages without speleothems. Due to the operation of the waterworks, the cave is filled up with water at present.

The utilization of the cave as waterworks for almost 130 years has also brought about many scientific achievements. The longest spring discharge data series in Hungary is available from here and this provided basic or control data for a range of karst infiltration models (e.g. Kessler 1958; Maucha 1990). Here Vass (2012) elaborated and estimation method which is able to predict annual utilizable water reserves for every year with high accuracy at the end of the first quarter of the year. The yield fluctuations of the spring are clearly illustrated by the data from more than a century-long measurement series that its annual discharge ranges from 175,000 to 4,000,000 m^3, while its monthly discharge from 7,000 to 800,000 m^3 (Kessler 1958; Aujeszky and Scheuer 1972; Vass 1997; 2012).

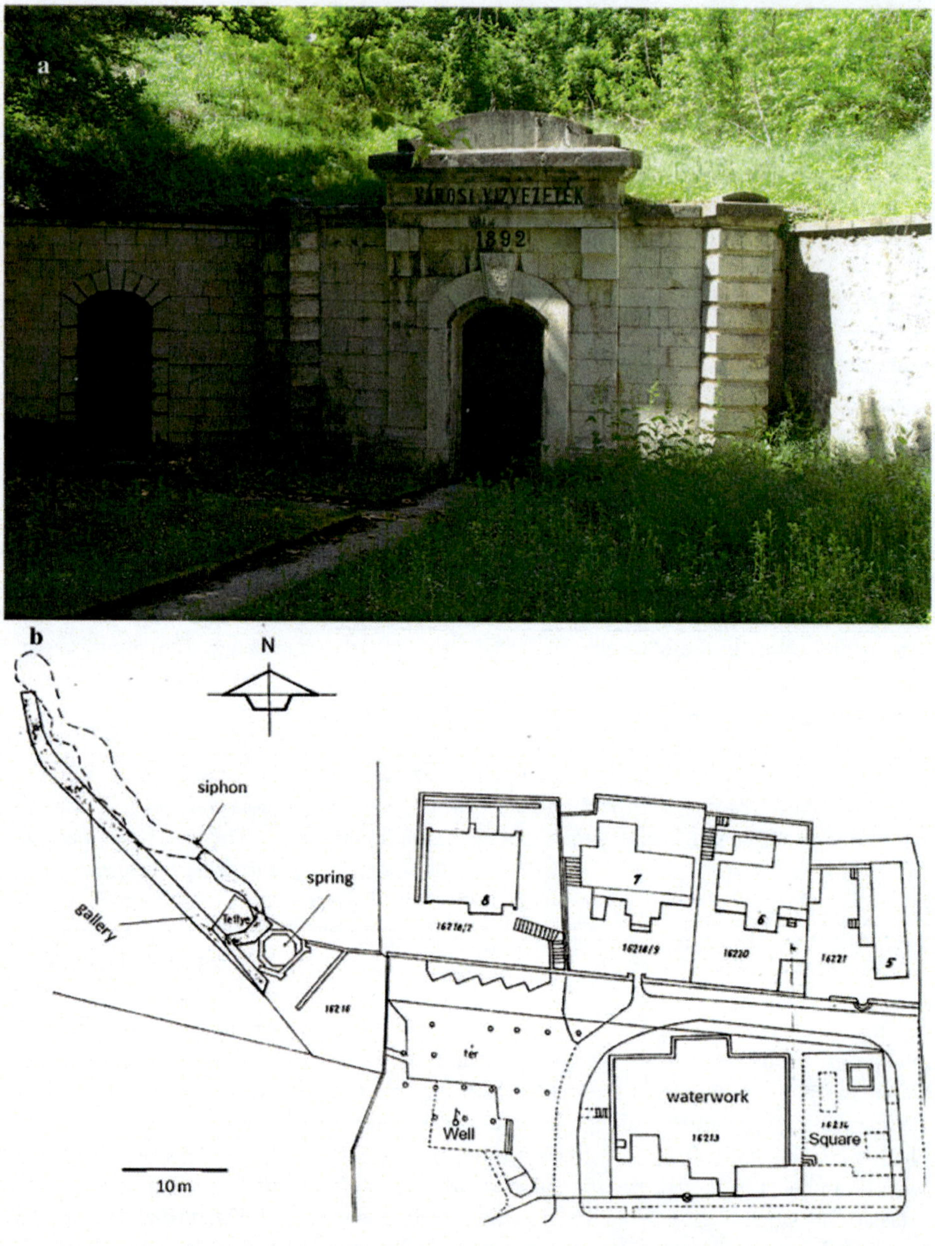

Fig. 14 The waterworks built in 1892 and the Tettye Cave today (*source of map* National Cave Registry)

Interestingly, in spite of the Tettye Spring being well explored, there is no agreement of opinions on the extension of its catchment area. In the available literary sources areas between 8 and 13 km^2 are found. The reason for this is probably the multiple outflow of underground water reserves on the margins of the catchment depending on karstwater level (Szabó 1953; Vass 1997; 2012; Böcker et al. 2003; Szőke 2012). In the case of the Tettye Spring precise data on the catchment would be even more important than for other karst springs since, on the one hand, it is a drinking water reservoir and, on the other, the southern zone of the catchment is a densely built-up urban area, which is a potential hazard to water quality. In addition to preserving good water quality, the infiltration of rainwater also has to be ensured. The remaining portions of the catchment are mostly continuous forest areas.

Given its autogenic character, from the water of the Tettye Spring huge amounts of freshwater limestone are deposited. Its formation has been continuous since the Pleistocene, but the intensity of the process has been considerably reduced with the growing utilization of the spring water. The volume of tufa accumulation is estimated at more than 300,000 m^3 (Aujeszky and Scheuer 1972; Rónaki 1976; Havasi and Parrag 1999; Kele 2009). It has a typical morphology of subhorizontal surface in the foreland of the Tettye Spring (partly due to anthropogenic planation of the terrain) and a steep drop towards the valley in its continuation to the south. In the tufa mound there are many syngenetic cavities and man-made cellars, out of which the Tettye Tufa Cave has the largest dimensions. The cave consists of natural cavities linked by carved galleries in more than 200 m length. It is used for tourism, along with the Abaliget Cave this is the only show cave in the Mecsek. In entire Hungary the only other tufa cave open for tourists is in the Anna Cave of the Bükk Mountains (North-Hungary).

7 Sinkholes

While the spring caves and chimneys of the Mecsek had been all explored before World War II, explorations of cave sections beyond active swallow holes only began in the 1990s. As it was mentioned above, all the swallow holes with independent catchment are associated with valleys arriving from sandstone terrain and, with the exception of the Nyárás Valley swallow hole, all of them feed the Vízfő Spring. Main properties of the valleys of swallow holes are listed in Table 1.

Among the explored swallow caves only the two largest, the Spirál Cave and the Trió Cave are treated here.

7.1 Spirál Cave

With its 86 m depth, Spirál Cave is the deepest in the Mecsek and with its 1400 m length it is the second longest after the Abaliget cave. Its water recharge derives from the same sandstone imbrication which is also exposed on the catchment of the Kőlyuk of Mánfa (Fig. 2). In the Büdöskút Valley, which leads to the swallow hole, there are three further holes, two of them completely filled up and the third (the Büdöskút Sinkhole) occasionally fills up and then opens up again in function of floodwater inflow and the explorers' activities. Its depth is only 5–7 m and its 13-m long passage follows a single tectonic crack. The flood overflow of this cave at valley floor position is the Spirál Swallow Hole. Although it is registered as a swallow hole, its dimensions are not primarily due to the solutional activity of the inflowing stream but to the fact that in the tectonically strongly affected area the submergent stream reached a vertical joint of several metres' width, perpendicular to the course of the stream (N to S aligned) where stream water sinks to more than 70 m depth. The enclosing rock of the cave is Triassic limestone and dolomite and its morphology is governed by a tectonic joint system and active collapses dissecting the cracks. Among the cracks the already mentioned "Szakadék" is outstanding (Fig. 15), here, beyond a narrow entrance and labyrinth, a 60-m long single gorge leads to the depth. At the bottom of the"Szakadék" the character of the cave changes: stream passages run in both directions dissected by narrow sections and an upper branch leads several ten metres higher, approaching the surface. Tectonic preformation plays a decisive part in the location of major chambers (e.g. Vízesés, Verecke and Spirálszíve Chambers), the chimneys and cracks reaching high and dangerous collapses are equally found in the internal and external sections of the cave. At the same time, dripstones, draperies, flowstones and other speleothems appear in great numbers and the manganese crustings, typical of other caves in the Mecsek, which cover large surfaces are worth mentioning (MKCs 2008; Rónaki 2011).

The entrance section of the Spirál Cave was explored by members of the Mecsek Karst Exploration Group as early as 1977, but only to 30 m depth and 60 m length. Further work was seriously hindered by the recurrent events of floodwater inflow. The major breakthrough happened in 1996, when they penetrated into the "Szakadék" and increased the depth of the cave to 80 m. In the same year the cave was closed and the entrance was secured by well rings. In the stream passages reached, however, further advance was not easy. The next milestone was in 2008, when the cave was enriched with four chambers and 281 m passage length (Fig. 16). The most remarkable success in exploration was that of the Spirálszíve Chamber, which, along with the Big Chamber of the Abaliget cave, is one of the largest chambers of the Mecsek. Groundfloor dimensions are 26 × 13 m, estimated height is 25 m and it is rich in speleothems.

Considerable explorations followed in 2010 and 2012, when 200 and 155 m new sections, respectively, were

Table 1 Characteristics of valleys with major swallow holes (modified after Barta and Tarnai 1999 and Gila 2012)

Valley with number of sinkholes	a (km^2)	b (m^3/min)	c
Nyárás Valley (1)	2.4	5–10	1.7 km from Abaliget Cave
Sózó Valley (1)	0.15	1–2	3.1 km from Vízfő
Körtvélyes (2)	1	4–5	1.8–2.7 km from Vízfő
Szuadó Valley (3)	3.4	5–10	2.5–2.8 km from Vízfő
Büdöskút Valley (4)	0.33	4–5	1.8–2 km from Vízfő

a—non-karstic catchment area of the sinkhole(s)
b—estimated flood yield (2006, 2010, 2018)
c—distance of swallow hole(s) from spring

Fig. 15 The tectonic fault of the "Szakadék" (ravine)

reported. In addition to the new sections expressed in metres, it is important to note that in 2012 a new passageway was opened by the members of the Mecsek Karst Exploration Group which highly shortens and alleviates the way to the points where obstacles could be demolished in the internal parts of the cave. Another important event of the year was the installation of the new rust-free closure, the exchange and enlargement of the old ladders increasing the length of ladders to 90 m. At present, cave length exceeds 1400 m (MKCs 1977, 1996, 2008, 2010, 2012, 2013, 2017; Rónaki 2011).

The cave is strictly protected since 2005 and its catchment is part of the Western Mecsek Protected Landscape since 2009 (by the ministerial decree 4/2009 (IV.10.)). The entire area is covered by deciduous forests and the only hazard to the cave is water inburst from the floods of the neighbouring stream but, relying on the experience of the past 25 years, the closure of the cave provides sufficient protection against floodwater inflows.

7.2 Trió Cave

The Trió Cave is one of the most important swallow hole caves in the Mecsek. Its length is nearly 300 m and depth is 58 m. It is a typical case for the above detailed hydromorphological situation when the streams arriving from the sandstone surface of the 500–600 m high Jakab Hill, which borders the Western Mecsek Karst in the south cross the boundary of karstic rocks and submerge underground

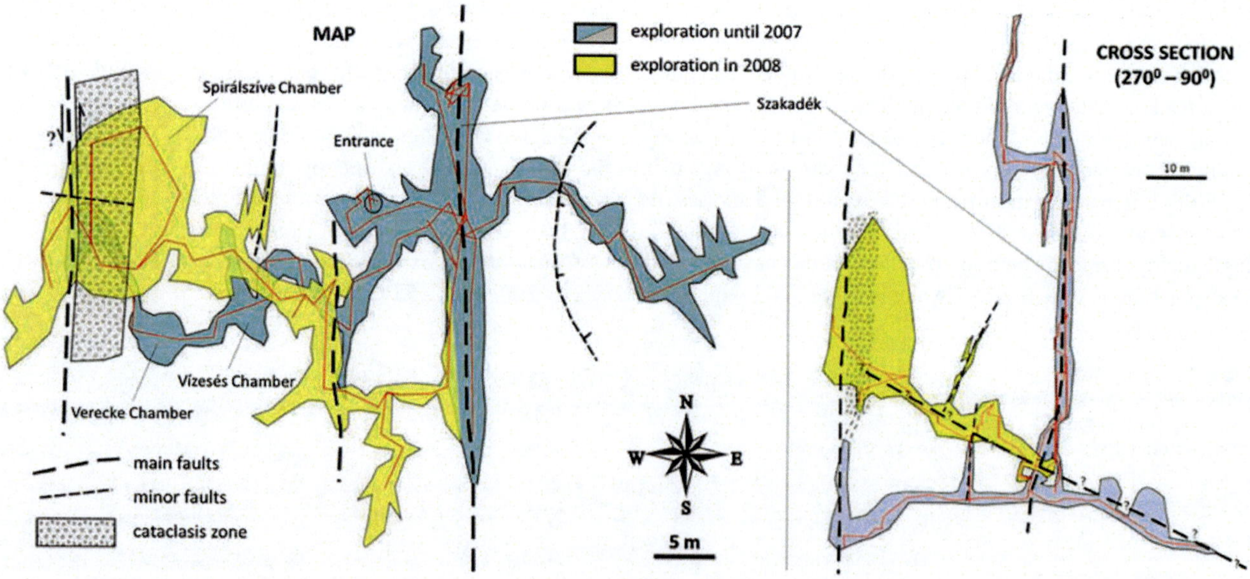

Fig. 16 Sketch of the explorations in 2008, which well demonstrates the role of tectonic preformation in cave development (modified after Gál and Glöckler 2009)

(Fig. 2; Table 1) and only re-emerge in the karst springs of the northern margin (Barta and Tarnai 1999). In the Szuadó Valley, where the cave is found, there are three swallow hole caves, each of which is deeper than 50 m and reach or approach the karstwater table. Under natural conditions, during low water only the uppermost Szuadó Swallow Hole was active, but during floods even the joint capacity of the three swallow holes was not sufficient to swallow the full discharge and the stream flowed further down to Orfű. The cross section area of the passages in itself would suffice to conduct floodwaters, but the 3–8-m thick sediment plugs at the entrance hinder the transport of greater masses of water. The mean discharge of the stream is some 100 l/min but during the dry summers and autumns it can dry out and rise to 5–10 m^3/min during floods.

The Trió Cave opens at 296 m above sea level, ca. 100 m higher than the Vízfő Spring and at 2.5 km distance from that. The water disappearing in the swallow hole re-emerges in the Vízfő Spring 90–210 h later (Szabó 1961; Gila 2012). The almost 60-m deep Trió Cave lies close to the karst water table and the water table of V shape cutting into the karstwater table, much higher under the neighbouring plateaus, levels off during humid periods. The most extreme example of this situation can be cited from 2010, when the water level of the cave rose above 290 m (more than 50 m rise) and approached the level of the entrance. Such events can be unambiguously distinguished from impoundment due to the plugging of end points, because well after rainfall events high levels were retained in all the three swallow holes and lower rather slowly. High floods are generally observed in the Szuadó Valley in every 2–3 years.

The enclosing rock of the cave is Middle Triassic (Anisian) limestone (Szabó 1961), which has a N dip on the northern limb of the Jakab Hill anticline and this controlled cave formation. In addition to solution and erosion along bedding planes, another important factor of preformation was the presence of N to S and E to W fracture zones, typical of the whole Western Mecsek Karst. Cave morphology is jointly controlled by both factors: sections along bedding planes are generally narrow, their cross sections are maximum 1 × 1 m, while the sections along tectonic lines appear as spectacular cracks (Fig. 17). The largest chambers and shafts of the cave are associated with fracture zones where, as a consequence of the brecciated rock, hydraulic conductivity increases to extremely high values. During rainy periods and snowmelt, in such zones the amount of dripping water increases tenfold within a matter of hours.

In addition to the classic solutional and erosional features typical of swallow holes, the richness in dripstones should be mentioned. Regarding cave sediments sandstone gravels with manganese coatings and locally deep clay fills are typical. A particular feature is the protrusion of thin marl layers intercalated into thicker limestone strata. Due to solution by percolating and precipitating waters on the passage walls the carbonate contents of the limestone and marl are removed. From the marl of high clay content, soft clayey layers of 1–2 cm thickness are left behind as residue and stand out from the wall even to 4–5 cm (Fig. 18). The protruding bedding planes present in masses during the exploration are highly vulnerable and, unfortunately, they are heavily destroyed during visits.

At very early times the position of the Trió Swallow Hole, the entrance to the cave, on valley floor, remote from the spring, attracted the attention of the explorers of the cave system behind the Vízfő Spring. Spring water was dyed as early as in the 1960s and fluorescein appeared at the spring after 210 h (Szabó 1961). In 1969 a group of three cavers cleared the entrance. The Trió name refers to them, who reached to 3.5 m depth (Rónaki 1970), but the pit was filled by floodwaters within some years. New clearing followed after almost 30 years. In October 1997 the Szeged Karst and Cave Exploration Association and the Cave Exploration Department of the Mecsek Association cleared the throat of the swallow hole filled with debris and sediment to the level of the valley floor, which presented a spectacular bedrock swallow hole in cleared condition. Further exploration was considerably hindered by the large amounts of clay fill and cave length was not more than ca 30 m in early 2001. The series of shafts which followed and the horizontal passages beyond that were only explored in 2001–2002, increasing total length to more than 200 m. No breakthrough has happened since that date. Two narrow sections in bedrock close to the karstwater level and some higher-lying, upward pointing passages filled with clay were cleared.

In addition to explorations, research with other purposes have to be mentioned in a non-exhaustive overview. In 2001 the"Búboskemence" dripstone feature was drilled for dating and environmental geochemistry investigations. The feature of ca. 1 m diameter was found to be 4,800 year old. Through rare earth metals the Thíra volcanic eruption (ca. 3650 years ago) and the impact of uranium ore mining in the Mecsek in the second half of the twentieth century, indicated by higher ^{238}U concentration in the dripstone through inwash of settling dust, could be identified (Siklósy et al. 2011; web4). In 2006 radon measurements were carried out led by Gabriella Koltai as part of a measurement campaign affecting more than 10 caves in the Mecsek (Koltai et al 2010). In spring 2012 12 temperature sensors were placed in the cave for two months. The objective of the measurements was the detection of the alternation of winter and summer air circulation as well as the impact of visiting groups on the cave climate (Muladi et al. 2013; Csépe-Muladi 2016). The relationships of microfractures and karstification processes were also studied in the Trió Cave within large-scale scientific programme (Bauer and Tóth 2015).

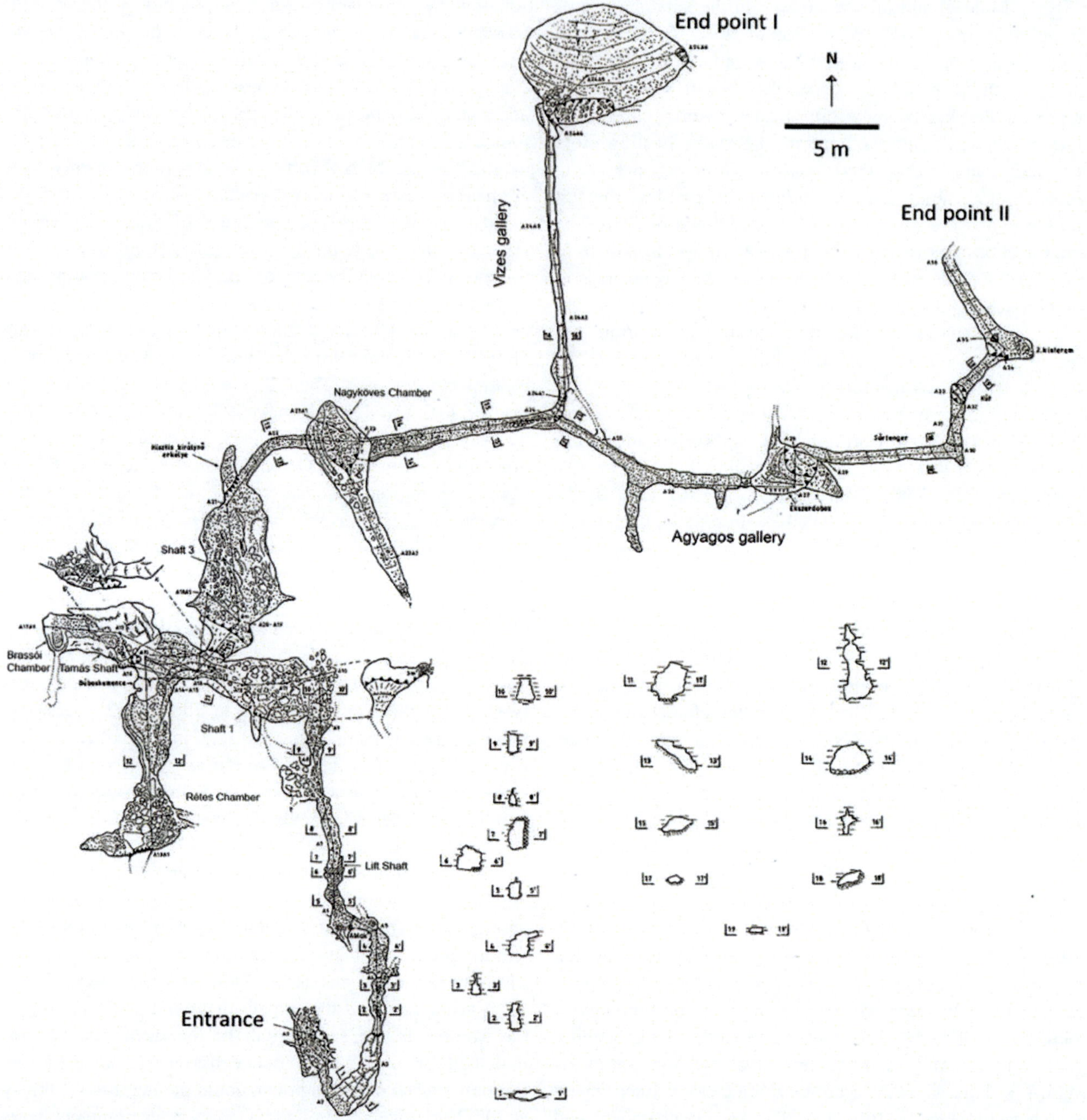

Fig. 17 Map of the Trió Cave from 2002 (*source* National Cave Registry)

From the cave fauna the stock of bats (150–200 specimens), populous in relation to the Mecsek Mountains, and the blind cave-dwelling crustacean *Stenasellus* are worth mentioning.

At present the cave is closed. In its shafts rust-free ladders make movement easier. Against water inbursts and infilling a 50–60-m high stone wall protects the entrance to the Trió Cave. Naturally, through the sediments on the floor can receive large amounts of water, but excluding sediment inwash the preservation of its state is ensured. The catchment is covered by deciduous forests and since 2009 it is part of the Western Mecsek Protected Landscape (by the ministerial decree 4/2009 (IV.10.)). The cave is used for tourism and adventure tours in overall are allowed.

Fig. 18 Speleothems in the Trió Cave and the protruding marl layers at the bottom of shaft 1

8 Conclusions

Western Mecsek is a typical allogenic karst in Southwest Hungary with more than 100 chimneys and with two dozens of spring caves and sinkholes. The area is built up mainly from Triassic limestone and dolomite. Its caves rarely exceed the 300 m length and the 50 m depth. The longest cave winds underground for ca. 2,000 m length. The area is under nature protection and it is covered mainly by deciduous forests but there are a few catchment areas where the anthropogenic activities (agriculture, settlement, tourism, traffic) can cause environmental conflicts, e.g. water pollution.

References

Angyal D (2012) A Mánfai-kőlyuk gerinctelen faunájának alakulása a vízmű általi hasznosítás tükrében—előzetes eredmények (Invertebrate fauna of the Mánfai-kőlyuk Cave (Mecsek, SW Hungary) in the light of utilization by waterworks—previous results). Természetvédelmi Közlemények 18:24–33 (in Hungarian)

Angyal D (2015) Integrative taxonomic revision of Niphargus spp. and other rare and endemic troglobiont macroinvertebrates from the caves of the Western Mecsek Mts. (South Hungary). Doctoral (PhD) Thesis, Keszthely, 2015; p 136. http://real-phd.mtak.hu/607/1/Angyal_Dorottya_dissertation.pdf. Accessed 4 Aug 2020

Angyal D, Korsós Z (2013) Millipedes (Diplopoda) of twelve caves in Western Mecsek, Southwest Hungary. Opuscula Zoologica. Budapest 44(2):99–106. http://epa.oszk.hu/02300/02340/00058/pdf/EPA02340_opuscula_zoologica_2013_tom44_2_099-106.pdf. Accessed 4 Aug 2020

Aujeszky G, Scheuer Gy (1972) A pécsi Tettye-forrás hidrológiai vizsgálata (Hydrology of the Tettye-Spring at Pécs). Hidrológiai Közlöny 1–2:6–15 (in Hungarian)

Barta K, Tarnai T (1997) Karsztkutatás az orfűi Vízfő-forrás vízgyűjtő területén (Karst research on the catchment area of the Vízfő Spring, Orfű, Mecsek Mountains). Karszt és Barlang 1–2:12–19 (in Hungarian)

Barta K, Tarnai T (1999) Karstmorphological research in the Mecsek Mountains South Hungary. Acta Carsologica Ljubljana 28(1):13–25

Barta K, Tarnai T, Bauer M (2012) A Nyugat-mecseki-karszt általános jellemzése (General characterization of the Western Mecsek Karst). In: Barta K, Tarnai T (eds) A Nyugat-Mecsek karsztja. GeoLitera, Szeged, pp 25–40 (in Hungarian)

Bauer M, Tóth TM (2011) A mikrotöréshálózat szerepe a barlangok fejlődésében az orfűi Vízfő-forrás vízgyűjtőjén (The role of microfracture network to cave development on the catchment area of Vízfő Spring, Orfű). Karsztfejlődés 16:103–122 (in Hungarian)

Bauer M, Tóth TM (2015) Modeling microfracture geometry to asses the function of a karst system (Vízfő spring catchment area, Western Mecsek Mountains, Hungary). Geologica Croatica 68 (1):11–23. http://www.geologia-croatica.hr/index.php/GC/article/view/GC.2015.02. Accessed 4 Aug 2020

Bencsik A, László N, Bauer M, Bozsó G (2015) Orfűi barlangi üledékek geokémiai és geológiai tulajdonságai (Geochemical and geological properties of cave sediments around Orfű). In: Pál-Molnár E, Raucsik B, Varga A (eds) Meddig ér a takarónk? A magmaképződéstől a regionális litoszféra formáló folyamatokig. 6. Kőzettani és geokémiai vándorgyűlés, p 149 (in Hungarian)

Berényi Üveges I (2012) Karsztvizek hasznosítása és védelme a Közép- és Nyugat-mecseki karsztvidéken (Water supply and protection on middle and western Mecsek Karst). In: Barta K, Tarnai T (eds) A Nyugat-Mecsek karsztja. GeoLitera, Szeged, pp 45–53 (in Hungarian)

Böcker T, Hidasi J, Hidasi G, Deák FJ, Székvölgyi K, Vönöczky Á, Kraft J, Majoros Gy, Rónaki L, Vass B (2003) Üzemelő, sérülékeny földtani környezetben levő ivóvízbázisok biztonságba helyezése a Pécsi Tettye Vízmű területén (Protection of water resources in vulnerable geological environment at Tettye Water Company, Pécs). Envicom Mérnöki és Számítástechnikai Kft. Mérnöki Irodája, Budapest, p 153 (in Hungarian)

Csépe-Muladi B (2016) A barlangklíma tér- és időbeli változásainak vizsgálata különböző magyarországi karsztterületeken (Investigation of spatial and temporal changes of cave climate in different karst areas of Hungary). Doctoral (PhD) Thesis. Szeged, p 137 (in Hungarian)

Czigány Sz, Parrag T, Szűcs T (1999) Vízminőség vizsgálatok az Orfű környéki karsztvidéken (Water quality studies around Orfű). Karsztfejlődés 3:139–151 (in Hungarian)

Demeter Gy (1994) A Paladilhia hungarica (Gastropoda, Prosobranchia) elterjedése a Mecsek szubterrán vizeiben és adatok a csiga biológiájához (Presence of Paladilhia hungarica (Gastropoda, Prosobranchia) in subterraneous water of Mecsek Mountains and data about the biology of the snail). MSc Thesis. Pécs, p 33 (in Hungarian)

Dezső J, Ország J, Leél-Őssy Sz, Zalán B (2012) Az Abaligeti-barlang kitöltésviszonyai (Sedimentation in the Abaliget Cave). In: Barta K, Tarnai T (eds) A Nyugat-Mecsek karsztja. GeoLitera, Szeged, pp 109–122 (in Hungarian)

Gál B, Glöckler G (2009) A Spirál-víznyelő barlang (The Spirál Sinkhole Cave). In: Glöckler G, Gál Gyné (eds) MKCs (Mecseki Karsztkutató Csoport) 2008. évi jelentése. Pécs, pp 26–37 (in Hungarian)

Gila Cs (2012) Vízfestések az orfűi Vízfő-forrás vízgyűjtő területén (Water tracing on the catchment area of Vízfő Spring, Orfű). In: Barta K, Tarnai T (eds) A Nyugat-Mecsek karsztja. GeoLitera, Szeged, pp 55–66 (in Hungarian)

Havasi I, Parrag T (1999) The planned landscape protection area in the Western Mecsek Hills, Hungary. In: Bárány-Kevei I, Gunn, J (eds) Essays in the ecology and conservation of karst. Szeged, pp 171–173. http://acta.bibl.u-szeged.hu/37204/1/geographica_036.pdf. Accessed 4 Aug 2020

Havasi I (2003) Mánfai-kőlyuk (Mánfa Cave). In: Székely K (ed) Magyarország fokozottan védett barlangjai. Mezőgazda Kiadó, Budapest, pp 393–394 (in Hungarian)

Haas J (2004) Magyarország geológiája. Triász (Geology of Hungary. Trias). ELTE Eötvös Kiadó, Budapest, p 350 (in Hungarian)

Havasi I, Székely K, Salamon G (2003) Abaligeti-barlang (Abaliget Cave). In: Székely K (ed) Magyarország fokozottan védett barlangjai. Mezőgazda Kiadó, Budapest, pp 388–392 (in Hungarian)

Illés A, Feketéné Kurtán A (2011) "Mikro utakon" a Mészégető-barlang rendszerében ("On Micro-ways" in the system of Mészégető Cave). In: A Pro Natura Karszt- és Barlangkutató Egyesület jelentése a 2010. évi kutatási tevékenységéről. Pécs, pp 32–45 (in Hungarian). http://www.termeszetvedelem.hu/_user/browser/File/barlangkutat%C3%A1si%20jelent%C3%A9sek/2010/pro_natura_2010_01.pdf. Accessed 4 Aug 2020

Jakucs L (1977) Morphogenetics of Karst Regions. Akadémiai Kiadó, Budapest, p 283

Kele S (2009) Édesvízi mészkövek vizsgálata a Kárpát-medencéből: paleoklimatológiai és szedimentológiai elemzések (Investigations on freshwater limestones from the Carpathian-Basin: palaeoclimatological and sedimentological studies). Doctoral (PhD) Thesis, Budapest, p 176 (in Hungarian)

Kéki A, Havasi I (2010) A fokozottan védett Abaligeti-barlang kezelési terve (Management plan of strictly protected Abaliget Cave). KvVm Budapest, p 36 (in Hungarian)

Kessler H (1958) Estimation of subsurface water resources in karstic regions. General Assembly of Toronto, vol II. Groundwater. 3-14 Sept 1957. International Association of Scientific Hydrology, Gentbrugge, pp 199–206

Koltai G, Ország J, Tegzes Z and Bárány-Kevei I (2010) Comprehensive radon concentration measurements in caves located in the area of Mecsek mountains. Acta Carsologica. Ljubljana 39(3):513–522. https://ojs.zrc-sazu.si/carsologica/article/view/79. Accessed 4 Aug 2020

Kordos L (1984) Magyarország barlangjai (Caves of Hungary). Gondolat Kiadó, Budapest, p 326 (in Hungarian)

Lovász Gy (1971) Adatok az Abaligeti-karszt geomorfológiai és hidrológiai jellemzéséhez (Data about geomorphological and hydrological characterisation of Abaliget Karst). Földrajzi Értesítő 20(3):283–296 (in Hungarian)

Maucha L (1990) A karsztos beszivárgás számítása (Estimation of infiltration in karst areas). Hidrológiai Közlöny (3):153–161 (in Hungarian)

Miklós G (1980) A Remény-zsomboly klímaelemzése (Climatic analysis of the Remény Karst Pit). Karszt és Barlang (2):81–86 (in Hungarian)

MKCs (Mecseki Karsztkutató Csoport) éves jelentései (Annual Reports of Mecsek Karst Research Group) 1977, 1996, 2008, 2010, 2012, 2013, 2017 (in Hungarian)

Muladi B, Csépe Z, Mucsi L, Puskás I, Koltai G, Bauer M (2013) Climatic features of different karst caves in Hungary. In: Filippi M, Bosák P (eds) 16th international congress of speleology: proceedings, vol 2. Czech Speleological Society, Prague, Czechia, pp 432–438

Nagy G (1975) A Mánfai Kőlyuk szifonjának kutatása (Exploration of the Siphon of Kőlyuk at Mánfa). Karszt és Barlang 1–2:19–22 (in Hungarian)

Parrag T (1998) A karsztforrások szerepe a Nyugat-Mecsek társadalmi-gazdasági életében (The impact on Karst springs on the socio-economic life of settlements in the Western Mecsek Hills). Földrajzi Értesítő 47(2):149–156 (in Hungarian)

Parrag T (2003a) Orfűi Vízfő-barlang (Vízfő Cave in Orfű). In: Székely K (ed) Magyarország fokozottan védett barlangjai. Mezőgazda Kiadó, Budapest, pp 396–398 (in Hungarian)

Parrag T (2003b) Mészégető-források barlangja (Cave of Mészégető Springs). In: Székely K (ed) Magyarország fokozottan védett barlangjai. Mezőgazda Kiadó, Budapest, pp 394–396 (in Hungarian)

Rónaki L (1962a) Az orfűi Vízfőforrás-barlang feltárt szakaszának földtani viszonyai (Geological conditions of known sections of Vízfő Spring Cave, Orfű). Karszt és Barlang (2):51–55 (in Hungarian)

Rónaki L (1962b) Újabb vízfestés a mecseki karszton (Further water tracing in the Mecsekian Karst). Karszt- és Barlangkutatási Tájékoztató 6–7:104–105 (in Hungarian)

Rónaki L (1970) Vízfő-forrás és barlangjának kutatása (Exploration of Vízfő Spring and its Cave at Orfű). Karszt és Barlang (1):25–30 (in Hungarian)

Rónaki L (1976) Pincebarlang a pécsi Tettye-karsztforrás mésztufájában (Cellar-cave in the Travertine of the Tettye Spring at Pécs, South Hungary). Karszt és Barlang 1–2:25–28 (in Hungarian)

Rónaki L (2011) Az orfűi Vízfő-forrásbarlang vízgyűjtő területén történt vizsgálatok és azok eredményeinek áttekintése (Investigations on the catchment area of Vízfő Spring Cave in Orfű). Hidrológiai Közlöny 91(3):33–45 (in Hungarian)

Rőth L (1997) A pécsi vízellátás története a római kortól 1892-ig (The history of water supply in Pécs from Roman times up to 1892). In: Szirtes G (ed) Víz, ember, természet. A százötéves pécsi vízmű története. Pécs, pp 11–28 (in Hungarian)

Sebe K, Csillag G, Konrád Gy (2008) The role of neotectonics in fluvial landscape development in the Western Mecsek Mountains and related foreland basins (SE Transdanubia, Hungary). Geomorphology 102(1):55–67

Siklosy Z, Kern Z, Demeny A, Pilet S, Leel-Ossy S, Lin K, Shen CC, Szeles E, Breitner D (2011) Speleothems and pine trees as sensitive indicators of environmental pollution—a case study of the effect of uranium-ore mining in Hungary. Appl Geochem 26(5):666–678

Szabó PZ (1953) Két mecseki karsztforrás vizsgálata Komló és Pécs vízellátása szempontjából (Study of Two Karst Springs in the Mecsek Mountains for Water Supply of Pécs). Földrajzi Közlemények 3–4:161–190 (in Hungarian)
Szabó PZ (1961) A Mecsek és a Villányi-hegység barlangjai (Caves of Mecsek and Villány mountains) Karszt és Barlang (1):3–19 (in Hungarian)
Szőke E (2012) A Tettye-forrás szökevényvizei vízhozamának és hőmérsékletének kapcsolata a csapadék mennyiségével és a beépítettséggel (Connection between precipitation, land use temperature and discharge of springs around Tettye spring). In: Barta K, Tarnai T (eds) A Nyugat-Mecsek karsztja. GeoLitera, Szeged, pp 75–83 (in Hungarian)
Vass B (1997) A Tettye forrásvíz vízműves hasznosításának története (Utilization of the Tettye source—a historical overview). In: Szirtes G (ed) Víz, ember, természet. A százötéves pécsi vízmű története. Pécs, pp 29–38 (in Hungarian)
Vass B (2012) Visszaemlékezések a mecseki barlangkutatás hőskorára (Memoirs about the beginnings of the cave exploration in Mecsek mountains). In: Barta K, Tarnai T (eds) A Nyugat-Mecsek karsztja. GeoLitera, Szeged, pp 25–40 (in Hungarian)
Vértes L (1952) A Mélyvölgyi-kőfülke és néhány más mecseki barlang kutatásáról (About Exploration of Mélyvölgy cave and other caves in Mecsek mountains). Földtani Közlöny 82(7–9):270–278. (in Hungarian). (http://epa.oszk.hu/01600/01635/00443/pdf/EPA01635_foldtani_kozlony_1952_82_07-09_270-278.pdf). Accessed 4 Aug 2020
Web1: http://www.abaligetibarlang.hu/tortenet/ Accessed 4 Aug 2020
Web2: https://map.mbfsz.gov.hu/fdt100/ Accessed 4 Aug 2020
Web3: http://www.termeszetvedelem.hu/index.php?pg=cave_4120-2 Accessed 4 Aug 2020
Web4: http://real.mtak.hu/11646/1/49713_ZJ1.pdf. Accessed 4 Aug 2020

The Caves of the Villány Mountains

Szabolcs Leél-Őssy

Abstract

Two big and a few smaller hypogene caves dissolved mainly in Cretaceous limestone in the Villány Mountains measuring 30 × 70 km. During stone mining, at the end of the twentieth century, the 850 m long Beremend Crystal Cave and the 1500 m long Nagyharsány Crystal Cave were discovered. The Beremend Crystal Cave is very rich in popcorn and in aragonite. This cave reaches the karst water table: there is a small lake at the bottom of the cave. The Nagyharsány Crystal Cave is famous for the most spectacular logomites, and there are lots of dripstones too. This cave has a built-in pavement of stainless steel. These caves are opening in a recently active mine. Mining is prohibited at a safe distance around these caves.

The smaller surfaces with karst phenomena are only on the southern side of Harsány Hill.

Keywords

Villány Mountains • Beremend Crystal Cave • Nagyharsány Crystal Cave

1 Introduction

The southernmost part of Hungary is the Villány region (24 km south of the Mecsek Mountains). These mountains are forming a southwest–northeast direction mountain range. This area is famous for its red wine and the medicinal bath of Harkány (there are no springs: the 62 °C hot water is coming across the 50–70 m deep artificial wells from the lower Triassic limestone and dolomite).

The geology of this region is fascinating. Based on the lower Triassic colourful shale and sandstone (with anhydrite and gypsum interbedding), the whole of the Villány Mountains is built up of Mesozoic rocks, which are made of Triassic, Jurassic, and Cretaceous limestones. Mostly, thick loess sediment covers the hills—mainly on the southern sides. These slopes enjoy a strong Mediterranean influence, which is most favourable for the red wine!

The Villány range has a typical schuppen structure: it consists of six slivers and the block of Beremend and the block of Siklós Castle Hill (these are a few kilometres away from the Villány range in southern direction). The block of Kistapolca (where there is a lukewarm spring) is not a hill: there the Triassic rock only reaches the surface. The southern side of the highest peak, Harsány Hill (also known as Szársomlyó) (442 m asl) is free of loess sediment: it shows impressive bedding (for details, see chapter "An Interpretation of Karstification and Karst Processes by Hungarian Researchers"). On the hill there are steep-walled deep dolines, which were filled up with bauxite at the Jurassic–Cretaceous boundary. This area is a very beautiful palaeokarstic region. Sometimes, the dolines are filled with yellowish-red clay. These are very rich in Pleistocene Vertebrata fossils. These finds are known over the world (Kubinyi 1854; Kordos 1991).

There are favourable conditions for mixture corrosion on the rims of big basins covered by water-tight strata and of the mountains consisting of karstifiable rocks. (You can read details about the water circulation and the hydrodynamical conditions in chapter "Hydrogeology of the Karst Regions in Hungary"). There are no cold cliff springs in the Villány Mountains: the annual precipitation on the karstic blocs travels a long way, and rises up further as a lukewarm or hot spring (and dissolves hypogene caves). The first article about the caves of the Villány Mountains was written by Kevi (1956) and Szabó (1961). In 2000, Rónaki wrote about 28 caves (about the mined down caves too).

Sz. Leél-Őssy (✉)
Department of Physical and Applied Geology, Eötvös Loránd University, Pázmány Péter sétány 1/C, 1117 Budapest, Hungary
e-mail: szabolcs.leel-Ossy@ttk.elte.hu

M. Veress and Sz. Leél-Őssy (eds.), *Cave and Karst Systems of Hungary*,
Cave and Karst Systems of the World, https://doi.org/10.1007/978-3-030-92960-2_15

2 The Caves of the Villány Mountains

The Villány Mountains have two big hypogene caves: the Beremend Crystal Cave and the Nagyharsány Crystal Cave. Both of them developed in very clean Cretaceous Nagyharsány Limestone Formation and were discovered during limestone mining at the end of the twentieth century.

Only a few other caves are known in the Villány Mountains. On the southern slope of Tenkes Hill, there is a natural, beautiful, 3 × 4 m cave mouth of a 10 m deep shaft: this is the Máriagyűd Cave, which is 45 m long and 20 m deep (Fig. 1). All the other caves in the Villány Mountains were discovered in an artificial way: mainly the limestone quarries opened their entrances. Caves were found during well-digging: we can see three small caves in a 30 m deep well in the 700-year-old fortress of Siklós. In the slope of the small fortress-hill, there is 60 m long and 24 m deep Váraljai Cave, which is reaching the karst water level. In the cave, there is an almost full spherical cupola; its diameter is 3 m.

Fig. 1 The entrance-shaft of the Máriagyűd Cave (Hegedűs)

It was a surprise when a beautiful natural cave (the Borpince Cave) was found in the big wine cellar. Its entrance is at the deepest point of the cave, on the level of the wine cellar. The walls of this 55 m long and 14 m deep cave are adorned by popcorn and few-centimetre long aragonite bushes. The crystals are absolutely intact!

Near Máriagyűd village, in the Rózsa (Rose) mine (where beautiful Jurassic limestone is mined, which is very rich in Ammonites), a 9 m deep pothole was found: unfortunately, it was mined out.

Templom (Church) Hill in Villány is well-known among palaeontological researchers. Ten small caves were found in the Jurassic limestone quarry: the bigger part of these caves was destroyed by the mine, but the protected shaft cave on Templom Hill is 16 m deep. Extremely rich Pleistocene fauna was found in the tectonic fissure in this mine and on the top of the neighbouring Somssich Hill (Pazonyi et al. 2018). The bones of small mammalian animals (mainly rodents: mouse, hamster, etc.) are very frequent here. Many vertebras of various snakes can also be found here.

The very small hill of Beremend (lower than 50 m, the steeple of the church is almost higher!) on the lowlands along the Drava abounds in caves. Near to the village, there is a big cement plant, and one of its mines is on this hill. Middle Cretaceous limestone is mined here. The mine "ate" the bigger part of the hill by now. During many decades of activity of the quarry, more than a dozen caves were annihilated. (Many palaeontological finds were collected in these caves. The bottom of several caves reached the lukewarm-water lakes). However, the biggest and most beautiful cave, the Beremend Crystal Cave is strictly protected. Mining around the cave has been prohibited. Besides, here are two other (35 and 10 m deep) wet caves. These three caves reach the karst water. (The difference between the bottom of the quarry and the karst water level is only 3 m!). There is a spa near to the quarry. One million-year-old Vertebrata fossils (more than 70 species!) were also found on the hill (Jánossy 1986, 1992).

The other big mine, which is rich in caves is the Nagyharsány quarry. Lower Middle Cretaceous limestone is mined here too, which is transported to the Beremend cement plant. Probably, mining destroyed several small caves also here. However, the small, 9 m deep, and the longest cave of the Villány range, the Crystal Cave of Nagyharsány is protected against mining. There are two natural cave entrances on the hill (originally named Szársomlyó). The length of these small caves is 8 and 3 m.

2.1 Beremend Crystal Cave

The entrance of this cave is at an unlucky place in the middle of the northern wing of the mine. Therefore, the prohibition of mining around the cave hinders the quarrying of limestone, the storage of which would be enough for several years (Fig. 2).

The 850 m long cave is on a 60 × 40 m area only! The absolute vertical difference in the cave is 53 m (Fig. 3).

According to Takácsné Bolner (2003a), based on Ford and Williams (2007), and on the basis of our observations, the Beremend Crystal Cave is a typical hydrothermal (thermal karst) cave system originated by mixture corrosion with a labyrinth-type network of passages characterized by sudden changes of size, by frequency of niches and by extraordinary rich and a variegated ensemble of minerals and forms, originated partly by hydrothermal effects. The crystal cave covers a small area and has no big chambers. Based on Ford (1988), it is a three-dimensional passage-system hydrothermal (hypogene) cave. The passages reach the recent karst water level in the Tavas-terem (Lake Chamber). The traces of the thermal influence were proved by stable isotopes measurements.

The evaporation of the warm karst water offers the possibility for the creation of corrosional pockets also recently at the upper sections of the cave. The active corrosional effect can be observed at the upper part of the cave, close to the surface, where weathered crust and lack of formations on the cave walls are typical. In the upper part of the cave the collapse processes are characteristic.

It is remarkable that some upper parts of the cave dissolved into the old karstic cave fill. Maybe, it was a palaeokarstic hole, which filled up with red clay. The red clay contains many bones of bats and big vertebrates. The "bedrock" of these places was the cave fill (Takácsné Bolner 2003a).

A one-metre wide cave entrance was opened at the base of the wall of the Beremend limestone quarry at 116 m amsl on 12 November 1984. Mihály Magusics, the leader of the shift descended into the first room the Előcsarnok (Entrance Hall). (As it was unsafe due to accidental collapse, it was filled up with stones during the construction of a new entrance). József Kárpát and Katalin Takács-Bolner from the Hungarian Nature Protection Office examined the first part of the new cave (around the Crystal Room). More Hungarian cave researchers—Péter Adamkó and Szabolcs Leél-Őssy among others—exposed the bigger part of the cave with the leading of Katalin Takács-Bolner on 6 December (the Washing Powder Brunch, which reaches the water level, and the Dripstone Hall, Bone Branch). They discovered the deepest point of the cave, the Tavas-terem (Lake Chamber) in January 1985, where crystal-clean water is filling up the bottom of the chamber on a 10–20 m^2 surface (Fig. 4). Katalin Csondor (in her PhD thesis) registered more than 1 m water level changing between 2018 and 2020. The divers sank 10 m down in the lake. The temperature of the

Fig. 2 Limestone quarry of Beremend: the protected cliff (in the middle) containing the Beremend Crystal Cave (Hegedűs)

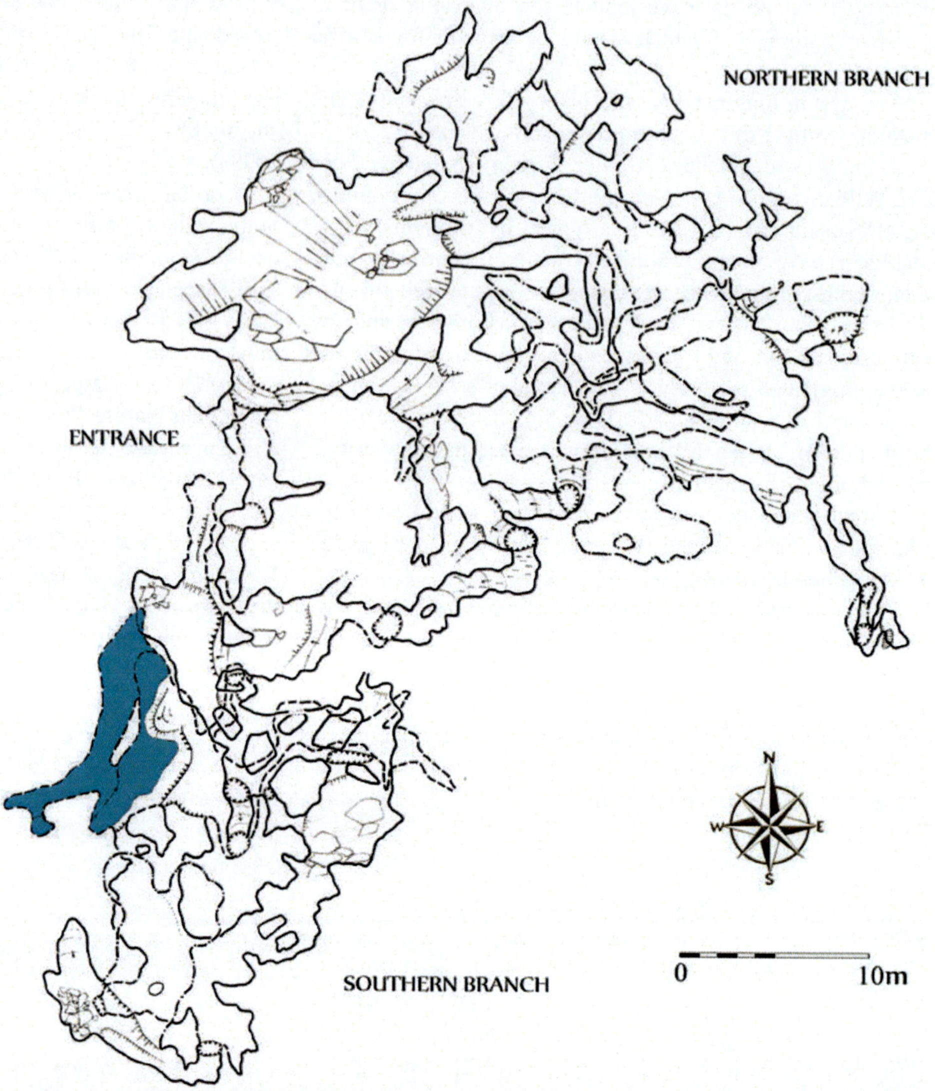

Fig. 3 Ground-plan of the Beremend Crystal Cave

lake is changing between 17.4 and 18.7 °C. She placed 18 pieces of limestone sheet (2 × 5 × 0.5 cm) 0.5 m under the water, in the air 2–3 m above the water level, and on the water–air boundary for 125 weeks. The most intensive precipitation processes were observed on the sheets that were placed on the boundary: rhombohedral calcite crystals and 10–100 micro-millimetre-long crystal needles (maybe aragonites) were precipitated on the surface of the sheets (Csondor 2021).

In the room there was no difference in the crystal precipitations at the surface of the lake: the level of the lake is changeable, which means 1–2 m in a year (see above)! White popcorn can be observed on the wall both above and under the water level.

On the way leading to the Tavas-terem, the 2–3 cm thick soft, grey-white altered rock is covering the wall of the cave, which is wet, so it is easy to smear (Fig. 5). Some calcite veins in the rock were harder and less soluble. Therefore, the surface of the wall looks coffered.

The thick shell of Requienia is visible on the wall of the cave in many places. There is thick soft, cream-colour infilling on the bottom of the Mosóporos ("washing powder") corridor. The temperature in the cave is not uniform: at different points of the caves it is alternating between 15 °C and 19 °C.

There are many beautiful, 1–3 m diameter cupolas in the caves, mainly on the upper part of the cave. (One of them is only 1.5 m from the base of the mine!) The cupolas are free

Fig. 4 Pond in the bottom of Beremend Crystal Cave (photo by Sz. Leél-Őssy)

Fig. 5 Grey-white altered rock wall in Beremend Crystal Cave (Hegedűs)

of precipitations sometimes on the upper part of the cave, but there are rich crystal precipitations sometimes in the middle part. Generally, a third of the ball-like cavities developed/dissolved. Sometimes the spherical cavities are creating a cluster of grapes-like forms.

The most frequent precipitation in the Beremend Crystal Cave is the popcorn. The common botryoid (Leél-Őssy et al. 2011), which is coralloid in Hill and Forti 1997, is everywhere on the walls in the middle and lower parts of the cave, but sometimes in the upper part, too. They often cover the whole wall and appear as 5–10 cm long, ramifying, sheaf-like clusters that stand out perpendicularly from the wall. It forms an agglomeration of spheroidal forms, usually like a green pea (Lake Chamber, Dripstone Chamber, and Crystal Chamber). The colour of the popcorn is always snow-white. The lower, recently developed part of the cave is abounding in precipitations. Here the popcorn is not only on the walls but on the floor too. It is interesting that the wall is red sometimes between the independent, separated peas because of the weathering that originated clay minerals (Washing Powder Corridor) (Fig. 6). Further-growing like a wart (Crystal Room) can be visible on the surface of the peas sometimes (Fig. 7). We can meet the angular botryoids (mentioned by Takácsné Bolner 1981), which look like a ball, but examined closely it is bounded by rhombus faces with an edge length of less than 1 cm. There are many dripstone-forms, logomite and logotite in the Lake Chamber (they consist of popcorns). A few dripstones are covered by popcorn. We did not find aragonite, only calcite here. The most common cave form of calcite, the dripstone, is not too frequent in this cave: it is visible only in the Lake Chamber and in the Dripstone Chamber. Here, 10–20 cm long stalactite is hanging down from the ceiling. Flowstones are also frequent here. Recently, only small stalactites are growing: the big dripstones developed earlier when the surface received more precipitation.

A few, corroded and 1–2 cm thick calcite veins are visible in the upper-middle part of the cave on the wall.

The most valuable and most spectacular mineral of the Beremend Crystal Cave is aragonite, mainly in the middle, sometimes in the upper part of the cave. Aragonite precipitated from the aerosol. The fragile, 1–3 cm long,

Fig. 6 Separated peas on the red wall (Kovács)

Fig. 7 Further-growing popcorns like warts in the Beremend Crystal Cave (Kovács)

Fig. 9 Huntite on the dripstones in Beremend Crystal Cave (photo by Sz. Leél-Őssy)

semi-transparent crystal needles are usually on the popcorns, but sometimes on the walls, or on the clay-fills too (Fig. 8). We also found aragonite crystal needles in the big, tube-form vertebrata bones. (In the Beremend Crystal Cave many bones of Pleistocene mammals were found: Pazonyi et al. 2019). Aragonite needles are always snow-white, like the popcorn. The most beautiful and the biggest aragonite crystals are in the Crystal Room and in the Washing Powder Branch.

Huntites are rare in Hungary, but they are common minerals in this cave (Fig. 9). In 1986, László Bognár detected it in Eötvös Loránd University (Bognár and Kiss 1985) from the Dripstone Room. It was the second description of huntite in Hungary. This mineral is also a snow-white, soft material with high water content: it is mountain milk. In this cave, the huntite appears together with aragonite on the vertical wall and on the surface of popcorns.

The huntites as the last products of the hydrothermal crystallization are typical of the cave system. The X-ray powder diffraction analysis showed magnesite, dolomite, gibbsite, kaolinite, and goethite content too (Vigassy et al. 2010).

Fig. 8 Aragonite crystals in Beremend Crystal Cave (Hegedűs)

The Duna–Dráva National Park gives the entry permission into the cave only for research purpose.

2.2 Nagyharsány Crystal Cave

The big and beautiful hill of Szársomlyó (climbing to the top of the hill needs to have the permission from the Duna-Dráva National Park) is a little independent hill of the Villány Mountains, on the eastern side of the range. Here is somewhat younger Cretaceous bauxite too besides the well-bedded Lower-Middle Cretaceous limestone, but this deposit is not pay ore.

This cave opens in the Nagyharsány limestone quarry, in the western part of the hill. This entrance is on a lucky place, at the southwestern edge of the mine. Therefore, the prohibition of quarrying around the cave (to protect the cave) is not a big problem for the mine. Today, the known length of the cave is 1,500 m, and the depth is 64 m on a 130 × 30 m area (Fig. 10).

The tectonical preformation is most striking in this cave: the bedding of the limestone is between 170° and 180°/55° and 60°. The east–west extension of the cave is parallel to the line of strike. The chambers and the chasms are following the dipping of the limestone. The ceiling and the upper part of the passages are rich in cupolas and in other solution forms (Fig. 11). There is a big difference between the eastern part and the western part of the cave. In the western part, the bedding plane creates the ceiling wall: this is absolutely unique! The section of this part has the form of a triangle. There are many striae on the 30–40 m^2 fault plane. The angular, fracture forms are characteristic in this part (Vigassy et al. 2010). Here the series of the chambers is 20 m wide and 10–15 m high!

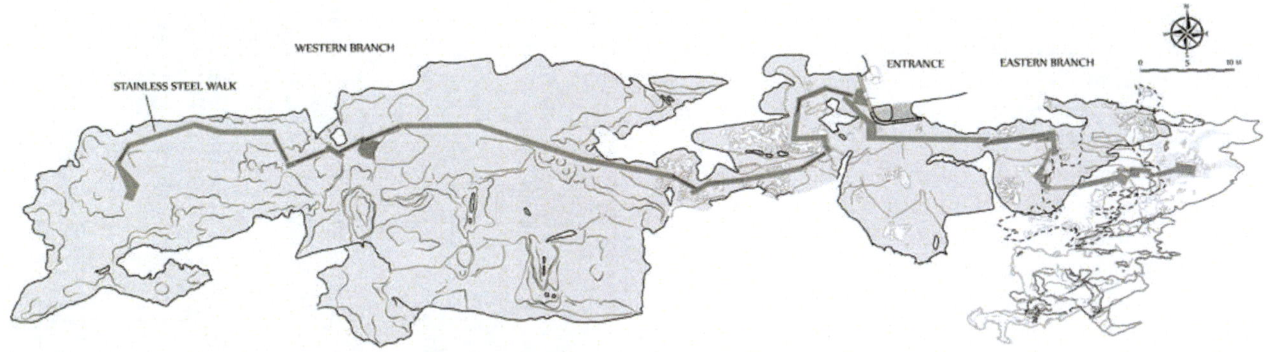

Fig. 10 Ground-plan of the Nagyharsány Crystal Cave

Fig. 11 The upper part of the chasms in Nagyharsány Crystal Cave is rich in solution forms (Hegedűs)

Fig. 12 Thick calcite vein is Nagyharsány Crystal cave (Hegedűs)

The eastern part of the cave is smaller than the western part. Here the extension of the cave is also determined by the dipping of limestone. The passages are following the steep bedding. The lower part is a cupola-labyrinth. The corroded wall surface, the prepared calcite veins are characteristic here. The width of these veins is sometimes 20–30 cm! (Fig. 12). Many cupolas dissolved into the big old veins. The prepared veins make a boxwork net. The cupolas and the situation of the calcite veins show the strong hydrothermal effect. The soft wall surface is very characteristic of the cave (mainly in the eastern part): there are 1–2 cm solution residual deposits on the wall: this is an altered rock. In the eastern part of the cave the hydrothermal minerals are richer than in the western part.

A 2 × 2 m entrance to the cave was opened after an explosion on 17 April 1994 on the 164 m level amsl, but only a few metres were visible from the cave. As every natural cave longer than 2 m is protected in Hungary since 1961, in the next year the leader of the mine asked for permission to mine this part of the mine together with the small cave. When Katalin Takácsné Bolner and Gábor Salamon from the Hungarian Nature Protection Office went to see the object, they were surprised: there was the continuation of the small hole! The new part of the cave is imposing indeed: there are several huge chambers with large chasms. In the next month, further five visits discovered the bigger part of the cave (Takácsné-Bolner 2003b).

There are many bones of mammalian in the western part of the cave, mainly in the Csontos (Bone) Chamber at the western end of the cave. Dénes Jánossy found here the fossils of Pleistocene animals—mainly bats and small rodents. Nowadays, Piroska Pazonyi prepares the rich findings, which are very young (about 40,000 years old—oral communication, 2021).

The Nagyharsány Crystal Cave is very rich in dripstones too. There are big (1–3 m high) stalagmites, and a very big (100 m^2) flowstone in the western part of the cave. The surfaces of the stalagmites and the flowstone are covered by white-yellowish popcorn (Fig. 13). The popcorns are usually on one side of the stalagmites only. There are very long, 1–2 m straw stalactites hanging from the ceiling. The stalagmites belong to two generations: there are some old, very much corroded and some living samples. Seeing two generations of precipitations is unique: popcorn on the old stalagmites, and new dripstones on the popcorn!

There are more and smaller dripstones in the eastern part of the cave. There are no big flowstones, but many one-metre long stalactites, and there are one to one-and-half-metre high thin stalagmites in the lower part. It is difficult to move between the stalagmites and the stalactites in the narrow place without destroying precipitations (Fig. 14). Here the passages are one to three-metre wide and one to two-metre high only. There is no clay sediment at the bottom. The temperature in the cave is around 14 °C, but it is a little bit higher in the lower part. Maybe, there is an air communication with the lukewarm karstic water, which is a few 10-m further down. However, the known passages of the cave did not reach the karstic water level.

The most valuable crystals of the cave are in the western part, in the Csoda (Miracle) Chamber. Here not only many dripstones, but many popcorns occur too. The popcorns create stalagmite-like forms: 30–40 cm high logomites, which consist of pure popcorn, with the hole in the middle (Fig. 15). There are beautiful aragonite crystals on the ceiling at some places. The size of the hemisphere created from aragonite needles is 1–3 cm, and the length of the independent needles is 1 cm.

Fig. 13 Several generations of dripstones and popcorns in Nagyharsány Crystal cave (Hegedűs)

Fig. 14 The lower passage in Nagyharsány Crystal Cave very rich in dripstones (Hegedűs)

Fig. 15 Logomite in Nagyharsány Crystal Cave (Hegedűs)

Also in the Miracle Chamber, one can observe the interesting edge at the mouth of the small passages leading down. These 20–30 cm high edges are consisting of calcite and aragonite. We also found huntite on the surface of the popcorn at some places, and 1–2 mm thick gypsum crust of 1–2 m^2 at one place in the eastern part.

By the X-ray powder diffraction analyses we found 15 minerals in the cave: not only calcite, aragonite, huntite and gypsum, but also ankerite, dolomite, magnesite, goethite, illite, smectite, gibbsite, kaolinite, chlorite, quartz and anorthite (Vigassy et al. 2010). Several minerals did not originate in the cave, but these are terrigenous sediments.

A pavement and ladders from stainless steel were built for protecting the precipitations from the visitors in the biggest part of the Nagyharsány Crystal Cave a few years ago, but there is no electricity. It is not a show-cave: everybody has to ask for permission to the visit every time from the Duna-Dráva National Park and from the active limestone quarry.

3 Conclusion

The lithological and hydrodynamical fundamentals in the Villány Mountains are most favourable for hypogene cave diagenesis. These caves (which are very rich in miracle white aragonite, calcite, gypsum and huntite crystals) are dissolved by rising lukewarm-warm thermal water: these hidden caves are without natural entrance. We could only find these caves with luck (and with the aid of stone mining).

We can preserve the wonderful and extremely sensitive precipitations of these caves only with rigorous rules and with the restriction of the visits.

References

Bognár L and Kiss J. (1985) Kutatási jelentés a Beremendi-kristálybarlang ásványtani vizsgálatáról (Research report on the mineralogical examination). Manuscript, ELTE, Budapest, p 37 (in Hungarian)

Csondor K (2021) A Villányi-termálkarszt hidrogeológiai vizsgálata (Hydrogeological investigation of Villány Thermal Karst). PhD thesis, R. Eötvös University, p 156 (in Hungarian)

Ford DC and Williams P. (2007) Karst Geomorphology and Hydrology. Wiley J West Sussex p 563

Hill CA and Forti P (1997) Cave minerals. Huntswill, Alabama, USA p 463

Jánossy D (1986) Pleistocene vertebrata faunas of Hungary. Akadémiai Kiadó et Elsevier Budapest-Amsterdam, p 208.

Jánossy D (1992) Lower Pleistocene Bord Remains from Beremend (S-Hungary, Lopc. 15. and 16.) Aquila, 99 9–25

Kevi L. (1956) Dél-dunántúli barlangok (The caves of South Transdanubia). Manuscript, Pécs, 20. (in Hungarian)

Kordos L (1991) Villányi-hegység, Csarnóta, pliocén ősgerinces lelőhelyek (Villány Mountains, Csarnóta, Pliocene palaeovertebrata localities) Magyarország Geológiai Alapszelvényei, MÁFI. Budapest 148:6 (in Hungarian)

Kubinyi F (1854) A beremendi mészképlet és az abbani csonttárlat (The limestone of Beremend and the bones in the limestone). Magyar És Erdélyország Képekben II:56–58 (in Hungarian)

Leél-Őssy Sz, Szanyi G, Surányi G (2011) Minerals and speleothems of the Józsefhegy Cave (Budapest, Hungary). Int J Speleol 40(2):191–203

Pazonyi P, Trembeczki M, Mészáros L, Szentesi Z (2019) Preliminary report on the early Pleistocene vertebrate sites of Beremend Crystal Cave (Beremend 16, South Hungary) and on their palaeo-ecological importance. Freagmenta Palaeontologica Hungary C. A. Volume 36 Budapest, 2019:115–140. https://doi.org/10.17111/FragmPalHung.2019.36.115

Pazonyi P, Virág A, Gereb K, Botfalvai G, Sebe K, Szentesi Z, Mészáros L (2018) Sedimentological, taphonomical and palaeoecological aspects of the late early Pleistocene vertebrate fauna from the Somssich Hill 2 site (South Hungary). C. R. Palevol 17:296–309

Rónaki L (2000) A Villányi-hegység és a beremendi rög barlangjai (The caves of Villány Mountains and the block of Beremend). Manuscript, Pécs, p 118. (in Hungarian)

Szabó PZ (1961) 1961) A Mecsek és a Villányi-hegység barlangjai (The caves of the Mecsek and Villány Mountains. Karszt És Barlangkutatás I:3–20 (in Hungarian)

Takácsné-Bolner K (1981) Új feltárások a Pál-völgyi-barlangban (New discoveries in Pál-völgy Cave). Karszt És Barlang 1980:87–92 (in Hungarian)

Takácsné-Bolner K (2003a) Beremendi-kristálybarlang (Beremend Crystal Cave). In: Székely K (ed) Magyarország fokozottan védett barlangjai (Highly protected caves of Hungary). Mezőgazda, Budapest pp 399–403 (in Hungarian)

Takácsné Bolner K (2003b) Nagyharsányi-kristálybarlang (Nagyharsány Crystal Cave). In: Székely K (ed) Magyarország fokozottan védett barlangjai (Highly protected caves of Hungary). Mezőgazda, Budapest pp 404–406 (in Hungarian)

Vigassy T, Leél-Őssy Sz, Demény A (2010) A Nagyharsányi és a Beremendi kristálybarlang és képződményeinek genetikai vizsgálata (Genetic investigations on the precipitations of the Beremend and Nagyharsány Caves). Földtani Közlöny 140(1):73 85 (in Hungarian)

Monitoring and Geochemical Investigations of Caves in Hungary: Implications for Climatological, Hydrological, and Speleothem Formation Processes

György Czuppon, Attila Demény, Szabolcs Leél-Őssy, József Stieber, Mihály Óvári, Péter Dobosy, Ágnes Berentés, and Richard Kovács

Abstract

Cave monitoring was conducted in the Béke and Baradla caves in the Aggtelek Karst, in the Ajándék-Ariadne cave system in the Pilis Mountain, as well as in the Szemlőhegy and Pálvölgy caves in the Buda Mountain. Climatological data outside of the cave (e.g., temperature, amount of precipitation), within the cave (temperature, humidity, CO_2 concentration), as well as the geochemical data of drip water (chemical and isotopic compositions) and carbonates (stable isotope compositions) were compiled in order to constrain the processes that influence the cave air, drip waters, and the precipitated carbonates. The monitoring activities in the Béke and Baradla caves revealed that the speleothems formed in these caves are able to record temperature and hydrological changes, and hence the measured trace element and isotopic compositions together with petrographic characteristics can be used to reconstruct past climate and environmental fluctuations. In addition, the equations that best describe the carbonate-water oxygen isotope fractionation and best reflect the temperature dependence of the isotopic composition of precipitations were identified for paleotemperature reconstruction. The investigations conducted in the Ajándék-Ariadne cave system indicate that the isotope fractionation during carbonate formation in the Legény and Leány caves is kinetically altered. Most promising sites are assumed to be in the inner part of Vacska Cave within this system, where speleothems can be selected for future paleoclimate studies. The monitoring in the caves (Szemlőhegy and Pálvölgy caves) located under urban areas revealed a complex influence of both natural and anthropogenic factors on chemical and isotopic compositions of the drip water.

G. Czuppon (✉) · A. Demény · Á. Berentés
Institute for Geological and Geochemical Research, Research Centre for Astronomy and Earth Sciences, Eötvös Loránd Research Network, Budaörsi út 45., Budapest, H-1112, Hungary
e-mail: czuppon.gyorgy@csfk.org

A. Demény
e-mail: demeny.attila@csfk.org

S. Leél-Őssy
Department of Physical and Applied Geology, Eötvös Loránd University, Pázmány Péter sétány 1/C, Budapest, H-1117, Hungary
e-mail: leel-ossy.szabolcs@ttk.elte.hu

J. Stieber
Stieber Environmental Ltd, Nyerges u. 6, H-1181 Budapest, Hungary
e-mail: stieber@stieber.hu

M. Óvári
Nuclear Security Department, Centre of Energy Research, Eötvös Loránd Research Network, Konkoly-Thege Miklós út 29-33, Budapest, H-1121, Hungary
e-mail: ovari.mihaly@ek-cer.hu

P. Dobosy
Institute of Aquatic Ecology, Centre for Ecological Research, Eötvös Loránd Research Network, Karolina str. 29., Budapest, H-1113, Hungary
e-mail: dobosy.peter@ecolres.hu

R. Kovács
Duna-Ipoly National Park Directorate, Költő u. 21, Budapest, H-1121, Hungary

Keywords

Cave • Speleothem • Monitoring • Drip water • Stable isotopes • Chemical compositions (paleo)climate • Hydrological processes

1 Introduction

Caves represent a complex environment in which speleothems can form and preserve information relating to the past climate and environmental changes (e.g., Fairchild and Baker 2012). The way this information is stored depends on many factors that may vary from cave to cave, or even within the same cave (e.g., Riechelmann et al. 2011). Therefore, it is essential to determine these factors in the caves from where speleothems are taken to be able to

M. Veress and Sz. Leél-Őssy (eds.), *Cave and Karst Systems of Hungary*,
Cave and Karst Systems of the World, https://doi.org/10.1007/978-3-030-92960-2_16

flawlessly reconstruct paleoclimate changes by utilizing the speleothem-based proxies. These kinds of investigation include monitoring of climatological parameters (e.g., precipitation, temperature, relative humidity) within and outside the cave, the measurement of various physical parameters (e.g., drip rate, speed of carbonate precipitation), and complex geochemical analyses (e.g., chemical and stable isotope composition) of regularly taken water samples (precipitation, drip water, soil water), and farmed carbonates. Such monitoring activities allow us to disentangle the complex interactions between environmental/climatological parameters and the response inside the cave, as well as identify the "transformation" of this information to the petrographic, chemical, and isotopic signal of the precipitating carbonates.

In addition, the results from these relatively long-term monitoring activities can be utilized to characterize the hydrological conditions of the (epi) karst and the recent changes in the climatological and/or environmental parameters (e.g., temperature).

Here we present a compilation of the monitoring results (Czuppon et al., 2018, 2021) from different caves in Hungary aiming to depict specific features of them that can help to better understand the climatological, hydrological, and speleothem formation processes.

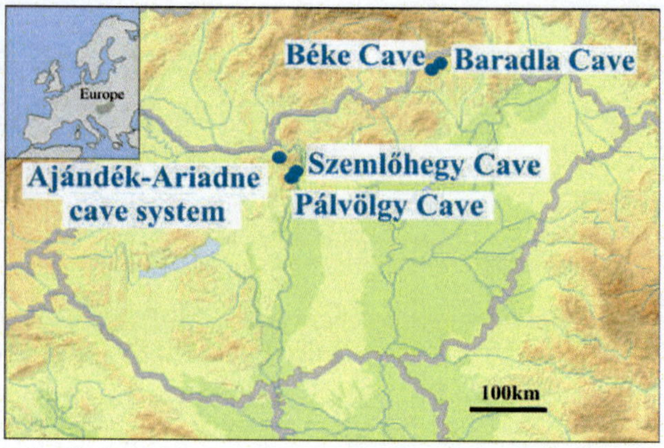

Fig. 1 The monitored caves in Hungary presented in this chapter

2 Studied Caves

Monitoring investigations are being conducted in several caves in Hungary (Fig. 1): the Baradla and Béke caves in the Aggtelek Karst, in Northeast Hungary; the Ajándék-Ariadne cave system in Pilis Mountain, in North Hungary; the Szemlőhegy and Pálvölgy caves in the Buda Mountain located in the city of Budapest. The monitoring activities involve collecting (1) the climatological parameters (temperature, amount of precipitation) outside the cave generally by utilizing the data of the nearest located meteorological station, (2) climatological parameters within the cave (temperature, relative humidity, cave air pCO_2), and (3) precipitation and drip waters for stable hydrogen and oxygen isotope measurement, in order to perform chemical analyses and trace element measurements.

3 Methods

Air temperature and relative humidity were measured manually using a Testo 610 device (temp.: ±0.5 °C in the range between −10 and +50 °C, humid.: ±2.5% at 0–100 °C). Cave air pCO_2 was measured using a portable CO_2 meter (Testo 535; ±3% of the value) in the Baradla and Béke caves and using another portable CO_2 meter (ENVIRO 100 by STIEBER Ltd.; range between 0 and 30,000 ppm; ±3% of the value) in Vacska Cave. Both electrical conductivity and pH value of the water samples were always measured at the site. The electrical conductivity (EC) and pH were determined using a Combo device (HI98129; pH: ±0.05 in the range between 0 and 14; electrical conductivity: ±2% in the range between 0 and 3,999 μS/cm) in the Baradla and Béke caves, while in other caves (Vacska, Szemlőhegy and Pálvölgy caves) using the WTW pH/cond. 3,320 device (temp.: ±0.1 °C in the range between −5 and +105 °C; pH: ±0.01 between −2 and +20; conductivity: ± 1 μS/cm between 0 and 2,000). The results of electrical conductivity and pH are given following temperature correction.

All stable isotope measurements were conducted at the Institute for Geological and Geochemical Research (IGGR) in Budapest, Hungary. Stable carbon and oxygen isotope compositions of approximately 150–200 μg carbonate samples were determined applying the carbonate-orthophosphoric acid reaction at 72 °C (Spötl and Vennemann 2003) and using an automated GASBENCH II sample preparation device attached to a Thermo Finnigan Delta Plus XP mass spectrometer. The isotopic compositions of carbonate samples are expressed as $\delta^{13}C$ and $\delta^{18}O$ in ‰ relative to V-PDB (Vienna Pee Dee Belemnite), with a precision better than ±0.1‰. Stable hydrogen and oxygen isotope analyses of the water samples were carried using the Liquid–Water Isotope Analyser-24d by Los Gatos Research. Precision is better than 1.0 and 0.15‰ for hydrogen and oxygen isotope ratios, respectively. The isotopic compositions of the water samples are expressed as δD and $\delta^{18}O$ in ‰ relative to V-SMOW (Vienna Standard Mean Ocean Water; Coplen 1996). More details of the stable isotope analyses are described inDemény et al. (2019) and Czuppon et al. (2018).

Trace element concentration measurements were conducted at the Institute of Chemistry, Eötvös Loránd University, Budapest, Hungary. About 10 mg of powdered sample was measured into high purity 15 mL polypropylene centrifuge vials using an ultra-micro balance, then dissolved in 12 mL, 0.5 mol/dm^3 nitric acid (Suprapure grade, E. Merck, Darmstadt, Germany). For the quantification 120 µL of 1 mg/dm^3 indium internal standard solution was added. The analysis was carried out using an Element 2 inductively coupled plasma sector field mass spectrometer (Thermo-Finnigan, Bremen, Germany) (see Demény et al. 2017 for more details).

The concentrations of anions (chloride, bromide, nitrate) and cations (ammonium, calcium, magnesium, sodium, potassium) were measured using a Dionex ICS 5000+ dual channel ion chromatography system (Thermo Fischer Scientific, USA) at Danube Research Institute. Alkalinity was determined by the standard titrimetric method (Eaton et al. 2005). Total organic carbon (TOC) and total nitrogen (TN) concentrations were measured by using a Multi N/C 3100 TC–TN analyzer (Analytik Jena, Germany).

4 Béke and Baradla Caves in the Aggtelek Karst

4.1 Background

The Baradla (N 48°28′ E 20°30′) and Béke (N 48°29′, E 20° 31′) caves are mainly located in the Gömör-Torna karst, northeastern Hungary and partly in Slovakia (Fig. 2). A detailed description of these cave can be found in Chap. 10; therefore, only a number of relevant characteristics are summarized briefly here based on Czuppon et al. (2018). Both caves developed in Guttenstein Limestone (with dolomitic layers), Steinalm Limestone, and Wetterstein Limestone formations (Triassic lagoon facies carbonates) (e.g., Hips 2003). The distance between the caves is about 2–3 km. Béke Cave is about 7 km long (in total) involving a main passage (4,300 m long) and two side branches (Felfedező ág and Kavicsos ág; Fig. 2). It has three entrances; the main entrance is situated at about 338 m a.s.l. The thickness of the overburden is around 30–60 m. The sites monitored in the period between 2013 and 2016 are situated along the main passage at different distances from the main entrance: (1) Nagy-tufa (B-NT, 120 m), (2) Felfedező ág (B-FA, 250 m), (3) point B560 (B-560 m), and (4) Buzogány Hall (B-BZ, 1,560 m) (Fig. 18.2). Baradla Cave has a total length of >25 km and five entrances as well as several side branches along the main passage. In this cave one site (Nehéz-út: BAR-NU) was selected for monitoring, approximately 1,100 m from the main entrance along the main passage (Fig. 2). The thickness of the overburden above the studied site is around 90–100 m. In addition to the monitoring campaign of 2013–2016, this site was visited in subsequent periods as well. Intensive monitoring started again at the end of 2019.

4.2 Reflection of Climatological and Hydrological Processes in Cave Climate and Drip Waters

The isotopic, chemical, and trace element compositions and temperature of the drip waters as well as the climatological parameters within the cave represent an intermediate stage between the observed climatological/environmental parameters outside of the cave and proxies preserved in cave carbonates (e.g., speleothems, flowstones). In the Béke and Baradla caves, these parameters were monitored intensively in the period 2013–2016 (Czuppon et al. 2018). Following this campaign only the BAR-NU site in Baradla Cave was monitored for cave air temperature and CO_2 concentration and occasionally for drip water isotope composition (Czuppon et al., 2021).

Temperature

The air temperature recorded in the Béke and Baradla caves varies between 9.3 and 11.4 °C and between 9.6 and 10.6 °C, respectively. The mean annual air surface temperature measured at the Jósvafő meteorological station in the period of 2013–2019 lies within both these ranges (10.2 °C; data source https://www.ogimet.com/). Despite the narrow range at the BAR-NU site clear seasonality can be observed (Fig. 3), while in Béke Cave no seasonality can be detected (Fig. 4). Moreover, in Béke Cave there is a small, but systematic temperature change as a function of distance from the main entrance: the annual average temperatures at the B-NT, B-FA, B-560, and B-BZ sites were 9.8, 10.1, 10.2, and 10.4 ° C, respectively, for the period 2013–2014. At the BAR-NU site in Baradla Cave much longer record of air temperature is available, extending from 1994 to 2020. In this long temperature time series, a weak increasing trend with time seems to appear (Fig. 3a). The temperature increase became more convincing by plotting only the temperatures measured in winter (Fig. 3b). This observation indicates that the apparent increasing trend in Fig. 3a is mainly modulated by the increase of winter temperature (in which case a stronger relationship can be detected, $R^2 = 0.62$).

The mean annual air temperatures measured at the Jósvafő meteorological station in the period 2013–2019 (data source https://www.ogimet.com/) also show higher values than the long term mean annual temperature (8.3 °C) determined based on the period 1961–2010 using the Carpatclim database (Szalai et al. 2013). It seems that the cave air temperature responds to and follows the long-term temperature increase of surface air.

The high resolution temperature measurements of drip water (conducted between the beginning of December 2019 and the end of May 2020 at the BAR-NU site) appear also to

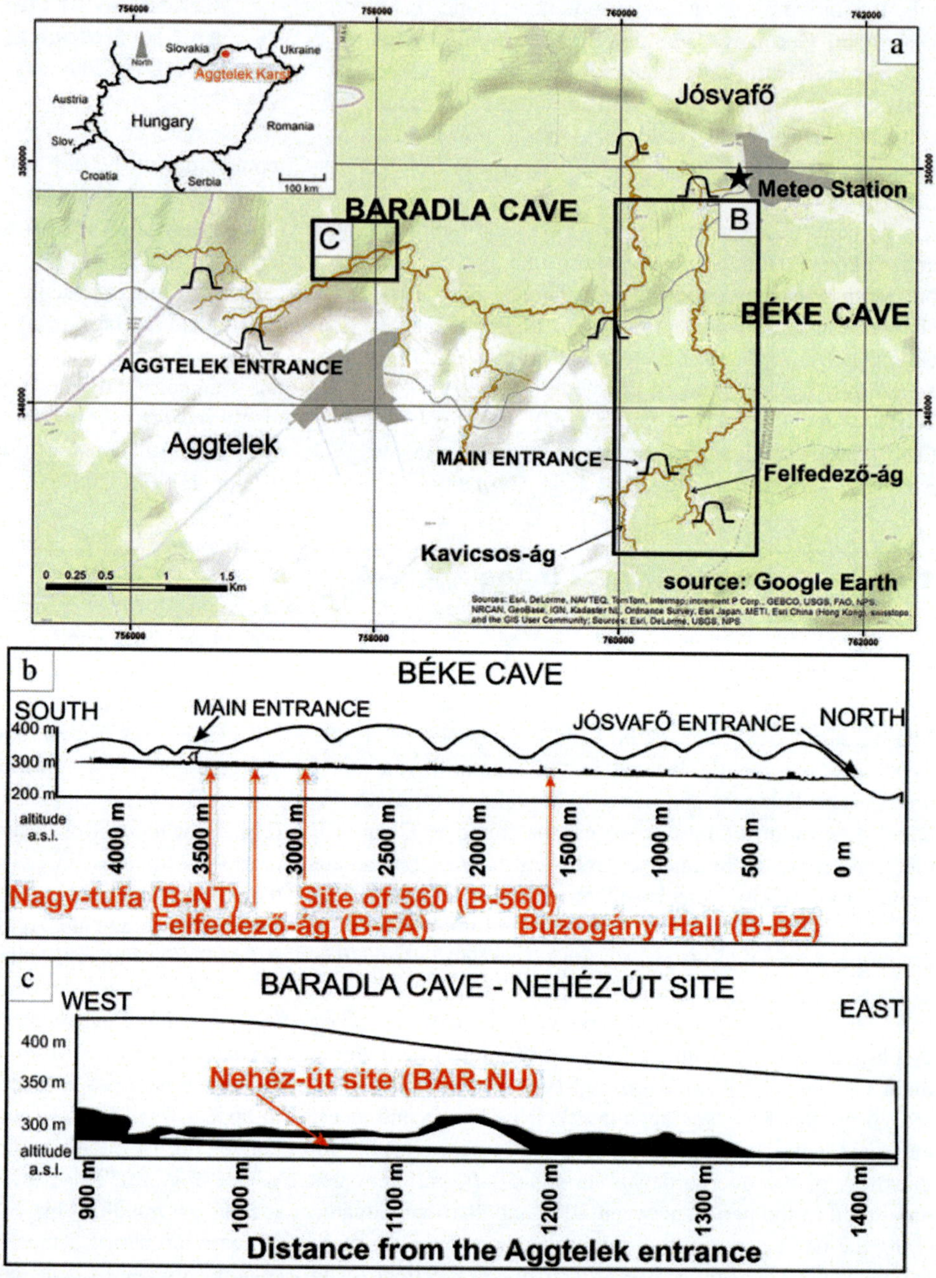

Fig. 2 The locations of Béke (**b**) and Baradla (**c**) caves in the Aggtelek Karst (**a**) showing the sampling sites. Modified after Czuppon et al. (2018)

generally follow the surface air temperature trend but with a much smaller amplitude (Fig. 5, $T_{dripwater}$ varies between 9.6 and 11.4 °C). Interestingly, the smooth temperature curve of drip water is interrupted by sudden temperature drops which tend to emerge following larger precipitation events. This observation suggests that the precipitation events have a significant influence on the temperature of the drip water by lowering it. Moreover, the effect of the precipitation depends strongly on the intensity and type of precipitation. Therefore, the time for the response (i.e., time delay) to the precipitation to appear in the drip water varies between one and six days in the studied period. This temperature fluctuation might also influence the isotope fractionation.

CO_2

The concentration of CO_2 in cave air is one of the most important variables that can influence calcite precipitation within the cave and hence chemical and isotopic composition, and the fabric of the forming speleothems (Spötl et al. 2005; Mattey et al. 2008, Boch et al. 2011; Sherwin and Baldini 2011; Breitenbach et al. 2015). The CO_2

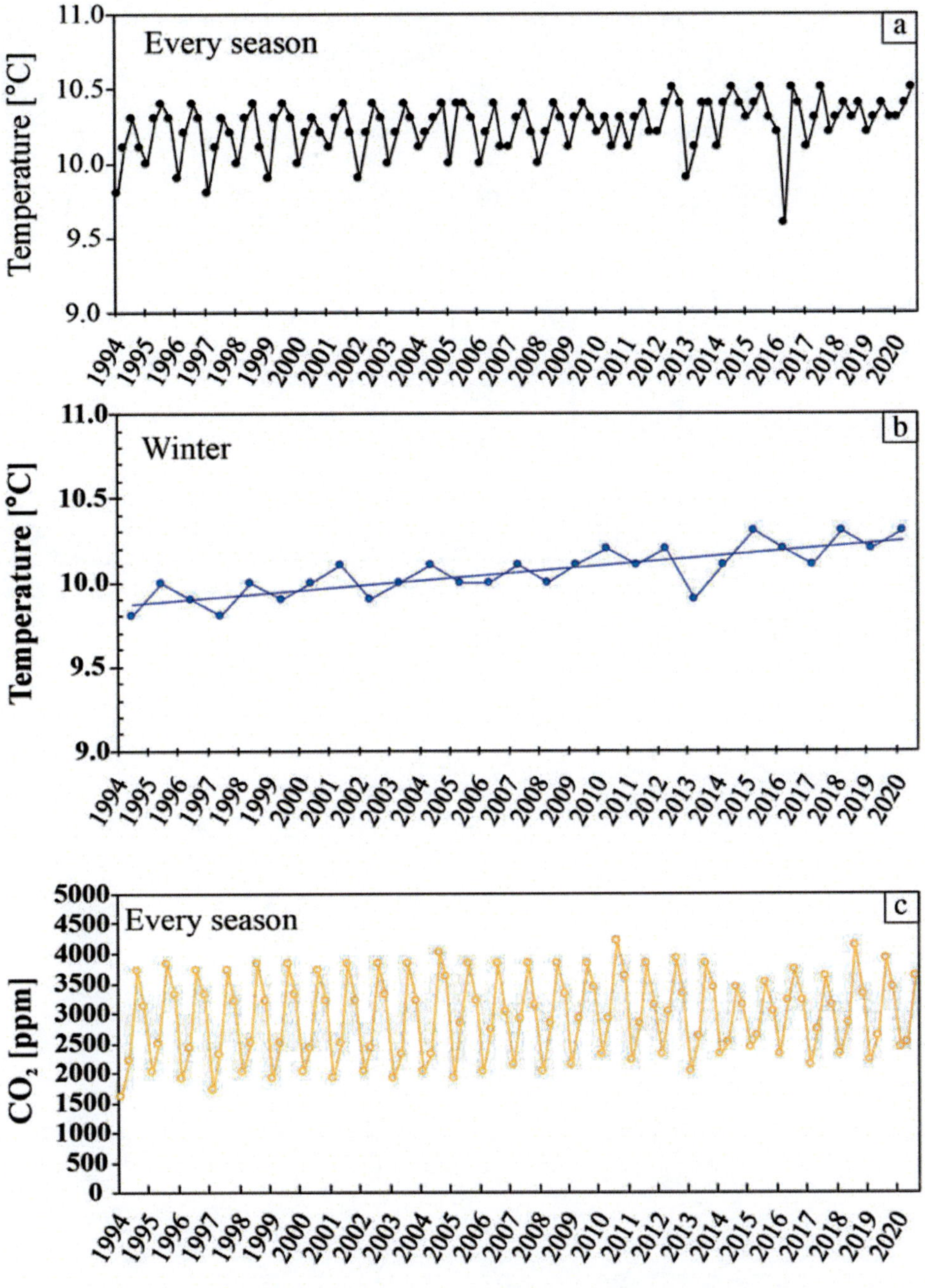

Fig. 3 Temperature (**a**, **b**) and CO_2 (**c**) variations at the Nehéz-út site (BAR-NU) in Baradla Cave. Modified after Czuppon et al. (2018)

concentration and its variability in a cave strongly depends on the ventilation dynamics. The strength and the timing of the ventilation are controlled by the temperature contrast between the cave interior and the outside, synoptic scale atmospheric pressure and wind conditions as well as cave morphology (Mattey et al. 2008). In both the Béke and Baradla caves, pCO_2 was extremely high during summer (Figs. 3 and 4, 30,000 ppm in Béke Cave and 4,200 ppm in Baradla Cave) and moderately high during winter (1,500 ppm in Béke Cave and 1,600 ppm in Baradla Cave). This seasonal variation of pCO_2 has a significant effect on calcite precipitation rate (and hence its stable isotope composition) as greater CO_2 concentrations may hold back carbonate formation. Thus, it is expected that calcite precipitation is enhanced during winter when cave air pCO_2 is drastically lowered (Czuppon et al. 2018).

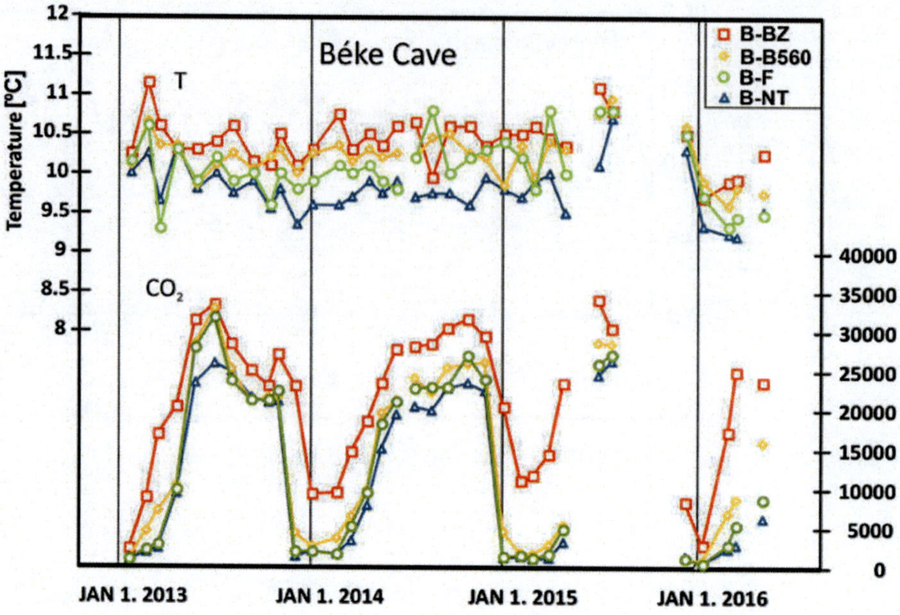

Fig. 4 Temperature (**a**) and CO_2 (**b**) variations in Béke Cave. Modified after Czuppon et al. (2018)

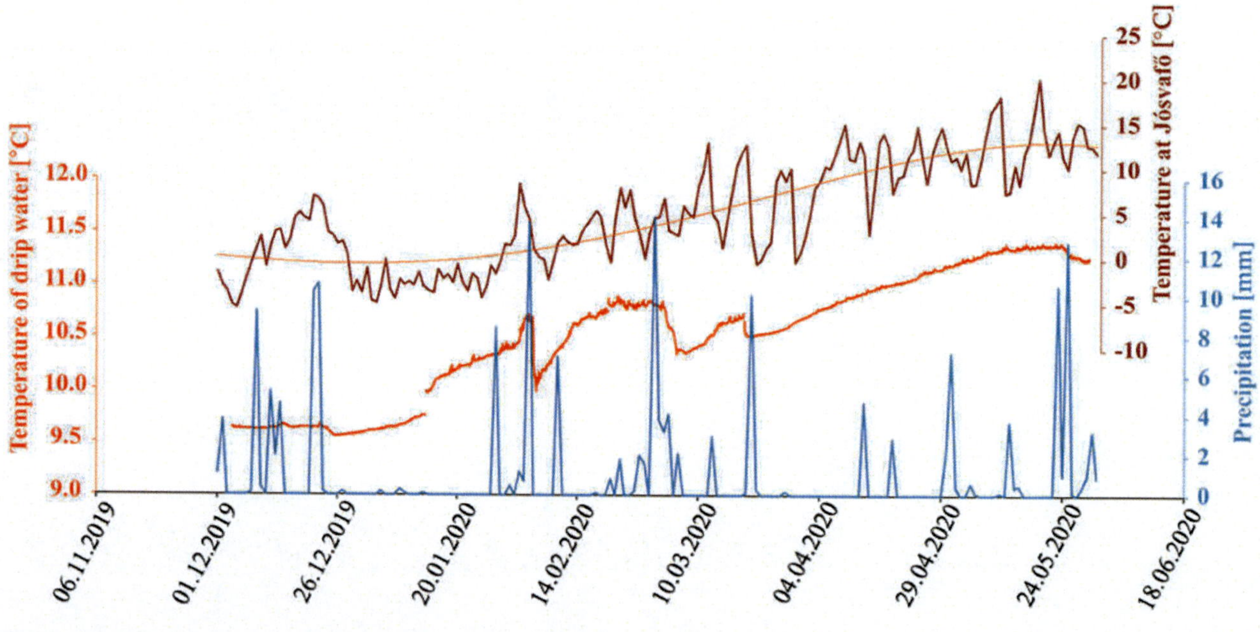

Fig. 5 Temperature variations of drip water (red curve) at the Nehéz-út site (BAR-NU) in Baradla Cave plotted along with air temperature (dark red curve) and precipitation amount (blue curve) measured at the settlement of Jósvafő (*source of the meteorological data* https://www.ogimet.com/)

Stable isotopes

Understanding the relationship between the stable isotope compositions (δD and $\delta^{18}O$) of precipitation and drip water is essential to interpret the hydrogen and oxygen isotope signal preserved in fluid inclusion and carbonates of speleothem. The isotopic composition of precipitation is mainly controlled by the following factors: (1) temperature, (2) moisture source, (3) seasonal effects, (4) continental effect, (5) altitude, and (6) amount effects (Dansgaard 1964; Rozanski et al. 1993; Lachniet 2009). Empirical correlations between these controls and the stable isotope compositions of precipitation need to be established (e.g., Mangini et al. 2005); especially the temperature, seasonal distribution of precipitation and moisture source, and their changes have the most significant impact on the δD and $\delta^{18}O$ values of the infiltrating water and hence the drip water in the studied area.

The monitoring conducted in the period 2013–2016 determined the temperature dependence of the hydrogen and oxygen isotope composition of precipitation observed at the Jósvafő station. Czuppon et al. (2018) obtained 2.5 and 0.33‰/°C for δD-T and $\delta^{18}O$-T annual gradients. The latter agrees well with those calculated for the northern hemisphere (0.31‰/°C; Rozanski et al. 1993).

Moreover, in such calculations and in applications where δD-T and $\delta^{18}O$-T gradients are used, it is also necessary to consider the period when the infiltration takes place. A comparison of the stable isotope composition of annual precipitation and the drip water can reveal the dominant infiltration time. The isotopic compositions of drip waters show constant and similar values ($\delta D_{AVERAGE} = -64.6 \pm 1.4‰$ and $\delta^{18}O_{AVERAGE} = -9.40 \pm 0.20‰$) at each site during 2013–2016 in both the Béke and Baradla caves (Fig. 6; Czuppon et al. 2018). These values are more negative than the amount-weighted stable isotope composition of annual precipitation ($\delta D_{AVERAGE} = -62.3‰$ and $\delta^{18}O_{AVERAGE} = -9.09‰$) suggesting a greater contribution from precipitation in cold months (winter). Although the stable isotope compositions of cave waters during 2013–2016 do not reveal clear seasonality, indicating a relatively well-mixed aquifer above the caves, the δD and $\delta^{18}O$ values of drip water collected at the BAR-NU site in subsequent years show elevated values (Fig. 7) forming an increasing trend from 2013 to 2020. In addition, the amount-weighted stable isotope composition of annual precipitation in 2019 also shows higher values than those in drip water suggesting that even if the isotopic composition of the precipitation has changed, the main infiltration period is still the cold half year (winter). Considering this observation, it is reasonable to separately define the temperature dependence of isotopic composition for the winter half year. Utilizing a relative longer data set from GNIP stations located close to the Aggtelek Karst and following the approach of Kern et al. (2019), Demény et al. 2020a found δD-T gradients of 5.0‰ for the winter half year.

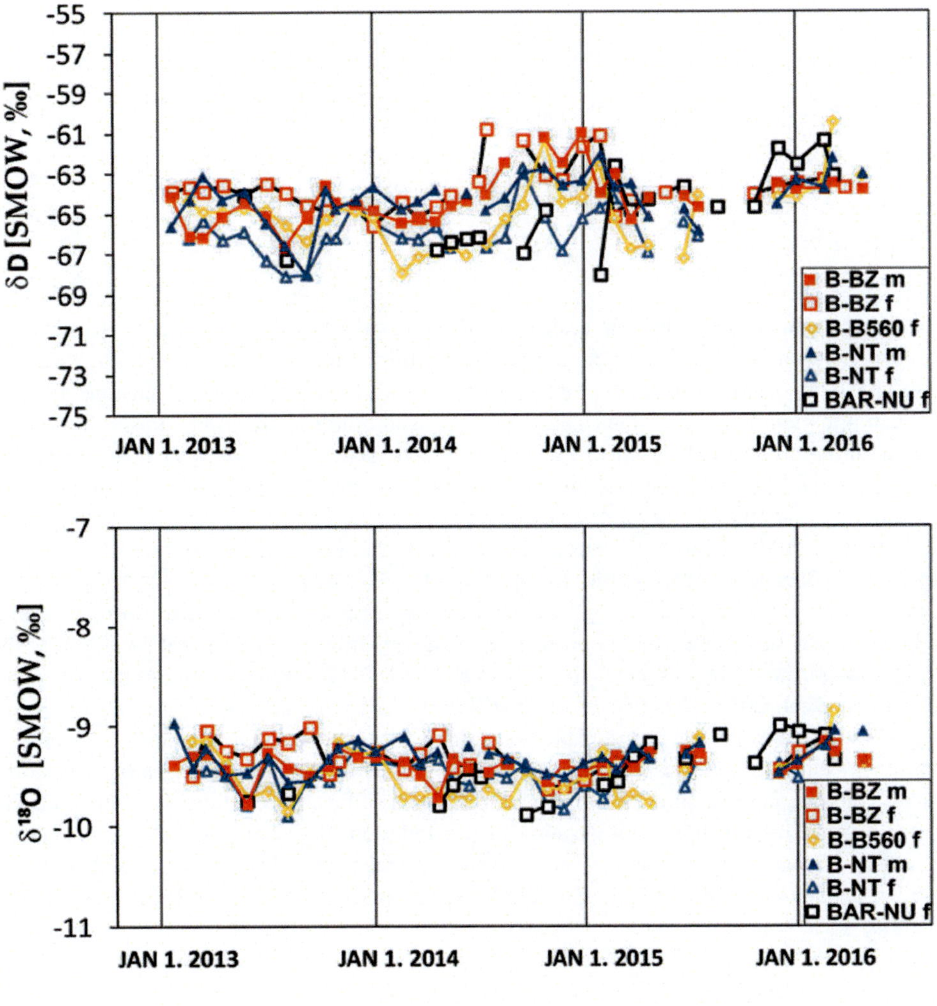

Fig. 6 Stable isotope composition (δD and $\delta^{18}O$) of drip waters in Béke Cave during the monitoring campaign between 2013–2016 (B-NT m: monthly drip water at Nagy-tufa; B-NT f: fresh drip water at Nagy-tufa; B-560 f: fresh drip water at the 560 site, B-BZ m: monthly drip water at Buzogány Hall; B-BZ f: fresh drip water at Buzogány Hall) and in Baradla Cave (BAR-NU f: fresh drip water at Nehéz-út). Modified after Czuppon et al. (2018)

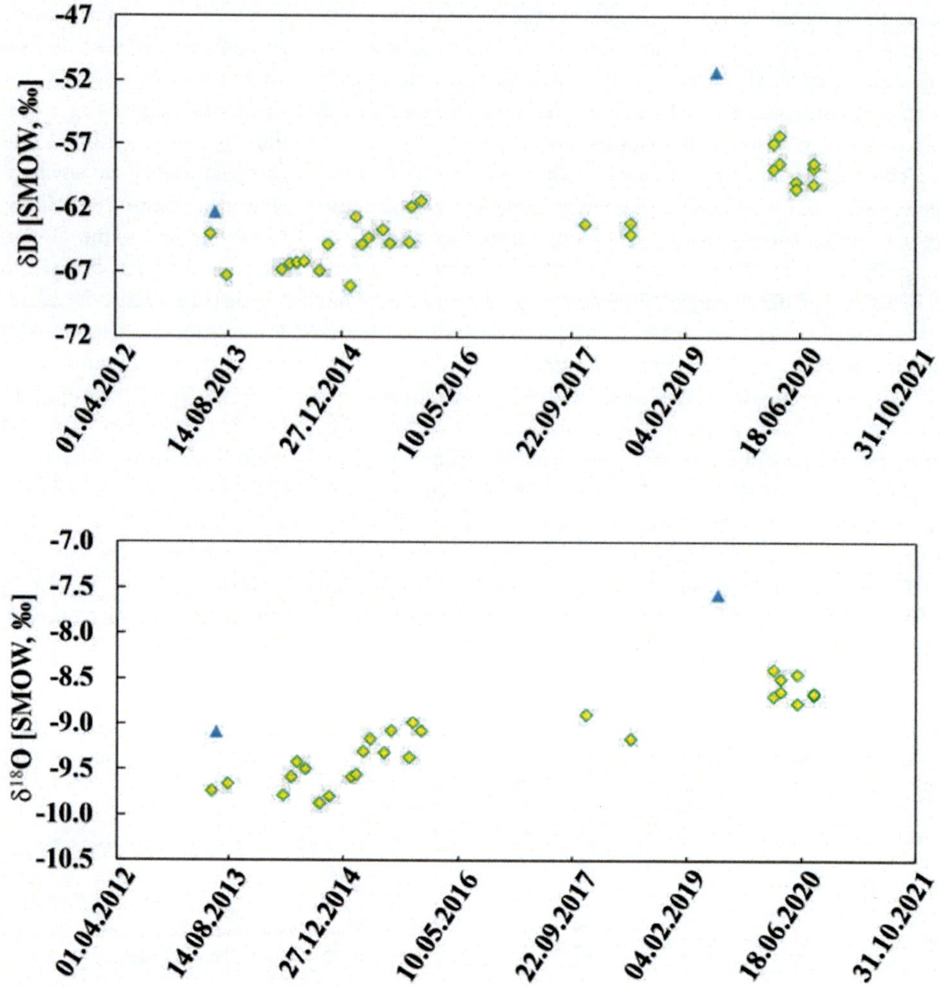

Fig. 7 Stable isotope compositions of drip water collected at the Nehéz-út site (BAR-NU) in Baradla Cave (green rhombus) and the amount-weighted stable isotope compositions of annual precipitation (blue triangle)

Thus, it can be concluded that besides temperature, which is definitely one of the major factors that controls the stable isotopic composition of precipitation and hence drip water, it is also important to consider the changes in the season of the main infiltration period during the interpretation of stable isotope variability of the speleothems collected at these sites.

Finally, it should also be noted that changes in the contribution of different moisture sources could also have significant influence on the stable isotope composition of precipitation and speleothems (e.g., Breitenbach et al. 2010). Hungary and hence the studied sites receive precipitation originating from different sources (Bottyán et al. 2013, 2017; Czuppon et al. 2017). Bottyán et al. (2017) and Czuppon et al. (2017) showed that the precipitation originating from different moisture sources is characterized by a distinct stable isotope composition, resulting in a detectable influence on the stable isotopic composition of precipitation in Hungary. Thus, it is also necessary to consider moisture source changes for the interpretation of the stable isotope data of speleothems.

In summary, it can be concluded that the stable isotope compositions of drip water respond to and follow the longer, intra-annual changes of the δD and $\delta^{18}O$ values of the precipitation, thus the precipitating carbonate can record long-term changes rather than seasonal variability of isotopic composition in precipitation.

Chemistry

Trace element composition of drip water is influenced by the hydrological condition of the epikarst, and thus it can reflect the intensity of the infiltration of rain water and the processes that take place before reaching the drip point. The Mg/Ca and Sr/Ca ratios are commonly used to detect such processes as the changes (i.e., concomitant increase) of these ratios can be indicative of Prior Calcite Precipitation (PCP) (Fairchild et al. 2000, 2006; Tremaine and Froelich 2013; Sherwin and Baldini 2011, and Wong et al. 2011) and of prolonged water–rock interaction in the karst (e.g., McGillen and Fairchild 2005; Borsato et al. 2016). PCP can generally occur during drier periods when the karst is less filled with percolating water, resulting in a greater possibility for degassing, which leads to the precipitation of some carbonate from the water before it is able to reach the cave chamber.

The drip water composition at the Buzogány Hall (B-BZ) and Nagy-tufa sites (B-NT) in Béke Cave revealed a negative relationship between Ca^{2+} concentration and the Mg/Ca ratio, as well as a positive correlation between the Mg/Ca and Sr/Ca ratios (Fig. 8, Czuppon et al. 2018). These findings can be explained by both PCP and prolonged water–rock interaction. The PCP preferentially removes Ca^{2+} with respect to other cations, and thus the Mg/Ca and Sr/Ca ratios will show an increase in the residual water (Borsato et al. 2016). Changes of the length of the water–rock interaction can also affect these ratios, as a shorter water–rock interaction favors preferential dissolution of Ca with respect to Mg and Sr, while longer residence times induce more dissolution of Mg and Sr. As the host rock is not pure calcite but also contains dolomitic layers (i.e., Guttenstein Limestone; e.g., Hips 2003), the different residence times (i.e., different degrees of water–rock interaction) may lead to incongruent carbonate dissolution. To find out which processes are responsible for the correlation between Mg/Ca and Sr/Ca, Borsato et al. (2016) proposed the use of additional element ratios which are not sourced from the carbonate host rock. As Si and Na are derived from siliciclastic minerals from the soil zone or anthrophonic activity (e.g., feeding animals salt during winter time) rather than from carbonate host rock, the Si/Ca and Na/Ca in drip waters can be used to complement Mg/Ca and Sr/Ca. The correlation between Mg/Ca and both Si/Ca and Na/Ca indicates that PCP plays a major role in producing the observed systematics rather than incongruent carbonate dissolution (Czuppon et al. 2018).

4.3 Cave Processes and Climatic Relevance of Isotopic Compositions of Recent Carbonates

Understanding the stable isotope variation of recently forming carbonates in terms of climatic and environmental variability is crucial for the interpretation of the speleothem data. Indeed, constraining the relationship between the $\delta^{13}C$ and $\delta^{18}O$ values measured in the carbonates and all of the monitored factors within and outside of the cave can help to reveal the main variables that govern the isotope systematics in speleothem. In such investigations, carbonates formed on glass plates/bulbs placed in the cave during the monitoring campaign (e.g., Czuppon et al. 2018; Demény et al. 2020b) and actively growing stalagmites with a well constrained age model covering the last decades or centuries (Demény et al. 2017) can be utilized.

The isotopic composition of the recent carbonates formed on glass bulbs in Béke Cave indicates that kinetic fractionation played a significant role during carbonate precipitation. Czuppon et al. (2018) argued that the observed positive shift in both $\delta^{18}O$ and $\delta^{13}C$ values results from the prolonged residence time of the thin solution film on the bulb, which is likely related to a reduced drip rate. Similar observations relating to the relationship between drip rate and isotope composition have been made in Bunker Cave (Riechelmann et al. 2013). Although even kinetically altered speleothem-based isotope records can be used to reconstruct past climate and environmental changes (e.g., Kennett et al. 2012; Tan et al. 2015; Cheng et al. 2016), for calculation of the past temperature, samples less affected by kinetic processes should be selected. Therefore, Czuppon et al. (2018) selected the most negative isotope values of the recently formed carbonates to determine the equation which best describes the isotope fractionation. It was found that the equations determined by Tremaine et al. (2011) and Coplen (2007) are the most adequate to describe the isotope fractionation during the carbonate formation in the Baradla and Béke caves (Fig. 9). Therefore, these equations are recommended for use in speleothem-based climate reconstruction.

Recently, actively forming speleothems with a well-defined age model, that cover the period when instrumental measurements of meteorological parameters are available (e.g., temperature, precipitation amount), provide an opportunity to constrain the factors, even on a relative longer timescale (several decades), that control the isotopic and chemical compositions as well as the petrographic features of the speleothems. Indeed, the investigation of recently formed speleothems from Baradla Cave which cover the last 100 years revealed a close relationship between $\delta^{13}C$ values and the amount of precipitation (Demény et al. 2017). In addition, Demény et al. (2017) suggested the observed variations in $\delta^{18}O$ values are related to the long-term temperature changes. These findings indicate that the carbon and oxygen isotope composition measured in speleothems derived from these caves can be used to reconstruct past changes in the precipitations and temperatures.

5 The Ajándék-Ariadne Cave System in Pilis Mt.

5.1 Backgrounds

The Ajándék-Ariadne cave system is located in Pilis Mountain, north of Budapest (Fig. 10). It is composed of several caves: Legény, Leány, Vacska, Ariadne, Rejtekút, Indikációs, Ősi, and Ajándék. The main goal of this study was to find the location best suited for the collection of

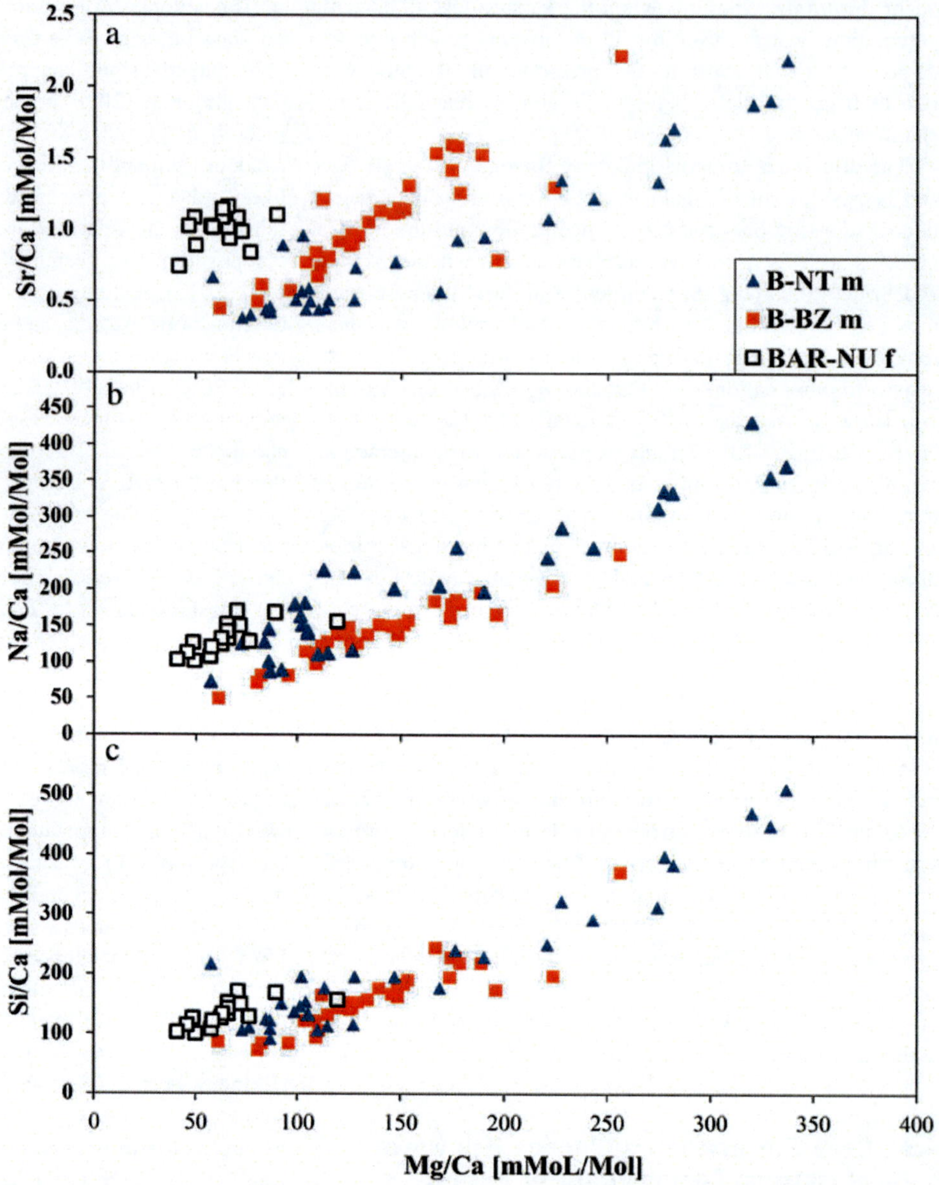

Fig. 8 The Mg/Ca ratios are plotted against Sr/Ca ratios (**a**), Na/Ca ratios (**b**) and Si/Ca for the different drip sites (B-NT m: monthly drip water at Nagy-tufa, Béke Cave; B-BZ m: monthly drip water at Buzogány Hall, Béke Cave; BARNU f: fresh drip water at Nehéz-út, Baradla Cave). Modified after Czuppon et al. (2018)

speleothems for paleoclimate analyses. The main criteria of the site selection are that the site should be closed, stable without large fluctuations in environmental parameters, free of kinetic fractionation, relatively well known with respect to the governing factors that determine speleothem compositions, have an almost constant supply of drip water, be rich in speleothem formations, and should be clean, i.e., free of detrital material. Among these criteria, knowledge of the factors controlling local environmental parameters can be obtained through monitoring. Hence before speleothem collection the potential sites were monitored during a 2-year campaign in Vacska Cave.

Physico-chemical parameters like temperature, drip water pH, cave air CO_2 concentration, solution chemistry, and stable isotope compositions of drip waters were measured with a sampling frequency of 2–4 weeks in Vacska Cave (Czuppon et al., 2021). Six sites were selected for monitoring, with solution chemistry measurements at three sites. CO_2 concentrations were measured at one site with a 3-h measurement frequency with breaks for battery replacement.

Fractionation processes during carbonate formation were investigated using two approaches: (1) calcite deposition on glass slides and collection every two months, and (2) sampling of speleothem surfaces. The main goal is to determine

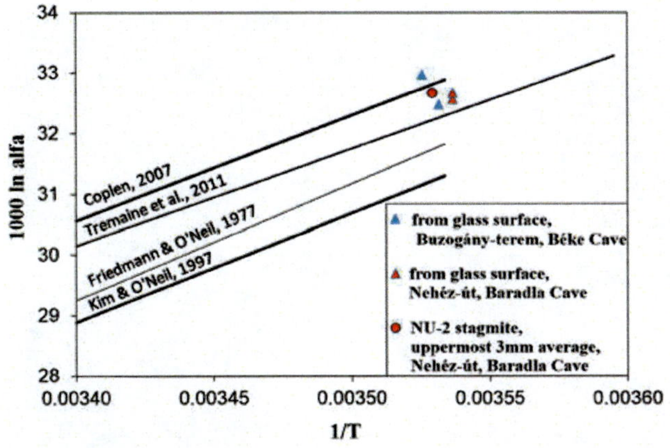

Fig. 9 Calcite-water 1000 lnα values versus 1/T (in °K) measured for carbonate precipitates collected in the Baradla and Béke caves and selected published curves (Coplen 2007; Tremaine et al. 2011; Friedmann and O'Neil 1977; Kim and O'Neil 1997). Modified after Czuppon et al. (2018)

whether the calcite precipitates are in isotopic equilibrium, or if kinetic fractionation affects the isotopic compositions. A constant drip water supply is needed for calcite farming on glass slides, which was observed only in one chamber (Rózsát-terem) in Vacska Cave. At the other cave sites carbonate was collected by scratching the surfaces. Where the speleothem size allowed, several samples were gathered from the stalagmite top to bottom (see Fig. 11). Altogether 23 speleothems locations were sampled this way, covering most of the Ajándék-Ariadne cave system (Fig. 10). Carbonate collection along the drip water path allowed us to gather presumably coeval samples on the stalagmite surface, producing a type of Hendy-test (Hendy 1971) that can detect strong degassing-evaporation processes that lead to kinetic isotope fractionation.

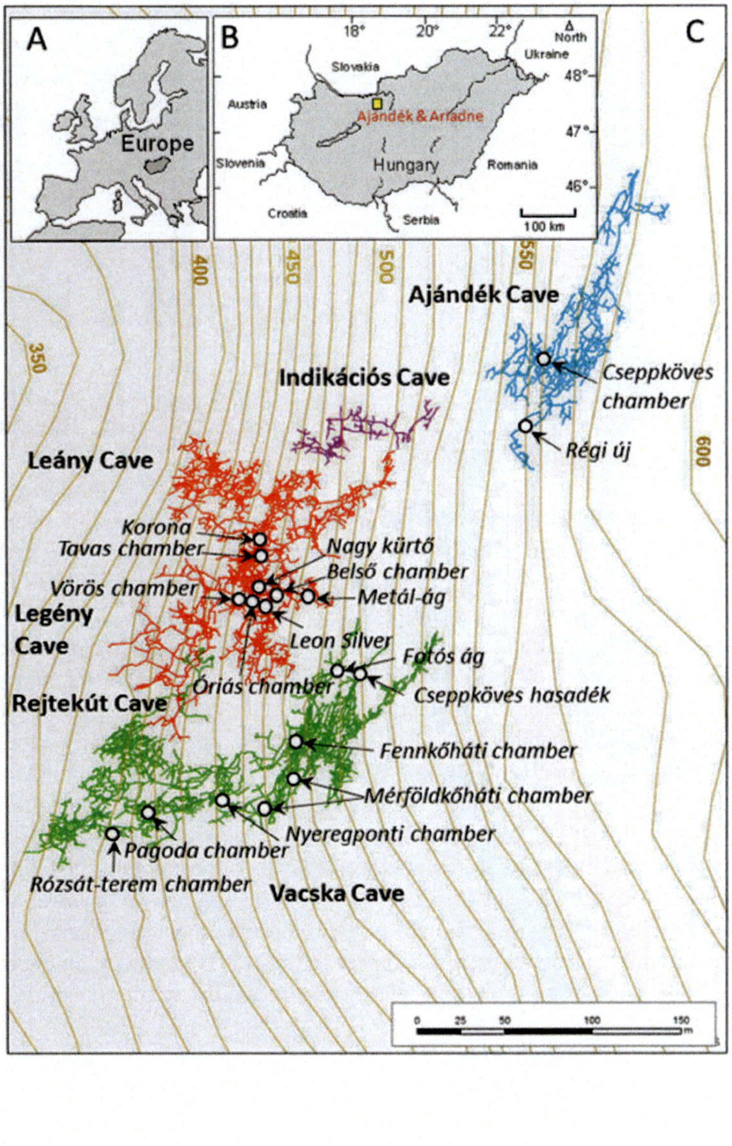

Fig. 10 Map of the Ajándék-Ariadne cave system showing the sampling sites and cave chambers from where the samples were taken

Fig. 11 Sampling positions of a stalagmite surface at the Nyeregponti site, Vacska Cave

5.2 Observations

Physico-chemical parameters

CO_2 concentration was measured at one location in Vacska Cave (Rózsát-terem) and showed well-expressed seasonality with values as low as $\sim$1,500 ppm in winter months and higher concentrations reaching 9,200 ppm in summer months (Fig. 12). Temperature was monitored at three locations. The most stable temperature was found in the Fennkőháti chamber, slight fluctuations were detected in the Mérföldkőháti chamber, and significant changes appeared in the Rózsát-terem chamber (Fig. 12). This last site is situated closest to the cave entrance and the first is the innermost, as expected. Solution pH values show clear seasonal changes at all of the sites, related to the CO_2 concentration variations.

Although strong temperature and pH fluctuations were found in the Rózsát-terem chamber, the stable H and O isotope compositions of drip waters are generally stable, only the Nyeregponti chamber indicates seasonal variations. This seasonality can originate from temporal changes in the winter and summer precipitation contribution, or from a varying degree of evaporation. Increasing ventilation results in stronger evaporation and elevated δD and $\delta^{18}O$ values. The hydrogen and oxygen isotope compositions plot slightly above the Global Meteoric Water Line (GMWL, Craig 1961) (Fig. 13), which may be related to Mediterranean moisture contribution. Although this theory has to be explored further, the recent studies by Bottyán et al. (2017) and by Czuppon et al. (2017) showed that among the marine moisture sources the Mediterranean is the most pronounced in the Pannonian Basin. In addition, the dominantly Mediterranean-originated moisture has characteristically higher d-excess values compared with others. The data of the Nyeregponti chamber define a different correlation with lower slope, typical for evaporated waters. Hence the seasonal variation is not related to the muted transfer of precipitation water δD and $\delta^{18}O$ values, but rather to temporal variations in ventilation.

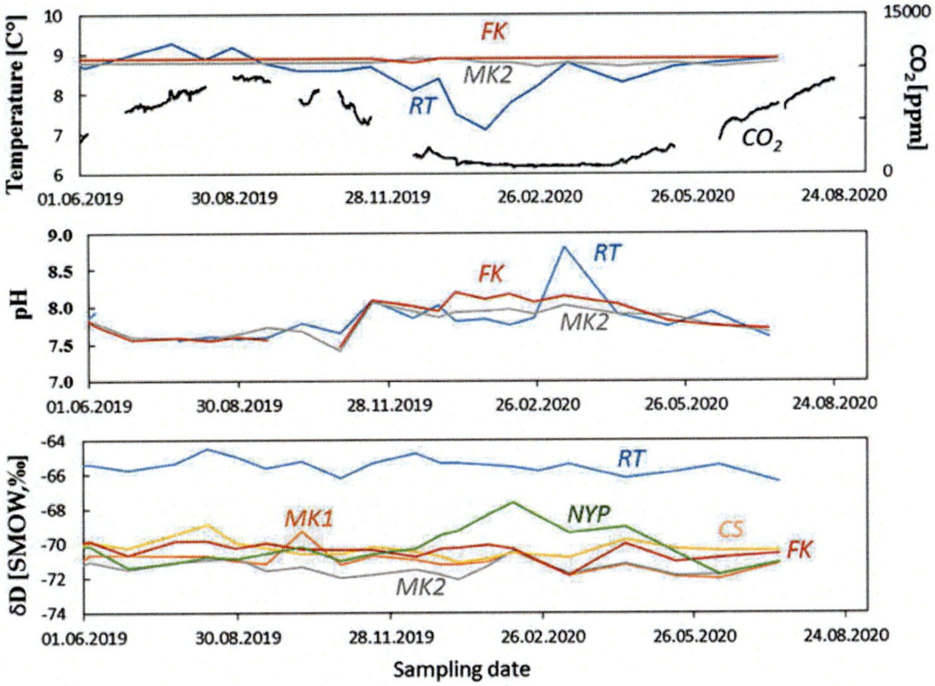

Fig. 12 Temperatures (in °C), CO_2 concentrations in cave air (in ppm), pH values and δD values (in ‰ relative to V-SMOW) of drip waters. RT: Rózsát-terem chamber, FK: Fennkőháti chamber, MK: Mérföldkőháti chamber, CS: Cseppköves hasadék, NYP: Nyeregponti chamber

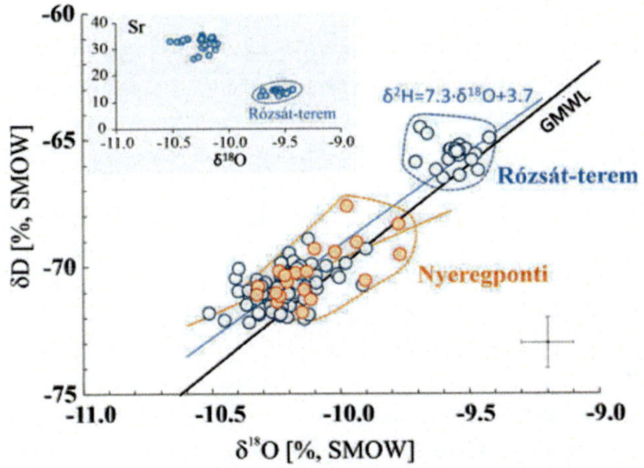

Fig. 13 Stable H and O isotope compositions (in ‰ relative to V-SMOW) of drip waters. GMWL: Global Meteoric Water Line (Craig 1961). Inserted figure: Sr contents (in ppm) and $\delta^{18}O$ values of drip waters

Another important aspect of the water isotope analyses is that the hydrogen and oxygen isotope compositions of the Rózsát-terem chamber (RT) are significantly higher than those of the other sites (Fig. 13). The RT data correspond to the annual average of amount-weighted precipitation compositions, while the other sites indicate a higher degree of winter precipitation contribution. This difference may stem from the seepage route and residence time. Faster infiltration would result in the infiltration of summer precipitation that otherwise evaporates in the soil zone. Residence time differences may be indicated by the chemical composition of the drip water, especially Sr, as longer residence time promotes both rock dissolution and evaporation along the flow route, leading to a Sr concentration increase. Calcite precipitation along the flow route due to evaporation ("Prior Calcite Precipitation", PCP) can further increase the Sr content of the water, and consequently the precipitated carbonate. In Fig. 13 the inserted plot shows that the Sr concentrations are indeed lower at the RT site, suggesting a shorter residence time compared to the other locations.

Among other chemical compounds, phosphorus can provide important environmental information as its concentration is related to $\delta^{13}C$ changes indicating variations in soil activity (Demény et al. 2017). In Vacska Cave, P contents are correlated with Fe and Th concentrations (Fig. 14), suggesting that these elements are related to an influx of detrital material, most probably due to rainfall events. This relationship has to be considered during later paleoclimate/paleoenvironmental reconstruction.

Carbonate formation and stable isotope compositions

The scratched samples cover almost the entire Ajándék-Ariadne cave system, and hence these analyses are discussed first. Figure 15 shows the carbon and oxygen isotope compositions of speleothem surface samples. The Leány-Legény caves (Fig. 15a) display large $\delta^{13}C$-$\delta^{18}O$ scatters compared to the other locations. Based on the on-site observations, the areas of the Óriás chamber, the Nagy Kürtő and Korona sites

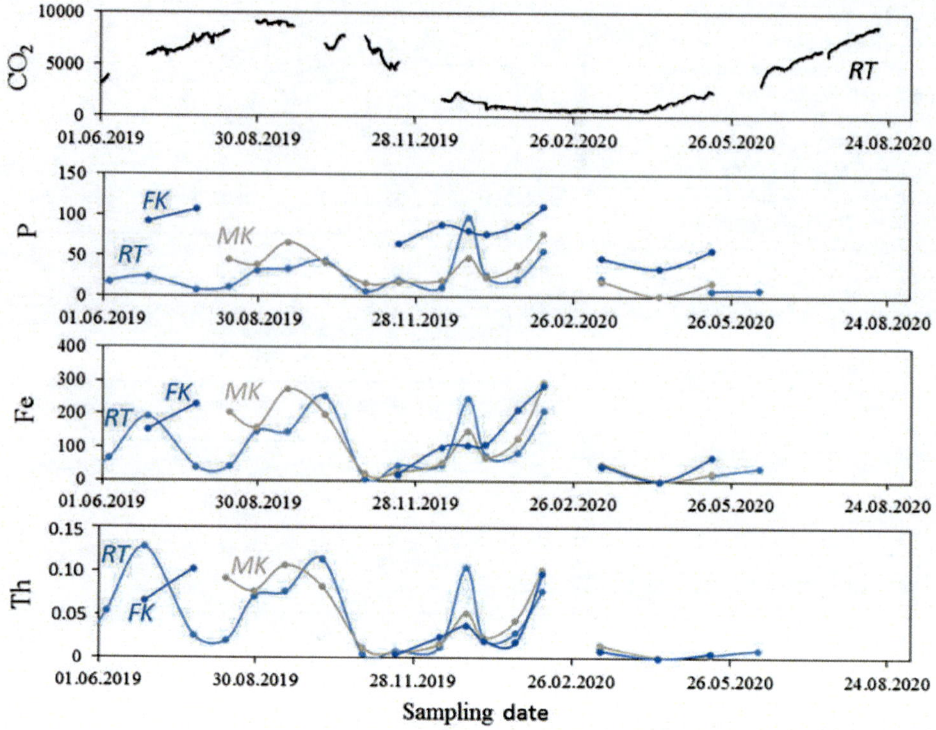

Fig. 14 CO_2 concentrations (ppm) in cave air, as well as P (ppb), Fe (ppb) and Th (ppb) contents of drip waters in Vacska Cave

are well ventilated, which appears in the general positive $\delta^{13}C$-$\delta^{18}O$ correlation within these sites. Additionally, these data are shifted toward less negative $\delta^{13}C$ values, which suggests a higher degree of rock carbonate contribution and decreased soil-derived carbon addition. The other sites with lower $\delta^{13}C$ values define a positive $\delta^{13}C$-$\delta^{18}O$ trend, again indicating ventilation-related kinetic fractionation. In conclusion, the Leány-Legény caves do not seem to be appropriate for further detailed paleoclimate investigations.

Only two speleothems were studied in Ajándék Cave (Fig. 15b). With one outlier, the data define a positive $\delta^{13}C$-$\delta^{18}O$ trend, similarly to the Leány-Legény cave samples. As earlier monitoring provided water $\delta^{18}O$ values and cave temperatures were measured, the expected calcite compositions could be calculated using the Tremaine et al. (2011) equation established for speleothem carbonates. The expected $\delta^{18}O$ value is lower than the lowest oxygen isotope composition obtained for Ajándék Cave, which also suggests that the sampled carbonates are characterized by kinetically affected compositions.

In contrast to the former locations, the samples of Vacska Cave indicate no kinetic fractionation with a general lack of significant positive $\delta^{13}C$-$\delta^{18}O$ correlation (Fig. 15c). Several sites were sampled twice during the monitoring campaign (e.g., the flowstone surface of the Nyeregponti chamber, shown in Fig. 15c), the data are similar; see the Nyeregponti site data marked with different colors and the mixed carbonate sampling locations (#1 and #2 in Fig. 15c). Similarly to Ajándék Cave, expected calcite composition was calculated using the $\delta^{18}O$ values of the drip waters, local cave temperatures, and the calcite-water oxygen isotope fractionation equation of Tremaine et al. (2011). The expected $\delta^{18}O$ (calcite) value is in good agreement with the Nyeregponti chamber data. The large $\delta^{18}O$ scatter observed for the other samples can be due to combined variations in water composition and temperature.

The lowest $\delta^{13}C$ value was obtained for the Rózsát-terem (RT) chamber, where only one stalagmite point was sampled due to preservation rules. As this chamber was characterized by several continuous dripping points, 6 glass slides were placed under the drip waters and collected after ~2 months. The C and O isotope compositions are well correlated (Fig. 15d), indicating that this site is affected by ventilation-driven kinetic fractionation. The lowest $\delta^{18}O$ values closely match the expected calcite O isotope composition calculated as above, but from this composition many of the farmed calcite samples show a strong shift. The detected kinetic effect is in agreement with the conclusions inferred from physico-chemical measurements at the same site (see above).

In conclusion, the potentially most appropriate sites for future paleoclimate and paleoenvironmental investigations are in Vacska Cave, far from the entrance. This does not mean that stalagmites from earlier sampling may not provide

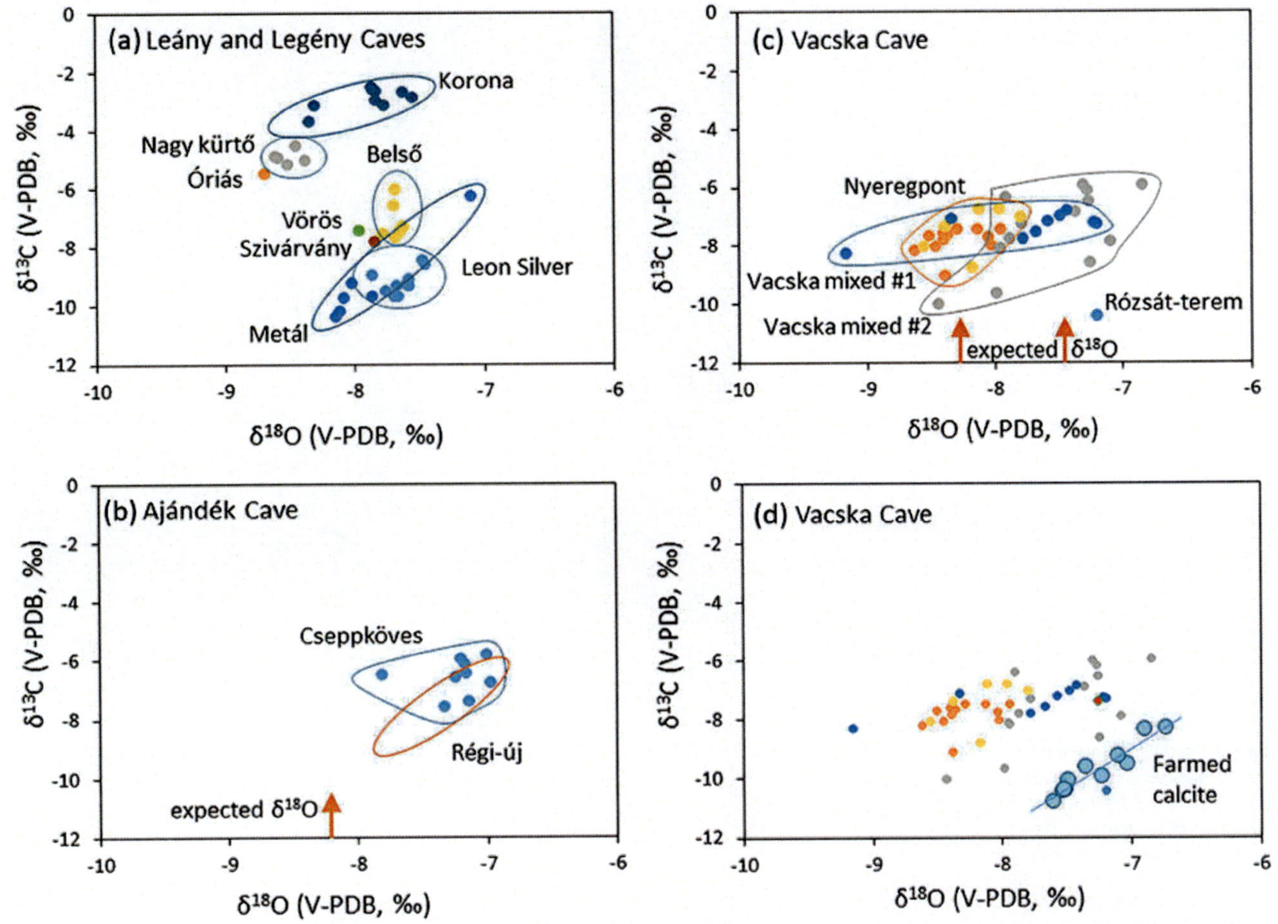

Fig. 15 Stable carbon and oxygen isotope compositions (in ‰ relative to V-PDB) of scratched carbonate samples from speleothem surfaces

valuable paleoclimate information, but future sampling will be done on the basis of the present observations.

5.3 Paleoclimate Perspectives

Several broken stalagmites have been collected from the Ajándék-Ariadne cave system that were sampled with a hand drill and the powder samples were assessed for U-Th age dating. Those stalagmites that showed high-enough U contents and low Th contents were cut in half and one half was polished. Samples were drilled from the polished surfaces, and preliminary U-Th age determinations were conducted at the Institute for Nuclear Research (Debrecen). A stalagmite (V-03) collected near the Fennkőháti chamber with continuous deposition (lack of visible hiatuses), fine and well visible lamination, and with clear calcite crystal appearance (Fig. 16) was selected for detailed investigations. Three preliminary U-Th ages indicate that the stalagmite was deposited from ~10 to ~7 ka ago, potentially covering the intriguing 8.2 ka climate change event (Fig. 16).

The V-03 stalagmite was sampled with 1-mm resolution for stable carbon and oxygen isotope analyses. Plotting the $\delta^{13}C$ values against $\delta^{18}O$ data (Fig. 17), a weak correlation appears, whose slope is much lower than the correlation slope of the farmed calcite. Although complex evaluation of climate-related and site-specific processes is generally needed (Dorale and Liu 2009), the lack of $\delta^{13}C$-$\delta^{18}O$ correlation or a weak relationship may indicate that kinetic fractionation did not significantly affect the stalagmite's isotopic compositions. Detailed radiocarbon and U-Th dating, measurements of stable carbon and oxygen isotope compositions of carbonate at high spatial resolution, stable H isotope analyses of inclusion hosted water as well as trace element concentrations of the stalagmite calcite matrix are planned to determine how the cave system behaved in a major climate change event.

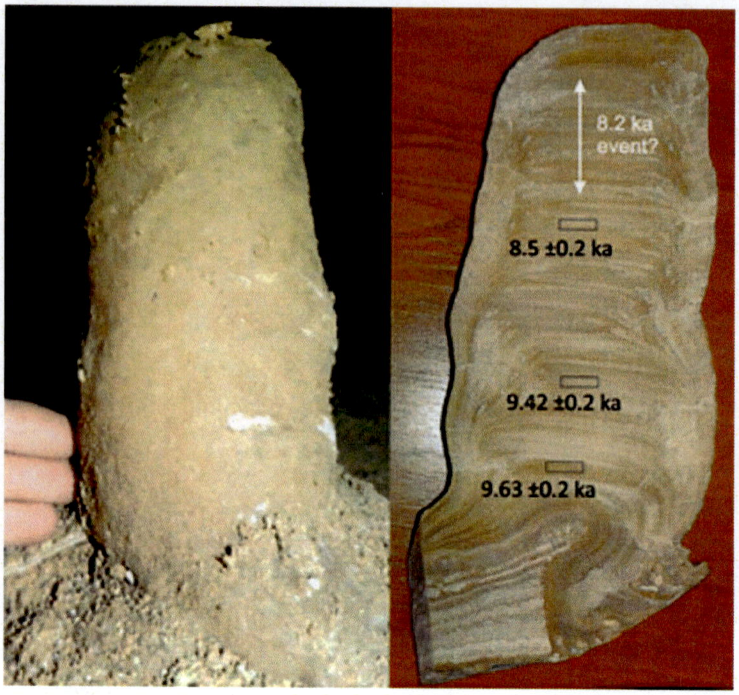

Fig. 16 Stalagmite (V-03) collected near the Fennkőháti chamber (Vacska Cave). Preliminary U-Th ages are shown with their sampling locations

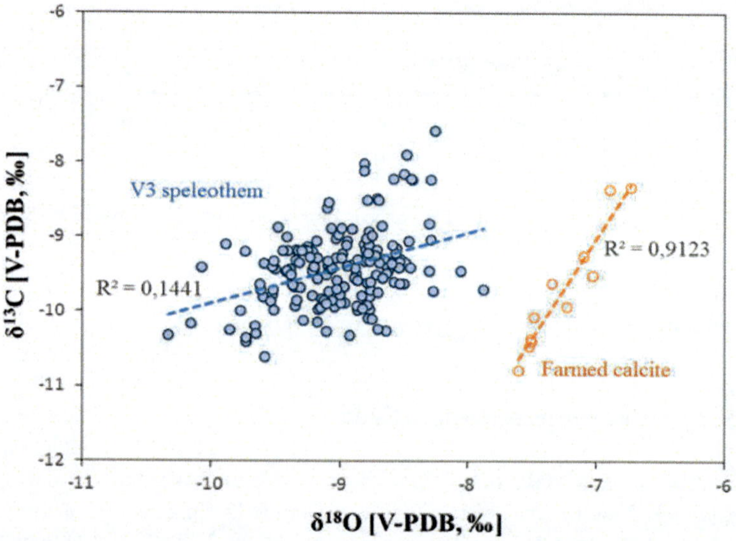

Fig. 17 Stable carbon and oxygen isotope compositions (in ‰ relative to V-PDB) of the V-03 stalagmite collected near the Fennkőháti chamber. Isotope compositions of farmed calcites collected at the Rózsát-terem chamber are shown for comparison

6 Pálvölgy and Szemlőhegy Caves in the Buda Mountains

6.1 Backgrounds

The Pálvölgy and Szemlőhegy caves are located in the Buda Mountains, within the capital city of Hungary, Budapest (Fig. 18). The main part of these caves developed in the Szépvölgy Limestone Formation; only the uppermost channels and chambers extend into the Buda Marl Formation, which overlies the limestone (e.g., Benkovics et al. 1995; Fehér 2011, Virág and Mádl-Szőnyi 2013). The Szépvölgy Limestone Formation deposited during the Late Eocene as the beginning part of a transgressive sequence (Győri et al. 2011). The thickness of the limestone varies between 50 and 80 m. The overlying marl contains various amounts of lime and mud representing somewhat deeper environments (Báldi

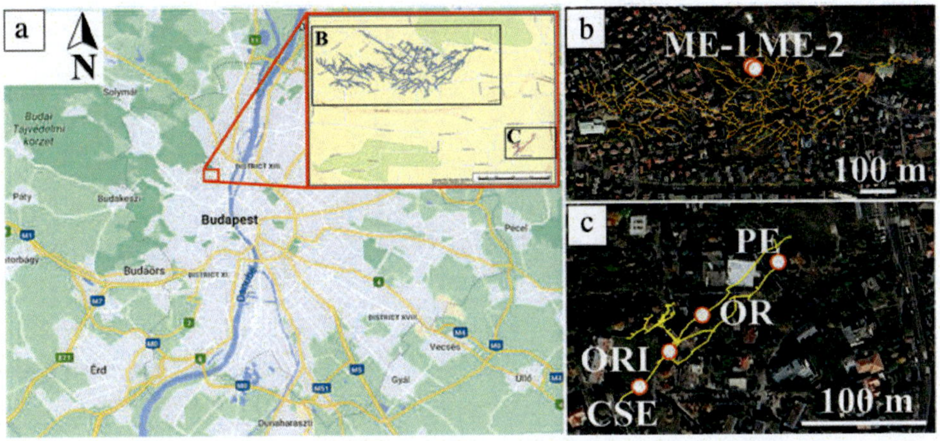

Fig. 18 Location of the Szemlőhegy and Pálvölgy caves in Budapest (**a**) and the sampling sites on the schematic map of the Pálvölgy (**b**) and Szemlőhegy caves (**c**)

1983). The thickness of the marl can reach 60–120 m in some parts in the region. A special characteristic of this marl is that it often contains pyrite (Győri et al. 2011).

Drip waters were monitored along the main passages (Fig. 18) of Szemlőhegy Cave at four sites (Pettyes-terem = PE, Csengő-terem = CSE), Örvény-folyosó = OR, Óriás-folyosó = ORI) and at two sites in Meseország within Pálvölgy Cave (ME-1 and ME-2) from 12 December 2018 to 12 March 2020 with a frequency of two weeks and from 12 March 2020 to 2 July 2020 with a frequency of one month. In addition to these sites, tap water was also sampled every time. At all of these sites, water samples were collected for chemical, trace element as well as stable hydrogen and oxygen isotope analyses. pH, electric conductivity, and temperature were measured inside the caves (Győri et al. 2011).

As both caves (along with some others) are located within the city, several studies have investigated the drip water compositions in order to trace anthropogenic influences (e.g., Fehér 2011; Fehér et al. 2009, 2016; Fehér and Borbás 2014; Virág et al. 2009, Virág and Mádl-Szőnyi 2013; Takácsné-Bolner 2013). Utilizing the results of the most recent monitoring campaign we further explore the natural and anthropogenic factors that affect not just the chemical but also the stable isotope composition of drip waters.

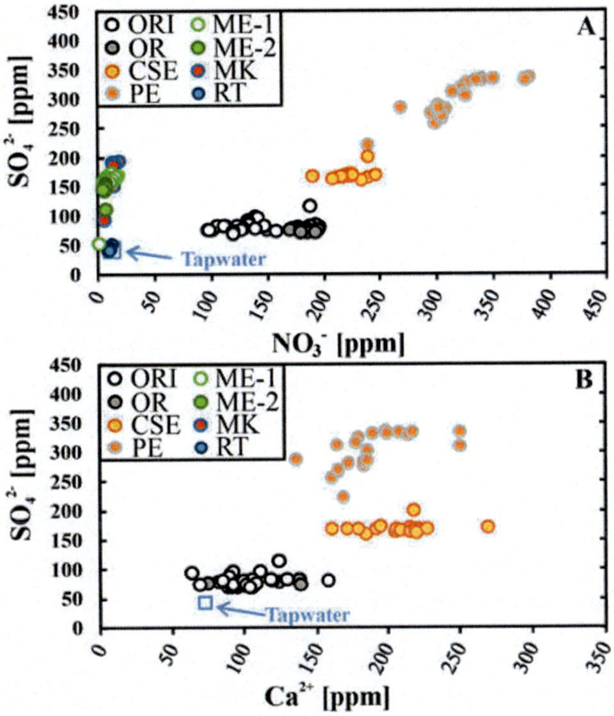

Fig. 19 Sulfate versus nitrate (**a**) and sulfate versus Ca concentrations (**b**) measured in drip water in Szemlőhegy Cave (ORI: Óriás-terem; OR: Örvény-folyosó; CSE: Csengő-terem; PE: Pettyes-terem), in Pálvölgy Cave (ME-1: Meseország 1 site; ME-2: Meseország 2 site) and in Vacska Cave (MK: Mérföldkőháti chamber; RT: Rózsát-terem chamber)

6.2 Observations

Physical and chemical parameters

The studied drip waters in Széplőhegy Cave are characterized by elevated sulfate (SO_4^{2-}), nitrate (NO_3^-), chloride (Cl^-), and sodium (Na^+) contents. The total phosphor (P) shows relatively high values at the OR and ORI sites. The amount of these components varies among the sites; the highest sulfate and nitrate contents have been observed at PE while the lowest at OR. In addition, the measured sulfate content shows a positive linear correlation with nitrate (Fig. 19), while its correlation with Ca is less pronounced. The highest Sr content was found in CSE, which represent the site deepest from the surface (~45 m; Fig. 20) while the lowest was observed at the sites closest to the surface (OR and ORI). SA similar trend can be recognized in case of electric conductivity (Fig. 20). The temperature at the PE site

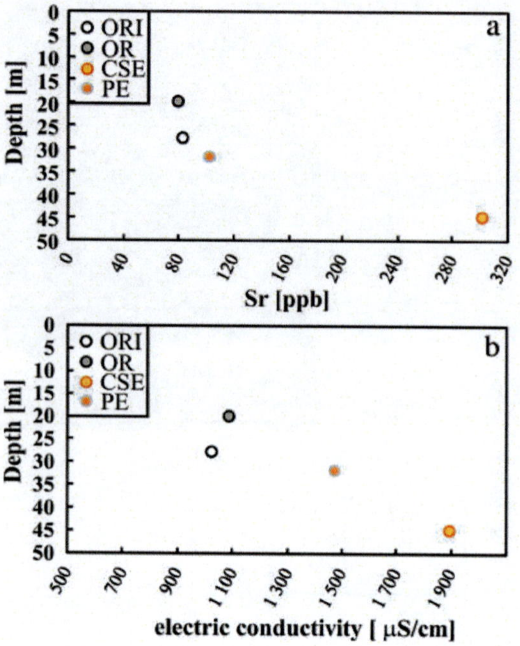

Fig. 20 Sr concentration and electric conductivity in drip water in Szemlőhegy Cave (ORI: Óriás-terem; OR: Örvény-folyosó; CSE: Csengő-terem; PE: Pettyes-terem) as a function of depth

shows somewhat lower values (12.5 ± 0.1 °C) relative to others (CSE = 13.6 ± 0.1 °C; OR = 13.4 ± 0.1 °C; ORI = 13.4 ± 0.1 °C). In addition, no seasonal variations were found, the temperatures at all sites are characterized by stable values during the studied periods.

Stable isotopes

The stable isotope compositions of drip waters display various patterns (Fig. 21A) in Szemlőhegy Cave. The time series of hydrogen and oxygen isotope compositions observed at the OR and CSE sites show no significant fluctuation characterized by very stable values (δD_{OR} = −71.0 ± 0.5‰ and $\delta^{18}O_{OR}$ = −9. ± 0.1‰). In contrast, the δD and $\delta^{18}O$ values measured at the PE site reveal clear seasonality, the greatest amplitudes between the lowest (summer) and highest values (winter) exceed 10 and 1.4‰ for hydrogen and oxygen, respectively. Beside this strong fluctuation, the PE site forms an isotopically distinct group from the others having more positive values (Fig. 22a) plotting closest to the isotopic composition of the amount weighted precipitation. The isotope compositions observed at the ORI site reveal somewhat lower fluctuations, the amplitudes for hydrogen and oxygen are 4 and 0.7‰, respectively. In addition, the stable isotope compositions of tap water also show strong seasonality characterized by lower values during summer and higher values during winter. This pattern actually follows the seasonal variation of the isotopic composition of the Danube because the tap water in Budapest is supplied by river bank filtration.

Similar to Szemlőhegy Cave, the drip sites monitored in Pálvölgy Cave revealed different patterns; one site (ME-1) showed very stable δD (−70.0 ± 1.0‰) and $\delta^{18}O$ (9.9 ± 0.1‰) values during the whole studied period, while the other one (ME-2) is characterized by strong seasonality: δD values range from −64.9 to −47.3‰, while the $\delta^{18}O$ values vary between −9.3 and 7.2‰.

6.3 Drip Waters: Natural and Anthropogenic Influences

In contrast to other caves monitored in Hungary (e.g., Baradla, Béke, Vacska caves; see previous sections), the caves located beneath the capital city of Hungary, which has undergone extensive urbanization in the last 150 years, have been exposed to significant anthropogenic activities. Tracing and determining the possible anthropogenic influences as well as evaluating their magnitude is crucial to protect the sub-surface water system including karst and thermal waters. Only a minor part of Pálvölgy Cave is located beneath woody land, the majority of the studied caves underlie a built-up area including garden suburbs, blocks of flats, pavements and roads (Virág et al. 2009; Fehér 2011; Fehér et al. 2016). Previous monitoring activities suspected significant anthropogenic pollution from water pipes, the sewage system, road salt (which is used as a de-icing chemical during winter) and fertilizer (e.g., Virág et al. 2009; Fehér 2011; Takácsné-Bolner 2013).

In the following, utilizing the results of the most recent monitoring of the chemical and isotopic composition of drip waters, an attempt has been made to constrain the factors that influence or contaminate the infiltrating waters into the caves. The sulfate content displays relatively high values in both Pálvölgy and Szemlőhegy caves partly overlapping the values observed in Vacska Cave (Pilis Mountain) (Fig. 19), which is located in a forest and hence it is lack of urban-influence. The highest values were observed at the PE site in Szemlőhegy Cave. This relatively high sulfate content can originate from both natural and anthropogenic sources. As the overlying marl contains pyrite, and the sulfate can originate from this marl. In addition, gypsum can be often found in the cave, which may also serve as source for sulfate.

The nitrate in Szemlőhegy Cave shows relative good correlation with this sulfate (Fig. 19) implying that some part of the sulfate at least has a common origin with the nitrate. Thus, it can be assumed that the nitrate also derives from the overlying marl. The clay minerals (smektite, illite) can contain ammonia (NH_4^+) , which may be released by thermal decomposition or water–rock interaction and oxidized

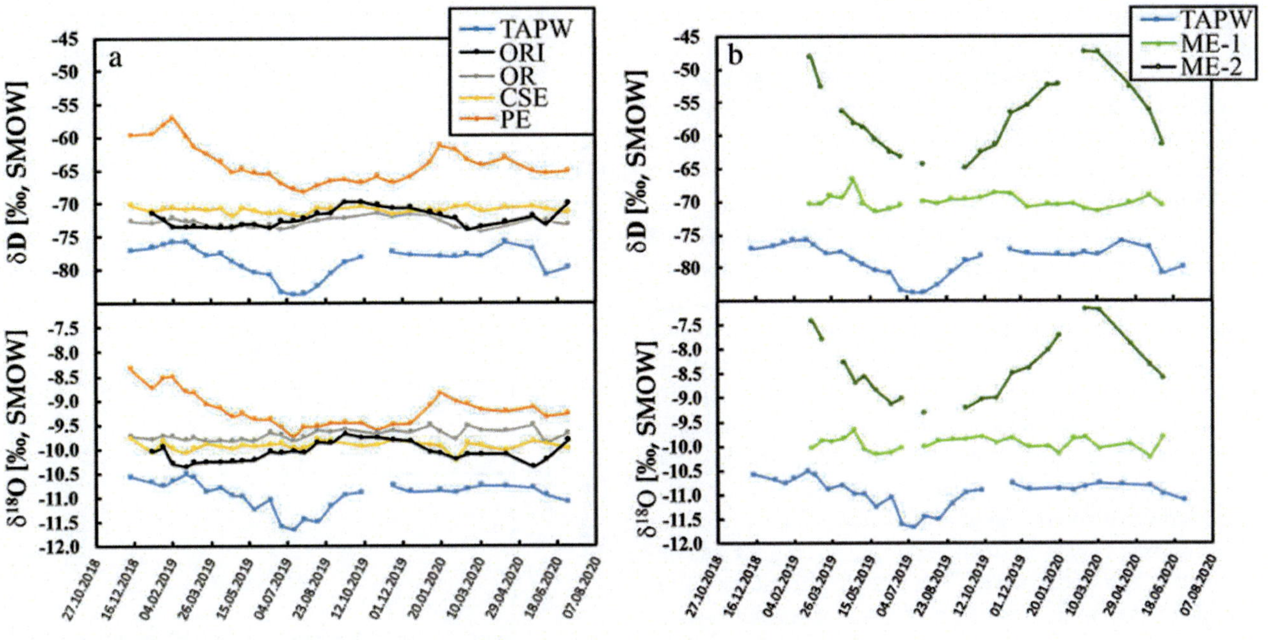

Fig. 21 Stable isotope variations of drip waters as a function of time observed in Szemlőhegy Cave (**a**) (ORI: Óriás-terem; OR: Örvény--folyosó; CSE: Csengő-terem; PE: Pettyes-terem) and in Pálvölgy Cave (**b**) (ME-1: Meseország 1 site; ME-2: Meseország 2 site). Tap water compositions (TAPW) are also shown

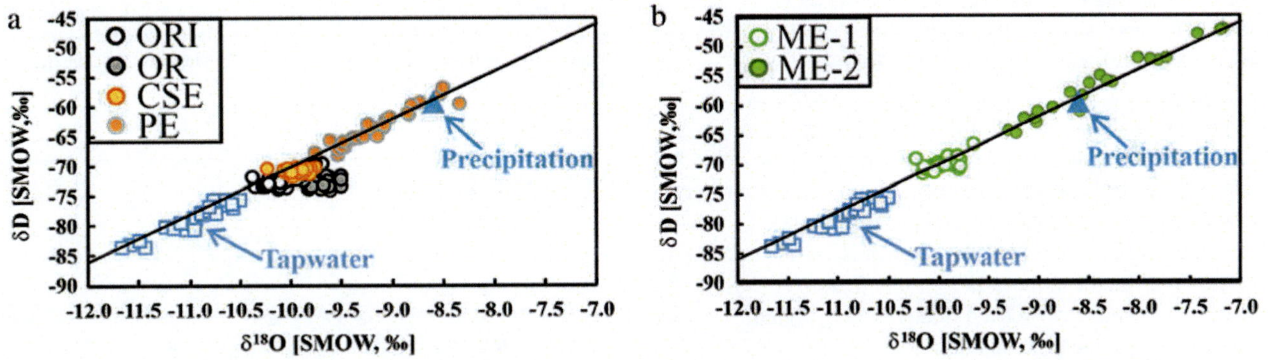

Fig. 22 δ^{18}O versus δD of drip waters from Szemlőhegy Cave (**a**) (ORI: Óriás-terem; OR: Örvény-folyosó; CSE: Csengő-terem; PE: Pettyes-terem), and Pálvölgy Cave (**b**) (ME-1: Meseország 1 site; ME-2: Meseország 2 site). The isotopic composition of tap water and the Global Meteoric Water Line (GMWL, Craig 1961) are also shown

by microbes (e.g., Holloway and Dahlgren 2002; Mingram et al. 2005). Although the Buda Marl contains illite, the ammonium content of this mineral is not known. In addition, the absence of a similarly high nitrate content in Pálvölgy Cave, where the Buda Marl is also the overburden and is not in line with this explanation.

Previous studies have invoked possible contamination from the sewage system to explain the observed characteristics (e.g., Virág et al. 2009; Fehér 2011; Takácsné-Bolner 2013). However, the stable isotope composition found in the drip waters during the 2019–2020 monitoring campaign does not support this explanation. Although the time series of δD and δ^{18}O of the PE site in Szemlőhegy Cave, where the highest sulfate and nitrate contents were found, show a similar pattern to those in tap water, the absolute values differ the most as compared to the other sites (Fig. 22). Therefore, the observed seasonality in PE cannot be explained by contamination from the water and/or sewage pipelines. Instead, it likely reflects the seasonal variation of precipitation at a smaller amplitude and shifted by six months. The amplitude of the seasonal variation of δD and δ^{18}O generally decreases and disappears with the increase of the infiltration (residence) time of the percolating water due to dispersion. The Sr content and the electric conductivity can be also a useful tool to trace this effect. These parameters apparently increase with depth (Fig. 20). Moreover, it should be emphasized that the dispersion of the stable isotope signal of the precipitation, and the increase of electrical conductivity depends on many factors; therefore, they do not always change systematically with depth.

The use of fertilizer in the area above the cave can serve as an alternative explanation, at least partly for the relative high nitrate and sulfate content, as well as for the observed correlation between them. As the PE site of Szemlőhegy Cave has probably the fastest response to the hydrological changes and processes on the surface (based on the stable isotope variation), it may also have the greatest sensitivity for indicating contamination from above. This feature is in agreement with the observation that revealed that this site is characterized by the highest sulfate and nitrate concentrations. Although correlation of sulfate and nitrate with other possible components of fertilizer (e.g., P and K) can be expected, it was not observed. The lack of correlation with P is probably due to the reactions between inorganic phosphate and calcium ions in the water, which results in calcium phosphate. The subsurface water even in the marl should be rich in calcium therefore this reaction is likely to already take place at the beginning of infiltration preventing the fertilizer-derived P to display a systematic signal in the drip water. The potassium in the water migrating downward can be fixed (or released) in the clay minerals through cation exchange (e.g., Simonsson et al. 2009), which may result in unsystematic variation with nitrate in the drip water.

P display a higher concentration only at the OR and ORI sites, which are located relatively close to each other within Szemlőhegy Cave (Fig. 23). As the other sites, especially PE, which are characterized by the greatest nitrate content do no show elevated P concentration, this feature is most likely related to local, cave-related features (e.g., guano) rather than anthropogenic contamination.

Although the drip sites in Pálvölgy Cave are also characterized by high sulfate content, the nitrate does not show similar elevated values to Szemlőhegy Cave or correlation with sulfate. As the overburden in this case also is the Buda Marl Formation, which contains pyrite, no anthropogenic source is required to explain the presence of sulfate in the drip water.

The sodium and chloride ions show elevated values both in the Szemlőhegy and Pálvölgy caves compared to Vacska Cave (Fig. 23). Especially, the ME-1 site in Pálvölgy Cave has an extremely high concentration of these ions (Cl^- up to 1,500 ppm, Na^+ up to 650 ppm). This high concentration is surely related to anthropogenic activity, most likely road salt is main source (e.g., Fehér 2011; Takácsné-Bolner 2013).

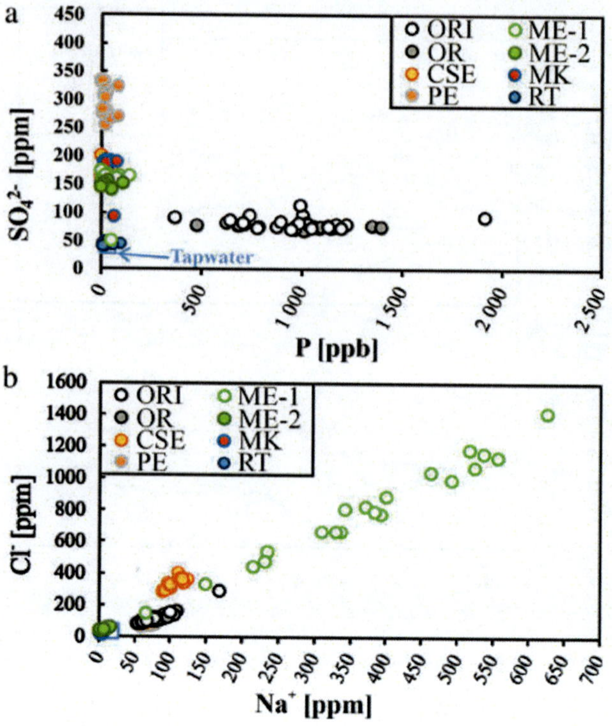

Fig. 23 P versus SO_4^{2-} and Na^+ and Cl^- of drip waters in Szemlőhegy Cave (ORI: Óriás-terem; OR: Örvény-folyosó; CSE: Csengő-terem; PE: Pettyes-terem), in Pálvölgy Cave (ME-1: Meseország 1 site; ME-2: Meseország 2 site) and in Vacska Cave (MK: Mérföldkőháti chamber; RT: Rózsát-terem chamber)

7 Concluding Remarks

(1) Complex monitoring investigations (including regular measurements of climatological parameters within and outside of the cave, chemical and isotopic analyses of drip water and cave carbonates) are essential:
   - to reveal cave and site specific characteristics
   - to constrain the hydrological processes in the relationships of meteorological parameters
   - to reveal and constrain anthropogenic contamination or influences
   - to determine the factors that influence the carbonate precipitation and the observed chemical and isotopic variations in speleothems.

(2) Multiannual monitoring activities in the Béke and Baradla caves revealed that the speleothems from these caves are suitable for recording multiannual winter biased climate trends.

(3) Complex investigations in the Ajandék-Ariadne cave system helped to identify the caves and sites where the carbonate formations are less influenced by kinetic processes and hence provide the most promising location to collect speleothems for reconstruction of past climate and environmental changes.
(4) The monitoring investigations in the Szemlőhegy and Pálvölgy caves revealed natural and anthropogenic factors that influenced the isotopic and chemical compositions of drip waters.

Acknowledgements The study was financially supported by the National Research, Development and Innovation Office, Hungary (OTKA 101664, OTKA 123871 and PD 121387), the Hungarian Academy of Sciences (NANOMIN project, KEP-8/2018), and the Eötvös Loránd Research Network (KEP-1/2020). We are also thankful for the support and permission to the Aggtelek National Park Directorate and Duna-Ipoly National Park Directorate. The János Bolyai Research Scholarship of the Hungarian Academy of Sciences financially supported György Czuppon's work. The work was language-edited by Iain Coulthard.

References

Báldi T (1983) Magyarországi oligocén és alsó-miocén formációk. (Oligocen and lower Miocen Formation in Hungary) Akadémia Kiadó, Budapest, p 293. (in Hungarian)
Benkovics L, Török Á, Nádor A (1995) A Ferenc-hegyi vonulat barlangjainak geológiája. (Geology of caves in Ferenc Hill) Karszt- és barlangkutatás 10:193–207. (in Hungarian)
Boch R, Spötl C, Frisia S (2011) Origin and palaeoenvironmental significance of lamination in stalagmites from Katerloch Cave, Austria. Sedimentology 58:508–531
Borsato A, Johnston VE, Frisia S, Miorandi R, Corradini F (2016) Temperature and altitudinal influence on karst dripwater chemistry: implications for regional-scale palaeoclimate reconstructions from speleothems. Geochim Cosmochim Acta 177:275–297
Bottyán E, Gy C, Weidinger T, Haszpra L, Kármán K (2013) Determination of air moisture source region for precipitation in Hungary. Cent Eur Geol 56:273
Bottyán E, Czuppon G, Weidinger T, Haszpra L, Kármán K (2017) Moisture source diagnostics and isotope characteristics for precipitation in east Hungary: implications for their relationship. Hydrol Sci J 62(12):2049–2060. https://doi.org/10.1080/02626667.2017.1358450
Breitenbach SFM, Lechleitner FA, Meyer H, Diengdoh G, Mattey D, Marwan N (2015) Cave ventilation and rainfall signals in dripwater in a monsoonal setting—a monitoring study from NE India. Chem Geol 402:111–124
Cheng H, Spötl C, Breitenbach SFM, Sinha A, Wassenburg JA, Jochum KP, Scholz D, Li X, Yi L, Peng Y, Lv Y, Zhang P, Votintseva A, Loginov V, Ning Y, Kathayat G, Edwards RL (2016) Climate variations of Central Asia on orbital to millennial timescales. Sci Rep 6:36975. https://doi.org/10.1038/srep36975
Coplen TB (1996) New guidelines for reporting stable hydrogen, carbon, and oxygen isotope-ratio data. Geochim Cosmochim Acta 60:3359–3360
Coplen TB (2007) Calibration of the calcite-water oxygen-isotope geothermometer at Devils Hole, Nevada, a natural laboratory. Geochim Cosmochim Acta 71:3948–3957
Craig H (1961) Isotopic variations in meteoric waters. Science 133:1702–1703
Czuppon G, Bottyán E, Haszpra L, Weidinger T, Kármán K (2017) Significance of the air moisture source on the stable isotope composition of the precipitation in Hungary. EGU General Assembly, Geophysical Research Abstracts, vol 19: EGU2017-13458
Czuppon G, Demény A, Leél-Őssy S, Óvari M, Molnár M, Stieber J, Kármán K, Kiss K, Haszpra L (2018) Cave monitoring in the Béke and Baradla caves (Northeastern Hungary): implications for the conditions for the formation cave carbonates. Int J Speleol 47 (1):13–28
Czuppon Gy, Demény A, Leél-Őssy Sz, Stieber J, Óvari M, Dobosy P, Berentés Á, Kovács R (2021) Cave monitoring in Hungary: An overview. Cent Eur Geol, https://doi.org/10.1556/24.2021.00109
Dansgaard W (1964) Stable isotopes in precipitation. Tellus 16:436–468
Demény A, Németh A, Kern Z, Czuppon Gy, Molnár M, Leél-Őssy S, Óvári M, Stieber J (2017) Recently forming stalagmites from the Baradla Cave and their suitability assessment for climate-proxy relationships. Cent Eur Geol 30(1):1–34. https://doi.org/10.1556/24.60.2017.001
Demény A, Kern Z, Németh A, Frisia, S, Hatvani IG, Czuppon G, Leél-Őssy S, Molnár M, Óvári M, Surányi G, Gilli A, Wu CC, Shen CC (2019) North Atlantic influences on climate conditions in East-Central Europe in the late Holocene reflected by flowstone compositions. Quat Int 512:99–112
Demény A, Rinyu L, Kern Z, Hatvani I, Czuppon G, Surányi G, Leél-Őssy S, Shen C-C, Koltai G (2020a) Paleotemperature reconstructions using speleothem fluid inclusion analyses from Hungary. Chem Geol (revised)
Demény A, Rinyu L, Németh P, Czuppon G, Enyedi N, Makk J, Leél-Őssy S, Kesjár D, Kovács I (2020b) Bacterial and abiogenic carbonates formed in caves—no vital effect on clumped isotope compositions. PLoS One (revised)
Dorale JA, Liu Z (2009) Limitations of Hendy test criteria in judging the paleoclimatic suitability of speleothems and the need for replication. J Cave Karst Stud 71:73–80
Eaton AD, Clesceri LS, Rice EW, Greenberg AE (eds) (2005) Standard methods for the examination of water and wastewater, 21st edn. APHA, AWWA, WEF, Washington, DC USA
Fairchild IJ, Borsato A, Tooth AF, Frisia S, Hawkesworth CJ, Huang YM, McDermott F, Spiro B (2000) Controls on trace element (Sr–Mg) compositions of carbonate cave waters: implications for speleothem climatic records. Chem Geol 166:255–269
Fairchild IJ, Smith CL, Baker A, Fuller L, Spötl C, Mattey D, McDermott F, EIMF (2006) Modification and preservation of environmental signals in speleothems. Earth Sci Rev 75:105–153
Fairchild IJ, Baker A (2012) Speleothem science. Wiley-Blackwell
Fehér K, Kovács J, Márkus L, Borbás E, Tanos P, Hatvani IG (2016) Analysis of drip water in an urban karst cave beneath the Hungarian Capital (Budapest). Acta Carsologica 45(3):213–231
Fehér K, Borbás E (2014) Long-term monitoring of cave drip water at two sampling locations in Pálvölgy Cave. Karsztfejlődés XIX:51–62
Fehér K, Kiss K, Kovács J, Kiss A (2009) Beszivárgás vizsgálatok a rózsadombi termálkarsztron. (Investigations of infiltrating waters in Rózsadomb thermalkarst) Karsztfejlődés XVI:203–221. (in Hungarian)
Fehér K (2011) Újabb adatok a budai termálkarszt beszivárgó vizeinek minőségéről. (New data on the quality of the infiltrating waters in Buda thermalkarst) Karsztfejlődés XIV:45–55. (in Hungarian)

Friedman I, O'Neil JR (1977) Compilation of stable isotope fractionation factors of geochemical interest. In: Fleisher M, Chap KK (eds) Data of geochemistry, 6th edn. US Geological Survey Professional Paper, vol 440, pp 1–12
Győri O, Poros Zs, Mindszenty A, Molnár F, Fodor L, Szabó R (2011) Budai-hegységi paleogén karbonátos kőzetek diagenezistörténete. (Diagenetic history of the Palaeogene carbonates, Buda Hills, Hungary) Földtani Közlöny 141(1):341–361. (in Hungarian)
Hendy CH (1971) The isotopic geochemistry of speleothems 1. The calculation of the effects of the different modes of formation on the isotopic composition of speleothems and their applicability as palaeoclimatic indicators. Geochim Cosmochim Acta 35:801–824
Hips K (2003) Gutenstein formation in the Aggtelek facies of the Silica nappe. Földtani Közlöny 133(4):445–468
Holloway JM, Dahlgren RA (2002) Nitrogen in rock: occurrences and biogeochemical implications. Glob Biogeochem Cycles 16(4):1118. https://doi.org/10.1029/2002GB001862
Kennett DJ, Breitenbach SFM, Aquino VV, Asmerom Y, Awe J, Baldini JUL, Bartlein P, Culleton NJ, Ebert C, Jazwa C, Macri MJ, Marwan N, Polyak V, Prufer KM, Ridley HR, Sodemann H, Winterhalder B, Haug GH (2012) Development and disintegration of maya political systems in response to climate change. Science 338:788–791
Kern Z, Demény A, Persoiu A, Hatvani IG (2019) Speleothem records from the Eastern part of Europe and Turkey—discussion on stable oxygen and carbon isotopes. Quaternary 2:31
Kim ST, O'Neil JR (1997) Equilibrium and nonequilibrium oxygen isotope effects in synthetic carbonates. Geochim Cosmochim Acta 61:3461–3475
Lachniet MS (2009) Climatic and environmental controls on speleothem oxygen isotope values. Quat Sci Rev 28:412–432
Mangini A, Spötl C, Verdes P (2005) Reconstruction of temperature in the Central Alps during the past 2000 yr from a $\delta^{18}O$ stalagmite record. Earth Planet Sci Lett 234:741–751
Mattey D, Lowry D, Duffet J, Fisher R, Hodge E, Frisia S (2008) A 53 year seasonally resolved oxygen and carbon isotope record from a modern Gibraltar speleothem: reconstructed drip water and relationship to local precipitation. Earth Planet Sci Lett 269:80–95
McGillen MR, Fairchild IJ (2005) An experimental study of incongruent dissolution of $CaCO_3$ under analogue glacial conditions. J Glaciol 51(174):383–390
Mingram B, Hoth P, Lüders V, Harlov D (2005) The significance of fixed ammonium in Palaeozoic sediments for the generation of nitrogen-rich natural gases in the North German Basin. Int J Earth Sci 94:1010–1022
Riechelmann DFC, Schröder-Ritzrau A, Scholz D, Fohlmeister J, Spötl C, Richter DK, Mangini A (2011) Monitoring Bunker Cave (NW Germany): a prerequisite to interpret geochemical proxy data of speleothems from this site. J Hydrol 409:682–695
Riechelmann DFC, Deininger M, Scholz D, Riechelmann S, Schröder-Ritzrau A, Spötl C, Richter DK, Mangini A, Immenhauser A (2013) Disequilibrium carbon and oxygen isotope fractionation in recent cave calcite: comparison of cave precipitates and model data. Geochim Cosmochim Acta 103:232–244
Rozanski K, Araguás-Araguás L, Gonfiantini R (1993) Isotopic patterns in modern global precipitation, in Climate Change. In: Swart PK, Lohmann KC, McKenzie J, Savin S (eds) Continental isotopic records, geophysical monograph series 78:1–36. American Geophysical Union, Washington D.C.
Sebastian FM, Breitenbach Jess F, Adkins H, Meyer N, Marwan KK, Kumar GH, Haug (2010) Strong influence of water vapor source dynamics on stable isotopes in precipitation observed in Southern Meghalaya NE India. Earth and Planetary Science Letters 292(1–2):212–220. https://doi.org/10.1016/j.epsl.2010.01.038
Sherwin CM, Baldini JUL (2011) Cave air and hydrological controls on prior calcite precipitation and stalagmite growth rates: implications for palaeoclimate reconstructions using speleothems. Geochim Cosmochim Acta 75:3915–3929. https://doi.org/10.1016/j.gca.2011.04.020
Simonsson M, Hillier S, Öborn I (2009) Changes in clay minerals and potassium fixation capacity as a result of release and fixation of potassium in long-term field experiments. Geoderma 151:109–120
Spötl C, Vennemann TW (2003) Continuous-flow isotope ratio mass spectrometric analysis of carbonate minerals. Rapid Commun Mass Spectrom 19(9):1004–1006
Spötl C, Fairchild I, Tooth AF (2005) Cave air control on dripwater geochemistry, Obir Caves (Austria): implications for speleothem deposition in dynamically ventilated caves. Geochim Cosmochim Acta 69(10):2451–2468
Szalai S, Auer I, Hiebl J, Milkovich J, Radim T, Stepanek P, Zahradnicek P, Bihari Z, Lakatos M, Szentimrey T, Limanowka D, Kilar P, Cheval S, Deak Gy, Mihic D, Antolovic I, Mihajlovic V, Nejedlik P, Stastny P, Mikulova K, Nabyvanets I, Skyryk O, Krakovskaya S, Vogt J, Antofie T. Spinoni J (2013) Climate of the Greater Carpathian region. Final Technical Report. www.carpatclim-eu.org
Takácsné Bolner K (2013) Az 1987–1994 közötti időszak csepegővíz-vizsgálatainak tanulságai. (Investigation of drip waters in the period of 1987–1994) In: Mindszenty A (ed) Budapest: földtani értékek és az ember (Budapest: geoheritage and humans - urban geological studies). ELTE Eötvös Kiadó Budapest 234–236. (in Hungarian)
Tan L, Cai Y, An Y, Cheng H, Shen CC, Breitenbach SFM, Gao Y, Edwards RL, Zhang H, Du (2015) A Chinese cave links climate change, social impacts, and human adaptation over the last 500 years. Sci Rep 5:12284. https://doi.org/10.1038/srep12284
Tremaine DM, Froelich PN (2013) Speleothem trace element signatures: a hydrologic geochemical study of modern cave dripwaters and farmed calcite. Geochim Cosmochim Acta 121:522–545
Tremaine DM, Froelich PN, Wang Y (2011) Speleothem calcite farmed in situ: modern calibration of $\delta^{18}O$ and $\delta^{13}C$ paleoclimate proxies in a continuously-monitored natural cave system. Geochim Cosmochim Acta 75:4929–4950
Virág M, Mádl-Szőnyi J (2013) A Szemlő-hegyi-barlang csepegővizeinek vizsgálata a rózsadombi törmeléktakaró beszivárgásmódosító hatásának értékelése céljából. (Investigation of drip waters in Szemlőhegy Cave in order to evaluate the influence of the overburden on the infiltration). In: Mindszenty A (ed) Budapest: földtani értékek és az ember (Budapest: geoheritage and humans—urban geological studies). ELTE Eötvös Kiadó Budapest: 239–243. (in Hungarian)
Virág M, Mádl-Szőnyi J, Mindszenty A, Zihné Perényi K, Leél-Őssy Sz, Erőss A, Siklósy Z (2009) Az urbanizáció hatása a budai barlangok csepegő vizeire a Szemlő-hegyi-barlang példáján. (Urban influence on the drip waters of Szemlőhegy Cave). Karsztfejlődés XIV:57–81. (in Hungarian)
Wong CI, Banner JL, Musgrove M (2011) Seasonal dripwater Mg/Ca and Sr/Ca variations driven by cave ventilation: implications for and modeling of speleothem paleoclimate records. Geochim Cosmochim Acta 75:3514–3529. https://doi.org/10.1016/j.gca.2011.03.025

Sedimentologic, Taphonomic Processes and Paleontological Values of Hungarian Caves

László Kordos

Abstract

The study is an updated summary of the paleontological values of the caves in Hungary in the light of sedimentologic and taphonomic conditions. The localities discussed in chronological order contain their locations, history of excavation, important sedimentologic and taphonomic characteristics, typical paleontologic and/or archaeologic remains, the age and biochronologic positions. Most of the illustrations have historical value, first published here. The reference is containing the basic works, the most important priority papers and the current publications.

Keywords

Hungary • Caves • Paleontology • Cave deposits • Taphonomy • Research history • Paleoenvironment • Chronology • Archeology • Cenozoic • Quaternary

1 Introduction

They are longer than two meters and have enough space for human to go in. Consequently the karst fissures, the completely filled fossil caves, are not part of investigations. According to Ford and Culllinford (1976) it is distinguished into two main types of cave sediment. The autochthonous or endogenetic sediments are derived by internal processes, i.e. those operating within the cave. The second type is the allochthonous or exogenetic sediments, derives from external sources and is carried into the cave by the agents of transportation. The definition of caves in Hungary including those natural origin underground cavities, which are differentiated two types of sedimentation: (1) entrance deposits—where the actual entrance zone(s) of caves, including rock shelters (abris), cave system fragments and several other possibilities determined by the dynamic contact between the external and internal cave climate (Kordos 1970; Ford and Williams 1989), (2) the cave interior sediments contains the basic division into clastic, organic and precipitated types and into sediments formed outside of a cave and transported into it (allochthonous) and that created within a cave, (Ford and Williams 1989). In Hungary mostly Vértes (1959) studied the sediments of archaeological cave sites.

1.1 Sediments in Caves

To understand the paleontological values of caves one has to study the sedimentation and taphonomical processes of the imbedded cave deposits. Otherwise the conclusions frequently have mistakes, and the misinterpretations for long time remained in the scientific community. For example in Hungary the traditional cave excavations (Kadić 1915, 1916) distinguished the layers by the visible sedimentologic characters (colour, grain size, proportions of clastic materials, etc.) in an arfical quadratic grid system. Jánossy et al. (1986) compared in detail the sedimentological characters and the vertebrate faunal changes in the sequence of Upony I. rockshelter (NE Hungary). They found that the faunal changes were faster than the sedimentological characters. Recognizing of this situation the cave excavation practise was modified, sampled the deposit sections in every 20–10 cms. It happened also on this case where the cave deposits formed slopes—the results: the fossils come to light as from distinguished layers, but really from the same time deposited sediments. After it the excavations in Hungarian caves followed this method (Hír 1991) or tried to use the open air archaeological system in caves (Ringer and Mester 2001; Kordos and Szolyák 2009).

L. Kordos (✉)
Savaria University Centre, ELTE SEK Eötvös Lóránd University, Szombathely, Hungary

M. Veress and Sz. Leél-Őssy (eds.), *Cave and Karst Systems of Hungary*, Cave and Karst Systems of the World, https://doi.org/10.1007/978-3-030-92960-2_17

1.2 Caves and Sediments Developed in Travertine and in Volcanic Rocks in Hungary

Primary travertino caves. The travertine is a form of limestone deposited by mineral springs. Travertine is a terrestrial sedimentary rock, formed by the precipitation of carbonate minerals. The word travertine is derived from the Italian travertino, itself a derivation of the Latin tiburtinus "of Tibur". The freshwater limestone of tufa caves is preliminary caves formed during the deposition of the tufa itself. The travertine and primary caves developing in a very various way worldwide and in Hungary too (Scheuer and Schweitzer 1973, 1988).

The most famous primary traverino cave is developed in the Bükk Mountain, close to town city of Miskolc at Lillafüred, by waterfall of cold karstic origin water of Szinva streem. The artificially made corridors join to each other the separate travertine chambers. The travertine rich in calcareus plant remains (branches, leafs, roots), molluscs but without any vertebrate remains. The cave discovered in 1833 and its shape later several times artificially modified. The cave is open for public. Its name is changed from Anna Cave to Petőfi Cave and in the scientific literature remained as Lillafüred Forrás-mésztufabarlang. Cave Cataster No is 5391–1 (Kordos 1984a).

Another travertine caves contains shell and bone remains opened in 1976 when the wall of the Dobó-bástya (pillar) collapsed in Eger (West Hungary), (Cave Cataster No is 5383–1, 9 and 16). The background of the stony wall is Middle Miocene rhyolithic tuff, covered by gravel and by freshwater limestone layers. The syngenetic caverns developed by the activity of thermal waters. The shelters behind the tetarata waterflows remained dry and the sediments of them contains rich Mollusca and Vertebrata fauna. The age of the fossil assemblages is in s.l. the Last Interglacial (Kordos and Krolopp 1980).

Among the syngenetic travertine caves in Hungary locating in the heart of the capital, Budapest (Cave Cataster No is 4762–1). The Castle Hill and the Castle caves are developed in uniform geological processes. The base is Oligocene marl covered by gravel of the Late Quaternary floodplain terasse. The temperate springs produced an 8–12 m thick travertin layer. The spring developed the travertine and the primary caves, which are simultaneously formed and partly filled by calcareous sinters (Figs. 17.1 and 17.2). Sometimes horizontal soil layers are visible. The several km long cave system mostly artificial, the cellars are locating two or three layers under each other (Krolopp et al. 1977; Jánossy 1986). The Castle Cave sinters have a lot of middle Pleistocene vertebrate and Molluscs remains and pebble tooth of Early Man, named as "Buda culture" (Vértes 1965; Kordos 2006).

Fig. 17.1 Syngenetic travertine cave forms fom the Castle Hill of Buda. The vertebrate microfaunas studied by Dénes Jánossy (in front of the picture)

In Hungary there a lot of primary caves which were developed mostly without any surface connections to reservoirs of thermal or mixed waters. They are generally very rich in different minerals (gypsum, aragonite, calcite, barite etc.) generations. Among these caves the largest and most famous is the József-hegy Cave at Buda, Cave Cataster No is 4762–6 (Leél-Őssy et al. 2011). The fossils in this case came from the cave walls by solutions. Close to the actual surfaces, through the fissures, the soil contains faunal remains, these are redepositing and if there is a cavity between open air and the cave the bats penetrate into the cave.

The caves developed in volcanic rocks are rare in Hungary (Eszterhás 1977; Takács-Bolner et al. 1989), the sediments are local origin, perhaps contains biological remains but not yet studied. Sometimes the non-karstic caverns have rich fossiliferous layers.

The Függő-kő Cave located near-by village Mátraszőlős, North Hungary. On the wall of the 20–12 m high andesite tuff and agglomerate cliff the cavern made by the meandering stream of the valley or by gaseous volcanic bulb and

Fig. 17.2 Stratified calcareous soft deposits on the top of the cave of Castle Hill of Buda

filled by detritic and clastic sediments (Cave Cataster No is 5223–2). The excavation was in two sections of the deposits in 1979–1982. In the upper series distinguished 10, and in the lower one 9 (numerated I-IX) layers. The Mollusca fauna is represent two different units. The upper part is deposited mostly forested and temperated environment like during the Holocene. The lower part looks like a „glacial" feature of the younger end of the last cold phase. The ecologic character of the vertebrate fauna is similar, the lower unit is steppe like and its possible equivalent with the loess period between the Dunaújváros Soil and Mende Upper Soil complex, in age of 25 and 21 kyrs (Jánossy et al. 1982–1983).

The Seybold (or Kenyér) Hill Quarry at Kőszeg (Western Hungary) contains several fissures and cavities with fossil vertebrates developed in carbonate type of metamorphic rocks. The 25 ms long and 1.0–1.5 ms average diameter labyrinth lie corridors developed by the solution of carbonate part of the bedrock. The bone remains collected several times between 1904 and 1947 and summarized and re-evaluated by Kordos (1979).The vertebrates deposited after the Last Glacial Maximum (LGM) during the steppe conditions of the "postglacial". It is including among other microvertebrates *Microtus oeconomus*, *Microtus gregalis*, *Microtus nivalis*, *Dicrostonyx torquatus*, *Ochotona pusilla*, *Lepus timidus* species.

2 Taphonomic Processes

The term "taphonomy" was originally defined by palaeontologist Efremov in 1940 as "the study of the transition (in all details) of animal remains from the biosphere into lithosphere". The etymology of the term resides in the Greek word *taphos* for burial and *nomos* for laws (Lymann 1994, 2010). The concept and the term were adopted by zooarchaeologists and paleobiologists. Behrensmayer and Kidwell (1985) indicated that taphonomic research involved „the study of processes of preservation and how they affect information in the fossil record". Taphonomy had become in a very wide range of paleontological objects, and cave research, too (Coles et al. 1989; Andrews 1990; Rounds 2004). The vertebrate fossil taphonomic papers and investigations of Hungarian caves are very poor (Kretzoi 1941; Mészáros 1996; Gál 2008; Sóron and Virág 2009; Lengyel et al. 2008–2009; Patou-Mathis et al. 2017). Kretzoi summarized in 1941 the main types of bone accumulation in caves. He distinguished eight vertebrate assemblages: (1) Bear faunas, generally large caves, among the bones the cave bear accumulation almost 100%, sometimes the carnivore, bird, ruminant and bat bones are frequent. (2) Large carnivore faunas, where the bones of cave lions, hyenas and wolf bones along the cave bear and ungulates are dominating; the entrance of these caves generally well defended. (3) Early man faunas, generally same as the former ones, but the man carry and deposit bones into the cave. The 1–3 types reflecting closed forest environment. (4) Small carnivore faunas (wolverine, badger, fox). Small size caves, foxhole ("rókalyuk" in Hungarian), with one or more well defended entrances. Aside the bone of carnivore the rabbit, hamster and the land dwelling bird remains are frequent. The small mammals, ungulates and carnivorous bird remains are missing; sometimes the trace of herpetofauna is present. From ecological point of view the accumulated assemblages shows a steppe environment than forested one. (5) Predatory bird faunas depositing where the cave entrance laying in open, well identifying caves, like rock shelters and abris. These contain only vertebrate microfaunas, more bird bones and sporadic meso- and macromammals, amphibians and reptiles mostly from the steppe species. (6) Snake faunas. In narrow, horizontal fissures the herpetofaunal species are dominating, and the small mammal bones are rare. (7) Shaft faunas contains the fall down animal remains of ungulates, frogs and sometimes birds lived in forest environment.

The first real vertebrate taphonomical investigations made in Hungary during the Rudabánya Multidisciplinary Research project by Andrews and Armour-Chelu (Bernor et al. 2005).

3 Paleontological Values

Several cavities and larger caverns were discovered and totally or partly destroyed by the underground coal, bauxite, fireclay mining activity without any information about of fossils. The open air quarries (limestone, dolomite, manganese, iron etc.) frequently explored caves and paleokarst phenomena.

This section has an overview of the Hungarian caves which deposits containing mostly vertebrate remains and having paleontological value. The importance selected by historical, biochronological, unic taxa and archaeological content and its importance on international interests. The cased studied listing in geochrological sequence.

3.1 Oligo-Miocene Cave Deposits

In Hungary, Oligo-Miocene sediments were deposited in the NE-SW ranges, mostly covering the area of the present Mid-Hungarian Mountains. The sediments, of epicontinental-terrigenous and marine origin and whose deposition began during the Eocene, extended to cover most of the carbonate rocks.

Surviving Oligocene emerged areas were unsuitable for karstification, which is why there are hardly any Latest Oligocene karst forms in Hungary (Bárdossy and Kordos 1989).

3.1.1 Bodajk–Kajmát Quarry

In the Triassic dolostone in Bodajk-Kajmát Quarry (Bakony Mts) 4–5 dolines filled with laminated terrigenous sediments were found. Their diameter is 3–13 m, and their depth is 2–6 m. From some of them, vertical solution channels lead downwards. At about 2 m, there are man-sized cavities apparently created by phreatic processes that are today preserved as cave fragments. In one of them, supposedly the richest Middle-Late Oligocene vertebrate assemblage of Central Europe was found. First collected the bones by the young university teacher Barnabás Géczy and presented to the specialist Miklós Kretzoi as "chicken bones" (personal information of late Kretzoi). Later Kretzoi (in Jámbor et al. 1971) and others mentioned the fossil locality (Korpás 1981). In 1976 Kordos identified the locality and several times collected bones from the grey colour fine sediment from the original cave. Cavers, looking for a longer cave destroyed the locality. The following short overview contains the characteristic vertebrata taxa from Bodajk (Kordos 1992). Among the rodents Cricetiae (Eucricetodon I–II), Eomyidae (*Eomys*, *Pseudotheridomys*) are dominant, and *Paracricetodon*, *Pseudocricetodon*, *Plesiosminthus* and *Melissiodon* are present (Fig. 17.3). The larger fauna contains *Anthracotherium*, *Bachitherium*, *Amphitragulus* and Rhinocerotidae remnants.

3.1.2 Other Oligocene Cave Fragments

At Csákberény in the Vértes Mts was a wider fissure with solutional walls in Middle Eocene limestone. It is filled by greyish and yellowish sandy clay in which Rupelian bone remain (*Protacerathirium albigense*) were found (Kretzoi 1940). The locality later is built up by a house. At the top of the Tatabanya Keselő-hegy quarry from the karstified Eocene karst surface depression or cave (?) infilling the variegated clay also produced the signal of vertebrates (*Brachyodus* sp.) of Oligocene age (Kordos 1992). Both of these karst deposits are parts of the Late Oligocene and Early Miocene Csatka and Szápár Formations. This sediment, of

Fig. 17.3 Upper molar of *Melissiodon* sp. from the Oligo-Miocene cave locality of Bodajk (*Photo* Katona)

mainly fluvial origin, covers the surface over a large area, the total carbonate area of Transdanubia (1 600 km^2) apart from some more elevated plateau areas in the Keszthely and Bakony Mts (Korpás 1981).

3.2 Late Miocene Fossiliferous Caves

During the Late Miocene Pannonian (11–5 Ma) in the territory of Hungary there was geomorphologic inversion in the Carpathian Basin. The early (since cca. 15 Ma) elevated surface subsided and were covered with various sediments of the Pannonian Lake, including several caves with vertebrate fossils.

3.2.1 The Csákvár Cave

The southern part of the Vértes Mountains the Csákvár cave developed among Triassic dolostone cliffs of an escarpment, probably due to tectonical causes, Cave Cataster No is 4521–1. In front of the present cave entrance in some hundred meters the shore in the Pannonian Lake was a mosaic of wet open and dry savanna-like habitats. The scientific exploration of the Csákvár Cave (or Báraczháza Cave) started in 1926 (Kadić and Kretzoi 1930) and in 1951 and 1952 finished by Kretzoi (1951, 1954). The lower and upper part of the cave corridor in cross-section are different (Fig. 17.4.). The lower is fissure like, and the upper is widening by a shoulder like erosion step. The lower part the cave stream caused a meander form channel. On the original cave walls of the upper part contains symmetrical scallop forms developed under slow energy water body. Between the stream and "lake" period of the cave was a dry interval, when the fossiliferous grey colour, with high phosphate concentration clay deposited. The phosphate contains came from the high density of bones. The extremely divers vertebrate assemblage contains a typical "Hipparion fauna". Along the dominance of *Hipparion*, large mammals appeared, as *Tapiriscus pannonicus*, *Gomphotherium longirostre*, *Zygolophodon*, *Stephanorhinus*, *Brachipotherium*. Rich in carnivores like *Allohyaena kadici*, *Adrcocuta advena*, *Xenohyaena csakvarensis*, *Protictitherium*

Fig. 17.4 Entrance and cross-section of Csákvár cave (*Photo* Mészáros)

Fig. 17.5 Hyaena tootmarks on the *Hipparion* bone from the Csákvár cave (*Photo* Mészáros)

csakvarense, *Machaerodus* and *Paramachaerodus* species. The artiodactyles represented a divers open habitat environment with *Csakvarotherium hungaricum*, *Palaeotragocerus*, *Lagomeryx celer*, *Tragoceras oryxoides*, *Miotragocerus csakvarensis*, *Pikermicerus platyceros*, *Dystychoceras foveatus*, *Graecoryx esterhazyi*, *Cervavitus esterhazyi*, *Cervavitulus mimus*, *Cervaviscus hylae taxa.* Taphonomical point of view the cave sometimes was a hyaena shelter (Figs. 17.5 and 17.6).

Because the 7–8 Ma (BP) years old fauna represent a typical Turolian open air habitat opposite of the former wet and mostly forested (Vallesian) one, Kretzoi established a new biochronoligical unit based the Csákvár fauna, as Csakvarium which Mediterranean counterpart is Pikermi in Greece.

The subcorridors and the cover sediments of the Csákvár cave contains Late Pleistocene and Holocene vertebrate faunas, too.

3.2.2 Caves of Polgárdi

From the Devonian limestone quarries of Somlyó Hill and Kőszár Hill near the village of Polgárdi;several caves (Pávai-Vajna 1911; Halász 1964) and cave deposits with vertebrate remains were discovered during the twentieth century (Kormos 1911; Kretzoi 1952; Mihály 1975; Kordos 1991). The localities numbered accordingly: Polgárdi 1 is the first place where fossil bones were found; Polgárdi 2 contains the classical "Polgárdi Fauna" of Late Miocene (="Pliocene"), described by Kormos (1911), Kretzoi (1952); Polgárdi 3 is an underground travertine bone breccia (Mihály 1975); Polgárdi 4 and 5 located on the East wall of the quarry in the year of 1984 and between 1985 and 1991 (Kordos 1991; Freudenthal and Kordos 1989). The classical locality (No. 2) is generally cited as "Polgárdi" in the literature. During the excavation job performed by Kormos in 1910, bone remains were collected from the subsurface hall of a large sinkhole cave related by mining completely filled deposits. In the profile taken from the cave, five beds were distinguished: 1—the lowermost bed containing testudinacean bones, 2—overlain by an *Ictitherium* bearing bed, 3—followed, upwards, by a bed rich in *Hipparion* bones, 4—overlain by a bed characterized by *Gazella* and "*Capreolus*" remains, and finally 5—the uppermost be characterized by the small mammals (Fig. 17.7). Exploring the "Polgárdi" locality this was the first time washing and sieving techniques were used in Hungary in order to obtain fossil vertebrate remains. Taphonomical point of view the Polgárdi 4 and 5 localities are similar and very unique. The morphology of these caves not bordered by sharp surfaces but contains a spongious structure. The sediments are laminating in the cavities, consequently deposited in water. Mostly small size bones are not broken, remained in high concentration and in excellent conditions. Looking for the general geological history of the Polgárdi Hill (Kőszárhegy), the postvolcanic hydrothermal activity was present and developed the cavities like a present active geyser (Vendl 1926). The limestone body covered and partly still covering the Late Pannonian *Congeria balatonica* layer (after Kókay, J. and Kordos, L. experiences see Jánossy 1991). Eroding the Pannonian sediment or the hydrothermal pressure penetrated through to the former surface, the cavities opened and probably because the warm water and gaseous exhalations was deadly for the animals. Among the main fossiliferous deposits all are in the MN13 Zone (latest Miocene), the Polgárdi 4 fauna is the older (early MN13), the Polgárdi 5 is a little bit younger and the classic Polgárdi 2 is belonged to the latest part of MN13 Zone.

Fig. 17.6 Holotype of *Allohyaena kadici* Kretzoi from the Csákvár Cave (*Photo* Mészáros)

Fig. 17.7 Section of classic Polgárdi locality (No2) after Kormos (1911) 1–4 = bone layers, 5 = rodent layer, pm = paleozoic limestone, pak = pannonian clay and sand, a = pliocene, l = loess, h = humus

3.3 From Pliocene to Middle Pleistocene Fossils in Caves

The Plio-Pleistocene border according to the Hungarian, the scientific community, is about 2.4 Ma instead of the proposal of the International Commission on Stratigraphy in 1.81 Ma. The border between the Early and Middle Pleistocene is at 0.6–0.7 Kyrs (Villanyian and Early Biharian biochronological stages). The end of the Middle Pleistocene is marked by the last occurrence of vole taxon *Mimomys savini* at about 0.3 Kyrs (Late Biharian).

3.3.1 Cavities and Karst Fissures at the Villány Region

The Villany Mountains (S Hungary) are a classical type-region of European terrestrial biochronology and chronostratigraphy. Numerous data were collected in the outcrops of Villány, Nagyharsány, Csarnóta and Beremens villages. The more widespread karstification took place here from the Middle Pliocene to the Middle Pleistocene. Most of the localities are cave deposits, and some of the caves have open air part (Nagyharsány 6 at the top of the Szársomlyó Hill; 'Sculptors' Quarry; Beremend Quarry, Villány 8 Locality; Figs. 17.8, 17.9, 17.10). According to several excavations (Petényi 1864; Kormos 1937; Kretzoi 1956; Jánossy 1986). The Csarnóta 2 locality is the type locality of Csarnotian (Upper Pliocene), Beremendium (Beremend 5 locality, Early Pleistocene while Villány 3(Villány Kalkberg) is that of the Villanyian (Early Pleistocene). As substages distinguished Nagyharsányhegy and Templomhegy units are in the Lower Biharian (Figs. 17.11, 17.12, 17.13). The cave filling of Villány 8 Locality is important from the point of view of global environmental changes (from warmer into dryer) because the older and deeper deposited red clay penetrates into the younger upper loess like sediment contains vertebrate remains.

3.3.2 Caves of Osztramos

The most complete chronological sequence of karst cave development is known at Osztramos Hill (NE Hungary), an isolated part of the Rudabánya Range. More than 20 localities yielding vertebrate remains have been covered by Triassic limestone quarrying (Figs. 17.14 and 17.15). There are two fossil cave layers at 312 m and 305 m a.s.l., and an

Fig. 17.8 The limestone quarry of Beremend explored several fossiliferous cave localities

Fig. 17.9 Shaft entrance from the bottom of Beremend quarry. The present thermal water level is about 15–20 m below

active cave at the modern karstwater level. On the basis of age determination of rich vertebrate finds the cavity at 312 ms a.s.l was largely filled with Middle Pliocene sediments. Though there is even a biostratotype, Estramontium. The cave floor, originally formed as symmetrical scallops (this morphological character first time distinguished here) at just under the actual karstwater level was covered by 3–4 m thick, redeposited sandy clay and then by several meters of laminated calcite. Its age indicated by the included vertebrate remains. This means that the open cavities can be continuously traced back to that period. The cave level at 305 ms a.s.l is filled by Lower and Middle Pleistocene deposits (Jánossy 1986; Jánossy and Kordos 1977; Kordos 1975a). Osztramos Hill lies at the southern margin of the Slovak Karst and, hence together with sites (Včelare–Méhész) is very important for the reconstruction of karst history in the whole region (Horaček and Kordos 1989). The Osztramos (=Esztramos) limestone quarry is closed during the early 1990s, and there only one cave is remained visible (Földváry cave). The mining tunnels at the base of the hill opened two large, mostly by karst water covered deep caves (Rákóczi I and II Caves).

3.3.3 Hajnóczy Cave

The Hajnóczy cave is situated in the SW Bükk Mts at 457 m a.s.l. and was formed in Triassic limestone by corrosion and erosion processes. The total length is 4250 m, and the maximum vertical distance is 135 m. Number of Cave Cataster: 5382–2. The cave contains red clay deposits of Early Pleistocene belong to the Templomhegy faunal unit (cca. 700 kyr). Characteristic old taxa: *Ursus deningeri*, *Mimomys savini*, *Allocricetus éhiki*, *Mimomys pusillus*, *Glis sackdillingensis* mammal taxa. At the Early Late Pleistocene, considerable amounts of red clay redeposited in different places of the cave. At the end of the Riss-Wurm Interglacial and in the Holocene shattered debris accumulated (Hír 1985).

3.3.4 Tar-Kő Rock Shelter

The small cave was formed in Triassic limestone below the peak of the Tar-kő, on the edge of the Bükk Plateau (N Hungary), at an altitude of 850 m a.s.l. Its original size before the excavations was 20–22 m wide by Fig. 17.16, 3–4 m long. Cave Cataster No is 5343–1 (Fig. 17.16). Behind entrance which filled up by Late Pleistocene clastic sediments and the lower layers (from No. 2 to 18) deposits varied from dark reddish (at the bottom) to yellowish (at the top) colour clay, with rocks split by a dripstone level at layer 8. Jánossy (1986) made six times excavations to discover the unique fauna, which contains more than 15 000 determined items and distinguished over 85 vertebrata taxa. According to Jánossy the relative frequencies of voles and species markedly reacting to climatic changes (layers 1–15) a forest phase may be discerned (layers 5–11). Apart from this, there is a slight (layers 12–15) and more pronounced (layers 1–4) arid phase shown by a repeated expansions of the remains of Celtis stony endocarps and lizards. The chronological position of Tar-kő Rock Shelter represents a separate biochronological unit, the Tarkő Phase of the Late Middle Pleistocene and *Mimomys savini–Lagurus transiens*, *Lagurus transiens–Arvicola cantiana* Biozones.

3.3.5 Uppony Rock Shelter No. 1

The rock shelter is situated close to the Uppony village (N Hungary) at the north limestone wall of a gorge, 47 m above the bottom of the valley, at an altitude 287 m a.s.l. Cave Cataster No is 5320–1. The excavations in the rock shelter opened 8 layers. The top 1–5 layers are rich in

Fig. 17.10 Holotype of the genus *Prospalax* described from the Pliocene karst sediments of Beremend

micromammals and larger (*Ursus deningeri*, *Leo gombaszögensis*, *Ovis ammon-polii*) animal remains (Jánossy 1986). Near the bottom of the rock shelter deposit layer No. 6 may suspect erosional discordance and signs of a calcite trust. Underlying No. 7 and 8 layers were 40–60 cms in width, reddish-brown with a few corroded rocks. The vertebrate fauna of the upper part of the deposits reflecting a steppe-tundra animal assemblages, while in the lower layers the forest elements are dominant. Because in the vertebrate remains of upper layers are absent in Late Biharian faunal elements but contains mostly Late Pleistocene taxa. For this transitional situation Jánossy (1986) proposed the Upponyian biochronological substage, and according to Kordos (1994a, b) it is belonging into the *Lagurus lagurus* Partial Range Biozone.

3.3.6 Pongor-Lyuk at Répáshuta

The Pongor-lyuk is a small cave close to the village of Répáshuta in the Bükk Mts., N Hungary. The length is 35.6 m, and vertical extension is 6.4 m. Cave Cataster No is 5381–20. The age of the excavated profile in the upper 1.6 m section is Holocene with two cultural level: Bronze age at 60–80 cms, and Upper Palaeolithic at 120–140 cms. Under the Holocene sediment has a long time gap and following the formation with Solymarian biochronolgic units (cca. 300–400 kyrs). The characteristic fossils are *Ursus stehlini*, *Talpa fossilis*, *Sorex runtonensis*, and *Cricetus runtonensis solymarensis*. The upper part (layer 3) of the Late Middle Pleistocene is poor in fauna and were formed under cool climate. The lower (4th) layer is rich in vertebrate remains on its bottom, and indicating a similar climate condition that of today but more drier than that (Hír 1986).

3.4 Important Late Pleistocene Cave Deposits

More than 90% of the fossil studied cave deposits sediment during the Late Pleistocene, between 140 and 10 kyrs. They were frequently contain palaeolithic cultural finds and large mammal bones. These caves used sporadically by the

Fig. 17.11 Remain of the classic Nagyharsány fossil site at Villány, today a "sculpturs quarry"

humans, bears, hyenas or remained open for the high variety of species of the actual biome.

3.4.1 Ördöglyuk Cave of Solymár

The Ördöglyuk Cave (Devils' Hole at Solymár) is situated close to Buda(pest) west from the village of Solymár on the NE slope of the Zsíros-hegy, appr. 348 m a.s.l. The cave was formed in Triassic limestone, the length is 5550 m, and vertical extension is 78 m. Cave Cataster No is 4773–1. The main entrance of the cave is opened in the top of on a former quarry but more part of the cave corridors opened to the hill slope making sub-entrances. The original hydrothermal cave after it partly filled by reddish color clay and debris sediments contains rich mollusca and vertebrata remains. The quarry of Solymár in a vertical shaft infilling with reddish-braun color clay and clastic sediments some bones (sometimes associated) in a good quality were collected in 1905 and 1906. According to Kubacskas' paper (1926) *Ursus spelaeus*, *Hyaena spelaea*, *Felis spelaea* and some others were determined. The faunal remains perhaps originally belonged into sediment complex of the Ördöglyuk Cave.

In the cave deposits there were no regular and detailed excavations because the cave sediments regularly deposited in dangerous places, like the "Kiskörút"locality is under the entrance shaft where the sediments is the pillar of the entrance built-in a staircase. At the top of the huge underground even at the Denevér Hall found vertebrate assemblage contains *Panthera leo fossilis* (Hankó 2007). Only occasionally collected mollusc and vertebrate remains (Rotaridesz 1943; Vértes 1950; Jánossy 1986; Krolopp 1973; Topál in Demeter et al. 2016). The Ördöglyuk Cave system in several places close to the hill surface shafts and rooms contains same type of sediments and fossil remains, like *Alces brevirostris*, *Dicerorhinus kirchbergensis*, *Bison* (*Urus*) *hungaricus*, and several taxa of small mammals like *Rhinolophus solymarensis*, *Myotis oxygnathoides*, *Spalax mehelyanus*, *Mus* (*Budamys*) *solymarensis*, *Lagurus lagurus* etc., determined by Kretzoi, Vértes, Jánossy, Topál, Vörös. The Ördöglyuk Cave paleovertebrate association is the type of the Solymarian substage (Kretzoi 1953). Among small mammals characteristic for the substage the *Sorex*, also a water vole (*Arvicola* sp. III) with a thin layer enamel, a few relict species in the rodent fauna (*Pitymys gregaloides*), the

Fig. 17.12 The Villány quarry contains several classic early Pleistocene vertebate localities. View of the quarry from the Somssich Hill 2nd locality

appearance of the true mice (*Mus* or *Budamys*), (Jánossy 1986). The biostratigraphy and its correlation is still dubious. From the Biostratigraphic point of view the fauna belongs to the *Arvicola terrestris* taxon Range Zone and it is possible to correlate the Eemian, between 125 and 140 kyrs (Fejfar and Heinrich 1990; Kordos 1994a, b).

3.4.2 Horváti-Lik, Uppony

The Horváti lik (Horváti Hole), named after a local legend, is a secret tunnel which is goes to the nearby village of Bánhorváti, is located close the village of Uppony (NE Hungary), on the left side of the Uppony Gorge.The explored cave lies at 270 m a.s.l. the bottom of the valley. The explored length is 17 m, and the vertical extension is 7.6 m (Cave Cataster No is 5320–23). The Uppony Gorge contains several small caves, including the Uppony Rock Shelter No. 1 (discussed earlier in this chapter). In the Horváti-lik Fűköh and Kordos in 1977 and 1978 excavated the debris deposits of the small cave (Fűköh and Kordos 1977, 1980). The total thickness of the excavated sediment, down to the bedrock, is 410 cm and that is divided into 25 layers.The uppermost layers 1–4 are Holocene, layers 5–6 are mixed, and layers 7–21 yielded faunal content of early Late Pleistocene (Figs. 17.17 and 17.18). The sequence showed a transition from temperate woodland in Layers 21–16 to cold-climate open steppe in Layers 15–7. The rodents are predominant, *Microtus arvalis* is the most abundant species. The steppe elements are present in high percentage (e.g. *Sicista*, *Ochotona*, *Allocricetus*, *Spalax*, *Allactaga*, Cricetus and *Lagurus*), and *Distroxtonyx* also occurs. Using cluster analysis, biochronological methods and correlation with the Marine Isotope Stages (MIS) and the continental stratigraphy units, the Pleistocene faunas of the Horváti-lik represent a significant steppe phases at the end of the Eemian, correlated with Marine Isotope Stage 5d to MIS 5a, or Herning Stadial, dates as 120–110 kyr BP (Pazonyi and Kordos 2004; Ruddy 2011).

3.4.3 Por-Lyuk Cave at Jósvafő

The entrance of Por-lyuk cave at Jósvafő has at the slope of Magas Galya at the Nagyoldal in the middle of the Aggtelek-Jósvafő karstic region, in the latter part of Lófej valley at 53 m height of the bottom of the valley, in the altitude at 404 m a.s.l. The total length of the cave is 90 m, and the depth is 16 m. Cave Cataster No is 5440. The cave developed in Triassic limestone ("Wetterstein" facies) and in the solution residue the kaolinite is predominant as well as in the bedrock. The cave sediments excavated for vertebrate

Fig. 17.13 The Villány 3 (=Kalkberg) is the type of the Villanyian bichronologic stage

remains in the year of 1967 by Jánossy and studied for pollen and molluscs (Jánossy et al. 1973). The lowermost reddish-braun color sediment contains compact vertebrate fauna. Typical of the last interglacial (Eemian) animal assemblages are the red-toothed shrews (*Sorex* sp.), the largest form of white-toothed shrews (*Crocidura robusta*), the mole rat (*Spalax*), the dwarf hamster (*Allocticetus*), the giant hamster (*Cricetus cricetus major*) and the large horse (*Equus abeli* group), and absolute predominance of the common vole (*Microtus arvalis*). All these features imply a closer relation with the Varbóian substage (see Lambrecht Kálmán Cave). The presence of the large *Crocidura*, small dormouse (cf. *Glis sackdillingensis*) and the sheep remains (*Ovis* sp.) point to a unique stratigraphically position (Jánossy 1986).

3.4.4 Lambrecht Kálmán Cave

The Lambrecht Kálmán cave is formed in Triassic limestone, 410 m a.s.l on the Nagygallya Hill, about 5 km south-southwest from village Varbó in the Bükk Mountains. The length of the cave is only 14 m; without the sediment removed during excavations in 1952 and 1960 by Vértes and Jánossy (Vértes 1965; Jánossy 1963–1964). The cave is named after Dr. Lambrecht Kálmán (1889–1936) paleornithologist, ethnographer and science writer. Cave Cataster No is 5363–18. The most fossiliferous layer (V/a) is 40-80 cm thick reddish-braun color clay, with corroded stones. Here Vértes found a few chipped flint implements of palaeolithic man, which he determined as "pre-Mousterian" (Vértes 1965; Jánossy 1986). This layer is the stereotype of the Varbóian substage of the biochronologic system. The Varbóian substage may be characterized by the presence of a small bear (*Ursus sractos* aff. *taubachensis*) and other carnivores (*Canis lupus*, Leo *spelaeus*); the giant hamster (*Cricetus cricetus major*) was still present, and the dwarf hamster (*Allocricetus*) appeared for the last time; a small crested procupine (*Hystrix vinogradovi*). The steppe and semi desert animals (*Allactaga*, *Spalax*, *Ochotona*, *Lagurus*) are present. The horizon-marker of the Varbóian substage is the *Hystrix* and the wild ass (*Asinus hydruntinus*). In correlation the V/a layer of the Lambrecht Kálmán Cave is belonging into the MIS 5d zone and into the interval of 116–103 kyr (Kordos 2013) or into the age range between MIS 5d and MIS 4 on the basis of *Arvicola* studies (Ruddy 2011).

Fig. 17.14 Az Osztramos (=Esztramos) limestone quarry in NE Hungary opened several important fossiliferous and crystal caves

3.4.5 Kálvária Cave No. 4

At the foot of the Kálvária Hill at the city of Tatabánya (Gerecse Mts) a small cave fragment developed in Eocene limestone (Cave Cataster No is 4610–37. The lower entrance is like a cellar, and the upper is a small shaft. Two type of sediments deposited in the cave, the first upper unit is a rocky debris, and the underlying second is a well stratified sandy clay and small size rock fragments. The upper unit contains steppe micromammals (*Citellus citellus*, *Lagurus lagurus*, *Ochotona*), and the *Microtus arvalis* is predominant. *Bison* and *Equus* associated skeletal remains also were found. The lower layer has a very similar micromammal assemblage. The whole sediments belongs to the same age in the Early Wurmian sometimes between 110 and 100 kyrs (Kordos 1994a, b). Both deposit units are very rich (1150 bone specimens) in bird remains. Each of the two assemblages (upper and lower layers) had distinct avian characteristics. The high number of kestrel (*Falco tinnuculus*) long bones in Assemblage I generally included complete skeletal parts from subadult birds, suggesting a colony which bred in the cave, whose chicks died of natural causes (Gál 2008).

3.4.6 Suba-Lyuk, the Neanderthal Man and the Steppe Paleobiome

Suba-lyuk cave locality is in NE Hungary at the south foot of the Bükk Mts. Cave Cataster No is 5382–1. Archaeological excavations started in 1932, and in that year bone remains of neanderthals were found by amateur caver János Dancza (Kadić 1940). Altogether 18 layers have been distinguished in the cave. According to the original interpretation the three lowermost layers belong to the last interglacial. Layers 7–16 were deposited next as „upper layers" and were determined to be Wurmian I. Beside the Mousterian type stone tools (Vértes 1965; Mester 1989) in the lower layers (1–3) are found *Capra ibex*, while in the upper ones (7–16), the cave-bear was the dominant fauna, accompanied by steppe-type small mammal faunas. In the layer 11 among 14 small mammal species with *Lagurus lagurus* steppe species as dominant. In the same layer—remained in the fissures of the cave wall—together with *Lagurus* an M_2 fragment of neanderthal man also found (Kordos 1993). The Lagurus migrated to Central Europe from East and is found there by the end of the Middle Pleistocene. Its sporadic occurrence

Fig. 17.15 The early Pleistocene shrew *Beremendia fissidens* from Osztramos

can be traced in the Carpathian Basin and in the territory of the present Poland up to the last Glacial Maximum. During this period its dominant presence can be correlated at about 60 kyr when they invaded Europe from two routes. One of them was from the southern part of the Russian Plain arriving at the Carpathian Basin following the Danube line, and they stopped here in the „dead end street". The other route led to the North, and across the German-Polish Plain until they got to the territories of Belgium, France and England (Kordos 1990, 2006, 2013). Biochronological point of view the vertebrate fauna of this cave is the type of Subalyuk Phase, and using the biostratigraphic classification it is the *Microtus gregalis–Lagurus lagurus* Zone (Jánossy 1986; Kordos 1990).

3.4.7 Szelim Cave

The Szelim Cave is located in north western Hungary at the western margin of the Gerecse Mountains, 289 m above the Által-ér valley near Tatabánya city, close to the huge Turul Bird statue. The cave interior is 45 m long and 14 m high. Cave Cataster No is 4630–8 (Fig. 17.19).The very thick-bedded Upper Triassic Dachstene Limestone is of a cyclic structure and was developed on the shallow-marine carbonate platform in the Tethys Ocean. The Szelim Cave opens some hundred meters North of the Turul monuments. It has two entrances looking out onto the town and shaft in the celling. The corridor of the cave runs in a meander pattern, formed by an underground river. The cave was recognized as an archaeological site in 1932 when the outstanding speleologist Hubert Kessler began excavations and István Gaál palaeontologist followed him. The excavated sediments are more than 12 m deep and rich in archaeological finds and caused a plenty of interpretations (Mester 2006). During the 1930s, researchers dug deeper in the cave sediments and found some mousterian like stone tools and Late Pleistocene cave bear dominant vertebrate fauna (*Ursus spelaeus*, *Hyaene spelaea*, *Canis lupus*, *Coelodonta antiquitatis* and *Mammuthus primigenius*). The re-interpretation of the sediments deposited in the shaft and compared with the former data (Bradák and Markó 2006). During the beginning of the 21th century local cavers working together with archaeologist and palaeontologist (by M. Juhász and L. Kordos) cleaned the deposit wall of the shaft and made four meters deep trench from the cave sediment surface to the bedrock (Fig. 17.20). The old and new section opened a very complicated, several times eroded and redeposited cave debris section (Bradák and Markó 2006; Kordos pers. investigations).

3.4.8 Istállós-Kő Cave

The Istállós-kő Cave is formed in Triassic Felsőtárkány Limestone Formation on the southwestern slope of

Fig. 17.16 Entrance of the Tarkő rockshelter

Istállós-kő peak (958.4 m a.s.l.), at village Szilvásvárad in the Bükk Mountains, at an altitude of 546 m a.s.l. (N48° 04′ 17, 88″; E20° 25′ 03, 62″. The entrance is 10 m wide and 7.5 m high, and the cave is 57 m long with ceiling 16 m high. The evolution of the cave was influenced by inclined limestone layers (Jánossy 1986; Patou-Mathis et al. 2017). Cave Cataster No is 5341–1. Istállós-kő Cave was investigated during twenty field campaigns between 1912 and 2005.The cave is considered as an important Aurignacian site in the Middle Danube basin. Traditionally two layer-complex were distinguished in the excavated sections, as Aurignacian I, in the lower layer, and Aurignacian II in the upper layer (Vértes 1965). The fauna from the excavations of Istállós-kő Cave first studied by Mottl (1942) who found a forest-steppe character with alpine and arctic elements. Faunal material from Vértes' excavations were studied by Jánossy (1952, 1986), who mentioned that the number of bones reached about 30,000 pieces. Half of this amount was cave bear remains. According to Vörös' (2003–2004) revision 19.242 mammal bones (of which 11,465 pieces of micromammals) and 765 bird bones were collected in total. The excavation campaigns of Ringer (2002) in 2000 opened a new section and the vertebrate fauna is evaluated by Kordos (2001). The large mammal remains (*Ursus spelaeus* 74–92%, *Canis lupus*, *Crocotta crocuta*, *Rangifer tarandus*, *Rupicapra rupicapra*) are typical for a "cave bear" shelter. The taphonomical analysis suggests that the bears lived in the cave before people. The humans occupied the cave only seasonally, in-between very different animals used the cave and diesd in it (Kordos 2001; Patou-Mathis et al. 2017). The radiometric age determinations in the whole section have a very wide range from 44,300 to 25,500 years (Markó 2015).

3.4.9 Szeleta Cave

On the eastern margin of the Bükk Mountains, in a cliff above the Garadna valley, close to Hotel Palota at Lillafüred has the Szeleta Cave. Cave Cataster No is 6363–1. Nevertheless, the cave, which was formed in the Middle Triassic shallow-marine, very thick-bedded limestone (Fehérkő Formation) is famous the first traces of the „caverman" of the Upper Palaeolithic age. It was Ottokár Kadić (proposed him to excavate by Ottó Herman) who found the first remains in 1906 and who laid the foundations for an organized

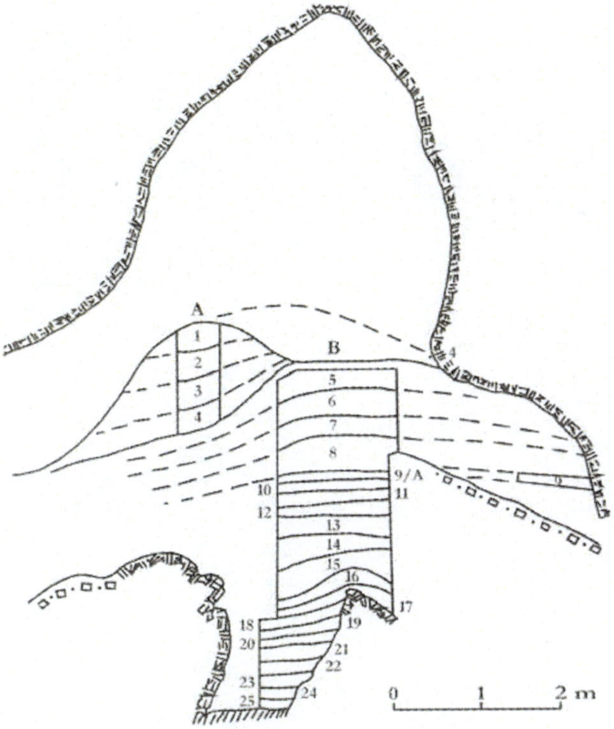

Fig. 17.17 Early Late Pleistocene (5–25 layers) and Holocene (1–4) layers in the Uppony Horváti-lik

Fig. 17.18 Excavation trench in the Uppony Horváti-lik deposits

approach to Stone Age archaeology and its related speleology. The cave 98 m above the valley consists of a large hall and two passages. The age of the findings derived from the cave sediments during the excavations, ranges from the Middle Stone Age—which started 30 000–40 000 years ago and was characterized by hunter-gatherer lifestyle up to the Early Iron Age. Within the succession of cave-filling sediments—of an average thickness of 4–5 m, but reaching 12 m in the hall—at least 9 layers could be separated; 6 of them were deposited in the Pleistocene and 3 in the Holocene. The lower section of the cave-filling succession is made up of stream load and red clay. It is overlain by the series of cave clays containing rock debris and is characterized by stone tools of different human cultures; furthermore, it yields the remains of cave bears in a large quantities. In the Szeleta Cave the earliest human visitors, on the basis of the Middle Palaeolithic tool types, were neanderthals. The first appearance at Szeleta Cave may be dated to MIS 4 (ca. 75–85 kyears BP) earliest. During its entire Pleistocene existence Szeleta Cave was occupied by cave bears during winters and humans were able to take over the cave for short durations in the summer, while bears were away (Lengyel et al. 2008–2009). In the lowermost part of the series, the so-called "Taubach" stone tools can be found, overlain by layers containing the "Bábony" and the "Early Szeletian" stone tools. Layers containing the Early Szeletian artefacts are covered by a bed which is famous for its characteristic "stone spearheads" and contains the tools of the advanced Szeleta Culture (Ringer 2008–2009; Kordos 1979, 2001).

3.4.10 Herman Ottó Cave

The Herman Ottó Cave (formerly Puskaporos Cave) is located between Alsó-Hámor and Felső-Hámor villages in western direction from Miskolc, in the eastern mouth of the Bükk Mts (NE Hungary). The cave has got an Upper and a Lower Entrance. The Lower was called Herman Ottó Cave, and the Upper Entrance was entered to the registration as the Herman Ottó Rock Shelter. The strata of the Lower Entrance and Hall thick of the 17 layers was about 2–3 m. The

Fig. 17.19 The corridor of the Szelim Cave (*Photo* Mészáros)

Fig. 17.20 Several times eroded and re-deposited four meters deep Early Late Pleistocene sediments in the bottom trench of the Szelim cave (*Photo* Mészáros)

No. 2 layer (yellow clay with limestone debris) served remains of *Ursus spelaeus*, *Canis lupus*, *Hyaena crocuta*, *Castor fiber*, *Cervus*, *Alces*, *Megaloceros*, *Rupicapra* etc. The No. 3 layer (limestone breccia) contained *Alopex* and *Alces* bones. The excavators found several associated remain of the skeleton of a *Megaloceros giganteus* in the 5th layer (yellow clay), (Kadić 1916). In 2005 and 2006, Szolyák directed excavations in the Hall 15 square metres area. The found undisturbed and average 30 cm thick Pleistocene deposits, which was served remain of vertebrate fauna and five knapped stones. This deposit corresponds to the fifth layer of Kadić. The vertebrate fauna is in the layers of the Upper Entrance and the Lower Entrance, and the Hall is few species. The Pleistocene layers of the cave and the rock-shelter are contemporaneous and accumulated in the same place but more times. Both of the fauna lived in an ecosystem with cold and temperate climate in the Late Pleistocene. The palaeolithic knapped stones (700 pieces) found in the No. 2, No. 3 and No. 5 layers. During the twentieth century the researchers defined the assemblage as to be Aurignacian or Solutréan or Chattelperonian or Cave Gravettian. The AMS measuring (Beta Analytic Laboratory, Miami, USA) data were 35,410 ± 660 BP and 35,630 ± 630 BP (Kordos and Szolyák 2009).

3.4.11 Peskő Cave

Peskő is a classical one chamber larger size cave fragment (approx. 33 m long, 10 m wide) in the bottom of a characteristic cliff of Pes-kő (860 m a.s.l.) at the Bükk Mountain (coordinates 48.05° N, 20.43° E, altitude 745 m a.s.l.), formed within Triassic limestone. Cave Cataster No is 5343–1. Several excavations were carried between 1912 and 1987 (see in more Vértes 1965; Jánossy 1986; Hír 1991). The litostratigraphic interpretations, and they vertebrate faunal elements in different excavations are controversial. The rich fossil assemblages were found from two horizons, which termed a "brick-red" and a "light-yellow cave loam. The characteristic rodents are *Microtus arvalis*, *Microtus gregalis*, *Microtus oeconomus*, *Dicrostonyx torquatus*. A small number of lithic artefacts classified as Aurignacian I. The faunal remains suggest a partially wooded steppe environment. Radiocarbon dating of Aurignacian bone artefacts from Layer 1 gave an age of 35 200 BP (Ruddy 2011).

3.4.12 Remete Felső Cave

The cave is locating at the NW border of Budapest, close to the former village of Máriaremete, among the Triassic cliffs of the slope of the Remete valley, 70 ms above the bottom and 350 ms a.s.l. Cave Cataster No is 4772–2. The small cave excavated in the years of 1969 and 1970 for archaeological reasons. The entrance hall contains five layers. The sediments on the top (layers 1–2) deposited during the younger Holocene, and the cryoturbate mixed layer 3 is covering a loess-like layer with Pleistocene vertebrate faunal remains (Layer 4). The lowermost 5th layer full of large block of rocks and showing cold Pleistocene period. In the upper hall of the cave similar lithological sequence found. In both sections the 4th layer have a same vertebrate remains: *Ursus spelaeus*, *Leo spelaeus*, Lepus *timidus*, *Coelodonta antiquitatis*, *Megaloceros*, and among others *Ovibos pallantis*. From the same layer (?) three specimens attributed to the *Homo neanderthalensis*. The paleolithic tools are indicating "Jankovichian" or trans-danubian Szeletien typology from the beginning of the last glacial (Wurm I), (Gáboriné 1984). The Remete valley opened several other rock shelters, cavities and shafts, too.

3.4.13 Caves of the Bajót Cliffs

The Öreg-kő ("Old Cliff"), which rises above the village of Bajót, belongs to the north eastern foothills of the Gerecse Mountains. It is 375 m high and made up of Late Triassic Dachstein Limestone. The side of the the hill reveals the famous caves of the Öreg-kő, which are registered as being among the most important archaeological localities in the Transdanubian region. The caves were dissolve thermal water; presumably occurred during the Pliocene when the limestone body of the hill had not yet risen above the surrounding areas (Albert 2010).

Jankovich Cave. The largest cave of the Cliffs of Bajót opening to North at relative height of 80 m in 330 m a.s.l. The entrance of cave including sediments is mostly destroyed, and the main hall has two skapt open to the sky. Cave Cataster No is 4661–1. Between 1923 and 1956 about ten excavations took place here and several interpretations of the archaeological materials. The best-known and most important cave locality in Transdanubia named after Béla Jankovich, the minister of religion and education of the first excavation period (1918–1925), who visited and financially supported the field works. Since the eighties the site was looked as the eponymous locality of a Middle Palaeolithic industry dated to the early Wurm or to the first Pleniglacial period and characterized by the presence of Middle Palaeolithic tools. On the other hand, the Upper Palaeolithic lithic are practically absent. The osseous tools generally interpreted as the effect of natural mixing or imprecise methods of excavations. The litho-stratigraphical observations of cave sediments show quite clearer picture. Below the unusually thick Holocene layer (the importance of it described later), yellow or red coloured clay with lithic and osseous artefacts was found, overlying a sterile plastic yellow clay (Markó 2013). Among the lithic tools separated some of it are represents the Transdanubian Szeletians, the "Jankovichian civilisation" (Vértes 1965; Gábori-Csánk 1968). The vertebrate fauna of the older layer is mixed in age and in taphoniomical point of view the *Equus*, *Coelodonta*, *Ursus spelaeus*, *Dicrostonyx*, *Ochotona*, *Putorius* are abundant (complicated by Kretzoi, in Vértes 1965).

Bajót No. 3 Rockshelter. A small cave West from the Jankovich cave of the Cliffs of Bajót. An excavation trench from the entrance opened 11 m long profile. The archaeological excavation in 1985 and 1986 made by traditional methods, and distinguished several layers in the "Entrance", in the "Entrance Hall", and in the "Chamber". From the most compact and in vertebrate remains the richest 5a layer is washed. For the vertebrate assemblage characteristic among the large mammals the *Ursus spelaeus*, *Rangifer tarandus* and *Canis lupus*. The micromammals (*Sciurus*, *Apodemus*, *Myodes*, *Spalax*, *Ochotona*, *Microtus arvalis*, *Microtus gregalis*) showing a mixed forest-steppe environment. The faunal picture and the age of the layer similar to the Tata mousterian site, cca 100 kyr (Kordos 1994a, b).

3.4.14 Pilisszántó Rock Shelter No. 1

The rock shelter is situated in Late Triassic limestone of the Pilis Hill, in the vicinaty of Pilisszántó, amount north-northeast of Budapest, an altitude of 423 m a.s.l. Cave Cataster No is 4840–3. The deposits of the rock shelter excavated since 1912 with interruptions till 1951 by Kadic, Kormos, Gábori (Jánossy 1986; Ruddy 2011). They distinguished seven Pleistocene layers below the humus, with the following designations: D_1 light-yellow cave loess; D_2 reddish dark-yellow cave loess; D_3 greenish-grey layer; D_4 "greyish-brown" layer; D_5 "rusty"; D_6 "brownish ash-gray"; and D_7 "greyish-brown" layer. Strata D_1 and D_2 were referred to as "upper", D_3 to D_5 as middle and D_6 and D_7 as "lower Diluvium". The various layers originally separated were so much different lithological that it would have been distinguish them faunistically, too (Vértes 1965). The rock shelter was the place of human settlement from time to time also for birds of prey and carnivorous mammals. The vertebrate assemblage is reminiscent of the Asiatic tundras today, the place of the cave bear (*Ursus spelaeus*) in large-mammal fauna being taken by the reindeer (*Rangifer*) (over 1,500 remains of the latter found here. Among the micromammals the *Dicrostonyx* also frequent. The whole series of faunal assemblages from Hungary have been placed in the Piliszántóian substage as a type locality. The chronological position is close to Last Glacial Maximum (LGM) cc. 17 kyr B.P (Kretzoi 1953; Jánossy 1986; Dobosi and Vörös 1987).

3.5 Holocene Vertebrate Cave Localities

From the second half of the nineteenth century intensive and detailed excavations were made in the Holocene cave sediments of Hungary. The studies focused to discover the fine layered sediments, their paleontologic contents, connections with archaeological levels, and on the base of vertebrate evaluate the faunal changes, the terrestrial stratigraphy and the climatic changes developed the „vole thermometer" and "Arvicola humidity" methods (Hermann et al. 1957; Jánossy and Kordos 1976; Kordos 1978a, b). From the vertebrate faunal evolution Kretzoi (1965, 1969) distinguished four Holocene phases (Bajót, Körös, Bükk and Alföld) to which Kordos (1978a) added the Kőhát phase. On the basis of ten vole species the July average temperature has been calculated (Kordos 1978b, c) which for the individual geochronological units have been the following: Pre-Boreal 17.9 °C, Boreal 18.8 °C, Atlantic 15.9 °C, Sub-Boreal 16.8 ° C, Sub-Atlantic 16.9 °C.

3.5.1 Jankovich Cave

In the Jankovich Cave the Late Pleistocene to Holocene period is represented by Beds 1 to 11 of Block II penetrated by excavations in 1956 (Hermann et al. 1957). Studied by Kretzoi in detail, the vertebrate finds from these beds constitute the first Hungarian biostratotype sequence of the Holocene. In this paper (in Hermann et al. 1957), Kretzoi did not give any detailed documentation, any faunal description, his information being restricted to frequency diagrams and concise faunistic, stratigraphy and palaeoecologic evaluations. The Bed 5are the represents the Pleistocene/Holocene boundary, while the fossils starting from Beds 5 to 4 standard fauna of the Bajót Stage, those from Beds 3 to 2 that of the Körös Stage, and Bed 1 has been identified with the Bükk Stage (Bácskay and Kordos 1984). This vertebrate biostratigraphic classification had already been revise, generalized and territorially extended (Kordos 1978a, b).

3.5.2 Petényi Cave and Rejtek I. Rock Shelter

Petényi Cave (otherwise Peskő II. Rock Shelter) is situated within one of the Triassic rocky formations being at the southwestern part of the Bükk Plateau, in the wall of Peskő. It is lying at a height of 735 m a.s.l. at the left side of Peskő Valley and at the eastern end of the rocks of Peskő, near Peskő Cave. The 13 m long, 8 m wide and 3.0–3.5 m high Rock Shelter is the remain of a former more spacious cave. It is named after the Hungarian pioneer of birds and vertebrate fossils, Salamon János Petényi. Rejtek I. Rock Shelter is in the middle of the Bükk Mountains East of the village Répáshuta, and locating in Triassic limestone 500 m a.s.l. Both small shelters excavated by Vértes és Jánossy have an unique Latest Pleistocene—Holocene sequence of vertebrates. The Holocene biochronologic phases established by Kretzoi (1965, 1969) after the investigations of the two shelter deposits Jánossy and Kordos (1976) re-defined.

3.5.3 Kis-Kőhát Shaft

The Kis-kőhát shaft opening on the southern edge of the Nagy-fennsík (Bükk Mts) is 114 m deep, between the spacious double entrance pits only one of which opens to the surface, and 50 m shaft with dripstone leading to the lowest point there is a huge hall. The cave opening is relatively high (920 m) above sea level and having an annual temperature of 4–6 °C, gives winter shelter to thousands of bats (Takács-Bolner 1989). Cave Cataster No is 5343–3. In the years of 1971 and 1972 Kordos and cavers collected a great number of *Rhinolophus hipposideros* skeletal remains (the number of individual 483) from the aven of the Óriás-terem. The chronology of the small mammal fauna of the bat population deposited during the Atlantic period, and named Kőhát phase as a new terrestrial biochronological unit (Kordos 1975b, 1978a).

3.5.4 Aggtelek Cave

There were several excavations and rescue activity in the Hungarian part (Baradla, Cave Cataster No is 5430–1) of the 25 km long Aggtelek Cave and also an enormous quantity of stray finds has been found. The oldest fossil vertebrates (*Ursus spelaeus*, *Coelodonta antiquitatis*, *Mammuthus primigenius* etc.) came to light from different parts of the reddish-brown yellow coloured clay and from the underground streams transported gravels. It is indicating the minimum age of the cave at the younger end of the Middle Pleistocene. In the Aggtelek part of the cave the oldest finds of human activity is from the Neolithic Bükk Culture. Similar situation discovered in the Hillebrand Jenő Cave, Bükk Mts (Fig. 17.21). In most part of the cave there is a sterile yellow clay below the cultural level. The next younger significant human settlement in the cave was in Late Bronze–Early Iron Age. The excavation in 1976 by Kordos made in the Csontház and in the Denevér-ág. The Csontház contained mostly mixed sediments but in one of the place the original Neolithic layer closed by a dripstone level (Fig. 17.22). It is contains historico-zoogeographic point of view highly significant rodents as *Microtus gregalis*, *Microtus oeconomus*, *Microtus agrestis* and first time from the Atlantic Phase from Europe, an *Ochotona* species (Kordos 1978d).

3.5.5 Nagyoldal Shaft

Nagyoldal Shaft opens in Nagyoldal 420 NE from the highest peak of the Aggtelek Karst, named Oltárkő (604 m a.s.l.) and at the SE slope of a do line. Cave Cataster No is 5440–26. The

Fig. 17.21 Neolithic fire layers in the Hillebrand Jenő cave (=Kőlyuk II)

original total depth was 21 m. In 1970 in the southern end of the shaft-level, speleologists started a large-scale exploration. Until 1972, they dug a 10 m deep shaft and Kordos sampled for paleontological investigations from the sediment remained in their original position after timbering the shaft the dripstone curtains. The samples were numerated downwards from 1 to 6. At about 3 m above the shaft there is a small niche used by owl. The samples taken from the shallow filling of the niche were marked as 0. The oldest 6th layer on the base of the single presence of cool indicating rodent *Ochotona* and morphological and chemical characteristics of the human remains represent Late Bronze Age, younger Sub-Boreal Phase. This stratigraphic column reveals an uninterrupted evolution of the fauna for roughly 2700 years. The vertebrate faunal changes in the deposit of the Nagyoldal Shaft are the key to reconstruct the deforestation processes in the Aggtelek Karst, forced by human activity (Kordos 1978d).

3.5.6 Hosszú-Hegy Shaft

The Hosszú-hegy Shaft or, as it is also called, The Három-lyukú Shaft, opens at 492 m altitude near the dummit of the Hosszú-hegy hill between Pilisszántó and Csobánka. During the excavations between 1974 and 1976 by Kordos with help of the cavers, recovered 9 layers deposited as shaft debris. The total thickness of the sampled section is 6.5 m, numerated from the top to down as layers 1–9. (Fig. 17.23). Reviewing the vertebrate fauna as a whole, and one can observe a sharp change in composition. In the older samples *Pelobates*, *Rana*, *Glis*, *Cricetus*, *Microtus arvalis* and *Arvicola* are frequent but gradually lesser in the younger ones. The chronological position of the sequence is determined by the constant presence of a low number of Pleistocene species and by animal species that became widespread and populous in the Holocene *Ochotona*, *Sicista* and *Spalax* are present, in a decreasing number though. Accordingly, the sample series 1

Fig. 17.22 Under the flowstone intact Neolithic sediments were excavated at the Csontház section of the Aggtelek Cave

to 9 from the Hosszú-hegy shaft can be assigned to the Körös Phase of the vertebrate biostratigraphic scale (Kordos 1983).

3.5.7 Rigó-Lyuk

The Rigó-lyuk locates several other rock shelters (160 m a.s.l.) in the gorge of Gaja-brook of village Bodajk at the Eastern margin of the Bakony Mountain in Trans-Danubia. During the excavation in 1976 nine layers can be distinguished in the 1.5 m thick sediments. These layers represent two sediment cycles. Below there was a 85 cms thick reddish brown clay with rock-debris connected by an erosion surface. On the basis of its vertebrate fauna of modern characters as well as of the distortion of this fauna resulted by cultural effects the sedimentary filling up to the Rigó rock-shelter can be dated by the beginning of our era. Within the vertebrate biochronologic system it mean the end of the Kőhát phase and the beginning of the Alföld phase (the end of the subboreal and beginning of subatlantic). The vertebrate faunal contains very similar those of the Nagyoldal Shaft younger layers (Kordos 1984b).

4 Conclusions

- The origin of fossiliferous cave sediments is different, sometimes within one cave even. The syngenetic deposits mostly developed by the processes of travertine formation. The caves of volcanic origin not contains in situ fossils but the erosional cavities of volcanic rocks frequently forming fossil contains shelters. Several caves deposited fossils from the bedrock, and on the wall surface of the caves the imbedded original fossils are prepared by natural ways (nummulits from the Eocene, *Megalodus* shells from the Triassic, *Congeria* shells from the Late Miocene), and for the geologists are distinguished stratotype sections (Aggtelek Cave), too. In the fluviatile deposits the fossils are rare, regularly eroded and mixed. The classical fossiliferous cave deposits in various way, by debris moving, shaft collapsing, ice breaked rocks, wind blowing dusts etc. Most of these types of sediments often redeposited and mixed.

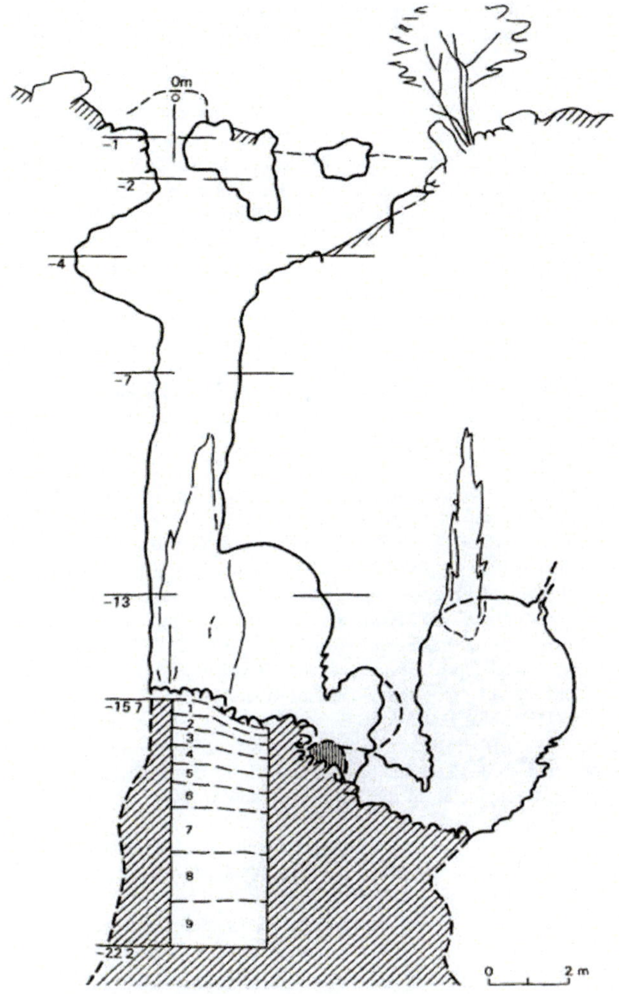

Fig. 17.23 Cross section of the Holocene sediments of the Hosszú-hegy shaft

- Determine the geological ages of the cave deposits have several methods. The traditional way is the interpretation from the bones of the animal taxa. The molluscs, pollens and charcoals are also available for age determinations. The radiometric kronological methods (C-14, uranium series, thermoluminescence, ESR, fission-track etc.). Stable isotopes (O^{16}/O^{18}) engage the correlations with the MIS (Marine Isotope Stages). Studying the characteristic artefacts and their interactions with the human activity and natural environmental changes producing an archaeological time scale. All of these methods—and others—used in Hungary.
- The evolution of the historical review of the vertebrate terrestrial biochronologic system is an important issue in Hungary and of course other territories of Europe, too. During the second half of the 19th and beginning of the twentieth century only one glacial period distinguished, before it was the preglaciale and after it is the postglacial (monoglacial theory). Sometimes used the diluvial and the alluvial categories. The poliglacialist based vertebrate biochronology and biostratigraphy established by Kretzoi and Jánossy, re-juvenated by Kordos. Some of the, i.e. Villanyian and Biharian stages built into the internationally accepted systems. Since 1906, the first trace of the presence of palaeolith man at Szeleta Cave started the hierarchy of the archaeological chronologic system which advances versions are still exist. The most stratigraphic units base on a separated layer or sequence of fossiliferous cave deposits and the units made a continuous time-line. Re-interpretation of the sedimentation and taphonomic processes correlating with the stable isotopes (MIS) units and different radiometric data clarified that the traditional units sometime mixed, has a gap in between and not covering the whole Quaternary. In this paper the author collected the most important and relevant stratigraphic and chronologic data from the cave layers.
- From a taphonomic point of view the most typical bone concentrations made by the large bat colonies and the owl pellets. Through the vertical chimneys and shafts from the frog-size to large mammal (sometimes human) bones a great amount of the vertebrates deposited in the cave floors. The bird bodies are also common. In several caves there are snake bones indicating the process that of the living snakes, captured in the bell form cave chambers, and they died by over-hybernation. The shallow rock shelters regularly used by small and medium size carnivores where accumulated they feeding remains. The taphonomic investigations in the fossils of cave deposits just during the last decades started. For taphonomic reasons need to develop and apply new excavation technics and making special investigations and interpretations.
- Among the some thousands the caves of Hungary cca. one hundred contains real paleontological values. Some of them—due to the traditional excavations and investigations—contains locus typicus of new taxa and chronologic, stratigraphic units. Cave deposits in Hungary are the basic pillars of the Quaternary researches.

References

Albert G (2010) Bajót, Öreg-kő; Bajót Öreg Stone. In: Budai T, Gyalog L (eds) Geological map of Hungary for tourists 1:200 000. Published by the Geological Institute of Hungary, Budapest, p 155 (in Hungarian)

Andrews P (1990) Owls, caves and fossils. The University of Chicago Press, p 231

Báccskay E, Kordos L (1984) Major finds scattered fossils in the Palaeovertebrate Collection of the Hungarian Geological Institute

(Communication No. 9). Magyar Állami Földtani Intézet Évi Jelentése az 1982. évről, pp 355–361

Bárdossy GY, Kordos L (1989) Paleokarst of Hungary. In: Bosak et al. (eds) Paleokarst. A systematic and regional review. Elsevier and Academica, Amsterdam-Praha, pp 137–153

Behrensmayer AK, Kidwell SM (1985) Taphonomy's contribution to paleobiology. Paleobiology 11:105–119

Bernor RL, Kordos L, Rook L (eds) (2005) Multidisciplinary research at Rudabánya. Palaeontographica Italica 90:1–313

Bradák B, Markó A (2006) A Szelim-barlang (Tatabánya) üledéksorának vizsgálata és fejlődéstörténetének vázlata (The study and historic reconstruction of the sedimentary sequence, in the Szelim Cave (Tatabánya). Földtani Közlöny 136(2):233–248 (in Hungarian)

Coles GM, Gilberston DD, Hunt CO, Jenkinson RDS (1989) Taphonomy and the palynology of Cave deposits. Cave Sci 16(3): 83–89

Demeter A, Sipos G, Csorba G (2016) In memoriam dr. György Topál. Annlas Mus Hist Nat Hung 108:10–32

Dobosi V, Vörös I (1987) The chronological revision of the Pilisszántó Rockshelter I. Folia Archaeologica 38:25–45

Efremov RL (1940) Taphonomy: a new branch of paleontology. Pan-Am Geol 74:81–93

Eszterhás I (1977) Az MKBT Vulkánszpeleológiai Kollektívájának Évkönyve 1977 [Yearbook of the Volcanspeleological Group of the Hungarian Karst and Cave Society, only in Hungarian]. Manuscript, Budapest, p 288 (in Hungarian)

Fejfar O, Heinrich W-D (1990) Proposed biostratigraphical division of the European continental Neogene and Quaternary based on muroid rodenta (Rodentia: Mammalia). Int Symp Evol Phyl Biostr Arvicolids. Pfeil-Verlag, Münich, pp 115–124

Ford TD, Cullingford CHD (1976) The science od speleology. Academic Press, London, New York, San Francisco, p 593

Ford TD, Williams PW (1989) Karst geomorphology and hydrology. Unwin Hyman, London, p 601

Freudenthal M, Kordos L (1989) Cricetus polgardiensis sp. nov. and Cricetus kormosi Schaub, 1931 from the Late Miocene Polgárdi localities (Hungary). Scripta Geologica 89:71–100

Fűköh L, Kordos L (1977) Jelentés az Uppony Horváti-lik 1977. évi őslénytani ásatásáról (Report on the paleontological excavation of the Uppony Horváti-lik, from the year of 1977). Az Egri Múzeum Évkönyve Eger 15: 21–32 (in Hungarian)

Fűköh L, Kordos L (1980) Jelentés az Uppony Horváti-lik 1978. évi őslénytani ásatásáról (Report on the paleontological excavation of the Uppony Horváti-lik, from the year of 1978). Az Egri Múzeum Évkönyve, Eger 16–17:21–43 (in Hungarian)

Gábori-Csánk V (1968) Az őskori Jankovichien kultúra Nyugat-Magyarországon [The palaeolithic Jankovichian Culture in Western Hungary]. Ph.D. Dissertation, Budapest, p 360 (in Hungarian)

Gáboriné Csánk V (1984) A Remete felső-barlang és a "dunántúli Szeletian" [The Remete Upper Cave and the "Transdanubian Szeletian"]. Budapest régiségei 26:5–32 (in Hungarian)

Gál E (2008) Faunal and taphonomic analyses of a Late Pleistocene bird-bone assemblage from cave deposits in north-west Hungary. Geobios 41(1):79–90

Halász Á (1964) Jelentés a somlyóhegyi 1.sz. barlangról és őskarszt jelenségektől (Report on the Somlyó-hegy No. 1 cave and paleokarst phenomena). Manuscipt, in Hungarian, pp 1–23, Budapest (in Hungarian)

Hankó ES (2007) A revision of three Pleistocene sebspewcies of Panthera, based on mandible and teeth remains, stored in Hungarian Collections. Fagmenta Palaeontologica Hungarica 24–25:25–43

Hermann M, Kretzoi M, Vértes L (1957) Neuere Forshungen in der Jankovich-Höhle. Folia Archaeologica 9:3–23

Hír J (1986) A répáshutai Pongor-lyuk ásatásának eddigi eredményei (Results of Excavation of the Répáshuta's Pongor-Hole). Nehézipari Műszaki Egyetem Közleményei, Miskolc, I. sorozat, Bányászat 33:75–85 (in Hungarian)

Hír J (1985) A Hajnóczy-barlang üledékei (Sediments in the Hajnóczy cave). Karszt és Barlang 1985(I–II):17–24 (in Hungarian)

Hír J (1991) Layer identifying excavation in the Pes-kő Cave. Folia Historico-naturalia Musei Matraensis, Gyöngyös 16:5–12

Horaček I, Kordos L (1989) Biostratigraphic Investigations in Paleokarst. In: Bosák et al. (eds) Paleokarst. A systematic and regional review. Elsevier and Academica, Amsterdam-Praha, pp 599–612

Jámbor Á, Korpás L, Kretzoi M, Pálfalvy I, és Rákosi L (1971) A dunántúli oligocén képződmények rétegtani problémái (Stratigraphical problems of the Transdanubian Oligocene rocks). Földtani Intézet Évi Jelentése 1964-ről, pp 249–264 (in Hungarian)

Jánossy D (1991) Late miocene bird remains from Polgárdi (W-Hungary). Aquila 98:13–35

Jánossy D, Kordos L (1976): Pleistocene-Holocene Mollusc and Vertebrate Fauna of Two Caves in Hungary. Ann Hist Nat Mus Nat Hung 68:5–29

Jánossy D, Kordos L (1977) Az Osztramos gerinces lelőhelyeinek faunisztikai és karsztmorfológiai áttekintése (1975-ig), (The faunistical and karst-morphological review of paleontological localities for vertebrates at Osztramos (Northern Hungary). Fragmenta Mineralogica et Palaeontologica 1977(8):39–72 (in Hungarian)

Jánossy D, Krolopp E, Brunnacker K (1968) Die Felsnische Uppony I (Nordungarn). Eiszeitalter und Gegenwart 19:31–47

Jánossy D, Kordos L, Krolopp E, Topál GY (1973) The Porlyuk Cave of Jósvafő. Karszt- és Barlangkutatás 8:15–59

Jánossy D, Kordos L, Krolopp E (1982–1983) Upper Pleistocene and Holocene fauna from the Függő-kő Cave (Mátraszőlős). Fol Hist Nat Mus Matr 8:47–61

Jánossy D (1952) Az Istállóskői-barlang aurignaci faunája (Aurignacian fauna of the Istállóskő cave). Földtani Közlöny 82(4–6):181–203 (in Hungarian)

Jánossy D (1963–1964) Letzintertglaziale Vertebratenfauna aus der Kálmán Lambrecht Höhle (Bükk-gebirge, Nordost Ungarn) I–II. Acta Zool Hung 9–10:139–197, 293–331

Jánossy D (1986) Pleistocene vertebrate faunas of Hungary. Akadémiai Kiadó, Budapest, p 208

Kadić O, Kretzoi M (1930) Ergebnisse der weiteren Grabungen in der Esterházy-Höhle (Csákvárer Höhle). Mitteilungen über Höhlen- und Karstforschung, Berlin, pp 45–49

Kadić O (1915) Über das kubizienrende und staffelweise Graben in Höhlen.- Barlangkutatás – Höhlenforschung 3(2):123–126

Kadić O (1916) Die Ergebnisse der Erforschung der Szeletahöhle .- Földtani Intézet Évkönyve 23:151–278

Kadić O (1940) A cserépfalui Mussolini barlang (The Mussolini Cave of Cserépfalu). Geologica Hungarica ser. Palaeontologica 14. (in Hungarian)

Kordos L (1984a) Magyarország barlangjai [Caves of Hungary]. Gondolat, Budapest, p 326 (in Hungarian)

Kordos L (1984b) Investigations on the young Holocene sedimentary infilling of Rigó rock-shelter near Bodajk. Folia Musei Historico-Naturalis Bakonyensis Zirc 3:31–42

Kordos L (2006) Stratigraphy of the Middle Pleistocene "Buda Culture" of Castle Hill, Budapest (Hungary). Praehistoria Miskolc 4–5:9–32

Kordos L, Krolopp E (1980) The Mollusca and Vertebrate fauna of the spring limestone deposit from Upper Pleistocene int he area of the Dobó-bástya at Eger. Folis Historico-Naturalia usei Matraensis Gyöngyös 6:5–12

Kordos L, Szolyák P (2009) A hámori Herman Ottó-barlang pleisztocén gerinces maradványai és a "középső felső-pleisztocén" faunája a Kárpát-medencében (The Pleistocene Vertebrata Fauna of Herman Ottó Cave and the Mid-Upper Pleistocene Funas in the Carpathian Basin). A Herman Ottó Múzeum Évkönyve, Miskolc 2008:199–220 (in Hungarian)

Kordos L (1970) Klímamegfigyelések a barlangok bejárati szakaszában (Climatic investigations of the entrance part of caves). Karszt és Barlang 1970(I):31–34 (in Hungarian)

Kordos L (1975a) Az Esztramos barlanggenetikai, hegységszerkezeti és üledékföldtani vizsgálata (Speleogenetical, tectonical and sedimentological investigations of Esztramos). Karszt és Barlang 1974 (I):21–26 (in Hungarian)

Kordos L (1975b) Untersuchung der subfossilen Fledermaus-population der Kis-kőháter Höhle. A Miskolci Herman Ottó Múzeum Évkönyve (23–14):567–586

Kordos L (1978a) A magyarországi holocén képződmények gerinces biosztratigráfiájának vázlata (A Sketch of the Vertebrate Biostratigraphy of the Hungarian Holocene). Földrajzi Közlemények 25[CI] (1–3):144–160

Kordos L (1978b) Holocén klímaváltozások kimutatása Magyarországon a "pocok hőmérő" segítségével (Changes in the Holocene Climate of Hungary Reflected by the "Vole-Thermometer" Method). Földrajzi Közlemények 25[CI](1–3):222–229

Kordos L (1978c) Historico-zoogeographical and ecological investigations of the subfossil vertebrate fauna of Aggtelek Karst. Vertebrata Hungarica 18:85–100

Kordos L (1978d) Historico-zoogeographical and ecological investigations of the subfossil vertebrate fauna of the Aggtelek Karst. Vertebrata Hungarica 28:85–100

Kordos L (1979) The revision of the upper pleistocene vertebrate fauna of seybold stone quarry at Kőszeg. Fragmenta Mineralogica et Palaeontologica 9:11–20

Kordos L (1983) Holocene Vertebrate Fauna of the Hosszú-hegy Shaft. Magyar Állami Földtani Intézet Évi Jelentése az 1981 évről, pp 425–437

Kordos L (1990) The evolution of Upper Pleistocene voles in Central Europe. Int Symp Evol Phyl Biostrat Arvicolids. Pfeil-Verlag, Münich, pp 275–284

Kordos L (1991) Late miocene palaeovertebrate localities, Polgárdi, Mezőföld. Magyarország Geológiai Alapszelvényei, pp 1–6. Geological Institute of Hungary, Budapest

Kordos L (1992) Magyarország harmad- és negyedidőszaki emlősfaunájának fejlődése és biokronológiája (The development and biocronology of the Hungarian Tertiary and Quaternary mammal fauna).Akadémiai Doktori Értekezés, Budapest, p 103 (in Hungarian)

Kordos L (1993) Paleoanthropology in Hungary. In: Sigmon BA (ed) Before the wall fell. Canadian Scholars Press, Toronto, pp 101–121

Kordos L (1994a) Revised biostratigraphy of the early man site at Vértesszőlős, Hungary. Courier Forschungs-Institut Senckenberg 171:225–236

Kordos L (1994b) A gerecsei barlangok ősgerinces kutatásának újabb eredményei (1970–1994b) [New Records of Paleovertebrate Research at the caves of the Gerecse Mts (1970–1994b)]. Limes, Komárom-Esztergom megyei Tudományos Szemle, Tatabánya 1994b(2):93–111. (in Hungarian)

Kordos L (2001) Az Istállóskői-barlang 2000. évi ásatásának gerinces maradványai [Vertebrate remains from the excavation of the year of 2000 in the Istállós-kő cave], Manuscript, Budapest—Miskolc, p 10. (in Hungarian)

Kordos L (2013) A sztyepplakó rágcsálók európai honfoglalásai (The European spread of grassland rodents). Természet Világa 144 (3):110–113. (in Hungarian)

Kormos T (1911) Der pliozäne Knoschenfund bei Polgárdi. Annales Instituti Geologici Hungarici 41(1–2):171–189

Kormos T (1937) Zur Geschichte und Geologie der oberliozänen Knochenbreccien des Villányer Gebirges. Mathematikai és Természettudományi Értesítő 56:1061–1110, Budapest

Korpás L (1981) Oligocene—lower miocene formations of the transdanubian central mountain sin Hungary. Földtani Intézet Évkönyve 64(1):81–140

Kretzoi M (1956) A Villányi-hegység alsó-pleisztocén gerinces faunái (The Lower Pleistocene vertebrate fauna of the Villány Mountains). Geologica Hungarica ser. Palaeontologica 27:1–264 (in Hungarian)

Kretzoi M (1940) Alttertiäre Perissodactylen aus Ungarn. Annales Museum Naturale Hungariae 33:87–98

Kretzoi M (1941) Ősemlősmaradványok Betfiáról (Die unterpleistozäne Säugetierfauna von Betfia bei Nagyvárad). Földtani Közlöny 71:235–261, 308–335. (in Hungarian)

Kretzoi M (1951) A csákvári Hipparion-fauna (The Hipparion fauna of Csákvár). Földtani Közlöny 81(10–12):384–401. (in Hungarian)

Kretzoi M (1952) Die Raubtiere der Hipparionfauna von Polgárdi. Annales Instituti Geologici Hungarici 40(3):5–42

Kretzoi M (1953) Quaternary geology and vertebrate fauna. Acta Geologica 2(1–2):67–77

Kretzoi M (1954) Befejező jelentés a Csákvári barlang őslénytani feltárásáról (Final report on the paleontological excavation of the Csákvár Cave). Magyar Állami Földtani Intézet Jelentése az 1952. évről, pp 37–55, Budapest (in Hungarian)

Kretzoi M (1965) The role of Vertebrate faunae and Palaeolithic industries of Hungary in Quaternary stratigraphy and chronology. Acta Geologica Hungarica 9:125–143

Kretzoi M (1969) Sketch of the late cenozoic (pliocene and quaternary) terrestrial Stratigraphy of Hungary. Földrajzi Közlemények (43)17 (3):179–204

Krolopp E (1973) Quaternary malacology in Hungary. Földr Közlem 2:181–171

Krolopp E, Schweitzer F, Scheuer F, Dénes GY, Kordos L, Skoflek I, Jánossy D (1977) A budai Várhegy negyedkori képződményei (Quaternary formations of Castle Hill in Buda). Földtani Közlöny 106(3):193–228. (in Hungarian)

Kubacska A (1926) A Solymári sziklaüreg pleisztocén csontlelete [Pleistocene bone finds from the Solymár rock shelter). Barlangvilág 1(1):17–26. (in Hungarian)

Leél-Őssy SZ, Szanyi GY, Surányi G (2011) Minerals and speleothems of the József-hegy Cave (Budapest, Hungary). Int J Speleol 40 (2):191–203

Lengyel GY, Szolyák P, Pacher M (2008–2009) Szeleta cave earliest occupation reconsidered. Praehistoria, Miskolc 9–10:9–20

Lymann RL (1994) Taphonomy. Cambridge Manuals in Archaeology, Cambridge University Press, p 524

Lymann RL (2010) What taphonomy is, what it isn't, and why taphonomists should care about the difference. J Taphonomy 8 (1):1–16

Markó A (2013) Ont he middle palaeolithic industry of the Jankovich Cave (Northeastern Transdanubia). Archaeologiai Értesítő 138:7–28

Markó A (2015) Istállóskő revisited: Lithic artefacts and assemblages, sixty years after. Acta Archaeologica Acadeniae Scientiarum Hungaricae 66:5–38

Mester ZS (1989) A subalyuk-barlang középső paleolitikus iparának újraértékelése [Re-examination of the middle palaeolithic industry at the Subalyuk Cave]. Folia Archaeologica 40:11–35

Mester ZS (2006) Mousterian industries in Hungary—15 years later. In: Kulakovska LV (ed) The European middle palaeolithic. Kiev, pp 170–179. (in Hungarian)

Mészáros LG (1996) Soricidae (Mammalia, Insectivora) remains from three Late Miocene localities in western Hungary. Annales Univ Sci Budapest Sect Geol 31:5–25

Mihály S (1975) A new occurrence of Pikermian-type Vertebrata int he Big Quarry of Polgárdi-Ipartelepek, Transdanubia, Hungary. Discuss Palaeont 22:89–93

Mottl M (1942) Das Aurignacian in Ungarn. Quartär 4:82–108

Patou-Mathis M, Vercoutére C, Lengyel GY, Szolyák P, Mester ZS (2017) New interpretation of the Upper Palaeolithic human occupations at Istállóskő cave (Bükk Mountains, Hungary). Eurasian Prehistory 13(1–21):77–90

Pávai-Vajna F (1911) Néhány újabb barlang ismertetése (Description of some new caves) [Polgárdi]. Földtani Közlöny 41(11–12):786–787. (in Hungarian)

Pazonyi P, Kordos L (2004) Late Eemian (Late Pleistocene) vertebrate fauna from the Horváti-lik (Uppony, NE Hungarty). Fragmenta Mineralogica Hungarica, Budapest 22:107–117

Petényi SJ (1864) Hátrahagyott munkái [Posthumous works]. Pest, p 130. (in Hungarian)

Ringer Á, Mester ZS (2001) Die Ergebnisse der Revision der Szeleta-Höhle der Jahren 1999–2000. A Herman Ottó Múzeum Évkönyve, Miskolc 40:5–19

Ringer Á (2002) The new image of Szeleta and Istállós-kő Caves in the Bükk Mountains: a revision project between 1999–2002. Praehistoria 3:47–52

Ringer Á (2008–2009) Nouvelles données sur le Szélétien du Bükk. Praehistoria (Miskolc) 9–10:21–34

Rotaridesz M (1943) Preglaziale Schnecken aus der Höhle von Solymár. Ann Hist Nat Mus Nat Hung Pars Min Geol Pal 36:85–90

Rounds A (2004) Cave taphonomy. Rev J Undergraduate Stud Res 7:7–11

Ruddy M (2011) The Western Palaearctic evolution of the water vole Arvicola. Ph.D. Dissertation, Rpyal Holloway College, University of London, p 397

Scheuer GY, Schweitzer F (1973) The development of the Hungarian travertine sequence in the Quaternary. Földrajzi közlemények 21 (2):133–141

Scheuer GY, Schweitzer F (1988) A Gerecse- és a Budai-hegység édesvízi mészkőösszletei (Freshwater limestone beds of the Gerecse Mountains and the Buda Hills). Földrajzi tanulmányok (20), Akadémiai kiadó, Budapest, p 129. (in Hungarian)

Sóron A, Virág A (2009) Detailed quantitative method in microvertebrate taphonomy int he case of Pleistocene filling of the Vaskapu II rock shelter. Central European Geology (Acta Geologica Hungarica) 52(2):185–198

Takács-Bolner IE, Juhász M, Kraus S (1989) The Caves of Hungary. Karszt és Barlang, Special Issue, pp 17–30

Vendl A (1926) A Somlyó és Szárhegy geológiája s egykori hévforrásai (Geology and ancient thermal springs of the Somlyó and Szárhegy). Hidrológiai Közlöny 4–5 (1924–1925):1–10. (in Hungarian)

Vértes L (1950) A Solymári-barlang rétegviszonyairól [On the stratigraphic conditions of the Solymár Cave]. Földtani Közlöny 80(4–6):199–203. (in Hungarian)

Vértes L (1959) Untersuchungen an Höhlensedimenten, Methode und Ergebnisse. Régészeti Füzetek, Ser. II., 7., Magyar Nemzeti Múzeum-Történeti Múzeum, Budapest, p 176

Vértes L (1965) Az őskőkor és az átmeneti kőkor emlékei Magyarországon (Paleolithic and Mesolithic Remnants of Hungary). A Magyar Régészet Kézikönyve I. Akadémiai Kiadó, Budapest, p 385 (in Hungarian)

Vörös I (2003–2004) Stratigraphy and biostratigraphy of Istállóskő Cave. Praehistoria 4–5:33–76

Age, Depositional Environment, and Geochemistry of Freshwater Carbonates (Travertine, Tufa) from Hungary

Sándor Kele and Barbara Bódai

Abstract

Secondary continental carbonate accumulations are quite frequent deposits in Hungary, indicating strong hydrothermal activity during the Pliocene and Quaternary. They form from $CaCO_3$-rich waters either at ambient or at higher temperatures due to CO_2 degassing and/or biological activity. Pliocene-Pleistocene travertines formed from hydrothermal fluids are characteristic for the area of the Gerecse and Buda Mts, while recent travertine accumulations form mostly at thermal wells as scalings. Freshwater tufa carbonates in Hungary are connected to karstic springs of the the Bükk Mts, Aggtelek Mts, Mecsek Mts, Bakony Mts, Balaton-felvidék. Sedimentology, geochemistry and dating of travertine and tufa can provide important information about palaeoclimatic, tectonic and palaeoenvironmental conditions and their petrophysical and microfacies study can contribute to the understanding of microbialite oil reservoir rocks. Furthermore, the recent deposits can be used for the calibration of the oxygen- and clumped isotope paleothermometers. The aim of this chapter is to provide a short review of the most important Hungarian travertine and tufa deposits including the achievements of most recent researches focusing on their age, depositional environment and geochemistry.

Keywords

Travertine • Tufa • Depositional environment • Isotope geochemistry • U–Th dating

S. Kele (✉)
Institute for Geological and Geochemical Research, Research Centre for Astronomy and Earth Sciences, Budaörsi út 45, 1112 Budapest, Hungary
e-mail: kele.sandor@csfk.org

B. Bódai
Department of Physical Geography, Eötvös Loránd University, Pázmány Péter sétány 1/c, 1117 Budapest, Hungary

1 Introduction

Freshwater carbonates were also called calcareous sinters, however, Pedley (1990) suggested the use of the terms tufa and travertine, in order to distinguish them from siliceous sinters, which are also common at high-temperature hydrothermal areas (e.g. Fournier and Rowe 1966). Freshwater carbonates form from $CaCO_3$-rich waters either at ambient or at higher temperatures due to CO_2 degassing and/or biological activity in a high variety of depositional environments such as hill slopes, streams, lakes and caves (Ford and Pedley 1996; Pentecost 2005; Pedley 2014). The term 'tufa' is used for calcium carbonate formed from cool, ambient temperature freshwater due to biologically induced processes and typically contain remains of micro- and macrophytes, invertebrates and bacteria (Pedley 1990; Ford and Pedley 1996; Chafetz and Folk 1984; Golubic et al. 1993; Violante et al. 1994), whereas travertines are spring-related continental carbonates forming from calcium bicarbonate rich hydrothermal fluids, mainly due to CO_2 degassing controlled by physical–chemical processes (Capezzuoli et al. 2014).

Sedimentology, geochemistry and dating of these carbonates can provide important information about palaeoclimatic, tectonic and palaeoenvironmental conditions (Chafetz and Folk 1984; Ford and Pedley 1996; Hancock et al. 1999; Arenas et al. 2000; Andrews 2006; Arenas-Abad et al. 2010; Capezzuoli et al. 2010; Özkul et al. 2013; Török et al. 2017; Della Porta et al. 2017a, b; Toker 2017). The aim of this chapter is to introduce the reader to the most important Hungarian travertine and tufa deposits including the achievements of most recent researches focusing on their age, depositional environment and geochemistry.

M. Veress and Sz. Leél-Őssy (eds.), *Cave and Karst Systems of Hungary*,
Cave and Karst Systems of the World, https://doi.org/10.1007/978-3-030-92960-2_18

2 History of Travertine Research in Hungary

Detailed review on Hungarian travertine and tufa research can be found in Kele (2009). This chapter is restricted to the most important works with special focus on papers published during the last decades and/or using new and novel methods. Hungarian travertines were already mentioned from the late eighteenth century (e.g. Townson 1797). The geological map of Horusitzky (1939) indicated several travertine sites, but the first detailed work on travertines was the one of Schréter (1953) providing a detailed review on "freshwater limestones" of the Gerecse and Buda Mts. Later on, based mainly on geomorphological, sedimentological and biostratigraphic considerations, the travertine deposits of the Gerecse and Buda Mts were correlated with the terrace levels of the Danube River and its tributaries (Pécsi 1959; Kretzoi and Pécsi 1982). Vértes (1964) published a monography about the travertine of the Tata, Porhanyó quarry and middle paleolithic archeological site. The discovery of the Vértesszőlős Early Human site—an important Lower Palaeolithic locality of Central Europe—within a slope tetarata type travertine deposit drew new attention to travertines and several papers were published about the site mostly focusing on its age (e.g. Kretzoi and Vértes 1965; Cherdintsev et al. 1965; Pécsi 1973; Járai-Komlódi 1973; Hennig et al. 1983a; Schwarcz and Latham 1984). One of the most important work is the monography of Scheuer and Schweitzer (1988), which is a detailed overview of travertines of the Gerecse and Buda Mts mainly from a geomorphologist's and hydrogeologist's point of view. Although further detailed studies have been conducted (e.g. Bakacsi 1993; Korpás et al. 2004; Kele et al. 2003; etc.), geochemical and stable isotope analyses of travertines were almost completely missing (except Opauszky et al. 1964; Szöőr et al. 1992) in Hungary until the early 2000's (Korpás 2003; Kele et al. 2003). Földvári et al. (2003) applied thermoanalytical analysis on the thermal lacustrine travertines of the Buda Castle Hill, Budakalász and Szomód, Les Hill as an alternative way to assess the temperature of carbonate deposition. Kovács-Pálffy and Földvári (2004) focussed on the mineralogy of seven sites from north-eastern Transdanubia, while Lantos (2004) attempted their magnetostratigraphic correlation. Török (2003) performed the facies analysis and genetic interpretation of the travertine of the Buda Castle Hill, but also studied travertine buildings of Budapest (e.g. Török 2004, 2008).

The IAS sponsored travertine field course "Climatic and tectonic controls on travertine formation: the case of the Pannonian Basin (4–9 July 2004, Tata and Egerszalók, Hungary)" was a cornerstone in the Hungarian travertine research generating a couple of new case studies. Kele et al. (2006) reconstructed the paleoenvironment of the Tata travertine complex using petrographic observations and stable isotope geochemistry. Kele (2009) provided an updated review of the travertines of the Gerecse and Buda Mts (Fig. 1) using sedimentological and geochemical methods, including stable isotope geochemistry and U–Th dating, thus providing a new chronological frame for these paleospring deposits. U–Th age data and stable isotope data were published from the sites of the Gellért Hill (Kele et al. 2009), Süttő (Sierralta et al. 2010), Rózsadomb (Kele et al. 2011), Tokod (Gasparik and Kele 2010) and other smaller sites from Budapest (Kele 2011; Kele and Scheuer 2012) . The detailed study of the recently forming Egerszalók travertine (Kele et al. 2008, see later in detail), was an important work about stable isotope fractionation processes accompanying travertine deposition. The detailed study of the "indoor travertine" deposit within the Rudas Spa included radiocarbon analyses as well (Virág et al. 2020).

Studies from the last decade have started to apply geophysical methods to reconstruct the dimensions and depositional environment of the travertines, e.g. at Kápolna Hill (Buda Mts) (Török et al. 2013). Recently, travertine deposits gained new interest due to the discovery of major hydrocarbon reservoirs in lacustrine basins of the South Atlantic rift-sag on the Sao Paulo Plateau (Wright 2012) and offshore Angola (Greenhalgh et al. 2011). Since core samples from the original microbialite reservoir rocks are limited and not easily accessible, travertines can serve as their field analogues. In the frame of the TraRAS (Travertine Research Analogue Studies), Hungarian sites at Budakalász and Süttő were also insvestigated mostly from sedimentological and petrophysical point of view (e.g. Degros 2013; Smekens 2013; Soete et al. 2015; Claes 2015; Claes et al. 2017; Török 2018; Török et al. 2017, 2019). These studies used a series of analytical methods and research approaches like stable isotope geochemistry, $^{87}Sr/^{86}Sr$ isotope ratio, U–Th dating, petrophysical measurements, cathodoluminescence microscopy, effective porosity, permeability, compressional-wave velocity (Vp), shear-wave velocity (Vs), sonic velocity measurements in combination with petrography, X-Ray Diffraction (XRD), micro-Computer Tomography (mCT) analyses as well as 3D reconstruction of the quarry outcrops using LiDaR (Laser Imaging Detection And Ranging) photogrammetry.

3 Depositional Environment of Hungarian Travertines

The reconstruction of the depositional environment of Hungarian travertines is in the focus of studies since the beginning of the twentieth century (Schréter 1912, 1953). These early works distinguished spring travertines and

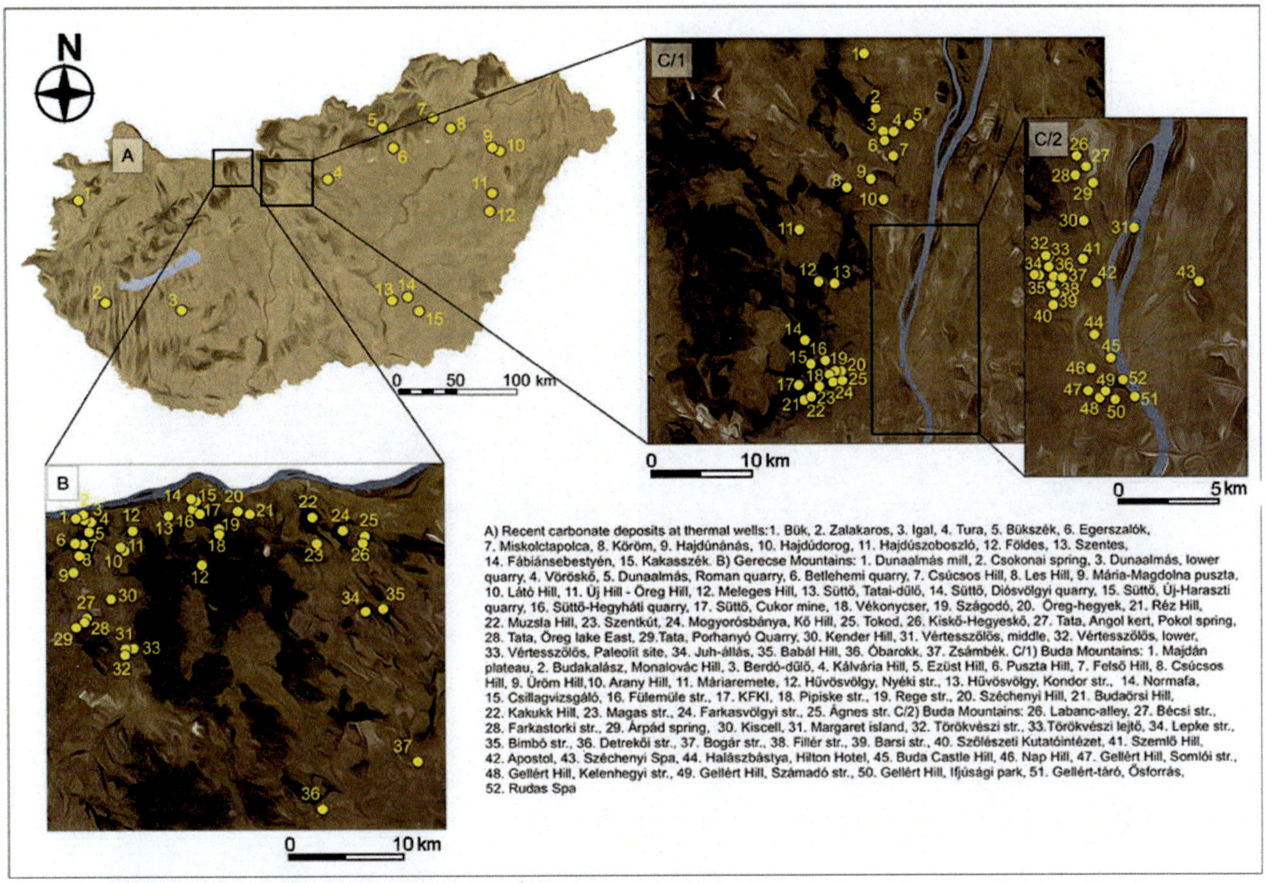

Fig. 1 Map of the Hungarian travertine occurrences including recent carbonate deposits of active thermal wells, and Pliocene-Pleistocene travertines of the Gerecse and Buda Mts (based on Scheuer and Schweitzer 1988; Kele 2009)

lacustrine-paludal travertines. Scheuer and Schweitzer (1988) classified the travertines as lacustrine-paludal, tetarata barrier, spring-cones and their combination. The first detailed sedimentological studies aiming to reconstruct travertine depositional environments are dated back to the middle of the 90's resulted mainly lacustrine environments (e.g. Bakacsi 1993; Kovács 1995; Török 2003; Kele et al. 2003). Török (2003) reconstructed a shallow lake environment for the travertine of the Buda Castle Hill. Kele (2009) studied all travertine occurrences from the Gerecse and Buda Mts and based on sedimentological and geochemical analyses concluded that travertines of the Gerecse Mts formed in characteristically low temperature lakes (e.g. Les Hill, Süttő, Dunaszentmiklós), whereas the travertines of Buda Mts precipitated from rather higher temperature, warm or hot thermal water, reflecting hydrothermal effects (Fig. 2). Difference in the deposition temperatures of the travertines of the Gerecse and Buda Mts were later confirmed by recent clumped isotope analyses (Kele et al. in prep, in prep, in prep).

Although remnants of spring cones (e.g. Gellért Hill, Ifjúsági-park, Törökvész lejtő, Kele et al. 2009, 2011) are common in the Buda Mts, the majority of the travertine occurrences formed in shallow lakes, and on slopes (e.g. Buda Castle Hill, Budakalász, Üröm-Hill, etc.) (Fig. 2). According to Török et al. (2013) the travertine of the Kápolna Hill (Buda Mts) is a thermogene-thermometeogene lacustrine sediment and can be regarded as a transition between a spring cone and a lake. The most significant and still being quarried travertine bodies are located at Budakalász (Buda Mts) and at Süttő and Dunaalmás (Gerecse Mts) and formed in lacustrine-palustrine depositional environments (Kele 2009; Claes et al. 2017; Török 2018). The multi-methodological study of Török et al. (2019) reconstructed a travertine fissure ridge sytem in the Cukor quarry (Süttő) and so far it is the only known fissure ridge in Hungary. Comparing the Hungarian travertine sites to the ones from the neighbouring countries, it can be stated that the Hungarian travertines are mostly lacustrine deposits, while travertine mounds of low temperature and saline (high TDS) springs are more characteristic for Slovakia (e.g. Dudince, Santovka) and Romania (e.g. Kalán, Olasztelek, Feredőgyógy).

Fig. 2 Travertines from the Gerecse and Buda Mts. **a** The largest (active) travertine quarry of the Buda Mts is located at Budakalász on the Monalovác Hill. **b** Abandoned travertine quarry at Kálvária tető (Kápolna Hill, Buda Mts), north of Budapest. **c** Remnant of a spring cone from the Gellért Hill, Ifjúsági park (Buda Mts) (Kele et al. 2009). **d** Compact travertine beds in the southern part of the Buda Castle Hill (Buda Mts). **e** Travertine walls being quarried in the Újharaszti quarry (Süttő, Gerecse Mts). **f** Location of the Heidelberg-type Early Human remnant within the travertine at Vértesszőlős (Gerecse Mts). **g** Lacustrine travertine exposed in the Római quarry at Dunaalmás (Gerecse Mts). **h** Well-preserved paleospring vent inside the travertine of Csonkás-hát, Alsóvadács (Central Gerecse Mts)

4 Age of Hungarian Travertines

During the Pliocene and Pleistocene, the area of the Gerecse and Buda Mts were characterised by strong hydrological activity (Fig. 1), depositing more than hundred individual travertine occurrences, in many cases on the former fluvial terrace deposits (Scheuer and Schweitzer 1988; Kele 2009; Budai et al. 2018). These travertines are important terrestrial archives of climate and environment change and have played a key role in correlating fluvial terraces along the Danube Valley, being an area of strong regional differences in neotectonic activity (e.g. Ruszkiczay-Rüdiger et al. 2005a, b; Kele 2009; Kele et al. 2009, 2011; Ruszkiczay-Rüdiger et al. 2018). The first chronological studies were based on paleontology (e.g. Jánossy and Krolopp 1981; Jánossy 1986). The ^{14}C method was used to date the "culture layer" and charcoal from the overlying travertine in the Porhanyó quarry (Tata, Gerecse Mts) by de Vries and de Waard (1964) followed by ESR (Electron Spin Resonance) and U–Th datings (Schwarcz and Skoflek 1982; Hennig et al. 1983a; Scheuer and Schweitzer 1988). Application of ESR and U–Th methods on the travertine of the Vértesszőlős archeological site (Gerecse Mts) resulted ages ranging from 202 ky to >350 ky (Cherdintsev et al. 1965; Pécsi 1973; Hennig et al. 1983a; Schwarz and Latham 1990). At the same time application of paleomagnetic studies in travertine dating has also started (Latham and Schwarcz 1990; Scheuer and Schweitzer 1988; Lantos 2004).

The detailed sedimentological and geochemical study of Hungarian travertines including their U–Th dating using the MC–ICPMS technique was performed by Kele (2009), who demonstrated that the travertines of the Gerecse and Buda Mts are significantly younger than considered before based mainly on the "traditional terrace-system" and geomorphological considerations (e.g. Pécsi 1959; Scheuer and Schweitzer 1988) and paleomagnetic age data. The new U–Th analyses showed that intensive spring activity characterised both mountains during mainly the Middle-Pleistocene, showing frequency peaks in the Mindel/Riss interglacial and in the Riss glacial (Fig. 3). The formation of travertines was not restricted to the Mindel/Riss, Riss/Würm interglacials, but occurred during glacials (e.g. Riss) as well, and travertines formed, when precipitation provided enough water for the springs (Kele 2009).

In the Buda Mts travertines at high (400–500 m asl) elevation are the oldest, while the travertines covering the tI–tIII terrace levels are the youngest. Age of the travertines located on the same elevation above sea level can differ significantly, whereas in the Gerecse Mts there is a closer relationship between altitude and travertine age (Kele 2009). No age data proves travertine formation in the Buda Mts from 180 ky until the Holocene. However, in the Western-Gerecse travertines have been formed also in the past 100 ky (e.g. Tata, Porhanyó quarry, Öreg lake, Angol-kert, Kele et al. 2004, 2006; Kele 2009). Within the Gerecse Mts, the most frequently dated travertine bodies are located at Vértesszőlős and Süttő. New dating studies at the Vértesszőlős Early Human site (Kele et al. 2016, in submission) indicated the presence of the Heidelberg-type human during the MIS 9. At Süttő, similarly to the previous results (422 to 273 ky, Kele 2009; Sierralta et al. 2010) published Middle Pleistocene age data ranging between 310 and 240 ky from the Újharaszti and Hegyhát quarries, while Török et al. (2019) analysed samples of the Cukor quarry (554–318 ky).

There are however difficulties in the U–Th dating of travertines. The upper limit of the method is $\sim$600 ky, thus applicable only for Middle Pleistocene or younger deposits. The oldest travertines are Upper Pannonian (e.g. Szabadság Hill area, Széchenyi Hill, Normafa in the Buda Mts) and several sites of the Gerecse Mts (e.g. Dunaalmás római quarry, Kőpite, Új Hill-Öreg Hill, Látó Hill, Mogyorósbánya Kő Hill) falls outside this range (Kele 2009), as a consequence their exact age is not known. Furthermore, recent study of Claes et al. (2020) at Süttő (Hungary) and Ballik (Turkey) travertine quarries resulted in U–Th ages (at Süttő: 553 to 120 ky) inconsistent with their stratigraphic position, calling the attention to the diagenetic susceptibility of travertines that can behave as an open system for uranium after deposition through recrystallisation (e.g. De Boever et al. 2017). Due to the above problems, U–Th ages of diagenetically altered samples can only be considered as minimum ages (Ruszkiczay-Rüdiger et al. 2018; Török 2018), and the use of independent dating methods and dating of surrounding deposits are suggested (Claes et al. 2020). Pleistocene vertebrate faunas of the Süttő travertine complex can be used as an alternative way of dating. Pazonyi et al. (2014) reviewed the fossil content of the Süttő travertine and concluded that the fauna of the Cukor quarry indicated Pliocene/Lower Pleistocene age (older than the ones published by Török et al. (2019), while travertines in the Hegyhát quarry indicates a Lower to Middle Pleistocene age. To conclude, although U–Th dating can provide a lot of new numerical age data for the Hungarian travertines, recrystallised samples and micritic and mixed micritic-sparitic travertines should be avoided as they can behave as an open system for uranium after deposition and their ages can be considered as minimum ages, while fully sparitic samples (e.g. vein samples) are more suited for dating and analysis of multiple samples in combination with other independent methods and fossil content is strongly recommended (Claes et al. 2020).

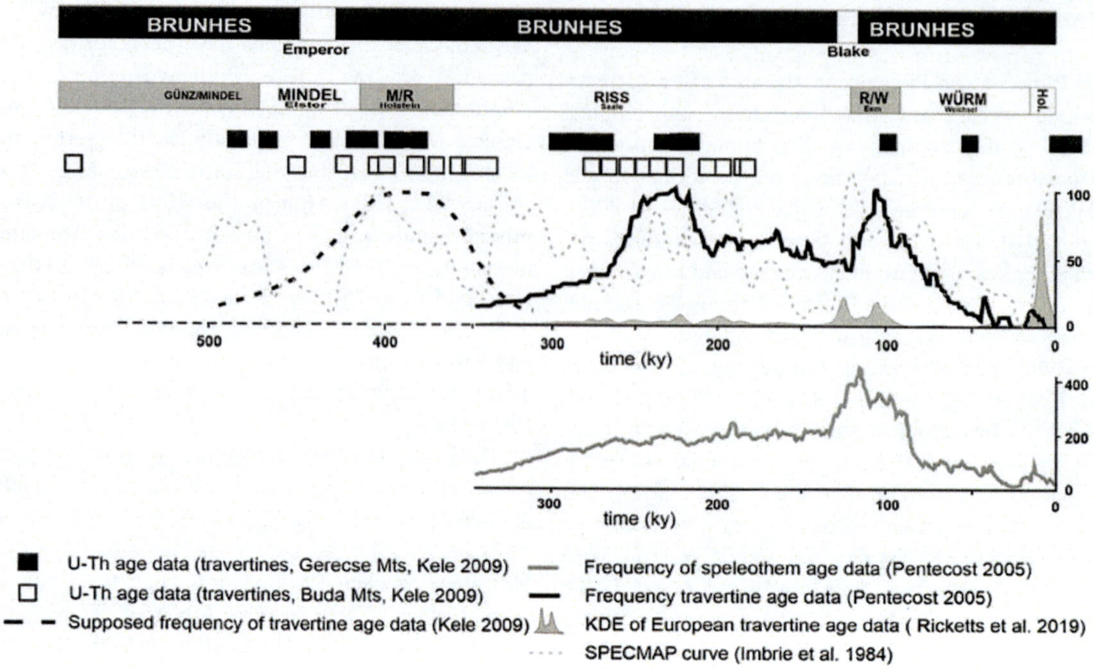

Fig. 3 Comparison of travertine ages from the Gerecse and Buda Mts (Kele 2009), with (1) magnetostratigraphy; (2) glacial stratigraphy; (3) frequency of U-series error-weighted dates of epigean travertine and speleothems (Hennig et al. 1983b; Pentecost 2005), (2) kernel density estimation (KDE) of age data from European travertines (Ricketts et al. 2019) and 3) the SPECMAP curve (Imbrie et al. 1984)

4.1 Calculated Uplift Rates and Relation with Terrace Stratigraphy

Age data of travertines covering terraces were used first by Ruszkiczay-Rüdiger et al. (2005a, b) in the quantification of uplift/river incision rate for the Danube Bend. Ruszkiczay-Rüdiger et al. (2005b) reconsidered the existing published age data for three segments of the Danube Valley and calculated incision rates between 0.14 and 0.41 mm/year for the last 360 ky. Based on several U–Th age data and using the altitude of the travertine localities above sea level Kele (2009) estimated average incision rates around 0.3–0.4 mm/year in the Gerecse and Buda Mts, and concluded that the incision rate of the individual sub-areas in both mountains may have varied in a certain degree (e.g. Gellért Hill: 0.47–0.52 mm/year; Buda Castle Hill: 0.11–0.14 mm/year, Kele et al. 2009; Kele et al. 2011). The difference in the incision rates of the individual sub-areas points to the fact that the elevation of the travertines does not definitely show their relative age, which means that it is possible to make only rough estimations regarding the changes of paleokarst water-level using the age and elevation of travertines.

Travertines along the fluvial terraces and tributaries of the Danube were used in the correlation and dating of terrace levels (Wein 1977; Scheuer and Schweitzer 1988; Gábris et al. 2014; Kele and Mindszenty 2014a) (Fig. 4). Published U –Th data (<360 ky) served as a good base for the dating of the tI–tIV terrace levels situated at lower elevations (e.g. Kretzoi and Pécsi 1982; Schwarcz and Skoflek 1982; Hennig et al. 1983a) but the age of the tV–tVII levels was previously determined older than 780 ky (Upper-Pliocene—Lower Pleistocene), based on mainly geomorphological considerations, sporadic paleontological findings and paleomagnetic measurement. Based on new U–Th age data Kele (2009) proposed that the travertines covering the tV–tVII terraces are much younger than it was believed before and were formed mostly during the Middle-Pleistocene. It must be mentioned however that travertine depositing paleosprings can discharge above the erosion base, thus their age do not necessarily useful for the dating of terrace levels and the above mentioned possible open system behaviour of travertines could also hinder their reliable dating. Ruszkiczay-Rüdiger et al. (2016, 2018) focussed on the western Gerecse Mts area as it is a suitable location for the revision of the terrace system of the Danube Valley in Hungary. The new cosmogenic radionuclide (CRN)- and luminescence (OSL)-based terrace ages were then compared to revised published paleontological, U–Th and magnetostratigraphic data in order to establish a robust chronology of the Danube Valley. Ruszkiczay-Rüdiger et al. (2018) found that the CRN, OSL, revised paleontological ages and the magnetostratigraphic position of the terraces are internally consistent, but U–Th ages of travertines covering the

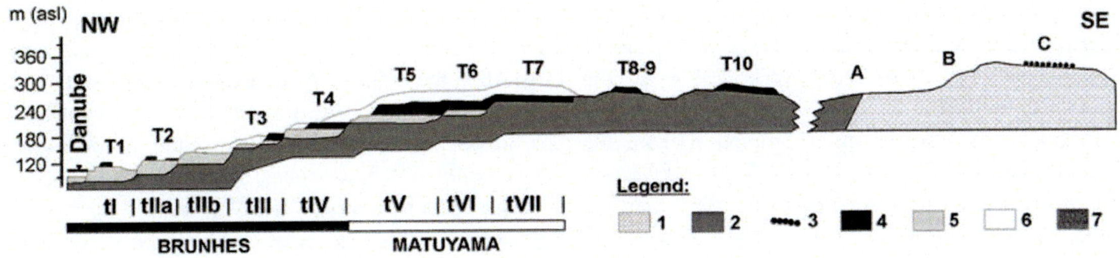

Fig. 4 Geomorphic horizons in the Western Gerecse in the Almásneszmély-Dunaszentmiklós section (modified after Pécsi et al. 1982; Ruszkiczay-Rüdiger et al. 2007; Kele 2009). 1. Triassic limestone, 2. Upper Miocene sand and clay, 3. Oligocene-Miocene terrestrial gravel, 4. Upper Miocene to Quaternary travertine, 5. Quaternary alluvial sand and gravel, 6. Loess, slope loess, 7. eolian sand. tI-tVII: terrace horizons, T1-10: travertine horizons. A: Pliocene pediment, B: late Miocene abrasion horizon, C: Cretaceous-Paleogene etchplain

terraces are scattered and significantly offset towards younger ages. To conclude, age data of travertines should be used carefully in terrace dating and in calculation of incision/uplift rates.

4.2 Spring Relocations Based on the U–Th Data

4.2.1 Buda Mountains

With the help of the radiometric age of travertines, the migration and development of paleokarst springs was reconstructed (Kele 2009; Kele and Mindszenty 2014b). In the Upper-Pannonian the travertine formation in the Buda Mts was characteristic of the Széchenyi Hill—János Hill area (>400 m asl) (Fig. 1). These travertines at Hármaskút-tető-Normafa, Csillagvizsgáló u., Budaörsi Hill, Kakukk Hill, Széchenyi Hill are the largest considering both their presence at the surface and also their volume. After the Upper-Pannonian, the spring orifices moved eastwards and in the Middle-Pleistocene significant spring-migration and thermal spring activity started in the Buda Mts, while the northern part of Budapest, and the territory between the Solymár Valley and Ős-Dera creek became the centre of water discharge. The biggest Middle-Pleistocene travertine sites are located north of Budapest at Budakalász, Üröm Hill, Ezüst Hill and Kálvária Hill. Travertine formation in the Budakalász area (210–240 m asl) started about 600–550 ky ago, while in the Ördögárok Valley the Máriaremete travertine occurrence (280 m asl) was formed only 400 ky ago. During the Middle-Pleistocene travertine was formed on the Buda Castle Hill, and 350 ky ago the Rózsadomb area (160–190 m asl) was also characterised with intensive thermal spring activity (Kele et al. 2011). About 250 ky ago there was a decrease in the discharge level of the springs which moved towards the Danube, while the individual areas (e.g. Rózsadomb, Gellért Hill, Buda Castle Hill) were uplifted differently. Travertines of the Rózsadomb, Gellért Hill and Ördögárok Valley are rather small and sporadic, compared to the Upper Pannonian sites and the sites at the Budakalász area. Travertines of the same age located at different elevations are the evidences of the different uplift rate. The springs from which the Máriaremete travertine deposited discharged to the surface in E-SE direction at lower elevation in the Ördögárok Valley about 200–250 ky ago. Due to the further relative decrease in the water level, about 240 ky ago travertine formation started at the Farkastorki u. (150–160 m asl) and 180 ky ago on the Kiscelli plateau (140 m asl) and the springs moved in EW direction from the Rózsadomb area. Travertines at the Castle Hill and at the Kiscelli plateau are the largest ones in the area of Budapest. The presence of travertines from the Upper-Pleistocene until the Holocene is not evidenced in the Buda Mts. The present discharge level of the thermal springs might have been evolved around the end of Pleistocene and early Holocene.

4.2.2 Gerecse Mountains

In the Gerecse Mts almost 40 travertine occurrences are known, located mostly on the northern part and along the marginal (eastern, western) areas (Scheuer and Schweitzer 1988; Sierralta et al. 2010) (Fig. 1). According to Scheuer (1999a, b) 41 palaeothermal springs were active since the Pliocene, but nowadays only some lukewarm springs (e.g. Fényes spring, Tata; Csokonai and Lilla springs, Dunaalmás) discharge in this area. Between 250–350 m (asl) travertines older than the upper limit (600 ky) of the U–Th method can be found. The oldest travertine occurrences are located in the Western, Central and Eastern Gerecse, towards the centre of the Gerecse Mts. One of the largest travertine deposit is located at Dunaalmás, Les Hill and Kőpite in Western Gerecse. The Dunaalmás travertine is older than the upper limit of the U–Th method (>600 ky) as well as the largest travertine body of the Eastern-Gerecse, at Mogyorósbánya, Kő Hill (>600 ky, Kele 2009).

During the Middle-Pleistocene, the discharge points of thermal springs moved to the marginal areas and the spring activity started at lower (150–250 m asl) elevations (e.g. Vértesszőlős). After the deposition of the Middle-Pleistocene travertines of the central parts of Gerecse Mts (e.g. Alsóvadács: 329 m asl; Vékonycser: 238 m asl) the travertines of Réz Hill (466 ky) and Öreg Hill (391 ky) at 225–235 m (asl) were formed in NE direction from the Middle-Pleistocene occurrences. The Süttő travertine occurrence, which is the largest travertine site of the Gerecse Mts, located on the nothern part of Central-Gerecse (220–250 m asl), was considered previously as Upper-Pleistocene-Lower Pleistocene, but according to Kele (2009) and Sierralta et al. (2010) it was formed during the Middle-Pleistocene as well. During the Middle Pleistocene in the Eastern-Gerecse, the spring orifices also moved eastwards. During the Upper-Pleistocene springs were active only in the area of Tata (Porhanyó quarry). Travertines at the eastern side of the Öreg Lake (Tata) were formed 50 ky ago, and further Holocene and recent occurrences can be found in the Angol-kert at Cseke Lake, at the Fényes springs and at the Csokonai spring (Dunaalmás, 110–140 m asl).

5 Stable and Clumped Isotope Geochemistry of Hungarian Travertines

During the past decades both recent and fossil travertines of Hungary were studied with the stable isotope method (e.g. Szöőr et al. 1992; Kele et al. 2003, 2006, 2008, 2009, 2011; Korpás et al. 2004; Kele 2009; Sierralta et al. 2010; Claes et al. 2017; Török 2018; Török et al. 2013, 2017, 2019). Most of these studies determined the $\delta^{13}C$ and $\delta^{18}O$ values, and the stable isotope data were mainly used to (1) reconstruct the depositional environment and paleoclimatic conditions during deposition, (2) water temperature based on the oxygen isotope composition (and on assumptions regarding the $\delta^{18}O$ of the depositing water) and (3) the origin of dissolved CO_2. Besides, stable isotope studies on recent carbonates were used for calibration purposes (e.g. Kele et al. 2008, 2015, see later). In some studies, however, Sr isotope geochemistry ($^{87}Rb/^{86}Sr$ and $^{87}Sr/^{86}Sr$) was also applied to provide information about the rocks, which could have acted as sources for the travertine depositing fluids (Claes et al. 2017; Török 2018; Török et al. 2019). For instance, according to Török (2018) the $^{87}Sr/^{86}Sr$ ratios of the Süttő travertine vary between 0.70835 and 0.70860 with an average value of 0.70848, inferring contact between Oligocene siliciclastic sequences in the Süttő area and downward percolating fluids.

Kele (2009) studied more than six hundred travertine samples of about hundred travertine sites of the Gerecse and Buda Mts and found significant difference between their stable carbon and oxygen isotope compositions (Fig. 5). The $\delta^{13}C$ values of most of the travertines of the Buda Mts are higher than 0‰ (in general between 1 to 3‰, V-PDB), while most of the $\delta^{18}O$ values are in the 14–24‰ (V-SMOW) range. In case of the travertines of the Gerecse Mts, the $\delta^{13}C$ values are generally lower than 1‰ and most of the data range between –7 and 0‰ (V-PDB), while the $\delta^{18}O$ values range mainly from 17 to 23‰ (V-SMOW) (Kele 2009). Spatial variations in the $\delta^{13}C$ values can be observed within the Gerecse Mts as well. Low $\delta^{13}C$ values (<4‰) in the area of Tata might be explained with the contribution of the organic CO_2, but the role of magmatic CO_2 can not be excluded, while towards the northern part of the Gerecse Mts these values approach 0‰ (e.g. at Süttő)

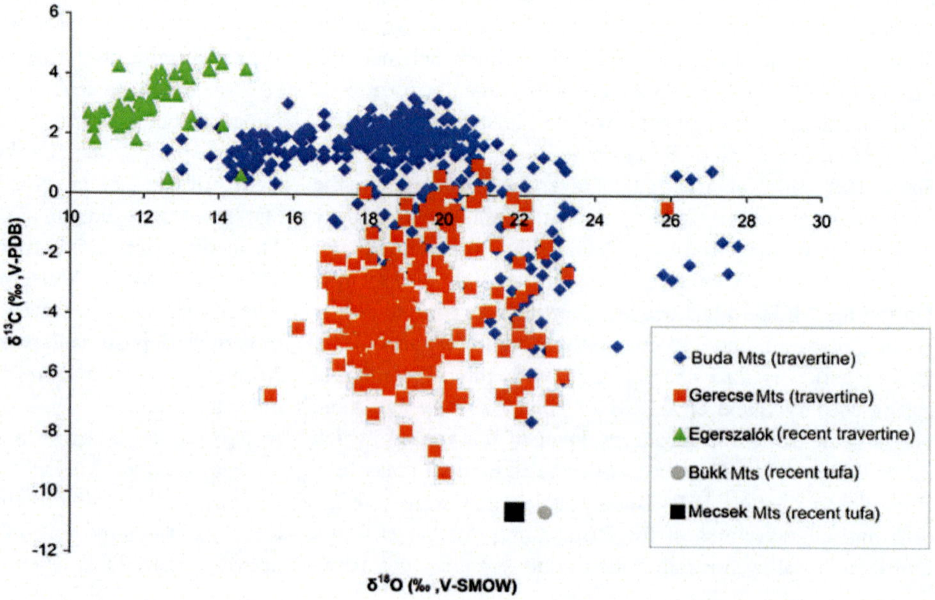

Fig. 5 Stable isotope geochemistry of travertines of the Gerecse and Buda Mts and comparison with values of recent travertine deposits from Egerszalók and recent tufa from the Bükk Mts and Mecsek Mts (Kele 2009)

(Kele 2009). In the $\delta^{18}O$ values, no trends can be observed according to the different areas (i.e. $\delta^{18}O$ the values of the Western-, Central- and Eastern-Gerecse Mts overlap).

In the Buda Mts, the $\delta^{13}C$ and $\delta^{18}O$ values are mainly connected to the age and genetics of the occurrences. The Upper-Pannonian travertines near the Szabadság Hill have a more positive $\delta^{18}O$ and negative $\delta^{13}C$ values (+19.2 to +27.8‰ V-SMOW and −7.6 to +3.2‰ V-PDB, respectively) compared to the Pleistocene and Holocene travertines of the Buda Mts situated at lower altitudes. This can be explained by differences in age and depositional environment as the Upper Pannonian limestones formed in lagoons with periodic connections to the sea, while the Pleistocene-Holocene travertines formed from thermal springs of higher temperature, and the $\delta^{18}O$ values of the travertine precipitating fluid could have been different as well. The $\delta^{13}C$ values of the Pleistocene and Holocene travertines are less negative than the values of the Upper-Pannonian ones, which can be related to the dominance of carbon deriving from dissolution of $\delta^{13}C$ rich marine limestones (Kele 2009). Since the individual mountains and their sub-areas can be characterised by distinct stable isotope values, their stable isotope compostions are useful to define their location in archaeological studies (Djurić et al. 2018).

In Hungary, application of clumped isotope thermometry has started first on recent travertines and tufa, because of calibration purposes (Kele et al. 2015). Compared to the conventional oxygen isotope-based carbonate-water paleothermometry, the biggest advantage of the clumped isotope thermometer is that it requires no assumptions on the $\delta^{18}O$ of the precipitating water (Ghosh et al. 2006; Eiler 2007). Among others, recent travertine samples of thermal wells of Köröm, Tura, Igal, Bük and Széchenyi spa (Budapest) and tufa deposits from the Szalajka Valley (Bükk Mts), Havasok (Baradla cave, Aggtelek Mts) were used for the calibration of the conventional oxygen isotope- and for the novel clumped isotope thermometer (Kele et al. 2015; Bernasconi et al. 2018) (Figs. 6 and 7).

The calibrated clumped isotope thermometer was then used to calculate the temperature of deposition (and the $\delta^{18}O$ of the depositing fluids) of Pleistocene travertines from several localities from both the Gerecse and Buda Mts. The first calculated temperature values of these ongoing studies provided reliable temperature data and emphasised the temperature differences between the travertine depositing paleosprings of the Gerecse and Buda Mts and provided data to assess the temperature of deposition at important sites such as the ones at Süttő and Vértesszőlős (Kele et al. 2016; Török 2018; Török et al. 2019; Kele in prep).

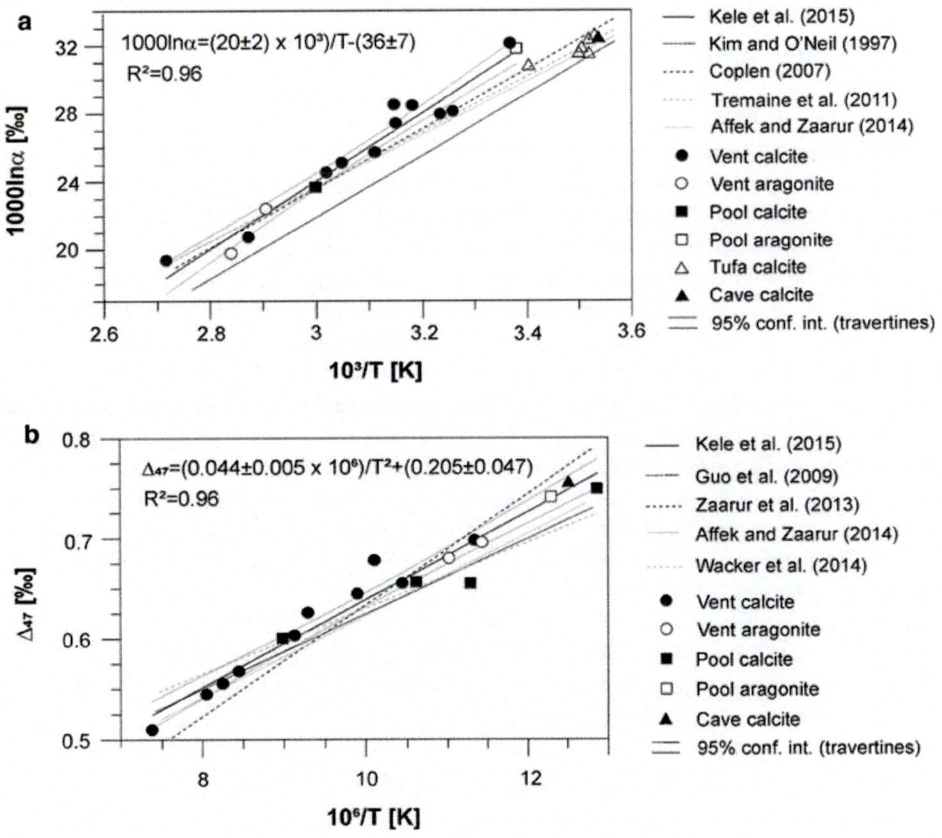

Fig. 6 Empirical calibration of the conventional oxygen isotope thermometer **a** and the clumped isotope (Δ_{47}) thermometer **b** using recently forming travertines, including carbonate deposits of the thermal wells of Köröm, Tura, Igal, Bük and Széchenyi spa (Budapest) and tufa deposits from the Szalajka Valley (Bükk Mts), Havasok (Baradla cave, Aggtelek Mts) as well as other foreign active travertine and tufa sites (based on Kele et al. 2015)

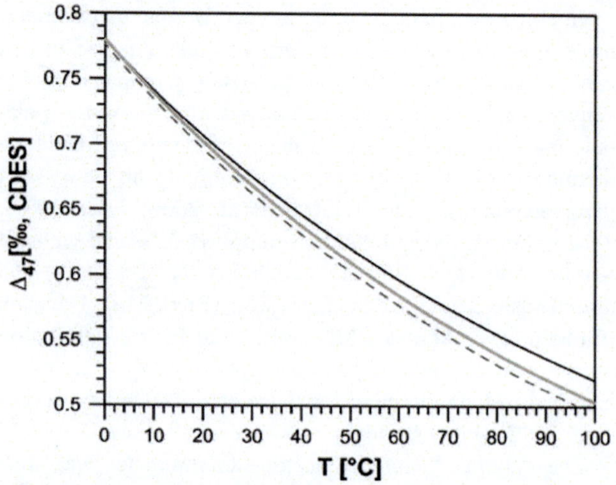

Fig. 7 Comparison of the original, travertine based Kele et al. (2015) calibration (gray dashed line) of the clumped isotope thermometer with its recalculated version (gray line) (Bernasconi et al. 2018) using the Brand parameters and projection of the data to the reference frame using the University of Washington (Seattle, USA) values. The experimental calibration of Kelson et al. (2017) (black line) is for comparison. CDES: carbon dioxide equilibrium scale (based on Bernasconi et al. 2018)

6 Recent Travertine Carbonate Accumulations in Hungary

6.1 Natural Springs

The average geothermal gradient (40–50 K/km) and heat flow density values are quite high in the Carpathian Basin (Dövényi and Horváth 1988; Lenkey et al. 2002; Békési et al. 2020) and as a consequence, Hungary is rich in geothermal water. More than hundred thermal springs discharge in the capital, Budapest, along the river Danube, where a geological fault separates the Buda Mts from the Great Plain. The Gellért Hill belongs to the Buda Thermal Karst containing thermal springs (e.g. Mátyás spring, Juventus spring, Gül Baba spring, Ősforrás spring, Sárosfürdő) with elevated temperatures (35–43 °C) (Virág et al. 2013a; Kovácsné Bodor 2020). The main aquifers of the Buda Thermal Karst are Mesozoic carbonates and Eocene limestone. The Buda Thermal Karst is in subsurface connection with the several thousand metres thick Mesozoic carbonate sequence of the Transdanubian Central Range, which serves as an aquifer and karstic recharge area for the Buda Thermal Karst. The cold water infiltrated in the Transdanubian Central Range is heated up by thermal convection during its long regional-scale flow and discharge as thermal water on the right side of the Danube, and in the riverbed along the step-faulted boundary of the uplifted hills and the subsided basin to the east (Kovács and Müller 1980; Kele et al. 2009). Several medium to high-temperature thermal springs are close to the current level of the Danube River (e.g. Békásmegyer, Attila spring, Óbuda, Árpád spring, Pünkösdfürdő, Lukács-Császár Spa, etc.) (Kele et al. 2009) and some of them (e.g. springs of the Gellért Spa and Rudas Spa) discharge directly from the lowest karstic blocks, while others (e.g. Római Spa and Attila spring at Békásmegyer) reach the surface breaking through permeable deposits (Scheuer and Schweitzer 1989). Although the number of active springs is high, their capacity for carbonate deposition is very low compared to the Pleistocene times. Probably the most well-known recent travertine accumulation in Budapest is the one in the Rudas Spa, where a flowstone-like historical indoor-travertine has precipitated during the past c. 120 years (Veres 2007; Virág et al. 2013b, 2020) and impressive carbonate deposits formed at the discharge points of some fountains inside the Gellért Spa as well.

Outside of the area of Budapest, the recent thermal spring activity and natural travertine deposition is negligible. Probably one of the exceptions is the cave bath of Miskolctapolca, where recent carbonate accumulation is significant from the thermal water of 29.4 °C (Fig. 8). Lukewarm natural springs in the Gerecse Mts discharge at Dunaalmás (Csokonai and Lilla springs, Fig. 8), and in Tata, Fényes Lake, however the carbonate deposition around these springs is minimal (Kele 2009).

6.2 Thermal Wells

Most of the recent carbonate deposits are restricted to thermal wells and spas (Fig. 8) and can be easily found in Budapest (e.g. Béke well, Dagály; Zsigmondi-5 well, Margitsziget) (Kele et al. 2009). These carbonate accumulations are located at the discharge point of fountains, inside tubes as scalings or in the water tank of the spa as in the Széchenyi thermal spa (70.9 °C), where the groundwater is extracted from a 1257 m deep well from the top of the Triassic aquifer (Virág et al. 2013c, d; Kele et al. 2015).

Probably the most famous travertine deposit is the one located at Egerszalók, referred also as the "Hungarian Pamukkale" (Kele et al. 2008; Kele 2009) (Fig. 8). This spectacular terraced-slope travertine formed since 1961, when the first oil-exploratory boreholes (De–42, De–42a) were drilled. Here, the thermal water discharges from the confined Mesozoic-Tertiary reservoir under artesian conditions and travertine accumulates with high (~1 mm/day) precipitation rate. Kele et al. (2008) performed detailed seasonal field, petrographic, mineralogical, chemical and stable isotope analyses along downstream sections in order to study how the different processes influencing the travertine depositions are preserved in the travertines

Fig. 8 Actively forming travertine carbonates at thermal springs, wells and spas in Hungary. **a** The spectacular terraced-slope travertine deposit at Egerszalók, also called the Hungarian Pamukkale (or as "Sódomb" by the locals). Right from the travertine mound artificial concrete terrace pools were created; **b** Fast travertine precipitation from the discharging hot (~65–68 °C) water at the De–42 thermal well at Egerszalók; **c** Travertine encrustation at the Köröm thermal well from a water of 83.9 °C; **d** Reddish-brownish travertine along the downstream channel at Köröm thermal well; **e** Recent travertine deposition on the fountains and walls, and terraced sloped forms **f** inside the Cave Bath of Miskolctapolca (Hungary); **g** Csokonai spring and **h** Lilla spring at Dunaalmás at the right bank of the Danube (N-Gerecse Mts, Hungary) (*Photos* Eva de Boever)

precipitated. Kele et al. (2008) observed that the fractionation of oxygen isotopes between water and the precipitated carbonate at Egerszalók shows offset from the previously determined "equilibrium" curves, and values of the vent samples define an empirical travertine curve which fits to the lowe-temperature calcite-water fractionation described by Coplen (2007) at Devils Hole (Nevada, USA) (Fig. 6). The vent carbonate samples collected at Egerszalók (completed with other recent vent travertine samples from Hungary, Italy, Turkey) were later used by Kluge et al. (2018) to assess the factors controlling clumped isotopes (Δ_{47}) and $\delta^{18}O$ values of hydrothermal vent calcites. They concluded that fast-growing hydrothermal vent calcites tend towards a small ($\sim$0.01‰) effect of DIC speciation on Δ_{47} values at the studied pH 6–8. Thus, pH-related effects are mostly small for Δ_{47} in the subsurface environment at low pH, and high growth rate control the magnitude of disequilibrium (Kluge et al. 2018).

Between Egerszalók and Demjén villages two other thermal wells (K–10, K–11) provided water with high dissolved carbonate content depositing small yellowish-reddish travertine accumulations (Bányai 2009) and there are other deep thermal wells providing hot water with carbonate precipitations at or near well-head pipes, as scalings. Nice examples occur at Bükfürdő Thermal Spa (54.9 °C, $\sim$1282 m depth), Köröm (79.2 °C, 1880 m, Fig. 8), Tura (95 °C, 2092 m), Igal (75 °C, 651 m) and they are all described in (Kele et al. 2015) and in SE-Hungary at Fábiánsebestyén (92 °C, 2012 m) and Kakasszék ($\sim$100 °C, $\sim$1800 m) by Kluge et al. (2018). Besides, Deák et al. (in submission) presents data from thermal wells at Bükkszék (38 °C, 1057 m), Szentes (96.1 °C, 2323 m), Földes (58.3 and 66.2 °C, 1236 m), Hajdúszoboszló (4 wells, 67.3–70.5 °C, 1001–1880 m), Hajdúnánás (66.3 °C, 1013 m, Deák 1986), Hajdúdorog (64 °C, 1037 m, Deák 1986) and Zalakaros (96 °C, 2000 m).

7 History of Freshwater Tufa Research in Hungary

The exact starting point of research of freshwater tufas in Hungary is difficult to establish. The term 'tufa' in previous studies was not clearly separated, and the 'freshwater lime' was the commonly used term for the previously studied formations. Cholnoky (1940)'s name is associated with, among other things, the introduction of the 'tetarata' term in the Hungarian language. The carbonate occurrences of the Bakony region and the geyser cones of the Tihany peninsula were described by Cholnoky (1932, 1935) and by Hoffer (1943). Jaskó (1959) studied the relationship between the karst water and the geological buildup in the Transdanubian Mountains. Veress (2002) described freshwater tufa occurrences deposited in watercourses of the Bakony Mountain, with special regard to their morphology. Bódai et al. (2015) and Bódai (2017a) examined the tufa occurrences of the Csurgó spring and Judit spring (Bakony Mountain) in detail using stable isotopes. Gyula Scheuer and Ferenc Schweitzer published articles about freshwater tufa mainly from hydrogeological and geomorphological point of view (e.g. Scheuer 1975, 2002, 2003; Scheuer and Schweitzer 1970, 1972, 1978, 1981, 1985, 1986). The tufa occurrences of the Bükk Mts were summarised by Hevesi (1972), who described the causes of tufa precipitation, tufa morphology, the role of the vegetation in tufa deposition and listed the major tufa occurrences. The famous and rare freshwater tufa cave (Anna Cave) in Lillafüred was first described in detail by Kadić (1921) and later more detailed descriptions were published by Lénárt (1981), Kessler (1953), Kordos (1984). Kraft et al. (1986) studied the freshwater tufa linked to springs around and in the Mecsek Mts. Molnár et al. (2004, 2006) applied the ^{14}C method in the dating of young karstic rocks. Koltai et al. (2012a, b) examined the tufas and karstic waters in the Mecsek Mts with stable isotope geochemical studies and started monitoring in the area. Scheuer et al. (2013) published macro- and micro element data from the thermal and karstic springs of the Wester-Mecsek. Kele (2009), Bódai et al. (2016) and Bódai (2017b) investigated the stable isotope composition of the tufa deposits of Sebesvíz and Szalajka Valley. Kele et al. (2015) used freshly precipitated tufa samples from the Szalajka Valley for the calibration of the clumped isotope method. Study of the Hungarian tufa sites and karstic water, including monitoring study of water geochemistry is still ongoing (Bódai et al. 2016, in prep).

7.1 Freshwater Tufa Accumulations in Hungary

Freshwater tufa carbonates in Hungary are connected to karstic springs of the the Bükk Mts, Aggtelek Mts, Mecsek Mts, Bakony Mts, and Balatonfelvidék (Figs. 9, 10). Rocks in the Bükk Mts are characterised by different water conductivity (Aujeszky and Scheuer 1974; Pelikán [edt.] 2005) and hydrological conditions, altitude, and residence time of water have created springs with different characteristics. Springs of karstic plateaus at higher altitude above sea level discharge from Carboniferous-Permian limestone and show higher runoff fluctuation and lower temperature compared to other springs discharging at the marginal areas of the Bükk Mts (Aujeszky et al. 1974; Aujeszky and Scheuer 1974). In contrast, springs of the Southern Bükk discharging from Upper Eocene limestone at lower altitude show the highest runoff, and characterised by balanced water regime and higher water temperature (Aujeszky et al. 1974; Aujeszky and Scheuer 1974). Tufa deposition is still active in this area.

The Szalajka spring discharges at the contact of Triassic limestone and impermeable clay shale and the temperature of spring water varies between 8.5 and 9.4 °C. Bódai (2017b) performed seasonal monitoring studies in the Szalajka Valley (Fig. 10), including in situ measurement of T, pH, EC in the field and stable isotope analyses of karstic water and freshly precipitated tufa in order to study the causes of seasonal changes in these parameters. Stable isotope analysis of recent tufa carbonates can help to reveal the relationship between past climatic events and tufa geochemistry and provides an opportunity for temperature estimation based on oxygen isotope fractionation between tufa carbonate and water. Measured clumped isotope data (Δ_{47}) of freshwater tufa of Szalajka creek and Havasok cave tufa (Baradla Cave, Aggtelek Mts, Hungary, Fig. 10) have also been published (Kele et al. 2015). Similar studies have been conducted at the recently depositing tufa sites at Dobrica spring (9–11 °C) and Sebesvíz (8–9 °C) (Bükk Mts) as well (Kele 2009) (Fig. 10). Beside the above-mentioned locations, many other small freshwater tufa site is still active in the Bükk Mts. Hevesi (1972) summarised the most significant ones as follows: Mónosbéli-vízfő (13–14 °C), Pacsirta spring, Vízfő-tető (13–14 °C). Intermittent active springs can be found at the southern part of the Great Plateau (e.g. Vöröskő-, Imó- and Fekete-len springs).

The touristic Anna tufa cave (Fig. 10) is located at Lillafüred (Bükk Mts) with a total length of 600 m, and it is constituted by the connection of 34 larger chambers. The age of the tufa, which hosts the cave, is thought to be Pleistocene. The cave chambers were formed at the site of huge tree trunks which were first encrusted by carbonate and then rotted with time. Within the cave, many encrusted plant remains, roots, moss fibres can be seen nowadays. The average temperature of the cave water is 6–14 °C (Lénárt and Hevesi 2003). The erosion of the cave is still ongoing due to the water of the Szinva and Garadna creeks and this is manifested by cracks, fractures and sliding of blocks inside the cave.

The Bakony Mountain consists mostly of dolomite and Dachstein limestone formed during the Late Triassic (Haas 2001). At some places, the subsurface thickness of the Main Dolomite Formation exceeds 800 m (Bihari 1981) and it is covered by karst. Due to the deep dwelling karstic rocks, many currently active tufa deposits can be seen in the streams and they associate with springs and smaller waterfalls as well. The Csurgó spring (9–10 °C) located in the Northern Bakony is perhaps one of the most spectacular tufa deposit in this region. It discharges on a slope, and builds a fan-shaped freshwater tufa mound, forming several metres high waterfalls. The spring was investigated by monthly monitoring studies between 2012 and 2014 by Bódai et al. (2015) and the stable isotope composition of water was almost stable during the study period. Fresh tufa deposits were also analysed by the stable isotope method, and there was no change in the stable isotope values of water during the study period. Besides, stable isotope studies were performed on recent tufa accumulations at Judit spring, Szömörke Valley, Erdei spring, Marhaállási spring, Vadászház 22. spring, Németbánya 3 spring, Németbánya 8 spring in Bakony Mountain (Bódai 2017a). In many cases, however, tufa deposition is no longer active near the karst springs and only old tufa deposits (dams, ramparts) can be recognised (e.g. Mésztufalejtő spring, Csöpögő spring).

The basement of Balatonfelvidék is also formed by karstic rocks, but these are covered by Pannonian sediments and a debris of Pliocene-Pleistocene basalt volcanoes. Because of the karstic building rock, there are many surface and deep karst forms in this area. The studied riverbed of Malom stream (Fig. 10) digs into a covered karst area, and its water is derived from the karst system. There are 18 wells in the area supplying drinking water, and the main body of water in Malom Valley discharges to the surface from an artificially constructed pipe. The values of measured stable isotope composition of water (δD, $\delta^{18}O$) and tufa ($\delta^{13}C$, $\delta^{18}O$) do not show significant changes downstream (Bódai et al. in prep).

Recent freshwater tufa precipitations are common in the Mecsek Mts (Kraft et al. 1986). Koltai et al. (2012a, b) and Koltai et al. (2013) examined the geochemistry of the most important tufa-bearing karstic springs in the Eastern and Western Mecsek. The karstic rocks of Western Mecsek belong to a single block which can be separated into 3 parts on the surface (Barta and Tarnai 1999). The studied Anyák spring, Melegmányi spring, Kánya spring, Mariska spring, Dagonyászó spring, Zsolnay spring are located on the first karstic block of nearly 40 km^2 (Fig. 10). The studied area is built up from Triassic karstic rocks, where numerous small and large caves, dolines and karst springs have formed (Barta and Tarnai 1999). The carbonate rocks of the Eastern Mecsek were formed during the Jurassic period and they are less suitable for karstification and cave formation. From this area, the Csurgó spring and Pásztor spring are studied in detail (Koltai et al. 2012a, b).

Stable isotope composition ($\delta^{18}O$, δD) of freshwater tufa bearing karstic springs (Csurgó spring, Szalajka spring, Dobrica spring, Sebesvíz, Malom Valley) in the Bakony Mts and Bükk Mts have not shown significant downstream change. The stable isotope compositions of samples collected from these two mountains differ from each other and were not influenced by 'altitude effect', but rather by the residence time and atmospheric effects (Bódai 2017a, b). The stable isotope analysis of springs in Mecsek Mts revealed a difference between the western and eastern part of the mountains which may be due to the 'altitude effect' (Koltai et al. 2012a, b). The stable carbon and oxygen isotope composition of the Hungarian freshwater tufa are

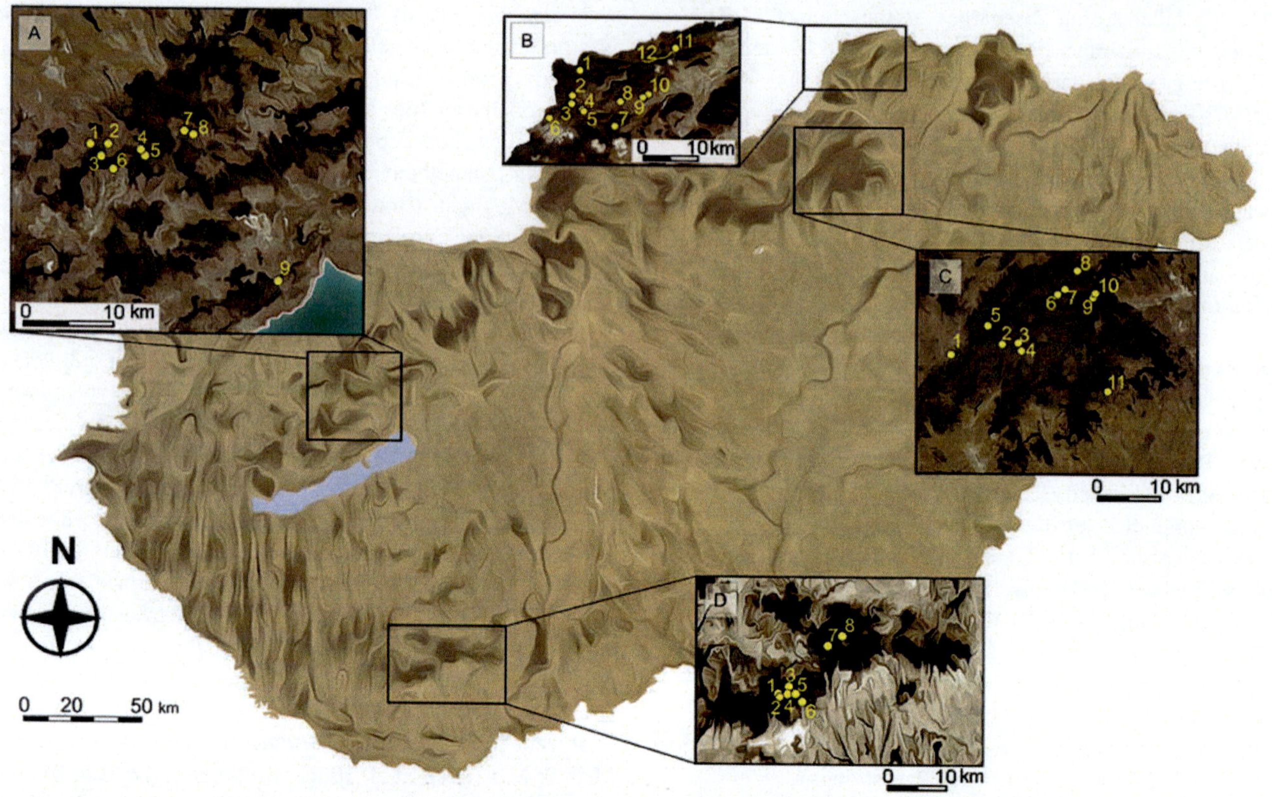

A) Bakony Mountains: 1. Erdei spring, 2. Marhaállási spring, 3. Vadászház 22. spring, 4. Németbánya 3. spring, 5. Németbánya 8. spring, 6. Csurgó spring, 7. Szömörke valley, 8. Judit spring, 9. Malom valley. B) Aggtelek 1. Lófej spring, 2. Tohonya spring, 3. Babot-kút spring, 4. Alsó-barlang spring, 5. Jósva spring, 6. Baradla cave, 7. Kecskekút spring, 8. Bolyamér spring, 9. Kis-Kopolya spring, 10. Kopolya spring, 11. Csurgó spring, 12. Ménes-valley and its tributary valley. C) Bükk Mountains: 1. Mónosbéli-vizfő, 2. Vöröskő spring, 3. Imó-kő, 4. Feketelen spring, 5. Szalajka spring, 6. Sebesviz, 7. Pacsirta spring, 8. Dobrica spring, 9. Lillafüredi waterfall, 10. Anna freshwater tufa cave, 11. Latorút. D) Mecsek 1. Kánya spring, 2. Mariska spring, 3. Melegmányi spring, 4. Anyák spring, 5. Dagonyászó spring, 6. Zsolnay spring, 7. Csurgó spring, 8. Pásztor spring.

Fig. 9 Map of the Hungarian tufa deposits

similar to each other and – compared to the stable isotope values of other tufas from European countries (Andrews 2006) – they reflect the continental effect moving from west to east (Fig. 11). The $\delta^{18}O$ values of Hungarian tufa are much more negative than those found in Western Europe, and closer to the values of samples collected in Poland and the Dinaric Kast. The $\delta^{13}C$ values of freshwater tufa indicate strong soil activity.

There are many recent tufa sediments in the Aggtelek Mts too, but these sites have not been studied in detail (Sásdi 2005). The most important tufa-bearing springs are: Kopolya- and Little-Kopolya spring, Bolyamér spring, Kecskekút spring, Kecső spring, Nagy spring, Babot spring, Jósfa spring, e.g. at the entrance of Rövid Alsó-barlangi and in the riverbed too, Tohonya spring, Lófej spring, and the spring in Ménes Valley and its tributary valleys, e.g. Sárogkerti spring group (Sárogkerti Valley), Csurgó spring (Bába Valley) (Fig. 9).

7.2 Most Common Morphology of Freshwater Tufas in Hungary

Freshwater tufa occurrences in Hungary are mainly related to karstic areas. In general, tufa deposition is not concentrated near the spring outflow, but may extend further downstream. This is because cold water contains more carbon dioxide and thus can keep more carbonate in the solution. Moving away from the spring orifice, the amount of dissolved CO_2 decreases in parallel with the downstream increase of water temperature, leading to gradual precipitation of freshwater tufa, as it can be seen in the Szalajka Valley in Szilvásvárad (Bükk Mts), where precipitation begins at a distance of ~500 m from the spring.

The tufa most commonly forms in stream bed as tufa barrages, dams (Fig. 10). One of the most spectacular forms of tufa precipitations are dams or dam (tetarata) systems which can also be observed along the Szalajka stream. Tufa

Fig. 10 Tufa deposits from karstic creeks and caves in Hungary. **a** Tufa dams at the Fátyol waterfall (Szalajka creek, Bükk Mts) and **b** at Sebesvíz (Bükk Mts); **c** Small tufa mound at Dobrica spring (Bükk Mts); **d** Tufa dams of the Malom Valley (Balatonfelvidék); **e** Tufa waterfalls in the Melegmányi Valley (Centreal Mecsek Mts); **f** Core sampling of fan-shaped tufa deposits formed in the riverbed (Mecsek Mts); **g** Periodically active snowwhite, clean tufa deposit (Havasok tufa) from the Baradla cave (Aggtelek Mts); H. Core sampling inside the Anna cave, formed within the tufa deposit (Lillafüred, Bükk Mts)

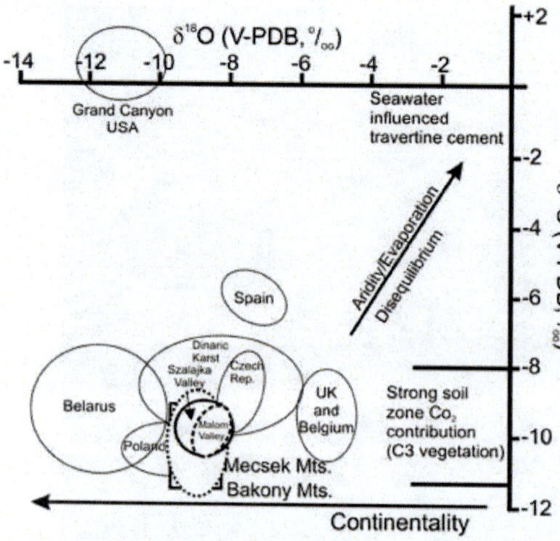

Fig. 11 Stable isotope composition of Hungarian tufa deposits and comparison with European tufa sites (based on Andrews 2006).

dams can be formed in connection with steeper slopes, topographic fractures, various obstacles in the riverbed (e.g. tree trunks, roots) (e.g. Alonso-Zarza and Tanner [edt.] 2010; Pentecost 2005). They can develop stepped river morphology in a longitudinal section of a stream creating small cascades, or high waterfalls. Dams are able to retain water during their vertical accumulation, thus creating basins, smaller or large lakes (Pentecost and Viles 1994; Pentecost 2005). There are many forms of appearance of tufa deposits that can be distinguished based on the height and distance of the dams (Pentecost 2005). In Central Europe, the term 'tetarata' became widespread (Cholnoky 1940). Probably the most spectacular example of the country is the Fátyol Waterfall, which is 17 m long, and consists of 18 consecutive tufa dams (Fig. 10).

Significant waterfalls occur where there is a significant difference in the topography, but carbonate precipitation itself can reach a certain height over time, which can work as a barrier fostering a waterfall formation. Such sites usually form on steep slopes below the erosion base (Scheuer 2002). Moss curtains, various plant remains can be incorporated into the tufa, creating a porous structure that allows water to leak through the sediment (e.g. the 20 m high Szinva waterfall in Lillafüred, Bükk Mts). Caves are often formed behind the waterfalls and tufa curtains, and the leaking water can form small stalactites. Due to the waterfalls, the Anna cave in Lillafüred should be mentioned which is a syngenetic cave, as its formation was parallel with the deposition of the host rock. Freshwater tufa encrustations in water beds can frequently be observed and tufa deposition around pebbles, smaller or larger branches, or anthropogenic objects.

8 Conclusions

Hungary is rich in both travertine and tufa deposits indicating strong hydrothermal and karstic spring activity in the past. Most of the travertine deposits formed in the Pliocene and Pleistocene times in the area of the Gerecse Mts and Buda Mts and despite the high geothermal gradient and available geothermal water, the amount of the actively precipitating travertine carbonate is negligible and restricted only to thermal wells. Travertines are important building and ornamental material and being quarried since the Roman times in Hungary. Their scientific research has started in the middle of the twentieth century and in the past decades it became clear that they preserve important information about the paleoenvironment, paleoclimate, geomorphology and they can be used in reservoir geology as well as in the calibration of the conventional oxygen- and the novel clumped paleothermometers. Reconstruction of travertine depositional environments was based on detailed microfacies studies, while their U–Th dating analyses, together with stable isotope and clumped isotope geochemistry provided important information about temperature, position and age of paleosprings.

Actively forming freshwater tufa carbonates in Hungary are connected to karstic springs and creeks of the the Bükk Mts, Aggtelek Mts, Mecsek Mts, Bakony Mts, and Balatonfelvidék. As they form in close association with fresh meteoric water and climate, their stable isotope geochemical study has the potential to reconstruct past climatic events, although for this precise numerical age data are necessary, which is one of the biggest challenges knowing their contaminated nature and uncertainties related to the available dating methods. Tufa researches in Hungary have started in the early '70 s and during the last decade detailed and still ongoing seasonal monitoring studies have started at the most important sites including the Szalajka Valley. However, there is still a lot to discover within the Hungarian tufa deposits and hopefully future studies will reveal and use these records to complete the paleoclimatic picture of the Carpathian Basin.

References

Alonso-Zarza AM, Tanner LH (2010) Carbonates in continental settings: facies, environments and processes. Elsevier. pp 152–175, 183–187, 269–274

Andrews JE (2006) Palaeoclimatic records from stable isotopes in riverine tufas: synthesis and review. Earth-Sci Rev 75:85–104

Arenas C, Gutierrez F, Osácar C, Sancho C (2000) Sedimentology and geochemistry of fluvio-lacustrine tufa deposits controlled by evaporite solution subsidence in the central Ebro Depression, NE Spain. Sedimentology 47:883–909

Arenas-Abad C, Vázquez-Urbez M, Pardo-Tirapu G, Sancho-Marcén C (2010) Fluvial and associated carbonate deposits. In: Alonso-Zarza A, Tanner LH (eds) Carbonates in continental settings: processes, facies and application. Dev Sedimentol 61:133–175

Aujeszky G, Karácsonyi S, Scheuer G (1974) A DNY-i Bükk karsztvízföldtani viszonyai (Karst hydrogeology of the SW Bükk Mts). Hidr Közl 54(10):465–476 [in Hungarian]

Aujeszky G, Scheuer G (1974) Adatok a Bükk-hegység karsztvízföldtani viszonyaihoz (Additional data to the karst hydrogeology of the Bükk Mts). Hidr Közl 54(3–4):173–183 [in Hungarian]

Bakacsi Z (1993) A süttői édesvízi mészkő szedimentológiai vizsgálata (Sedimentological study of the Süttő travertine). MSc thesis, ELTE, Alkalmazott és Környezetföldtani Tanszék, Budapest, 108 pp. [in Hungarian]

Barta K, Tarnai T (1999) Karstmorphological research in the Mecsek Mountains. South Hungary. Acta Carsol 28(1):13–26

Bányai A (2009) A Demjén K-11 számú hévízkút vízéből történő karbonátos kiválások vizsgálata (Study of the carbonate precipitations from Demjén K-11 thermal well). Karsztfejlődés XIV. Szombathely 2009:83–101 [in Hungarian]

Bernasconi SM, Muller IA, Bergmann KD, Breitenbach SFM, Fernandez A, Hodell DA, Jaggi M, Meckler AN, Millan I, Ziegler M (2018) Reducing uncertainties in carbonate clumped isotope analysis through consistent carbonate-based standardization. Geochem Geophys Geosyst 19. https://doi.org/10.1029/2017GC007385

Békési E, Lenkey L, Limberger J, Porkoláb K, Balázs A, Bonté D, Vrijlandt M, Horváth F, Cloetingh S, van Wees J-D (2020) Subsurface temperature model of the Hungarian part of the Pannonian Basin. Glob and Plan Change 171:48–64

Bihari D (1981) Magyarázó a Bakony hegység 20000–es földtani térképsorozatához: Magyarpolány (Explanatory book to the geological map of the Bakony Mts: Magyarpolány). Magyar Állami Földtani Intézet. Budapest. 65 p [in Hungarian]

Bódai B, Kele S, Kármán K, Czuppon Gy, Móga J (2015) A Bakonyi Csurgó-kút vizének és forrásmészkövének stabilizotóp-geokémiai vizsgálata (Stable isotope geochemical study of water and freshwater tufa of Csurgó Spring (Bakony Mts., Hungary). Karsztfejlődés XX., Szombathely, 49–62 [in Hungarian]

Bódai B, Móga J, Bartha A, Holló S, Kele S (2016) A Szilvásváradi Szalajka-patak mésztufáinak szezonális geokémiai vizsgálata (Seasonal study of the calcareous tufa of the Szalajka-creek, Szilvásvárad). Karsztfejlődés XXI, Szombathely, 45–64 [in Hungarian]

Bódai B, Czuppon G, Fórizs I, Kele S (2022) Seasonal study of calcite-water oxygen isotope fractionation at recent freshwater tufa sites from Hungary. Geologica Carpathica (in press)

Bódai B (2017a) Bakonyi mésztufa-előfordulások geokémiai és geomorfológiai vizsgálata (Geochemical and geomorphological study of the calcareous tufa of the Bakony Mts.) OTDK. Eötvös Loránd Tudományegyetem. Természetföldrajzi Tanszék. Budapes, 42 pp. [in Hungarian]

Bódai B (2017b) A szilvásváradi Szalajka-patak mésztufáinak vizsgálata a stabilizotópos összetételen alapuló paleohőmérsékleti számítások pontosítása céljából (Study of the calcareous tufa of the Szalajka-creek, Szilvásvárad for the refienement of the stable isotope based paleothermometry). MSc thesis, Eötvös Loránd Tudományegyetem. Természetföldrajzi Tanszék. Budapest, 42 pp. [in Hungarian]

Budai T, Fodor L, Sztanó O, Kercsmár Z, Császár G, Csillag G, Gál N, Kele S, Kiszely M, Selmeczi I, Babinszki E, Thamóné Bozsó E, Lantos Z (2018) A Gerecse hegység földtana. Magyarázó a Gerecse hegység tájegységi földtani térképéhez (1: 50 000) (Geology of the Gerecse Mountains. Explanatory book to the geological map of the Gerecse Mountains (1:50 000). Budapest, MBFSZ, 2018. 491 p. (Magyarország tájegységi térképsorozata) [in Hungarian]

Capezzuoli E, Gandin A, Sandrelli F (2010) Calcareous tufa as indicators of climatic variability: a case from the Southern Tuscany (Italy). In: Pedley M, Rogerson M (eds) Tufas, speleothems and stromatolites: unravelling the physical and microbial controls. Geol Soc Lond, Special Publications 336:263–281

Capezzuoli E, Gandin A, Pedley M (2014) Decoding tufa and travertine (fresh water carbonates) in the sedimentary record: the state of the art. Sedimentology 61:1–21

Chafetz HS, Folk RL (1984) Travertines—depositional morphology and the bacterially constructed constituents. J Sed Petrol 54:289–316

Cherdintsev VV, Kazachevskiy IV, Kuzmina EA (1965) Age of Pleistocene carbonate formation according to thorium and uranium isotopes. Geokhimia 9:1085–1092 [in Russian]

Cholnoky J (1932) Tihany. Morfológiai megfigyelések (Tihany. Morphological observations). Mathematikai és Természettudományi Értesítő 48/1:214–236 [in Hungarian]

Cholnoky J (1935) A tihanyi gejzírkúpok (Geyser cones of Tihany). Földgömb 6(2):41–49 [in Hungarian]

Cholnoky J (1940) A mésztufa vagy travertino képződéséről (About the formation of travertines). Mathematikai És Természettudományi Értesítő. 59(3):1023–1038 [in Hungarian]

Claes H (2015) Sediment-petrological and geochemical characterization of travertine. Travertine as a potential Pre-Salt carbonate analogue. PhD thesis, KU Leuven, 300 p.

Claes H, Degros M, Soete J, Claes S, Kele S, Mindszenty A, Török Á, El Desouky H, Vanhaecke F, Swennen R (2017) Geobody architecture, genesis and petrophysical characteristics of the Budakalász continental spring-related carbonates, Buda Hills (Hungary). Quat Int 437:107–128

Claes H, Török Á, Soete J, Mohammadi Z, Vassilieva E, Hamaekers H, Marques Erthal M, Aratman C, Cheng H, Edwards RL, Shen C-C, Özkul M, Kele S, Mindszenty A, Swennen R (2020) U/Th dating and open system behavior: implications for travertines based on the study of Süttő (Hungary) and Ballik (Türkey) sites. Quaternaire 30 (2):117–132

Coplen TB (2007) Calibration of the calcite-water oxygen-isotope geothermometer at Devils Hole, Nevada, a natural laboratory. Geochim Cosmochim Acta 71:3948–3957

Deák J (1986) Paleoklíma vizsgálatok stabil izotóp elemzésekkel (Paleoclimate studies using stable isotope analyses, in Hungarian). VITUKI Final Research Report, Budapest, 1986 [in Hungarian]

Deák J, Fórizs I (in submission) Paleoclimate information obtained from isotope data of travertines

De Boever E, Brasier AT, Foubert A, Kele S (2017) What do we really know about early diagenesis of non-marine carbonates? Sed Geol 361:25–51

Degros M (2013) Sedimentological-Petrological Study and Reservoir Characterization of the Budakalász Travertine Carbonates (Buda Mountains, Hungary). MSc Theses, KU Leuven, 157 pp

Della Porta G, Capezzuoli E, De Bernardo A (2017a) Facies character and depositional architecture of hydrothermal travertine slope aprons (Pleistocene, Acquasanta Terme, Central Italy). Mar and Petr Geol 87:171–187

Della Porta G, Croci A, Marini M, Kele S (2017b) Depositional architecture, facies character and geochemical signature of the Tivoli travertines (Pleistocene, Acque Albule Basin, Central Italy). Riv Ital Paleontol Stratigr 123(3):487–540

Dövényi P, Horváth F (1988) A review of temperature, thermal conductivity, and heat flow data for the Pannonian Basin, In: Royden LH and Horváth F (editors), The Pannonian Basin, a study in basin evolution: American Association of Petroleum Geologists - Memorie 45:195–233

Djurić B, Kele S, Rižnar I (2018) The Budakalász travertine production —*ASMOSIA Acts*. In: Poljak DM, Marasovic K (eds) Proceedings of the XI ASMOSIA conference (Split 18–22 May, 2015), pp 545–556

Eiler JM (2007) Clumped-isotope geochemistry—the study of naturally-occurring, multiply-substituted isotopologues. Earth Planet Sci Lett 262:309–327

Ford TD, Pedley HM (1996) A review of tufa and travertine deposits of the world. Earth Sci Rev 41:117–175

Fournier RO, Rowe JJ (1966) Estimation of underground temperatures from the silica content of water from hot springs and wet-steam wells. Am J Sci 264:685–697

Földvári M, Berner Z, Stüben D (2003) Thermoanalytical study of Quaternary thermal lacustrine travertine occurrences in Hungary (Buda-Vár-hegy, Budakalász, Szomód-Les- hegy). Acta Geol Hung 46:195–204

Gasparik M, Kele S (2010) A tokodi (Keleti-Gerecse) édesvízi mészkő és gerinces fauna lelőhely geokémiai, szedimentológiai és őslénytani vizsgálata, valamint U/Th sorozatos kormeghatározása (Sedimentological and paleontological study and Uranium-series dating of the Tokod travertine paleovertebrate site (Eastern Gerecse)). A Tatabányai Múzeum Évkönyve i. Tatabánya 2010(1):29–39 [in Hungarian]

Gábris G, Mindszenty A, Kele S (2014) Folyóteraszok és travertínók Budapesten (River terraces and travertines in Budapest). In: Mindszenty A (ed) Budapest: földtani értékek és az ember. Városgeológiai tanulmányok (Budapest: geoheritage and humans – urban geological studies). 311 pp, ELTE - Eötvös Kiadó, Budapest, 54–57 [in Hungarian]

Ghosh P, Adkins J, Affek H, Balta B, Guo W, Schauble EA, Schrag D, Eiler JM (2006) 13C–18O bonds in carbonate minerals: a new kind of paleothermometer. Geochim Cosmochim Acta 70:1439–1456

Golubic S, Violante C, Ferreri V, D'Argenio B (1993) Algal control and early diagenesis in Quaternary travertine formation (Rocchetta a Volturno, Central Apennines). Studies on Fossil Benthic Algae. In: Barattolo F et al. (eds) Boll Soc Paleont Ital Spec. vol 1, Mucchi, Modena, pp 231–247

Greenhalgh J, Wells S, Borsato R, Pratt D, Martin M, Roberson R, Fontes C, Obaje WA (2011) A fresh look at prospectivity of the equatorial conjugate margin of Brazil and Africa. First Break 29:67–72

Haas J (2001) Geology of Hungary. Eötvös University Press. Budapest, pp 35–78

Hancock PL, Chalmers RML, Altunel E, Çakir Z (1999) Travitonics: using travertines in active fault studies. J Struct Geol 21:903–916

Hevesi A (1972) Forrásmészkő-képződés a Bükkben (Spring carbonate formation in the Bükk Mts). Földr Ért XXI. 2–3:187–205 [in Hungarian]

Hennig GJ, Grün R, Brunnacker K, Pécsi M (1983a) Th^{230}/U^{234} sowie ESR Alterbestimmungen einiger Travertine in Ungarn. Eiszeitalter and Gegenwart 33:7–19

Hennig GJ, Grün R, Brunnacker K (1983b) Speleothems, travertines and paleoclimates. Quat Res 20:1–29

Hoffer A (1943) A Tihanyi–félsziget vulkáni képződményei (Volcanic formations of the Tihany peninsula). Földt Közl 73(4–9):375–429 [in Hungarian]

Horusitzky H (1939) Budapest Duna jobbparti részének hidrogeológiája (Hydrogeology of the areas right from the Danube). Hidr Közl 18:1–404 [in Hungarian]

Imbrie J, Hays JD, Martinson DG, McIntyre A, Mix AC, Morley JJ, Pisias NG, Prell WL, Shackleton NJ (1984) The orbitall theory of Pleistocene climate: support from a revised chronology of the marine $\delta^{18}O$ record. In: Berger et al. (Eds.), Milankovitch and Climate, Part 1, Riedel D, Hingham MA, pp 269–305

Járai-Komlódi M (1973) Pollenstatisztikai vizsgálatok a vértesszőlősi őstelepülés mésztufa rétegeiből. Földr Közlemények 97:127–132 [in Hungarian]

Jánossy D, Krolopp E (1981) Die pleistozanen Schnecken und Vertebraten-Faunen von Süttő. Fragm Mineral Palaeontol 10:31–58

Jánossy D (1986) Pleistocene Vertebrate Faunas of Hungary. Akadémiai Kiadó, Budapest 208 pp.

Jaskó S (1959) A földtani felépítés és a karsztvíz elterjedésének kapcsolata a Dunántúli középhegységben (The connection of the geological buildup and the karstic water in the Transdanubian Mountains). Hidr Közl 39(4):289–297 [in Hungarian]

Kadić O (1921) A Hámori barlang Borsodmegyében (The Hámori cave in Borsod county). A Természet 17:56–58 [in Hungarian]

Kele S (2009) Édesvízi mészkövek vizsgálata a Kárpát-medencéből: paleoklimatológiai és szedimentológiai elemzések (Study of Freshwater Limestones from the Carpathian Basin: Paleoclimatological and Sedimentological Implications). PhD Thesis. Eötvös Loránd University, Budapest, Hungary, 176 pp [in Hungarian]

Kele S (2011) A máriaremetei édesvízi mészkő U-soros kormeghatározása és paleokörnyezeti rekonstrukciója (Uranium-series dating and paleoenvironmental reconstruction of the freshwater limestone of Máriaremete). Hidr Táj 50(1):87–90 [in Hungarian]

Kele S, Vaselli O, Szabó C, Minissale A (2003) Stable isotope geochemistry of Pleistocene travertine from Budakalász (Buda Mts., Hungary). Acta Geol Hung 46(2):161–175

Kele S, Korpás L, Kovács-Pálffy P, Lantos M (2004) Sedimentology, mineralogy, lake evolution and chronology of the Quaternary Tata thermal lacustrine travertine (Hungary). Topical issues of the research of Middle Palaeolithic period in Central Europe", Tud füzetek (ISSN 0866-2908), vol 12, pp 33–35

Kele S, Korpás L, Demény A, Kovács-Pálffy P, Bajnóczi B, Medzihradszky Z (2006) Paleoenvironmental evaluation of the Tata Travertine Complex (Hungary), based on stable isotopic and petrographic studies. Acta Geol Hung 49(1):1–31

Kele S, Demény A, Siklósy Z, Németh T, Tóth M, Kovács MB (2008) Chemical and stable isotope composition of recent hot-water travertines and associated thermal waters, from Egerszalók, Hungary: Depositional facies and non-equilibrium fractionation. Sed Geol 211:53–72

Kele S, Gy S, Demény A, Shen C-C, Chiang H-W (2009) U-series dating and isotope geochemical study of the Gellért Hill (Budapest) travertine. Centr Eur Geol 52(3–4):199–224

Kele S, Gy S, Demény A, Shen C-C, Hong-Wei C (2011) A Rózsadomb (Budapest) édesvízi mészköveinek U/Th sorozatos kormeghatározása és stabilizotóp-geokémiai vizsgálata (Uranium-series dating and geochemical study of the travertines located on the Rózsadomb Hill (Budapest). Földt Közl 141(2):445–468 [in Hungarian]

Kele S, Scheuer G (2012) Az Ördögárok-völgyéhez kapcsolódó Kondor úti és Nyéki úti édesvízi mészkő előfordulások U-soros kormeghatározása és képződési környezetük rekonstrukciója (Uranium-series dating and paleoenvironmental reconstruction of the freshwater limestones of the Kondor and Nyéki streets, Ördögárok valley). Hidr Táj 1(1):53–55 [in Hungarian]

Kele S, Mindszenty A (2014a) A terasz szintek és az édesvízi mészkövek kapcsolata a Budai-hegységben (Relationship between terrace levels and freshwater limestones in the Buda Mts). In: Mindszenty A (ed) Budapest: földtani értékek és az ember. Városgeológiai tanulmányok. (Budapest: geoheritage and humans - urban geological studies), ELTE Eötvös Kiadó, Budapest, pp 102 [in Hungarian]

Kele S, Mindszenty A (2014b) A forráskilépési pontok térbeli átrendeződésének története az édesvízi mészkövek korának tükrében (History of rearrangement of the discharge points of springs in relation to the age of freshwater limestones). In: Mindszenty A (ed.) Budapest: földtani értékek és az ember. Városgeológiai tanulmányok. (Budapest: geoheritage and humans - urban geological studies). Budapest: ELTE - Eötvös Kiadó, pp 102–103 [in Hungarian]

Kele S, Breitenbach SFM, Capezzuoli E, Meckler AN, Ziegler M, Millan IM, Kluge T, Deák J, Hanselmann K, John CM, Yan H, Liu Z, Bernasconi SM (2015) Temperature dependence of oxygen- and clumped isotope fractionation in carbonates: a study of travertines and tufas in the 6–95°C temperature range. Geochim Et Cosm Acta 168:172–192

Kele S, Markó A, Cseh J, Shen C-C, Wu C-C, Bernasconi SM (2016) Dating and clumped isotope-based temperature of a paleo-jacuzzi (Vértesszőlős Early Man site, Hungary). In: 5th international clumped isotope workshop, pp 6–9, St. Petersburg, Florida, USA

Kele S, Markó A, Kisné Cseh J, Shen C–C, Wu C–C, Bernasconi SM (in preparation) Samuel's paleo–jacuzzi: dating and clumped isotope–based temperature of the Vértesszőlős Early Human site (Hungary)

Kele S, Ziegler M, Rinyu L, Hong-Wei C, Shen C-C, Scheuer G (in preparation) Paleotemperature and $\delta^{18}O$ reconstruction of Pleistocene spring activity in the Gerecse Mts (Hungary) based on clumped isotopes from travertines

Kele S, Scheuer G, Hong-Wei C, Shen C-C, Schauer A, Huntington KW (in preparation) Reconstruction of paleohydrothermal changes in the Buda Mts

Kelson JR, Huntington KW, Schauer AJ, Saenger C, Lechler AR (2017) Toward a universal carbonate clumped isotope calibration: Diverse synthesis and preparatory methods suggest a single temperature relationship. Geochim Et Cosmochim Acta 197: 104–131

Kessler H (1953) A lillafüredi Anna-barlang forrásai (Springs of the Anna cave, Lillafüred). Hidr Közl 33:60–65 [in Hungarian]

Kluge T, John CM, Boch R, Kele S (2018) Assessment of factors controlling clumped isotopes and $\delta^{18}O$ values of hydrothermal calcites. Geochem Geophys Geosys

Koltai G, Kele S, Kármán K, Keveiné Bárány I (2012a) Édesvízi mészkövet lerakó források geokémiai vizsgálata a Mecsekben (Geochemical study of springs depositing freshwater limestones in the Mecsek Mts.). Karsztfejlődés XVII, Szombathely 35–46 [in Hungarian]

Koltai G, Kele S, Keveiné I (2012b) Preliminary studies of freshwater tufa deposits in Mecsek Mts. Hungary Acta Clim Et Chorol 46: 143–151

Koltai G, Palcsu L, Kele S, Futó I, Heim E, Bárány-Kevei I (2013) Downstream section changes of freshwater tufa depositing streams, Budapest. Hungary. Cent Eur Geol 56(2–3):229–230

Kordos L (1984) Magyarország barlangjai (The caves of Hungary). Gondolat Könyvkiadó, Budapest. 178–180(276):285 [in Hungarian]

Korpás L (2003) Basic pattern of quaternary travertine: a review with special regard to the Hungarian deposits. Acta Geol Hung 46(2): 131–148

Korpás L, Kovács-Pálffy P, Lantos M, Földvári M, Kordos L, Krolopp E, Stüben D, Berner Z (2004) Sedimentology, geochemistry, chronology and palaeokarst evolution of Quaternary thermal lacustrine travertine. An integrated case study from Vár-hegy, Budapest, Hungary. Földt Közl 134/4:541–562

Kovács A (1995) A budakalászi édesvízi mészkő mikrofácies vizsgálata (Microfacies study of the freshwater limestone of Budakalász). Diplomamunka, ELTE, Alkalmazott és Környezetföldtani Tanszék, 98 pp [in Hungarian]

Kovács J, Müller P (1980) A Budai-hegyek hévizes tevékenységének kialakulása és nyomai (Development and traces of the hydrothermal activity of the Buda Mts.). Karszt és Barlang, II. füzet, pp 93–98 [in Hungarian]

Kovács-Pálffy P, Földvári M (2004) Mineralogy of the travertines in NE Transdanubia (Hungary). Földt Közl 134(4):105–126

Kovácsné Bodor P (2020) Termálvíz megcsapolódáshoz köthető vízkémiai és kiválási folyamatok kísérleti tanulmányozása (Experimental study of processes related to water chemistry and deposition connected to thermal water discharge). PhD thesis, Általános és Alkalmazott Földtani Tanszék, ELTE, Budapest, 200 p [in Hungarian]

Kraft J, Scheuer G, Schweitzer F (1986) A Pécs környéki forráslerakódások vizsgálata (Study of spring deposits around Pécs). Pécsi Műszaki Szemle. 31(3):13–18 [in Hungarian]

Kretzoi M, Vértes L (1965) The role of Vertebrate fauna and Palaeolithic industries of Hungary in Quaternary stratigraphy and chronology. Acta Geol 9:125–143

Kretzoi M, Pécsi M (1982) Pliocene and quaternary chronostratigraphy and continental surface development of the Pannonian Basin. Quat Stud Hungary, INQUA Hungarian, National Committee, pp 11–42

Lantos M (2004) Magnetostratigraphic correlation of Quaternary travertine sequences in NE Transdanubia. Földt Közl 134/2: 227–236

Latham AG, Schwarcz HP (1990) Magnetic polarity of travertine samples from Vertesszőlős. In: Kretzoi M and Dobosi VT (Eds.), Vertesszőlős - Site, Man and Culture. Akademiai Kiadó, Budapest, pp 553–555

Lenkey L, Dövényi P, Horváth F, Cloetingh S (2002) Geothermics of the Pannonian basin and its bearing on the neotectonics. EGU Stephan Mueller Spec Publ Series 3:29–40

Lénárt L (1981) Adalékok a lillafüredi mésztufabarlangok kutatásához (Contributions to the study of the tufa caves of Lillafüred). Karszt és Barlang, 1–2. félév. pp 1–8 [in Hungarian]

Lénárt L, Hevesi A (2003) Anna-barlang (Anna-cave). In: Székely K (ed) Magyarország fokozottan védett barlangjai (Highly protected caves of Hungary). Mezőgazda Kiadó, Budapest, pp 194–197 [in Hungarian]

Molnár M, Dezső Z, Futó I, Rinyu L, Svingor É (2006) Fiatal karsztos kőzetek 14C korának mérése és értelmezése (Measurement and interpretation of 14C age of young karstic rocks). In: Veress M (ed.) Karsztfejlődés XI. Szombathely, Magyarország, Berzsenyi Dániel Főiskola Természetföldrajzi Tanszék, p 37 [in Hungarian]

Molnár M, Dezső Z, Palcsu L, Szántó Z, Futó I, Svingor É, Rinyu L (2004) 14C mérés alkalmazása a karsztkutatásban (Application of the 14C method in karst research). In: Veress M (ed) Karsztfejlődés IX. Szombathely, Magyarország, Berzsenyi Dániel Főiskola Természetföldrajzi Tanszék, p 9 [in Hungarian]

Opauszky L, Matus L, Lukács J, Kiss I (1964) Die Bestimmung der Paleotemperatur durch Messung des $^{18}O/^{16}O$ Isotopverhältnisses von Karbonaten. In: Vértes L (Ed.): Tata, eine mittelpaläolithische Travertin-Siedlung in Ungarn. Akadémiai Kiadó, Budapest, pp 19–29

Özkul M, Kele S, Gökgöz A, Shen C-C, Jones B, Baykara MO, Fórizs I, Németh T, Chang Y-W, Alçiçek MC (2013) Comparison of the Quaternary travertine sites in the Denizli extensional basin based on their depositional and geochemical data. Sed Geol 294:179–204

Pazonyi P, Kordos L, Magyari E, Marinova E, Füköh L, Venczel M (2014) Pleistocene vertebrate faunas of the Süttő Travertine Complex. Quat Int 319:50–63

Pedley HM (1990) Classification and environmental models of cool freshwater tufas. Sed Geol 68:143–154

Pedley M (2014) The morphology and function of thrombolytic calcite precipitating biofilms: a universal model derived from freshwater mesocosm experiments. Sedimentol 61:22–40

Pelikán P [ed.] (2005) A Bükk hegység földtana. Magyar Állami Földtani Intézet. Budapest. pp 147–177 [in Hungarian]

Pentecost A (2005) Travertine. Springer, Berlin Heidelberg, p 445

Pentecost A, Viles H (1994) A review and reassessment of travertine classification. Géograph Phys Et Quat. 48:305–314
Pécsi M (1959) Formation and geomorphology of the Danube valley in Hungary. Akadémiai Kiadó 346, Budapest
Pécsi M (1973) Geomorphological position and absolute age of the Lower Palaeolithic site at Vértesszőlős. Hungary. Földr Közl 21 (2):109–115
Ricketts JW, Ma L, Wagler AE, Garcia VH (2019) Global travertine deposition modulated by oscillations in climate. J Quat Sci 1–11
Ruszkiczay-Rüdiger Z (2007) Tectonic and climatic forcing in Quaternary landscape evolution in the Central Pannonian Basin: a quantitative, geomorphological, geochronological and structural analysis. Department of Isotope Geochemistry, Vrije Universiteit, Amsterdam, PhD-thesis, p 149
Ruszkiczay-Rüdiger Z, Fodor L, Csillag G, Braucher R, Kele S, Novothny Á, Thamó-Bozsó E, Virág A, Molnár G, Madarász B, Team A (2016) Spatially and temporally varying Quaternary uplift rates of the Gerecse Hills, Northern Pannonian Basin, using dated geomorphological horizons in the Danube valley. Geophysical Research Abstracts, vol 18, EGU2016-6463, EGU General Assembly
Ruszkiczay-Rüdiger Z, Fodor L, Bada G, Leél-Őssy S, Horváth E, Dunai TJ (2005a) Quantification of Quaternary vertical movements in the central Pannonian Basin: a review of chronologic data along the Danube River. Hungary Tectonophys 410(1–4):157–172
Ruszkiczay-Rüdiger Z, Dunai TJ, Bada G, Fodor L, Horváth E (2005b) Middle to late pleistocene uplift rate of the Hungarian mountain range at the Danube bend (Pannonian Basin) using in situ produced 3He. Tectonophysics 410(1–4):173–187
Ruszkiczay-Rüdiger Z, Csillag G, Fodor L, Braucher R, Novothny Á, Thamó-Bozsó E, Virág A, Pazonyi P, Tímár G, ASTER Team (2018) Integration of new and revised chronological data to constrain the terrace evolution of the Danube River (Gerecse Hills, Pannonian Basin). Quat Geochron 48:148–170
Sásdi L (2005) Az Aggtelek-Rudabányai-hegység édesvízi mészkő előfordulásai (Travertine deposits of the Aggtelek-Rudabánya Mountains). Karsztfejlődés X. 137–151 [in Hungarian]
Scheuer G (1975) Kiegészítő adatok a Bükk hegységi édesvízi mészkövek előfordulásaihoz (Additional data to the freshwater limestones of the Bükk Mts). Földr Ért XXIV. 1:75–78 [in Hungarian]
Scheuer G (1999a) A gerecsei mészképző hévforrások összehasonlító vizsgálata (Comparative study of carbonate depositing thermal springs of the Gerecse). Hidr Közl 79(3):152–156 [in Hungarian]
Scheuer G (1999b) A gerecsei alsó pleisztocén és felső neogén hévforrástavi édesvízi összletek összehasonlító vizsgálata (Comparative study of the Lower Pleistocene and Upper Neogene freshwater thermal lake deposits). Hidr Közl 79(1):11–14 [in Hungarian]
Scheuer G (2002) A hideg karsztvizek mésztufáinak vizsgálata és főbb típusaik (Study of the main types of calcareous tufa formed from cold karstic waters). Hidr Közl 82(4):225–232 [in Hungarian]
Scheuer G (2003) Mészképző ásványvízforrások és kiválásaikat befolyásoló tényezők (Carbonate depositing mineral springs and their influencing factors). Hidr Közl 83(6):339–347 [in Hungarian]
Scheuer G, Schweitzer F (1970) A karsztvíz eredetű édesvízi mészkövek csoportosítása (Classification of freshwater limestones of karstic origin). Földr Ért XIX. Évf. 1970. 3 füzet, 356–360 [in Hungarian]
Scheuer G, Schweitzer F (1972) Az édesvízi mészkövet lerakó karsztforrások paleogeográfiai viszonyai és osztályozásuk (Paleogeographycal features of freshwater limestone depositing karstic springs and their classification). Földr Ért XXI. Évf. 1972(2–3):385–390 [in Hungarian]
Scheuer G, Schweitzer F (1978) Az édesvízi mészkövet lerakó források sajátosságai (Features of freshwater limestone depositing springs). Földr Ért XXVII. Évf. 3–4:475–486 [in Hungarian]
Scheuer G, Schweitzer F (1981) A hazai édesvízi mészkövek származása és összehasonlító vizsgálatuk (Origin and comparative study of freshwater limestones of Hungary). Földt Közl 111(1):67–97 [in Hungarian]
Scheuer G, Schweitzer F (1985) Az édesvízi mészkőkúpok típusai és alakulati formái (Types and forms of freshwater limestone cones). Földt Közl 115:385–398 [in Hungarian]
Scheuer G, Schweitzer F (1986) A lejtői édesvízi mészkőképződés formái és típusai (Forms and types of freshwater limestone formation on slopes). Földt Közl 116:363–376 [in Hungarian]
Scheuer G, Schweitzer F (1988) A Gerecse és a Budai-hegység édesvízi mészkőösszletei (Freshwater Limestones of the Gerecse and Buda Hills). Földr Tan vol. 20. Akad. Kiadó, Budapest, pp 129 [in Hungarian]
Scheuer G, Schweitzer F (1989) Újabb adatok a Budai-hegység pannóniai hévforrás tevékenységéhez (New data to the Pannonian thermal spring activity of the Buda Mts.). Hidr Táj, pp 41–44 [in Hungarian]
Scheuer G, Koltai G, Kele S (2013) A nyugat-mecseki hévíz és karsztforrások makro- és mikroelem adottságainak előzetes vizsgálata (Preliminary study of the macro and micro elements of the karst and thermal waters of the Western Gerecse). Hidr Közl 2:27–40 [in Hungarian]
Schréter Z (1912) Harmadkori pleisztocén hévforrások tevékenységének nyomai a Budai hegységben (Traces of Pleistocene thermal springs in the Buda Mts.). MÁFI Évkönyv 19:179–231 [in Hungarian]
Schréter Z (1953) A Budai- és a Gerecse-hegység peremi édesvízi mészkő előfordulásai. (Travertine deposits along the border of the Buda Gerecse Mts). Földtani Intézet Évi Jelentése 1949-ről, pp 111–150 [in Hungarian]
Schwarcz HP, Skoflek I (1982) New dates for the Tata, Hungary archaeological site. Nature 295:590–591
Schwarcz HP, Latham AG (1984) Uranium series age determination of travertines from the site of Vértesszőlős, Hungary. J Archaeol Sci 11:327–336
Schwarcz HP, Latham AG (1990) Absolute age determination of travertines from the Vértesszőlős site. In: Kretzoi M, Dobosi V (eds) Vértesszőlős—site, man and culture. Budapest, pp 549–552
Sierralta M, Kele S, Melcher F, Hambach U, Reinders J, van Geldern R, Frechen M (2010) Characterisation and Uranium-series dating of Travertine from Süttő in Hungary. Quat Int 222:178–193
Smekens T (2013) Sedimentological study and reservoir characterization of travertine carbonates near Süttő (Gerecse Mountains, Hungary). MSc Theses, KU Leuven, 102 pp.
Soete J, Kleipool LM, Claes H, Claes S, Hamaekers H, Kele S, Özkul M, Foubert A, Reijmer JJG, Swennen R (2015) Acoustic properties in travertines and their relation to porosity and pore types. Mar Pet Geol 59:320–335
Szöőr G, Balázs É, Sümeghy P, Scheuer G, Schweitzer F, Hertelendi E (1992) A magyarországi édesvízi mészkövek termoanalitikai és izotópgeokémiai elemzése fáciestani és rétegtani értékeléssel (Thermoanalytical and isotope geochemical study and facies and stratigraphical evaluation of Hungarian freshwater limestones). In: Szöőr G (ed) Fáciesanalitikai, paleobiogeokémiai és paleoökológiai kutatások (Facies analytical, paleobiogeochemical and paleoecological researches) Debrecen, pp 93–107 [in Hungarian]
Toker E (2017) Quaternary fluvial tufas of Sarıkavak area, southwestern Turkey: facies and depositional systems. Quat Int 437:37–50

Townson R (1797) Travels in Hungary, with a Short Account of Vienna in the Year 1797, 506 pp. In: Vértes L 1964: Die Geschichte der Freilegung, pp 9–11. In: Vértes L (ed) 1964: Tata, eine mittelpaläolithische Travertin-Siedlung in Ungarn. – Akadémiai Kiadó, Budapest

Török Á (2003) Facies analysis and genetic interpretation of travertine, Buda Vár-hegy. Hungary Acta Geol Hung 46(2):177–192

Török Á (2004) Comparison of the process of decay of two limestones in a polluted urban environment. Land Reconst Manag 3:3–92

Török Á (2008) Black crusts on travertine: factors controlling development and stability. Environ Geol. https://doi.org/10.1007/s00254-008-1297-x

Török Á, Ünnep V, Balázs A, Mindszenty A, Kele S (2013) A kápolna-hegyi édesvízi mészkőkúp complex szedimentológiai, geokémiai és geofizikai vizsgálata (Budai-hegység) (Complex sedimentological, geochemical and geophysical study of the Kápolna-hegy spring-cone (Buda Mts, Hungary). Földt Közl 143 (3):251–264 [in Hungarian]

Török Á, Mindszenty A, Claes H, Kele S, Fodor L, Swennen R (2017) Geobody architecture of continental carbonates: "Gazda" travertine quarry (Süttő, Gerecse Hills, Hungary). Quat Int 437:164–185

Török Á (2018) The anatomy of a Pleistocene travertine complex—sedimentological, diagenetic and reservoir studies on the Süttő travertine System (Gerecse Hills, Hungary). PhD Theses, KU Leuven and ELTE Budapest, 150 pp

Török Á, Claes H, Brogi A, Liotta D, Tóth Á, Mindszenty A, Kudó I, Kele S, Huntington KW, Shen C-C, Swennen R (2019) A multi-methodological approach to reconstruct a dismantled travertine fissure ridge: the case of the Cukor quarry (Süttő, Gerecse Hills, Hungary). Geomorph 345:106836

Veres V (2007) A Rudas-fürdő travertínója (The travertine of the Rudas Spa). MSc thesis, ELTE, Természettudományi Kar, Alkalmazott és Környezetföldtani Tanszék, 116 p. [in Hungarian]

Veress M (2002) Néhány bakonyi patakmeder mésztufa kiválásának morfogenetikai típusai (Morphogenetic types of calcareous tufa deposits of some watercourses of the Bakony Mts.). Hidr Közl 82/1:15–22 [in Hungarian]

Vértes L (ed) (1964) Tata, eine mittelpaläolithische Travertinsiedlung in Ungarn. Akadémiai Kiadó, Budapest, 253 pp.

Violante C, Ferreri V, D'Argenio B, Golubic S, (1994) Quaternary Travertines at (1994) IAS—15th regional meeting—pre meeting fieldtrip guidebook. Defrede, Napoli, pp 5–23

Virág M, Mindszenty A, Bendő Z (2013a) Az Ősforrás kalcitlemezei (The calcite rafts in Gellért Ősforrás). In: Mindszenty A (ed) Budapest: földtani értékek és az ember (Budapest: geoheritage and humans - urban geological studies), ELTE Eötvös Kiadó, Budapest, pp 188–191 [in Hungarian]

Virág M, Mindszenty A, Weidinger T, Molnár M, Bendő Z, Tanos P, Mádl-Szőnyi J (2013b) A Rudas fürdő travertínója (The travertine of Rudas Spa, Budapest). In: Mindszenty A (ed) Budapest: földtani értékek és az ember (Budapest: geoheritage and humans - urban geological studies), ELTE Eötvös Kiadó, Budapest, pp 191–199 [in Hungarian]

Virág M, Mindszenty A, Bendő Z (2013c) A Városliget-II. fúrás (Széchenyi fürdő) ásványkiválásainak tanulságai a Budai Termálkarszt barlangjai számára. (Unusual speleothems from a non-spelean environment. Mineral precipitates of the Széchenyi Spa, Buda Thermal Karst, Budapest, Hungary). In Mindszenty A (Ed.): Budapest: földtani értékek és az ember (Budapest: geoheritage and humans - urban geological studies), ELTE Eötvös Kiadó, Budapest, pp 270–274 [in Hungarian]

Virág M, Mindszenty A, Bendő Z, Hegedűs A (2013d) Unusual speleothems from a non-spelean environment—mineral precipitates of the Széchenyi Spa (Buda Thermal Karst, Budapest, Hungary). In: Mádl-Szőnyi J, Erőss A, Mindszenty A, Tóth Á (Eds) International symposium on hierarchical flow systems in karst regions—in honour of Professor József Tóth in celebration of his 80th birthday, 4–7 September 2013d, ELTE, Budapest, Hungary, p 140 [in Hungarian]

Virág M, Molnár M, Braun M, Mindszenty A (2020) Investigation of a flowstone-like historical indoor-travertine (Rudas Spa, Budapest, Hungary) using the ^{14}C bomb- peak". Radiocarbon 62(5):1419–1435. https://doi.org/10.1017/RDC.2020.24

de Vries H, de Waard H (1964) Die Untersuchungen des C14 Laboratoriums zu Groningen. In: Vértes L (Ed.): Tata, eine mittelpaläolithische Travertin-Siedlung in Ungarn – Akadémiai Kiadó, Budapest, pp 35–36

Wein G (1977) A Budai-hegység tektonikája (Tectonics of the Buda Mts.). MÁFI Alkalmi kiadvány, 76 p. [in Hungarian]

Wright V (2012) Lacustrine carbonates in rift settings: the interaction of volcanic and microbial processes on carbonate deposition. Geol Soc, London 1–20. Spec. Publ

The Vegetation of Karsts in Hungary

Zoltán Bátori, Norbert Bauer, László Erdős, and András Vojtkó

Abstract

Karst systems in Hungary are topographically complex environments supporting habitats for a number of zonal and edaphic plant communities that harbour many endangered, endemic and relict species, indicating past evolutionary dynamics and the alternation of colder and warmer climate periods during the Quaternary. Vegetation structure, species richness, floristic composition and functional traits vary along steep environmental gradients, resulting in a high alpha, beta and gamma diversity over relatively short distances. Different human activities, such as agriculture, mining, construction, grazing, forestry and tourism have been shaping the vegetation of these karst systems for centuries, although there are regional differences concerning the level and type of anthropogenic impacts. Increasing temperatures and longer drought periods have been detected for the area of the Carpathian Basin, and this trend will probably continue in coming decades. The capacity of karstic refugia to buffer against the negative effects of climate oscillations on vegetation is likely to be exceeded if the current management practices will remain unchanged. In this chapter, the phytogeography, community structure and threatening factors of the different karst areas in Hungary were discussed by reviewing the most relevant literature, primarily focusing on the vascular vegetation of these karst areas.

Z. Bátori (✉)
Department of Ecology, University of Szeged, Közép fasor 52, 6726 Szeged, Hungary

N. Bauer
Department of Botany, Hungarian Natural History Museum, Könyves Kálmán krt. 40, 1087 Budapest, Hungary

L. Erdős
Department of Ecology, University of Debrecen, Egyetem tér 1, 4032 Debrecen, Hungary
e-mail: erdos.laszlo@okologia.mta.hu

A. Vojtkó
Department of Botany, Eszterházy Károly University of Applied Sciences, Eszterházy tér 1, 3300 Eger, Hungary

Keywords

Biodiversity hotspot • Climate change • Endemic species • Refugium • Threatening factor

1 Introduction

1.1 Macroclimate and Vegetation

The natural vegetation of the Hungarian karst areas (ca. 1350 km^2, Jakucs 1977) is extremely diverse and shows considerable differences among the different regions. This is true not only for the vegetation of the foothill (<400 m asl), submontane (400–750 m asl) and montane (>750 m asl) zones, but also for the area and cover of the different vegetation types. The main reason for these patterns is associated with the geographic location of the Carpathian Basin (Fig. 1).

The inner part of the Carpathian Basin is characterised by different climatic conditions. The influence of Central European–sub-oceanic climate effects is most evident in the western third of the basin, while continental climate effects are more pronounced in the eastern parts (combined with the effects of montane climate in northern Hungary). The southernmost karst areas of Hungary (Mecsek and Villány Mts) and the south-facing slopes of the Transdanubian Mts are affected by a strong sub-Mediterranean climate influence, which is manifested through warmer average temperatures and a double rainfall maximum (Borbás 1900; Zólyomi 1942; Soó 1962, 1967; Barina 2004; Somlyay 2009; Borhidi et al. 2012; Bauer 2014). The main axis of the Transdanubian and North Hungarian Mts is oriented in a roughly southwest–northeast direction (Fig. 1), and the amount of precipitation falling on the western–northwestern parts and plateaus of the mountain ranges is much higher than on the other parts (partially because of the Föhn-type winds). As a result, great differences can be observed in the amount of

M. Veress and Sz. Leél-Őssy (eds.), *Cave and Karst Systems of Hungary*,
Cave and Karst Systems of the World, https://doi.org/10.1007/978-3-030-92960-2_19

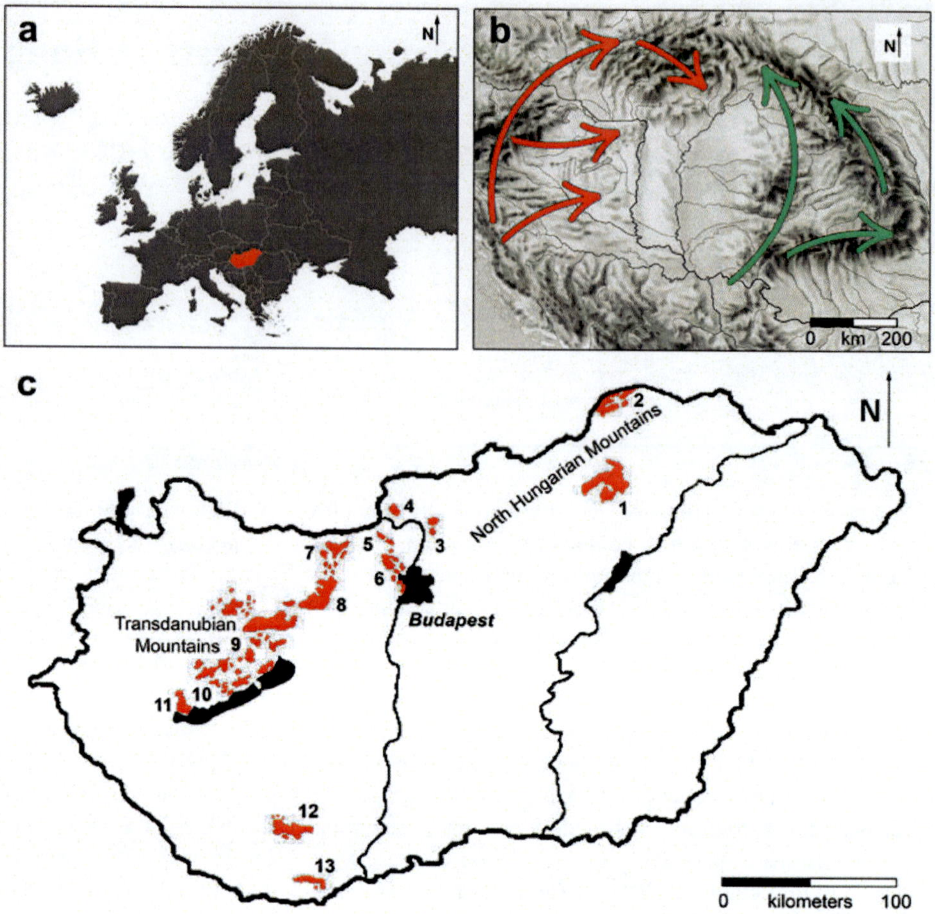

Fig. 1 **a** The location of Hungary in Europe; **b** Potential migration routes (green and red arrows) for karst-dwelling species during the Pleistocene and Holocene climate oscillations (after Borhidi 2011); **c** Karst areas in Hungary. North Hungarian Mts: Bükk Mts (1) and Aggtelek Karst (2); Transdanubian Mts: southwestern Cserhát Mts (3), southern Börzsöny Mts (4), Pilis Mts (5), Buda Mts (6), Gerecse Mts (7), Vértes Mts (8), Bakony Mts (9), Balaton Upland (10) and Keszthely Mts (11); Mecsek (12) and Villány Mts (13) (after Jakucs 1977)

precipitation within relatively short distances. For instance, some areas in the western part of the Bakony Mts (Kőris Hill and Hárskút plateau) receive much more precipitation (800–900 mm) than the eastern parts (600 mm), although the distance between them is only around 30–40 km (Kakas 1960; Mersich et al. 2000; Dövényi 2010). The average precipitation is about 30% less on the southern side of the Bükk Mts (550–600 mm) than on the Bükk plateau (800–900 mm). According to the measurements between 1981 and 2010, the highest average precipitation in Hungary was recorded on the Bükk plateau (839 mm; climatological station: Jávorkút); additionally, the highest yearly precipitation (1555 mm in 2010) was also recorded here.

Mesic and xeric forests are the most common vegetation types in the Hungarian karst areas (Zólyomi 1967; Fekete 1988). Turkey oak–sessile oak forest is the main zonal vegetation type of the foothills, but in some regions (Balaton Upland, eastern Bakony, Mecsek and Villány Mts) pubescent oak also forms a vegetation zone (Fekete and Zólyomi 1966). Zonal oak–hornbeam and beech forests dominate the submontane regions, while zonal beech forest is the main forest type in the montane regions. However, in some karst areas, oak–hornbeam (e.g. in the north western part of the Bakony Mts) and beech forests (e.g. in the Mecsek Mts) are mainly found as extrazonal communities (Fekete 1964; Kevey 2008).

1.2 Geomorphology and Vegetation

The complex topography of the Hungarian karst areas contributes to the maintenance of various edaphic communities, such as limestone/dolomite beech forests, rocky grasslands, scree forests, shrub forests and slope steppes on stony soils. Vegetation pattern changes along steep environmental gradients, but the different bedrock types (limestone and dolomite) also play an important role in structuring vegetation. Extensive dolomite areas can be found in the Transdanubian Mts, but dolomite also occurs in most of the other karst areas to a lower extent. These surfaces are characterised by a diverse geomorphology (i.e. rocks of various sizes) resulting in a wide range of microhabitats over relatively short distances. Extremely cool and warm microhabitats occur close

Fig. 2 Plant species of high conservation value in the Hungarian karst areas. **a** *Linum dolomiticum* (Buda Mts), **b** *Viola biflora* (Bükk Mts), **c** *Primula auricula* (Transdanubian Mts), **d** *Himantoglossum calcaratum* subsp. *jankae* (North Hungarian, Transdanubian, Mecsek and Villány Mts), **e** *Stachys alpina* (Mecsek Mts), **f** *Allium victorialis* (Bakony and Bükk Mts), **g** *Primula farinosa* (Balaton Upland), **h** *Dracocephalum ruyschiana* (Bükk Mts), **i** *Colchicum hungaricum* (Villány Mts), **j** *Vincetoxicum pannonicum* (Buda and Villány Mts), **k** *Clinopodium thymifolium* (Bükk Mts), **l** *Seseli leucospermum* (Transdanubian Mts). Photos **a**, **c**, **d**, **f** and **l** courtesy of Norbert Bauer, **b** and **e** courtesy of Zoltán Bátori, and **g–k** courtesy of Attila Molnár V.

to each other, where species could potentially track suitable microclimates over a wide range of glacial and postglacial climatic changes with minimal movement (Gams 1930; Zólyomi 1942; Jakucs 1955, 1961; Fekete et al. 1989; Kun et al. 2005). As a result, these heterogeneous landscapes maintain a number of relict species (Figs. 2 and 3). Similar patterns can be observed on the limestone surfaces (Jakucs 1955, 1956, 1961), where relict populations tend to be more concentrated in unique karstic microhabitats (e.g. dolines, sinkholes, limestone pavements, ravines and rocky outcrops); here, the local abiotic factors (e.g. humidity, moisture, temperature and nutrients) significantly differ from those found on the surrounding plateaus. Vegetation and karst interact with each other to some extent. For instance, climatic factors strongly influence soil thickness and nutrient content, which in turn influence vegetation patterns, while the root system of plant species contributes to the formation of karst landforms (Jakucs 1956).

Fig. 3 Natural vegetation and different types of disturbances in the Hungarian karst areas. **a** Typical karst landscape in the Pilis Mts, **b** Mosaics of shrub forests and rocky grasslands in the Transdanubian Mts, **c** Ravine forest in the Mecsek Mts with the dominance of *Allium ursinum* and *Lunaria rediviva,* **d** Early spring vegetation in a karst doline in the Mecsek Mts (the north-facing slope is covered by *Cardamine enneaphyllos*), **e** *Calamagrostis epigejos* dominated doline in the Bükk Mts, **f** Motocross tracks on dolomite slopes in the Bakony Mts, **g** Environmental legacy of past mining in the Aggtelek Karst (Esztramos Hill), **h** Intensive logging in the Mecsek Mts. Photos **a**, **b** and **f** courtesy of Norbert Bauer, **c**–**e** and **g** courtesy of Zoltán Bátori, and **h** courtesy of Imola Bóni

1.3 Biogeographic Features

The karst areas of Hungary belong to the Pannonian biogeographic region, bordering on the Alpine biogeographic region to the north and west, and the Continental biogeographic region to the east and south. In the interglacial periods (e.g. in the Riss/Würm interglacial) and especially after the last glacial period (ca. 12,000 yr BP) karst-dwelling species have migrated from the Balkans to the north (Borhidi 2011). A migration route following the eastern part and borderland of the Alps appears to have been open, providing an opportunity for many species to migrate across the White Carpathians and Fatra Mts toward the karst areas in eastern Slovakia (e.g. Slovak Karst) and the North Hungarian Mts (e.g. Aggtelek Karst and Bükk Mts) (Fig. 1). Meanwhile, species have colonised the karst areas in the Transdanubian Mts (from the Keszthely Mts to the Naszály Mts) and southern Hungary (Mecsek and Villány Mts). The eastern part of the Carpathian Basin and its vicinity (e.g. Western Romanian Carpathians) have served as another migration route for many karst-dwelling species now inhabiting the northeastern Carpathians and the North Hungarian Mts (Vojtkó 2016). Abrupt climate oscillations both in the Pleistocene and Holocene may have facilitated the migration of species with various ecological requirements from the local refugial centres (Willis et al. 2000).

The relatively high number of climate relicts indicates the alternation of colder (e.g. *Allium victorialis* in the Bükk and Bakony Mts, *Arabis alpina* in the Bükk Mts, and *Stachys alpina* in the Mecsek Mts) and warmer periods (e.g. *Carpinus orientalis* in the Vértes Mts, *Clinopodium thymifolium* in the Bükk Mts, *Colchicum hungaricum* in the Villány Mts), and the presence of endemic species (e.g. *Linum dolomiticum* in the Buda Mts, *Seseli leucospermum* in the Transdanubian Mts, and *Sesleria heufleriana* in the Aggtelek Karst and Bükk Mts) indicates the unique characteristics and isolated development of the vegetation in the different Hungarian karst areas (Fig. 2; see further examples in Sect. 5.1).

In the present chapter, we primarily focus on the vascular vegetation of the different karst areas in Hungary. Our main aim is to discuss the flora and vegetation of these karst areas in three subsections, by reviewing the relevant literature (Jávorka 1926; Máthé 1940, 1941; Zólyomi 1942, 1958; Boros 1954a, b; Jakucs 1955, 1961; Fekete 1959; Kárpáti 1960; Soó 1962; Horvát 1972; Vojtkó 1993, 1997, 1998, 2001, 2002, 2010, 2014, 2016; Kevey and Borhidi 1998, 2002; Kevey 1999, 2008; Kun et al. 2000; Somlyay and Pifkó 2002; Barina 2004; Németh 2006, 2007; Nagy 2007; Somlyay 2009; 2010, 2011; Erdős and Morschhauser, 2010; Erdős et al. 2014, 2015, 2017; Pintér et al. 2010; Bátori et al. 2012; Borhidi et al. 2012; Kevey and Csiky 2012; Bauer 2014; Bauer and Somlyay 2015; Virók et al. 2016; Varga et al. 2018), taking into account the species composition, vegetation history and location of these areas as follows: (1) North Hungarian Mts: Aggtelek Karst and Bükk Mts; (2) Transdanubian Mts: from the Keszthely Mts to the karst areas of the southwestern Cserhát Mts (e.g. Naszály Mts and Vas Hill); and (3) Mecsek and Villány Mts (Fig. 1). Nomenclature follows the 'AlgaeBase' (www.algaebase.org) and 'The Plant List' (www.theplantlist.org).

2 North Hungarian Mountains

2.1 Floristic Composition

Karst surfaces in the Bükk Mts are generally located at higher altitudes compared to the Aggtelek Karst, which is reflected both in their zonal and edaphic community structure. Montane beech forests cover large areas in the island-like Bükk Mts, while deep valleys and steep slopes provide heterogeneous environments for a number of edaphic communities. In contrast, karst surfaces in the Aggtelek Karst are geologically connected to the Slovak Karst and the distribution of many species is strongly influenced by the proximity of the Carpathians.

In this region with considerable continental climatic influences, the species composition of the south-facing and north-facing slopes is remarkably different. The warm south-facing slopes provide habitats for a number of Pannonian endemic and continental species. Many of them reach their northernmost distribution in these two karst areas and are absent from southeastern Slovakia. These endemic species are *Astragalus vesicarius* subsp. *albidus* (Aggtelek Karst), *Carduus collinus*, *Crepis pannonica*, *Dianthus giganteiformis* subsp. *pontederae* and *Thlaspi jankae* (Aggtelek Karst and Bükk Mts). Typical continental species include *Aconitum anthora*, *Adonis vernalis*, *Asyneuma canescens*, *Bromus inermis*, *Chamaecytisus albus*, *Clinopodium menthifolium*, *Jurinea mollis*, *Klasea radiata*, *Linum flavum*, *Linum hirsutum*, *Linum tenuifolium*, *Piptatherum virescens*, *Pontechium maculatum*, *Prunus fruticosa*, *Pulsatilla grandis* and *Vinca herbacea*. In contrast, many montane species occur on the north-facing slopes and valley bottoms, reaching their southernmost distribution in northern Hungary, such as the Carpathian endemic species *Aconitum moldavicum* (Bükk Mts), *Campanula xylocarpa*, *Cardamine glanduligera* (Aggtelek Karst), *Dianthus plumarius* subsp. *praecox*, *Poa pannonica*, *Sesleria heufleriana* and *Sorbus hazslinszkyana* (Aggtelek Karst and Bükk Mts).

The deep ravines, north-facing slopes and dolines in the Aggtelek Karst and Bükk Mts have several characteristics (e.g. cool microclimate) that make them important microrefugia for cool-adapted species (e.g. for Carpathian and boreal species), such as *Actaea europaea*, *Allium victorialis*, *Arabis alpina*, *Cirsium erisithales*, *Clematis alpina*, *Moehringia muscosa*, *Pleurospermum austriacum*, *Valeriana tripteris*, *Viola biflora* (Bükk Mts), *Aconitum variegatum*, *Anthriscus nitida*, *Asplenium viride*, *Calamagrostis varia*, *Primula elatior* (Aggtelek Karst and Bükk Mts). Some cool-adapted species prefer treeless habitats (e.g. rocky grasslands) and/or karst forests (e.g. dolomite beech forests). Notable examples include *Biscutella laevigata*, *Dracocephalum ruyschiana*, *Hieracium bupleuroides* subsp. *glaberrimum*, *Phyteuma orbiculare*, *Saxifraga adscendens* (Bükk Mts), *Saxifraga paniculata* and *Sesleria albicans* (Aggtelek Karst and Bükk Mts). The warm and dry south-facing slopes in the Bükk Mts are important microrefugia for some Atlantic–sub-Mediterranean and sub-Mediterranean species, such as *Asplenium haussknechtii*, *Clinopodium thymifolium* and *Cotinus coggygria*. One of the most vulnerable species of the Aggtelek Karst is *Onosma viridis*, which inhabits dry and warm grasslands and open forests.

2.2 Vegetation Characteristics

Altitudinal and climatic gradients in the North Hungarian Mts are among the most pronounced in Hungary, resulting in typical zonation of the vegetation. Extrazonal forest-steppe mosaics (grasslands and shrublands) are frequent on the river terraces and southern foothills, where agricultural intensification has generally been the lowest. Typical species include *Oxytropis pilosa*, *Prunus fruticosa*, *Prunus tenella*, *Pulsatilla grandis*, *Rosa gallica*, *Stipa dasyphylla* and *Stipa tirsa*. Turkey oak–sessile oak and oak–hornbeam forests cover large areas at higher altitudes. The turkey oak–sessile oak forest belt in the Bükk Mts occurs at an altitude between 300 and 400 m. *Quercus cerris* and *Quercus petraea* agg. are the dominant tree species in the canopy, while *Ligustrum vulgare* is frequent in the shrub layer. *Lapsana communis*, *Lathyrus niger*, *Melica uniflora*, *Poa nemoralis*, *Tanacetum corymbosum* and *Trifolium medium* are the typical herbs of this community. Most of the zonal turkey oak–sessile oak forests have been transformed by human management in the Aggtelek Karst. Typical stands can only be found far from the settlements, where they grow on south-facing slopes. *Acer campestre*, *Quercus petraea* agg. and *Sorbus torminalis* are frequent, while *Quercus cerris* is rare in the canopy. The shrub layer is mostly dense, with shrub species such as *Cornus mas* and *Crataegus monogyna*. Frequent species in the herb layer include *Clinopodium vulgare*, *Festuca heterophylla*, *Geum urbanum* and *Symphytum tuberosum*. A unique oak–hornbeam forest community covers the limestone pavements and rocky surfaces at an altitude between 500 and 600 m, which shares similarities with the scree forests. The typical species of these species-rich forests are *Erythronium dens-canis* (Aggtelek Karst), *Helleborus purpurascens* (Bükk Mts), *Carex brevicollis*, *Lamium galeobdolon* s.l., *Mercurialis perennis*, *Scilla bifolia* and *Waldsteinia geoides*. Beech forest is the main zonal vegetation type of the higher altitudes in the Bükk Mts, while it grows only on north-facing slopes and covers smaller areas in the Aggtelek Karst. These forests in the Bükk Mts contain many montane species, such as *Daphne mezereum*, *Rosa pendulina* and *Sambucus racemosa* in the shrub layer, and *Aconitum moldavicum*, *Astrantia major*, *Hesperis matronalis*, *Lathyrus transsylvanicus*, *Polygonatum verticillatum*, *Prenanthes purpurea*, *Scilla bifolia* and *Senecio ovatus* in the herb layer.

Shrub forests cover large areas on the warm and dry south-facing slopes. These low forests form mosaics with rocky grasslands and slope steppes, containing a relatively high number of sub-Mediterranean and continental species, delineating the northern border of the Pannonian biogeographic region. *Quercus pubescens*, *Cornus mas*, *Prunus mahaleb*, *Sorbus aria* s.l. and *Sorbus torminalis* are frequent in the canopy and/or shrub layers. There is a strong biogeographic gradient between the western and eastern parts of the southern Bükk Mts, which is manifested in their different floristic composition. More sub-Mediterranean species (e.g. *Cotinus coggygria*, *Medicago rigidula*, *Orlaya grandiflora* and *Vicia sparsiflora*) occur in the western part, while the eastern part provides habitats for a number of continental and/or Pontic species (e.g. *Cerastium arvense* subsp. *molle*, *Inula oculus-christi*, *Klasea radiata*, *Ranunculus illyricus* and *Seseli peucedanoides*). Slope steppes cover large areas on the south-facing slopes and are dominated by *Festuca rupicola* and *Festuca valesiaca*. Typical species in the Bükk Mts include *Adonis vernalis*, *Allium flavum*, *Allium senescens*, *Alyssum montanum*, *Bupleurum falcatum*, *Carex humilis*, *Dianthus giganteiformis* subsp. *pontederae*, *Erysimum odoratum* and *Inula ensifolia*, while *Daphne cneorum* and *Jurinea glycacantha* are rare. The proportion of this vegetation type is larger in the Aggtelek Karst than in the Bükk Mts. Typical species in the Aggtelek Karst include *Campanula sibirica*, *Carex humilis*, *Cyanus triumfettii*, *Filipendula vulgaris*, *Helianthemum ovatum*, *Peucedanum cervaria*, *Potentilla incana*, *Pulsatilla grandis*, *Teucrium chamaedrys*, *Teucrium montanum* and *Trifolium rubens*. Unique species are *Dracocephalum austriacum* and *Himantoglossum calcaratum* subsp. *jankae*. Rocky outcrops and rock clefts in the Bükk Mts serve as important habitats for diverse pioneer vegetation, where two unique species (*Asplenium haussknechtii* and *Clinopodium thymifolium*) have maintained viable populations. The treeless rocky grasslands of the north-facing slopes are dominated by *Sesleria* species (*Sesleria albicans*, *Sesleria heufleriana* and *Sesleria hungarica*). Typical species of these grasslands are *Asperula tinctoria*, *Campanula rotundifolia*, *Carex humilis*, *Dianthus plumarius*, *Draba lasiocarpa*, *Helianthemum oelandicum* subsp. *incanum*, *Ophrys insectifera*, *Phyteuma orbiculare*, *Poa badensis*, *Saxifraga paniculata*, *Scabiosa columbaria* and *Thalictrum foetidum*.

The slopes and bottoms of deep ravines and valleys in the Bükk Mts provide suitable conditions for the development of typical scree and ravine forests. Their canopy is diverse and often dominated by *Acer*, *Tilia* and *Fraxinus* species. The occurrence of *Clematis alpina*, *Daphne mezereum* and *Ribes alpinum* in the shrub layer, and *Hesperis matronalis*, *Lunaria rediviva* and *Silene dioica* in the herb layer indicates cool and humid microclimate. *Asplenium scolopendrium*, *Omphalodes scorpioides*, *Petasites albus*, *Polystichum aculeatum*, *Rubus saxatilis* and *Scopolia carniolica* frequently occur on the rocky slopes of these ravines and valleys. Riverine ash-alder forests have developed along the permanent creeks. The canopy is dominated by *Alnus glutinosa*. These forests contain some unique species such as *Valeriana dioica* subsp. *simplicifolia*, *Telekia speciosa* (Bükk Mts) and *Poa remota* (Aggtelek Karst). The flowers of *Cardamine amara* form a white carpet on the woodland floor during springtime. *Cardamine amara* may also be dominant close to the karst springs. Dolomite beech forests—and, to a lesser degree, limestone beech forests—grow on the north-facing slopes of the submontane and montane zones in the Bükk Mts. These forests contain many high montane and glacial relict species. The canopy is similar to that of zonal beech forests, with the main difference that *Taxus baccata* is typical in the lower canopy. The herb layer is especially rich in dicots and orchids and is often referred to as 'wildflower gardens'. Characteristic species include *Calamagrostis varia*, *Sesleria albicans*, *Sesleria hungarica* and *Carex humilis*. The unique distribution of *Allium victorialis*, *Cypripedium calceolus*, *Lathyrus transsylvanicus*, *Polystichum lonchitis*, *Sorbus hazslinszkyana* and a hybrid of *Sorbus aria* and *Sorbus hazslinszkyana* contributes to the conservation value of this community. Other typical species include *Aquilegia vulgaris*, *Asplenium viride*, *Cephalanthera damasonium*, *Cephalanthera longifolia*, *Cephalanthera rubra*, *Cirsium erisithales*, *Corallorhiza trifida*, *Epipactis atrorubens*, *Moehringia muscosa*, *Orthilia secunda*, *Phyteuma orbiculare*, *Polystichum aculeatum*, *Rubus saxatilis*, *Valeriana tripteris* and *Viola collina*. On steeper slopes the growth of *Fagus sylvatica* is suppressed, and *Tilia cordata*, *Tilia platyphyllos*, *Sorbus aria* s.l. and *Sorbus graeca* s.l have become dominant in the canopy. The cover of *Sesleria hungarica*, *Carex humilis* and *Calamagrostis varia* is much lower in this forest type. The herb layer is composed of a mixture of ravine forest, dolomite beech forest and rocky grassland species, such as *Actaea europaea*, *Asplenium viride*, *Carex digitata*, *Clematis alpina*, *Primula veris* subsp. *canescens*, *Saxifraga paniculata*, *Scabiosa columbaria* and *Valeriana tripteris*. Yet another scree forest type occurs on the limestone pavements, ridges, plateaus and doline rims, consisting of *Acer campestre*, *Acer platanoides*, *Acer pseudoplatanus*, *Carpinus betulus*, *Fraxinus excelsior*, *Tilia platyphyllos*, *Tilia cordata* and *Ulmus glabra* in its canopy. This is the climax vegetation of these limestone surfaces, which contains a number of rare and endangered species. *Corylus avellana*, *Cornus mas*, *Spiraea media* and *Viburnum lantana* are typical in the shrub layer. The herb layer is composed of a mixture of xeric oak forest and mesic forest species, such as *Aconitum anthora*, *Buglossoides purpurocaerulea*, *Laser trilobum*, *Primula veris*, *Lamium galeobdolon* s.l., *Mercurialis perennis*, *Scilla bifolia* and *Waldsteinia geoides*. Other typical species include *Erythronium dens-canis* (Aggtelek Karst), *Hesperis matronalis* (Bükk Mts) and *Carex brevicollis*. *Spiraea* scrubs have developed on the limestone pavements. Characteristic shrubs are *Rosa spinosissima*, *Sorbus aria* s.l. and *Spiraea media*, while *Sesleria hungarica* (Bükk Mts), *Sesleria heufleriana*, *Aconitum anthora* and *Waldsteinia geoides* are typical in the herb layer.

Nardus swards (Bükk Mts), *Arrhenatherum* hay meadows and *Festuca rubra* hay meadows can be found on the karst plateaus where forests were previously cut to convert them to pastures and hay meadows. These species-rich grasslands contain a number of montane species, such as *Lilium bulbiferum* (Bükk Mts), *Geranium sylvaticum* (Aggtelek Karst), *Alchemilla* spp., *Carlina acaulis*, *Gentiana pneumonanthe*, *Gymnadenia conopsea*, *Hypericum maculatum* and *Traunsteinera globosa*. *Nardus* swards are less species-rich during spring and early summer than in late summer. *Antennaria dioica*, *Botrychium multifidum*, *Dianthus deltoides*, *Gentianella austriaca*, *Gentianella livonica*, *Lycopodium clavatum*, *Potentilla erecta*, *Vaccinium myrtillus* and *Viola canina* subsp. *montana* are typical during summer. In the bottom of larger dolines in the Bükk Mts, where sinkholes do occur, a special tall-herb community has developed. These small patches contain several cool-adapted species, such as *Aconitum variegatum*, *Euphorbia lucida*, *Filipendula ulmaria*, *Geranium palustre*, *Iris sibirica* and *Primula elatior*.

3 Transdanubian Mountains

3.1 Floristic Composition

One of the most important endemic plant species of this karst region is *Seseli leucospermum*, which is characteristic of the rocky dolomite surfaces from the Keszthely Mts to the Naszály Mts. *Linum dolomiticum* is an endemic species restricted exclusively to the Buda Mts, while one of the two known populations of *Vincetoxicum pannonicum* worldwide also occurs in this mountain. Shrub forests and dolomite beech forests maintain the populations of many apomictic *Sorbus* taxa with a narrow distribution, such as *Sorbus semiincisa* (Buda and Pilis Mts), *Sorbus gerecseensis* (Gerecse and Vértes Mts), *Sorbus pseudolatifolia*, *Sorbus vertesensis* (Vértes Mts), *Sorbus degenii*, *Sorbus eugenii-kelleri* (eastern Bakony and Vértes Mts), *Sorbus redliana*, *Sorbus barthae*, *Sorbus tobani* (eastern Bakony Mts), *Sorbus veszpremensis* (southern Bakony Mts), *Sorbus balatonica* (Balaton Upland and Keszthely Mts), *Sorbus gayerana* (southern Bakony and Keszthely Mts), *Sorbus andreanszkyana* and *Sorbus decipientiformis* (Keszthely Mts). Some other endemic species with wider distributions occur in the eastern parts of the karst region, such as *Carduus collinus* (Buda, Gerecse, Pilis and southwestern Cserhát Mts), *Sempervivum matricum* (Gerecse, Naszály and Pilis Mts) and *Erysimum witmannii* (Naszály Mts).

Many relict species (i.e. boreal and high-montane species) have maintained viable populations in the cool habitats of dolomite outcrops, rocky grasslands and karst forests. Notable examples include *Arabidopsis lyrata* subsp. *petraea*,

Biscutella laevigata, *Calamagrostis varia*, *Carduus defloratus* subsp. *glaucus*, *Festuca amethystina*, *Moehringia muscosa*, *Phyteuma orbiculare*, *Polygala amara*, *Primula auricula*, *Sesleria sadleriana* and *Viola collina*. Beech, ravine and scree forests also provide important habitats for a number of relict species, such as *Actaea europaea* (Keszthely Mts, †—extinct), *Allium victorialis* (eastern Bakony Mts), *Festuca amethystina* (Buda and Keszthely Mts), *Moehringia muscosa* (Bakony and Vértes Mts) and *Ribes alpinum* (Bakony Mts). Other montane species of the karst region are *Corydalis intermedia*, *Aquilegia vulgaris*, *Anthriscus nitida* and *Veronica montana*.

The highest number of continental, Pontic and Pontic–Pannonian elements occurs in the eastern and southeastern parts of the karst region, where they primarily inhabit dry habitats (grasslands, shrublands and forests), and sometimes mesic forests. Although some of these species (e.g. *Artemisia austriaca*, *Astragalus vesicarius* subsp. *albidus*, *Carduus hamulosus*, *Euphorbia glareosa*, *Hypericum elegans*, *Klasea radiata*, *Phlomoides tuberosa* and *Prunus tenella*) occur in various parts of this karst region, most of them reach their western distribution in the eastern Bakony Mts and in the eastern third of the Balaton Upland. One of the most valuable steppe species of the karst region is *Klasea lycopifolia*, which has maintained viable populations on the dolomite surfaces in the eastern Bakony and Vértes Mts. *Ajuga laxmannii* occurs in some parts of the Balaton Upland and the eastern Bakony and Vértes Mts, where rocks are covered by a considerable layer of loess. The Pontic–Pannonian species *Lathyrus pallescens* only occurs in the Buda Mts. Some species with a southerly distribution occur only in the eastern parts of the karst region (e.g. *Conringia austriaca* in the Buda and Pilis Mts, and *Waldsteinia geoides* in the Buda, Pilis and southwestern Cserhát Mts). This is also true for some continental elements, such as *Aconitum anthora*, while others (e.g. *Spiraea media*) sporadically occur in the western parts of the karst region as well. Limestone surfaces in the Börzsöny Mts provide refuges for some locally rare steppe species, such as *Astragalus austriacus*.

The distribution of some sub-Atlantic–sub-Mediterranean species (e.g. *Cytisus scoparius*, *Daphne laureola*, *Dioscorea communis*, *Primula vulgaris* and *Ruscus aculeatus*) is primarily restricted to the western or central parts of the karst region (where the amount of precipitation is higher), while many sub-Mediterranean species are widespread throughout the Transdanubian Mts (e.g. *Convolvulus cantabrica*, *Chrysopogon gryllus*, *Piptatherum virescens* and *Scutellaria columnae*). The abundance of sub-Mediterranen species may be exceptionally high in the xeric grasslands and oak forests, and many of these species (e.g. *Aethionema saxatile*, *Chrysopogon gryllus*, *Cotinus coggygria*, *Coronilla coronata*, *Hornungia petraea*, *Ononis pusilla*, *Plantago argentea*, *Stipa eriocaulis* and *Sternbergia colchiciflora*) play an important role in structuring local plant communities. Some species, such as *Hippocrepis emerus* (Balaton Upland and Keszthely Mts), *Prospero autumnale* (Balaton Upland, southern and eastern Bakony Mts), *Ruscus hypoglossum* (Bakony and Keszthely Mts) have a narrower distribution. The Mediterranean (relict) species *Stipa bromoides* (Bakony Mts) and *Carpinus orientalis* (Vértes Mts) are represented with only one population, respectively.

3.2 Vegetation Characteristics

Beech forest is the main zonal vegetation type of the higher altitudes and moister areas in the Transdanubian Mts. The dominating tree species is *Fagus sylvatica*. Typical species of the herb layer are *Actaea spicata*, *Anemone ranunculoides*, *Asarum europaeum*, *Carex pilosa*, *Cardamine bulbifera*, *Corydalis cava*, *Euphorbia amygdaloides*, *Galanthus nivalis*, *Galium odoratum*, *Lamium galeobdolon* subsp. *montanum*, *Lathyrus vernus*, *Melica uniflora*, *Mercurialis perennis*, *Oxalis acetosella* and *Sanicula europaea*. *Corydalis intermedia*, *Daphne laureola* and *Scutellaria columnae* are the diagnostic species of these beech forests, and all three are absent from the flora of the other Hungarian karst areas. Some species such as *Allium ursinum*, *Galium sylvaticum*, *Knautia drymeia*, *Lathyrus venetus* and *Luzula forsteri* are more typical for the central and/or western parts, while others, such as *Helleborus purpurascens* and *Waldsteinia geoides*, for the eastern parts of the karst region. Oak–hornbeam and beech forests have similar species composition, but the canopy of the oak–hornbeam forests is composed of more tree species (e.g. *Quercus petraea* agg. and *Carpinus betulus*). Zonal oak–hornbeam forests are absent from the northwestern part of the Bakony Mts, while extrazonal oak–hornbeam forests occur in all parts of the Transdanubian Mts (e.g. in valleys). The main zonal vegetation type of the foothill region is the turkey oak–sessile oak forest, which primarily occurs on drier and warmer plateaus and slopes and has a sub-Mediterranean–sub-Atlantic character. The canopy is composed of *Acer campestre*, *Quercus petraea* agg., *Quercus cerris* and *Pyrus pyraster*. Both the shrub and herb layers are species-rich. The most characteristic herbaceous plants are *Astragalus glycyphyllos*, *Campanula persicifolia*, *Clinopodium vulgare*, *Festuca heterophylla*, *Melittis melissophyllum* subsp. *carpatica*, *Poa nemoralis*, *Primula veris*, *Silene viridiflora*, *Silene viscaria*, *Stachys officinalis* and *Tanacetum corymbosum*. The distribution of *Asphodelus albus* and *Muscari botryoides* is primarily restricted to the western part of the mountain range, while *Galium intermedium* is more abundant in the eastern part. *Veratrum nigrum* is a typical element of the warmer oak forests across the Transdanubian Mts.

Sub-Mediterranean type pubescent oak forests occupy large areas on the warm and dry south-facing slopes. These species-rich forests form a vegetation zone in some parts of the mountain range, such as the Balaton Upland and the Veszprém–Várpalota plateau. On the south-facing slopes, pubescent oak forests often form a mosaic with shrub forests and rocky grasslands and provide refuges for a number of warm-adapted species, such as *Anacamptis pyramidalis*, *Buglossoides purpurocaerulea*, *Cornus mas*, *Cotinus coggygria*, *Dictamnus albus*, *Fraxinus ornus*, *Geranium sanguineum*, *Laser trilobum*, *Limodorum abortivum*, *Piptatherum virescens*, *Polygonatum odoratum*, *Silene nutans* and *Viburnum lantana*. One of the main characteristic species of these forests in the Keszthely Mts and the Balaton Upland is *Hippocrepis emerus*. *Mercurialis ovata* has a predominantly southern distribution, while *Lathyrus lacteus* and *Vicia sparsiflora* have a predominantly eastern distribution within the mountain range. *Sternbergia colchiciflora* and *Vinca herbacea* are relatively common in the pubescent oak forests and dry lowland oak forests.

Dry grasslands occupy large areas on the limestone and dolomite areas in the Transdanubian Mts and contain many rare and endangered species. Dominant graminoids in the slope steppes are *Carex humilis*, *Chrysopogon gryllus*, *Festuca rupicola*, *Festuca valesiaca*, *Stipa capillata*, *Stipa pennata* and *Stipa pulcherrima*, while *Bromus pannonicus*, *Carex humilis*, *Festuca pallens*, *Stipa eriocaulis* and *Stipa pulcherrima* dominate the rocky grasslands. Typical species include *Fumana procumbens*, *Hornungia petraea*, *Koeleria macrantha* agg., *Phleum phleoides*, *Teucrium montanum*, *Teucrium chamaedrys* and *Thymus praecox*. *Seseli leucospermum* is an endemic species of these rocky grasslands. The distribution of *Arabidopsis lyrata* subsp. *petraea* and *Leontodon incanus* is restricted to the western part of the mountain range, while *Allium moschatum*, *Artemisia alba*, *Carduus collinus*, *Helianthemum oelandicum* subsp. *incanum*, *Lappula heteracantha*, *Plantago argentea* and *Sempervivum matricum* occur primarily in the central and/or eastern parts.

Scree forests have developed on the periglacial debris of steep northern slopes and outcrops. The canopy is composed of *Acer*, *Fraxinus* and *Tilia* trees, while *Corylus avellana*, *Sambucus nigra* and *Staphylea pinnata* are typical in the shrub layer. Characteristic species in the herb layer include *Aconitum vulparia*, *Cardamine enneaphyllos*, *Geranium lucidum*, *Lilium martagon*, *Mercurialis perennis*, *Omphalodes scorpioides* and *Parietaria officinalis*, but they share many other species with the adjacent zonal forests. *Waldsteinia geoides* is especially abundant in the Buda and Pilis Mts. The habitat structure, microclimate and species composition of the stands on limestone and dolomite differ from each other, due to the different denudation and erosion processes. As a result, scree forests on dolomite contain more xeric species, such as *Cornus mas*, *Primula veris* and *Tanacetum corymbosum*. Small scree forest patches also occur in the dolines of the Bakony Mts. Ravine forests occur in the lower third of larger valleys, where the microclimate is cooler and more humid than in the scree forests. *Acer pseudoplatanus* and *Fraxinus excelsior* are typical in the canopy, and the shrub and herb layers contain many beech and scree forest species. Characteristic species include *Asplenium scolopendrium*, *Geranium phaeum*, *Lunaria rediviva*, *Polystichum aculeatum*, *Ranunculus lanuginosus* and *Silene dioica*. Other typical ravine forest (i.e. cool-adapted) species are *Anthriscus nitida* (Bakony Mts), *Festuca altissima* (Bakony, Gerecse, Naszály and Pilis Mts) and *Scrophularia vernalis* (Bakony, Gerecse, Pilis and Vértes Mts). Riverine ash-alder forests usually cover small areas along the fast-flowing permanent creeks in the Transdanubian Mts. The most diverse stands grow in the Bakony Mts. The dominant herbaceous plants are *Aegopodium podagraria*, *Caltha palustris*, *Carex remota*, *Chrysosplenium alternifolium* and *Circaea lutetiana*, but *Cirsium oleraceum*, *Petasites hybridus* and *Scirpus sylvaticus* are also frequent. Some montane species (e.g. *Aruncus dioicus* and *Leucojum vernum*) also occur in these forests. *Cardamine amara*, *Crepis paludosa* and *Veratrum album* are typical close to the karst springs. Dolomite beech forests primarily occur on the north-facing ridges, edges of the plateaus and convex slopes. These forests represent a unique combination of microhabitats with widely different environmental conditions (e.g. dry microhabitats and cool microhabitats occur close to each other). *Fagus sylvatica* is the dominant tree species in the submontane zone, while many tree species typical for xeric (e.g. *Fraxinus ornus*) and mesic rocky forests (e.g. *Tilia cordata*) can be found in the canopy at lower altitudes. Apomictic *Sorbus* taxa with narrow distributions are also typical in this forest type. Many montane, sub-alpine and relict species have maintained viable populations in the cool microhabitats. Some of them (e.g. *Carex alba*, *Calamagrostis varia*, *Laserpitium latifolium*, *Phyteuma orbiculare*, *Polygala amara* and *Viola collina*) are frequent species, while others (e.g. *Allium victorialis*, *Bupleurum longifolium*, *Carduus defloratus* subsp. *glaucus*, *Moehringia muscosa*, *Primula auricula*, *Ranunculus serpens* subsp. *nemorosus*, *Rubus saxatilis*, *Taxus baccata* and *Valeriana tripteris*) occur sporadically. On the ridges of the Bakony Mts and the Balaton Upland (in the contact zone of the mesic communities of the north-facing and the xeric vegetation of the south-facing slopes), top-forests have developed. The canopy is dominated by *Fraxinus ornus* and *Quercus pubescens*, while *Cornus mas* is the most abundant shrub. The herb layer displays a strange mixture of mesic (e.g. *Allium ursinum*, *Anemone ranunculoides* and *Corydalis cava*) and xeric species (e.g. *Mercurialis ovata*, *Veratrum nigrum* and

Vincetoxicum hirundinaria). Tiny and fragmented stands of *Spiraea* scrubs grow in the Gerecse, Naszály and Pilis Mts.

In the Bakony Mts (Hárskút and Tés Plateaus) and sporadically in the Pilis Mts (above 450 m asl), where forests were previously cut, different types of hay meadows have developed. Today, many of these meadows have been abandoned and became overgrown by shrub and tree species. The dominant grasses are *Agrostis capillaris*, *Anthoxanthum odoratum*, *Helictotrichon pubescens* and *Trisetum flavescens*. Characteristic species include *Gentianopsis ciliata*, *Dianthus deltoides*, *Polygala vulgaris* and *Viola canina*. Alkaline fens and fen meadows occur at the foot of the Transdanubian Mts, containing many boreal, montane relict and fen species such as *Allium suaveolens*, *Carex hartmanii*, *Carex lasiocarpa*, *Crepis mollis* subsp. *succisifolia*, *Gladiolus palustris*, *Lathyrus pannonicus*, *Primula farinosa*, *Senecio umbrosus* and *Trollius europaeus*.

4 Mecsek and Villány Mountains

4.1 Floristic Composition

In this region with considerable sub-Mediterranean climatic influences, the difference between the flora of the south-facing and north-facing slopes is conspicuous. The warm and dry south-facing slopes are rich in sub-Mediterranean, Illyrian, Pannonian-Balkanic, Balkanic and Mediterranean elements. Species of the xeric forests and xeric grasslands include *Asplenium javorkaeanum*, *Convolvulus cantabrica*, *Festuca dalmatica*, *Doronicum hungaricum*, *Himantoglossum calcaratum* subsp. *jankae*, *Inula spiraeifolia*, *Ophrys scolopax*, *Orchis simia*, *Plantago argentea* and *Sedum acre*. The populations of *Colchicum hungaricum*, *Sempervivum tectorum* and *Trigonella gladiata* in the Villány Mts (they do not occur elsewhere in Hungary) are considered to be relicts from warmer periods. In the Villány Mts, many other species with a southern distribution occur, such as *Medicago orbicularis* and *Pisum sativum* subsp. *elatius*. An especially valuable endemic species of the Villány Mts is *Vincetoxicum pannonicum*. Rare plant species of the xeric forest–grassland mosaics in the Mecsek Mts include *Cytisus austriacus* subsp. *heuffelii*, *Coronilla coronata* and *Klasea lycopifolia*.

In the flora of top-forests that occupy the mountain tops and ridges, there is a mixture of mesic and dry forest species. The mesic species of these top-forests are, among others, *Asperula taurina*, *Doronicum orientale*, *Helleborus odorus*, *Lonicera caprifolium*, *Ruscus hypoglossum*, *Knautia drymeia*, *Lathyrus venetus*, *Limodorum abortivum*, *Polystichum setiferum* and *Primula vulgaris*, as well as *Waldsteinia geoides* (Mecsek Mts) and *Orobanche nana* (Villány Mts). In spring, *Allium ursinum* is abundant, *Anemone ranunculoides*, *Cardamine bulbifera* and *Galanthus nivalis* are frequent in all mesic forest types. *Cardamine enneaphyllos* only reaches higher cover values on the coolest slopes (e.g. in ravines, valleys and dolines). *Aremonia agrimonoides* and *Chaerophyllum aureum* are rarities of the beech forests in the Mecsek Mts, while *Stachys alpina* is a montane species occurring mainly in cool and rocky forests. In the bottom of deeper valleys and larger dolines, there is a special vegetation type similar to springs, with occurrences of *Chrysosplenium alternifolium*, *Cardamine flexuosa* and *Carex strigosa*. The bottoms and north-facing slopes of ravines, valleys and dolines harbour numerous species associated with cool and moist conditions. Notable examples include *Actaea spicata*, *Aconitum vulparia*, *Asplenium scolopendrium*, *Daphne mezereum*, *Dryopteris affinis*, *Dryopteris dilatata*, *Dryopteris expansa* and *Polystichum aculeatum*. *Aruncus dioicus*, *Lunaria rediviva* and *Silene dioica* are restricted to ravines and deep valleys.

4.2 Vegetation Characteristics

The Mecsek and Villány Mts have a considerable diversity of plant communities. Oak–hornbeam forests cover the largest areas in the two mountains. They can be found in different exposures in the higher parts of the Mecsek Mts, while they are confined to north-facing slopes in the Villány Mts. The canopy is dominated by *Quercus petraea* agg. and *Carpinus betulus*, but *Tilia tomentosa* also plays an important role. This latter even becomes dominant in some parts of the Villány Mts. *Staphylea pinnata* and *Ligustrum vulgare* are frequent in the shrub layer. *Ruscus aculeatus* is typical, in places even abundant. Typical species of the herb layer are *Allium ursinum*, *Anemone ranunculoides*, *Cardamine bulbifera*, *Carex pilosa*, *Galanthus nivalis*, *Galium odoratum*, *Helleborus odorus* and *Melica uniflora*, while *Asperula taurina* is also relatively widespread. Submontane beech forests can be found in higher altitudes in the Mecsek Mts as well as on the north-facing slopes and bottom sections of the deepest valleys and dolines. In the Villány Mts, beech forests form only small fragmentary stands, mainly in valley bottoms and at the foothills of the north-facing slopes. The dominating tree species is *Fagus sylvatica*. The shrub layer is sparse or absent. In addition to *Ruscus aculeatus*, *Ruscus hypoglossum* is also typical. The typical species of the herb layer are similar to those of the oak-hornbeam forests. Turkey oak–sessile oak forests are found mainly near mountain tops and ridges, occasionally also at the foot of the mountains. The main tree species are *Quercus petraea* agg., *Quercus cerris* and *Tilia tomentosa*. The shrub layer is mostly dense, with shrub species such as *Cornus mas*, *Crataegus monogyna* and *Ligustrum vulgare*. *Ruscus aculeatus* is also frequent. The most important species of the herb layer are, among others, *Brachypodium pinnatum*, *Brachypodium sylvaticum*, *Buglossoides purpurocaerulea*, *Convallaria majalis*, *Doronicum orientale*, *Helleborus odorus*, *Melica uniflora* and *Potentilla micrantha*.

On the south-facing slopes of the Villány Mts and in similar exposures in the karst areas of the Mecsek Mts, there is a mosaic of xeric forests and grasslands. The small patches of shrub forests are formed by short individuals of *Quercus pubescens* and *Fraxinus ornus*. The most typical species of the shrub layer are *Euonymus verrucosus*, *Ligustrum vulgare* and *Viburnum lantana*. In striking contrast to the shrub forests of the Transdanubian and North Hungarian Mts, *Cotinus coggygria* is very rare in the Mecsek Mts and completely absent from the Villány Mts. *Ruscus aculeatus* is quite frequent. Widespread species of the herb layer are *Dictamnus albus*, *Galium lucidum*, *Geranium sanguineum*, *Inula spiraeifolia*, *Iris variegata*, *Polygonatum odoratum*, *Sedum maximum* and *Tanacetum corymbosum*. Where the soil is a bit deeper, pubescent oak forests have developed. Their canopy, dominated by *Fraxinus ornus*, *Quercus pubescens* and *Tilia tomentosa*, is considerably higher than that of the shrub forests. Their shrub layer is mostly dense, with species such as *Cornus mas*, *Crataegus monogyna*, *Ligustrum vulgare* and *Viburnum lantana*. *Ruscus aculeatus* is frequent and the lianas *Dioscorea communis* and *Lonicera caprifolium* may be abundant. The most frequent species of the herb layer are, among others, *Brachypodium sylvaticum*, *Buglossoides purpurocaerulea*, *Carex flacca*, *Melica uniflora* and *Viola alba*.

Open rocky grasslands are typical plant communities on the south-facing slopes in the Villány Mts, however, they only cover small areas in the Mecsek Mts. Bare rocks cover up to 50% of these grasslands. The dominant species is *Festuca dalmatica*. Frequent species include *Allium flavum*, *Artemisia alba*, *Asplenium javorkaeanum*, *Helianthemum ovatum*, *Linaria genistifolia*, *Melica ciliata*, *Orlaya grandiflora*, *Potentilla incana*, *Ranunculus illyricus*, *Stachys recta* and *Teucrium chamaedrys*. Species with southerly distributions as well as Mediterranean and sub-Mediterranean species are frequent and usually reach high cover values, such as *Colchicum hungaricum*, *Pisum sativum* subsp. *elatius*, *Ranunculus psilostachys* (Villány Mts), *Convolvulus cantabrica*, *Fumana procumbens*, *Lathyrus sphaericus* and *Plantago argentea*. Early spring ephemerals are also typical (e.g. *Arabis nova*, *Arenaria serpyllifolia*, *Cerastium brachypetalum*, *Saxifraga tridactylites* and *Thlaspi perfoliatum*). Closed rocky grasslands in the Villány Mts cover smaller areas than open rocky grasslands. These grasslands may be dominated by *Festuca dalmatica*, *Festuca rupicola*, *Festuca valesiaca*, *Carex humilis*, *Chrysopogon gryllus* or *Bromus pannonicus*. Other typical species include *Allium montanum*, *Convolvulus cantabrica*, *Inula spiraeifolia*, *Iris variegata*, *Jurinea mollis*, *Klasea radiata* and *Plantago argentea*, as well as some orchids (e.g. *Orchis simia* and *Anacamptis pyramidalis*). Rocky and slope steppes cover larger areas in the Mecsek Mts and on the southern foot of the Villány Mts (where rocks are covered by a considerable layer of loess). The dominant grasses are *Cleistogenes serotina*, *Festuca rupicola*, *Festuca valesiaca* and *Stipa pulcherrima*, but *Elymus hispidus* or *Bothriochloa ischaemum* may also be abundant. Other typical species are *Adonis vernalis*, *Allium sphaerocephalon*, *Campanula sibirica*, *Dianthus giganteiformis*, *Filipendula vulgaris*, *Galium lucidum*, *Jurinea mollis*, *Melica ciliata*, *Orlaya grandiflora*, *Sanguisorba minor*, *Teucrium chamaedrys* and *Verbascum phoeniceum*, as well as some orchids, including *Himantoglossum calcaratum* subsp. *jankae*, *Neotinea tridentata* and *Orchis simia*.

Scree forests occupy areas with shallow and stony soils on the north-facing parts of mountain tops and ridges. In the Villány Mts, only small and not really typical stands are present. The dominant species of the canopy are *Fraxinus ornus* and *Tilia tomentosa*, but *Tilia platyphyllos* is also typical. Characteristic species of the herb layer include *Waldsteinia geoides* (Mecsek Mts), *Allium ursinum*, *Aconitum vulparia*, *Doronicum orientale*, *Lilium martagon* and *Mercurialis perennis*. Ravine forests or their fragments have formed in the ravines, valleys and the bottoms of larger dolines. Typical species of the canopy layer are *Acer platanoides*, *Acer pseudoplatanus* (although in the Villány Mts this species is rare and only occurs as saplings), *Fagus sylvatica*, *Fraxinus excelsior* and *Ulmus glabra*. *Sambucus nigra* is frequent in the shrub layer. Characteristic species in the herb layer include *Dryopteris affinis* (Mecsek Mts), *Actaea spicata*, *Asplenium scolopendrium*, *Cystopteris fragilis*, *Daphne mezereum*, *Lunaria rediviva*, *Polystichum aculeatum*, *Polystichum setiferum* and *Scutellaria altissima*. Riverine ash-alder forests grow along the fast-flowing creeks in the Mecsek Mts. The main component of the canopy is *Alnus glutinosa*, while *Aegopodium podagraria*, *Allium ursinum*, *Carex pendula*, *Carex strigosa*, *Chrysosplenium alternifolium*, *Cornus sanguinea* and *Frangula alnus* are frequent in the understorey. *Cirsium oleraceum*, *Equisetum telmateia*, *Mentha aquatica* and *Scrophularia umbrosa* are somewhat rarer but nevertheless typical. On the mountain tops and ridges of the Mecsek and Villány Mts, top-forests have developed. The canopy is formed by *Fraxinus ornus*, *Quercus pubescens* and *Tilia tomentosa*. *Cornus mas* is the most abundant shrub. The herb layer contains a mixture of mesic (e.g. *Adoxa moschatellina*, *Chaerophyllum aureum*, *Corydalis cava*, *Galanthus nivalis* and *Scilla vindobonensis*) and xeric species (e.g. *Aconitum anthora*, *Doronicum hungaricum* and *Mercurialis ovata*). On the northern sides of ridges, tiny stands of *Spiraea* scrubs can be found, where *Spiraea media* is the dominant species, forming 1–1.5 m high thickets. *Waldsteinia geoides* is a typical species of the community in the Mecsek Mts, while the species is absent from the Villány Mts. The herb layer of the *Spiraea* scrub is largely restricted to its edges and is formed by species of the neighbouring forests and grasslands.

Arrhenatherum hay meadows can be found in areas where forests were previously cut to convert them to pastures and hay meadows. They only cover larger areas in the Mecsek Mts. These grasslands may be dominated by *Arrhenatherum elatius*, *Briza media*, *Dactylis glomerata*, *Holcus lanatus*, *Ranunculus acris* and *Stellaria graminea*. The number of open water habitats (e.g. small lakes and water-filled ruts) is low both in the Mecsek and Villány Mts, with species such as *Lemna minor*, *Lemna trisulca*, *Lythrum hyssopifolia* and *Lythrum portula*.

5 Biodiversity Features and Conservation Status

5.1 Biogeographic Importance and Biodiversity Hotspots

Karst systems are topographically complex environments supporting unique microclimates in several microhabitats, such as alkaline fens, local depressions, limestone pavements, ravines, rocky outcrops, screes, steep slopes and valleys (Zólyomi 1942; Jakucs 1955; Vojtkó 2001; Bauer et al. 2008; Bátori et al. 2017). These microhabitats have the capacity to harbour 'trailing edge' or 'stable' climate relicts (cf. Kiedrzyński et al. 2017), indicating the alternation of colder (C) and warmer (W) periods in the Carpathian Basin during the Quaternary (Gáyer 1924; Fekete 1964; Lájer 1998; Farkas 1999; Kevey 1999; Vojtkó 2001; Barina 2006; Somlyay 2009; Bauer and Somlyay 2015; Virók et al. 2016). Notable examples include *Allium victorialis*, *Arabis alpina*, *Asplenium viride*, *Calamagrostis varia*, *Carduus defloratus* subsp. *glaucus*, *Carex hartmanii*, *Crepis mollis* subsp. *succisifolia*, *Dracocephalum ruyschiana*, *Moehringia muscosa*, *Primula auricula*, *Primula farinosa*, *Ranunculus serpens* subsp. *nemorosus*, *Senecio umbrosus*, *Stachys alpina*, *Taxus baccata* and *Viola biflora* (C), and *Clinopodium thymifolium*, *Convolvulus cantabrica*, *Carpinus orientalis*, *Colchicum hungaricum*, *Onosma viridis*, *Orobanche nana*, *Paronychia cephalotes*, *Silene flavescens* and *Trigonella gladiata* (W) (Fig. 2). The stable microclimate of cave entrances supports the persistence of rare bryophytes (e.g. *Amblystegium jungermannioides*, *Orthothecium intricatum* and *Timmia megapolitana* subsp. *bavarica*) (cf. Boros 1971; Buczkó and Rajczy 1989), while karst springs and streams are important habitats for algae (e.g. *Chara vulgaris*) and rare hydrophytic plants (e.g. *Nasturtium officinale* and *Potamogeton coloratus*) (Boros 1937; Bauer 2006). The island-like dolomite and limestone areas also provide habitats for a number of endemic and subendemic taxa, such as *Dianthus plumarius* agg., *Erysimum witmannii*, *Ferula sadleriana*, *Linum dolomiticum*, *Seseli leucospermum*, *Thalictrum minus* subsp. *pseudominus* and *Vincetoxicum pannonicum*, and for many other plants of high conservation value (e.g. *Astragalus vesicarius*, *Biscutella laevigata*, *Cypripedium calceolus*, *Himantoglossum adriaticum*, *Himantoglossum calcaratum* subsp. *jankae*, *Leontodon incanus*, *Phyteuma orbiculare*, *Poa badensis* and *Sesleria albicans*) (Király 2009; Molnár et al. 2012). In addition, some Illyrian, Mediterranean and sub-Mediterranean elements (e.g. *Colchicum hungaricum*, *Pisum sativum* subsp. *elatius* and *Trigonella gladiata*) reach their northernmost distribution in the karst areas of Hungary. Almost all of these plants are classified as red-listed species in Hungary (Király 2007), and some of them are listed on the IUCN Red List (e.g. *Himantoglossum adriaticum*, *Ferula sadleriana*, *Linum dolomiticum*, *Seseli leucospermum* and *Vincetoxicum pannonicum*) (Fig. 2).

Several karst systems in Hungary have been recognised as important biodiversity hotspots. Vegetation structure, plant species richness, floristic composition and functional traits may vary along steep environmental gradients (e.g. humidity, moisture, nutrient and temperature), resulting in a high diversity of grassland, shrub and forest communities over relatively short distances (see Sects. 2–4; Fig. 3). For instance, Erdős et al. (2012) investigated the plant species composition along 200-m-long transects that were established perpendicular to the ridges of the Villány Mts, and found that mesic forest communities (oak–hornbeam forests and scree forests) of the north-facing slopes are replaced by dry shrub forests and closed rocky grasslands near the ridge, while mosaics of shrub forests and open rocky grasslands occur on the south-facing slopes. These dry grasslands are remarkably species-rich, containing on average 39 plant species per 2 m^2 (Erdős et al. 2011). Extremely high fine-scale plant species richness was observed in the dry grasslands and mountain hay meadows of the dolines in the Aggtelek Karst and Bükk Mts as well, where the most species-rich 1 m^2 vegetation plots contained 31 and 38 plant species, respectively (Bátori et al. 2017). Previous investigations in the Bakony region indicated that karst habitats may harbour more plant species than adjacent non-karstic habitats (e.g. igneous rocks) (Bauer 2014).

Patterns of species richness and composition in forests on karst substrate are strongly influenced by variations in canopy cover and soil properties. If the canopy is relatively open and the soil is dry, a high number of xeric plant species may survive in the understory, while denser canopies and moister soils support the growth of plants adapted to cooler and moister conditions (Jakucs 1961; Kevey 2008; Borhidi et al. 2012). In some cases, however, karst systems represent a unique combination of plant species (i.e. species adapted to widely different environmental conditions), forming particular communities. For instance, limestone/dolomite beech forests, scree forests and top-forests provide key habitats not only for cool-adapted plants (e.g. glacial relicts), but also for plants associated with warm and/or dry conditions (Zólyomi 1987; Kevey and Borhidi 1998).

5.2 Capacity of Karst Dolines as Microrefugia

Dolines, small to large-sized bowl-shaped depressions, are the most typical landforms in limestone karst surfaces. They have several characteristics that make them primary candidates to serve as microrefugia. Microrefugia are small areas with favourable environmental conditions that enable the persistence of viable populations of species during regional climate changes (Keppel et al. 2012). The high geomorphological diversity associated with dolines is reflected in the high variation in microclimate, soil characteristics and microhabitat diversity (Bacsó and Zólyomi 1934; Jakucs 1971; Bárány-Kevei 1999; Bátori et al. 2012, 2019). Thermal inversion maintains cooler and moister conditions on the bottom and north-facing slopes of dolines, while the south-facing slopes receive much more insolation and therefore provide warmer microclimates. As a result, more microhabitats and cool-adapted species can usually be found within dolines than in their surroundings (Fig. 3). For instance, dolines in the Aggtelek Karst (between 400 and 600 m asl) maintain beech forest, ravine forest and scree forest patches in the 'ocean' of oak–hornbeam forests and provide important microhabitats for the populations of montane and ravine forest species such as *Geranium sylvaticum* and *Lunaria rediviva* (Vojtkó et al. 2018). Similar patterns can also be observed in the Mecsek Mts (between 250 and 500 m asl) (Bátori et al. 2012) and the Bakony Mts (between 400 and 450 m asl), where beech forests and ravine forests in larger dolines have an important role in the maintenance of montane species (e.g. *Aconitum vulparia* and *Dryopteris affinis* in the Mecsek Mts, and *Anthriscus nitida* and *Festuca altissima* in the Bakony Mts). The south-facing slopes of the non-forested dolines in the Aggtelek Karst (between 400 and 600 m asl) and the Bükk Mts (between 700 and 900 m asl) are covered by dry and/or semi-dry grasslands, while the north-facing slopes and bottoms are dominated by mountain hay meadows, wet meadows or tall-herb communities (Vojtkó 2001; Virók et al. 2016). These non-forested dolines have the capacity to support the populations of montane and/or relict species, such as *Primula elatior* and *Rubus saxatilis* in the Aggtelek Karst, and *Bupleurum longifolium* and *Dracocephalum ruyschiana* in the Bükk Mts. Dolines may also provide refuges for rare bryophytes (e.g. *Calliergon cordifolium* in the Mecsek Mts) (Csiky et al. 2014). Recent investigations revealed that not only the number of plant communities and cool-adapted species but also the total number of plant species and plant families (i.e. taxonomic diversity) may be significantly higher in dolines than on the surrounding plateaus (Fig. 4).

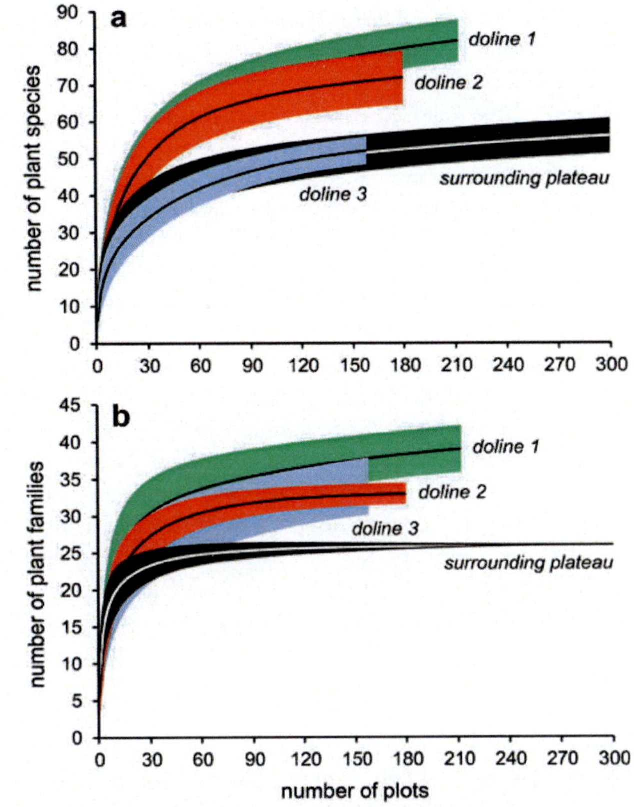

Fig. 4 **a** Plant species accumulation curves for three large dolines and the surrounding plateau in the Mecsek Mts. **b** Plant family accumulation curves for the same sites. Transects were established with north to south orientation across each doline and consisted of 1 m × 1 m contiguous plots. Similar sampling design was used for the plateau. Data were collected by Zoltán Bátori. Analysis was performed using the Estimates software (Colwell 2004)

5.3 Karst-Related Protected Areas

Out of the ten national parks established in Hungary between 1973 and 2002, five (Aggtelek National Park, Balaton Upland National Park, Bükk National Park, Danube-Dráva National Park and Danube-Ipoly National Park) play an important role in the protection of karst ecosystems. The most significant values of the Bükk National Park (1977) and the Aggtelek National Park (1985) (northern Hungary) are the special surface formations (e.g. dolines, sinkholes and limestone pavements) and caves, and the unique habitats, flora and fauna associated with these karst systems. The extensive limestone and dolomite karst areas in the Transdanubian Mts (central Hungary and central Transdanubia) are managed by the Balaton Upland National Park (1997) and the Danube-Ipoly National Park (1997), while the Danube-Dráva National Park is responsible for the management of the karst areas in southern Transdanubia. The system of national protected areas largely overlaps with the network of the nature protection areas of the European Union (Natura 2000) (www.termeszetvedelem.hu).

5.4 Threatening Factors

Karst systems in Hungary have a long history of human presence. Agriculture, mining, lime burning, cultivation, construction, grazing, forestry and tourism have been shaping the vegetation for centuries, although there are regional differences concerning the types and levels of disturbances (Fig. 3; Table 1). Extensive dolomite and limestone habitats have been destroyed or severely damaged by mining (e.g. Bél-kő, Esztramos, Malom and Szársomlyó Hills, and Naszály Mts), resulting in changes in the population numbers of endangered species. For instance, one population of *Orobanche nana* and one population of *Vincetoxicum pannonicum* have become extinct in the Villány Mts due to limestone mining. Water level fluctuations in karst aquifers —associated with bauxite mining—have contributed to the extinction of rare plant species, such as *Pinguicula alpina* (Király 2007). Open and closed rocky grasslands, slope steppes and mosaics of dry grasslands and shrublands are degraded by game pressure (e.g. mouflon and wild boar), heavy trampling (e.g. tourism and motorcycling), flower picking (e.g. *Cypripedium calceolus* and *Pulsatilla grandis*), littering (e.g. car accumulators and plastic bags), expansion of settlements (e.g. on the hills in the northern shore of Lake Balaton) and plant invasion (e.g. *Ailanthus altissima* and *Robinia pseudoacacia*) (Borhidi and Sánta 1999; Bölöni et al. 2011). The abandonment of traditional land use (e.g. grazing and mowing) also has important ecological consequences. For instance, Erdős et al. (2013) found that shrub forests in the Villány Mts began to expand rapidly at the expense of grasslands after grazing had come to an end during the 1970's.

Forestry activity may greatly affect stand structural characteristics and therefore the diversity of habitats. A large proportion of the semi-dry and mesic forests of the Hungarian karst areas have been converted into plantations of non-native trees (e.g. *Picea abies*, *Pinus nigra* and *Quercus rubra*), decreasing the diversity of the herb and shrub layers and increasing the possibility of extreme fire events (Borhidi and Sánta 1999; Cseresnyés et al. 2006). Intensive logging may have a great direct or indirect influence on species composition through its effects on climatic conditions, light availability and population interactions (Bátori et al. 2020). For instance, Kiss et al. (2020) found that the cover and number of oak forest species and disturbance-tolerant species increased, while the cover and number of beech and ravine forest species decreased within the dolines of the Mecsek Mts 10 years after logging.

In recent decades, many karstic caves have experienced intensified tourist traffic. As a result of artificial illumination, special communities of cyanobacteria, algae, mosses and ferns called lampenflora (i.e. lamp flora) develop in close and remote proximity to lights. Such communities are not only unsightly in appearance, but also deteriorate the substrata on which they grow (Mulec 2019; Rajczy 2019).

Increasing temperatures and longer drought periods have been detected for Hungary, and this trend will likely

Table 1 Types and levels (low: *, moderately high: **, high: ***, and not typical: –) of disturbances in the karst areas of Hungary, which have a key role in shaping vegetation patterns. Bükk Mts (1), Aggtelek Karst (2), southwestern Cserhát Mts (3), southern Börzsöny Mts (4), Pilis Mts (5), Buda Mts (6), Gerecse Mts (7), Vértes Mts (8), Bakony Mts (9), Balaton Upland (10), Keszthely Mts (11), Mecsek (12) and Villány Mts (13)

Disturbance	Karst areas in Hungary												
Types of disturbance	1	2	3	4	5	6	7	8	9	10	11	12	13
Abandonment of grazing/mowing	*	*	–	*	–	*	–	**	**	**	*	*	*
Adventive species	*	*	*	*	*	**	*	*	*	**	***	*	***
Agriculture	–	–	*	*	–	*	*	*	**	*	–	–	**
Expansion of settlements	*	–	**	**	*	**	*	*	*	**	**	*	**
Forestry	***	***	***	*	*	*	**	**	***	*	***	**	**
Game pressure	***	*	*	*	**	*	**	***	**	**	**	**	*
Human-induced fire	*	*	–	–	–	–	–	–	*	*	*	–	*
Illegal waste deposit	–	–	*	*	*	**	*	*	*	**	*	*	*
Mining	*	*	***	**	*	*	***	***	***	*	**	*	***
Tourism	**	*	*	**	**	**	*	**	**	**	**	**	**

continue in coming decades (Bartholy et al. 2008). As a result, the main vegetation zones will undergo range shifts and many cool-adapted species are likely to disappear from lower elevations (Bátori et al. 2017). The capacity of current and potential karstic microrefugia (e.g. dolines, north-facing rocky outcrops and ravines) to buffer against the negative effects of climate change on vegetation is likely to be exceeded if the current management practices (e.g. intensive logging) will remain unchanged.

6 Conclusions

The flora and vegetation of the Hungarian karst areas show considerable differences from north to south and east to west. The main reason for these patterns relates to the climatic variability, geology, geomorphology and vegetation history of these areas. The relatively high number of climate relicts indicates the alternation of colder and warmer periods in the Quaternary, and the presence of endemic species indicates the unique characteristics and isolated development of the vegetation. Several karst systems in Hungary have been recognised as important biodiversity hotspots. Vegetation pattern changes along steep temperature and moisture gradients, but the different bedrock types (limestone and dolomite) also play a crucial role in structuring vegetation. Turkey oak–sessile oak forest is the main zonal vegetation type of the foothills, oak–hornbeam and beech forests dominate the submontane regions, while beech forest is the main zonal vegetation type of the montane regions. The complex topography of these karst areas contributes to the maintenance of various edaphic communities, such as ravine forests, scree forests, shrub forests and rocky grasslands. Unfortunately, extensive dolomite and limestone forest and grassland areas have been destroyed or severely damaged in Hungary by various human activities, such as agriculture, mining, lime burning, cultivation, construction, grazing, forestry and tourism. Currently, climate change is also considered to be a serious threat to the vegetation of these areas. Therefore, there is an urgent need to assess the combined effects of different human activities and climate change, providing solutions for planning and management with the aim of minimising the negative effects of rising temperatures on karst ecosystems.

Acknowledgements The prepara\on of this book chapter was supported by the NKFIH K 124796 grant. The authors would like to acknowledge Prof. Acla Molnár V. and Imola Bóni for providing photos of the respec\ve plants and habitats presented in this chapter. Zoltán Bátori was supported by the János Bolyai Research Scholarship of the Hungarian Academy of Sciences.

References

Bacsó N, Zólyomi B (1934) Mikroklíma és növényzet a Bükkfennsíkon (Microclimate and vegetation on the Bükk plateau). Az Időjárás 38:177–196 [in Hungarian]

Bárány-Kevei I (1999) Microclimate of karstic dolines. Acta Climatologica Universitatis Szegediensis 32–33:19–27

Barina Z (2004) A Dunántúli-középhegység növényföldrajzának főbb jellemzői (Characteristics of the phytogeography of Transdanubian Mountains in Hungary). Flora Pannonica 2:37–55 [in Hungarian]

Barina Z (2006) A Gerecse hegység flórájának katalógusa (Flora of the Gerecse Mountains). Magyar Természettudományi Múzeum és a Duna–Ipoly Nemzeti Park Igazgatóság, Budapest, p 612 [in Hungarian]

Bartholy J, Pongrácz R, Gelybó G, Szabó P (2008) Analysis of expected climate change in the Carpathian Basin using the PRUDENCE results. Időjárás 112:249–264

Bátori Z, Körmöczi L, Erdős L, Zalatnai M, Csiky J (2012) Importance of karst sinkholes in preserving relict, mountain and wet woodland plant species under sub-Mediterranean climate: a case study from southern Hungary. J Cave Karst Stud 74:127–144

Bátori Z, Vojtkó A, Farkas T, Szabó A, Havadtői K, E-Vojtkó A, Tölgyesi C, Cseh V, Erdős L, Maák I, Keppel G (2017) Large- and small-scale environmental factors drive distributions of cool-adapted plants in karstic microrefugia. Ann Bot Lond 119:301–309

Bátori Z, Vojtkó A, Maák IE, Lőrinczi G, Farkas T, Kántor N, Tanács E, Kiss PJ, Juhász O, Módra G, Tölgyesi C, Erdős L, Aguilon DJ, Keppel G (2019) Karst dolines provide diverse microhabitats for different functional groups in multiple phyla. Sci Rep-UK 9:7176

Bátori Z, Vojtkó A, Keppel G, Tölgyesi C, Čarni A, Zorn M, Farkas T, Erdős L, Kiss PJ, Módra G (2020) Anthropogenic disturbances alter the conservation value of karst dolines. Biodivers Conserv 29:503–525

Bauer N (2006) A *Potamogeton coloratus* Hornem. Magyarországon (*Potamogeton coloratus* Hornem. in Hungary). Flora Pannonica 4:111–119 [in Hungarian]

Bauer N, Somlyay L (2015) A *Crepis mollis* (Jacq.) Asch. subsp. *hieracioides* (Waldst. & Kit.) Domin újrafelfedezése Magyarországon (Rediscovery of *Crepis mollis* (Jacq.) Asch. subsp. *hieracioides* (Waldst. & Kit.) Domin in Hungary). Kitaibelia 20:150–156 [in Hungarian]

Bauer N, Szabó I, Futó J (2008) Distribution of *Primula auricula* L. in the Transdanubian Mountains (Hungary)—with a new record: Rezi: Meleg-hegy (Keszthely Mts). Acta Bot Hung 50:237–256

Bauer N (2014) A Bakony-vidék szárazgyepjei. Sztyeprétek és sziklagyepek osztályozása és növényföldrajzi karaktere (Dry grasslands of the Bakony Region—classification and phytogeographical character of dry and rocky grasslands). A Bakony természettudományi kutatásának eredményei 33. Magyar Természettudományi Múzeum Bakonyi Természettudományi Múzeuma, Zirc, p 336 [in Hungarian]

Bölöni J, Molnár Z, Kun A (eds) (2011) Magyarország élőhelyei. A hazai vegetációtípusok leírása és határozója. ÁNÉR 2011. (Habitats of Hungary. Descriptions and identification keys of vegetation types of Hungary). MTA Ökológiai és Botanikai Kutatóintézete, Vácrátót, p 341 [in Hungarian]

Borbás V (1900) A Balaton tavának és partmellékének növényföldrajza és edényes növényzete (The vascular flora and phytogeography of Lake Balaton and its surroundings). A Balaton tudományos tanulmányozásának eredményei 1, Budapest, p 432 [in Hungarian]

Borhidi A, Sánta A (1999) Vörös Könyv Magyarország növénytársulásairól 1–2. (Red book on the plant associations of Hungary 1–2).

TermészetBÚVÁR Alapítvány Kiadó, Budapest, p 362 and 404 [in Hungarian]

Borhidi A, Kevey B, Lendvai G (2012) Plant communities of Hungary. Akadémiai Kiadó, Budapest, p 544

Borhidi A (2011) Növényföldrajz (Phytogeography). In: Schweitzer F (ed) Magyarország térképekben (Hungary in Maps). Magyar Tudományos Akadémia Földrajztudományi Kutatóintézet, Budapest, pp 65–70 [in Hungarian]

Boros Á (1937) A magyarországi hévizek felsőbbrendű növényzete (Macrophyte vegetation of thermal waters in Hungary). Bot Közl 34:89–118 [in Hungarian]

Boros Á (1954a) Florisztikai közlemények IV (Floristic records IV). Bot Közl 45:247–250 [in Hungarian]

Boros Á (1954b) A Vértes, Velencei-hegység, a Velencei-tó és környékük növényföldrajza (Phytogeography of the Vértes and Velence Mts, and the lake Velence). Földrajzi Értesítő 3:280–300 [in Hungarian]

Boros Á (1971) A barlangok felsőbbrendű növényvilága (Superior vegetation in caves). Karszt és Barlang 1971:49–52 [in Hungarian]

Buczkó K, Rajczy M (1989) Contributions to the flora of the Hungarian caves II. Flora of three caves near Beremend, Hungary. Studia Botanica Hungarica 21:13–26

Colwell RK (2004) Estimates: statistical estimation of species richness and shared species from samples. Version 9. http://viceroy.eeb.uconn.edu/estimates/

Cseresnyés I, Csontos P, Bózsing E (2006) Stand age influence on litter mass of *Pinus nigra* plantations on dolomite hills in Hungary. Can J Botany 84:363–370

Csiky J, Atkári B, Deme J, Radnai É (2014) Mohaflorisztikai érdekességek a Nyugat-Mecsekből (Interesting moss and liverwort records from the Western Mecsek Mts, South Transdanubia, Hungary). Kitaibelia 19:29–38 [in Hungarian]

Dövényi Z (ed) (2010) Magyarország kistájainak katasztere (Inventory of microregions in Hungary). MTA Földrajztudományi Kutatóintézet, Budapest, p 876 [in Hungarian]

Erdős L, Morschhauser T (2010) The rock-heath association Helleboro odori-Spiraeetum mediae in the Villány Mts. (South Hungary). Natura Somogyiensis 10:7–14

Erdős L, Gallé R, Bátori Z, Papp M, Körmöczi L (2011) Properties of shrubforest edges: a case study from South Hungary. Cent Eur J Biol 6:639–658

Erdős L, Méri Á, Bátori Z, Gallé R, Körmöczi L (2012) North-south facing vegetation gradients in the Villány Mountains: a case study on the population and the community level. Pak J Bot 44:927–932

Erdős L, Cserhalmi D, Bátori Z, Kiss T, Morschhauser T, Benyhe B, Dénes A (2013) Shrub encroachment in a wooded-steppe mosaic: combining GIS methods with landscape historical analysis. Appl Ecol Env Res 11:371–384

Erdős L, Dénes A, Tolnay D, Magnes M, Bátori Z, Tölgyesi C, Kevey B (2015) Phytosociology and nature conservation value of thermophilous oak forests in a South Hungarian karst area. Austrian J for Sci 132:145–168

Erdős L, Bátori Z, Tölgyesi C, Kevey B (2017) The Illyrian mesic forests of the Villány Mts: phytosociology and conservation importance. Biologia 72:510–519

Erdős L, Tölgyesi C, Dénes A, Darányi N, Fodor A, Bátori Z, Tolnay D (2014) Comparative analysis of the natural and semi-natural plant communities of Mt Nagy and other parts of the Villány Mts (south Hungary). Thaiszia - J Bot 24:1–21

Farkas S (ed) (1999) Magyarország védett növényei (Protected plants of Hungary). Mezőgazda Kiadó, Budapest, p 416 [in Hungarian]

Fekete G (1959) *Stipa bromoides* (L.) Dörfl., eine neuentdeckte Pflanze in Ungarn. Acta Bot Hung 5:349–356

Fekete G, Zólyomi B (1966) Über die Vegetationszonen und pflanzengeographischen Charakteristik des Bakony-Gebirges. Annales Historico-Naturales Musei Nationalis Hungarici 58:197–205

Fekete G, Tölgyesi G, Horánszky A (1989) Dolomite versus limestone habitats: a study of ionic accumulation on a broader floristic basis. Flora 183:337–348

Fekete G (1964) A Bakony növénytakarója. A Bakony természettudományi kutatásának eredményei I. (Vegetation of the Bakony Mountains. Results of the investigations of the Bakony Mountains I.). Bakonyi Múzeum, Veszprém, p 53 [in Hungarian]

Fekete G (1988) Természetes növénytakaró (Natural vegetation). In: Ádám L, Marosi S, Szilárd J (eds) A Dunántúli-középhegység, B) Regionális tájföldrajz (The Transdanubian Mountains, B) Regional geography). Akadémiai Kiadó, Budapest, pp 149–174 [in Hungarian]

Gams H (1930) Über Reliktföhrenwälder und das Dolomitphänomen. In: Rübel E (ed) Ergebnisse der Internationalen Pflanzengeographischen Exkursion durch die Tschechoslovakei und Polen 1928. Veröffentlichungen des Geobotanischen Institutes Rübel in Zürich 6, pp 32–80

Gáyer G (1924) Die alpinen Moorpflanzen des Balatongebietes. Magyar Botanikai Lapok 23:57–61

Horvát AO (1972) Die Vegetation des Mecsek Gebirges und seiner Umgebung. Akadémiai Kiadó, Budapest, p 376

Jakucs P (1955) Geobotanische Untersuchungen und Karstaufforstung in Nordungarn. Acta Bot Hung 2:89–131

Jakucs P (1956) Karrosodás és növényzet (Limestone pavements and vegetation). Földrajzi Közlemények 4:241–249 [in Hungarian]

Jakucs L (1977) A magyarországi karsztok fejlődéstörténeti típusai (Genetic types of the Hungarian karst). Karszt és Barlang 1–2:1–16 [in Hungarian]

Jakucs P (1961) Die phytozönologischen Verhältnisse der Flaumeichen-Buschwälder Südostmitteleuropas. Akadémiai Kiadó, Budapest, p 313

Jakucs L (1971) A karsztok morfogenetikája. A karsztfejlődés varienciái (Morphogenetics of karst regions. Variants of karst evolution). Akadémiai Kiadó, Budapest, p 311 [in Hungarian]

Jávorka S (1926) A *Sorbus torminalis* (L.) CR. magyar keverékfajai (Natural hybrids of *Sorbus torminalis* (L.) CR. in Hungary). Magyar Botanikai Lapok 25:83–90 [in Hungarian]

Kakas J (1960) Magyarország éghajlati atlasza (Climatic atlas of Hungary). Akadémiai Kiadó, Budapest, p 78 [in Hungarian]

Kárpáti Z (1960) Die Sorbus-Arten Ungarns und der angrenzenden Gebiete. Feddes Repertorium 62:71–334

Keppel G, Van Niel KP, Wardell-Johnson GW, Yates CJ, Byrne M, Mucina L, Schut AGT, Hopper SD, Franklin SE (2012) Refugia: identifying and understanding safe havens for biodiversity under climate change. Global Ecol Biogeogr 21:393–404

Kevey B (1999) Újabb montán reliktum a magyar flórában: *Ranunculus nemorosus* DC. (*Ranunculus nemorosus* DC.: a new montane relict species in the Hungarian flora). Kitaibelia 4:271–272 [in Hungarian]

Kevey B (2008) Magyarország erdőtársulásai (Forest associations of Hungary). Tilia 14:1–488 [in Hungarian]

Kevey B, Borhidi A (1998) Top-forest (Aconito anthorae-Fraxinetum orni) a special ecotonal case in the phytosociological system (Mecsek Mts, South Hungary). Acta Bot Hung 41:27–121

Kevey B, Borhidi A (2002) Top-forests (Veratro nigri-Fraxinetum orni) of the Bakony Mountains. Acta Bot Hung 44:85–115

Kevey B, Csiky J (2012) A Nyugat-Mecsek növényvilága (Vegetation of the western Mecsek Mountains). In: Barta K, Tarnai T (eds) A Nyugat-Mecsek karsztja (Karst areas of the western Mecsek Mountains). GeoLitera, Szeged, pp. 127–141 [in Hungarian]

Kiedrzyński M, Zielińska KM, Kiedrzyńska E, Rewicz A (2017) Refugial debate: on small sites according to they function and capacity. Evol Ecol 31:815–827

Király G (ed) (2007) Vörös Lista. A magyarországi edényes flóra veszélyeztetett fajai (Red list of the vascular flora of Hungary). Sopron, p 73 [in Hungarian]

Király G (ed) (2009) Új magyar füvészkönyv. Magyarország hajtásos növényei. Határozókulcsok (New Hungarian herbal. The vascular plants of Hungary. Identification key). Aggteleki Nemzeti Park Igazgatóság, Jósvafő, p 616 [in Hungarian]

Kiss PJ, Tölgyesi C, Bóni I, Erdős L, Vojtkó A, Maák IE, Bátori Z (2020) The effects of intensive logging on the capacity of karst dolines to provide potential microrefugia for cool-adapted plants. Acta Geogr Slov 60:37–48

Kun A, Ittzés P, Facsar G, Höhn M (2000) Sziklagyepek és lejtősztyeppek a Középdunai Flóraválasztó környékén II. Mészkő- és dolomitvegetáció a Cserhát-hegységben (Rocky grasslands and steppe slopes in the region of the Middle-Danubian Flora Boundary II. Dolomite and limestone vegetation in the Cserhát Hills). Kitaibelia 5:209–215 [in Hungarian]

Kun A, Tóth T, Szabó B, Koncz J (2005) A dolomitjelenség: közettani, talajtani és növényzeti összefüggések (The dolomite phenomenon: relations among rocks, soils and vegetation). Bot Közl 92:1–25 [in Hungarian]

Lájer K (1998) Bevezetés a magyarországi lápok vegetáció-ökológiájába (Introduction to the vegetation ecology of Hungarian mires). Tilia 6:84–238 [in Hungarian]

Máthé I (1940) Magyarország növényzetének flóraelemei (Geoelements of the Hungarian flora). Acta Geobot Hung 3:117–147 [in Hungarian]

Máthé I (1941) Magyarország növényzetének flóraelemei II. (Geoelements of the Hungarian flora II.). Acta Geobot Hung 4:84–108 [in Hungarian]

Mersich I, Práger T, Ambrózy P, Hunkár M, Dunkel Z (eds) (2000) Magyarország éghajlati atlasza (Climatological atlas of Hungary). Országos Meteorológiai Szolgálat, Budapest, p 107 [in Hungarian]

Molnár VA, Kreutz CAJ, Óvári M, Sennikoov A, Bateman RM, Takács A, Somlyay L, Sramkó G (2012) *Himantoglossum jankae* (Orchidaceae: Orchideae), a new name for a longmisnamed lizard orchid. Phytotaxa 73:8–12

Mulec J (2019) Lampenflora. In: White WB, Culver DC, Pipan T (eds) Encyclopedia of caves. Academic Press, Third Edition, pp 635–641

Nagy J (2007) A Börzsöny hegység edényes flórája (Vascular flora of the Börzsöny Mountains). Duna–Ipoly Nemzeti Park Igazgatóság, Budapest, p 376 [in Hungarian]

Németh C (2006) Hibrid eredetű, bennszülött Sorbus taxonok elterjedése a Vértesben és környékén. Flora Pannonica 4:17–33

Németh C (2007) Új berkenye kisfaj a magyar flórában *Sorbus tobani* Németh (*Sorbus tobani* Németh, a new hybrid species in the Hungarian flora). Flora Pannonica 5:173–184 [in Hungarian]

Pintér B, Vojtkó A, Tímár G (2010) A Naszály edényes flórája (Vascular flora of the Mt Naszály). In: Pintér B, Tímár G (eds) A Naszály természetrajza (A natural history of Mt Naszály, Hungary). Duna-Ipoly Nemzeti Park Igazgatóság, Budapest, pp 217–444 [in Hungarian]

Rajczy M (2019) A kivilágított barlangokat veszélyeztető növényesedés és visszaszorítása hazánkban. Történeti áttekintés az első 40 évről (Lampenflora of caves and its suppression: historical perspectives in Hungary over a 40-year period). Karszt és Barlang (különszám):37–40 [in Hungarian]

Somlyay L (2009) A Budai-hegység florisztikai növényföldrajzának fő vonásai (The main features of floristical phytogeography of the Buda Mts). Kitaibela 14:35–68 [in Hungarian]

Somlyay L (2010) Distribution of *Crepis pannonica* in Hungary. Studia Botanica Hungarica 41:113–128

Somlyay L (2011) Distribution of *Conringia austriaca* (Brassicaceae) in Hungary and its phytogeographical significance in Central Europe. Annales Historico-Naturales Musei Nationalis Hungarici 103:5–22

Somlyay L, Pifkó D (2002) A *Lathyrus pallescens* (Bieb.) C. Koch Magyarországon, és más adatok a Budai-hegység flórájának ismeretéhez (The *Lathyrus pallescens* (Bieb.) C. Koch in Hungary and other data to the flora of the Buda Mts). Kitaibelia 7:237–245 [in Hungarian]

Soó R (1967) Submediterran-illyrische Einflüsse in der Flora und Vegetaion Ungarns. Sonderdruck Aus Mitteilungen Der Ostalpin-Dinarischen Pflanzensoziologischen Arbaitsgemeinschaft 7:119–122

Soó R (1962) Növényföldrajz (Phytogeography). Tankönyvkiadó, Budapest, p 159 [in Hungarian]

Varga S, Sinigla M, Bauer N (2018) A *Spiraea media* Fr. Schm. Balaton-felvidéki előfordulásáról (About the occurrence of *Spiraea media* Fr. Schm. on the Balaton Uplands). Kitaibelia 23:151–154 [in Hungarian]

Virók V, Farkas R, Farkas T, Šuvada R, Vojtkó A (eds) (2016) A Gömör–Tornai-karszt flórája. Enumeráció (Flora of the Gömör–Torna Karst. Enumeration). ANP Füzetek XIV. Aggteleki Nemzeti Park Igazgatóság, Jósvafő, p 922 [in Hungarian]

Vojtkó A (1993) A váci Naszály vegetációtérképe. (Vegetation map of the Naszály at Vác]). Bot Közl 80:103–110 [in Hungarian]

Vojtkó A (1997) The vegetation of the Bükk plateau (NE Hungary) II. The grassland communities of the limestone and dolomite rocks. Acta Bot Hung 40:239–270

Vojtkó A, E-Vojtkó A, Dulai S, Farkas T (2018) A fás vegetáció jellegzetességei az Alsó-hegy (Gömör–Tornai-karszt) karsztfennsíkján (Characteristics of forest vegetation on the karst plateau of Alsó-hegy (Gömör–Torna karst, Hungary)). Bot Közl 105:97–108 [in Hungarian]

Vojtkó A (1998) A Bükk hegység sziklagyepjeinek és sztyeprétjeinek jellemzése (Rocky grasslands and steppe communities of the Bükk Mountains, Hungary). In: Csontos P (ed) Sziklagyepek szünbotanikai kutatása (Synbotanical studies of rock grasslands). Scientia Kiadó, Budapest, pp 133–155 [in Hungarian]

Vojtkó A (ed) (2001) A Bükk hegység flórája (Flora of the Bükk Mountains). Sorbus Kiadó, Eger, p 340 [in Hungarian]

Vojtkó A (2002) A hegység növénytakarója (Vegetation of the mountains). In: Baráz C (ed) A Bükki Nemzeti Park. Hegyek, erdők, emberek (Bükk National Park. Mountains, forests, people). Bükki Nemzeti Park Igazgatóság, Eger, pp 237–26 [in Hungarian]

Vojtkó A (2010) A Naszály vegetációja (Vegetation of the Mt Naszály). In: Pintér B, Tímár G (eds) A Naszály természetrajza (A natural history of Mt Naszály, Hungary). Duna-Ipoly Nemzeti Park Igazgatóság, Budapest, pp 445–489 [in Hungarian]

Vojtkó A (2014) Vegetáció (Vegetation). In: Virók V, Farkas R, Farkas T, Boldoghné Szűts F, Vojtkó A (eds) A Gömör–Tornai-karszt flórája. Általános rész (Flora of the Gömör–Torna Karst. General concepts). ANP Füzetek XIII. Aggteleki Nemzeti Park Igazgatóság, Jósvafő, pp 55–224 [in Hungarian]

Vojtkó A (2016) A Gömör–Tornai-karszt növényföldrajza (Phytogeography of the Gömör–Torna Karst). In: Virók V, Farkas R, Farkas T, Šuvada R, Vojtkó A (eds) A Gömör–Tornai-karszt flórája. Enumeráció (Flora of the Gömör–Torna Karst. Enumeration). ANP Füzetek XIV. Aggteleki Nemzeti Park Igazgatóság, Jósvafő, pp 9–70 [in Hungarian]

Willis KJ, Rudner E, Sümegi P (2000) The full-glacial forests of central and southeastern Europe. Quaternary Res 53:203–213

Zólyomi B (1942) A középdunai flóraválasztó és a dolomitjelenség (Middle-Danubian Flora Boundary and the dolomite phenomenon). Bot Közl 39:209–231 [in Hungarian]

Zólyomi B (1987) Coenotone, ecotone and their role in preserving relic species. Acta Bot Hung 33:3–18

Zólyomi B (1958) Budapest és környékének természetes növénytakarója (Natural vegetation of Budapest and its environs). In: Pécsi M, Marosi S, Szilárd J (eds) Budapest természeti képe (Natural portray of Budapest). Akadémiai Kiadó, Budapest, pp 509–642 [in Hungarian]

Zólyomi B (1967) Rekonstruált növénytakaró, 1: 1.500.000 (map) (Reconstructed vegetetaion cover). In: Radó S (ed) Magyarország Nemzeti Atlasza (National Atlas of Hungary). Kartográfiai vállalat, Budapest, p 31 [in Hungarian]

The Fauna of Karsts in Hungary

Csaba Szinetár, Péter Paulovics, Gábor Csorba, and Gergely Balázs

Abstract

This chapter outlines the development and current state of the surface and cave fauna of Hungary's karst regions. Concerning the cave fauna, the country's geographical situation is truly unique because it lies north from the mid-latitude biodiversity ridge. In the detailed discussion of the three main karst regions, we present the most important endemic species of Hungary's cave fauna including, for example, *Duvalius hungaricus, Eukoenenia spelaea vagvoelgyii, Helodrilus mozsaryorum, Pseudosinella aggtelekiensis, and Niphargus molnari.* When presenting the surface-dwelling fauna of karst regions, we focus on protected species of high environmental value. Both the cave and the surface-dwelling karst fauna reflect Hungary's exceptional biodiversity through their unique mix of various fauna elements.

Keywords

Pannonicum • Subterranean habitats • Troglobiont • Stygobiont fauna • Mid-latitude biodiversity ridge

C. Szinetár (✉)
Department of Biology, Savaria University Centre, ELTE SEK, Eötvös Loránd University, Károlyi Gáspár tér 4., 9700 Szombathely, Hungary
e-mail: szinetar.csaba@sek.elte.hu

P. Paulovics
Juhász Gyula Faculty of Education, University of Szeged, Boldogasszony sugárút 6., 6725 Szeged, Hungary
e-mail: paulovicsp@student.elte.hu

G. Csorba
Hungarian Natural History Museum, Baross u. 13, 1088 Budapest, Hungary
e-mail: csorba.gabor@nhmus.hu

G. Balázs
Behavioural Ecology Group, Department of Systematic Zoology and Ecology, Eötvös Loránd University, Pázmány Péter sétány 1/C, 1117 Budapest, Hungary

1 Introduction

1.1 Biogeographic Features of the Surface-Dwelling Karst Fauna

Hungary's karst fauna is fundamentally determined by its biogeographic location. That is, the Central-European deciduous forest zone meets the continental forest steppe zone at the Carpathian Basin. The whole Hungarian karst regions belong to the Pannonicum biogeographic region (Roekaerts 2002). The North Hungarian Karst, lying close to the Carpathians (which itself is part of the Alpine BR), hosts a fair share of Carpathian fauna elements. As a result, the latter karst area also hosts a fair share of Carpathian fauna elements. A large part (up to 50–80 per cent) of the total species pool is made up of European, Palearctic, and Holarctic species with wide distribution ranges (Varga 2019). A smaller part is consisted of "colouring elements" defined by edaphic and topographic factors the latter ones playing the more important role due to their great heterogeneity. After the last glacial period, the area was colonized from the borderlands of the Carpathian Basin. As a result, the highest biogeographic diversity can be found in the Mecsek Mountains and Villány Hills and in the North Hungarian Karst regions. The Mecsek Mountains and Villány Hills show a strong Mediterranean and West-Balcanic (Illyrian) influence, whereas the Gömör-Torna Karst Region is under considerable North and East Carpathian, Boreo-Continental, and arid Xeromontane influences. The latter ones are also present in the Bükk Mountains. Endemic species in the Carpathian Basin primarily belong to the soil-dwelling and cave-dwelling fauna. The surface-dwelling fauna chiefly exhibits only younger quaternary endemisms at subspecies level. Noctuids on dolomite areas (e.g. Buda and Vértes Mountains) are especially rich in endemic species. Additionally, our karst regions preserved relict species too. Especially, rocky grasslands and downy oak (*Quercus*

M. Veress and Sz. Leél-Őssy (eds.), *Cave and Karst Systems of Hungary*, Cave and Karst Systems of the World, https://doi.org/10.1007/978-3-030-92960-2_20

pubescens) scrub woodlands host a number of species from the major postglacial warm periods. Relict lepidoptera species of these habitats include *Pieris ergane* (Geyer 1828) and *Iolana iolas* (Ochsenheimer 1816) (Fig. 1) and the early spring geometrid, *Erannis ankeraria* (Staudinger 1861) (Fig. 2). The fauna of karst woodlands (beech and sessile oak-hornbeam forests) is dominated by Palearctic and Holarctic species with large distribution range that generally occur in Hungary's other habitats too. The "colouring elements" take their largest share in the south-facing xerothermic karst habitats. They come from Holomediterranean, Pontomediterranean, and West-Balcanic (Illyrian) species. Alpine (Arctic) species only occur in our coldest microclimate karst habitats. Such occurrences include the isolated populations of the newt *Ichthyosaura alpestris* (Laurenti 1768) in the Bakony and North Hungarian Mountains or a snail species, *Vertigo alpestris* Alder, 1838 in the Aggtelek Karst Region. The typical Carpathian species can be found on the Bükk Plateau and in the Aggtelek Karst. Xeromontane fauna elements also occur in dolomite habitats of the Transdanubian Mountains in addition to the Bükk Plateau and the Aggtelek Karst (Varga 2019). Due to the relatively good environmental condition of Hungary's karst regions, they only host a smaller number of adventive species. The spreading of these can be attributed to intensive land management practices, the effect of which is meant to be mitigated by environment protection measures.

1.2 Biogeographic Features of the Caves

The cave fauna of Hungary has a very complex history and from this history still a lot of pieces are missing. The geographical setup of the country is quite special in a sense. Hungary lies north from the mid-latitude biodiversity ridge (Culver and Pipan 2009). This expression was created to describe the experienced pattern where caves with very high biodiversity are situated between the 42° and 46° N latitudes in Europe and between 33° and 35° in North America. The diversity ridge is explained by the high surface productivity, big continuous karst areas, and by the fact that these areas were not affected by the Pleistocene permanent ice sheets (Culver et al. 2009). Although the karst areas of Hungary do not fall into the defined ridge, yet they lie right at its northern boundary and they were also not affected by permanent glaciations. It seems like that the only ingredient which is missing is the presence of the big continuous karst areas resulting in less diverse cave communities. On the other hand, the fragmented karst areas with less diversity theoretically provide unique opportunity to examine general biological phenomena which are often hard to detect because of the disturbing complexity in big karst areas of the mid-latitude biodiversity ridge. For instance, the cave biota of Hungary has elements from the surrounding karst areas and the exact origin of the elements is often possible to trace back which helps to understand the driving forces behind distribution patterns.

Fig. 1 *Iolana iolas* (Ochsenheimer 1816)

Fig. 2 *Erannis ankeraria* (Staudinger 1861)

The areas where the highest cave biodiversity can be found are the Aggtelek Karst and the Mecsek Mountains. The caves of the Aggtelek Karst are the home to fauna elements with southern origin like the aquatic annelid *Helodrilus mozsaryorum* (Zicsi 1974) (described as *Allolobophora mozsaryorum*) and the palpigrade *Eukoenenia spelaea vagvoelgyii* (Szalay 1956) (Fig. 3) the latter of which represents the northernmost occurrence of the genus. Other species such as the amphipod *Niphargus aggetlekiensis* Dudich 1932 (Fig. 4) or the coleopteran *Duvalius hungaricus* (Csiki 1903) are from Northern-Carpathian clades.

The cave fauna of the Mecsek Mountains shows similar mosaic-like pattern but due to its separated location and stabile climate on a geological scale it also served as a refuge for some species. The aquatic isopoda *Protelsonia hungarica hungarica* (Méhelÿ 1924) is a fine example for relict species as it is a representative of an ancient isopod lineage. Yet the pattern can be better seen by comparing the two stygobiont amphipod species endemic to the caves of the

Fig. 3 *Eukoenenia spelaea vagvoelgyii* (Szalay 1956)

Fig. 4 *Niphargus aggetlekiensis* Dudich 1932

Mecsek Mountains. *Niphargus molnari* (Méhelÿ 1927) is an ancient species which has its closest relatives in Central Europe and believed to colonize the caves of the area long time ago. The other Niphargus species, the *Niphargus gebhardti* Schellenberg, 1934 (described as *N. foreli gebhardti*), colonized the caves more recently and belongs to a clade that contains species mostly with Western Carpathian distribution (Angyal et al. 2015; Balázs et al. 2015). Naturally, the picture is not always this clear and there are questions to be answered. There are taxa like the genus *Niphargus* with big scale phylogenetic studies, where a species can easily be placed in its clade using available genetic data. In case of other groups, like the springtails, the genetic information is scarce, therefore the phylogenetic position and the origin of lineages are hard if not impossible to detect.

It is always easier to explain the presence of taxa than to find the reason behind the absence of a taxonomic group. For instance, amphipods have many representatives in European caves and this fact is especially true for the caves of the Dinaric Karst. The northern part of the Dinaric Karst lies close to the Hungarian karst areas, yet the only troglobiont amphipod genus to be found in Hungary is *Niphargus*. Surprisingly, even this genus is missing from certain karst areas like the Bakony Mountains where an otherwise diverse cave fauna can be found (Tóth 2015). Besides the taxonomic and biodiversity importance of species inhabiting Hungarian caves they provide a good opportunity to answer general colonization, speciation, and adaptation questions as in some occasions they represent replicates on a population, species, or higher taxonomic level. The possibilities provided by the fragmented karst areas of Hungary and the mosaic-like assembly of the cave fauna are already giving us insights to better understand big questions of cave biology. The research potential will certainly improve with the help of the rapidly developing research tools.

1.3 Relationships Between Surface and Cave Fauna

When examining a cave fauna one of the main questions is what determines the composition of the given fauna. Most of the caves are permanently in connection with the surface and the surface fauna by cracks, fissures, and entrances. Theoretically this means that all surface species have the chance to colonize the subterranean realm and evolve into a true troglobiont. In reality, this is rarely the case since not all species are capable of surviving in a subterranean environment and even if they are, then the permanent gene flow between the surface and the subterranean populations acts against speciation. Moreover, there are species like the spiders *Meta menardi* (Latreille 1804) (Fig. 5) and *Metellina merianae* Scopoli, 1763 which are common in the entrance zones of the Hungarian caves (Mammola et al. 2019) but never occur in the deeper parts. Their presence in the twilight zone is because they prefer cooler, shady, humid environments.

In cases when the colonizer surface source population gone extinct, the information available on the driving forces

Fig. 5 *Meta menardi* (Latreille 1804)

or the origin of the cave species is limited (Culver et al. 2009), and therefore the most exciting elements of the Hungarian cave fauna are probably the eutroglophile (sensu Sket 2008) species which have surface and cave populations as well. Sometimes these eutroglophile species look identical regardless of the environment they live like the diplopods *Hungarosoma bokori* Verhoeff, 1928 and *Haasea hungarica* (Verhoeff 1928) (described as *Orobainosoma hungaricum*). On other occasions, troglomorphic adaptations can be detected in the cave populations making these populations valuable for scientific examinations. This is the situation with the isopod *Asellus aquaticus* (Linnaeus 1758) (Fig. 6) population living in the Molnár János Cave (Pérez-Moreno et al. 2017) or the ostracod *Cryptocandona dudichi* (Klie 1930) (described as *Candona dudichi*) living in the Baradla Cave and in the nearby springs (Namiotko et al. 2001).

2 North Hungarian Mountains

2.1 Surface Fauna of Karst Regions in the North Hungarian Mountains

The karst regions of Bükk and Aggtelek-Rudabánya Mountains are within the North Hungarian Mountains, a hill range of mainly volcanic origin. Due to their elevation and direct connection to the Carpathians, the North Hungarian Mountains harbour the highest number of mountain species in Hungary. The region's fauna is determined by the large size of the area, its high forest coverage, and its above-mentioned connection to the bordering higher elevation mountain regions. Even with that in mind, the two karst regions host an extremely rich flora and fauna thanks to their diverse surface formations and mosaic-like habitat structure. Although the two karst regions are situated close to each other and thus have their similarities, they differ significantly concerning their formation, current geographic features, and fauna.

The Aggtelek Karst Region is mainly known for its magnificent caves; however, its surface also has its specialities. Its elevation of 300–600 m is lower than that of the Bükk Mountains. The south-facing slopes are covered by xerothermic vegetation types such as xerothermic oak forests, dry steppic grasslands, and rocky grasslands. Some of the latter are secondary habitats. The ecological highlight of the region is the area around the Nagyoldal Mountain whose fauna and flora have been extensively researched. It is unique in the sense that warm-loving species and East Carpathian ones requiring cooler conditions with more moisture live in close proximity. The nearness of the Carpathians is shown by such mollusc species as *Cocldina cerata* (Rossmässler 1836) and *Chilostoma faustinus* (Rossmässler 1835) (Varga 1999, 2019). The region's rocky grasslands are home to *Stenobothrus eurasius* Zubovski, 1898 (Fig. 7), a locust species with a distribution ranging from the Mongolian steppes to the Carpathian Basin (Kenyeres et al. 2020). This habitat type also harbours some rare xeromontane noctuids, including *Euxoa decora* (Denis and Schiffermüller 1775) and *E. birivia* (Denis and Schiffermüller 1775). The deforested plateaus also exhibit unique climatic conditions. Shrubby dolina rims and closed depressions are inhabited by the rare grasshopper *Pholidoptera transsylvanica* (Fischer-Waldheim 1853) and the endemic butterfly subspecies *Aricia artaxerxes issekutzi*

Fig. 6 *Asellus aquaticus* (Linnaeus 1758)

Fig. 7 *Stenobothrus eurasius* Zubovski 189

(Fabricius 1793) (Fig. 8). A typical species of the region's grassland habitats is the protected owfly *Libelloides macaronius* (Scopoli 1763) (Fig. 9) *whose strongest populations live in the Central Transdanubian Region, in the Bakony Mountains* (Ábrahám et al. 2016). *Otiorhynchus roubali* (Penecke 1931) *is an endemic weevil (Coleoptera) species of the xerothermic rocky forest habitats.* The endemic lamprey species of the Tisza water system *Eudontomyzon danfordi* (Rega 1911) (Fig. 10) lives in several streams in the Aggtelek Karst Region (Sallai 2014). *The diverse insect fauna is reflected in the abundant vertebrate fauna, including Ablepharus kitaibelii fitzingeri* (Mertens 1952) (Fig. 11), *a protected lizard species. The rich birdlife includes such rare species as Emberiza cia* (Linnaeus 1766) (Fig. 12),

Fig. 8 *Aricia artaxerxes* (Fabricius 1793)

Fig. 9 *Libelloides macaronius* (Scopoli 1763)

Circaetus gallicus (Gmelin 1788) (Fig. 13), and the Hazel grouse, *Bonasia bonasia* (Linnaeus 1758), which is probably the highest value resident bird of this karst region. Of the mammals, our two large predators, *Canis lupus* (Linnaeus 1758) and *Lynx lynx* (Linnaeus 1758) have permanently recolonized the region since the 1980s and 1990s, respectively.

Bükk Mountains is Hungary's highest elevation karst region with a unique fauna. The beech forests of the large plateau had been cut down from the eighteenth century onwards to provide charcoal for the local smelters, creating an open karst grassland used as hay meadows and pastures. The diversity of the area can be attributed to the special climate of the plateau's dolina and depressions and, along

Fig. 10 *Eudontomyzon danfordi* (Regan 1911)

Fig. 11 *Ablepharus kitaibelii fitzingeri* (Mertens 1952)

the southern edge of the plateau, of the rock walls and rocky mountain slopes. Calcareous base rocks are an important factor for snails and the unique climate sustains a unique fauna here. Rare mollusc species include *Spelaeodiscus triatus* (Rossmässler 1839), a snail occurring in the Tatra Mountains and in the high elevation areas of Transylvania, and *Phenacolimax annularis* (Studer 1820), whose relatives inhabit Mediterranean high mountain areas (Varga 2019). *Bielzia coerulescens* (Bielz 1851) (Fig. 14) is a Carpathian fauna element of the mountain beech forests. *Bithynella pannonica* (von Frauenfeld 1865) is an endemic snail species of the karst springs and cold-water creeks of the Bükk Mountains. *Entephria cyanata gerennae* Gyulai, 1984, an endemic geometrid feeding on *Arabis alpina*, favours the uniquely cold microclimate of valley bases. The Southern part of the Bükk Mountains is the home of some sub-Mediterranean noctuids as *Eutelia adulatrix* (Hübner 1813) and *Pyrrhia purpurina* (Esper 1804) (Varga 2019).

Fig. 12 *Emberiza cia* (Linnaeus 1766)

Fig. 13 Circaetus gallicus (Gmelin 1788)

A valuable part of the vertebrate fauna is consisted of the predatory birds requiring extensive undisturbed woodlands, including *Aquila heliaca* (Savigny 1809) and *Falco cherrug* (Gray 1834), and the large predatory mammals *Lynx lynx* and *Canis lupus* that have been recolonized this area during the recent decades similar to the Aggtelek Karst Region. The mouflon (*Ovis aries*, Linnaeus 1758), introduced to Central Europe 120 years ago, behaves as an invasive species and causes significant damages to the fauna and flora of these shallow soil habitats.

2.2 The Fauna of the Caves in the Aggtelek Karst and in the Bükk Mountains

In the North Hungarian Mountains—in accordance with the geology—two well-defined units can be found regarding the subterranean fauna: the Aggtelek Karst and the Bükk Mountains. From a biological point of view, two basic morphological cave types exist within the Aggtelek Karst. These types can be simplified as shaft caves with vertical entrance typical to the Alsó-hegy Region and the mostly

Fig. 14 *Bielzia coerulescens* (Bielz 1851)

horizontal multi-level caves with sinking rivers. The morphology of the caves affects the available food for its inhabitants and therefore profoundly influences the fauna they sustain. The large horizontal caves with sinking rivers and numerous swallow holes provide plenty of food sources brought by water to support diverse subterranean communities, while the caves with vertical entrance get moderate amount of nutrients entering the cave from the surface via gravity and to a lesser extent via temporary surface streamlets. The difference is not just in the quantity but also in quality. The nutrients of epigean origin are more evenly distributed in time and space in the horizontal caves, whereas they are concentrated at the entrance part and sporadic in the vertical shafts. By far the most important cave in the area is the Baradla-Domica Cave System. This cave has two main historical parts as it is crossing the Hungarian-Slovakian border. The Slovakian part is called Domica whereas the Hungarian part is called Baradla. This reflects not only the geopolitical situation but the exploration history as the two parts were only connected in 1932. The

Fig. 15 *Mesoniscus graniger* (Frivaldszky 1865)

Fig. 16 *Pseudosinella aggtelekiensis* (Stach 1929)

Baradla-Domica Cave System is the most diverse subterranean habitat in Hungary and as such it has attracted the interest of biologists as early as the middle of the nineteenth century (Frivaldszky 1865). It has been the centre of the Hungarian cave biology ever since with numerous outstanding findings (e.g. Dudich 1932). The active research of the cave was facilitated by the foundation of a cave laboratory in 1958 (Dudich 1960) which helped the research until the 80s.

As a result of decades of effort, up until now 597 invertebrate species have been found in the cave system (Gruber and Gaál 2014) and the Baradla section is a type locality of more than 30 species and subspecies (Salamon et al. 2014). It is worth to note that while out of the 597 species 397 species known from the Baradla part and 271 species from the Domica part, the overlap is only 12%. This surprising data can only be weakly explained by presence of different microhabitats in the two parts but is most probably the result of the different taxon specialists involved in the researches on the two sides of the border. The most common terrestrial species are the eutroglophile (sensu Sket 2008) isopod *Mesoniscus graniger* (Frivaldszky 1865) (Fig. 15) and the springtail *Pseudosinella aggtelekiensis* (Stach 1929) (described as *Lepidocyrtus* (*Pseudosinella*) *aggtelekiensis*) (Fig. 16). Both species can be found in almost all caves in the area. The minute predator beetle *Duvalius hungaricus* is a rare endemic troglobiont of the Aggtelek Karst. The stygobiont amphipod *Niphargus aggtelekiensis* is endemic to the Baradla and the surrounding caves. The endemic aquatic annelid *Helodrilus mozsaryorum* (Fig. 17) can only be found in the Baradla Rövid-Alsó Cave, and its closest relative *Helodrilus kratochvili* (Černosvitov 1937) lives in the Žira Cave in Western-Herczegovina (Szederjesi et al. 2014). The vertical shafts of Alsó-hegy are home to a range of cave-dwelling springtail (Collembola) species most of them known only from this region (Loksa 1967, 1969; Dányi 2011).

The characteristic caves of the Bükk Mountains can be described as deep vertical shafts connected by phreatic horizontal passages. Although these caves are forming a spacious network of subterranean habitats and the surface bio-productivity is considerable, the cave fauna in the area is less diverse than of the Aggtelek Karst. There are two possible explanations for this stark contrast. One is that although the productivity is high on the surface, the nutrients cannot reach the caves in large quantity due to the vertical entrances and the lack of permanent inflows. The second possible reason behind the pattern is that in these big cave systems the water level is extremely variable from dry streambeds with stagnant siphons to high water floods which even partially filling up the vertical shafts. These extreme, stochastic, and rapid water level changes can be unfavourable for both terrestrial and aquatic species. These ideas are supported by the fact that the most diverse cave in the area is the Kecske-lyuk Cave which is a relatively small horizontal cave with permanent stream.

The characteristic terrestrial cave-dwelling species of the area are *Mesoniscus graniger* and the endemic *Duvalius gebhardti* (Bokor 1926) (Fig. 18). The most typical representatives of the aquatic stygofauna are *Niphargus forroi*

Fig. 17 *Helodrilus mozsaryorum* Helodrilus mozsariorum (Zicsi 1974)

Fig. 18 *Duvalius gebhardti* (Bokor 1926)

(Karaman 1985), which was described from the Diabáz Cave and can only be found in a handful of caves in the area, and the widespread *Niphargus tatrensis* Wrzesniowsky, 1888. The name *N. tatrensis* probably covers a species group with numerous morphologically similar but genetically distinct cryptic species (Fišer et al. 2010; Balázs et al. 2015), and therefore the geographically isolated Bükk population potentially deserve separate taxonomic rank.

The caves of the two major karst areas of the Northern Mountain Range are home to the most species-rich bat communities of present-day Hungary. The diversity of these subterranean habitats is mirrored by the community richness of bats dwelling in trees and buildings. Along with the biotic and environmental attributes of the surface, the size, the characteristics, and the microclimate of warmer caves allow the formation of bat communities matching those in the

Fig. 19 *Rhinolophus euryale* (Blasius 1853)

Balkans or the Mediterranean. Typical species of these communities are *Rhinolophus* (Lacépède 1799) species, *Myotis emarginatus* (Geoffroy 1806), *Miniopterus schreibersii* (Kuhl 1819), *Myotis myotis* (Borkhausen 1797), *Myotis blythii* Tomes, 1857 (collectively known as large *Myotis*) species), as well as *Pipistrellus pipistrellus* (Schreber 1774). These species form mixed maternal colonies in the centre of their range (often with various other species), and at the beginning of the 1990s, this phenomenon could still be observed this far north (e.g. in the Herman Ottó Cave). The first Hungarian record of *Hypsugo savii* Bonaparte, 1837—originally distributed across the Mediterranean—was also from this region (Dobrosi 1993). *Rhinolophus euryale* (Blasius 1853) is present here with a ca. 10,000-strong population, located along both sides of the Hungarian-Slovakian border (Paulovics and Juhász 2008) (Fig. 19). *Miniopterus schreibersii* also reaches its northernmost occurrence here. The caves also serve as wintering as well as swarming sites for several other species.

3 Transdanubian Mountains

3.1 Surface Fauna of Karst Regions in the Transdanubian Mountains

The karst areas stretching from the Keszthely Mountains to the Buda Mountains show significant sub-Mediterranean and Atlantic climatic influences. Its calcareous base rocks are mainly consisting of dolomite, creating habitats that are typical examples of the so-called "dolomite phenomenon". Possibly the best and most species-rich example of this phenomenon is the Southern part of the Vértes Mountains which is the only known occurrence of the Holomediterranean *Scolopendra cingulata* (Latreille 1829) in Hungary (Fig. 20). Because the extent of open treeless habitats is relatively high in the Keszthely, Bakony, and South Vértes Mountains, they also host some Submediterranean and South European species with larger populations than those in the more southern Mecsek Mountains. It is the main reason for certain species occurring only in these areas. Higher elevation areas of the Transdanubian Mountains lie in the oak-hornbeam and beech forest zones. Their fauna is essentially no different from that of similar habitats in other parts of Hungary. The unique species of the area are connected to the xerothermic vegetation types, especially to scrub woodlands. The current look of the region, especially of the calcareous areas, is strongly shaped by prior land use practices. The areas most heavily affected by the former deforestations and the subsequent pine forestations, the quarrying, the creation of fruit orchards and vineyards, and the growth of built-up areas were the Buda and Keszthely Mountains. A way for small animals to adapt to surfaces heating up extremely is by taking shelter under the rocks or in the cracks. That is a typical behaviour of Hungary's only protected myriapod species, *Scolopendra cingulata*, only known from the warm scrub woodlands and rocky grasslands of the Vértes Mountains. Its strongest populations live in the Balkans. The karst areas of the Transdanubian Mountain Range are also the single known home to the geometrid *Charissa intermedia budensis* (Kovács 1954). During daytime, the adult moths hide in the cracks of

Fig. 20 *Scolopendra cingulata* (Latreille 1829)

dolomite rocks. *Lignyoptera fumidaria* (Hübner 1825), a steppic relict species is a characteristic geometrid of dry dolomite grasslands in the Buda, Vértes, and Bakony Mountains. The species reaches the western borders of its distribution range in the Carpathian Basin. The noctuid *Cucullia mixta lorica* (Ronkay and Ronkay 1987) is another steppic relict subspecies, so far recorded only in the Buda and Vértes Mountains. *Euxoa vitta* (Esper 1789) also occurs only in these Transdanubian dolomite grasslands. *Pieris ergane* is a specialist butterfly of the dolomite grasslands. Its larvae feed on *Aethionema saxatile*. Of necessity, we can find the lepidoptera species in common with the South Transdanubian karst areas, including *Iolana iolas*. During the summer, the southern slopes of the mountain range echoes the chirping of cicadas (*Cicada orni* (Linnaeus 1758), *Tibicen haematodes* (Scopoli 1763)). *Marumba quercus* (Denis and Schiffermüller 1775) (Fig. 21) is an oak-feeding hawk moth with a Holomediterranean distribution range. The rare geometrid *Erannis ankeraria*, also occurring in the South Transdanubian karst regions, feeds solely on downy oak. Calcareous fen meadows such as the Sásd Meadows in the Balaton Uplands constitute a unique habitat type, harbouring the wetland ecotype of *Euphydryas aurinia* (Rottemburg 1775) (Fig. 22), whose larvae feed on *Succisa pratensis* (Ábrahám 2014). The Buda Mountains constitute the geologically and topographically most diverse area formed on calcareous base rock in the Transdanubian Mountain Range. Its fauna is extensively researched on a number of sites covered by rocky dolomite grasslands, including the Sas Hill of Buda, the Tétény Plateau, and the hills around Budaörs. The karst areas of the Transdanubian Mountains host isolated populations of *Saga pedo* (Pallas 1771) (Fig. 23), Hungary's largest insect. *Arcyptera microptera* (Fischer de Waldheim 1833) is another valuable species, inhabiting the remaining fragments of Eurasian forest steppes in the Carpathian Basin. A high number of its occurrences fall to 250–350 melevation sites in the Bakony and Buda Mountains (Kenyeres et al. 2008). *Isophya kraussii* (Brunner von Wattenwyl 1878) (Fig. 24) is a typical species of any karst forest-grassland mosaic in the Transdanubian Mountains (Kenyeres et al. 2009). Among the coleoptera, the highly protected *Carabus hungaricus* (Fabricius 1792) is a typical species of the continental forest steppe zone, which besides the Pannonic sand steppes in the Hungarian Plain also present in the dolomite grasslands of Buda, Gerecse, and East Bakony Mountains (Bérces et al. 2007). Beetles can also be specialists, i.e. *Chlorophorus hungaricus* Seidlitz, 1791 (Fig. 25)—its host plant is *Dorycnium germanicum* and thus the species inhabits dolomite and limestone grasslands in the Transdanubian Mountains (Merkl and Szél 2012). *Atypus affinis* (Eichwald 1830) is a characteristic spider species of scrub woodlands. *Nemesia pannonica* (Herman 1879) is a typical spider of hill base loess grasslands with deeper soil, its

Fig. 21 *Marumba quercus* (Denis and Schiffermüller 1775)

Fig. 22 *Euphydryas aurinia* (Rottemburg 1775)

largest populations in Hungary live in the karst regions (Szinetár 2006). All three Hungarian spider species of the genus *Eresus* are present in the Transdanubian Mountains. *Araneus grossus* (Koch 1844) is a protected, large-sized orb-weaver spider inhabiting scrub woodlands with abundant *Cotinus coggygria* (Szinetár et al. 2013). The region also hosts a rich reptilian fauna, including two highly protected species, *Dolichophis caspius* (Gmelin 1789) (Fig. 26) and *Ablepharus kitaibelii fitzingeri* (Vörös and Halpern 2012).

Fig. 23 *Saga pedo* (Pallas 1771)

Fig. 24 *Isophya kraussii* (Brunner von Wattenwyl 1878)

3.2 The Fauna of the Caves in the Karst Regions of the Transdanubian Mountains

Regarding the cave fauna three characteristic regions can be defined within the Transdanubian Mountains from the east to the west.

The easternmost region includes the Pilis Mountains and the Buda Mountains. Most of the caves here are of thermal karst origin, and therefore they have restricted connection with the surface which results in very limited food availability for the cave fauna. In accordance with this, the cave fauna is rather poor, thus only a few fauna elements can be

Fig. 25 *Chlorophorus hungaricus* (Seidlitz 1791)

Fig. 26 *Dolichophis caspius* (Gmelin 1789)

found per cave and these populations are typically representatives of eutroglophile and subtroglophile (sensu Sket 2008) species. The only exception from this general pattern is the Molnár János Cave in the Buda Mountains with its relatively diverse stygobiont fauna (Balázs et al. 2015). The outstanding fauna of this cave can be explained by the possibility of chemoautotrophy-based food source (Herczeg et al. 2020). Since these caves are easily reachable from Budapest the fauna is well studied resulting in noticeable findings. The lithobiid, *Lithobius lucifugus infernus* (Loksa 1948), subspecies (Loksa 1948; Novák and Dányi 2011), two bat parasite nematoda species, the *Pterothominx neopulchra* (Babos 1954) (as *Capillaria neopulchra*), and the *Physaloptera myotis* (Babos 1954) were described from caves of the Buda Mountains, while the bat tick species *Ixodes ariadnae* (Hornok et al. 2016) and the beetle subspecies *Choleva lederina pilisensis* (Růžička 1998) were described from the Ariadne Caves System in the Pilis Mountains.

The middle region consists of the North Bakony, the South Bakony, and the Balaton Uplands. Since these mountains are built up from various kind and partially covered limestone, continuous big cave areas cannot be found here. The biggest caves of the area are concentrated in the North Bakony on the Tési Plateau. The cave fauna is relatively rich although highly specialized troglobionts are scarce (Tóth 2015; Loksa 1960a; Eszterhás 1984). The most well-examined cave of the area is the only 154-m-long relict hydrothermal Lóczy Cave near Balatonfüred. This cave is the type locality of two springtail species *Arrhopalites loczy* Loksa 1960a b and *Pseudosinella argentea* Loksa 1960a, b) and two dipluran species (*Plusiocampa breviantenna* Loksa 1960a, b and *Eutricocampa paurociliata* Loksa 1960a, b).

The westernmost region consists of the Keszthely Mountains and the Tapolca Basin. Both geologically and biologically the most important cave of the Keszthely Mountains is the 6400-m-long tectonic Csodabogyós Cave at Balatonederics (Szabó et al. 2013; Novák and Kutasi 2014). According to the literature, the almost 15-km-long, partially flooded hydrothermal Tapolcai Cave System in the Tapolca Basin is home to various eutroglophile and subtroglophile taxa (Geyer and Mann 1940; Loksa 1960b). Unfortunately, this cave temporarily run dry during the 80s because of the nearby mining activities, and therefore its fauna should be re-evaluated.

Due to its diverse geology, the Transdanubian Mountain Range harbours a rich bat community. As opposed to the eastern region with its large caves housing thermophilous bat populations (from the Danube Bend to Mt. Vértes), the middle and western regions (Bakony, Balaton Uplands, Keszthely Mountain Range) show a different characteristic. Until recently, one of the largest mixed colonies of *Myotis myotis* and *Myotis blythii* (with 8,000–10,000 individuals) used to overwinter in the Szopláki-Ördöglyuk of Mt. Pilis (Topál 1989). By today, their numbers have dropped to about a few hundred. In the Jankovich Cave of Mt. Gerecse, these species may form maternal colonies exceeding a thousand females (Kováts et al. 2011). Once *Miniopterus schreibersii* was present across the region with several maternal colonies of hundreds of individuals. By now, the species can be considered virtually extinct from these mountains. A hundred years ago, *Rhinolophus euryale*, another thermophilous species, was also characterized by a contiguous distribution in the eastern parts of the mountain range and a 1,000-strong maternal colony was still known from the Pisznice Cave of Gerecse at the beginning of the 1970s (Topál 1989). Today, a species conservation action targets the last Transdanubian population of the species. The caves of the Bakony are characterized with much smaller populations, too, but many forest-dwelling bat species use these sites during the periods of swarming and winter hibernation. The first Hungarian record of *Eptesicus nilssonii* (Keyserling and Blasius 1839)—the most northernly distributed European bat—is also from here (Paulovics 1998). The mountain range is a stronghold for other cold-tolerant species such as *Myotis nattereri* (Kuhl 1817) and *Barbastella barbastellus* (Schreiber 1774) (Fig. 27). Caves of the Keszthely Mountain Range and the Tapolca Basin are again warmer. The most significant bat community can be found in the Csodabogyós-Széllik Cave System with

Fig. 27 *Barbastella barbastellus* (Schreiber 1774)

species such as *M. natterei*, *M. emarginatus*, *Plecotus auritus* (Linnaeus 1758), and large *Myotis* species.

4 Mecsek and Villányi Mountains

4.1 Surface Fauna of Karst Regions in the Mecsek and Villány Mountains

Hungary's southernmost karst region is situated in the Mecsek Mountains and Villány Hills. Due to its location, submediterranean influence can be felt more strongly here, especially on the south-facing slopes. The region hosts the highest number of submediterranean-pontomediterranean and Balkanic fauna elements. The three main karst areas (Abaliget, Orfű, and the Misina-Tubes) of Mecsek Mountains fall in the northwestern part of the mountain range. These areas are mostly forested, grassland coverage is rather small and is mainly confined to the Misina-Tubes area. The woodland (beech and oak-hornbeam forests, scree, and ravine forests) fauna of this karst region is like that of other hill forest habitats in the country. It is the streams and caves that constitute area-specific habitats. The Melegmány Creek harbours a stable population of *Cordulegaster heros* (Theischinger 1979), a rare and protected dragonfly (Ambrus et al. 2018) and is also home to an endemic and highly protected flightless caddisfly, *Chaetopteryx mecsekensis* (Nógrádi 1986; Szivák et al. 2017). *Rosalia alpina* (Linnaeus 1758) (Fig. 28) a typical longicorn species of beech forests across the country, which also occurs here. The decoratively coloured millipede *Polydesmus collari* (Koch 1847), a Balkanic (Illyrian) fauna element, lives in the forest litter of moist woodlands.

The dominant habitat types of the south-facing slopes are the downy oak (*Quercus pubescens*) scrub forests, steppic limestone slopes, and rocky grasslands. They harbour several specialist lepidoptera species, among them are *Cuculia formosa* (Rogenhofer 1860), known from the Mecsek and Villány Mountains only, and *Polymixis rufocincta isolate* (Geyer 1828) (Fig. 29), a Pontomediterranean noctuid, recorded only from the Szársomlyó Hill (Ábrahám and Uherkovich 2000). Other Mediterranean butterfly and moths species include *Iolana iolas* (also in the Keszthely Mountains), *Paraboarmia viertilii* (Bohatsch 1883), and *Erannis ankeraria*, an early spring geometrid inhabiting warm downy oak scrub woodlands. The leaf beetle *Eupales ulema* (Germar 1813) as a typical Balkanic fauna element occurs on the south slopes of the Mecsek (Vig 2009). The Palaearctic red cicada (*Tibicina haematodes*), occurring on the south-facing slopes of the region, is another Mediterranean element in the fauna. The region hosts valuable specialist spider species too. One of them is *Pelecopsis loksai* Szinetár and Samu 2003, which was described from the Szársomlyó Hill but has been discovered around Budapest and other karst areas in the Balaton Uplands since then. *Cryptodrassus hungaricus* (Balogh 1935) (Fig. 30), a secretive rare spider living in the cracks of rocky slopes, was also discovered on Szársomlyó in 2020. *Eresus moravicus* Řezáč 2008 (Fig. 31)

Fig. 28 *Rosalia alpina* (Linnaeus 1758)

Fig. 29 *Polymixis rufocincta* isolata (Geyer 1828)

is one of the most eye-candy Hungarian spiders whose males can be seen during May in the Mecsek and Villány Mountains (it also occurs in the other two karst regions of the country). The region is also home to valuable Orthoptera species such as *Saga pedo*. The amphibian fauna includes *Bombina variegata* (Linnaeus 1758), living in the small ponds of beech forests at higher elevations. The southern-facing slopes host strong populations of reptiles, including *Podarcis muralis* (Laurenti 1768), *Lacerta viridis* (Laurenti 1768), and *Dolichophis caspius*, a large whipsnake species occurring on the Szársomlyó Hill. The region is also rich in birds, hosting the largest population of *Crex crex* (Linnaeus 1758) in the South Transdanubian area, in the hay meadows of Bükkösd Valley.

4.2 The Fauna of the Caves in the Karst Regions of the Mecsek and Villány Mountains

In spite of their close proximity, the Mecsek and the Villány Mountains differ greatly in their geological background and therefore in the subterranean environment they provide. The cave systems of the Mecsek are classic epigean karst features concentrated on the western part of the mountain range. These caves are isolated from any other karst area and are home to a unique subterranean fauna with ancient and often endemic elements. Although the cave systems of the Mecsek are typically small formations, yet as they are old and rich in nutrients, sustain complex communities. The two most important caves are the Abaligeti Cave and the

Fig. 30 *Cryptodrassus hungaricus* (Balogh 1935)

Fig. 31 *Eresus moravicus* Řezáč 2008

Mánfai-kőlyuk Cave where most of the biospeleological investigations were carried out (Bokor 1924; Daday 1889; Dudich 1928; Gebhardt 1931, 1933, 1934; Méhelÿ 1924, 1925) making these two caves the best studied ones in Hungary after Baradla. As a result of the intense zoological surveys—according to the latest comprehensive publication (Gebhardt 1967)—in the Mánfai-kőlyuk Cave and the Abaligeti Cave, 159 and 286 species—including protozoans—are known, respectively.

The biggest cave of the area is the 2000-m-long Abaligeti Cave which is the type locality of numerous species and subspecies, namely, the minute gastropod *Paladilhiopsis hungarica* (Soós 1927) (described as *Lartetia hungarica*); the isopod *Protelsonia hungarica hungarica* (Fig. 32); the amphipod *Niphargus gebhardti*; the nematode *Aulolaimoides phoxodorus* Andrássy 1964; the bat-parasite trematoda *Prosotocus vespertilionis* Mődlinger 1930; and the diplopods *Brachydesmus troglobius* Daday 1889 , *H. bokori*, and *H. hungarica*.

Fig. 32 *Protelsonia hungarica hungarica* Méhelÿ 1924

Fig. 33 *Rhinolophus ferrumequinum* (Schreber 1774)

The Mánfai-kölyuk Cave is the type locality of the isopod *Protelsonia hungarica robusta* Méhelÿ 1927, the amphipod *Niphargus molnari* Méhelÿ 1927, the gastropod *Paladilhiopsis gebhardti* (Wagner 1931), and the triclads *Polycelis tothi* Méhelÿ 1927 and *Dendrocoelum pannonicum* Méhelÿ 1927. While most of the above-quoted early researches solely concentrated on the two caves, recent studies tried to involve more locations and find solutions for some long-lasting taxonomic and distribution questions using integrative taxonomic approach (Angyal and Korsós 2013; Angyal et al. 2015, 2018).

The hydrothermal hypogean caves of the Villány Mountains with their unique speleothems are highly valuable geological features. These caves do not have direct connection with the surface's primer production and therefore they are poor in nutrients and as a result no macroscopic fauna have been reported from any caves in the area yet.

As it was characteristic of other parts of the country, a few decades ago the local caves were inhabited by several maternal and wintering colonies—the most renowned of them being the Abaliget Cave that served as a maternal site for about 600 of *Miniopterus schreibersii* and as a wintering site for 1,500–2,000 *Myotis blythii* (Topál 1989). Despite the drastic reduction of population sizes, the cave is still an important swarming site for a dozen bat species as well as a significant winter hibernaculum for *Rhinolophus ferrumequinum* (Schreber 1774) (Fig. 33) along with the *Myotis dasycneme* (Boie 1825) and *Myotis emarginatus* (Boldogh and Estók 2007). Except for three species, every species known from Hungary has been recorded either from the cave itself or from its immediate surroundings. Other significant bat habitats of the region are the Mánfai-Kőlyuk (currently the only local winter hibernaculum and swarming site of *Miniopterus schreibersii*), the Trió Cave, the Vásárosúti-shaft, and the Büdöskúti-shaft.

The caves of the Villányi Mountain Range and especially those of the Szársomlyó are far less explored. Bats have moved in following the cessation of mining activities that had opened up many natural cavities. Their populations may well be connected to those in Croatia, consisting of the characteristic species communities of Southern Europe (*Miniopterus schreibersii*, *Rhinolophus ferrumequinum*, *Myotis myotis*, *Myotis blythii*, *Myotis emarginatus*) that are sometimes present in colonies of several thousand individuals (Dombi 1995).

5 Conservation Status

In terms of conservation, the karst regions in Hungary are in a relatively favourable situation. All of them are parts of National Parks and their protection and management are administered by five separate National Park Directorates (see Sect. 6.3 in "The Vegetation of Karsts in Hungary" chapter). The most valuable sites have a highly protected status. A large part of the protected animals in Hungary occur in the karst regions because a large share of the specialist species in xerothermic mountain and hill habitats is protected in every group (e.g. insects, snails, spiders). Karst regions also play an important role in the conservation of the Hungarian populations of several vertebrate species. Two of our highly protected reptiles, *Ablepharus kitaibelii* and *Dolichophis caspius,* occur almost exclusively in karst areas. In terms of mammals, a number of bat species are strongly connected to karst cave habitats. The Aggtelek and Bükk Mountains karst regions harbour strong populations of the country's many highly protected predatory birds, providing a breeding area for them.

The first Hungarian cave protection document dates back as early as 1839. This local regulation intended to save the speleothems of Baradla Cave from collection by the visitors. Ever since the Hungarian cave protection regulations focus mainly on the geological aspects and only to a lesser extent treating caves as ecosystems.

The Forest Act in 1935—which was the first national-scale decree on cave protection—states that if a cave has a unique character and scientific value then the cave and its catchment area should be protected. It meant that the researchers had to prove whether a cave deserve protection. As a result of this requirement, this act only provided protection for the most outstanding caves where the geologists, palaeontologists, archaeologists, and biologists could join forces to convince the authorities. The turning point in cave protection was the Nature Conservation Act in 1961. This act states that all the discovered caves are automatically, *ex lege* protected in Hungary. The conservation act in 1982 gave possibility to upgrade the protection status of certain caves to strictly protected based on special scientific, touristic, medical, or water management importance. This act also declares that all activities, including even a visit in a cave not open to the public, require permission. The *ex lege* protection proved to be a good general tool to conserve the subterranean values but it has some unplanned and unwanted consequences as well.

Since the caves are automatically protected detailed evaluations are not necessary, therefore often missing. This is especially true when it comes to biological evaluations, and since it has been unnecessary for decades, experts are scarce, as well. As a consequence, we have very limited data on species distributions and population sizes. The protection can only be maintained if the visits of a cave are regulated and therefore almost all caves in Hungary are locked with gates. These gated entrances alter the microclimate and the nutrient circulation of a cave. This influences the fragile ecosystems and the negative effects can only be minimized but cannot be avoided even with the most careful planning and installation. Since the caves are protected the cave-dwelling animals are protected by law, too. This is the reason why only three cave species have legal protection on their own right (*Duvalius hungaricus, Duvalius gebhardti, Paladilhiopsis hungarica*) while dozens of other species would deserve it as they are only known from a single location.

The legal background of bat protection in Hungary developed independently from that of the caves or the invertebrates. The first national act on protecting "advantageous" animals in 1894 (this part came into force in 1901) listed 34 vertebrate species as "protected" including all 24 bat species known in the country at that time. In accordance with the EU legislation, all bat species (the recent count is 28 species) occurring in the country are protected.

The Hungarian cave biota is threatened by many factors, but the general protection serves as a strong base. In the past, most severe problem was the mining industry as some caves were completely destroyed in open pit mines (e.g. caves in Villányi Mts) (Buczkó and Rajczy 1989) while others suffered from the artificially lowered karst water level (e.g. Tapolcai Cave System). Since nowadays the mining industry in the country is less active this threat is of less importance. Karst spring caves are often used for public water supply and this utilization surely changes the hydrology of the area and therefore affects the biota (Angyal 2012). The caves under settlements like the Molnár János Cave in Budapest are obviously highly endangered by anthropogenic pollutions (Leél-Őssy et al. 2011). Whatever is the problem, effective protection can only be planned and executed when enough information is available for the stakeholders. Therefore, regular biological evaluation of caves would be the most important step to keep our cave ecosystems safe.

6 Conclusions

The surface-dwelling fauna of Hungary's karst regions is not significantly dissimilar to those found in the country's geologically different mountain regions with different base rocks. Dolomite habitats, especially in the Transdanubian Mountains, host the most distinctive species mix. A special feature of dolomite grasslands is that they are also home to a number of relict steppic species. This surface-dwelling karst fauna consisting of the typical species of the Pannonian region is enriched by mostly Carpathian colouring elements in the Aggtelek Karst and Bükk Plateau of the North Hungarian Mountains) and Mediterranean and West-Balcanic (Illyrian) elements in the Mecsek and Villány Mountains of South Transdanubia, making it truly exceptional and diverse.

Besides the taxonomic and biodiversity importance of species inhabiting Hungarian caves they provide a good opportunity to answer general colonization, speciation, and adaptation questions as in some occasions they represent replicates on a population, species or higher taxonomic level. The possibilities provided by the fragmented karst areas of Hungary and the mosaic-like assembly of the cave fauna are already giving us insights to better understand big questions of cave biology.

Acknowledgements The authors are grateful to many colleagues for providing photos and for particularly important suggestions and comments: Ambrus András, Ábrahám Levente, Bauer Norbert, Boldogh Sándor, Deli Tamás, Dobrosi Dénes, Dombi Imre, Dudás György, Estók Péter, Fehér Csaba Endre, Hadarics Tibor, Hudák Tamás, Kenyeres Zoltán, Korsós Zoltán, Kováts Dávid, Merkl Ottó, Varga András, Halpern Bálint, Sallai Zoltán, Szabadfalvi András, Uherkovich Ákos.Our project was supported by *Hungarian National Research, Developmental and Innovation Fund for international cooperation* grant (#SNN-125627).

References

Ábrahám L (2014) Lápi tarkalepke (*Euphydryas aurinia*) (Rottemburg, 1775) In. Haraszthy L (szerk) Natura 2000 fajok és élőhelyek Magyarországon (Natura 2000 species and habitats in Hungary). Pro Vértes Közalapítvány, Csákvár, pp 323–326 (in Hungarian)

Ábrahám L, Uherkovich Á (2000) A nagylepke (Lepidoptera) fauna kutatásának eddigi eredményei a Villányi-hegységben (Updated results of studies on the butterflies and larger moths (Lepidoptera) of the Villány Hills, South Hungary). Dunántúli Dolgozatok Természettudományi Sorozat 10:309–339 (in Hungarian)

Ábrahám L, Deli T, Jakab G, Danyik T (2016) A keleti rablópille *Libelloides macaronius* (Scopoli, 1763) az Alföldön (*Libelloides macaronius* (Scopoli, 1763) on the Great Hungarian Plain). Crisicum 9:165–170 (in Hungarian)

Ambrus A, Danyik T, Kovács T, Olajos P (2018) Magyarország szitakötőinek kézikönyve (Handbook of Hungarian dragonflies). Herman Ottó Intézet Budapest, p 290 (in Hungarian)

Angyal D (2012) Invertebrate fauna of the Mánfai-Kőlyuk Cave (Mecsek, SW Hungary) in the light of utilization by waterworks.—Previous results. Term Közl 18:24–33

Angyal D, Korsós Z (2013) Millipedes (Diplopoda) of twelve caves in Western Mecsek, South-west Hungary. Opusc Zool Budapest 44 (2):85–106

Angyal D, Balázs G, Zakšek V, Krízsik V, Fišer C (2015) Redescription of two subterranean amphipods *Niphargus molnari* Méhely, 1827 and *Niphargus gebhardti* Schellenberg, 1934 (Amphipoda, Niphargidae) and their phylogenic position. Zookeys 509:53–65

Angyal D, Balázs G, Krízsik V, Herczeg G, Fehér Z (2018) Molecular and morphological divergence in a stygobiont gastropod lineage (Truncatelloidea, Moitessieriidae, Paladilhiopsis) within an isolated karstic area in the Mecsek Mountains (Hungary). J Zool Syst Evol Res 56(4):493–504

Balázs G, Angyal D, Kondorosy E (2015) *Niphargus* (Crustacea: Amphipoda) species in Hungary: literature review, current taxonomy and updated distribution of valid taxa. Zootaxa 3974(3):361–376

Balogh JI (1935) A Sashegy pókfaunája (Spider fauna of Sas Mountain). Budapest, p 60 (in Hungarian)

Bérces S, Szél G, Ködöböcz V, Kutasi C, Szabó K, Fülöp D, Pénzers Z, Peregovits L (2007) A magyar futrinka (*Carabus hungaricus*) In: Forró L (ed) A Kárpát-medence állatvilágának kialakulása (The origin of the fauna of the Carpathian Basin). Magyar Természettudományi Múzeum, Budapest, pp 107–124 (In Hungarian)

Bokor E (1924) Beitrage zur rezenten Fauna der Abaligeter Höhle. Zool Anz 61:111–121

Boldogh S, és Estók P (ed) (2007) Földalatti denevérszállások katasztere I (Catalogue of underground bat roost in Hungary I). ANP Füzetek III. Aggteleki Nemzeti Park Igazgatóság (in Hungarian)

Buczkó K, Rajczy M (1989) Contributions to the flora of Hungarian caves II. Flora of three caves near Beremend, Hungary. Stud Bot Hung 21:13–25

Culver DC, Pipan T (2009) Biology of caves and other subterranean habitats. Oxford University Press, Oxford, p 254

Culver DC, Pipan T, Schneider K (2009) Vicariance, dispersal and scale in the aquatic subterranean fauna of Karst regions. Freshw Biol 54:918–929

Daday J (1889) A magyarországi Myriapodák magánrajza (Monograph on the Hungarian Myriapods). Kir M Természettud Társ, Budapest, 126 pp

Dányi L (2011) Cave dwelling springtails (Collembola) of Hungary: a review. Soil Org 83(3):419–432

Dobrosi D (1993) *Adatok a Bükk denevérfaunájához* (Data to the bat fauna of the Bükk Mountains). Folia Historico-naturalia Musei Matraensis 18:191–197 (in Hungarian)

Dombi I (1995) Dél-Magyarország gyöngyszeme—a *Szársomlyó* denevérfaunája (The pearl of Southern Hungary—the bat fauna of the Szársomlyó). Denevérkutatás Hung Bat Res News 1:7–10 (in Hungarian)

Dudich E (1928) Faunisztikai jegyzetek. III (Faunistic notes III). Áll Közl 24:38–45

Dudich E (1932) Biologie der Aggteleker Tropfstein-Höhle "Baradla" in Ungarn. Wien Spaläol Monog 12:246

Dudich E (1960) Das höhlenbiologische Laboratorium der Eötvös Lorand Universität. Biospeologica Hungarica X. Ann Univ Sci Bud De Rol Eötv Nom Sec Biol 3:131–135

Eszterhás I (1984) Az Alba Regia-barlang állatvilága (The animals of the Alba Regia Cave). Karszt És Barlang 1:25–30

Fišer C, Coleman CO, Zagmajster M, Zwittnig B, Gerecke R, Sket B (2010) Old museum samples and recent taxonomy: a taxonomic, biogeographic and conservation perspective of the *Niphargus tatrensis* species complex (Crustacea: Amphipoda). Org Divers Evol 10:5–22

Frivaldszky J (1865) Adatok a magyarhoni barlangok faunájához (Data on the cave fauna of Hungary). Math És Term-Tud Közl 3:17–53

Gebhardt A (1931) Az Abaligeti-barlang életvilága (The flora nad fauna of the Abaligeti Cave). A Természet 15–16:1–7

Gebhardt A (1933) Vergleichung der Tierwelt der Abaligeter – und Manfaer Höhlen. Áll Közl 30(1–2):36–44

Gebhardt A (1934) Az Abaligeti-barlang élővilága (The wildlife of the Abaligeti Cave). Math És Term-Tud Közl 37(4):264

Geyer F, Mann H (1940) Studien an Höhle und Thermalteich von Tapolca am Plattensee in Ungarn. Arch Hydrobiol 36:359–385

Gruber P, Gaál Ľ (eds) (2014) A Baradla-Domica-barlangrendszer. A Barlang, mely összeköt (The Baradla-Domica Cave System. The cave which connects). Aggteleki Nemzeti Park Igazgatóság, Jósvafő, p 512

Herczeg G, Hafenscher VP, Balázs G, Fišer Ž, Kralj-Fišer S, Horváth G (2020) The evolution of foraging innovation following colonisation of a less variable environment: a case study in surface- vs. cave-dwelling *Asellus aquaticus*. Ecol Evol 00:1–9

Hornok S, Kováts D, Angyal D, Dányi L, Kovács R, Kontschán J (2016) Description of the male and the larva of Ixodes ariadnae Hornok, 2014. Ticks Tick Borne Dis 7:1252–1255

Kenyeres Z, Nagy B, Bauer N (2008) Distribution and habitat requirements of *Arcyptera microptera* (Fischer de Waldheim, 1833) in Hungary. Articulata 23(2):25–36

Kenyeres Z, Rácz IA, Varga Z (2009) Endemism hot spots, core areas and disjunctions in European Orthoptera. Acta Zool Cracoviensia 52B(1–2):189–211

Kenyeres Z, Szabó S, Bauer N (2020) Conservation possibilities of the rare grasshopper *Stenobothrus eurasius* Zubovski, 1898 are hampered by wild game in its fragmented western outposts. J Insect Conserv 24:115–124

Kováts D, Kurali A, Wizl V, Kukoda O (2011) A Dunazugi Denevérkutató Program első eredményei (First results of the Dunazug Bat Research Program). In: Penksza K et al (eds) VII.

Kárpát-medencei Biológiai Szimpózium (Budapest, 2011. október 13–14). Magyar Biológiai Társaság, Budapest, pp 123–126

Leél-Őssy S, Bergmann C, Bognár C (2011) A Budapesti Molnár János-barlang termálvizének veszélyeztetettsége (The threats to the termal water of the Molnár János Cave). A Miskolci Egyetem Közleménye Bányászat 81:91–102

Loksa I (1960a) Faunistisch-systematische und ökologische Untersuchungen in der Lóczy-Höhle bei Balatonfüred (Biospeologica Hungarica, XI). Opusc Zool Budapest 6:253–266

Loksa I (1960b) Über die Landarthropoden der Teichhöhle von Tapolca (Ungarn). Opusc Zool Budapest 4:39–51

Loksa I (1967) Vier neue Höhlencollembolen aus Ungarn. Opusc Zool Budapest 6(2):289–296

Loksa I (1969) Zwei neue *Arrhopalites* Unterarten (Collembola) aus Höhlen in Ungarn. Opusc Zool Budapest 9(2):357–361

Mammola S, Cardoso P, Angyal D, Balázs G, Blick T et al (2019) Local-versus broad-scale environmental drivers of continental β-diversity patterns in subterranean spider communities across Europe. Proc Royal Soc B Biol Sci. 286:20191579

Méhelÿ L (1924) *Protelsonia hungarica*, nov. gen. Einer blinder Isopode aus Ungarn. Zool Anz 58:353–357

Méhelÿ L (1925) A magyar középhegység, jelesen a Bükk, a Bakony és a Mecsek planáriái (Planarias of the Bükk, Bakony and the Mecsek Mountains). Math És Term-Tud Ért 41:178–184

Merkl O, Szél G (2012) A Sas-hegy bogárfaunája (Coleoptera). (Beetles (Coleoptera) of Mt Sas-hegy (Budapest, Hungary)). Rosalia 8:373–458 (in Hungarian)

Namiotko T, Meisch C, Gidó ZS, Danielopol DL (2001) Redescription, taxonomy, distribution and ecology of Cryptocandona dudichi (Klie, 1930) (Crustacea, Ostracoda). Bull Soc Nat Luxemb 102:109–130

Novák J, Kutasi C (2014) New data on the Pseudoscorpion fauna of the caves of the Bakony Mountains, Hungary. Opusc Zool Budapest 45 (2):189–194

Paulovics P, Juhász M (2008) Egy lehetséges módszer a kereknyergű patkósdenevér (Rhinolophus euryale Blasius, 1853) dunántúli maradványállományának megmentésére: megerősítés áttelepítéssel (protokoll tervezet) (One possible way to save the remaining Transdanubian population of the Mediterranean horseshoe bat (Rhinolophus euryale) from extinction: strengthening with resettlement (action plan). Denevérkutatás – Hung Bat Res News 4:18–37 (in Hungarian)

Paulovics P (1998) Az északi denevér (*Eptesicus nilssonii* (Keyserling et Blasius, 1839) első hazai előfordulása (First record of the Northern bat (Eptesicus nilssonii) in Hungary). Denevérkutatás – Hung Bat Res News 3:25–28 (in Hungarian)

Pérez-Moreno JL, Balázs G, Wilkins B, Herczeg G, Bracken-Grissom HD (2017) The role of isolation on contrasting phylogeographic patterns in two cave crustaceans. BMC Evol Biol 17:247

Roekaerts M (2002) The biogeographical regions map of Europe. Basic principles of its creation and overview of its development. European Environment Agency, Copenhagen, p 15

Salamon G, Dányi L, Angyal D, Balázs G, Forró L (2014) A Baradla gerinctelen faunája (The ivertebrate fauna of the Baradla). In: Gruber P, Gaál L' (eds) A Baradla-Domica-barlangrendszer. A Barlang, mely összeköt (The Baradla-Domica cave system. The cave which connects). Aggteleki Nemzeti Park Igazgatóság, Jósvafő, pp 279–306

Sallai Z (2014) Tiszai ingola. (*Eudontomyzon danfordi* Regan, 1911) In: Haraszthy L (szerk.) Natura 2000 fajok és élőhelyek Magyarországon (Natura 2000 species and habitats in Hungary). Pro Vértes Közalapítvány, Csákvár, pp 402–404 (in Hungarian)

Sket B (2008) Can we agree on an ecological classification of subterranean animals? J Nat Hist 42(21–22):1549–1563

Szabó Á, Ripka G, Hajdú Z, Tempfli B, Varga M, Mészáros I, Kutasi C, Németh T, Pénzes B (2013) New data on the Mesostigmatid Mite Fauna of Hungary (Acari: Mesostigmata). Acta Phytopathol Entomol Hung 48(1):149–154

Szederjesi T, Angyal D, Balázs G, Dányi L (2014) Remarks on the earthworm genus *Helodrilus* Hoffmeister, 1845 with new epigean and subterranean records (Oligochaeta, Lumbricidae). Opusc Zool Budapest 45(2):181–188

Szinetár C (2006) Pókok. Keresztespókok, farkaspókok, ugrópókok és rokonaik a Kárpát-medencében (Spiders. Cross spiders, wolf spiders, jumping spiders and their relatives in the Carpathian Basin). Élővilág Könyvtár, Kossuth Kiadó, Budapest, p 112 (in Hungarian)

Szinetár C, Samu F (2003) *Pelecopsis loksai* sp. n., a new erigonine spider from Hungary (Araneae: Linyphiidae). Bull Br Arachnol Soc 12:412–414

Szinetár C, Rákóczi AM, Bleicher K, Botos E, Kovács P, Samu F (2013) A Sas-hegy pókfaunája II. A Sas-hegy faunakutatásának 80 éve – a hegyről kimutatott pókfajok kommentált listája. (Spider fauna of Mt Sas-hegy II. 80 yeaers of of fauna research on Mt Sas-hegy, with the annotated list of spiders). Rosalia 8:333–362 (in Hungarian)

Sziválc I, Mikes T, Szalontai B, Kučinić M, Vučković I, Vadkerti E, Kisfalvi P, Pauls SU, Bálint M (2017) Ecological divergence of Chaetopteryx rugulosa species complex (Insecta, Trichoptera) linked to climatic niche diversification. Hydrobiologia 794:31–47

Topál G (1989) A barlangi denevérek magyarországi kutatásának áttekintése (An overview of research on cave bats in Hungary). Karszt és Barlang 1989(1–2):85–86 (in Hungarian)

Tóth S (2015) Adatok a Bakony-vidék barlangjainak faunájához (Data on the cave fauna of the Bakony area). Fol Mus Hist Nat Bakonyiensis 32:121–163

Varga Z (1999) Az Aggteleki-karszt gerinctelen állatai (Invertebrates of the Aggtelek Karst) Aggteleki Nemzeti Park Igazgatóság, Szeged, p 20 (in Hungarian)

Varga Z (2019) Biogeográfia. Az élet földrajza (Biogeography. The geography of life). Pars Kft, Nagykovácsi, p 610 (in Hungarian)

Vig K (2009) Leaf beetles from the Mecsek Hills (Southern Hungary). History of entomological research on the territory and some remarks on the Hungarian Chrysomelid Fauna (Coleoptera: Chrysomelidae). Res Chrysomelidae Koninklienje Brille Leiden 2:235–254

Vörös J, Halpern B (2012) A Sas-hegy herpetofaunája (Herpetofauna of Mt Sas-hegy). Rosalia 8:533–548 (in Hungarian)

Geoecology of Hungarian Karsts

Ilona Bárány Kevei and Márton Kiss

Abstract

The three-dimensional geoecosystem of karsts is very fragile, vulnerable, and it responds rather sensitively to any type of pollution, anthropogenic influence and also to direct and indirect effects of climate change. Population growth and therefore primarily the increasing agricultural demand, as well as the quarrying of carbonate rocks and mining of embedded minerals have become more important than the protection of these landscapes. Nowadays, these processes are studied in a complex way, both in space and time, by the interdisciplinary science of geoecology. The geoecological system of karsts is complex and diverse. It is a structural and dynamic system in which rock, water, soil, micro- and macroclimate constitute the abiogenic elements, while micro- and macroflora and fauna as well as people themselves constitute the biogenic elements. Our study presents the geoecological status of Hungarian karsts based on the evaluation of some geoecological parameters and some elements of their interaction.

Keywords

Karst-geoecosystem • Climate-soil-vegetation system of karst • Ecosystem service • Hungarian karsts

I. Bárány Kevei (✉)
Department of Climatology and Landscape Ecology, University of Szeged, Egyetem u 2, 6722 Szeged, Hungary
e-mail: keveibar@geo.u-szeged.hu

M. Kiss
Centre for Ecological Research, Institute of Ecology and Botany, Alkotmány u. 2-4., 2163 Vácrátót, Hungary
e-mail: kiss.marton@geo.u-szeged.hu

1 Introduction

Since the middle of the last century, research on the climate-specific interpretation of karst development has shown that, in addition to geological features, climate, soil and vegetation play an important role in the functioning of karst geoecosystems. Nowadays, these processes are studied in a complex way, both in space and time, by the interdisciplinary science of geoecology. It analyses geoedaphic processes, the geoscientific background of biodiversity, and the relationship between geo- and biodiversity (Keveiné Bárány 2008; Knoll et al. 2012; Rajakaruna and Boyd 2014; Efe 2019; Bátori et al. 2019 Luque-Espinar et al. 2020). Today, biogeomorphological feedbacks also form a part of karst research (Viles 2011; Corenblit et al. 2011; Phillips 2016; Bárány Kevei and Kiss 2016). The three-dimensional system of karsts is very fragile: erosion induced barrenness, grazing, deforestation, soil acidification, agricultural use, the modifying effect of industrial production resulted in a significant change in the condition of karsts (Bárány Kevei 1987; Bognár 1987; Sauro 1993, Bárány Kevei et al. 2006; Kaszala and Bárány Kevei 2004). The karst research of the 2000s focused on the analysis of environmental effects (Keveiné Bárány 2004; Ballut and Faivre 2012; Telbisz et al. 2015; Ramsey 2015 Keveiné Bárány 2018). Extraneous karst materials damage cave stalactites formed over several millennia, endanger the quality of karst water and change the aesthetic and recreational values of karst landscapes. The geoecological research of epikarst and karst water came to the fore (Williams 2008; Lu 2007; Bonacci et al. 2009; Ekmekci and Tezcan 2011; Ramos et al. 2018; Gunn 2020). Human activity has significantly disturbed the natural geoecological status of karsts (Bárány Kevei and Gunn 1999; Bárány Kevei 1999a; Gams and Gabrovec 1999; Gines 1999; van Beynen et al. 2011; Keveiné Bárány 2014; Telbisz et al. 2015; Zapletalová et al. 2016). The study of the parameters used for the prediction of geohazards in karsts has started with the help of GIS (Day 2007, 2010; Gao 2008;

M. Veress and Sz. Leél-Őssy (eds.), *Cave and Karst Systems of Hungary*,
Cave and Karst Systems of the World, https://doi.org/10.1007/978-3-030-92960-2_21

Galve et al. 2009; Gutierrez et al. 2014; Parise and Gunn 2007; Siska et al. 2016; Mauro et al. 2019). The study of karst ecosystem services and that of significant differences in the spatial pattern of karsts are also important parts of contemporary karst research (Goldscheider 2012; Kiss et al. 2011; Bátori et al. 2011; Phillips 2016; Keveiné Bárány and Kiss 2018). The INTERREG Danube Transnational Programme ECO Karst (2017) research project explores the possibilities of both the sustainable use and the conservation of protected karst areas and their ecosystem services in Central Europe. Our study evaluates the geoecological status of Hungarian karsts on the examples of environmentally sensitive karst areas in the mountains of Aggtelek, Bükk and Mecsek.

The geoecological system of karsts is complex and diverse (Pfeffer 1990; Bárány Kevei 1998a; Daoxian 2001; Wenhua and Jianhua 2015; Phillips 2016). It is a structural and dynamic system in which rock, water, soil, micro- and macroclimate constitute the abiogenic elements, while micro- and macroflora and fauna as well as people themselves constitute the biogenic elements. The operation of the system is ensured by the interaction of abiogenic and biogenic factors, as well as by the material and energy flow generated during the interaction, while its structure is determined by the three-dimensional arrangements of the elements (Fig. 1). Without wishing to be exhaustive, our study presents the geoecological status of Hungarian karsts based on the evaluation of some geoecological parameters and the study of their interaction.

2 The Role of Climate in Shaping the Geoecological Character of Karsts

Climate is a morphogenetic factor in the process of karst rock degradation/erosion, an eco-factor ensuring the dynamism of the karst system. Changes in rainwater affect the solution process and the hydrological system of karsts (Kresic and Stevanovic 2010; Meixian et al. 2016). Climate change is a lot more significant stress factor in karsts than in non-karstic areas. While being under the influence of macroclimate, the spatial and temporal differences of energy intake in the specific microclimate spaces of karstic landscapes also have an important effect on the geoecological processes of karsts (Keveiné Bárány 2009). The karsts of Hungary are located in the real temperate zone (on three climate borders). As a result of global climate change, extreme meteorological events have been influencing the geoecological changes of Hungarian karsts in recent decades.

The karstic mountains of Hungary are dominated by cool-moderately humid and cool-moderately dry climates. The mean annual temperature is 6–7 °C, the average annual amount of sunny hours reaches 1950 h per year, and the mean annual amount of precipitation is 700–800 mm (Péczely 1979).

Researchers have been studying the macro- and microclimatic processes of karsts since the 1950s (Schmidt 1930; Bacsó and Zólyomi 1934; Reicher 1936; Horvát 1953;

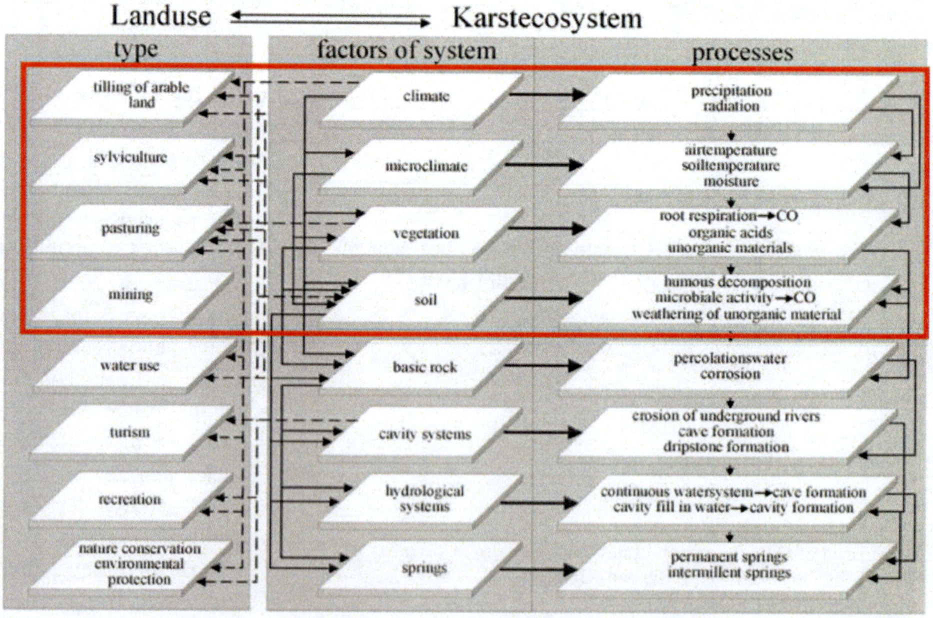

Fig. 1 The relationship of climate-soil-vegetation and their utilization in the karst system (Keveiné Bárány 2009)

Geiger 1961; Sauberer Dirmhirn 1954; Wagner 1964; Yoshino 1984; Bárány Kevei 1967, 1985). The formation of "cold air lakes" filling the dolines at night is a unique microclimatic phenomenon in karstic areas (Bárány Kevei 1999b; Clements et al. 2003; Whiteman et al. 2004; Keveiné Bárány 2011; Caputa 2016; Vilović et al. 2019; Quintero-Ruiz et al. 2019a, b). On the bottom and the cooler slopes of these depressions, colder air buffers the effects of climate change, so these local spatial units can act as microrefugia for cold-tolerant species (Bátori et al. 2009, 2014, 2019; Maclean et al. 2015).

In the independent microclimate spaces of the depressions of Hungarian karsts, the warming and the cooling of the different slopes result in a difference in energy intake and release. In the early morning hours, the air temperature at the bottom of the dolines is several degrees (4–5 °C) lower than at the rim of the dolines. During the day, the dolines are filled with very high temperature air, which is mixed by local, turbulent movements. As is known, the daily amount of radiation is the highest on the southern slopes (slopes with southern exposure), and it is the lowest on northern slopes (Bárány Kevei 1999b). A significant radiation deficit occurs at the bottom of the doline due to self-shadow early in the afternoon. In the eastern-western cross section, both slopes receive the same amount of radition, however, the slope facing east receives more radiation in the morning, while the slope facing west receives more radiation in the afternoon (the daily maximum radiation occurs between 8 and 9 in the morning on the slope facing east, while it occurs between 2 and 3 in the afternoon on the slope facing west), which influences the ecological rhythm of biogenic factors (Fig. 2).

The temperature stratification of the air filling the doline changes significantly during the day. While temperature rise on the valley floor where the doline is located reaches 10 °C between 5 and 6 a.m. (i.e. in one hour), the doline itself is still in self-shadow (fast temperature rise is delayed by the cold air that accumulates in the doline at night). Temperature stratification in the doline is straight during the day, but an inversion occurs between 6 and 7 p.m., the air at the bottom is the coldest then, and temperature rises upwards the slope. Figure 3 shows the temperature stratification of a "cold air lake" in one of the plateau dolines in the Bükk Mountain (the lowest air temperature measure at the bottom was −3.5 °C at a summer night).

The daily course of the soil temperature follows the course of the air temperature with a phase delay according to the exposure conditions and the shadow effect. Warming starts on the southern slope facing north and on the western slope facing east at the earliest, then the northern slope facing south reaches higher temperature, while the eastern slope facing west warms with a lower temperature maximum. During the day, the near-surface and the deeper soil layers of the southern slope with northern exposure are the warmest. The southern slope with northern exposure is the coldest, and it here that the phase delay is the largest towards the deeper layers of the soil. The analysis of the soil temperature amplitudes (Fig. 4) shows how large the differences between soil temperature minima and maxima are within only a few 10 m distances. The soil temperature amplitudes of the southern and western slopes differ by some orders of magnitude from the amplitudes of the northern and eastern slopes within a small distance (those of the southern and western slopes are smaller). The amplitude is three times greater on the eastern slope than on the southern slope in topsoil samples. The daily temperature fluctuation is smaller, the air temperature is lower, and snow cover lasts longer on the southern and western slopes. This results in the survival of shade-tolerant and cold-tolerant plants on these slopes.

The local microclimate in the soils of the different slopes also affects microbial activity. The soil temperature phase shift results in different bioactivity on the slopes, moreover, different microclimates develop on karst surfaces covered by

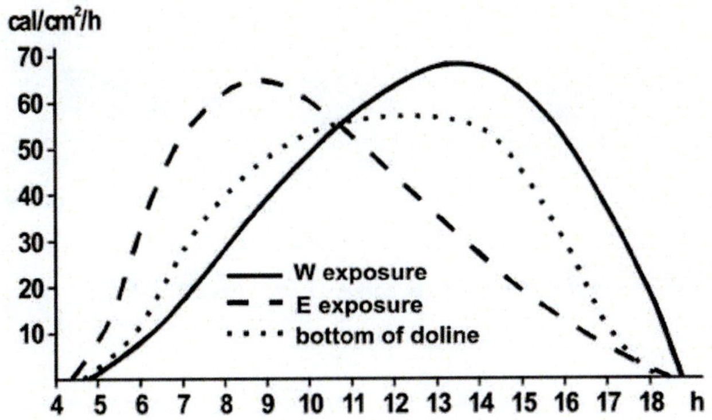

Fig. 2 The daily course of global radiation in eastern and western exposure and at the bottom of the doline (Keveiné Bárány 2009)

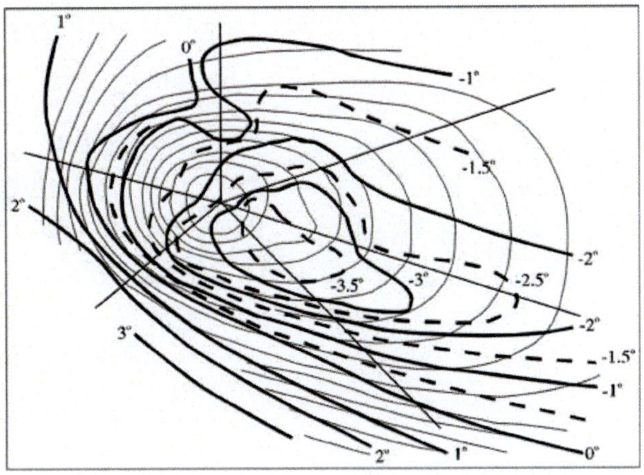

Fig. 3 Cold air lake in a doline of the Bükk Mountains (Keveiné Bárány 2009)

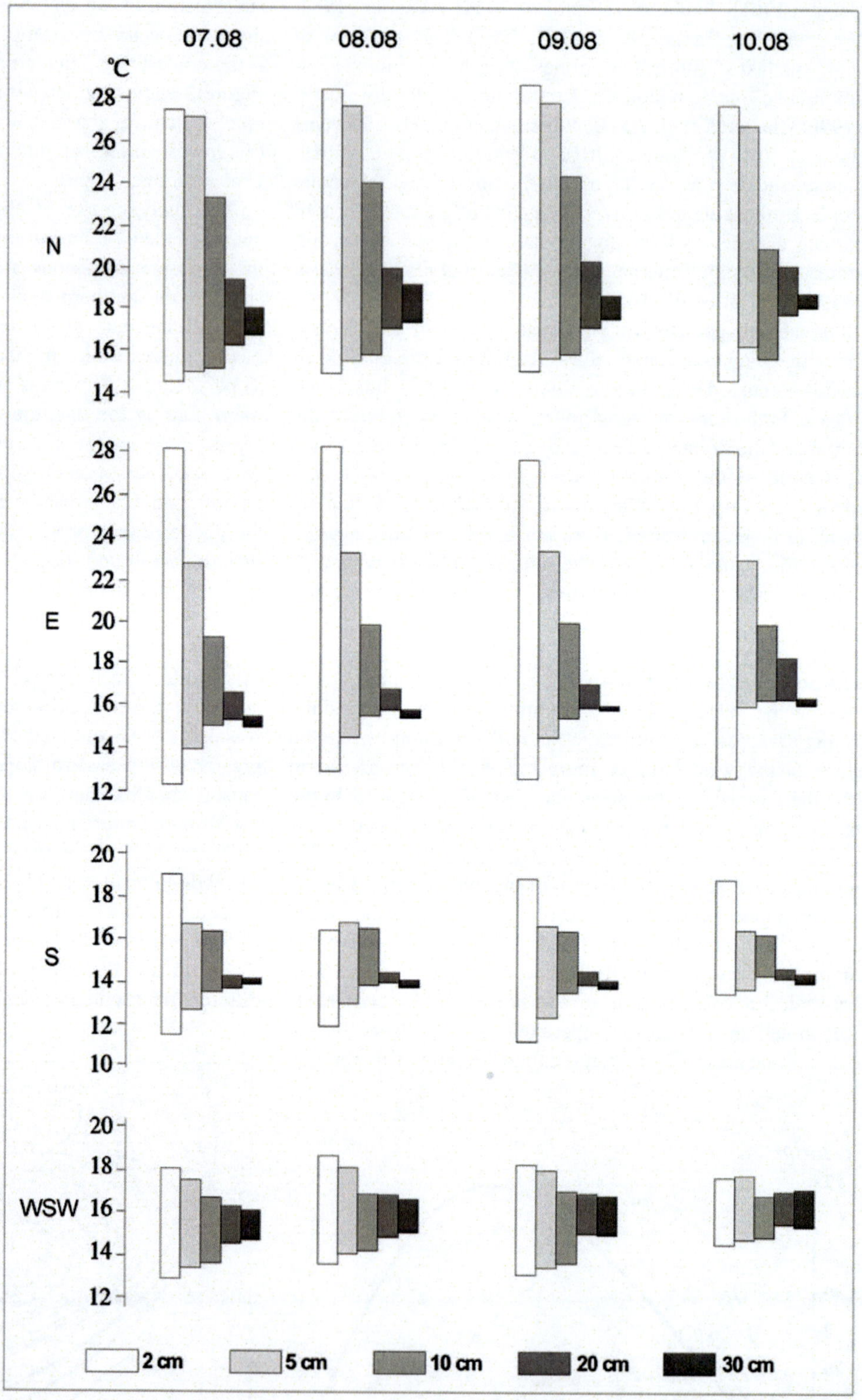

Fig. 4 Differences in soil temperature amplitudes on the slopes of the doline (Bükk Mountain, on 4 August days) N = northern slope; E = eastern slope; S = southern slope; WSW = west-south-western slope (Keveiné Bárány 2009)

different vegetation. In open, grassy areas, air temperature is higher during the day, and it is lower at night than in the forest. In the forests, the shadow effect prevails all day, the heat conduction is slow. At night, on the other hand, a double cold level forms, one at the top of the canopy level, which is the primary active level, and one on the ground since the colder, high density air descends towards the ground.

3 The Role of Soils in Karstic Geoecosystems

The near-surface, water-permeable and vulnerable transitional epikarst zone (Jones et al. 2004; Williams 2008; Bakalowicz 2014; Móga et al. 2017; Phillips 2018), of which soil is an important part, plays an important role in karst development. The epikarst connects the surface to the hydrological network of the karst system through the network of cracks and fissures in the rock. Due to external abiogenic environmental influences (such as climate, acid rain, chemical and physical changes, heavy metal pollution, etc.), the dynamics of the epicarst (including the soil) also changes. Biogenic CO_2in soil atmosphere plays in an important role in karst corrosion (rock dissolution), which was recognized early on (Jakucs 1971, 1980; Trudgill 1976, 1986; Ford and Williams 2007; Zámbó 1970; Kevei and Zámbó 1986). The CO_2 content of soil gas comes from the root respiration of higher plants and the decomposition of organic matter (Kuzyakov 2006).

In Central Europe (including Hungary), 8.1 g NO/NO_2 forms 11.7 g HNO_5/HNO_3, and 20.2 g SO_2 forms 30.6 g H_2SO_4 in the soil, and these acidic solutions dissolve 10.6 g of calcium carbonate per square meter (Arnold 1984). However, the filtering and buffering capacity of a soil can also reduce adverse external impacts (Zámbó 1970). In this sense, soil is the indicator sphere of the geoecosystem of karsts (Bárány Kevei 1992); it indicates and, to an extent, balances unfavourable external effects. In Hungary, the study of karstic soils, which is even further justified due to present-day environmental loads, has become more important in the last few decades (Zámbó 1986; Jakucs 1971; Bárány Kevei 1998b; Zámbó and Telbisz 1999). Soil processes, due to causing changes in soil properties, affect the whole system either directly or indirectly. From the point of view of changes in the system, we consider the role and interaction of climate-soil and vegetation to be of key importance (Fig. 5).

The operation and interactions of the above three system factors can be excellently studied in dolines that are the "hot spots" of karsts (Keveiné Bárány 2018). The Hungarian karst soils formed on Triassic-Jurassic Ladinian and Carnian limestone in the Bükk Mountains, on Triassic Gutenstein and Wetterstein Ladinian limestone and dolomite in the Aggtelek Karst Region, on Middle Triassic-Jurassic limestone in the Mecsek Hills. All of these three areas have different types of rendzina soils, some of which are particularly shallow on the slopes. The figure shows the relationship of soils with morphology and vegetation in the Aggtelek Karst Region (Fig. 6).

3.1 Physical and Chemical Properties of Karst Soils

The physical and chemical properties of soil determine the functioning of the entire karst ecological system. The structure and tissue of soils determine the air, water and heat management of a soil, they affect their chemical properties, and they also play a significant role in soil biological processes. Dark or black rendzina of shallow depth (some 10 cm), mixed with rock debris, and with A and C horizons, occur in the studied karsts (Table 1). According to the international WRB system (World Soil Reference Base System), these soils are black rendzina soils (FR2: Mollic Rendzic Leptosol, Humic, Eutric) soils (Barta et al. 2009), brown rendzina soils (BR2: Leptic Phaeozem) and Leptic Luvisol or red loamy soils (Humic, Hypereutric, Clayic, Chromic), which indicate the dynamics of forest soils. In addition, brown earth, rust-brown, clayey, acidic and podzolic brown forest soils occur in smaller patches on our karsts. Unconsolidated soils of karsts include slope alluvium type of soils and rocky-stony soils with little or no profile differentiation.

Karst soils are not very well classified, the dominant fraction is clay, clayey loam, with a smaller proportion of clay (Fig. 7). In the soils of dolines in the Aggtelek Karst Region, the Bükk Mountains and the Mecsek Hills, loam (i.e. silt, according to geological nomenclature) occurs in the highest proportion (%).

Clay appears in larger quantities at the deeper levels (50 cm), and it can form an impermeable layer at the deepest points of the dolines. In this case, subsoil corrosion may shift towards the edge of the dolines in time, and they widen instead of deepening (Zámbó 1970, 1986). An increase in the proportion of sand results in faster drainage, which facilitates the entry of possible external contaminants into cave passages and karst water.

Soil pH plays an important role in karstic dissolution. The pH values of the soils of the Aggtelek Karst Region, the Bükk Mountains and the Mecsek Hills indicate acidic, weakly acidic and neutral pH (Bárány 1980, 1985). The soils of the Mecsek Hills and the Aggtelek Karst Region are slightly more acidic than those of the Bükk Mountains (Zseni and Keveiné Bárány 2009). The pH of soils formed on limestone is generally not acidic, which is in line with the properties of the bedrock. The acidification tendency of soils

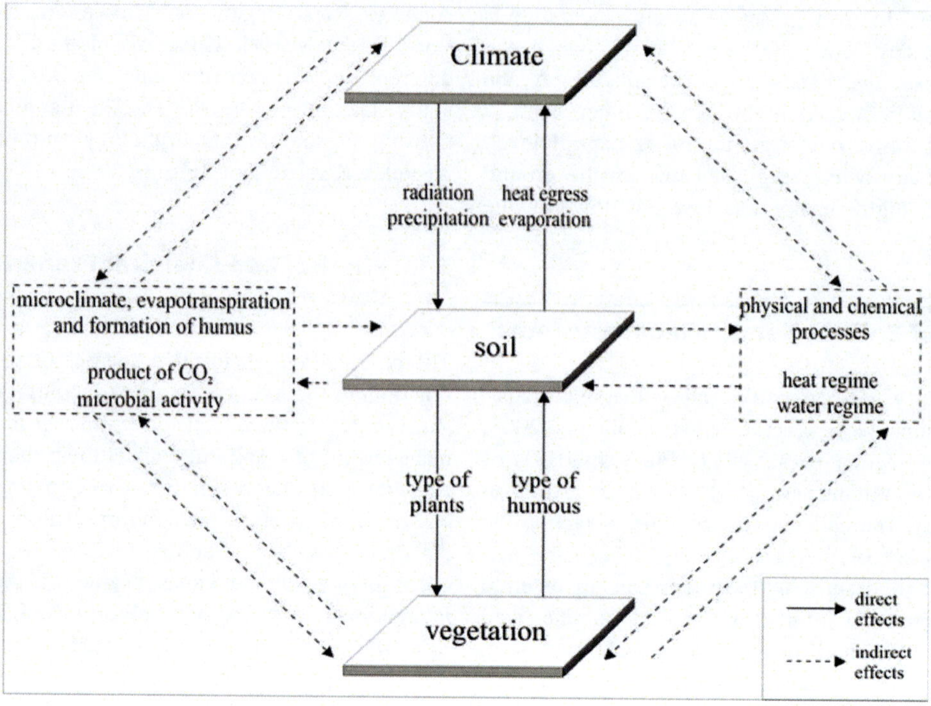

Fig. 5 Interactions within the climate-soil-vegetation systems on karsts (Keveiné Bárány 2009)

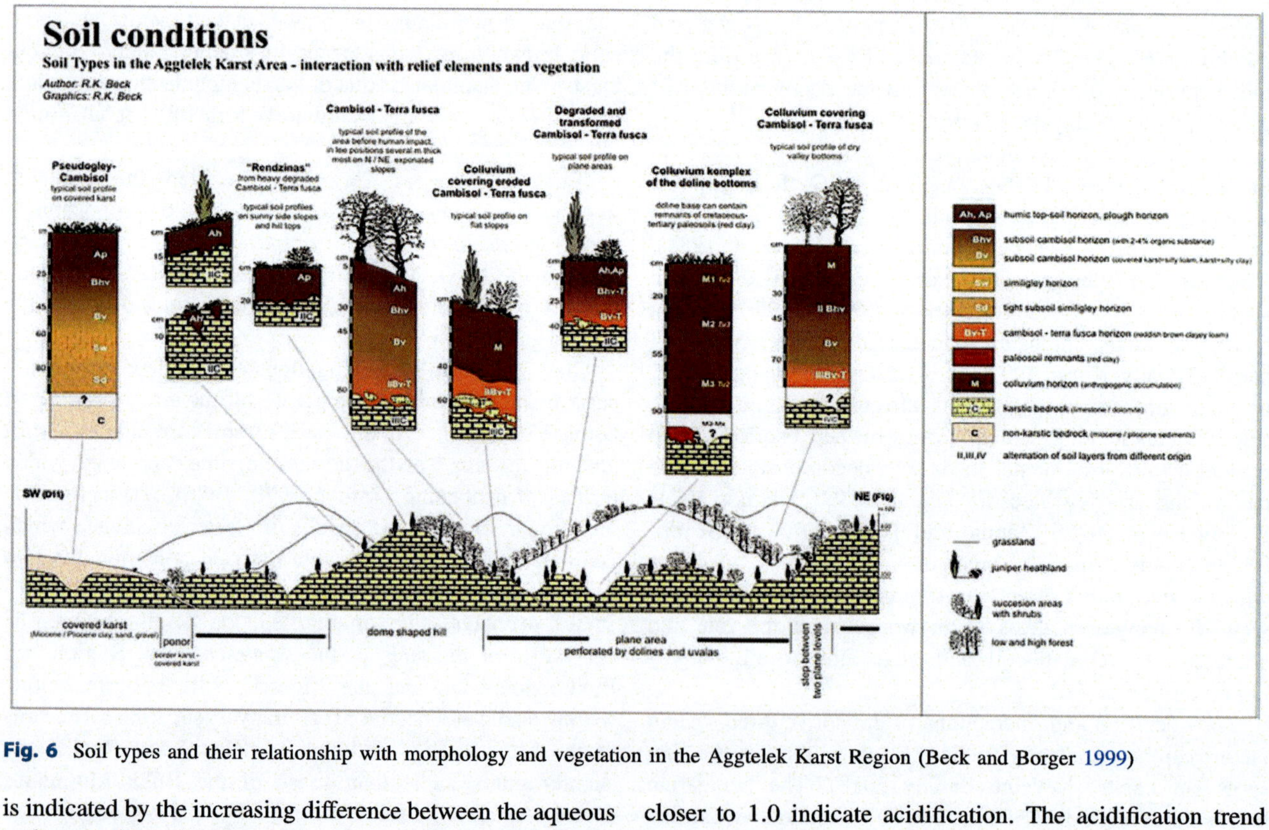

Fig. 6 Soil types and their relationship with morphology and vegetation in the Aggtelek Karst Region (Beck and Borger 1999)

is indicated by the increasing difference between the aqueous and potassium chloride soil solution (ΔpH) (Stefanovits 1992). The value of ΔpH is normal at 0.2–0.5, while values closer to 1.0 indicate acidification. The acidification trend has been observed in the dolines of all three areas (Bükk Mountains, Aggtelek Karst Region, Mecsek Hills) since the

Table 1 Characteristics of two black rendzina soils (FR1 and FR2)(based on Bartaet al. 2009)

Characteristics of soils	FR1		FR2	
	A-horizon	R-horizon	A-horizon	R-horizon
	0–30 cm	30 cm ≤	0–20 cm	20cm ≤
(a) KA and phys. fraction	60, clay	Solid rock	66, clay	Solid rock
(b) Clay content (<0,002 mm)	47.3 %	–	45.1 %	–
(c) Rough frame (%)	<5%	>90 %	<5%	>90 %
(d) Structure	cramb.grain	(Limestone)	grain	(Dolomite)
(e) Dry colour	10YR3-4/2	Off-white	10YR3/2	Off-white
(f) Wet colour	10YR3/1	Off-white	10YR3/1	Off-white
pH (H_2O)	7.06	–	6.96	–
pH (KCl)	6.35	–	6.70	–
(g) Carbonate cont. (m/m%)	0.31	–	0.97	–
(h) Humus content (m/m%)	11.95	–	10.64	–
(i) Base saturation (S/T, %)	92.1	–	91.5	–
(j) Roots	Medium	Craks	Many	Craks
(k) Transition	Sharp	–	Sharp	–
(l) Hydrol. conduct.($mm{\cdot}h^{-1}$)	45–50	–	22–23	–
(m) Max. water capac. (v/v%)	52.4	–	48.3	–
(n) Min. water capacity (v/v%)	45.4	–	37.8	–
(o) Volume weight ($\cdot cm^{-3}$)	1.1	–	1.03	–

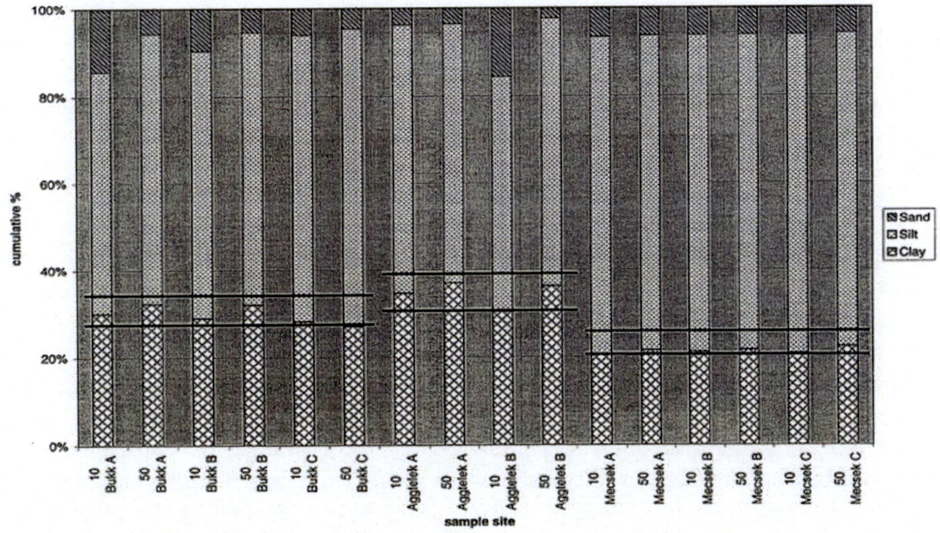

Fig. 7 The average particle size (cumulative%) of the analysed soils (based on 109 soil samples) at the depths of 10 and 50 cm of the examined doline (Charnok 2000)

early 1980s. It is worth mentioning that not only are there significant differences in space (appearing on different slopes or surfaces having different land cover), but the difference between the two pH values also increases over time. In 1982, the ΔpH value was 0.2–0.6, it increased to 0.3–1.3 in 5–6 years, but it was mostly above 0.7. In the 2000s, this value ranged from 0.47 to 1.16 (Table 2).

The carbonated lime content of the soils can only be detected in traces in neutral, weakly alkaline soils, which can be attributed to the stronger leaching effect of acidic sediments, in addition to acidic humic substances. Although the buffering capacity of soils decreases due to the low lime content, the high organic matter content compensates for the lack of carbonated lime (Zámbó and Telbisz 2000). In the black rendzina soils of the Bükk dolines, the proportion of organic matter accumulated near the surface varies between 70 and 80%.

Table 2 Average pH values on the three sample areas (The table shows the average of 288 soil samples)

Location of dolines	pH H_2O	pH KCL	pH H_2O–KCL
Bükk A	6.40	5.93	0.47
Bükk B	6.50	5.45	1,04
Bükk C	6.60	5.54	1.06
Average	*6,50*	*5.64*	*0.86*
Aggtelek A	5.35	4.36	0.99
Aggtelek B	6.13	5.49	0.64
Average	5.74	4.93	0.81
Mecsek A	5.48	4.33	1.16
Mecsek B	6.04	5.16	0.88
Mecsek C	5.83	4.67	1.15
Average	*5.78*	*4.72*	*1.06*

Nowadays, heavy metals play a key role in the changes of the condition of soils; therefore their knowledge is especially important in karsts. Heavy metals are present in a soil in several forms (bound, bioavailable). When analysing the heavy metal retention capacity of a soil, it is not sufficient to examine the "total" (acid-soluble) metal content of the soil, but it is also necessary to determine the amount that plants are potentially able to absorb (bioavailable). As metals can enter the water leaking through the soil, and they can also leak into the karst water, as a result, they damage its quality, and thus the supply of potable water to the population of the given area is endangered. The great amount of heavy metals that can be taken up by plants may also pose a serious threat (Keveiné Bárány et al. 2002a, b; Kaszala and Bárány Kevei 2015). When grazing, animals may consume too much polluted plants, which may end in a serious health problem. The metal retention of soils is influenced by the interaction of different metals; but the pH and organic matter content of a soil also affect the mobility of metals.

Concerning heavy metals, Cd oxides affect the retention of Mn oxides in soils, while Fe oxides affect the retention of Pb and Mn oxides (Arenas-Lago 2014). In general, there are more Mn oxides in soils formed on limestone than it would result from the natural state of the rock (Mulwa et al. 2012). Heavy metals are mobilized in soils with low pH values (in acidic soils). The heavy metal load of the soils of the Aggtelek and Bükk dolines (Keveiné Bárány et al. 2002; Zseni 2003; Bárány Kevei et al. 2006; Kaszala and Bárány Kevei 2004, 2015) is higher than expected in several places. The Aggtelek soil samples have higher Co and Cr contents, and lower Cd content, while the Cd content is higher in Bükk samples (Cd is mobilized at near-neutral pH, while most metals are mobilized at pH 5.5). Ni contamination is high in only a few places (1.5 times higher than the limit value). The Aggtelek and Bükk karsts have higher heavy metal load than the Western Mecsek karst. The sources of environmental contamination in the Aggtelek Karst Region and the Bükk Mountains were probably the Sajó Valley chemical industry (e.g. ore concentrator) and the nearby Slovakian industrial area. The buffering effect of soils is also manifested in the reduction of the mobility of heavy metals in karst soils.

3.2 The Geoecological Significance of Soil Biota: Microbes in Karst Soils

The pH value of soils influences soil life, i.e. the activity of microbial life (Bárány Kevei and Mezősi 1978; KeveiI and Zámbó 1986; Darabos 1999; Gaozhong et al. 2019). The populations of soil microbes (bacteria, *Actinomyces* or "ray fungi", microbial fungi) range to several millions in a gram. In recent years, more and more microbiological studies have contributed to the understanding and assessment of the diversity of bacterial populations in karst soils (Borsodi et al. 2012; Knáb et al. 2012, 2018). More than two-thirds of the CO_2 produced in soil is released through the decomposition of organic matter in these microbial populations. More intensive CO_2 production takes place in the upper 20–30 cm of soil. The amount of soil microbes present in forest soils ranges to 15–20 million, while they are present in significantly smaller numbers in soils under herbaceous vegetation. Examining the relationships among soil temperature, soil moisture and the amount of bacteria, bacteria count in the dolines near the surface was found to correlate significantly with soil temperature, while it correlated strongly with soil moisture in the deeper soil layer (below 30 cm). The amount of carbon dioxide produced by microbes is thus a real ecological. Under Hungarian climatic conditions, temperature ranging from 22.2 to 24.6 °C and the calculated 14–25% dry weight percent of soil moisture in karst areas give the optimum range for microbial activity, where microbial activity develops favourably (Fig. 8).

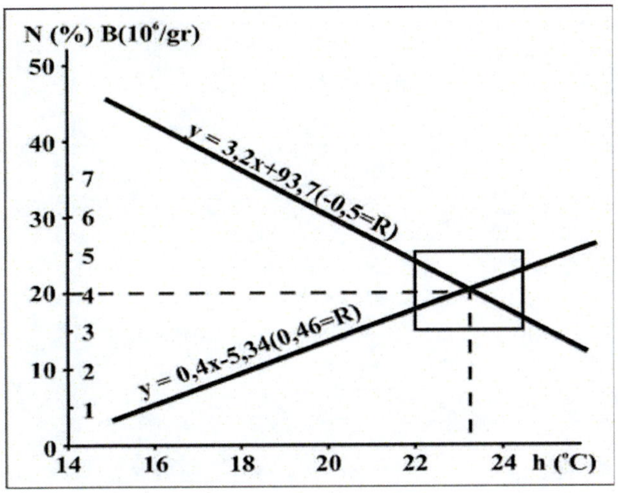

Fig. 8 Correlations among soil temperature (°C), soil moisture (N%), and soil bacteria (B10^6/gr) in Hungarian karsts (Bárány Kevei 2009)

As bacterial processes take place depending on soil moisture and temperature, the amount of carbon dioxide produced by microbes is an ecological quantity in the true sense. The biogenic processes that determine soil dynamics also take place in the upper part of the epicarst, only material and energy flows take place at greater soil depth, which are induced by physical, chemical and biological processes near the surface. It also means that interfering with the system at this upper level can cause significant changes. Biological processes affect the chemical properties of the soil through the decomposition of humic substances, so it is desirable to maintain a natural bacterial population.

The vegetation covering the surface of karsts (Efe 2014, 2019; Donato et al. 2018; Run Liu et al. 2019; Öztürk and Savran 2020), its composition and related habitats significantly influence soil processes (Bárány Kevei and Horváth 1996). Deforestation in our mountains was significant in the 1800s and early 1900s, and species diversity decreased in new, planted forests compared to the original composition. Although recent research on dolines has shown that Abies species, which have replaced natural karst forests, support the survival of species that prefer colder and moister environments (Bátori et al. 2020), the species diversity of Hungarian karsts has decreased considerably. Vegetation strongly influences microclimate and soil function. As the vegetation of the karsts is analysed in another chapter of this volume, we present the currently more important and already changed association types and habitats on the example of the Aggtelek Karst Region in this chapter (Fig. 9).

Vegetation has a special composition due to rendzina and brown forest soils (with forest soil dynamics) formed on karst bedrock. Species that adapt to extreme soil water management form an association here. Through root respiration, the macroflora plays a significant role in the production of carbon dioxide, which determines the aggressiveness of the dissolving water, but it also influences the size of microbial populations.

4 Ecosystem Services in Hungarian Karsts

At the landscape scale, the operation of a karst ecological system extending over different geospheres and the protection of its special natural values can be helped best by the application of special environmental assessment and management methods. One of these is the assessment of ecosystem services, an essential element of which is an integrated, interdisciplinary approach, which is also reflected in the karst ecological research presented above. Ecosystem services are those products or services of ecosystems that contribute directly or indirectly to human well-being. Temperate karst areas, which also include Hungary, play a significant role in one of the supply services, timber production (Kiss et al. 2011). It is related to an important regulatory service, the presence and magnitude of carbon sequestration. The importance of carbon sequestration in karst forests is underlined by the fact that these landscapes, which include areas with diverse topography, often with low productivity, are often less suitable for silviculture, especially for its intensive form. Old, untreated forests of diverse age-composition or naturally managed forests are of great importance for carbon sequestration and storage (Tanács et al. 2016). Carbon sequestration during photosynthesis is manifested in an increase in biomass, the long-term effect of which is an increase in the intensity of karst corrosion, which can be considered as a positive biogeomorphological feedback (Bárány Kevei and Kiss 2016). Finally, karst areas have traditional and significant tourist values, i.e. important cultural services. It is not only due to caves and some special karst morphological phenomena, but also to the structure and functioning of the karst ecosystem: the entire functioning of the climate-soil-vegetation relationship system can create the complexity of a diverse landscape, surface forms and natural vegetation that tourism is based on.

The geoecological features and diversity of karst areas, the large number of inanimate natural values can also be characterized by the concept of geodiversity and some of its indicators. The concept characterizes the diversity of parent rock, hydrological factors, surface forms and soils within a territorial unit. The specific karstic micro-topography, the diversity of the soils and vegetation found on it make lots of karstic territorial units "hot spots' (Butorac Buzjak 2020). According to the results of large-scale assessments, geodiversity is statistically positively related to the amount of ecosystem services provided. That is, geodiversity increases

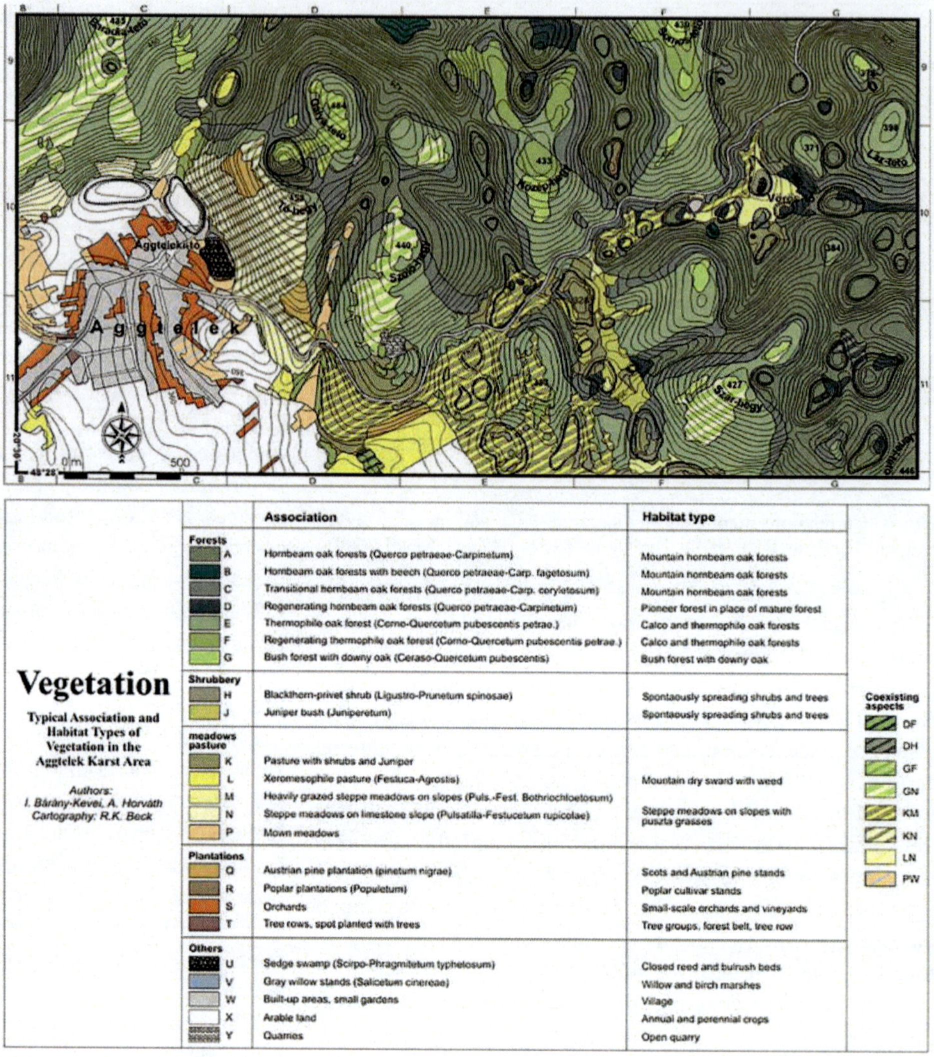

Fig. 9 Vegetation of Aggtelek Karst Region: associations and habitat types (Bárány Kevei et al. 2005)

the determining role of biodiversity as an influencing factor in the creation of services in several cases. For some services, it is true for up to a certain threshold (Alahuhta et al. 2018). Taking geodiversity into account in the process of evaluating ecosystem services is most effective when the concept and its related aspects can be incorporated into theoretical frameworks. One of the most widely used evaluation frameworks worldwide, which also determines the policy of the European Union, is the cascade model (Potschin Haines-Young 2011). The essence of this model is to systematically link the state variables of ecosystems to human well-being through the evaluation of potential and used services, with indicators applied separately for each evaluation level. In other words, one of the main goals of the framework is to make the primary ecological and natural factors of the spatial and temporal patterns of ecosystem services as identifiable as possible. Therefore, the impact assessment of geodiversity as a characteristic significantly influencing biodiversity and other ecological status indicators is an important task of ongoing research on the topic (Fox et al. 2020).

The maintenance of karst ecosystem services, and the land use and environmental management based on them can be helped by several relevant policy processes. An important environmental policy objective of the European Union is the assessment and mapping of Member States' ecosystems, their condition, and the services they provide (EC 2013). Its indirect goal is to visualize ecosystem services in spatial planning and impact assessment processes (Vogiatzakis et al. 2020). In particular, it may strengthen the conservation status of protected landscape types of extensive land use, as these areas provide a large number of services that are only partly taken into account in current environmental assessment and development processes. It includes karsts through the significant amount of regulatory and cultural services they provide. Based on the mapping and evaluation of ecosystem services, another policy objective is the display of natural capital in statistical and accounting systems (Hein et al. 2020). It can also help record the ecosystem services of karsts and use them economically for planning purposes.

5 Conclusions

Climate, soil and vegetation have an important role in the functioning of the geoecosystem of karsts. Climate is a morphogenetic factor in the process of karst rock degradation, an eco-factor that ensures the dynamism of the karst system. Climate change is a much more significant stress factor in karsts than in non-karst areas. In the independent microclimate spaces of karst depressions, the warming and cooling of different slopes results in a difference in energy intake and release, which influences the management of soil biogenic activity and thus also the dissolution processes. A particular phenomenon in karst depressions is the formation of nocturnal cold air lakes, which buffer the effects of climate change, so these spaces can act as microrefugia for cold-tolerant species.

Soil is the indicator sphere of the karst ecological system; soil changes demonstrate various environmental effects. The pH of Hungarian karst soils varies from weakly acidic, acidic to neutral, showing acidification tendency in time, which reduces the buffering of environmental effects. The process of acidification has been taking place on our karsts since the early 1980s, which is unfavourable for future karst ecological processes. Soils contain heavy metals around the limit value, the content of Ni and Cd exceeds the permissible value. Currently, heavy metal pollution in karsts has reached or exceeded the limit value for only a few elements, but the growing tendency is not negligible from an environmental point of view.

More than two-thirds of the CO_2 produced in the soil is released by the decomposition of organic matter by microbial populations. In the topsoil of dolines, the number of bacteria is significantly related to soil temperature, and in the deeper soil layer to moisture content. The amount of carbon dioxide produced by microbes depends on soil moisture and temperature, so it is an ecological size in the true sense. Under Hungarian climatic conditions, the temperature between 22.2 and 24.6 °C and the soil moisture calculated in 14–25% dry weight percent in the karst areas give the optimum range for microbial activity, where microbial activity develops favourably.

The preservation of the characteristic geo-ecosystem of karsts and its values can be supported by the methodology of evaluating ecosystem services. Karsts provide a wide range of provisioning, regulating and cultural services, which is a result of the outstanding level of geodiversity of these landscapes. The sustainable use of these values can be helped by a number of recent, relevant policy processes and tools (e.g. mapping and assessment of ecosystem services, impact assessment applications, ecosystem accounting).

References

Alahuhta J, Ala-Hulkko T, Tukiainen H, Purola L, Akujärvi A, Lampinen R, Hjort J (2018) The role of geodiversity in providing ecosystem services at broad scales. Ecol Ind 91:47–56

Arenas-Lago D, Lago-Vila M, Rodríguez-Seijo A, Andrade ML, (2014) Risk of metal mobility in soils from a Pb/Zn depleted mine (Lugo, Spain). Environ Earth Sci 72:2541–2556

Arnold A (1984) Teoretische Betrachtungen zur antropogenen Beschleunigung des Verkarstungprozesses durch Immissionen saurer Gase, Aerosole und Staube. Mitteil Verb Deutsch Höhlen U Karstforscher 30(2):24–26

Bacsó N, Zólyomi B (1934) Mikroklíma és növényzet a Bükk-fennsíkon (Microclimate and vegetation on the Bükk Plateau) Időjárás 38 (9–10):177–196 (in Hungarian)

Bakalowicz M (2014) The Epikarst, the Skin of Karst. 16 Epikarst. Karst Waters Inst Special Publ 9:16–22

Ballut Ch, Faivre S (2012) New data on the dolines of Velebit Mountain: an evaluation of their sedimentary arhive potential in the reconstruction of landscape evolution. Acta Carsol 41(1):59–74

Barta K, Tanács E, Samu A, Keveiné Bárány I (2009) Hazai rendzinák megfeleltetése a WBR nemzetközi talajosztályozási rendszerben. (Correspondance of Hungarian rendzinas to the international soil classification system WBR) Agrokémia és Talajtan 581:7–18 (in Hungarian)

Bárány I (1967) Der Einfluss des Niveauunterschiedes und der Exposition auf die Luftttemperatur in einer Doline im Bükk-Gebirge. Acta Climatologica Univ. Szegediensis Tomus VII Fasc 1–4:85–109

Bárány I (1980) Some data about the physical and chemical properties of the soil of karst dolines. Acta Geogr Univ Szegediensis Tomus XX:37–49

Bárány I (1985) Ökologische Untersuchung der Karstdolinen unter besonderer Berücksichtigung des Mikroklimas. Acta Geogr Univ Szegediensis Tom XXV:109–130

Bárány Kevei I (1987) Man's impact in karst areas Hungary. In: Karst and Man (Ed.) Kunaver J. Ljubljana, pp 207–219

Bárány Kevei I (1992) Karst soil as indicators of karst development in Hungarian karsts. Zeitschrift Für Geomorphologie N. F. Suppl 85:101–110

Bárány Kevei I (1998a) Geoecological system of Karsts. Acta Carsologica. Krasoslovni Zbornik, XXVII/1:13–25

Bárány Kevei I (1998b) The geo-ecology of three Hungarian karsts. Cave und Karst Science. Transaction of the British Cave Research Association. Num 3:113–117

Bárány Kevei I (1999a) Impact of agricultural land use on some Hungarian Karst region. Int J Speleol 28B(1)4:89–98

Bárány Kevei I (1999b) Microclimate of karstic dolines. Acta Climatologica Universitatis Szegediensis. Tom. 32–33:19–27

Bárány Kevei I (2003) Human impact on Hungarian karst terrains, with special regard to sylviculture. Acta Carsol 32(2):175–187

Bárány Kevei I, Botos C (2001) Landscape ecological problem in Aggtelek National Park with special regard to sustainable silviculture. Ecológia (Bratislava). Supplement 4:151–156

Bárány Kevei I, Gunn J (ed) (1999) Essay in the ecology conservation of Karsts. Acta Ggeograhica Tom. XXXVI, Special Issue of Acta Geographica Szegediensis, p 218

Bárány Kevei I, Horváth A (1996) Survey of the interaction between soil and vegetation in karstecological system a Aggtelek (Hngary). Acta Geographica Szegediensis, Tom. XXXV:81–87

Bárány Kevei I, Horváth A, Beck RK (2005) Geo Ecological Aspects of the Aggtelek Karst Area Hungary. International Atlas of Karst Phenomene. Union Internationale Speleologie Shert 18. Zeitschrift für Geomorphologie Suppl, vol 139

Bárány Kevei I, Kiss M (2016) Biogeomorphological feedback in karst areas. Landsc Environ 10(3–4):101–108. https://doi.org/10.21120/LE/10/3-4/1

Bárány I, Mezősi G (1978) Adatok a karsztos dolinák talajökológiai viszonyaihoz. (Data for the soil ecological condition in karst dolines Földrajzi Értesítő. XXVII.1:65–73 (in Hungarian)

Bárány Kevei I, Zseni A, Hoyk E, Kaszala R (2006) Investigations of heavy metal pollution in the karst regions of Hungary. In: Halasi-Kun GJ (ed) Pollution and water resources. Columbia Univ. 2004–2006. Sustainable development in Central Europe, pp 181–191

Bátori Z, Csiky J, Erdős L, Morschhauser T, Török P, Körmöczi L (2009) Vegetation of the dolines in Mecsek Monuntains (South Hungary) in relation to the local plant communities. Acta Carsol 38/2–3:237–252

Bátori Z, Gallé R, Erdős L, Körmöczi L (2011) Ecological conditions, flora and vegetation of a large doline in the Mecsek Mountains (South Hungary)Acta Bot. Croat 70:147–155. https://doi.org/10.2478/v10184-010-0018-1

Bátori Z, Lengyel A, Maróti M, Körmöczi L, Tölgyesi C, Bíró A, Tóth M, Kincses Z (2014) Microclimate-vegetation relationships in natural habitat islands: species preservation and conservation perspectives. Időjárás 118:257–281

Bátori Z, Vojtkó A, Maák IE, Lőrinczi G, Farkas T, Kántor N, Tanács E, Kiss PJ, Juhász O, Módra G, Tölgyesi C, Erdős L, Aguilon DJ, Keppel G (2019) Karst dolines provide diverse microhabitats for different functional groups in multiple phyla. Sci Rep 9(1). https://doi.org/10.1038/s41598-019-43603-x

Bátori Z, Vojtkó A, Keppel G (2020) Anthropogenic disturbances alter the conservation value of karst dolines. Biodivers Conserv 29:503–525. https://doi.org/10.1007/s10531-019-01896-4)

Beck RK, Borger H (1999) Soils and relief of the Aggtelek karst (NE Hungary): a record of the ecological impact of palaeoweathering effects and human activity. In: Bárány Kevei I, Gunn J (eds) Essay in the ecology and conservation of Karst: Special Issue of Acta Universitatis Szegediensis for the International Geoographical Union: commission on sustainable development and management of Karst terrains. Acta Geogr Tomus XXXVI:54–59

van Beynen Philip E, van Beynen Kaya M (2011) Human disturbance of karst environments. Karst Manage 379–397

Bognár A (1987) The exploitation of bauxite ore deposition in the area of Obrovac and its influence on karst landscape. In: Karst and Man (Kunaver J ed) Proceeding of the internatioonal symposium on human influence in Karst, Postojna, pp 169–179

Bonacci O, Pipan T, Culver DC (2009) A framework for karstecohydrology. Environ Geol 56(5):891–900. https://doi.org/10.1007/s00254-008-1189-0

Borsodi AK, Knáb M, Krett G, Makk J, Márialigeti K, Erőss A, Mádl-Szőnyi J (2012) Biofilm bacterial communities inhabiting the cave walls of the Buda Thermal Karst System. Hungary. Geomicrobiol J 29(7):611–627

Butorac V, Buzjak N (2020) Geodiversity and Landscape Services in the Region of Ogulinsko-Plašćanska Zavala, Croatia. Ekológia (bratislava) 39:130–144

Caputa Z (2016) The impact of solar radiation on the temperature of the exposed rocks of the karst canyon (the Kraków-Częstochowa Upland, Poland). Bull Geogr Phys Geogr Ser 10:19–30. https://doi.org/10.1515/bgeo-2016-0002

Charnok SE (2000) Geo-ecological Sensitivity of Karst Doline Systems, Hungary MSc Dissertation (Manuskript). Collingwood College, Durham, p 115

Clements CB, Whiteman CD, Horel JD (2003) Cold-airpool structure and evolution in a mountain basin: Peter Sinks. Utah J Appl Meteo 42:752–776

Corenblit D, Baas ACW, Bornette G, Delmotte S, Francis RA, Gurnell A, Julien F, Naiman RJ, Steiger J (2011) Feedbacks between geomorphology and biota controlling Earth surface processes and landforms. Earth-Sci Rev 106(3):307–331

Daoxian Y (2001) On the Karst ecosystem. Acta Gelogica Sinica 75/3:336–339 http://www.geojournals.cn/dzxbcn/ch/index.aspx

Darabos G (1999) The role of soil microorganisms in karst corrosion. In: Bárány Kevei I, Gunn J (eds) Essay in the ecology and conservation of Karst: Special Issue of Acta Universitatis Szegediensis, Acta Geographica Tomus XXXVI:54–59

Day MJ (2007) Natural and anthropogenic hazards in the karst of Jamaica. Geol Soc Lond, Special Publications 279:173–164–184

Day MJ (2010) Human interaction with Caaribbean karst landscapes: past, present and future. Acta Carsol 39(1):137–146

Donato CR, Ribeiro AS, Oliveira DG, Souto LS (2018) Análise de aspectos ecológicos da vegetação como ferramenta indicadora da conservação do entorno de cavernas em Laranjeiras, Sergipe. Scientia Plena. vol 14. Num 12:1–13

EC—European Commission (2013) Mapping and assessment of ecosystems and their services, an analytical framework for ecosystem assessments under Action 5 of the EU Biodiversity Strategy to 2020. Discussion paper - Final, April 2013

Efe R (2019) Geomorphology, soil and vegetation relationships in Karst Terrains: a case study of the central Taurus Mountains (Southern Turkey). In: Krystev V, Efe R, Atasoy E (eds) Theory and practice in social sciences. ST. Kliment Ohridski University Press

Ekmekci M, Tezcan L (2011) Management of Karst aquifers under climate change: implication for sustinable use. In: Baba A et al (eds) Climate change and its effects on water resources. NATO Secur Sci Ser C: Environ Secur 3:77–85 https://doi.org/10.1007/978-94-007-1143-3_9

Espinar JAL, Igúzquiza EP, Galiano VFR, Olmo MC, Mario Chica-Olmo, Panizo RV (2018) Karst and vegetation: biodiversity and geobotany in the Sierra de las Nieves Karst Aquifer (Málaga, Spain) In: Bertrand C et al (eds) Eurokarst Besançon, Advances in Karst Science, pp 11–22. https://doi.org/10.1007/978-3-030-14015-1_2

Faimon J, Ličbinská M (2010) Carbon dioxide in the soils and adjacent caves of the Moravian Karst December 2010 Acta Carsologica/Karsoslovni Zbornik 39(3):463–475

Ford D, Williams P (2007) Karst hydrogeology and geomorphology. John Wiley and Sons, Chichester, p 562

Fox N, Lj G, Eigenbrod F, Bullock JM, Parks KE (2020) Incorporating geodiversity in ecosystem servicedecisions. Ecosyst People 16 (1):151–159

Galve JPF, Gutiérrez P, Lucha J, Bonachea J, Remondo A, Cendrero M, Gutiérrez M, Gimeno J, Pardo G (2009) Sinkholes in the salt-bearing evaporite karst of the Ebro River valley upstream of Zaragoza city (NE Spain): geomorphological mapping and analysis as a basis for risk management. Geomorphology 108(3–4):145–158

Gams I, Gabrovec M (1999) Land use and human impact in the Dinaric Karst. Int J Speleol 28 B(1/4):55–70. ISSN 0392-6672

Gao Y (2008) Spatial operations in a GIS-based karst feature database. Environ Geol 54(5):1017–1027

Gaozhong P, Yanna L, Lina D, Longwu Z, Kechao H, Danjuan Z, Ling M, Guangping X (2019) Profiling the bacterial diversity in a typical Karst Tiankeng of China. Biomolecules 9:1–18. https://doi.org/10.3390/biom9050187

Geiger R (1961) Das Klima der bodennahen Luftschicht. Braunschweg p. 646

Gines A (1999) Agriculture, grazing, and land use changes at the Serra de Tramuntana karstic Mountains. Int J Speleol 28B(1/4):5–14

Goldscheider N (2012) A holistic approach to groundwater protection and ecosystem services in karst terrain. AQUA Mundi 117–124. https://doi.org/10.4409/Am-046-12-0047

Gunn J (2020) Karst groundwater in UNESCO protected areas: global verwiew. Hydrogeol J 1–18.https://doi.org/10.1007/s10040-020-022c-x

Gutiérrez F, Parise M, Waele JD, Jourde H (2014) A review on natural and human-induced geohazards and impacts in karst. Earth Sci Rev 138:61–88

Hein L, Bagstad KJ, Edens B, Schenau S, Castillo G, Soulard F, Brown C, Driver A, Bordt M, Steurer A, Harris R, Caparrós A (2020) Progress in natural capital accounting for ecosystems. Science 367:514–515

Horvát S (1953) Die Vegetation der Karstdolinen. Geografski Glasnik. Nos. 14–15.

Hoyk E (2001) A Nyugat-Mecsek növényzetének értékelése optimalizációs térkép alapján. (Evaluation of the vegetation of the Western Mecsek based on an optimization map) Földrajzi Konferencia, Szeged 1–8 (in Hungarian)

INTERREG Danube Transnational Progrmme ECO Karst (2017). Mapping of ecosystem services at the regional level. A practical guide 1-174. Project co-funded by European Union funds (ERDF and IPA). http://www.interreg-danube.eu/approved-projects/eco-karst

Jakucs L (1971) Karsztok morfogenetikája. A karsztfejlődés varienciái. (Morphogenetics of karsts. Variations in karst development) Akadémiai Kiadó, p 310 (in Hungarian)

Jakucs L (1980) Karszt biológiai produktum (Karst biological product) Földrajzi Közlemények 28(4):331–339 (in Hungarian)

Jones WK, Culver DC, Herman JS (2004) Epikarst Vol. Special publication 9. Leesburg, Virginia: Karst Waters Institute, p. 160

Kaszala R, Bárány Kevei I (2004) Heavy metal content of soils in the karstic area of North Hungary. In: Calabrese EJ, Kostecki PT, Dragun J (eds) Contaminated soils, sediments and water. Science in the real world, vol 9, pp 167–175. Springer

Kaszala R, Bárány Kevei I (2015) Heavy metal concentrations in the soils and vegetation of Béke-cave watershead (Aggtelek Karst, Hungary. Acta Geogr Debrecina Landsc Environ 9(2):51–58

Keveiné BI (1985) A karsztdolinák talajainak és növényzetének sajátosságai. (Characteristics of the soils and vegetation of karst dolines) Földrajzi Értesítő. XXXIV. Évf. 3:195–207 (in Hungarian)

Kevei Bárány I (1987) Comparative karstic soil researches in the Bükk Mountain and on the Aggtelek Karst, Hungary. In: Gams I, Kunaver J (eds) Karst and Man. Ljubljana, p 265

Keveiné Bárány I, Zseni A, Kaszala R (2002a) A talaj és növényzet nehézfém taratlmának vizsgálata karsztos mintaterületen (Investigation of heavy metal content of soil and vegetation in a karst sample area) Karsztfejlődés VII.1:297–315 (in Hungarian)

Keveiné Bárány I, Zseni A, Kaszala R (2002b) A talaj és a növényzet nehézfém-tartalmának vizsgálata karsztos területen (Research of heavy metal content of soils and plants in a karst area) Karsztfejlődés VII.297–315) (in Hungarian)

Keveiné Bárány I (2008) Geodiverzitás és tájdiverzitás. (Geodiversity and landscape diversity) Földrajzi Közlemények 132/4:31–439 (in Hungarian)

Keveiné Bárány I (2009) A karsztok ökológiai rendszere: Szerkezet, működés, környezeti hatások a rendszerben (Ecological system of karsts: Structure, activity, environmental effects in the system) Földrajzi tanulmányok, vol 4, p 121 (in Hungarian)

Keveiné Bárány I (2014) Tájváltozások a karsztokon különös tekintettel az emberi használatra.(Landscape changes in karsts with special regard to human use) Karsztfejlődés XIX:5–21 (in Hungarian)

Keveiné Bárány I (2018) Karsztdolinákról másképpen (Geoökológiai értelmezés) (From karst dolinas otherweise (Geoecological interpretation) Karsztfejlődés XXIII:163–185 (in Hungarian)

Keveiné Bárány I, Kiss M (2018) A karsztok ökoszisztéma-szolgáltatásai integrált geográfiai megközelítésben.(Karst ecosystem services in an integrated geographic approach). In: Pap N, Szalai G (eds.) Táj geográfus ecsettel (.Landscape with geographer brush), pp 31–44 (in Hungarian)

Kevei I, Zámbó L (1986) Study of the relationship between bacteria activity in karstic soils and corrosion. Annales Universitatis Scientiarum Budapestinensis de Rolando Eötvös Nominatae. Tomus XX–XXI:325–333

Keveiné Bárány I, Hoyk E, Zseni A (1999) Karsztökológiai egyensúly megbomlások néhány hazai karsztterületen. (Disturbance of the karstecological balance on some Hungarian karsts) Karsztfejlődés III:79–91 (in Hungarian)

Kiss M, Tanács E, Kevei Bárány I (2011) Ecosystem services in Hungsrian karst areas. Acta Climatologica Et Chorologica 44–45 (1):41–49

Knáb M, Szili-Kovács T, Kiss K, Palatinszky M, Márialigeti K, Móga J, Borsodi AK (2012) Comparison of soil microbial communities from two distinct karst areas in Hungary. Acta Microbiol Immunol Hung 59(1):91–105

Knáb M, Szili-Kovács T, Márialigeti K, Móga J, Borsodi AK (2018) Bacterial diversity in soils of different Hungarian karst areas. Acta Microbiol Immunol Hung 1:1–20

Knoll AH, Canfield DE, Kurt O (2012) What is the geobiology. In: Konhauser et al. (eds) Fundamentals of geobiology. Wiley-Blackwell, Oxford, p 456

Kresic N, Stevanovic Z (2010) Groundwater hydrology of springs: engineering, theory, management, and sustainability. Butterworth-Heinemann, Amsterdam, ISBN 978-1-85617-502-9, p 592

Kuzyakov Y (2006) Sources of CO_2 efflux from soil and review of partitioning methods. Soil Biol Biochem 38:425–448

Liu R, Zhaohui Z, Jiachen S, Zhihui W (2019) Bryophyte diversity in karst sinkholes affected by different degrees of human disturbance. Acta Soc Bot Pol 1–14. https://doi.org/10.5586/asbp.3620

Lu YR (2007) Karst water resources and geo-ecology in typical regions of China. Environ Geol 51(5):695–699

Luque-Espinar JA, Pardo-Igúzquiza E, Rodríguez-Galiano VF, Chica-Olmo M, de la Vega-Panizo R (2020) Karst and Vegetation: Biodiversity and Geobotany in the Sierra de las Nieves Karst Aquifer (Málaga, Spain). In: Bertrand C, Denimal S, Steinmann M, Renard P (eds) Eurokarst 2018, Besançon. Advances in Karst Science. Springer, pp 11–22 Cham. https://doi.org/10.1007/978-3-030-14015-1_2.OnlineISBN978-3-030-140151

Maclean IMD, Hopkins JJ, Bennie J, Lawson CR, Wilson RJ (2015) Microclimates buffer the responses of plant communities to climate change (Global Ecol. Biogeogr.) A J Macroecol. https://doi.org/10.1111/geb.12359. http://wileyonlinelibrary.com/journal/geb:1–11

Mauro G, de Azevedo RU, Ferreira R, Goulart FF (2019) Landscape Fragmentation around Ferruginous Caves of the Iron Quadrangle, Minas Gerais, Brazil. Cuadernos De Geografía: Revista Colombiana De Geografía 28(1):89–101. https://doi.org/10.15446/rcdg.v28n1.67506

Meixian L, Xianli X, Dingbao W, Ke-Lin W (2016) Karst catchments exhibited higher degradation stress from climate change than the non-karst catchments in southwest China: an ecohydrological perspective. J Hydrol 535:173–180

Móga J, Strat D, Iván V, Mari L, Kiss K, Szabó M, Borsodi A, Csüllög G (2017) Changes of the karst landscape and epikarst system in the area of the Tapolca karst terrains, North–West Balaton Highlands, Hungary. Forum geografic. Studii şi cercetări de geografie şi protecţia mediului XVI/1:12–25. https://doi.org/10.5775/fg.2017.074.i

Mulwa BM, Maina DM, Patel JP (2012) Multielelemental analysis of limestone and soil samples Kitui South (Kenya) limestone deposits. Int J Fund Phys Sci 2(4):48–51

Öztürk MZ, Savran A (2020) An oasis in the Central Anatolian steppe: the ecology of a collapse doline. Acta BiologicaTurcica 33(2):100–113

Parise M, Gunn J (2007) Natural and anthropogenic hazards in karst areas: an introduction. Geol Soc Lond. Special Publications 279 (1):1–3. https://doi.org/10.1144/SP279.1

Péczely G (1979) Éghajlattan (Climatology.in Hungarian) Tankönyvkiadó, Budapest, p 336 (Hungarian)

Pfeffer KH (1990) Wissenschaftliche Informationen zu Karst-Ökosystemen-eine wichtige Aufgabe für praxisorientierte Forschungen und Planungen. Tübinger Geographische Studien pp 1–35

Phillips D (2016) Biogeomorphology and contingent ecosystem engineering in karst landscapes. Prog Phys Geogr 40(4):503–526

Phillips DJ (2018) Self-limited biogeomorphic ecosystem engineering in epikarst soils. Phys Geogr 39(4):304–328

Potschin M, Haines-Young R (2011) Ecosystem services: exploring a geographical perspective. Prog Phys Geogr 35:575–594

Quintero-Ruiz Joab R, Yáñez-Espinosa L, Flores J, Fortanelli J, De-Nova A, Reyes-Hernandez H, Rodas-Ortíz JP (2019a) Analysis of the soil and microclimate relationship in two dolines of Carso Huasteco, Mexico. J Nat Resour Dev 09:25–33

Quintero-Ruiz Joab R, Yáñez-Espinosa L, Flores J, Fortanelli J, De-Nova A, Reyes-Hernandez H, Rodas-Ortíz JP (2019b) Analysis of the soil and microclimate relationship in two dolines of Carso Huasteco, Mexico. J Nat Resour Dev 09:25–33. https://doi.org/10.5027/jnrd.v9i0.03

Ramos J, Gracia JS, Ortiz J, González-Chávez R Iturbe (2018) Water quality in aguadas within a protected karstic rain forest: The role of the vegetation-soil water interaction. 121/1:2–11. https://doi.org/10.1016/jecoleng.2017.12.006

Ramsey LC (2015) Morphometry and basic ecological characteristics of dolines in unlogged temperate rainforest karstlandscapes of Northern Vancouver Island, British Columbia, Canada. PhD dissertation. University Nova Gorica Slovenia, p 315

Reicher M (1936) Naechtlicher Kaltluftfluss an Hindernissen. Biokl. B.3. 152–163. In: Geiger R (ed) 1961 Das Klima der bodennachen Luftshicht, p 646

Rajakaruna NR, Boyd S (2014) Geoecology. In: Gibson D (ed) Oxford bibliographies in ecology. Oxford University Press, NewYork

Sauberer F, Dirmhirn I (1954) Über die Entstehung der extremen Temperatur minima in der Doline Gstettner-Alm (On the occurrence of extreme temperature minimums in the Gstettner-Alm Doline). Arch Meteor Geophys Bioklimatol 5B:307–326

Sauro U (1993) Human impact on the karst of the Venetian Fore-Alps. Italy Environ Geol 21(3):115–121. https://doi.org/10.1007/BF00775294

Siska PP, Kemmerly PR (2007) Doline feature database (DFD) development and assessment of hazards using a co-regionalization model. Papers Appl Geogr Conf 31:34–42

Siska PP, Goovaerts P, Hung IK (2016) Evaluating susceptibility of Karst dolines (sinkholes) for collapse in Sango, Tennessee, USA. Prog Phys Geogr 40:579–597

Stefanovits P (1992) Talajtan (Soil Science) (in Hungarian) Mezőgaszda Kiadó, pp 369

Tanács E, Krizsán T, Kiss M (2016) Carbon sequestration modeling in a temperate near-natural beech forest using CO2Fix. Acta Climatol Et Chorol 49–50:81–90

Telbisz T, Bottlik Z, Mari L, Petrvalska A (2015) Exploring relationships between Karst terrains and social features by the Example of Gömör-Torna Karst (Hungary-Slovakia). Acta Carsol 44(1):121–137

Trudgill ST (1976) The erosion of limestone under soil and the long term stability of soil-vegetation systems on limestone. Earth Surf Process 1:31–42

Trudgill ST (1986) Limestone weathering under a soil cover and the evolution of limestone pavements, Malham District north Yorkshire UK. New Directions in karst. Proceeding of the Anglo-French Karst Symposium, pp 461–471

Viles HA (2011) Biogeomorphology. In: Gregory KJ, Goudie AS (eds) The handbook of geomorphology. SAGE, London, pp 246–259

Vilovci T, Buzjak S, Buzjak N (2019) Floristic and microclimatic features of the Sovljak doline (Mt. Velika Kapela, Croatia). Article in Sumarski List, February 2019. Šumarski List 1–2:35–43. https://doi.org/10.31298/sl.143.1-2.4

Vrynioty PK, Alexakis D, Bathrellos GD, Vryniotis SHD, D, Vasiliades E (2014) Environmental research and evalution of agricultural soil of Arta plain West Hellas. J Geochem Explor 136:84–92

Vogiatzakis IN, Zotos S, Litskas VD, Manolaki P, Sarris D, Stavrinides MC (2020) Towards implementing mapping and assessment of ecosystems and their services in cyprus: a first set of indicators for ecosystem management. One Ecosyst 5:e47715

Wenhua LI, Jianhua CAO (eds) (2015) Karst ecosystem: an ecosystem on carbonate rocks. J Resour Ecol 6(4):197–198. https://doi.org/10.5814/j.issn.1674-764x.2015.04.001;www.jorae.cn

Wagner R (1964) Lufttemperatur Messungen in einer Doline des Bükkgebirges. Zeitschrift Für Angewandte Meteorologie 3(5):3–4

Whiteman CD, Haiden T, Pospichal B, Eisenbch S, and. Steinacker DR, (2004) Minimum temperatures, diurnal temperature ranges, and temperature inversions in limestone sinkholes of different sizes and shapes. Am Meteorol Soc 43:1224–1236

Williams PW (2008) The role of the epikarst in karst and cave hydrogeology: a review. Int J Speleol 37/1:1–10 Bologna (Italy)

Worthington SRH (2011) Management of Carbonate Aquifers. In: van Beynen P (eds) Karst management, pp 243–261. https://doi.org/10.1007/978-94-007-1207-2_11

Yoshino MM (1984) Thermal belt and cold air drainage on the mountain slope and cold air lake in the basin at quiet, clear night. Geo J 8(3):235–250

Zapletalová J, Stefanova D, Vaishar A, Stefanov P, Dvořák P, Tcherkezova E (2016) Social development of ecologycally sensitive rural areas case studies of Moravian Karst (Czech Republic) and the Devetashko Plato (Buégaria). Bul Acad Sci Prob Geogr 3–4:88–109

Zámbó L (1970) A vörösagyagok és a felszíni karsztosodás kapcsolata az Aggteleki-karszt délnyugati részén (Relationship between red clays and surface karstification in the southwestern part of the Aggtelek Karst) Földrajzi Közlemények 4:281–293 (in Hungarian)

Zámbó (1986) A talajhatás jelentősége a karszt korróziós fejlődésben (Significance of soil influence in karst corrosion development) Kandidátusi értekezés (PhD Dissertation. Budapest, Eötvös Lóránd University), p.170) (in Hungarian)

Zámbó L, Telbisz T (1999) The influence of the soil zone on karst corrosion and karren development, pp 185–192. In: Bárány-Kevei I,

Gunn J (eds) Essays in the ecology and conservation of Karst. Szeged – Budapest – Miskolc, p 218

Zámbó L, Telbisz T (2000) A mikrobiális befolyásoltságú karsztkorrózió vizsgálata magyarországi karsztok talajaiból származó kismintákon (Investigation of microbially influenced karst corrosion on small samples from karst soils in Hungary) Karsztfejlődés V:21–39 (in Hungarian)

Zseni A (2003) Karsztos mintaterületek talajainak tápanyag gazdálkodása és nehézfém terhelése.(Nutrient management and heavy metal contaminaion of soils in karst sample areas). PhD Dissertation, Szeged. p. 118 (in Hungrian)

Zseni A, Keveiné Bárány I (2001) Talajtulajdonságok változása és jellemzői a dolinákban.(Changes in soil properties and characteristics in the dolines). Földrajzi Konferencia Szeged. pp 1–16. http://geography.hu/mfk2001/cikkek/ZseniKeveine.pdf (in Hungrian)

Index

M. Veress and Sz. Leél-Őssy (eds.), *Cave and Karst Systems of Hungary*,
Cave and Karst Systems of the World, https://doi.org/10.1007/978-3-030-92960-2

E

F

M

N

Q

R